Table A.3 Standard Normal Curve Areas *(cont.)*

z	.00	.01	.02	.03	.04	.05	.06	.07	.08	.09
0.0	.5000	.5040	.5080	.5120	.5160	.5199	.5239	.5279	.5319	.5359
0.1	.5398	.5438	.5478	.5517	.5557	.5596	.5636	.5675	.5714	.5753
0.2	.5793	.5832	.5871	.5910	.5948	.5987	.6026	.6064	.6103	.6141
0.3	.6179	.6217	.6255	.6293	.6331	.6368	.6406	.6443	.6480	.6517
0.4	.6554	.6591	.6628	.6664	.6700	.6736	.6772	.6808	.6844	.6879
0.5	.6915	.6950	.6985	.7019	.7054	.7088	.7123	.7157	.7190	.7224
0.6	.7257	.7291	.7324	.7357	.7389	.7422	.7454	.7486	.7517	.7549
0.7	.7580	.7611	.7642	.7673	.7704	.7734	.7764	.7794	.7823	.7852
0.8	.7881	.7910	.7939	.7967	.7995	.8023	.8051	.8078	.8106	.8133
0.9	.8159	.8186	.8212	.8238	.8264	.8289	.8315	.8340	.8365	.8389
1.0	.8413	.8438	.8461	.8485	.8508	.8531	.8554	.8577	.8599	.8621
1.1	.8643	.8665	.8686	.8708	.8729	.8749	.8770	.8790	.8810	.8830
1.2	.8849	.8869	.8888	.8907	.8925	.8944	.8962	.8980	.8997	.9015
1.3	.9032	.9049	.9066	.9082	.9099	.9115	.9131	.9147	.9162	.9177
1.4	.9192	.9207	.9222	.9236	.9251	.9265	.9278	.9292	.9306	.9319
1.5	.9332	.9345	.9357	.9370	.9382	.9394	.9406	.9418	.9429	.9441
1.6	.9452	.9463	.9474	.9484	.9495	.9505	.9515	.9525	.9535	.9545
1.7	.9554	.9564	.9573	.9582	.9591	.9599	.9608	.9616	.9625	.9633
1.8	.9641	.9649	.9656	.9664	.9671	.9678	.9686	.9693	.9699	.9706
1.9	.9713	.9719	.9726	.9732	.9738	.9744	.9750	.9756	.9761	.9767
2.0	.9772	.9778	.9783	.9788	.9793	.9798	.9803	.9808	.9812	.9817
2.1	.9821	.9826	.9830	.9834	.9838	.9842	.9846	.9850	.9854	.9857
2.2	.9861	.9864	.9868	.9871	.9875	.9878	.9881	.9884	.9887	.9890
2.3	.9893	.9896	.9898	.9901	.9904	.9906	.9909	.9911	.9913	.9916
2.4	.9918	.9920	.9922	.9925	.9927	.9929	.9931	.9932	.9934	.9936
2.5	.9938	.9940	.9941	.9943	.9945	.9946	.9948	.9949	.9951	.9952
2.6	.9953	.9955	.9956	.9957	.9959	.9960	.9961	.9962	.9963	.9964
2.7	.9965	.9966	.9967	.9968	.9969	.9970	.9971	.9972	.9973	.9974
2.8	.9974	.9975	.9976	.9977	.9977	.9978	.9979	.9979	.9980	.9981
2.9	.9981	.9982	.9982	.9983	.9984	.9984	.9985	.9985	.9986	.9986
3.0	.9987	.9987	.9987	.9988	.9988	.9989	.9989	.9989	.9990	.9990
3.1	.9990	.9991	.9991	.9991	.9992	.9992	.9992	.9992	.9993	.9993
3.2	.9993	.9993	.9994	.9994	.9994	.9994	.9994	.9995	.9995	.9995
3.3	.9995	.9995	.9995	.9996	.9996	.9996	.9996	.9996	.9996	.9997
3.4	.9997	.9997	.9997	.9997	.9997	.9997	.9997	.9997	.9997	.9998

NINTH EDITION

Probability and Statistics for Engineering and the Sciences

JAY DEVORE

California Polytechnic State University, San Luis Obispo

CENGAGE
Learning

Australia • Brazil • Mexico • Singapore • United Kingdom • United States

CENGAGE
Learning

Probability and Statistics for Engineering and the Sciences, Ninth Edition

Jay L. Devore

Senior Product Team Manager: Richard Stratton

Senior Product Manager: Molly Taylor

Senior Content Developer: Jay Campbell

Product Assistant: Spencer Arritt

Media Developer: Andrew Coppola

Marketing Manager: Julie Schuster

Content Project Manager: Cathy Brooks

Art Director: Linda May

Manufacturing Planner: Sandee Milewski

IP Analyst: Christina Ciaramella

IP Project Manager: Farah Fard

Production Service and Compositor: MPS Limited

Text and Cover Designer: C Miller Design

For product information and technology assistance, contact us at
Cengage Learning Customer & Sales Support, 1-800-354-9706

For permission to use material from this text or product, submit all requests online at **www.cengage.com/permissions**
Further permissions questions can be emailed to
permissionrequest@cengage.com

Unless otherwise noted, all items © Cengage Learning

Library of Congress Control Number: 2014946237

ISBN: 978-1-305-25180-9

Cengage Learning
20 Channel Center Street
Boston, MA 02210
USA

Cengage Learning is a leading provider of customized learning solutions with office locations around the globe, including Singapore, the United Kingdom, Australia, Mexico, Brazil, and Japan. Locate your local office at **www.cengage.com/global**.

Cengage Learning products are represented in Canada by Nelson Education, Ltd.

To learn more about Cengage Learning Solutions, **visit www.cengage.com**.

Purchase any of our products at your local college store or at our preferred online store **www.cengagebrain.com**.

Printed in the United States of America
Print Number: 02   Print Year: 2015

To my beloved grandsons
Philip and Elliot, who are
highly statistically significant.

Contents

8 Tests of Hypotheses Based on a Single Sample

9 Inferences Based on Two Samples

10 The Analysis of Variance

Preface

Purpose

The use of probability models and statistical methods for analyzing data has become common practice in virtually all scientific disciplines. This book attempts to provide a comprehensive introduction to those models and methods most likely to be encountered and used by students in their careers in engineering and the natural sciences. Although the examples and exercises have been designed with scientists and engineers in mind, most of the methods covered are basic to statistical analyses in many other disciplines, so that students of business and the social sciences will also profit from reading the book.

Approach

Students in a statistics course designed to serve other majors may be initially skeptical of the value and relevance of the subject matter, but my experience is that students *can* be turned on to statistics by the use of good examples and exercises that blend their everyday experiences with their scientific interests. Consequently, I have worked hard to find examples of real, rather than artificial, data—data that someone thought was worth collecting and analyzing. Many of the methods presented, especially in the later chapters on statistical inference, are illustrated by analyzing data taken from published sources, and many of the exercises also involve working with such data. Sometimes the reader may be unfamiliar with the context of a particular problem (as indeed I often was), but I have found that students are more attracted by real problems with a somewhat strange context than by patently artificial problems in a familiar setting.

Mathematical Level

The exposition is relatively modest in terms of mathematical development. Substantial use of the calculus is made only in Chapter 4 and parts of Chapters 5 and 6. In particular, with the exception of an occasional remark or aside, calculus appears in the inference part of the book only—in the second section of Chapter 6. Matrix algebra is not used at all. Thus almost all the exposition should be accessible to those whose mathematical background includes one semester or two quarters of differential and integral calculus.

Content

Chapter 1 begins with some basic concepts and terminology—population, sample, descriptive and inferential statistics, enumerative versus analytic studies, and so on—and continues with a survey of important graphical and numerical descriptive methods. A rather traditional development of probability is given in Chapter 2, followed by probability distributions of discrete and continuous random variables in Chapters 3 and 4, respectively. Joint distributions and their properties are discussed in the first part of Chapter 5. The latter part of this chapter introduces statistics and their sampling distributions, which form the bridge between probability and inference. The next three chapters cover point estimation, statistical intervals, and hypothesis testing based on a single sample. Methods of inference involving two independent samples and paired data are presented in Chapter 9. The analysis of variance is the subject of Chapters 10 and 11 (single-factor and multifactor, respectively). Regression makes its initial appearance in Chapter 12 (the simple linear regression model and correlation) and returns for an extensive encore in Chapter 13. The last three chapters develop chi-squared methods, distribution-free (nonparametric) procedures, and techniques from statistical quality control.

Helping Students Learn

Although the book's mathematical level should give most science and engineering students little difficulty, working toward an understanding of the concepts and gaining an appreciation for the logical development of the methodology may sometimes require substantial effort. To help students gain such an understanding and appreciation, I have provided numerous exercises ranging in difficulty from many that involve routine application of text material to some that ask the reader to extend concepts discussed in the text to somewhat new situations. There are many more exercises than most instructors would want to assign during any particular course, but I recommend that students be required to work a substantial number of them. In a problem-solving discipline, active involvement of this sort is the surest way to identify and close the gaps in understanding that inevitably arise. Answers to most odd-numbered exercises appear in the answer section at the back of the text. In addition, a Student Solutions Manual, consisting of worked-out solutions to virtually all the odd-numbered exercises, is available.

To access additional course materials and companion resources, please visit www.cengagebrain.com. At the CengageBrain.com home page, search for the ISBN of your title (from the back cover of your book) using the search box at the top of the page. This will take you to the product page where free companion resources can be found.

New for This Edition

- The major change for this edition is the elimination of the rejection region approach to hypothesis testing. Conclusions from a hypothesis-testing analysis are now based entirely on P-values. This has necessitated completely rewriting Section 8.1, which now introduces hypotheses and then test procedures based on P-values. Substantial revision of the remaining sections of Chapter 8 was then required, and this in turn has been propagated through the hypothesis-testing sections and subsections of Chapters 9–15.
- Many new examples and exercises, almost all based on real data or actual problems. Some of these scenarios are less technical or broader in scope than what has been included in previous editions—for example, investigating the nocebo effect (the inclination of those told about a drug's side effects to experience them), comparing sodium contents of cereals produced by three different manufacturers, predicting patient height from an easy-to-measure anatomical characteristic, modeling the relationship between an adolescent mother's age and the birth weight of her baby, assessing the effect of smokers' short-term abstinence on the accurate perception of elapsed time, and exploring the impact of phrasing in a quantitative literacy test.
- More examples and exercises in the probability material (Chapters 2–5) are based on information from published sources.

• The exposition has been polished whenever possible to help students gain a better intuitive understanding of various concepts.

Acknowledgments

My colleagues at Cal Poly have provided me with invaluable support and feedback over the years. I am also grateful to the many users of previous editions who have made suggestions for improvement (and on occasion identified errors). A special note of thanks goes to Jimmy Doi for his accuracy checking and to Matt Carlton for his work on the two solutions manuals, one for instructors and the other for students.

The generous feedback provided by the following reviewers of this and previous editions has been of great benefit in improving the book: Robert L. Armacost, University of Central Florida; Bill Bade, Lincoln Land Community College; Douglas M. Bates, University of Wisconsin–Madison; Michael Berry, West Virginia Wesleyan College; Brian Bowman, Auburn University; Linda Boyle, University of Iowa; Ralph Bravaco, Stonehill College; Linfield C. Brown, Tufts University; Karen M. Bursic, University of Pittsburgh; Lynne Butler, Haverford College; Troy Butler, Colorado State University; Barrett Caldwell, Purdue University; Kyle Caudle, South Dakota School of Mines & Technology; Raj S. Chhikara, University of Houston–Clear Lake; Edwin Chong, Colorado State University; David Clark, California State Polytechnic University at Pomona; Ken Constantine, Taylor University; Bradford Crain, Portland State University; David M. Cresap, University of Portland; Savas Dayanik, Princeton University; Don E. Deal, University of Houston; Annjanette M. Dodd, Humboldt State University; Jimmy Doi, California Polytechnic State University–San Luis Obispo; Charles E. Donaghey, University of Houston; Patrick J. Driscoll, U.S. Military Academy; Mark Duva, University of Virginia; Nassir Eltinay, Lincoln Land Community College; Thomas English, College of the Mainland; Nasser S. Fard, Northeastern University; Ronald Fricker, Naval Postgraduate School; Steven T. Garren, James Madison University; Mark Gebert, University of Kentucky; Harland Glaz, University of Maryland; Ken Grace, Anoka-Ramsey Community College; Celso Grebogi, University of Maryland; Veronica Webster Griffis, Michigan Technological University; Jose Guardiola, Texas A&M University–Corpus Christi; K. L. D. Gunawardena, University of Wisconsin–Oshkosh; James J. Halavin, Rochester Institute of Technology; James Hartman, Marymount University; Tyler Haynes, Saginaw Valley State University; Jennifer Hoeting, Colorado State University; Wei-Min Huang, Lehigh University; Aridaman Jain, New Jersey Institute of Technology; Roger W. Johnson, South Dakota School of Mines & Technology; Chihwa Kao, Syracuse University; Saleem A. Kassam, University of Pennsylvania; Mohammad T. Khasawneh, State University of NewYork–Binghamton; Kyungduk Ko, Boise State University; Stephen Kokoska, Colgate University; Hillel J. Kumin, University of Oklahoma; Sarah Lam, Binghamton University; M. Louise Lawson, Kennesaw State University; Jialiang Li, University of Wisconsin–Madison; Wooi K. Lim, William Paterson University; Aquila Lipscomb, The Citadel; Manuel Lladser, University of Colorado at Boulder; Graham Lord, University of California–Los Angeles; Joseph L. Macaluso, DeSales University; Ranjan Maitra, Iowa State University; David Mathiason, Rochester Institute of Technology; Arnold R. Miller, University of Denver; John J. Millson, University of Maryland; Pamela Kay Miltenberger, West Virginia Wesleyan College; Monica Molsee, Portland State University; Thomas Moore, Naval Postgraduate School; Robert M. Norton, College of Charleston; Steven Pilnick, Naval Postgraduate School; Robi Polikar, Rowan University; Justin Post, North Carolina State University; Ernest Pyle, Houston Baptist University;

Xianggui Qu, Oakland University; Kingsley Reeves, University of South Florida; Steve Rein, California Polytechnic State University–San Luis Obispo; Tony Richardson, University of Evansville; Don Ridgeway, North Carolina State University; Larry J. Ringer, Texas A&M University; Nabin Sapkota, University of Central Florida; Robert M. Schumacher, Cedarville University; Ron Schwartz, Florida Atlantic University; Kevan Shafizadeh, California State University–Sacramento; Mohammed Shayib, Prairie View A&M; Alice E. Smith, Auburn University; James MacGregor Smith, University of Massachusetts; Paul J. Smith, University of Maryland; Richard M. Soland, The George Washington University; Clifford Spiegelman, Texas A&M University; Jery Stedinger, Cornell University; David Steinberg, Tel Aviv University; William Thistleton, State University of New York Institute of Technology; J A Stephen Viggiano, Rochester Institute of Technology; G. Geoffrey Vining, University of Florida; Bhutan Wadhwa, Cleveland State University; Gary Wasserman, Wayne State University; Elaine Wenderholm, State University of New York–Oswego; Samuel P. Wilcock, Messiah College; Michael G. Zabetakis, University of Pittsburgh; and Maria Zack, Point Loma Nazarene University.

Preeti Longia Sinha of MPS Limited has done a terrific job of supervising the book's production. Once again I am compelled to express my gratitude to all those people at Cengage who have made important contributions over the course of my textbook writing career. For this most recent edition, special thanks go to Jay Campbell (for his timely and informed feedback throughout the project), Molly Taylor, Ryan Ahern, Spencer Arritt, Cathy Brooks, and Andrew Coppola. I also greatly appreciate the stellar work of all those Cengage Learning sales representatives who have labored to make my books more visible to the statistical community. Last but by no means least, a heartfelt thanks to my wife Carol for her decades of support, and to my daughters for providing inspiration through their own achievements.

Jay Devore

Overview and Descriptive Statistics

> *"I took statistics at business school, and it was a transformative experience. Analytical training gives you a skill set that differentiates you from most people in the labor market."*
>
> —Laszlo Bock, Senior Vice President of People Operations (in charge of all hiring) at Google
>
> April 20, 2014, *The New York Times*, interview with columnist Thomas Friedman

> *"I am not much given to regret, so I puzzled over this one a while. Should have taken much more statistics in college, I think."*
>
> —Max Levchin, Paypal Co-founder, Slide Founder
>
> Quote of the week from the Web site of the American Statistical Association on November 23, 2010

> *"I keep saying that the sexy job in the next 10 years will be statisticians, and I'm not kidding."*
>
> —Hal Varian, Chief Economist at Google
>
> August 6, 2009, *The New York Times*

INTRODUCTION

Statistical concepts and methods are not only useful but indeed often indispensable in understanding the world around us. They provide ways of gaining new insights into the behavior of many phenomena that you will encounter in your chosen field of specialization in engineering or science.

The discipline of statistics teaches us how to make intelligent judgments and informed decisions in the presence of uncertainty and variation. Without uncertainty or variation, there would be little need for statistical methods or statisticians. If every component of a particular type had exactly the same lifetime, if all resistors produced by a certain manufacturer had the same resistance value, if pH determinations for soil specimens from a particular locale gave identical results, and so on, then a single observation would reveal all desired information.

An interesting manifestation of variation appeared in connection with determining the "greenest" way to travel. The article **"Carbon Conundrum" (*Consumer Reports*, 2008: 9)** identified organizations that help consumers calculate carbon output. The following results on output for a flight from New York to Los Angeles were reported:

Carbon Calculator	CO_2 (lb)
Terra Pass	1924
Conservation International	3000
Cool It	3049
World Resources Institute/Safe Climate	3163
National Wildlife Federation	3465
Sustainable Travel International	3577
Native Energy	3960
Environmental Defense	4000
Carbonfund.org	4820
The Climate Trust/CarbonCounter.org	5860
Bonneville Environmental Foundation	6732

There is clearly rather substantial disagreement among these calculators as to exactly how much carbon is emitted, characterized in the article as "from a ballerina's to Bigfoot's." A website address was provided where readers could learn more about how the various calculators work.

How can statistical techniques be used to gather information and draw conclusions? Suppose, for example, that a materials engineer has developed a coating for retarding corrosion in metal pipe under specified circumstances. If this coating is applied to different segments of pipe, variation in environmental conditions and in the segments themselves will result in more substantial corrosion on some segments than on others. Methods of statistical analysis could be used on data from such an experiment to decide whether the *average* amount of corrosion exceeds an upper specification limit of some sort or to predict how much corrosion will occur on a single piece of pipe.

Alternatively, suppose the engineer has developed the coating in the belief that it will be superior to the currently used coating. A comparative experiment could be carried out to investigate this issue by applying the current coating to some segments of pipe and the new coating to other segments. This must be done with care lest the wrong conclusion emerge. For example, perhaps the average amount of corrosion is identical for the two coatings. However, the new coating may be applied to segments that have superior ability to resist corrosion and under less stressful environmental conditions compared to the segments and conditions for the current coating. The investigator would then likely observe a difference between the two coatings attributable not to the coatings themselves, but just to extraneous variation. Statistics offers not only methods for analyzing the results of experiments once they have been carried out but also suggestions for how experiments can be performed in an efficient manner to mitigate the effects of variation and have a better chance of producing correct conclusions.

1.1 Populations, Samples, and Processes

Engineers and scientists are constantly exposed to collections of facts, or **data**, both in their professional capacities and in everyday activities. The discipline of statistics provides methods for organizing and summarizing data and for drawing conclusions based on information contained in the data.

An investigation will typically focus on a well-defined collection of objects constituting a **population** of interest. In one study, the population might consist of all gelatin capsules of a particular type produced during a specified period. Another investigation might involve the population consisting of all individuals who received a B.S. in engineering during the most recent academic year. When desired information is available for all objects in the population, we have what is called a **census**. Constraints on time, money, and other scarce resources usually make a census impractical or infeasible. Instead, a subset of the population—**a sample**—is selected in some prescribed manner. Thus we might obtain a sample of bearings from a particular production run as a basis for investigating whether bearings are conforming to manufacturing specifications, or we might select a sample of last year's engineering graduates to obtain feedback about the quality of the engineering curricula.

We are usually interested only in certain characteristics of the objects in a population: the number of flaws on the surface of each casing, the thickness of each capsule wall, the gender of an engineering graduate, the age at which the individual graduated, and so on. A characteristic may be categorical, such as gender or type of malfunction, or it may be numerical in nature. In the former case, the *value* of the characteristic is a category (e.g., female or insufficient solder), whereas in the latter case, the value is a number (e.g., age = 23 years or diameter = .502 cm). A **variable** is any characteristic whose value may change from one object to another in the population. We shall initially denote variables by lowercase letters from the end of our alphabet. Examples include

   x = brand of calculator owned by a student

   y = number of visits to a particular Web site during a specified period

   z = braking distance of an automobile under specified conditions

Data results from making observations either on a single variable or simultaneously on two or more variables. A **univariate** data set consists of observations on a single variable. For example, we might determine the type of transmission, automatic (A) or manual (M), on each of ten automobiles recently purchased at a certain dealership, resulting in the categorical data set

   M A A A M A A M A A

The following sample of pulse rates (beats per minute) for patients recently admitted to an adult intensive care unit is a numerical univariate data set:

   88 80 71 103 154 132 67 110 60 105

We have **bivariate** data when observations are made on each of two variables. Our data set might consist of a (height, weight) pair for each basketball player on a team, with the first observation as (72, 168), the second as (75, 212), and so on. If an engineer determines the value of both $x =$ component lifetime and $y =$ reason for component failure, the resulting data set is bivariate with one variable numerical and the other categorical. **Multivariate** data arises when observations are made on more than one variable (so bivariate is a special case of multivariate). For example, a research physician might determine the systolic blood pressure, diastolic blood pressure, and serum cholesterol level for each patient participating in a study. Each observation would be a triple of numbers, such as (120, 80, 146). In many multivariate data sets, some variables are numerical and others are categorical. Thus the annual automobile issue of *Consumer Reports* gives values of such variables as type of vehicle (small, sporty, compact, mid-size, large), city fuel efficiency (mpg), highway fuel efficiency (mpg), drivetrain type (rear wheel, front wheel, four wheel), and so on.

Branches of Statistics

An investigator who has collected data may wish simply to summarize and describe important features of the data. This entails using methods from **descriptive statistics**. Some of these methods are graphical in nature; the construction of histograms, boxplots, and scatter plots are primary examples. Other descriptive methods involve calculation of numerical summary measures, such as means, standard deviations, and correlation coefficients. The wide availability of statistical computer software packages has made these tasks much easier to carry out than they used to be. Computers are much more efficient than human beings at calculation and the creation of pictures (once they have received appropriate instructions from the user!). This means that the investigator doesn't have to expend much effort on "grunt work" and will have more time to study the data and extract important messages. Throughout this book, we will present output from various packages such as Minitab, SAS, JMP, and R. The R software can be downloaded without charge from the site **http://www.r-project.org**. It has achieved great popularity in the statistical community, and many books describing its various uses are available (it does entail programming as opposed to the pull-down menus of Minitab and JMP).

EXAMPLE 1.1 Charity is a big business in the United States. The Web site **charitynavigator.com** gives information on roughly 6000 charitable organizations, and there are many smaller charities that fly below the navigator's radar screen. Some charities operate very efficiently, with fundraising and administrative expenses that are only a small percentage of total expenses, whereas others spend a high percentage of what they take in on such activities. Here is data on fundraising expenses as a percentage of total expenditures for a random sample of 60 charities:

6.1	12.6	34.7	1.6	18.8	2.2	3.0	2.2	5.6	3.8
2.2	3.1	1.3	1.1	14.1	4.0	21.0	6.1	1.3	20.4
7.5	3.9	10.1	8.1	19.5	5.2	12.0	15.8	10.4	5.2
6.4	10.8	83.1	3.6	6.2	6.3	16.3	12.7	1.3	0.8
8.8	5.1	3.7	26.3	6.0	48.0	8.2	11.7	7.2	3.9
15.3	16.6	8.8	12.0	4.7	14.7	6.4	17.0	2.5	16.2

Without any organization, it is difficult to get a sense of the data's most prominent features—what a typical (i.e., representative) value might be, whether values are highly concentrated about a typical value or quite dispersed, whether there are any gaps in the data, what fraction of the values are less than 20%, and so on. Figure 1.1

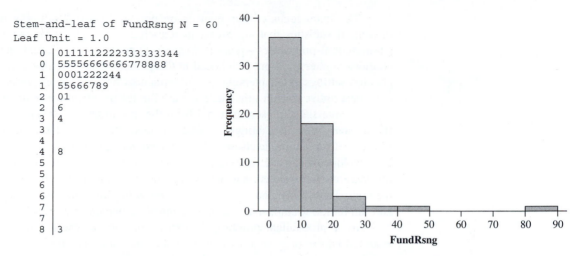

```
Stem-and-leaf of FundRsng N = 60
Leaf Unit = 1.0
    0 | 0111112222333333344
    0 | 55556666666778888
    1 | 0001222244
    1 | 55666789
    2 | 01
    2 | 6
    3 | 4
    3 |
    4 |
    4 | 8
    5 |
    5 |
    6 |
    6 |
    7 |
    7 |
    8 | 3
```

Figure 1.1 A Minitab stem-and-leaf display (tenths digit truncated) and histogram for the charity fundraising percentage data

shows what is called a *stem-and-leaf display* as well as a *histogram.* In Section 1.2 we will discuss construction and interpretation of these data summaries. For the moment, we hope you see how they begin to describe how the percentages are distributed over the range of possible values from 0 to 100. Clearly a substantial majority of the charities in the sample spend less than 20% on fundraising, and only a few percentages might be viewed as beyond the bounds of sensible practice. ■

Having obtained a sample from a population, an investigator would frequently like to use sample information to draw some type of conclusion (make an inference of some sort) about the population. That is, the sample is a means to an end rather than an end in itself. Techniques for generalizing from a sample to a population are gathered within the branch of our discipline called **inferential statistics**.

EXAMPLE 1.2 Material strength investigations provide a rich area of application for statistical methods. The article **"Effects of Aggregates and Microfillers on the Flexural Properties of Concrete"** (*Magazine of Concrete Research*, **1997: 81–98**) reported on a study of strength properties of high-performance concrete obtained by using superplasticizers and certain binders. The compressive strength of such concrete had previously been investigated, but not much was known about flexural strength (a measure of ability to resist failure in bending). The accompanying data on flexural strength (in MegaPascal, MPa, where 1 Pa (Pascal) = 1.45×10^{-4} psi) appeared in the article cited:

| 5.9 | 7.2 | 7.3 | 6.3 | 8.1 | 6.8 | 7.0 | 7.6 | 6.8 | 6.5 | 7.0 | 6.3 | 7.9 | 9.0 |

| 8.2 | 8.7 | 7.8 | 9.7 | 7.4 | 7.7 | 9.7 | 7.8 | 7.7 | 11.6 | 11.3 | 11.8 | 10.7 |

Suppose we want an *estimate* of the average value of flexural strength for all beams that could be made in this way (if we conceptualize a population of all such beams, we are trying to estimate the population mean). It can be shown that, with a high degree of confidence, the population mean strength is between 7.48 MPa and 8.80 MPa; we call this a *confidence interval* or *interval estimate.* Alternatively, this data could be used to predict the flexural strength of a *single* beam of this type. With a high degree of confidence, the strength of a single such beam will exceed 7.35 MPa; the number 7.35 is called a *lower prediction bound.* ■

The main focus of this book is on presenting and illustrating methods of inferential statistics that are useful in scientific work. The most important types of inferential procedures—point estimation, hypothesis testing, and estimation by confidence intervals—are introduced in Chapters 6–8 and then used in more complicated settings in Chapters 9–16. The remainder of this chapter presents methods from descriptive statistics that are most used in the development of inference.

Chapters 2–5 present material from the discipline of probability. This material ultimately forms a bridge between the descriptive and inferential techniques. Mastery of probability leads to a better understanding of how inferential procedures are developed and used, how statistical conclusions can be translated into everyday language and interpreted, and when and where pitfalls can occur in applying the methods. Probability and statistics both deal with questions involving populations and samples, but do so in an "inverse manner" to one another.

In a probability problem, properties of the population under study are assumed known (e.g., in a numerical population, some specified distribution of the population values may be assumed), and questions regarding a sample taken from the population are posed and answered. In a statistics problem, characteristics of a sample are available to the experimenter, and this information enables the experimenter to draw conclusions about the population. The relationship between the two disciplines can be summarized by saying that probability reasons from the population to the sample (deductive reasoning), whereas inferential statistics reasons from the sample to the population (inductive reasoning). This is illustrated in Figure 1.2.

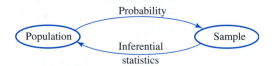

Figure 1.2 The relationship between probability and inferential statistics

Before we can understand what a particular sample can tell us about the population, we should first understand the uncertainty associated with taking a sample from a given population. This is why we study probability before statistics.

EXAMPLE 1.3 As an example of the contrasting focus of probability and inferential statistics, consider drivers' use of manual lap belts in cars equipped with automatic shoulder belt systems. (The article **"Automobile Seat Belts: Usage Patterns in Automatic Belt Systems," *Human Factors*, 1998: 126–135,** summarizes usage data.) In probability, we might assume that 50% of all drivers of cars equipped in this way in a certain metropolitan area regularly use their lap belt (an assumption about the population), so we might ask, "How likely is it that a sample of 100 such drivers will include at least 70 who regularly use their lap belt?" or "How many of the drivers in a sample of size 100 can we expect to regularly use their lap belt?" On the other hand, in inferential statistics, we have sample information available; for example, a sample of 100 drivers of such cars revealed that 65 regularly use their lap belt. We might then ask, "Does this provide substantial evidence for concluding that more than 50% of all such drivers in this area regularly use their lap belt?" In this latter scenario, we are attempting to use sample information to answer a question about the structure of the entire population from which the sample was selected. ∎

In the foregoing lap belt example, the population is well defined and concrete: all drivers of cars equipped in a certain way in a particular metropolitan area. In Example 1.2, however, the strength measurements came from a sample of prototype beams that had not been selected from an existing population. Instead, it is convenient to think of the population as consisting of all possible strength measurements that might be made under similar experimental conditions. Such a population is referred to as a **conceptual** or **hypothetical population**. There are a number of problem situations in which we fit questions into the framework of inferential statistics by conceptualizing a population.

The Scope of Modern Statistics

These days statistical methodology is employed by investigators in virtually all disciplines, including such areas as

- molecular biology (analysis of microarray data)
- ecology (describing quantitatively how individuals in various animal and plant populations are spatially distributed)
- materials engineering (studying properties of various treatments to retard corrosion)
- marketing (developing market surveys and strategies for marketing new products)
- public health (identifying sources of diseases and ways to treat them)
- civil engineering (assessing the effects of stress on structural elements and the impacts of traffic flows on communities)

As you progress through the book, you'll encounter a wide spectrum of different scenarios in the examples and exercises that illustrate the application of techniques from probability and statistics. Many of these scenarios involve data or other material extracted from articles in engineering and science journals. The methods presented herein have become established and trusted tools in the arsenal of those who work with data. Meanwhile, statisticians continue to develop new models for describing randomness, and uncertainty and new methodology for analyzing data. As evidence of the continuing creative efforts in the statistical community, here are titles and capsule descriptions of some articles that have recently appeared in statistics journals (*Journal of the American Statistical Association* is abbreviated *JASA*, and *AAS* is short for the *Annals of Applied Statistics,* two of the many prominent journals in the discipline):

- **"How Many People Do You Know? Efficiently Estimating Personal Network Size"** (*JASA,* **2010: 59–70):** How many of the N individuals at your college do you know? You could select a random sample of students from the population and use an estimate based on the fraction of people in this sample that you know. Unfortunately this is very inefficient for large populations because the fraction of the population someone knows is typically very small. A "latent mixing model" was proposed that the authors asserted remedied deficiencies in previously used techniques. A simulation study of the method's effectiveness based on groups consisting of first names ("How many people named Michael do you know?") was included as well as an application of the method to actual survey data. The article concluded with some practical guidelines for the construction of future surveys designed to estimate social network size.

- **"Active Learning Through Sequential Design, with Applications to the Detection of Money Laundering"** (*JASA,* **2009: 969–981):** Money laundering involves concealing the origin of funds obtained through illegal activities. The huge number of transactions occurring daily at financial institutions makes detection of money laundering difficult. The standard approach has been to extract various summary quantities from the transaction history and conduct a time-consuming investigation of suspicious activities. The article proposes a more efficient statistical method and illustrates its use in a case study.

- **"Robust Internal Benchmarking and False Discovery Rates for Detecting Racial Bias in Police Stops"** (*JASA*, 2009: 661–668): Allegations of police actions that are attributable at least in part to racial bias have become a contentious issue in many communities. This article proposes a new method that is designed to reduce the risk of flagging a substantial number of "false positives" (individuals falsely identified as manifesting bias). The method was applied to data on 500,000 pedestrian stops in New York City in 2006; of the 3000 officers regularly involved in pedestrian stops, 15 were identified as having stopped a substantially greater fraction of Black and Hispanic people than what would be predicted were bias absent.

- **"Records in Athletics Through Extreme Value Theory"** (*JASA*, 2008: 1382–1391): The focus here is on the modeling of extremes related to world records in athletics. The authors start by posing two questions: (1) What is the ultimate world record within a specific event (e.g., the high jump for women)? and (2) How "good" is the current world record, and how does the quality of current world records compare across different events? A total of 28 events (8 running, 3 throwing, and 3 jumping for both men and women) are considered. For example, one conclusion is that only about 20 seconds can be shaved off the men's marathon record, but that the current women's marathon record is almost 5 minutes longer than what can ultimately be achieved. The methodology also has applications to such issues as ensuring airport runways are long enough and that dikes in Holland are high enough.

- **"Self-Exciting Hurdle Models for Terrorist Activity"** (*AAS*, 2012: 106–124): The authors developed a predictive model of terrorist activity by considering the daily number of terrorist attacks in Indonesia from 1994 through 2007. The model estimates the chance of future attacks as a function of the times since past attacks. One feature of the model considers the excess of nonattack days coupled with the presence of multiple coordinated attacks on the same day. The article provides an interpretation of various model characteristics and assesses its predictive performance.

- **"Prediction of Remaining Life of Power Transformers Based on Left Truncated and Right Censored Lifetime Data"** (*AAS*, 2009: 857–879): There are roughly 150,000 high-voltage power transmission transformers in the United States. Unexpected failures can cause substantial economic losses, so it is important to have predictions for remaining lifetimes. Relevant data can be complicated because lifetimes of some transformers extend over several decades during which records were not necessarily complete. In particular, the authors of the article use data from a certain energy company that began keeping careful records in 1980. But some transformers had been installed before January 1, 1980, and were still in service after that date ("left truncated" data), whereas other units were still in service at the time of the investigation, so their complete lifetimes are not available ("right censored" data). The article describes various procedures for obtaining an interval of plausible values (a *prediction interval*) for a remaining lifetime and for the cumulative number of failures over a specified time period.

- **"The BARISTA: A Model for Bid Arrivals in Online Auctions"** (*AAS*, 2007: 412–441): Online auctions such as those on eBay and uBid often have characteristics that differentiate them from traditional auctions. One particularly important difference is that the number of bidders at the outset of many traditional auctions is fixed, whereas in online auctions this number and the number of resulting bids are not predetermined. The article proposes a new BARISTA (for Bid ARrivals In STAges) model for describing the way in which bids arrive online. The model allows for higher bidding intensity at the outset of the auction and also as the auction comes to a close. Various properties of the model are investigated and then validated using data from eBay.com on auctions for Palm M515 personal assistants, Microsoft Xbox games, and Cartier watches.

- **"Statistical Challenges in the Analysis of Cosmic Microwave Background Radiation"** (*AAS*, 2009: 61–95): The cosmic microwave background (CMB) is a significant source of information about the early history of the universe. Its radiation level is uniform, so extremely delicate instruments have been developed to measure fluctuations. The authors provide a review of statistical issues with CMB data analysis; they also give many examples of the application of statistical procedures to data obtained from a recent NASA satellite mission, the *Wilkinson Microwave Anisotropy Probe.*

Statistical information now appears with increasing frequency in the popular media, and occasionally the spotlight is even turned on statisticians. For example, the **Nov. 23, 2009,** *New York Times* reported in an article "Behind Cancer Guidelines, Quest for Data" that the new science for cancer investigations and more sophisticated methods for data analysis spurred the U.S. Preventive Services task force to re-examine guidelines for how frequently middle-aged and older women should have mammograms. The panel commissioned six independent groups to do statistical modeling. The result was a new set of conclusions, including an assertion that mammograms every two years are nearly as beneficial to patients as annual mammograms, but confer only half the risk of harms. Donald Berry, a very prominent biostatistician, was quoted as saying he was pleasantly surprised that the task force took the new research to heart in making its recommendations. The task force's report has generated much controversy among cancer organizations, politicians, and women themselves.

It is our hope that you will become increasingly convinced of the importance and relevance of the discipline of statistics as you dig more deeply into the book and the subject. Hopefully you'll be turned on enough to want to continue your statistical education beyond your current course.

Enumerative Versus Analytic Studies

W. E. Deming, a very influential American statistician who was a moving force in Japan's quality revolution during the 1950s and 1960s, introduced the distinction between *enumerative studies* and *analytic studies.* In the former, interest is focused on a finite, identifiable, unchanging collection of individuals or objects that make up a population. A *sampling frame*—that is, a listing of the individuals or objects to be sampled—is either available to an investigator or else can be constructed. For example, the frame might consist of all signatures on a petition to qualify a certain initiative for the ballot in an upcoming election; a sample is usually selected to ascertain whether the number of *valid* signatures exceeds a specified value. As another example, the frame may contain serial numbers of all furnaces manufactured by a particular company during a certain time period; a sample may be selected to infer something about the average lifetime of these units. The use of inferential methods to be developed in this book is reasonably noncontroversial in such settings (though statisticians may still argue over which particular methods should be used).

An analytic study is broadly defined as one that is not enumerative in nature. Such studies are often carried out with the objective of improving a future product by taking action on a process of some sort (e.g., recalibrating equipment or adjusting the level of some input such as the amount of a catalyst). Data can often be obtained only on an existing process, one that may differ in important respects from the future process. There is thus no sampling frame listing the individuals or objects of interest. For example, a sample of five turbines with a new design may be experimentally manufactured and tested to investigate efficiency. These five could be viewed as a sample from the conceptual population of all prototypes that could be manufactured under similar conditions, but *not* necessarily as representative of the population of units manufactured once regular production gets underway. Methods for using sample information to draw conclusions about future production units may be problematic. Someone with expertise in the area of turbine design and engineering (or whatever other subject area is relevant) should be called upon to judge whether such extrapolation is sensible. A good exposition of these issues is contained in the article **"Assumptions for Statistical Inference" by Gerald Hahn and William Meeker (*The American Statistician*, 1993: 1–11).**

Collecting Data

Statistics deals not only with the organization and analysis of data once it has been collected but also with the development of techniques for collecting the data. If data is not properly collected, an investigator may not be able to answer the questions under consideration with a reasonable degree of confidence. One common problem is that the target population—the one about which conclusions are to be drawn—may be different from the population actually sampled. For example, advertisers would like various kinds of information about the television-viewing habits of potential customers. The most systematic information of this sort comes from placing monitoring devices in a small number of homes across the United States. It has been conjectured that placement of such devices in and of itself alters viewing behavior, so that characteristics of the sample may be different from those of the target population.

When data collection entails selecting individuals or objects from a frame, the simplest method for ensuring a representative selection is to take a *simple random sample*. This is one for which any particular subset of the specified size (e.g., a sample of size 100) has the same chance of being selected. For example, if the frame consists of 1,000,000 serial numbers, the numbers 1, 2, . . . , up to 1,000,000 could be placed on identical slips of paper. After placing these slips in a box and thoroughly mixing, slips could be drawn one by one until the requisite sample size has been obtained. Alternatively (and much to be preferred), a table of random numbers or a software package's random number generator could be employed.

Sometimes alternative sampling methods can be used to make the selection process easier, to obtain extra information, or to increase the degree of confidence in conclusions. One such method, *stratified sampling,* entails separating the population units into nonoverlapping groups and taking a sample from each one. For example, a study of how physicians feel about the Affordable Care Act might proceed by stratifying according to specialty: select a sample of surgeons, another sample of radiologists, yet another sample of psychiatrists, and so on. This would result in information separately from each specialty and ensure that no one specialty is over- or underrepresented in the entire sample.

Frequently a "convenience" sample is obtained by selecting individuals or objects without systematic randomization. As an example, a collection of bricks may be stacked in such a way that it is extremely difficult for those in the center to be selected. If the bricks on the top and sides of the stack were somehow different from the others, resulting sample data would not be representative of the population. Often an investigator will assume that such a convenience sample approximates a random sample, in which case a statistician's repertoire of inferential methods can be used; however, this is a judgment call. Most of the methods discussed herein are based on a variation of simple random sampling described in Chapter 5.

Engineers and scientists often collect data by carrying out some sort of designed experiment. This may involve deciding how to allocate several different treatments (such as fertilizers or coatings for corrosion protection) to the various experimental units (plots of land or pieces of pipe). Alternatively, an investigator may systematically vary the levels or categories of certain factors (e.g., pressure or type of insulating material) and observe the effect on some response variable (such as yield from a production process).

EXAMPLE 1.4   An article in the *New York Times* **(Jan. 27, 1987)** reported that heart attack risk could be reduced by taking aspirin. This conclusion was based on a designed experiment involving both a control group of individuals that took a placebo having the appearance of aspirin but known to be inert and a treatment group that took aspirin according to a specified regimen. Subjects were randomly assigned to the groups to protect against any biases and so that probability-based methods could be used to analyze the data. Of the 11,034 individuals in the control group, 189 subsequently experienced heart attacks, whereas only 104 of the 11,037 in the aspirin group had a heart attack. The incidence rate of heart attacks in the treatment group was only about half that in the control group. One possible explanation for this result is chance variation—that aspirin really doesn't have the desired effect and the observed difference is just typical variation in the same way that tossing two identical coins would usually produce different numbers of heads. However, in this case, inferential methods suggest that chance variation by itself cannot adequately explain the magnitude of the observed difference. ∎

EXAMPLE 1.5   An engineer wishes to investigate the effects of both adhesive type and conductor material on bond strength when mounting an integrated circuit (IC) on a certain substrate. Two adhesive types and two conductor materials are under consideration. Two observations are made for each adhesive-type/conductor-material combination, resulting in the accompanying data:

Adhesive Type	Conductor Material	Observed Bond Strength	Average
1	1	82, 77	79.5
1	2	75, 87	81.0
2	1	84, 80	82.0
2	2	78, 90	84.0

The resulting average bond strengths are pictured in Figure 1.3. It appears that adhesive type 2 improves bond strength as compared with type 1 by about the same amount whichever one of the conducting materials is used, with the 2, 2 combination being best. Inferential methods can again be used to judge whether these effects are real or simply due to chance variation.

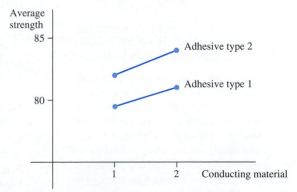

Figure 1.3   Average bond strengths in Example 1.5

Suppose additionally that there are two cure times under consideration and also two types of IC post coating. There are then $2 \cdot 2 \cdot 2 \cdot 2 = 16$ combinations of these four factors, and our engineer may not have enough resources to make even a single observation for each of these combinations. In Chapter 11, we will see how the careful selection of a fraction of these possibilities will usually yield the desired information. ∎

EXERCISES Section 1.1 (1–9)

1. Give one possible sample of size 4 from each of the following populations:
   a. All daily newspapers published in the United States
   b. All companies listed on the New York Stock Exchange
   c. All students at your college or university
   d. All grade point averages of students at your college or university

2. For each of the following hypothetical populations, give a plausible sample of size 4:
   a. All distances that might result when you throw a football
   b. Page lengths of books published 5 years from now
   c. All possible earthquake-strength measurements (Richter scale) that might be recorded in California during the next year
   d. All possible yields (in grams) from a certain chemical reaction carried out in a laboratory

3. Consider the population consisting of all computers of a certain brand and model, and focus on whether a computer needs service while under warranty.
   a. Pose several probability questions based on selecting a sample of 100 such computers.
   b. What inferential statistics question might be answered by determining the number of such computers in a sample of size 100 that need warranty service?

4. a. Give three different examples of concrete populations and three different examples of hypothetical populations.
   b. For one each of your concrete and your hypothetical populations, give an example of a probability question and an example of an inferential statistics question.

5. Many universities and colleges have instituted supplemental instruction (SI) programs, in which a student facilitator meets regularly with a small group of students enrolled in the course to promote discussion of course material and enhance subject mastery. Suppose that students in a large statistics course (what else?) are randomly divided into a control group that will not participate in SI and a treatment group that will participate. At the end of the term, each student's total score in the course is determined.
   a. Are the scores from the SI group a sample from an existing population? If so, what is it? If not, what is the relevant conceptual population?

   b. What do you think is the advantage of randomly dividing the students into the two groups rather than letting each student choose which group to join?
   c. Why didn't the investigators put all students in the treatment group? [*Note:* The article **"Supplemental Instruction: An Effective Component of Student Affairs Programming" (*J. of College Student Devel.*, 1997: 577–586)** discusses the analysis of data from several SI programs.]

6. The California State University (CSU) system consists of 23 campuses, from San Diego State in the south to Humboldt State near the Oregon border. A CSU administrator wishes to make an inference about the average distance between the hometowns of students and their campuses. Describe and discuss several different sampling methods that might be employed. Would this be an enumerative or an analytic study? Explain your reasoning.

7. A certain city divides naturally into ten district neighborhoods. How might a real estate appraiser select a sample of single-family homes that could be used as a basis for developing an equation to predict appraised value from characteristics such as age, size, number of bathrooms, distance to the nearest school, and so on? Is the study enumerative or analytic?

8. The amount of flow through a solenoid valve in an automobile's pollution-control system is an important characteristic. An experiment was carried out to study how flow rate depended on three factors: armature length, spring load, and bobbin depth. Two different levels (low and high) of each factor were chosen, and a single observation on flow was made for each combination of levels.
   a. The resulting data set consisted of how many observations?
   b. Is this an enumerative or analytic study? Explain your reasoning.

9. In a famous experiment carried out in 1882, Michelson and Newcomb obtained 66 observations on the time it took for light to travel between two locations in Washington, D.C. A few of the measurements (coded in a certain manner) were 31, 23, 32, 36, −2, 26, 27, and 31.
   a. Why are these measurements not identical?
   b. Is this an enumerative study? Why or why not?

1.2 Pictorial and Tabular Methods in Descriptive Statistics

Descriptive statistics can be divided into two general subject areas. In this section, we consider representing a data set using visual displays. In Sections 1.3 and 1.4, we will develop some numerical summary measures for data sets. Many visual techniques may already be familiar to you: frequency tables, tally sheets, histograms, pie charts, bar graphs, scatter diagrams, and the like. Here we focus on a selected few of these techniques that are most useful and relevant to probability and inferential statistics.

Notation

Some general notation will make it easier to apply our methods and formulas to a wide variety of practical problems. The number of observations in a single sample, that is, the *sample size,* will often be denoted by n, so that $n = 4$ for the sample of universities {Stanford, Iowa State, Wyoming, Rochester} and also for the sample of pH measurements {6.3, 6.2, 5.9, 6.5}. If two samples are simultaneously under consideration, either m and n or n_1 and n_2 can be used to denote the numbers of observations. An experiment to compare thermal efficiencies for two different types of diesel engines might result in samples {29.7, 31.6, 30.9} and {28.7, 29.5, 29.4, 30.3}, in which case $m = 3$ and $n = 4$.

Given a data set consisting of n observations on some variable x, the individual observations will be denoted by $x_1, x_2, x_3, \ldots, x_n$. The subscript bears no relation to the magnitude of a particular observation. Thus x_1 will not in general be the smallest observation in the set, nor will x_n typically be the largest. In many applications, x_1 will be the first observation gathered by the experimenter, x_2 the second, and so on. The ith observation in the data set will be denoted by x_i.

Stem-and-Leaf Displays

Consider a numerical data set $x_1, x_2, \ldots, x_n$ for which each x_i consists of at least two digits. A quick way to obtain an informative visual representation of the data set is to construct a *stem-and-leaf display.*

Constructing a Stem-and-Leaf Display

1. Select one or more leading digits for the stem values. The trailing digits become the leaves.
2. List possible stem values in a vertical column.
3. Record the leaf for each observation beside the corresponding stem value.
4. Indicate the units for stems and leaves someplace in the display.

For a data set consisting of exam scores, each between 0 and 100, the score of 83 would have a stem of 8 and a leaf of 3. If all exam scores are in the 90s, 80s, and 70s (an instructor's dream!), use of the tens digit as the stem would give a display with only three rows. In this case, it is desirable to stretch the display by repeating each stem value twice—9H, 9L, 8H, . . . ,7L—once for high leaves 9, . . . , 5 and again for low leaves 4, . . . , 0. Then a score of 93 would have a stem of 9L and leaf of 3. In general, a display based on between 5 and 20 stems is recommended.

EXAMPLE 1.6　A common complaint among college students is that they are getting less sleep than they need. The article **"Class Start Times, Sleep, and Academic Performance in College: A Path Analysis" (*Chronobiology Intl.*, 2012: 318–335)** investigated factors that impact sleep time. The stem-and-leaf display in Figure 1.4 shows the average number of hours of sleep per day over a two-week period for a sample of 253 students.

5L	00
5H	6889
6L	000111123444444
6H	55556778899999
7L	000011111112222223333333344444444
7H	555555555666666666666677777788888888889999999999999999
8L	000000000000011111122222222222222222222333333333333344444444444444
8H	5555555556666666666777777788888888899999999999
9L	00001111111222223334
9H	666678999
10L	00
10H	56

Stem: ones digit
Leaf: tenths digit

Figure 1.4　Stem-and-leaf display for average sleep time per day

The first observation in the top row of the display is 5.0, corresponding to a stem of 5 and leaf of 0, and the last observation at the bottom of the display is 10.6. Note that in the absence of a context, without the identification of stem and leaf digits in the display, we wouldn't know whether the observation with stem 7 and leaf 9 was .79, 7.9, or 79. The leaves in each row are ordered from smallest to largest; this is commonly done by software packages but is not necessary if a display is created by hand.

The display suggests that a typical or representative sleep time is in the stem 8L row, perhaps 8.1 or 8.2. The data is not highly concentrated about this typical value as would be the case if almost all students were getting between 7.5 and 9.5 hours of sleep on average. The display appears to rise rather smoothly to a peak in the 8L row and then decline smoothly (we conjecture that the minor peak in the 6L row would disappear if more data was available). The general shape of the display is rather symmetric, bearing strong resemblance to a bell-shaped curve; it does not stretch out more in one direction than the other. The two smallest and two largest values seem a bit separated from the remainder of the data—perhaps they are very mild, but certainly not extreme, "outliers". A reference in the cited article suggests that individuals in this age group need about 8.4 hours of sleep per day. So it appears that a substantial percentage of students in the sample are sleep deprived. ■

A stem-and-leaf display conveys information about the following aspects of the data:

- identification of a typical or representative value
- extent of spread about the typical value
- presence of any gaps in the data

- extent of symmetry in the distribution of values
- number and locations of peaks
- presence of any *outliers*—values far from the rest of the data

EXAMPLE 1.7 Figure 1.5 presents stem-and-leaf displays for a random sample of lengths of golf courses (yards) that have been designated by *Golf Magazine* as among the most challenging in the United States. Among the sample of 40 courses, the shortest is 6433 yards long, and the longest is 7280 yards. The lengths appear to be distributed in a roughly uniform fashion over the range of values in the sample. Notice that a stem choice here of either a single digit (6 or 7) or three digits (643, ... , 728) would yield an uninformative display, the first because of too few stems and the latter because of too many.

64	35	64	33	70			Stem: Thousands and hundreds digits
65	26	27	06	83			Leaf: Tens and ones digits
66	05	94	14				
67	90	70	00	98	70	45	13
68	90	70	73	50			
69	00	27	36	04			
70	51	05	11	40	50	22	
71	31	69	68	05	13	65	
72	80	09					

```
Stem-and-leaf of yardage   N = 40
Leaf Unit = 10
    4    64  3367
    8    65  0228
   11    66  019
   18    67  0147799
   (4)   68  5779
   18    69  0023
   14    70  012455
    8    71  013666
    2    72  08
```

(a)

(b)

Figure 1.5 Stem-and-leaf displays of golf course lengths: (a) two-digit leaves; (b) display from Minitab with truncated one-digit leaves

Statistical software packages do not generally produce displays with multiple-digit stems. The Minitab display in Figure 1.5(b) results from *truncating* each observation by deleting the ones digit. ∎

Dotplots

A dotplot is an attractive summary of numerical data when the data set is reasonably small or there are relatively few distinct data values. Each observation is represented by a dot above the corresponding location on a horizontal measurement scale. When a value occurs more than once, there is a dot for each occurrence, and these dots are stacked vertically. As with a stem-and-leaf display, a dotplot gives information about location, spread, extremes, and gaps.

EXAMPLE 1.8 There is growing concern in the U.S. that not enough students are graduating from college. America used to be number 1 in the world for the percentage of adults with college degrees, but it has recently dropped to 16th. Here is data on the percentage of 25- to 34-year-olds in each state who had some type of postsecondary degree as of 2010 (listed in alphabetical order, with the District of Columbia included):

31.5	32.9	33.0	28.6	37.9	43.3	45.9	37.2	68.8	36.2	35.5
40.5	37.2	45.3	36.1	45.5	42.3	33.3	30.3	37.2	45.5	54.3
37.2	49.8	32.1	39.3	40.3	44.2	28.4	46.0	47.2	28.7	49.6
37.6	50.8	38.0	30.8	37.6	43.9	42.5	35.2	42.2	32.8	32.2
38.5	44.5	44.6	40.9	29.5	41.3	35.4				

Figure 1.6 shows a dotplot of the data. Dots corresponding to some values close together (e.g., 28.6 and 28.7) have been vertically stacked to prevent crowding. There is clearly a great deal of state-to-state variability. The largest value, for D.C., is obviously an extreme outlier, and four other values on the upper end of the data are candidates for mild outliers (MA, MN, NY, and ND). There is also a cluster of states at the low end, primarily located in the South and Southwest. The overall percentage for the entire country is 39.3%; this is not a simple average of the 51 numbers but an average weighted by population sizes.

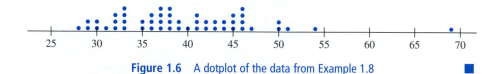

Figure 1.6　A dotplot of the data from Example 1.8　■

A dotplot can be quite cumbersome to construct and look crowded when the number of observations is large. Our next technique is well suited to such situations.

Histograms

Some numerical data is obtained by counting to determine the value of a variable (the number of traffic citations a person received during the last year, the number of customers arriving for service during a particular period), whereas other data is obtained by taking measurements (weight of an individual, reaction time to a particular stimulus). The prescription for drawing a histogram is generally different for these two cases.

DEFINITION

> A numerical variable is **discrete** if its set of possible values either is finite or else can be listed in an infinite sequence (one in which there is a first number, a second number, and so on). A numerical variable is **continuous** if its possible values consist of an entire interval on the number line.

A discrete variable x almost always results from counting, in which case possible values are 0, 1, 2, 3, … or some subset of these integers. Continuous variables arise from making measurements. For example, if x is the pH of a chemical substance, then in theory x could be any number between 0 and 14: 7.0, 7.03, 7.032, and so on. Of course, in practice there are limitations on the degree of accuracy of any measuring instrument, so we may not be able to determine pH, reaction time, height, and concentration to an arbitrarily large number of decimal places. However, from the point of view of creating mathematical models for distributions of data, it is helpful to imagine an entire continuum of possible values.

Consider data consisting of observations on a discrete variable x. The **frequency** of any particular x value is the number of times that value occurs in the data set. The **relative frequency** of a value is the fraction or proportion of times the value occurs:

$$\text{relative frequency of a value} = \frac{\text{number of times the value occurs}}{\text{number of observations in the data set}}$$

Suppose, for example, that our data set consists of 200 observations on $x =$ the number of courses a college student is taking this term. If 70 of these x values are 3, then

frequency of the x value 3:　70

relative frequency of the x value 3:　$\dfrac{70}{200} = .35$

Multiplying a relative frequency by 100 gives a percentage; in the college-course example, 35% of the students in the sample are taking three courses. The relative frequencies, or percentages, are usually of more interest than the frequencies themselves. In theory, the relative frequencies should sum to 1, but in practice the sum may differ slightly from 1 because of rounding. A **frequency distribution** is a tabulation of the frequencies and/or relative frequencies.

> ### Constructing a Histogram for Discrete Data
>
> First, determine the frequency and relative frequency of each x value. Then mark possible x values on a horizontal scale. Above each value, draw a rectangle whose height is the relative frequency (or alternatively, the frequency) of that value; the rectangles should have equal widths.

This construction ensures that the *area* of each rectangle is proportional to the relative frequency of the value. Thus if the relative frequencies of $x = 1$ and $x = 5$ are .35 and .07, respectively, then the area of the rectangle above 1 is five times the area of the rectangle above 5.

EXAMPLE 1.9 How unusual is a no-hitter or a one-hitter in a major league baseball game, and how frequently does a team get more than 10, 15, or even 20 hits? Table 1.1 is a frequency distribution for the number of hits per team per game for all nine-inning games that were played between 1989 and 1993.

Table 1.1 Frequency Distribution for Hits in Nine-Inning Games

Hits/Game	Number of Games	Relative Frequency	Hits/Game	Number of Games	Relative Frequency
0	20	.0010	14	569	.0294
1	72	.0037	15	393	.0203
2	209	.0108	16	253	.0131
3	527	.0272	17	171	.0088
4	1048	.0541	18	97	.0050
5	1457	.0752	19	53	.0027
6	1988	.1026	20	31	.0016
7	2256	.1164	21	19	.0010
8	2403	.1240	22	13	.0007
9	2256	.1164	23	5	.0003
10	1967	.1015	24	1	.0001
11	1509	.0779	25	0	.0000
12	1230	.0635	26	1	.0001
13	834	.0430	27	1	.0001
				19,383	1.0005

The corresponding histogram in Figure 1.7 rises rather smoothly to a single peak and then declines. The histogram extends a bit more on the right (toward large values) than it does on the left—a slight "positive skew."

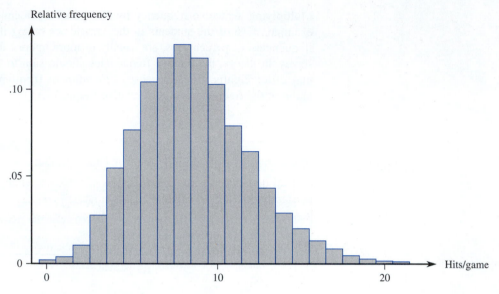

Figure 1.7 Histogram of number of hits per nine-inning game

Either from the tabulated information or from the histogram itself, we can determine the following:

$$
\begin{array}{l}
\text{proportion of games with} \\
\text{at most two hits}
\end{array}
=
\begin{array}{c}
\text{relative} \\
\text{frequency} \\
\text{for } x = 0
\end{array}
+
\begin{array}{c}
\text{relative} \\
\text{frequency} \\
\text{for } x = 1
\end{array}
+
\begin{array}{c}
\text{relative} \\
\text{frequency} \\
\text{for } x = 2
\end{array}
$$

$$= .0010 + .0037 + .0108 = .0155$$

Similarly,

$$
\begin{array}{l}
\text{proportion of games with} \\
\text{between 5 and 10 hits (inclusive)}
\end{array}
= .0752 + .1026 + \cdots + .1015 = .6361
$$

That is, roughly 64% of all these games resulted in between 5 and 10 (inclusive) hits. ■

Constructing a histogram for continuous data (measurements) entails subdividing the measurement axis into a suitable number of **class intervals** or **classes**, such that each observation is contained in exactly one class. Suppose, for example, that we have 50 observations on x = fuel efficiency of an automobile (mpg), the smallest of which is 27.8 and the largest of which is 31.4. Then we could use the class boundaries 27.5, 28.0, 28.5, … , and 31.5 as shown here:

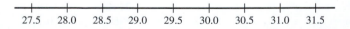

One potential difficulty is that occasionally an observation lies on a class boundary so therefore does not fall in exactly one interval, for example, 29.0. One way to deal with this problem is to use boundaries like 27.55, 28.05, … , 31.55. Adding a hundredths digit to the class boundaries prevents observations from falling on the resulting boundaries. Another approach is to use the classes $27.5{-}<28.0$, $28.0{-}<28.5$, … , $31.0{-}<31.5$. Then 29.0 falls in the class $29.0{-}<29.5$ rather than in the class $28.5{-}<29.0$. In other words, with this convention, an observation on a boundary is placed in the interval to the *right* of the boundary. This is how Minitab constructs a histogram.

> ### Constructing a Histogram for Continuous Data: Equal Class Widths
>
> Determine the frequency and relative frequency for each class. Mark the class boundaries on a horizontal measurement axis. Above each class interval, draw a rectangle whose height is the corresponding relative frequency (or frequency).

EXAMPLE 1.10 Power companies need information about customer usage to obtain accurate forecasts of demands. Investigators from Wisconsin Power and Light determined energy consumption (BTUs) during a particular period for a sample of 90 gas-heated homes. An adjusted consumption value was calculated as follows:

$$\text{adjusted consumption} = \frac{\text{consumption}}{(\text{weather, in degree days})(\text{house area})}$$

This resulted in the accompanying data (part of the stored data set FURNACE.MTW available in Minitab), which we have ordered from smallest to largest.

2.97	4.00	5.20	5.56	5.94	5.98	6.35	6.62	6.72	6.78
6.80	6.85	6.94	7.15	7.16	7.23	7.29	7.62	7.62	7.69
7.73	7.87	7.93	8.00	8.26	8.29	8.37	8.47	8.54	8.58
8.61	8.67	8.69	8.81	9.07	9.27	9.37	9.43	9.52	9.58
9.60	9.76	9.82	9.83	9.83	9.84	9.96	10.04	10.21	10.28
10.28	10.30	10.35	10.36	10.40	10.49	10.50	10.64	10.95	11.09
11.12	11.21	11.29	11.43	11.62	11.70	11.70	12.16	12.19	12.28
12.31	12.62	12.69	12.71	12.91	12.92	13.11	13.38	13.42	13.43
13.47	13.60	13.96	14.24	14.35	15.12	15.24	16.06	16.90	18.26

The most striking feature of the histogram in Figure 1.8 is its resemblance to a bell-shaped curve, with the point of symmetry roughly at 10.

Class	1–<3	3–<5	5–<7	7–<9	9–<11	11–<13	13–<15	15–<17	17–<19
Frequency	1	1	11	21	25	17	9	4	1
Relative frequency	.011	.011	.122	.233	.278	.189	.100	.044	.011

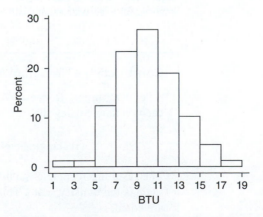

Figure 1.8 Histogram of the energy consumption data from Example 1.10

From the histogram,

$$\begin{array}{l} \text{proportion of} \\ \text{observations} \\ \text{less than 9} \end{array} \approx .01 + .01 + .12 + .23 = .37 \quad \left(\text{exact value} = \frac{34}{90} = .378\right)$$

The relative frequency for the $9 - {<}11$ class is about .27, so we estimate that roughly half of this, or .135, is between 9 and 10. Thus

$$\begin{array}{l} \text{proportion of observations} \\ \text{less than 10} \end{array} \approx .37 + .135 = .505 \text{ (slightly more than 50\%)}$$

The exact value of this proportion is $47/90 = .522$. ∎

There are no hard-and-fast rules concerning either the number of classes or the choice of classes themselves. Between 5 and 20 classes will be satisfactory for most data sets. Generally, the larger the number of observations in a data set, the more classes should be used. A reasonable rule of thumb is

$$\text{number of classes} \approx \sqrt{\text{number of observations}}$$

Equal-width classes may not be a sensible choice if there are some regions of the measurement scale that have a high concentration of data values and other parts where data is quite sparse. Figure 1.9 shows a dotplot of such a data set; there is high concentration in the middle, and relatively few observations stretched out to either side. Using a small number of equal-width classes results in almost all observations falling in just one or two of the classes. If a large number of equal-width classes are used, many classes will have zero frequency. A sound choice is to use a few wider intervals near extreme observations and narrower intervals in the region of high concentration.

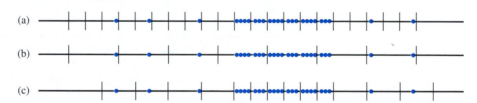

Figure 1.9 Selecting class intervals for "varying density" data: (a) many short equal-width intervals; (b) a few wide equal-width intervals; (c) unequal-width intervals

Constructing a Histogram for Continuous Data: Unequal Class Widths

After determining frequencies and relative frequencies, calculate the height of each rectangle using the formula

$$\text{rectangle height} = \frac{\text{relative frequency of the class}}{\text{class width}}$$

The resulting rectangle heights are usually called *densities,* and the vertical scale is the **density scale.** This prescription will also work when class widths are equal.

EXAMPLE 1.11 Corrosion of reinforcing steel is a serious problem in concrete structures located in environments affected by severe weather conditions. For this reason, researchers have been investigating the use of reinforcing bars made of composite material. One study was carried out to develop guidelines for bonding glass-fiber-reinforced plastic rebars to concrete **("Design Recommendations for Bond of GFRP Rebars to Concrete," *J. of Structural Engr.*, 1996: 247–254).** Consider the following 48 observations on measured bond strength:

11.5	12.1	9.9	9.3	7.8	6.2	6.6	7.0	13.4	17.1	9.3	5.6
5.7	5.4	5.2	5.1	4.9	10.7	15.2	8.5	4.2	4.0	3.9	3.8
3.6	3.4	20.6	25.5	13.8	12.6	13.1	8.9	8.2	10.7	14.2	7.6
5.2	5.5	5.1	5.0	5.2	4.8	4.1	3.8	3.7	3.6	3.6	3.6

Class	2–<4	4–<6	6–<8	8–<12	12–<20	20–<30
Frequency	9	15	5	9	8	2
Relative frequency	.1875	.3125	.1042	.1875	.1667	.0417
Density	.094	.156	.052	.047	.021	.004

The resulting histogram appears in Figure 1.10. The right or upper tail stretches out much farther than does the left or lower tail—a substantial departure from symmetry.

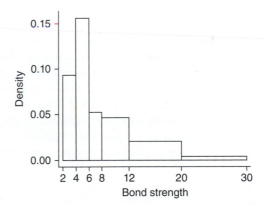

Figure 1.10 A Minitab density histogram for the bond strength data of Example 1.11 ■

When class widths are unequal, not using a density scale will give a picture with distorted areas. For equal-class widths, the divisor is the same in each density calculation, and the extra arithmetic simply results in a rescaling of the vertical axis (i.e., the histogram using relative frequency and the one using density will have exactly the same appearance). A density histogram does have one interesting property. Multiplying both sides of the formula for density by the class width gives

relative frequency = (class width)(density) = (rectangle width)(rectangle height)

= rectangle area

That is, *the area of each rectangle is the relative frequency of the corresponding class.* Furthermore, since the sum of relative frequencies should be 1, *the total area of all rectangles in a density histogram is* 1. It is always possible to draw a histogram so that the area equals the relative frequency (this is true also for a histogram of discrete data)—just use the density scale. This property will play an important role in motivating models for distributions in Chapter 4.

Histogram Shapes

Histograms come in a variety of shapes. A **unimodal** histogram is one that rises to a single peak and then declines. A **bimodal** histogram has two different peaks. Bimodality can occur when the data set consists of observations on two quite different kinds of individuals or objects. For example, consider a large data set consisting of driving times for automobiles traveling between San Luis Obispo, California, and Monterey, California (exclusive of stopping time for sightseeing, eating, etc.). This histogram would show two peaks: one for those cars that took the inland route (roughly 2.5 hours) and another for those cars traveling up the coast (3.5–4 hours). However, bimodality does not automatically follow in such situations. Only if the two separate histograms are "far apart" relative to their spreads will bimodality occur in the histogram of combined data. Thus a large data set consisting of heights of college students should not result in a bimodal histogram because the typical male height of about 69 inches is not far enough above the typical female height of about 64–65 inches. A histogram with more than two peaks is said to be **multimodal**. Of course, the number of peaks may well depend on the choice of class intervals, particularly with a small number of observations. The larger the number of classes, the more likely it is that bimodality or multimodality will manifest itself.

EXAMPLE 1.12 Figure 1.11(a) shows a Minitab histogram of the weights (lb) of the 124 players listed on the rosters of the San Francisco 49ers and the New England Patriots (teams the author would like to see meet in the Super Bowl) as of Nov. 20, 2009. Figure 1.11(b) is a smoothed histogram (actually what is called a *density estimate*) of the data from the R software package. Both the histogram and the smoothed histogram show three distinct peaks; the one on the right is for linemen, the middle peak corresponds to linebacker weights, and the peak on the left is for all other players (wide receivers, quarterbacks, etc.).

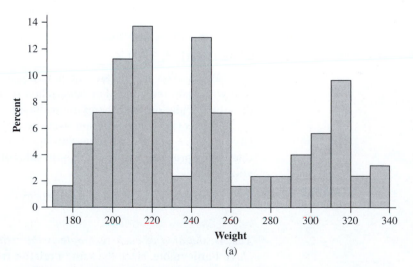

Figure 1.11 NFL player weights (a) Histogram (b) Smoothed histogram

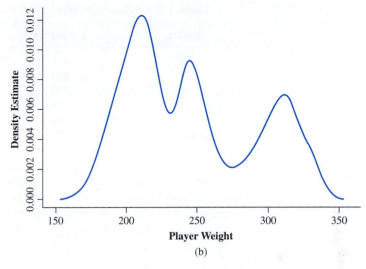

Figure 1.11 (continued)

A histogram is **symmetric** if the left half is a mirror image of the right half. A unimodal histogram is **positively skewed** if the right or upper tail is stretched out compared with the left or lower tail and **negatively skewed** if the stretching is to the left. Figure 1.12 shows "smoothed" histograms, obtained by superimposing a smooth curve on the rectangles, that illustrate the various possibilities.

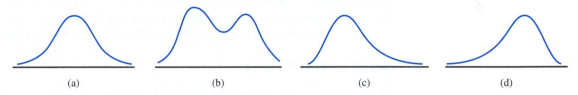

Figure 1.12 Smoothed histograms: (a) symmetric unimodal; (b) bimodal; (c) positively skewed; and (d) negatively skewed

Qualitative Data

Both a frequency distribution and a histogram can be constructed when the data set is *qualitative* (categorical) in nature. In some cases, there will be a natural ordering of classes—for example, freshmen, sophomores, juniors, seniors, graduate students—whereas in other cases the order will be arbitrary—for example, Catholic, Jewish, Protestant, and the like. With such categorical data, the intervals above which rectangles are constructed should have equal width.

EXAMPLE 1.13 **The Public Policy Institute of California** carried out a telephone survey of 2501 California adult residents during April 2006 to ascertain how they felt about various aspects of K–12 public education. One question asked was "Overall, how would you rate the quality of public schools in your neighborhood today?" Table 1.2 displays the frequencies and relative frequencies, and Figure 1.13 shows the corresponding histogram (bar chart).

Table 1.2 Frequency Distribution for the School Rating Data

Rating	Frequency	Relative Frequency
A	478	.191
B	893	.357
C	680	.272
D	178	.071
F	100	.040
Don't know	172	.069
	2501	1.000

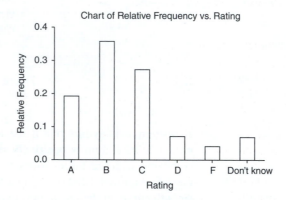

Figure 1.13   Histogram of the school rating data from Minitab

More than half the respondents gave an A or B rating, and only slightly more than 10% gave a D or F rating. The percentages for parents of public school children were somewhat more favorable to schools: 24%, 40%, 24%, 6%, 4%, and 2%.   ■

Multivariate Data

Multivariate data is generally rather difficult to describe visually. Several methods for doing so appear later in the book, notably scatterplots for bivariate numerical data.

EXERCISES Section 1.2 (10–32)

10. Consider the strength data for beams given in Example 1.2.

  a. Construct a stem-and-leaf display of the data. What appears to be a representative strength value? Do the observations appear to be highly concentrated about the representative value or rather spread out?

  b. Does the display appear to be reasonably symmetric about a representative value, or would you describe its shape in some other way?

  c. Do there appear to be any outlying strength values?

  d. What proportion of strength observations in this sample exceed 10 MPa?

11. The accompanying specific gravity values for various wood types used in construction appeared in the article **"Bolted Connection Design Values Based on European Yield Model"** (*J. of Structural Engr.*, **1993: 2169–2186**):

.31 .35 .36 .36 .37 .38 .40 .40 .40
.41 .41 .42 .42 .42 .42 .42 .43 .44
.45 .46 .46 .47 .48 .48 .48 .51 .54
.54 .55 .58 .62 .66 .66 .67 .68 .75

Construct a stem-and-leaf display using repeated stems, and comment on any interesting features of the display.

12. The accompanying summary data on CeO_2 particle sizes (nm) under certain experimental conditions was read from a graph in the article **"Nanoceria— Energetics of Surfaces, Interfaces and Water Adsorption"** (*J. of the Amer. Ceramic Soc.*, **2011: 3992–3999**):

3.0–<3.5	3.5–<4.0	4.0–<4.5	4.5–<5.0	5.0–<5.5
5	15	27	34	22

5.5–<6.0	6.0–<6.5	6.5–<7.0	7.0–<7.5	7.5–<8.0
14	7	2	4	1

a. What proportion of the observations are less than 5?
b. What proportion of the observations are at least 6?
c. Construct a histogram with relative frequency on the vertical axis and comment on interesting features. In particular, does the distribution of particle sizes appear to be reasonably symmetric or somewhat skewed? [*Note:* The investigators fit a lognormal distribution to the data; this is discussed in Chapter 4.]
d. Construct a histogram with density on the vertical axis and compare to the histogram in (c).

13. Allowable mechanical properties for structural design of metallic aerospace vehicles requires an approved method for statistically analyzing empirical test data. The article **"Establishing Mechanical Property Allowables for Metals"** (*J. of Testing and Evaluation*, **1998: 293–299**) used the accompanying data on tensile ultimate strength (ksi) as a basis for addressing the difficulties in developing such a method.

```
122.2 124.2 124.3 125.6 126.3 126.5 126.5 127.2 127.3
127.5 127.9 128.6 128.8 129.0 129.2 129.4 129.6 130.2
130.4 130.8 131.3 131.4 131.4 131.5 131.6 131.6 131.8
131.8 132.3 132.4 132.4 132.5 132.5 132.5 132.5 132.6
132.7 132.9 133.0 133.1 133.1 133.1 133.1 133.2 133.2
133.2 133.3 133.3 133.5 133.5 133.5 133.8 133.9 134.0
134.0 134.0 134.0 134.1 134.2 134.3 134.4 134.4 134.6
134.7 134.7 134.7 134.8 134.8 134.8 134.9 134.9 135.2
135.2 135.2 135.3 135.3 135.4 135.5 135.5 135.6 135.6
135.7 135.8 135.8 135.8 135.8 135.8 135.9 135.9 135.9
135.9 136.0 136.0 136.1 136.2 136.2 136.3 136.4 136.4
136.6 136.8 136.9 136.9 137.0 137.1 137.2 137.6 137.6
137.8 137.8 137.8 137.9 137.9 138.2 138.2 138.3 138.3
138.4 138.4 138.4 138.5 138.5 138.6 138.7 138.7 139.0
139.1 139.5 139.6 139.8 139.8 140.0 140.0 140.7 140.7
140.9 140.9 141.2 141.4 141.5 141.6 142.9 143.4 143.5
143.6 143.8 143.8 143.9 144.1 144.5 144.5 147.7 147.7
```

a. Construct a stem-and-leaf display of the data by first deleting (truncating) the tenths digit and then repeating each stem value five times (once for leaves 1 and 2, a second time for leaves 3 and 4, etc.). Why is it relatively easy to identify a representative strength value?
b. Construct a histogram using equal-width classes with the first class having a lower limit of 122 and an upper limit of 124. Then comment on any interesting features of the histogram.

14. The accompanying data set consists of observations on shower-flow rate (L/min) for a sample of $n = 129$ houses in Perth, Australia (**"An Application of Bayes Methodology to the Analysis of Diary Records in a Water Use Study,"** *J. Amer. Stat. Assoc.*, **1987: 705–711**):

```
4.6  12.3  7.1   7.0   4.0   9.2  6.7   6.9  11.5   5.1
11.2 10.5  14.3  8.0   8.8   6.4  5.1   5.6   9.6   7.5
7.5   6.2  5.8   2.3   3.4  10.4  9.8   6.6   3.7   6.4
8.3   6.5  7.6   9.3   9.2   7.3  5.0   6.3  13.8   6.2
5.4   4.8  7.5   6.0   6.9  10.8  7.5   6.6   5.0   3.3
7.6   3.9 11.9   2.2  15.0   7.2  6.1  15.3  18.9   7.2
5.4   5.5  4.3   9.0  12.7  11.3  7.4   5.0   3.5   8.2
8.4   7.3 10.3  11.9   6.0   5.6  9.5   9.3  10.4   9.7
5.1   6.7 10.2   6.2   8.4   7.0  4.8   5.6  10.5  14.6
10.8 15.5  7.5   6.4   3.4   5.5  6.6   5.9  15.0   9.6
7.8   7.0  6.9   4.1   3.6  11.9  3.7   5.7   6.8  11.3
9.3   9.6 10.4   9.3   6.9   9.8  9.1  10.6   4.5   6.2
8.3   3.2  4.9   5.0   6.0   8.2  6.3   3.8   6.0
```

a. Construct a stem-and-leaf display of the data.
b. What is a typical, or representative, flow rate?
c. Does the display appear to be highly concentrated or spread out?
d. Does the distribution of values appear to be reasonably symmetric? If not, how would you describe the departure from symmetry?
e. Would you describe any observation as being far from the rest of the data (an outlier)?

15. Do running times of American movies differ somehow from running times of French movies? The author investigated this question by randomly selecting 25 recent movies of each type, resulting in the following running times:

Am: 94 90 95 93 128 95 125 91 104 116 162 102 90
     110 92 113 116 90 97 103 95 120 109 91 138
Fr: 123 116 90 158 122 119 125 90 96 94 137 102
     105 106 95 125 122 103 96 111 81 113 128 93 92

Construct a *comparative* stem-and-leaf display by listing stems in the middle of your paper and then placing the Am leaves out to the left and the Fr leaves out to the right. Then comment on interesting features of the display.

16. The article cited in Example 1.2 also gave the accompanying strength observations for cylinders:

6.1	5.8	7.8	7.1	7.2	9.2	6.6	8.3	7.0	8.3
7.8	8.1	7.4	8.5	8.9	9.8	9.7	14.1	12.6	11.2

a. Construct a comparative stem-and-leaf display (see the previous exercise) of the beam and cylinder data, and then answer the questions in parts (b)–(d) of Exercise 10 for the observations on cylinders.

b. In what ways are the two sides of the display similar? Are there any obvious differences between the beam observations and the cylinder observations?

c. Construct a dotplot of the cylinder data.

17. The accompanying data came from a study of collusion in bidding within the construction industry (**"Detection of Collusive Behavior,"** *J. of Construction Engr. and Mgmnt*, **2012: 1251–1258**).

No. Bidders	No. Contracts
2	7
3	20
4	26
5	16
6	11
7	9
8	6
9	8
10	3
11	2

a. What proportion of the contracts involved at most five bidders? At least five bidders?

b. What proportion of the contracts involved between five and 10 bidders, inclusive? Strictly between five and 10 bidders?

c. Construct a histogram and comment on interesting features.

18. Every corporation has a governing board of directors. The number of individuals on a board varies from one corporation to another. One of the authors of the article **"Does Optimal Corporate Board Size Exist? An Empirical Analysis"** (*J. of Applied Finance*, **2010: 57–69**) provided the accompanying data on the number of directors on each board in a random sample of 204 corporations.

No. directors:	4	5	6	7	8	9
Frequency:	3	12	13	25	24	42

No. directors:	10	11	12	13	14	15
Frequency:	23	19	16	11	5	4

No. directors:	16	17	21	24	32
Frequency:	1	3	1	1	1

a. Construct a histogram of the data based on relative frequencies and comment on any interesting features.

b. Construct a frequency distribution in which the last row includes all boards with at least 18 directors. If this distribution had appeared in the cited article, would you be able to draw a histogram? Explain.

c. What proportion of these corporations have at most 10 directors?

d. What proportion of these corporations have more than 15 directors?

19. The number of contaminating particles on a silicon wafer prior to a certain rinsing process was determined for each wafer in a sample of size 100, resulting in the following frequencies:

Number of particles	0	1	2	3	4	5	6	7
Frequency	1	2	3	12	11	15	18	10
Number of particles	8	9	10	11	12	13	14	
Frequency	12	4	5	3	1	2	1	

a. What proportion of the sampled wafers had at least one particle? At least five particles?

b. What proportion of the sampled wafers had between five and ten particles, inclusive? Strictly between five and ten particles?

c. Draw a histogram using relative frequency on the vertical axis. How would you describe the shape of the histogram?

20. The article **"Determination of Most Representative Subdivision"** (*J. of Energy Engr.*, **1993: 43–55**) gave data on various characteristics of subdivisions that could be used in deciding whether to provide electrical power using overhead lines or underground lines. Here are the values of the variable x = total length of streets within a subdivision:

1280	5320	4390	2100	1240	3060	4770
1050	360	3330	3380	340	1000	960
1320	530	3350	540	3870	1250	2400
960	1120	2120	450	2250	2320	2400
3150	5700	5220	500	1850	2460	5850
2700	2730	1670	100	5770	3150	1890
510	240	396	1419	2109		

a. Construct a stem-and-leaf display using the thousands digit as the stem and the hundreds digit as the leaf, and comment on the various features of the display.

b. Construct a histogram using class boundaries 0, 1000, 2000, 3000, 4000, 5000, and 6000. What proportion of subdivisions have total length less than 2000? Between 2000 and 4000? How would you describe the shape of the histogram?

21. The article cited in Exercise 20 also gave the following values of the variables y = number of culs-de-sac and z = number of intersections:

```
y 1 0 1 0 0 2 0 1 1 1 2 1 0 0 1 1 0 1 1
z 1 8 6 1 1 5 3 0 0 4 4 0 0 1 2 1 4 0 4
y 1 1 0 0 0 1 1 2 0 1 2 2 1 1 0 2 1 1 0
z 0 3 0 1 1 0 1 3 2 4 6 6 0 1 1 8 3 3 5
y 1 5 0 3 0 1 1 0 0
z 0 5 2 3 1 0 0 0 3
```

a. Construct a histogram for the y data. What proportion of these subdivisions had no culs-de-sac? At least one cul-de-sac?

b. Construct a histogram for the z data. What proportion of these subdivisions had at most five intersections? Fewer than five intersections?

22. How does the speed of a runner vary over the course of a marathon (a distance of 42.195 km)? Consider determining both the time to run the first 5 km and the time to run between the 35-km and 40-km points, and then subtracting the former time from the latter time. A positive value of this difference corresponds to a runner slowing down toward the end of the race. The accompanying histogram is based on times of runners who participated in several different Japanese marathons (**"Factors Affecting Runners' Marathon Performance,"** *Chance,* *Fall,* **1993: 24–30).**

What are some interesting features of this histogram? What is a typical difference value? Roughly what proportion of the runners ran the late distance more quickly than the early distance?

Histogram for Exercise 22

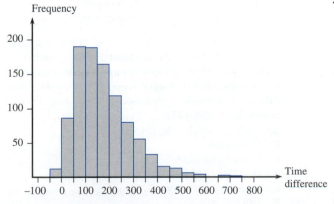

23. The article **"Statistical Modeling of the Time Course of Tantrum Anger"** (*Annals of Applied Stats,* **2009: 1013–1034)** discussed how anger intensity in children's tantrums could be related to tantrum duration as well as behavioral indicators such as shouting, stamping, and pushing or pulling. The following frequency distribution was given (and also the corresponding histogram):

0–<2 :	136	2–<4:	92	4–<11:	71
11–<20:	26	20–<30:	7	30–<40:	3

Draw the histogram and then comment on any interesting features.

24. The accompanying data set consists of observations on shear strength (lb) of ultrasonic spot welds made on a certain type of alclad sheet. Construct a relative frequency histogram based on ten equal-width classes with boundaries 4000, 4200, …. [The histogram will agree with the one in **"Comparison of Properties of Joints Prepared by Ultrasonic Welding and Other Means"** (*J. of Aircraft,* **1983: 552–556).**] Comment on its features.

5434	4948	4521	4570	4990	5702	5241
5112	5015	4659	4806	4637	5670	4381
4820	5043	4886	4599	5288	5299	4848
5378	5260	5055	5828	5218	4859	4780
5027	5008	4609	4772	5133	5095	4618
4848	5089	5518	5333	5164	5342	5069
4755	4925	5001	4803	4951	5679	5256
5207	5621	4918	5138	4786	4500	5461
5049	4974	4592	4173	5296	4965	5170
4740	5173	4568	5653	5078	4900	4968
5248	5245	4723	5275	5419	5205	4452
5227	5555	5388	5498	4681	5076	4774
4931	4493	5309	5582	4308	4823	4417
5364	5640	5069	5188	5764	5273	5042
5189	4986					

25. A transformation of data values by means of some mathematical function, such as $\sqrt{x}$ or $1/x$, can often yield a set of numbers that has "nicer" statistical properties than the original data. In particular, it may be possible to find a function for which the histogram of transformed values is more symmetric (or, even better, more like a bell-shaped curve) than the original data. As an example, the article **"Time Lapse Cinematographic Analysis of Beryllium–Lung Fibroblast Interactions"** (*Environ. Research,* **1983: 34–43)** reported the results of experiments designed to study the behavior of certain individual cells that had been exposed to beryllium. An important characteristic of such an individual cell is its interdivision time (IDT). IDTs were determined for a large number of cells, both in exposed (treatment) and unexposed (control) conditions. The authors of the article used a logarithmic transformation, that is, transformed value = log(original value). Consider the following representative IDT data:

IDT	$\log_{10}$(IDT)	IDT	$\log_{10}$(IDT)	IDT	$\log_{10}$(IDT)
28.1	1.45	60.1	1.78	21.0	1.32
31.2	1.49	23.7	1.37	22.3	1.35
13.7	1.14	18.6	1.27	15.5	1.19
46.0	1.66	21.4	1.33	36.3	1.56
25.8	1.41	26.6	1.42	19.1	1.28
16.8	1.23	26.2	1.42	38.4	1.58
34.8	1.54	32.0	1.51	72.8	1.86
62.3	1.79	43.5	1.64	48.9	1.69
28.0	1.45	17.4	1.24	21.4	1.33
17.9	1.25	38.8	1.59	20.7	1.32
19.5	1.29	30.6	1.49	57.3	1.76
21.1	1.32	55.6	1.75	40.9	1.61
31.9	1.50	25.5	1.41		
28.9	1.46	52.1	1.72		

Use class intervals 10−<20, 20−<30,… to construct a histogram of the original data. Use intervals 1.1−<1.2, 1.2−<1.3,… to do the same for the transformed data. What is the effect of the transformation?

26. Automated electron backscattered diffraction is now being used in the study of fracture phenomena. The following information on misorientation angle (degrees) was extracted from the article **"Observations on the Faceted Initiation Site in the Dwell-Fatigue Tested Ti-6242 Alloy: Crystallographic Orientation and Size Effects"** (*Metallurgical and Materials Trans.*, 2006: 1507–1518).

Class:	0−<5	5−<10	10−<15	15−<20
Rel freq:	.177	.166	.175	.136
Class:	20−<30	30−<40	40−<60	60−<90
Rel freq:	.194	.078	.044	.030

   a. Is it true that more than 50% of the sampled angles are smaller than 15°, as asserted in the paper?
   b. What proportion of the sampled angles are at least 30°?
   c. Roughly what proportion of angles are between 10° and 25°?
   d. Construct a histogram and comment on any interesting features.

27. The article **"Study on the Life Distribution of Microdrills"** (*J. of Engr. Manufacture*, 2002: 301–305) reported the following observations, listed in increasing order, on drill lifetime (number of holes that a drill machines before it breaks) when holes were drilled in a certain brass alloy.

11	14	20	23	31	36	39	44	47	50
59	61	65	67	68	71	74	76	78	79
81	84	85	89	91	93	96	99	101	104
105	105	112	118	123	136	139	141	148	158
161	168	184	206	248	263	289	322	388	513

   a. Why can a frequency distribution not be based on the class intervals 0–50, 50–100, 100–150, and so on?
   b. Construct a frequency distribution and histogram of the data using class boundaries 0, 50, 100,…, and then comment on interesting characteristics.
   c. Construct a frequency distribution and histogram of the natural logarithms of the lifetime observations, and comment on interesting characteristics.
   d. What proportion of the lifetime observations in this sample are less than 100? What proportion of the observations are at least 200?

28. The accompanying frequency distribution on deposited energy (mJ) was extracted from the article **"Experimental Analysis of Laser-Induced Spark Ignition of Lean Turbulent Premixed Flames"** (*Combustion and Flame*, 2013: 1414–1427).

1.0−<2.0	5	2.0−<2.4	11
2.4−<2.6	13	2.6−<2.8	30
2.8−<3.0	46	3.0−<3.2	66
3.2−<3.4	133	3.4−<3.6	141
3.6−<3.8	126	3.8−<4.0	92
4.0−<4.2	73	4.2−<4.4	38
4.4−<4.6	19	4.6−<5.0	11

   a. What proportion of these ignition trials resulted in a deposited energy of less than 3 mJ?
   b. What proportion of these ignition trials resulted in a deposited energy of at least 4 mJ?
   c. Roughly what proportion of the trials resulted in a deposited energy of at least 3.5 mJ?
   d. Construct a histogram and comment on its shape.

29. The following categories for type of physical activity involved when an industrial accident occurred appeared in the article **"Finding Occupational Accident Patterns in the Extractive Industry Using a Systematic Data Mining Approach"** (*Reliability Engr. and System Safety*, 2012: 108–122):
   A. Working with handheld tools
   B. Movement
   C. Carrying by hand
   D. Handling of objects
   E. Operating a machine
   F. Other

Construct a frequency distribution, including relative frequencies, and histogram for the accompanying data from 100 accidents (the percentages agree with those in the cited article):

```
A  B  D  A  A  F  C  A  C  B  E  B  A  C
F  D  B  C  D  A  A  C  B  E  B  B  C  E  A
B  A  A  A  B  C  C  D  F  D  B  B  A  F
C  B  A  C  B  E  E  D  A  B  C  E  A  A
F  C  B  D  D  D  B  D  C  A  F  A  A  B
D  E  A  E  D  B  C  A  F  A  C  D  D  A
A  B  A  F  D  C  A  C  B  F  D  A  E  A
C  D
```

30. A **Pareto diagram** is a variation of a histogram for categorical data resulting from a quality control study. Each category represents a different type of product nonconformity or production problem. The categories are ordered so that the one with the largest frequency appears on the far left, then the category with the second largest frequency, and so on. Suppose the following information on nonconformities in circuit packs is obtained: failed component, 126; incorrect component, 210; insufficient solder, 67; excess solder, 54; missing component, 131. Construct a Pareto diagram.

31. The **cumulative frequency** and cumulative relative frequency for a particular class interval are the sum of frequencies and relative frequencies, respectively, for that interval and all intervals lying below it. If, for example, there are four intervals with frequencies 9, 16, 13, and 12, then the cumulative frequencies are 9,

25, 38, and 50, and the cumulative relative frequencies are .18, .50, .76, and 1.00. Compute the cumulative frequencies and cumulative relative frequencies for the data of Exercise 24.

32. Fire load (MJ/m^2) is the heat energy that could be released per square meter of floor area by combustion of contents and the structure itself. The article **"Fire Loads in Office Buildings" (*J. of Structural Engr.*, 1997: 365–368)** gave the following cumulative percentages (read from a graph) for fire loads in a sample of 388 rooms:

Value	0	150	300	450	600
Cumulative %	0	19.3	37.6	62.7	77.5

Value	750	900	1050	1200	1350
Cumulative %	87.2	93.8	95.7	98.6	99.1

Value	1500	1650	1800	1950
Cumulative %	99.5	99.6	99.8	100.0

a. Construct a relative frequency histogram and comment on interesting features.

b. What proportion of fire loads are less than 600? At least 1200?

c. What proportion of the loads are between 600 and 1200?

1.3 Measures of Location

Visual summaries of data are excellent tools for obtaining preliminary impressions and insights. More formal data analysis often requires the calculation and interpretation of numerical summary measures. That is, from the data we try to extract several summarizing quantities that might serve to characterize the data set and convey some of its prominent features. Our primary concern will be with numerical data; some comments regarding categorical data appear at the end of the section.

Suppose, then, that our data set is of the form $x_1, x_2, \ldots, x_n$, where each x_i is a number. What features of such a set of numbers are of most interest and deserve emphasis? One important characteristic of a set of numbers is its location, and in particular its center. This section presents methods for describing the location of a data set; in Section 1.4 we will turn to methods for assessing variability in a set of numbers.

The Mean

For a given set of numbers $x_1, x_2, \ldots, x_n$, the most familiar and useful measure of the center is the *mean,* or arithmetic average of the set. Because we will almost always think of the x_i's as constituting a sample, we will often refer to the arithmetic average as the *sample mean* and denote it by $\bar{x}$.

DEFINITION

The **sample mean** $\bar{x}$ of observations $x_1, x_2, \ldots, x_n$ is given by

$$\bar{x} = \frac{x_1 + x_2 + \cdots + x_n}{n} = \frac{\sum_{i=1}^{n} x_i}{n}$$

The numerator of $\bar{x}$ can be written more informally as Σx_i, where the summation is over all sample observations.

For reporting $\bar{x}$, we recommend using decimal accuracy of one digit more than the accuracy of the x_i's. Thus if observations are stopping distances with $x_1 = 125$, $x_2 = 131$, and so on, we might have $\bar{x} = 127.3$ ft.

EXAMPLE 1.14   Recent years have seen growing commercial interest in the use of what is known as *internally cured concrete*. This concrete contains porous inclusions most commonly in the form of lightweight aggregate (LWA). The article **"Characterizing Lightweight Aggregate Desorption at High Relative Humidities Using a Pressure Plate Apparatus" (*J. of Materials in Civil Engr*, 2012: 961–969)** reported on a study in which researchers examined various physical properties of 14 LWA specimens. Here are the 24-hour water-absorption percentages for the specimens:

$x_1 = 16.0$	$x_2 = 30.5$	$x_3 = 17.7$	$x_4 = 17.5$	$x_5 = 14.1$
$x_6 = 10.0$	$x_7 = 15.6$	$x_8 = 15.0$	$x_9 = 19.1$	$x_{10} = 17.9$
$x_{11} = 18.9$	$x_{12} = 18.5$	$x_{13} = 12.2$	$x_{14} = 6.0$	

Figure 1.14 shows a dotplot of the data; a water-absorption percentage in the mid-teens appears to be "typical." With $\Sigma x_i = 229.0$, the sample mean is

$$\bar{x} = \frac{229.0}{14} = 16.36$$

A physical interpretation of the sample mean demonstrates how it assesses the center of a sample. Think of each dot in the dotplot as representing a 1-lb weight. Then a fulcrum placed with its tip on the horizontal axis will balance precisely when it is located at $\bar{x}$. So the sample mean can be regarded as the balance point of the distribution of observations.

Figure 1.14   Dotplot of the data from Example 1.14   ■

Just as $\bar{x}$ represents the average value of the observations in a sample, the average of all values in the population can be calculated. This average is called the **population mean** and is denoted by the Greek letter μ. When there are N values in the population (a finite population), then $\mu = $ (sum of the N population values)$/N$. We will give a more general definition for μ in Chapters 3 and 4 that applies to both finite and (conceptually) infinite populations. Just as $\bar{x}$ is an interesting and important measure of sample location, μ is an interesting and important (often the most important) characteristic of a population. One of our first tasks in statistical inference will be to present methods based on the sample mean for drawing conclusions about a population mean. For example, we might use the sample mean $\bar{x} = 16.36$ computed in Example 1.14 as a *point estimate* (a single number that is our "best" guess) of μ = the true average water-absorption percentage for all specimens treated as described.

The mean suffers from one deficiency that makes it an inappropriate measure of center under some circumstances: Its value can be greatly affected by the presence of even a single outlier (unusually large or small observation). For example, if a sample of employees contains nine who earn \$50,000 per year and one whose yearly salary is \$150,000, the sample mean salary is \$60,000; this value certainly does not seem representative of the data. In such situations, it is desirable to employ a measure that is less sensitive to outlying values than $\bar{x}$, and we will momentarily propose one. However, although $\bar{x}$ does have this potential defect, it is still the most widely used measure, largely because there are many populations for which an extreme outlier in the sample would be highly unlikely. When sampling from such a population (a normal or bell-shaped population being the most important example), the sample mean will tend to be stable and quite representative of the sample.

The Median

The word *median* is synonymous with "middle," and the sample median is indeed the middle value once the observations are ordered from smallest to largest. When the observations are denoted by $x_1, \ldots, x_n$, we will use the symbol $\tilde{x}$ to represent the sample median.

DEFINITION

The **sample median** is obtained by first ordering the n observations from smallest to largest (with any repeated values included so that every sample observation appears in the ordered list). Then,

$$
\tilde{x} = \begin{cases} \text{The single middle value if } n \text{ is odd} & = \left(\dfrac{n+1}{2}\right)^{\text{th}} \text{ordered value} \\[2em] \text{The average of the two middle values if } n \text{ is even} & = \text{average of } \left(\dfrac{n}{2}\right)^{\text{th}} \text{and } \left(\dfrac{n}{2}+1\right)^{\text{th}} \text{ordered values} \end{cases}
$$

EXAMPLE 1.15 People not familiar with classical music might tend to believe that a composer's instructions for playing a particular piece are so specific that the duration would not depend at all on the performer(s). However, there is typically plenty of room for interpretation, and orchestral conductors and musicians take full advantage of this. The author went to the Web site **ArkivMusic.com** and selected a sample of 12 recordings of Beethoven's Symphony No. 9 (the "Choral," a stunningly beautiful work), yielding the following durations (min) listed in increasing order:

62.3 62.8 63.6 65.2 65.7 66.4 67.4 68.4 68.8 70.8 75.7 79.0

Here is a dotplot of the data:

Figure 1.15 Dotplot of the data from Example 1.15

Since $n = 12$ is even, the sample median is the average of the $n/2 = 6^{th}$ and $(n/2 + 1) = 7^{th}$ values from the ordered list:

$$\widetilde{x} = \frac{66.4 + 67.4}{2} = 66.90$$

Note that if the largest observation 79.0 had not been included in the sample, the resulting sample median for the $n = 11$ remaining observations would have been the single middle value 66.4 (the $[n + 1]/2 = 6^{th}$ ordered value, i.e., the 6th value in from either end of the ordered list). The sample mean is $\bar{x} = \Sigma x_i = 816.1/12 = 68.01$, roughly a minute larger than the median. The mean is pulled out relative to the median because the sample "stretches out" somewhat more on the upper end than on the lower end. ■

The data in Example 1.15 illustrates an important property of $\widetilde{x}$ in contrast to $\bar{x}$: The sample median is very insensitive to outliers. If, for example, the two largest x_is are increased from 75.7 and 79.0 to 85.7 and 89.0, respectively, $\widetilde{x}$ would be unaffected. Thus, in the treatment of outlying data values, $\bar{x}$ and $\widetilde{x}$ are at opposite ends of a spectrum. Both quantities describe where the data is centered, but they will not in general be equal because they focus on different aspects of the sample.

Analogous to $\widetilde{x}$ as the middle value in the sample is a middle value in the population, the **population median**, denoted by $\widetilde{\mu}$. As with $\bar{x}$ and μ, we can think of using the sample median $\widetilde{x}$ to make an inference about $\widetilde{\mu}$. In Example 1.15, we might use $\widetilde{x} = 66.90$ as an estimate of the median time for the population of all recordings.

The population mean μ and median $\widetilde{\mu}$ will not generally be identical. If the population distribution is positively or negatively skewed, as pictured in Figure 1.16, then $\mu \neq \widetilde{\mu}$. When this is the case, in making inferences we must first decide which of the two population characteristics is of greater interest and then proceed accordingly.

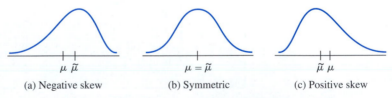

(a) Negative skew   (b) Symmetric   (c) Positive skew

Figure 1.16 Three different shapes for a population distribution

Other Measures of Location: Quartiles, Percentiles, and Trimmed Means

The median (population or sample) divides the data set into two parts of equal size. To obtain finer measures of location, we could divide the data into more than two such parts. Roughly speaking, quartiles divide the data set into four equal parts, with the observations above the third quartile constituting the upper quarter of the data set, the second quartile being identical to the median, and the first quartile separating the lower quarter from the upper three-quarters.

Similarly, a data set (sample or population) can be even more finely divided using percentiles; the 99th percentile separates the highest 1% from the bottom 99%, and so on. Unless the number of observations is a multiple of 100, care must be exercised in obtaining percentiles. We will revisit percentiles in Chapter 4 in connection with certain models for infinite populations.

The mean is quite sensitive to a single outlier, whereas the median is impervious to many outliers. Since extreme behavior of either type might be undesirable, we briefly consider alternative measures that are neither as sensitive as $\bar{x}$ nor as insensitive as $\tilde{x}$. To motivate these alternatives, note that $\bar{x}$ and $\tilde{x}$ are at opposite extremes of the same "family" of measures. The mean is the average of all the data, whereas the median results from eliminating all but the middle one or two values and then averaging. To paraphrase, the mean involves trimming 0% from each end of the sample, whereas for the median the maximum possible amount is trimmed from each end. A **trimmed mean** is a compromise between $\bar{x}$ and $\tilde{x}$. A 10% trimmed mean, for example, would be computed by eliminating the smallest 10% and the largest 10% of the sample and then averaging what remains.

EXAMPLE 1.16 The production of Bidri is a traditional craft of India. Bidri wares (bowls, vessels, and so on) are cast from an alloy containing primarily zinc along with some copper. Consider the following observations on copper content (%) for a sample of Bidri artifacts in London's Victoria and Albert Museum **("Enigmas of Bidri,"** *Surface Engr.*, **2005: 333–339),** listed in increasing order:

> 2.0   2.4   2.5   2.6   2.6   2.7   2.7   2.8   3.0   3.1   3.2   3.3   3.3
> 3.4   3.4   3.6   3.6   3.6   3.6   3.7   4.4   4.6   4.7   4.8   5.3   10.1

Figure 1.17 is a dotplot of the data. A prominent feature is the single outlier at the upper end; the distribution is somewhat sparser in the region of larger values than is the case for smaller values. The sample mean and median are 3.65 and 3.35, respectively. A trimmed mean with a trimming percentage of $100(2/26) = 7.7\%$ results from eliminating the two smallest and two largest observations; this gives $\bar{x}_{tr(7.7)} = 3.42$. Trimming here eliminates the larger outlier and so pulls the trimmed mean toward the median.

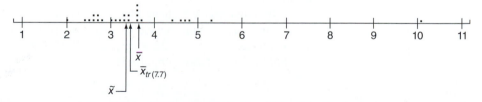

Figure 1.17   Dotplot of copper contents from Example 1.16   ∎

A trimmed mean with a moderate trimming percentage—someplace between 5% and 25%—will yield a measure of center that is neither as sensitive to outliers as is the mean nor as insensitive as the median. If the desired trimming percentage is $100\alpha\%$ and $n\alpha$ is not an integer, the trimmed mean must be calculated by interpolation. For example, consider $\alpha = .10$ for a 10% trimming percentage and $n = 26$ as in Example 1.16. Then $\bar{x}_{tr(10)}$ would be the appropriate weighted average of the 7.7% trimmed mean calculated there and the 11.5% trimmed mean resulting from trimming three observations from each end.

Categorical Data and Sample Proportions

When the data is categorical, a frequency distribution or relative frequency distribution provides an effective tabular summary of the data. The natural numerical summary quantities in this situation are the individual frequencies and the relative frequencies. For example, if a survey of individuals who own digital cameras is undertaken to study brand preference, then each individual in the sample would identify the brand of camera that he or she owned, from which we could count the number owning Canon, Sony, Kodak, and so on. Consider sampling a dichotomous population—one that consists of only two categories (such as voted or did not vote in the last election, does or does not own a digital camera, etc.). If we let x denote the number in the sample falling in category 1, then the number in category 2 is $n - x$. The relative frequency or *sample proportion* in category 1 is x/n and the sample proportion in category 2 is $1 - x/n$. Let's denote a response that falls in category 1 by a 1 and a response that falls in category 2 by a 0. A sample size of $n = 10$ might then yield the responses 1, 1, 0, 1, 1, 1, 0, 0, 1, 1. The sample mean for this numerical sample is (since number of 1s = $x = 7$)

$$\frac{x_1 + \cdots + x_n}{n} = \frac{1 + 1 + 0 + \cdots + 1 + 1}{10} = \frac{7}{10} = \frac{x}{n} = \text{sample proportion}$$

More generally, *focus attention on a particular category and code the sample results so that a 1 is recorded for an observation in the category and a 0 for an observation not in the category. Then the sample proportion of observations in the category is the sample mean of the sequence of 1's and 0's.* Thus a sample mean can be used to summarize the results of a categorical sample. These remarks also apply to situations in which categories are defined by grouping values in a numerical sample or population (e.g., we might be interested in knowing whether individuals have owned their present automobile for at least 5 years, rather than studying the exact length of ownership).

Analogous to the sample proportion x/n of individuals or objects falling in a particular category, let p represent the proportion of those in the entire population falling in the category. As with x/n, p is a quantity between 0 and 1, and while x/n is a sample characteristic, p is a characteristic of the population. The relationship between the two parallels the relationship between $\tilde{x}$ and $\tilde{\mu}$ and between $\bar{x}$ and μ. In particular, we will subsequently use x/n to make inferences about p. If a sample of 100 students from a large university reveals that 38 have Macintosh computers, then we could use $38/100 = .38$ as a point estimate of the proportion of all students at the university who have Macs. Or we might ask whether this sample provides strong evidence for concluding that at least 1/3 of all students are Mac owners. With k categories ($k > 2$), we can use the k sample proportions to answer questions about the population proportions $p_1, \ldots, p_k$.

EXERCISES Section 1.3 (33–43)

33. The **May 1, 2009, issue of *The Montclarian*** reported the following home sale amounts for a sample of homes in Alameda, CA that were sold the previous month (1000s of $):

590 815 575 608 350 1285 408 540 555 679

a. Calculate and interpret the sample mean and median.

b. Suppose the 6th observation had been 985 rather than 1285. How would the mean and median change?

c. Calculate a 20% trimmed mean by first trimming the two smallest and two largest observations.

d. Calculate a 15% trimmed mean.

34. Exposure to microbial products, especially endotoxin, may have an impact on vulnerability to allergic diseases.

The article **"Dust Sampling Methods for Endotoxin— An Essential, But Underestimated Issue"** (**Indoor Air, 2006: 20–27**) considered various issues associated with determining endotoxin concentration. The following data on concentration (EU/mg) in settled dust for one sample of urban homes and another of farm homes was kindly supplied by the authors of the cited article.

U: 6.0 5.0 11.0 33.0 4.0 5.0 80.0 18.0 35.0 17.0 23.0
F: 4.0 14.0 11.0 9.0 9.0 8.0 4.0 20.0 5.0 8.9 21.0
9.2 3.0 2.0 0.3

a. Determine the sample mean for each sample. How do they compare?
b. Determine the sample median for each sample. How do they compare? Why is the median for the urban sample so different from the mean for that sample?
c. Calculate the trimmed mean for each sample by deleting the smallest and largest observation. What are the corresponding trimming percentages? How do the values of these trimmed means compare to the corresponding means and medians?

35. Mercury is a persistent and dispersive environmental contaminant found in many ecosystems around the world. When released as an industrial by-product, it often finds its way into aquatic systems where it can have deleterious effects on various avian and aquatic species. The accompanying data on blood mercury concentration (μg/g) for adult females near contaminated rivers in Virginia was read from a graph in the article **"Mercury Exposure Effects the Reproductive Success of a Free-Living Terrestrial Songbird, the Carolina Wren"** (**The Auk, 2011: 759–769;** this is a publication of the American Ornithologists' Union).

.20 .22 .25 .30 .34 .41 .55 .56
1.42 1.70 1.83 2.20 2.25 3.07 3.25

a. Determine the values of the sample mean and sample median and explain why they are different. [Hint: $\Sigma x_1 = 18.55$.]
b. Determine the value of the 10% trimmed mean and compare to the mean and median.
c. By how much could the observation .20 be increased without impacting the value of the sample median?

36. A sample of 26 offshore oil workers took part in a simulated escape exercise, resulting in the accompanying data on time (sec) to complete the escape (**"Oxygen Consumption and Ventilation During Escape from an Offshore Platform,"** **Ergonomics, 1997: 281–292**):

389 356 359 363 375 424 325 394 402
373 373 370 364 366 364 325 339 393
392 369 374 359 356 403 334 397

a. Construct a stem-and-leaf display of the data. How does it suggest that the sample mean and median will compare?

b. Calculate the values of the sample mean and median. [Hint: $\Sigma x_i = 9638$.]
c. By how much could the largest time, currently 424, be increased without affecting the value of the sample median? By how much could this value be decreased without affecting the value of the sample median?
d. What are the values of $\bar{x}$ and $\tilde{x}$ when the observations are reexpressed in minutes?

37. The article **"Snow Cover and Temperature Relationships in North America and Eurasia"** (**J. Climate and Applied Meteorology, 1983: 460–469**) used statistical techniques to relate the amount of snow cover on each continent to average continental temperature. Data presented there included the following ten observations on October snow cover for Eurasia during the years 1970–1979 (in million km^2):

6.5 12.0 14.9 10.0 10.7 7.9 21.9 12.5 14.5 9.2

What would you report as a representative, or typical, value of October snow cover for this period, and what prompted your choice?

38. Blood pressure values are often reported to the nearest 5 mmHg (100, 105, 110, etc.). Suppose the actual blood pressure values for nine randomly selected individuals are

118.6 127.4 138.4 130.0 113.7 122.0 108.3
131.5 133.2

a. What is the median of the *reported* blood pressure values?
b. Suppose the blood pressure of the second individual is 127.6 rather than 127.4 (a small change in a single value). How does this affect the median of the reported values? What does this say about the sensitivity of the median to rounding or grouping in the data?

39. The propagation of fatigue cracks in various aircraft parts has been the subject of extensive study in recent years. The accompanying data consists of propagation lives (flight hours/10^4) to reach a given crack size in fastener holes intended for use in military aircraft (**"Statistical Crack Propagation in Fastener Holes Under Spectrum Loading,"** **J. Aircraft, 1983: 1028–1032**):

.736 .863 .865 .913 .915 .937 .983 1.007
1.011 1.064 1.109 1.132 1.140 1.153 1.253 1.394

a. Compute and compare the values of the sample mean and median.
b. By how much could the largest sample observation be decreased without affecting the value of the median?

40. Compute the sample median, 25% trimmed mean, 10% trimmed mean, and sample mean for the lifetime data given in Exercise 27, and compare these measures.

41. A sample of $n = 10$ automobiles was selected, and each was subjected to a 5-mph crash test. Denoting a car with no visible damage by S (for success) and a car with such damage by F, results were as follows:

S S F S S S F F S S

a. What is the value of the sample proportion of successes x/n?

b. Replace each S with a 1 and each F with a 0. Then calculate $\bar{x}$ for this numerically coded sample. How does $\bar{x}$ compare to x/n?

c. Suppose it is decided to include 15 more cars in the experiment. How many of these would have to be S's to give $x/n = .80$ for the entire sample of 25 cars?

42. a. If a constant c is added to each x_i in a sample, yielding $y_i = x_i + c$, how do the sample mean and median of the y_is relate to the mean and median of the x_is? Verify your conjectures.

b. If each x_i is multiplied by a constant c, yielding $y_i = cx_i$, answer the question of part (a). Again, verify your conjectures.

43. An experiment to study the lifetime (in hours) for a certain type of component involved putting ten components into operation and observing them for 100 hours. Eight of the components failed during that period, and those lifetimes were recorded. Denote the lifetimes of the two components still functioning after 100 hours by 100+. The resulting sample observations were

48   79   100+   35   92   86   57   100+   17   29

Which of the measures of center discussed in this section can be calculated, and what are the values of those measures? [*Note:* The data from this experiment is said to be "censored on the right."]

1.4 Measures of Variability

Reporting a measure of center gives only partial information about a data set or distribution. Several samples or populations may have identical measures of center yet differ from one another in other important ways. Figure 1.18 shows dotplots of three samples with the same mean and median, yet the extent of spread about the center is different for all three samples. The first sample has the largest amount of variability, the third has the smallest amount, and the second is intermediate to the other two in this respect.

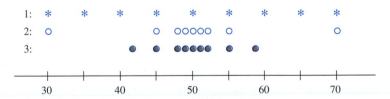

Figure 1.18 Samples with identical measures of center but different amounts of variability

Measures of Variability for Sample Data

The simplest measure of variability in a sample is the **range**, which is the difference between the largest and smallest sample values. The value of the range for sample 1 in Figure 1.18 is much larger than it is for sample 3, reflecting more variability in the first sample than in the third. A defect of the range, though, is that it depends on only the two most extreme observations and disregards the positions of the remaining values. Samples 1 and 2 in Figure 1.18 have identical ranges, yet when the observations between the two extremes are taken into account, there is much less variability or dispersion in the second sample than in the first.

Our primary measures of variability involve the **deviations from the mean**, $x_1 - \bar{x}, x_2 - \bar{x}, \ldots, x_n - \bar{x}$. That is, the deviations from the mean are obtained by subtracting $\bar{x}$ from each of the n sample observations. A deviation will be positive if the observation is larger than the mean (to the right of the mean on the measurement axis) and negative if the observation is smaller than the mean. If all the deviations are small in magnitude, then all x_i's are close to the mean and there is little variability. Alternatively, if some of the deviations are large in magnitude, then some x_i's lie far from $\bar{x}$, suggesting a greater amount of variability. A simple way to combine the deviations into a single quantity is to average them. Unfortunately, this is a bad idea:

$$\text{sum of deviations} = \sum_{i=1}^{n}(x_i - \bar{x}) = 0$$

so that the average deviation is always zero. The verification uses several standard rules of summation and the fact that $\Sigma \bar{x} = \bar{x} + \bar{x} + \cdots + \bar{x} = n\bar{x}$:

$$\sum(x_i - \bar{x}) = \sum x_i - \sum \bar{x} = \sum x_i - n\bar{x} = \sum x_i - n\left(\frac{1}{n}\sum x_i\right) = 0$$

There are several ways to prevent negative and positive deviations from counteracting one another when they are combined. One possibility is to work with the absolute values of the deviations and calculate the average absolute deviation $\Sigma|x_i - \bar{x}|/n$. Because the absolute value operation leads to a number of theoretical difficulties, consider instead the squared deviations $(x_1 - \bar{x})^2, (x_2 - \bar{x})^2, \ldots, (x_n - \bar{x})^2$. Rather than use the average squared deviation $\Sigma(x_i - \bar{x})^2/n$, for several reasons we divide the sum of squared deviations by $n - 1$ instead of n.

DEFINITION

> The **sample variance,** denoted by s^2, is given by
>
> $$s^2 = \frac{\sum(x_i - \bar{x})^2}{n - 1} = \frac{S_{xx}}{n - 1}$$
>
> The **sample standard deviation,** denoted by s, is the (positive) square root of the variance:
>
> $$s = \sqrt{s^2}$$

Note that s^2 and s are both nonnegative. One appealing property of the standard deviation is that the unit for s is the same as the unit for each of the x_i's. If, for example, the observations are fuel efficiencies in miles per gallon, then we might have $s = 2.0$ mpg. The sample standard deviation can be interpreted as roughly the size of a typical or representative deviation from the sample mean within the given sample. Thus if $s = 2.0$ mpg, then some x_i's in the sample are closer than 2.0 to $\bar{x}$, whereas others are farther away; 2.0 is a representative (or "standard") deviation from the mean fuel efficiency. If $s = 3.0$ for a second sample of cars of another type, a typical deviation in this sample is roughly 1.5 times what it is in the first sample, an indication of more variability in the second sample.

EXAMPLE 1.17   The Web site **www.fueleconomy.gov** contains a wealth of information about fuel characteristics of various vehicles. In addition to EPA mileage ratings, there are many vehicles for which users have reported their own values of fuel efficiency (mpg). Consider the following sample of $n = 11$ efficiencies for the 2009 Ford

Focus equipped with an automatic transmission (for this model, EPA reports an overall rating of 27 mpg–24 mpg for city driving and 33 mpg for highway driving):

Car	x_i	$x_i - \bar{x}$	$(x_i - \bar{x})^2$
1	27.3	−5.96	35.522
2	27.9	−5.36	28.730
3	32.9	−0.36	0.130
4	35.2	1.94	3.764
5	44.9	11.64	135.490
6	39.9	6.64	44.090
7	30.0	−3.26	10.628
8	29.7	−3.56	12.674
9	28.5	−4.76	22.658
10	32.0	−1.26	1.588
11	37.6	4.34	18.836
	$\Sigma x_i = 365.9$	$\Sigma(x_i - \bar{x}) = .04$	$\Sigma(x_i - \bar{x})^2 = 314.110$	$\bar{x} = 33.26$

Effects of rounding account for the sum of deviations not being exactly zero. The numerator of s^2 is $S_{xx} = 314.110$, from which

$$s^2 = \frac{S_{xx}}{n - 1} = \frac{314.110}{11 - 1} = 31.41, \qquad s = 5.60$$

The size of a representative deviation from the sample mean 33.26 is roughly 5.6 mpg. Note: Of the nine people who also reported driving behavior, only three did more than 80% of their driving in highway mode; we bet you can guess which cars they drove. We haven't a clue why all 11 reported values exceed the EPA figure—maybe only drivers with really good fuel efficiencies communicate their results. ∎

Motivation for s^2

To explain the rationale for the divisor $n - 1$ in s^2, note first that whereas s^2 measures sample variability, there is a measure of variability in the population called the **population variance**. We will use σ^2 (the square of the lowercase Greek letter sigma) to denote the population variance and σ to denote the **population standard deviation** (the square root of σ^2). The value of σ can be interpreted as roughly the size of a typical deviation from μ within the entire population of x values. When the population is finite and consists of N values,

$$\sigma^2 = \sum_{i=1}^{N} (x_i - \mu)^2 / N$$

which is the average of all squared deviations from the population mean (for the population, the divisor is N and not $N - 1$). More general definitions of σ^2 appear in Chapters 3 and 4.

Just as $\bar{x}$ will be used to make inferences about the population mean μ, we should define the sample variance so that it can be used to make inferences about σ^2. Now note that σ^2 involves squared deviations about the population mean μ. If we actually knew the value of μ, then we could define the sample variance as the average squared deviation of the sample x_i's about μ. However, the value of μ is almost never known, so the sum of squared deviations about $\bar{x}$ must be used. But *the x_i's tend to be closer to their average $\bar{x}$ than to the population average μ. To compensate for this,* the divisor $n - 1$ is used rather than the sample size n. In other words, if we used a divisor n in the sample variance, then the resulting quantity would tend to underestimate σ^2 (produce estimated values that are too small on the average), whereas dividing by the slightly smaller $n - 1$ corrects this underestimating.

It is customary to refer to s^2 as being based on $n - 1$ **degrees of freedom** (df). This terminology reflects the fact that although s^2 is based on the n quantities $x_1 - \bar{x}, x_2 - \bar{x}, \ldots, x_n - \bar{x}$, these sum to 0, so specifying the values of any $n - 1$ of the quantities determines the remaining value. For example, if $n = 4$ and $x_1 - \bar{x} = 8$, $x_2 - \bar{x} = -6$, and $x_4 - \bar{x} = -4$, then automatically $x_3 - \bar{x} = 2$, so only three of the four values of $x_i - \bar{x}$ are freely determined (3 df).

A Computing Formula for s^2

It is best to obtain s^2 from statistical software or else use a calculator that allows you to enter data into memory and then view s^2 with a single keystroke. If your calculator does not have this capability, there is an alternative formula that avoids calculating the deviations. The formula involves both $(\Sigma x_i)^2$, summing and then squaring, and Σx_i^2, squaring and then summing.

An alternative expression for the numerator of s^2 is

$$S_{xx} = \sum (x_i - \bar{x})^2 = \sum x_i^2 - \frac{\left(\sum x_i\right)^2}{n}$$

Proof Because $\bar{x} = \Sigma x_i / n$, $n(\bar{x})^2 = (\Sigma x_i)^2 / n$. Then,

$$\sum (x_i - \bar{x})^2 = \sum (x_i^2 - 2\bar{x} \cdot x_i + \bar{x}^2) = \sum x_i^2 - 2\bar{x} \sum x_i + \sum (\bar{x})^2$$
$$= \sum x_i^2 - 2\bar{x} \cdot n\bar{x} + n(\bar{x})^2 = \sum x_i^2 - n(\bar{x})^2 \quad \blacksquare$$

EXAMPLE 1.18 Traumatic knee dislocation often requires surgery to repair ruptured ligaments. One measure of recovery is range of motion (measured as the angle formed when, starting with the leg straight, the knee is bent as far as possible). The given data on post-surgical range of motion appeared in the article **"Reconstruction of the Anterior and Posterior Cruciate Ligaments After Knee Dislocation"** (*Amer. J. Sports Med.*, **1999: 189–197**):

154 142 137 133 122 126 135 135 108 120 127 134 122

The sum of these 13 sample observations is $\Sigma x_i = 1695$, and the sum of their squares is

$$\sum x_i^2 = (154)^2 + (142)^2 + \cdots + (122)^2 = 222{,}581$$

Thus the numerator of the sample variance is

$$S_{xx} = \sum x_i^2 - \left[\left(\sum x_i\right)^2\right]/n = 222{,}581 - (1695)^2/13 = 1579.0769$$

from which $s^2 = 1579.0769/12 = 131.59$ and $s = 11.47$. $\blacksquare$

Both the defining formula and the computational formula for s^2 can be sensitive to rounding, so as much decimal accuracy as possible should be used in intermediate calculations.

Several other properties of s^2 enhance understanding and facilitate computation.

PROPOSITION

Let $x_1, x_2, \ldots, x_n$ be a sample and c be any nonzero constant.

1. If $y_1 = x_1 + c, y_2 = x_2 + c, \ldots, y_n = x_n + c$, then $s_y^2 = s_x^2$, and
2. If $y_1 = cx_1, \ldots, y_n = cx_n$, then $s_y^2 = c^2 s_x^2$, $s_y = |c| s_x$

where s_x^2 is the sample variance of the x's and s_y^2 is the sample variance of the y's.

In words, Result 1 says that the variance is unchanged when a constant c is added to (or subtracted from) each data value. This is intuitive, since adding or subtracting c shifts the location of the data set but leaves distances between data values unchanged. According to Result 2, multiplication of each x_i by c results in s^2 being multiplied by a factor of c^2. These properties can be proved by noting in Result 1 that $\bar{y} = \bar{x} + c$ and in Result 2 that $\bar{y} = c\bar{x}$.

Boxplots

Stem-and-leaf displays and histograms convey rather general impressions about a data set, whereas a single summary such as the mean or standard deviation focuses on just one aspect of the data. In recent years, a pictorial summary called a *boxplot* has been used successfully to describe several of a data set's most prominent features. These features include (1) center, (2) spread, (3) the extent and nature of any departure from symmetry, and (4) identification of "outliers," observations that lie unusually far from the main body of the data. Because even a single outlier can drastically affect the values of $\bar{x}$ and s, a boxplot is based on measures that are "resistant" to the presence of a few outliers—the median and a measure of variability called the *fourth spread*.

DEFINITION

Order the n observations from smallest to largest and separate the smallest half from the largest half; the median $\tilde{x}$ is included in both halves if n is odd. Then the **lower fourth** is the median of the smallest half and the **upper fourth** is the median of the largest half. A measure of spread that is resistant to outliers is the **fourth spread** f_s, given by

$$f_s = \text{upper fourth} - \text{lower fourth}$$

Roughly speaking, the fourth spread is unaffected by the positions of those observations in the smallest 25% or the largest 25% of the data. Hence it is resistant to outliers. The fourths are very similar to quartiles, and the fourth spread is similar to the *interquartile range*, the difference between the upper and lower quartiles. But quartiles are a bit more cumbersome than fourths to calculate by hand, and there are several different sensible ways to compute the quartiles (so values may vary from one software package to another).

The simplest boxplot is based on the following five-number summary:

$$\text{smallest } x_i \quad \text{lower fourth} \quad \text{median} \quad \text{upper fourth} \quad \text{largest } x_i$$

First, draw a rectangle above a horizontal measurement scale; the left edge of the rectangle is above the lower fourth, and the right edge is above the upper fourth (so box width $= f_s$). Place a vertical line segment or some other symbol inside the rectangle at the location of the median; the position of the median symbol relative to the two edges conveys information about skewness in the middle 50% of the data. Finally, draw "whiskers" out from either end of the rectangle to the smallest and largest observations. A boxplot with a vertical orientation can also be drawn by making obvious modifications in the construction process.

EXAMPLE 1.19   The accompanying data consists of observations on the time until failure (1000s of hours) for a sample of turbochargers from one type of engine (from **"The Beta Generalized Weibull Distribution: Properties and Applications,"** *Reliability Engr. and System Safety*, **2012: 5–15**).

1.6	2.0	2.6	3.0	3.5	3.9	4.5	4.6	4.8	5.0
5.1	5.3	5.4	5.6	5.8	6.0	6.0	6.1	6.3	6.5
6.5	6.7	7.0	7.1	7.3	7.3	7.3	7.7	7.7	7.8
7.9	8.0	8.1	8.3	8.4	8.4	8.5	8.7	8.8	9.0

The five-number summary is as follows.

> smallest: 1.6 lower fourth: 5.05 median: 6.5 upper fourth: 7.85 largest: 9.0

Figure 1.19 shows Minitab output from a request to describe the data. Q1 and Q3 are the lower and upper quartiles, respectively, and IQR (interquartile range) is the difference between these quartiles. SE Mean is $s/\sqrt{n}$, the "standard error of the mean"; it will be important in our subsequent development of several widely used procedures for making inferences about the population mean μ.

Variable	Count	Mean	SE Mean	StDev	Minimum	Q1	Median	Q3	Maximum	IQR
lifetime	40	6.253	0.309	1.956	1.600	5.025	6.500	7.875	9.000	2.850

Figure 1.19 Minitab description of the turbocharger lifetime data

Figure 1.20 shows both a dotplot of the data and a boxplot. Both plots indicate that there is a reasonable amount of symmetry in the middle 50% of the data, but overall values stretch out more toward the low end than toward the high end—a negative skew. The box itself is not very narrow, indicating a fair amount of variability in the middle half of the data, and the lower whisker is especially long.

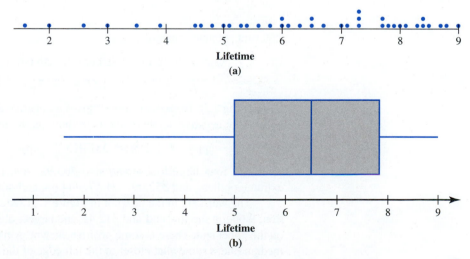

Figure 1.20 (a) Dotplot and (b) Boxplot for the lifetime data

Boxplots That Show Outliers

A boxplot can be embellished to indicate explicitly the presence of outliers. Many inferential procedures are based on the assumption that the population distribution is normal (a certain type of bell curve). Even a single extreme outlier in the sample warns the investigator that such procedures may be unreliable, and the presence of several mild outliers conveys the same message.

DEFINITION

> Any observation farther than $1.5f_s$ from the closest fourth is an **outlier.** An outlier is **extreme** if it is more than $3f_s$ from the nearest fourth, and it is **mild** otherwise.

Let's modify our previous construction of a boxplot by drawing a whisker out from each end of the box to the smallest and largest observations that are *not* outliers. Now represent each mild outlier by a closed circle and each extreme outlier by an open circle. Some statistical computer packages do not distinguish between mild and extreme outliers.

EXAMPLE 1.20 The Clean Water Act and subsequent amendments require that all waters in the United States meet specific pollution reduction goals to ensure that water is "fishable and swimmable." The article **"Spurious Correlation in the USEPA Rating Curve Method for Estimating Pollutant Loads" (*J. of Environ. Engr.*, 2008: 610–618)** investigated various techniques for estimating pollutant loads in watersheds; the authors "discuss the imperative need to use sound statistical methods" for this purpose. Among the data considered is the following sample of TN (total nitrogen) loads (kg N/day) from a particular Chesapeake Bay location, displayed here in increasing order.

9.69	13.16	17.09	18.12	23.70	24.07	24.29	26.43
30.75	31.54	35.07	36.99	40.32	42.51	45.64	48.22
49.98	50.06	55.02	57.00	58.41	61.31	64.25	65.24
66.14	67.68	81.40	90.80	92.17	92.42	100.82	101.94
103.61	106.28	106.80	108.69	114.61	120.86	124.54	143.27
143.75	149.64	167.79	182.50	192.55	193.53	271.57	292.61
312.45	352.09	371.47	444.68	460.86	563.92	690.11	826.54
1529.35							

Relevant summary quantities are

$$\tilde{x} = 92.17 \qquad \text{lower } 4^{th} = 45.64 \qquad \text{upper } 4^{th} = 167.79$$

$$f_s = 122.15 \qquad 1.5f_s = 183.225 \qquad 3f_s = 366.45$$

Subtracting $1.5f_s$ from the lower 4th gives a negative number, and none of the observations are negative, so there are no outliers on the lower end of the data. However,

$$\text{upper } 4^{th} + 1.5f_s = 351.015 \qquad \text{upper } 4^{th} + 3f_s = 534.24$$

Thus the four largest observations—563.92, 690.11, 826.54, and 1529.35—are extreme outliers, and 352.09, 371.47, 444.68, and 460.86 are mild outliers.

The whiskers in the boxplot in Figure 1.21 extend out to the smallest observation, 9.69, on the low end and 312.45, the largest observation that is not an outlier, on the upper end. There is some positive skewness in the middle half of the data (the median line is somewhat closer to the left edge of the box than to the right edge) and a great deal of positive skewness overall.

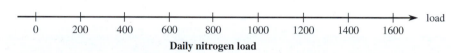

Figure 1.21 A boxplot of the nitrogen load data showing mild and extreme outliers ■

Comparative Boxplots

A comparative or side-by-side boxplot is a very effective way of revealing similarities and differences between two or more data sets consisting of observations on the same variable—fuel efficiency observations for four different types of automobiles, crop yields for three different varieties, and so on.

EXAMPLE 1.21   High levels of sodium in food products represent a growing health concern. The accompanying data consists of values of sodium content in one serving of cereal for one sample of cereals manufactured by General Mills, another sample manufactured by Kellogg, and a third sample produced by Post (see the website **http://www .nutritionresource.com/foodcomp2.cfm?id=0800** rather than visiting your neighborhood grocery store!).

G:	211	408	171	178	359	249	205	203	201	223	234	256	218
K:	143	202	120	229	150	5	207	362	252	275	224		
P:	253	220	212	41	140	215	266	3	214	280			

Figure 1.22 shows a comparative boxplot of the data from the software package R. The typical sodium content (median) is roughly the same for all three companies. But the distributions differ markedly in other respects. The Kellogg data shows a substantial

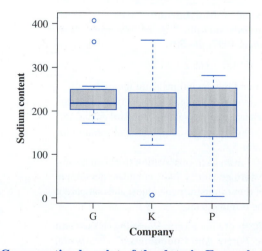

Figure 1.22   Comparative boxplot of the data in Example 1.21, from R

positive skew both in the middle 50% and overall, with two outliers at the upper end. The Kellogg data exhibits a negative skew in the middle 50% and a positive skew overall, except for the outlier at the low end (this outlier is not identified by Minitab). And the Post data is negatively skewed both in the middle 50% and overall with no outliers. Variability as assessed by the box length (here the interquartile range rather than the fourth spread) is smallest for the G brand and largest for the P brand, with the K brand intermediate to the other two; looking instead at standard deviations, s_K and s_P are roughly the same and both much larger than s_G. ■

EXERCISES Section 1.4 (44–61)

44. Poly(3-hydroxybutyrate) (PHB), a semicrystalline polymer that is fully biodegradable and biocompatible, is obtained from renewable resources. From a sustainability perspective, PHB offers many attractive properties though it is more expensive to produce than standard plastics. The accompanying data on melting point (°C) for each of 12 specimens of the polymer using a differential scanning calorimeter appeared in the article **"The Melting Behaviour of Poly(3-Hydroxybutyrate) by DSC. Reproducibility Study" (Polymer Testing, 2013: 215–220).**

180.5 181.7 180.9 181.6 182.6 181.6
181.3 182.1 182.1 180.3 181.7 180.5

Compute the following:

a. The sample range
b. The sample variance s^2 from the definition [*Hint:* First subtract 180 from each observation.]
c. The sample standard deviation
d. s^2 using the shortcut method

45. The value of Young's modulus (GPa) was determined for cast plates consisting of certain intermetallic substrates, resulting in the following sample observations (**"Strength and Modulus of a Molybdenum-Coated Ti-25Al-10Nb-3U-1Mo Intermetallic," J. of Materials Engr. and Performance, 1997: 46–50):**

116.4 115.9 114.6 115.2 115.8

a. Calculate $\bar{x}$ and the deviations from the mean.
b. Use the deviations calculated in part (a) to obtain the sample variance and the sample standard deviation.
c. Calculate s^2 by using the computational formula for the numerator S_{xx}.
d. Subtract 100 from each observation to obtain a sample of transformed values. Now calculate the sample variance of these transformed values, and compare it to s^2 for the original data.

46. The article **"Effects of Short-Term Warming on Low and High Latitude Forest Ant Communities" (Ecoshpere, May 2011, Article 62)** described an experiment in which observations on various characteristics were made using minichambers of three different types: (1) cooler (PVC frames covered with shade cloth), (2) control (PVC frames only), and (3) warmer (PVC frames covered with plastic). One of the article's authors kindly supplied the accompanying data on the difference between air and soil temperatures (°C).

Cooler	Control	Warmer
1.59	1.92	2.57
1.43	2.00	2.60
1.88	2.19	1.93
1.26	1.12	1.58
1.91	1.78	2.30
1.86	1.84	0.84
1.90	2.45	2.65
1.57	2.03	0.12
1.79	1.52	2.74
1.72	0.53	2.53
2.41	1.90	2.13
2.34		2.86
0.83		2.31
1.34		1.91
1.76		

a. Compare measures of center for the three different samples.
b. Calculate, interpret, and compare the standard deviations for the three different samples.
c. Do the fourth spreads for the three samples convey the same message as do the standard deviations about relative variability?
d. Construct a comparative boxplot (which was included in the cited article) and comment on any interesting features.

47. Zinfandel is a popular red wine varietal produced almost exclusively in California. It is rather controversial among wine connoisseurs because its alcohol content varies quite substantially from one producer to another. In May 2013, the author went to the website **klwines .com,** randomly selected 10 zinfandels from among the

325 available, and obtained the following values of alcohol content (%):

14.8   14.5   16.1   14.2   15.9
13.7   16.2   14.6   13.8   15.0

a. Calculate and interpret several measures of center.
b. Calculate the sample variance using the defining formula.
c. Calculate the sample variance using the shortcut formula after subtracting 13 from each observation.

48. Exercise 34 presented the following data on endotoxin concentration in settled dust both for a sample of urban homes and for a sample of farm homes:

U: 6.0 5.0 11.0 33.0 4.0 5.0 80.0 18.0 35.0 17.0 23.0
F: 4.0 14.0 11.0 9.0 9.0 8.0 4.0 20.0 5.0 8.9 21.0
    9.2 3.0 2.0 0.3

a. Determine the value of the sample standard deviation for each sample, interpret these values, and then contrast variability in the two samples. [*Hint:* $\Sigma x_i = 237.0$ for the urban sample and $= 128.4$ for the farm sample, and $\Sigma x_i^2 = 10,079$ for the urban sample and 1617.94 for the farm sample.]
b. Compute the fourth spread for each sample and compare. Do the fourth spreads convey the same message about variability that the standard deviations do? Explain.
c. The authors of the cited article also provided endotoxin concentrations in dust bag dust:

U: 34.0 49.0 13.0 33.0 24.0 24.0 35.0 104.0 34.0 40.0 38.0 1.0
F: 2.0 64.0 6.0 17.0 35.0 11.0 17.0 13.0 5.0 27.0 23.0
   28.0 10.0 13.0 0.2

Construct a comparative boxplot (as did the cited paper) and compare and contrast the four samples.

49. A study of the relationship between age and various visual functions (such as acuity and depth perception) reported the following observations on the area of scleral lamina (mm^2) from human optic nerve heads (**"Morphometry of Nerve Fiber Bundle Pores in the Optic Nerve Head of the Human,"** *Experimental Eye Research*, 1988: 559–568):

2.75 2.62 2.74 3.85 2.34 2.74 3.93 4.21 3.88
4.33 3.46 4.52 2.43 3.65 2.78 3.56 3.01

a. Calculate Σx_i and Σx_i^2.
b. Use the values calculated in part (a) to compute the sample variance s^2 and then the sample standard deviation s.

50. In 1997 a woman sued a computer keyboard manufacturer, charging that her repetitive stress injuries were caused by the keyboard (*Genessy v. Digital Equipment Corp.*). The injury awarded about $3.5 million for pain and suffering, but the court then set aside that award as being unreasonable compensation. In making this

determination, the court identified a "normative" group of 27 similar cases and specified a reasonable award as one within two standard deviations of the mean of the awards in the 27 cases. The 27 awards were (in $1000s) 37, 60, 75, 115, 135, 140, 149, 150, 238, 290, 340, 410, 600, 750, 750, 750, 1050, 1100, 1139, 1150, 1200, 1200, 1250, 1576, 1700, 1825, and 2000, from which $\Sigma x_i = 20,179$, $\Sigma x_i^2 = 24,657,511$. What is the maximum possible amount that could be awarded under the two-standard-deviation rule?

51. The article **"A Thin-Film Oxygen Uptake Test for the Evaluation of Automotive Crankcase Lubricants"** (*Lubric. Engr.*, 1984: 75–83) reported the following data on oxidation-induction time (min) for various commercial oils:

87 103 130 160 180 195 132 145 211 105 145
153 152 138 87 99 93 119 129

a. Calculate the sample variance and standard deviation.
b. If the observations were reexpressed in hours, what would be the resulting values of the sample variance and sample standard deviation? Answer without actually performing the reexpression.

52. The first four deviations from the mean in a sample of $n = 5$ reaction times were .3, .9, 1.0, and 1.3. What is the fifth deviation from the mean? Give a sample for which these are the five deviations from the mean.

53. A mutual fund is a professionally managed investment scheme that pools money from many investors and invests in a variety of securities. Growth funds focus primarily on increasing the value of investments, whereas blended funds seek a balance between current income and growth. Here is data on the expense ratio (expenses as a % of assets, from **www .morningstar.com**) for samples of 20 large-cap balanced funds and 20 large-cap growth funds ("large-cap" refers to the sizes of companies in which the funds invest; the population sizes are 825 and 762, respectively):

Bl	1.03	1.23	1.10	1.64	1.30
	1.27	1.25	0.78	1.05	0.64
	0.94	2.86	1.05	0.75	0.09
	0.79	1.61	1.26	0.93	0.84

Gr	0.52	1.06	1.26	2.17	1.55
	0.99	1.10	1.07	1.81	2.05
	0.91	0.79	1.39	0.62	1.52
	1.02	1.10	1.78	1.01	1.15

a. Calculate and compare the values of $\bar{x}$, $\tilde{x}$, and s for the two types of funds.
b. Construct a comparative boxplot for the two types of funds, and comment on interesting features.

54. Grip is applied to produce normal surface forces that compress the object being gripped. Examples include two people shaking hands, or a nurse squeezing a patient's forearm to stop bleeding. The article **"Investigation of Grip Force, Normal Force, Contact Area, Hand Size, and Handle Size for Cylindrical Handles"** (*Human Factors*, **2008: 734–744**) included the following data on grip strength (N) for a sample of 42 individuals:

```
 16   18   18   26   33   41   54   56   66   68   87   91   95
 98  106  109  111  118  127  127  135  145  147  149  151  168
172  183  189  190  200  210  220  229  230  233  238  244  259
294  329  403
```

   a. Construct a stem-and-leaf display based on repeating each stem value twice, and comment on interesting features.

   b. Determine the values of the fourths and the fourthspread.

   c. Construct a boxplot based on the five-number summary, and comment on its features.

   d. How large or small does an observation have to be to qualify as an outlier? An extreme outlier? Are there any outliers?

   e. By how much could the observation 403, currently the largest, be decreased without affecting f_s?

55. Here is a stem-and-leaf display of the escape time data introduced in Exercise 36 of this chapter.

32	55
33	49
34	
35	6699
36	34469
37	03345
38	9
39	2347
40	23
41	
42	4

   a. Determine the value of the fourth spread.

   b. Are there any outliers in the sample? Any extreme outliers?

   c. Construct a boxplot and comment on its features.

   d. By how much could the largest observation, currently 424, be decreased without affecting the value of the fourth spread?

56. The following data on distilled alcohol content (%) for a sample of 35 port wines was extracted from the article **"A Method for the Estimation of Alcohol in Fortified Wines Using Hydrometer Baumé and Refractometer Brix"** (*Amer. J. Enol. Vitic.*, **2006: 486–490**). Each value is an average of two duplicate measurements.

```
16.35 18.85 16.20 17.75 19.58 17.73 22.75 23.78 23.25
19.08 19.62 19.20 20.05 17.85 19.17 19.48 20.00 19.97
17.48 17.15 19.07 19.90 18.68 18.82 19.03 19.45 19.37
19.20 18.00 19.60 19.33 21.22 19.50 15.30 22.25
```

Use methods from this chapter, including a boxplot that shows outliers, to describe and summarize the data.

57. A sample of 20 glass bottles of a particular type was selected, and the internal pressure strength of each bottle was determined. Consider the following partial sample information:

median = 202.2   lower fourth = 196.0
upper fourth = 216.8

| *Three smallest observations* | 125.8 | 188.1 | 193.7 |
| *Three largest observations* | 221.3 | 230.5 | 250.2 |

   a. Are there any outliers in the sample? Any extreme outliers?

   b. Construct a boxplot that shows outliers, and comment on any interesting features.

58. A company utilizes two different machines to manufacture parts of a certain type. During a single shift, a sample of $n = 20$ parts produced by each machine is obtained, and the value of a particular critical dimension for each part is determined. The comparative boxplot at the bottom of this page is constructed from the resulting data. Compare and contrast the two samples.

Comparative boxplot for Exercise 58

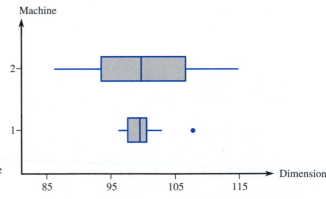

59. Blood cocaine concentration (mg/L) was determined both for a sample of individuals who had died from cocaine-induced excited delirium (ED) and for a sample of those who had died from a cocaine overdose without excited delirium; survival time for people in both groups was at most 6 hours. The accompanying data was read from a comparative boxplot in the article **"Fatal Excited Delirium Following Cocaine Use"** (*J. of Forensic Sciences*, **1997: 25–31**).

ED	0	0	0	0	.1	.1	.1	.1	.2	.2	.3	.3
		.3	.4	.5	.7	.8	1.0	1.5	2.7	2.8		
		3.5	4.0	8.9	9.2	11.7	21.0					
Non-ED	0	0	0	0	0	.1	.1	.1	.1	.2	.2	.2
		.3	.3	.3	.4	.5	.5	.6	.8	.9	1.0	
		1.2	1.4	1.5	1.7	2.0	3.2	3.5	4.1			
		4.3	4.8	5.0	5.6	5.9	6.0	6.4	7.9			
		8.3	8.7	9.1	9.6	9.9	11.0	11.5				
		12.2	12.7	14.0	16.6	17.8						

a. Determine the medians, fourths, and fourth spreads for the two samples.

b. Are there any outliers in either sample? Any extreme outliers?

c. Construct a comparative boxplot, and use it as a basis for comparing and contrasting the ED and non-ED samples.

60. Observations on burst strength (lb/in²) were obtained both for test nozzle closure welds and for production cannister nozzle welds (**"Proper Procedures Are the Key to Welding Radioactive Waste Cannisters,"** *Welding J.,* **Aug. 1997: 61–67**).

Test	7200	6100	7300	7300	8000	7400
	7300	7300	8000	6700	8300	
Cannister	5250	5625	5900	5900	5700	6050
	5800	6000	5875	6100	5850	6600

Construct a comparative boxplot and comment on interesting features (the cited article did not include such a picture, but the authors commented that they had looked at one).

61. The accompanying comparative boxplot of gasoline vapor coefficients for vehicles in Detroit appeared in the article **"Receptor Modeling Approach to VOC Emission Inventory Validation"** (*J. of Envir. Engr.,* **1995: 483–490**). Discuss any interesting features.

Comparative boxplot for Exercise 61

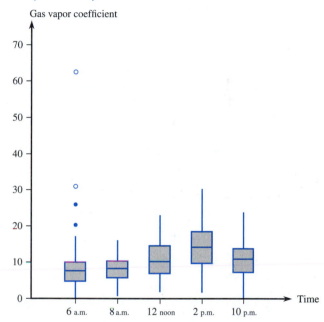

SUPPLEMENTARY EXERCISES (62–83)

62. Consider the following information on ultimate tensile strength (lb/in) for a sample of $n = 4$ hard zirconium copper wire specimens (from **"Characterization Methods for Fine Copper Wire,"** *Wire J. Intl.,* **Aug., 1997: 74–80**):

$\bar{x} = 76{,}831 \quad s = 180 \quad$ smallest $x_i = 76{,}683$
largest $x_i = 77{,}048$

Determine the values of the two middle sample observations (and don't do it by successive guessing!).

63. A sample of 77 individuals working at a particular office was selected and the noise level (dBA) experienced by each individual was determined, yielding the following data (**"Acceptable Noise Levels for Construction Site Offices,"** *Building Serv. Engr. Research and Technology,* **2009: 87–94**).

55.3 55.3 55.3 55.9 55.9 55.9 55.9 56.1 56.1 56.1 56.1
56.1 56.1 56.8 56.8 57.0 57.0 57.0 57.8 57.8 57.8 57.9
57.9 57.9 58.8 58.8 58.8 59.8 59.8 59.8 62.2 62.2 63.8
63.8 63.8 63.9 63.9 63.9 64.7 64.7 64.7 65.1 65.1 65.1
65.3 65.3 65.3 65.3 67.4 67.4 67.4 67.4 68.7 68.7 68.7
68.7 69.0 70.4 70.4 71.2 71.2 71.2 73.0 73.0 73.1 73.1
74.6 74.6 74.6 74.6 79.3 79.3 79.3 79.3 83.0 83.0 83.0

Use various techniques discussed in this chapter to organize, summarize, and describe the data.

64. Fretting is a wear process that results from tangential oscillatory movements of small amplitude in machine parts. The article **"Grease Effect on Fretting Wear of Mild Steel"** (*Industrial Lubrication and Tribology,* **2008: 67–78**) included the following data on volume

wear (10^{-4}mm^3) for base oils having four different viscosities.

Viscosity			Wear			
20.4	58.8	30.8	27.3	29.9	17.7	76.5
30.2	44.5	47.1	48.7	41.6	32.8	18.3
89.4	73.3	57.1	66.0	93.8	133.2	81.1
252.6	30.6	24.2	16.6	38.9	28.7	23.6

a. The *sample coefficient of variation* $100s/\bar{x}$ assesses the extent of variability relative to the mean (specifically, the standard deviation as a percentage of the mean). Calculate the coefficient of variation for the sample at each viscosity. Then compare the results and comment.

b. Construct a comparative boxplot of the data and comment on interesting features.

65. The accompanying frequency distribution of fracture strength (MPa) observations for ceramic bars fired in a particular kiln appeared in the article **"Evaluating Tunnel Kiln Performance" (*Amer. Ceramic Soc. Bull.*, Aug. 1997: 59–63).**

Class	81–<83	83–<85	85–<87	87–<89	89–<91
Frequency	6	7	17	30	43
Class	91–<93	93–<95	95–<97	97–<99	
Frequency	28	22	13	3	

a. Construct a histogram based on relative frequencies, and comment on any interesting features.

b. What proportion of the strength observations are at least 85? Less than 95?

c. Roughly what proportion of the observations are less than 90?

66. A deficiency of the trace element selenium in the diet can negatively impact growth, immunity, muscle and neuromuscular function, and fertility. The introduction of selenium supplements to dairy cows is justified when pastures have low selenium levels. Authors of the article **"Effects of Short-Term Supplementation with Selenised Yeast on Milk Production and Composition of Lactating Cows" (*Australian J. of Dairy Tech.*, 2004: 199–203)** supplied the following data on milk selenium concentration (mg/L) for a sample of cows given a selenium supplement and a control sample given no supplement, both initially and after a 9-day period.

Obs	Init Se	Init Cont	Final Se	Final Cont
1	11.4	9.1	138.3	9.3
2	9.6	8.7	104.0	8.8
3	10.1	9.7	96.4	8.8
4	8.5	10.8	89.0	10.1
5	10.3	10.9	88.0	9.6
6	10.6	10.6	103.8	8.6

Obs	Init Se	Init Cont	Final Se	Final Cont
7	11.8	10.1	147.3	10.4
8	9.8	12.3	97.1	12.4
9	10.9	8.8	172.6	9.3
10	10.3	10.4	146.3	9.5
11	10.2	10.9	99.0	8.4
12	11.4	10.4	122.3	8.7
13	9.2	11.6	103.0	12.5
14	10.6	10.9	117.8	9.1
15	10.8		121.5	
16	8.2		93.0	

a. Do the initial Se concentrations for the supplement and control samples appear to be similar? Use various techniques from this chapter to summarize the data and answer the question posed.

b. Again use methods from this chapter to summarize the data and then describe how the final Se concentration values in the treatment group differ from those in the control group.

67. *Aortic stenosis* refers to a narrowing of the aortic valve in the heart. The article **"Correlation Analysis of Stenotic Aortic Valve Flow Patterns Using Phase Contrast MRI" (*Annals of Biomed. Engr.*, 2005: 878–887)** gave the following data on aortic root diameter (cm) and gender for a sample of patients having various degrees of aortic stenosis:

M: 3.7 3.4 3.7 4.0 3.9 3.8 3.4 3.6 3.1 4.0 3.4 3.8 3.5
F: 3.8 2.6 3.2 3.0 4.3 3.5 3.1 3.1 3.2 3.0

a. Compare and contrast the diameter observations for the two genders.

b. Calculate a 10% trimmed mean for each of the two samples, and compare to other measures of center (for the male sample, the interpolation method mentioned in Section 1.3 must be used).

68. a. For what value of c is the quantity $\Sigma(x_i - c)^2$ minimized? [*Hint:* Take the derivative with respect to c, set equal to 0, and solve.]

b. Using the result of part (a), which of the two quantities $\Sigma(x_i - \bar{x})^2$ and $\Sigma(x_i - \mu)^2$ will be smaller than the other (assuming that $\bar{x} \neq \mu$)?

69. a. Let a and b be constants and let $y_i = ax_i + b$ for $i = 1, 2, \ldots, n$. What are the relationships between $\bar{x}$ and $\bar{y}$ and between s_x^2 and s_y^2?

b. A sample of temperatures for initiating a certain chemical reaction yielded a sample average (°C) of 87.3 and a sample standard deviation of 1.04. What are the sample average and standard deviation measured in °F? [*Hint:* $F = \frac{9}{5}C + 32$.]

70. Elevated energy consumption during exercise continues after the workout ends. Because calories burned after exercise contribute to weight loss and have other consequences, it is important to understand this process. The article **"Effect of Weight Training Exercise and**

Treadmill Exercise on Post-Exercise Oxygen Consumption" (*Medicine and Science in Sports and Exercise*, **1998: 518–522**) reported the accompanying data from a study in which oxygen consumption (liters) was measured continuously for 30 minutes for each of 15 subjects both after a weight training exercise and after a treadmill exercise.

Subject	1	2	3	4	5	6	7
Weight (x)	14.6	14.4	19.5	24.3	16.3	22.1	23.0
Treadmill (y)	11.3	5.3	9.1	15.2	10.1	19.6	20.8

Subject	8	9	10	11	12	13	14	15
Weight (x)	18.7	19.0	17.0	19.1	19.6	23.2	18.5	15.9
Treadmill (y)	10.3	10.3	2.6	16.6	22.4	23.6	12.6	4.4

a. Construct a comparative boxplot of the weight and treadmill observations, and comment on what you see.

b. The data is in the form of (x, y) pairs, with x and y measurements on the same variable under two different conditions, so it is natural to focus on the differences within pairs: $d_1 = x_1 - y_1, \ldots, d_n = x_n - y_n$. Construct a boxplot of the sample differences. What does it suggest?

71. Here is a description from Minitab of the strength data given in Exercise 13.

```
Variable   N   Mean Median TrMean StDev SE Mean
strength 153 135.39 135.40 135.41  4.59    0.37

Variable   Minimum   Maximum        Q1        Q3
strength    122.20    147.70    132.95    138.25
```

a. Comment on any interesting features (the quartiles and fourths are virtually identical here).

b. Construct a boxplot of the data based on the quartiles, and comment on what you see.

72. Anxiety disorders and symptoms can often be effectively treated with benzodiazepine medications. It is known that animals exposed to stress exhibit a decrease in benzodiazepine receptor binding in the frontal cortex. The article **"Decreased Benzodiazepine Receptor Binding in Prefrontal Cortex in Combat-Related Posttraumatic Stress Disorder"** (*Amer. J. of Psychiatry*, **2000: 1120–1126**) described the first study of benzodiazepine receptor binding in individuals suffering from PTSD. The accompanying data on a receptor binding measure (adjusted distribution volume) was read from a graph in the article.

PTSD: 10, 20, 25, 28, 31, 35, 37, 38, 38, 39, 39, 42, 46

Healthy: 23, 39, 40, 41, 43, 47, 51, 58, 63, 66, 67, 69, 72

Use various methods from this chapter to describe and summarize the data.

73. The article **"Can We Really Walk Straight?"** (*Amer. J. of Physical Anthropology*, **1992: 19–27**) reported on an experiment in which each of 20 healthy men was asked to walk as straight as possible to a target 60 m away at normal speed. Consider the following observations on cadence (number of strides per second):

.95 .85 .92 .95 .93 .86 1.00 .92 .85 .81
.78 .93 .93 1.05 .93 1.06 1.06 .96 .81 .96

Use the methods developed in this chapter to summarize the data; include an interpretation or discussion wherever appropriate. [*Note:* The author of the article used a rather sophisticated statistical analysis to conclude that people cannot walk in a straight line and suggested several explanations for this.]

74. The **mode** of a numerical data set is the value that occurs most frequently in the set.

a. Determine the mode for the cadence data given in Exercise 73.

b. For a categorical sample, how would you define the modal category?

75. Specimens of three different types of rope wire were selected, and the fatigue limit (MPa) was determined for each specimen, resulting in the accompanying data.

Type 1	350	350	350	358	370	370	370	371
	371	372	372	384	391	391	392	
Type 2	350	354	359	363	365	368	369	371
	373	374	376	380	383	388	392	
Type 3	350	361	362	364	364	365	366	371
	377	377	377	379	380	380	392	

a. Construct a comparative boxplot, and comment on similarities and differences.

b. Construct a comparative dotplot (a dotplot for each sample with a common scale). Comment on similarities and differences.

c. Does the comparative boxplot of part (a) give an informative assessment of similarities and differences? Explain your reasoning.

76. The three measures of center introduced in this chapter are the mean, median, and trimmed mean. Two additional measures of center that are occasionally used are the *midrange,* which is the average of the smallest and largest observations, and the *midfourth,* which is the average of the two fourths. Which of these five measures of center are resistant to the effects of outliers and which are not? Explain your reasoning.

77. The authors of the article **"Predictive Model for Pitting Corrosion in Buried Oil and Gas Pipelines"** (*Corrosion*, **2009: 332–342**) provided the data on which their investigation was based.

a. Consider the following sample of 61 observations on maximum pitting depth (mm) of pipeline specimens buried in clay loam soil.

```
0.41  0.41   0.41  0.41  0.43   0.43   0.43  0.48  0.48
0.58  0.79   0.79  0.81  0.81   0.81   0.91  0.94  0.94
1.02  1.04   1.04  1.17  1.17   1.17   1.17  1.17  1.17
1.17  1.19   1.19  1.27  1.40   1.40   1.59  1.59  1.60
1.68  1.91   1.96  1.96  1.96   2.10   2.21  2.31  2.46
2.49  2.57   2.74  3.10  3.18   3.30   3.58  3.58  4.15
4.75  5.33   7.65  7.70  8.13  10.41  13.44
```

Construct a stem-and-leaf display in which the two largest values are shown in a last row labeled HI.

b. Refer back to (a), and create a histogram based on eight classes with 0 as the lower limit of the first class and class widths of .5, .5, .5, .5, 1, 2, 5, and 5, respectively.

c. The accompanying comparative boxplot from Minitab shows plots of pitting depth for four different types of soils. Describe its important features.

78. Consider a sample $x_1, x_2, \ldots, x_n$ and suppose that the values of $\bar{x}$, s^2, and s have been calculated.

a. Let $y_i = x_i - \bar{x}$ for $i = 1, \ldots, n$. How do the values of s^2 and s for the y_i's compare to the corresponding values for the x_i's? Explain.

b. Let $z_i = (x_i - \bar{x})/s$ for $i = 1, \ldots, n$. What are the values of the sample variance and sample standard deviation for the z_i's?

79. Let $\bar{x}_n$ and s_n^2 denote the sample mean and variance for the sample $x_1, \ldots, x_n$ and let $\bar{x}_{n+1}$ and s_{n+1}^2 denote these quantities when an additional observation x_{n+1} is added to the sample.

a. Show how $\bar{x}_{n+1}$ can be computed from $\bar{x}_n$ and x_{n+1}.

b. Show that

$$ns_{n+1}^2 = (n-1)s_n^2 + \frac{n}{n+1}(x_{n+1} - \bar{x}_n)^2$$

so that s_{n+1}^2 can be computed from x_{n+1}, $\bar{x}_n$, and s_n^2.

c. Suppose that a sample of 15 strands of drapery yarn has resulted in a sample mean thread elongation of 12.58 mm and a sample standard deviation of .512 mm. A 16th strand results in an elongation value of

11.8. What are the values of the sample mean and sample standard deviation for all 16 elongation observations?

80. Lengths of bus routes for any particular transit system will typically vary from one route to another. The article **"Planning of City Bus Routes"** (*J. of the Institution of Engineers*, **1995: 211–215**) gives the following information on lengths (km) for one particular system:

Length	6–<8	8–<10	10–<12	12–<14	14–<16
Frequency	6	23	30	35	32
Length	16–<18	18–<20	20–<22	22–<24	24–<26
Frequency	48	42	40	28	27
Length	26–<28	28–<30	30–<35	35–<40	40–<45
Frequency	26	14	27	11	2

a. Draw a histogram corresponding to these frequencies.

b. What proportion of these route lengths are less than 20? What proportion of these routes have lengths of at least 30?

c. Roughly what is the value of the 90th percentile of the route length distribution?

d. Roughly what is the median route length?

81. A study carried out to investigate the distribution of total braking time (reaction time plus accelerator-to-brake movement time, in ms) during real driving conditions at 60 km/hr gave the following summary information on the distribution of times ("A Field Study on Braking Responses During Driving," *Ergonomics*, **1995: 1903–1910**):

mean = 535   median = 500   mode = 500
sd = 96   minimum = 220   maximum = 925
5th percentile = 400   10th percentile = 430
90th percentile = 640   95th percentile = 720

What can you conclude about the shape of a histogram of this data? Explain your reasoning.

82. The sample data $x_1, x_2, \ldots, x_n$ sometimes represents a **time series,** where $x_t =$ the observed value of a response variable x at time t. Often the observed series shows a great deal of random variation, which makes it difficult to study longer-term behavior. In such situations, it is desirable to produce a smoothed version of the series. One technique for doing so involves **exponential smoothing**. The value of a smoothing constant α is chosen ($0 < \alpha < 1$). Then with $\bar{x}_t =$ smoothed value at time t, we set $\bar{x}_1 = x_1$, and for $t = 2, 3, \ldots, n, \bar{x}_t = \alpha x_t + (1 - \alpha)\bar{x}_{t-1}$.

a. Consider the following time series in which $x_t =$ temperature (°F) of effluent at a sewage treatment plant on day t: 47, 54, 53, 50, 46, 46, 47, 50, 51, 50, 46, 52, 50, 50. Plot each x_t against t on a two-dimensional coordinate system (a time-series plot). Does there appear to be any pattern?

b. Calculate the $\bar{x}_t$s using $\alpha = .1$. Repeat using $\alpha = .5$. Which value of α gives a smoother $\bar{x}_t$ series?

Comparative boxplot for Exercise 77

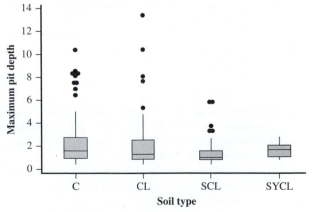

c. Substitute $\bar{x}_{t-1} = \alpha x_{t-1} + (1 - \alpha)\bar{x}_{t-2}$ on the right-hand side of the expression for $\bar{x}_t$, then substitute $\bar{x}_{t-2}$ in terms of x_{t-2} and $\bar{x}_{t-3}$, and so on. On how many of the values $x_t, x_{t-1}, \ldots, x_1$ does $\bar{x}_t$ depend? What happens to the coefficient on x_{t-k} as k increases?

d. Refer to part (c). If t is large, how sensitive is $\bar{x}_t$ to the initialization $\bar{x}_1 = x_1$? Explain.

[*Note:* A relevant reference is the article **"Simple Statistics for Interpreting Environmental Data,"** *Water Pollution Control Fed. J.,* **1981: 167–175.**]

83. Consider numerical observations $x_1, \ldots, x_n$. It is frequently of interest to know whether the x_i s are (at least approximately) symmetrically distributed about some value. If n is at least moderately large, the extent of symmetry can be assessed from a stem-and-leaf display or histogram. However, if n is not very large, such pictures are not particularly informative. Consider the following alternative. Let y_1 denote the smallest x_i, y_2 the second smallest x_i, and so on. Then plot the following pairs as points on a two-dimensional coordinate system: $(y_n - \tilde{x}, \tilde{x} - y_1)$, $(y_{n-1} - \tilde{x}, \tilde{x} - y_2)$, $(y_{n-2} - \tilde{x}, \tilde{x} - y_3), \ldots$ There are $n/2$ points when n is even and $(n - 1)/2$ when n is odd.

a. What does this plot look like when there is perfect symmetry in the data? What does it look like when observations stretch out more above the median than below it (a long upper tail)?

b. The accompanying data on rainfall (acre-feet) from 26 seeded clouds is taken from the article **"A Bayesian Analysis of a Multiplicative Treatment Effect in Weather Modification"** (***Technometrics,*** **1975: 161–166)**. Construct the plot and comment on the extent of symmetry or nature of departure from symmetry.

4.1	7.7	17.5	31.4	32.7	40.6	92.4
115.3	118.3	119.0	129.6	198.6	200.7	242.5
255.0	274.7	274.7	302.8	334.1	430.0	489.1
703.4	978.0	1656.0	1697.8	2745.6		

84. Consider a sample $x_1, \ldots, x_n$ with n even. Let $\bar{x}_L$ and $\bar{x}_U$ denote the average of the smallest $n/2$ and the largest $n/2$ observations, respectively. Show that the mean absolute deviation from the median for this sample satisfies

$$\Sigma|x_i - \tilde{x}|/n = (\bar{x}_U - \bar{x}_L)/2$$

Then show that if n is odd and the two averages are calculated after excluding the median from each half, replacing n on the left with $n - 1$ gives the correct result. [*Hint*: Break the sum into two parts, the first involving observations less than or equal to the median and the second involving observations greater than or equal to the median.]

BIBLIOGRAPHY

Albert, Jim and Maria Rizzo, *R by Example, Springer*, New York, 2012. An up-to-date introduction whose focus is on applying statistical techniques rather than on details of the R programming language.

Chambers, John, William Cleveland, Beat Kleiner, and Paul Tukey, *Graphical Methods for Data Analysis,* Brooks/Cole, Pacific Grove, CA, 1983. A highly recommended presentation of various graphical and pictorial methodology in statistics.

Cleveland, William, *Visualizing Data,* Hobart Press, Summit, NJ, 1993. An entertaining tour of pictorial techniques.

Freedman, David, Robert Pisani, and Roger Purves, *Statistics* (4th ed.), Norton, New York, 2007. An excellent, very nonmathematical survey of basic statistical reasoning and methodology.

Hoaglin, David, Frederick Mosteller, and John Tukey, *Understanding Robust and Exploratory Data Analysis,* Wiley, New York, 1983. Discusses why, as well as how, exploratory methods should be employed; it is good on details of stem-and-leaf displays and boxplots.

Moore, David, and William Notz, *Statistics: Concepts and Controversies* (7th ed.), Freeman, San Francisco, 2009. An extremely readable and entertaining paperback that contains an intuitive discussion of problems connected with sampling and designed experiments.

Peck, Roxy, and Jay Devore, *Statistics: The Exploration and Analysis of Data* (7th ed.), Cengage Brooks/Cole, Belmont, CA, 2012. The first few chapters give a very nonmathematical survey of methods for describing and summarizing data.

Peck, Roxy, et al. (eds.), *Statistics: A Guide to the Unknown* (4th ed.), Cengage Learning, Belmont, CA, 2006. Contains many short nontechnical articles describing various applications of statistics.

Verzani, John, *Using R for Introductory Statistics,* Chapman and Hall/CRC, Boca Raton, FL, 2005. A very nice introduction to the R software package.

2

Probability

INTRODUCTION

The term **probability** refers to the study of randomness and uncertainty. In any situation in which one of a number of possible outcomes may occur, the discipline of probability provides methods for quantifying the chances, or likelihoods, associated with the various outcomes. The language of probability is constantly used in an informal manner in both written and spoken contexts. Examples include such statements as "It is likely that the Dow Jones average will increase by the end of the year," "There is a 50–50 chance that the incumbent will seek reelection," "There will probably be at least one section of that course offered next year," "The odds favor a quick settlement of the strike," and "It is expected that at least 20,000 concert tickets will be sold." In this chapter, we introduce some elementary probability concepts, indicate how probabilities can be interpreted, and show how the rules of probability can be applied to compute the probabilities of many interesting events. The methodology of probability will then permit us to express in precise language such informal statements as those given above.

The study of probability as a branch of mathematics goes back over 300 years, where it had its genesis in connection with questions involving games of chance. Many books are devoted exclusively to probability, but our objective here is to cover only that part of the subject that has the most direct bearing on problems of statistical inference.

2.1 Sample Spaces and Events

An **experiment** is any activity or process whose outcome is subject to uncertainty. Although the word *experiment* generally suggests a planned or carefully controlled laboratory testing situation, we use it here in a much wider sense. Thus experiments that may be of interest include tossing a coin once or several times, selecting a card or cards from a deck, weighing a loaf of bread, ascertaining the commuting time from home to work on a particular morning, obtaining blood types from a group of individuals, or measuring the compressive strengths of different steel beams.

The Sample Space of an Experiment

DEFINITION

> The **sample space** of an experiment, denoted by $\mathscr{S}$, is the set of all possible outcomes of that experiment.

EXAMPLE 2.1 The simplest experiment to which probability applies is one with two possible outcomes. One such experiment consists of examining a single weld to see whether it is defective. The sample space for this experiment can be abbreviated as $\mathscr{S} = \{N, D\}$, where N represents not defective, D represents defective, and the braces are used to enclose the elements of a set. Another such experiment would involve tossing a thumbtack and noting whether it landed point up or point down, with sample space $\mathscr{S} = \{U, D\}$, and yet another would consist of observing the gender of the next child born at the local hospital, with $\mathscr{S} = \{M, F\}$. ∎

EXAMPLE 2.2 If we examine three welds in sequence and note the result of each examination, then an outcome for the entire experiment is any sequence of N's and D's of length 3, so

$$\mathscr{S} = \{NNN, NND, NDN, NDD, DNN, DND, DDN, DDD\}$$

If we had tossed a thumbtack three times, the sample space would be obtained by replacing N by U in $\mathscr{S}$ above, with a similar notational change yielding the sample space for the experiment in which the genders of three newborn children are observed. ∎

EXAMPLE 2.3 Two gas stations are located at a certain intersection. Each one has six gas pumps. Consider the experiment in which the number of pumps in use at a particular time of day is determined for each of the stations. An experimental outcome specifies how many pumps are in use at the first station and how many are in use at the second one. One possible outcome is (2, 2), another is (4, 1), and yet another is (1, 4). The 49 outcomes in $\mathscr{S}$ are displayed in the accompanying table. The sample space for the experiment in which a six-sided die is thrown twice results from deleting the 0 row and 0 column from the table, giving 36 outcomes.

		Second Station						
		0	**1**	**2**	**3**	**4**	**5**	**6**
	0	(0, 0)	(0, 1)	(0, 2)	(0, 3)	(0, 4)	(0, 5)	(0, 6)
	1	(1, 0)	(1, 1)	(1, 2)	(1, 3)	(1, 4)	(1, 5)	(1, 6)
	2	(2, 0)	(2, 1)	(2, 2)	(2, 3)	(2, 4)	(2, 5)	(2, 6)
First Station	**3**	(3, 0)	(3, 1)	(3, 2)	(3, 3)	(3, 4)	(3, 5)	(3, 6)
	4	(4, 0)	(4, 1)	(4, 2)	(4, 3)	(4, 4)	(4, 5)	(4, 6)
	5	(5, 0)	(5, 1)	(5, 2)	(5, 3)	(5, 4)	(5, 5)	(5, 6)
	6	(6, 0)	(6, 1)	(6, 2)	(6, 3)	(6, 4)	(6, 5)	(6, 6)

∎

EXAMPLE 2.4 A reasonably large percentage of C++ programs written at a particular company compile on the first run, but some do not (a *compiler* is a program that translates source code, in this case C++ programs, into machine language so programs can be executed). Suppose an experiment consists of selecting and compiling C++ programs at this location one by one until encountering a program that compiles on the first run. Denote a program that compiles on the first run by *S* (for success) and one that doesn't do so by *F* (for failure). Although it may not be very likely, a possible outcome of this experiment is that the first 5 (or 10 or 20 or …) are *F*'s and the next one is an *S*. That is, for any positive integer *n*, we may have to examine *n* programs before seeing the first *S*. The sample space is $\mathscr{S} = \{S, FS, FFS, FFFS, \dots\}$, which contains an infinite number of possible outcomes. The same abbreviated form of the sample space is appropriate for an experiment in which, starting at a specified time, the gender of each newborn infant is recorded until the birth of a male is observed. ■

Events

In our study of probability, we will be interested not only in the individual outcomes of $\mathscr{S}$ but also in various collections of outcomes from $\mathscr{S}$.

DEFINITION

> An **event** is any collection (subset) of outcomes contained in the sample space $\mathscr{S}$. An event is **simple** if it consists of exactly one outcome and **compound** if it consists of more than one outcome.

When an experiment is performed, a particular event *A* is said to occur if the resulting experimental outcome is contained in *A*. In general, exactly one simple event will occur, but many compound events will occur simultaneously.

EXAMPLE 2.5 Consider an experiment in which each of three vehicles taking a particular freeway exit turns left (*L*) or right (*R*) at the end of the exit ramp. The eight possible outcomes that comprise the sample space are *LLL, RLL, LRL, LLR, LRR, RLR, RRL,* and *RRR*. Thus there are eight simple events, among which are $E_1 = \{LLL\}$ and $E_5 = \{LRR\}$. Some compound events include

$$A = \{RLL, LRL, LLR\} = \text{the event that exactly one of the three}$$
$$\text{vehicles turns right}$$

$$B = \{LLL, RLL, LRL, LLR\} = \text{the event that at most one of the}$$
$$\text{vehicles turns right}$$

$$C = \{LLL, RRR\} = \text{the event that all three vehicles turn in the}$$
$$\text{same direction}$$

Suppose that when the experiment is performed, the outcome is *LLL*. Then the simple event E_1 has occurred and so also have the events *B* and *C* (but not *A*). ■

EXAMPLE 2.6
(Example 2.3
continued)
When the number of pumps in use at each of two six-pump gas stations is observed, there are 49 possible outcomes, so there are 49 simple events: $E_1 = \{(0, 0)\}$, $E_2 = \{(0, 1)\}, \dots, E_{49} = \{(6, 6)\}$. Examples of compound events are

$$A = \{(0, 0), (1, 1), (2, 2), (3, 3), (4, 4), (5, 5), (6, 6)\} = \text{the event that}$$
the number of pumps in use is the same for both stations

$$B = \{(0, 4), (1, 3), (2, 2), (3, 1), (4, 0)\} = \text{the event that}$$
the total number of pumps in use is four

$$C = \{(0, 0), (0, 1), (1, 0), (1, 1)\} = \text{the event that}$$
at most one pump is in use at each station ■

EXAMPLE 2.7
(Example 2.4
continued)

The sample space for the program compilation experiment contains an infinite number of outcomes, so there are an infinite number of simple events. Compound events include

$$A = \{S, FS, FFS\} = \text{the event that at most three programs are examined}$$

$$E = \{FS, FFFS, FFFFFS, \ldots\} = \text{the event that an even number of}$$
programs are examined ■

Some Relations from Set Theory

An event is just a set, so relationships and results from elementary set theory can be used to study events. The following operations will be used to create new events from given events.

DEFINITION

1. The **complement** of an event A, denoted by A', is the set of all outcomes in $\mathscr{S}$ that are not contained in A.

2. The **union** of two events A and B, denoted by $A \cup B$ and read "A or B," is the event consisting of all outcomes that are *either in A or in B or in both events* (so that the union includes outcomes for which both A and B occur as well as outcomes for which exactly one occurs)—that is, all outcomes in at least one of the events.

3. The **intersection** of two events A and B, denoted by $A \cap B$ and read "A and B," is the event consisting of all outcomes that are in *both A and B.*

EXAMPLE 2.8
(Example 2.3
continued)

For the experiment in which the number of pumps in use at a single six-pump gas station is observed, let $A = \{0, 1, 2, 3, 4\}$, $B = \{3, 4, 5, 6\}$, and $C = \{1, 3, 5\}$. Then

$$A' = \{5, 6\}, \quad A \cup B = \{0, 1, 2, 3, 4, 5, 6\} = \mathscr{S}, \quad A \cup C = \{0, 1, 2, 3, 4, 5\},$$

$$A \cap B = \{3, 4\}, \quad A \cap C = \{1, 3\}, \quad (A \cap C)' = \{0, 2, 4, 5, 6\}$$ ■

EXAMPLE 2.9
(Example 2.4
continued)

In the program compilation experiment, define A, B, and C by

$$A = \{S, FS, FFS\}, \quad B = \{S, FFS, FFFFS\}, \quad C = \{FS, FFFS, FFFFFS, \ldots\}$$

Then

$$A' = \{FFFS, FFFFS, FFFFFS, \ldots\}, \quad C' = \{S, FFS, FFFFS, \ldots\}$$

$$A \cup B = \{S, FS, FFS, FFFFS\}, \quad A \cap B = \{S, FFS\}$$ ■

Sometimes A and B have no outcomes in common, so that the intersection of A and B contains no outcomes.

DEFINITION

Let $\emptyset$ denote the *null event* (the event consisting of no outcomes whatsoever). When $A \cap B = \emptyset$, A and B are said to be **mutually exclusive** or **disjoint** events.

EXAMPLE 2.10 A small city has three automobile dealerships: a GM dealer selling Chevrolets and Buicks; a Ford dealer selling Fords and Lincolns; and a Toyota dealer. If an experiment consists of observing the brand of the next car sold, then the events $A = \{\text{Chevrolet, Buick}\}$ and $B = \{\text{Ford, Lincoln}\}$ are mutually exclusive because the next car sold cannot be both a GM product and a Ford product (at least until the two companies merge!). ∎

The operations of union and intersection can be extended to more than two events. For any three events A, B, and C, the event $A \cup B \cup C$ is the set of outcomes contained in at least one of the three events, whereas $A \cap B \cap C$ is the set of outcomes contained in all three events. Given events $A_1, A_2, A_3, \ldots$, these events are said to be mutually exclusive (or pairwise disjoint) if no two events have any outcomes in common.

A pictorial representation of events and manipulations with events is obtained by using Venn diagrams. To construct a Venn diagram, draw a rectangle whose interior will represent the sample space $\mathcal{S}$. Then any event A is represented as the interior of a closed curve (often a circle) contained in $\mathcal{S}$. Figure 2.1 shows examples of Venn diagrams.

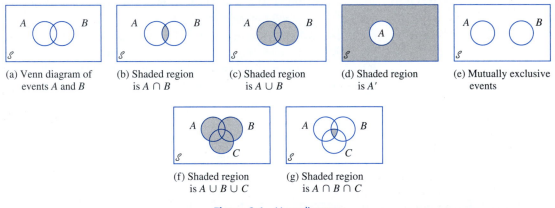

(a) Venn diagram of events A and B

(b) Shaded region is $A \cap B$

(c) Shaded region is $A \cup B$

(d) Shaded region is A'

(e) Mutually exclusive events

(f) Shaded region is $A \cup B \cup C$

(g) Shaded region is $A \cap B \cap C$

Figure 2.1 Venn diagrams

EXERCISES Section 2.1 (1–10)

1. Four universities—1, 2, 3, and 4—are participating in a holiday basketball tournament. In the first round, 1 will play 2 and 3 will play 4. Then the two winners will play for the championship, and the two losers will also play. One possible outcome can be denoted by 1324 (1 beats 2 and 3 beats 4 in first-round games, and then 1 beats 3 and 2 beats 4).
  a. List all outcomes in $\mathcal{S}$.
  b. Let A denote the event that 1 wins the tournament. List outcomes in A.
  c. Let B denote the event that 2 gets into the championship game. List outcomes in B.
  d. What are the outcomes in $A \cup B$ and in $A \cap B$? What are the outcomes in A'?

2. Suppose that vehicles taking a particular freeway exit can turn right (R), turn left (L), or go straight (S). Consider observing the direction for each of three successive vehicles.

   a. List all outcomes in the event A that all three vehicles go in the same direction.

   b. List all outcomes in the event B that all three vehicles take different directions.

   c. List all outcomes in the event C that exactly two of the three vehicles turn right.

   d. List all outcomes in the event D that exactly two vehicles go in the same direction.

   e. List outcomes in D', $C \cup D$, and $C \cap D$.

3. Three components are connected to form a system as shown in the accompanying diagram. Because the components in the 2–3 subsystem are connected in parallel, that subsystem will function if at least one of the two individual components functions. For the entire system to function, component 1 must function and so must the 2–3 subsystem.

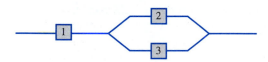

The experiment consists of determining the condition of each component [S (success) for a functioning component and F (failure) for a nonfunctioning component].

   a. Which outcomes are contained in the event A that exactly two out of the three components function?

   b. Which outcomes are contained in the event B that at least two of the components function?

   c. Which outcomes are contained in the event C that the system functions?

   d. List outcomes in C', $A \cup C$, $A \cap C$, $B \cup C$, and $B \cap C$.

4. Each of a sample of four home mortgages is classified as fixed rate (F) or variable rate (V).

   a. What are the 16 outcomes in $\mathcal{S}$?

   b. Which outcomes are in the event that exactly three of the selected mortgages are fixed rate?

   c. Which outcomes are in the event that all four mortgages are of the same type?

   d. Which outcomes are in the event that at most one of the four is a variable-rate mortgage?

   e. What is the union of the events in parts (c) and (d), and what is the intersection of these two events?

   f. What are the union and intersection of the two events in parts (b) and (c)?

5. A family consisting of three persons—A, B, and C—goes to a medical clinic that always has a doctor at each of stations 1, 2, and 3. During a certain week, each member of the family visits the clinic once and is assigned at random to a station. The experiment consists of recording the station number for each member. One outcome is (1, 2, 1) for A to station 1, B to station 2, and C to station 1.

   a. List the 27 outcomes in the sample space.

   b. List all outcomes in the event that all three members go to the same station.

   c. List all outcomes in the event that all members go to different stations.

   d. List all outcomes in the event that no one goes to station 2.

6. A college library has five copies of a certain text on reserve. Two copies (1 and 2) are first printings, and the other three (3, 4, and 5) are second printings. A student examines these books in random order, stopping only when a second printing has been selected. One possible outcome is 5, and another is 213.

   a. List the outcomes in $\mathcal{S}$.

   b. Let A denote the event that exactly one book must be examined. What outcomes are in A?

   c. Let B be the event that book 5 is the one selected. What outcomes are in B?

   d. Let C be the event that book 1 is not examined. What outcomes are in C?

7. An academic department has just completed voting by secret ballot for a department head. The ballot box contains four slips with votes for candidate A and three slips with votes for candidate B. Suppose these slips are removed from the box one by one.

   a. List all possible outcomes.

   b. Suppose a running tally is kept as slips are removed. For what outcomes does A remain ahead of B throughout the tally?

8. An engineering construction firm is currently working on power plants at three different sites. Let A_i denote the event that the plant at site i is completed by the contract date. Use the operations of union, intersection, and complementation to describe each of the following events in terms of A_1, A_2, and A_3, draw a Venn diagram, and shade the region corresponding to each one.

   a. At least one plant is completed by the contract date.

   b. All plants are completed by the contract date.

   c. Only the plant at site 1 is completed by the contract date.

   d. Exactly one plant is completed by the contract date.

   e. Either the plant at site 1 or both of the other two plants are completed by the contract date.

9. Use Venn diagrams to verify the following two relationships for any events A and B (these are called De Morgan's laws):

   a. $(A \cup B)' = A' \cap B'$

   b. $(A \cap B)' = A' \cup B'$

[*Hint:* In each part, draw a diagram corresponding to the left side and another corresponding to the right side.]

10. **a.** In Example 2.10, identify three events that are mutually exclusive.

   b. Suppose there is no outcome common to all three of the events A, B, and C. Are these three events necessarily mutually exclusive? If your answer is yes, explain why; if your answer is no, give a counterexample using the experiment of Example 2.10.

2.2 Axioms, Interpretations, and Properties of Probability

Given an experiment and a sample space $\mathcal{S}$, *the objective of probability is to assign to each event A a number P(A), called the probability of the event A, which will give a precise measure of the chance that A will occur.* To ensure that the probability assignments will be consistent with our intuitive notions of probability, all assignments should satisfy the following axioms (basic properties) of probability.

AXIOM 1   For any event A, $P(A) \geq 0$.

AXIOM 2 $P(\mathcal{S}) = 1$.

AXIOM 3   If $A_1, A_2, A_3, \ldots$ is an infinite collection of disjoint events, then

$$P(A_1 \cup A_2 \cup A_3 \cup \cdots) = \sum_{i=1}^{\infty} P(A_i)$$

You might wonder why the third axiom contains no reference to a *finite* collection of disjoint events. It is because the corresponding property for a finite collection can be derived from our three axioms. We want the axiom list to be as short as possible and not contain any property that can be derived from others on the list. Axiom 1 reflects the intuitive notion that the chance of A occurring should be nonnegative. The sample space is by definition the event that must occur when the experiment is performed ($\mathcal{S}$ contains all possible outcomes), so Axiom 2 says that the maximum possible probability of 1 is assigned to $\mathcal{S}$. The third axiom formalizes the idea that if we wish the probability that at least one of a number of events will occur and no two of the events can occur simultaneously, then the chance of at least one occurring is the sum of the chances of the individual events.

PROPOSITION $P(\emptyset) = 0$ where $\emptyset$ is the null event (the event containing no outcomes whatsoever). This in turn implies that the property contained in Axiom 3 is valid for a *finite* collection of disjoint events.

Proof   First consider the infinite collection $A_1 = \emptyset, A_2 = \emptyset, A_3 = \emptyset, \ldots$. Since $\emptyset \cap \emptyset = \emptyset$, the events in this collection are disjoint and $\cup A_i = \emptyset$. The third axiom then gives

$$P(\emptyset) = \Sigma P(\emptyset)$$

This can happen only if $P(\emptyset) = 0$.

Now suppose that $A_1, A_2, \ldots, A_k$ are disjoint events, and append to these the infinite collection $A_{k+1} = \emptyset, A_{k+2} = \emptyset, A_{k+3} = \emptyset, \ldots$. Again invoking the third axiom,

$$P\left(\bigcup_{i=1}^{k} A_i\right) = P\left(\bigcup_{i=1}^{\infty} A_i\right) = \sum_{i=1}^{\infty} P(A_i) = \sum_{i=1}^{k} P(A_i)$$

as desired.   ∎

EXAMPLE 2.11 Consider tossing a thumbtack in the air. When it comes to rest on the ground, either its point will be up (the outcome U) or down (the outcome D). The sample space for this event is therefore $\mathscr{S} = \{U, D\}$. The axioms specify $P(\mathscr{S}) = 1$, so the probability assignment will be completed by determining $P(U)$ and $P(D)$. Since U and D are disjoint and their union is $\mathscr{S}$, the foregoing proposition implies that

$$1 = P(\mathscr{S}) = P(U) + P(D)$$

It follows that $P(D) = 1 - P(U)$. One possible assignment of probabilities is $P(U) = .5$, $P(D) = .5$, whereas another possible assignment is $P(U) = .75$, $P(D) = .25$. In fact, letting p represent any fixed number between 0 and 1, $P(U) = p$, $P(D) = 1 - p$ is an assignment consistent with the axioms. ■

EXAMPLE 2.12 Consider testing batteries coming off an assembly line one by one until one having a voltage within prescribed limits is found. The simple events are $E_1 = \{S\}$, $E_2 = \{FS\}$, $E_3 = \{FFS\}$, $E_4 = \{FFFS\}$,.... Suppose the probability of any particular battery being satisfactory is .99. Then it can be shown that $P(E_1) = .99$, $P(E_2) = (.01)(.99)$, $P(E_3) = (.01)^2(.99)$,...is an assignment of probabilities to the simple events that satisfies the axioms. In particular, because the E_i's are disjoint and $\mathscr{S} = E_1 \cup E_2 \cup E_3 \cup \ldots$, it must be the case that

$$1 = P(\mathscr{S}) = P(E_1) + P(E_2) + P(E_3) + \cdots$$
$$= .99[1 + .01 + (.01)^2 + (.01)^3 + \cdots]$$

Here we have used the formula for the sum of a geometric series:

$$a + ar + ar^2 + ar^3 + \cdots = \frac{a}{1 - r}$$

However, another legitimate (according to the axioms) probability assignment of the same "geometric" type is obtained by replacing .99 by any other number p between 0 and 1 (and .01 by $1 - p$). ■

Interpreting Probability

Examples 2.11 and 2.12 show that the axioms do not completely determine an assignment of probabilities to events. The axioms serve only to rule out assignments inconsistent with our intuitive notions of probability. In the tack-tossing experiment of Example 2.11, two particular assignments were suggested. The appropriate or correct assignment depends on the nature of the thumbtack and also on one's interpretation of probability. The interpretation that is most frequently used and most easily understood is based on the notion of relative frequencies.

Consider an experiment that can be repeatedly performed in an identical and independent fashion, and let A be an event consisting of a fixed set of outcomes of the experiment. Simple examples of such repeatable experiments include the tack-tossing and die-tossing experiments previously discussed. If the experiment is performed n times, on some of the replications the event A will occur (the outcome will be in the set A), and on others, A will not occur. Let $n(A)$ denote the number of replications on which A does occur. Then the ratio $n(A)/n$ is called the *relative frequency* of occurrence of the event A in the sequence of n replications.

For example, let A be the event that a package sent within the state of California for 2nd day delivery actually arrives within one day. The results from sending 10 such packages (the first 10 replications) are as follows:

Package #	1	2	3	4	5	6	7	8	9	10
Did A occur?	N	Y	Y	Y	N	N	Y	Y	N	N
Relative frequency of A	0	.5	.667	.75	.6	.5	.571	.625	.556	.5

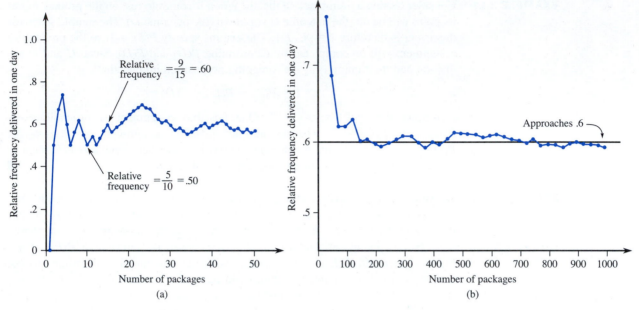

Figure 2.2 Behavior of relative frequency (a) Initial fluctuation (b) Long-run stabilization

Figure 2.2(a) shows how the relative frequency $n(A)/n$ fluctuates rather substantially over the course of the first 50 replications. But as the number of replications continues to increase, Figure 2.2(b) illustrates how the relative frequency stabilizes.

More generally, empirical evidence, based on the results of many such repeatable experiments, indicates that any relative frequency of this sort will stabilize as the number of replications n increases. That is, as n gets arbitrarily large, $n(A)/n$ approaches a limiting value referred to as the *limiting* (or *long-run*) *relative frequency* of the event A. The *objective interpretation of probability* identifies this limiting relative frequency with $P(A)$. Suppose that probabilities are assigned to events in accordance with their limiting relative frequencies. Then a statement such as "the probability of a package being delivered within one day of mailing is .6" means that of a large number of mailed packages, roughly 60% will arrive within one day. Similarly, if B is the event that an appliance of a particular type will need service while under warranty, then $P(B) = .1$ is interpreted to mean that in the long run 10% of such appliances will need warranty service. This doesn't mean that exactly 1 out of 10 will need service, or that exactly 10 out of 100 will need service, because 10 and 100 are not the long run.

This relative frequency interpretation of probability is said to be objective because it rests on a property of the experiment rather than on any particular individual concerned with the experiment. For example, two different observers of a sequence of coin tosses should both use the same probability assignments since the observers have nothing to do with limiting relative frequency. In practice, this interpretation is not as objective as it might seem, since the limiting relative frequency of an event will not be known. Thus we will have to assign probabilities based on our beliefs about the limiting relative frequency of events under study. Fortunately, there are many experiments for which there will be a consensus with respect to probability assignments. When we speak of a fair coin, we shall mean $P(H) = P(T) = .5$, and a fair die is one for which limiting relative frequencies of the six outcomes are all $1/6$, suggesting probability assignments $P(\{1\}) = \cdots = P(\{6\}) = 1/6$.

Because the objective interpretation of probability is based on the notion of limiting frequency, its applicability is limited to experimental situations that are repeatable. Yet the language of probability is often used in connection with situations that are inherently unrepeatable. Examples include: "The chances are good for a peace agreement"; "It is likely that our company will be awarded the contract"; and "Because their best quarterback is injured, I expect them to score no more than 10 points against us." In such situations we would like, as before, to assign numerical probabilities to various outcomes and events (e.g., the probability is .9 that we will get the contract). This necessitates adopting an alternative interpretation of these probabilities. Because different observers may have different prior information and opinions concerning such experimental situations, probability assignments may now differ from individual to individual. Interpretations in such situations are thus referred to as *subjective*. The book by Robert Winkler listed in the chapter references gives a very readable survey of several subjective interpretations.

More Probability Properties

PROPOSITION

For any event A, $P(A) + P(A') = 1$, from which $P(A) = 1 - P(A')$.

Proof In Axiom 3, let $k = 2$, $A_1 = A$, and $A_2 = A'$. Since by definition of A', $A \cup A' = \mathcal{S}$ while A and A' are disjoint, $1 = P(\mathcal{S}) = P(A \cup A') = P(A) + P(A')$. ■

This proposition is surprisingly useful because there are many situations in which $P(A')$ is more easily obtained by direct methods than is $P(A)$.

EXAMPLE 2.13 Consider a system of five identical components connected in series, as illustrated in Figure 2.3.

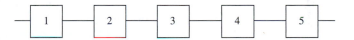

Figure 2.3 A system of five components connected in a series

Denote a component that fails by F and one that doesn't fail by S (for success). Let A be the event that the *system* fails. For A to occur, at least one of the individual components must fail. Outcomes in A include *SSFSS* (1, 2, 4, and 5 all work, but 3 does not), *FFSSS*, and so on. There are in fact 31 different outcomes in A. However, A', the event that the system works, consists of the single outcome *SSSSS*. We will see in Section 2.5 that if 90% of all such components do not fail and different components fail independently of one another, then $P(A') = P(SSSSS) = .9^5 = .59$. Thus $P(A) = 1 - .59 = .41$; so among a large number of such systems, roughly 41% will fail. ■

In general, the foregoing proposition is useful when the event of interest can be expressed as "at least …," since then the complement "less than …" may be easier to work with (in some problems, "more than …" is easier to deal with than "at most …"). When you are having difficulty calculating $P(A)$ directly, think of determining $P(A')$.

PROPOSITION

For any event A, $P(A) \leq 1$.

This is because $1 = P(A) + P(A') \geq P(A)$ since $P(A') \geq 0$.

When events A and B are mutually exclusive, $P(A \cup B) = P(A) + P(B)$. For events that are not mutually exclusive, adding $P(A)$ and $P(B)$ results in "double-counting" outcomes in the intersection. The next result, the *addition rule* for a double union probability, shows how to correct for this.

PROPOSITION

> For any two events *A* and *B*,
>
> $$P(A \cup B) = P(A) + P(B) - P(A \cap B)$$

Proof Note first that $A \cup B$ can be decomposed into two *disjoint* events, *A* and $B \cap A'$; the latter is the part of *B* that lies outside *A* (see Figure 2.4). Furthermore, *B* itself is the union of the two disjoint events $A \cap B$ and $A' \cap B$, so $P(B) = P(A \cap B) + P(A' \cap B)$. Thus

$$P(A \cup B) = P(A) + P(B \cap A') = P(A) + [P(B) - P(A \cap B)]$$
$$= P(A) + P(B) - P(A \cap B)$$

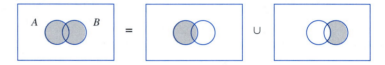

Figure 2.4   Representing A ∪ B as a union of disjoint events   ∎

EXAMPLE 2.14 In a certain residential suburb, 60% of all households get Internet service from the local cable company, 80% get television service from that company, and 50% get both services from that company. If a household is randomly selected, what is the probability that it gets at least one of these two services from the company, and what is the probability that it gets exactly one of these services from the company?

With $A = \{$gets Internet service$\}$ and $B = \{$gets TV service$\}$, the given information implies that $P(A) = .6$, $P(B) = .8$, and $P(A \cap B) = .5$. The foregoing proposition now yields

P(subscribes to at least one of the two services)

$$= P(A \cup B) = P(A) + P(B) - P(A \cap B) = .6 + .8 - .5 = .9$$

The event that a household subscribes only to tv service can be written as $A' \cap B$ [(not Internet) and TV]. Now Figure 2.4 implies that

$$.9 = P(A \cup B) = P(A) + P(A' \cap B) = .6 + P(A' \cap B)$$

from which $P(A' \cap B) = .3$. Similarly, $P(A \cap B') = P(A \cup B) - P(B) = .1$. This is all illustrated in Figure 2.5, from which we see that

$$P(\text{exactly one}) = P(A \cap B') + P(A' \cap B) = .1 + .3 = .4$$

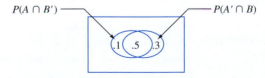

Figure 2.5   Probabilities for Example 2.14   ∎

The addition rule for a triple union probability is similar to the foregoing rule.

PROPOSITION

For any three events A, B, and C,
$$P(A \cup B \cup C) = P(A) + P(B) + P(C) - P(A \cap B) - P(A \cap C)$$
$$-P(B \cap C) + P(A \cap B \cap C)$$

This can be verified by examining the Venn diagram of $A \cup B \cup C$, shown in Figure 2.6. When $P(A)$, $P(B)$, and $P(C)$ are added, the intersection probabilities $P(A \cap B)$, $P(A \cap C)$, and $P(B \cap C)$ are all counted twice. Each one must therefore be subtracted. But then $P(A \cap B \cap C)$ has been added in three times and subtracted out three times, so it must be added back. In general, the probability of a union of k events is obtained by summing individual event probabilities, subtracting double intersection probabilities, adding triple intersection probabilities, subtracting quadruple intersection probabilities, and so on.

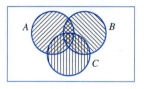

Figure 2.6 $A \cup B \cup C$

Determining Probabilities Systematically

Consider a sample space that is either finite or "countably infinite" (the latter means that outcomes can be listed in an infinite sequence, so there is a first outcome, a second outcome, a third outcome, and so on—for example, the battery testing scenario of Example 2.12). Let E_1, E_2, $E_3, \ldots$ denote the corresponding simple events, each consisting of a single outcome. A sensible strategy for probability computation is to first determine each simple event probability, with the requirement that $\Sigma P(E_i) = 1$. Then the probability of any compound event A is computed by adding together the $P(E_i)$'s for all E_i's in A:

$$P(A) = \sum_{\text{all } E_i\text{'s in } A} P(E_i)$$

EXAMPLE 2.15

During off-peak hours a commuter train has five cars. Suppose a commuter is twice as likely to select the middle car (#3) as to select either adjacent car (#2 or #4), and is twice as likely to select either adjacent car as to select either end car (#1 or #5). Let $p_i = P(\text{car } i \text{ is selected}) = P(E_i)$. Then we have $p_3 = 2p_2 = 2p_4$ and $p_2 = 2p_1 = 2p_5 = p_4$. This gives

$$1 = \sum P(E_i) = p_1 + 2p_1 + 4p_1 + 2p_1 + p_1 = 10p_1$$

implying $p_1 = p_5 = .1$, $p_2 = p_4 = .2$, $p_3 = .4$. The probability that one of the three middle cars is selected (a compound event) is then $p_2 + p_3 + p_4 = .8$. ∎

Equally Likely Outcomes

In many experiments consisting of N outcomes, it is reasonable to assign equal probabilities to all N simple events. These include such obvious examples as tossing a fair coin or fair die once or twice (or any fixed number of times), or selecting one or several cards from a well-shuffled deck of 52. With $p = P(E_i)$ for every i,

$$1 = \sum_{i=1}^{N} P(E_i) = \sum_{i=1}^{N} p = p \cdot N \quad \text{so } p = \frac{1}{N}$$

That is, if there are N equally likely outcomes, the probability for each is $1/N$.

Now consider an event A, with $N(A)$ denoting the number of outcomes contained in A. Then

$$P(A) = \sum_{E_i \text{ in } A} P(E_i) = \sum_{E_i \text{ in } A} \frac{1}{N} = \frac{N(A)}{N}$$

Thus when outcomes are equally likely, computing probabilities reduces to counting: determine both the number of outcomes $N(A)$ in A and the number of outcomes N in $\mathscr{S}$, and form their ratio.

EXAMPLE 2.16 You have six unread mysteries and six unread science fiction books on your bookshelf. The first three of each type are hardcover, and the last three are paperback. Consider randomly selecting one of the six mysteries and then randomly selecting one of the six science fiction books to take on a post-finals vacation to Acapulco (after all, you need something to read on the beach). Number the mysteries 1, 2,..., 6, and do the same for the science fiction books. Then each outcome is a pair of numbers such as (4, 1), and there are $N = 36$ possible outcomes (For a visual of this situation, refer to the table in Example 2.3 and delete the first row and column). With random selection as described, the 36 outcomes are equally likely. Nine of these outcomes are such that both selected books are paperbacks (those in the lower right-hand corner of the referenced table): (4,4), (4,5),..., (6,6). So the probability of the event A that both selected books are paperbacks is

$$P(A) = \frac{N(A)}{N} = \frac{9}{36} = .25 \qquad \blacksquare$$

EXERCISES Section 2.2 (11–28)

11. A mutual fund company offers its customers a variety of funds: a money-market fund, three different bond funds (short, intermediate, and long-term), two stock funds (moderate and high-risk), and a balanced fund. Among customers who own shares in just one fund, the percentages of customers in the different funds are as follows:

Money-market	20%	High-risk stock	18%
Short bond	15%	Moderate-risk	
		stock	25%
Intermediate		Balanced	7%
bond	10%		
Long bond	5%		

A customer who owns shares in just one fund is randomly selected.

a. What is the probability that the selected individual owns shares in the balanced fund?

b. What is the probability that the individual owns shares in a bond fund?

c. What is the probability that the selected individual does not own shares in a stock fund?

12. Consider randomly selecting a student at a large university, and let A be the event that the selected student has a Visa card and B be the analogous event for MasterCard. Suppose that $P(A) = .6$ and $P(B) = .4$.

a. Could it be the case that $P(A \cap B) = .5$? Why or why not? [*Hint:* See Exercise 24.]

b. From now on, suppose that $P(A \cap B) = .3$. What is the probability that the selected student has at least one of these two types of cards?

c. What is the probability that the selected student has neither type of card?

d. Describe, in terms of A and B, the event that the selected student has a Visa card but not a MasterCard, and then calculate the probability of this event.

e. Calculate the probability that the selected student has exactly one of the two types of cards.

13. A computer consulting firm presently has bids out on three projects. Let $A_i = \{$awarded project $i\}$, for $i = 1, 2, 3$, and suppose that $P(A_1) = .22$, $P(A_2) = .25$, $P(A_3) = .28$, $P(A_1 \cap A_2) = .11$, $P(A_1 \cap A_3) = .05$, $P(A_2 \cap A_3) = .07$, $P(A_1 \cap A_2 \cap A_3) = .01$. Express in words each of the following events, and compute the probability of each event:

a. $A_1 \cup A_2$

b. $A_1' \cap A_2'$ [*Hint:* $(A_1 \cup A_2)' = A_1' \cap A_2'$]

c. $A_1 \cup A_2 \cup A_3$ **d.** $A_1' \cap A_2' \cap A_3'$

e. $A_1' \cap A_2' \cap A_3$ **f.** $(A_1' \cap A_2') \cup A_3$

14. Suppose that 55% of all adults regularly consume coffee, 45% regularly consume carbonated soda, and 70% regularly consume at least one of these two products.

  a. What is the probability that a randomly selected adult regularly consumes both coffee and soda?

  b. What is the probability that a randomly selected adult doesn't regularly consume at least one of these two products?

15. Consider the type of clothes dryer (gas or electric) purchased by each of five different customers at a certain store.

  a. If the probability that at most one of these purchases an electric dryer is .428, what is the probability that at least two purchase an electric dryer?

  b. If P(all five purchase gas) $= .116$ and P(all five purchase electric) $= .005$, what is the probability that at least one of each type is purchased?

16. An individual is presented with three different glasses of cola, labeled C, D, and P. He is asked to taste all three and then list them in order of preference. Suppose the same cola has actually been put into all three glasses.

  a. What are the simple events in this ranking experiment, and what probability would you assign to each one?

  b. What is the probability that C is ranked first?

  c. What is the probability that C is ranked first and D is ranked last?

17. Let A denote the event that the next request for assistance from a statistical software consultant relates to the SPSS package, and let B be the event that the next request is for help with SAS. Suppose that $P(A) = .30$ and $P(B) = .50$.

  a. Why is it not the case that $P(A) + P(B) = 1$?

  b. Calculate $P(A')$.

  c. Calculate $P(A \cup B)$.

  d. Calculate $P(A' \cap B')$.

18. A wallet contains five \$10 bills, four \$5 bills, and six \$1 bills (nothing larger). If the bills are selected one by one in random order, what is the probability that at least two bills must be selected to obtain a first \$10 bill?

19. Human visual inspection of solder joints on printed circuit boards can be very subjective. Part of the problem stems from the numerous types of solder defects (e.g., pad non-wetting, knee visibility, voids) and even the degree to which a joint possesses one or more of these defects. Consequently, even highly trained inspectors can disagree on the disposition of a particular joint. In one batch of 10,000 joints, inspector A found 724 that were judged defective, inspector B found 751 such joints, and 1159 of the joints were judged defective by at least one of the inspectors. Suppose that one of the 10,000 joints is randomly selected.

  a. What is the probability that the selected joint was judged to be defective by neither of the two inspectors?

  b. What is the probability that the selected joint was judged to be defective by inspector B but not by inspector A?

20. A certain factory operates three different shifts. Over the last year, 200 accidents have occurred at the factory. Some of these can be attributed at least in part to unsafe working conditions, whereas the others are unrelated to working conditions. The accompanying table gives the percentage of accidents falling in each type of accident–shift category.

		Unsafe Conditions	Unrelated to Conditions
	Day	10%	35%
Shift	*Swing*	8%	20%
	Night	5%	22%

Suppose one of the 200 accident reports is randomly selected from a file of reports, and the shift and type of accident are determined.

  a. What are the simple events?

  b. What is the probability that the selected accident was attributed to unsafe conditions?

  c. What is the probability that the selected accident did not occur on the day shift?

21. An insurance company offers four different deductible levels—none, low, medium, and high—for its homeowner's policyholders and three different levels—low, medium, and high—for its automobile policyholders. The accompanying table gives proportions for the various categories of policyholders who have both types of insurance. For example, the proportion of individuals with both low homeowner's deductible and low auto deductible is .06 (6% of all such individuals).

Auto	Homeowner's			
	N	L	M	H
L	.04	.06	.05	.03
M	.07	.10	.20	.10
H	.02	.03	.15	.15

Suppose an individual having both types of policies is randomly selected.

  a. What is the probability that the individual has a medium auto deductible and a high homeowner's deductible?

  b. What is the probability that the individual has a low auto deductible? A low homeowner's deductible?

  c. What is the probability that the individual is in the same category for both auto and homeowner's deductibles?

  d. Based on your answer in part (c), what is the probability that the two categories are different?

  e. What is the probability that the individual has at least one low deductible level?

  f. Using the answer in part (e), what is the probability that neither deductible level is low?

22. The route used by a certain motorist in commuting to work contains two intersections with traffic signals. The probability that he must stop at the first signal is .4, the analogous probability for the second signal is .5, and the probability that he must stop at at least one of the two signals is .7. What is the probability that he must stop

a. At both signals?
b. At the first signal but not at the second one?
c. At exactly one signal?

23. The computers of six faculty members in a certain department are to be replaced. Two of the faculty members have selected laptop machines and the other four have chosen desktop machines. Suppose that only two of the setups can be done on a particular day, and the two computers to be set up are randomly selected from the six (implying 15 equally likely outcomes; if the computers are numbered 1, 2,…, 6, then one outcome consists of computers 1 and 2, another consists of computers 1 and 3, and so on).

a. What is the probability that both selected setups are for laptop computers?
b. What is the probability that both selected setups are desktop machines?
c. What is the probability that at least one selected setup is for a desktop computer?
d. What is the probability that at least one computer of each type is chosen for setup?

24. Show that if one event A is contained in another event B (i.e., A is a subset of B), then $P(A) \leq P(B)$. [*Hint:* For such A and B, A and $B \cap A'$ are disjoint and $B = A \cup (B \cap A')$, as can be seen from a Venn diagram.] For general A and B, what does this imply about the relationship among $P(A \cap B)$, $P(A)$ and $P(A \cup B)$?

25. The three most popular options on a certain type of new car are a built-in GPS (A), a sunroof (B), and an automatic transmission (C). If 40% of all purchasers request A, 55% request B, 70% request C, 63% request A or B, 77% request A or C, 80% request B or C, and 85% request A or B or C, determine the probabilities of the following events. [*Hint:* "A or B" is the event that at least one of the two options is requested; try drawing a Venn diagram and labeling all regions.]

a. The next purchaser will request at least one of the three options.
b. The next purchaser will select none of the three options.

c. The next purchaser will request only an automatic transmission and not either of the other two options.
d. The next purchaser will select exactly one of these three options.

26. A certain system can experience three different types of defects. Let $A_i (i = 1,2,3)$ denote the event that the system has a defect of type i. Suppose that

$$P(A_1) = .12 \quad P(A_2) = .07 \quad P(A_3) = .05$$
$$P(A_1 \cup A_2) = .13 \quad P(A_1 \cup A_3) = .14$$
$$P(A_2 \cup A_3) = .10 \quad P(A_1 \cap A_2 \cap A_3) = .01$$

a. What is the probability that the system does not have a type 1 defect?
b. What is the probability that the system has both type 1 and type 2 defects?
c. What is the probability that the system has both type 1 and type 2 defects but not a type 3 defect?
d. What is the probability that the system has at most two of these defects?

27. An academic department with five faculty members—Anderson, Box, Cox, Cramer, and Fisher—must select two of its members to serve on a personnel review committee. Because the work will be time-consuming, no one is anxious to serve, so it is decided that the representatives will be selected by putting the names on identical pieces of paper and then randomly selecting two.

a. What is the probability that both Anderson and Box will be selected? [*Hint:* List the equally likely outcomes.]
b. What is the probability that at least one of the two members whose name begins with C is selected?
c. If the five faculty members have taught for 3, 6, 7, 10, and 14 years, respectively, at the university, what is the probability that the two chosen representatives have a total of at least 15 years' teaching experience there?

28. In Exercise 5, suppose that any incoming individual is equally likely to be assigned to any of the three stations irrespective of where other individuals have been assigned. What is the probability that

a. All three family members are assigned to the same station?
b. At most two family members are assigned to the same station?
c. Every family member is assigned to a different station?

2.3 Counting Techniques

When the various outcomes of an experiment are equally likely (the same probability is assigned to each simple event), the task of computing probabilities reduces to counting. Letting N denote the number of outcomes in a sample space and $N(A)$ represent the number of outcomes contained in an event A,

$$P(A) = \frac{N(A)}{N} \tag{2.1}$$

If a list of the outcomes is easily obtained and N is small, then N and $N(A)$ can be determined without the benefit of any general counting principles.

There are, however, many experiments for which the effort involved in constructing such a list is prohibitive because N is quite large. By exploiting some general counting rules, it is possible to compute probabilities of the form (2.1) without out a listing of outcomes. These rules are also useful in many problems involving outcomes that are not equally likely. Several of the rules developed here will be used in studying probability distributions in the next chapter.

The Product Rule for Ordered Pairs

Our first counting rule applies to any situation in which a set (event) consists of ordered pairs of objects and we wish to count the number of such pairs. By an ordered pair, we mean that, if O_1 and O_2 are objects, then the pair (O_1, O_2) is different from the pair (O_2, O_1). For example, if an individual selects one airline for a trip from Los Angeles to Chicago and (after transacting business in Chicago) a second one for continuing on to New York, one possibility is (American, United), another is (United, American), and still another is (United, United).

PROPOSITION

> If the first element or object of an ordered pair can be selected in n_1 ways, and for each of these n_1 ways the second element of the pair can be selected in n_2 ways, then the number of pairs is $n_1 n_2$.

An alternative interpretation involves carrying out an operation that consists of two stages. If the first stage can be performed in any one of n_1 ways, and for each such way there are n_2 ways to perform the second stage, then $n_1 n_2$ is the number of ways of carrying out the two stages in sequence.

EXAMPLE 2.17

A homeowner doing some remodeling requires the services of both a plumbing contractor and an electrical contractor. If there are 12 plumbing contractors and 9 electrical contractors available in the area, in how many ways can the contractors be chosen? If we denote the plumbers by $P_1, \ldots, P_{12}$ and the electricians by $Q_1, \ldots, Q_9$, then we wish the number of pairs of the form (P_i, Q_j). With $n_1 = 12$ and $n_2 = 9$, the product rule yields $N = (12)(9) = 108$ possible ways of choosing the two types of contractors. ∎

In Example 2.17, the choice of the second element of the pair did not depend on which first element was chosen or occurred. As long as there is the same number of choices of the second element for each first element, the product rule is valid even when the set of possible second elements depends on the first element.

EXAMPLE 2.18

A family has just moved to a new city and requires the services of both an obstetrician and a pediatrician. There are two easily accessible medical clinics, each having two obstetricians and three pediatricians. The family will obtain maximum health insurance benefits by joining a clinic and selecting both doctors from that clinic. In how many ways can this be done? Denote the obstetricians by O_1, O_2, O_3, and O_4 and the pediatricians by $P_1, \ldots, P_6$. Then we wish the number of pairs (O_i, P_j) for which O_i and P_j are associated with the same clinic. Because there are four obstetricians, $n_1 = 4$, and for each there are three choices of pediatrician, so $n_2 = 3$. Applying the product rule gives $N = n_1 n_2 = 12$ possible choices. ∎

In many counting and probability problems, a configuration called a **tree diagram** can be used to represent pictorially all the possibilities. The tree diagram associated with

Example 2.18 appears in Figure 2.7. Starting from a point on the left side of the diagram, for each possible first element of a pair a straight-line segment emanates rightward. Each of these lines is referred to as a first-generation branch. Now for any given first-generation branch we construct another line segment emanating from the tip of the branch for each possible choice of a second element of the pair. Each such line segment is a second-generation branch. Because there are four obstetricians, there are four first-generation branches, and three pediatricians for each obstetrician yields three second-generation branches emanating from each first-generation branch.

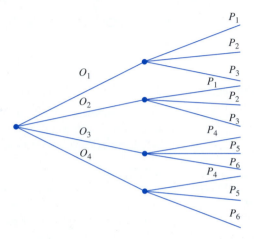

Figure 2.7 Tree diagram for Example 2.18

Generalizing, suppose there are n_1 first-generation branches, and for each first-generation branch there are n_2 second-generation branches. The total number of second-generation branches is then $n_1 n_2$. Since the end of each second-generation branch corresponds to exactly one possible pair (choosing a first element and then a second puts us at the end of exactly one second-generation branch), there are $n_1 n_2$ pairs, verifying the product rule.

The construction of a tree diagram does not depend on having the same number of second-generation branches emanating from each first-generation branch. If the second clinic had four pediatricians, then there would be only three branches emanating from two of the first-generation branches and four emanating from each of the other two first-generation branches. A tree diagram can thus be used to represent pictorially experiments other than those to which the product rule applies.

A More General Product Rule

If a six-sided die is tossed five times in succession rather than just twice, then each possible outcome is an ordered collection of five numbers such as $(1, 3, 1, 2, 4)$ or $(6, 5, 2, 2, 2)$. We will call an ordered collection of k objects a **k-tuple** (so a pair is a 2-tuple and a triple is a 3-tuple). Each outcome of the die-tossing experiment is then a 5-tuple.

Product Rule for k-Tuples

Suppose a set consists of ordered collections of k elements (k-tuples) and that there are n_1 possible choices for the first element; for each choice of the first element, there are n_2 possible choices of the second element;...; for each possible choice of the first $k - 1$ elements, there are n_k choices of the kth element. Then there are $n_1 n_2 \cdot \cdots \cdot n_k$ possible k-tuples.

An alternative interpretation involves carrying out an operation in k stages. If the first stage can be performed in any one of n_1 ways, and for each such way there are n_2 ways to perform the second stage, and for each way of performing the first two stages there are n_3 ways to perform the 3rd stage, and so on, then $n_1 n_2 \cdots n_k$ is the number of ways to carry out the entire k-stage operation in sequence. This more general rule can also be visualized with a tree diagram. For the case $k = 3$, simply add an appropriate number of 3rd generation branches to the tip of each 2nd generation branch. If, for example, a college town has four pizza places, a theater complex with six screens, and three places to go dancing, then there would be four 1st generation branches, six 2nd generation branches emanating from the tip of each 1st generation branch, and three 3rd generation branches leading off each 2nd generation branch. Each possible 3-tuple corresponds to the tip of a 3rd generation branch.

EXAMPLE 2.19
(Example 2.17 continued)

Suppose the home remodeling job involves first purchasing several kitchen appliances. They will all be purchased from the same dealer, and there are five dealers in the area. With the dealers denoted by $D_1, \ldots, D_5$, there are $N = n_1 n_2 n_3 = (5)(12)(9) = 540$ 3-tuples of the form (D_i, P_j, Q_k), so there are 540 ways to choose first an appliance dealer, then a plumbing contractor, and finally an electrical contractor. ∎

EXAMPLE 2.20
(Example 2.18 continued)

If each clinic has both three specialists in internal medicine and two general surgeons, there are $n_1 n_2 n_3 n_4 = (4)(3)(3)(2) = 72$ ways to select one doctor of each type such that all doctors practice at the same clinic. ∎

Permutations and Combinations

Consider a group of n distinct individuals or objects ("distinct" means that there is some characteristic that differentiates any particular individual or object from any other). How many ways are there to select a subset of size k from the group? For example, if a Little League team has 15 players on its roster, how many ways are there to select 9 players to form a starting lineup? Or if a university bookstore sells ten different laptop computers but has room to display only three of them, in how many ways can the three be chosen?

An answer to the general question just posed requires that we distinguish between two cases. In some situations, such as the baseball scenario, the order of selection is important. For example, Angela being the pitcher and Ben the catcher gives a different lineup from the one in which Angela is catcher and Ben is pitcher. Often, though, order is not important and one is interested only in which individuals or objects are selected, as would be the case in the laptop display scenario.

DEFINITION

> An ordered subset is called a **permutation.** The number of permutations of size k that can be formed from the n individuals or objects in a group will be denoted by $P_{k,n}$. An unordered subset is called a **combination.** One way to denote the number of combinations is $C_{k,n}$, but we shall instead use notation that is quite common in probability books: $\binom{n}{k}$, read "n choose k."

The number of permutations can be determined by using our earlier counting rule for k-tuples. Suppose, for example, that a college of engineering has seven departments, which we denote by $a, b, c, d, e, f,$ and g. Each department has one representative on the college's student council. From these seven representatives, one is to be chosen chair, another is to be selected vice-chair, and a third will be secretary. How many ways are there to select the three officers? That is, how many permutations of size 3 can be formed from the 7 representatives? To answer this question, think of forming a triple (3-tuple) in which the first element is the chair, the second is the vice-chair, and the third is the secretary. One such triple is (*a, g, b*), another is (*b, g, a*), and yet another is (*d, f, b*). Now the chair can be selected in any of $n_1 = 7$ ways. For each way of selecting the chair, there are $n_2 = 6$ ways to select the vice-chair, and hence $7 \times 6 = 42$ (chair, vice-chair) pairs. Finally, for each way of selecting a chair and vice-chair, there are $n_3 = 5$ ways of choosing the secretary. This gives

$$P_{3,7} = (7)(6)(5) = 210$$

as the number of permutations of size 3 that can be formed from 7 distinct individuals. A tree diagram representation would show three generations of branches.

The expression for $P_{3,7}$ can be rewritten with the aid of *factorial notation*. Recall that 7! (read "7 factorial") is compact notation for the descending product of integers (7)(6)(5)(4)(3)(2)(1). More generally, for any positive integer m, $m! = m(m-1)(m-2) \cdot \cdots \cdot (2)(1)$. This gives $1! = 1$, and we also define $0! = 1$. Then

$$P_{3,7} = (7)(6)(5) = \frac{(7)(6)(5)(4!)}{(4!)} = \frac{7!}{4!}$$

Generalizing to arbitrary group size n and subset size k yields

$$P_{k,n} = n(n-1)(n-2) \cdot \cdots \cdot (n-(k-2))(n-(k-1))$$

Multiplying and dividing this by $(n-k)!$ gives a compact expression for the number of permutations.

PROPOSITION

$$P_{k,n} = \frac{n!}{(n-k)!}$$

EXAMPLE 2.21 There are ten teaching assistants available for grading papers in a calculus course at a large university. The first exam consists of four questions in increasing order of difficulty, and the professor wishes to select a different assistant to grade each question (only one assistant per question). In how many ways can the assistants be chosen for grading? Here n = group size = 10 and k = subset size = 4. The number of permutations is

$$P_{4,10} = \frac{10!}{(10-4)!} = \frac{10!}{6!} = 10(9)(8)(7) = 5040$$

That is, the professor could give 5040 different four-question exams without using the same assignment of graders to questions, by which time all the teaching assistants would hopefully have finished their degree programs! ∎

Now let's move on to combinations (i.e., unordered subsets). Again refer to the student council scenario, and suppose that three of the seven representatives are to be selected to attend a statewide convention. The order of selection is not important; all that matters is which three get selected. So we are looking for $\binom{7}{3}$, the number of combinations of size 3 that can be formed from the 7 individuals. Consider for a moment the combination a,c,g. These three individuals can be ordered in $3! = 6$ ways to produce permutations:

$$a, c, g \quad a, g, c \quad c, a, g \quad c, g, a \quad g, a, c \quad g, c, a$$

Similarly, there are $3! = 6$ ways to order the combination b,c,e to produce permutations, and in fact $3!$ ways to order any particular combination of size 3 to produce permutations. This implies the following relationship between the number of combinations and the number of permutations:

$$P_{3,7} = (3!) \cdot \binom{7}{3} \Rightarrow \binom{7}{3} = \frac{P_{3,7}}{3!} = \frac{7!}{(3!)(4!)} = \frac{(7)(6)(5)}{(3)(2)(1)} = 35$$

It would not be too difficult to list the 35 combinations, but there is no need to do so if we are interested only in how many there are. Notice that the number of permutations 210 far exceeds the number of combinations; the former is larger than the latter by a factor of $3!$ since that is how many ways each combination can be ordered.

Generalizing the foregoing line of reasoning gives a simple relationship between the number of permutations and the number of combinations that yields a concise expression for the latter quantity.

PROPOSITION

$$\binom{n}{k} = \frac{P_{k,n}}{k!} = \frac{n!}{k!(n-k)!}$$

Notice that $\binom{n}{n} = 1$ and $\binom{n}{0} = 1$ since there is only one way to choose a set of (all) n elements or of no elements, and $\binom{n}{1} = n$ since there are n subsets of size 1.

EXAMPLE 2.22 A particular iPod playlist contains 100 songs, 10 of which are by the Beatles. Suppose the shuffle feature is used to play the songs in random order (the randomness of the shuffling process is investigated in **"Does Your iPod *Really* Play Favorites?"** (***The Amer. Statistician, 2009: 263–268***). What is the probability that the first Beatles song heard is the fifth song played?

In order for this event to occur, it must be the case that the first four songs played are not Beatles' songs (NBs) and that the fifth song is by the Beatles (B). The number of ways to select the first five songs is $100(99)(98)(97)(96)$. The number of ways to select these five songs so that the first four are NBs and the next is a B is $90(89)(88)(87)(10)$. The random shuffle assumption implies that any particular set of 5 songs from amongst the 100 has the same chance of being selected as the first five played as does any other set of five songs; each outcome is equally likely. Therefore the desired probability is the ratio of the number of outcomes for which the event of interest occurs to the number of possible outcomes:

$$P(1^{\text{st}} \text{ B is the } 5^{\text{th}} \text{ song played}) = \frac{90 \cdot 89 \cdot 88 \cdot 87 \cdot 10}{100 \cdot 99 \cdot 98 \cdot 97 \cdot 96} = \frac{P_{4,90} \cdot (10)}{P_{5,100}} = .0679$$

Here is an alternative line of reasoning involving combinations. Rather than focusing on selecting just the first five songs, think of playing all 100 songs in random order. The number of ways of choosing 10 of these songs to be the Bs (without regard to the order in which they are then played) is $\binom{100}{10}$. Now if we choose 9 of the last 95 songs to be Bs, which can be done in $\binom{95}{9}$ ways, that leaves four NBs and one B for the first five songs. There is only one further way for these five to start with four NBs and then follow with a B (remember that we are considering *unordered* subsets). Thus

$$P(1^{st} \text{ B is the } 5^{th} \text{ song played}) = \frac{\binom{95}{9}}{\binom{100}{10}}$$

It is easily verified that this latter expression is in fact identical to the first expression for the desired probability, so the numerical result is again .0679.

The probability that one of the first five songs played is a Beatles' song is

$P(1^{st} \text{ B is the } 1^{st} \text{ or } 2^{nd} \text{ or } 3^{rd} \text{ or } 4^{th} \text{ or } 5^{th} \text{ song played})$

$$= \frac{\binom{99}{9}}{\binom{100}{10}} + \frac{\binom{98}{9}}{\binom{100}{10}} + \frac{\binom{97}{9}}{\binom{100}{10}} + \frac{\binom{96}{9}}{\binom{100}{10}} + \frac{\binom{95}{9}}{\binom{100}{10}} = .4162$$

It is thus rather likely that a Beatles' song will be one of the first five songs played. Such a "coincidence" is not as surprising as might first appear to be the case. ∎

EXAMPLE 2.23 A university warehouse has received a shipment of 25 printers, of which 10 are laser printers and 15 are inkjet models. If 6 of these 25 are selected at random to be checked by a particular technician, what is the probability that exactly 3 of those selected are laser printers (so that the other 3 are inkjets)?

Let $D_3 = \{$exactly 3 of the 6 selected are inkjet printers$\}$. Assuming that any particular set of 6 printers is as likely to be chosen as is any other set of 6, we have equally likely outcomes, so $P(D_3) = N(D_3)/N$, where N is the number of ways of choosing 6 printers from the 25 and $N(D_3)$ is the number of ways of choosing 3 laser printers and 3 inkjet models. Thus $N = \binom{25}{6}$. To obtain $N(D_3)$, think of first choosing 3 of the 15 inkjet models and then 3 of the laser printers. There are $\binom{15}{3}$ ways of choosing the 3 inkjet models, and there are $\binom{10}{3}$ ways of choosing the 3 laser printers; $N(D_3)$ is now the product of these two numbers (visualize a tree diagram—we are really using a product rule argument here), so

$$P(D_3) = \frac{N(D_3)}{N} = \frac{\binom{15}{3}\binom{10}{3}}{\binom{25}{6}} = \frac{\frac{15!}{3!12!} \cdot \frac{10!}{3!7!}}{\frac{25!}{6!19!}} = .3083$$

Let $D_4 = \{$exactly 4 of the 6 printers selected are inkjet models$\}$ and define D_5 and D_6 in an analogous manner. Then the probability that at least 3 inkjet printers are selected is

$$P(D_3 \cup D_4 \cup D_5 \cup D_6) = P(D_3) + P(D_4) + P(D_5) + P(D_6)$$

$$= \frac{\binom{15}{3}\binom{10}{3}}{\binom{25}{6}} + \frac{\binom{15}{4}\binom{10}{2}}{\binom{25}{6}} + \frac{\binom{15}{5}\binom{10}{1}}{\binom{25}{6}} + \frac{\binom{15}{6}\binom{10}{0}}{\binom{25}{6}} = .8530$$

∎

EXERCISES Section 2.3 (29–44)

29. As of April 2006, roughly 50 million .com web domain names were registered (e.g., yahoo.com).

    a. How many domain names consisting of just two letters in sequence can be formed? How many domain names of length two are there if digits as well as letters are permitted as characters? [*Note:* A character length of three or more is now mandated.]

    b. How many domain names are there consisting of three letters in sequence? How many of this length are there if either letters or digits are permitted? [*Note:* All are currently taken.]

    c. Answer the questions posed in (b) for four-character sequences.

    d. As of April 2006, 97,786 of the four-character sequences using either letters or digits had not yet been claimed. If a four-character name is randomly selected, what is the probability that it is already owned?

30. A friend of mine is giving a dinner party. His current wine supply includes 8 bottles of zinfandel, 10 of merlot, and 12 of cabernet (he only drinks red wine), all from different wineries.

    a. If he wants to serve 3 bottles of zinfandel and serving order is important, how many ways are there to do this?

    b. If 6 bottles of wine are to be randomly selected from the 30 for serving, how many ways are there to do this?

    c. If 6 bottles are randomly selected, how many ways are there to obtain two bottles of each variety?

    d. If 6 bottles are randomly selected, what is the probability that this results in two bottles of each variety being chosen?

    e. If 6 bottles are randomly selected, what is the probability that all of them are the same variety?

31. The composer Beethoven wrote 9 symphonies, 5 piano concertos (music for piano and orchestra), and 32 piano sonatas (music for solo piano).

    a. How many ways are there to play first a Beethoven symphony and then a Beethoven piano concerto?

    b. The manager of a radio station decides that on each successive evening (7 days per week), a Beethoven symphony will be played followed by a Beethoven piano concerto followed by a Beethoven piano sonata. For how many years could this policy be continued before exactly the same program would have to be repeated?

32. An electronics store is offering a special price on a complete set of components (receiver, compact disc player, speakers, turntable). A purchaser is offered a choice of manufacturer for each component:

    Receiver: Kenwood, Onkyo, Pioneer, Sony, Sherwood
    Compact disc player: Onkyo, Pioneer, Sony, Technics
    Speakers: Boston, Infinity, Polk
    Turntable: Onkyo, Sony, Teac, Technics

    A switchboard display in the store allows a customer to hook together any selection of components (consisting of one of each type). Use the product rules to answer the following questions:

    a. In how many ways can one component of each type be selected?

    b. In how many ways can components be selected if both the receiver and the compact disc player are to be Sony?

    c. In how many ways can components be selected if none is to be Sony?

    d. In how many ways can a selection be made if at least one Sony component is to be included?

    e. If someone flips switches on the selection in a completely random fashion, what is the probability that the system selected contains at least one Sony component? Exactly one Sony component?

33. Again consider a Little League team that has 15 players on its roster.

    a. How many ways are there to select 9 players for the starting lineup?

    b. How many ways are there to select 9 players for the starting lineup and a batting order for the 9 starters?

    c. Suppose 5 of the 15 players are left-handed. How many ways are there to select 3 left-handed outfielders and have all 6 other positions occupied by right-handed players?

34. Computer keyboard failures can be attributed to electrical defects or mechanical defects. A repair facility currently has 25 failed keyboards, 6 of which have electrical defects and 19 of which have mechanical defects.

    a. How many ways are there to randomly select 5 of these keyboards for a thorough inspection (without regard to order)?

    b. In how many ways can a sample of 5 keyboards be selected so that exactly two have an electrical defect?

    c. If a sample of 5 keyboards is randomly selected, what is the probability that at least 4 of these will have a mechanical defect?

35. A production facility employs 10 workers on the day shift, 8 workers on the swing shift, and 6 workers on the graveyard shift. A quality control consultant is to select 5 of these workers for in-depth interviews. Suppose the selection is made in such a way that any particular group of 5 workers has the same chance of being selected as

does any other group (drawing 5 slips without replacement from among 24).

a. How many selections result in all 5 workers coming from the day shift? What is the probability that all 5 selected workers will be from the day shift?

b. What is the probability that all 5 selected workers will be from the same shift?

c. What is the probability that at least two different shifts will be represented among the selected workers?

d. What is the probability that at least one of the shifts will be unrepresented in the sample of workers?

36. An academic department with five faculty members narrowed its choice for department head to either candidate *A* or candidate *B*. Each member then voted on a slip of paper for one of the candidates. Suppose there are actually three votes for *A* and two for *B*. If the slips are selected for tallying in random order, what is the probability that *A* remains ahead of *B* throughout the vote count (e.g., this event occurs if the selected ordering is *AABAB*, but not for *ABBAA*)?

37. An experimenter is studying the effects of temperature, pressure, and type of catalyst on yield from a certain chemical reaction. Three different temperatures, four different pressures, and five different catalysts are under consideration.

a. If any particular experimental run involves the use of a single temperature, pressure, and catalyst, how many experimental runs are possible?

b. How many experimental runs are there that involve use of the lowest temperature and two lowest pressures?

c. Suppose that five different experimental runs are to be made on the first day of experimentation. If the five are randomly selected from among all the possibilities, so that any group of five has the same probability of selection, what is the probability that a different catalyst is used on each run?

38. A sonnet is a 14-line poem in which certain rhyming patterns are followed. The writer Raymond Queneau published a book containing just 10 sonnets, each on a different page. However, these were structured such that other sonnets could be created as follows: the first line of a sonnet could come from the first line on any of the 10 pages, the second line could come from the second line on any of the 10 pages, and so on (successive lines were perforated for this purpose).

a. How many sonnets can be created from the 10 in the book?

b. If one of the sonnets counted in part (a) is selected at random, what is the probability that none of its lines came from either the first or the last sonnet in the book?

39. A box in a supply room contains 15 compact fluorescent lightbulbs, of which 5 are rated 13-watt, 6 are rated 18-watt, and 4 are rated 23-watt. Suppose that three of these bulbs are randomly selected.

a. What is the probability that exactly two of the selected bulbs are rated 23-watt?

b. What is the probability that all three of the bulbs have the same rating?

c. What is the probability that one bulb of each type is selected?

d. If bulbs are selected one by one until a 23-watt bulb is obtained, what is the probability that it is necessary to examine at least 6 bulbs?

40. Three molecules of type *A*, three of type *B*, three of type *C*, and three of type *D* are to be linked together to form a chain molecule. One such chain molecule is *ABCDABCDABCD*, and another is *BCDDAAABDBCC*.

a. How many such chain molecules are there? [*Hint:* If the three *A*'s were distinguishable from one another— A_1, A_2, A_3—and the *B*'s, *C*'s, and *D*'s were also, how many molecules would there be? How is this number reduced when the subscripts are removed from the *A*'s?]

b. Suppose a chain molecule of the type described is randomly selected. What is the probability that all three molecules of each type end up next to one another (such as in *BBBAAADDDCCC*)?

41. An ATM personal identification number (PIN) consists of four digits, each a 0, 1, 2,... 8, or 9, in succession.

a. How many different possible PINs are there if there are no restrictions on the choice of digits?

b. According to a representative at the author's local branch of Chase Bank, there are in fact restrictions on the choice of digits. The following choices are prohibited: (i) all four digits identical (ii) sequences of consecutive ascending or descending digits, such as 6543 (iii) any sequence starting with 19 (birth years are too easy to guess). So if one of the PINs in (a) is randomly selected, what is the probability that it will be a legitimate PIN (that is, not be one of the prohibited sequences)?

c. Someone has stolen an ATM card and knows that the first and last digits of the PIN are 8 and 1, respectively. He has three tries before the card is retained by the ATM (but does not realize that). So he randomly selects the 2nd and 3rd digits for the first try, then randomly selects a different pair of digits for the second try, and yet another randomly selected pair of digits for the third try (the individual knows about the restrictions described in (b) so selects only from the legitimate possibilities). What is the probability that the individual gains access to the account?

d. Recalculate the probability in (c) if the first and last digits are 1 and 1, respectively.

42. A starting lineup in basketball consists of two guards, two forwards, and a center.

a. A certain college team has on its roster three centers, four guards, four forwards, and one individual (X) who can play either guard or forward. How many different starting lineups can be created? [*Hint:* Consider lineups without X, then lineups with X as guard, then lineups with X as forward.]

b. Now suppose the roster has 5 guards, 5 forwards, 3 centers, and 2 "swing players" (X and Y) who can play either guard or forward. If 5 of the 15 players are randomly selected, what is the probability that they constitute a legitimate starting lineup?

43. In five-card poker, a straight consists of five cards with adjacent denominations (e.g., 9 of clubs, 10 of hearts, jack of hearts, queen of spades, and king of clubs). Assuming that aces can be high or low, if you are dealt a five-card hand, what is the probability that it will be a straight with high card 10? What is the probability that it will be a straight? What is the probability that it will be a straight flush (all cards in the same suit)?

44. Show that $\binom{n}{k} = \binom{n}{n-k}$. Give an interpretation involving subsets.

2.4 Conditional Probability

The probabilities assigned to various events depend on what is known about the experimental situation when the assignment is made. Subsequent to the initial assignment, partial information relevant to the outcome of the experiment may become available. Such information may cause us to revise some of our probability assignments. For a particular event A, we have used $P(A)$ to represent the probability, assigned to A; we now think of $P(A)$ as the original, or unconditional probability, of the event A.

In this section, we examine how the information "an event B has occurred" affects the probability assigned to A. For example, A might refer to an individual having a particular disease in the presence of certain symptoms. If a blood test is performed on the individual and the result is negative (B = negative blood test), then the probability of having the disease will change (it should decrease, but not usually to zero, since blood tests are not infallible). We will use the notation $P(A|B)$ to represent the **conditional probability of A given that the event B has occurred**. B is the "conditioning event."

As an example, consider the event A that a randomly selected student at your university obtained all desired classes during the previous term's registration cycle. Presumably $P(A)$ is not very large. However, suppose the selected student is an athlete who gets special registration priority (the event B). Then $P(A|B)$ should be substantially larger than $P(A)$, although perhaps still not close to 1.

EXAMPLE 2.24   Complex components are assembled in a plant that uses two different assembly lines, A and A'. Line A uses older equipment than A', so it is somewhat slower and less reliable. Suppose on a given day line A has assembled 8 components, of which 2 have been identified as defective (B) and 6 as nondefective (B'), whereas A' has produced 1 defective and 9 nondefective components. This information is summarized in the accompanying table.

		Condition	
		B	B'
Line	A	2	6
	A'	1	9

Unaware of this information, the sales manager randomly selects 1 of these 18 components for a demonstration. Prior to the demonstration

$$P(\text{line } A \text{ component selected}) = P(A) = \frac{N(A)}{N} = \frac{8}{18} = .44$$

However, if the chosen component turns out to be defective, then the event B has occurred, so the component must have been 1 of the 3 in the B column of the table. Since these 3 components are equally likely among themselves after B has occurred,

$$P(A|B) = \frac{2}{3} = \frac{2/18}{3/18} = \frac{P(A \cap B)}{P(B)} \tag{2.2}$$

∎

In Equation (2.2), the conditional probability is expressed as a ratio of unconditional probabilities: The numerator is the probability of the intersection of the two events, whereas the denominator is the probability of the conditioning event B. A Venn diagram illuminates this relationship (Figure 2.8).

Figure 2.8 Motivating the definition of conditional probability

Given that B has occurred, the relevant sample space is no longer $\mathcal{S}$ but consists of outcomes in B; A has occurred if and only if one of the outcomes in the intersection occurred, so the conditional probability of A given B is proportional to $P(A \cap B)$. The proportionality constant $1/P(B)$ is used to ensure that the probability $P(B|B)$ of the new sample space B equals 1.

The Definition of Conditional Probability

Example 2.24 demonstrates that when outcomes are equally likely, computation of conditional probabilities can be based on intuition. When experiments are more complicated, though, intuition may fail us, so a general definition of conditional probability is needed that will yield intuitive answers in simple problems. The Venn diagram and Equation (2.2) suggest how to proceed.

DEFINITION

> For any two events A and B with $P(B) > 0$, the **conditional probability of A given that B has occurred** is defined by
>
> $$P(A|B) = \frac{P(A \cap B)}{P(B)} \tag{2.3}$$

EXAMPLE 2.25 Suppose that of all individuals buying a certain digital camera, 60% include an optional memory card in their purchase, 40% include an extra battery, and 30% include both a card and battery. Consider randomly selecting a buyer and let $A = \{\text{memory card purchased}\}$ and $B = \{\text{battery purchased}\}$. Then $P(A) = .60$, $P(B) = .40$, and $P(\text{both purchased}) = P(A \cap B) = .30$. Given that the selected individual purchased an extra battery, the probability that an optional card was also purchased is

$$P(A|B) = \frac{P(A \cap B)}{P(B)} = \frac{.30}{.40} = .75$$

That is, of all those purchasing an extra battery, 75% purchased an optional memory card. Similarly,

$$P(\text{battery}|\text{memory card}) = P(B|A) = \frac{P(A \cap B)}{P(A)} = \frac{.30}{.60} = .50$$

Notice that $P(A|B) \neq P(A)$ and $P(B|A) \neq P(B)$. ■

The event whose probability is desired might be a union or intersection of other events, and the same could be true of the conditioning event.

EXAMPLE 2.26 A news magazine publishes three columns entitled "Art" (A), "Books" (B), and "Cinema" (C). Reading habits of a randomly selected reader with respect to these columns are

Read regularly	A	B	C	$A \cap B$	$A \cap C$	$B \cap C$	$A \cap B \cap C$
Probability	.14	.23	.37	.08	.09	.13	.05

Figure 2.9 illustrates relevant probabilities.

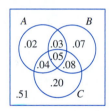

Figure 2.9 Venn diagram for Example 2.26

Consider the following four conditional probabilities:

(i) $P(A|B) = \dfrac{P(A \cap B)}{P(B)} = \dfrac{.08}{.23} = .348$

(ii) The probability that the selected individual regularly reads the Art column given that he or she regularly reads at least one of the other two columns is

$$P(A|B \cup C) = \frac{P(A \cap (B \cup C))}{P(B \cup C)} = \frac{.04 + .05 + .03}{.47} = \frac{.12}{.47} = .255$$

(iii) $P(A|\text{reads at least one}) = P(A|A \cup B \cup C) = \dfrac{P(A \cap (A \cup B \cup C))}{P(A \cup B \cup C)}$

$$= \frac{P(A)}{P(A \cup B \cup C)} = \frac{.14}{.49} = .286$$

(iv) The probability that the selected individual reads at least one of the first two columns given that he or she reads the Cinema column is

$$P(A \cup B|C) = \frac{P((A \cup B) \cap C)}{P(C)} = \frac{.04 + .05 + .08}{.37} = .459$$ ■

The Multiplication Rule for $P(A \cap B)$

The definition of conditional probability yields the following result, obtained by multiplying both sides of Equation (2.3) by $P(B)$.

The Multiplication Rule

$$P(A \cap B) = P(A|B) \cdot P(B)$$

This rule is important because it is often the case that $P(A \cap B)$ is desired, whereas both $P(B)$ and $P(A|B)$ can be obtained from the problem description. Consideration of $P(B|A)$ gives $P(A \cap B) = P(B|A) \cdot P(A)$.

EXAMPLE 2.27   Four individuals have responded to a request by a blood bank for blood donations. None of them has donated before, so their blood types are unknown. Suppose only type O+ is desired and only one of the four actually has this type. If the potential donors are selected in random order for typing, what is the probability that at least three individuals must be typed to obtain the desired type?

Making the identification $B = \{$first type not O+$\}$ and $A = \{$second type not O+$\}$, $P(B) = 3/4$. Given that the first type is not O+, two of the three individuals left are not O+, so $P(A|B) = 2/3$. The multiplication rule now gives

$$
\begin{aligned}
P(\text{at least three individuals are typed}) &= P(A \cap B) \\
&= P(A|B) \cdot P(B) \\
&= \frac{2}{3} \cdot \frac{3}{4} = \frac{6}{12} \\
&= .5
\end{aligned}
$$
■

The multiplication rule is most useful when the experiment consists of several stages in succession. The conditioning event B then describes the outcome of the first stage and A the outcome of the second, so that $P(A|B)$—conditioning on what occurs first—will often be known. The rule is easily extended to experiments involving more than two stages. For example, consider three events A_1, A_2, and A_3. The triple intersection of these events can be represented as the double intersection $(A_1 \cap A_2) \cap A_3$. Applying our previous multiplication rule to this intersection and then to $A_1 \cap A_2$ gives

$$
\begin{aligned}
P(A_1 \cap A_2 \cap A_3) &= P(A_3|A_1 \cap A_2) \cdot P(A_1 \cap A_2) \\
&= P(A_3|A_1 \cap A_2) \cdot P(A_2|A_1) \cdot P(A_1)
\end{aligned}
\tag{2.4}
$$

Thus the triple intersection probability is a product of three probabilities, two of which are conditional.

EXAMPLE 2.28   For the blood-typing experiment of Example 2.27,

$$
\begin{aligned}
P(\text{third type is O+}) &= P(\text{third is}|\text{first isn't} \cap \text{second isn't}) \\
&\quad \cdot P(\text{second isn't}|\text{first isn't}) \cdot P(\text{first isn't}) \\
&= \frac{1}{2} \cdot \frac{2}{3} \cdot \frac{3}{4} = \frac{1}{4} = .25
\end{aligned}
$$
■

When the experiment of interest consists of a sequence of several stages, it is convenient to represent these with a tree diagram. Once we have an appropriate tree diagram, probabilities and conditional probabilities can be entered on the various branches; this will make repeated use of the multiplication rule quite straightforward.

EXAMPLE 2.29   An electronics store sells three different brands of DVD players. Of its DVD player sales, 50% are brand 1 (the least expensive), 30% are brand 2, and 20% are brand 3. Each manufacturer offers a 1-year warranty on parts and labor. It is known that 25% of brand 1's DVD players require warranty repair work, whereas the corresponding percentages for brands 2 and 3 are 20% and 10%, respectively.

1. What is the probability that a randomly selected purchaser has bought a brand 1 DVD player that will need repair while under warranty?

2. What is the probability that a randomly selected purchaser has a DVD player that will need repair while under warranty?

3. If a customer returns to the store with a DVD player that needs warranty repair work, what is the probability that it is a brand 1 DVD player? A brand 2 DVD player? A brand 3 DVD player?

The first stage of the problem involves a customer selecting one of the three brands of DVD player. Let A_i = {brand i is purchased}, for i = 1, 2, and 3. Then $P(A_1)$ = .50, $P(A_2)$ = .30, and $P(A_3)$ = .20. Once a brand of DVD player is selected, the second stage involves observing whether the selected DVD player needs warranty repair. With B = {needs repair} and B' = {doesn't need repair}, the given information implies that $P(B|A_1)$ = .25, $P(B|A_2)$ = .20, and $P(B|A_3)$ = .10.

The tree diagram representing this experimental situation is shown in Figure 2.10. The initial branches correspond to different brands of DVD players; there are two second-generation branches emanating from the tip of each initial branch, one for "needs repair" and the other for "doesn't need repair." The probability $P(A_i)$ appears on the ith initial branch, whereas the conditional probabilities $P(B|A_i)$ and $P(B'|A_i)$ appear on the second-generation branches. To the right of each second-generation branch corresponding to the occurrence of B, we display the product of probabilities on the branches leading out to that point. This is simply the multiplication rule in action. The answer to the question posed in 1 is thus $P(A_1 \cap B) = P(B|A_1) \cdot P(A_1)$ = .125. The answer to question 2 is

$$P(B) = P[(\text{brand 1 and repair}) \text{ or } (\text{brand 2 and repair}) \text{ or } (\text{brand 3 and repair})]$$
$$= P(A_1 \cap B) + P(A_2 \cap B) + P(A_3 \cap B)$$
$$= .125 + .060 + .020 = .205$$

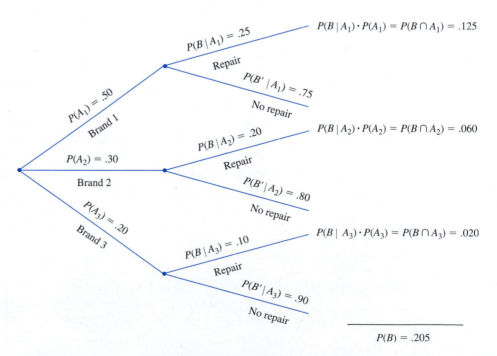

Figure 2.10 Tree diagram for Example 2.29

Finally,

$$P(A_1|B) = \frac{P(A_1 \cap B)}{P(B)} = \frac{.125}{.205} = .61$$

$$P(A_2|B) = \frac{P(A_2 \cap B)}{P(B)} = \frac{.060}{.205} = .29$$

and

$$P(A_3|B) = 1 - P(A_1|B) - P(A_2|B) = .10$$

The initial or *prior probability* of brand 1 is .50. Once it is known that the selected DVD player needed repair, the *posterior probability* of brand 1 increases to .61. This is because brand 1 DVD players are more likely to need warranty repair than are the other brands. The posterior probability of brand 3 is $P(A_3|B) = .10$, which is much less than the prior probability $P(A_3) = .20$. ∎

Bayes' Theorem

The computation of a posterior probability $P(A_j|B)$ from given prior probabilities $P(A_i)$ and conditional probabilities $P(B|A_i)$ occupies a central position in elementary probability. The general rule for such computations, which is really just a simple application of the multiplication rule, goes back to Reverend Thomas Bayes, who lived in the eighteenth century. To state it we first need another result. Recall that events $A_1,\ldots, A_k$ are mutually exclusive if no two have any common outcomes. The events are *exhaustive* if one A_i must occur, so that $A_1 \cup \ldots \cup A_k = \mathcal{S}$.

The Law of Total Probability

Let $A_1,\ldots, A_k$ be mutually exclusive and exhaustive events. Then for any other event B,

$$P(B) = P(B|A_1)P(A_1) + \cdots + P(B|A_k)P(A_k)$$

$$= \sum_{i=1}^{k} P(B|A_i)P(A_i) \qquad (2.5)$$

Proof Because the A_i's are mutually exclusive and exhaustive, if B occurs it must be in conjunction with exactly one of the A_i's. That is, $B = (A_1 \cap B) \cup \ldots \cup (A_k \cap B)$, where the events $(A_i \cap B)$ are mutually exclusive. This "partitioning of B" is illustrated in Figure 2.11. Thus

$$P(B) = \sum_{i=1}^{k} P(A_i \cap B) = \sum_{i=1}^{k} P(B|A_i)P(A_i)$$

as desired.

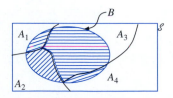

Figure 2.11 Partition of B by mutually exclusive and exhaustive A_i's ∎

EXAMPLE 2.30   An individual has 3 different email accounts. Most of her messages, in fact 70%, come into account #1, whereas 20% come into account #2 and the remaining 10% into account #3. Of the messages into account #1, only 1% are spam, whereas the corresponding percentages for accounts #2 and #3 are 2% and 5%, respectively. What is the probability that a randomly selected message is spam?

To answer this question, let's first establish some notation:

$$A_i = \{\text{message is from account} \ \# \ i\} \ \text{for} \ i = 1, 2, 3, \quad B = \{\text{message is spam}\}$$

Then the given percentages imply that

$$P(A_1) = .70, P(A_2) = .20, P(A_3) = .10$$
$$P(B|A_1) = .01, P(B|A_2) = .02, P(B|A_3) = .05$$

Now it is simply a matter of substituting into the equation for the law of total probability:

$$P(B) = (.01)(.70) + (.02)(.20) + (.05)(.10) = .016$$

In the long run, 1.6% of this individual's messages will be spam.   ■

Bayes' Theorem

Let $A_1, A_2,..., A_k$ be a collection of k mutually exclusive and exhaustive events with *prior* probabilities $P(A_i)$ $(i = 1,..., k)$. Then for any other event B for which $P(B) > 0$, the *posterior* probability of A_j given that B has occurred is

$$P(A_j|B) = \frac{P(A_j \cap B)}{P(B)} = \frac{P(B|A_j)P(A_j)}{\sum\limits_{i=1}^{k} P(B|A_i) \cdot P(A_i)} \quad j = 1,..., k \quad (2.6)$$

The transition from the second to the third expression in (2.6) rests on using the multiplication rule in the numerator and the law of total probability in the denominator. The proliferation of events and subscripts in (2.6) can be a bit intimidating to probability newcomers. As long as there are relatively few events in the partition, a tree diagram (as in Example 2.29) can be used as a basis for calculating posterior probabilities without ever referring explicitly to Bayes' theorem.

EXAMPLE 2.31   *Incidence of a rare disease.* Only 1 in 1000 adults is afflicted with a rare disease for which a diagnostic test has been developed. The test is such that when an individual actually has the disease, a positive result will occur 99% of the time, whereas an individual without the disease will show a positive test result only 2% of the time (the *sensitivity* of this test is 99% and the *specificity* is 98%; in contrast, the **Sept. 22, 2012 issue of *The Lancet*** reports that the first at-home HIV test has a sensitivity of only 92% and a specificity of 99.98%). If a randomly selected individual is tested and the result is positive, what is the probability that the individual has the disease?

To use Bayes' theorem, let A_1 = individual has the disease, A_2 = individual does not have the disease, and B = positive test result. Then $P(A_1) = .001$, $P(A_2) = .999$, $P(B|A_1) = .99$, and $P(B|A_2) = .02$. The tree diagram for this problem is in Figure 2.12.

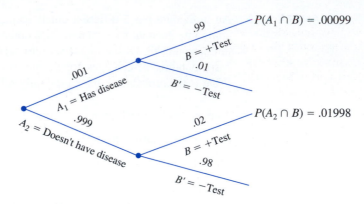

$$P(A_1 \cap B) = .00099$$

$$P(A_2 \cap B) = .01998$$

Figure 2.12 Tree diagram for the rare-disease problem

Next to each branch corresponding to a positive test result, the multiplication rule yields the recorded probabilities. Therefore, $P(B) = .00099 + .01998 = .02097$, from which we have

$$P(A_1 | B) = \frac{P(A_1 \cap B)}{P(B)} = \frac{.00099}{.02097} = .047$$

This result seems counterintuitive; the diagnostic test appears so accurate that we expect someone with a positive test result to be highly likely to have the disease, whereas the computed conditional probability is only .047. However, the rarity of the disease implies that most positive test results arise from errors rather than from diseased individuals. The probability of having the disease has increased by a multiplicative factor of 47 (from prior .001 to posterior .047); but to get a further increase in the posterior probability, a diagnostic test with much smaller error rates is needed. ∎

EXERCISES Section 2.4 (45–69)

45. The population of a particular country consists of three ethnic groups. Each individual belongs to one of the four major blood groups. The accompanying *joint probability table* gives the proportions of individuals in the various ethnic group–blood group combinations.

		Blood Group			
		O	**A**	**B**	**AB**
Ethnic Group	**1**	.082	.106	.008	.004
	2	.135	.141	.018	.006
	3	.215	.200	.065	.020

Suppose that an individual is randomly selected from the population, and define events by $A = \{$type A selected$\}$, $B = \{$type B selected$\}$, and $C = \{$ethnic group 3 selected$\}$.

a. Calculate $P(A)$, $P(C)$, and $P(A \cap C)$.

b. Calculate both $P(A|C)$ and $P(C|A)$, and explain in context what each of these probabilities represents.

c. If the selected individual does not have type B blood, what is the probability that he or she is from ethnic group 1?

46. Suppose an individual is randomly selected from the population of all adult males living in the United States. Let A be the event that the selected individual is over 6 ft in height, and let B be the event that the selected individual is a professional basketball player. Which do you think is larger, $P(A|B)$ or $P(B|A)$? Why?

47. Return to the credit card scenario of Exercise 12 (Section 2.2), and let C be the event that the selected student has an American Express card. In addition to $P(A) = .6$, $P(B) = .4$, and $P(A \cap B) = .3$, suppose that $P(C) = .2$, $P(A \cap C) = .15$, $P(B \cap C) = .1$, and $P(A \cap B \cap C) = .08$.

a. What is the probability that the selected student has at least one of the three types of cards?

b. What is the probability that the selected student has both a Visa card and a MasterCard but not an American Express card?

c. Calculate and interpret $P(B|A)$ and also $P(A|B)$.

d. If we learn that the selected student has an American Express card, what is the probability that she or he also has both a Visa card and a MasterCard?

e. Given that the selected student has an American Express card, what is the probability that she or he has at least one of the other two types of cards?

48. Reconsider the system defect situation described in Exercise 26 (Section 2.2).

a. Given that the system has a type 1 defect, what is the probability that it has a type 2 defect?

b. Given that the system has a type 1 defect, what is the probability that it has all three types of defects?

c. Given that the system has at least one type of defect, what is the probability that it has exactly one type of defect?

d. Given that the system has both of the first two types of defects, what is the probability that it does not have the third type of defect?

49. The accompanying table gives information on the type of coffee selected by someone purchasing a single cup at a particular airport kiosk.

	Small	**Medium**	**Large**
Regular	14%	20%	26%
Decaf	20%	10%	10%

Consider randomly selecting such a coffee purchaser.

a. What is the probability that the individual purchased a small cup? A cup of decaf coffee?

b. If we learn that the selected individual purchased a small cup, what now is the probability that he/she chose decaf coffee, and how would you interpret this probability?

c. If we learn that the selected individual purchased decaf, what now is the probability that a small size was selected, and how does this compare to the corresponding unconditional probability of (a)?

50. A department store sells sport shirts in three sizes (small, medium, and large), three patterns (plaid, print, and stripe), and two sleeve lengths (long and short). The accompanying tables give the proportions of shirts sold in the various category combinations.

Short-sleeved

	Pattern		
Size	**Pl**	**Pr**	**St**
S	.04	.02	.05
M	.08	.07	.12
L	.03	.07	.08

Long-sleeved

	Pattern		
Size	**Pl**	**Pr**	**St**
S	.03	.02	.03
M	.10	.05	.07
L	.04	.02	.08

a. What is the probability that the next shirt sold is a medium, long-sleeved, print shirt?

b. What is the probability that the next shirt sold is a medium print shirt?

c. What is the probability that the next shirt sold is a short-sleeved shirt? A long-sleeved shirt?

d. What is the probability that the size of the next shirt sold is medium? That the pattern of the next shirt sold is a print?

e. Given that the shirt just sold was a short-sleeved plaid, what is the probability that its size was medium?

f. Given that the shirt just sold was a medium plaid, what is the probability that it was short-sleeved? Long-sleeved?

51. According to a July 31, 2013, posting on **cnn.com** subsequent to the death of a child who bit into a peanut, a 2010 study in the journal ***Pediatrics*** found that 8% of children younger than 18 in the United States have at least one food allergy. Among those with food allergies, about 39% had a history of severe reaction.

a. If a child younger than 18 is randomly selected, what is the probability that he or she has at least one food allergy and a history of severe reaction?

b. It was also reported that 30% of those with an allergy in fact are allergic to multiple foods. If a child younger than 18 is randomly selected, what is the probability that he or she is allergic to multiple foods?

52. A system consists of two identical pumps, #1 and #2. If one pump fails, the system will still operate. However, because of the added strain, the remaining pump is now more likely to fail than was originally the case. That is, $r = P(\#2 \text{ fails} \mid \#1 \text{ fails}) > P(\#2 \text{ fails}) = q$. If at least one pump fails by the end of the pump design life in 7% of all systems and both pumps fail during that period in only 1%, what is the probability that pump #1 will fail during the pump design life?

53. A certain shop repairs both audio and video components. Let *A* denote the event that the next component brought in for repair is an audio component, and let *B* be the event that the next component is a compact disc player (so the event *B* is contained in *A*). Suppose that $P(A) = .6$ and $P(B) = .05$. What is $P(B|A)$?

54. In Exercise 13, $A_i = \{$awarded project $i\}$, for $i = 1, 2, 3$. Use the probabilities given there to compute the following probabilities, and explain in words the meaning of each one.

a. $P(A_2|A_1)$ b. $P(A_2 \cap A_3|A_1)$
c. $P(A_2 \cup A_3|A_1)$ d. $P(A_1 \cap A_2 \cap A_3|A_1 \cup A_2 \cup A_3)$

55. Deer ticks can be carriers of either Lyme disease or human granulocytic ehrlichiosis (HGE). Based on a recent study, suppose that 16% of all ticks in a certain location carry Lyme disease, 10% carry HGE, and 10% of the ticks that carry at least one of these diseases in fact carry both of them. If a randomly selected tick is found to have carried HGE, what is the probability that the selected tick is also a carrier of Lyme disease?

56. For any events A and B with $P(B) > 0$, show that $P(A|B) + P(A'|B) = 1$.

57. If $P(B|A) > P(B)$, show that $P(B'|A) < P(B')$. [*Hint:* Add $P(B'|A)$ to both sides of the given inequality and then use the result of Exercise 56.]

58. Show that for any three events A, B, and C with $P(C) > 0$, $P(A \cup B|C) = P(A|C) + P(B|C) - P(A \cap B|C)$.

59. At a certain gas station, 40% of the customers use regular gas (A_1), 35% use plus gas (A_2), and 25% use premium (A_3). Of those customers using regular gas, only 30% fill their tanks (event B). Of those customers using plus, 60% fill their tanks, whereas of those using premium, 50% fill their tanks.

a. What is the probability that the next customer will request plus gas and fill the tank ($A_2 \cap B$)?
b. What is the probability that the next customer fills the tank?
c. If the next customer fills the tank, what is the probability that regular gas is requested? Plus? Premium?

60. Seventy percent of the light aircraft that disappear while in flight in a certain country are subsequently discovered. Of the aircraft that are discovered, 60% have an emergency locator, whereas 90% of the aircraft not discovered do not have such a locator. Suppose a light aircraft has disappeared.

a. If it has an emergency locator, what is the probability that it will not be discovered?
b. If it does not have an emergency locator, what is the probability that it will be discovered?

61. Components of a certain type are shipped to a supplier in batches of ten. Suppose that 50% of all such batches contain no defective components, 30% contain one defective component, and 20% contain two defective components. Two components from a batch are randomly selected and tested. What are the probabilities associated with 0, 1, and 2 defective components being in the batch under each of the following conditions?

a. Neither tested component is defective.
b. One of the two tested components is defective. [*Hint:* Draw a tree diagram with three first-generation branches for the three different types of batches.]

62. Blue Cab operates 15% of the taxis in a certain city, and Green Cab operates the other 85%. After a nighttime hit-and-run accident involving a taxi, an eyewitness said the vehicle was blue. Suppose, though, that under night vision conditions, only 80% of individuals can correctly distinguish between a blue and a green vehicle. What is the (posterior) probability that the taxi at fault was blue? In answering, be sure to indicate which probability rules you are using. [*Hint:* A tree diagram might help. *Note:* This is based on an actual incident.]

63. For customers purchasing a refrigerator at a certain appliance store, let A be the event that the refrigerator was manufactured in the U.S., B be the event that the refrigerator had an icemaker, and C be the event that the customer purchased an extended warranty. Relevant probabilities are

$P(A) = .75$ $P(B|A) = .9$ $P(B|A') = .8$
$P(C|A \cap B) = .8$ $P(C|A \cap B') = .6$
$P(C|A' \cap B) = .7$ $P(C|A' \cap B') = .3$

a. Construct a tree diagram consisting of first-, second-, and third-generation branches, and place an event label and appropriate probability next to each branch.
b. Compute $P(A \cap B \cap C)$.
c. Compute $P(B \cap C)$.
d. Compute $P(C)$.
e. Compute $P(A|B \cap C)$, the probability of a U.S. purchase given that an icemaker and extended warranty are also purchased.

64. The Reviews editor for a certain scientific journal decides whether the review for any particular book should be short (1–2 pages), medium (3–4 pages), or long (5–6 pages). Data on recent reviews indicates that 60% of them are short, 30% are medium, and the other 10% are long. Reviews are submitted in either Word or LaTeX. For short reviews, 80% are in Word, whereas 50% of medium reviews are in Word and 30% of long reviews are in Word. Suppose a recent review is randomly selected.

a. What is the probability that the selected review was submitted in Word format?
b. If the selected review was submitted in Word format, what are the posterior probabilities of it being short, medium, or long?

65. A large operator of timeshare complexes requires anyone interested in making a purchase to first visit the site of interest. Historical data indicates that 20% of all potential purchasers select a day visit, 50% choose a one-night visit, and 30% opt for a two-night visit. In addition, 10% of day visitors ultimately make a purchase, 30% of one-night visitors buy a unit, and 20% of those visiting for two nights decide to buy. Suppose a visitor is randomly selected and is found to have made a purchase. How likely is it that this person made a day visit? A one-night visit? A two-night visit?

66. Consider the following information about travelers on vacation (based partly on a recent Travelocity poll): 40% check work email, 30% use a cell phone to stay connected to work, 25% bring a laptop with them, 23% both check work email and use a cell phone to stay connected, and 51% neither check work email nor use a cell phone to stay connected nor bring a laptop. In addition, 88 out of every 100 who bring a laptop also check work email, and 70 out of every 100 who use a cell phone to stay connected also bring a laptop.

    a. What is the probability that a randomly selected traveler who checks work email also uses a cell phone to stay connected?
    b. What is the probability that someone who brings a laptop on vacation also uses a cell phone to stay connected?
    c. If the randomly selected traveler checked work email and brought a laptop, what is the probability that he/she uses a cell phone to stay connected?

67. There has been a great deal of controversy over the last several years regarding what types of surveillance are appropriate to prevent terrorism. Suppose a particular surveillance system has a 99% chance of correctly identifying a future terrorist and a 99.9% chance of correctly identifying someone who is not a future terrorist. If there are 1000 future terrorists in a population of 300 million, and one of these 300 million is randomly selected, scrutinized by the system, and identified as a future terrorist, what is the probability that he/she actually is a future terrorist? Does the value of this probability make you uneasy about using the surveillance system? Explain.

68. A friend who lives in Los Angeles makes frequent consulting trips to Washington, D.C.; 50% of the time she travels on airline #1, 30% of the time on airline #2, and the remaining 20% of the time on airline #3. For airline #1, flights are late into D.C. 30% of the time and late into L.A. 10% of the time. For airline #2, these percentages are 25% and 20%, whereas for airline #3 the percentages are 40% and 25%. If we learn that on a particular trip she arrived late at exactly one of the two destinations, what are the posterior probabilities of having flown on airlines #1, #2, and #3? Assume that the chance of a late arrival in L.A. is unaffected by what happens on the flight to D.C. [*Hint:* From the tip of each first-generation branch on a tree diagram, draw three second-generation branches labeled, respectively, 0 late, 1 late, and 2 late.]

69. In Exercise 59, consider the following additional information on credit card usage:

    70% of all regular fill-up customers use a credit card.
    50% of all regular non-fill-up customers use a credit card.
    60% of all plus fill-up customers use a credit card.
    50% of all plus non-fill-up customers use a credit card.
    50% of all premium fill-up customers use a credit card.
    40% of all premium non-fill-up customers use a credit card.

    Compute the probability of each of the following events for the next customer to arrive (a tree diagram might help).

    a. {plus and fill-up and credit card}
    b. {premium and non-fill-up and credit card}
    c. {premium and credit card}
    d. {fill-up and credit card}
    e. {credit card}
    f. If the next customer uses a credit card, what is the probability that premium was requested?

2.5 Independence

The definition of conditional probability enables us to revise the probability $P(A)$ originally assigned to A when we are subsequently informed that another event B has occurred; the new probability of A is $P(A|B)$. In our examples, it frequently happened that $P(A|B)$ differed from the unconditional probability $P(A)$. Then the information "B has occurred" resulted in a change in the likelihood of A occurring. Often the chance that A will occur or has occurred is not affected by knowledge that B has occurred, so that $P(A|B) = P(A)$. It is then natural to regard A and B as independent events, meaning that the occurrence or nonoccurrence of one event has no bearing on the chance that the other will occur.

DEFINITION

Two events A and B are **independent** if $P(A|B) = P(A)$ and are **dependent** otherwise.

The definition of independence might seem "unsymmetric" because we do not also demand that $P(B|A) = P(B)$. However, using the definition of conditional probability and the multiplication rule,

$$P(B|A) = \frac{P(A \cap B)}{P(A)} = \frac{P(A|B)P(B)}{P(A)} \tag{2.7}$$

The right-hand side of Equation (2.7) is $P(B)$ if and only if $P(A|B) = P(A)$ (independence). Thus the equality in the definition implies the other equality (and vice versa). It is also straightforward to show that if A and B are independent, then so are the following pairs of events: (1) A' and B, (2) A and B', and (3) A' and B'.

EXAMPLE 2.32   Consider a gas station with six pumps numbered $1, 2, \ldots, 6$, and let E_i denote the simple event that a randomly selected customer uses pump i $(i = 1, \ldots, 6)$. Suppose that

$$P(E_1) = P(E_6) = .10, \quad P(E_2) = P(E_5) = .15, \quad P(E_3) = P(E_4) = .25$$

Define events A, B, C by

$$A = \{2, 4, 6\}, B = \{1, 2, 3\}, C = \{2, 3, 4, 5\}.$$

We then have $P(A) = .50$, $P(A|B) = .30$, and $P(A|C) = .50$. That is, events A and B are dependent, whereas events A and C are independent. Intuitively, A and C are independent because the relative division of probability among even- and odd-numbered pumps is the same among pumps 2, 3, 4, 5 as it is among all six pumps. ∎

EXAMPLE 2.33   Let A and B be any two mutually exclusive events with $P(A) > 0$. For example, for a randomly chosen automobile, let $A = \{$the car has a four cylinder engine$\}$ and $B = \{$the car has a six cylinder engine$\}$. Since the events are mutually exclusive, if B occurs, then A cannot possibly have occurred, so $P(A|B) = 0 \neq P(A)$. The message here is that *if two events are mutually exclusive, they cannot be independent.* When A and B are mutually exclusive, the information that A occurred says something about B (it cannot have occurred), so independence is precluded. ∎

The Multiplication Rule for $P(A \cap B)$

Frequently the nature of an experiment suggests that two events A and B should be assumed independent. This is the case, for example, if a manufacturer receives a circuit board from each of two different suppliers, each board is tested on arrival, and $A = \{$first is defective$\}$ and $B = \{$second is defective$\}$. If $P(A) = .1$, it should also be the case that $P(A|B) = .1$; knowing the condition of the second board shouldn't provide information about the condition of the first. The probability that both events will occur is easily calculated from the individual event probabilities when the events are independent.

PROPOSITION

> A and B are independent if and only if (iff)
>
> $$P(A \cap B) = P(A) \cdot P(B) \tag{2.8}$$

The verification of this multiplication rule is as follows:

$$P(A \cap B) = P(A|B) \cdot P(B) = P(A) \cdot P(B) \tag{2.9}$$

where the second equality in Equation (2.9) is valid iff A and B are independent. Equivalence of independence and Equation (2.8) imply that the latter can be used as a definition of independence.

EXAMPLE 2.34 It is known that 30% of a certain company's washing machines require service while under warranty, whereas only 10% of its dryers need such service. If someone purchases both a washer and a dryer made by this company, what is the probability that both machines will need warranty service?

Let A denote the event that the washer needs service while under warranty, and let B be defined analogously for the dryer. Then $P(A) = .30$ and $P(B) = .10$. Assuming that the two machines will function independently of one another, the desired probability is

$$P(A \cap B) = P(A) \cdot P(B) = (.30)(.10) = .03 \qquad \blacksquare$$

It is straightforward to show that A and B are independent iff A' and B are independent, iff A and B' are independent, and iff A' and B' are independent. Thus in Example 2.34, the probability that neither machine needs service is

$$P(A' \cap B') = P(A') \cdot P(B') = (.70)(.90) = .63$$

EXAMPLE 2.35 Each day, Monday through Friday, a batch of components sent by a first supplier arrives at a certain inspection facility. Two days a week, a batch also arrives from a second supplier. Eighty percent of all supplier 1's batches pass inspection, and 90% of supplier 2's do likewise. What is the probability that, on a randomly selected day, two batches pass inspection? We will answer this assuming that on days when two batches are tested, whether the first batch passes is independent of whether the second batch does so. Figure 2.13 displays the relevant information.

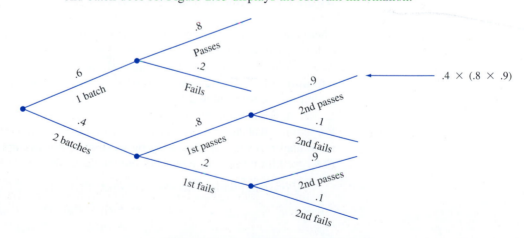

Figure 2.13 Tree diagram for Example 2.35

$$P(\text{two pass}) = P(\text{two received} \cap \text{both pass})$$
$$= P(\text{both pass} \mid \text{two received}) \cdot P(\text{two received})$$
$$= [(.8)(.9)](.4) = .288 \qquad \blacksquare$$

Independence of More Than Two Events

The notion of independence of two events can be generalized to collections of more than two events. Although it is possible to extend the definition for two independent events by working in terms of conditional and unconditional probabilities, it is more direct and less cumbersome to proceed along the lines of the last proposition.

DEFINITION

> Events $A_1, \ldots, A_n$ are **mutually independent** if for every k ($k = 2, 3, \ldots, n$) and every subset of indices $i_1, i_2, \ldots, i_k$,
>
> $$P(A_{i_1} \cap A_{i_2} \cap \ldots \cap A_{i_k}) = P(A_{i_1}) \cdot P(A_{i_2}) \cdots P(A_{i_k})$$

To paraphrase the definition, the events are mutually independent if the probability of the intersection of any subset of the n events is equal to the product of the individual probabilities. In using the multiplication property for more than two independent events, it is legitimate to replace one or more of the A_i's by their complements (e.g., if A_1, A_2, and A_3 are independent events, so are A_1', A_2', and A_3'). As was the case with two events, we frequently specify at the outset of a problem the independence of certain events. The probability of an intersection can then be calculated via multiplication.

EXAMPLE 2.36 The article **"Reliability Evaluation of Solar Photovoltaic Arrays"** (*Solar Energy, 2002: 129–141*) presents various configurations of solar photovoltaic arrays consisting of crystalline silicon solar cells. Consider first the system illustrated in Figure 2.14(a).

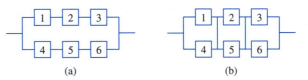

(a)   (b)

Figure 2.14 System configurations for Example 2.36: (a) series-parallel; (b) total-cross-tied

There are two subsystems connected in parallel, each one containing three cells. In order for the system to function, at least one of the two parallel subsystems must work. Within each subsystem, the three cells are connected in series, so a subsystem will work only if all cells in the subsystem work. Consider a particular lifetime value t_0, and suppose we want to determine the probability that the system lifetime exceeds t_0. Let A_i denote the event that the lifetime of cell i exceeds t_0 ($i = 1, 2, \ldots, 6$). We assume that the $A_i's$ are independent events (whether any particular cell lasts more than t_0 hours has no bearing on whether or not any other cell does) and that $P(A_i) = .9$ for every i since the cells are identical. Then

$$P(\text{system lifetime exceeds } t_0) = P[(A_1 \cap A_2 \cap A_3) \cup (A_4 \cap A_5 \cap A_6)]$$
$$= P(A_1 \cap A_2 \cap A_3) + P(A_4 \cap A_5 \cap A_6)$$
$$- P[(A_1 \cap A_2 \cap A_3) \cap (A_4 \cap A_5 \cap A_6)]$$
$$= (.9)(.9)(.9) + (.9)(.9)(.9) - (.9)(.9)(.9)(.9)(.9)(.9) = .927$$

Alternatively,

$$P(\text{system lifetime exceeds } t_0) = 1 - P(\text{both subsystem lives are} \le t_0)$$
$$= 1 - [P(\text{subsystem life is} \le t_0)]^2$$
$$= 1 - [1 - P(\text{subsystem life is} > t_0)]^2$$
$$= 1 - [1 - (.9)^3]^2 = .927$$

Next consider the total-cross-tied system shown in Figure 2.14(b), obtained from the series-parallel array by connecting ties across each column of junctions. Now the system fails as soon as an entire column fails, and system lifetime exceeds t_0 only if the life of every column does so. For this configuration,

$$P(\text{system lifetime is at least } t_0) = [P(\text{column lifetime exceeds } t_0)]^3$$
$$= [1 - P(\text{column lifetime is} \le t_0)]^3$$
$$= [1 - P(\text{both cells in a column have lifetime} \le t_0)]^3$$
$$= [1 - (1 - .9)^2]^3 = .970$$

■

EXERCISES Section 2.5 (70–89)

70. Reconsider the credit card scenario of Exercise 47 (Section 2.4), and show that A and B are dependent first by using the definition of independence and then by verifying that the multiplication property does not hold.

71. An oil exploration company currently has two active projects, one in Asia and the other in Europe. Let A be the event that the Asian project is successful and B be the event that the European project is successful. Suppose that A and B are independent events with $P(A) = .4$ and $P(B) = .7$.
   a. If the Asian project is not successful, what is the probability that the European project is also not successful? Explain your reasoning.
   b. What is the probability that at least one of the two projects will be successful?
   c. Given that at least one of the two projects is successful, what is the probability that only the Asian project is successful?

72. In Exercise 13, is any A_i independent of any other A_j? Answer using the multiplication property for independent events.

73. If A and B are independent events, show that A' and B are also independent. [*Hint:* First establish a relationship between $P(A' \cap B)$, $P(B)$, and $P(A \cap B)$.]

74. The proportions of blood phenotypes in the U.S. population are as follows:

A	B	AB	O
.40	.11	.04	.45

Assuming that the phenotypes of two randomly selected individuals are independent of one another, what is the probability that both phenotypes are O? What is the probability that the phenotypes of two randomly selected individuals match?

75. One of the assumptions underlying the theory of control charting (see Chapter 16) is that successive plotted points are independent of one another. Each plotted point can signal either that a manufacturing process is operating correctly or that there is some sort of malfunction.

Even when a process is running correctly, there is a small probability that a particular point will signal a problem with the process. Suppose that this probability is .05. What is the probability that at least one of 10 successive points indicates a problem when in fact the process is operating correctly? Answer this question for 25 successive points.

76. In October, 1994, a flaw in a certain Pentium chip installed in computers was discovered that could result in a wrong answer when performing a division. The manufacturer initially claimed that the chance of any particular division being incorrect was only 1 in 9 billion, so that it would take thousands of years before a typical user encountered a mistake. However, statisticians are not typical users; some modern statistical techniques are so computationally intensive that a billion divisions over a short time period is not outside the realm of possibility. Assuming that the 1 in 9 billion figure is correct and that results of different divisions are independent of one another, what is the probability that at least one error occurs in one billion divisions with this chip?

77. An aircraft seam requires 25 rivets. The seam will have to be reworked if any of these rivets is defective. Suppose rivets are defective independently of one another, each with the same probability.
   a. If 15% of all seams need reworking, what is the probability that a rivet is defective?
   b. How small should the probability of a defective rivet be to ensure that only 10% of all seams need reworking?

78. A boiler has five identical relief valves. The probability that any particular valve will open on demand is .96. Assuming independent operation of the valves, calculate P(at least one valve opens) and P(at least one valve fails to open).

79. Two pumps connected in parallel fail independently of one another on any given day. The probability that only the older pump will fail is .10, and the probability that only the newer pump will fail is .05. What is the probability that the pumping system will fail on any given day (which happens if both pumps fail)?

80. Consider the system of components connected as in the accompanying picture. Components 1 and 2 are connected in parallel, so that subsystem works iff either 1 or 2 works; since 3 and 4 are connected in series, that subsystem works iff both 3 and 4 work. If components work independently of one another and P(component i works) = .9 for $i = 1,2$ and = .8 for $i = 3,4$, calculate P(system works).

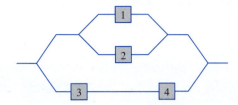

81. Refer back to the series-parallel system configuration introduced in Example 2.36, and suppose that there are only two cells rather than three in each parallel subsystem [in Figure 2.14(a), eliminate cells 3 and 6, and renumber cells 4 and 5 as 3 and 4]. Using $P(A_i) = .9$, the probability that system lifetime exceeds t_0 is easily seen to be .9639. To what value would .9 have to be changed in order to increase the system lifetime reliability from .9639 to .99? [*Hint:* Let $P(A_i) = p$, express system reliability in terms of p, and then let $x = p^2$.]

82. Consider independently rolling two fair dice, one red and the other green. Let A be the event that the red die shows 3 dots, B be the event that the green die shows 4 dots, and C be the event that the total number of dots showing on the two dice is 7. Are these events pairwise independent (i.e., are A and B independent events, are A and C independent, and are B and C independent)? Are the three events mutually independent?

83. Components arriving at a distributor are checked for defects by two different inspectors (each component is checked by both inspectors). The first inspector detects 90% of all defectives that are present, and the second inspector does likewise. At least one inspector does not detect a defect on 20% of all defective components. What is the probability that the following occur?
   a. A defective component will be detected only by the first inspector? By exactly one of the two inspectors?
   b. All three defective components in a batch escape detection by both inspectors (assuming inspections of different components are independent of one another)?

84. Consider purchasing a system of audio components consisting of a receiver, a pair of speakers, and a CD player. Let A_1 be the event that the receiver functions properly throughout the warranty period, A_2 be the event that the speakers function properly throughout the warranty period, and A_3 be the event that the CD player functions properly throughout the warranty period. Suppose that these events are (mutually) independent with $P(A_1) = .95$, $P(A_2) = .98$, and $P(A_3) = .80$.

   a. What is the probability that all three components function properly throughout the warranty period?
   b. What is the probability that at least one component needs service during the warranty period?
   c. What is the probability that all three components need service during the warranty period?
   d. What is the probability that only the receiver needs service during the warranty period?
   e. What is the probability that exactly one of the three components needs service during the warranty period?
   f. What is the probability that all three components function properly throughout the warranty period but that at least one fails within a month after the warranty expires?

85. A quality control inspector is examining newly produced items for faults. The inspector searches an item for faults in a series of independent fixations, each of a fixed duration. Given that a flaw is actually present, let p denote the probability that the flaw is detected during any one fixation (this model is discussed in **"Human Performance in Sampling Inspection," *Human Factors*, 1979: 99–105**).

   a. Assuming that an item has a flaw, what is the probability that it is detected by the end of the second fixation (once a flaw has been detected, the sequence of fixations terminates)?
   b. Give an expression for the probability that a flaw will be detected by the end of the nth fixation.
   c. If when a flaw has not been detected in three fixations, the item is passed, what is the probability that a flawed item will pass inspection?
   d. Suppose 10% of all items contain a flaw [P(randomly chosen item is flawed) = .1]. With the assumption of part (c), what is the probability that a randomly chosen item will pass inspection (it will automatically pass if it is not flawed, but could also pass if it is flawed)?
   e. Given that an item has passed inspection (no flaws in three fixations), what is the probability that it is actually flawed? Calculate for $p = .5$.

86. a. A lumber company has just taken delivery on a shipment of 10,000 2 × 4 boards. Suppose that 20% of these boards (2000) are actually too green to be used in first-quality construction. Two boards are selected at random, one after the other. Let $A = \{$the first board is green$\}$ and $B = \{$the second board is green$\}$. Compute $P(A)$, $P(B)$, and $P(A \cap B)$ (a tree diagram might help). Are A and B independent?
   b. With A and B independent and $P(A) = P(B) = .2$, what is $P(A \cap B)$? How much difference is there between this answer and $P(A \cap B)$ in part (a)? For purposes of calculating $P(A \cap B)$, can we assume that A and B of part (a) are independent to obtain essentially the correct probability?
   c. Suppose the shipment consists of ten boards, of which two are green. Does the assumption of independence

now yield approximately the correct answer for $P(A \cap B)$? What is the critical difference between the situation here and that of part (a)? When do you think an independence assumption would be valid in obtaining an approximately correct answer to $P(A \cap B)$?

87. Consider randomly selecting a single individual and having that person test drive 3 different vehicles. Define events A_1, A_2, and A_3 by

$A_1 =$ likes vehicle #1 $A_2 =$ likes vehicle #2
$A_3 =$ likes vehicle #3

Suppose that $P(A_1) = .55$, $P(A_2) = .65$, $P(A_3) = .70$, $P(A_1 \cup A_2) = .80$, $P(A_2 \cap A_3) = .40$, and $P(A_1 \cup A_2 \cup A_3) = .88$.
   a. What is the probability that the individual likes both vehicle #1 and vehicle #2?
   b. Determine and interpret $P(A_2 | A_3)$.
   c. Are A_2 and A_3 independent events? Answer in two different ways.
   d. If you learn that the individual did not like vehicle #1, what now is the probability that he/she liked at least one of the other two vehicles?

88. The probability that an individual randomly selected from a particular population has a certain disease is .05. A diagnostic test correctly detects the presence of the disease 98% of the time and correctly detects the absence of the disease 99% of the time. If the test is applied twice, the two test results are independent, and both are positive, what is the (posterior) probability that the selected individual has the disease? [*Hint:* Tree diagram with first-generation branches corresponding to Disease and No Disease, and second- and third-generation branches corresponding to results of the two tests.]

89. Suppose identical tags are placed on both the left ear and the right ear of a fox. The fox is then let loose for a period of time. Consider the two events $C_1 = \{$left ear tag is lost$\}$ and $C_2 = \{$right ear tag is lost$\}$. Let $\pi = P(C_1) = P(C_2)$, and assume C_1 and C_2 are independent events. Derive an expression (involving π) for the probability that exactly one tag is lost, given that at most one is lost (**"Ear Tag Loss in Red Foxes," *J. Wildlife Mgmt.*, 1976: 164–167**). [*Hint:* Draw a tree diagram in which the two initial branches refer to whether the left ear tag was lost.]

SUPPLEMENTARY EXERCISES (90–114)

90. A certain legislative committee consists of 10 senators. A subcommittee of 3 senators is to be randomly selected.
   a. How many different such subcommittees are there?
   b. If the senators are ranked 1, 2,…, 10 in order of seniority, how many different subcommittees would include the most senior senator?
   c. What is the probability that the selected subcommittee has at least 1 of the 5 most senior senators?
   d. What is the probability that the subcommittee includes neither of the two most senior senators?

91. A factory uses three production lines to manufacture cans of a certain type. The accompanying table gives percentages of nonconforming cans, categorized by type of nonconformance, for each of the three lines during a particular time period.

	Line 1	Line 2	Line 3
Blemish	15	12	20
Crack	50	44	40
Pull-Tab Problem	21	28	24
Surface Defect	10	8	15
Other	4	8	2

During this period, line 1 produced 500 nonconforming cans, line 2 produced 400 such cans, and line 3 was responsible for 600 nonconforming cans. Suppose that one of these 1500 cans is randomly selected.

   a. What is the probability that the can was produced by line 1? That the reason for nonconformance is a crack?
   b. If the selected can came from line 1, what is the probability that it had a blemish?
   c. Given that the selected can had a surface defect, what is the probability that it came from line 1?

92. An employee of the records office at a certain university currently has ten forms on his desk awaiting processing. Six of these are withdrawal petitions and the other four are course substitution requests.
   a. If he randomly selects six of these forms to give to a subordinate, what is the probability that only one of the two types of forms remains on his desk?
   b. Suppose he has time to process only four of these forms before leaving for the day. If these four are randomly selected one by one, what is the probability that each succeeding form is of a different type from its predecessor?

93. One satellite is scheduled to be launched from Cape Canaveral in Florida, and another launching is scheduled for Vandenberg Air Force Base in California. Let A denote the event that the Vandenberg launch goes off on schedule, and let B represent the event that the Cape Canaveral launch goes off on schedule. If A and B are independent events with $P(A) > P(B)$, $P(A \cup B) = .626$, and $P(A \cap B) = .144$, determine the values of $P(A)$ and $P(B)$.

94. A transmitter is sending a message by using a binary code, namely, a sequence of 0's and 1's. Each transmitted bit (0 or 1) must pass through three relays to reach the receiver. At each relay, the probability is .20 that the bit sent will be different from the bit received (a reversal). Assume that the relays operate independently of one another.

Transmitter → Relay 1 → Relay 2 → Relay 3 → Receiver

 a. If a 1 is sent from the transmitter, what is the probability that a 1 is sent by all three relays?

 b. If a 1 is sent from the transmitter, what is the probability that a 1 is received by the receiver? [*Hint:* The eight experimental outcomes can be displayed on a tree diagram with three generations of branches, one generation for each relay.]

 c. Suppose 70% of all bits sent from the transmitter are 1s. If a 1 is received by the receiver, what is the probability that a 1 was sent?

95. Individual A has a circle of five close friends (B, C, D, E, and F). A has heard a certain rumor from outside the circle and has invited the five friends to a party to circulate the rumor. To begin, A selects one of the five at random and tells the rumor to the chosen individual. That individual then selects at random one of the four remaining individuals and repeats the rumor. Continuing, a new individual is selected from those not already having heard the rumor by the individual who has just heard it, until everyone has been told.

 a. What is the probability that the rumor is repeated in the order B, C, D, E, and F?

 b. What is the probability that F is the third person at the party to be told the rumor?

 c. What is the probability that F is the last person to hear the rumor?

 d. If at each stage the person who currently "has" the rumor does not know who has already heard it and selects the next recipient at random from all five possible individuals, what is the probability that F has still not heard the rumor after it has been told ten times at the party?

96. According to the article **"Optimization of Distribution Parameters for Estimating Probability of Crack Detection"** (*J. of Aircraft*, **2009: 2090–2097),** the following "Palmberg" equation is commonly used to determine the probability $P_d(c)$ of detecting a crack of size c in an aircraft structure:

$$P_d(c) = \frac{(c/c^*)^\beta}{1 + (c/c^*)^\beta}$$

where c^* is the crack size that corresponds to a .5 detection probability (and thus is an assessment of the quality of the inspection process).

 a. Verify that $P_d(c^*) = .5$

 b. What is $P_d(2c^*)$ when $\beta = 4$?

 c. Suppose an inspector inspects two different panels, one with a crack size of c^* and the other with a crack

size of $2c^*$. Again assuming $\beta = 4$ and also that the results of the two inspections are independent of one another, what is the probability that exactly one of the two cracks will be detected?

 d. What happens to $P_d(c)$ as $\beta \to \infty$?

97. A chemical engineer is interested in determining whether a certain trace impurity is present in a product. An experiment has a probability of .80 of detecting the impurity if it is present. The probability of not detecting the impurity if it is absent is .90. The prior probabilities of the impurity being present and being absent are .40 and .60, respectively. Three separate experiments result in only two detections. What is the posterior probability that the impurity is present?

98. Five friends—Allison, Beth, Carol, Diane, and Evelyn—have identical calculators and are studying for a statistics exam. They set their calculators down in a pile before taking a study break and then pick them up in random order when they return from the break. What is the probability that at least one of the five gets her own calculator? [*Hint:* Let A be the event that Alice gets her own calculator, and define events B, C, D, and E analogously for the other four students.] How can the event {at least one gets her own calculator} be expressed in terms of the five events A, B, C, D, and E? Now use a general law of probability. [*Note:* This is called the *matching problem*. Its solution is easily extended to n individuals. Can you recognize the result when n is large (the approximation to the resulting series)?]

99. Fasteners used in aircraft manufacturing are slightly crimped so that they lock enough to avoid loosening during vibration. Suppose that 95% of all fasteners pass an initial inspection. Of the 5% that fail, 20% are so seriously defective that they must be scrapped. The remaining fasteners are sent to a recrimping operation, where 40% cannot be salvaged and are discarded. The other 60% of these fasteners are corrected by the recrimping process and subsequently pass inspection.

 a. What is the probability that a randomly selected incoming fastener will pass inspection either initially or after recrimping?

 b. Given that a fastener passed inspection, what is the probability that it passed the initial inspection and did not need recrimping?

100. Jay and Maurice are playing a tennis match. In one particular game, they have reached deuce, which means each player has won three points. To finish the game, one of the two players must get two points ahead of the other. For example, Jay will win if he wins the next two points (JJ), or if Maurice wins the next point and Jay the three points after that (MJJJ), or if the result of the next six points is JMMJJJ, and so on.

 a. Suppose that the probability of Jay winning a point is .6 and outcomes of successive points are independent of one another. What is the probability that Jay wins the game? [*Hint:* In the law of total probability, let A_1 = Jay wins each of the next two points, A_2 = Maurice wins each of the next two points, and A_3 = each player wins one of the next two points. Also let $p = P$(Jay wins the game). How does p compare to P(Jay wins the game $| A_3$)?]

b. If Jay wins the game, what is the probability that he needed only two points to do so?

101. A system consists of two components. The probability that the second component functions in a satisfactory manner during its design life is .9, the probability that at least one of the two components does so is .96, and the probability that both components do so is .75. Given that the first component functions in a satisfactory manner throughout its design life, what is the probability that the second one does also?

102. The accompanying table categorizing each student in a sample according to gender and eye color appeared in the article **"Does Eye Color Depend on Gender? It Might Depend on Who or How You Ask"** (**J. of Statistics Educ., 2013, Vol. 21, Num. 2**).

Eye Color

Gender	Blue	Brown	Green	Hazel	Total
Male	370	352	198	187	1107
Female	359	290	110	160	919
Total	729	642	308	347	2026

Suppose that one of these 2026 students is randomly selected. Let F denote the event that the selected individual is a female, and A, B, C, and D represent the events that he or she has blue, brown, green, and hazel eyes, respectively.

a. Calculate both $P(F)$ and $P(C)$.

b. Calculate $P(F \cap C)$. Are the events F and C independent? Why or why not?

c. If the selected individual has green eyes, what is the probability that he or she is a female?

d. If the selected individual is female, what is the probability that she has green eyes?

e. What is the "conditional distribution" of eye color for females (i.e., $P(A|F)$, $P(B|F)$, $P(C|F)$, and $P(D|F)$), and what is it for males? Compare the two distributions.

103. a. A certain company sends 40% of its overnight mail parcels via express mail service E_1. Of these parcels, 2% arrive after the guaranteed delivery time (denote the event "late delivery" by L). If a record of an overnight mailing is randomly selected from the company's file, what is the probability that the parcel went via E_1 and was late?

b. Suppose that 50% of the overnight parcels are sent via express mail service E_2 and the remaining 10% are sent via E_3. Of those sent via E_2, only 1% arrive late, whereas 5% of the parcels handled by E_3 arrive late. What is the probability that a randomly selected parcel arrived late?

c. If a randomly selected parcel has arrived on time, what is the probability that it was not sent via E_1?

104. A company uses three different assembly lines—A_1, A_2, and A_3—to manufacture a particular component. Of those manufactured by line A_1, 5% need rework to remedy a defect, whereas 8% of A_2's components need rework and 10% of A_3's need rework. Suppose that 50% of all components are produced by line A_1, 30% are produced by line A_2, and 20% come from line A_3. If a randomly selected component needs rework, what is the probability that it came from line A_1? From line A_2? From line A_3?

105. Disregarding the possibility of a February 29 birthday, suppose a randomly selected individual is equally likely to have been born on any one of the other 365 days.

a. If ten people are randomly selected, what is the probability that all have different birthdays? That at least two have the same birthday?

b. With k replacing ten in part (a), what is the smallest k for which there is at least a 50-50 chance that two or more people will have the same birthday?

c. If ten people are randomly selected, what is the probability that either at least two have the same birthday or at least two have the same last three digits of their Social Security numbers? [*Note:* The article **"Methods for Studying Coincidences"** (**F. Mosteller and P. Diaconis, J. Amer. Stat. Assoc., 1989: 853–861**) discusses problems of this type.]

106. One method used to distinguish between granitic (G) and basaltic (B) rocks is to examine a portion of the infrared spectrum of the sun's energy reflected from the rock surface. Let R_1, R_2, and R_3 denote measured spectrum intensities at three different wavelengths; typically, for granite $R_1 < R_2 < R_3$, whereas for basalt $R_3 < R_1 < R_2$. When measurements are made remotely (using aircraft), various orderings of the R_is may arise whether the rock is basalt or granite. Flights over regions of known composition have yielded the following information:

	Granite	Basalt
$R_1 < R_2 < R_3$	60%	10%
$R_1 < R_3 < R_2$	25%	20%
$R_3 < R_1 < R_2$	15%	70%

Suppose that for a randomly selected rock in a certain region, P(granite) = .25 and P(basalt) = .75.

a. Show that P(granite $| R_1 < R_2 < R_3$) > P(basalt $| R_1 < R_2 < R_3$). If measurements yielded $R_1 < R_2 < R_3$, would you classify the rock as granite or basalt?

b. If measurements yielded $R_1 < R_3 < R_2$, how would you classify the rock? Answer the same question for $R_3 < R_1 < R_2$.

c. Using the classification rules indicated in parts (a) and (b), when selecting a rock from this region, what is the probability of an erroneous classification? [*Hint:* Either G could be classified as B or B as G, and $P(B)$ and $P(G)$ are known.]

d. If $P(\text{granite}) = p$ rather than .25, are there values of p (other than 1) for which one would always classify a rock as granite?

107. A subject is allowed a sequence of glimpses to detect a target. Let $G_i = \{\text{the target is detected on the }i\text{th glimpse}\}$, with $p_i = P(G_i)$. Suppose the G_i's are independent events, and write an expression for the probability that the target has been detected by the end of the nth glimpse. [*Note:* This model is discussed in **"Predicting Aircraft Detectability," *Human Factors*, 1979: 277–291.**]

108. In a Little League baseball game, team A's pitcher throws a strike 50% of the time and a ball 50% of the time, successive pitches are independent of one another, and the pitcher never hits a batter. Knowing this, team B's manager has instructed the first batter not to swing at anything. Calculate the probability that

a. The batter walks on the fourth pitch

b. The batter walks on the sixth pitch (so two of the first five must be strikes), using a counting argument or constructing a tree diagram

c. The batter walks

d. The first batter up scores while no one is out (assuming that each batter pursues a no-swing strategy)

109. Four engineers, A, B, C, and D, have been scheduled for job interviews at 10 A.M. on Friday, January 13, at Random Sampling, Inc. The personnel manager has scheduled the four for interview rooms 1, 2, 3, and 4, respectively. However, the manager's secretary does not know this, so assigns them to the four rooms in a completely random fashion (what else!). What is the probability that

a. All four end up in the correct rooms?

b. None of the four ends up in the correct room?

110. A particular airline has 10 A.M. flights from Chicago to New York, Atlanta, and Los Angeles. Let A denote the event that the New York flight is full and define events B and C analogously for the other two flights. Suppose $P(A) = .9$, $P(B) = .7$, $P(C) = .8$ and the three events are independent. What is the probability that

a. All three flights are full? That at least one flight is not full?

b. Only the New York flight is full? That exactly one of the three flights is full?

111. A personnel manager is to interview four candidates for a job. These are ranked 1, 2, 3, and 4 in order of preference and will be interviewed in random order. However, at the conclusion of each interview, the manager will know only how the current candidate compares to those previously interviewed. For example, the interview order 3, 4, 1, 2 generates no information after the first interview, shows that the second candidate is worse than the first, and that the third is better than the first two. However, the order 3, 4, 2, 1 would generate the same information after each of the first three interviews. The manager wants to hire the best candidate but must make an irrevocable hire/no hire decision after each interview. Consider the following strategy: Automatically reject the first s candidates and then hire the first subsequent candidate who is best among those already interviewed (if no such candidate appears, the last one interviewed is hired).

For example, with $s = 2$, the order 3, 4, 1, 2 would result in the best being hired, whereas the order 3, 1, 2, 4 would not. Of the four possible s values (0, 1, 2, and 3), which one maximizes $P(\text{best is hired})$? [*Hint:* Write out the 24 equally likely interview orderings: $s = 0$ means that the first candidate is automatically hired.]

112. Consider four independent events A_1, A_2, A_3, and A_4, and let $p_i = P(A_i)$ for $i = 1,2,3,4$. Express the probability that at least one of these four events occurs in terms of the p_is, and do the same for the probability that at least two of the events occur.

113. A box contains the following four slips of paper, each having exactly the same dimensions: (1) win prize 1; (2) win prize 2; (3) win prize 3; (4) win prizes 1, 2, and 3. One slip will be randomly selected. Let $A_1 = \{\text{win prize 1}\}$, $A_2 = \{\text{win prize 2}\}$, and $A_3 = \{\text{win prize 3}\}$. Show that A_1 and A_2 are independent, that A_1 and A_3 are independent, and that A_2 and A_3 are also independent (this is *pairwise* independence). However, show that $P(A_1 \cap A_2 \cap A_3) \neq P(A_1) \cdot P(A_2) \cdot P(A_3)$, so the three events are *not* mutually independent.

114. Show that if A_1, A_2, and A_3 are independent events, then $P(A_1 \mid A_2 \cap A_3) = P(A_1)$.

BIBLIOGRAPHY

Carlton, Matthew and Jay Devore, *Probability with Applications in Engineering, Science, and Technology*, Springer, New York, 2014. A comprehensive introduction to probability, written at a slightly higher mathematical level than this text but containing many good examples.

Durrett, Richard, *Elementary Probability for Applications*, Cambridge Univ. Press, London, England, 2009. A concise yet readable presentation at a slightly higher level than this text.

Mosteller, Frederick, Robert Rourke, and George Thomas, *Probability with Statistical Applications* (2nd ed.), Addison-Wesley, Reading, MA, 1970. A very good precalculus introduction to probability, with many entertaining examples; especially good on counting rules and their application.

Ross, Sheldon, *A First Course in Probability* (8th ed.), Macmillan, New York, 2009. Rather tightly written and more mathematically sophisticated than this text but contains a wealth of interesting examples and exercises.

Winkler, Robert, *Introduction to Bayesian Inference and Decision*, Holt, Rinehart & Winston, New York, 1972. A very good introduction to subjective probability.

Discrete Random Variables and Probability Distributions

3

INTRODUCTION

Whether an experiment yields qualitative or quantitative outcomes, methods of statistical analysis require that we focus on certain numerical aspects of the data (such as a sample proportion x/n, mean $\bar{x}$, or standard deviation s). The concept of a random variable allows us to pass from the experimental outcomes themselves to a numerical function of the outcomes. There are two fundamentally different types of random variables—discrete random variables and continuous random variables. In this chapter, we examine the basic properties and discuss the most important examples of discrete variables. Chapter 4 focuses on continuous random variables.

3.1 Random Variables

In any experiment, there are numerous characteristics that can be observed or measured, but in most cases an experimenter will focus on some specific aspect or aspects of a sample. For example, in a study of commuting patterns in a metropolitan area, each individual in a sample might be asked about commuting distance and the number of people commuting in the same vehicle, but not about IQ, income, family size, and other such characteristics. Alternatively, a researcher may test a sample of components and record only the number that have failed within 1000 hours, rather than record the individual failure times.

In general, each outcome of an experiment can be associated with a number by specifying a rule of association (e.g., the number among the sample of ten components that fail to last 1000 hours or the total weight of baggage for a sample of 25 airline passengers). Such a rule of association is called a **random variable**—a variable because different numerical values are possible and random because the observed value depends on which of the possible experimental outcomes results (Figure 3.1).

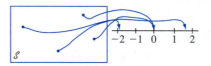

Figure 3.1   A random variable

DEFINITION

> For a given sample space $\mathscr{S}$ of some experiment, a **random variable (rv)** is any rule that associates a number with each outcome in $\mathscr{S}$. In mathematical language, a random variable is a function whose domain is the sample space and whose range is the set of real numbers.

Random variables are customarily denoted by uppercase letters, such as X and Y, near the end of our alphabet. In contrast to our previous use of a lowercase letter, such as x, to denote a variable, we will now use lowercase letters to represent some particular value of the corresponding random variable. The notation $X(\omega) = x$ means that x is the value associated with the outcome ω by the rv X.

EXAMPLE 3.1   When a student calls a university help desk for technical support, he/she will either immediately be able to speak to someone (S, for success) or will be placed on hold (F, for failure). With $\mathscr{S} = \{S, F\}$, define an rv X by

$$X(S) = 1 \quad X(F) = 0$$

The rv X indicates whether (1) or not (0) the student can immediately speak to someone.   ■

The rv X in Example 3.1 was specified by explicitly listing each element of $\mathscr{S}$ and the associated number. Such a listing is tedious if $\mathscr{S}$ contains more than a few outcomes, but it can frequently be avoided.

EXAMPLE 3.2   Consider the experiment in which a telephone number in a certain area code is dialed using a random number dialer (such devices are used extensively by polling organizations), and define an rv Y by

$$Y = \begin{cases} 1 & \text{if the selected number is unlisted} \\ 0 & \text{if the selected number is listed in the directory} \end{cases}$$

For example, if 5282966 appears in the telephone directory, then $Y(5282966) = 0$, whereas $Y(7727350) = 1$ tells us that the number 7727350 is unlisted. A word description of this sort is more economical than a complete listing, so we will use such a description whenever possible.   ■

In Examples 3.1 and 3.2, the only possible values of the random variable were 0 and 1. Such a random variable arises frequently enough to be given a special name, after the individual who first studied it.

DEFINITION

> Any random variable whose only possible values are 0 and 1 is called a **Bernoulli random variable.**

We will sometimes want to consider several different random variables from the same sample space.

EXAMPLE 3.3   Example 2.3 described an experiment in which the number of pumps in use at each of two six-pump gas stations was determined. Define rv's X, Y, and U by

X = the total number of pumps in use at the two stations

Y = the difference between the number of pumps in use at station 1 and the number in use at station 2

U = the maximum of the numbers of pumps in use at the two stations

If this experiment is performed and $\omega = (2, 3)$ results, then $X((2, 3)) = 2 + 3 = 5$, so we say that the observed value of X was $x = 5$. Similarly, the observed value of Y would be $y = 2 - 3 = -1$, and the observed value of U would be $u = \max(2, 3) = 3$.   ■

Each of the random variables of Examples 3.1–3.3 can assume only a finite number of possible values. This need not be the case.

EXAMPLE 3.4   Consider an experiment in which 9-volt batteries are tested until one with an acceptable voltage (S) is obtained. The sample space is $\mathscr{S} = \{S, FS, FFS,\dots\}$. Define an rv X by

X = the number of batteries tested before the experiment terminates

Then $X(S) = 1$, $X(FS) = 2$, $X(FFS) = 3,\dots, X(FFFFFFS) = 7$, and so on. Any positive integer is a possible value of X, so the set of possible values is infinite.   ■

EXAMPLE 3.5   Suppose that in some random fashion, a location (latitude and longitude) in the continental United States is selected. Define an rv Y by

Y = the height above sea level at the selected location

For example, if the selected location were (39°50′N, 98°35′W), then we might have $Y((39°50′N, 98°35′W)) = 1748.26$ ft. The largest possible value of Y is 14,494 (Mt. Whitney), and the smallest possible value is -282 (Death Valley). The set of all possible values of Y is the set of all numbers in the interval between -282 and 14,494— that is,

$$\{y : y \text{ is a number}, -282 \le y \le 14{,}494\}$$

and there are an infinite number of numbers in this interval.   ■

Two Types of Random Variables

In Section 1.2, we distinguished between data resulting from observations on a counting variable and data obtained by observing values of a measurement variable. A slightly more formal distinction characterizes two different types of random variables.

DEFINITION

A **discrete** random variable is an rv whose possible values either constitute a finite set or else can be listed in an infinite sequence in which there is a first element, a second element, and so on ("countably" infinite).

A random variable is **continuous** if *both* of the following apply:

1. Its set of possible values consists either of all numbers in a single interval on the number line (possibly infinite in extent, e.g., from $-\infty$ to ∞) or all numbers in a disjoint union of such intervals (e.g., $[0, 10] \cup [20, 30]$).

2. No possible value of the variable has positive probability, that is, $P(X = c) = 0$ for any possible value c.

Although any interval on the number line contains an infinite number of numbers, it can be shown that there is no way to create an infinite listing of all these values—there are just too many of them. The second condition describing a continuous random variable is perhaps counterintuitive, since it would seem to imply a total probability of zero for all possible values. But we shall see in Chapter 4 that *intervals* of values have positive probability; the probability of an interval will decrease to zero as the width of the interval shrinks to zero.

EXAMPLE 3.6   All random variables in Examples 3.1–3.4 are discrete. As another example, suppose we select married couples at random and do a blood test on each person until we find a husband and wife who both have the same Rh factor. With $X =$ the number of blood tests to be performed, possible values of X are $D = \{2, 4, 6, 8, \dots\}$. Since the possible values have been listed in sequence, X is a discrete rv. ■

To study basic properties of discrete rv's, only the tools of discrete mathematics—summation and differences—are required. The study of continuous variables requires the continuous mathematics of the calculus—integrals and derivatives.

EXERCISES Section 3.1 (1–10)

1. A concrete beam may fail either by shear (S) or flexure (F). Suppose that three failed beams are randomly selected and the type of failure is determined for each one. Let $X =$ the number of beams among the three selected that failed by shear. List each outcome in the sample space along with the associated value of X.

2. Give three examples of Bernoulli rv's (other than those in the text).

3. Using the experiment in Example 3.3, define two more random variables and list the possible values of each.

4. Let $X =$ the number of nonzero digits in a randomly selected 4-digit PIN that has no restriction on the digits.

   What are the possible values of X? Give three possible outcomes and their associated X values.

5. If the sample space $\mathscr{S}$ is an infinite set, does this necessarily imply that any rv X defined from $\mathscr{S}$ will have an infinite set of possible values? If yes, say why. If no, give an example.

6. Starting at a fixed time, each car entering an intersection is observed to see whether it turns left (L), right (R), or goes straight ahead (A). The experiment terminates as soon as a car is observed to turn left. Let $X =$ the number of cars observed. What are possible X values? List five outcomes and their associated X values.

7. For each random variable defined here, describe the set of possible values for the variable, and state whether the variable is discrete.

  a. X = the number of unbroken eggs in a randomly chosen standard egg carton

  b. Y = the number of students on a class list for a particular course who are absent on the first day of classes

  c. U = the number of times a duffer has to swing at a golf ball before hitting it

  d. X = the length of a randomly selected rattlesnake

  e. Z = the sales tax percentage for a randomly selected amazon.com purchase

  f. Y = the pH of a randomly chosen soil sample

  g. X = the tension (psi) at which a randomly selected tennis racket has been strung

  h. X = the total number of times three tennis players must spin their rackets to obtain something other than *UUU* or *DDD* (to determine which two play next)

8. Each time a component is tested, the trial is a success (S) or failure (F). Suppose the component is tested repeatedly until a success occurs on three *consecutive* trials. Let Y denote the number of trials necessary to achieve this. List all outcomes corresponding to the five smallest possible values of Y, and state which Y value is associated with each one.

9. An individual named Claudius is located at the point 0 in the accompanying diagram.

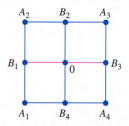

Using an appropriate randomization device (such as a tetrahedral die, one having four sides), Claudius first moves to one of the four locations B_1, B_2, B_3, B_4. Once at one of these locations, another randomization device is used to decide whether Claudius next returns to 0 or next visits one of the other two adjacent points. This process then continues; after each move, another move to one of the (new) adjacent points is determined by tossing an appropriate die or coin.

  a. Let X = the number of moves that Claudius makes before first returning to 0. What are possible values of X? Is X discrete or continuous?

  b. If moves are allowed also along the diagonal paths connecting 0 to A_1, A_2, A_3, and A_4, respectively, answer the questions in part (a).

10. The number of pumps in use at both a six-pump station and a four-pump station will be determined. Give the possible values for each of the following random variables:

  a. T = the total number of pumps in use

  b. X = the difference between the numbers in use at stations 1 and 2

  c. U = the maximum number of pumps in use at either station

  d. Z = the number of stations having exactly two pumps in use

3.2 Probability Distributions for Discrete Random Variables

Probabilities assigned to various outcomes in $\mathcal{S}$ in turn determine probabilities associated with the values of any particular rv X. The *probability distribution of X* says how the total probability of 1 is distributed among (allocated to) the various possible X values. Suppose, for example, that a business has just purchased four laser printers, and let X be the number among these that require service during the warranty period. Possible X values are then 0, 1, 2, 3, and 4. The probability distribution will tell us how the probability of 1 is subdivided among these five possible values—how much probability is associated with the X value 0, how much is apportioned to the X value 1, and so on. We will use the following notation for the probabilities in the distribution:

$$p(0) = \text{the probability of the } X \text{ value } 0 = P(X = 0)$$
$$p(1) = \text{the probability of the } X \text{ value } 1 = P(X = 1)$$

and so on. In general, $p(x)$ will denote the probability assigned to the value x.

EXAMPLE 3.7 The Cal Poly Department of Statistics has a lab with six computers reserved for statistics majors. Let X denote the number of these computers that are in use at a particular time of day. Suppose that the probability distribution of X is as given in the following table; the first row of the table lists the possible X values and the second row gives the probability of each such value.

x	0	1	2	3	4	5	6
$p(x)$	.05	.10	.15	.25	.20	.15	.10

We can now use elementary probability properties to calculate other probabilities of interest. For example, the probability that at most 2 computers are in use is

$$P(X \leq 2) = P(X = 0 \text{ or } 1 \text{ or } 2) = p(0) + p(1) + p(2) = .05 + .10 + .15 = .30$$

Since the event *at least 3 computers are in use* is complementary to *at most 2 computers are in use,*

$$P(X \geq 3) = 1 - P(X \leq 2) = 1 - .30 = .70$$

which can, of course, also be obtained by adding together probabilities for the values 3, 4, 5, and 6. The probability that between 2 and 5 computers *inclusive* are in use is

$$P(2 \leq X \leq 5) = P(X = 2, 3, 4, \text{ or } 5) = .15 + .25 + .20 + .15 = .75$$

whereas the probability that the number of computers in use is *strictly between* 2 and 5 is

$$P(2 < X < 5) = P(X = 3 \text{ or } 4) = .25 + .20 = .45 \quad \blacksquare$$

DEFINITION

> The **probability distribution** or **probability mass function** (pmf) of a discrete rv is defined for every number x by $p(x) = P(X = x) = P(\text{all } \omega \in \mathcal{S}: X(\omega) = x)$.

In words, for every possible value x of the random variable, the pmf specifies the probability of observing that value when the experiment is performed. The conditions $p(x) \geq 0$ and $\Sigma_{\text{all possible } x}\, p(x) = 1$ are required of any pmf.

The pmf of X in the previous example was simply given in the problem description. We now consider several examples in which various probability properties are exploited to obtain the desired distribution.

EXAMPLE 3.8 Six boxes of components are ready to be shipped by a certain supplier. The number of defective components in each box is as follows:

Box	1	2	3	4	5	6
Number of defectives	0	2	0	1	2	0

One of these boxes is to be randomly selected for shipment to a particular customer. Let X be the number of defectives in the selected box. The three possible X values are 0, 1, and 2. Of the six equally likely simple events, three result in $X = 0$, one in $X = 1$, and the other two in $X = 2$. Then

$$p(0) = P(X = 0) = P(\text{box 1 or 3 or 6 is sent}) = \frac{3}{6} = .500$$

$$p(1) = P(X = 1) = P(\text{box 4 is sent}) = \frac{1}{6} = .167$$

$$p(2) = P(X = 2) = P(\text{box 2 or 5 is sent}) = \frac{2}{6} = .333$$

That is, a probability of .500 is distributed to the X value 0, a probability of .167 is placed on the X value 1, and the remaining probability, .333, is associated with the X value 2. The values of X along with their probabilities collectively specify the pmf. If this experiment were repeated over and over again, in the long run $X = 0$ would occur one-half of the time, $X = 1$ one-sixth of the time, and $X = 2$ one-third of the time. ■

EXAMPLE 3.9 Consider whether the next person buying a computer at a certain electronics store buys a laptop or a desktop model. Let

$$X = \begin{cases} 1 & \text{if the customer purchases a desktop computer} \\ 0 & \text{if the customer purchases a laptop computer} \end{cases}$$

If 20% of all purchasers during that week select a desktop, the pmf for X is

$$p(0) = P(X = 0) = P(\text{next customer purchases a laptop model}) = .8$$
$$p(1) = P(X = 1) = P(\text{next customer purchases a desktop model}) = .2$$
$$p(x) = P(X = x) = 0 \text{ for } x \neq 0 \text{ or } 1$$

An equivalent description is

$$p(x) = \begin{cases} .8 & \text{if } x = 0 \\ .2 & \text{if } x = 1 \\ 0 & \text{if } x \neq 0 \text{ or } 1 \end{cases}$$

Figure 3.2 is a picture of this pmf, called a *line graph*. X is, of course, a Bernoulli rv and $p(x)$ is a Bernoulli pmf.

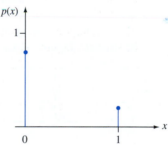

Figure 3.2 The line graph for the pmf in Example 3.9 ■

EXAMPLE 3.10 In a group of five potential blood donors—a, b, c, d, and e—only a and b have type O-positive blood. Five blood samples, one from each individual, will be typed in random order until an O+ individual is identified. Let the rv Y = the number of typings necessary to identify an O+ individual. Then the pmf of Y is

$$p(1) = P(Y = 1) = P(a \text{ or } b \text{ typed first}) = \frac{2}{5} = .4$$

$$p(2) = P(Y = 2) = P(c, d, \text{ or } e \text{ first, and then } a \text{ or } b)$$

$$= P(c, d, \text{ or } e \text{ first}) \cdot P(a \text{ or } b \text{ next} \mid c, d, \text{ or } e \text{ first}) = \frac{3}{5} \cdot \frac{2}{4} = .3$$

$$p(3) = P(Y = 3) = P(c, d, \text{ or } e \text{ first and second, and then } a \text{ or } b)$$

$$= \left(\frac{3}{5}\right)\left(\frac{2}{4}\right)\left(\frac{2}{3}\right) = .2$$

$$p(4) = P(Y = 4) = P(c, d, \text{ and } e \text{ all done first}) = \left(\frac{3}{5}\right)\left(\frac{2}{4}\right)\left(\frac{1}{3}\right) = .1$$

$$p(y) = 0 \quad \text{if } y \neq 1, 2, 3, 4$$

In tabular form, the pmf is

y	1	2	3	4
$p(y)$	.4	.3	.2	.1

where any y value not listed receives zero probability. Figure 3.3 shows a line graph of the pmf.

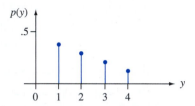

Figure 3.3 The line graph for the pmf in Example 3.10 ∎

The name "probability mass function" is suggested by a model used in physics for a system of "point masses." In this model, masses are distributed at various locations x along a one-dimensional axis. Our pmf describes how the total probability mass of 1 is distributed at various points along the axis of possible values of the random variable (where and how much mass at each x).

Another useful pictorial representation of a pmf, called a **probability histogram,** is similar to histograms discussed in Chapter 1. Above each y with $p(y) > 0$, construct a rectangle centered at y. The height of each rectangle is proportional to $p(y)$, and the base width is the same for all rectangles. When possible values are equally spaced, the base width is frequently chosen as the distance between successive y values (though it could be smaller). Figure 3.4 shows two probability histograms.

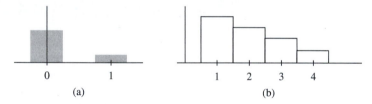

Figure 3.4 Probability histograms: (a) Example 3.9; (b) Example 3.10

It is often helpful to think of a pmf as specifying a mathematical model for a discrete population.

EXAMPLE 3.11 Consider selecting a household in a certain region at random and let $X = $ the number of individuals in the selected household. Suppose the pmf of X is as follows:

x	1	2	3	4	5	6	7	8	9	10
$p(x)$	.140	.175	.220	.260	.155	.025	.015	.005	.004	.001

[this is very close to the household size distribution for rural Thailand given in the article **"The Probability of Containment for Multitype Branching Process Models for Emerging Epidemics"** (**J. of Applied Probability**, **2011: 173–188**), which modeled influenza transmission.]

Suppose this is based on 1 million households. One way to view this situation is to think of the population as consisting of 1 million households, each with its own X value; the proportion with each X value is given by $p(x)$ in the above table. An alternative viewpoint is to forget about the households and think of the population itself as consisting of X values: 14% of these values are 1, 17.5% are 2, and so on. The pmf then describes the distribution of the possible population values $1, 2, \ldots, 10$. ■

Once we have such a population model, we will use it to compute values of population characteristics (e.g., the mean μ) and make inferences about such characteristics.

A Parameter of a Probability Distribution

The pmf of the Bernoulli rv X in Example 3.9 was $p(0) = .8$ and $p(1) = .2$ because 20% of all purchasers selected a desktop computer. At another store, it may be the case that $p(0) = .9$ and $p(1) = .1$. More generally, the pmf of any Bernoulli rv can be expressed in the form $p(1) = \alpha$ and $p(0) = 1 - \alpha$, where $0 < \alpha < 1$. Because the pmf depends on the particular value of α, we often write $p(x; \alpha)$ rather than just $p(x)$:

$$p(x; \alpha) = \begin{cases} 1 - \alpha & \text{if } x = 0 \\ \alpha & \text{if } x = 1 \\ 0 & \text{otherwise} \end{cases} \tag{3.1}$$

Then each choice of α in Expression (3.1) yields a different pmf.

DEFINITION

> Suppose $p(x)$ depends on a quantity that can be assigned any one of a number of possible values, with each different value determining a different probability distribution. Such a quantity is called a **parameter** of the distribution. The collection of all probability distributions for different values of the parameter is called a **family** of probability distributions.

The quantity α in Expression (3.1) is a parameter. Each different number α between 0 and 1 determines a different member of the Bernoulli family of distributions.

EXAMPLE 3.12   Starting at a fixed time, we observe the gender of each newborn child at a certain hospital until a boy (B) is born. Let $p = P(B)$, assume that successive births are independent, and define the rv X by $x = $ number of births observed. Then

$$p(1) = P(X = 1) = P(B) = p$$
$$p(2) = P(X = 2) = P(GB) = P(G) \cdot P(B) = (1 - p)p$$

and

$$p(3) = P(X = 3) = P(GGB) = P(G) \cdot P(G) \cdot P(B) = (1 - p)^2 p$$

Continuing in this way, a general formula emerges:

$$p(x) = \begin{cases} (1 - p)^{x-1} p & x = 1, 2, 3, \ldots \\ 0 & \text{otherwise} \end{cases} \tag{3.2}$$

The parameter p can assume any value between 0 and 1. Expression (3.2) describes the family of *geometric* distributions. In the gender scenario, $p = .51$ might be appropriate, but if we were looking for the first child with Rh-positive blood, then it might be the case that $p = .85$. ■

The Cumulative Distribution Function

For some fixed value x, we often wish to compute the probability that the observed value of X will be at most x. For example, let X be the number of number of beds occupied in a hospital's emergency room at a certain time of day; suppose the pmf of X is given by

x	0	1	2	3	4
$p(x)$	.20	.25	.30	.15	.10

Then the probability that at most two beds are occupied is

$$P(X \le 2) = p(0) + p(1) + p(2) = .75$$

Furthermore, since $X \le 2.7$ if and only if $X \le 2$, we also have $P(X \le 2.7) = .75$, and similarly $P(X \le 2.999) = .75$. Since 0 is the smallest possible X value, $P(X \le -1.5) = 0$, $P(X \le -10) = 0$, and in fact for any negative number x, $P(X \le x) = 0$. And because 4 is the largest possible value of X, $P(X \le 4) = 1$, $P(X \le 9.8) = 1$, and so on.

Very importantly,

$$P(X < 2) = p(0) + p(1) = .45 < .75 = P(X \le 2)$$

because the latter probability includes the probability mass at the x value 2 whereas the former probability does not. More generally, $P(X < x) < P(X \le x)$ whenever x is a possible value of X. Furthermore, $P(X \le x)$ is a well-defined and computable probability for *any* number x.

DEFINITION

> The **cumulative distribution function** (cdf) $F(x)$ of a discrete rv variable X with pmf $p(x)$ is defined for every number x by
>
> $$F(x) = P(X \le x) = \sum_{y:y \le x} p(y) \tag{3.3}$$
>
> For any number x, $F(x)$ is the probability that the observed value of X will be at most x.

EXAMPLE 3.13 A store carries flash drives with either 1 GB, 2 GB, 4 GB, 8 GB, or 16 GB of memory. The accompanying table gives the distribution of Y = the amount of memory in a purchased drive:

y	1	2	4	8	16
$p(y)$	.05	.10	.35	.40	.10

Let's first determine $F(y)$ for each of the five possible values of Y:

$$F(1) = P(Y \le 1) = P(Y = 1) = p(1) = .05$$
$$F(2) = P(Y \le 2) = P(Y = 1 \text{ or } 2) = p(1) + p(2) = .15$$
$$F(4) = P(Y \le 4) = P(Y = 1 \text{ or } 2 \text{ or } 4) = p(1) + p(2) + p(4) = .50$$
$$F(8) = P(Y \le 8) = p(1) + p(2) + p(4) + p(8) = .90$$
$$F(16) = P(Y \le 16) = 1$$

Now for any other number y, $F(y)$ will equal the value of F at the closest possible value of Y to the left of y. For example,

$$F(2.7) = P(Y \le 2.7) = P(Y \le 2) = F(2) = .15$$
$$F(7.999) = P(Y \le 7.999) = P(Y \le 4) = F(4) = .50$$

If y is less than 1, $F(y) = 0$ [e.g. $F(.58) = 0$], and if y is at least 16, $F(y) = 1$ [e.g., $F(25) = 1$]. The cdf is thus

$$F(y) = \begin{cases} 0 & y < 1 \\ .05 & 1 \le y < 2 \\ .15 & 2 \le y < 4 \\ .50 & 4 \le y < 8 \\ .90 & 8 \le y < 16 \\ 1 & 16 \le y \end{cases}$$

A graph of this cdf is shown in Figure 3.5.

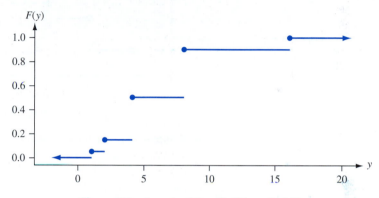

Figure 3.5   A graph of the cdf of Example 3.13

For X a discrete rv, the graph of $F(x)$ will have a jump at every possible value of X and will be flat between possible values. Such a graph is called a **step function.**

EXAMPLE 3.14
(Example 3.12 continued)

The pmf of $X =$ *the number of births up to and including that of the first boy* had the form

$$p(x) = \begin{cases} (1 - p)^{x-1}p & x = 1, 2, 3,\dots \\ 0 & \text{otherwise} \end{cases}$$

For any positive integer x,

$$F(x) = \sum_{y \le x} p(y) = \sum_{y=1}^{x} (1 - p)^{y-1}p = p\sum_{y=0}^{x-1} (1 - p)^y \tag{3.4}$$

To evaluate this sum, recall that the partial sum of a geometric series is

$$\sum_{y=0}^{k} a^y = \frac{1 - a^{k+1}}{1 - a}$$

Using this in Equation (3.4), with $a = 1 - p$ and $k = x - 1$, gives

$$F(x) = p \cdot \frac{1 - (1 - p)^x}{1 - (1 - p)} = 1 - (1 - p)^x \quad x \text{ a positive integer}$$

Since F is constant in between positive integers,

$$F(x) = \begin{cases} 0 & x < 1 \\ 1 - (1 - p)^{[x]} & x \ge 1 \end{cases} \tag{3.5}$$

where $[x]$ is the largest integer $\leq x$ (e.g., $[2.7] = 2$). Thus if $p = .51$ as in the birth example, then the probability of having to examine at most five births to see the first boy is $F(5) = 1 - (.49)^5 = 1 - .0282 = .9718$, whereas $F(10) \approx 1.0000$. This cdf is graphed in Figure 3.6.

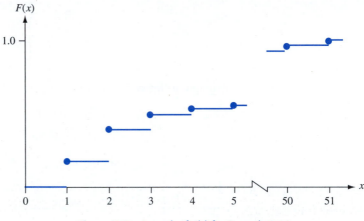

Figure 3.6 A graph of $F(x)$ for Example 3.14 ■

In examples thus far, the cdf has been derived from the pmf. This process can be reversed to obtain the pmf from the cdf whenever the latter function is available. For example, consider again the rv of Example 3.7 (the number of computers being used in a lab); possible X values are $0, 1, \ldots, 6$. Then

$$
\begin{aligned}
p(3) &= P(X = 3) \\
&= [p(0) + p(1) + p(2) + p(3)] - [p(0) + p(1) + p(2)] \\
&= P(X \leq 3) - P(X \leq 2) \\
&= F(3) - F(2)
\end{aligned}
$$

More generally, the probability that X falls in a specified interval is easily obtained from the cdf. For example,

$$
\begin{aligned}
P(2 \leq X \leq 4) &= p(2) + p(3) + p(4) \\
&= [p(0) + \cdots + p(4)] - [p(0) + p(1)] \\
&= P(X \leq 4) - P(X \leq 1) \\
&= F(4) - F(1)
\end{aligned}
$$

Notice that $P(2 \leq X \leq 4) \neq F(4) - F(2)$. This is because the X value 2 is included in $2 \leq X \leq 4$, so we do not want to subtract out its probability. However, $P(2 < X \leq 4) = F(4) - F(2)$ because $X = 2$ is not in the interval $2 < X \leq 4$.

PROPOSITION

For any two numbers a and b with $a \leq b$,

$$P(a \leq X \leq b) = F(b) - F(a-)$$

where "$a-$" represents the largest possible X value that is strictly less than a. In particular, if the only possible values are integers and if a and b are integers, then

$$
\begin{aligned}
P(a \leq X \leq b) &= P(X = a \text{ or } a + 1 \text{ or} \ldots \text{ or } b) \\
&= F(b) - F(a - 1)
\end{aligned}
$$

Taking $a = b$ yields $P(X = a) = F(a) - F(a - 1)$ in this case.

The reason for subtracting $F(a-)$ rather than $F(a)$ is that we want to include $P(X = a)$; $F(b) - F(a)$ gives $P(a < X \le b)$. This proposition will be used extensively when computing binomial and Poisson probabilities in Sections 3.4 and 3.6.

EXAMPLE 3.15 Let $X =$ the number of days of sick leave taken by a randomly selected employee of a large company during a particular year. If the maximum number of allowable sick days per year is 14, possible values of X are $0, 1, \ldots, 14$. With $F(0) = .58$, $F(1) = .72$, $F(2) = .76$, $F(3) = .81$, $F(4) = .88$, and $F(5) = .94$,

$$P(2 \le X \le 5) = P(X = 2, 3, 4, \text{ or } 5) = F(5) - F(1) = .22$$

and

$$P(X = 3) = F(3) - F(2) = .05 \qquad \blacksquare$$

EXERCISES Section 3.2 (11–28)

11. Let X be the number of students who show up for a professor's office hour on a particular day. Suppose that the pmf of X is $p(0) = .20$, $p(1) = .25$, $p(2) = .30$, $p(3) = .15$, and $p(4) = .10$.
  a. Draw the corresponding probability histogram.
  b. What is the probability that at least two students show up? More than two students show up?
  c. What is the probability that between one and three students, inclusive, show up?
  d. What is the probability that the professor shows up?

12. Airlines sometimes overbook flights. Suppose that for a plane with 50 seats, 55 passengers have tickets. Define the random variable Y as the number of ticketed passengers who actually show up for the flight. The probability mass function of Y appears in the accompanying table.

y	45	46	47	48	49	50	51	52	53	54	55
$p(y)$	.05	.10	.12	.14	.25	.17	.06	.05	.03	.02	.01

  a. What is the probability that the flight will accommodate all ticketed passengers who show up?
  b. What is the probability that not all ticketed passengers who show up can be accommodated?
  c. If you are the first person on the standby list (which means you will be the first one to get on the plane if there are any seats available after all ticketed passengers have been accommodated), what is the probability that you will be able to take the flight? What is this probability if you are the third person on the standby list?

13. A mail-order computer business has six telephone lines. Let X denote the number of lines in use at a specified time. Suppose the pmf of X is as given in the accompanying table.

x	0	1	2	3	4	5	6
$p(x)$	.10	.15	.20	.25	.20	.06	.04

Calculate the probability of each of the following events.
  a. {at most three lines are in use}
  b. {fewer than three lines are in use}
  c. {at least three lines are in use}
  d. {between two and five lines, inclusive, are in use}
  e. {between two and four lines, inclusive, are not in use}
  f. {at least four lines are not in use}

14. A contractor is required by a county planning department to submit one, two, three, four, or five forms (depending on the nature of the project) in applying for a building permit. Let $Y =$ the number of forms required of the next applicant. The probability that y forms are required is known to be proportional to y—that is, $p(y) = ky$ for $y = 1, \ldots, 5$.
  a. What is the value of k? [*Hint:* $\sum_{y=1}^{5} p(y) = 1$]
  b. What is the probability that at most three forms are required?
  c. What is the probability that between two and four forms (inclusive) are required?
  d. Could $p(y) = y^2/50$ for $y = 1, \ldots, 5$ be the pmf of Y?

15. Many manufacturers have quality control programs that include inspection of incoming materials for defects. Suppose a computer manufacturer receives circuit boards in batches of five. Two boards are selected from each batch for inspection. We can represent possible outcomes of the selection process by pairs. For example, the pair $(1, 2)$ represents the selection of boards 1 and 2 for inspection.
  a. List the ten different possible outcomes.
  b. Suppose that boards 1 and 2 are the only defective boards in a batch. Two boards are to be chosen at

random. Define X to be the number of defective boards observed among those inspected. Find the probability distribution of X.

c. Let $F(x)$ denote the cdf of X. First determine $F(0) = P(X \leq 0)$, $F(1)$, and $F(2)$; then obtain $F(x)$ for all other x.

16. Some parts of California are particularly earthquake-prone. Suppose that in one metropolitan area, 25% of all homeowners are insured against earthquake damage. Four homeowners are to be selected at random; let X denote the number among the four who have earthquake insurance.

a. Find the probability distribution of X. [*Hint:* Let S denote a homeowner who has insurance and F one who does not. Then one possible outcome is $SFSS$, with probability $(.25)(.75)(.25)(.25)$ and associated X value 3. There are 15 other outcomes.]

b. Draw the corresponding probability histogram.

c. What is the most likely value for X?

d. What is the probability that at least two of the four selected have earthquake insurance?

17. A new battery's voltage may be acceptable (A) or unacceptable (U). A certain flashlight requires two batteries, so batteries will be independently selected and tested until two acceptable ones have been found. Suppose that 90% of all batteries have acceptable voltages. Let Y denote the number of batteries that must be tested.

a. What is $p(2)$, that is, $P(Y = 2)$?

b. What is $p(3)$? [*Hint:* There are two different outcomes that result in $Y = 3$.]

c. To have $Y = 5$, what must be true of the fifth battery selected? List the four outcomes for which $Y = 5$ and then determine $p(5)$.

d. Use the pattern in your answers for parts (a)–(c) to obtain a general formula for $p(y)$.

18. Two fair six-sided dice are tossed independently. Let M = the maximum of the two tosses (so $M(1,5) = 5$, $M(3,3) = 3$, etc.).

a. What is the pmf of M? [*Hint:* First determine $p(1)$, then $p(2)$, and so on.]

b. Determine the cdf of M and graph it.

19. A library subscribes to two different weekly news magazines, each of which is supposed to arrive in Wednesday's mail. In actuality, each one may arrive on Wednesday, Thursday, Friday, or Saturday. Suppose the two arrive independently of one another, and for each one $P(\text{Wed.}) = .3$, $P(\text{Thurs.}) = .4$, $P(\text{Fri.}) = .2$, and $P(\text{Sat.}) = .1$. Let Y = the number of days beyond Wednesday that it takes for both magazines to arrive (so possible Y values are 0, 1, 2, or 3). Compute the pmf of Y. [*Hint:* There are 16 possible outcomes; $Y(W,W) = 0$, $Y(F,Th) = 2$, and so on.]

20. Three couples and two single individuals have been invited to an investment seminar and have agreed to attend. Suppose the probability that any particular

couple or individual arrives late is .4 (a couple will travel together in the same vehicle, so either both people will be on time or else both will arrive late). Assume that different couples and individuals are on time or late independently of one another. Let X = the number of people who arrive late for the seminar.

a. Determine the probability mass function of X. [*Hint:* label the three couples #1, #2, and #3 and the two individuals #4 and #5.]

b. Obtain the cumulative distribution function of X, and use it to calculate $P(2 \leq X \leq 6)$.

21. Suppose that you read through this year's issues of the *New York Times* and record each number that appears in a news article—the income of a CEO, the number of cases of wine produced by a winery, the total charitable contribution of a politician during the previous tax year, the age of a celebrity, and so on. Now focus on the leading digit of each number, which could be 1, 2, ..., 8, or 9. Your first thought might be that the leading digit X of a randomly selected number would be equally likely to be one of the nine possibilities (a discrete uniform distribution). However, much empirical evidence as well as some theoretical arguments suggest an alternative probability distribution called *Benford's law:*

$$p(x) = P(\text{1st digit is } x) = \log_{10}\left(\frac{x + 1}{x}\right) \quad x = 1, 2, \ldots, 9$$

a. Without computing individual probabilities from this formula, show that it specifies a legitimate pmf.

b. Now compute the individual probabilities and compare to the corresponding discrete uniform distribution.

c. Obtain the cdf of X.

d. Using the cdf, what is the probability that the leading digit is at most 3? At least 5?

[*Note:* Benford's law is the basis for some auditing procedures used to detect fraud in financial reporting—for example, by the Internal Revenue Service.]

22. Refer to Exercise 13, and calculate and graph the cdf $F(x)$. Then use it to calculate the probabilities of the events given in parts (a)–(d) of that problem.

23. A branch of a certain bank in New York City has six ATMs. Let X represent the number of machines in use at a particular time of day. The cdf of X is as follows:

$$F(x) = \begin{cases} 0 & x < 0 \\ .06 & 0 \leq x < 1 \\ .19 & 1 \leq x < 2 \\ .39 & 2 \leq x < 3 \\ .67 & 3 \leq x < 4 \\ .92 & 4 \leq x < 5 \\ .97 & 5 \leq x < 6 \\ 1 & 6 \leq x \end{cases}$$

Calculate the following probabilities directly from the cdf:

a. $p(2)$, that is, $P(X = 2)$ **b.** $P(X > 3)$

c. $P(2 \le X \le 5)$ **d.** $P(2 < X < 5)$

24. An insurance company offers its policyholders a number of different premium payment options. For a randomly selected policyholder, let $X =$ the number of months between successive payments. The cdf of X is as follows:

$$F(x) = \begin{cases} 0 & x < 1 \\ .30 & 1 \le x < 3 \\ .40 & 3 \le x < 4 \\ .45 & 4 \le x < 6 \\ .60 & 6 \le x < 12 \\ 1 & 12 \le x \end{cases}$$

a. What is the pmf of X?

b. Using just the cdf, compute $P(3 \le X \le 6)$ and $P(4 \le X)$.

25. In Example 3.12, let $Y =$ the number of girls born before the experiment terminates. With $p = P(B)$ and $1 - p = P(G)$, what is the pmf of Y? [*Hint:* First list the possible values of Y, starting with the smallest, and proceed until you see a general formula.]

26. Alvie Singer lives at 0 in the accompanying diagram and has four friends who live at A, B, C, and D. One day Alvie decides to go visiting, so he tosses a fair coin twice to decide which of the four to visit. Once at a friend's house, he will either return home or else proceed to one of the two adjacent houses (such as 0, A, or C when at B), with each of the three possibilities having probability

1/3. In this way, Alvie continues to visit friends until he returns home.

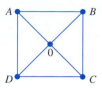

a. Let $X =$ the number of times that Alvie visits a friend. Derive the pmf of X.

b. Let $Y =$ the number of straight-line segments that Alvie traverses (including those leading to and from 0). What is the pmf of Y?

c. Suppose that female friends live at A and C and male friends at B and D. If $Z =$ the number of visits to female friends, what is the pmf of Z?

27. After all students have left the classroom, a statistics professor notices that four copies of the text were left under desks. At the beginning of the next lecture, the professor distributes the four books in a completely random fashion to each of the four students (1, 2, 3, and 4) who claim to have left books. One possible outcome is that 1 receives 2's book, 2 receives 4's book, 3 receives his or her own book, and 4 receives 1's book. This outcome can be abbreviated as (2, 4, 3, 1).

a. List the other 23 possible outcomes.

b. Let X denote the number of students who receive their own book. Determine the pmf of X.

28. Show that the cdf $F(x)$ is a nondecreasing function; that is, $x_1 < x_2$ implies that $F(x_1) \le F(x_2)$. Under what condition will $F(x_1) = F(x_2)$?

3.3 Expected Values

Consider a university having 15,000 students and let $X =$ the number of courses for which a randomly selected student is registered. The pmf of X follows. Since $p(1) = .01$, we know that $(.01) \cdot (15,000) = 150$ of the students are registered for one course, and similarly for the other x values.

x	1	2	3	4	5	6	7	
$p(x)$	.01	.03	.13	.25	.39	.17	.02	(3.6)
Number registered	150	450	1950	3750	5850	2550	300	

The average number of courses per student, or the average value of X in the population, results from computing the total number of courses taken by all students and dividing by the total number of students. Since each of 150 students is taking one course, these 150 contribute 150 courses to the total. Similarly, 450 students contribute 2(450) courses, and so on. The population average value of X is then

$$\frac{1(150) + 2(450) + 3(1950) + \cdots + 7(300)}{15,000} = 4.57 \quad (3.7)$$

Since $150/15,000 = .01 = p(1)$, $450/15,000 = .03 = p(2)$, and so on, an alternative expression for (3.7) is

$$1 \cdot p(1) + 2 \cdot p(2) + \cdots + 7 \cdot p(7) \tag{3.8}$$

Expression (3.8) shows that to compute the population average value of X, we need only the possible values of X along with their probabilities (proportions). In particular, the population size is irrelevant as long as the pmf is given by (3.6). The average or mean value of X is then a *weighted* average of the possible values $1, \ldots, 7$, where the weights are the probabilities of those values.

The Expected Value of X

DEFINITION

Let X be a discrete rv with set of possible values D and pmf $p(x)$. The **expected value** or **mean value** of X, denoted by $E(X)$ or μ_X or just μ, is

$$E(X) = \mu_X = \sum_{x \in D} x \cdot p(x)$$

EXAMPLE 3.16   For the pmf of X = *number of courses* in (3.6),

$$\mu = 1 \cdot p(1) + 2 \cdot p(2) + \cdots + 7 \cdot p(7)$$
$$= (1)(.01) + 2(.03) + \cdots + (7)(.02)$$
$$= .01 + .06 + .39 + 1.00 + 1.95 + 1.02 + .14 = 4.57$$

If we think of the population as consisting of the X values $1, 2, \ldots, 7$, then $\mu = 4.57$ is the population mean. In the sequel, we will often refer to μ as the *population mean* rather than the mean of X in the population. Notice that μ here is not 4, the ordinary average of $1, \ldots, 7$, because the distribution puts more weight on 4, 5, and 6 than on other X values. ∎

In Example 3.16, the expected value μ was 4.57, which is not a possible value of X. The word *expected* should be interpreted with caution because one would not expect to see an X value of 4.57 when a single student is selected.

EXAMPLE 3.17   Just after birth, each newborn child is rated on a scale called the Apgar scale. The possible ratings are $0, 1, \ldots, 10$, with the child's rating determined by color, muscle tone, respiratory effort, heartbeat, and reflex irritability (the best possible score is 10). Let X be the Apgar score of a randomly selected child born at a certain hospital during the next year, and suppose that the pmf of X is

x	0	1	2	3	4	5	6	7	8	9	10
$p(x)$	.002	.001	.002	.005	.02	.04	.18	.37	.25	.12	.01

Then the mean value of X is

$$E(X) = \mu = 0(.002) + 1(.001) + 2(.002)$$
$$+ \cdots + 8(.25) + 9(.12) + 10(.01)$$
$$= 7.15$$

Again, μ is not a possible value of the variable X. Also, because the variable relates to a future child, there is no concrete existing population to which μ refers. Instead, we think of the pmf as a model for a conceptual population consisting of the values $0, 1, 2, \ldots, 10$. The mean value of this conceptual population is then $\mu = 7.15$. ∎

EXAMPLE 3.18   Let $X = 1$ if a randomly selected vehicle passes an emissions test and $X = 0$ otherwise. Then X is a Bernoulli rv with pmf $p(1) = p$ and $p(0) = 1 - p$, from which $E(X) = 0 \cdot p(0) + 1 \cdot p(1) = 0(1 - p) + 1(p) = p$. That is, the expected value of X is just the probability that X takes on the value 1. If we conceptualize a population consisting of 0s in proportion $1 - p$ and 1s in proportion p, then the population average is $\mu = p$. ∎

EXAMPLE 3.19   The general form for the pmf of $X = $ *the number of children born up to and including the first boy* is

$$p(x) = \begin{cases} p(1 - p)^{x-1} & x = 1, 2, 3, \ldots \\ 0 & \text{otherwise} \end{cases}$$

From the definition,

$$E(X) = \sum_D x \cdot p(x) = \sum_{x=1}^{\infty} xp(1 - p)^{x-1} = p \sum_{x=1}^{\infty} \left[-\frac{d}{dp}(1 - p)^x \right] \qquad (3.9)$$

Interchanging the order of taking the derivative and the summation, the sum is that of a geometric series. After the sum is computed, the derivative is taken, resulting in $E(X) = 1/p$. If p is near 1, we expect to see a boy very soon, whereas if p is near 0, we expect many births before the first boy. For $p = .5$, $E(X) = 2$. ∎

There is another frequently used interpretation of μ. Consider observing a first value x_1 of X, then a second value x_2, a third value x_3, and so on. After doing this a large number of times, calculate the sample average of the observed x_i's. This average will typically be quite close to μ. That is, μ can be interpreted as the long-run average observed value of X when the experiment is performed repeatedly. In Example 3.17, the long-run average Apgar score is $\mu = 7.15$.

EXAMPLE 3.20   Let X, the number of interviews a student has prior to getting a job, have pmf

$$p(x) = \begin{cases} k/x^2 & x = 1, 2, 3, \ldots \\ 0 & \text{otherwise} \end{cases}$$

where $k = \pi^2/6$ insures that $\Sigma p(x) = 1$ (the value of k comes from a result in Fourier series). The expected value of X is

$$\mu = E(X) = \sum_{x=1}^{\infty} x \cdot \frac{k}{x^2} = k \sum_{x=1}^{\infty} \frac{1}{x} \qquad (3.10)$$

The sum on the right of Equation (3.10) is the famous harmonic series of mathematics and can be shown to equal ∞. $E(X)$ is not finite here because $p(x)$ does not decrease sufficiently fast as x increases; statisticians say that the probability distribution of X has "a heavy tail." If a sequence of X values is chosen using this distribution, the sample average will not settle down to some finite number but will tend to grow without bound.

Statisticians use the phrase "heavy tails" in connection with any distribution having a large amount of probability far from μ (so heavy tails do not require $\mu = \infty$). Such heavy tails make it difficult to make inferences about μ. ∎

The Expected Value of a Function

Sometimes interest will focus on the expected value of some function $h(X)$ rather than on just $E(X)$.

EXAMPLE 3.21   Suppose a bookstore purchases ten copies of a book at $6.00 each to sell at $12.00 with the understanding that at the end of a 3-month period any unsold copies can be redeemed for $2.00. If $X =$ the number of copies sold, then net revenue $= h(X) = 12X + 2(10 - X) - 60 = 10X - 40$. In this situation, we might be interested not only in the expected number of copies sold [i.e., $E(X)$] but also in the expected net revenue—that is, the expected value of a particular function of X. ∎

An easy way of computing the expected value of $h(X)$ is suggested by the following example.

EXAMPLE 3.22   The cost of a certain vehicle diagnostic test depends on the number of cylinders X in the vehicle's engine. Suppose the cost function is given by $h(X) = 20 + 3X + .5X^2$. Since X is a random variable, so is $Y = h(X)$. The pmf of X and derived pmf of Y are as follows:

x	4	6	8		y	40	56	76
$p(x)$	.5	.3	.2	$\Rightarrow$	$p(y)$	.5	.3	.2

With D^* denoting possible values of Y,

$$
\begin{aligned}
E(Y) = E[h(X)] &= \sum_{D^*} y \cdot p(y) \\
&= (40)(.5) + (56)(.3) + (76)(.2) \\
&= h(4) \cdot (.5) + h(6) \cdot (.3) + h(8) \cdot (.2) \\
&= \sum_{D} h(x) \cdot p(x)
\end{aligned}
\tag{3.11}
$$

According to Equation (3.11), it was not necessary to determine the pmf of Y to obtain $E(Y)$; instead, the desired expected value is a weighted average of the possible $h(x)$ (rather than x) values. ∎

PROPOSITION
> If the rv X has a set of possible values D and pmf $p(x)$, then the expected value of any function $h(X)$, denoted by $E[h(X)]$ or $\mu_{h(X)}$, is computed by
>
> $$ E[h(X)] = \sum_{D} h(x) \cdot p(x) $$

That is, $E[h(X)]$ is computed in the same way that $E(X)$ itself is, except that $h(x)$ is substituted in place of x.

EXAMPLE 3.23   A computer store has purchased three computers of a certain type at $500 apiece. It will sell them for $1000 apiece. The manufacturer has agreed to repurchase any computers still unsold after a specified period at $200 apiece. Let X denote the number of computers sold, and suppose that $p(0) = .1$, $p(1) = .2$, $p(2) = .3$, and $p(3) = .4$. With $h(X)$ denoting the profit associated with selling X units, the given information implies that $h(X) =$ revenue $-$ cost $=$ $1000X + 200(3 - X) - 1500 = 800X - 900$. The expected profit is then

$$
\begin{aligned}
E[h(X)] &= h(0) \cdot p(0) + h(1) \cdot p(1) + h(2) \cdot p(2) + h(3) \cdot p(3) \\
&= (-900)(.1) + (-100)(.2) + (700)(.3) + (1500)(.4) \\
&= \$700
\end{aligned}
$$
∎

Expected Value of a Linear Function

The $h(X)$ function of interest is quite frequently a linear function $aX + b$. In this case, $E[h(X)]$ is easily computed from $E(X)$ without the need for additional summation.

PROPOSITION

$$
E(aX + b) = a \cdot E(X) + b
$$

(Or, using alternative notation, $\mu_{aX+b} = a \cdot \mu_X + b$)

*[handwritten: *only linear]*

To paraphrase, the expected value of a linear function equals the linear function evaluated at the expected value $E(X)$. Since $h(X)$ in Example 3.23 is linear and $E(X) = 2$, $E[h(X)] = 800(2) - 900 = \$700$, as before.

Proof

$$
E(aX + b) = \sum_D (ax + b) \cdot p(x) = a \sum_D x \cdot p(x) + b \sum_D p(x)
$$
$$
= aE(X) + b
$$
∎

Two special cases of the proposition yield two important rules of expected value.

1. For any constant a, $E(aX) = a \cdot E(X)$ (take $b = 0$).
2. For any constant b, $E(X + b) = E(X) + b$ (take $a = 1$).   (3.12)

Multiplication of X by a constant a typically changes the unit of measurement, for example, from inches to cm, where $a = 2.54$. Rule 1 says that the expected value in the new units equals the expected value in the old units multiplied by the conversion factor a. Similarly, if a constant b is added to each possible value of X, then the expected value will be shifted by that same constant amount.

The Variance of X

The expected value of X describes where the probability distribution is centered. Using the physical analogy of placing point mass $p(x)$ at the value x on a one-dimensional axis, if the axis were then supported by a fulcrum placed at μ, there would be no tendency for the axis to tilt. This is illustrated for two different distributions in Figure 3.7.

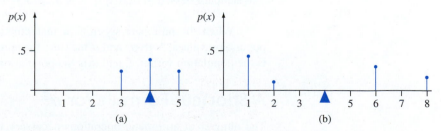

Figure 3.7 Two different probability distributions with $\mu = 4$

Although both distributions pictured in Figure 3.7 have the same center μ, the distribution of Figure 3.7(b) has greater spread (i.e., variability or dispersion) than does that of Figure 3.7(a). We will use the variance of X to assess the amount of variability in (the distribution of) X, just as s^2 was used in Chapter 1 to measure variability in a sample.

DEFINITION

> Let X have pmf $p(x)$ and expected value μ. Then the **variance** of X, denoted by $V(X)$ or σ_X^2, or just σ^2, is
>
> $$V(X) = \sum_D (x - \mu)^2 \cdot p(x) = E[(X - \mu)^2]$$
>
> The **standard deviation** (SD) of X is
>
> $$\sigma_X = \sqrt{\sigma_X^2}$$

The quantity $h(X) = (X - \mu)^2$ is the squared deviation of X from its mean, and σ^2 is the expected squared deviation—i.e., the weighted average of squared deviations, where the weights are probabilities from the distribution. If most of the probability distribution is close to μ, then σ^2 will be relatively small. However, if there are x values far from μ that have large $p(x)$, then σ^2 will be quite large. Very roughly, σ can be interpreted as the size of a representative deviation from the mean value μ. So if $\sigma = 10$, then in a long sequence of observed X values, some will deviate from μ by more than 10 while others will be closer to the mean than that—a typical deviation from the mean will be something on the order of 10.

EXAMPLE 3.24 A library has an upper limit of 6 on the number of DVDs that can be checked out to an individual at one time. Consider only those who currently have DVDs checked out, and let X denote the number of DVDs checked out to a randomly selected individual. The pmf of X is as follows:

x	1	2	3	4	5	6
$p(x)$	.30	.25	.15	.05	.10	.15

The expected value of X is easily seen to be $\mu = 2.85$. The variance of X is then

$$V(X) = \sigma^2 = \sum_{x=1}^{6} (x - 2.85)^2 \cdot p(x)$$

$$= (1 - 2.85)^2(.30) + (2 - 2.85)^2(.25) + \cdots + (6 - 2.85)^2(.15) = 3.2275$$

The standard deviation of X is $\sigma = \sqrt{3.2275} = 1.800$. ∎

When the pmf $p(x)$ specifies a mathematical model for the distribution of population values, both σ^2 and σ measure the spread of values in the population; σ^2 is the population variance, and σ is the population standard deviation.

A Shortcut Formula for σ^2

The number of arithmetic operations necessary to compute σ^2 can be reduced by using an alternative formula.

Shortcut

PROPOSITION

$$V(X) = \sigma^2 = \left[\sum_D x^2 \cdot p(x)\right] - \mu^2 = E(X^2) - [E(X)]^2$$

In using this formula, $E(X^2)$ is computed first without any subtraction; then $E(X)$ is computed, squared, and subtracted (once) from $E(X^2)$.

EXAMPLE 3.25
(Example 3.24 continued)

The pmf of the number X of DVDs checked out was given as $p(1) = .30$, $p(2) = .25$, $p(3) = .15$, $p(4) = .05$, $p(5) = .10$, and $p(6) = .15$, from which $\mu = 2.85$ and

$$E(X^2) = \sum_{x=1}^{6} x^2 \cdot p(x) = (1^2)(.30) + (2^2)(.25) + \cdots + (6^2)(.15) = 11.35$$

Thus $\sigma^2 = 11.35 - (2.85)^2 = 3.2275$ as obtained previously from the definition. ∎

Proof of the Shortcut Formula   Expand $(x - \mu)^2$ in the definition of σ^2 to obtain $x^2 - 2\mu x + \mu^2$, and then carry Σ through to each of the three terms:

$$\sigma^2 = \sum_D x^2 \cdot p(x) - 2\mu \cdot \sum_D x \cdot p(x) + \mu^2 \sum_D p(x)$$
$$= E(X^2) - 2\mu \cdot \mu + \mu^2 = E(X^2) - \mu^2$$
∎

Variance of a Linear Function

The variance of $h(X)$ is the expected value of the squared difference between $h(X)$ and its expected value:

$$V[h(X)] = \sigma^2_{h(X)} = \sum_D \{h(x) - E[h(X)]\}^2 \cdot p(x) \qquad (3.13)$$

Challenge problem

When $h(X) = aX + b$, a linear function,

$$h(x) - E[h(X)] = ax + b - (a\mu + b) = a(x - \mu)$$

Substituting this into (3.13) gives a simple relationship between $V[h(X)]$ and $V(X)$:

PROPOSITION

$$V(aX + b) = \sigma^2_{aX+b} = a^2 \cdot \sigma^2_X \quad \text{and} \quad \sigma_{aX+b} = |a| \cdot \sigma_X$$

In particular,

$$\sigma_{aX} = |a| \cdot \sigma_X, \quad \sigma_{X+b} = \sigma_X \qquad (3.14)$$

The absolute value is necessary because a might be negative, yet a standard deviation cannot be. Usually multiplication by a corresponds to a change in the unit of measurement (e.g., kg to lb or dollars to euros). According to the first relation in (3.14), the sd in the new unit is the original sd multiplied by the conversion factor. The second relation says that adding or subtracting a constant does not impact variability; it just rigidly shifts the distribution to the right or left.

EXAMPLE 3.26 In the computer sales scenario of Example 3.23, $E(X) = 2$ and

$$E(X^2) = (0)^2(.1) + (1)^2(.2) + (2)^2(.3) + (3)^2(.4) = 5$$

so $V(X) = 5 - (2)^2 = 1$. The profit function $h(X) = 800X - 900$ then has variance $(800)^2 \cdot V(X) = (640,000)(1) = 640,000$ and standard deviation 800. ∎

EXERCISES Section 3.3 (29–45)

29. The pmf of the amount of memory X (GB) in a purchased flash drive was given in Example 3.13 as

x	1	2	4	8	16
$p(x)$	.05	.10	.35	.40	.10

Compute the following:
   a. $E(X)$
   b. $V(X)$ directly from the definition
   c. The standard deviation of X
   d. $V(X)$ using the shortcut formula

30. An individual who has automobile insurance from a certain company is randomly selected. Let Y be the number of moving violations for which the individual was cited during the last 3 years. The pmf of Y is

y	0	1	2	3
$p(y)$	.60	.25	.10	.05

   a. Compute $E(Y)$.
   b. Suppose an individual with Y violations incurs a surcharge of $\$100Y^2$. Calculate the expected amount of the surcharge.

31. Refer to Exercise 12 and calculate $V(Y)$ and σ_Y. Then determine the probability that Y is within 1 standard deviation of its mean value.

32. A certain brand of upright freezer is available in three different rated capacities: 16 ft³, 18 ft³, and 20 ft³. Let $X =$ the rated capacity of a freezer of this brand sold at a certain store. Suppose that X has pmf

x	16	18	20
$p(x)$	.2	.5	.3

   a. Compute $E(X)$, $E(X^2)$, and $V(X)$.
   b. If the price of a freezer having capacity X is $70X - 650$, what is the expected price paid by the next customer to buy a freezer?
   c. What is the variance of the price paid by the next customer?

   d. Suppose that although the rated capacity of a freezer is X, the actual capacity is $h(X) = X - .008X^2$. What is the expected actual capacity of the freezer purchased by the next customer?

33. Let X be a Bernoulli rv with pmf as in Example 3.18.
   a. Compute $E(X^2)$.
   b. Show that $V(X) = p(1 - p)$.
   c. Compute $E(X^{79})$.

34. Suppose that the number of plants of a particular type found in a rectangular sampling region (called a quadrat by ecologists) in a certain geographic area is an rv X with pmf

$$p(x) = \begin{cases} c/x^3 & x = 1, 2, 3, \dots \\ 0 & \text{otherwise} \end{cases}$$

Is $E(X)$ finite? Justify your answer (this is another distribution that statisticians would call heavy-tailed).

35. A small market orders copies of a certain magazine for its magazine rack each week. Let $X =$ demand for the magazine, with pmf

x	1	2	3	4	5	6
$p(x)$	$\frac{1}{15}$	$\frac{2}{15}$	$\frac{3}{15}$	$\frac{4}{15}$	$\frac{3}{15}$	$\frac{2}{15}$

Suppose the store owner actually pays $\$2.00$ for each copy of the magazine and the price to customers is $\$4.00$. If magazines left at the end of the week have no salvage value, is it better to order three or four copies of the magazine? [*Hint:* For both three and four copies ordered, express net revenue as a function of demand X, and then compute the expected revenue.]

36. Let X be the damage incurred (in $\$$) in a certain type of accident during a given year. Possible X values are 0, 1000, 5000, and 10000, with probabilities .8, .1, .08, and .02, respectively. A particular company offers a $\$500$ deductible policy. If the company wishes its expected profit to be $\$100$, what premium amount should it charge?

37. The n candidates for a job have been ranked 1, 2, 3,..., n. Let X = the rank of a randomly selected candidate, so that X has pmf

$$p(x) = \begin{cases} 1/n & x = 1, 2, 3,\ldots, n \\ 0 & \text{otherwise} \end{cases}$$

(this is called the *discrete uniform distribution*). Compute $E(X)$ and $V(X)$ using the shortcut formula. [*Hint:* The sum of the first n positive integers is $n(n + 1)/2$, whereas the sum of their squares is $n(n + 1)(2n + 1)/6$.]

38. Possible values of X, the number of components in a system submitted for repair that must be replaced, are 1, 2, 3, and 4 with corresponding probabilities .15, .35, .35, and .15, respectively.
   a. Calculate $E(X)$ and then $E(5 - X)$.
   b. Would the repair facility be better off charging a flat fee of \$75 or else the amount \$[150/(5 − X)]? [*Note:* It is not generally true that $E(c/Y) = c/E(Y)$.]

39. A chemical supply company currently has in stock 100 lb of a certain chemical, which it sells to customers in 5-lb batches. Let X = the number of batches ordered by a randomly chosen customer, and suppose that X has pmf

x	1	2	3	4
$p(x)$	.2	.4	.3	.1

Compute $E(X)$ and $V(X)$. Then compute the expected number of pounds left after the next customer's order is shipped and the variance of the number of pounds left. [*Hint:* The number of pounds left is a linear function of X.]

40. a. Draw a line graph of the pmf of X in Exercise 35. Then determine the pmf of $-X$ and draw its line

$$V(X) = V(-X)$$

graph. From these two pictures, what can you say about $V(X)$ and $V(-X)$?
   b. Use the proposition involving $V(aX + b)$ to establish a general relationship between $V(X)$ and $V(-X)$.

41. Use the definition in Expression (3.13) to prove that $V(aX + b) = a^2 \cdot \sigma_X^2$. [*Hint:* With $h(X) = aX + b$, $E[h(X)] = a\mu + b$ where $\mu = E(X)$.]

42. Suppose $E(X) = 5$ and $E[X(X - 1)] = 27.5$. What is
   a. $E(X^2)$? [*Hint:* First verify that $E[X(X - 1)] = E(X^2) - E(X)$?]
   b. $V(X)$?
   c. The general relationship among the quantities $E(X)$, $E[X(X - 1)]$, and $V(X)$?

43. Write a general rule for $E(X - c)$ where c is a constant. What happens when $c = \mu$, the expected value of X?

44. A result called **Chebyshev's inequality** states that for any probability distribution of an rv X and any number k that is at least 1, $P(|X - \mu| \geq k\sigma) \leq 1/k^2$. In words, the probability that the value of X lies at least k standard deviations from its mean is at most $1/k^2$.
   a. What is the value of the upper bound for $k = 2$? $k = 3$? $k = 4$? $k = 5$? $k = 10$?
   b. Compute μ and σ for the distribution of Exercise 13. Then evaluate $P(|X - \mu| \geq k\sigma)$ for the values of k given in part (a). What does this suggest about the upper bound relative to the corresponding probability?
   c. Let X have possible values -1, 0, and 1, with probabilities $\frac{1}{18}, \frac{8}{9}$, and $\frac{1}{18}$, respectively. What is $P(|X - \mu| \geq 3\sigma)$, and how does it compare to the corresponding bound?
   d. Give a distribution for which $P(|X - \mu| \geq 5\sigma) = .04$.

45. If $a \leq X \leq b$, show that $a \leq E(X) \leq b$.

3.4 The Binomial Probability Distribution

① finite amount of trials

② boolean outcome

③ trials are independent

④ P(true) is constant

There are many experiments that conform either exactly or approximately to the following list of requirements:

1. The experiment consists of a sequence of n smaller experiments called *trials,* where n is fixed in advance of the experiment.
2. Each trial can result in one of the same two possible outcomes (dichotomous trials), which we generically denote by success (S) and failure (F). The assignment of the S and F labels to the two sides of the dichotomy is arbitrary.
3. The trials are independent, so that the outcome on any particular trial does not influence the outcome on any other trial.
4. The probability of success $P(S)$ is constant from trial to trial; we denote this probability by p.

DEFINITION

An experiment for which Conditions 1–4 (a fixed number of dichotomous, independent, homogenous trials) are satisfied is called a **binomial experiment.**

EXAMPLE 3.27 Consider each of the next n vehicles undergoing an emissions test, and let S denote a vehicle that passes the test and F denote one that fails to pass. Then this experiment satisfies Conditions 1–4. Tossing a thumbtack n times, with S = point up and F = point down, also results in a binomial experiment, as would the experiment in which the gender (S for female and F for male) is determined for each of the next n children born at a particular hospital. ∎

Many experiments involve a sequence of independent trials for which there are more than two possible outcomes on any one trial. A binomial experiment can then be created by dividing the possible outcomes into two groups.

EXAMPLE 3.28 The color of pea seeds is determined by a single genetic locus. If the two alleles at this locus are AA or Aa (the genotype), then the pea will be yellow (the phenotype), and if the allele is aa, the pea will be green. Suppose we pair off 20 Aa seeds and cross the two seeds in each of the ten pairs to obtain ten new genotypes. Call each new genotype a success S if it is aa and a failure otherwise. Then with this identification of S and $F,$ the experiment is binomial with $n = 10$ and $p = P(\text{aa genotype})$. If each member of the pair is equally likely to contribute a or A, then $p = P(a) \cdot P(a) = (.5)(.5) = .25$. ∎

EXAMPLE 3.29 The pool of prospective jurors for a certain case consists of 50 individuals, of whom 35 are employed. Suppose that 6 of these individuals are randomly selected one by one to sit in the jury box for initial questioning by lawyers for the defense and the prosecution. Label the ith person selected (the ith trial) as a success S if he or she is employed and a failure F otherwise. Then

$$P(S \text{ on first trial}) = \frac{35}{50} = .70$$

and

$$P(S \text{ on second trial}) = P(SS) + P(FS)$$
$$= P(\text{second } S \mid \text{first } S)P(\text{first } S)$$
$$+ P(\text{second } S \mid \text{first } F)P(\text{first } F)$$
$$= \frac{34}{49} \cdot \frac{35}{50} + \frac{35}{49} \cdot \frac{15}{50} = \frac{35}{50}\left(\frac{34}{49} + \frac{15}{49}\right) = \frac{35}{50} = .70$$

Similarly, it can be shown that $P(S \text{ on } i\text{th trial}) = .70$ for $i = 3, 4, 5, 6$. However, if the first five individuals selected are all S, then only 30 Ss remain for the sixth selection. Thus,

$$P(S \text{ on sixth trial} \mid SSSSS) = 30/45 = .667$$

whereas

$$P(S \text{ on sixth trial} \mid FFFFF) = 35/45 = .778$$

The experiment is not binomial because the trials are not independent. In general, if sampling is without replacement, the experiment will not yield independent trials.

Now consider a large county that has 500,000 individuals in its jury pool, of whom 400,000 are employed. A sample of 10 individuals from the pool is chosen without replacement. Again the ith trial is regarded as a success S if the ith individual is employed. The important difference between this and the previous scenario is that the size of the population being sampled is very large relative to the sample size. In this case

$$P(S \text{ on } 2 \mid S \text{ on } 1) = \frac{399{,}999}{499{,}999} = .8000$$

and

$$P(S \text{ on } 10 \mid S \text{ on first } 9) = \frac{399{,}991}{499{,}991} = .799996 \approx .8000$$

$$P(S \text{ on } 10 \mid F \text{ on first } 9) = \frac{400{,}000}{499{,}991} = .800014 \approx .8000$$

These calculations suggest that although the trials are not exactly independent, the conditional probabilities differ so slightly from one another that for practical purposes the trials can be regarded as independent with constant $P(S) = .8$. Thus, to a very good approximation, the experiment is binomial with $n = 10$ and $p = .8$. ■

We will use the following rule of thumb in deciding whether a "without-replacement" experiment can be treated as being binomial.

RULE

> Consider sampling without replacement from a dichotomous population of size N. If the sample size (number of trials) n is at most 5% of the population size, the experiment can be analyzed as though it were a binomial experiment.

By "analyzed," we mean that probabilities based on the binomial experiment assumptions will be quite close to the actual "without-replacement" probabilities, which are typically more difficult to calculate. In the first scenario of Example 3.29, $n/N = 6/50 = .12 > .05$, so the binomial experiment is not a good approximation, but in the second scenario, $n/N = 10/500{,}000 \ll .05$.

The Binomial Random Variable and Distribution

In most binomial experiments, it is the total number of S's, rather than knowledge of exactly which trials yielded S's, that is of interest.

DEFINITION

> The **binomial random variable** X associated with a binomial experiment consisting of n trials is defined as
>
> $$X = \text{the number of } S\text{'s among the } n \text{ trials}$$

Suppose, for example, that $n = 3$. Then there are eight possible outcomes for the experiment:

$$SSS \quad SSF \quad SFS \quad SFF \quad FSS \quad FSF \quad FFS \quad FFF$$

From the definition of X, $X(SSF) = 2$, $X(SFF) = 1$, and so on. Possible values for X in an n-trial experiment are $x = 0, 1, 2, \ldots, n$. We will often write $X \sim \text{Bin}(n, p)$ to indicate that X is a binomial rv based on n trials with success probability p.

NOTATION

> Because the pmf of a binomial rv X depends on the two parameters n and p, we denote the pmf by $b(x; n, p)$.

Consider first the case $n = 4$ for which each outcome, its probability, and corresponding x value are displayed in Table 3.1. For example,

$$P(SSFS) = P(S) \cdot P(S) \cdot P(F) \cdot P(S) \quad \text{(independent trials)}$$
$$= p \cdot p \cdot (1 - p) \cdot p \quad \text{(constant } P(S)\text{)}$$
$$= p^3 \cdot (1 - p)$$

Table 3.1 Outcomes and Probabilities for a Binomial Experiment with Four Trials

Outcome	x	Probability	Outcome	x	Probability
SSSS	4	p^4	FSSS	3	$p^3(1-p)$
SSSF	3	$p^3(1-p)$	FSSF	2	$p^2(1-p)^2$
SSFS	3	$p^3(1-p)$	FSFS	2	$p^2(1-p)^2$
SSFF	2	$p^2(1-p)^2$	FSFF	1	$p(1-p)^3$
SFSS	3	$p^3(1-p)$	FFSS	2	$p^2(1-p)^2$
SFSF	2	$p^2(1-p)^2$	FFSF	1	$p(1-p)^3$
SFFS	2	$p^2(1-p)^2$	FFFS	1	$p(1-p)^3$
SFFF	1	$p(1-p)^3$	FFFF	0	$(1-p)^4$

In this special case, we wish $b(x; 4, p)$ for $x = 0, 1, 2, 3$, and 4. For $b(3; 4, p)$, let's identify which of the 16 outcomes yield an x value of 3 and sum the probabilities associated with each such outcome:

$$b(3; 4, p) = P(FSSS) + P(SFSS) + P(SSFS) + P(SSSF) = 4p^3(1 - p)$$

There are four outcomes with $X = 3$ and each has probability $p^3(1 - p)$ (the order of S's and F's is not important, only the number of S's), so

$$b(3; 4, p) = \left\{ \begin{matrix} \text{number of outcomes} \\ \text{with } X = 3 \end{matrix} \right\} \cdot \left\{ \begin{matrix} \text{probability of any particular} \\ \text{outcome with } X = 3 \end{matrix} \right\}$$

Similarly, $b(2; 4, p) = 6p^2(1 - p)^2$, which is also the product of the number of outcomes with $X = 2$ and the probability of any such outcome.

In general,

$$b(x; n, p) = \left\{ \begin{matrix} \text{number of sequences of} \\ \text{length } n \text{ consisting of } x \text{ } S\text{'s} \end{matrix} \right\} \cdot \left\{ \begin{matrix} \text{probability of any} \\ \text{particular such sequence} \end{matrix} \right\}$$

Since the ordering of S's and F's is not important, the second factor in the previous equation is $p^x(1 - p)^{n-x}$ (e.g., the first x trials resulting in S and the last $n - x$ resulting in F). The first factor is the number of ways of choosing x of the n trials to be S's—that is, the number of combinations of size x that can be constructed from n distinct objects (trials here).

THEOREM

$$b(x; n, p) = \begin{cases} \dbinom{n}{x} p^x(1 - p)^{n-x} & x = 0, 1, 2, \ldots, n \\ 0 & \text{otherwise} \end{cases}$$

EXAMPLE 3.30 Each of six randomly selected cola drinkers is given a glass containing cola S and one containing cola F. The glasses are identical in appearance except for a code on the bottom to identify the cola. Suppose there is actually no tendency among cola drinkers to prefer one cola to the other. Then $p = P(\text{a selected individual prefers } S) = .5$, so with $X =$ the number among the six who prefer S, $X \sim \text{Bin}(6,.5)$.

Thus

$$P(X = 3) = b(3; 6, .5) = \binom{6}{3}(.5)^3(.5)^3 = 20(.5)^6 = .313$$

The probability that at least three prefer S is

$$P(3 \le X) = \sum_{x=3}^{6} b(x; 6, .5) = \sum_{x=3}^{6} \binom{6}{x}(.5)^x(.5)^{6-x} = .656$$

and the probability that at most one prefers S is

$$P(X \le 1) = \sum_{x=0}^{1} b(x; 6, .5) = .109$$ ∎

Using Binomial Tables*

Even for a relatively small value of n, the computation of binomial probabilities can be tedious. Appendix Table A.1 tabulates the cdf $F(x) = P(X \le x)$ for $n = 5, 10, 15, 20, 25$ in combination with selected values of p corresponding to different columns of the table. Various other probabilities can then be calculated using the proposition on cdf's from Section 3.2. A table entry of 0 signifies only that the probability is 0 to three significant digits since all table entries are actually positive.

NOTATION

For $X \sim \text{Bin}(n, p)$, the cdf will be denoted by

$$B(x; n, p) = P(X \le x) = \sum_{y=0}^{x} b(y; n, p) \quad x = 0, 1, \ldots, n$$

EXAMPLE 3.31 Suppose that 20% of all copies of a particular textbook fail a certain binding strength test. Let X denote the number among 15 randomly selected copies that fail the test. Then X has a binomial distribution with $n = 15$ and $p = .2$.

cdf given
use Complement

1. The probability that at most 8 fail the test is

$$P(X \le 8) = \sum_{y=0}^{8} b(y; 15, .2) = B(8; 15, .2)$$

which is the entry in the $x = 8$ row and the $p = .2$ column of the $n = 15$ binomial table. From Appendix Table A.1, the probability is $B(8; 15, .2) = .999$.

2. The probability that exactly 8 fail is

$$P(X = 8) = P(X \le 8) - P(X \le 7) = B(8; 15, .2) - B(7; 15, .2)$$

which is the difference between two consecutive entries in the $p = .2$ column. The result is $.999 - .996 = .003$.

---

* Statistical software packages such as Minitab and R will provide the pmf or cdf almost instantaneously upon request for any value of p and n ranging from 2 up into the millions. There is also an R command for calculating the probability that X lies in some interval.

3. The probability that at least 8 fail is

$$P(X \geq 8) = 1 - P(X \leq 7) = 1 - B(7; 15, .2)$$

$$= 1 - \begin{pmatrix} \text{entry in } x = 7 \\ \text{row of } p = .2 \text{ column} \end{pmatrix}$$

$$= 1 - .996 = .004$$

4. Finally, the probability that between 4 and 7, inclusive, fail is

$$P(4 \leq X \leq 7) = P(X = 4, 5, 6, \text{ or } 7) = P(X \leq 7) - P(X \leq 3)$$

$$= B(7; 15, .2) - B(3; 15, .2) = .996 - .648 = .348$$

Notice that this latter probability is the difference between entries in the $x = 7$ and $x = 3$ rows, *not* the $x = 7$ and $x = 4$ rows. ∎

EXAMPLE 3.32 An electronics manufacturer claims that at most 10% of its power supply units need service during the warranty period. To investigate this claim, technicians at a testing laboratory purchase 20 units and subject each one to accelerated testing to simulate use during the warranty period. Let p denote the probability that a power supply unit needs repair during the period (the proportion of all such units that need repair). The laboratory technicians must decide whether the data resulting from the experiment supports the claim that $p \leq .10$. Let X denote the number among the 20 sampled that need repair, so $X \sim \text{Bin}(20, p)$. Consider the decision rule:

Reject the claim that $p \leq .10$ in favor of the conclusion that $p > .10$ if $x \geq 5$

(where x is the observed value of X), and consider the claim plausible if $x \leq 4$.

The probability that the claim is rejected when $p = .10$ (an incorrect conclusion) is

$$P(X \geq 5 \text{ when } p = .10) = 1 - B(4; 20, .1) = 1 - .957 = .043$$

The probability that the claim is not rejected when $p = .20$ (a different type of incorrect conclusion) is

$$P(X \leq 4 \text{ when } p = .2) = B(4; 20, .2) = .630$$

The first probability is rather small, but the second is intolerably large. When $p = .20$, so that the manufacturer has grossly understated the percentage of units that need service, and the stated decision rule is used, 63% of all samples will result in the manufacturer's claim being judged plausible!

One might think that the probability of this second type of erroneous conclusion could be made smaller by changing the cutoff value 5 in the decision rule to something else. However, although replacing 5 by a smaller number would yield a probability smaller than .630, the other probability would then increase. The only way to make both "error probabilities" small is to base the decision rule on an experiment involving many more units. ∎

The Mean and Variance of X

For $n = 1$, the binomial distribution becomes the Bernoulli distribution. From Example 3.18, the mean value of a Bernoulli variable is $\mu = p$, so the expected number of S's on any single trial is p. Since a binomial experiment consists of n trials, intuition suggests that for $X \sim \text{Bin}(n, p)$, $E(X) = np$, the product of the number of trials and the probability of success on a single trial. The expression for $V(X)$ is not so intuitive.

PROPOSITION

> If $X \sim \text{Bin}(n, p)$, then $E(X) = np$, $V(X) = np(1 - p) = npq$, and $\sigma_X = \sqrt{npq}$ (where $q = 1 - p$).

Thus, calculating the mean and variance of a binomial rv does not necessitate evaluating summations. The proof of the result for $E(X)$ is sketched in Exercise 64.

EXAMPLE 3.33 If 75% of all purchases at a certain store are made with a credit card and X is the number among ten randomly selected purchases made with a credit card, then $X \sim \text{Bin}(10, .75)$. Thus $E(X) = np = (10)(.75) = 7.5$, $V(X) = npq = 10(.75)(.25) = 1.875$, and $\sigma = \sqrt{1.875} = 1.37$. Again, even though X can take on only integer values, $E(X)$ need not be an integer. If we perform a large number of independent binomial experiments, each with $n = 10$ trials and $p = .75$, then the average number of S's per experiment will be close to 7.5.

The probability that X is within 1 standard deviation of its mean value is $P(7.5 - 1.37 \le X \le 7.5 + 1.37) = P(6.13 \le X \le 8.87) = P(X = 7 \text{ or } 8) = .532.$ ∎

EXERCISES Section 3.4 (46–67)

46. Compute the following binomial probabilities directly from the formula for $b(x; n, p)$:

   a. $b(3; 8, .35)$
   b. $b(5; 8, .6)$
   c. $P(3 \le X \le 5)$ when $n = 7$ and $p = .6$
   d. $P(1 \le X)$ when $n = 9$ and $p = .1$

47. The article **"Should You Report That Fender-Bender?"** (*Consumer Reports*, **Sept. 2013: 15**) reported that 7 in 10 auto accidents involve a single vehicle (the article recommended always reporting to the insurance company an accident involving multiple vehicles). Suppose 15 accidents are randomly selected. Use Appendix Table A.1 to answer each of the following questions.

   a. What is the probability that at most 4 involve a single vehicle?
   b. What is the probability that exactly 4 involve a single vehicle?
   c. What is the probability that exactly 6 involve multiple vehicles?
   d. What is the probability that between 2 and 4, inclusive, involve a single vehicle?
   e. What is the probability that at least 2 involve a single vehicle?
   f. What is the probability that exactly 4 involve a single vehicle and the other 11 involve multiple vehicles?

48. NBC News reported on May 2, 2013, that 1 in 20 children in the United States have a food allergy of some sort. Consider selecting a random sample of 25 children and let X be the number in the sample who have a food allergy. Then $X \sim \text{Bin}(25, .05)$.

   a. Determine both $P(X \le 3)$ and $P(X < 3)$.
   b. Determine $P(X \ge 4)$.
   c. Determine $P(1 \le X \le 3)$.
   d. What are $E(X)$ and σ_X?
   e. In a sample of 50 children, what is the probability that none has a food allergy?

49. A company that produces fine crystal knows from experience that 10% of its goblets have cosmetic flaws and must be classified as "seconds."

   a. Among six randomly selected goblets, how likely is it that only one is a second?
   b. Among six randomly selected goblets, what is the probability that at least two are seconds?
   c. If goblets are examined one by one, what is the probability that at most five must be selected to find four that are not seconds?

50. A particular telephone number is used to receive both voice calls and fax messages. Suppose that 25% of the incoming calls involve fax messages, and consider a sample of 25 incoming calls. What is the probability that

a. At most 6 of the calls involve a fax message?

b. Exactly 6 of the calls involve a fax message?

c. At least 6 of the calls involve a fax message?

d. More than 6 of the calls involve a fax message?

51. Refer to the previous exercise.

a. What is the expected number of calls among the 25 that involve a fax message?

b. What is the standard deviation of the number among the 25 calls that involve a fax message?

c. What is the probability that the number of calls among the 25 that involve a fax transmission exceeds the expected number by more than 2 standard deviations?

52. Suppose that 30% of all students who have to buy a text for a particular course want a new copy (the successes!), whereas the other 70% want a used copy. Consider randomly selecting 25 purchasers.

a. What are the mean value and standard deviation of the number who want a new copy of the book?

b. What is the probability that the number who want new copies is more than two standard deviations away from the mean value?

c. The bookstore has 15 new copies and 15 used copies in stock. If 25 people come in one by one to purchase this text, what is the probability that all 25 will get the type of book they want from current stock? [*Hint:* Let X = the number who want a new copy. For what values of X will all 25 get what they want?]

d. Suppose that new copies cost $100 and used copies cost $70. Assume the bookstore currently has 50 new copies and 50 used copies. What is the expected value of total revenue from the sale of the next 25 copies purchased? Be sure to indicate what rule of expected value you are using. [*Hint:* Let $h(X)$ = the revenue when X of the 25 purchasers want new copies. Express this as a linear function.]

53. Exercise 30 (Section 3.3) gave the pmf of Y, the number of traffic citations for a randomly selected individual insured by a particular company. What is the probability that among 15 randomly chosen such individuals

a. At least 10 have no citations?

b. Fewer than half have at least one citation?

c. The number that have at least one citation is between 5 and 10, inclusive?*

54. A particular type of tennis racket comes in a midsize version and an oversize version. Sixty percent of all customers at a certain store want the oversize version.

a. Among ten randomly selected customers who want this type of racket, what is the probability that at least six want the oversize version?

b. Among ten randomly selected customers, what is the probability that the number who want the oversize

* "Between a and b, inclusive" is equivalent to ($a \le X \le b$).

version is within 1 standard deviation of the mean value?

c. The store currently has seven rackets of each version. What is the probability that all of the next ten customers who want this racket can get the version they want from current stock?

55. Twenty percent of all telephones of a certain type are submitted for service while under warranty. Of these, 60% can be repaired, whereas the other 40% must be replaced with new units. If a company purchases ten of these telephones, what is the probability that exactly two will end up being replaced under warranty?

56. The College Board reports that 2% of the 2 million high school students who take the SAT each year receive special accommodations because of documented disabilities (*Los Angeles Times*, **July 16, 2002**). Consider a random sample of 25 students who have recently taken the test.

a. What is the probability that exactly 1 received a special accommodation?

b. What is the probability that at least 1 received a special accommodation?

c. What is the probability that at least 2 received a special accommodation?

d. What is the probability that the number among the 25 who received a special accommodation is within 2 standard deviations of the number you would expect to be accommodated?

e. Suppose that a student who does not receive a special accommodation is allowed 3 hours for the exam, whereas an accommodated student is allowed 4.5 hours. What would you expect the average time allowed the 25 selected students to be?

57. A certain type of flashlight requires two type-D batteries, and the flashlight will work only if both its batteries have acceptable voltages. Suppose that 90% of all batteries from a certain supplier have acceptable voltages. Among ten randomly selected flashlights, what is the probability that at least nine will work? What assumptions did you make in the course of answering the question posed?

58. A very large batch of components has arrived at a distributor. The batch can be characterized as acceptable only if the proportion of defective components is at most .10. The distributor decides to randomly select 10 components and to accept the batch only if the number of defective components in the sample is at most 2.

a. What is the probability that the batch will be accepted when the actual proportion of defectives is .01? .05? .10? .20? .25?

b. Let p denote the actual proportion of defectives in the batch. A graph of P(batch is accepted) as a function of p, with p on the horizontal axis and P(batch is accepted) on the vertical axis, is called the *operating characteristic curve* for the acceptance sampling plan. Use the results of part (a) to sketch this curve for $0 \leq p \leq 1$.

c. Repeat parts (a) and (b) with "1" replacing "2" in the acceptance sampling plan.

d. Repeat parts (a) and (b) with "15" replacing "10" in the acceptance sampling plan.

e. Which of the three sampling plans, that of part (a), (c), or (d), appears most satisfactory, and why?

59. An ordinance requiring that a smoke detector be installed in all previously constructed houses has been in effect in a particular city for 1 year. The fire department is concerned that many houses remain without detectors. Let $p =$ the true proportion of such houses having detectors, and suppose that a random sample of 25 homes is inspected. If the sample strongly indicates that fewer than 80% of all houses have a detector, the fire department will campaign for a mandatory inspection program. Because of the costliness of the program, the department prefers not to call for such inspections unless sample evidence strongly argues for their necessity. Let X denote the number of homes with detectors among the 25 sampled. Consider rejecting the claim that $p \geq .8$ if $x \leq 15$.

a. What is the probability that the claim is rejected when the actual value of p is .8?

b. What is the probability of not rejecting the claim when $p = .7$? When $p = .6$?

c. How do the "error probabilities" of parts (a) and (b) change if the value 15 in the decision rule is replaced by 14?

60. A toll bridge charges $1.00 for passenger cars and $2.50 for other vehicles. Suppose that during daytime hours, 60% of all vehicles are passenger cars. If 25 vehicles cross the bridge during a particular daytime period, what is the resulting expected toll revenue? [*Hint:* Let $X =$ the number of passenger cars; then the toll revenue $h(X)$ is a linear function of X.]

61. A student who is trying to write a paper for a course has a choice of two topics, A and B. If topic A is chosen, the student will order two books through interlibrary loan, whereas if topic B is chosen, the student will order four books. The student believes that a good paper necessitates receiving and using at least half the books ordered for either topic chosen. If the probability that a book ordered through interlibrary loan actually arrives in time is .9 and books arrive independently of one another, which topic should the student choose to maximize the probability of writing a good paper? What if the arrival probability is only .5 instead of .9?

62. a. For fixed n, are there values of $p(0 \leq p \leq 1)$ for which $V(X) = 0$? Explain why this is so.

b. For what value of p is $V(X)$ maximized? [*Hint:* Either graph $V(X)$ as a function of p or else take a derivative.]

63. a. Show that $b(x; n, 1 - p) = b(n - x; n, p)$.

b. Show that $B(x; n, 1 - p) = 1 - B(n - x - 1; n, p)$. [*Hint:* At most x S's is equivalent to at least $(n - x)$ F's.]

c. What do parts (a) and (b) imply about the necessity of including values of p greater than .5 in Appendix Table A.1?

64. Show that $E(X) = np$ when X is a binomial random variable. [*Hint:* First express $E(X)$ as a sum with lower limit $x = 1$. Then factor out np, let $y = x - 1$ so that the sum is from $y = 0$ to $y = n - 1$, and show that the sum equals 1.]

65. Customers at a gas station pay with a credit card (A), debit card (B), or cash (C). Assume that successive customers make independent choices, with $P(A) = .5$, $P(B) = .2$, and $P(C) = .3$.

a. Among the next 100 customers, what are the mean and variance of the number who pay with a debit card? Explain your reasoning.

b. Answer part (a) for the number among the 100 who don't pay with cash.

66. An airport limousine can accommodate up to four passengers on any one trip. The company will accept a maximum of six reservations for a trip, and a passenger must have a reservation. From previous records, 20% of all those making reservations do not appear for the trip. Answer the following questions, assuming independence wherever appropriate.

a. If six reservations are made, what is the probability that at least one individual with a reservation cannot be accommodated on the trip?

b. If six reservations are made, what is the expected number of available places when the limousine departs?

c. Suppose the probability distribution of the number of reservations made is given in the accompanying table.

Number of reservations	3	4	5	6
Probability	.1	.2	.3	.4

Let X denote the number of passengers on a randomly selected trip. Obtain the probability mass function of X.

67. Refer to Chebyshev's inequality given in Exercise 44. Calculate $P(|X - \mu| \geq k\sigma)$ for $k = 2$ and $k = 3$ when $X \sim \text{Bin}(20, .5)$, and compare to the corresponding upper bound. Repeat for $X \sim \text{Bin}(20, .75)$.

3.5 Hypergeometric and Negative Binomial Distributions

The hypergeometric and negative binomial distributions are both related to the binomial distribution. The binomial distribution is the approximate probability model for sampling without replacement from a finite dichotomous (S–F) population provided the sample size n is small relative to the population size N; the hypergeometric distribution is the exact probability model for the number of S's in the sample. The binomial rv X is the number of S's when the number n of trials is fixed, whereas the negative binomial distribution arises from fixing the number of S's desired and letting the number of trials be random.

The Hypergeometric Distribution

The assumptions leading to the hypergeometric distribution are as follows:

1. The population or set to be sampled consists of N individuals, objects, or elements (a *finite* population).

2. Each individual can be characterized as a success (S) or a failure (F), and there are M successes in the population.

3. A sample of n individuals is selected without replacement in such a way that each subset of size n is equally likely to be chosen.

The random variable of interest is X = the number of S's in the sample. The probability distribution of X depends on the parameters n, M, and N, so we wish to obtain $P(X = x) = h(x; n, M, N)$.

X out of n

M out of N

EXAMPLE 3.34 During a particular period a university's information technology office received 20 service orders for problems with printers, of which 8 were laser printers and 12 were inkjet models. A sample of 5 of these service orders is to be selected for inclusion in a customer satisfaction survey. Suppose that the 5 are selected in a completely random fashion, so that any particular subset of size 5 has the same chance of being selected as does any other subset. What then is the probability that exactly x (x = 0, 1, 2, 3, 4, or 5) of the selected service orders were for inkjet printers?

Here, the population size is N = 20, the sample size is n = 5, and the number of S's (inkjet = S) and F's in the population are M = 12 and N − M = 8, respectively. Consider the value x = 2. Because all outcomes (each consisting of 5 particular orders) are equally likely,

$$P(X = 2) = h(2; 5, 12, 20) = \frac{\text{number of outcomes having } X = 2}{\text{number of possible outcomes}}$$

The number of possible outcomes in the experiment is the number of ways of selecting 5 from the 20 service orders without regard to order—that is, $\binom{20}{5}$. To count the number of outcomes having X = 2, note that there are $\binom{12}{2}$ ways of selecting 2 of the inkjet orders, and for each such way there are $\binom{8}{3}$ ways of selecting the 3 laser orders to fill out the sample. The product rule from Chapter 2 then gives $\binom{12}{2}\binom{8}{3}$ as the number of outcomes with X = 2, so

$$h(2; 5, 12, 20) = \frac{\binom{12}{2}\binom{8}{3}}{\binom{20}{5}} = \frac{77}{323} = .238$$

In general, if the sample size n is smaller than the number of successes in the population (M), then the largest possible X value is n. However, if $M < n$ (e.g., a sample size of 25 and only 15 successes in the population), then X can be at most M. Similarly, whenever the number of population failures ($N - M$) exceeds the sample size, the smallest possible X value is 0 (since all sampled individuals might then be failures). However, if $N - M < n$, the smallest possible X value is $n - (N - M)$. Thus, the possible values of X satisfy the restriction $\max(0, n - (N - M)) \leq x \leq \min(n, M)$. An argument parallel to that of the previous example gives the pmf of X.

PROPOSITION

If X is the number of S's in a completely random sample of size n drawn from a population consisting of M S's and ($N - M$) F's, then the probability distribution of X, called the **hypergeometric distribution,** is given by

$$P(X = x) = h(x; n, M, N) = \frac{\binom{M}{x}\binom{N - M}{n - x}}{\binom{N}{n}} \tag{3.15}$$

for x an integer satisfying $\max(0, n - N + M) \leq x \leq \min(n, M)$.

In Example 3.34, $n = 5$, $M = 12$, and $N = 20$, so $h(x; 5, 12, 20)$ for $x = 0, 1, 2,$ 3, 4, 5 can be obtained by substituting these numbers into Equation (3.15).

EXAMPLE 3.35

Five individuals from an animal population thought to be near extinction in a certain region have been caught, tagged, and released to mix into the population. After they have had an opportunity to mix, a random sample of 10 of these animals is selected. Let $X =$ the number of tagged animals in the second sample. Suppose there are actually 25 animals of this type in the region.

The parameter values are $n = 10$, $M = 5$ (5 tagged animals in the population), and $N = 25$, so the pmf of X is

$$h(x; 10, 5, 25) = \frac{\binom{5}{x}\binom{20}{10 - x}}{\binom{25}{10}} \qquad x = 0, 1, 2, 3, 4, 5$$

The probability that exactly two of the animals in the second sample are tagged is

$$P(X = 2) = h(2; 10, 5, 25) = \frac{\binom{5}{2}\binom{20}{8}}{\binom{25}{10}} = .385$$

The probability that at most two of the animals in the recapture sample are tagged is

$$P(X \leq 2) = P(X = 0, 1, \text{ or } 2) = \sum_{x=0}^{2} h(x; 10, 5, 25)$$

$$= .057 + .257 + .385 = .699 \qquad \blacksquare$$

Various statistical software packages will easily generate hypergeometric probabilities (tabulation is cumbersome because of the three parameters).

As in the binomial case, there are simple expressions for $E(X)$ and $V(X)$ for hypergeometric rv's.

PROPOSITION

> The mean and variance of the hypergeometric rv X having pmf $h(x; n, M, N)$ are
>
> Mean $\quad E(X) = n \cdot \dfrac{M}{N} \qquad V(X) = \left(\dfrac{N-n}{N-1}\right) \cdot n \cdot \dfrac{M}{N} \cdot \left(1 - \dfrac{M}{N}\right)$ Variance

The ratio M/N is the proportion of S's in the population. Replacing M/N by p in $E(X)$ and $V(X)$ gives

$$E(X) = np$$

$$V(X) = \left(\frac{N-n}{N-1}\right) \cdot np(1-p) \tag{3.16}$$

Expression (3.16) shows that the means of the binomial and hypergeometric rv's are equal, whereas the variances of the two rv's differ by the factor $(N - n)/(N - 1)$, often called the **finite population correction factor**. This factor is less than 1, so the hypergeometric variable has smaller variance than does the binomial rv. The correction factor can be written as $(1 - n/N)/(1 - 1/N)$, which is approximately 1 when n is small relative to N.

EXAMPLE 3.36
(Example 3.35 continued)

In the animal-tagging example, $n = 10$, $M = 5$, and $N = 25$, so $p = 5/25 = .2$ and

$$E(X) = 10(.2) = 2$$

$$V(X) = \frac{15}{24}(10)(.2)(.8) = (.625)(1.6) = 1$$

If the sampling had been carried out with replacement, $V(X) = 1.6$.

Suppose the population size N is not actually known, so the value x is observed and we wish to estimate N. It is reasonable to equate the observed sample proportion of S's, x/n, with the population proportion, M/N, giving the estimate

estimate N $\quad \hat{N} = \dfrac{M \cdot n}{x}$

If $M = 100$, $n = 40$, and $x = 16$, then $\hat{N} = 250$. ∎

Our general rule of thumb in Section 3.4 stated that if sampling was without replacement but n/N was at most .05, then the binomial distribution could be used to compute approximate probabilities involving the number of S's in the sample. A more precise statement is as follows: Let the population size, N, and number of population S's, M, get large with the ratio M/N approaching p. Then $h(x; n, M, N)$ approaches $b(x; n, p)$; so for n/N small, the two are approximately equal provided that p is not too near either 0 or 1. This is the rationale for the rule.

The Negative Binomial Distribution

The negative binomial rv and distribution are based on an experiment satisfying the following conditions:

1. The experiment consists of a sequence of independent trials.
2. Each trial can result in either a success (S) or a failure (F).

3. The probability of success is constant from trial to trial, so $P(S$ on trial $i) = p$ for $i = 1, 2, 3, \ldots$.

4. The experiment continues (trials are performed) until a total of r successes have been observed, where r is a specified positive integer.

The random variable of interest is $X =$ the number of failures that precede the rth success; X is called a **negative binomial random variable** because, in contrast to the binomial rv, the number of successes is fixed and the number of trials is random.

Possible values of X are $0, 1, 2, \ldots$. Let $nb(x; r, p)$ denote the pmf of X. Consider $nb(7; 3, p) = P(X = 7)$, the probability that exactly 7 F's occur before the 3rd S. In order for this to happen, the 10th trial must be an S and there must be exactly 2 S's among the first 9 trials. Thus

$$nb(7; 3, p) = \left\{ \binom{9}{2} \cdot p^2(1-p)^7 \right\} \cdot p = \binom{9}{2} \cdot p^3(1-p)^7$$

Generalizing this line of reasoning gives the following formula for the negative binomial pmf.

PROPOSITION

The pmf of the negative binomial rv X with parameters $r =$ number of S's and $p = P(S)$ is

$$nb(x; r, p) = \binom{x + r - 1}{r - 1} p^r(1-p)^x \quad x = 0, 1, 2, \ldots$$

EXAMPLE 3.37

A pediatrician wishes to recruit 5 couples, each of whom is expecting their first child, to participate in a new natural childbirth regimen. Let $p = P($a randomly selected couple agrees to participate$)$. If $p = .2$, what is the probability that 15 couples must be asked before 5 are found who agree to participate? That is, with $S = \{$agrees to participate$\}$, what is the probability that 10 F's occur before the fifth S? Substituting $r = 5$, $p = .2$, and $x = 10$ into $nb(x; r, p)$ gives

$$nb(10; 5, .2) = \binom{14}{4}(.2)^5(.8)^{10} = .034$$

The probability that at most 10 F's are observed (at most 15 couples are asked) is

$$P(X \leq 10) = \sum_{x=0}^{10} nb(x; 5, .2) = (.2)^5 \sum_{x=0}^{10} \binom{x+4}{4}(.8)^x = .164 \qquad \blacksquare$$

In some sources, the negative binomial rv is taken to be the number of trials $X + r$ rather than the number of failures.

In the special case $r = 1$, the pmf is

$$nb(x; 1, p) = (1-p)^x p \quad x = 0, 1, 2, \ldots \tag{3.17}$$

In Example 3.12, we derived the pmf for the number of trials necessary to obtain the first S, and the pmf there is similar to Expression (3.17). Both $X =$ number of F's and $Y =$ number of trials $(= 1 + X)$ are referred to in the literature as **geometric random variables**, and the pmf in Expression (3.17) is called the **geometric distribution**.

The expected number of trials until the first S was shown in Example 3.19 to be $1/p$, so that the expected number of F's until the first S is $(1/p) - 1 = (1 - p)/p$. Intuitively, we would expect to see $r \cdot (1 - p)/p$ F's before the rth S, and this is indeed $E(X)$. There is also a simple formula for $V(X)$.

PROPOSITION

If X is a negative binomial rv with pmf $nb(x; r, p)$, then

$$E(X) = \frac{r(1 - p)}{p} \qquad V(X) = \frac{r(1 - p)}{p^2}$$

Finally, by expanding the binomial coefficient in front of $p^r(1 - p)^x$ and doing some cancellation, it can be seen that $nb(x; r, p)$ is well defined even when r is not an integer. This *generalized negative binomial distribution* has been found to fit observed data quite well in a wide variety of applications.

EXERCISES Section 3.5 (68–78)

68. Eighteen individuals are scheduled to take a driving test at a particular DMV office on a certain day, eight of whom will be taking the test for the first time. Suppose that six of these individuals are randomly assigned to a particular examiner, and let X be the number among the six who are taking the test for the first time.
  a. What kind of a distribution does X have (name and values of all parameters)?
  b. Compute $P(X = 2)$, $P(X \le 2)$, and $P(X \ge 2)$.
  c. Calculate the mean value and standard deviation of X.

69. Each of 12 refrigerators of a certain type has been returned to a distributor because of an audible, high-pitched, oscillating noise when the refrigerators are running. Suppose that 7 of these refrigerators have a defective compressor and the other 5 have less serious problems. If the refrigerators are examined in random order, let X be the number among the first 6 examined that have a defective compressor.
  a. Calculate $P(X = 4)$ and $P(X \le 4)$
  b. Determine the probability that X exceeds its mean value by more than 1 standard deviation.
  c. Consider a large shipment of 400 refrigerators, of which 40 have defective compressors. If X is the number among 15 randomly selected refrigerators that have defective compressors, describe a less tedious way to calculate (at least approximately) $P(X \le 5)$ than to use the hypergeometric pmf.

70. An instructor who taught two sections of engineering statistics last term, the first with 20 students and the second with 30, decided to assign a term project. After all projects had been turned in, the instructor randomly ordered them before grading. Consider the first 15 graded projects.
  a. What is the probability that exactly 10 of these are from the second section?
  b. What is the probability that at least 10 of these are from the second section?
  c. What is the probability that at least 10 of these are from the same section?
  d. What are the mean value and standard deviation of the number among these 15 that are from the second section?
  e. What are the mean value and standard deviation of the number of projects not among these first 15 that are from the second section?

71. A geologist has collected 10 specimens of basaltic rock and 10 specimens of granite. The geologist instructs a laboratory assistant to randomly select 15 of the specimens for analysis.
  a. What is the pmf of the number of granite specimens selected for analysis?
  b. What is the probability that all specimens of one of the two types of rock are selected for analysis?
  c. What is the probability that the number of granite specimens selected for analysis is within 1 standard deviation of its mean value?

72. A personnel director interviewing 11 senior engineers for four job openings has scheduled six interviews for the first day and five for the second day of interviewing. Assume that the candidates are interviewed in random order.

a. What is the probability that x of the top four candidates are interviewed on the first day?

b. How many of the top four candidates can be expected to be interviewed on the first day?

73. Twenty pairs of individuals playing in a bridge tournament have been seeded $1, \ldots, 20$. In the first part of the tournament, the 20 are randomly divided into 10 east–west pairs and 10 north–south pairs.

a. What is the probability that x of the top 10 pairs end up playing east–west?

b. What is the probability that all of the top five pairs end up playing the same direction?

c. If there are $2n$ pairs, what is the pmf of X = the number among the top n pairs who end up playing east–west? What are $E(X)$ and $V(X)$?

74. A second-stage smog alert has been called in a certain area of Los Angeles County in which there are 50 industrial firms. An inspector will visit 10 randomly selected firms to check for violations of regulations.

a. If 15 of the firms are actually violating at least one regulation, what is the pmf of the number of firms visited by the inspector that are in violation of at least one regulation?

b. If there are 500 firms in the area, of which 150 are in violation, approximate the pmf of part (a) by a simpler pmf.

c. For X = the number among the 10 visited that are in violation, compute $E(X)$ and $V(X)$ both for the exact pmf and the approximating pmf in part (b).

75. The probability that a randomly selected box of a certain type of cereal has a particular prize is .2. Suppose you purchase box after box until you have obtained two of these prizes.

a. What is the probability that you purchase x boxes that do not have the desired prize?

b. What is the probability that you purchase four boxes?

c. What is the probability that you purchase at most four boxes?

d. How many boxes without the desired prize do you expect to purchase? How many boxes do you expect to purchase?

76. A family decides to have children until it has three children of the same gender. Assuming $P(B) = P(G) = .5$, what is the pmf of X = the number of children in the family?

77. Three brothers and their wives decide to have children until each family has two female children. What is the pmf of X = the total number of male children born to the brothers? What is $E(X)$, and how does it compare to the expected number of male children born to each brother?

78. According to the article **"Characterizing the Severity and Risk of Drought in the Poudre River, Colorado"** (*J. of Water Res. Planning and Mgmnt.*, **2005: 383–393),** the drought length Y is the number of consecutive time intervals in which the water supply remains below a critical value y_0 (a deficit), preceded by and followed by periods in which the supply exceeds this critical value (a surplus). The cited paper proposes a geometric distribution with $p = .409$ for this random variable.

a. What is the probability that a drought lasts exactly 3 intervals? At most 3 intervals?

b. What is the probability that the length of a drought exceeds its mean value by at least one standard deviation?

3.6 The Poisson Probability Distribution

The binomial, hypergeometric, and negative binomial distributions were all derived by starting with an experiment consisting of trials or draws and applying the laws of probability to various outcomes of the experiment. There is no simple experiment on which the Poisson distribution is based, though we will shortly describe how it can be obtained by certain limiting operations.

DEFINITION

A discrete random variable X is said to have a **Poisson distribution** with parameter μ ($\mu > 0$) if the pmf of X is

$$p(x; \mu) = \frac{e^{-\mu} \cdot \mu^x}{x!} \qquad x = 0, 1, 2, 3, \ldots$$

It is no accident that we are using the symbol μ for the Poisson parameter; we shall see shortly that μ is in fact the expected value of X. The letter e in the pmf represents the base of the natural logarithm system; its numerical value is approximately 2.71828. In contrast to the binomial and hypergeometric distributions, the Poisson distribution spreads probability over *all* non-negative integers, an infinite number of possibilities.

It is not obvious by inspection that $p(x; \mu)$ specifies a legitimate pmf, let alone that this distribution is useful. First of all, $p(x; \mu) > 0$ for every possible x value because of the requirement that $\mu > 0$. The fact that $\Sigma p(x; \mu) = 1$ is a consequence of the Maclaurin series expansion of e^{μ} (check your calculus book for this result):

$$e^{\mu} = 1 + \mu + \frac{\mu^2}{2!} + \frac{\mu^3}{3!} + \cdots = \sum_{x=0}^{\infty} \frac{\mu^x}{x!} \tag{3.18}$$

If the two extreme terms in (3.18) are multiplied by $e^{-\mu}$ and then this quantity is moved inside the summation on the far right, the result is

$$1 = \sum_{x=0}^{\infty} \frac{e^{-\mu} \cdot \mu^x}{x!}$$

Appendix Table A.2 contains the Poisson cdf $F(x; \mu)$ for $\mu = .1, .2, \ldots, 1, 2, \ldots, 10$, 15, and 20. Alternatively, many software packages will provide $F(x; \mu)$ and $p(x; \mu)$ upon request.

EXAMPLE 3.38 Let X denote the number of traps (defects of a certain kind) in a particular type of metal oxide semiconductor transistor, and suppose it has a Poisson distribution with $\mu = 2$ (the Poisson model is suggested in the article **"Analysis of Random Telegraph Noise in 45-nm CMOS Using On-Chip Characterization System "**(*IEEE Trans. on Electron Devices*, **2013: 1716–1722);** we changed the value of the parameter for computational ease).

The probability that there are exactly three traps is

$$P(X = 3) = p(3;2) = \frac{e^{-2}2^3}{3!} = .180,$$

and the probability that there are at most three traps is

$$P(X \leq 3) = F(3; 2) = \sum_{x=0}^{3} \frac{e^{-2}2^x}{x!} = .135 + .271 + .271 + .180 = .857$$

This latter cumulative probability is found at the intersection of the $\mu = 2$ column and the $x = 3$ row of Appendix Table A.2, whereas $p(3;2) = F(3;2) - F(2;2) = .857 - .677 = .180$, the difference between two consecutive entries in the $\mu = 2$ column of the cumulative Poisson table. ∎

The Poisson Distribution as a Limit

The rationale for using the Poisson distribution in many situations is provided by the following proposition.

PROPOSITION

Suppose that in the binomial pmf $b(x; n, p)$, we let $n \to \infty$ and $p \to 0$ in such a way that np approaches a value $\mu > 0$. Then $b(x; n, p) \to p(x; \mu)$.

According to this result, *in any binomial experiment in which n is large and p is small, $b(x; n, p) \approx p(x; \mu)$, where $\mu = np$.* As a rule of thumb, this approximation can safely be applied if $n > 50$ and $np < 5$.

EXAMPLE 3.39   If a publisher of nontechnical books takes great pains to ensure that its books are free of typographical errors, so that the probability of any given page containing at least one such error is .005 and errors are independent from page to page, what is the probability that one of its 600-page novels will contain exactly one page with errors? At most three pages with errors?

With S denoting a page containing at least one error and F an error-free page, the number X of pages containing at least one error is a binomial rv with $n = 600$ and $p = .005$, so $np = 3$. We wish

$$P(X = 1) = b(1; 600, .005) \approx p(1; 3) = \frac{e^{-3}(3)^1}{1!} = .14936$$

The binomial value is $b(1; 600, .005) = .14899$, so the approximation is very good. Similarly,

$$P(X \le 3) \approx \sum_{x=0}^{3} p(x; 3) = F(3;3) = .647$$

which to three-decimal-place accuracy is identical to $B(3; 600, .005)$.   ■

Table 3.2 shows the Poisson distribution for $\mu = 3$ along with three binomial distributions with $np = 3$, and Figure 3.8 plots the Poisson along with the first two binomial distributions. The approximation is of limited use for $n = 30$, but of course the accuracy is better for $n = 100$ and much better for $n = 300$.

Table 3.2 Comparing the Poisson and Three Binomial Distributions

x	$n = 30, p = .1$	$n = 100, p = .03$	$n = 300, p = .01$	Poisson, $\mu = 3$
0	0.042391	0.047553	0.049041	0.049787
1	0.141304	0.147070	0.148609	0.149361
2	0.227656	0.225153	0.224414	0.224042
3	0.236088	0.227474	0.225170	0.224042
4	0.177066	0.170606	0.168877	0.168031
5	0.102305	0.101308	0.100985	0.100819
6	0.047363	0.049610	0.050153	0.050409
7	0.018043	0.020604	0.021277	0.021604
8	0.005764	0.007408	0.007871	0.008102
9	0.001565	0.002342	0.002580	0.002701
10	0.000365	0.000659	0.000758	0.000810

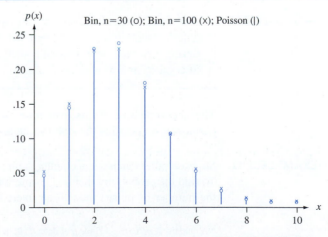

Figure 3.8 Comparing a Poisson and two binomial distributions

The Mean and Variance of X

Since $b(x; n, p) \rightarrow p(x; \mu)$ as $n \rightarrow \infty$, $p \rightarrow 0$, $np \rightarrow \mu$, the mean and variance of a binomial variable should approach those of a Poisson variable. These limits are $np \rightarrow \mu$ and $np(1 - p) \rightarrow \mu$.

PROPOSITION

> If X has a Poisson distribution with parameter μ, then $E(X) = V(X) = \mu$.

These results can also be derived directly from the definitions of mean and variance.

EXAMPLE 3.40
(Example 3.38 continued)

Both the expected number of traps and the variance of the number of traps equal 2, and $\sigma_X = \sqrt{\mu} = \sqrt{2} = 1.414$. ■

The Poisson Process

A very important application of the Poisson distribution arises in connection with the occurrence of events of some type over time. Events of interest might be visits to a particular Web site, pulses of some sort recorded by a counter, email messages sent to a particular address, accidents in an industrial facility, or cosmic ray showers observed by astronomers at a particular observatory. We make the following assumptions about the way in which the events of interest occur:

1. There exists a parameter $\alpha > 0$ such that for any short time interval of length Δt, the probability that exactly one event occurs is $\alpha \cdot \Delta t + o(\Delta t)$.*

2. The probability of more than one event occurring during Δt is $o(\Delta t)$ [which, along with Assumption 1, implies that the probability of no events during Δt is $1 - \alpha \cdot \Delta t - o(\Delta t)$].

3. The number of events occurring during the time interval Δt is independent of the number that occur prior to this time interval.

Informally, Assumption 1 says that for a short interval of time, the probability of a single event occurring is approximately proportional to the length of the time interval, where α is the constant of proportionality. Now let $Pk(t)$ denote the probability that k events will be observed during any particular time interval of length t.

PROPOSITION

> $P_k(t) = e^{-at} \cdot (\alpha t)^k / k!$, so that the number of events during a time interval of length t is a Poisson rv with parameter $\mu = \alpha t$. The expected number of events during any such time interval is then αt, so the expected number during a unit interval of time is α.

The occurrence of events over time as described is called a *Poisson process;* the parameter α specifies the *rate* for the process.

EXAMPLE 3.41

Suppose pulses arrive at a counter at an average rate of six per minute, so that $\alpha = 6$. To find the probability that in a .5-min interval at least one pulse is received, note that the number of pulses in such an interval has a Poisson distribution with parameter

---

* A quantity is $o(\Delta t)$ (read "little o of delta t") if, as Δt approaches 0, so does $o(\Delta t)/\Delta t$. That is, $o(\Delta t)$ is even more negligible (approaches 0 faster) than Δt itself. The quantity $(\Delta t)^2$ has this property, but $\sin(\Delta t)$ does not.

$\alpha t = 6(.5) = 3$ (.5 min is used because α is expressed as a rate per minute). Then with $X =$ the number of pulses received in the 30-sec interval,

$$P(1 \le X) = 1 - P(X = 0) = 1 - \frac{e^{-3}(3)^0}{0!} = .950 \qquad \blacksquare$$

Instead of observing events over time, consider observing events of some type that occur in a two- or three-dimensional region. For example, we might select on a map a certain region R of a forest, go to that region, and count the number of trees. Each tree would represent an event occurring at a particular point in space. Under assumptions similar to 1–3, it can be shown that the number of events occurring in a region R has a Poisson distribution with parameter $\alpha \cdot a(R)$, where $a(R)$ is the area of R. The quantity α is the expected number of events per unit area or volume.

EXERCISES Section 3.6 (79–93)

79. The article **"Expectation Analysis of the Probability of Failure for Water Supply Pipes"** (*J. of Pipeline Systems Engr. and Practice*, **May 2012: 36–46**) proposed using the Poisson distribution to model the number of failures in pipelines of various types. Suppose that for cast-iron pipe of a particular length, the expected number of failures is 1 (very close to one of the cases considered in the article). Then X, the number of failures, has a Poisson distribution with $\mu = 1$.
    a. Obtain $P(X \le 5)$ by using Appendix Table A.2.
    b. Determine $P(X = 2)$ first from the pmf formula and then from Appendix Table A.2.
    c. Determine $P(2 \le X \le 4)$.
    d. What is the probability that X exceeds its mean value by more than one standard deviation?

80. Let X be the number of material anomalies occurring in a particular region of an aircraft gas-turbine disk. The article **"Methodology for Probabilistic Life Prediction of Multiple-Anomaly Materials"** (*Amer. Inst. of Aeronautics and Astronautics J.*, **2006: 787–793**) proposes a Poisson distribution for X. Suppose that $\mu = 4$.
    a. Compute both $P(X \le 4)$ and $P(X < 4)$.
    b. Compute $P(4 \le X \le 8)$.
    c. Compute $P(8 \le X)$.
    d. What is the probability that the number of anomalies exceeds its mean value by no more than one standard deviation?

81. Suppose that the number of drivers who travel between a particular origin and destination during a designated time period has a Poisson distribution with parameter $\mu = 20$ (suggested in the article **"Dynamic Ride Sharing: Theory and Practice,"** *J. of Transp. Engr.*, **1997: 308–312**). What is the probability that the number of drivers will
    a. Be at most 10?
    b. Exceed 20?

    c. Be between 10 and 20, inclusive? Be strictly between 10 and 20?
    d. Be within 2 standard deviations of the mean value?

82. Consider writing onto a computer disk and then sending it through a certifier that counts the number of missing pulses. Suppose this number X has a Poisson distribution with parameter $\mu = .2$. (Suggested in **"Average Sample Number for Semi-Curtailed Sampling Using the Poisson Distribution,"** *J. Quality Technology*, **1983: 126–129.**)
    a. What is the probability that a disk has exactly one missing pulse?
    b. What is the probability that a disk has at least two missing pulses?
    c. If two disks are independently selected, what is the probability that neither contains a missing pulse?

83. An article in the *Los Angeles Times* (**Dec. 3, 1993**) reports that 1 in 200 people carry the defective gene that causes inherited colon cancer. In a sample of 1000 individuals, what is the approximate distribution of the number who carry this gene? Use this distribution to calculate the approximate probability that
    a. Between 5 and 8 (inclusive) carry the gene.
    b. At least 8 carry the gene.

84. The **Centers for Disease Control and Prevention** reported in 2012 that 1 in 88 American children had been diagnosed with an autism spectrum disorder (ASD).
    a. If a random sample of 200 American children is selected, what are the expected value and standard deviation of the number who have been diagnosed with ASD?
    b. Referring back to (a), calculate the approximate probability that at least 2 children in the sample have been diagnosed with ASD?

c. If the sample size is 352, what is the approximate probability that fewer than 5 of the selected children have been diagnosed with ASD?

85. Suppose small aircraft arrive at a certain airport according to a Poisson process with rate $\alpha = 8$ per hour, so that the number of arrivals during a time period of t hours is a Poisson rv with parameter $\mu = 8t$.

   a. What is the probability that exactly 6 small aircraft arrive during a 1-hour period? At least 6? At least 10?

   b. What are the expected value and standard deviation of the number of small aircraft that arrive during a 90-min period?

   c. What is the probability that at least 20 small aircraft arrive during a 2.5-hour period? That at most 10 arrive during this period?

86. Organisms are present in ballast water discharged from a ship according to a Poisson process with a concentration of 10 organisms/m³ [the article **"Counting at Low Concentrations: The Statistical Challenges of Verifying Ballast Water Discharge Standards"** (*Ecological Applications*, 2013: 339–351) considers using the Poisson process for this purpose].

   a. What is the probability that one cubic meter of discharge contains at least 8 organisms?

   b. What is the probability that the number of organisms in 1.5 m³ of discharge exceeds its mean value by more than one standard deviation?

   c. For what amount of discharge would the probability of containing at least 1 organism be .999?

87. The number of requests for assistance received by a towing service is a Poisson process with rate $\alpha = 4$ per hour.

   a. Compute the probability that exactly ten requests are received during a particular 2-hour period.

   b. If the operators of the towing service take a 30-min break for lunch, what is the probability that they do not miss any calls for assistance?

   c. How many calls would you expect during their break?

88. In proof testing of circuit boards, the probability that any particular diode will fail is .01. Suppose a circuit board contains 200 diodes.

   a. How many diodes would you expect to fail, and what is the standard deviation of the number that are expected to fail?

   b. What is the (approximate) probability that at least four diodes will fail on a randomly selected board?

   c. If five boards are shipped to a particular customer, how likely is it that at least four of them will work properly? (A board works properly only if all its diodes work.)

89. The article **"Reliability-Based Service-Life Assessment of Aging Concrete Structures"** (*J. Structural Engr.*, 1993: 1600–1621) suggests that a Poisson process can be used to represent the occurrence of structural loads over time. Suppose the mean time between occurrences of loads is .5 year.

   a. How many loads can be expected to occur during a 2-year period?

   b. What is the probability that more than five loads occur during a 2-year period?

   c. How long must a time period be so that the probability of no loads occurring during that period is at most .1?

90. Let X have a Poisson distribution with parameter μ. Show that $E(X) = \mu$ directly from the definition of expected value. [*Hint:* The first term in the sum equals 0, and then x can be canceled. Now factor out μ and show that what is left sums to 1.]

91. Suppose that trees are distributed in a forest according to a two-dimensional Poisson process with parameter α, the expected number of trees per acre, equal to 80.

   a. What is the probability that in a certain quarter-acre plot, there will be at most 16 trees?

   b. If the forest covers 85,000 acres, what is the expected number of trees in the forest?

   c. Suppose you select a point in the forest and construct a circle of radius .1 mile. Let $X =$ the number of trees within that circular region. What is the pmf of X? [*Hint:* 1 sq mile = 640 acres.]

92. Automobiles arrive at a vehicle equipment inspection station according to a Poisson process with rate $\alpha = 10$ per hour. Suppose that with probability .5 an arriving vehicle will have no equipment violations.

   a. What is the probability that exactly ten arrive during the hour and all ten have no violations?

   b. For any fixed $y \geq 10$, what is the probability that y arrive during the hour, of which ten have no violations?

   c. What is the probability that ten "no-violation" cars arrive during the next hour? [*Hint:* Sum the probabilities in part (b) from $y = 10$ to ∞.]

93. **a.** In a Poisson process, what has to happen in both the time interval $(0, t)$ and the interval $(t, t + \Delta t)$ so that no events occur in the entire interval $(0, t + \Delta t)$? Use this and Assumptions 1–3 to write a relationship between $P_0(t + \Delta t)$ and $P_0(t)$.

   b. Use the result of part (a) to write an expression for the difference $P_0(t + \Delta t) - P_0(t)$. Then divide by Δt and let $\Delta t \to 0$ to obtain an equation involving $(d/dt)P_0(t)$, the derivative of $P_0(t)$ with respect to t.

   c. Verify that $P_0(t) = e^{-\alpha t}$ satisfies the equation of part (b).

   d. It can be shown in a manner similar to parts (a) and (b) that the $P_k(t)$s must satisfy the system of differential equations

$$\frac{d}{dt} P_k(t) = \alpha P_{k-1}(t) - \alpha P_k(t)$$
$$k = 1, 2, 3, \ldots$$

Verify that $P_k(t) = e^{-\alpha t}(\alpha t)^k/k!$ satisfies the system. (This is actually the only solution.)

SUPPLEMENTARY EXERCISES (94–122)

94. Consider a deck consisting of seven cards, marked 1, 2,…, 7. Three of these cards are selected at random. Define an rv W by $W = $ the sum of the resulting numbers, and compute the pmf of W. Then compute μ and σ^2. [*Hint:* Consider outcomes as unordered, so that (1, 3, 7) and (3, 1, 7) are not different outcomes. Then there are 35 outcomes, and they can be listed. (This type of rv actually arises in connection with a statistical procedure called Wilcoxon's rank-sum test, in which there is an x sample and a y sample and W is the sum of the ranks of the x's in the combined sample; see Section 15.2.)

95. After shuffling a deck of 52 cards, a dealer deals out 5. Let $X = $ the number of suits represented in the five-card hand.

   a. Show that the pmf of X is

x	1	2	3	4
$p(x)$	.002	.146	.588	.264

   [*Hint:* $p(1) = 4P$(all are spades), $p(2) = 6P$(only spades and hearts with at least one of each suit), and $p(4) = 4P$(2 spades $\cap$ one of each other suit).]

   b. Compute μ, σ^2, and σ.

96. The negative binomial rv X was defined as the number of F's preceding the rth S. Let $Y = $ the number of trials necessary to obtain the rth S. In the same manner in which the pmf of X was derived, derive the pmf of Y.

97. Of all customers purchasing automatic garage-door openers, 75% purchase a chain-driven model. Let $X = $ the number among the next 15 purchasers who select the chain-driven model.

   a. What is the pmf of X?
   b. Compute $P(X > 10)$.
   c. Compute $P(6 \le X \le 10)$.
   d. Compute μ and σ^2.
   e. If the store currently has in stock 10 chain-driven models and 8 shaft-driven models, what is the probability that the requests of these 15 customers can all be met from existing stock?

98. In some applications the distribution of a discrete rv X resembles the Poisson distribution except that zero is not a possible value of X. For example, let $X = $ the number of tattoos that an individual wants removed when she or he arrives at a tattoo-removal facility. Suppose the pmf of X is

$$p(x) = k\frac{e^{-\theta}\theta^x}{x} \quad x = 1, 2, 3,\ldots$$

   a. Determine the value of k. *Hint:* The sum of all probabilities in the Poisson pmf is 1, and this pmf must also sum to 1.

   b. If the mean value of X is 2.313035, what is the probability that an individual wants at most 5 tattoos removed?

   c. Determine the standard deviation of X when the mean value is as given in (b).

   [*Note:* The article **"An Exploratory Investigation of Identity Negotiation and Tattoo Removal"** (*Academy of Marketing Science Review*, **vol. 12, no. 6, 2008**) gave a sample of 22 observations on the number of tattoos people wanted removed; *estimates* of μ and σ calculated from the data were 2.318182 and 1.249242, respectively.]

99. A *k-out-of-n system* is one that will function if and only if at least k of the n individual components in the system function. If individual components function independently of one another, each with probability .9, what is the probability that a 3-out-of-5 system functions?

100. A manufacturer of integrated circuit chips wishes to control the quality of its product by rejecting any batch in which the proportion of defective chips is too high. To this end, out of each batch (10,000 chips), 25 will be selected and tested. If at least 5 of these 25 are defective, the entire batch will be rejected.

   a. What is the probability that a batch will be rejected if 5% of the chips in the batch are in fact defective?
   b. Answer the question posed in (a) if the percentage of defective chips in the batch is 10%.
   c. Answer the question posed in (a) if the percentage of defective chips in the batch is 20%.
   d. What happens to the probabilities in (a)–(c) if the critical rejection number is increased from 5 to 6?

101. Of the people passing through an airport metal detector, .5% activate it; let $X = $ the number among a randomly selected group of 500 who activate the detector.

   a. What is the (approximate) pmf of X?
   b. Compute $P(X = 5)$.
   c. Compute $P(5 \le X)$.

102. An educational consulting firm is trying to decide whether high school students who have never before used a hand-held calculator can solve a certain type of problem more easily with a calculator that uses reverse Polish logic or one that does not use this logic. A sample of 25 students is selected and allowed to practice on both calculators. Then each student is asked to work one problem on the reverse Polish calculator and a similar problem on the other. Let $p = P(S)$, where S indicates that a student worked the problem more quickly using reverse Polish logic than without, and let $X = $ number of S's.

   a. If $p = .5$, what is $P(7 \le X \le 18)$?
   b. If $p = .8$, what is $P(7 \le X \le 18)$?

c. If the claim that $p = .5$ is to be rejected when either $x \leq 7$ or $x \geq 18$, what is the probability of rejecting the claim when it is actually correct?

d. If the decision to reject the claim $p = .5$ is made as in part (c), what is the probability that the claim is not rejected when $p = .6$? When $p = .8$?

e. What decision rule would you choose for rejecting the claim $p = .5$ if you wanted the probability in part (c) to be at most .01?

103. Consider a disease whose presence can be identified by carrying out a blood test. Let p denote the probability that a randomly selected individual has the disease. Suppose n individuals are independently selected for testing. One way to proceed is to carry out a separate test on each of the n blood samples. A potentially more economical approach, group testing, was introduced during World War II to identify syphilitic men among army inductees. First, take a part of each blood sample, combine these specimens, and carry out a single test. If no one has the disease, the result will be negative, and only the one test is required. If at least one individual is diseased, the test on the combined sample will yield a positive result, in which case the n individual tests are then carried out. If $p = .1$ and $n = 3$, what is the expected number of tests using this procedure? What is the expected number when $n = 5$? [The article **"Random Multiple-Access Communication and Group Testing" (*IEEE Trans. on Commun.*, 1984: 769–774)** applied these ideas to a communication system in which the dichotomy was active/idle user rather than diseased/nondiseased.]

104. Let p_1 denote the probability that any particular code symbol is erroneously transmitted through a communication system. Assume that on different symbols, errors occur independently of one another. Suppose also that with probability p_2 an erroneous symbol is corrected upon receipt. Let X denote the number of correct symbols in a message block consisting of n symbols (after the correction process has ended). What is the probability distribution of X?

105. The purchaser of a power-generating unit requires c consecutive successful start-ups before the unit will be accepted. Assume that the outcomes of individual start-ups are independent of one another. Let p denote the probability that any particular start-up is successful. The random variable of interest is $X =$ the number of start-ups that must be made prior to acceptance. Give the pmf of X for the case $c = 2$. If $p = .9$, what is $P(X \leq 8)$? [*Hint:* For $x \geq 5$, express $p(x)$ "recursively" in terms of the pmf evaluated at the smaller values $x - 3, x - 4, \ldots, 2$.] (This problem was suggested by the article **"Evaluation of a Start-Up Demonstration Test," *J. Quality Technology*, 1983: 103–106.**)

106. A plan for an executive travelers' club has been developed by an airline on the premise that 10% of its current customers would qualify for membership.

a. Assuming the validity of this premise, among 25 randomly selected current customers, what is the probability that between 2 and 6 (inclusive) qualify for membership?

b. Again assuming the validity of the premise, what are the expected number of customers who qualify and the standard deviation of the number who qualify in a random sample of 100 current customers?

c. Let X denote the number in a random sample of 25 current customers who qualify for membership. Consider rejecting the company's premise in favor of the claim that $p > .10$ if $x \geq 7$. What is the probability that the company's premise is rejected when it is actually valid?

d. Refer to the decision rule introduced in part (c). What is the probability that the company's premise is not rejected even though $p = .20$ (i.e., 20% qualify)?

107. Forty percent of seeds from maize (modern-day corn) ears carry single spikelets, and the other 60% carry paired spikelets. A seed with single spikelets will produce an ear with single spikelets 29% of the time, whereas a seed with paired spikelets will produce an ear with single spikelets 26% of the time. Consider randomly selecting ten seeds.

a. What is the probability that exactly five of these seeds carry a single spikelet and produce an ear with a single spikelet?

b. What is the probability that exactly five of the ears produced by these seeds have single spikelets? What is the probability that at most five ears have single spikelets?

108. A trial has just resulted in a hung jury because eight members of the jury were in favor of a guilty verdict and the other four were for acquittal. If the jurors leave the jury room in random order and each of the first four leaving the room is accosted by a reporter in quest of an interview, what is the pmf of $X =$ the number of jurors favoring acquittal among those interviewed? How many of those favoring acquittal do you expect to be interviewed?

109. A reservation service employs five information operators who receive requests for information independently of one another, each according to a Poisson process with rate $\alpha = 2$ per minute.

a. What is the probability that during a given 1-min period, the first operator receives no requests?

b. What is the probability that during a given 1-min period, exactly four of the five operators receive no requests?

c. Write an expression for the probability that during a given 1-min period, all of the operators receive exactly the same number of requests.

110. Grasshoppers are distributed at random in a large field according to a Poisson process with parameter $\alpha = 2$ per square yard. How large should the radius R of a circular sampling region be taken so that the probability of finding at least one in the region equals .99?

111. A newsstand has ordered five copies of a certain issue of a photography magazine. Let X = the number of individuals who come in to purchase this magazine. If X has a Poisson distribution with parameter $\mu = 4$, what is the expected number of copies that are sold?

112. Individuals A and B begin to play a sequence of chess games. Let S = {A wins a game}, and suppose that outcomes of successive games are independent with $P(S) = p$ and $P(F) = 1 - p$ (they never draw). They will play until one of them wins ten games. Let X = the number of games played (with possible values 10, 11,..., 19).

  a. For $x = 10, 11, \ldots, 19$, obtain an expression for $p(x) = P(X = x)$.

  b. If a draw is possible, with $p = P(S)$, $q = P(F)$, $1 - p - q = P(\text{draw})$, what are the possible values of X? What is $P(20 \leq X)$? [*Hint:* $P(20 \leq X) = 1 - P(X < 20)$.]

113. A test for the presence of a certain disease has probability .20 of giving a false-positive reading (indicating that an individual has the disease when this is not the case) and probability .10 of giving a false-negative result. Suppose that ten individuals are tested, five of whom have the disease and five of whom do not. Let X = the number of positive readings that result.

  a. Does X have a binomial distribution? Explain your reasoning.

  b. What is the probability that exactly three of the ten test results are positive?

114. The generalized negative binomial pmf is given by

$$nb(x; r, p) = k(r, x) \cdot p^r (1 - p)^x$$

$$x = 0, 1, 2, \ldots$$

Let X, the number of plants of a certain species found in a particular region, have this distribution with $p = .3$ and $r = 2.5$. What is $P(X = 4)$? What is the probability that at least one plant is found?

115. There are two Certified Public Accountants in a particular office who prepare tax returns for clients. Suppose that for a particular type of complex form, the number of errors made by the first preparer has a Poisson distribution with mean value μ_1, the number of errors made by the second preparer has a Poisson distribution with mean value μ_2, and that each CPA prepares the same number of forms of this type. Then if a form of this type is randomly selected, the function

$$p(x; \mu_1, \mu_2) = .5 \frac{e^{-\mu_1}\mu_1^x}{x!} + .5 \frac{e^{-\mu_2}\mu_2^x}{x!} \quad x = 0, 1, 2, \ldots$$

gives the pmf of X = the number of errors on the selected form.

  a. Verify that $p(x; \mu_1, \mu_2)$ is in fact a legitimate pmf (≥ 0 and sums to 1).

  b. What is the expected number of errors on the selected form?

  c. What is the variance of the number of errors on the selected form?

  d. How does the pmf change if the first CPA prepares 60% of all such forms and the second prepares 40%?

116. The *mode* of a discrete random variable X with pmf $p(x)$ is that value x^* for which $p(x)$ is largest (the most probable x value).

  a. Let $X \sim \text{Bin}(n, p)$. By considering the ratio $b(x + 1; n, p)/b(x; n, p)$, show that $b(x; n, p)$ increases with x as long as $x < np - (1 - p)$. Conclude that the mode x^* is the integer satisfying $(n + 1)p - 1 \leq x^* \leq (n + 1)p$.

  b. Show that if X has a Poisson distribution with parameter μ, the mode is the largest integer less than μ. If μ is an integer, show that both $\mu - 1$ and μ are modes.

117. A computer disk storage device has ten concentric tracks, numbered 1, 2,..., 10 from outermost to innermost, and a single access arm. Let p_i = the probability that any particular request for data will take the arm to track $i (i = 1, \ldots, 10)$. Assume that the tracks accessed in successive seeks are independent. Let X = the number of tracks over which the access arm passes during two successive requests (excluding the track that the arm has just left, so possible X values are $x = 0, 1, \ldots, 9$). Compute the pmf of X. [*Hint:* $P(\text{the arm is now on track } i \text{ and } X = j) = P(X = j | \text{arm now on } i) \cdot p_i$. After the conditional probability is written in terms of $p_1, \ldots, p_{10}$, by the law of total probability, the desired probability is obtained by summing over i.]

118. If X is a hypergeometric rv, show directly from the definition that $E(X) = nM/N$ (consider only the case $n < M$). [*Hint:* Factor nM/N out of the sum for $E(X)$, and show that the terms inside the sum are of the form $h(y; n - 1, M - 1, N - 1)$, where $y = x - 1$.]

119. Use the fact that

$$\sum_{\text{all } x} (x - \mu)^2 p(x) \geq \sum_{x: |x - \mu| \geq k\sigma} (x - \mu)^2 p(x)$$

to prove Chebyshev's inequality given in Exercise 44.

120. The simple Poisson process of Section 3.6 is characterized by a constant rate α at which events occur per unit time. A generalization of this is to suppose that the probability of exactly one event occurring in the interval $[t, t + \Delta t]$ is $\alpha(t) \cdot \Delta t + o(\Delta t)$. It can then be shown that the number of events occurring during an interval $[t_1, t_2]$ has a Poisson distribution with parameter

$$\mu = \int_{t_2}^{t_1} \alpha(t) \, dt$$

The occurrence of events over time in this situation is called a *nonhomogeneous Poisson process*. The article **"Inference Based on Retrospective Ascertainment,"**

(*J. Amer. Stat. Assoc.*, **1989: 360–372),** considers the intensity function

$$\alpha(t) = e^{a+bt}$$

as appropriate for events involving transmission of HIV (the AIDS virus) via blood transfusions. Suppose that $a = 2$ and $b = .6$ (close to values suggested in the paper), with time in years.

a. What is the expected number of events in the interval [0, 4]? In [2, 6]?

b. What is the probability that at most 15 events occur in the interval [0, .9907]?

121. Consider a collection $A_1, \ldots, A_k$ of mutually exclusive and exhaustive events, and a random variable X whose distribution depends on which of the A_i's occurs (e.g., a commuter might select one of three possible routes from home to work, with X representing the commute time). Let $E(X|A_i)$ denote the expected value of X given that the event A_i occurs. Then it can be shown that $E(X) = \Sigma E(X|A_i) \cdot P(A_i)$, the weighted average of the individual "conditional expectations" where the weights are the probabilities of the partitioning events.

a. The expected duration of a voice call to a particular telephone number is 3 minutes, whereas the expected duration of a data call to that same number is 1 minute. If 75% of all calls are voice calls, what is the expected duration of the next call?

b. A deli sells three different types of chocolate chip cookies. The number of chocolate chips in a type i cookie has a Poisson distribution with parameter $\mu_i = i + 1$ $(i = 1, 2, 3)$. If 20% of all customers purchasing a chocolate chip cookie select the first type, 50% choose the second type, and the remaining 30% opt for the third type, what is the expected number of chips in a cookie purchased by the next customer?

122. Consider a communication source that transmits packets containing digitized speech. After each transmission, the receiver sends a message indicating whether the transmission was successful or unsuccessful. If a transmission is unsuccessful, the packet is re-sent. Suppose a voice packet can be transmitted a maximum of 10 times. Assuming that the results of successive transmissions are independent of one another and that the probability of any particular transmission being successful is p, determine the probability mass function of the rv $X =$ the number of times a packet is transmitted. Then obtain an expression for the expected number of times a packet is transmitted.

BIBLIOGRAPHY

Johnson, Norman, Samuel Kotz, and Adrienne Kemp, *Discrete Univariate Distributions,* Wiley, New York, 1992. An encyclopedia of information on discrete distributions.

Olkin, Ingram, Cyrus Derman, and Leon Gleser, *Probability Models and Applications* (2nd ed.), Macmillan, New York, 1994. Contains an in-depth discussion of both general properties of discrete and continuous distributions and results for specific distributions.

Ross, Sheldon, *Introduction to Probability Models* (10th ed.), Academic Press, New York, 2010. A good source of material on the Poisson process and generalizations, and a nice introduction to other topics in applied probability.

Continuous Random Variables and Probability Distributions

4

INTRODUCTION

Chapter 3 concentrated on the development of probability distributions for discrete random variables. In this chapter, we consider the second general type of random variable that arises in many applied problems. Sections 4.1 and 4.2 present the basic definitions and properties of continuous random variables and their probability distributions. In Section 4.3, we study in detail the normal random variable and distribution, unquestionably the most important and useful in probability and statistics. Sections 4.4 and 4.5 discuss some other continuous distributions that are often used in applied work. In Section 4.6, we introduce a method for assessing whether given sample data is consistent with a specified distribution.

4.1 Probability Density Functions

A discrete random variable (rv) is one whose possible values either constitute a finite set or else can be listed in an infinite sequence (a list in which there is a first element, a second element, etc.). A random variable whose set of possible values is an entire interval of numbers is not discrete.

Recall from Chapter 3 that a random variable X is continuous if (1) possible values comprise either a single interval on the number line (for some $A < B$, any number x between A and B is a possible value) or a union of disjoint intervals, and (2) $P(X = c) = 0$ for any number c that is a possible value of X.

EXAMPLE 4.1 If in the study of the ecology of a lake, we make depth measurements at randomly chosen locations, then $X =$ the depth at such a location is a continuous rv. Here A is the minimum depth in the region being sampled, and B is the maximum depth. ∎

EXAMPLE 4.2 If a chemical compound is randomly selected and its pH X is determined, then X is a continuous rv because any pH value between 0 and 14 is possible. If more is known about the compound selected for analysis, then the set of possible values might be a subinterval of [0, 14], such as $5.5 \le x \le 6.5$, but X would still be continuous. ∎

EXAMPLE 4.3 Let X represent the amount of time a randomly selected customer spends waiting for a haircut before his/her haircut commences. Your first thought might be that X is a continuous random variable, since a measurement is required to determine its value. However, there are customers lucky enough to have no wait whatsoever before climbing into the barber's chair. So it must be the case that $P(X = 0) > 0$. Conditional on no chairs being empty, though, the waiting time will be continuous since X could then assume any value between some minimum possible time A and a maximum possible time B. This random variable is neither purely discrete nor purely continuous but instead is a mixture of the two types. ∎

One might argue that although in principle variables such as height, weight, and temperature are continuous, in practice the limitations of our measuring instruments restrict us to a discrete (though sometimes very finely subdivided) world. However, continuous models often approximate real-world situations very well, and continuous mathematics (the calculus) is frequently easier to work with than mathematics of discrete variables and distributions.

Probability Distributions for Continuous Variables

Suppose the variable X of interest is the depth of a lake at a randomly chosen point on the surface. Let $M =$ the maximum depth (in meters), so that any number in the interval $[0, M]$ is a possible value of X. If we "discretize" X by measuring depth to the nearest meter, then possible values are nonnegative integers less than or equal to M. The resulting discrete distribution of depth can be pictured using a probability histogram. If we draw the histogram so that the area of the rectangle above any possible integer k is the proportion of the lake whose depth is (to the nearest meter) k, then the total area of all rectangles is 1. A possible histogram appears in Figure 4.1(a).

If depth is measured much more accurately and the same measurement axis as in Figure 4.1(a) is used, each rectangle in the resulting probability histogram is much narrower, though the total area of all rectangles is still 1. A possible histogram is pictured in Figure 4.1(b); it has a much smoother appearance than the histogram in Figure 4.1(a). If we continue in this way to measure depth more and more finely, the resulting sequence of histograms approaches a smooth curve, such as is pictured in Figure 4.1(c). Because for each histogram the total area of all rectangles equals 1, the total area under the smooth curve is also 1. The probability that the depth at a randomly chosen point is between a and b is just the area under the smooth curve between a and b. It is exactly a smooth curve of the type pictured in Figure 4.1(c) that specifies a continuous probability distribution.

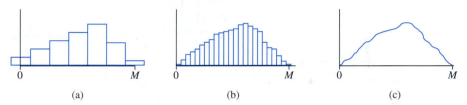

| (a) | (b) | (c) |

Figure 4.1   (a) Probability histogram of depth measured to the nearest meter; (b) probability histogram of depth measured to the nearest centimeter; (c) a limit of a sequence of discrete histograms

DEFINITION

Let X be a continuous rv. Then a **probability distribution** or **probability density function** (pdf) of X is a function $f(x)$ such that for any two numbers a and b with $a \leq b$,

$$P(a \leq X \leq b) = \int_a^b f(x)\,dx$$

That is, the probability that X takes on a value in the interval $[a, b]$ is the area above this interval and under the graph of the density function, as illustrated in Figure 4.2. The graph of $f(x)$ is often referred to as the *density curve*.

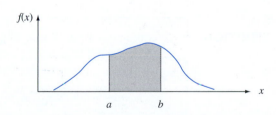

Figure 4.2 $P(a \leq X \leq b)$ = the area under the density curve between a and b

For $f(x)$ to be a legitimate pdf, it must satisfy the following two conditions:

1. $f(x) \geq 0$ for all x

2. $\displaystyle\int_{-\infty}^{\infty} f(x)\,dx$ = area under the entire graph of $f(x)$
   $= 1$

EXAMPLE 4.4   The direction of an imperfection with respect to a reference line on a circular object such as a tire, brake rotor, or flywheel is, in general, subject to uncertainty. Consider the reference line connecting the valve stem on a tire to the center point, and let X

be the angle measured clockwise to the location of an imperfection. One possible pdf for X is

$$f(x) = \begin{cases} \dfrac{1}{360} & 0 \le x < 360 \\ 0 & \text{otherwise} \end{cases}$$

The pdf is graphed in Figure 4.3. Clearly $f(x) \ge 0$. The area under the density curve is just the area of a rectangle: (height)(base) = (1/360)(360) = 1. The probability that the angle is between 90° and 180° is

$$P(90 \le X \le 180) = \int_{90}^{180} \frac{1}{360}\, dx = \frac{x}{360} \Big|_{x=90}^{x=180} = \frac{1}{4} = .25$$

The probability that the angle of occurrence is within 90° of the reference line is

$$P(0 \le X \le 90) + P(270 \le X < 360) = .25 + .25 = .50$$

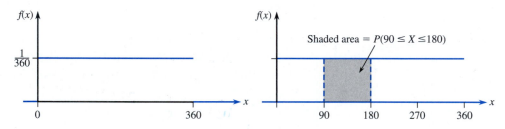

Figure 4.3 The pdf and probability from Example 4.4 ■

Because whenever $0 \le a \le b \le 360$ in Example 4.4, $P(a \le X \le b)$ depends only on the width $b - a$ of the interval, X is said to have a uniform distribution.

DEFINITION

A continuous rv X is said to have a **uniform distribution** on the interval $[A, B]$ if the pdf of X is

$$f(x; A, B) = \begin{cases} \dfrac{1}{B - A} & A \le x \le B \\ 0 & \text{otherwise} \end{cases}$$

The graph of any uniform pdf looks like the graph in Figure 4.3 except that the interval of positive density is $[A, B]$ rather than $[0, 360]$.

In the discrete case, a probability mass function (pmf) tells us how little "blobs" of probability mass of various magnitudes are distributed along the measurement axis. In the continuous case, probability density is "smeared" in a continuous fashion along the interval of possible values. When density is smeared uniformly over the interval, a uniform pdf, as in Figure 4.3, results.

When X is a discrete random variable, each possible value is assigned positive probability. This is not true of a continuous random variable (that is, the second condition of the definition is satisfied) because the area under a density curve that lies above any single value is zero:

$$P(X = c) = \int_c^c f(x)\,dx = \lim_{\varepsilon \to 0} \int_{c-\varepsilon}^{c+\varepsilon} f(x)\,dx = 0$$

The fact that $P(X = c) = 0$ when X is continuous has an important practical consequence: The probability that X lies in some interval between a and b does not depend on whether the lower limit a or the upper limit b is included in the probability calculation:

$$P(a \le X \le b) = P(a < X < b) = P(a < X \le b) = P(a \le X < b) \qquad (4.1)$$

If X is discrete and both a and b are possible values (e.g., X is binomial with $n = 20$ and $a = 5, b = 10$), then all four of the probabilities in (4.1) are different.

The zero probability condition has a physical analog. Consider a solid circular rod with cross-sectional area $= 1\ \text{in}^2$. Place the rod alongside a measurement axis and suppose that the density of the rod at any point x is given by the value $f(x)$ of a density function. Then if the rod is sliced at points a and b and this segment is removed, the amount of mass removed is $\int_a^b f(x)\,dx$; if the rod is sliced just at the point c, no mass is removed. Mass is assigned to interval segments of the rod but not to individual points.

EXAMPLE 4.5 "Time headway" in traffic flow is the elapsed time between the time that one car finishes passing a fixed point and the instant that the next car begins to pass that point. Let $X =$ the time headway for two randomly chosen consecutive cars on a freeway during a period of heavy flow. The following pdf of X is essentially the one suggested in **"The Statistical Properties of Freeway Traffic" (*Transp. Res.*, vol. 11: 221–228):**

$$f(x) = \begin{cases} .15e^{-.15(x-.5)} & x \ge .5 \\ 0 & \text{otherwise} \end{cases}$$

The graph of $f(x)$ is given in Figure 4.4; there is no density associated with headway times less than .5, and headway density decreases rapidly (exponentially fast) as x increases from .5. Clearly, $f(x) \ge 0$; to show that $\int_{-\infty}^{\infty} f(x)\,dx = 1$, we use the calculus result $\int_a^{\infty} e^{-kx}\,dx = (1/k)e^{-k \cdot a}$. Then

$$\int_{-\infty}^{\infty} f(x)\,dx = \int_{.5}^{\infty} .15e^{-.15(x-.5)}\,dx = .15e^{.075} \int_{.5}^{\infty} e^{-.15x}\,dx$$

$$= .15e^{.075} \cdot \frac{1}{.15} e^{-(.15)(.5)} = 1$$

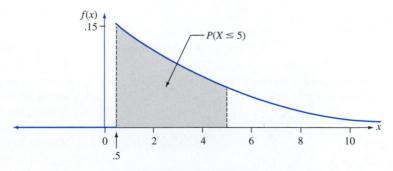

Figure 4.4   The density curve for time headway in Example 4.5

The probability that headway time is at most 5 sec is

$$P(X \le 5) = \int_{-\infty}^{5} f(x)\, dx = \int_{.5}^{5} .15e^{-.15(x-.5)}\, dx$$

$$= .15e^{.075} \int_{.5}^{5} e^{-.15x}\, dx = .15e^{.075} \cdot \left(-\frac{1}{.15} e^{-.15x} \Big|_{x=.5}^{x=5} \right)$$

$$= e^{.075}(-e^{-.75} + e^{-.075}) = 1.078(-.472 + .928) = .491$$

$$= P(\text{less than 5 sec}) = P(X < 5) \qquad \blacksquare$$

Unlike discrete distributions such as the binomial, hypergeometric, and negative binomial, the distribution of any given continuous rv cannot usually be derived using simple probabilistic arguments. Instead, one must make a judicious choice of pdf based on prior knowledge and available data. Fortunately, there are some general families of pdf's that have been found to be sensible candidates in a wide variety of experimental situations; several of these are discussed later in the chapter.

Just as in the discrete case, it is often helpful to think of the population of interest as consisting of X values rather than individuals or objects. The pdf is then a model for the distribution of values in this numerical population, and from this model various population characteristics (such as the mean) can be calculated.

EXERCISES Section 4.1 (1–10)

1. The current in a certain circuit as measured by an ammeter is a continuous random variable X with the following density function:

$$f(x) = \begin{cases} .075x + .2 & 3 \le x \le 5 \\ 0 & \text{otherwise} \end{cases}$$

   a. Graph the pdf and verify that the total area under the density curve is indeed 1.
   b. Calculate $P(X \le 4)$. How does this probability compare to $P(X < 4)$?
   c. Calculate $P(3.5 \le X \le 4.5)$ and also $P(4.5 < X)$.

2. Suppose the reaction temperature X (in °C) in a certain chemical process has a uniform distribution with $A = -5$ and $B = 5$.
   a. Compute $P(X < 0)$.
   b. Compute $P(-2.5 < X < 2.5)$.
   c. Compute $P(-2 \le X \le 3)$.
   d. For k satisfying $-5 < k < k + 4 < 5$, compute $P(k < X < k + 4)$.

3. The error involved in making a certain measurement is a continuous rv X with pdf

$$f(x) = \begin{cases} .09375(4 - x^2) & -2 \le x \le 2 \\ 0 & \text{otherwise} \end{cases}$$

   a. Sketch the graph of $f(x)$.
   b. Compute $P(X > 0)$.

   c. Compute $P(-1 < X < 1)$.
   d. Compute $P(X < -.5$ or $X > .5)$.

4. Let X denote the vibratory stress (psi) on a wind turbine blade at a particular wind speed in a wind tunnel. The article **"Blade Fatigue Life Assessment with Application to VAWTS"** (*J. of Solar Energy Engr.*, 1982: 107–111) proposes the Rayleigh distribution, with pdf

$$f(x; \theta) = \begin{cases} \dfrac{x}{\theta^2} \cdot e^{-x^2/(2\theta^2)} & x > 0 \\ 0 & \text{otherwise} \end{cases}$$

   as a model for the X distribution.
   a. Verify that $f(x; \theta)$ is a legitimate pdf.
   b. Suppose $\theta = 100$ (a value suggested by a graph in the article). What is the probability that X is at most 200? Less than 200? At least 200?
   c. What is the probability that X is between 100 and 200 (again assuming $\theta = 100$)?
   d. Give an expression for $P(X \le x)$.

5. A college professor never finishes his lecture before the end of the hour and always finishes his lectures within 2 min after the hour. Let $X =$ the time that elapses between the end of the hour and the end of the lecture and suppose the pdf of X is

$$f(x) = \begin{cases} kx^2 & 0 \le x \le 2 \\ 0 & \text{otherwise} \end{cases}$$

a. Find the value of k and draw the corresponding density curve. [*Hint*: Total area under the graph of $f(x)$ is 1.]

b. What is the probability that the lecture ends within 1 min of the end of the hour?

c. What is the probability that the lecture continues beyond the hour for between 60 and 90 sec?

d. What is the probability that the lecture continues for at least 90 sec beyond the end of the hour?

6. The actual tracking weight of a stereo cartridge that is set to track at 3 g on a particular changer can be regarded as a continuous rv X with pdf

$$f(x) = \begin{cases} k[1 - (x - 3)^2] & 2 \le x \le 4 \\ 0 & \text{otherwise} \end{cases}$$

a. Sketch the graph of $f(x)$.

b. Find the value of k.

c. What is the probability that the actual tracking weight is greater than the prescribed weight?

d. What is the probability that the actual weight is within .25 g of the prescribed weight?

e. What is the probability that the actual weight differs from the prescribed weight by more than .5 g?

7. The article **"Second Moment Reliability Evaluation vs. Monte Carlo Simulations for Weld Fatigue Strength"** (*Quality and Reliability Engr. Intl.*, **2012: 887–896**) considered the use of a uniform distribution with $A = .20$ and $B = 4.25$ for the diameter X of a certain type of weld (mm).

a. Determine the pdf of X and graph it.

b. What is the probability that diameter exceeds 3 mm?

c. What is the probability that diameter is within 1 mm of the mean diameter?

d. For any value a satisfying $.20 < a < a + 1 < 4.25$, what is $P(a < X < a + 1)$?

8. In commuting to work, a professor must first get on a bus near her house and then transfer to a second bus. If the waiting time (in minutes) at each stop has a uniform distribution with $A = 0$ and $B = 5$, then it can be shown that the total waiting time Y has the pdf

$$f(y) = \begin{cases} \dfrac{1}{25}y & 0 \le y < 5 \\[2ex] \dfrac{2}{5} - \dfrac{1}{25}y & 5 \le y \le 10 \\[2ex] 0 & y < 0 \text{ or } y > 10 \end{cases}$$

a. Sketch a graph of the pdf of Y.

b. Verify that $\int_{-\infty}^{\infty} f(y)\, dy = 1$.

c. What is the probability that total waiting time is at most 3 min?

d. What is the probability that total waiting time is at most 8 min?

e. What is the probability that total waiting time is between 3 and 8 min?

f. What is the probability that total waiting time is either less than 2 min or more than 6 min?

9. Based on an analysis of sample data, the article **"Pedestrians' Crossing Behaviors and Safety at Unmarked Roadways in China"** (*Accident Analysis and Prevention*, **2011: 1927–1936**) proposed the pdf $f(x) = .15e^{-.15(x-1)}$ when $x \ge 1$ as a model for the distribution of $X =$ time (sec) spent at the median line.

a. What is the probability that waiting time is at most 5 sec? More than 5 sec?

b. What is the probability that waiting time is between 2 and 5 sec?

10. A family of pdf's that has been used to approximate the distribution of income, city population size, and size of firms is the Pareto family. The family has two parameters, k and θ, both > 0, and the pdf is

$$f(x; k, \theta) = \begin{cases} \dfrac{k \cdot \theta^k}{x^{k+1}} & x \ge \theta \\[2ex] 0 & x < \theta \end{cases}$$

a. Sketch the graph of $f(x; k, \theta)$.

b. Verify that the total area under the graph equals 1.

c. If the rv X has pdf $f(x; k, \theta)$, for any fixed $b > \theta$, obtain an expression for $P(X \le b)$.

d. For $\theta < a < b$, obtain an expression for the probability $P(a \le X \le b)$.

4.2 Cumulative Distribution Functions and Expected Values

Several of the most important concepts introduced in the study of discrete distributions also play an important role for continuous distributions. Definitions analogous to those in Chapter 3 involve replacing summation by integration.

The Cumulative Distribution Function

The cumulative distribution function (cdf) $F(x)$ for a discrete rv X gives, for any specified number x, the probability $P(X \le x)$. It is obtained by summing the pmf $p(y)$ over all possible values y satisfying $y \le x$. The cdf of a continuous rv gives the same probabilities $P(X \le x)$ and is obtained by integrating the pdf $f(y)$ between the limits $-\infty$ and x.

DEFINITION

The **cumulative distribution function** $F(x)$ for a continuous rv X is defined for every number x by

$$F(x) = P(X \le x) = \int_{-\infty}^{x} f(y)\,dy$$

For each x, $F(x)$ is the area under the density curve to the left of x. This is illustrated in Figure 4.5, where $F(x)$ increases smoothly as x increases.

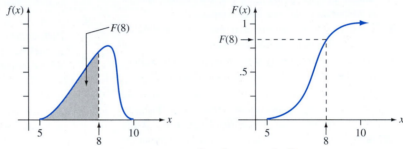

Figure 4.5 A pdf and associated cdf

EXAMPLE 4.6   Let X, the thickness of a certain metal sheet, have a uniform distribution on $[A, B]$. The density function is shown in Figure 4.6. For $x < A$, $F(x) = 0$, since there is no area under the graph of the density function to the left of such an x. For $x \ge B$, $F(x) = 1$, since all the area is accumulated to the left of such an x. Finally, for $A \le x \le B$,

$$F(x) = \int_{-\infty}^{x} f(y)\,dy = \int_{A}^{x} \frac{1}{B - A}\,dy = \frac{1}{B - A} \cdot y \,\Big|_{y=A}^{y=x} = \frac{x - A}{B - A}$$

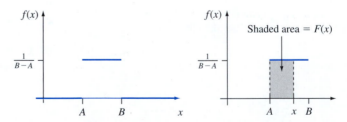

Figure 4.6 The pdf for a uniform distribution

The entire cdf is

$$F(x) = \begin{cases} 0 & x < A \\ \dfrac{x - A}{B - A} & A \le x < B \\ 1 & x \ge B \end{cases}$$

The graph of this cdf appears in Figure 4.7.

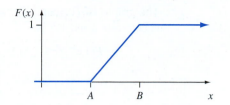

Figure 4.7 The cdf for a uniform distribution ■

Using $F(x)$ to Compute Probabilities

The importance of the cdf here, just as for discrete rv's, is that probabilities of various intervals can be computed from a formula for or table of $F(x)$.

PROPOSITION

Let X be a continuous rv with pdf $f(x)$ and cdf $F(x)$. Then for any number a,

$$P(X > a) = 1 - F(a)$$

and for any two numbers a and b with $a < b$,

$$P(a \leq X \leq b) = F(b) - F(a)$$

Figure 4.8 illustrates the second part of this proposition; the desired probability is the shaded area under the density curve between a and b, and it equals the difference between the two shaded cumulative areas. This is different from what is appropriate for a discrete integer-valued random variable (e.g., binomial or Poisson): $P(a \leq X \leq b) = F(b) - F(a - 1)$ when a and b are integers.

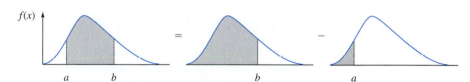

Figure 4.8 Computing $P(a \leq X \leq b)$ from cumulative probabilities

EXAMPLE 4.7 Suppose the pdf of the magnitude X of a dynamic load on a bridge (in newtons) is given by

$$f(x) = \begin{cases} \dfrac{1}{8} + \dfrac{3}{8}x & 0 \leq x \leq 2 \\ 0 & \text{otherwise} \end{cases}$$

For any number x between 0 and 2,

$$F(x) = \int_{-\infty}^{x} f(y)\,dy = \int_{0}^{x}\left(\frac{1}{8} + \frac{3}{8}y\right)dy = \frac{x}{8} + \frac{3}{16}x^2$$

Thus

$$F(x) = \begin{cases} 0 & x < 0 \\ \dfrac{x}{8} + \dfrac{3}{16}x^2 & 0 \leq x \leq 2 \\ 1 & 2 < x \end{cases}$$

The graphs of $f(x)$ and $F(x)$ are shown in Figure 4.9. The probability that the load is between 1 and 1.5 is

$$P(1 \le X \le 1.5) = F(1.5) - F(1)$$

$$= \left[\frac{1}{8}(1.5) + \frac{3}{16}(1.5)^2 \right] - \left[\frac{1}{8}(1) + \frac{3}{16}(1)^2 \right]$$

$$= \frac{19}{64} = .297$$

The probability that the load exceeds 1 is

$$P(X > 1) = 1 - P(X \le 1) = 1 - F(1) = 1 - \left[\frac{1}{8}(1) + \frac{3}{16}(1)^2 \right]$$

$$= \frac{11}{16} = .688$$

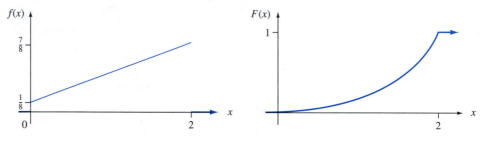

Figure 4.9 The pdf and cdf for Example 4.7 ■

Once the cdf has been obtained, any probability involving X can easily be calculated without any further integration.

Obtaining $f(x)$ from $F(x)$

For X discrete, the pmf is obtained from the cdf by taking the difference between two $F(x)$ values. The continuous analog of a difference is a derivative. The following result is a consequence of the Fundamental Theorem of Calculus.

PROPOSITION

If X is a continuous rv with pdf $f(x)$ and cdf $F(x)$, then at every x at which the derivative $F'(x)$ exists, $F'(x) = f(x)$.

EXAMPLE 4.8
(Example 4.6 continued)

When X has a uniform distribution, $F(x)$ is differentiable except at $x = A$ and $x = B$, where the graph of $F(x)$ has sharp corners. Since $F(x) = 0$ for $x < A$ and $F(x) = 1$ for $x > B$, $F'(x) = 0 = f(x)$ for such x. For $A < x < B$,

$$F'(x) = \frac{d}{dx}\left(\frac{x - A}{B - A} \right) = \frac{1}{B - A} = f(x) \qquad ■$$

Percentiles of a Continuous Distribution

When we say that an individual's test score was at the 85th percentile of the population, we mean that 85% of all population scores were below that score and 15% were above. Similarly, the 40th percentile is the score that exceeds 40% of all scores and is exceeded by 60% of all scores (having a value corresponding to a high percentile is not necessarily good; e.g., you would not want to be at the 99th percentile for blood alcohol content).

DEFINITION

Secondtest + final

Let p be a number between 0 and 1. The **(100p)th percentile** of the distribution of a continuous rv X, denoted by $\eta(p)$, is defined by

$$p = F(\eta(p)) = \int_{-\infty}^{\eta(p)} f(y)\,dy \qquad (4.2)$$

According to Expression (4.2), $\eta(p)$ is that value on the measurement axis such that $100p\%$ of the area under the graph of $f(x)$ lies to the left of $\eta(p)$ and $100(1 - p)\%$ lies to the right. Thus $\eta(.75)$, the 75th percentile, is such that the area under the graph of $f(x)$ to the left of $\eta(.75)$ is .75. Figure 4.10 illustrates the definition.

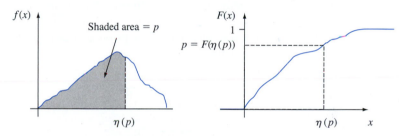

Figure 4.10   The (100p)th percentile of a continuous distribution

EXAMPLE 4.9   The distribution of the amount of gravel (in tons) sold by a particular construction supply company in a given week is a continuous rv X with pdf

$$f(x) = \begin{cases} \dfrac{3}{2}(1 - x^2) & 0 \le x \le 1 \\[2mm] 0 & \text{otherwise} \end{cases}$$

The cdf of sales for any x between 0 and 1 is

$$F(x) = \int_0^x \frac{3}{2}(1 - y^2)\,dy = \frac{3}{2}\left(y - \frac{y^3}{3}\right)\Bigg|_{y=0}^{y=x} = \frac{3}{2}\left(x - \frac{x^3}{3}\right)$$

The graphs of both $f(x)$ and $F(x)$ appear in Figure 4.11. The (100p)th percentile of this distribution satisfies the equation

$$p = F(\eta(p)) = \frac{3}{2}\left[\eta(p) - \frac{(\eta(p))^3}{3}\right]$$

that is,

$$(\eta(p))^3 - 3\eta(p) + 2p = 0$$

For the 50th percentile, $p = .5$, and the equation to be solved is $\eta^3 - 3\eta + 1 = 0$; the solution is $\eta = \eta(.5) = .347$. If the distribution remains the same from week to week, then in the long run 50% of all weeks will result in sales of less than .347 ton and 50% in more than .347 ton.

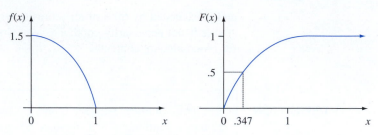

Figure 4.11 The pdf and cdf for Example 4.9 ■

DEFINITION

The **median** of a continuous distribution, denoted by $\tilde{\mu}$, is the 50th percentile, so $\tilde{\mu}$ satisfies $.5 = F(\tilde{\mu})$. That is, half the area under the density curve is to the left of $\tilde{\mu}$ and half is to the right of $\tilde{\mu}$.

A continuous distribution whose pdf is **symmetric**—the graph of the pdf to the left of some point is a mirror image of the graph to the right of that point—has median $\tilde{\mu}$ equal to the point of symmetry, since half the area under the curve lies to either side of this point. Figure 4.12 gives several examples. The error in a measurement of a physical quantity is often assumed to have a symmetric distribution.

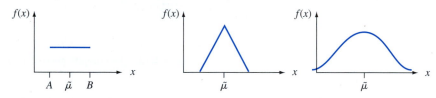

Figure 4.12 Medians of symmetric distributions

Expected Values

For a discrete random variable X, $E(X)$ was obtained by summing $x \cdot p(x)$ over possible X values. Here we replace summation by integration and the pmf by the pdf to get a continuous weighted average.

DEFINITION

The **expected** or **mean value** of a continuous rv X with pdf $f(x)$ is

$$\mu_X = E(X) = \int_{-\infty}^{\infty} x \cdot f(x)\,dx$$

EXAMPLE 4.10
(Example 4.9
continued)

The pdf of weekly gravel sales X was

$$f(x) = \begin{cases} \dfrac{3}{2}(1 - x^2) & 0 \le x \le 1 \\ 0 & \text{otherwise} \end{cases}$$

so

$$E(X) = \int_{-\infty}^{\infty} x \cdot f(x)\,dx = \int_{0}^{1} x \cdot \frac{3}{2}(1 - x^2)\,dx$$

$$= \frac{3}{2} \int_{0}^{1} (x - x^3)\,dx = \frac{3}{2}\left(\frac{x^2}{2} - \frac{x^4}{4}\right)\Bigg|_{x=0}^{x=1} = \frac{3}{8} \quad\blacksquare$$

When the pdf $f(x)$ specifies a model for the distribution of values in a numerical population, then μ is the population mean, which is the most frequently used measure of population location or center.

Often we wish to compute the expected value of some function $h(X)$ of the rv X. If we think of $h(X)$ as a new rv Y, techniques from mathematical statistics can be used to derive the pdf of Y, and $E(Y)$ can then be computed from the definition. Fortunately, as in the discrete case, there is an easier way to compute $E[h(X)]$.

PROPOSITION

weighted mean

> If X is a continuous rv with pdf $f(x)$ and $h(X)$ is any function of X, then
>
> $$E[h(X)] = \mu_{h(X)} = \int_{-\infty}^{\infty} h(x) \cdot f(x)\,dx$$

That is, just as $E(X)$ is a weighted average of possible X values, where the weighting function is the pdf $f(x)$, $E[h(X)]$ is a weighted average of $h(X)$ values.

EXAMPLE 4.11 Two species are competing in a region for control of a limited amount of a certain resource. Let $X =$ the proportion of the resource controlled by species 1 and suppose X has pdf

$$f(x) = \begin{cases} 1 & 0 \le x \le 1 \\ 0 & \text{otherwise} \end{cases}$$

which is a uniform distribution on [0, 1]. (In her book ***Ecological Diversity***, E. C. Pielou calls this the "broken-stick" model for resource allocation, since it is analogous to breaking a stick at a randomly chosen point.) Then the species that controls the majority of this resource controls the amount

$$h(X) = \max(X, 1 - X) = \begin{cases} 1 - X & \text{if } 0 \le X < \dfrac{1}{2} \\[2mm] X & \text{if } \dfrac{1}{2} \le X \le 1 \end{cases}$$

The expected amount controlled by the species having majority control is then

$$E[h(X)] = \int_{-\infty}^{\infty} \max(x, 1 - x) \cdot f(x)\,dx = \int_{0}^{1} \max(x, 1 - x) \cdot 1\,dx$$

$$= \int_{0}^{1/2} (1 - x) \cdot 1\,dx + \int_{1/2}^{1} x \cdot 1\,dx = \frac{3}{4} \quad\blacksquare$$

In the discrete case, the variance of X was defined as the expected squared deviation from μ and was calculated by summation. Here again integration replaces summation.

DEFINITION

The **variance** of a continuous random variable X with pdf $f(x)$ and mean value μ is

$$\sigma_X^2 = V(X) = \int_{-\infty}^{\infty} (x - \mu)^2 \cdot f(x)dx = E[(X - \mu)^2]$$

The **standard deviation** (SD) of X is $\sigma_X = \sqrt{V(X)}$.

The variance and standard deviation give quantitative measures of how much spread there is in the distribution or population of x values. Again σ is roughly the size of a typical deviation from μ. Computation of σ^2 is facilitated by using the same short-cut formula employed in the discrete case.

PROPOSITION

$$V(X) = E(X^2) - [E(X)]^2$$

EXAMPLE 4.12
(Example 4.10 continued)

For X = weekly gravel sales, we computed $E(X) = \frac{3}{8}$. Since

$$E(X^2) = \int_{-\infty}^{\infty} x^2 \cdot f(x)\, dx = \int_0^1 x^2 \cdot \frac{3}{2}(1 - x^2)\, dx$$

$$= \int_0^1 \frac{3}{2}(x^2 - x^4)\, dx = \frac{1}{5}$$

$$V(X) = \frac{1}{5} - \left(\frac{3}{8}\right)^2 = \frac{19}{320} = .059 \quad \text{and} \quad \sigma_X = .244 \qquad\blacksquare$$

When $h(X) = aX + b$, the expected value and variance of $h(X)$ satisfy the same properties as in the discrete case: $E[h(X)] = a\mu + b$ and $V[h(X)] = a^2 \cdot \sigma^2$.

EXERCISES Section 4.2 (11–27)

11. Let X denote the amount of time a book on two-hour reserve is actually checked out, and suppose the cdf is

$$F(x) = \begin{cases} 0 & x < 0 \\ \dfrac{x^2}{4} & 0 \le x < 2 \\ 1 & 2 \le x \end{cases}$$

  a. Calculate $P(X \le 1)$.
  b. Calculate $P(.5 \le X \le 1)$.
  c. Calculate $P(X > 1.5)$.
  d. What is the median checkout duration $\tilde{\mu}$? [solve $.5 = F(\tilde{\mu})$].
  e. Obtain the density function $f(x)$.
  f. Calculate $E(X)$.

  g. Calculate $V(X)$ and σ_X.
  h. If the borrower is charged an amount $h(X) = X^2$ when checkout duration is X, compute the expected charge $E[h(X)]$.

12. The cdf for X (= measurement error) of Exercise 3 is

$$F(x) = \begin{cases} 0 & x < -2 \\ \dfrac{1}{2} + \dfrac{3}{32}\left(4x - \dfrac{x^3}{3}\right) & -2 \le x < 2 \\ 1 & 2 \le x \end{cases}$$

  a. Compute $P(X < 0)$.
  b. Compute $P(-1 < X < 1)$.
  c. Compute $P(.5 < X)$.

d. Verify that $f(x)$ is as given in Exercise 3 by obtaining $F'(x)$.

e. Verify that $\tilde{\mu} = 0$.

13. Example 4.5 introduced the concept of time headway in traffic flow and proposed a particular distribution for $X =$ the headway between two randomly selected consecutive cars (sec). Suppose that in a different traffic environment, the distribution of time headway has the form

$$f(x) = \begin{cases} \dfrac{k}{x^4} & x > 1 \\ 0 & x \leq 1 \end{cases}$$

a. Determine the value of k for which $f(x)$ is a legitimate pdf.

b. Obtain the cumulative distribution function.

c. Use the cdf from (b) to determine the probability that headway exceeds 2 sec and also the probability that headway is between 2 and 3 sec.

d. Obtain the mean value of headway and the standard deviation of headway.

e. What is the probability that headway is within 1 standard deviation of the mean value?

14. The article **"Modeling Sediment and Water Column Interactions for Hydrophobic Pollutants"** (**Water Research**, 1984: 1169–1174) suggests the uniform distribution on the interval (7.5, 20) as a model for depth (cm) of the bioturbation layer in sediment in a certain region.

a. What are the mean and variance of depth?

b. What is the cdf of depth?

c. What is the probability that observed depth is at most 10? Between 10 and 15?

d. What is the probability that the observed depth is within 1 standard deviation of the mean value? Within 2 standard deviations?

15. Let X denote the amount of space occupied by an article placed in a 1-ft^3 packing container. The pdf of X is

$$f(x) = \begin{cases} 90x^8(1 - x) & 0 < x < 1 \\ 0 & \text{otherwise} \end{cases}$$

a. Graph the pdf. Then obtain the cdf of X and graph it.

b. What is $P(X \leq .5)$ [i.e., $F(.5)$]?

c. Using the cdf from (a), what is $P(.25 < X \leq .5)$? What is $P(.25 \leq X \leq .5)$?

d. What is the 75th percentile of the distribution?

e. Compute $E(X)$ and σ_X.

f. What is the probability that X is more than 1 standard deviation from its mean value?

16. The article **"A Model of Pedestrians' Waiting Times for Street Crossings at Signalized Intersections"** (**Transportation Research**, 2013: 17–28) suggested that under some circumstances the distribution of waiting time X could be modeled with the following pdf:

$$f(x; \theta, \tau) = \begin{cases} \dfrac{\theta}{\tau}(1 - x/\tau)^{\theta - 1} & 0 \leq x < \tau \\ 0 & \text{otherwise} \end{cases}$$

a. Graph $f(x; \theta, 80)$ for the three cases $\theta = 4$, 1, and .5 (these graphs appear in the cited article) and comment on their shapes.

b. Obtain the cumulative distribution function of X.

c. Obtain an expression for the median of the waiting time distribution.

d. For the case $\theta = 4$, $\tau = 80$, calculate $P(50 \leq X \leq 70)$ without at this point doing any additional integration.

17. Let X have a uniform distribution on the interval $[A, B]$.

a. Obtain an expression for the $(100p)$th percentile.

b. Compute $E(X)$, $V(X)$, and σ_X.

c. For n, a positive integer, compute $E(X^n)$.

18. Let X denote the voltage at the output of a microphone, and suppose that X has a uniform distribution on the interval from -1 to 1. The voltage is processed by a "hard limiter" with cutoff values $-.5$ and .5, so the limiter output is a random variable Y related to X by $Y = X$ if $|X| \leq .5$, $Y = .5$ if $X > .5$, and $Y = -.5$ if $X < -.5$.

a. What is $P(Y = .5)$?

b. Obtain the cumulative distribution function of Y and graph it.

19. Let X be a continuous rv with cdf

$$F(x) = \begin{cases} 0 & x \leq 0 \\ \dfrac{x}{4}\left[1 + \ln\left(\dfrac{4}{x}\right)\right] & 0 < x \leq 4 \\ 1 & x > 4 \end{cases}$$

[This type of cdf is suggested in the article **"Variability in Measured Bedload-Transport Rates"** (**Water Resources Bull.**, 1985: 39–48) as a model for a certain hydrologic variable.] What is

a. $P(X \leq 1)$?

b. $P(1 \leq X \leq 3)$?

c. The pdf of X?

20. Consider the pdf for total waiting time Y for two buses

$$f(y) = \begin{cases} \dfrac{1}{25}y & 0 \leq y < 5 \\ \dfrac{2}{5} - \dfrac{1}{25}y & 5 \leq y \leq 10 \\ 0 & \text{otherwise} \end{cases}$$

introduced in Exercise 8.

a. Compute and sketch the cdf of Y. [*Hint*: Consider separately $0 \leq y < 5$ and $5 \leq y \leq 10$ in computing $F(y)$. A graph of the pdf should be helpful.]

b. Obtain an expression for the $(100p)$th percentile. [*Hint*: Consider separately $0 < p < .5$ and $.5 < p < 1$.]

c. Compute $E(Y)$ and $V(Y)$. How do these compare with the expected waiting time and variance for a single bus when the time is uniformly distributed on [0, 5]?

21. An ecologist wishes to mark off a circular sampling region having radius 10 m. However, the radius of the resulting region is actually a random variable R with pdf

$$f(r) = \begin{cases} \dfrac{3}{4}[1 - (10 - r)^2] & 9 \leq r \leq 11 \\ 0 & \text{otherwise} \end{cases}$$

What is the expected area of the resulting circular region?

22. The weekly demand for propane gas (in 1000s of gallons) from a particular facility is an rv X with pdf

$$f(x) = \begin{cases} 2\left(1 - \dfrac{1}{x^2}\right) & 1 \leq x \leq 2 \\ 0 & \text{otherwise} \end{cases}$$

a. Compute the cdf of X.
b. Obtain an expression for the $(100p)$th percentile. What is the value of $\tilde{\mu}$?
c. Compute $E(X)$ and $V(X)$.
d. If 1.5 thousand gallons are in stock at the beginning of the week and no new supply is due in during the week, how much of the 1.5 thousand gallons is expected to be left at the end of the week? [*Hint*: Let $h(x) = $ amount left when demand $= x$.]

23. If the temperature at which a certain compound melts is a random variable with mean value 120°C and standard deviation 2°C, what are the mean temperature and standard deviation measured in °F? [*Hint*: °F $= 1.8°C + 32$.]

24. Let X have the Pareto pdf

$$f(x; k, \theta) = \begin{cases} \dfrac{k \cdot \theta^k}{x^{k+1}} & x \geq \theta \\ 0 & x < \theta \end{cases}$$

introduced in Exercise 10.

a. If $k > 1$, compute $E(X)$.
b. What can you say about $E(X)$ if $k = 1$?
c. If $k > 2$, show that $V(X) = k\theta^2(k-1)^{-2}(k-2)^{-1}$.
d. If $k = 2$, what can you say about $V(X)$?
e. What conditions on k are necessary to ensure that $E(X^n)$ is finite?

25. Let X be the temperature in °C at which a certain chemical reaction takes place, and let Y be the temperature in °F (so $Y = 1.8X + 32$).

a. If the median of the X distribution is $\tilde{\mu}$, show that $1.8\tilde{\mu} + 32$ is the median of the Y distribution.
b. How is the 90th percentile of the Y distribution related to the 90th percentile of the X distribution? Verify your conjecture.
c. More generally, if $Y = aX + b$, how is any particular percentile of the Y distribution related to the corresponding percentile of the X distribution?

26. Let X be the total medical expenses (in 1000s of dollars) incurred by a particular individual during a given year. Although X is a discrete random variable, suppose its distribution is quite well approximated by a continuous distribution with pdf $f(x) = k(1 + x/2.5)^{-7}$ for $x \geq 0$.

a. What is the value of k?
b. Graph the pdf of X.
c. What are the expected value and standard deviation of total medical expenses?
d. This individual is covered by an insurance plan that entails a $500 deductible provision (so the first $500 worth of expenses are paid by the individual). Then the plan will pay 80% of any additional expenses exceeding $500, and the maximum payment by the individual (including the deductible amount) is $2500. Let Y denote the amount of this individual's medical expenses paid by the insurance company. What is the expected value of Y?
[*Hint*: First figure out what value of X corresponds to the maximum out-of-pocket expense of $2500. Then write an expression for Y as a function of X (which involves several different pieces) and calculate the expected value of this function.]

27. When a dart is thrown at a circular target, consider the location of the landing point relative to the bull's eye. Let X be the angle in degrees measured from the horizontal, and assume that X is uniformly distributed on [0, 360]. Define Y to be the transformed variable $Y = h(X) = (2\pi/360)X - \pi$, so Y is the angle measured in radians and Y is between $-\pi$ and π. Obtain $E(Y)$ and σ_Y by first obtaining $E(X)$ and σ_X, and then using the fact that $h(X)$ is a linear function of X.

4.3 The Normal Distribution

The normal distribution is the most important one in all of probability and statistics. Many numerical populations have distributions that can be fit very closely by an appropriate normal curve. Examples include heights, weights, and other physical characteristics (the famous **1903 *Biometrika* article "On the Laws of Inheritance in Man"** discussed many examples of this sort), measurement errors in scientific experiments, anthropometric measurements on fossils, reaction times in psychological experiments, measurements of intelligence and aptitude, scores on various tests, and numerous economic measures and indicators. In addition, even when individual variables themselves are not normally distributed, sums and averages of the variables will under suitable conditions have approximately a normal distribution; this is the content of the Central Limit Theorem discussed in the next chapter.

DEFINITION

> A continuous rv X is said to have a **normal distribution** with parameters μ and σ (or μ and σ^2), where $-\infty < \mu < \infty$ and $0 < \sigma$, if the pdf of X is
>
> $$f(x; \mu, \sigma) = \frac{1}{\sqrt{2\pi}\sigma} e^{-(x-\mu)^2/(2\sigma^2)} \qquad -\infty < x < \infty \qquad (4.3)$$

[handwritten margin notes: pdf; Mean μ E(X) ✶; Variance σ = V(X)]

Again e denotes the base of the natural logarithm system and equals approximately 2.71828, and π represents the familiar mathematical constant with approximate value 3.14159. The statement that X is normally distributed with parameters μ and σ^2 is often abbreviated $X \sim N(\mu, \sigma^2)$.

Clearly $f(x; \mu, \sigma) \geq 0$, but a somewhat complicated calculus argument must be used to verify that $\int_{-\infty}^{\infty} f(x; \mu, \sigma)\,dx = 1$. It can be shown that $E(X) = \mu$ and $V(X) = \sigma^2$, so the parameters are the mean and the standard deviation of X. Figure 4.13 presents graphs of $f(x; \mu, \sigma)$ for several different (μ, σ) pairs. Each density curve is symmetric about μ and bell-shaped, so the center of the bell (point of symmetry) is both the mean of the distribution and the median. The mean μ is a *location parameter*, since changing its value rigidly shifts the density curve to one side or the other; σ is referred to as a *scale parameter*, because changing its value stretches or compresses the curve horizontally without changing the basic shape. The inflection points of a normal curve (points at which the curve changes from turning downward to turning upward) occur at $\mu - \sigma$ and $\mu + \sigma$. Thus the value of σ can be visualized as the distance from the mean to these inflection points. A large value of σ corresponds to a density curve that is quite spread out about μ, whereas a small value yields a highly concentrated curve. The larger the value of σ, the more likely it is that a value of X far from the mean may be observed.

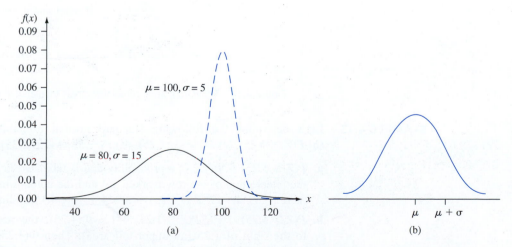

Figure 4.13 (a) Two different normal density curves (b) Visualizing μ and σ for a normal distribution

The Standard Normal Distribution

The computation of $P(a \leq X \leq b)$ when X is a normal rv with parameters μ and σ requires evaluating

$$\int_a^b \frac{1}{\sqrt{2\pi}\sigma} e^{-(x-\mu)^2/(2\sigma^2)} dx \qquad (4.4)$$

None of the standard integration techniques can be used to accomplish this. Instead, for $\mu = 0$ and $\sigma = 1$, Expression (4.4) has been calculated using numerical techniques and tabulated for certain values of a and b. This table can also be used to compute probabilities for any other values of μ and σ under consideration.

DEFINITION

The normal distribution with parameter values $\mu = 0$ and $\sigma = 1$ is called the **standard normal distribution.** A random variable having a standard normal distribution is called a **standard normal random variable** and will be denoted by Z. The pdf of Z is

$$f(z; 0, 1) = \frac{1}{\sqrt{2\pi}} e^{-z^2/2} \qquad -\infty < z < \infty$$

The graph of $f(z; 0, 1)$ is called the *standard normal* (or z) curve. Its inflection points are at 1 and -1. The cdf of Z is $P(Z \leq z) = \int_{-\infty}^z f(y; 0, 1) \, dy$, which we will denote by $\Phi(z)$.

The standard normal distribution almost never serves as a model for a naturally arising population. Instead, it is a reference distribution from which information about other normal distributions can be obtained. Appendix Table A.3 gives $\Phi(z) = P(Z \leq z)$, the area under the standard normal density curve to the left of z, for $z = -3.49, -3.48, \ldots, 3.48, 3.49$. Figure 4.14 illustrates the type of cumulative area (probability) tabulated in Table A.3. From this table, various other probabilities involving Z can be calculated.

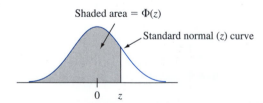

Figure 4.14 Standard normal cumulative areas tabulated in Appendix Table A.3

EXAMPLE 4.13 Let's determine the following standard normal probabilities: (a) $P(Z \leq 1.25)$, (b) $P(Z > 1.25)$, (c) $P(Z \leq -1.25)$, (d) $P(-.38 \leq Z \leq 1.25)$, and (e) $P(Z \leq 5)$.

a. $P(Z \leq 1.25) = \Phi(1.25)$, a probability that is tabulated in Appendix Table A.3 at the intersection of the row marked 1.2 and the column marked .05. The number there is .8944, so $P(Z \leq 1.25) = .8944$. Figure 4.15(a) illustrates this probability.

b. $P(Z > 1.25) = 1 - P(Z \leq 1.25) = 1 - \Phi(1.25)$, the area under the z curve to the right of 1.25 (an upper-tail area). Then $\Phi(1.25) = .8944$ implies that $P(Z > 1.25) = .1056$. Since Z is a continuous rv, $P(Z \geq 1.25) = .1056$. See Figure 4.15(b).

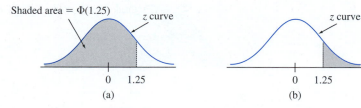

Shaded area = $\Phi(1.25)$ z curve

z curve

0 1.25
(a)

0 1.25
(b)

Figure 4.15 Normal curve areas (probabilities) for Example 4.13

c. $P(Z \leq -1.25) = \Phi(-1.25)$, a lower-tail area. Directly from Appendix Table A.3, $\Phi(-1.25) = .1056$. By symmetry of the z curve, this is the same answer as in part (b).

d. $P(-.38 \leq Z \leq 1.25)$ is the area under the standard normal curve above the interval whose left endpoint is $-.38$ and whose right endpoint is 1.25. From Section 4.2, if X is a continuous rv with cdf $F(x)$, then $P(a \leq X \leq b) = F(b) - F(a)$. Thus $P(-.38 \leq Z \leq 1.25) = \Phi(1.25) - \Phi(-.38) = .8944 - .3520 = .5424$. (See Figure 4.16.)

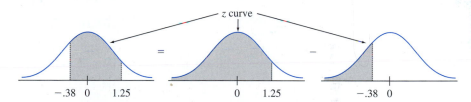

z curve

$-.38$ 0 1.25

0 1.25

$-.38$ 0

Figure 4.16 $P(-.38 \leq Z \leq 1.25)$ as the difference between two cumulative areas

e. $P(Z \leq 5) = \Phi(5)$, the cumulative area under the z curve to the left of 5. This probability does not appear in the table because the last row is labeled 3.4. However, the last entry in that row is $\Phi(3.49) = .9998$. That is, essentially all of the area under the curve lies to the left of 3.49 (at most 3.49 standard deviations to the right of the mean). Therefore we conclude that $P(Z \leq 5) \approx 1$. ■

Percentiles of the Standard Normal Distribution

For any p between 0 and 1, Appendix Table A.3 can be used to obtain the $(100p)$th percentile of the standard normal distribution.

EXAMPLE 4.14 The 99th percentile of the standard normal distribution is that value on the horizontal axis such that the area under the z curve to the left of the value is .9900. Appendix Table A.3 gives for fixed z the area under the standard normal curve to the left of z, whereas here we have the area and want the value of z. This is the "inverse" problem to $P(Z \leq z) = ?$ so the table is used in an inverse fashion: Find in the middle of the table .9900; the row and column in which it lies identify the 99th z percentile. Here .9901 lies at the intersection of the row marked 2.3 and column marked .03, so the 99th percentile is (approximately) $z = 2.33$. (See Figure 4.17.) By symmetry, the first percentile is as far below 0 as the 99th is above 0, so equals -2.33 (1% lies below the first and also above the 99th). (See Figure 4.18.)

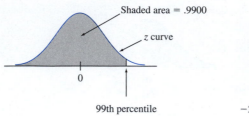

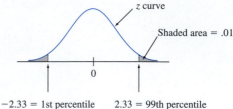

Figure 4.17 Finding the 99th percentile

Figure 4.18 The relationship between the 1st and 99th percentiles

In general, the $(100p)$th percentile is identified by the row and column of Appendix Table A.3 in which the entry p is found (e.g., the 67th percentile is obtained by finding .6700 in the body of the table, which gives $z = .44$). If p does not appear, the number closest to it is typically used, although linear interpolation gives a more accurate answer. For example, to find the 95th percentile, look for .9500 inside the table. Although it does not appear, both .9495 and .9505 do, corresponding to $z = 1.64$ and 1.65, respectively. Since .9500 is halfway between the two probabilities that do appear, we will use 1.645 as the 95th percentile and -1.645 as the 5th percentile.

z_α Notation for z Critical Values

In statistical inference, we will need the values on the horizontal z axis that capture certain small tail areas under the standard normal curve.

> **Notation**
>
> z_α will denote the value on the z axis for which α of the area under the z curve lies to the right of z_α. (See Figure 4.19.)

For example, $z_{.10}$ captures upper-tail area .10, and $z_{.01}$ captures upper-tail area .01.

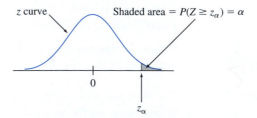

Figure 4.19 z_α notation Illustrated

Since α of the area under the z curve lies to the right of $z_\alpha, 1 - \alpha$ of the area lies to its left. Thus z_α *is the* $100(1 - \alpha)$*th percentile of the standard normal distribution.* By symmetry the area under the standard normal curve to the left of $-z_\alpha$ is also α. The z_α's are usually referred to as **z critical values.** Table 4.1 lists the most useful z percentiles and z_α values.

Table 4.1 Standard Normal Percentiles and Critical Values

Percentile	90	95	97.5	99	99.5	99.9	99.95
α (upper-tail area)	.1	.05	.025	.01	.005	.001	.0005
$z_\alpha = 100(1 - \alpha)$th percentile	1.28	1.645	1.96	2.33	2.58	3.08	3.27

EXAMPLE 4.15 $z_{.05}$ is the $100(1 - .05)$th $= 95$th percentile of the standard normal distribution, so $z_{.05} = 1.645$. The area under the standard normal curve to the left of $-z_{.05}$ is also .05. (See Figure 4.20.)

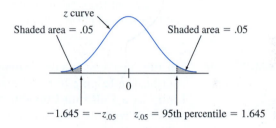

Shaded area = .05   Shaded area = .05

$-1.645 = -z_{.05}$ $z_{.05} = $ 95th percentile $= 1.645$

Figure 4.20 Finding $z_{.05}$

Nonstandard Normal Distributions

When $X \sim N(\mu, \sigma^2)$, probabilities involving X are computed by "standardizing." The **standardized variable** is $(X - \mu)/\sigma$. Subtracting μ shifts the mean from μ to zero, and then dividing by σ scales the variable so that the standard deviation is 1 rather than σ.

PROPOSITION

If X has a normal distribution with mean μ and standard deviation σ, then

$$Z = \frac{X - \mu}{\sigma}$$

has a standard normal distribution. Thus

$$P(a \le X \le b) = P\left(\frac{a - \mu}{\sigma} \le Z \le \frac{b - \mu}{\sigma}\right)$$

$$= \Phi\left(\frac{b - \mu}{\sigma}\right) - \Phi\left(\frac{a - \mu}{\sigma}\right)$$

$$P(X \le a) = \Phi\left(\frac{a - \mu}{\sigma}\right) \qquad P(X \ge b) = 1 - \Phi\left(\frac{b - \mu}{\sigma}\right)$$

According to the first part of the proposition, the area under the normal (μ, σ^2) curve that lies above the interval $[a, b]$ is identical to the area under the standard normal curve that lies above the interval from the standardized lower limit $(a - \mu)/\sigma$ to the standardized upper limit $(b - \mu)/\sigma$. An illustration of the second part appears in Figure 4.21. The key idea is that by standardizing, any probability involving X can be expressed as a probability involving a standard normal rv Z, so that Appendix Table A.3 can be used. The proposition can be proved by writing the cdf of $Z = (X - \mu)/\sigma$ as

$$P(Z \le z) = P(X \le \sigma z + \mu) = \int_{-\infty}^{\sigma z + \mu} f(x; \mu, \sigma)\, dx$$

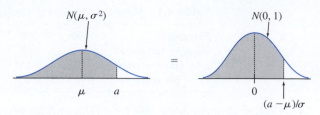

Figure 4.21 Equality of nonstandard and standard normal curve areas

Using a result from calculus, this integral can be differentiated with respect to z to yield the desired pdf $f(z; 0, 1)$.

EXAMPLE 4.16 The time that it takes a driver to react to the brake lights on a decelerating vehicle is critical in helping to avoid rear-end collisions. The article **"Fast-Rise Brake Lamp as a Collision-Prevention Device"** (***Ergonomics, 1993: 391–395***) suggests that reaction time for an in-traffic response to a brake signal from standard brake lights can be modeled with a normal distribution having mean value 1.25 sec and standard deviation of .46 sec. What is the probability that reaction time is between 1.00 sec and 1.75 sec? If we let X denote reaction time, then standardizing gives

$$1.00 \leq X \leq 1.75$$

if and only if

$$\frac{1.00 - 1.25}{.46} \leq \frac{X - 1.25}{.46} \leq \frac{1.75 - 1.25}{.46}$$

Thus

$$P(1.00 \leq X \leq 1.75) = P\left(\frac{1.00 - 1.25}{.46} \leq Z \leq \frac{1.75 - 1.25}{.46}\right)$$

$$= P(-.54 \leq Z \leq 1.09) = \Phi(1.09) - \Phi(-.54)$$

$$= .8621 - .2946 = .5675$$

This is illustrated in Figure 4.22. Similarly, if we view 2 sec as a critically long reaction time, the probability that actual reaction time will exceed this value is

$$P(X > 2) = P\left(Z > \frac{2 - 1.25}{.46}\right) = P(Z > 1.63) = 1 - \Phi(1.63) = .0516$$

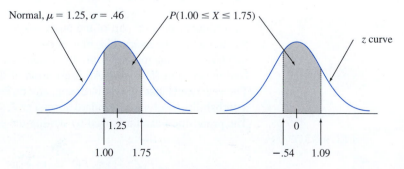

Figure 4.22 Normal curves for Example 4.16

Standardizing amounts to nothing more than calculating a distance from the mean value and then reexpressing the distance as some number of standard deviations. Thus, if $\mu = 100$ and $\sigma = 15$, then $x = 130$ corresponds to $z = (130 - 100)/15 = 30/15 = 2.00$. That is, 130 is 2 standard deviations above (to the right of) the mean value. Similarly, standardizing 85 gives $(85 - 100)/15 = -1.00$, so 85 is 1 standard deviation below the mean. The z table applies to *any* normal distribution provided that we think in terms of number of standard deviations away from the mean value.

EXAMPLE 4.17 The breakdown voltage of a randomly chosen diode of a particular type is known to be normally distributed. What is the probability that a diode's breakdown voltage is within 1 standard deviation of its mean value? This question can be answered without knowing either μ or σ, as long as the distribution is known to be normal; the answer is the same for *any* normal distribution:

$$P(X \text{ is within 1 standard deviation of its mean}) = P(\mu - \sigma \le X \le \mu + \sigma)$$

$$= P\left(\frac{\mu - \sigma - \mu}{\sigma} \le Z \le \frac{\mu + \sigma - \mu}{\sigma}\right)$$

$$= P(-1.00 \le Z \le 1.00)$$

$$= \Phi(1.00) - \Phi(-1.00) = .6826$$

The probability that X is within 2 standard deviations of its mean is $P(-2.00 \le Z \le 2.00) = .9544$ and within 3 standard deviations of the mean is $P(-3.00 \le Z \le 3.00) = .9974$. ∎

The results of Example 4.17 are often reported in percentage form and referred to as the *empirical rule* (because empirical evidence has shown that histograms of real data can very frequently be approximated by normal curves).

> If the population distribution of a variable is (approximately) normal, then
>
> **1.** Roughly 68% of the values are within 1 SD of the mean.
> **2.** Roughly 95% of the values are within 2 SDs of the mean.
> **3.** Roughly 99.7% of the values are within 3 SDs of the mean.

It is indeed unusual to observe a value from a normal population that is much farther than 2 standard deviations from μ. These results will be important in the development of hypothesis-testing procedures in later chapters.

Percentiles of an Arbitrary Normal Distribution

The $(100p)$th percentile of a normal distribution with mean μ and standard deviation σ is easily related to the $(100p)$th percentile of the standard normal distribution.

PROPOSITION

$$\begin{pmatrix} (100p)\text{th percentile} \\ \text{for normal } (\mu, \sigma) \end{pmatrix} = \mu + \begin{bmatrix} (100p)\text{th for} \\ \text{standard normal} \end{bmatrix} \cdot \sigma$$

Another way of saying this is that if z is the desired percentile for the standard normal distribution, then the desired percentile for the normal (μ, σ) distribution is z standard deviations from μ.

EXAMPLE 4.18   The authors of **"Assessment of Lifetime of Railway Axle"** (**_Intl. J. of Fatigue,_** **2013: 40–46**) used data collected from an experiment with a specified initial crack length and number of loading cycles to propose a normal distribution with mean value 5.496 mm and standard deviation .067 mm for the rv X = final crack depth. For this model, what value of final crack depth would be exceeded by only .5% of all cracks under these circumstances? Let c denote the requested value. Then the desired condition is that $P(X > c) = .005$, or, equivalently, that $P(X \leq c) = .995$. Thus c is the 99.5th percentile of the normal distribution with $\mu = 5.496$ and $\sigma = .067$. The 99.5th percentile of the standard normal distribution is 2.58, so

$$c = \eta(.995) = 5.496 + (2.58)(.067) = 5.496 + .173 = 5.669 \text{ mm}$$

This is illustrated in Figure 4.23.

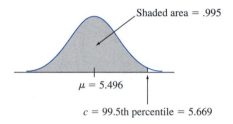

Shaded area = .995

$\mu = 5.496$

c = 99.5th percentile = 5.669

Figure 4.23   Distribution of final crack depth for Example 4.18   ■

The Normal Distribution and Discrete Populations

The normal distribution is often used as an approximation to the distribution of values in a discrete population. In such situations, extra care should be taken to ensure that probabilities are computed in an accurate manner.

EXAMPLE 4.19   IQ in a particular population (as measured by a standard test) is known to be approximately normally distributed with $\mu = 100$ and $\sigma = 15$. What is the probability that a randomly selected individual has an IQ of at least 125? Letting X = the IQ of a randomly chosen person, we wish $P(X \geq 125)$. The temptation here is to standardize $X \geq 125$ as in previous examples. However, the IQ population distribution is actually discrete, since IQs are integer-valued. So the normal curve is an approximation to a discrete probability histogram, as pictured in Figure 4.24.

The rectangles of the histogram are _centered_ at integers. IQs of at least 125 correspond to rectangles beginning at 124.5, as shaded in Figure 4.24. Thus we really want the area under the approximating normal curve to the right of 124.5. Standardizing this value gives $P(Z \geq 1.63) = .0516$, whereas standardizing 125 results in $P(Z \geq 1.67) = .0475$. The difference is not great, but the answer .0516 is more accurate. Similarly, $P(X = 125)$ would be approximated by the area between 124.5 and 125.5, since the area under the normal curve above the single value 125 is zero.

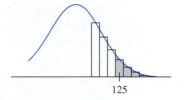

125

Figure 4.24 A normal approximation to a discrete distribution

The correction for discreteness of the underlying distribution in Example 4.19—that is, the addition or subtraction of .5 before standardizing—is often called a **continuity correction**. It is useful in the following application of the normal distribution to the computation of binomial probabilities.

Approximating the Binomial Distribution

Recall that the mean value and standard deviation of a binomial random variable X are $\mu_X = np$ and $\sigma_X = \sqrt{npq}$, respectively. Figure 4.25 displays a binomial probability histogram for the binomial distribution with $n = 25, p = .6$, for which $\mu = 25(.6) = 15$ and $\sigma = \sqrt{25(.6)(.4)} = 2.449$. A normal curve with this μ and σ has been superimposed on the probability histogram. Although the probability histogram is a bit skewed (because $p \neq .5$), the normal curve gives a very good approximation, especially in the middle part of the picture. The area of any rectangle (probability of any particular X value) except those in the extreme tails can be accurately approximated by the corresponding normal curve area. For example, $P(X = 10) = B(10; 25, .6) - B(9; 25, .6) = .021$, whereas the area under the normal curve between 9.5 and 10.5 is $P(-2.25 \leq Z \leq -1.84) = .0207$.

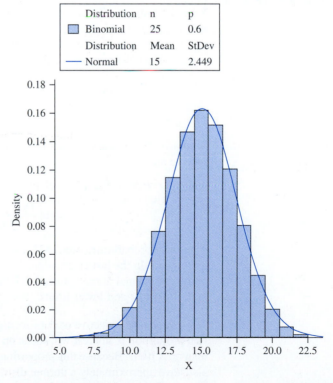

Figure 4.25 Binomial probability histogram for $n = 25, p = .6$ with normal approximation curve superimposed

More generally, as long as the binomial probability histogram is not too skewed, binomial probabilities can be well approximated by normal curve areas. It is then customary to say that X has approximately a normal distribution.

PROPOSITION

> Let X be a binomial rv based on n trials with success probability p. Then if the binomial probability histogram is not too skewed, X has approximately a normal distribution with $\mu = np$ and $\sigma = \sqrt{npq}$. In particular, for $x =$ a possible value of X,
>
> $$P(X \le x) = B(x, n, p) \approx \left(\begin{array}{c} \text{area under the normal curve} \\ \text{to the left of } x + .5 \end{array} \right)$$
>
> $$= \Phi\left(\frac{x + .5 - np}{\sqrt{npq}} \right)$$
>
> In practice, the approximation is adequate provided that both $np \ge 10$ and $nq \ge 10$ (i.e., the expected number of successes and the expected number of failures are both at least 10), since there is then enough symmetry in the underlying binomial distribution.

A direct proof of the approximation's validity is quite difficult. In the next chapter we'll see that it is a consequence of a more general result called the Central Limit Theorem. In all honesty, the approximation is not so important for probability calculation as it once was. This is because software can now calculate binomial probabilities exactly for quite large values of n.

EXAMPLE 4.20

Suppose that 25% of all students at a large public university receive financial aid. Let X be the number of students in a random sample of size 50 who receive financial aid, so that $p = .25$. Then $\mu = 12.5$ and $\sigma = 3.06$. Since $np = 50(.25) = 12.5 \ge 10$ and $nq = 37.5 \ge 10$, the approximation can safely be applied. The probability that at most 10 students receive aid is

$$P(X \le 10) = B(10; 50, .25) \approx \Phi\left(\frac{10 + .5 - 12.5}{3.06} \right)$$

$$= \Phi(-.65) = .2578$$

Similarly, the probability that between 5 and 15 (inclusive) of the selected students receive aid is

$$P(5 \le X \le 15) = B(15; 50, .25) - B(4; 50, .25)$$

$$\approx \Phi\left(\frac{15.5 - 12.5}{3.06} \right) - \Phi\left(\frac{4.5 - 12.5}{3.06} \right) = .8320$$

The exact probabilities are .2622 and .8348, respectively, so the approximations are quite good. In the last calculation, $P(5 \le X \le 15)$ is being approximated by the area under the normal curve between 4.5 and 15.5—the continuity correction is used for both the upper and lower limits. ∎

When the objective of our investigation is to make an inference about a population proportion p, interest will focus on the sample proportion of successes X/n rather than on X itself. Because this proportion is just X multiplied by the constant $1/n$, it will also have approximately a normal distribution (with mean $\mu = p$ and standard deviation $\sigma = \sqrt{pq/n}$) provided that both $np \ge 10$ and $nq \ge 10$. This normal approximation is the basis for several inferential procedures to be discussed in later chapters.

EXERCISES Section 4.3 (28–58)

28. Let Z be a standard normal random variable and calculate the following probabilities, drawing pictures wherever appropriate.

   a. $P(0 \leq Z \leq 2.17)$ **b.** $P(0 \leq Z \leq 1)$

   c. $P(-2.50 \leq Z \leq 0)$ **d.** $P(-2.50 \leq Z \leq 2.50)$

   e. $P(Z \leq 1.37)$ **f.** $P(-1.75 \leq Z)$

   g. $P(-1.50 \leq Z \leq 2.00)$ **h.** $P(1.37 \leq Z \leq 2.50)$

   i. $P(1.50 \leq Z)$ **j.** $P(|Z| \leq 2.50)$

29. In each case, determine the value of the constant c that makes the probability statement correct.

   a. $\Phi(c) = .9838$ **b.** $P(0 \leq Z \leq c) = .291$

   c. $P(c \leq Z) = .121$ **d.** $P(-c \leq Z \leq c) = .668$

   e. $P(c \leq |Z|) = .016$

30. Find the following percentiles for the standard normal distribution. Interpolate where appropriate.

   a. 91st   **b.** 9th   **c.** 75th

   d. 25th   **e.** 6th

31. Determine z_α for the following values of α:

   a. $\alpha = .0055$ **b.** $\alpha = .09$

   c. $\alpha = .663$

32. Suppose the force acting on a column that helps to support a building is a normally distributed random variable X with mean value 15.0 kips and standard deviation 1.25 kips. Compute the following probabilities by standardizing and then using Table A.3.

   a. $P(X \leq 15)$ **b.** $P(X \leq 17.5)$

   c. $P(X \geq 10)$ **d.** $P(14 \leq X \leq 18)$

   e. $P(|X - 15| \leq 3)$

33. Mopeds (small motorcycles with an engine capacity below 50 cm³) are very popular in Europe because of their mobility, ease of operation, and low cost. The article **"Procedure to Verify the Maximum Speed of Automatic Transmission Mopeds in Periodic Motor Vehicle Inspections"** (*J. of Automobile Engr.*, **2008: 1615–1623**) described a rolling bench test for determining maximum vehicle speed. A normal distribution with mean value 46.8 km/h and standard deviation 1.75 km/h is postulated. Consider randomly selecting a single such moped.

   a. What is the probability that maximum speed is at most 50 km/h?

   b. What is the probability that maximum speed is at least 48 km/h?

   c. What is the probability that maximum speed differs from the mean value by at most 1.5 standard deviations?

34. The article **"Reliability of Domestic-Waste Biofilm Reactors"** (*J. of Envir. Engr.*, **1995: 785–790**) suggests that substrate concentration (mg/cm³) of influent to a reactor is normally distributed with $\mu = .30$ and $\sigma = .06$.

   a. What is the probability that the concentration exceeds .50?

   b. What is the probability that the concentration is at most .20?

   c. How would you characterize the largest 5% of all concentration values?

35. In a road-paving process, asphalt mix is delivered to the hopper of the paver by trucks that haul the material from the batching plant. The article **"Modeling of Simultaneously Continuous and Stochastic Construction Activities for Simulation"** (*J. of Construction Engr. and Mgmnt.*, **2013: 1037–1045**) proposed a normal distribution with mean value 8.46 min and standard deviation .913 min for the rv $X = $ truck haul time.

   a. What is the probability that haul time will be at least 10 min? Will exceed 10 min?

   b. What is the probability that haul time will exceed 15 min?

   c. What is the probability that haul time will be between 8 and 10 min?

   d. What value c is such that 98% of all haul times are in the interval from $8.46 - c$ to $8.46 + c$?

   e. If four haul times are independently selected, what is the probability that at least one of them exceeds 10 min?

36. Spray drift is a constant concern for pesticide applicators and agricultural producers. The inverse relationship between droplet size and drift potential is well known. The paper **"Effects of 2,4-D Formulation and Quinclorac on Spray Droplet Size and Deposition"** (*Weed Technology*, **2005: 1030–1036**) investigated the effects of herbicide formulation on spray atomization. A figure in the paper suggested the normal distribution with mean 1050 μm and standard deviation 150 μm was a reasonable model for droplet size for water (the "control treatment") sprayed through a 760 ml/min nozzle.

   a. What is the probability that the size of a single droplet is less than 1500 μm? At least 1000 μm?

   b. What is the probability that the size of a single droplet is between 1000 and 1500 μm?

   c. How would you characterize the smallest 2% of all droplets?

   d. If the sizes of five independently selected droplets are measured, what is the probability that exactly two of them exceed 1500 μm?

37. Suppose that blood chloride concentration (mmol/L) has a normal distribution with mean 104 and standard deviation 5 (information in the article **"Mathematical Model of Chloride Concentration in Human Blood,"** *J. of Med. Engr. and Tech.*, **2006: 25–30**, including a normal probability plot as described in Section 4.6, supports this assumption).

a. What is the probability that chloride concentration equals 105? Is less than 105? Is at most 105?

b. What is the probability that chloride concentration differs from the mean by more than 1 standard deviation? Does this probability depend on the values of μ and σ?

c. How would you characterize the most extreme .1% of chloride concentration values?

38. There are two machines available for cutting corks intended for use in wine bottles. The first produces corks with diameters that are normally distributed with mean 3 cm and standard deviation .1 cm. The second machine produces corks with diameters that have a normal distribution with mean 3.04 cm and standard deviation .02 cm. Acceptable corks have diameters between 2.9 cm and 3.1 cm. Which machine is more likely to produce an acceptable cork?

39. The defect length of a corrosion defect in a pressurized steel pipe is normally distributed with mean value 30 mm and standard deviation 7.8 mm [suggested in the article **"Reliability Evaluation of Corroding Pipelines Considering Multiple Failure Modes and Time-Dependent Internal Pressure"** (*J. of Infrastructure Systems*, 2011: 216–224)].

a. What is the probability that defect length is at most 20 mm? Less than 20 mm?

b. What is the 75th percentile of the defect length distribution—that is, the value that separates the smallest 75% of all lengths from the largest 25%?

c. What is the 15th percentile of the defect length distribution?

d. What values separate the middle 80% of the defect length distribution from the smallest 10% and the largest 10%?

40. The article **"Monte Carlo Simulation—Tool for Better Understanding of LRFD"** (*J. of Structural Engr.*, 1993: 1586–1599) suggests that yield strength (ksi) for A36 grade steel is normally distributed with $\mu = 43$ and $\sigma = 4.5$.

a. What is the probability that yield strength is at most 40? Greater than 60?

b. What yield strength value separates the strongest 75% from the others?

41. The automatic opening device of a military cargo parachute has been designed to open when the parachute is 200 m above the ground. Suppose opening altitude actually has a normal distribution with mean value 200 m and standard deviation 30 m. Equipment damage will occur if the parachute opens at an altitude of less than 100 m. What is the probability that there is equipment damage to the payload of at least one of five independently dropped parachutes?

42. The temperature reading from a thermocouple placed in a constant-temperature medium is normally distributed

with mean μ, the actual temperature of the medium, and standard deviation σ. What would the value of σ have to be to ensure that 95% of all readings are within .1° of μ?

43. Vehicle speed on a particular bridge in China can be modeled as normally distributed (**"Fatigue Reliability Assessment for Long-Span Bridges under Combined Dynamic Loads from Winds and Vehicles," *J. of Bridge Engr.*, 2013: 735–747**).

a. If 5% of all vehicles travel less than 39.12 m/h and 10% travel more than 73.24 m/h, what are the mean and standard deviation of vehicle speed? [*Note:* The resulting values should agree with those given in the cited article.]

b. What is the probability that a randomly selected vehicle's speed is between 50 and 65 m/h?

c. What is the probability that a randomly selected vehicle's speed exceeds the speed limit of 70 m/h?

44. If bolt thread length is normally distributed, what is the probability that the thread length of a randomly selected bolt is

a. Within 1.5 SDs of its mean value?

b. Farther than 2.5 SDs from its mean value?

c. Between 1 and 2 SDs from its mean value?

45. A machine that produces ball bearings has initially been set so that the true average diameter of the bearings it produces is .500 in. A bearing is acceptable if its diameter is within .004 in. of this target value. Suppose, however, that the setting has changed during the course of production, so that the bearings have normally distributed diameters with mean value .499 in. and standard deviation .002 in. What percentage of the bearings produced will not be acceptable?

46. The Rockwell hardness of a metal is determined by impressing a hardened point into the surface of the metal and then measuring the depth of penetration of the point. Suppose the Rockwell hardness of a particular alloy is normally distributed with mean 70 and standard deviation 3.

a. If a specimen is acceptable only if its hardness is between 67 and 75, what is the probability that a randomly chosen specimen has an acceptable hardness?

b. If the acceptable range of hardness is $(70 - c, 70 + c)$, for what value of c would 95% of all specimens have acceptable hardness?

c. If the acceptable range is as in part (a) and the hardness of each of ten randomly selected specimens is indepen-dently determined, what is the expected number of acceptable specimens among the ten?

d. What is the probability that at most eight of ten independently selected specimens have a hardness of less than 73.84? [*Hint:* Y = the number among the ten specimens with hardness less than 73.84 is a binomial variable; what is p?]

47. The weight distribution of parcels sent in a certain manner is normal with mean value 12 lb and standard deviation 3.5 lb. The parcel service wishes to establish a weight value c beyond which there will be a surcharge. What value of c is such that 99% of all parcels are at least 1 lb under the surcharge weight?

48. Suppose Appendix Table A.3 contained $\Phi(z)$ only for $z \geq 0$. Explain how you could still compute
    a. $P(-1.72 \leq Z \leq -.55)$
    b. $P(-1.72 \leq Z \leq .55)$

    Is it necessary to tabulate $\Phi(z)$ for z negative? What property of the standard normal curve justifies your answer?

49. Consider babies born in the "normal" range of 37–43 weeks gestational age. Extensive data supports the assumption that for such babies born in the United States, birth weight is normally distributed with mean 3432 g and standard deviation 482 g. [The article "**Are Babies Normal?**" (**The American Statistician, 1999: 298–302**) analyzed data from a particular year; for a sensible choice of class intervals, a histogram did not look at all normal, but after further investigations it was determined that this was due to some hospitals measuring weight in grams and others measuring to the nearest ounce and then converting to grams. A modified choice of class intervals that allowed for this gave a histogram that was well described by a normal distribution.]
    a. What is the probability that the birth weight of a randomly selected baby of this type exceeds 4000 g? Is between 3000 and 4000 g?
    b. What is the probability that the birth weight of a randomly selected baby of this type is either less than 2000 g or greater than 5000 g?
    c. What is the probability that the birth weight of a randomly selected baby of this type exceeds 7 lb?
    d. How would you characterize the most extreme .1% of all birth weights?
    e. If X is a random variable with a normal distribution and a is a numerical constant ($a \neq 0$), then $Y = aX$ also has a normal distribution. Use this to determine the distribution of birth weight expressed in pounds (shape, mean, and standard deviation), and then recalculate the probability from part (c). How does this compare to your previous answer?

50. In response to concerns about nutritional contents of fast foods, McDonald's has announced that it will use a new cooking oil for its french fries that will decrease substantially trans fatty acid levels and increase the amount of more beneficial polyunsaturated fat. The company claims that 97 out of 100 people cannot detect a difference in taste between the new and old oils. Assuming that this figure is correct (as a long-run proportion), what is the approximate probability that in a random sample of 1000 individuals who have purchased fries at McDonald's,

    a. At least 40 can taste the difference between the two oils?
    b. At most 5% can taste the difference between the two oils?

51. Chebyshev's inequality, (see Exercise 44, Chapter 3), is valid for continuous as well as discrete distributions. It states that for any number k satisfying $k \geq 1$, $P(|X - \mu| \geq k\sigma) \leq 1/k^2$ (see Exercise 44 in Chapter 3 for an interpretation). Obtain this probability in the case of a normal distribution for $k = 1, 2$, and 3, and compare to the upper bound.

52. Let X denote the number of flaws along a 100-m reel of magnetic tape (an integer-valued variable). Suppose X has approximately a normal distribution with $\mu = 25$ and $\sigma = 5$. Use the continuity correction to calculate the probability that the number of flaws is
    a. Between 20 and 30, inclusive.
    b. At most 30. Less than 30.

53. Let X have a binomial distribution with parameters $n = 25$ and p. Calculate each of the following probabilities using the normal approximation (with the continuity correction) for the cases $p = .5, .6$, and .8 and compare to the exact probabilities calculated from Appendix Table A.1.
    a. $P(15 \leq X \leq 20)$
    b. $P(X \leq 15)$
    c. $P(20 \leq X)$

54. Suppose that 10% of all steel shafts produced by a certain process are nonconforming but can be reworked (rather than having to be scrapped). Consider a random sample of 200 shafts, and let X denote the number among these that are nonconforming and can be reworked. What is the (approximate) probability that X is
    a. At most 30?
    b. Less than 30?
    c. Between 15 and 25 (inclusive)?

55. Suppose only 75% of all drivers in a certain state regularly wear a seat belt. A random sample of 500 drivers is selected. What is the probability that
    a. Between 360 and 400 (inclusive) of the drivers in the sample regularly wear a seat belt?
    b. Fewer than 400 of those in the sample regularly wear a seat belt?

56. Show that the relationship between a general normal percentile and the corresponding z percentile is as stated in this section.

57. a. Show that if X has a normal distribution with parameters μ and σ, then $Y = aX + b$ (a linear function of X) also has a normal distribution. What are the parameters of the distribution of Y [i.e., $E(Y)$ and $V(Y)$]? [*Hint*: Write the cdf of Y, $P(Y \leq y)$, as an integral involving the pdf of X, and then differentiate with respect to y to get the pdf of Y.]

b. If, when measured in °C, temperature is normally distributed with mean 115 and standard deviation 2, what can be said about the distribution of temperature measured in °F?

58. There is no nice formula for the standard normal cdf $\Phi(z)$, but several good approximations have been published in articles. The following is from **"Approximations for Hand Calculators Using Small Integer Coefficients"** (*Mathematics of Computation*, 1977: 214–222). For $0 < z \leq 5.5$,

$$P(Z \geq z) = 1 - \Phi(z)$$

$$\approx .5 \exp\left\{-\left[\frac{(83z + 351)z + 562}{703/z + 165}\right]\right\}$$

The relative error of this approximation is less than .042%. Use this to calculate approximations to the following probabilities, and compare whenever possible to the probabilities obtained from Appendix Table A.3.

a. $P(Z \geq 1)$ **b.** $P(Z < -3)$

c. $P(-4 < Z < 4)$ **d.** $P(Z > 5)$

4.4 The Exponential and Gamma Distributions

The density curve corresponding to any normal distribution is bell-shaped and therefore symmetric. There are many practical situations in which the variable of interest to an investigator might have a skewed distribution. One family of distributions that has this property is the gamma family. We first consider a special case, the exponential distribution, and then generalize later in the section.

The Exponential Distribution

The family of exponential distributions provides probability models that are very widely used in engineering and science disciplines.

DEFINITION

> X is said to have an **exponential distribution** with (scale) parameter λ ($\lambda > 0$) if the pdf of X is
>
> $$f(x; \lambda) = \begin{cases} \lambda e^{-\lambda x} & x \geq 0 \\ 0 & \text{otherwise} \end{cases} \qquad (4.5)$$

Some sources write the exponential pdf in the form $(1/\beta)e^{-x/\beta}$, so that $\beta = 1/\lambda$. The expected value of an exponentially distributed random variable X is

$$\mu = E(X) = \int_0^\infty x\lambda e^{-\lambda x}\, dx$$

Obtaining this expected value necessitates doing an integration by parts. The variance of X can be computed using the fact that $V(X) = E(X^2) - [E(X)]^2$. The determination of $E(X^2)$ requires integrating by parts twice in succession. The results of these integrations are as follows:

$$\mu = \frac{1}{\lambda} \qquad \sigma^2 = \frac{1}{\lambda^2}$$

Both the mean and standard deviation of the exponential distribution equal $1/\lambda$. Graphs of several exponential pdf's are illustrated in Figure 4.26.

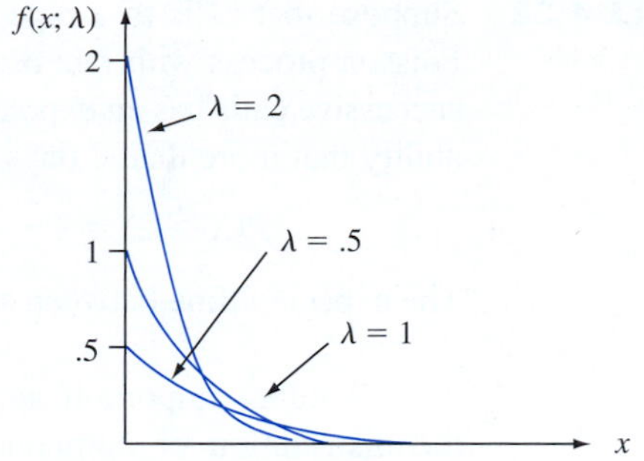

$f(x; \lambda)$

2

1

.5

$\lambda = 2$

$\lambda = .5$

$\lambda = 1$

x

Figure 4.26   Exponential density curves

The exponential pdf is easily integrated to obtain the cdf.

$$F(x; \lambda) = \begin{cases} 0 & x < 0 \\ 1 - e^{-\lambda x} & x \geq 0 \end{cases}$$

EXAMPLE 4.21   The article **"Probabilistic Fatigue Evaluation of Riveted Railway Bridges"** (*J. of Bridge Engr.*, **2008: 237–244**) suggested the exponential distribution with mean value 6 MPa as a model for the distribution of stress range in certain bridge connections. Let's assume that this is in fact the true model. Then $E(X) = 1/\lambda = 6$ implies that $\lambda = .1667$. The probability that stress range is at most 10 MPa is

$$P(X \leq 10) = F(10; .1667) = 1 - e^{-(.1667)(10)} = 1 - .189 = .811$$

The probability that stress range is between 5 and 10 MPa is

$$P(5 \leq X \leq 10) = F(10; .1667) - F(5; .1667) = (1 - e^{-1.667}) - (1 - e^{-.8335})$$
$$= .246 \qquad \blacksquare$$

The exponential distribution is frequently used as a model for the distribution of times between the occurrence of successive events, such as customers arriving at a service facility or calls coming in to a switchboard. The reason for this is that the exponential distribution is closely related to the Poisson process discussed in Chapter 3.

PROPOSITION

Suppose that the number of events occurring in any time interval of length t has a Poisson distribution with parameter αt (where α, the rate of the event process, is the expected number of events occurring in 1 unit of time) and that numbers of occurrences in nonoverlapping intervals are independent of one another. Then the distribution of elapsed time between the occurrence of two successive events is exponential with parameter $\lambda = \alpha$.

Although a complete proof is beyond the scope of the text, the result is easily verified for the time X_1 until the first event occurs:

$$P(X_1 \leq t) = 1 - P(X_1 > t) = 1 - P[\text{no events in } (0, t)]$$

$$= 1 - \frac{e^{-\alpha t} \cdot (\alpha t)^0}{0!} = 1 - e^{-\alpha t}$$

which is exactly the cdf of the exponential distribution.

EXAMPLE 4.22 Suppose that calls to a rape crisis center in a certain county occur according to a Poisson process with rate $\alpha = .5$ call per day. Then the number of days X between successive calls has an exponential distribution with parameter value .5, so the probability that more than 2 days elapse between calls is

$$P(X > 2) = 1 - P(X \leq 2) = 1 - F(2; .5) = e^{-(.5)(2)} = .368$$

The expected time between successive calls is $1/.5 = 2$ days. ∎

Another important application of the exponential distribution is to model the distribution of component lifetime. A partial reason for the popularity of such applications is the **"memoryless" property** of the exponential distribution. Suppose component lifetime is exponentially distributed with parameter λ. After putting the component into service, we leave for a period of t_0 hours and then return to find the component still working; what now is the probability that it lasts at least an additional t hours? In symbols, we wish $P(X \geq t + t_0 | X \geq t_0)$. By the definition of conditional probability,

$$P(X \geq t + t_0 | X \geq t_0) = \frac{P[(X \geq t + t_0) \cap (X \geq t_0)]}{P(X \geq t_0)}$$

But the event $X \geq t_0$ in the numerator is redundant, since both events can occur if and only if $X \geq t + t_0$. Therefore,

$$P(X \geq t + t_0 | X \geq t_0) = \frac{P(X \geq t + t_0)}{P(X \geq t_0)} = \frac{1 - F(t + t_0; \lambda)}{1 - F(t_0; \lambda)} = e^{-\lambda t}$$

This conditional probability is identical to the original probability $P(X \geq t)$ that the component lasted t hours. Thus *the distribution of additional lifetime is exactly the same as the original distribution of lifetime,* so at each point in time the component shows no effect of wear. In other words, the distribution of remaining lifetime is independent of current age.

Although the memoryless property can be justified at least approximately in many applied problems, in other situations components deteriorate with age or occasionally improve with age (at least up to a certain point). More general lifetime models are then furnished by the gamma, Weibull, and lognormal distributions (the latter two are discussed in the next section).

The Gamma Function

To define the family of gamma distributions, we first need to introduce a function that plays an important role in many branches of mathematics.

DEFINITION

For $\alpha > 0$, the **gamma function** $\Gamma(\alpha)$ is defined by

$$\Gamma(\alpha) = \int_0^{\infty} x^{\alpha - 1} e^{-x} \, dx \tag{4.6}$$

The most important properties of the gamma function are the following:

1. For any $\alpha > 1$, $\Gamma(\alpha) = (\alpha - 1) \cdot \Gamma(\alpha - 1)$ [via integration by parts]

2. For any positive integer, n, $\Gamma(n) = (n - 1)!$

3. $\Gamma(1/2) = \sqrt{\pi}$

Now let

$$f(x; \alpha) = \begin{cases} \dfrac{x^{\alpha-1}e^{-x}}{\Gamma(\alpha)} & x \geq 0 \\ 0 & \text{otherwise} \end{cases} \tag{4.7}$$

Then $f(x; \alpha) \geq 0$. Expression (4.6) implies that $\int_0^\infty f(x; \alpha)\, dx = \Gamma(\alpha)/\Gamma(\alpha) = 1$. Thus $f(x; \alpha)$ satisfies the two basic properties of a pdf.

The Gamma Distribution

DEFINITION

A continuous random variable X is said to have a **gamma distribution** if the pdf of X is

$$f(x; \alpha, \beta) = \begin{cases} \dfrac{1}{\beta^\alpha \Gamma(\alpha)} x^{\alpha-1} e^{-x/\beta} & x \geq 0 \\ 0 & \text{otherwise} \end{cases} \tag{4.8}$$

where the parameters α and β satisfy $\alpha > 0$, $\beta > 0$. The **standard gamma distribution** has $\beta = 1$, so the pdf of a standard gamma rv is given by (4.7).

The exponential distribution results from taking $\alpha = 1$ and $\beta = 1/\lambda$.

Figure 4.27(a) illustrates the graphs of the gamma pdf $f(x; \alpha, \beta)$ (4.8) for several (α, β) pairs, whereas Figure 4.27(b) presents graphs of the standard gamma pdf. For the standard pdf, when $\alpha \leq 1$, $f(x; \alpha)$ is strictly decreasing as x increases from 0; when $\alpha > 1$, $f(x; \alpha)$ rises from 0 at $x = 0$ to a maximum and then decreases. The parameter β in (4.8) is a scale parameter, and α is referred to as a *shape parameter* because changing its value alters the basic shape of the density curve.

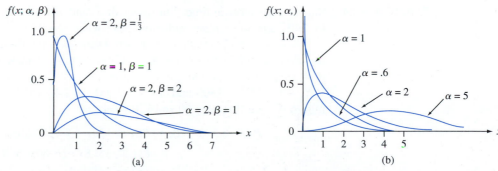

Figure 4.27 (a) Gamma density curves; (b) standard gamma density curves

The mean and variance of a random variable X having the gamma distribution $f(x; \alpha, \beta)$ are

$$E(X) = \mu = \alpha\beta \qquad V(X) = \sigma^2 = \alpha\beta^2$$

When X is a standard gamma rv, the cdf of X,

$$F(x; \alpha) = \int_0^x \frac{y^{\alpha-1}e^{-y}}{\Gamma(\alpha)}\, dy \qquad x > 0 \tag{4.9}$$

is called the **incomplete gamma function** [sometimes the incomplete gamma function refers to Expression (4.9) without the denominator $\Gamma(\alpha)$ in the integrand]. There are extensive tables of $F(x; \alpha)$ available; in Appendix Table A.4, we present a small tabulation for $\alpha = 1, 2, \ldots, 10$ and $x = 1, 2, \ldots, 15$.

EXAMPLE 4.23   The article **"The Probability Distribution of Maintenance Cost of a System Affected by the Gamma Process of Degradation"** (**Reliability Engr. and System Safety, 2012: 65–76**) notes that the gamma distribution is widely used to model the extent of degradation such as corrosion, creep, or wear. Let X represent the amount of degradation of a certain type, and suppose that it has a standard gamma distribution with $\alpha = 2$. Since

$$P(a \le X \le b) = F(b) - F(a)$$

when X is continuous,

$$P(3 \le X \le 5) = F(5; 2) - F(3; 2) = .960 - .801 = .159$$

The probability that the amount of degradation exceeds 4 is

$$P(X > 4) = 1 - P(X \le 4) = 1 - F(4; 2) = 1 - .908 = .092 \qquad \blacksquare$$

The incomplete gamma function can also be used to compute probabilities involving nonstandard gamma distributions. These probabilities can also be obtained almost instantaneously from various software packages.

PROPOSITION

> Let X have a gamma distribution with parameters α and β. Then for any $x > 0$, the cdf of X is given by
>
> $$P(X \le x) = F(x; \alpha, \beta) = F\left(\frac{x}{\beta}; \alpha\right)$$
>
> where $F(\cdot; \alpha)$ is the incomplete gamma function.

EXAMPLE 4.24   Suppose the survival time X in weeks of a randomly selected male mouse exposed to 240 rads of gamma radiation has (what else!) a gamma distribution with $\alpha = 8$ and $\beta = 15$. (Data in **Survival Distributions: Reliability Applications in the Biomedical Services, by A. J. Gross and V. Clark,** suggests $\alpha \approx 8.5$ and $\beta \approx 13.3$.) The expected survival time is $E(X) = (8)(15) = 120$ weeks, whereas $V(X) = (8)(15)^2 = 1800$ and $\sigma_X = \sqrt{1800} = 42.43$ weeks. The probability that a mouse survives between 60 and 120 weeks is

$$
\begin{aligned}
P(60 \le X \le 120) &= P(X \le 120) - P(X \le 60) \\
&= F(120/15; 8) - F(60/15; 8) \\
&= F(8; 8) - F(4; 8) = .547 - .051 = .496
\end{aligned}
$$

The probability that a mouse survives at least 30 weeks is

$$
\begin{aligned}
P(X \ge 30) &= 1 - P(X < 30) = 1 - P(X \le 30) \\
&= 1 - F(30/15; 8) = .999 \qquad \blacksquare
\end{aligned}
$$

The Chi-Squared Distribution

The chi-squared distribution is important because it is the basis for a number of procedures in statistical inference. The central role played by the chi-squared distribution in inference springs from its relationship to normal distributions (see Exercise 71). We'll discuss this distribution in more detail in later chapters.

DEFINITION

Let ν be a positive integer. Then a random variable X is said to have a **chi-squared distribution** with parameter ν if the pdf of X is the gamma density with $\alpha = \nu/2$ and $\beta = 2$. The pdf of a chi-squared rv is thus

$$f(x; \nu) = \begin{cases} \dfrac{1}{2^{\nu/2}\Gamma(\nu/2)} x^{(\nu/2)-1} e^{-x/2} & x \geq 0 \\ 0 & x < 0 \end{cases} \qquad (4.10)$$

The parameter ν is called the **number of degrees of freedom** (df) of X. The symbol χ^2 is often used in place of "chi-squared."

EXERCISES Section 4.4 (59–71)

59. Let $X = $ the time between two successive arrivals at the drive-up window of a local bank. If X has an exponential distribution with $\lambda = 1$ (which is identical to a standard gamma distribution with $\alpha = 1$), compute the following:
   a. The expected time between two successive arrivals
   b. The standard deviation of the time between successive arrivals
   c. $P(X \leq 4)$ **d.** $P(2 \leq X \leq 5)$

60. Let X denote the distance (m) that an animal moves from its birth site to the first territorial vacancy it encounters. Suppose that for banner-tailed kangaroo rats, X has an exponential distribution with parameter $\lambda = .01386$ (as suggested in the article **"Competition and Dispersal from Multiple Nests,"** *Ecology,* **1997: 873–883**).
   a. What is the probability that the distance is at most 100 m? At most 200 m? Between 100 and 200 m?
   b. What is the probability that distance exceeds the mean distance by more than 2 standard deviations?
   c. What is the value of the median distance?

61. Data collected at Toronto Pearson International Airport suggests that an exponential distribution with mean value 2.725 hours is a good model for rainfall duration (*Urban Stormwater Management Planning with Analytical Probabilistic Models,* **2000, p. 69**).
   a. What is the probability that the duration of a particular rainfall event at this location is at least 2 hours? At most 3 hours? Between 2 and 3 hours?
   b. What is the probability that rainfall duration exceeds the mean value by more than 2 standard deviations? What is the probability that it is less than the mean value by more than one standard deviation?

62. The article **"Microwave Observations of Daily Antarctic Sea-Ice Edge Expansion and Contribution Rates"** (*IEEE Geosci. and Remote Sensing Letters,* **2006: 54–58**) states that "The distribution of the daily

sea-ice advance/retreat from each sensor is similar and is approximately double exponential." The proposed double exponential distribution has density function $f(x) = .5\lambda e^{-\lambda|x|}$ for $-\infty < x < \infty$. The standard deviation is given as 40.9 km.
   a. What is the value of the parameter λ?
   b. What is the probability that the extent of daily sea-ice change is within 1 standard deviation of the mean value?

63. A consumer is trying to decide between two long-distance calling plans. The first one charges a flat rate of 10¢ per minute, whereas the second charges a flat rate of 99¢ for calls up to 20 minutes in duration and then 10¢ for each additional minute exceeding 20 (assume that calls lasting a noninteger number of minutes are charged proportionately to a whole-minute's charge). Suppose the consumer's distribution of call duration is exponential with parameter λ.
   a. Explain intuitively how the choice of calling plan should depend on what the expected call duration is.
   b. Which plan is better if expected call duration is 10 minutes? 15 minutes? [*Hint:* Let $h_1(x)$ denote the cost for the first plan when call duration is x minutes and let $h_2(x)$ be the cost function for the second plan. Give expressions for these two cost functions, and then determine the expected cost for each plan.]

64. Evaluate the following:
   a. $\Gamma(6)$ **b.** $\Gamma(5/2)$
   c. $F(4; 5)$ (the incomplete gamma function) and $F(5; 4)$
   d. $P(X \leq 5)$ when X has a standard gamma distribution with $\alpha = 7$.
   e. $P(3 < X < 8)$ when X has the distribution specified in (d).

65. Let X denote the data transfer time (ms) in a grid computing system (the time required for data transfer between a "worker" computer and a "master" computer. Suppose that X has a gamma distribution with mean value 37.5 ms and standard deviation 21.6 (suggested by the article **"Computation Time of Grid Computing with Data Transfer Times that Follow a Gamma Distribution,"** *Proceedings of the First International Conference on Semantics, Knowledge, and Grid,* **2005**).

a. What are the values of α and β?

b. What is the probability that data transfer time exceeds 50 ms?

c. What is the probability that data transfer time is between 50 and 75 ms?

66. The two-parameter gamma distribution can be generalized by introducing a third parameter γ, called a *threshold* or *location* parameter: replace x in (4.8) by $x - \gamma$ and $x \geq 0$ by $x \geq \gamma$. This amounts to shifting the density curves in Figure 4.27 so that they begin their ascent or descent at γ rather than 0. The article **"Bivariate Flood Frequency Analysis with Historical Information Based on Copulas"** (*J. of Hydrologic Engr.,* **2013: 1018–1030**) employs this distribution to model $X =$ 3-day flood volume (10^8 m^3). Suppose that values of the parameters are $\alpha = 12$, $\beta = 7$, $\gamma = 40$ (very close to estimates in the cited article based on past data).

a. What are the mean value and standard deviation of X?

b. What is the probability that flood volume is between 100 and 150?

c. What is the probability that flood volume exceeds its mean value by more than one standard deviation?

d. What is the 95th percentile of the flood volume distribution?

67. Suppose that when a transistor of a certain type is subjected to an accelerated life test, the lifetime X (in weeks) has a gamma distribution with mean 24 weeks and standard deviation 12 weeks.

a. What is the probability that a transistor will last between 12 and 24 weeks?

b. What is the probability that a transistor will last at most 24 weeks? Is the median of the lifetime distribution less than 24? Why or why not?

c. What is the 99th percentile of the lifetime distribution?

d. Suppose the test will actually be terminated after t weeks. What value of t is such that only .5% of all transistors would still be operating at termination?

68. The special case of the gamma distribution in which α is a positive integer n is called an Erlang distribution. If we replace β by $1/\lambda$ in Expression (4.8), the Erlang pdf is

$$f(x; \lambda, n) = \begin{cases} \dfrac{\lambda(\lambda x)^{n-1} e^{-\lambda x}}{(n-1)!} & x \geq 0 \\ 0 & x < 0 \end{cases}$$

It can be shown that if the times between successive events are independent, each with an exponential distribution with parameter λ, then the total time X that

elapses before all of the next n events occur has pdf $f(x; \lambda, n)$.

a. What is the expected value of X? If the time (in minutes) between arrivals of successive customers is exponentially distributed with $\lambda = .5$, how much time can be expected to elapse before the tenth customer arrives?

b. If customer interarrival time is exponentially distributed with $\lambda = .5$, what is the probability that the tenth customer (after the one who has just arrived) will arrive within the next 30 min?

c. The event $\{X \leq t\}$ occurs iff at least n events occur in the next t units of time. Use the fact that the number of events occurring in an interval of length t has a Poisson distribution with parameter λt to write an expression (involving Poisson probabilities) for the Erlang cdf $F(t; \lambda, n) = P(X \leq t)$.

69. A system consists of five identical components connected in series as shown:

As soon as one component fails, the entire system will fail. Suppose each component has a lifetime that is exponentially distributed with $\lambda = .01$ and that components fail independently of one another. Define events $A_i = \{i\text{th component lasts at least } t \text{ hours}\}$, $i = 1,\ldots, 5$, so that the A_is are independent events. Let $X =$ the time at which the system fails—that is, the shortest (minimum) lifetime among the five components.

a. The event $\{X \geq t\}$ is equivalent to what event involving $A_1,\ldots, A_5$?

b. Using the independence of the A_i's, compute $P(X \geq t)$. Then obtain $F(t) = P(X \leq t)$ and the pdf of X. What type of distribution does X have?

c. Suppose there are n components, each having exponential lifetime with parameter λ. What type of distribution does X have?

70. If X has an exponential distribution with parameter λ, derive a general expression for the $(100p)$th percentile of the distribution. Then specialize to obtain the median.

71. **a.** The event $\{X^2 \leq y\}$ is equivalent to what event involving X itself?

b. If X has a standard normal distribution, use part (a) to write the integral that equals $P(X^2 \leq y)$. Then differentiate this with respect to y to obtain the pdf of X^2 [the square of a $N(0, 1)$ variable]. Finally, show that X^2 has a chi-squared distribution with $\nu = 1$ df [see (4.10)]. [*Hint*: Use the following identity.]

$$\frac{d}{dy}\left\{\int_{a(y)}^{b(y)} f(x)\, dx\right\} = f[b(y)] \cdot b'(y) - f[a(y)] \cdot a'(y)$$

4.5 Other Continuous Distributions

The normal, gamma (including exponential), and uniform families of distributions provide a wide variety of probability models for continuous variables, but there are many practical situations in which no member of these families fits a set of observed data very well. Statisticians and other investigators have developed other families of distributions that are often appropriate in practice.

The Weibull Distribution

The family of Weibull distributions was introduced by the Swedish physicist Waloddi Weibull in 1939; his 1951 article **"A Statistical Distribution Function of Wide Applicability"** (*J. of Applied Mechanics*, **vol. 18: 293–297**) discusses a number of applications.

DEFINITION

> A random variable X said to have a **Weibull distribution** with shape parameter α and scale parameter β ($\alpha > 0, \beta > 0$) if the pdf of X is
>
> $$f(x; \alpha, \beta) = \begin{cases} \dfrac{\alpha}{\beta^\alpha} x^{\alpha-1} e^{-(x/\beta)^\alpha} & x \geq 0 \\ \\ 0 & x < 0 \end{cases} \qquad (4.11)$$

In some situations, there are theoretical justifications for the appropriateness of the Weibull distribution, but in many applications $f(x; \alpha, \beta)$ simply provides a good fit to observed data for particular values of α and β. When $\alpha = 1$, the pdf reduces to the exponential distribution (with $\lambda = 1/\beta$), so the exponential distribution is a special case of both the gamma and Weibull distributions. However, there are gamma distributions that are not Weibull distributions and vice versa, so one family is not a subset of the other. Both α and β can be varied to obtain a number of different-looking density curves, as illustrated in Figure 4.28.

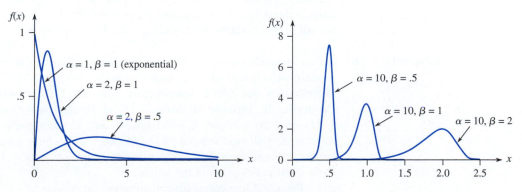

Figure 4.28 Weibull density curves

Integrating to obtain $E(X)$ and $E(X^2)$ yields

$$\mu = \beta\Gamma\left(1 + \frac{1}{\alpha}\right) \qquad \sigma^2 = \beta^2\left\{\Gamma\left(1 + \frac{2}{\alpha}\right) - \left[\Gamma\left(1 + \frac{1}{\alpha}\right)\right]^2\right\}$$

The computation of μ and σ^2 thus necessitates using the gamma function.

The integration $\int_0^x f(y; \alpha, \beta)\, dy$ is easily carried out to obtain the cdf of X.

The cdf of a Weibull rv having parameters α and β is

$$F(x; \alpha, \beta) = \begin{cases} 0 & x < 0 \\ 1 - e^{-(x/\beta)^\alpha} & x \geq 0 \end{cases} \tag{4.12}$$

EXAMPLE 4.25   In recent years the Weibull distribution has been used to model engine emissions of various pollutants. Let X denote the amount of NO_x emission (g/gal) from a randomly selected four-stroke engine of a certain type, and suppose that X has a Weibull distribution with $\alpha = 2$ and $\beta = 10$ (suggested by information in the article **"Quantification of Variability and Uncertainty in Lawn and Garden Equipment NO_x and Total Hydrocarbon Emission Factors," *J. of the Air and Waste Management Assoc.*, 2002: 435–448).** The corresponding density curve looks exactly like the one in Figure 4.28 for $\alpha = 2$, $\beta = 1$ except that now the values 50 and 100 replace 5 and 10 on the horizontal axis. Then

$$P(X \leq 10) = F(10; 2, 10) = 1 - e^{-(10/10)^2} = 1 - e^{-1} = .632$$

Similarly, $P(X \leq 25) = .998$, so the distribution is almost entirely concentrated on values between 0 and 25. The value c which separates the 5% of all engines having the largest amounts of NO_x emissions from the remaining 95% satisfies

$$.95 = 1 - e^{-(c/10)^2}$$

Isolating the exponential term on one side, taking logarithms, and solving the resulting equation gives $c \approx 17.3$ as the 95th percentile of the emission distribution. ∎

In practical situations, a Weibull model may be reasonable except that the smallest possible X value may be some value γ not assumed to be zero (this would also apply to a gamma model; see Exercise 66). The quantity γ can then be regarded as a third (threshold or location) parameter of the distribution, which is what Weibull did in his original work. For, say, $\gamma = 3$, all curves in Figure 4.28 would be shifted 3 units to the right. This is equivalent to saying that $X - \gamma$ has the pdf (4.11), so that the cdf of X is obtained by replacing x in (4.12) by $x - \gamma$.

EXAMPLE 4.26   An understanding of the volumetric properties of asphalt is important in designing mixtures which will result in high-durability pavement. The article **"Is a Normal Distribution the Most Appropriate Statistical Distribution for Volumetric Properties in Asphalt Mixtures?" (*J. of Testing and Evaluation*, Sept. 2009: 1–11)** used the analysis of some sample data to recommend that for a particular mixture, $X = $ air void volume (%) be modeled with a three-parameter Weibull distribution. Suppose the values of the parameters are $\gamma = 4$, $\alpha = 1.3$, and $\beta = .8$ (quite close to estimates given in the article).

For $x > 4$, the cumulative distribution function is

$$F(x; \alpha, \beta, \gamma) = F(x; 1.3, .8, 4) = 1 - e^{-[(x-4)/.8]^{1.3}}$$

The probability that the air void volume of a specimen is between 5% and 6% is

$$P(5 \le X \le 6) = F(6; 1.3, .8, 4) - F(5; 1.3, .8, 4) = e^{-[(5-4)/.8]^{1.3}} - e^{-[(6-4)/.8]^{1.3}}$$
$$= .263 - .037 = .226$$

Figure 4.29 shows a graph from Minitab of the corresponding Weibull density function in which the shaded area corresponds to the probability just calculated.

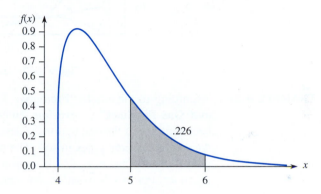

Figure 4.29 Weibull density curve with threshold = 4, shape = 1.3, scale = .8 ∎

The Lognormal Distribution

DEFINITION

> A nonnegative rv X is said to have a **lognormal distribution** if the rv $Y = \ln(X)$ has a normal distribution. The resulting pdf of a lognormal rv when $\ln(X)$ is normally distributed with parameters μ and σ is
>
> $$f(x; \mu, \sigma) = \begin{cases} \dfrac{1}{\sqrt{2\pi}\sigma x} e^{-[\ln(x) - \mu]^2/(2\sigma^2)} & x \ge 0 \\ 0 & x < 0 \end{cases}$$

Be careful here; the parameters μ and σ are not the mean and standard deviation of X but of $\ln(X)$. The mean and variance of X can be shown to be

$$E(X) = e^{\mu + \sigma^2/2} \qquad V(X) = e^{2\mu + \sigma^2} \cdot (e^{\sigma^2} - 1)$$

In Chapter 5, we will present a theoretical justification for this distribution in connection with the Central Limit Theorem. But as with other distributions, the lognormal can be used as a model even in the absence of such justification. Figure 4.30 illustrates graphs of the lognormal pdf; although a normal curve is symmetric, a lognormal curve has a positive skew.

Because $\ln(X)$ has a normal distribution, the cdf of X can be expressed in terms of the cdf $\Phi(z)$ of a standard normal rv Z.

$$F(x; \mu, \sigma) = P(X \le x) = P[\ln(X) \le \ln(x)]$$

$$= P\left(Z \le \frac{\ln(x) - \mu}{\sigma}\right) = \Phi\left(\frac{\ln(x) - \mu}{\sigma}\right) \qquad x \ge 0 \qquad (4.13)$$

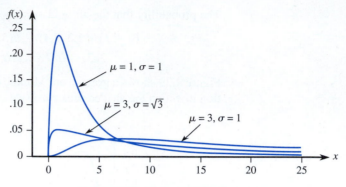

Figure 4.30 Lognormal density curves

EXAMPLE 4.27 According to the article **"Predictive Model for Pitting Corrosion in Buried Oil and Gas Pipelines"** (*Corrosion*, **2009: 332–342),** the lognormal distribution has been reported as the best option for describing the distribution of maximum pit depth data from cast iron pipes in soil. The authors suggest that a lognormal distribution with $\mu = .353$ and $\sigma = .754$ is appropriate for maximum pit depth (mm) of buried pipelines. For this distribution, the mean value and variance of pit depth are

$$E(X) = e^{.353 + (.754)^2/2} = e^{.6373} = 1.891$$
$$V(X) = e^{2(.353) + (.754)^2} \cdot (e^{(.754)^2} - 1) = (3.57697)(.765645) = 2.7387$$

The probability that maximum pit depth is between 1 and 2 mm is

$$P(1 \leq X \leq 2) = P(\ln(1) \leq \ln(X) \leq \ln(2)) = P(0 \leq \ln(X) \leq .693)$$
$$= P\left(\frac{0 - .353}{.754} \leq Z \leq \frac{.693 - .353}{.754}\right) = \Phi(.47) - \Phi(-.45) = .354$$

This probability is illustrated in Figure 4.31 (from Minitab).

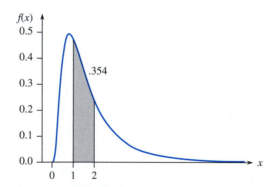

Figure 4.31 Lognormal density curve with $\mu = .353$ and $\sigma = .754$

What value c is such that only 1% of all specimens have a maximum pit depth exceeding c? The desired value satisfies

$$.99 = P(X \leq c) = P\left(Z \leq \frac{\ln(c) - .353}{.754}\right)$$

The z critical value 2.33 captures an upper-tail area of .01 ($z_{.01} = 2.33$), and thus a cumulative area of .99. This implies that

$$\frac{\ln(c) - .353}{.754} = 2.33$$

from which $\ln(c) = 2.1098$ and $c = 8.247$. Thus 8.247 is the 99th percentile of the maximum pit depth distribution. ∎

The Beta Distribution

All families of continuous distributions discussed so far except for the uniform distribution have positive density over an infinite interval (though typically the density function decreases rapidly to zero beyond a few standard deviations from the mean). The beta distribution provides positive density only for X in an interval of finite length.

DEFINITION

A random variable X is said to have a **beta distribution** with parameters α, β (both positive), A, and B if the pdf of X is

$$f(x; \alpha, \beta, A, B) = \begin{cases} \dfrac{1}{B - A} \cdot \dfrac{\Gamma(\alpha + \beta)}{\Gamma(\alpha) \cdot \Gamma(\beta)} \left(\dfrac{x - A}{B - A}\right)^{\alpha - 1} \left(\dfrac{B - x}{B - A}\right)^{\beta - 1} & A \le x \le B \\ 0 & \text{otherwise} \end{cases}$$

The case $A = 0, B = 1$ gives the **standard beta distribution.**

Figure 4.32 illustrates several standard beta pdf's. Graphs of the general pdf are similar, except they are shifted and then stretched or compressed to fit over $[A, B]$. Unless α and β are integers, integration of the pdf to calculate probabilities is difficult. Either a table of the incomplete beta function or appropriate software should be used. The mean and variance of X are

$$\mu = A + (B - A) \cdot \frac{\alpha}{\alpha + \beta} \qquad \sigma^2 = \frac{(B - A)^2 \alpha \beta}{(\alpha + \beta)^2 (\alpha + \beta + 1)}$$

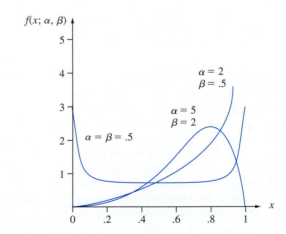

Figure 4.32   Standard beta density curves

EXAMPLE 4.28   Project managers often use a method labeled PERT—for program evaluation and review technique—to coordinate the various activities making up a large project. (One successful application was in the construction of the *Apollo* spacecraft.) A standard assumption in PERT analysis is that the time necessary to complete any particular activity once it has been started has a beta distribution with $A =$ the optimistic time (if everything goes well) and $B =$ the pessimistic time (if everything goes badly). Suppose that in constructing a single-family house, the time X (in days) necessary for laying the foundation has a beta distribution with $A = 2$, $B = 5$, $\alpha = 2$, and $\beta = 3$. Then $\alpha/(\alpha + \beta) = .4$, so $E(X) = 2 + (3)(.4) = 3.2$. For these values of α and β, the pdf of X is a simple polynomial function. The probability that it takes at most 3 days to lay the foundation is

$$P(X \le 3) = \int_2^3 \frac{1}{3} \cdot \frac{4!}{1!2!} \left(\frac{x-2}{3}\right) \left(\frac{5-x}{3}\right)^2 dx$$

$$= \frac{4}{27} \int_2^3 (x-2)(5-x)^2 dx = \frac{4}{27} \cdot \frac{11}{4} = \frac{11}{27} = .407 \quad \blacksquare$$

The standard beta distribution is commonly used to model variation in the proportion or percentage of a quantity occurring in different samples, such as the proportion of a 24-hour day that an individual is asleep or the proportion of a certain element in a chemical compound.

EXERCISES Section 4.5 (72–86)

72. The lifetime X (in hundreds of hours) of a certain type of vacuum tube has a Weibull distribution with parameters $\alpha = 2$ and $\beta = 3$. Compute the following:
   a. $E(X)$ and $V(X)$
   b. $P(X \le 6)$
   c. $P(1.5 \le X \le 6)$

   (This Weibull distribution is suggested as a model for time in service in **"On the Assessment of Equipment Reliability: Trading Data Collection Costs for Precision,"** *J. of Engr. Manuf.*, 1991: 105–109.)

73. The authors of the article **"A Probabilistic Insulation Life Model for Combined Thermal-Electrical Stresses"** (*IEEE Trans. on Elect. Insulation*, 1985: 519–522) state that "the Weibull distribution is widely used in statistical problems relating to aging of solid insulating materials subjected to aging and stress." They propose the use of the distribution as a model for time (in hours) to failure of solid insulating specimens subjected to AC voltage. The values of the parameters depend on the voltage and temperature; suppose $\alpha = 2.5$ and $\beta = 200$ (values suggested by data in the article).
   a. What is the probability that a specimen's lifetime is at most 250? Less than 250? More than 300?
   b. What is the probability that a specimen's lifetime is between 100 and 250?
   c. What value is such that exactly 50% of all specimens have lifetimes exceeding that value?

74. Once an individual has been infected with a certain disease, let X represent the time (days) that elapses before the individual becomes infectious. The article **"The Probability of Containment for Multitype Branching Process Models for Emerging Epidemics"** (*J. of Applied Probability*, 2011: 173–188) proposes a Weibull distribution with $\alpha = 2.2$, $\beta = 1.1$, and $\gamma = .5$ (refer to Example 4.26).
   a. Calculate $P(1 < X < 2)$.
   b. Calculate $P(X > 1.5)$.
   c. What is the 90th percentile of the distribution?
   d. What are the mean and standard deviation of X?

75. Let X have a Weibull distribution with the pdf from Expression (4.11). Verify that $\mu = \beta\Gamma(1 + 1/\alpha)$. [*Hint*: In the integral for $E(X)$, make the change of variable $y = (x/\beta)^\alpha$, so that $x = \beta y^{1/\alpha}$.]

76. The article **"The Statistics of Phytotoxic Air Pollutants"** (*J. of Royal Stat. Soc.*, 1989: 183–198) suggests the lognormal distribution as a model for SO_2 concentration above a certain forest. Suppose the parameter values are $\mu = 1.9$ and $\sigma = .9$.
   a. What are the mean value and standard deviation of concentration?
   b. What is the probability that concentration is at most 10? Between 5 and 10?

77. The authors of the article from which the data in Exercise 1.27 was extracted suggested that a reasonable probability model for drill lifetime was a lognormal distribution with $\mu = 4.5$ and $\sigma = .8$.
   a. What are the mean value and standard deviation of lifetime?
   b. What is the probability that lifetime is at most 100?

c. What is the probability that lifetime is at least 200? Greater than 200?

78. The article **"On Assessing the Accuracy of Offshore Wind Turbine Reliability-Based Design Loads from the Environmental Contour Method"** (*Intl. J. of Offshore and Polar Engr.*, 2005: 132–140) proposes the Weibull distribution with $\alpha = 1.817$ and $\beta = .863$ as a model for 1-hour significant wave height (m) at a certain site.

a. What is the probability that wave height is at most .5 m?

b. What is the probability that wave height exceeds its mean value by more than one standard deviation?

c. What is the median of the wave-height distribution?

d. For $0 < p < 1$, give a general expression for the $100p$th percentile of the wave-height distribution.

79. Nonpoint source loads are chemical masses that travel to the main stem of a river and its tributaries in flows that are distributed over relatively long stream reaches, in contrast to those that enter at well-defined and regulated points. The article **"Assessing Uncertainty in Mass Balance Calculation of River Nonpoint Source Loads"** (*J. of Envir. Engr.*, 2008: 247–258) suggested that for a certain time period and location, $X =$ nonpoint source load of total dissolved solids could be modeled with a lognormal distribution having mean value 10,281 kg/day/km and a coefficient of variation $CV = .40$ ($CV = \sigma_X/\mu_X$).

a. What are the mean value and standard deviation of $\ln(X)$?

b. What is the probability that X is at most 15,000 kg/day/km?

c. What is the probability that X exceeds its mean value, and why is this probability not .5?

d. Is 17,000 the 95th percentile of the distribution?

80. a. Use Equation (4.13) to write a formula for the median $\tilde{\mu}$ of the lognormal distribution. What is the median for the load distribution of Exercise 79?

b. Recalling that z_α is our notation for the $100(1 - \alpha)$ percentile of the standard normal distribution, write an expression for the $100(1 - \alpha)$ percentile of the lognormal distribution. In Exercise 79, what value will load exceed only 1% of the time?

81. Sales delay is the elapsed time between the manufacture of a product and its sale. According to the article **"Warranty Claims Data Analysis Considering Sales Delay"** (*Quality and Reliability Engr. Intl.*, 2013: 113–123), it is quite common for investigators to model sales delay using a lognormal distribution. For a particular product, the cited article proposes this distribution with parameter values $\mu = 2.05$ and $\sigma^2 = .06$ (here the unit for delay is months).

a. What are the variance and standard deviation of delay time?

b. What is the probability that delay time exceeds 12 months?

c. What is the probability that delay time is within one standard deviation of its mean value?

d. What is the median of the delay time distribution?

e. What is the 99th percentile of the delay time distribution?

f. Among 10 randomly selected such items, how many would you expect to have a delay time exceeding 8 months?

82. As in the case of the Weibull and Gamma distributions, the lognormal distribution can be modified by the introduction of a third parameter γ such that the pdf is shifted to be positive only for $x > \gamma$. The article cited in Exercise 4.39 suggested that a shifted lognormal distribution with shift (i.e., threshold) $= 1.0$, mean value $= 2.16$, and standard deviation $= 1.03$ would be an appropriate model for the rv $X = $ *maximum-to-average depth ratio* of a corrosion defect in pressurized steel.

a. What are the values of μ and σ for the proposed distribution?

b. What is the probability that depth ratio exceeds 2?

c. What is the median of the depth ratio distribution?

d. What is the 99th percentile of the depth ratio distribution?

83. What condition on α and β is necessary for the standard beta pdf to be symmetric?

84. Suppose the proportion X of surface area in a randomly selected quadrat that is covered by a certain plant has a standard beta distribution with $\alpha = 5$ and $\beta = 2$.

a. Compute $E(X)$ and $V(X)$.

b. Compute $P(X \le .2)$.

c. Compute $P(.2 \le X \le .4)$.

d. What is the expected proportion of the sampling region not covered by the plant?

85. Let X have a standard beta density with parameters α and β.

a. Verify the formula for $E(X)$ given in the section.

b. Compute $E[(1 - X)^m]$. If X represents the proportion of a substance consisting of a particular ingredient, what is the expected proportion that does not consist of this ingredient?

86. Stress is applied to a 20-in. steel bar that is clamped in a fixed position at each end. Let $Y = $ the distance from the left end at which the bar snaps. Suppose $Y/20$ has a standard beta distribution with $E(Y) = 10$ and $V(Y) = \frac{100}{7}$.

a. What are the parameters of the relevant standard beta distribution?

b. Compute $P(8 \le Y \le 12)$.

c. Compute the probability that the bar snaps more than 2 in. from where you expect it to.

4.6 Probability Plots

An investigator will often have obtained a numerical sample $x_1, x_2, \ldots, x_n$ and wish to know whether it is plausible that it came from a population distribution of some particular type (e.g., from a normal distribution). For one thing, many formal procedures from statistical inference are based on the assumption that the population distribution is of a specified type. The use of such a procedure is inappropriate if the actual underlying probability distribution differs greatly from the assumed type. For example, the article **"Toothpaste Detergents: A Potential Source of Oral Soft Tissue Damage" (*Intl. J. of Dental Hygiene*, 2008: 193–198)** contains the following statement: "Because the sample number for each experiment (replication) was limited to three wells per treatment type, the data were assumed to be normally distributed." As justification for this leap of faith, the authors wrote that "Descriptive statistics showed standard deviations that suggested a normal distribution to be highly likely." *Note*: This argument is not very persuasive.

Additionally, understanding the underlying distribution can sometimes give insight into the physical mechanisms involved in generating the data. An effective way to check a distributional assumption is to construct what is called a **probability plot**. The essence of such a plot is that if the distribution on which the plot is based is correct, the points in the plot should fall close to a straight line. If the actual distribution is quite different from the one used to construct the plot, the points will likely depart substantially from a linear pattern.

Sample Percentiles

The details involved in constructing probability plots differ a bit from source to source. The basis for our construction is a comparison between percentiles of the sample data and the corresponding percentiles of the distribution under consideration. Recall that the $(100p)$th percentile of a continuous distribution with cdf $F(\cdot)$ is the number $\eta(p)$ that satisfies $F(\eta(p)) = p$. That is, $\eta(p)$ is the number on the measurement scale such that the area under the density curve to the left of $\eta(p)$ is p. Thus the 50th percentile $\eta(.5)$ satisfies $F(\eta(.5)) = .5$, and the 90th percentile satisfies $F(\eta(.9)) = .9$. Consider as an example the standard normal distribution, for which we have denoted the cdf by $\Phi(\cdot)$. From Appendix Table A.3, we find the 20th percentile by locating the row and column in which .2000 (or a number as close to it as possible) appears inside the table. Since .2005 appears at the intersection of the $-.8$ row and the .04 column, the 20th percentile is approximately $-.84$. Similarly, the 25th percentile of the standard normal distribution is (using linear interpolation) approximately $-.675$.

Roughly speaking, sample percentiles are defined in the same way that percentiles of a population distribution are defined. The 50th-sample percentile should separate the smallest 50% of the sample from the largest 50%, the 90th percentile should be such that 90% of the sample lies below that value and 10% lies above, and so on. Unfortunately, we run into problems when we actually try to compute the sample percentiles for a particular sample of n observations. If, for example, $n = 10$, we can split off 20% of these values or 30% of the data, but there is no value that will split off exactly 23% of these ten observations. To proceed further, we need an operational definition of sample percentiles (this is one place where different people do slightly different things). Recall that when n is odd, the sample median or 50th-sample percentile is the middle value in the ordered list, for example, the sixth-largest value when $n = 11$. This amounts to regarding the middle observation as being half in the lower half of the data and half in the upper half. Similarly, suppose $n = 10$. Then if we call the third-smallest value the 25th percentile, we are regarding that value as being half in the lower group (consisting of the two smallest observations) and half in the upper group (the seven largest observations). This leads to the following general definition of sample percentiles.

DEFINITION

Order the n sample observations from smallest to largest. Then the ith smallest observation in the list is taken to be the **$[100(i - .5)/n]$th sample percentile.**

Once the percentage values $100(i - .5)/n$ $(i = 1, 2, \ldots, n)$ have been calculated, sample percentiles corresponding to intermediate percentages can be obtained by linear interpolation. For example, if $n = 10$, the percentages corresponding to the ordered sample observations are $100(1 - .5)/10 = 5\%$, $100(2 - .5)/10 = 15\%$, $25\%, \ldots$, and $100(10 - .5)/10 = 95\%$. The 10th percentile is then halfway between the 5th percentile (smallest sample observation) and the 15th percentile (second-smallest observation). For our purposes, such interpolation is not necessary because a probability plot will be based only on the percentages $100(i - .5)/n$ corresponding to the n sample observations.

A Probability Plot

Suppose now that for percentages $100(i - .5)/n$ $(i = 1, \ldots, n)$ the percentiles are determined for a specified population distribution whose plausibility is being investigated. If the sample was actually selected from the specified distribution, the sample percentiles (ordered sample observations) should be reasonably close to the corresponding population distribution percentiles. That is, for $i = 1, 2, \ldots, n$ there should be reasonable agreement between the ith smallest sample observation and the $[100(i - .5)/n]$th percentile for the specified distribution. Let's consider the (population percentile, sample percentile) pairs—that is, the pairs

$$\left(\begin{array}{cc} [100(i - .5)/n]\text{th percentile} & i\text{th smallest sample} \\ \text{of the distribution,} & \text{observation} \end{array} \right)$$

for $i = 1, \ldots, n$. Each such pair can be plotted as a point on a two-dimensional coordinate system. If the sample percentiles are close to the corresponding population distribution percentiles, the first number in each pair will be roughly equal to the second number. The plotted points will then fall close to a 45° line. Substantial deviations of the plotted points from a 45° line cast doubt on the assumption that the distribution under consideration is the correct one.

Example 4.29

The value of a certain physical constant is known to an experimenter. The experimenter makes $n = 10$ independent measurements of this value using a particular measurement device and records the resulting measurement errors (error = observed value − true value). These observations appear in the accompanying table.

Percentage	5	15	25	35	45
z percentile	−1.645	−1.037	−.675	−.385	−.126
Sample observation	−1.91	−1.25	−.75	−.53	.20
Percentage	55	65	75	85	95
z percentile	.126	.385	.675	1.037	1.645
Sample observation	.35	.72	.87	1.40	1.56

Is it plausible that the random variable *measurement error* has a standard normal distribution? The needed standard normal (z) percentiles are also displayed in the table. Thus the points in the probability plot are $(-1.645, -1.91)$, $(-1.037, -1.25),\ldots$, and $(1.645, 1.56)$. Figure 4.33 shows the resulting plot. Although the points deviate a bit from the 45° line, the predominant impression is that this line fits the points very well. The plot suggests that the standard normal distribution is a reasonable probability model for measurement error.

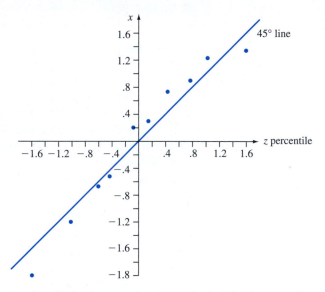

Figure 4.33   Plot of pairs (z percentile, observed value) for the data of Example 4.29

Figure 4.34 shows a plot of pairs (z percentile, observation) for a second sample of ten observations. The 45° line gives a good fit to the middle part of the sample but not to the extremes. The plot has a well-defined S-shaped appearance. The two smallest sample observations are considerably larger than the corresponding z percentiles (the points on the far left of the plot are well above the 45° line). Similarly, the two largest sample observations are much smaller than the associated z percentiles. This plot indicates that the standard normal distribution would not be a plausible choice for the probability model that gave rise to these observed measurement errors.

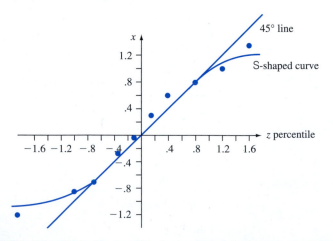

Figure 4.34   Plots of pairs (z percentile, observed value) for the scenario of Example 4.29: second sample

An investigator is typically not interested in knowing just whether a particular probability distribution, such as the standard normal distribution (normal with $\mu = 0$ and $\sigma = 1$) or the exponential distribution with $\lambda = .1$, is a plausible model for the population distribution from which the sample was selected. Instead, the issue is whether *some* member of a family of probability distributions specifies a plausible model—the family of normal distributions, the family of exponential distributions, the family of Weibull distributions, and so on. The values of the parameters of a distribution are usually not specified at the outset. If the family of Weibull distributions is under consideration as a model for lifetime data, are there *any* values of the parameters α and β for which the corresponding Weibull distribution gives a good fit to the data? Fortunately, it is frequently the case that just one probability plot will suffice for assessing the plausibility of an entire family. If the plot deviates substantially from a straight line, no member of the family is plausible. When the plot is quite straight, further work is necessary to estimate values of the parameters that yield the most reasonable distribution of the specified type.

Let's focus on a plot for checking normality. Such a plot is useful in applied work because many formal statistical procedures give accurate inferences only when the population distribution is at least approximately normal. These procedures should generally not be used if the normal probability plot shows a very pronounced departure from linearity. The key to constructing an omnibus normal probability plot is the relationship between standard normal (z) percentiles and those for any other normal distribution:

$$\text{normal } (\mu, \sigma) \text{ percentile} = \mu + \sigma \cdot (\text{corresponding } z \text{ percentile})$$

Consider first the case $\mu = 0$. If each observation is exactly equal to the corresponding normal percentile for some value of σ, the pairs $(\sigma \cdot [z \text{ percentile}], \text{observation})$ fall on a 45° line, which has slope 1. This then implies that the (z percentile, observation) pairs fall on a line passing through $(0, 0)$ (i.e., one with y-intercept 0) but having slope σ rather than 1. The effect of a nonzero value of μ is simply to change the y-intercept from 0 to μ.

A plot of the n pairs

$$([100(i - .5)/n]\text{th } z \text{ percentile, } i\text{th smallest observation})$$

is called a **normal probability plot.** If the sample observations are in fact drawn from a normal distribution with mean value μ and standard deviation σ, the points should fall close to a straight line with slope σ and intercept μ. Thus a plot for which the points fall close to some straight line suggests that the assumption of a normal population distribution is plausible.

EXAMPLE 4.30 There has been recent increased use of augered cast-in-place (ACIP) and drilled displacement (DD) piles in the foundations of buildings and transportation structures. In the article **"Design Methodology for Axially Loaded Auger Cast-in-Place and Drilled Displacement Piles"** (*J. of Geotech. Geoenviron. Engr.*, 2012: 1431–1441), researchers propose a design methodology to enhance the efficiency of these piles. Here are length-diameter ratio measurements based on 17 static pile load tests on

ACIP and DD piles from various construction sites. The values of p for which z percentiles are needed are $(1 - .5)/17 = .029, (2 - .5)/17 = .088, \ldots$ and $.971$.

$x_{(i)}$:	30.86	37.68	39.04	42.78	42.89	42.89	45.05	47.08	47.08
z percentile:	−1.89	−1.35	−1.05	−0.82	−0.63	−0.46	−0.30	−0.15	0.00

$x_{(i)}$:	48.79	48.79	52.56	52.56	54.80	55.17	56.31	59.94
z percentile:	0.15	0.30	0.46	0.63	0.82	1.05	1.35	1.89

Figure 4.35 shows the corresponding normal probability plot generated by the R software package. The pattern in the plot is quite straight, indicating it is plausible that the population distribution of length-diameter ratio is normal.

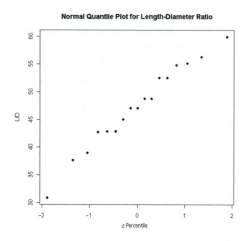

Figure 4.35 Normal probability plot from R for the Length-Diameter Ratio data ■

There is an alternative version of a normal probability plot in which the z percentile axis is replaced by a nonlinear percentage axis. The scaling on this axis is constructed so that plotted points should again fall close to a line when the sampled distribution is normal. Figure 4.36 shows such a plot from Minitab for the ratio data of Example 4.30. (The last two numbers in the small box on the right will be explained in Chapter 14.)

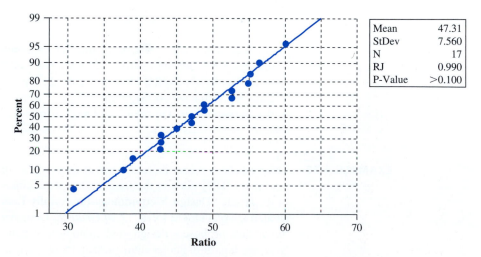

Figure 4.36 Normal probability plot of the ratio data from Minitab

A nonnormal population distribution can often be placed in one of the following three categories:

1. It is symmetric and has "lighter tails" than does a normal distribution; that is, the density curve declines more rapidly out in the tails than does a normal curve.

2. It is symmetric and heavy-tailed compared to a normal distribution.

3. It is skewed.

A uniform distribution is light-tailed, since its density function drops to zero outside a finite interval. The Cauchy density function $f(x) = 1/[\pi\beta(1 + ((x - \theta)/\beta)^2)]$ for $-\infty < x < \infty$ is heavy-tailed, since $1/(1 + x^2)$ declines much less rapidly than does $e^{-x^2/2}$. Lognormal and Weibull distributions are among those that are skewed. When the points in a normal probability plot do not adhere to a straight line, the pattern will frequently suggest that the population distribution is in a particular one of these three categories.

The largest and smallest observations in a sample from a light-tailed distribution are usually not as extreme as would be expected from a normal random sample. Visualize a straight line drawn through the middle part of the plot; points on the far right tend to be below the line (observed value $<$ z percentile), whereas points on the left end of the plot tend to fall above the straight line (observed value $>$ z percentile). The result is an **S**-shaped pattern of the type pictured in Figure 4.34.

A sample from a heavy-tailed distribution also tends to produce an **S**-shaped plot. However, in contrast to the light-tailed case, the left end of the plot curves downward (observed $<$ z percentile), as shown in Figure 4.37(a). If the underlying distribution is positively skewed (a short left tail and a long right tail), the smallest sample observations will be larger than expected from a normal sample and so will the largest observations. In this case, points on both ends of the plot will fall above a straight line through the middle part, yielding a curved pattern, as illustrated in Figure 4.37(b). A sample from a lognormal distribution will usually produce such a pattern. A plot of (z percentile, $\ln(x)$) pairs should then resemble a straight line.

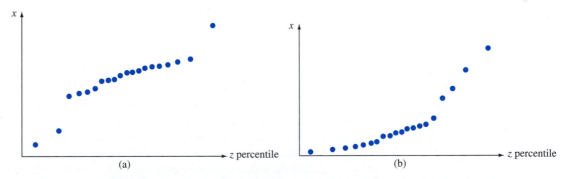

Figure 4.37 Probability plots that suggest a nonnormal distribution: (a) a plot consistent with a heavy-tailed distribution; (b) a plot consistent with a positively skewed distribution

Even when the population distribution is normal, the sample percentiles will not coincide exactly with the theoretical percentiles because of sampling variability. How much can the points in the probability plot deviate from a straight-line pattern before the assumption of population normality is no longer plausible? This is not an easy question to answer. Generally speaking, a small sample from a normal distribution is more likely to yield a plot with a nonlinear pattern than is a large sample. The book *Fitting Equations to Data* (see the Chapter 13 bibliography) presents the results of a simulation study in which numerous samples of different sizes were selected from normal distributions. The authors concluded that there is typically greater variation in the appearance of the probability plot for sample sizes smaller than 30, and only for much larger sample sizes does a linear pattern generally predominate. When a plot is based on a small sample size, only a very substantial departure from linearity should be taken as conclusive evidence of nonnormality. A similar comment applies to probability plots for checking the plausibility of other types of distributions.

Beyond Normality

Consider a family of probability distributions involving two parameters, θ_1 and θ_2, and let $F(x; \theta_1, \theta_2)$ denote the corresponding cdf's. The family of normal distributions is one such family, with $\theta_1 = \mu$, $\theta_2 = \sigma$, and $F(x; \mu, \sigma) = \Phi[(x - \mu)/\sigma]$. Another example is the Weibull family, with $\theta_1 = \alpha$, $\theta_2 = \beta$, and

$$F(x; \alpha, \beta) = 1 - e^{-(x/\beta)^{\alpha}}$$

Still another family of this type is the gamma family, for which the cdf is an integral involving the incomplete gamma function that cannot be expressed in any simpler form.

The parameters θ_1 and θ_2 are said to be **location** and **scale parameters,** respectively, if $F(x; \theta_1, \theta_2)$ is a function of $(x - \theta_1)/\theta_2$. The parameters μ and σ of the normal family are location and scale parameters, respectively. In general, changing θ_1 shifts the location of the corresponding density curve to the right or left, and changing θ_2 amounts to stretching or compressing the horizontal measurement scale. Another example is given by the cdf

$$F(x; \theta_1, \theta_2) = 1 - e^{-e^{(x - \theta_1)/\theta_2}} \quad -\infty < x < \infty$$

A random variable with this cdf is said to have an *extreme value distribution.* It is used in applications involving component lifetime and material strength.

Although the form of the extreme value cdf might at first glance suggest that θ_1 is the point of symmetry for the density function, and therefore the mean and median, this is not the case. Instead, $P(X \le \theta_1) = F(\theta_1; \theta_1, \theta_2) = 1 - e^{-1} = .632$, and the density function $f(x; \theta_1, \theta_2) = F'(x; \theta_1, \theta_2)$ is negatively skewed (a long lower tail). Similarly, the scale parameter θ_2 is not the standard deviation ($\mu = \theta_1 - .5772\theta_2$ and $\sigma = 1.283\theta_2$). However, changing the value of θ_1 does rigidly shift the density curve to the left or right, whereas a change in θ_2 rescales the measurement axis.

The parameter β of the Weibull distribution is a scale parameter, but α is not a location parameter. A similar comment applies to the parameters α and β of the gamma distribution. And for the lognormal distribution, μ is not a location parameter, nor is σ a scale parameter. In the usual form, the density function for any member of these families is positive for $x > 0$ and 0 otherwise. Examples and exercises in the two previous sections introduced a third location (i.e., threshold) parameter γ for these three distributions; this shifts the density function so that it is positive if $x > \gamma$ and zero otherwise.

When the family under consideration has only location and scale parameters, the issue of whether any member of the family is a plausible population distribution can be addressed via a single, easily constructed probability plot. One first obtains the percentiles of the *standard distribution,* the one with $\theta_1 = 0$ and $\theta_2 = 1$, for percentages $100(i - .5)/n$ $(i = 1, \ldots, n)$. The n (standardized percentile, observation) pairs give the points in the plot. This is exactly what we did to obtain an omnibus normal probability plot. Somewhat surprisingly, this methodology can be applied to yield an omnibus Weibull probability plot. The key result is that if X has a Weibull distribution with shape parameter α and scale parameter β, then the transformed variable $\ln(X)$ has an extreme value distribution with location parameter $\theta_1 = \ln(\beta)$ and scale parameter $1/\alpha$. Thus a plot of the (extreme value standardized percentile, $\ln(x)$) pairs showing a strong linear pattern provides support for choosing the Weibull distribution as a population model.

EXAMPLE 4.31 The accompanying observations are on lifetime (in hours) of power apparatus insulation when thermal and electrical stress acceleration were fixed at particular values (**"On the Estimation of Life of Power Apparatus Insulation Under Combined Electrical and Thermal Stress," *IEEE Trans. on Electrical Insulation*, 1985: 70–78).** A Weibull probability plot necessitates first computing the 5th, 15th, . . . , and 95th percentiles of the standard extreme value distribution. The $(100p)$th percentile $\eta(p)$ satisfies

$$p = F(\eta(p)) = 1 - e^{-e^{\eta(p)}}$$

from which $\eta(p) = \ln[-\ln(1 - p)]$.

Percentile	−2.97	−1.82	−1.25	−.84	−.51
x	282	501	741	851	1072
ln(x)	5.64	6.22	6.61	6.75	6.98
Percentile	−.23	.05	.33	.64	1.10
x	1122	1202	1585	1905	2138
ln(x)	7.02	7.09	7.37	7.55	7.67

The pairs $(-2.97, 5.64)$, $(-1.82, 6.22), \ldots, (1.10, 7.67)$ are plotted as points in Figure 4.38. The straightness of the plot argues strongly for using the Weibull distribution as a model for insulation life, a conclusion also reached by the author of the cited article.

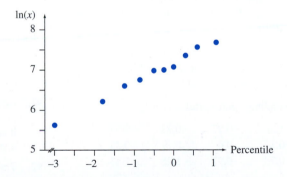

Figure 4.38 A Weibull probability plot of the insulation lifetime data ■

The gamma distribution is an example of a family involving a shape parameter for which there is no transformation $h(\ \cdot\ )$ such that $h(X)$ has a distribution that depends only on location and scale parameters. Construction of a probability plot necessitates first estimating the shape parameter from sample data (some methods for doing this are described in Chapter 6). Sometimes an investigator wishes to know whether the transformed variable X^θ has a normal distribution for some value of θ (by convention, $\theta = 0$ is identified with the logarithmic transformation, in which case X has a lognormal distribution). The book ***Graphical Methods for Data Analysis***, listed in the Chapter 1 bibliography, discusses this type of problem as well as other refinements of probability plotting. Fortunately, the wide availability of various probability plots with statistical software packages means that the user can often sidestep technical details.

EXERCISES Section 4.6 (87–97)

87. The accompanying normal probability plot was constructed from a sample of 30 readings on tension for mesh screens behind the surface of video display tubes used in computer monitors. Does it appear plausible that the tension distribution is normal?

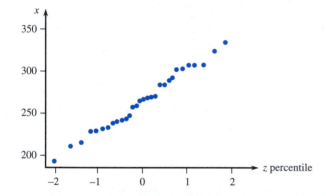

88. A sample of 15 female collegiate golfers was selected and the clubhead velocity (km/hr) while swinging a driver was determined for each one, resulting in the following data (**"Hip Rotational Velocities During the Full Golf Swing,"** *J. of Sports Science and Medicine,* **2009: 296–299):**

69.0	69.7	72.7	80.3	81.0
85.0	86.0	86.3	86.7	87.7
89.3	90.7	91.0	92.5	93.0

The corresponding z percentiles are

−1.83	−1.28	−0.97	−0.73	−0.52
−0.34	−0.17	0.0	0.17	0.34
0.52	0.73	0.97	1.28	1.83

Construct a normal probability plot and a dotplot. Is it plausible that the population distribution is normal?

89. The accompanying sample consisting of $n = 20$ observations on dielectric breakdown voltage of a piece of epoxy resin appeared in the article **"Maximum Likelihood Estimation in the 3-Parameter Weibull Distribution** (*IEEE Trans. on Dielectrics and Elec. Insul.,* **1996: 43–55).** The values of $(i − .5)/n$ for which z percentiles are needed are $(1 − .5)/20 = .025, (2 − .5)/20 = .075,…,$ and .975. Would you feel comfortable estimating population mean voltage using a method that assumed a normal population distribution?

Observation	24.46	25.61	26.25	26.42	26.66
z percentile	−1.96	−1.44	−1.15	−.93	−.76
Observation	27.15	27.31	27.54	27.74	27.94
z percentile	−.60	−.45	−.32	−.19	−.06
Observation	27.98	28.04	28.28	28.49	28.50
z percentile	.06	.19	.32	.45	.60
Observation	28.87	29.11	29.13	29.50	30.88
z percentile	.76	.93	1.15	1.44	1.96

90. The article **"A Probabilistic Model of Fracture in Concrete and Size Effects on Fracture Toughness"** (*Magazine of Concrete Res.,* **1996: 311–320)** gives arguments for why fracture toughness in concrete specimens should have a Weibull distribution and presents several histograms of data that appear well fit by superimposed Weibull curves. Consider the following sample of size $n = 18$ observations on toughness for high-strength concrete (consistent with one of the histograms); values of $p_i = (i − .5)/18$ are also given.

Observation	.47	.58	.65	.69	.72	.74
p_i	.0278	.0833	.1389	.1944	.2500	.3056
Observation	.77	.79	.80	.81	.82	.84
p_i	.3611	.4167	.4722	.5278	.5833	.6389
Observation	.86	.89	.91	.95	1.01	1.04
p_i	.6944	.7500	.8056	.8611	.9167	.9722

Construct a Weibull probability plot and comment.

91. Construct a normal probability plot for the fatigue-crack propagation data given in Exercise 39 (Chapter 1). Does it appear plausible that propagation life has a normal distribution? Explain.

92. The article **"The Load-Life Relationship for M50 Bearings with Silicon Nitride Ceramic Balls"** (*Lubrication Engr.,* **1984: 153–159)** reports the accompanying data on bearing load life (million revs.) for bearings tested at a 6.45 kN load.

47.1	68.1	68.1	90.8	103.6	106.0	115.0
126.0	146.6	229.0	240.0	240.0	278.0	278.0
289.0	289.0	367.0	385.9	392.0	505.0	

a. Construct a normal probability plot. Is normality plausible?

b. Construct a Weibull probability plot. Is the Weibull distribution family plausible?

93. Construct a probability plot that will allow you to assess the plausibility of the lognormal distribution as a model for the rainfall data of Exercise 83 in Chapter 1.

94. The accompanying observations are precipitation values during March over a 30-year period in Minneapolis-St. Paul.

.77	1.20	3.00	1.62	2.81	2.48
1.74	.47	3.09	1.31	1.87	.96
.81	1.43	1.51	.32	1.18	1.89
1.20	3.37	2.10	.59	1.35	.90
1.95	2.20	.52	.81	4.75	2.05

   a. Construct and interpret a normal probability plot for this data set.
   b. Calculate the square root of each value and then construct a normal probability plot based on this transformed data. Does it seem plausible that the square root of precipitation is normally distributed?
   c. Repeat part (b) after transforming by cube roots.

95. Use a statistical software package to construct a normal probability plot of the tensile ultimate-strength data given in Exercise 13 of Chapter 1, and comment.

96. Let the *ordered* sample observations be denoted by $y_1, y_2, \ldots, y_n$ (y_1 being the smallest and y_n the largest). Our suggested check for normality is to plot the $(\Phi^{-1}((i - .5)/n), y_i)$ pairs. Suppose we believe that the observations come from a distribution with mean 0, and let $w_1, \ldots, w_n$ be the *ordered absolute* values of the x_i's. A **half-normal** plot is a probability plot of the w_i's. More specifically, since $P(|Z| \leq w) = P(-w \leq Z \leq w) = 2\Phi(w) - 1$, a half-normal plot is a plot of the $(\Phi^{-1}(\{[(i - .5)/n + 1]/2\}, w_i)$ pairs. The virtue of this plot is that small or large outliers in the original sample will now appear only at the upper end of the plot rather than at both ends. Construct a half-normal plot for the following sample of measurement errors, and comment: $-3.78, -1.27, 1.44, -.39, 12.38, -43.40, 1.15, -3.96, -2.34, 30.84$.

97. The following failure time observations (1000s of hours) resulted from accelerated life testing of 16 integrated circuit chips of a certain type:

82.8	11.6	359.5	502.5	307.8	179.7
242.0	26.5	244.8	304.3	379.1	212.6
229.9	558.9	366.7	204.6		

   Use the corresponding percentiles of the exponential distribution with $\lambda = 1$ to construct a probability plot. Then explain why the plot assesses the plausibility of the sample having been generated from *any* exponential distribution.

SUPPLEMENTARY EXERCISES (98–128)

98. Let $X =$ the time it takes a read/write head to locate a desired record on a computer disk memory device once the head has been positioned over the correct track. If the disks rotate once every 25 millisec, a reasonable assumption is that X is uniformly distributed on the interval $[0, 25]$.
   a. Compute $P(10 \leq X \leq 20)$.
   b. Compute $P(X \geq 10)$.
   c. Obtain the cdf $F(X)$.
   d. Compute $E(X)$ and σ_X.

99. A 12-in. bar that is clamped at both ends is to be subjected to an increasing amount of stress until it snaps. Let $Y =$ the distance from the left end at which the break occurs. Suppose Y has pdf

$$f(y) = \begin{cases} \left(\dfrac{1}{24}\right)y\left(1 - \dfrac{y}{12}\right) & 0 \leq y \leq 12 \\ 0 & \text{otherwise} \end{cases}$$

   Compute the following:
   a. The cdf of Y, and graph it.
   b. $P(Y \leq 4)$, $P(Y > 6)$, and $P(4 \leq Y \leq 6)$
   c. $E(Y)$, $E(Y^2)$, and $V(Y)$
   d. The probability that the break point occurs more than 2 in. from the expected break point.

   e. The expected length of the shorter segment when the break occurs.

100. Let X denote the time to failure (in years) of a certain hydraulic component. Suppose the pdf of X is $f(x) = 32/(x + 4)^3$ for $x < 0$.
   a. Verify that $f(x)$ is a legitimate pdf.
   b. Determine the cdf.
   c. Use the result of part (b) to calculate the probability that time to failure is between 2 and 5 years.
   d. What is the expected time to failure?
   e. If the component has a salvage value equal to $100/(4 + x)$ when its time to failure is x, what is the expected salvage value?

101. The completion time X for a certain task has cdf $F(x)$ given by

$$\begin{cases} 0 & x < 0 \\ \dfrac{x^3}{3} & 0 \leq x < 1 \\ 1 - \dfrac{1}{2}\left(\dfrac{7}{3} - x\right)\left(\dfrac{7}{4} - \dfrac{3}{4}x\right) & 1 \leq x \leq \dfrac{7}{3} \\ 1 & x > \dfrac{7}{3} \end{cases}$$

a. Obtain the pdf $f(x)$ and sketch its graph.
b. Compute $P(.5 \leq X \leq 2)$.
c. Compute $E(X)$.

102. Let X represent the number of individuals who respond to a particular online coupon offer. Suppose that X has approximately a Weibull distribution with $\alpha = 10$ and $\beta = 20$. Calculate the best possible approximation to the probability that X is between 15 and 20, inclusive.

103. The article **"Computer Assisted Net Weight Control"** (*Quality Progress*, **1983: 22–25**) suggests a normal distribution with mean 137.2 oz and standard deviation 1.6 oz for the actual contents of jars of a certain type. The stated contents was 135 oz.

a. What is the probability that a single jar contains more than the stated contents?
b. Among ten randomly selected jars, what is the probability that at least eight contain more than the stated contents?
c. Assuming that the mean remains at 137.2, to what value would the standard deviation have to be changed so that 95% of all jars contain more than the stated contents?

104. When circuit boards used in the manufacture of compact disc players are tested, the long-run percentage of defectives is 5%. Suppose that a batch of 250 boards has been received and that the condition of any particular board is independent of that of any other board.

a. What is the approximate probability that at least 10% of the boards in the batch are defective?
b. What is the approximate probability that there are exactly 10 defectives in the batch?

105. Exercise 38 introduced two machines that produce wine corks, the first one having a normal diameter distribution with mean value 3 cm and standard deviation .1 cm, and the second having a normal diameter distribution with mean value 3.04 cm and standard deviation .02 cm. Acceptable corks have diameters between 2.9 and 3.1 cm. If 60% of all corks used come from the first machine and a randomly selected cork is found to be acceptable, what is the probability that it was produced by the first machine?

106. The reaction time (in seconds) to a certain stimulus is a continuous random variable with pdf

$$f(x) = \begin{cases} \dfrac{3}{2} \cdot \dfrac{1}{x^2} & 1 \leq x \leq 3 \\ 0 & \text{otherwise} \end{cases}$$

a. Obtain the cdf.
b. What is the probability that reaction time is at most 2.5 sec? Between 1.5 and 2.5 sec?
c. Compute the expected reaction time.
d. Compute the standard deviation of reaction time.
e. If an individual takes more than 1.5 sec to react, a light comes on and stays on either until one further second has elapsed or until the person reacts (whichever happens first). Determine the expected amount of time that the light remains lit. [*Hint:* Let $h(X) =$ the time that the light is on as a function of reaction time X.]

107. Let X denote the temperature at which a certain chemical reaction takes place. Suppose that X has pdf

$$f(x) = \begin{cases} \dfrac{1}{9}(4 - x^2) & -1 \leq x \leq 2 \\ 0 & \text{otherwise} \end{cases}$$

a. Sketch the graph of $f(x)$.
b. Determine the cdf and sketch it.
c. Is 0 the median temperature at which the reaction takes place? If not, is the median temperature smaller or larger than 0?
d. Suppose this reaction is independently carried out once in each of ten different labs and that the pdf of reaction time in each lab is as given. Let $Y =$ the number among the ten labs at which the temperature exceeds 1. What kind of distribution does Y have? (Give the names and values of any parameters.)

108. An oocyte is a female germ cell involved in reproduction. Based on analyses of a large sample, the article **"Reproductive Traits of Pioneer Gastropod Species Colonizing Deep-Sea Hydrothermal Vents After an Eruption"** (*Marine Biology*, **2011: 181–192**) proposed the following mixture of normal distributions as a model for the distribution of $X =$ oocyte diameter (μm):

$$f(x) = p f_1(x; \mu_1, \sigma) + (1 - p) f_2(x; \mu_2, \sigma)$$

where f_1 and f_2 are normal pdfs. Suggested parameter values were $p = .35$, $\mu_1 = 4.4$, $\mu_2 = 5.0$, and $\sigma = .27$.

a. What is the expected (i.e. mean) value of oocyte diameter?
b. What is the probability that oocyte diameter is between 4.4 μm and 5.0 μm? [*Hint:* Write an expression for the corresponding integral, carry the integral operation through to the two components, and then use the fact that each component is a normal pdf.]
c. What is the probability that oocyte diameter is smaller than its mean value? What does this imply about the shape of the density curve?

109. The article **"The Prediction of Corrosion by Statistical Analysis of Corrosion Profiles"** (*Corrosion Science*, **1985: 305–315**) suggests the following cdf for the depth X of the deepest pit in an experiment involving the exposure of carbon manganese steel to acidified seawater.

$$F(x; \alpha, \beta) = e^{-e^{-(x - \alpha)/\beta}} \quad -\infty < x < \infty$$

The authors propose the values $\alpha = 150$ and $\beta = 90$. Assume this to be the correct model.

a. What is the probability that the depth of the deepest pit is at most 150? At most 300? Between 150 and 300?

b. Below what value will the depth of the maximum pit be observed in 90% of all such experiments?

c. What is the density function of X?

d. The density function can be shown to be unimodal (a single peak). Above what value on the measurement axis does this peak occur? (This value is the mode.)

e. It can be shown that $E(X) \approx .5772\beta + \alpha$. What is the mean for the given values of α and β, and how does it compare to the median and mode? Sketch the graph of the density function. [*Note:* This is called the *largest extreme value distribution.*]

110. Let t = the amount of sales tax a retailer owes the government for a certain period. The article **"Statistical Sampling in Tax Audits"** (*Statistics and the Law*, **2008: 320–343**) proposes modeling the uncertainty in t by regarding it as a normally distributed random variable with mean value μ and standard deviation σ (in the article, these two parameters are estimated from the results of a tax audit involving n sampled transactions). If a represents the amount the retailer is assessed, then an under-assessment results if $t > a$ and an over-assessment results if $a > t$. The proposed penalty (i.e., loss) function for over- or under-assessment is $L(a, t) = t - a$ if $t > a$ and $= k(a - t)$ if $t \le a$ ($k > 1$ is suggested to incorporate the idea that over-assessment is more serious than under-assessment).

a. Show that $a^* = \mu + \sigma\Phi^{-1}(1/(k + 1))$ is the value of a that minimizes the expected loss, where Φ^{-1} is the inverse function of the standard normal cdf.

b. If $k = 2$ (suggested in the article), $\mu = \$100,000$, and $\sigma = \$10,000$, what is the optimal value of a, and what is the resulting probability of over-assessment?

111. The *mode* of a continuous distribution is the value x^* that maximizes $f(x)$.

a. What is the mode of a normal distribution with parameters μ and σ?

b. Does the uniform distribution with parameters A and B have a single mode? Why or why not?

c. What is the mode of an exponential distribution with parameter λ? (Draw a picture.)

d. If X has a gamma distribution with parameters α and β, and $\alpha > 1$, find the mode. [*Hint:* $\ln[f(x)]$ will be maximized iff $f(x)$ is, and it may be simpler to take the derivative of $\ln[f(x)]$.]

e. What is the mode of a chi-squared distribution having ν degrees of freedom?

112. The article **"Error Distribution in Navigation"** (*J. of the Institute of Navigation*, **1971: 429–442**) suggests that the frequency distribution of positive errors (magnitudes of errors) is well approximated by an exponential distribution. Let X = the lateral position error (nautical miles), which can be either negative or positive. Suppose the pdf of X is

$$f(x) = (.1)e^{-.2|x|} \quad -\infty < x < \infty$$

a. Sketch a graph of $f(x)$ and verify that $f(x)$ is a legitimate pdf (show that it integrates to 1).

b. Obtain the cdf of X and sketch it.

c. Compute $P(X \le 0)$, $P(X \le 2)$, $P(-1 \le X \le 2)$, and the probability that an error of more than 2 miles is made.

113. The article **"Statistical Behavior Modeling for Driver-Adaptive Precrash Systems"** (*IEEE Trans. on Intelligent Transp. Systems*, **2013: 1–9**) proposed the following mixture of two exponential distributions for modeling the behavior of what the authors called "the criticality level of a situation" X.

$$f(x; \lambda_1, \lambda_2, p) = \begin{cases} p\lambda_1 e^{-\lambda_1 x} + (1 - p)\lambda_2 e^{-\lambda_2 x} & x \ge 0 \\ 0 & \text{otherwise} \end{cases}$$

This is often called the hyperexponential or mixed exponential distribution. This distribution is also proposed as a model for rainfall amount in **"Modeling Monsoon Affected Rainfall of Pakistan by Point Processes"** (*J. of Water Resources Planning and Mgmnt.*, **1992: 671–688**).

a. Determine $E(X)$ and $V(X)$. *Hint:* For X distributed exponentially, $E(X) = 1/\lambda$ and $V(X) = 1/\lambda^2$; what does this imply about $E(X^2)$?

b. Determine the cdf of X.

c. If $p = .5$, $\lambda_1 = 40$, and $\lambda_2 = 200$ (values of the λ's suggested in the cited article), calculate $P(X > .01)$.

d. For the parameter values given in (c), what is the probability that X is within one standard deviation of its mean value?

e. The coefficient of variation of a random variable (or distribution) is $CV = \sigma/\mu$. What is CV for an exponential rv? What can you say about the value of CV when X has a hyperexponential distribution?

f. What is CV for an Erlang distribution with parameters λ and n as defined in Exercise 68? [*Note:* In applied work, the sample CV is used to decide which of the three distributions might be appropriate.]

114. Suppose a particular state allows individuals filing tax returns to itemize deductions only if the total of all itemized deductions is at least $5000. Let X (in 1000s of dollars) be the total of itemized deductions on a randomly chosen form. Assume that X has the pdf

$$f(x; \alpha) = \begin{cases} k/x^\alpha & x \ge 5 \\ 0 & \text{otherwise} \end{cases}$$

a. Find the value of k. What restriction on α is necessary?

b. What is the cdf of X?

c. What is the expected total deduction on a randomly chosen form? What restriction on α is necessary for $E(X)$ to be finite?

d. Show that $\ln(X/5)$ has an exponential distribution with parameter $\alpha - 1$.

115. Let I_i be the input current to a transistor and I_0 be the output current. Then the current gain is proportional to $\ln(I_0/I_i)$. Suppose the constant of proportionality is 1 (which amounts to choosing a particular unit of measurement), so that current gain $= X = \ln(I_0/I_i)$. Assume X is normally distributed with $\mu = 1$ and $\sigma = .05$.

a. What type of distribution does the ratio I_0/I_i have?

b. What is the probability that the output current is more than twice the input current?

c. What are the expected value and variance of the ratio of output to input current?

116. The article **"Response of SiC$_f$/Si$_3$N$_4$ Composites Under Static and Cyclic Loading—An Experimental and Statistical Analysis"** (*J. of Engr. Materials and Technology*, 1997: 186–193) suggests that tensile strength (MPa) of composites under specified conditions can be modeled by a Weibull distribution with $\alpha = 9$ and $\beta = 180$.

a. Sketch a graph of the density function.

b. What is the probability that the strength of a randomly selected specimen will exceed 175? Will be between 150 and 175?

c. If two randomly selected specimens are chosen and their strengths are independent of one another, what is the probability that at least one has a strength between 150 and 175?

d. What strength value separates the weakest 10% of all specimens from the remaining 90%?

117. Let Z have a standard normal distribution and define a new rv Y by $Y = \sigma Z + \mu$. Show that Y has a normal distribution with parameters μ and σ. [*Hint*: $Y \le y$ iff $Z \le$? Use this to find the cdf of Y and then differentiate it with respect to y.]

118. a. Suppose the lifetime X of a component, when measured in hours, has a gamma distribution with parameters α and β. Let $Y =$ the lifetime measured in minutes. Derive the pdf of Y. [*Hint*: $Y \le y$ iff $X \le y/60$. Use this to obtain the cdf of Y and then differentiate to obtain the pdf.]

b. If X has a gamma distribution with parameters α and β, what is the probability distribution of $Y = cX$?

119. In Exercises 117 and 118, as well as many other situations, one has the pdf $f(x)$ of X and wishes to know the pdf of $y = h(X)$. Assume that $h(\cdot)$ is an invertible function, so that $y = h(x)$ can be solved for x to yield $x = k(y)$. Then it can be shown that the pdf of Y is

$$g(y) = f[k(y)] \cdot |k'(y)|$$

a. If X has a uniform distribution with $A = 0$ and $B = 1$, derive the pdf of $Y = -\ln(X)$.

b. Work Exercise 117, using this result.

c. Work Exercise 118(b), using this result.

120. Based on data from a dart-throwing experiment, the article **"Shooting Darts"** (*Chance*, Summer 1997, 16–19) proposed that the horizontal and vertical errors from aiming at a point target should be independent of one another, each with a normal distribution having mean 0 and

variance σ^2. It can then be shown that the pdf of the distance V from the target to the landing point is

$$f(v) = \frac{v}{\sigma^2} \cdot e^{-v^2/2\sigma^2} \quad v > 0$$

a. This pdf is a member of what family introduced in this chapter?

b. If $\sigma = 20$ mm (close to the value suggested in the paper), what is the probability that a dart will land within 25 mm (roughly 1 in.) of the target?

121. The article **"Three Sisters Give Birth on the Same Day"** (*Chance*, Spring 2001, 23–25) used the fact that three Utah sisters had all given birth on March 11, 1998 as a basis for posing some interesting questions regarding birth coincidences.

a. Disregarding leap year and assuming that the other 365 days are equally likely, what is the probability that three randomly selected births all occur on March 11? Be sure to indicate what, if any, extra assumptions you are making.

b. With the assumptions used in part (a), what is the probability that three randomly selected births all occur on the same day?

c. The author suggested that, based on extensive data, the length of gestation (time between conception and birth) could be modeled as having a normal distribution with mean value 280 days and standard deviation 19.88 days. The due dates for the three Utah sisters were March 15, April 1, and April 4, respectively. Assuming that all three due dates are at the mean of the distribution, what is the probability that all births occurred on March 11? [*Hint*: The deviation of birth date from due date is normally distributed with mean 0.]

d. Explain how you would use the information in part (c) to calculate the probability of a common birth date.

122. Let X denote the lifetime of a component, with $f(x)$ and $F(x)$ the pdf and cdf of X. The probability that the component fails in the interval $(x, x + \Delta x)$ is approximately $f(x) \cdot \Delta x$. The conditional probability that it fails in $(x, x + \Delta x)$ given that it has lasted at least x is $f(x) \cdot \Delta x/[1 - F(x)]$. Dividing this by Δx produces the **failure rate function:**

$$r(x) = \frac{f(x)}{1 - F(x)}$$

An increasing failure rate function indicates that older components are increasingly likely to wear out, whereas a decreasing failure rate is evidence of increasing reliability with age. In practice, a "bathtub-shaped" failure is often assumed.

a. If X is exponentially distributed, what is $r(x)$?

b. If X has a Weibull distribution with parameters α and β, what is $r(x)$? For what parameter values will $r(x)$ be increasing? For what parameter values will $r(x)$ decrease with x?

c. Since $r(x) = -(d/dx)\ln[1 - F(x)]$, $\ln[1 - F(x)] = -\int r(x)dx$. Suppose

$$r(x) = \begin{cases} \alpha\left(1 - \dfrac{x}{\beta}\right) & 0 \le x \le \beta \\ 0 & \text{otherwise} \end{cases}$$

so that if a component lasts β hours, it will last forever (while seemingly unreasonable, this model can be used to study just "initial wearout"). What are the cdf and pdf of X?

123. Let U have a uniform distribution on the interval $[0, 1]$. Then observed values having this distribution can be obtained from a computer's random number generator. Let $X = -(1/\lambda)\ln(1 - U)$.
   a. Show that X has an exponential distribution with parameter λ. [*Hint*: The cdf of X is $F(x) = P(X \le x)$; $X \le x$ is equivalent to $U \le ?$]
   b. How would you use part (a) and a random number generator to obtain observed values from an exponential distribution with parameter $\lambda = 10$?

124. Consider an rv X with mean μ and standard deviation σ, and let $g(X)$ be a specified function of X. The first-order Taylor series approximation to $g(X)$ in the neighborhood of μ is

$$g(X) \approx g(\mu) + g'(\mu) \cdot (X - \mu)$$

The right-hand side of this equation is a linear function of X. If the distribution of X is concentrated in an interval over which $g(\cdot)$ is approximately linear [e.g., $\sqrt{x}$ is approximately linear in $(1, 2)$], then the equation yields approximations to $E(g(X))$ and $V(g(X))$.
   a. Give expressions for these approximations. [*Hint*: Use rules of expected value and variance for a linear function $aX + b$.]
   b. If the voltage v across a medium is fixed but current I is random, then resistance will also be a random variable related to I by $R = v/I$. If $\mu_I = 20$ and $\sigma_I = .5$, calculate approximations to μ_R and σ_R.

125. A function $g(x)$ is *convex* if the chord connecting any two points on the function's graph lies above the graph. When $g(x)$ is differentiable, an equivalent condition is that for every x, the tangent line at x lies entirely on or below the graph. (See the figure below.) How does $g(\mu) = g(E(X))$ compare to $E(g(X))$? [*Hint*: The equation of the tangent line at $x = \mu$ is $y = g(\mu) + g'(\mu) \cdot (x - \mu)$. Use the condition of convexity, substitute X for x, and take expected values. [*Note*: Unless $g(x)$ is linear, the resulting inequality (usually called Jensen's inequality) is strict ($<$ rather than $\le$); it is valid for both continuous and discrete rv's.]

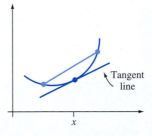

126. Let X have a Weibull distribution with parameters $\alpha = 2$ and β. Show that $Y = 2X^2/\beta^2$ has a chi-squared distribution with $\nu = 2$. [*Hint*: The cdf of Y is $P(Y \le y)$; express this probability in the form $P(X \le g(y))$, use the fact that X has a cdf of the form in Expression (4.12), and differentiate with respect to y to obtain the pdf of Y.]

127. An individual's credit score is a number calculated based on that person's credit history that helps a lender determine how much he/she should be loaned or what credit limit should be established for a credit card. An article in the **Los Angeles Times** gave data which suggested that a beta distribution with parameters $A = 150$, $B = 850$, $\alpha = 8$, $\beta = 2$ would provide a reasonable approximation to the distribution of American credit scores. [*Note*: credit scores are integer-valued].
   a. Let X represent a randomly selected American credit score. What are the mean value and standard deviation of this random variable? What is the probability that X is within 1 standard deviation of its mean value?
   b. What is the approximate probability that a randomly selected score will exceed 750 (which lenders consider a very good score)?

128. Let V denote rainfall volume and W denote runoff volume (both in mm). According to the article **"Runoff Quality Analysis of Urban Catchments with Analytical Probability Models"** (*J. of Water Resource Planning and Management*, 2006: 4–14), the runoff volume will be 0 if $V \le \nu_d$ and will be $k(V - \nu_d)$ if $V > \nu_d$. Here ν_d is the volume of depression storage (a constant), and k (also a constant) is the runoff coefficient. The cited article proposes an exponential distribution with parameter λ for V.
   a. Obtain an expression for the cdf of W. [*Note*: W is neither purely continuous nor purely discrete; instead it has a "mixed" distribution with a discrete component at 0 and is continuous for values $w > 0$.]
   b. What is the pdf of W for $w > 0$? Use this to obtain an expression for the expected value of runoff volume.

BIBLIOGRAPHY

Bury, Karl, *Statistical Distributions in Engineering,* Cambridge Univ. Press, Cambridge, England, 1999. A readable and informative survey of distributions and their properties.

Johnson, Norman, Samuel Kotz, and N. Balakrishnan, *Continuous Univariate Distributions,* vols. 1–2, Wiley, New York, 1994. These two volumes together present an exhaustive survey of various continuous distributions.

Nelson, Wayne, *Applied Life Data Analysis,* Wiley, New York, 1982. Gives a comprehensive discussion of distributions and methods that are used in the analysis of lifetime data.

Olkin, Ingram, Cyrus Derman, and Leon Gleser, *Probability Models and Applications* (2nd ed.), Macmillan, New York, 1994. Good coverage of general properties and specific distributions.

5

Joint Probability Distributions and Random Samples

INTRODUCTION

In Chapters 3 and 4 we developed probability models for a single random variable. Many problems in probability and statistics involve working simultaneously with two or more random variables. For example, X and Y might be the height and weight, respectively, of a randomly selected individual. Or X_1, X_2, and X_3 might be the number of purchases made with Visa, MasterCard, and American Express credit cards, respectively, at a store on a particular day. In Section 5.1 we discuss probability models for the joint (i.e., simultaneous) behavior of two or more random variables. The very important concept of independence of several random variables is then introduced and explored. Section 5.2 considers the expected value of a function of two or more random variables [e.g., the expected value of Y/X^2, the body mass index (BMI) when X is expressed in cm and Y is expressed in kg]. This leads to a discussion of covariance and correlation as measures of the degree of association between two variables. At the end of the section, the bivariate normal distribution is introduced as a generalization of the univariate normal distribution.

Sections 5.3 and 5.4 consider functions of the n variables $X_1, X_2, \ldots, X_n$ that constitute a sample from some population or distribution (for example, a sample of weights of newborn children). The most important function of this type is the sample average $(X_1 + X_2 + \cdots + X_n)/n$. We will call any such function, itself a random variable, a *statistic*. Methods from probability are used to obtain information about the distribution of a statistic. The premier result of this type is the Central Limit Theorem (CLT), the basis for many inferential procedures involving large sample sizes. The last section of the chapter deals with linear functions of the form $a_1X_1 + \cdots + a_nX_n$ where the a_i's are numerical constants.

5.1 Jointly Distributed Random Variables

There are many experimental situations in which more than one random variable (rv) will be of interest to an investigator. We first consider joint probability distributions for two random variables. The "pure" cases, in which both variables are discrete or both are continuous, are the ones most frequently encountered in practice.

Two Discrete Random Variables

The probability mass function (pmf) of a single discrete rv X specifies how much probability mass is placed on each possible X value. The joint pmf of two discrete rv's X and Y describes how much probability mass is placed on each possible pair of values (x, y).

DEFINITION

Let X and Y be two discrete rv's defined on the sample space $\mathcal{S}$ of an experiment. The **joint probability mass function** $p(x, y)$ is defined for each pair of numbers (x, y) by

$$p(x, y) = P(X = x \text{ and } Y = y)$$

It must be the case that $p(x, y) \geq 0$ and $\sum_x \sum_y p(x, y) = 1$.

Now let A be any particular set consisting of pairs of (x, y) values (e.g., $A = \{(x, y): x + y = 5\}$ or $\{(x, y): \max(x, y) \leq 3\}$). Then the probability $P[(X, Y) \in A]$ that the random pair (X, Y) lies in the set A is obtained by summing the joint pmf over pairs in A:

$$P[(X, Y) \in A] = \sum_{(x, y) \in A} \sum p(x, y)$$

EXAMPLE 5.1 Anyone who purchases an insurance policy for a home or automobile must specify a deductible amount, the amount of loss to be absorbed by the policyholder before the insurance company begins paying out. Suppose that a particular company offers auto deductible amounts of $100, $500, and $1000, and homeowner deductible amounts of $500, $1000, and $2000. Consider randomly selecting someone who has both auto and homeowner insurance with this company, and let $X =$ the amount of the auto policy deductible and $Y =$ the amount of the homeowner policy deductible. The joint pmf of these two variables appears in the accompanying *joint probability table*:

		y		
$p(x, y)$		500	1000	5000
	100	.30	.05	0
x	500	.15	.20	.05
	1000	.10	.10	.05

According to this joint pmf, there are nine possible (X, Y) pairs: (100, 500), (100, 1000), ..., and finally (1000, 5000). The probability of (100, 500) is $p(100, 500) = P(X = 100, Y = 500) = .30$. Clearly $p(x, y) \geq 0$, and it is easily confirmed that the sum of the nine displayed probabilities is 1. The probability $P(X = Y)$ is computed by summing $p(x, y)$ over the two (x, y) pairs for which the two deductible amounts are identical:

$$P(X = Y) = p(500, 500) + p(1000, 1000) = .15 + .10 = .25$$

Similarly, the probability that the auto deductible amount is at least \$500 is the sum of all probabilities corresponding to (x, y) pairs for which $x \geq 500$; this is the sum of the probabilities in the bottom two rows of the joint probability table:

$$P(X \geq 500) = .15 + .20 + .05 + .10 + .10 + .05 = .65 \qquad \blacksquare$$

Once the joint pmf of the two variables X and Y is available, it is in principle straightforward to obtain the distribution of just one of these variables. As an example, let X and Y be the number of statistics and mathematics courses, respectively, currently being taken by a randomly selected statistics major. Suppose that we wish the distribution of X, and that when $X = 2$, the only possible values of Y are 0, 1, and 2. Then

$$p_X(2) = P(X = 2) = P[(X, Y) = (2, 0) \text{ or } (2, 1) \text{ or } (2, 2)]$$
$$= p(2, 0) + p(2, 1) + p(2, 2)$$

That is, the joint pmf is summed over all pairs of the form $(2, y)$. More generally, for any possible value x of X, the probability $p_X(x)$ results from holding x fixed and summing the joint pmf $p(x, y)$ over all y for which the pair (x, y) has positive probability mass. The same strategy applies to obtaining the distribution of Y by itself.

DEFINITION

> The **marginal probability mass function of X,** denoted by $p_X(x)$, is given by
>
> $$p_X(x) = \sum_{y: p(x, y) > 0} p(x, y) \quad \text{for each possible value } x$$
>
> Similarly, the **marginal probability mass function of Y** is
>
> $$p_Y(y) = \sum_{x: p(x, y) > 0} p(x, y) \quad \text{for each possible value } y.$$

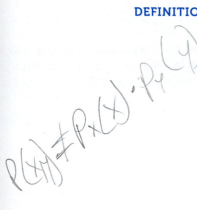

The use of the word *marginal* here is a consequence of the fact that if the joint pmf is displayed in a rectangular table as in Example 5.1, then the row totals give the marginal pmf of X and the column totals give the marginal pmf of Y. Once these marginal pmf's are available, the probability of any event involving only X or only Y can be calculated.

EXAMPLE 5.2
(Example 5.1 continued)

Possible X values are $x = 100$, 500, and 1000. Computing row totals from the joint probability table yields

$$p_X(100) = p(100, 500) + p(100, 1000) + p(100, 5000) = .30 + .05 + 0 = .35$$
$$p_X(500) = .15 + .20 + .05 = .40, \quad p_X(1000) = 1 - (.35 + .40) = .25$$

The marginal pmf of X is then

$$p_X(x) = \begin{cases} .35 & x = 100 \\ .40 & x = 500 \\ .25 & x = 1000 \\ 0 & \text{otherwise} \end{cases}$$

From this pmf, $P(X \geq 500) = .40 + .25 = .65$, which we already calculated in Example 5.1. Similarly, the marginal pmf of Y is obtained from the column totals as

$$p_Y(y) = \begin{cases} .55 & y = 500 \\ .35 & y = 1000 \\ .10 & y = 5000 \\ 0 & \text{otherwise} \end{cases}$$ ∎

Two Continuous Random Variables

The probability that the observed value of a continuous rv X lies in a one-dimensional set A (such as an interval) is obtained by integrating the pdf $f(x)$ over the set A. Similarly, the probability that the pair (X, Y) of continuous rv's falls in a two-dimensional set A (such as a rectangle) is obtained by integrating a function called the *joint density function*.

DEFINITION

Let X and Y be continuous rv's. A **joint probability density function** $f(x, y)$ for these two variables is a function satisfying $f(x, y) \geq 0$ and $\int_{-\infty}^{\infty} \int_{-\infty}^{\infty} f(x, y) \, dx \, dy = 1$. Then for any two-dimensional set A

$$P[(X, Y) \in A] = \iint_A f(x, y) \, dx \, dy$$

In particular, if A is the two-dimensional rectangle $\{(x, y): a \leq x \leq b, c \leq y \leq d\}$, then

$$P[(X,Y) \in A] = P(a \leq X \leq b, c \leq Y \leq d) = \int_a^b \int_c^d f(x, y) \, dy \, dx$$

We can visualize $f(x, y)$ as specifying a surface at height $f(x, y)$ above the point (x, y) in a three-dimensional coordinate system. Then $P[(X, Y) \in A]$ is the volume underneath this surface and above the region A, analogous to the area under a curve in the case of a single rv. This is illustrated in Figure 5.1.

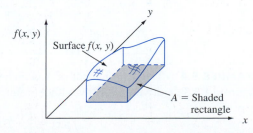

Figure 5.1 $P[(X, Y) \in A] =$ volume under density surface above A

EXAMPLE 5.3 A bank operates both a drive-up facility and a walk-up window. On a randomly selected day, let $X =$ the proportion of time that the drive-up facility is in use (at least one customer is being served or waiting to be served) and $Y =$ the proportion of time that the walk-up window is in use. Then the set of possible values for (X, Y) is the rectangle $D = \{(x, y): 0 \le x \le 1, 0 \le y \le 1\}$. Suppose the joint pdf of (X, Y) is given by

$$f(x, y) = \begin{cases} \dfrac{6}{5}(x + y^2) & 0 \le x \le 1, 0 \le y \le 1 \\ 0 & \text{otherwise} \end{cases}$$

To verify that this is a legitimate pdf, note that $f(x, y) \ge 0$ and

$$\int_{-\infty}^{\infty} \int_{-\infty}^{\infty} f(x, y)\, dx\, dy = \int_0^1 \int_0^1 \frac{6}{5}(x + y^2)\, dx\, dy$$

$$= \int_0^1 \int_0^1 \frac{6}{5} x\, dx\, dy + \int_0^1 \int_0^1 \frac{6}{5} y^2\, dx\, dy$$

$$= \int_0^1 \frac{6}{5} x\, dx + \int_0^1 \frac{6}{5} y^2\, dy = \frac{6}{10} + \frac{6}{15} = 1$$

The probability that neither facility is busy more than one-quarter of the time is

$$P\left(0 \le X \le \frac{1}{4}, 0 \le Y \le \frac{1}{4}\right) = \int_0^{1/4} \int_0^{1/4} \frac{6}{5}(x + y^2)\, dx\, dy$$

$$= \frac{6}{5} \int_0^{1/4} \int_0^{1/4} x\, dx\, dy + \frac{6}{5} \int_0^{1/4} \int_0^{1/4} y^2\, dx\, dy$$

$$= \frac{6}{20} \cdot \frac{x^2}{2} \Big|_{x=0}^{x=1/4} + \frac{6}{20} \cdot \frac{y^3}{3} \Big|_{y=0}^{y=1/4} = \frac{7}{640}$$

$$= .0109 \qquad \blacksquare$$

The marginal pdf of each variable can be obtained in a manner analogous to what we did in the case of two discrete variables. The marginal pdf of X at the value x results from holding x fixed in the pair (x, y) and *integrating* the joint pdf over y. Integrating the joint pdf with respect to x gives the marginal pdf of Y.

DEFINITION

The **marginal probability density functions** of X and Y, denoted by $f_X(x)$ and $f_Y(y)$, respectively, are given by

$$f_X(x) = \int_{-\infty}^{\infty} f(x, y)\, dy \qquad \text{for } -\infty < x < \infty$$

$$f_Y(y) = \int_{-\infty}^{\infty} f(x, y)\, dx \qquad \text{for } -\infty < y < \infty$$

EXAMPLE 5.4
(Example 5.3 continued)

The marginal pdf of X, which gives the probability distribution of busy time for the drive-up facility without reference to the walk-up window, is

$$f_X(x) = \int_{-\infty}^{\infty} f(x, y)\, dy = \int_0^1 \frac{6}{5}(x + y^2)\, dy = \frac{6}{5} x + \frac{2}{5}$$

for $0 \le x \le 1$ and 0 otherwise. The marginal pdf of Y is

$$f_Y(y) = \begin{cases} \dfrac{6}{5} y^2 + \dfrac{3}{5} & 0 \le y \le 1 \\ 0 & \text{otherwise} \end{cases}$$

Then

$$P(.25 \le Y \le .75) = \int_{.25}^{.75} f_Y(y)\,dy = \frac{37}{80} = .4625$$ ■

In Example 5.3, the region of positive joint density was a rectangle, which made computation of the marginal pdf's relatively easy. Consider now an example in which the region of positive density is more complicated.

EXAMPLE 5.5 A nut company markets cans of deluxe mixed nuts containing almonds, cashews, and peanuts. Suppose the net weight of each can is exactly 1 lb, but the weight contribution of each type of nut is random. Because the three weights sum to 1, a joint probability model for any two gives all necessary information about the weight of the third type. Let X = the weight of almonds in a selected can and Y = the weight of cashews. Then the region of positive density is $D = \{(x, y): 0 \le x \le 1, 0 \le y \le 1, x + y \le 1\}$, the shaded region pictured in Figure 5.2.

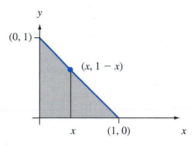

Figure 5.2 Region of positive density for Example 5.5

Now let the joint pdf for (X, Y) be

$$f(x, y) = \begin{cases} 24xy & 0 \le x \le 1, 0 \le y \le 1, x + y \le 1 \\ 0 & \text{otherwise} \end{cases}$$

For any fixed x, $f(x, y)$ increases with y; for fixed y, $f(x, y)$ increases with x. This is appropriate because the word *deluxe* implies that most of the can should consist of almonds and cashews rather than peanuts, so that the density function should be large near the upper boundary and small near the origin. The surface determined by $f(x, y)$ slopes upward from zero as (x, y) moves away from either axis.

Clearly, $f(x, y) \ge 0$. To verify the second condition on a joint pdf, recall that a double integral is computed as an iterated integral by holding one variable fixed (such as x as in Figure 5.2), integrating over values of the other variable lying along the straight line passing through the value of the fixed variable, and finally integrating over all possible values of the fixed variable. Thus

$$\int_{-\infty}^{\infty} \int_{-\infty}^{\infty} f(x, y)\,dy\,dx = \iint_D f(x, y)\,dy\,dx = \int_0^1 \left\{ \int_0^{1-x} 24xy\,dy \right\} dx$$

$$= \int_0^1 24x \left\{ \frac{y^2}{2} \Big|_{y=0}^{y=1-x} \right\} dx = \int_0^1 12x(1 - x)^2\,dx = 1$$

To compute the probability that the two types of nuts together make up at most 50% of the can, let $A = \{(x, y): 0 \le x \le 1, 0 \le y \le 1, \text{ and } x + y \le .5\}$, as shown in Figure 5.3. Then

$$P((X, Y) \in A) = \iint_A f(x, y) \, dx \, dy = \int_0^{.5} \int_0^{.5-x} 24xy \, dy \, dx = .0625$$

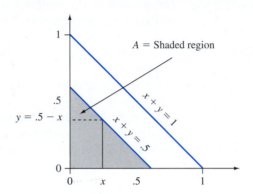

Figure 5.3 Computing $P[(X, Y) \in A]$ for Example 5.5

The marginal pdf for almonds is obtained by holding X fixed at x and integrating the joint pdf $f(x, y)$ along the vertical line through x:

$$f_X(x) = \int_{-\infty}^{\infty} f(x, y) \, dy = \begin{cases} \int_0^{1-x} 24xy \, dy = 12x(1 - x)^2 & 0 \le x \le 1 \\ 0 & \text{otherwise} \end{cases}$$

By symmetry of $f(x, y)$ and the region D, the marginal pdf of Y is obtained by replacing x and X in $f_X(x)$ by y and Y, respectively. ∎

Independent Random Variables

In many situations, information about the observed value of one of the two variables X and Y gives information about the value of the other variable. In Example 5.1, the marginal probability of X at $x = 100$ is .35 and at $x = 1000$ is .25. However, if we learn that $Y = 5000$, the last column of the joint probability table tells us that X can't possibly be 100 and the other two possibilities, 500 and 1000, are now equally likely. Thus knowing the value of Y changes the distribution of X; in such situations it is natural to say that there is a dependence between the two variables.

In Chapter 2, we pointed out that one way of defining independence of two events is via the condition $P(A \cap B) = P(A) \cdot P(B)$. Here is an analogous definition for the independence of two rv's.

DEFINITION

Two random variables X and Y are said to be **independent** if for every pair of x and y values

$$p(x, y) = p_X(x) \cdot p_Y(y) \qquad \text{when } X \text{ and } Y \text{ are discrete}$$

or   (5.1)

$$f(x, y) = f_X(x) \cdot f_Y(y) \qquad \text{when } X \text{ and } Y \text{ are continuous}$$

If (5.1) is not satisfied for all (x, y), then X and Y are said to be **dependent**.

The definition says that two variables are independent if their joint pmf or pdf is the product of the two marginal pmf's or pdf's. Intuitively, independence says that knowing the value of one of the variables does not provide additional information about what the value of the other variable might be. That is, the distribution of one variable does not depend on the value of the other variable.

EXAMPLE 5.6   In the insurance situation of Examples 5.1 and 5.2,

$$p(1000, 5000) = .05 \neq (.10)(.25) = p_X(1000) \cdot p_Y(5000)$$

so X and Y are not independent. In fact, the joint probability table has an entry which is 0, yet the corresponding row and column totals are both positive. Independence of X and Y requires that *every* entry in the joint probability table be the product of the corresponding row and column marginal probabilities.   ∎

EXAMPLE 5.7
(Example 5.5
continued)
Because $f(x, y)$ has the form of a product, X and Y would appear to be independent. However, although $f_X(3/4) = f_Y(3/4) = 9/16$, $f(3/4, 3/4) = 0 \neq 9/16 \cdot 9/16$, so the variables are not in fact independent. To be independent, $f(x, y)$ must have the form $g(x) \cdot h(y)$ *and* the region of positive density must be a rectangle whose sides are parallel to the coordinate axes.   ∎

Independence of two random variables is most useful when the description of the experiment under study suggests that X and Y have no effect on one another. Then once the marginal pmf's or pdf's have been specified, the joint pmf or pdf is simply the product of the two marginal functions. It follows that

$$P(a \leq X \leq b, c \leq Y \leq d) = P(a \leq X \leq b) \cdot P(c \leq Y \leq d)$$

EXAMPLE 5.8   Suppose that the lifetimes of two components are independent of one another and that the first lifetime, X_1, has an exponential distribution with parameter λ_1, whereas the second, X_2, has an exponential distribution with parameter λ_2. Then the joint pdf is

$$f(x_1, x_2) = f_{X_1}(x_1) \cdot f_{X_2}(x_2)$$
$$= \begin{cases} \lambda_1 e^{-\lambda_1 x_1} \cdot \lambda_2 e^{-\lambda_2 x_2} = \lambda_1 \lambda_2 e^{-\lambda_1 x_1 - \lambda_2 x_2} & x_1 > 0, x_2 > 0 \\ 0 & \text{otherwise} \end{cases}$$

Let $\lambda_1 = 1/1000$ and $\lambda_2 = 1/1200$, so that the expected lifetimes are 1000 hours and 1200 hours, respectively. The probability that both component lifetimes are at least 1500 hours is

$$P(1500 \leq X_1, 1500 \leq X_2) = P(1500 \leq X_1) \cdot P(1500 \leq X_2)$$
$$= e^{-\lambda_1(1500)} \cdot e^{-\lambda_2(1500)}$$
$$= (.2231)(.2865) = .0639 \qquad ∎$$

More Than Two Random Variables

To model the joint behavior of more than two random variables, we extend the concept of a joint distribution of two variables.

DEFINITION

> If $X_1, X_2, \ldots, X_n$ are all discrete random variables, the joint pmf of the variables is the function
>
> $$p(x_1, x_2, \ldots, x_n) = P(X_1 = x_1, X_2 = x_2, \ldots, X_n = x_n)$$
>
> If the variables are continuous, the joint pdf of $X_1, \ldots, X_n$ is the function $f(x_1, x_2, \ldots, x_n)$ such that for any n intervals $[a_1, b_1], \ldots, [a_n, b_n]$,
>
> $$P(a_1 \le X_1 \le b_1, \ldots, a_n \le X_n \le b_n) = \int_{a_1}^{b_1} \cdots \int_{a_n}^{b_n} f(x_1, \ldots, x_n) \, dx_n \ldots dx_1$$

EXAMPLE 5.9   A binomial experiment consists of n dichotomous (success–failure), homogenous (constant success probability) independent trials. Now consider a *trinomial* experiment in which each of the n trials can result in one of *three* possible outcomes. For example, each successive customer at a store might pay with cash, a credit card, or a debit card. The trials are assumed independent. Let $p_1 = P$(trial results in a type 1 outcome) and define p_2 and p_3 analogously for type 2 and type 3 outcomes. The random variables of interest here are $X_i =$ the number of trials that result in a type i outcome for $i = 1, 2, 3$.

In $n = 10$ trials, the probability that the first five are type 1 outcomes, the next three are type 2, and the last two are type 3—that is, the probability of the experimental outcome 1111122233—is $p_1^5 \cdot p_2^3 \cdot p_3^2$. This is also the probability of the outcome 1122311123, and in fact the probability of any outcome that has exactly five 1's, three 2's, and two 3's. Now to determine the probability $P(X_1 = 5, X_2 = 3,$ and $X_3 = 2)$, we have to count the number of outcomes that have exactly five 1's, three 2's, and two 3's. First, there are $\binom{10}{5}$ ways to choose five of the trials to be the type 1 outcomes. Now from the remaining five trials, we choose three to be the type 2 outcomes, which can be done in $\binom{5}{3}$ ways. This determines the remaining two trials, which consist of type 3 outcomes. So the total number of ways of choosing five 1's, three 2's, and two 3's is

$$\binom{10}{5} \cdot \binom{5}{3} = \frac{10!}{5!5!} \cdot \frac{5!}{3!2!} = \frac{10!}{5!3!2!} = 2520$$

Thus we see that $P(X_1 = 5, X_2 = 3, X_3 = 2) = 2520 \, p_1^5 \cdot p_2^3 \cdot p_3^2$. Generalizing this to n trials gives

$$p(x_1, x_2, x_3) = P(X_1 = x_1, X_2 = x_2, X_3 = x_3) = \frac{n!}{x_1! x_2! x_3!} p_1^{x_1} p_2^{x_2} p_3^{x_3}$$

for $x_1 = 0, 1, 2, \ldots$; $x_2 = 0, 1, 2, \ldots$; $x_3 = 0, 1, 2, \ldots$ such that $x_1 + x_2 + x_3 = n$. Notice that whereas there are three random variables here, the third variable X_3 is actually redundant. For example, in the case $n = 10$, having $X_1 = 5$ and $X_2 = 3$ implies that $X_3 = 2$ (just as in a binomial experiment there are actually two rv's—the number of successes and number of failures—but the latter is redundant).

As a specific example, the genetic allele of a pea section can be either AA, Aa, or aa. A simple genetic model specifies $P(\text{AA}) = .25$, $P(\text{Aa}) = .50$, and $P(\text{aa}) = .25$. If the alleles of 10 independently obtained sections are determined, the probability that exactly five of these are Aa and two are AA is

$$p(2, 5, 3) = \frac{10!}{2!5!3!} (.25)^2 (.50)^5 (.25)^3 = 0.769$$

A natural extension of the trinomial scenario is an experiment consisting of n independent and identical trials, in which each trial can result in any one of r possible outcomes. Let $p_i = P$(outcome i on any particular trial), and define random variables by $X_i =$ the number of trials resulting in outcome i ($i = 1, \ldots, r$). This is called a **multinomial experiment**, and the joint pmf of $X_1, \ldots, X_r$ is called the **multinomial distribution**. An argument analogous to the one used to derive the trinomial pmf gives the multinomial pmf as

$$p(x_1, \ldots, x_r)$$

$$= \begin{cases} \dfrac{n!}{(x_1!)(x_2!) \cdot \ldots \cdot (x_r!)} \cdot p_1^{x_1} \cdot \ldots \cdot p_r^{x_r} & x_i = 0, 1, 2, \ldots; \quad x_1 + \cdots + x_r = n \\ 0 & \text{otherwise} \end{cases}$$

■

EXAMPLE 5.10 When a certain method is used to collect a fixed volume of rock samples in a region, there are four resulting rock types. Let X_1, X_2, and X_3 denote the proportion by volume of rock types 1, 2, and 3 in a randomly selected sample (the proportion of rock type 4 is $1 - X_1 - X_2 - X_3$, so a variable X_4 would be redundant). If the joint pdf of X_1, X_2, X_3 is

$$f(x_1, x_2, x_3)$$

$$= \begin{cases} kx_1 x_2 (1 - x_3) & 0 \le x_1 \le 1, 0 \le x_2 \le 1, 0 \le x_3 \le 1, x_1 + x_2 + x_3 \le 1 \\ 0 & \text{otherwise} \end{cases}$$

then k is determined by

$$1 = \int_{-\infty}^{\infty} \int_{-\infty}^{\infty} \int_{-\infty}^{\infty} f(x_1, x_2, x_3) \, dx_3 \, dx_2 \, dx_1$$

$$= \int_0^1 \left\{ \int_0^{1-x_1} \left[\int_0^{1-x_1-x_2} kx_1 x_2 (1 - x_3) \, dx_3 \right] dx_2 \right\} dx_1$$

This iterated integral has value $k/144$, so $k = 144$. The probability that rocks of types 1 and 2 together account for at most 50% of the sample is

$$P(X_1 + X_2 \le .5) = \iiint_{\begin{cases} 0 \le x_i \le 1 \text{ for } i = 1, 2, 3 \\ x_1 + x_2 + x_3 \le 1, x_1 + x_2 \le .5 \end{cases}} f(x_1, x_2, x_3) \, dx_3 \, dx_2 \, dx_1$$

$$= \int_0^{.5} \left\{ \int_0^{.5-x_1} \left[\int_0^{1-x_1-x_2} 144x_1 x_2 (1 - x_3) \, dx_3 \right] dx_2 \right\} dx_1$$

$$= .6066$$

■

The notion of independence of more than two random variables is similar to the notion of independence of more than two events.

DEFINITION

The random variables $X_1, X_2, \ldots, X_n$ are said to be **independent** if for *every* subset $X_{i_1}, X_{i_2}, \ldots, X_{i_k}$ of the variables (each pair, each triple, and so on), the joint pmf or pdf of the subset is equal to the product of the marginal pmf's or pdf's.

Thus if the variables are independent with $n = 4$, then the joint pmf or pdf of any two variables is the product of the two marginals, and similarly for any three variables and all four variables together. Intuitively, independence means that learning the values of some variables doesn't change the distribution of the remaining variables. Most importantly, once we are told that n variables are independent, then the joint pmf or pdf is the product of the n marginals.

EXAMPLE 5.11 If $X_1, \ldots, X_n$ represent the lifetimes of n components, the components operate independently of one another, and each lifetime is exponentially distributed with parameter λ, then for $x_1 \geq 0, x_2 \geq 0, \ldots, x_n \geq 0$,

$$f(x_1, x_2, \ldots, x_n) = (\lambda e^{-\lambda x_1}) \cdot (\lambda e^{-\lambda x_2}) \cdot \cdots \cdot (\lambda e^{-\lambda x_n}) = \lambda^n e^{-\lambda \Sigma x_i}$$

Suppose a system consisting of these components will fail as soon as a single component fails. Let T represent system lifetime. Then the probability that the system lasts past time t is

$$P(T > t) = P(X_1 > t, \ldots, X_n > t) = \int_t^\infty \cdots \int_t^\infty f(x_1, \ldots, x_n) \, dx_1 \ldots dx_n$$

$$= \left(\int_t^\infty \lambda e^{-\lambda x_1} \, dx_1 \right) \ldots \left(\int_t^\infty \lambda e^{-\lambda x_n} \, dx_n \right)$$

$$= (e^{-\lambda t})^n = e^{-n\lambda t}$$

Therefore,

$$P(\text{system lifetime} \leq t) = 1 - e^{-n\lambda t} \qquad \text{for } t \geq 0$$

which shows that *system* lifetime has an exponential distribution with parameter $n\lambda$; the expected value of system lifetime is $1/n\lambda$.

A variation on the foregoing scenario appeared in the article **"A Method for Correlating Field Life Degradation with Reliability Prediction for Electronic Modules"** (*Quality and Reliability Engr. Intl.*, *2005: 715–726*). The investigators considered a circuit card with n soldered chip resistors. The failure time of a card is the minimum of the individual solder connection failure times (mileages here). It was assumed that the solder connection failure mileages were independent, that failure mileage would exceed t if and only if the shear strength of a connection exceeded a threshold d, and that each shear strength was normally distributed with a mean value and standard deviation that depended on the value of mileage t: $\mu(t) = a_1 - a_2 t$ and $\sigma(t) = a_3 + a_4 t$ (a weld's shear strength typically deteriorates and becomes more variable as mileage increases). Then the probability that the failure mileage of a card exceeds t is

$$P(T > t) = \left(1 - \Phi\left(\frac{d - (a_1 - a_2 t)}{a_3 + a_4 t} \right) \right)^n$$

The cited article suggested values for d and the a_i's based on data. In contrast to the exponential scenario, normality of individual lifetimes does not imply normality of system lifetime. ∎

In many experimental situations to be considered in this book, independence is a reasonable assumption, so that specifying the joint distribution reduces to deciding on appropriate marginal distributions.

Conditional Distributions

Suppose $X =$ the number of major defects in a randomly selected new automobile and $Y =$ the number of minor defects in that same auto. If we learn that the selected car has one major defect, what now is the probability that the car has at most three minor defects—that is, what is $P(Y \leq 3 \mid X = 1)$? Similarly, if X and Y denote the lifetimes of the front and rear tires on a motorcycle, and it happens that $X = 10,000$ miles, what now is the probability that Y is at most 15,000 miles, and what is the expected lifetime of the rear tire "conditional on" this value of X? Questions of this sort can be answered by studying conditional probability distributions.

DEFINITION

> Let X and Y be two continuous rv's with joint pdf $f(x, y)$ and marginal X pdf $f_X(x)$. Then for any X value x for which $f_X(x) > 0$, the **conditional probability density function of Y given that $X = x$** is
>
> $$f_{Y \mid X}(y \mid x) = \frac{f(x, y)}{f_X(x)} \qquad -\infty < y < \infty$$
>
> If X and Y are discrete, replacing pdf's by pmf's in this definition gives the **conditional probability mass function of Y when $X = x$.**

Notice that the definition of $f_{Y \mid X}(y \mid x)$ parallels that of $P(B \mid A)$, the conditional probability that B will occur, given that A has occurred. Once the conditional pdf or pmf has been determined, questions of the type posed at the outset of this subsection can be answered by integrating or summing over an appropriate set of Y values.

EXAMPLE 5.12

Reconsider the situation of Examples 5.3 and 5.4 involving $X =$ the proportion of time that a bank's drive-up facility is busy and $Y =$ the analogous proportion for the walk-up window. The conditional pdf of Y given that $X = .8$ is

$$f_{Y \mid X}(y \mid .8) = \frac{f(.8, y)}{f_X(.8)} = \frac{1.2(.8 + y^2)}{1.2(.8) + .4} = \frac{1}{34}(24 + 30y^2) \quad 0 < y < 1$$

The probability that the walk-up facility is busy at most half the time given that $X = .8$ is then

$$P(Y \leq .5 \mid X = .8) = \int_{-\infty}^{.5} f_{Y \mid X}(y \mid .8) \, dy = \int_{0}^{.5} \frac{1}{34}(24 + 30y^2) \, dy = .390$$

Using the marginal pdf of Y gives $P(Y \leq .5) = .350$. Also $E(Y) = .6$, whereas the expected proportion of time that the walk-up facility is busy given that $X = .8$ (a *conditional* expectation) is

$$E(Y \mid X = .8) = \int_{-\infty}^{\infty} y \cdot f_{Y \mid X}(y \mid .8) \, dy = \frac{1}{34} \int_{0}^{1} y(24 + 30y^2) \, dy = .574 \quad \blacksquare$$

If the two variables are independent, the marginal pmf or pdf in the denominator will cancel the corresponding factor in the numerator. The conditional distribution is then identical to the corresponding marginal distribution.

EXERCISES Section 5.1 (1–21)

1. A service station has both self-service and full-service islands. On each island, there is a single regular unleaded pump with two hoses. Let X denote the number of hoses being used on the self-service island at a particular time, and let Y denote the number of hoses on the full-service island in use at that time. The joint pmf of X and Y appears in the accompanying tabulation.

$p(x, y)$		y	
	0	1	2
0	.10	.04	.02
x 1	.08	.20	.06
2	.06	.14	.30

   a. What is $P(X = 1 \text{ and } Y = 1)$?
   b. Compute $P(X \le 1 \text{ and } Y \le 1)$.
   c. Give a word description of the event $\{X \ne 0 \text{ and } Y \ne 0\}$, and compute the probability of this event.
   d. Compute the marginal pmf of X and of Y. Using $p_X(x)$, what is $P(X \le 1)$?
   e. Are X and Y independent rv's? Explain.

2. A large but sparsely populated county has two small hospitals, one at the south end of the county and the other at the north end. The south hospital's emergency room has four beds, whereas the north hospital's emergency room has only three beds. Let X denote the number of south beds occupied at a particular time on a given day, and let Y denote the number of north beds occupied at the same time on the same day. Suppose that these two rv's are independent; that the pmf of X puts probability masses .1, .2, .3, .2, and .2 on the x values 0, 1, 2, 3, and 4, respectively; and that the pmf of Y distributes probabilities .1, .3, .4, and .2 on the y values 0, 1, 2, and 3, respectively.
   a. Display the joint pmf of X and Y in a joint probability table.
   b. Compute $P(X \le 1 \text{ and } Y \le 1)$ by adding probabilities from the joint pmf, and verify that this equals the product of $P(X \le 1)$ and $P(Y \le 1)$.
   c. Express the event that the total number of beds occupied at the two hospitals combined is at most 1 in terms of X and Y, and then calculate this probability.
   d. What is the probability that at least one of the two hospitals has no beds occupied?

3. A certain market has both an express checkout line and a superexpress checkout line. Let X_1 denote the number of customers in line at the express checkout at a particular time of day, and let X_2 denote the number of customers in line at the superexpress checkout at the same time. Suppose the joint pmf of X_1 and X_2 is as given in the accompanying table.

			x_2	
		0	1	2	3
	0	.08	.07	.04	.00
	1	.06	.15	.05	.04
x_1	2	.05	.04	.10	.06
	3	.00	.03	.04	.07
	4	.00	.01	.05	.06

   a. What is $P(X_1 = 1, X_2 = 1)$, that is, the probability that there is exactly one customer in each line?
   b. What is $P(X_1 = X_2)$, that is, the probability that the numbers of customers in the two lines are identical?
   c. Let A denote the event that there are at least two more customers in one line than in the other line. Express A in terms of X_1 and X_2, and calculate the probability of this event.
   d. What is the probability that the total number of customers in the two lines is exactly four? At least four?

4. Return to the situation described in Exercise 3.
   a. Determine the marginal pmf of X_1, and then calculate the expected number of customers in line at the express checkout.
   b. Determine the marginal pmf of X_2.
   c. By inspection of the probabilities $P(X_1 = 4)$, $P(X_2 = 0)$, and $P(X_1 = 4, X_2 = 0)$, are X_1 and X_2 independent random variables? Explain.

5. The number of customers waiting for gift-wrap service at a department store is an rv X with possible values 0, 1, 2, 3, 4 and corresponding probabilities .1, .2, .3, .25, .15. A randomly selected customer will have 1, 2, or 3 packages for wrapping with probabilities .6, .3, and .1, respectively. Let $Y =$ the total number of packages to be wrapped for the customers waiting in line (assume that the number of packages submitted by one customer is independent of the number submitted by any other customer).
   a. Determine $P(X = 3, Y = 3)$, i.e., $p(3, 3)$.
   b. Determine $p(4, 11)$.

6. Let X denote the number of Canon SLR cameras sold during a particular week by a certain store. The pmf of X is

x	0	1	2	3	4
$p_X(x)$	.1	.2	.3	.25	.15

   Sixty percent of all customers who purchase these cameras also buy an extended warranty. Let Y denote the number of purchasers during this week who buy an extended warranty.
   a. What is $P(X = 4, Y = 2)$? [*Hint*: This probability equals $P(Y = 2 \mid X = 4) \cdot P(X = 4)$; now think of the four purchases as four trials of a binomial

experiment, with success on a trial corresponding to buying an extended warranty.]

b. Calculate $P(X = Y)$.

c. Determine the joint pmf of X and Y and then the marginal pmf of Y.

7. The joint probability distribution of the number X of cars and the number Y of buses per signal cycle at a proposed left-turn lane is displayed in the accompanying joint probability table.

$p(x, y)$		$y = 0$	1	2
	0	.025	.015	.010
	1	.050	.030	.020
	2	.125	.075	.050
x	3	.150	.090	.060
	4	.100	.060	.040
	5	.050	.030	.020

a. What is the probability that there is exactly one car and exactly one bus during a cycle?

b. What is the probability that there is at most one car and at most one bus during a cycle?

c. What is the probability that there is exactly one car during a cycle? Exactly one bus?

d. Suppose the left-turn lane is to have a capacity of five cars, and that one bus is equivalent to three cars. What is the probability of an overflow during a cycle?

e. Are X and Y independent rv's? Explain.

8. A stockroom currently has 30 components of a certain type, of which 8 were provided by supplier 1, 10 by supplier 2, and 12 by supplier 3. Six of these are to be randomly selected for a particular assembly. Let X = the number of supplier 1's components selected, Y = the number of supplier 2's components selected, and $p(x, y)$ denote the joint pmf of X and Y.

a. What is $p(3, 2)$? [*Hint*: Each sample of size 6 is equally likely to be selected. Therefore, $p(3, 2)$ = (number of outcomes with $X = 3$ and $Y = 2$)/(total number of outcomes). Now use the product rule for counting to obtain the numerator and denominator.]

b. Using the logic of part (a), obtain $p(x, y)$. (This can be thought of as a multivariate hypergeometric distribution—sampling without replacement from a finite population consisting of more than two categories.)

9. Each front tire on a particular type of vehicle is supposed to be filled to a pressure of 26 psi. Suppose the actual air pressure in each tire is a random variable—X for the right tire and Y for the left tire, with joint pdf

$$f(x, y) = \begin{cases} K(x^2 + y^2) & 20 \le x \le 30,\ 20 \le y \le 30 \\ 0 & \text{otherwise} \end{cases}$$

a. What is the value of K?

b. What is the probability that both tires are underfilled?

c. What is the probability that the difference in air pressure between the two tires is at most 2 psi?

d. Determine the (marginal) distribution of air pressure in the right tire alone.

e. Are X and Y independent rv's?

10. Annie and Alvie have agreed to meet between 5:00 P.M. and 6:00 P.M. for dinner at a local health-food restaurant. Let X = Annie's arrival time and Y = Alvie's arrival time. Suppose X and Y are independent with each uniformly distributed on the interval [5, 6].

a. What is the joint pdf of X and Y?

b. What is the probability that they both arrive between 5:15 and 5:45?

c. If the first one to arrive will wait only 10 min before leaving to eat elsewhere, what is the probability that they have dinner at the health-food restaurant? [*Hint*: The event of interest is $A = \{(x, y): |x - y| \le 1/6\}$.]

11. Two different professors have just submitted final exams for duplication. Let X denote the number of typographical errors on the first professor's exam and Y denote the number of such errors on the second exam. Suppose X has a Poisson distribution with parameter μ_1, Y has a Poisson distribution with parameter μ_2, and X and Y are independent.

a. What is the joint pmf of X and Y?

b. What is the probability that at most one error is made on both exams combined?

c. Obtain a general expression for the probability that the total number of errors in the two exams is m (where m is a nonnegative integer). [*Hint*: $A = \{(x, y): x + y = m\} = \{(m, 0), (m - 1, 1), \ldots, (1, m - 1), (0, m)\}$. Now sum the joint pmf over $(x, y) \in A$ and use the binomial theorem, which says that

$$\sum_{k=0}^{m} \binom{m}{k} a^k b^{m-k} = (a + b)^m$$

for any a, b.]

12. Two components of a minicomputer have the following joint pdf for their useful lifetimes X and Y:

$$f(x, y) = \begin{cases} xe^{-x(1+y)} & x \ge 0 \text{ and } y \ge 0 \\ 0 & \text{otherwise} \end{cases}$$

a. What is the probability that the lifetime X of the first component exceeds 3?

b. What are the marginal pdf's of X and Y? Are the two lifetimes independent? Explain.

c. What is the probability that the lifetime of at least one component exceeds 3?

13. You have two lightbulbs for a particular lamp. Let X = the lifetime of the first bulb and Y = the lifetime of the second bulb (both in 1000s of hours). Suppose that X and Y are independent and that each has an exponential distribution with parameter $\lambda = 1$.

a. What is the joint pdf of X and Y?

b. What is the probability that each bulb lasts at most 1000 hours (i.e., $X \le 1$ and $Y \le 1$)?

c. What is the probability that the total lifetime of the two bulbs is at most 2? [*Hint*: Draw a picture of the region $A = \{(x, y): x \geq 0, y \geq 0, x + y \leq 2\}$ before integrating.]

d. What is the probability that the total lifetime is between 1 and 2?

14. Suppose that you have ten lightbulbs, that the lifetime of each is independent of all the other lifetimes, and that each lifetime has an exponential distribution with parameter λ.

   a. What is the probability that all ten bulbs fail before time t?

   b. What is the probability that exactly k of the ten bulbs fail before time t?

   c. Suppose that nine of the bulbs have lifetimes that are exponentially distributed with parameter λ and that the remaining bulb has a lifetime that is exponentially distributed with parameter θ (it is made by another manufacturer). What is the probability that exactly five of the ten bulbs fail before time t?

15. Consider a system consisting of three components as pictured. The system will continue to function as long as the first component functions and either component 2 or component 3 functions. Let X_1, X_2, and X_3 denote the lifetimes of components 1, 2, and 3, respectively. Suppose the X_i's are independent of one another and each X_i has an exponential distribution with parameter λ.

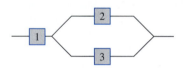

   a. Let Y denote the system lifetime. Obtain the cumulative distribution function of Y and differentiate to obtain the pdf. [*Hint*: $F(y) = P(Y \leq y)$; express the event $\{Y \leq y\}$ in terms of unions and/or intersections of the three events $\{X_1 \leq y\}$, $\{X_2 \leq y\}$, and $\{X_3 \leq y\}$.]

   b. Compute the expected system lifetime.

16. a. For $f(x_1, x_2, x_3)$ as given in Example 5.10, compute the *joint marginal density function* of X_1 and X_3 alone (by integrating over x_2).

   b. What is the probability that rocks of types 1 and 3 together make up at most 50% of the sample? [*Hint*: Use the result of part (a).]

   c. Compute the marginal pdf of X_1 alone. [*Hint*: Use the result of part (a).]

17. An ecologist wishes to select a point inside a circular sampling region according to a uniform distribution (in practice this could be done by first selecting a direction and then a distance from the center in that direction). Let $X =$ the x coordinate of the point selected and $Y =$ the y coordinate of the point selected. If the circle is centered at $(0, 0)$ and has radius R, then the joint pdf of X and Y is

$$f(x, y) = \begin{cases} \dfrac{1}{\pi R^2} & x^2 + y^2 \leq R^2 \\ 0 & \text{otherwise} \end{cases}$$

   a. What is the probability that the selected point is within $R/2$ of the center of the circular region? [*Hint*: Draw a picture of the region of positive density D. Because $f(x, y)$ is constant on D, computing a probability reduces to computing an area.]

   b. What is the probability that both X and Y differ from 0 by at most $R/2$?

   c. Answer part (b) for $R/\sqrt{2}$ replacing $R/2$.

   d. What is the marginal pdf of X? Of Y? Are X and Y independent?

18. Refer to Exercise 1 and answer the following questions:

   a. Given that $X = 1$, determine the conditional pmf of Y—i.e., $p_{Y|X}(0 \mid 1)$, $p_{Y|X}(1 \mid 1)$, and $p_{Y|X}(2 \mid 1)$.

   b. Given that two hoses are in use at the self-service island, what is the conditional pmf of the number of hoses in use on the full-service island?

   c. Use the result of part (b) to calculate the conditional probability $P(Y \leq 1 \mid X = 2)$.

   d. Given that two hoses are in use at the full-service island, what is the conditional pmf of the number in use at the self-service island?

19. The joint pdf of pressures for right and left front tires is given in Exercise 9.

   a. Determine the conditional pdf of Y given that $X = x$ and the conditional pdf of X given that $Y = y$.

   b. If the pressure in the right tire is found to be 22 psi, what is the probability that the left tire has a pressure of at least 25 psi? Compare this to $P(Y \geq 25)$.

   c. If the pressure in the right tire is found to be 22 psi, what is the expected pressure in the left tire, and what is the standard deviation of pressure in this tire?

20. Let X_1, X_2, X_3, X_4, X_5, and X_6 denote the numbers of blue, brown, green, orange, red, and yellow M&M candies, respectively, in a sample of size n. Then these X_i's have a multinomial distribution. According to the M&M Web site, the color proportions are $p_1 = .24$, $p_2 = .13$, $p_3 = .16$, $p_4 = .20$, $p_5 = .13$, and $p_6 = .14$.

   a. If $n = 12$, what is the probability that there are exactly two M&Ms of each color?

   b. For $n = 20$, what is the probability that there are at most five orange candies? [*Hint*: Think of an orange candy as a success and any other color as a failure.]

   c. In a sample of 20 M&Ms, what is the probability that the number of candies that are blue, green, or orange is at least 10?

21. Let X_1, X_2, and X_3 be the lifetimes of components 1, 2, and 3 in a three-component system.

   a. How would you define the conditional pdf of X_3 given that $X_1 = x_1$ and $X_2 = x_2$?

   b. How would you define the conditional joint pdf of X_2 and X_3 given that $X_1 = x_1$?

5.2 Expected Values, Covariance, and Correlation

Any function $h(X)$ of a single rv X is itself a random variable. However, we saw that to compute $E[h(X)]$, it is not necessary to obtain the probability distribution of $h(X)$. Instead, $E[h(X)]$ is computed as a weighted average of $h(x)$ values, where the weight function is the pmf $p(x)$ or pdf $f(x)$ of X. A similar result holds for a function $h(X, Y)$ of two jointly distributed random variables.

PROPOSITION

Let X and Y be jointly distributed rv's with pmf $p(x, y)$ or pdf $f(x, y)$ according to whether the variables are discrete or continuous. Then the expected value of a function $h(X, Y)$, denoted by $E[h(X, Y)]$ or $\mu_{h(X, Y)}$, is given by

$$E[h(X, Y)] = \begin{cases} \sum_x \sum_y h(x, y) \cdot p(x, y) & \text{if } X \text{ and } Y \text{ are discrete} \\ \int_{-\infty}^{\infty} \int_{-\infty}^{\infty} h(x, y) \cdot f(x, y) \, dx \, dy & \text{if } X \text{ and } Y \text{ are continuous} \end{cases}$$

EXAMPLE 5.13

Five friends have purchased tickets to a certain concert. If the tickets are for seats 1–5 in a particular row and the tickets are randomly distributed among the five, what is the expected number of seats separating any particular two of the five? Let X and Y denote the seat numbers of the first and second individuals, respectively. Possible (X, Y) pairs are $\{(1, 2), (1, 3), \ldots, (5, 4)\}$, and the joint pmf of (X, Y) is

$$p(x, y) = \begin{cases} \dfrac{1}{20} & x = 1, \ldots, 5; y = 1, \ldots, 5; x \neq y \\ 0 & \text{otherwise} \end{cases}$$

The number of seats separating the two individuals is $h(X, Y) = |X - Y| - 1$. The accompanying table gives $h(x, y)$ for each possible (x, y) pair.

$h(x, y)$		x			
	1	2	3	4	5
y 1	—	0	1	2	3
2	0	—	0	1	2
3	1	0	—	0	1
4	2	1	0	—	0
5	3	2	1	0	—

Thus

$$E[h(X, Y)] = \sum\sum_{(x, y)} h(x, y) \cdot p(x, y) = \sum_{x=1}^{5} \sum_{\substack{y=1 \\ x \neq y}}^{5} (|x - y| - 1) \cdot \frac{1}{20} = 1 \quad \blacksquare$$

EXAMPLE 5.14

In Example 5.5, the joint pdf of the amount X of almonds and amount Y of cashews in a 1-lb can of nuts was

$$f(x, y) = \begin{cases} 24xy & 0 \leq x \leq 1, 0 \leq y \leq 1, x + y \leq 1 \\ 0 & \text{otherwise} \end{cases}$$

If 1 lb of almonds costs the company $1.50, 1 lb of cashews costs $2.25, and 1 lb of peanuts costs $.75, then the total cost of the contents of a can is

$$h(X, Y) = (1.5)X + (2.25)Y + (.75)(1 - X - Y) = .75 + .75X + 1.5Y$$

(since $1 - X - Y$ of the weight consists of peanuts). The expected total cost is

$$E[h(X, Y)] = \int_{-\infty}^{\infty} \int_{-\infty}^{\infty} h(x, y) \cdot f(x, y)\, dx\, dy$$

$$= \int_{0}^{1} \int_{0}^{1-x} (.75 + .75x + 1.5y) \cdot 24xy\, dy\, dx = \$1.65 \qquad ■$$

The method of computing the expected value of a function $h(X_1, \ldots, X_n)$ of n random variables is similar to that for two random variables. If the X_i's are discrete, $E[h(X_1, \ldots, X_n)]$ is an n-dimensional sum; if the X_i's are continuous, it is an n-dimensional integral.

Covariance

When two random variables X and Y are not independent, it is frequently of interest to assess how strongly they are related to one another.

DEFINITION

The **covariance** between two rv's X and Y is

$$Cov(X, Y) = E[(X - \mu_X)(Y - \mu_Y)]$$

$$= \begin{cases} \displaystyle\sum_{x}\sum_{y}(x - \mu_X)(y - \mu_Y)p(x, y) & X, Y \text{ discrete} \\ \displaystyle\int_{-\infty}^{\infty}\int_{-\infty}^{\infty}(x - \mu_X)(y - \mu_Y)f(x, y)\, dx\, dy & X, Y \text{ continuous} \end{cases}$$

That is, since $X - \mu_X$ and $Y - \mu_Y$ are the deviations of the two variables from their respective mean values, the covariance is the expected product of deviations. Note that $Cov(X, X) = E[(X - \mu_X)^2] = V(X)$.

The rationale for the definition is as follows. Suppose X and Y have a strong positive relationship to one another, by which we mean that large values of X tend to occur with large values of Y and small values of X with small values of Y. Then most of the probability mass or density will be associated with $(x - \mu_X)$ and $(y - \mu_Y)$, either both positive (both X and Y above their respective means) or both negative, so the product $(x - \mu_X)(y - \mu_Y)$ will tend to be positive. Thus for a strong positive relationship, $Cov(X, Y)$ should be quite positive. For a strong negative relationship, the signs of $(x - \mu_X)$ and $(y - \mu_Y)$ will tend to be opposite, yielding a negative product. Thus for a strong negative relationship, $Cov(X, Y)$ should be quite negative. If X and Y are not strongly related, positive and negative products will tend to cancel one another, yielding a covariance near 0. Figure 5.4 illustrates the different possibilities. The covariance depends on *both* the set of possible pairs and the probabilities. In Figure 5.4, the probabilities could be changed without altering the set of possible pairs, and this could drastically change the value of $Cov(X, Y)$.

Figure 5.4 $p(x, y) = 1/10$ for each of ten pairs corresponding to indicated points: (a) positive covariance; (b) negative covariance; (c) covariance near zero

EXAMPLE 5.15 The joint and marginal pmf's for $X =$ automobile policy deductible amount and $Y =$ homeowner policy deductible amount in Example 5.1 were

		y			x	100	500	1000		y	500	1000	5000
$p(x, y)$	500	1000	5000		$p_X(x)$	.35	.40	.25		$p_Y(y)$	.55	.35	.10
	100	.30	.05	0									
x	500	.15	.20	.05									
	1000	.10	.10	.05									

from which $\mu_X = \Sigma x p_X(x) = 485$ and $\mu_Y = 1125$. Therefore,

$$\text{Cov}(X, Y) = \sum\sum_{(x, y)} (x - 485)(y - 1125)p(x, y)$$

$$= (100 - 485)(500 - 1125)(.30) + \dots$$

$$+ (1000 - 485)(5000 - 1125)(.05)$$

$$= 136{,}875$$

The following shortcut formula for $\text{Cov}(X, Y)$ simplifies the computations.

PROPOSITION

$$\text{Cov}(X, Y) = E(XY) - \mu_X \cdot \mu_Y$$

According to this formula, no intermediate subtractions are necessary; only at the end of the computation is $\mu_X \cdot \mu_Y$ subtracted from $E(XY)$. The proof involves expanding $(X - \mu_X)(Y - \mu_Y)$ and then carrying the summation or integration through to each individual term.

EXAMPLE 5.16
(Example 5.5 continued)

The joint and marginal pdf's of $X =$ amount of almonds and $Y =$ amount of cashews were

$$f(x, y) = \begin{cases} 24xy & 0 \le x \le 1, 0 \le y \le 1, x + y \le 1 \\ 0 & \text{otherwise} \end{cases}$$

$$f_X(x) = \begin{cases} 12x(1 - x)^2 & 0 \le x \le 1 \\ 0 & \text{otherwise} \end{cases}$$

with $f_Y(y)$ obtained by replacing x by y in $f_X(x)$. It is easily verified that $\mu_X = \mu_Y = \frac{2}{5}$, and

$$E(XY) = \int_{-\infty}^{\infty} \int_{-\infty}^{\infty} xy f(x, y)\, dx\, dy = \int_0^1 \int_0^{1-x} xy \cdot 24xy\, dy\, dx$$

$$= 8 \int_0^1 x^2 (1 - x)^3\, dx = \frac{2}{15}$$

Thus $\text{Cov}(X, Y) = 2/15 - (2/5)(2/5) = 2/15 - 4/25 = -2/75$. A negative covariance is reasonable here because more almonds in the can implies fewer cashews. ∎

It might appear that the relationship in the insurance example is quite strong since $\text{Cov}(X, Y) = 136{,}875$, whereas $\text{Cov}(X, Y) = -2/75$ in the nut example would seem to imply quite a weak relationship. Unfortunately, the covariance has a serious defect that makes it impossible to interpret a computed value. In the insurance example, suppose we had expressed the deductible amount in cents rather than in dollars. Then $100X$ would replace X, $100Y$ would replace Y, and the resulting covariance would be $\text{Cov}(100X, 100Y) = (100)(100)\text{Cov}(X, Y) = 1{,}368{,}750{,}000$. If, on the other hand, the deductible amount had been expressed in hundreds of dollars, the computed covariance would have been $(.01)(.01)(136{,}875) = 13.6875$. *The defect of covariance is that its computed value depends critically on the units of measurement.* Ideally, the choice of units should have no effect on a measure of strength of relationship. This is achieved by scaling the covariance.

Correlation

DEFINITION

> The **correlation coefficient** of X and Y, denoted by $\text{Corr}(X, Y)$, $\rho_{X,Y}$, or just ρ, is defined by
>
> $$\rho_{X, Y} = \frac{\text{Cov}(X, Y)}{\sigma_X \cdot \sigma_Y}$$

EXAMPLE 5.17 It is easily verified that in the insurance scenario of Example 5.15, $E(X^2) = 353{,}500$, $\sigma_X^2 = 353{,}500 - (485)^2 = 118{,}275$, $\sigma_X = 343.911$, $E(Y^2) = 2{,}987{,}500$, $\sigma_Y^2 = 1{,}721{,}875$, and $\sigma_Y = 1312.202$. This gives

$$\rho = \frac{136.875}{(343{,}911)(1312.202)} = .303 \qquad \blacksquare$$

The following proposition shows that ρ remedies the defect of $\text{Cov}(X, Y)$ and also suggests how to recognize the existence of a strong (linear) relationship.

PROPOSITION

> **1.** If a and c are either both positive or both negative,
>
> $$\text{Corr}(aX + b, cY + d) = \text{Corr}(X, Y)$$
>
> **2.** For any two rv's X and Y, $-1 \le \rho \le 1$. The two variables are said to be **uncorrelated** when $\rho = 0$.

Statement 1 says precisely that the correlation coefficient is not affected by a linear change in the units of measurement (if, say, X = temperature in °C, then $9X/5 + 32$ = temperature in °F). According to Statement 2, the strongest possible positive relationship is evidenced by $\rho = +1$, the strongest possible negative relationship corresponds to $\rho = -1$, and $\rho = 0$ indicates the absence of a relationship. The proof of the first statement is sketched in Exercise 35, and that of the second appears in Supplementary Exercise 87 at the end of the chapter. For descriptive purposes, the relationship will be described as strong if $|\rho| \geq .8$, moderate if $.5 < |\rho| < .8$, and weak if $|\rho| \leq .5$.

If we think of $p(x, y)$ or $f(x, y)$ as prescribing a mathematical model for how the two numerical variables X and Y are distributed in some population (height and weight, verbal SAT score and quantitative SAT score, etc.), then ρ is a population characteristic or parameter that measures how strongly X and Y are related in the population. In Chapter 12, we will consider taking a sample of pairs $(x_1, y_1), \ldots,$ (x_n, y_n) from the population. The sample correlation coefficient r will then be defined and used to make inferences about ρ.

The correlation coefficient ρ is actually not a completely general measure of the strength of a relationship.

PROPOSITION

> **1.** If X and Y are independent, then $\rho = 0$, but $\rho = 0$ does not imply independence.
>
> **2.** $\rho = 1$ or -1 iff $Y = aX + b$ for some numbers a and b with $a \neq 0$.

This proposition says that ρ is a measure of the degree of **linear** relationship between X and Y, and only when the two variables are perfectly related in a linear manner will ρ be as positive or negative as it can be. However, if $|\rho| << 1$, there may still be a strong relationship between the two variables, just one that is not linear. And even if $|\rho|$ is close to 1, it may be that the relationship is really nonlinear but can be well approximated by a straight line.

EXAMPLE 5.18   Let X and Y be discrete rv's with joint pmf

$$p(x, y) = \begin{cases} .25 & (x, y) = (-4, 1), (4, -1), (2, 2), (-2, -2) \\ 0 & \text{otherwise} \end{cases}$$

The points that receive positive probability mass are identified on the (x, y) coordinate system in Figure 5.5. It is evident from the figure that the value of X is completely determined by the value of Y and vice versa, so the two variables are completely dependent. However, by symmetry $\mu_X = \mu_Y = 0$ and $E(XY) = (-4)(.25) + (-4)(.25) + (4)(.25) + (4)(.25) = 0$. The covariance is then $\text{Cov}(X,Y) = E(XY) - \mu_X \cdot \mu_Y = 0$ and thus $\rho_{X,Y} = 0$. Although there is perfect dependence, there is also complete absence of any linear relationship!

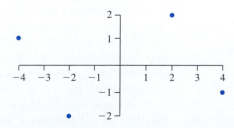

Figure 5.5 The population of pairs for Example 5.18

A value of ρ near 1 does not necessarily imply that increasing the value of X causes Y to increase. It implies only that large X values are *associated* with large Y values. For example, in the population of children, vocabulary size and number of cavities are quite positively correlated, but it is certainly not true that cavities cause vocabulary to grow. Instead, the values of both these variables tend to increase as the value of age, a third variable, increases. For children of a fixed age, there is probably a low correlation between number of cavities and vocabulary size. In summary, association (a high correlation) is not the same as causation.

The Bivariate Normal Distribution

Just as the most useful univariate distribution in statistical practice is the normal distribution, the most useful joint distribution for two rv's X and Y is the bivariate normal distribution. The pdf is somewhat complicated:

$$f(x, y) = \frac{1}{2\pi\sigma_1\sigma_2\sqrt{1 - \rho^2}}\exp$$

$$\left(-\frac{1}{2(1 - \rho^2)}\left[\left(\frac{x - \mu_1}{\sigma_1}\right)^2 - 2\rho\left(\frac{x - \mu_1}{\sigma_1}\right)\left(\frac{y - \mu_2}{\sigma_2}\right) + \left(\frac{y - \mu_2}{\sigma_2}\right)^2\right]\right)$$

A graph of this pdf, the density surface, appears in Figure 5.6. It follows (after some tricky integration) that the marginal distribution of X is normal with mean value μ_1 and standard deviation σ_1, and similarly the marginal distribution of Y is normal with mean μ_2 and standard deviation σ_2. The fifth parameter of the distribution is ρ, which can be shown to be the correlation coefficient between X and Y.

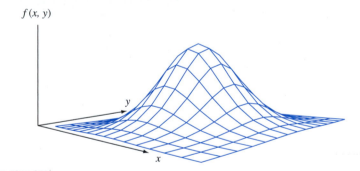

Figure 5.6   A graph of the bivariate normal pdf

It is not at all straightforward to integrate the bivariate normal pdf in order to calculate probabilities. Instead, selected software packages employ numerical integration techniques for this purpose.

EXAMPLE 5.19   Many students applying for college take the SAT, which for a few years consisted of three components: Critical Reading, Mathematics, and Writing. While some colleges used all three components to determine admission, many only looked at the first two (reading and math). Let X and Y denote the Critical Reading and Mathematics scores, respectively, for a randomly selected student. According to the College Board website, the population of students taking the exam in Fall 2012 had the following characteristics: $\mu_1 = 496$, $\sigma_1 = 114$, $\mu_2 = 514$, $\sigma_2 = 117$.

Suppose that X and Y have (approximately, since both variables are discrete) a bivariate normal distribution with correlation coefficient $\rho = .25$. The Matlab software package gives $P(X \leq 650, Y \leq 650) = P(\text{both scores are at most } 650) = .8097$. ■

It can also be shown that the conditional distribution of Y given that $X = x$ is normal. This can be seen geometrically by slicing the density surface with a plane perpendicular to the (x, y) passing through the value x on that axis; the result is a normal curve sketched out on the slicing plane. The conditional mean value is $\mu_{Y \cdot x} = (\mu_2 - \rho\mu_1\sigma_2/\sigma_1) + \rho\sigma_2 x/\sigma_1$, a linear function of x, and the conditional variance is $\sigma_{Y \cdot x}^2 = (1 - \rho^2)\sigma_2^2$. The closer the correlation coefficient is to 1 or -1, the less variability there is in the conditional distribution. Analogous results hold for the conditional distribution of X given that $Y = y$.

The bivariate normal distribution can be generalized to the *multivariate normal distribution*. Its density function is quite complicated, and the only way to write it compactly is to employ matrix notation. If a collection of variables has this distribution, then the marginal distribution of any single variable is normal, the conditional distribution of any single variable given values of the other variables is normal, the joint marginal distribution of any pair of variables is bivariate normal, and the joint marginal distribution of any subset of three or more of the variables is again multivariate normal.

EXERCISES Section 5.2 (22–36)

22. An instructor has given a short quiz consisting of two parts. For a randomly selected student, let $X =$ the number of points earned on the first part and $Y =$ the number of points earned on the second part. Suppose that the joint pmf of X and Y is given in the accompanying table.

				y	
$p(x, y)$		0	5	10	15
	0	.02	.06	.02	.10
x	5	.04	.15	.20	.10
	10	.01	.15	.14	.01

a. If the score recorded in the grade book is the total number of points earned on the two parts, what is the expected recorded score $E(X + Y)$?

b. If the maximum of the two scores is recorded, what is the expected recorded score?

23. The difference between the number of customers in line at the express checkout and the number in line at the super-express checkout in Exercise 3 is $X_1 - X_2$. Calculate the expected difference.

24. Six individuals, including A and B, take seats around a circular table in a completely random fashion. Suppose the seats are numbered 1, . . . , 6. Let $X =$ A's seat number and $Y =$ B's seat number. If A sends a written message around the table to B in the direction in which they are closest, how many individuals (including A and B) would you expect to handle the message?

25. A surveyor wishes to lay out a square region with each side having length L. However, because of a measurement error, he instead lays out a rectangle in which the north–south sides

both have length X and the east–west sides both have length Y. Suppose that X and Y are independent and that each is uniformly distributed on the interval $[L - A, L + A]$ (where $0 < A < L$). What is the expected area of the resulting rectangle?

26. Consider a small ferry that can accommodate cars and buses. The toll for cars is \$3, and the toll for buses is \$10. Let X and Y denote the number of cars and buses, respectively, carried on a single trip. Suppose the joint distribution of X and Y is as given in the table of Exercise 7. Compute the expected revenue from a single trip.

27. Annie and Alvie have agreed to meet for lunch between noon (0:00 P.M.) and 1:00 P.M. Denote Annie's arrival time by X, Alvie's by Y, and suppose X and Y are independent with pdf's

$$f_X(x) = \begin{cases} 3x^2 & 0 \le x \le 1 \\ 0 & \text{otherwise} \end{cases}$$

$$f_Y(y) = \begin{cases} 2y & 0 \le y \le 1 \\ 0 & \text{otherwise} \end{cases}$$

What is the expected amount of time that the one who arrives first must wait for the other person? [*Hint:* $h(X, Y) = |X - Y|$.]

28. Show that if X and Y are independent rv's, then $E(XY) = E(X) \cdot E(Y)$. Then apply this in Exercise 25. [*Hint:* Consider the continuous case with $f(x, y) = f_X(x) \cdot f_Y(y)$.]

29. Compute the correlation coefficient ρ for X and Y of Example 5.16 (the covariance has already been computed).

30. a. Compute the covariance for X and Y in Exercise 22.
  b. Compute ρ for X and Y in the same exercise.

31. a. Compute the covariance between X and Y in Exercise 9.
  b. Compute the correlation coefficient ρ for this X and Y.

32. Reconsider the minicomputer component lifetimes X and Y as described in Exercise 12. Determine $E(XY)$. What can be said about $\text{Cov}(X, Y)$ and ρ?

33. Use the result of Exercise 28 to show that when X and Y are independent, $\text{Cov}(X, Y) = \text{Corr}(X, Y) = 0$.

34. a. Recalling the definition of σ^2 for a single rv X, write a formula that would be appropriate for computing the variance of a function $h(X, Y)$ of two random variables. [*Hint*: Remember that variance is just a special expected value.]
  b. Use this formula to compute the variance of the recorded score $h(X, Y)$ [$= \max(X, Y)$] in part (b) of Exercise 22.

35. a. Use the rules of expected value to show that $\text{Cov}(aX + b, cY + d) = ac\, \text{Cov}(X, Y)$.
  b. Use part (a) along with the rules of variance and standard deviation to show that $\text{Corr}(aX + b, cY + d) = \text{Corr}(X, Y)$ when a and c have the same sign.
  c. What happens if a and c have opposite signs?

36. Show that if $Y = aX + b$ ($a \neq 0$), then $\text{Corr}(X, Y) = +1$ or -1. Under what conditions will $\rho = +1$?

5.3 Statistics and Their Distributions

The observations in a single sample were denoted in Chapter 1 by $x_1, x_2, \ldots, x_n$. Consider selecting two different samples of size n from the same population distribution. The x_i's in the second sample will virtually always differ at least a bit from those in the first sample. For example, a first sample of $n = 3$ cars of a particular type might result in fuel efficiencies $x_1 = 30.7$, $x_2 = 29.4$, $x_3 = 31.1$, whereas a second sample may give $x_1 = 28.8$, $x_2 = 30.0$, and $x_3 = 32.5$. Before we obtain data, there is uncertainty about the value of each x_i. Because of this uncertainty, *before* the data becomes available we now regard each observation as a random variable and denote the sample by $X_1, X_2, \ldots, X_n$ (uppercase letters for random variables).

This variation in observed values in turn implies that the value of any function of the sample observations—such as the sample mean, sample standard deviation, or sample fourth spread—also varies from sample to sample. That is, prior to obtaining $x_1, \ldots, x_n$, there is uncertainty as to the value of $\bar{x}$, the value of s, and so on.

EXAMPLE 5.20 Suppose that material strength for a randomly selected specimen of a particular type has a Weibull distribution with parameter values $\alpha = 2$ (shape) and $\beta = 5$ (scale). The corresponding density curve is shown in Figure 5.7. Formulas from Section 4.5 give

$$\mu = E(X) = 4.4311 \quad \tilde{\mu} = 4.1628 \quad \sigma^2 = V(X) = 5.365 \quad \sigma = 2.316$$

The mean exceeds the median because of the distribution's positive skew.

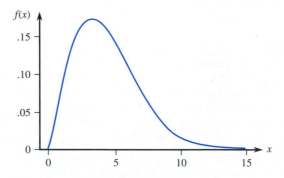

Figure 5.7 The Weibull density curve for Example 5.20

We used statistical software to generate six different samples, each with $n = 10$, from this distribution (material strengths for six different groups of ten specimens each). The results appear in Table 5.1, followed by the values of the sample mean, sample median, and sample standard deviation for each sample. Notice first that the ten observations in any particular sample are all different from those in any other sample. Second, the six values of the sample mean are all different from one another, as are the six values of the sample median and the six values of the sample standard deviation. The same is true of the sample 10% trimmed means, sample fourth spreads, and so on.

Table 5.1 Samples from the Weibull Distribution of Example 5.20

Sample	1	2	3	4	5	6
1	6.1171	5.07611	3.46710	1.55601	3.12372	8.93795
2	4.1600	6.79279	2.71938	4.56941	6.09685	3.92487
3	3.1950	4.43259	5.88129	4.79870	3.41181	8.76202
4	0.6694	8.55752	5.14915	2.49759	1.65409	7.05569
5	1.8552	6.82487	4.99635	2.33267	2.29512	2.30932
6	5.2316	7.39958	5.86887	4.01295	2.12583	5.94195
7	2.7609	2.14755	6.05918	9.08845	3.20938	6.74166
8	10.2185	8.50628	1.80119	3.25728	3.23209	1.75468
9	5.2438	5.49510	4.21994	3.70132	6.84426	4.91827
10	4.5590	4.04525	2.12934	5.50134	4.20694	7.26081
$\bar{x}$	4.401	5.928	4.229	4.132	3.620	5.761
$\tilde{x}$	4.360	6.144	4.608	3.857	3.221	6.342
s	2.642	2.062	1.611	2.124	1.678	2.496

Furthermore, the value of the sample mean from any particular sample can be regarded as a *point estimate* ("point" because it is a single number, corresponding to a single point on the number line) of the population mean μ, whose value is known to be 4.4311. None of the estimates from these six samples is identical to what is being estimated. The estimates from the second and sixth samples are much too large, whereas the fifth sample gives a substantial underestimate. Similarly, the sample standard deviation gives a point estimate of the population standard deviation. All six of the resulting estimates are in error by at least a small amount.

In summary, the values of the individual sample observations vary from sample to sample, so will in general the value of any quantity computed from sample data, and the value of a sample characteristic used as an estimate of the corresponding population characteristic will virtually never coincide with what is being estimated. ■

DEFINITION

> A **statistic** is any quantity whose value can be calculated from sample data. Prior to obtaining data, there is uncertainty as to what value of any particular statistic will result. Therefore, a statistic is a random variable and will be denoted by an uppercase letter; a lowercase letter is used to represent the calculated or observed value of the statistic.

Thus the sample mean, regarded as a statistic (before a sample has been selected or an experiment carried out), is denoted by $\overline{X}$; the calculated value of this statistic is $\bar{x}$. Similarly, S represents the sample standard deviation thought of as a statistic, and its computed value is s. If samples of two different types of bricks are selected and the individual compressive strengths are denoted by $X_1, \ldots, X_m$ and $Y_1, \ldots, Y_n$, respectively, then the statistic $\overline{X} - \overline{Y}$, the difference between the two sample mean compressive strengths, is often of great interest.

Any statistic, being a random variable, has a probability distribution. In particular, the sample mean $\overline{X}$ has a probability distribution. Suppose, for example, that $n = 2$ components are randomly selected and the number of breakdowns while under warranty is determined for each one. Possible values for the sample mean number of breakdowns $\overline{X}$ are 0 (if $X_1 = X_2 = 0$), .5 (if either $X_1 = 0$ and $X_2 = 1$ or $X_1 = 1$ and $X_2 = 0$), 1, 1.5, ... The probability distribution of $\overline{X}$ specifies $P(\overline{X} = 0)$, $P(\overline{X} = .5)$, and so on, from which other probabilities such as $P(1 \leq \overline{X} \leq 3)$ and $P(\overline{X} \geq 2.5)$ can be calculated. Similarly, if for a sample of size $n = 2$, the only possible values of the sample variance are 0, 12.5, and 50 (which is the case if X_1 and X_2 can each take on only the values 40, 45, or 50), then the probability distribution of S^2 gives $P(S^2 = 0)$, $P(S^2 = 12.5)$, and $P(S^2 = 50)$. The probability distribution of a statistic is sometimes referred to as its **sampling distribution** to emphasize that it describes how the statistic varies in value across all samples that might be selected.

Random Samples

The probability distribution of any particular statistic depends not only on the population distribution (normal, uniform, etc.) and the sample size n but also on the method of sampling. Consider selecting a sample of size $n = 2$ from a population consisting of just the three values 1, 5, and 10, and suppose that the statistic of interest is the sample variance. If sampling is done "with replacement," then $S^2 = 0$ will result if $X_1 = X_2$. However, S^2 cannot equal 0 if sampling is "without replacement." So $P(S^2 = 0) = 0$ for one sampling method, and this probability is positive for the other method. Our next definition describes a sampling method often encountered (at least approximately) in practice.

DEFINITION

> The rv's $X_1, X_2, \ldots, X_n$ are said to form a (simple) **random sample** of size n if
>
> 1. The X_i's are independent rv's.
> 2. Every X_i has the same probability distribution.

Conditions 1 and 2 can be paraphrased by saying that the X_i's are *independent and identically distributed* (iid). If sampling is either with replacement or from an infinite (conceptual) population, Conditions 1 and 2 are satisfied exactly. These conditions will be approximately satisfied if sampling is without replacement, yet the sample size n is much smaller than the population size N. In practice, if $n/N \leq .05$ (at most 5% of the population is sampled), we can proceed as if the X_i's form a random sample. The virtue of such random sampling is that the probability distribution of any statistic can be more easily obtained than for any other sampling procedure.

There are two general methods for obtaining information about a statistic's sampling distribution. One method involves calculations based on probability rules, and the other involves carrying out a simulation experiment.

Deriving a Sampling Distribution

Probability rules can be used to obtain the distribution of a statistic provided that it is a "fairly simple" function of the X_i's and either there are relatively few different X values in the population or else the population distribution has a "nice" form. Our next two examples illustrate such situations.

EXAMPLE 5.21 A certain brand of MP3 player comes in three configurations: a model with 2 GB of memory, costing $80, a 4 GB model priced at $100, and an 8 GB version with a price tag of $120. If 20% of all purchasers choose the 2 GB model, 30% choose the 4 GB model, and 50% choose the 8 GB model, then the probability distribution of the cost X of a single randomly selected MP3 player purchase is given by

x	80	100	120
$p(x)$	.2	.3	.5

with $\mu = 106$, $\sigma^2 = 244$ (5.2)

Suppose on a particular day only two MP3 players are sold. Let X_1 = the revenue from the first sale and X_2 = the revenue from the second. Suppose that X_1 and X_2 are independent, each with the probability distribution shown in (5.2) [so that X_1 and X_2 constitute a random sample from the distribution (5.2)]. Table 5.2 lists possible (x_1, x_2) pairs, the probability of each [computed using (5.2) and also the assumption of independence], and the resulting $\bar{x}$ and s^2 values. [Note that when $n = 2$, $s^2 = (x_1 - \bar{x})^2 + (x_2 - \bar{x})^2$.] Now to obtain the probability distribution of $\bar{X}$, the sample average revenue per sale, we must consider each possible value $\bar{x}$ and compute its probability. For example, $\bar{x} = 100$ occurs three times in the table with probabilities .10, .09, and .10, so

$$p_{\bar{X}}(100) = P(\bar{X} = 100) = .10 + .09 + .10 = .29$$

Similarly,

$$p_{S^2}(800) = P(S^2 = 800) = P(X_1 = 80, X_2 = 120 \text{ or } X_1 = 120, X_2 = 80)$$
$$= .10 + .10 = .20$$

Table 5.2 Outcomes, Probabilities, and Values of $\bar{x}$ and s^2 for Example 5.21

x_1	x_2	$p(x_1, x_2)$	$\bar{x}$	s^2
80	80	.04	80	0
80	100	.06	90	200
80	120	.10	100	800
100	80	.06	90	200
100	100	.09	100	0
100	120	.15	110	200
120	80	.10	100	800
120	100	.15	110	200
120	120	.25	120	0

The complete sampling distributions of $\bar{X}$ and S^2 appear in (5.3) and (5.4).

$\bar{x}$	80	90	100	110	120
$p_{\bar{X}}(\bar{x})$	.04	.12	.29	.30	.25

(5.3)

s^2	0	200	800
$p_{S^2}(s^2)$	.38	.42	.20

(5.4)

Figure 5.8 pictures a probability histogram for both the original distribution (5.2) and the $\bar{X}$ distribution (5.3). The figure suggests first that the mean (expected value) of the $\bar{X}$ distribution is equal to the mean 106 of the original distribution, since both histograms appear to be centered at the same place.

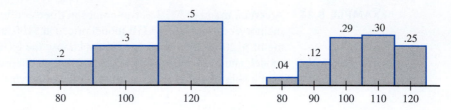

Figure 5.8 Probability histograms for the underlying distribution and $\overline{X}$ distribution in Example 5.21

From (5.3),

$$\mu_{\overline{X}} = E(\overline{X}) = \Sigma \overline{x} p_{\overline{X}}(\overline{x}) = (80)(.04) + \cdots + (120)(.25) = 106 = \mu$$

Second, it appears that the $\overline{X}$ distribution has smaller spread (variability) than the original distribution, since probability mass has moved in toward the mean. Again from (5.3),

$$\sigma_{\overline{X}}^2 = V(\overline{X}) = \Sigma \overline{x}^2 \cdot p_{\overline{X}}(\overline{x}) - \mu_{\overline{X}}^2$$
$$= (80^2)(.04) + \cdots + (120^2)(.25) - (106)^2$$
$$= 122 = 244/2 = \sigma^2/2$$

The variance of $\overline{X}$ is precisely half that of the original variance (because $n = 2$).

Using (5.4), the mean value of S^2 is

$$\mu_{S^2} = E(S^2) = \sum s^2 \cdot p_{S^2}(s^2)$$
$$= (0)(.38) + (200)(.42) + (800)(.20) = 244 = \sigma^2$$

That is, the $\overline{X}$ sampling distribution is centered at the population mean μ, and the S^2 sampling distribution is centered at the population variance σ^2.

If there had been four purchases on the day of interest, the sample average revenue $\overline{X}$ would be based on a random sample of four X_i's, each having the distribution (5.2). Mildly tedious calculations yield the pmf of $\overline{X}$ for $n = 4$ as

$\overline{x}$	80	85	90	95	100	105	110	115	120
$p_{\overline{X}}(\overline{x})$	.0016	.0096	.0376	.0936	.1761	.2340	.2350	.1500	.0625

From this, $\mu_{\overline{x}} = 106 = \mu$ and $\sigma_{\overline{X}}^2 = 61 = \sigma^2/4$. Figure 5.9 is a probability histogram of this pmf.

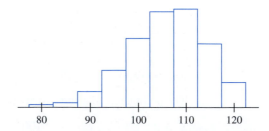

Figure 5.9 Probability histogram for $\overline{X}$ based on $n = 4$ in Example 5.21 ∎

Example 5.21 should suggest first of all that the computation of $p_{\overline{X}}(\overline{x})$ and $p_{S^2}(s^2)$ can be tedious. If the original distribution (5.2) had allowed for more than three possible values, then even for $n = 2$ the computations would have been more involved. The example should also suggest, however, that there are some general relationships between $E(\overline{X})$, $V(\overline{X})$, $E(S^2)$, and the mean μ and variance σ^2 of the original distribution. These are stated in the next section. Now consider an example in which the random sample is drawn from a continuous distribution.

EXAMPLE 5.22 Service time for a certain type of bank transaction is a random variable having an exponential distribution with parameter λ. Suppose X_1 and X_2 are service times for two different customers, assumed independent of each other. Consider the total service time $T_o = X_1 + X_2$ for the two customers, also a statistic. The cdf of T_o is, for $t \geq 0$,

$$F_{T_o}(t) = P(X_1 + X_2 \leq t) = \iint\limits_{\{(x_1, x_2):\, x_1 + x_2 \leq t\}} f(x_1, x_2)\, dx_1\, dx_2$$

$$= \int_0^t \int_0^{t-x_1} \lambda e^{-\lambda x_1} \cdot \lambda e^{-\lambda x_2}\, dx_2\, dx_1 = \int_0^t [\lambda e^{-\lambda x_1} - \lambda e^{-\lambda t}]\, dx_1$$

$$= 1 - e^{-\lambda t} - \lambda t e^{-\lambda t}$$

The region of integration is pictured in Figure 5.10.

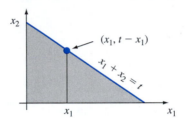

Figure 5.10 Region of integration to obtain cdf of T_o in Example 5.22

The pdf of T_o is obtained by differentiating $F_{T_o}(t)$:

$$f_{T_o}(t) = \begin{cases} \lambda^2 t e^{-\lambda t} & t \geq 0 \\ 0 & t < 0 \end{cases} \tag{5.5}$$

This is a gamma pdf ($\alpha = 2$ and $\beta = 1/\lambda$). The pdf of $\overline{X} = T_o/2$ is obtained from the relation $\{\overline{X} \leq \overline{x}\}$ iff $\{T_o \leq 2\overline{x}\}$ as

$$f_{\overline{X}}(\overline{x}) = \begin{cases} 4\lambda^2 \overline{x} e^{-2\lambda \overline{x}} & \overline{x} \geq 0 \\ 0 & \overline{x} < 0 \end{cases} \tag{5.6}$$

The mean and variance of the underlying exponential distribution are $\mu = 1/\lambda$ and $\sigma^2 = 1/\lambda^2$. From Expressions (5.5) and (5.6), it can be verified that $E(\overline{X}) = 1/\lambda$, $V(\overline{X}) = 1/(2\lambda^2)$, $E(T_o) = 2/\lambda$, and $V(T_o) = 2/\lambda^2$. These results again suggest some general relationships between means and variances of $\overline{X}$, T_o, and the underlying distribution. ∎

Simulation Experiments

The second method for obtaining information about a statistic's sampling distribution is to perform a simulation experiment. This method is usually used when a derivation via probability rules is very difficult or even impossible. Such an experiment is virtually always done with the aid of a computer. The following characteristics of an experiment must be specified:

1. The statistic of interest ($\overline{X}$, S, a particular trimmed mean, etc.)
2. The population distribution (normal with $\mu = 100$ and $\sigma = 15$, uniform with lower limit $A = 5$ and upper limit $B = 10$, etc.)
3. The sample size n (e.g., $n = 10$ or $n = 50$)
4. The number of replications k (number of samples to be obtained)

Then use appropriate software to obtain k different random samples, each of size n, from the designated population distribution. For each sample, calculate the value of the statistic and construct a histogram of the k values. This histogram gives the *approximate* sampling distribution of the statistic. The larger the value of k, the better the approximation will tend to be (the actual sampling distribution emerges as $k \rightarrow \infty$). In practice, $k = 500$ or 1000 is usually sufficient if the statistic is "fairly simple."

EXAMPLE 5.23   The population distribution for our first simulation study is normal with $\mu = 8.25$ and $\sigma = .75$, as pictured in Figure 5.11. [The article **"Platelet Size in Myocardial Infarction"** (**British Med. J., 1983: 449–451**) suggests this distribution for platelet volume in individuals with no history of serious heart problems.]

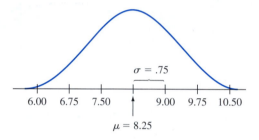

Figure 5.11   Normal distribution, with $\mu = 8.25$ and $\sigma = .75$

We actually performed four different experiments, with 500 replications for each one. In the first experiment, 500 samples of $n = 5$ observations each were generated using Minitab, and the sample sizes for the other three were $n = 10$, $n = 20$, and $n = 30$, respectively. The sample mean was calculated for each sample, and the resulting histograms of $\bar{x}$ values appear in Figure 5.12.

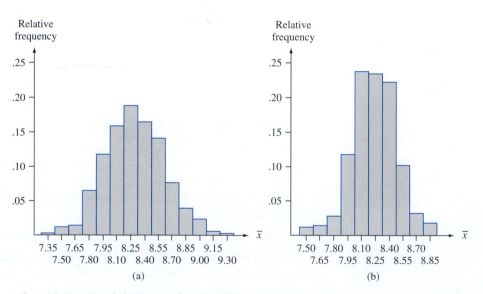

Figure 5.12   Sample histograms for $\bar{x}$ based on 500 samples, each consisting of n observations: (a) $n = 5$; (b) $n = 10$; (c) $n = 20$; (d) $n = 30$

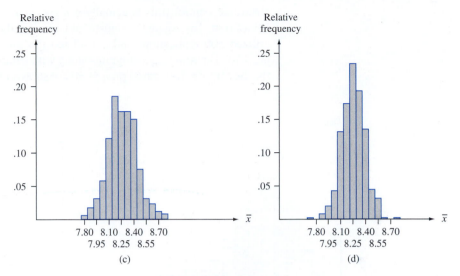

Figure 5.12 (continued)

The first thing to notice about the histograms is their shape. To a reasonable approximation, each of the four looks like a normal curve. The resemblance would be even more striking if each histogram had been based on many more than 500 $\bar{x}$ values. Second, each histogram is centered approximately at 8.25, the mean of the population being sampled. Had the histograms been based on an unending sequence of $\bar{x}$ values, their centers would have been exactly the population mean, 8.25.

The final aspect of the histograms to note is their spread relative to one another. The larger the value of n, the more concentrated is the sampling distribution about the mean value. This is why the histograms for $n = 20$ and $n = 30$ are based on narrower class intervals than those for the two smaller sample sizes. For the larger sample sizes, most of the $\bar{x}$ values are quite close to 8.25. This is the effect of averaging. When n is small, a single unusual x value can result in an $\bar{x}$ value far from the center. With a larger sample size, any unusual x values, when averaged in with the other sample values, still tend to yield an $\bar{x}$ value close to μ. Combining these insights yields a result that should appeal to your intuition: $\overline{X}$ **based on a large n tends to be closer to μ than does $\overline{X}$ based on a small n.** ∎

EXAMPLE 5.24   Consider a simulation experiment in which the population distribution is quite skewed. Figure 5.13 shows the density curve for lifetimes of a certain type of

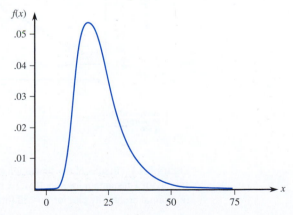

Figure 5.13 Density curve for the simulation experiment of Example 5.24 [$E(X) = 21.7584$, $V(X) = 82.1449$]

electronic control [this is actually a lognormal distribution with $E(\ln(X)) = 3$ and $V(\ln(X)) = .16$]. Again the statistic of interest is the sample mean $\overline{X}$. The experiment utilized 500 replications and considered the same four sample sizes as in Example 5.23. The resulting histograms along with a normal probability plot from Minitab for the 500 $\overline{x}$ values based on $n = 30$ are shown in Figure 5.14.

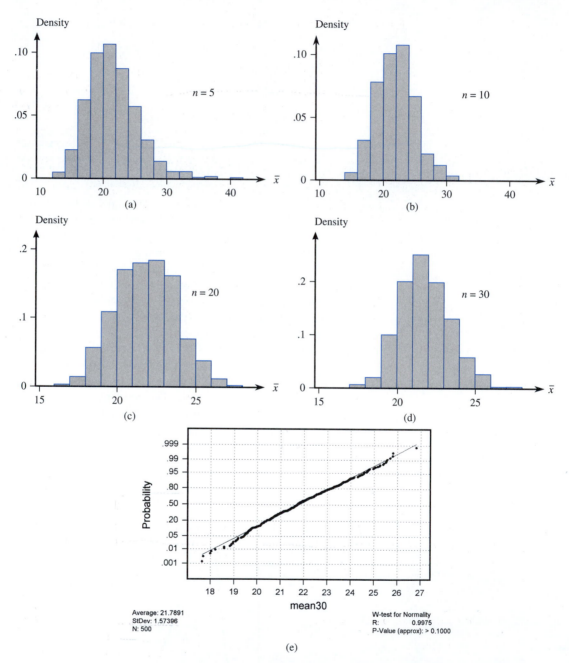

Figure 5.14 Results of the simulation experiment of Example 5.24: (a) $\overline{x}$ histogram for $n = 5$; (b) $\overline{x}$ histogram for $n = 10$; (c) $\overline{x}$ histogram for $n = 20$; (d) $\overline{x}$ histogram for $n = 30$; (e) normal probability plot for $n = 30$ (from Minitab)

Unlike the normal case, these histograms all differ in shape. In particular, they become progressively less skewed as the sample size n increases. The average of the 500 $\bar{x}$ values for the four different sample sizes are all quite close to the mean value of the population distribution. If each histogram had been based on an unending sequence of $\bar{x}$ values rather than just 500, all four would have been centered at exactly 21.7584. Thus different values of n change the shape but not the center of the sampling distribution of $\overline{X}$. Comparison of the four histograms in Figure 5.14 also shows that as n increases, the spread of the histograms decreases. Increasing n results in a greater degree of concentration about the population mean value and makes the histogram look more like a normal curve. The histogram of Figure 5.14(d) and the normal probability plot in Figure 5.14(e) provide convincing evidence that a sample size of $n = 30$ is sufficient to overcome the skewness of the population distribution and give an approximately normal $\overline{X}$ sampling distribution. ∎

EXERCISES Section 5.3 (37–45)

37. A particular brand of dishwasher soap is sold in three sizes: 25 oz, 40 oz, and 65 oz. Twenty percent of all purchasers select a 25-oz box, 50% select a 40-oz box, and the remaining 30% choose a 65-oz box. Let X_1 and X_2 denote the package sizes selected by two independently selected purchasers.

   a. Determine the sampling distribution of $\overline{X}$, calculate $E(\overline{X})$, and compare to μ.

   b. Determine the sampling distribution of the sample variance S^2, calculate $E(S^2)$, and compare to σ^2.

38. There are two traffic lights on a commuter's route to and from work. Let X_1 be the number of lights at which the commuter must stop on his way to work, and X_2 be the number of lights at which he must stop when returning from work. Suppose these two variables are independent, each with pmf given in the accompanying table (so X_1, X_2 is a random sample of size $n = 2$).

x_1	0	1	2
$p(x_1)$	.2	.5	.3

$\mu = 1.1, \sigma^2 = .49$

   a. Determine the pmf of $T_o = X_1 + X_2$.

   b. Calculate μ_{T_o}. How does it relate to μ, the population mean?

   c. Calculate $\sigma^2_{T_o}$. How does it relate to σ^2, the population variance?

   d. Let X_3 and X_4 be the number of lights at which a stop is required when driving to and from work on a second day assumed independent of the first day. With $T_o =$ the sum of all four X_i's, what now are the values of $E(T_o)$ and $V(T_o)$?

   e. Referring back to (d), what are the values of $P(T_o = 8)$ and $P(T_o \geq 7)$ [*Hint:* Don't even think of listing all possible outcomes!]

39. It is known that 80% of all brand A external hard drives work in a satisfactory manner throughout the warranty period (are "successes"). Suppose that $n = 15$ drives are randomly selected. Let $X =$ the number of successes in the sample. The statistic X/n is the sample proportion (fraction) of successes. Obtain the sampling distribution of this statistic. [*Hint:* One possible value of X/n is .2, corresponding to $X = 3$. What is the probability of this value (what kind of rv is X)?]

40. A box contains ten sealed envelopes numbered 1, . . . , 10. The first five contain no money, the next three each contains $5, and there is a $10 bill in each of the last two. A sample of size 3 is selected *with* replacement (so we have a random sample), and you get the largest amount in any of the envelopes selected. If X_1, X_2, and X_3 denote the amounts in the selected envelopes, the statistic of interest is $M =$ the maximum of X_1, X_2, and X_3.

   a. Obtain the probability distribution of this statistic.

   b. Describe how you would carry out a simulation experiment to compare the distributions of M for various sample sizes. How would you guess the distribution would change as n increases?

41. Let X be the number of packages being mailed by a randomly selected customer at a certain shipping facility. Suppose the distribution of X is as follows:

x	1	2	3	4
$p(x)$	.4	.3	.2	.1

   a. Consider a random sample of size $n = 2$ (two customers), and let $\overline{X}$ be the sample mean number of packages shipped. Obtain the probability distribution of $\overline{X}$.

   b. Refer to part (a) and calculate $P(\overline{X} \leq 2.5)$.

   c. Again consider a random sample of size $n = 2$, but now focus on the statistic $R =$ the sample range (difference between the largest and smallest values in the sample). Obtain the distribution of R. [*Hint:* Calculate the value of R for each outcome and use the probabilities from part (a).]

d. If a random sample of size $n = 4$ is selected, what is $P(\overline{X} \le 1.5)$? [Hint: You should not have to list all possible outcomes, only those for which $\bar{x} \le 1.5$.]

42. A company maintains three offices in a certain region, each staffed by two employees. Information concerning yearly salaries (1000s of dollars) is as follows:

Office	1	1	2	2	3	3
Employee	1	2	3	4	5	6
Salary	29.7	33.6	30.2	33.6	25.8	29.7

a. Suppose two of these employees are randomly selected from among the six (without replacement). Determine the sampling distribution of the sample mean salary $\overline{X}$.

b. Suppose one of the three offices is randomly selected. Let X_1 and X_2 denote the salaries of the two employees. Determine the sampling distribution of $\overline{X}$.

c. How does $E(\overline{X})$ from parts (a) and (b) compare to the population mean salary μ?

43. Suppose the amount of liquid dispensed by a certain machine is uniformly distributed with lower limit $A = 8$ oz and upper limit $B = 10$ oz. Describe how you would carry out simulation experiments to compare the sampling distribution of the (sample) fourth spread for sample sizes $n = 5, 10, 20,$ and 30.

44. Carry out a simulation experiment using a statistical computer package or other software to study the sampling distribution of $\overline{X}$ when the population distribution is Weibull with $\alpha = 2$ and $\beta = 5$, as in Example 5.20. Consider the four sample sizes $n = 5, 10, 20,$ and 30, and in each case use 1000 replications. For which of these sample sizes does the $\overline{X}$ sampling distribution appear to be approximately normal?

45. Carry out a simulation experiment using a statistical computer package or other software to study the sampling distribution of $\overline{X}$ when the population distribution is lognormal with $E(\ln(X)) = 3$ and $V(\ln(X)) = 1$. Consider the four sample sizes $n = 10, 20, 30,$ and 50, and in each case use 1000 replications. For which of these sample sizes does the $\overline{X}$ sampling distribution appear to be approximately normal?

5.4 The Distribution of the Sample Mean

The importance of the sample mean $\overline{X}$ springs from its use in drawing conclusions about the population mean μ. Some of the most frequently used inferential procedures are based on properties of the sampling distribution of $\overline{X}$. A preview of these properties appeared in the calculations and simulation experiments of the previous section, where we noted relationships between $E(\overline{X})$ and μ and also among $V(\overline{X})$, σ^2, and n.

PROPOSITION

Let $X_1, X_2, \ldots, X_n$ be a random sample from a distribution with mean value μ and standard deviation σ. Then

1. $E(\overline{X}) = \mu_{\overline{X}} = \mu$
2. $V(\overline{X}) = \sigma_{\overline{X}}^2 = \sigma^2/n$ and $\sigma_{\overline{X}} = \sigma/\sqrt{n}$

In addition, with $T_o = X_1 + \cdots + X_n$ (the sample total), $E(T_o) = n\mu$, $V(T_o) = n\sigma^2$, and $\sigma_{T_o} = \sqrt{n}\sigma$.

Proofs of these results are deferred to the next section. According to Result 1, the sampling (i.e., probability) distribution of $\overline{X}$ is centered precisely at the mean of the population from which the sample has been selected. Result 2 shows that the $\overline{X}$ distribution becomes more concentrated about μ as the sample size n increases. In marked contrast, the distribution of T_o becomes more spread out as n increases. Averaging moves probability in toward the middle, whereas totaling spreads probability out over a wider and wider range of values. The standard deviation $\sigma_{\overline{X}} = \sigma/\sqrt{n}$ is often called the **standard error of the mean**; it describes the magnitude of a typical or representative deviation of the sample mean from the population mean.

EXAMPLE 5.25 In a notched tensile fatigue test on a titanium specimen, the expected number of cycles to first acoustic emission (used to indicate crack initiation) is $\mu = 28{,}000$, and the standard deviation of the number of cycles is $\sigma = 5000$. Let $X_1, X_2, \ldots, X_{25}$ be a random sample of size 25, where each X_i is the number of cycles on a different randomly selected specimen. Then the expected value of the sample mean number of cycles until first emission is $E(\overline{X}) = \mu = 28{,}000$, and the expected total number of cycles for the 25 specimens is $E(T_0) = n\mu = 25(28{,}000) = 700{,}000$. The standard deviation of $\overline{X}$ (standard error of the mean) and of T_0 are

$$\sigma_{\overline{X}} = \sigma/\sqrt{n} = \frac{5000}{\sqrt{25}} = 1000$$

$$\sigma_{T_o} = \sqrt{n}\sigma = \sqrt{25}(5000) = 25{,}000$$

If the sample size increases to $n = 100$, $E(\overline{X})$ is unchanged, but $\sigma_{\overline{X}} = 500$, half of its previous value (the sample size must be quadrupled to halve the standard deviation of $\overline{X}$). ∎

The Case of a Normal Population Distribution

The simulation experiment of Example 5.23 indicated that when the population distribution is normal, a histogram of $\bar{x}$ values for any sample size n is well approximated by a normal curve.

PROPOSITION

> Let $X_1, X_2, \ldots, X_n$ be a random sample from a *normal* distribution with mean μ and standard deviation σ. Then for *any n*, $\overline{X}$ is normally distributed (with mean μ and standard deviation $\sigma/\sqrt{n}$), as is T_o (with mean $n\mu$ and standard deviation $\sqrt{n}\sigma$).*

We know everything there is to know about the $\overline{X}$ and T_o distributions when the population distribution is normal. In particular, probabilities such as $P(a \leq \overline{X} \leq b)$ and $P(c \leq T_o \leq d)$ can be obtained simply by standardizing. Figure 5.15 illustrates the $\overline{X}$ part of the proposition.

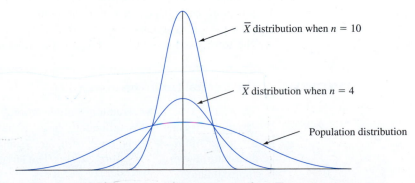

Figure 5.15 A normal population distribution and $\overline{X}$ sampling distributions

* A proof of the result for T_o when $n = 2$ is possible using the method in Example 5.22, but the details are messy. The general result is usually proved using a theoretical tool called a *moment generating function*. One of the chapter references can be consulted for more information.

EXAMPLE 5.26 The distribution of egg weights (g) of a certain type is normal with mean value 53 and standard deviation .3 (consistent with data in the article **"Evaluation of Egg Quality Traits of Chickens Reared under Backyard System in Western Uttar Pradesh" (*Indian J. of Poultry Sci.*, 2009: 261–262)**). Let $X_1, X_2, \ldots, X_{12}$ denote the weights of a dozen randomly selected eggs; these X_i's constitute a random sample of size 12 from the specified normal distribution.

The total weight of the 12 eggs is $T_o = X_1 + \ldots + X_{12}$; it is normally distributed with mean value $E(T_o) = n\mu = 12(53) = 636$ and variance $V(T_o) = n\sigma^2 = 12(.3)^2 = 1.08$. The probability that the total weight is between 635 and 640 is now obtained by standardizing and referring to Appendix Table A.3:

$$P(635 < T_o < 640) = \left(\frac{635 - 636}{\sqrt{1.08}} < Z < \frac{640 - 636}{\sqrt{1.08}} \right) = P(-.96 < Z < 3.85)$$

$$= \Phi(3.85) - \Phi(-.96) \approx 1 - .1685 = .8315$$

If cartons containing a dozen eggs are repeatedly selected, in the long run slightly more than 83% of the eggs in a carton will weigh in total between 635 g and 640 g. Notice that $635 < T_o < 640$ is equivalent to $52.9167 < \overline{X} < 53.3333$ (divide each term in the original system of inequalities by 12). Thus $P(52.9167 < \overline{X} < 53.3333) \approx .8315$. This latter probability can also be obtained by standardizing $\overline{X}$ directly.

Now consider randomly selecting just four of these eggs. The sample mean weight $\overline{X}$ is then normally distributed with mean value $\mu_{\overline{X}} = \mu = 53$ and standard deviation $\sigma_{\overline{X}} = \sigma/\sqrt{n} = .3/\sqrt{4} = .15$. The probability that the sample mean weight exceeds 53.5 g is then

$$P(\overline{X} > 53.5) = P\left(Z > \frac{53.5 - 53}{.15} \right)$$

$$= P(Z > 3.33) = 1 - \Phi(3.33) = 1 - .9996 = .0004$$

Because 53.5 is 3.33 standard deviations (of $\overline{X}$) larger than the mean value 53, it is exceedingly unlikely that the sample mean will exceed 53.5. ∎

The Central Limit Theorem

When the X_i's are normally distributed, so is $\overline{X}$ for every sample size n. The derivations in Example 5.21 and simulation experiment of Example 5.24 suggest that even when the population distribution is highly nonnormal, averaging produces a distribution more bell-shaped than the one being sampled. A reasonable conjecture is that if n is large, a suitable normal curve will approximate the actual distribution of $\overline{X}$. The formal statement of this result is the most important theorem of probability.

THEOREM

The Central Limit Theorem (CLT)

Let $X_1, X_2, \ldots, X_n$ be a random sample from a distribution with mean μ and variance σ^2. Then if n is sufficiently large, $\overline{X}$ has approximately a normal distribution with $\mu_{\overline{X}} = \mu$ and $\sigma_{\overline{X}}^2 = \sigma^2/n$, and T_o also has approximately a normal distribution with $\mu_{T_o} = n\mu$, $\sigma_{T_o}^2 = n\sigma^2$. The larger the value of n, the better the approximation.

Figure 5.16 illustrates the Central Limit Theorem. According to the CLT, when n is large and we wish to calculate a probability such as $P(a \leq \overline{X} \leq b)$, we need only "pretend" that $\overline{X}$ is normal, standardize it, and use the normal table. The resulting

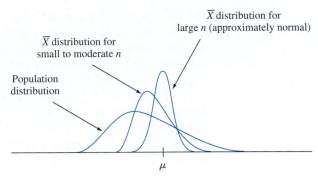

Figure 5.16 The Central Limit Theorem illustrated

answer will be approximately correct. The exact answer could be obtained only by first finding the distribution of $\overline{X}$, so the CLT provides a truly impressive shortcut. The proof of the theorem involves much advanced mathematics.

EXAMPLE 5.27

The amount of a particular impurity in a batch of a certain chemical product is a random variable with mean value 4.0 g and standard deviation 1.5 g. If 50 batches are independently prepared, what is the (approximate) probability that the sample average amount of impurity $\overline{X}$ is between 3.5 and 3.8 g? According to the rule of thumb to be stated shortly, $n = 50$ is large enough for the CLT to be applicable. $\overline{X}$ then has approximately a normal distribution with mean value $\mu_{\overline{X}} = 4.0$ and $\sigma_{\overline{X}} = 1.5/\sqrt{50} = .2121$, so

$$P(3.5 \le \overline{X} \le 3.8) \approx P\left(\frac{3.5 - 4.0}{.2121} \le Z \le \frac{3.8 - 4.0}{.2121}\right)$$
$$= \Phi(-.94) - \Phi(-2.36) = .1645$$

Now consider randomly selecting 100 batches, and let T_o represent the total amount of impurity in these batches. Then the mean value and standard deviation of T_o are $100(4) = 400$ and $\sqrt{100}(1.5) = 15$, respectively, and the CLT implies that T_0 has approximately a normal distribution. The probability that this total is at most 425 g is

$$P(T_0 \le 425) \approx P\left(Z \le \frac{425 - 400}{15}\right) = P(Z \le 1.67) = \Phi(1.67) = .9525 \quad \blacksquare$$

EXAMPLE 5.28

Let X = the number of different people sent text messages during a particular day by a randomly selected student at a large university. Suppose the mean value of X is 7 and the standard deviation is 6 (values very close to those reported in the article **"Cell Phone Use and Grade Point Average Among Undergraduate University Students"** (*College Student J.*, **2011: 544–551**). Among 100 randomly selected such students, how likely is it that the sample mean number of different people texted exceeds 5? Notice that the distribution being sampled is discrete, but the CLT is applicable whether the variable of interest is discrete or continuous. Also, although the fact that the standard deviation of this nonnegative variable is quite large relative to the mean value suggests that its distribution is positively skewed, the large sample size implies that $\overline{X}$ does have approximately a normal distribution. Using $\mu_{\overline{X}} = 7$ and $\sigma_{\overline{X}} = .6$, ·

$$P(\overline{X} > 5) \approx P\left(Z > \frac{5 - 7}{.6}\right) = 1 - \Phi(-3.33) = .9996$$

Note: The cited article stated that text messaging frequency was negatively correlated with GPA. $\blacksquare$

The CLT provides insight into why many random variables have probability distributions that are approximately normal. For example, the measurement error in a scientific experiment can be thought of as the sum of a number of underlying perturbations and errors of small magnitude.

A practical difficulty in applying the CLT is in knowing when n is sufficiently large. The problem is that the accuracy of the approximation for a particular n depends on the shape of the original underlying distribution being sampled. If the underlying distribution is close to a normal density curve, then the approximation will be good even for a small n, whereas if it is far from being normal, then a large n will be required.

Always Check first ✳✳

Rule of Thumb

The Central Limit Theorem can generally be used if $n > 30$.

✳✳

There are population distributions for which even an n of 40 or 50 does not suffice, but such distributions are rarely encountered in practice. On the other hand, the rule of thumb is often conservative; for many population distributions, an n much less than 30 would suffice. For example, in the case of a uniform population distribution, the CLT gives a good approximation for $n \geq 12$.

EXAMPLE 5.29   Consider the distribution shown in Figure 5.17 for the amount purchased (rounded to the nearest dollar) by a randomly selected customer at a particular gas station (a similar distribution for purchases in Britain (in £) appeared in the article **"Data Mining for Fun and Profit,"** *Statistical Science*, **2000: 111–131**; there were big spikes at the values, 10, 15, 20, 25, and 30). The distribution is obviously quite non-normal.

We asked Minitab to select 1000 different samples, each consisting of $n = 15$ observations, and calculate the value of the sample mean $\overline{X}$ for each one. Figure 5.18 is a histogram of the resulting 1000 values; this is the approximate

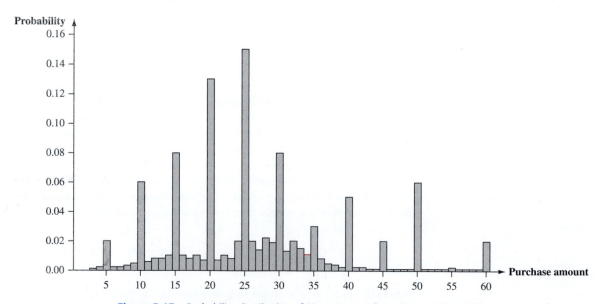

Figure 5.17   Probability distribution of $X =$ amount of gasoline purchased ($)

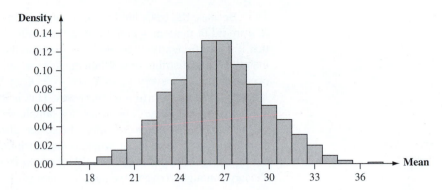

Figure 5.18 Approximate sampling distribution of the sample mean amount purchased when $n = 15$ and the population distribution is as shown in Figure 5.17

sampling distribution of $\overline{X}$ under the specified circumstances. This distribution is clearly approximately normal even though the sample size is actually much smaller than 30, our rule-of-thumb cutoff for invoking the Central Limit Theorem. As further evidence for normality, Figure 5.19 shows a normal probability plot of the 1000 $\overline{x}$ values; the linear pattern is very prominent. It is typically not nonnormality in the central part of the population distribution that causes the CLT to fail, but instead very substantial skewness.

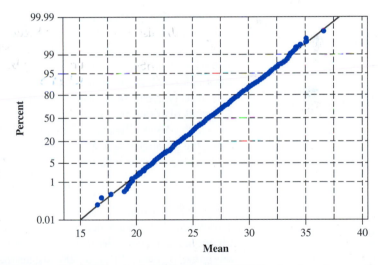

Figure 5.19 Normal probability plot from Minitab of the 1000 $\overline{x}$ values based on samples of size $n = 15$

Other Applications of the Central Limit Theorem

The CLT can be used to justify the normal approximation to the binomial distribution discussed in Chapter 4. Recall that a binomial variable X is the number of successes in a binomial experiment consisting of n independent success/failure trials with $p = P(S)$ for any particular trial. Define a new rv X_1 by

$$X_1 = \begin{cases} 1 & \text{if the 1st trial results in a success} \\ 0 & \text{if the 1st trial results in a failure} \end{cases}$$

and define X_2, X_3, . . . , X_n analogously for the other $n - 1$ trials. Each X_i indicates whether or not there is a success on the corresponding trial.

Because the trials are independent and $P(S)$ is constant from trial to trial, the X_i's are iid (a random sample from a Bernoulli distribution). The CLT then implies that if n is sufficiently large, both the sum and the average of the X_i's have approximately normal distributions. When the X_i's are summed, a 1 is added for every S that occurs and a 0 for every F, so $X_1 + \cdots + X_n = X$. The sample mean of the X_i's is X/n, the sample proportion of successes. That is, both X and X/n are approximately normal when n is large. The necessary sample size for this approximation depends on the value of p: When p is close to .5, the distribution of each X_i is reasonably symmetric (see Figure 5.20), whereas the distribution is quite skewed when p is near 0 or 1. Using the approximation only if both $np \geq 10$ and $n(1 - p) \geq 10$ ensures that n is large enough to overcome any skewness in the underlying Bernoulli distribution.

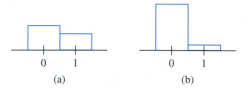

Figure 5.20 Two Bernoulli distributions: (a) $p = .4$ (reasonably symmetric); (b) $p = .1$ (very skewed)

Consider n independent Poisson rv's $X_1, \ldots, X_n$, each having mean value μ/n. It can be shown that $X = X_1 + \cdots + X_n$ has a Poisson distribution with mean value μ (because in general a sum of independent Poisson rv's has a Poisson distribution). The CLT then implies that a Poisson rv with sufficiently large μ has approximately a normal distribution. A common rule of thumb for this is $\mu > 20$.

Lastly, recall from Section 4.5 that X has a lognormal distribution if $\ln(X)$ has a normal distribution. Let $X_1, X_2, \ldots, X_n$ be a random sample from a distribution for which only positive values are possible $[P(X_i > 0) = 1]$. Then if n is sufficiently large, the product $Y = X_1 X_2 \cdot \cdots \cdot X_n$ has approximately a lognormal distribution. To verify this, note that

$$\ln(Y) = \ln(X_1) + \ln(X_2) + \cdots + \ln(X_n)$$

Since $\ln(Y)$ is a sum of independent and identically distributed rv's [the $\ln(X_i)$'s], it is approximately normal when n is large, so Y itself has approximately a lognormal distribution. As an example of the applicability of this result, Bury (*Statistical Models in Applied Science,* Wiley, p. 590) argues that the damage process in plastic flow and crack propagation is a multiplicative process, so that variables such as percentage elongation and rupture strength have approximately lognormal distributions.

EXERCISES Section 5.4 (46–57)

46. Young's modulus is a quantitative measure of stiffness of an elastic material. Suppose that for aluminum alloy sheets of a particular type, its mean value and standard deviation are 70 GPa and 1.6 GPa, respectively (values given in the article **"Influence of Material Properties Variability on Springback and Thinning in Sheet**

Stamping Processes: A Stochastic Analysis" (*Intl. J. of Advanced Manuf. Tech.,* **2010: 117–134**)).

a. If $\overline{X}$ is the sample mean Young's modulus for a random sample of $n = 16$ sheets, where is the sampling distribution of $\overline{X}$ centered, and what is the standard deviation of the $\overline{X}$ distribution?

b. Answer the questions posed in part (a) for a sample size of $n = 64$ sheets.

c. For which of the two random samples, the one of part (a) or the one of part (b), is $\overline{X}$ more likely to be within 1 GPa of 70 GPa? Explain your reasoning.

47. Refer to Exercise 46. Suppose the distribution is normal (the cited article makes that assumption and even includes the corresponding normal density curve).

a. Calculate $P(69 \leq \overline{X} \leq 71)$ when $n = 16$.

b. How likely is it that the sample mean diameter exceeds 71 when $n = 25$?

48. The **National Health Statistics Reports dated Oct. 22, 2008**, stated that for a sample size of 277 18-year-old American males, the sample mean waist circumference was 86.3 cm. A somewhat complicated method was used to *estimate* various population percentiles, resulting in the following values:

5th	10th	25th	50th	75th	90th	95th
69.6	70.9	75.2	81.3	95.4	107.1	116.4

a. Is it plausible that the waist size distribution is at least approximately normal? Explain your reasoning. If your answer is no, conjecture the shape of the population distribution.

b. Suppose that the population mean waist size is 85 cm and that the population standard deviation is 15 cm. How likely is it that a random sample of 277 individuals will result in a sample mean waist size of at least 86.3 cm?

c. Referring back to (b), suppose now that the population mean waist size in 82 cm. Now what is the (approximate) probability that the sample mean will be at least 86.3 cm? In light of this calculation, do you think that 82 cm is a reasonable value for μ?

49. There are 40 students in an elementary statistics class. On the basis of years of experience, the instructor knows that the time needed to grade a randomly chosen first examination paper is a random variable with an expected value of 6 min and a standard deviation of 6 min.

a. If grading times are independent and the instructor begins grading at 6:50 P.M. and grades continuously, what is the (approximate) probability that he is through grading before the 11:00 P.M. TV news begins?

b. If the sports report begins at 11:10, what is the probability that he misses part of the report if he waits until grading is done before turning on the TV?

50. Let X denote the courtship time for a randomly selected female–male pair of mating scorpion flies (time from the beginning of interaction until mating). Suppose the mean value of X is 120 min and the standard deviation of X is 110 min (suggested by data in the article **"Should I Stay or Should I Go? Condition- and Status-Dependent Courtship Decisions in the Scorpion Fly *Panorpa Cognate*"** (*Animal Behavior*, 2009: 491–497)).

a. Is it plausible that X is normally distributed?

b. For a random sample of 50 such pairs, what is the (approximate) probability that the sample mean courtship time is between 100 min and 125 min?

c. For a random sample of 50 such pairs, what is the (approximate) probability that the total courtship time exceeds 150 hr?

d. Could the probability requested in (b) be calculated from the given information if the sample size were 15 rather than 50? Explain.

51. The time taken by a randomly selected applicant for a mortgage to fill out a certain form has a normal distribution with mean value 10 min and standard deviation 2 min. If five individuals fill out a form on one day and six on another, what is the probability that the sample average amount of time taken on each day is at most 11 min?

52. The lifetime of a certain type of battery is normally distributed with mean value 10 hours and standard deviation 1 hour. There are four batteries in a package. What lifetime value is such that the total lifetime of all batteries in a package exceeds that value for only 5% of all packages?

53. Rockwell hardness of pins of a certain type is known to have a mean value of 50 and a standard deviation of 1.2.

a. If the distribution is normal, what is the probability that the sample mean hardness for a random sample of 9 pins is at least 51?

b. Without assuming population normality, what is the (approximate) probability that the sample mean hardness for a random sample of 40 pins is at least 51?

54. Suppose the sediment density (g/cm) of a randomly selected specimen from a certain region is normally distributed with mean 2.65 and standard deviation .85 (suggested in **"Modeling Sediment and Water Column Interactions for Hydrophobic Pollutants,"** *Water Research*, 1984: 1169–1174).

a. If a random sample of 25 specimens is selected, what is the probability that the sample average sediment density is at most 3.00? Between 2.65 and 3.00?

b. How large a sample size would be required to ensure that the first probability in part (a) is at least .99?

55. The number of parking tickets issued in a certain city on any given weekday has a Poisson distribution with parameter $\mu = 50$.

a. Calculate the approximate probability that between 35 and 70 tickets are given out on a particular day.

b. Calculate the approximate probability that the total number of tickets given out during a 5-day week is between 225 and 275.

c. Use software to obtain the exact probabilities in (a) and (b) and compare to their approximations.

56. A binary communication channel transmits a sequence of "bits" (0s and 1s). Suppose that for any particular bit transmitted, there is a 10% chance of a transmission error (a 0 becoming a 1 or a 1 becoming a 0). Assume that bit errors occur independently of one another.

a. Consider transmitting 1000 bits. What is the approximate probability that at most 125 transmission errors occur?

b. Suppose the same 1000-bit message is sent two different times independently of one another. What is the approximate probability that the number of errors in the first transmission is within 50 of the number of errors in the second?

57. Suppose the distribution of the time X (in hours) spent by students at a certain university on a particular project is gamma with parameters $\alpha = 50$ and $\beta = 2$. Because α is large, it can be shown that X has approximately a normal distribution. Use this fact to compute the approximate probability that a randomly selected student spends at most 125 hours on the project.

5.5 The Distribution of a Linear Combination

The sample mean $\overline{X}$ and sample total T_o are special cases of a type of random variable that arises very frequently in statistical applications.

[handwritten: not random sample]

DEFINITION

Given a collection of n random variables $X_1, \ldots, X_n$ and n numerical constants $a_1, \ldots, a_n$, the rv

$$Y = a_1 X_1 + \cdots + a_n X_n = \sum_{i=1}^{n} a_i X_i \qquad (5.7)$$

is called a **linear combination** of the X_i's.

For example, consider someone who owns 100 shares of stock A, 200 shares of stock B, and 500 shares of stock C. Denote the share prices of these three stocks at some particular time by X_1, X_2, and X_3, respectively. Then the value of this individual's stock holdings is the linear combination $Y = 100X_1 + 200X_2 + 500X_3$.

Taking $a_1 = a_2 = \cdots = a_n = 1$ gives $Y = X_1 + \cdots + X_n = T_o$, and $a_1 = a_2 = \cdots = a_n = 1/n$ yields

$$Y = \frac{1}{n}X_1 + \cdots + \frac{1}{n}X_n = \frac{1}{n}(X_1 + \cdots + X_n) = \frac{1}{n}T_o = \overline{X}$$

Notice that we are not requiring the X_i's to be independent or identically distributed. All the X_i's could have different distributions and therefore different mean values and variances. Our first result concerns the expected value and variance of a linear combination.

[handwritten: $E(ax+b) = aE(x)+b$; $V(ax+b) = a^2E(x)$]

PROPOSITION

Let $X_1, X_2, \ldots, X_n$ have mean values $\mu_1, \ldots, \mu_n$, respectively, and variances $\sigma_1^2, \ldots, \sigma_n^2$, respectively.

1. Whether or not the X_i's are independent,

$$E(a_1 X_1 + a_2 X_2 + \cdots + a_n X_n) = a_1 E(X_1) + a_2 E(X_2) + \cdots + a_n E(X_n)$$
$$= a_1 \mu_1 + \cdots + a_n \mu_n \qquad (5.8)$$

2. If $X_1, \ldots, X_n$ are independent,

$$V(a_1X_1 + a_2X_2 + \cdots + a_nX_n) = a_1^2 V(X_1) + a_2^2 V(X_2) + \cdots + a_n^2 V(X_n)$$
$$= a_1^2 \sigma_1^2 + \cdots + a_n^2 \sigma_n^2 \qquad (5.9)$$

and

$$\sigma_{a_1X_1 + \cdots + a_nX_n} = \sqrt{a_1^2 \sigma_1^2 + \cdots + a_n^2 \sigma_n^2} \qquad (5.10)$$

3. For any $X_1, \ldots, X_n$,

$$V(a_1X_1 + \cdots + a_nX_n) = \sum_{i=1}^{n} \sum_{j=1}^{n} a_i a_j \text{Cov}(X_i, X_j) \qquad (5.11)$$

Proofs are sketched out at the end of the section. A paraphrase of (5.8) is that the expected value of a linear combination is the same as the linear combination of the expected values—for example, $E(2X_1 + 5X_2) = 2\mu_1 + 5\mu_2$. The result (5.9) in Statement 2 is a special case of (5.11) in Statement 3; when the X_i's are independent, $\text{Cov}(X_i, X_j) = 0$ for $i \neq j$ and $= V(X_i)$ for $i = j$ (this simplification actually occurs when the X_i's are uncorrelated, a weaker condition than independence). Specializing to the case of a random sample (X_i's iid) with $a_i = 1/n$ for every i gives $E(\overline{X}) = \mu$ and $V(\overline{X}) = \sigma^2/n$, as discussed in Section 5.4. A similar comment applies to the rules for T_o.

EXAMPLE 5.30 A gas station sells three grades of gasoline: regular, extra, and super. These are priced at \$3.00, \$3.20, and \$3.40 per gallon, respectively. Let $X_1, X_2,$ and X_3 denote the amounts of these grades purchased (gallons) on a particular day. Suppose the X_i's are independent with $\mu_1 = 1000$, $\mu_2 = 500$, $\mu_3 = 300$, $\sigma_1 = 100$, $\sigma_2 = 80$, and $\sigma_3 = 50$. The revenue from sales is $Y = 3.0X_1 + 3.2X_2 + 3.4X_3$, and

$$E(Y) = 3.0\mu_1 + 3.2\mu_2 + 3.4\mu_3 = \$5620$$
$$V(Y) = (3.0)^2\sigma_1^2 + (3.2)^2\sigma_2^2 + (3.4)^2\sigma_3^2 = 184{,}436$$
$$\sigma_Y = \sqrt{184{,}436} = \$429.46 \qquad \blacksquare$$

The Difference Between Two Random Variables

An important special case of a linear combination results from taking $n = 2$, $a_1 = 1$, and $a_2 = -1$:

$$Y = a_1X_1 + a_2X_2 = X_1 - X_2$$

We then have the following corollary to the proposition.

COROLLARY

$E(X_1 - X_2) = E(X_1) - E(X_2)$ for any two rv's X_1 and X_2.

$V(X_1 - X_2) = V(X_1) + V(X_2)$ if X_1 and X_2 are independent rv's.

The expected value of a difference is the difference of the two expected values. However, the variance of a difference between two independent variables is the *sum*, *not* the difference, of the two variances. There is just as much variability in $X_1 - X_2$ as in $X_1 + X_2$ [writing $X_1 - X_2 = X_1 + (-1)X_2$, $(-1)X_2$ has the same amount of variability as X_2 itself].

EXAMPLE 5.31 A certain automobile manufacturer equips a particular model with either a six-cylinder engine or a four-cylinder engine. Let X_1 and X_2 be fuel efficiencies for independently and randomly selected six-cylinder and four-cylinder cars, respectively. With $\mu_1 = 22$, $\mu_2 = 26$, $\sigma_1 = 1.2$, and $\sigma_2 = 1.5$,

$$E(X_1 - X_2) = \mu_1 - \mu_2 = 22 - 26 = -4$$
$$V(X_1 - X_2) = \sigma_1^2 + \sigma_2^2 = (1.2)^2 + (1.5)^2 = 3.69$$
$$\sigma_{X_1 - X_2} = \sqrt{3.69} = 1.92$$

If we relabel so that X_1 refers to the four-cylinder car, then $E(X_1 - X_2) = 4$, but the variance of the difference is still 3.69. ∎

The Case of Normal Random Variables

When the X_i's form a random sample from a normal distribution, $\overline{X}$ and T_o are both normally distributed. Here is a more general result concerning linear combinations.

PROPOSITION

> If $X_1, X_2, \ldots, X_n$ are independent, normally distributed rv's (with possibly different means and/or variances), then any linear combination of the X_i's also has a normal distribution. In particular, the difference $X_1 - X_2$ between two independent, normally distributed variables is itself normally distributed.

EXAMPLE 5.32
(Example 5.30 continued)

The total revenue from the sale of the three grades of gasoline on a particular day was $Y = 3.0X_1 + 3.2X_2 + 3.4X_3$, and we calculated $\mu_Y = 5620$ and (assuming independence) $\sigma_Y = 429.46$. If the X_i's are normally distributed, the probability that revenue exceeds 4500 is

$$P(Y > 4500) = P\left(Z > \frac{4500 - 5620}{429.46}\right)$$
$$= P(Z > -2.61) = 1 - \Phi(-2.61) = .9955 \quad ∎$$

The CLT can also be generalized so it applies to certain linear combinations. Roughly speaking, if n is large and no individual term is likely to contribute too much to the overall value, then Y has approximately a normal distribution.

Proofs for the Case $n = 2$

For the result concerning expected values, suppose that X_1 and X_2 are continuous with joint pdf $f(x_1, x_2)$. Then

$$E(a_1X_1 + a_2X_2) = \int_{-\infty}^{\infty} \int_{-\infty}^{\infty} (a_1x_1 + a_2x_2)f(x_1, x_2)\, dx_1\, dx_2$$
$$= a_1 \int_{-\infty}^{\infty} \int_{-\infty}^{\infty} x_1 f(x_1, x_2)\, dx_2\, dx_1$$
$$+ a_2 \int_{-\infty}^{\infty} \int_{-\infty}^{\infty} x_2 f(x_1, x_2)\, dx_1\, dx_2$$
$$= a_1 \int_{-\infty}^{\infty} x_1 f_{X_1}(x_1)\, dx_1 + a_2 \int_{-\infty}^{\infty} x_2 f_{X_2}(x_2)\, dx_2$$
$$= a_1 E(X_1) + a_2 E(X_2)$$

Summation replaces integration in the discrete case. The argument for the variance result does not require specifying whether either variable is discrete or continuous. Recalling that $V(Y) = E[(Y - \mu_Y)^2]$,

$$V(a_1X_1 + a_2X_2) = E\{[a_1X_1 + a_2X_2 - (a_1\mu_1 + a_2\mu_2)]^2\}$$
$$= E\{a_1^2(X_1 - \mu_1)^2 + a_2^2(X_2 - \mu_2)^2 + 2a_1a_2(X_1 - \mu_1)(X_2 - \mu_2)\}$$

The expression inside the braces is a linear combination of the variables $Y_1 = (X_1 - \mu_1)^2$, $Y_2 = (X_2 - \mu_2)^2$, and $Y_3 = (X_1 - \mu_1)(X_2 - \mu_2)$, so carrying the E operation through to the three terms gives $a_1^2V(X_1) + a_2^2V(X_2) + 2a_1a_2 \text{Cov}(X_1, X_2)$ as required. ∎

EXERCISES Section 5.5 (58–74)

58. A shipping company handles containers in three different sizes: (1) 27 ft³ (3 × 3 × 3), (2) 125 ft³, and (3) 512 ft³. Let X_i ($i = 1, 2, 3$) denote the number of type i containers shipped during a given week. With $\mu_i = E(X_i)$ and $\sigma_i^2 = V(X_i)$, suppose that the mean values and standard deviations are as follows:

$$\mu_1 = 200 \qquad \mu_2 = 250 \qquad \mu_3 = 100$$
$$\sigma_1 = 10 \qquad \sigma_2 = 12 \qquad \sigma_3 = 8$$

a. Assuming that X_1, X_2, X_3 are independent, calculate the expected value and variance of the total volume shipped. [*Hint*: Volume = $27X_1 + 125X_2 + 512X_3$.]

b. Would your calculations necessarily be correct if the X_i's were not independent? Explain.

59. Let X_1, X_2, and X_3 represent the times necessary to perform three successive repair tasks at a certain service facility. Suppose they are independent, normal rv's with expected values μ_1, μ_2, and μ_3 and variances σ_1^2, σ_2^2, and σ_3^2, respectively.

a. If $\mu_1 = \mu_2 = \mu_3 = 60$ and $\sigma_1^2 = \sigma_2^2 = \sigma_3^2 = 15$, calculate $P(T_o \le 200)$ and $P(150 \le T_o \le 200)$?

b. Using the μ_i's and σ_i's given in part (a), calculate both $P(55 \le \bar{X})$ and $P(58 \le \bar{X} \le 62)$.

c. Using the μ_i's and σ_i's given in part (a), calculate and interpret $P(-10 \le X_1 - .5X_2 - .5X_3 \le 5)$.

d. If $\mu_1 = 40$, $\mu_2 = 50$, $\mu_3 = 60$, $\sigma_1^2 = 10$, $\sigma_2^2 = 12$, and $\sigma_3^2 = 14$, calculate $P(X_1 + X_2 + X_3 \le 160)$ and also $P(X_1 + X_2 \ge 2X_3)$.

60. Refer back to Example 5.31. Two cars with six-cylinder engines and three with four-cylinder engines are to be driven over a 300-mile course. Let $X_1, \ldots X_5$ denote the resulting fuel efficiencies (mpg). Consider the linear combination

$$Y = (X_1 + X_2)/2 - (X_3 + X_4 + X_5)/3$$

which is a measure of the difference between four-cylinder and six-cylinder vehicles. Compute $P(0 \le Y)$ and $P(Y > -2)$. [*Hint*: $Y = a_1X_1 + \cdots + a_5X_5$, with $a_1 = 1/2, \ldots, a_5 = -1/3$.]

61. Exercise 26 introduced random variables X and Y, the number of cars and buses, respectively, carried by a ferry on a single trip. The joint pmf of X and Y is given in the table in Exercise 7. It is readily verified that X and Y are independent.

a. Compute the expected value, variance, and standard deviation of the total number of vehicles on a single trip.

b. If each car is charged $3 and each bus $10, compute the expected value, variance, and standard deviation of the revenue resulting from a single trip.

62. Manufacture of a certain component requires three different machining operations. Machining time for each operation has a normal distribution, and the three times are independent of one another. The mean values are 15, 30, and 20 min, respectively, and the standard deviations are 1, 2, and 1.5 min, respectively. What is the probability that it takes at most 1 hour of machining time to produce a randomly selected component?

63. Refer to Exercise 3.

a. Calculate the covariance between X_1 = the number of customers in the express checkout and X_2 = the number of customers in the superexpress checkout.

b. Calculate $V(X_1 + X_2)$. How does this compare to $V(X_1) + V(X_2)$?

64. Suppose your waiting time for a bus in the morning is uniformly distributed on [0, 8], whereas waiting time in the evening is uniformly distributed on [0, 10] independent of morning waiting time.

a. If you take the bus each morning and evening for a week, what is your total expected waiting time? [*Hint*: Define rv's $X_1, \ldots, X_{10}$ and use a rule of expected value.]

b. What is the variance of your total waiting time?

c. What are the expected value and variance of the difference between morning and evening waiting times on a given day?

d. What are the expected value and variance of the difference between total morning waiting time and total evening waiting time for a particular week?

65. Suppose that when the pH of a certain chemical compound is 5.00, the pH measured by a randomly selected beginning chemistry student is a random variable with mean 5.00 and standard deviation .2. A large batch of the compound is subdivided and a sample given to each student in a morning lab and each student in an afternoon lab. Let $\overline{X}$ = the average pH as determined by the morning students and $\overline{Y}$ = the average pH as determined by the afternoon students.

a. If pH is a normal variable and there are 25 students in each lab, compute $P(-.1 \le \overline{X} - \overline{Y} \le .1)$. [*Hint:* $\overline{X} - \overline{Y}$ is a linear combination of normal variables, so is normally distributed. Compute $\mu_{\overline{X}-\overline{Y}}$ and $\sigma_{\overline{X}-\overline{Y}}$.]

b. If there are 36 students in each lab, but pH determinations are not assumed normal, calculate (approximately) $P(-.1 \le \overline{X} - \overline{Y} \le .1)$.

66. If two loads are applied to a cantilever beam as shown in the accompanying drawing, the bending moment at 0 due to the loads is $a_1X_1 + a_2X_2$.

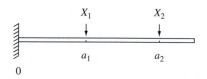

a. Suppose that X_1 and X_2 are independent rv's with means 2 and 4 kips, respectively, and standard deviations .5 and 1.0 kip, respectively. If a_1 = 5 ft and a_2 = 10 ft, what is the expected bending moment and what is the standard deviation of the bending moment?

b. If X_1 and X_2 are normally distributed, what is the probability that the bending moment will exceed 75 kip-ft?

c. Suppose the positions of the two loads are random variables. Denoting them by A_1 and A_2, assume that these variables have means of 5 and 10 ft, respectively, that each has a standard deviation of .5, and that all A_i's and X_i's are independent of one another. What is the expected moment now?

d. For the situation of part (c), what is the variance of the bending moment?

e. If the situation is as described in part (a) except that $\text{Corr}(X_1, X_2)$ = .5 (so that the two loads are not independent), what is the variance of the bending moment?

67. One piece of PVC pipe is to be inserted inside another piece. The length of the first piece is normally distributed with mean value 20 in. and standard deviation .5 in. The length of the second piece is a normal rv with mean and standard deviation 15 in. and .4 in., respectively. The amount of overlap is normally distributed with mean value 1 in. and standard deviation .1 in. Assuming that the lengths and amount of overlap are independent of one another, what is the probability that the total length after insertion is between 34.5 in. and 35 in.?

68. Two airplanes are flying in the same direction in adjacent parallel corridors. At time $t = 0$, the first airplane is 10 km ahead of the second one. Suppose the speed of the first plane (km/hr) is normally distributed with mean 520 and standard deviation 10 and the second plane's speed is also normally distributed with mean and standard deviation 500 and 10, respectively.

a. What is the probability that after 2 hr of flying, the second plane has not caught up to the first plane?

b. Determine the probability that the planes are separated by at most 10 km after 2 hr.

69. Three different roads feed into a particular freeway entrance. Suppose that during a fixed time period, the number of cars coming from each road onto the freeway is a random variable, with expected value and standard deviation as given in the table.

	Road 1	Road 2	Road 3
Expected value	800	1000	600
Standard deviation	16	25	18

a. What is the expected total number of cars entering the freeway at this point during the period? [*Hint:* Let X_i = the number from road i.]

b. What is the variance of the total number of entering cars? Have you made any assumptions about the relationship between the numbers of cars on the different roads?

c. With X_i denoting the number of cars entering from road i during the period, suppose that $\text{Cov}(X_1, X_2) = 80$, $\text{Cov}(X_1, X_3) = 90$, and $\text{Cov}(X_2, X_3) = 100$ (so that the three streams of traffic are not independent). Compute the expected total number of entering cars and the standard deviation of the total.

70. Consider a random sample of size n from a continuous distribution having median 0 so that the probability of any one observation being positive is .5. Disregarding the signs of the observations, rank them from smallest to largest in absolute value, and let W = the sum of the ranks of the observations having positive signs. For example, if the observations are $-.3$, $+.7$, $+2.1$, and -2.5, then the ranks of positive observations are 2 and 3, so $W = 5$. In Chapter 15, W will be called *Wilcoxon's signed-rank statistic*. W can be represented as follows:

$$W = 1 \cdot Y_1 + 2 \cdot Y_2 + 3 \cdot Y_3 + \cdots + n \cdot Y_n$$
$$= \sum_{i=1}^{n} i \cdot Y_i$$

where the Y_i's are independent Bernoulli rv's, each with $p = .5$ ($Y_i = 1$ corresponds to the observation with rank i being positive).

a. Determine $E(Y_i)$ and then $E(W)$ using the equation for W. [*Hint*: The first n positive integers sum to $n(n + 1)/2$.]

b. Determine $V(Y_i)$ and then $V(W)$. [*Hint*: The sum of the squares of the first n positive integers can be expressed as $n(n + 1)(2n + 1)/6$.]

71. In Exercise 66, the weight of the beam itself contributes to the bending moment. Assume that the beam is of uniform thickness and density so that the resulting load is uniformly distributed on the beam. If the weight of the beam is random, the resulting load from the weight is also random; denote this load by W (kip-ft).

a. If the beam is 12 ft long, W has mean 1.5 and standard deviation .25, and the fixed loads are as described in part (a) of Exercise 66, what are the expected value and variance of the bending moment? [*Hint*: If the load due to the beam were w kip-ft, the contribution to the bending moment would be $w \int_0^{12} x \, dx$.]

b. If all three variables (X_1, X_2, and W) are normally distributed, what is the probability that the bending moment will be at most 200 kip-ft?

72. I have three errands to take care of in the Administration Building. Let X_i = the time that it takes for the ith errand (i = 1, 2, 3), and let X_4 = the total time in minutes that I spend walking to and from the building and between each errand. Suppose the X_i's are independent, and normally distributed, with the following means and standard deviations: $\mu_1 = 15$, $\sigma_1 = 4$, $\mu_2 = 5$, $\sigma_2 = 1$, $\mu_3 = 8$,

$\sigma_3 = 2$, $\mu_4 = 12$, $\sigma_4 = 3$. I plan to leave my office at precisely 10:00 A.M. and wish to post a note on my door that reads, "I will return by t A.M." What time t should I write down if I want the probability of my arriving after t to be .01?

73. Suppose the expected tensile strength of type-A steel is 105 ksi and the standard deviation of tensile strength is 8 ksi. For type-B steel, suppose the expected tensile strength and standard deviation of tensile strength are 100 ksi and 6 ksi, respectively. Let $\overline{X}$ = the sample average tensile strength of a random sample of 40 type-A specimens, and let $\overline{Y}$ = the sample average tensile strength of a random sample of 35 type-B specimens.

a. What is the approximate distribution of $\overline{X}$? Of $\overline{Y}$?

b. What is the approximate distribution of $\overline{X} - \overline{Y}$? Justify your answer.

c. Calculate (approximately) $P(-1 \le \overline{X} - \overline{Y} \le 1)$.

d. Calculate $P(\overline{X} - \overline{Y} \ge 10)$. If you actually observed $\overline{X} - \overline{Y} \ge 10$, would you doubt that $\mu_1 - \mu_2 = 5$?

74. In an area having sandy soil, 50 small trees of a certain type were planted, and another 50 trees were planted in an area having clay soil. Let X = the number of trees planted in sandy soil that survive 1 year and Y = the number of trees planted in clay soil that survive 1 year. If the probability that a tree planted in sandy soil will survive 1 year is .7 and the probability of 1-year survival in clay soil is .6, compute an approximation to $P(-5 \le X - Y \le 5)$ (do not bother with the continuity correction).

SUPPLEMENTARY EXERCISES (75–96)

75. A restaurant serves three fixed-price dinners costing $12, $15, and $20. For a randomly selected couple dining at this restaurant, let X = the cost of the man's dinner and Y = the cost of the woman's dinner. The joint pmf of X and Y is given in the following table:

$p(x, y)$		y	
	12	15	20
x 12	.05	.05	.10
15	.05	.10	.35
20	0	.20	.10

a. Compute the marginal pmf's of X and Y.

b. What is the probability that the man's and the woman's dinner cost at most $15 each?

c. Are X and Y independent? Justify your answer.

d. What is the expected total cost of the dinner for the two people?

e. Suppose that when a couple opens fortune cookies at the conclusion of the meal, they find the message "You will receive as a refund the difference between the cost

of the more expensive and the less expensive meal that you have chosen." How much would the restaurant expect to refund?

76. In cost estimation, the total cost of a project is the sum of component task costs. Each of these costs is a random variable with a probability distribution. It is customary to obtain information about the total cost distribution by adding together characteristics of the individual component cost distributions—this is called the "roll-up" procedure. For example, $E(X_1 + \cdots + X_n) = E(X_1) + \cdots + E(X_n)$, so the roll-up procedure is valid for mean cost. Suppose that there are two component tasks and that X_1 and X_2 are independent, normally distributed random variables. Is the roll-up procedure valid for the 75th percentile? That is, is the 75th percentile of the distribution of $X_1 + X_2$ the same as the sum of the 75th percentiles of the two individual distributions? If not, what is the relationship between the percentile of the sum and the sum of percentiles? For what percentiles is the roll-up procedure valid in this case?

77. A health-food store stocks two different brands of a certain type of grain. Let X = the amount (lb) of brand A on hand and Y = the amount of brand B on hand. Suppose the joint pdf of X and Y is

$$f(x, y) = \begin{cases} kxy & x \geq 0, y \geq 0, 20 \leq x + y \leq 30 \\ 0 & \text{otherwise} \end{cases}$$

a. Draw the region of positive density and determine the value of k.

b. Are X and Y independent? Answer by first deriving the marginal pdf of each variable.

c. Compute $P(X + Y \leq 25)$.

d. What is the expected total amount of this grain on hand?

e. Compute $\text{Cov}(X, Y)$ and $\text{Corr}(X, Y)$.

f. What is the variance of the total amount of grain on hand?

78. According to the article **"Reliability Evaluation of Hard Disk Drive Failures Based on Counting Processes"** (*Reliability Engr. and System Safety*, **2013: 110–118**), particles accumulating on a disk drive come from two sources, one external and the other internal. The article proposed a model in which the internal source contains a number of loose particles W having a Poisson distribution with mean value μ; when a loose particle releases, it immediately enters the drive, and the release times are independent and identically distributed with cumulative distribution function $G(t)$. Let X denote the number of loose particles not yet released at a particular time t. Show that X has a Poisson distribution with parameter $\mu[1 - G(t)]$. [*Hint:* Let Y denote the number of particles accumulated on the drive from the internal source by time t so that $X + Y = W$. Obtain an expression for $P(X = x, Y = y)$ and then sum over y.]

79. Suppose that for a certain individual, calorie intake at breakfast is a random variable with expected value 500 and standard deviation 50, calorie intake at lunch is random with expected value 900 and standard deviation 100, and calorie intake at dinner is a random variable with expected value 2000 and standard deviation 180. Assuming that intakes at different meals are independent of one another, what is the probability that average calorie intake per day over the next (365-day) year is at most 3500? [*Hint:* Let X_i, Y_i, and Z_i denote the three calorie intakes on day i. Then total intake is given by $\Sigma(X_i + Y_i + Z_i)$.]

80. The mean weight of luggage checked by a randomly selected tourist-class passenger flying between two cities on a certain airline is 40 lb, and the standard deviation is 10 lb. The mean and standard deviation for a business-class passenger are 30 lb and 6 lb, respectively.

a. If there are 12 business-class passengers and 50 tourist-class passengers on a particular flight, what are the expected value of total luggage weight and the standard deviation of total luggage weight?

b. If individual luggage weights are independent, normally distributed rv's, what is the probability that total luggage weight is at most 2500 lb?

81. We have seen that if $E(X_1) = E(X_2) = \cdots = E(X_n) = \mu$, then $E(X_1 + \cdots + X_n) = n\mu$. In some applications, the number of X_i's under consideration is not a fixed number n but instead is an rv N. For example, let N = the number of components that are brought into a repair shop on a particular day, and let X_i denote the repair shop time for the ith component. Then the total repair time is $X_1 + X_2 + \cdots + X_N$, the sum of a *random* number of random variables. When N is independent of the X_i's, it can be shown that

$$E(X_1 + \cdots + X_N) = E(N) \cdot \mu$$

a. If the expected number of components brought in on a particularly day is 10 and expected repair time for a randomly submitted component is 40 min, what is the expected total repair time for components submitted on any particular day?

b. Suppose components of a certain type come in for repair according to a Poisson process with a rate of 5 per hour. The expected number of defects per component is 3.5. What is the expected value of the total number of defects on components submitted for repair during a 4-hour period? Be sure to indicate how your answer follows from the general result just given.

82. Suppose the proportion of rural voters in a certain state who favor a particular gubernatorial candidate is .45 and the proportion of suburban and urban voters favoring the candidate is .60. If a sample of 200 rural voters and 300 urban and suburban voters is obtained, what is the approximate probability that at least 250 of these voters favor this candidate?

83. Let μ denote the true pH of a chemical compound. A sequence of n independent sample pH determinations will be made. Suppose each sample pH is a random variable with expected value μ and standard deviation .1. How many determinations are required if we wish the probability that the sample average is within .02 of the true pH to be at least .95? What theorem justifies your probability calculation?

84. If the amount of soft drink that I consume on any given day is independent of consumption on any other day and is normally distributed with μ = 13 oz and σ = 2 and if I currently have two six-packs of 16-oz bottles, what is the probability that I still have some soft drink left at the end of 2 weeks (14 days)?

85. Refer to Exercise 58, and suppose that the X_i's are independent with each one having a normal distribution. What is the probability that the total volume shipped is at most 100,000 ft^3?

86. A student has a class that is supposed to end at 9:00 A.M. and another that is supposed to begin at 9:10 A.M.

Suppose the actual ending time of the 9 A.M. class is a normally distributed rv X_1 with mean 9:02 and standard deviation 1.5 min and that the starting time of the next class is also a normally distributed rv X_2 with mean 9:10 and standard deviation 1 min. Suppose also that the time necessary to get from one classroom to the other is a normally distributed rv X_3 with mean 6 min and standard deviation 1 min. What is the probability that the student makes it to the second class before the lecture starts? (Assume independence of X_1, X_2, and X_3, which is reasonable if the student pays no attention to the finishing time of the first class.)

87. Garbage trucks entering a particular waste-management facility are weighed prior to offloading their contents. Let X = the total processing time for a randomly selected truck at this facility (waiting, weighing, and offloading). The article **"Estimating Waste Transfer Station Delays Using GPS"** (*Waste Mgmt.*, 2008: 1742–1750) suggests the plausibility of a normal distribution with mean 13 min and standard deviation 4 min for X. Assume that this is in fact the correct distribution.

    a. What is the probability that a single truck's processing time is between 12 and 15 min?

    b. Consider a random sample of 16 trucks. What is the probability that the *sample mean* processing time is between 12 and 15 min?

    c. Why is the probability in (b) much larger than the probability in (a)?

    d. What is the probability that the sample mean processing time for a random sample of 16 trucks will be at least 20 min?

88. Each customer making a particular Internet purchase must pay with one of three types of credit cards (think Visa, MasterCard, AmEx). Let A_i ($i = 1, 2, 3$) be the event that a type i credit card is used, with $P(A_1) = .5$, $P(A_2) = .3$, and $P(A_3) = .2$. Suppose that the number of customers who make such a purchase on a given day is a Poisson rv with parameter λ. Define rv's X_1, X_2, X_3 by X_i = the number among the N customers who use a type i card ($i = 1, 2, 3$). Show that these three rv's are independent with Poisson distributions having parameters $.5\lambda$, $.3\lambda$, and $.2\lambda$, respectively. [*Hint:* For non-negative integers x_1, x_2, x_3, let $n = x_1 + x_2 + x_3$. Then $P(X_1 = x_1$, $X_2 = x_2, X_3 = x_3) = P(X_1 = x_1, X_2 = x_2, X_3 = x_3, N = n)$ [why is this?]. Now condition on $N = n$, in which case the three X_i's have a trinomial distribution (multinomial with three categories) with category probabilities .5, .3, and .2.]

89. a. Use the general formula for the variance of a linear combination to write an expression for $V(aX + Y)$. Then let $a = \sigma_Y/\sigma_X$, and show that $\rho \geq -1$. [*Hint:* Variance is always ≥ 0, and $\text{Cov}(X, Y) = \sigma_X \cdot \sigma_Y \cdot \rho$.]

    b. By considering $V(aX - Y)$, conclude that $\rho \leq 1$.

    c. Use the fact that $V(W) = 0$ only if W is a constant to show that $\rho = 1$ only if $Y = aX + b$.

90. Suppose a randomly chosen individual's verbal score X and quantitative score Y on a nationally administered aptitude examination have a joint pdf

$$f(x, y) = \begin{cases} \dfrac{2}{5}(2x + 3y) & 0 \leq x \leq 1, 0 \leq y \leq 1 \\ 0 & \text{otherwise} \end{cases}$$

You are asked to provide a prediction t of the individual's total score $X + Y$. The error of prediction is the mean squared error $E[(X + Y - t)^2]$. What value of t minimizes the error of prediction?

91. a. Let X_1 have a chi-squared distribution with parameter ν_1 (see Section 4.4), and let X_2 be independent of X_1 and have a chi-squared distribution with parameter ν_2. Use the technique of Example 5.22 to show that $X_1 + X_2$ has a chi-squared distribution with parameter $\nu_1 + \nu_2$.

    b. In Exercise 71 of Chapter 4, you were asked to show that if Z is a standard normal rv, then Z^2 has a chi-squared distribution with $\nu = 1$. Let $Z_1, Z_2, \ldots, Z_n$ be n independent standard normal rv's. What is the distribution of $Z_1^2 + \cdots + Z_n^2$? Justify your answer.

    c. Let $X_1, \ldots, X_n$ be a random sample from a normal distribution with mean μ and variance σ^2. What is the distribution of the sum $Y = \sum_{i=1}^{n}[(X_i - \mu)/\sigma]^2$? Justify your answer.

92. a. Show that $\text{Cov}(X, Y + Z) = \text{Cov}(X, Y) + \text{Cov}(X, Z)$.

    b. Let X_1 and X_2 be quantitative and verbal scores on one aptitude exam, and let Y_1 and Y_2 be corresponding scores on another exam. If $\text{Cov}(X_1, Y_1) = 5$, $\text{Cov}(X_1, Y_2) = 1$, $\text{Cov}(X_2, Y_1) = 2$, and $\text{Cov}(X_2, Y_2) = 8$, what is the covariance between the two total scores $X_1 + X_2$ and $Y_1 + Y_2$?

93. A rock specimen from a particular area is randomly selected and weighed two different times. Let W denote the actual weight and X_1 and X_2 the two measured weights. Then $X_1 = W + E_1$ and $X_2 = W + E_2$, where E_1 and E_2 are the two measurement errors. Suppose that the E_i's are independent of one another and of W and that $V(E_1) = V(E_2) = \sigma_E^2$.

    a. Express ρ, the correlation coefficient between the two measured weights X_1 and X_2, in terms of σ_W^2, the variance of actual weight, and σ_X^2, the variance of measured weight.

    b. Compute ρ when $\sigma_W = 1$ kg and $\sigma_E = .01$ kg.

94. Let A denote the percentage of one constituent in a randomly selected rock specimen, and let B denote the percentage of a second constituent in that same specimen. Suppose D and E are measurement errors in determining the values of A and B so that measured values are $X = A + D$ and $Y = B + E$, respectively. Assume that measurement errors are independent of one another and of actual values.

    a. Show that
    $$\text{Corr}(X, Y) = \text{Corr}(A, B) \cdot \sqrt{\text{Corr}(X_1, X_2)} \cdot \sqrt{\text{Corr}(Y_1, Y_2)}$$

where X_1 and X_2 are replicate measurements on the value of A, and Y_1 and Y_2 are defined analogously with respect to B. What effect does the presence of measurement error have on the correlation?

b. What is the maximum value of Corr(X, Y) when Corr(X_1, X_2) = .8100 and Corr(Y_1, Y_2) = .9025? Is this disturbing?

95. Let $X_1, \ldots, X_n$ be independent rv's with mean values $\mu_1, \ldots, \mu_n$ and variances $\sigma_1^2, \ldots, \sigma_n^2$. Consider a function $h(x_1, \ldots, x_n)$, and use it to define a rv $Y = h(X_1, \ldots, X_n)$. Under rather general conditions on the h function, if the σ_i's are all small relative to the corresponding μ_i's, it can be shown that $E(Y) \approx h(\mu_1, \ldots, \mu_n)$ and

$$V(Y) \approx \left(\frac{\partial h}{\partial x_1}\right)^2 \cdot \sigma_1^2 + \cdots + \left(\frac{\partial h}{\partial x_n}\right)^2 \cdot \sigma_n^2$$

where each partial derivative is evaluated at $(x_1, \ldots, x_n) = (\mu_1, \ldots, \mu_n)$. Suppose three resistors with resistances X_1, X_2, X_3 are connected in parallel across a battery with voltage X_4. Then by Ohm's law, the current is

$$Y = X_4\left[\frac{1}{X_1} + \frac{1}{X_2} + \frac{1}{X_3}\right]$$

Let $\mu_1 = 10$ ohms, $\sigma_1 = 1.0$ ohm, $\mu_2 = 15$ ohms, $\sigma_2 = 1.0$ ohm, $\mu_3 = 20$ ohms, $\sigma_3 = 1.5$ ohms, $\mu_4 = 120$ V,

$\sigma_4 = 4.0$ V. Calculate the approximate expected value and standard deviation of the current (suggested by **"Random Samplings," *CHEMTECH*, 1984: 696–697**).

96. A more accurate approximation to $E[h(X_1, \ldots, X_n)]$ in Exercise 95 is

$$h(\mu_1, \ldots, \mu_n) + \frac{1}{2}\sigma_1^2\left(\frac{\partial^2 h}{\partial x_1^2}\right) + \cdots + \frac{1}{2}\sigma_n^2\left(\frac{\partial^2 h}{\partial x_n^2}\right)$$

Compute this for $Y = h(X_1, X_2, X_3, X_4)$ given in Exercise 93, and compare it to the leading term $h(\mu_1, \ldots, \mu_n)$.

97. Let X and Y be independent standard normal random variables, and define a new rv by $U = .6X + .8Y$.
a. Determine Corr(X, U).
b. How would you alter U to obtain Corr(X, U) = ρ for a specified value of ρ?

98. Let $X_1, X_2, \ldots, X_n$ be random variables denoting n independent bids for an item that is for sale. Suppose each X_i is uniformly distributed on the interval [100, 200]. If the seller sells to the highest bidder, how much can he expect to earn on the sale? [*Hint*: Let $Y = \max(X_1, X_2, \ldots, X_n)$. First find $F_Y(y)$ by noting that $Y \le y$ iff each X_i is $\le y$. Then obtain the pdf and $E(Y)$.]

BIBLIOGRAPHY

Carlton, Matthew, and Jay Devore, *Probability with Applications in Engineering, Science, and Technology*, Springer, New York, 2015. A somewhat more sophisticated and extensive exposition of probability topics than in the present book.

Olkin, Ingram, Cyrus Derman, and Leon Gleser, *Probability Models and Applications* (2nd ed.), Macmillan, New York, 1994. Contains a careful and comprehensive exposition of joint distributions, rules of expectation, and limit theorems.

Point Estimation

6

INTRODUCTION

Given a parameter of interest, such as a population mean μ or population proportion p, the objective of point estimation is to use a sample to compute a number that represents in some sense an educated guess for the true value of the parameter. The resulting number is called a *point estimate*. Section 6.1 introduces some general concepts of point estimation. In Section 6.2, we describe and illustrate two important methods for obtaining point estimates: the method of moments and the method of maximum likelihood.

6.1 Some General Concepts of Point Estimation

Statistical inference is almost always directed toward drawing some type of conclusion about one or more parameters (population characteristics). To do so requires that an investigator obtain sample data from each of the populations under study. Conclusions can then be based on the computed values of various sample quantities. For example, let μ (a parameter) denote the true average breaking strength of wire connections used in bonding semiconductor wafers. A random sample of $n = 10$ connections might be made, and the breaking strength of each one determined, resulting in observed strengths $x_1, x_2, \ldots, x_{10}$. The sample mean breaking strength $\bar{x}$ could then be used to draw a conclusion about the value of μ. Similarly, if σ^2 is the variance of the breaking strength distribution (population variance, another parameter), the value of the sample variance s^2 can be used to infer something about σ^2.

When discussing general concepts and methods of inference, it is convenient to have a generic symbol for the parameter of interest. We will use the Greek letter θ for this purpose. In many investigations, θ will be a population mean μ, a difference $\mu_1 - \mu_2$ between two population means, or a population proportion of "successes" p. The objective of point estimation is to select a single number, based on sample data, that represents a sensible value for θ. As an example, the parameter of interest might be μ, the true average lifetime of batteries of a certain type. A random sample of $n = 3$ batteries might yield observed lifetimes (hours) $x_1 = 5.0$, $x_2 = 6.4, x_3 = 5.9$. The computed value of the sample mean lifetime is $\bar{x} = 5.77$, and it is reasonable to regard 5.77 as a very plausible value of μ—our "best guess" for the value of μ based on the available sample information.

Suppose we want to estimate a parameter of a single population (e.g., μ or σ) based on a random sample of size n. Recall from the previous chapter that before data is available, the sample observations must be considered random variables (rv's) X_1, $X_2, \ldots, X_n$. It follows that any function of the X_i's—that is, any statistic—such as the sample mean $\bar{X}$ or sample standard deviation S is also a random variable. The same is true if available data consists of more than one sample. For example, we can represent tensile strengths of m type 1 specimens and n type 2 specimens by $X_1, \ldots, X_m$ and $Y_1, \ldots, Y_n$, respectively. The difference between the two sample mean strengths is $\bar{X} - \bar{Y}$; this is the natural statistic for making inferences about $\mu_1 - \mu_2$, the difference between the population mean strengths.

DEFINITION

> A **point estimate** of a parameter θ is a single number that can be regarded as a sensible value for θ. It is obtained by selecting a suitable statistic and computing its value from the given sample data. The selected statistic is called the **point estimator** of θ.

In the foregoing battery example, the estimator used to obtain the point estimate of μ was $\bar{X}$, and the point estimate of μ was 5.77. If the three observed lifetimes had instead been $x_1 = 5.6$, $x_2 = 4.5$, and $x_3 = 6.1$, use of the estimator $\bar{X}$ would have resulted in the estimate $\bar{x} = (5.6 + 4.5 + 6.1)/3 = 5.40$. The symbol $\hat{\theta}$ ("theta hat") is customarily used to denote both the estimator of θ and the point estimate resulting from a given sample.* Thus $\hat{\mu} = \bar{X}$ is read as "the point estimator of μ is the sample

---

* Following earlier notation, we could use $\hat{\Theta}$ (an uppercase theta) for the estimator, but this is cumbersome to write.

mean $\overline{X}$." The statement "the point estimate of μ is 5.77" can be written concisely as $\hat{\mu} = 5.77$. Notice that in writing $\hat{\theta} = 72.5$, there is no indication of how this point estimate was obtained (what statistic was used). It is recommended that both the estimator and the resulting estimate be reported.

EXAMPLE 6.1 An automobile manufacturer has developed a new type of bumper, which is supposed to absorb impacts with less damage than previous bumpers. The manufacturer has used this bumper in a sequence of 25 controlled crashes against a wall, each at 10 mph, using one of its compact car models. Let X = the number of crashes that result in no visible damage to the automobile. The parameter to be estimated is p = the proportion of all such crashes that result in no damage [alternatively, $p = P$(no damage in a single crash)]. If X is observed to be $x = 15$, the most reasonable estimator and estimate are

$$\text{estimator } \hat{p} = \frac{X}{n} \qquad \text{estimate} = \frac{x}{n} = \frac{15}{25} = .60 \qquad \blacksquare$$

If for each parameter of interest there were only one reasonable point estimator, there would not be much to point estimation. In most problems, though, there will be more than one reasonable estimator.

EXAMPLE 6.2 Consider the accompanying 20 observations on dielectric breakdown voltage for pieces of epoxy resin first introduced in Exercise 4.89.

| 24.46 | 25.61 | 26.25 | 26.42 | 26.66 | 27.15 | 27.31 | 27.54 | 27.74 | 27.94 |
| 27.98 | 28.04 | 28.28 | 28.49 | 28.50 | 28.87 | 29.11 | 29.13 | 29.50 | 30.88 |

The pattern in the normal probability plot given there is quite straight, so we now assume that the distribution of breakdown voltage is normal with mean value μ. Because normal distributions are symmetric, μ is also the median lifetime of the distribution. The given observations are then assumed to be the result of a random sample $X_1, X_2, \dots, X_{20}$ from this normal distribution. Consider the following estimators and resulting estimates for μ:

a. Estimator = $\overline{X}$, estimate = $\bar{x} = \Sigma x_i/n = 555.86/20 = 27.793$
b. Estimator = $\widetilde{X}$, estimate = $\tilde{x} = (27.94 + 27.98)/2 = 27.960$
c. Estimator = $[\min(X_i) + \max(X_i)]/2$ = the average of the two extreme lifetimes, estimate = $[\min(x_i) + \max(x_i)]/2 = (24.46 + 30.88)/2 = 27.670$
d. Estimator = $\overline{X}_{tr(10)}$, the 10% trimmed mean (discard the smallest and largest 10% of the sample and then average),

$$\begin{aligned}
\text{estimate} &= \bar{x}_{tr(10)} \\
&= \frac{555.86 - 24.46 - 25.61 - 29.50 - 30.88}{16} \\
&= 27.838
\end{aligned}$$

Each one of the estimators (a)–(d) uses a different measure of the center of the sample to estimate μ. Which of the estimates is closest to the true value? This question cannot be answered without knowing the true value. A question that can be answered is, "Which estimator, when used on other samples of X_i's, will tend to produce estimates closest to the true value?" We will shortly address this issue. $\blacksquare$

EXAMPLE 6.3 The article **"Is a Normal Distribution the Most Appropriate Statistical Distribution for Volumetric Properties in Asphalt Mixtures?"** first cited in Example 4.26, reported the following observations on $X = $ *voids filled with asphalt* (%) for 52 specimens of a certain type of hot-mix asphalt:

74.33	71.07	73.82	77.42	79.35	82.27	77.75	78.65	77.19
74.69	77.25	74.84	60.90	60.75	74.09	65.36	67.84	69.97
68.83	75.09	62.54	67.47	72.00	66.51	68.21	64.46	64.34
64.93	67.33	66.08	67.31	74.87	69.40	70.83	81.73	82.50
79.87	81.96	79.51	84.12	80.61	79.89	79.70	78.74	77.28
79.97	75.09	74.38	77.67	83.73	80.39	76.90		

Let's estimate the variance σ^2 of the population distribution. A natural estimator is the sample variance:

$$\hat{\sigma}^2 = S^2 = \frac{\sum(X_i - \overline{X})^2}{n - 1}$$

Minitab gave the following output from a request to display descriptive statistics:

Variable	Count	Mean	SE Mean	StDev	Variance	Q1	Median	Q3
VFA(B)	52	73.880	0.889	6.413	41.126	67.933	74.855	79.470

Thus the point estimate of the population variance is

$$\hat{\sigma}^2 = s^2 = \frac{\sum(x_i - \overline{x})^2}{52 - 1} = 41.126$$

[alternatively, the computational formula for the numerator of s^2 gives

$$S_{xx} = \sum x_i^2 - \left(\sum x_i\right)^2/n = 285{,}929.5964 - (3841.78)^2/52 = 2097.4124].$$

A point estimate of the population standard deviation is then $\hat{\sigma} = s = \sqrt{41.126} = 6.413$.

An alternative estimator results from using the divisor n rather than $n - 1$:

$$\hat{\sigma}^2 = \frac{\sum(X_i - \overline{X})^2}{n}, \qquad \text{estimate} = \frac{2097.4124}{52} = 40.335$$

We will shortly indicate why many statisticians prefer S^2 to this latter estimator.

The cited article considered fitting four different distributions to the data: normal, lognormal, two-parameter Weibull, and three-parameter Weibull. Several different techniques were used to conclude that the two-parameter Weibull provided the best fit (a normal probability plot of the data shows some deviation from a linear pattern). From Section 4.5, the variance of a Weibull random variable is

$$\sigma^2 = \beta^2\{\Gamma(1 + 2/\alpha) - [\Gamma(1 + 1/\alpha)]^2\}$$

where α and β are the shape and scale parameters of the distribution. The authors of the article used the method of maximum likelihood (see Section 6.2) to estimate these parameters. The resulting estimates were $\hat{\alpha} = 11.9731$, $\hat{\beta} = 77.0153$. A sensible estimate of the population variance can now be obtained from substituting the estimates of the two parameters into the expression for σ^2; the result is $\hat{\sigma}^2 = 56.035$. This latter estimate is obviously quite different from the sample variance. Its validity depends on the population distribution being Weibull, whereas the sample variance is a sensible way to estimate σ^2 when there is uncertainty as to the specific form of the population distribution. ∎

In the best of all possible worlds, we could find an estimator $\hat{\theta}$ for which $\hat{\theta} = \theta$ always. However, $\hat{\theta}$ is a function of the sample X_i's, so it is a random variable. For some samples, $\hat{\theta}$ will yield a value larger than θ, whereas for other samples $\hat{\theta}$ will underestimate θ. If we write

$$\hat{\theta} = \theta + \text{error of estimation}$$

then an accurate estimator would be one resulting in small estimation errors, so that estimated values will be near the true value.

A sensible way to quantify the idea of $\hat{\theta}$ being close to θ is to consider the squared error $(\hat{\theta} - \theta)^2$. For some samples, $\hat{\theta}$ will be quite close to θ and the resulting squared error will be near 0. Other samples may give values of $\hat{\theta}$ far from θ, corresponding to very large squared errors. An omnibus measure of accuracy is the *expected* or *mean square error* MSE $= E[(\hat{\theta} - \theta)^2]$. If a first estimator has smaller MSE than does a second, it is natural to say that the first estimator is the better one. However, MSE will generally depend on the value of θ. What often happens is that one estimator will have a smaller MSE for some values of θ and a larger MSE for other values. Finding an estimator with the smallest MSE is typically not possible.

One way out of this dilemma is to restrict attention just to estimators that have some specified desirable property and then find the best estimator in this restricted group. A popular property of this sort in the statistical community is *unbiasedness*.

Unbiased Estimators

Suppose we have two measuring instruments; one instrument has been accurately calibrated, but the other systematically gives readings larger than the true value being measured. When each instrument is used repeatedly on the same object, because of measurement error, the observed measurements will not be identical. However, the measurements produced by the first instrument will be distributed about the true value in such a way that on average this instrument measures what it purports to measure, so it is called an unbiased instrument. The second instrument yields observations that have a systematic error component or bias. Figure 6.1 shows 10 measurements from both an unbiased and a biased instrument.

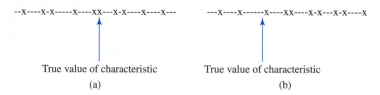

Figure 6.1 Measurements from (a) an unbiased instrument, and (b) a biased instrument

DEFINITION

A point estimator $\hat{\theta}$ is said to be an **unbiased estimator** of θ if $E(\hat{\theta}) = \theta$ for every possible value of θ. If $\hat{\theta}$ is not unbiased, the difference $E(\hat{\theta}) - \theta$ is called the **bias** of $\hat{\theta}$.

That is, $\hat{\theta}$ is unbiased if its probability (i.e., sampling) distribution is always "centered" at the true value of the parameter. Suppose $\hat{\theta}$ is an unbiased estimator; then if $\theta = 100$, the $\hat{\theta}$ sampling distribution is centered at 100; if $\theta = 27.5$, then the $\hat{\theta}$ sampling distribution is centered at 27.5, and so on. Figure 6.2 pictures the distributions of several biased and unbiased estimators. Note that "centered" here means that the expected value, not the median, of the distribution of $\hat{\theta}$ is equal to θ.

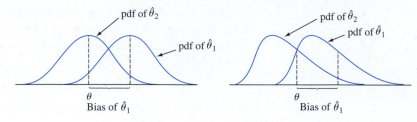

Figure 6.2   The pdf's of a biased estimator $\hat{\theta}_1$ and an unbiased estimator $\hat{\theta}_2$ for a parameter θ

It may seem as though it is necessary to know the value of θ (in which case estimation is unnecessary) to see whether $\hat{\theta}$ is unbiased. This is not usually the case, though, because unbiasedness is a general property of the estimator's sampling distribution—where it is centered—which is typically not dependent on any particular parameter value.

In Example 6.1, the sample proportion X/n was used as an estimator of p, where X, the number of sample successes, had a binomial distribution with parameters n and p. Thus

$$E(\hat{p}) = E\left(\frac{X}{n}\right) = \frac{1}{n} E(X) = \frac{1}{n}(np) = p$$

PROPOSITION

> When X is a binomial rv with parameters n and p, the sample proportion $\hat{p} = X/n$ is an unbiased estimator of p.

No matter what the true value of p is, the distribution of the estimator $\hat{p}$ will be centered at the true value.

EXAMPLE 6.4   Suppose that X, the reaction time to a certain stimulus, has a uniform distribution on the interval from 0 to an unknown upper limit θ (so the density function of X is rectangular in shape with height $1/\theta$ for $0 \le x \le \theta$). It is desired to estimate θ on the basis of a random sample $X_1, X_2, \ldots, X_n$ of reaction times. Since θ is the largest possible time in the entire population of reaction times, consider as a first estimator the largest sample reaction time: $\hat{\theta}_1 = \max(X_1, \ldots, X_n)$. If $n = 5$ and $x_1 = 4.2$, $x_2 = 1.7$, $x_3 = 2.4$, $x_4 = 3.9$, and $x_5 = 1.3$, the point estimate of θ is $\hat{\theta}_1 = \max(4.2, 1.7, 2.4, 3.9, 1.3) = 4.2$.

Unbiasedness implies that some samples will yield estimates that exceed θ and other samples will yield estimates smaller than θ—otherwise θ could not possibly be the center (balance point) of $\hat{\theta}_1$'s distribution. However, our proposed estimator will never overestimate θ (the largest sample value cannot exceed the largest population value) and will underestimate θ unless the largest sample value equals θ. This intuitive argument shows that $\hat{\theta}_1$ is a biased estimator. More precisely, it can be shown (see Exercise 32) that

$$E(\hat{\theta}_1) = \frac{n}{n+1} \cdot \theta < \theta \qquad \left(\text{since } \frac{n}{n+1} < 1\right)$$

The bias of $\hat{\theta}_1$ is given by $n\theta/(n+1) - \theta = -\theta/(n+1)$, which approaches 0 as n gets large.

It is easy to modify $\hat{\theta}_1$ to obtain an unbiased estimator of θ. Consider the estimator

$$\hat{\theta}_2 = \frac{n+1}{n} \cdot \max{(X_1, \ldots, X_n)}$$

Using this estimator on the data gives the estimate $(6/5)(4.2) = 5.04$. The fact that $(n + 1)/n > 1$ implies that $\hat{\theta}_2$ will overestimate θ for some samples and underestimate it for others. The mean value of this estimator is

$$E(\hat{\theta}_2) = E\left[\frac{n+1}{n} \max(X_1, \ldots, X_n)\right] = \frac{n+1}{n} \cdot E\left[\max(X_1, \ldots, X_n)\right]$$

$$= \frac{n+1}{n} \cdot \frac{n}{n+1}\theta = \theta$$

If $\hat{\theta}_2$ is used repeatedly on different samples to estimate θ, some estimates will be too large and others will be too small, but in the long run there will be no systematic tendency to underestimate or overestimate θ. ■

> ### Principle of Unbiased Estimation
>
> When choosing among several different estimators of θ, select one that is unbiased.

According to this principle, the unbiased estimator $\hat{\theta}_2$ in Example 6.4 should be preferred to the biased estimator $\hat{\theta}_1$. Consider now the problem of estimating σ^2.

PROPOSITION

> Let $X_1, X_2, \ldots, X_n$ be a random sample from a distribution with mean μ and variance σ^2. Then the estimator
>
> $$\hat{\sigma}^2 = S^2 = \frac{\sum(X_i - \overline{X})^2}{n-1}$$
>
> is unbiased for estimating σ^2.

Proof For any rv Y, $V(Y) = E(Y^2) - [E(Y)]^2$, so $E(Y^2) = V(Y) + [E(Y)]^2$. Applying this to

$$S^2 = \frac{1}{n-1}\left[\sum X_i^2 - \frac{(\sum X_i)^2}{n}\right]$$

gives

$$E(S^2) = \frac{1}{n-1}\left\{\sum E(X_i^2) - \frac{1}{n}E\left[\left(\sum X_i\right)^2\right]\right\}$$

$$= \frac{1}{n-1}\left\{\sum(\sigma^2 + \mu^2) - \frac{1}{n}\left\{V\left(\sum X_i\right) + \left[E\left(\sum X_i\right)\right]^2\right\}\right\}$$

$$= \frac{1}{n-1}\left\{n\sigma^2 + n\mu^2 - \frac{1}{n}n\sigma^2 - \frac{1}{n}(n\mu)^2\right\}$$

$$= \frac{1}{n-1}\{n\sigma^2 - \sigma^2\} = \sigma^2 \quad \text{(as desired)} \quad ■$$

The estimator that uses divisor n can be expressed as $(n-1)S^2/n$, so

$$E\left[\frac{(n-1)S^2}{n}\right] = \frac{n-1}{n}E(S^2) = \frac{n-1}{n}\sigma^2$$

This estimator is therefore not unbiased. The bias is $(n-1)\sigma^2/n - \sigma^2 = -\sigma^2/n$. Because the bias is negative, the estimator with divisor n tends to underestimate σ^2, and this is why the divisor $n-1$ is preferred by many statisticians (though when n is large, the bias is small and there is little difference between the two).

Unfortunately, the fact that S^2 is unbiased for estimating σ^2 does not imply that S is unbiased for estimating σ. Taking the square root invalidates the property of unbiasedness (the expected value of the square root is not the square root of the expected value). Fortunately, the bias of S is small unless n is quite small. There are other good reasons to use S as an estimator, especially when the population distribution is normal. These will become more apparent when we discuss confidence intervals and hypothesis testing in the next several chapters.

In Example 6.2, we proposed several different estimators for the mean μ of a normal distribution. If there were a unique unbiased estimator for μ, the estimation problem would be resolved by using that estimator. Unfortunately, this is not the case.

PROPOSITION

> If $X_1, X_2, \ldots, X_n$ is a random sample from a distribution with mean μ, then $\overline{X}$ is an unbiased estimator of μ. If in addition the distribution is continuous and symmetric, then $\widetilde{X}$ and any trimmed mean are also unbiased estimators of μ.

The fact that $\overline{X}$ is unbiased is just a restatement of one of our rules of expected value: $E(\overline{X}) = \mu$ for every possible value of μ (for discrete as well as continuous distributions). The unbiasedness of the other estimators is more difficult to verify.

The next example introduces another situation in which there are several unbiased estimators for a particular parameter.

EXAMPLE 6.5 Under certain circumstances organic contaminants adhere readily to wafer surfaces and cause deterioration in semiconductor manufacturing devices. The article **"Ceramic Chemical Filter for Removal of Organic Contaminants"** (*J. of the Institute of Envir. Sciences and Tech.*, **2003: 59–65**) discussed a recently developed alternative to conventional charcoal filters for removing organic airborne molecular contamination in cleanroom applications. One aspect of the investigation of filter performance involved studying how contaminant concentration in air related to concentration on a wafer surface after prolonged exposure. Consider the following representative data on $x =$ DBP concentration in air and $y =$ DBP concentration on a wafer surface after 4-hour exposure (both in $\mu g/m^3$, where DBP = dibutyl phthalate).

Obs. i:	1	2	3	4	5	6
x:	.8	1.3	1.5	3.0	11.6	26.6
y:	.6	1.1	4.5	3.5	14.4	29.1

The authors comment that "DBP adhesion on the wafer surface was roughly proportional to the DBP concentration in air." Figure 6.3 shows a plot of y versus x—i.e., of the (x, y) pairs.

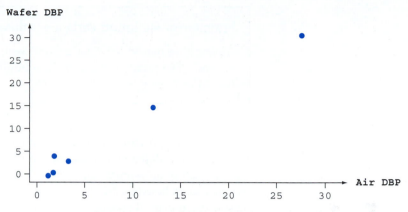

Figure 6.3   Plot of the DBP data from Example 6.5

If y were exactly proportional to x, then $y = \beta x$ for some value β, which says that the (x, y) points in the plot would lie exactly on a straight line with slope β passing through $(0, 0)$. But this is only approximately the case. So we now assume that for any fixed x, wafer DBP is a random variable Y having mean value βx. That is, we postulate that the *mean* value of Y is related to x by a line passing through $(0, 0)$ but that the observed value of Y will typically deviate from this line (this is referred to in the statistical literature as "regression through the origin").

Consider the following three estimators for the slope parameter β:

$$\#1: \hat{\beta} = \frac{1}{n}\sum\frac{Y_i}{x_i} \quad \#2: \hat{\beta} = \frac{\sum Y_i}{\sum x_i} \quad \#3: \hat{\beta} = \frac{\sum x_i Y_i}{\sum x_i^2}$$

The resulting estimates based on the given data are 1.3497, 1.1875, and 1.1222, respectively. So the estimate definitely depends on which estimator is used. If one of these three estimators were unbiased and the other two were biased, there would be a good case for using the unbiased one. But all three are unbiased; the argument relies on the fact that each one is a linear function of the Y_i's (we are assuming here that the x_i's are fixed, not random):

$$E\left(\frac{1}{n}\sum\frac{Y_i}{x_i}\right) = \frac{1}{n}\sum\frac{E(Y_i)}{x_i} = \frac{1}{n}\sum\frac{\beta x_i}{x_i} = \frac{1}{n}\sum\beta = \frac{n\beta}{n} = \beta$$

$$E\left(\frac{\sum Y_i}{\sum x_i}\right) = \frac{1}{\sum x_i}E\left(\sum Y_i\right) = \frac{1}{\sum x_i}\left(\sum\beta x_i\right) = \frac{1}{\sum x_i}\beta\left(\sum x_i\right) = \beta$$

$$E\left(\frac{\sum x_i Y_i}{\sum x_i^2}\right) = \frac{1}{\sum x_i^2}E\left(\sum x_i Y_i\right) = \frac{1}{\sum x_i^2}\left(\sum x_i\beta x_i\right) = \frac{1}{\sum x_i^2}\beta\left(\sum x_i^2\right) = \beta \quad ∎$$

In both the foregoing example and the situation involving estimating a normal population mean, the principle of unbiasedness (preferring an unbiased estimator to a biased one) cannot be invoked to select an estimator. What we now need is a criterion for choosing among unbiased estimators.

Estimators with Minimum Variance

Suppose $\hat{\theta}_1$ and $\hat{\theta}_2$ are two estimators of θ that are both unbiased. Then, although the distribution of each estimator is centered at the true value of θ, the spreads of the distributions about the true value may be different.

> ### Principle of Minimum Variance Unbiased Estimation
>
> Among all estimators of θ that are unbiased, choose the one that has minimum variance. The resulting $\hat{\theta}$ is called the **minimum variance unbiased estimator (MVUE)** of θ.

Figure 6.4(a) shows distributions of two different unbiased estimators. Use of the estimator with the more concentrated distribution is more likely than the other one to result in an estimate closer to θ. Figure 6.4(b) displays estimates from the two estimators based on 10 different samples. The MVUE is, in a certain sense, the most likely among all unbiased estimators to produce an estimate close to the true θ.

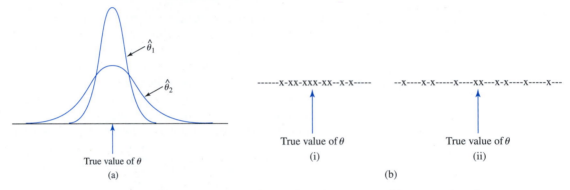

Figure 6.4 (a) Distributions of two unbiased estimators (b) Estimates based on 10 different samples

In Example 6.5, suppose each Y_i is normally distributed with mean βx_i and variance σ^2 (the assumption of constant variance). Then it can be shown that the third estimator $\hat{\beta} = \Sigma x_i Y_i / \Sigma x_i^2$ not only has smaller variance than either of the other two unbiased estimators, but in fact is the MVUE—it has smaller variance than *any* other unbiased estimator of β.

EXAMPLE 6.6 We argued in Example 6.4 that when $X_1, \ldots, X_n$ is a random sample from a uniform distribution on $[0, \theta]$, the estimator

$$\hat{\theta}_1 = \frac{n + 1}{n} \cdot \max (X_1, \ldots, X_n)$$

is unbiased for θ (we previously denoted this estimator by $\hat{\theta}_2$). This is not the only unbiased estimator of θ. The expected value of a uniformly distributed rv is just the midpoint of the interval of positive density, so $E(X_i) = \theta/2$. This implies that $E(\overline{X}) = \theta/2$, from which $E(2\overline{X}) = \theta$. That is, the estimator $\hat{\theta}_2 = 2\overline{X}$ is unbiased for θ.

If X is uniformly distributed on the interval from A to B, then $V(X) = \sigma^2 = (B - A)^2/12$. Thus, in our situation, $V(X_i) = \theta^2/12$, $V(\overline{X}) = \sigma^2/n = \theta^2/(12n)$, and $V(\hat{\theta}_2) = V(2\overline{X}) = 4V(\overline{X}) = \theta^2/(3n)$. The results of Exercise 32 can be used to show that $V(\hat{\theta}_1) = \theta^2/[n(n + 2)]$. The estimator $\hat{\theta}_1$ has smaller variance than does $\hat{\theta}_2$ if $3n < n(n + 2)$—that is, if $0 < n^2 - n = n(n - 1)$. As long as $n > 1$, $V(\hat{\theta}_1) < V(\hat{\theta}_2)$, so $\hat{\theta}_1$ is a better estimator than $\hat{\theta}_2$. More advanced methods can be used to show that $\hat{\theta}_1$ is the MVUE of θ—every other unbiased estimator of θ has variance that exceeds $\theta^2/[n(n + 2)]$. ∎

One of the triumphs of mathematical statistics has been the development of methodology for identifying the MVUE in a wide variety of situations. The most important result of this type for our purposes concerns estimating the mean μ of a normal distribution.

THEOREM

> Let $X_1, \ldots, X_n$ be a random sample from a normal distribution with parameters μ and σ. Then the estimator $\hat{\mu} = \overline{X}$ is the MVUE for μ.

Whenever we are convinced that the population being sampled is normal, the theorem says that $\bar{x}$ should be used to estimate μ. In Example 6.2, then, our estimate would be $\bar{x} = 27.793$.

In some situations, it is possible to obtain an estimator with small bias that would be preferred to the best unbiased estimator. This is illustrated in Figure 6.5. However, MVUEs are often easier to obtain than the type of biased estimator whose distribution is pictured.

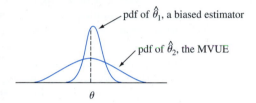

Figure 6.5 A biased estimator that is preferable to the MVUE

Some Complications

The last theorem does not say that in estimating a population mean μ, the estimator $\overline{X}$ should be used irrespective of the distribution being sampled.

EXAMPLE 6.7 Suppose we wish to estimate the thermal conductivity μ of a certain material. Using standard measurement techniques, we will obtain a random sample $X_1, \ldots, X_n$ of n thermal conductivity measurements. Let's assume that the population distribution is a member of one of the following three families:

$$f(x) = \frac{1}{\sqrt{2\pi\sigma^2}} e^{-(x-\mu)^2/(2\sigma^2)} \quad -\infty < x < \infty \tag{6.1}$$

$$f(x) = \frac{1}{\pi\beta[1 + ((x-\mu)/\beta)^2]} \quad -\infty < x < \infty \tag{6.2}$$

$$f(x) = \begin{cases} \dfrac{1}{2c} & \mu - c \le x \le \mu + c \\ 0 & \text{otherwise} \end{cases} \tag{6.3}$$

The pdf (6.1) is the normal distribution, (6.2) is called the Cauchy distribution, and (6.3) is a uniform distribution. All three distributions are symmetric about μ. The Cauchy density curve is bell-shaped but with much heavier tails (more probability farther out) than the normal curve. In fact, the tails are so heavy that the mean value does not exist, though μ is still the median and a location parameter for the distribution. The uniform distribution has no tails. The four estimators for μ considered earlier are $\overline{X}, \widetilde{X}, \overline{X}_e$ (the average of the two extreme observations), and $\overline{X}_{\text{tr}(10)}$, a trimmed mean.

The very important moral here is that the best estimator for μ depends crucially on which distribution is being sampled. In particular,

1. If the random sample comes from a normal distribution, then $\overline{X}$ is the best of the four estimators, since it has minimum variance among all unbiased estimators.

2. If the random sample comes from a Cauchy distribution, then $\overline{X}$ and $\overline{X}_e$ are terrible estimators for μ, whereas $\widetilde{X}$ is quite good (the MVUE is not known); $\overline{X}$ is bad because it is very sensitive to outlying observations, and the heavy tails of the Cauchy distribution make a few such observations likely to appear in any sample.

3. If the underlying distribution is uniform, the best estimator is $\overline{X}_e$; this estimator is greatly influenced by outlying observations, but the lack of tails makes such observations impossible.

4. *The trimmed mean is best in none of these three situations but works reasonably well in all three.* That is, $\overline{X}_{\text{tr}(10)}$ does not suffer too much in comparison with the best procedure in any of the three situations. ∎

More generally, recent research in statistics has established that when estimating a point of symmetry μ of a continuous probability distribution, a trimmed mean with trimming proportion 10% or 20% (from each end of the sample) produces reasonably behaved estimates over a very wide range of possible models. For this reason, a trimmed mean with small trimming percentage is said to be a **robust estimator**.

In some situations, the choice is not between two different estimators constructed from the same sample, but instead between estimators based on two different experiments.

EXAMPLE 6.8   Suppose a certain type of component has a lifetime distribution that is exponential with parameter λ so that expected lifetime is $\mu = 1/\lambda$. A sample of n such components is selected, and each is put into operation. If the experiment is continued until all n lifetimes, $X_1, \ldots, X_n$, have been observed, then $\overline{X}$ is an unbiased estimator of μ.

In some experiments, though, the components are left in operation only until the time of the rth failure, where $r < n$. This procedure is referred to as **censoring**. Let Y_1 denote the time of the first failure (the minimum lifetime among the n components), Y_2 denote the time at which the second failure occurs (the second smallest lifetime), and so on. Since the experiment terminates at time Y_r, the total accumulated lifetime at termination is

$$T_r = \sum_{i=1}^{r} Y_i + (n - r)Y_r$$

We now demonstrate that $\hat{\mu} = T_r/r$ is an unbiased estimator for μ. To do so, we need two properties of exponential variables:

1. The memoryless property (see Section 4.4), which says that at any time point, remaining lifetime has the same exponential distribution as original lifetime.

2. When $X_1, \ldots, X_k$ are independent, each exponentially distributed with parameter λ, $\min(X_1, \ldots, X_k)$, is exponential with parameter $k\lambda$.

Since all n components last until Y_1, $n - 1$ last an additional $Y_2 - Y_1$, $n - 2$ an additional $Y_3 - Y_2$ amount of time, and so on, another expression for T_r is

$$T_r = nY_1 + (n - 1)(Y_2 - Y_1) + (n - 2)(Y_3 - Y_2) + \cdots$$
$$+ (n - r + 1)(Y_r - Y_{r-1})$$

But Y_1 is the minimum of n exponential variables, so $E(Y_1) = 1/(n\lambda)$. Similarly, $Y_2 - Y_1$ is the smallest of the $n - 1$ remaining lifetimes, each exponential with parameter λ (by the memoryless property), so $E(Y_2 - Y_1) = 1/[(n - 1)\lambda]$. Continuing, $E(Y_{i+1} - Y_i) = 1/[(n - i)\lambda]$, so

$$E(T_r) = nE(Y_1) + (n - 1)E(Y_2 - Y_1) + \cdots + (n - r + 1)E(Y_r - Y_{r-1})$$

$$= n \cdot \frac{1}{n\lambda} + (n - 1) \cdot \frac{1}{(n - 1)\lambda} + \cdots + (n - r + 1) \cdot \frac{1}{(n - r + 1)\lambda}$$

$$= \frac{r}{\lambda}$$

Therefore, $E(T_r/r) = (1/r)E(T_r) = (1/r) \cdot (r/\lambda) = 1/\lambda = \mu$ as claimed.

As an example, suppose 20 components are tested and $r = 10$. Then if the first ten failure times are 11, 15, 29, 33, 35, 40, 47, 55, 58, and 72, the estimate of μ is

$$\hat{\mu} = \frac{11 + 15 + \cdots + 72 + (10)(72)}{10} = 111.5$$

The advantage of the experiment with censoring is that it terminates more quickly than the uncensored experiment. However, it can be shown that $V(T_r/r) = 1/(\lambda^2 r)$, which is larger than $1/(\lambda^2 n)$, the variance of $\overline{X}$ in the uncensored experiment. ■

Reporting a Point Estimate: The Standard Error

Besides reporting the value of a point estimate, some indication of its precision should be given. The usual measure of precision is the standard error of the estimator used.

DEFINITION

> The **standard error** of an estimator $\hat{\theta}$ is its standard deviation $\sigma_{\hat{\theta}} = \sqrt{V(\hat{\theta})}$. It is the magnitude of a typical or representative deviation between an estimate and the value of θ. If the standard error itself involves unknown parameters whose values can be estimated, substitution of these estimates into $\sigma_{\hat{\theta}}$ yields the **estimated standard error** (estimated standard deviation) of the estimator. The estimated standard error can be denoted either by $\hat{\sigma}_{\hat{\theta}}$ (the $\hat{}$ over σ emphasizes that $\sigma_{\hat{\theta}}$ is being estimated) or by $s_{\hat{\theta}}$.

EXAMPLE 6.9
(Example 6.2 continued)

Assuming that breakdown voltage is normally distributed, $\hat{\mu} = \overline{X}$ is the best estimator of μ. If the value of σ is known to be 1.5, the standard error of $\overline{X}$ is $\sigma_{\overline{X}} = \sigma/\sqrt{n} = 1.5/\sqrt{20} = .335$. If, as is usually the case, the value of σ is unknown, the estimate $\hat{\sigma} = s = 1.462$ is substituted into $\sigma_{\overline{X}}$ to obtain the estimated standard error $\hat{\sigma}_{\overline{X}} = s_{\overline{X}} = s/\sqrt{n} = 1.462/\sqrt{20} = .327$. ■

EXAMPLE 6.10
(Example 6.1 continued)

The standard error of $\hat{p} = X/n$ is

$$\sigma_{\hat{p}} = \sqrt{V(X/n)} = \sqrt{\frac{V(X)}{n^2}} = \sqrt{\frac{npq}{n^2}} = \sqrt{\frac{pq}{n}}$$

Since p and $q = 1 - p$ are unknown (else why estimate?), we substitute $\hat{p} = x/n$ and $\hat{q} = 1 - x/n$ into $\sigma_{\hat{p}}$, yielding the estimated standard error $\hat{\sigma}_{\hat{p}} = \sqrt{\hat{p}\hat{q}/n} = \sqrt{(.6)(.4)/25} = .098$. Alternatively, since the largest value of pq is attained when $p = q = .5$, an upper bound on the standard error is $\sqrt{1/(4n)} = .10$. ■

When the point estimator $\hat{\theta}$ has approximately a normal distribution, which will often be the case when n is large, then we can be reasonably confident that the true value of θ lies within approximately 2 standard errors (standard deviations) of $\hat{\theta}$. Thus if a sample of $n = 36$ component lifetimes gives $\hat{\mu} = \bar{x} = 28.50$ and $s = 3.60$, then $s/\sqrt{n} = .60$, so within 2 estimated standard errors, $\hat{\mu}$ translates to the interval $28.50 \pm (2)(.60) = (27.30, 29.70)$.

If $\hat{\theta}$ is not necessarily approximately normal but is unbiased, then it can be shown that the estimate will deviate from θ by as much as 4 standard errors at most 6% of the time. We would then expect the true value to lie within 4 standard errors of $\hat{\theta}$ (and this is a very conservative statement, since it applies to *any* unbiased $\hat{\theta}$). Summarizing, the standard error tells us roughly within what distance of $\hat{\theta}$ we can expect the true value of θ to lie.

The form of the estimator $\hat{\theta}$ may be sufficiently complicated so that standard statistical theory cannot be applied to obtain an expression for $\sigma_{\hat{\theta}}$. This is true, for example, in the case $\theta = \sigma$, $\hat{\theta} = S$; the standard deviation of the statistic S, σ_S, cannot in general be determined. In recent years, a new computer-intensive method called the **bootstrap** has been introduced to address this problem. Suppose that the population pdf is $f(x; \theta)$, a member of a particular parametric family, and that data $x_1, x_2, \ldots, x_n$ gives $\hat{\theta} = 21.7$. We now use statistical software to obtain "bootstrap samples" from the pdf $f(x; 21.7)$, and for each sample calculate a "bootstrap estimate" $\hat{\theta}^*$:

> First bootstrap sample: $x_1^*, x_2^*, \ldots, x_n^*$; estimate $= \hat{\theta}_1^*$
> Second bootstrap sample: $x_1^*, x_2^*, \ldots, x_n^*$; estimate $= \hat{\theta}_2^*$
> $\vdots$
> Bth bootstrap sample: $x_1^*, x_2^*, \ldots, x_n^*$; estimate $= \hat{\theta}_B^*$

$B = 100$ or 200 is often used. Now let $\bar{\theta}^* = \Sigma\hat{\theta}_i^*/B$, the sample mean of the bootstrap estimates. The **bootstrap estimate** of $\hat{\theta}$'s standard error is now just the sample standard deviation of the $\hat{\theta}_i^*$'s:

$$s_{\hat{\theta}} = \sqrt{\frac{1}{B-1}\Sigma(\hat{\theta}_i^* - \bar{\theta}^*)^2}$$

(In the bootstrap literature, B is often used in place of $B - 1$; for typical values of B, there is usually little difference between the resulting estimates.)

EXAMPLE 6.11   A theoretical model suggests that X, the time to breakdown of an insulating fluid between electrodes at a particular voltage, has $f(x; \lambda) = \lambda e^{-\lambda x}$, an exponential distribution. A random sample of $n = 10$ breakdown times (min) gives the following data:

$$41.53 \quad 18.73 \quad 2.99 \quad 30.34 \quad 12.33 \quad 117.52 \quad 73.02 \quad 223.63 \quad 4.00 \quad 26.78$$

Since $E(X) = 1/\lambda$, $E(\bar{X}) = 1/\lambda$, so a reasonable estimate of λ is $\hat{\lambda} = 1/\bar{x} = 1/55.087 = .018153$. We then used a statistical computer package to obtain $B = 100$ bootstrap samples, each of size 10, from $f(x; .018153)$. The first such sample was 41.00, 109.70, 16.78, 6.31, 6.76, 5.62, 60.96, 78.81, 192.25, 27.61, from which $\Sigma x_i^* = 545.8$ and $\hat{\lambda}_1^* = 1/54.58 = .01832$. The average of the 100 bootstrap estimates is $\bar{\lambda}^* = .02153$, and the sample standard deviation of these 100 estimates is $s_{\hat{\lambda}} = .0091$, the bootstrap estimate of $\hat{\lambda}$'s standard error. A histogram of the 100$\hat{\lambda}_i^*$'s was somewhat positively skewed, suggesting that the sampling distribution of $\hat{\lambda}$ also has this property. ∎

Sometimes an investigator wishes to estimate a population characteristic without assuming that the population distribution belongs to a particular parametric family. An instance of this occurred in Example 6.7, where a 10% trimmed mean was proposed for estimating a symmetric population distribution's center θ. The data of Example 6.2 gave $\hat{\theta} = \bar{x}_{tr(10)} = 27.838$, but now there is no assumed $f(x; \theta)$, so how can we obtain a bootstrap sample? The answer is to regard the sample itself as constituting the population (the $n = 20$ observations in Example 6.2) and take B different samples, each of size n, *with* replacement from this population. Several of the books listed in the chapter bibliography provide more information about bootstrapping.

EXERCISES Section 6.1 (1–19)

1. The accompanying data on flexural strength (MPa) for concrete beams of a certain type was introduced in Example 1.2.

5.9	7.2	7.3	6.3	8.1	6.8	7.0
7.6	6.8	6.5	7.0	6.3	7.9	9.0
8.2	8.7	7.8	9.7	7.4	7.7	9.7
7.8	7.7	11.6	11.3	11.8	10.7	

   a. Calculate a point estimate of the mean value of strength for the conceptual population of all beams manufactured in this fashion, and state which estimator you used. [*Hint*: $\Sigma x_i = 219.8$.]
   b. Calculate a point estimate of the strength value that separates the weakest 50% of all such beams from the strongest 50%, and state which estimator you used.
   c. Calculate and interpret a point estimate of the population standard deviation σ. Which estimator did you use? [*Hint*: $\Sigma x_i^2 = 1860.94$.]
   d. Calculate a point estimate of the proportion of all such beams whose flexural strength exceeds 10 MPa. [*Hint*: Think of an observation as a "success" if it exceeds 10.]
   e. Calculate a point estimate of the population coefficient of variation σ/μ, and state which estimator you used.

2. The **National Health and Nutrition Examination Survey (NHANES)** collects demographic, socioeconomic, dietary, and health-related information on an annual basis. Here is a sample of 20 observations on HDL cholesterol level (mg/dl) obtained from the 2009–2010 survey (HDL is "good" cholesterol; the higher its value, the lower the risk for heart disease):

35	49	52	54	65	51	51
47	86	36	46	33	39	45
39	63	95	35	30	48	

   a. Calculate a point estimate of the population mean HDL cholesterol level.
   b. Making no assumptions about the shape of the population distribution, calculate a point estimate of the value that separates the largest 50% of HDL levels from the smallest 50%.

   c. Calculate a point estimate of the population standard deviation.
   d. An HDL level of at least 60 is considered desirable as it corresponds to a significantly lower risk of heart disease. Making no assumptions about the shape of the population distribution, estimate the proportion p of the population having an HDL level of at least 60.

3. Consider the following sample of observations on coating thickness for low-viscosity paint (**"Achieving a Target Value for a Manufacturing Process: A Case Study," *J. of Quality Technology*, 1992: 22–26**):

.83	.88	.88	1.04	1.09	1.12	1.29	1.31
1.48	1.49	1.59	1.62	1.65	1.71	1.76	1.83

   Assume that the distribution of coating thickness is normal (a normal probability plot strongly supports this assumption).

   a. Calculate a point estimate of the mean value of coating thickness, and state which estimator you used.
   b. Calculate a point estimate of the median of the coating thickness distribution, and state which estimator you used.
   c. Calculate a point estimate of the value that separates the largest 10% of all values in the thickness distribution from the remaining 90%, and state which estimator you used. [*Hint*: Express what you are trying to estimate in terms of μ and σ.]
   d. Estimate $P(X < 1.5)$, i.e., the proportion of all thickness values less than 1.5. [*Hint*: If you knew the values of μ and σ, you could calculate this probability. These values are not available, but they can be estimated.]
   e. What is the estimated standard error of the estimator that you used in part (b)?

4. The article from which the data in Exercise 1 was extracted also gave the accompanying strength observations for cylinders:

6.1	5.8	7.8	7.1	7.2	9.2	6.6	8.3	7.0	8.3
7.8	8.1	7.4	8.5	8.9	9.8	9.7	14.1	12.6	11.2

   Prior to obtaining data, denote the beam strengths by $X_1, \ldots, X_m$ and the cylinder strengths by $Y_1, \ldots, Y_n$. Suppose that the X_i's constitute a random sample from a distribution with mean μ_1 and standard deviation σ_1 and that the Y_i's form a random sample (independent of the X_i's) from another distribution with mean μ_2 and standard deviation σ_2.

a. Use rules of expected value to show that $\overline{X} - \overline{Y}$ is an unbiased estimator of $\mu_1 - \mu_2$. Calculate the estimate for the given data.

b. Use rules of variance from Chapter 5 to obtain an expression for the variance and standard deviation (standard error) of the estimator in part (a), and then compute the estimated standard error.

c. Calculate a point estimate of the ratio σ_1/σ_2 of the two standard deviations.

d. Suppose a single beam and a single cylinder are randomly selected. Calculate a point estimate of the variance of the difference $X - Y$ between beam strength and cylinder strength.

5. As an example of a situation in which several different statistics could reasonably be used to calculate a point estimate, consider a population of N invoices. Associated with each invoice is its "book value," the recorded amount of that invoice. Let T denote the total book value, a known amount. Some of these book values are erroneous. An audit will be carried out by randomly selecting n invoices and determining the audited (correct) value for each one. Suppose that the sample gives the following results (in dollars).

Invoice

	1	2	3	4	5
Book value	300	720	526	200	127
Audited value	300	520	526	200	157
Error	0	200	0	0	−30

Let

$\overline{Y}$ = sample mean book value

$\overline{X}$ = sample mean audited value

$\overline{D}$ = sample mean error

Propose three different statistics for estimating the total audited (i.e., correct) value—one involving just N and $\overline{X}$, another involving T, N, and $\overline{D}$, and the last involving T and $\overline{X}/\overline{Y}$. If $N = 5000$ and $T = 1,761,300$, calculate the three corresponding point estimates. (The article **"Statistical Models and Analysis in Auditing,"** *Statistical Science,* **1989: 2–33** discusses properties of these estimators.)

6. Urinary angiotensinogen (AGT) level is one quantitative indicator of kidney function. The article **"Urinary Angiotensinogen as a Potential Biomarker of Chronic Kidney Diseases"** (*J. of the Amer. Society of Hypertension,* **2008: 349–354**) describes a study in which urinary AGT level (μg) was determined for a sample of adults with chronic kidney disease. Here is representative data (consistent with summary quantities and descriptions in the cited article):

2.6	6.2	7.4	9.6	11.5	13.5	14.5	17.0
20.0	28.8	29.5	29.5	41.7	45.7	56.2	56.2
66.1	66.1	67.6	74.1	97.7	141.3	147.9	177.8
186.2	186.2	190.6	208.9	229.1	229.1	288.4	288.4
346.7	407.4	426.6	575.4	616.6	724.4	812.8	1122.0

An appropriate probability plot supports the use of the lognormal distribution (see Section 4.5) as a reasonable model for urinary AGT level (this is what the investigators did).

a. Estimate the parameters of the distribution. [*Hint*: Remember that X has a lognormal distribution with parameters μ and σ^2 if $\ln(X)$ is normally distributed with mean μ and variance σ^2.]

b. Use the estimates of part (a) to calculate an estimate of the expected value of AGT level. [*Hint*: What is $E(X)$?]

7. **a.** A random sample of 10 houses in a particular area, each of which is heated with natural gas, is selected and the amount of gas (therms) used during the month of January is determined for each house. The resulting observations are 103, 156, 118, 89, 125, 147, 122, 109, 138, 99. Let μ denote the average gas usage during January by all houses in this area. Compute a point estimate of μ.

b. Suppose there are 10,000 houses in this area that use natural gas for heating. Let τ denote the total amount of gas used by all of these houses during January. Estimate τ using the data of part (a). What estimator did you use in computing your estimate?

c. Use the data in part (a) to estimate p, the proportion of all houses that used at least 100 therms.

d. Give a point estimate of the population median usage (the middle value in the population of all houses) based on the sample of part (a). What estimator did you use?

8. In a random sample of 80 components of a certain type, 12 are found to be defective.

a. Give a point estimate of the proportion of all such components that are *not* defective.

b. A system is to be constructed by randomly selecting two of these components and connecting them in series, as shown here.

The series connection implies that the system will function if and only if neither component is defective (i.e., both components work properly). Estimate the proportion of all such systems that work properly. [*Hint*: If p denotes the probability that a component works properly, how can P(system works) be expressed in terms of p?]

9. Each of 150 newly manufactured items is examined and the number of scratches per item is recorded (the items

are supposed to be free of scratches), yielding the following data:

Number of scratches per item	0	1	2	3	4	5	6	7
Observed frequency	18	37	42	30	13	7	2	1

Let X = the number of scratches on a randomly chosen item, and assume that X has a Poisson distribution with parameter μ.

a. Find an unbiased estimator of μ and compute the estimate for the data. [*Hint:* $E(X) = \mu$ for X Poisson, so $E(\overline{X}) = $?]

b. What is the standard deviation (standard error) of your estimator? Compute the estimated standard error. [*Hint:* $\sigma_X^2 = \mu$ for X Poisson.]

10. Using a long rod that has length μ, you are going to lay out a square plot in which the length of each side is μ. Thus the area of the plot will be μ^2. However, you do not know the value of μ, so you decide to make n independent measurements $X_1, X_2, \ldots, X_n$ of the length. Assume that each X_i has mean μ (unbiased measurements) and variance σ^2.

a. Show that $\overline{X}^2$ is not an unbiased estimator for μ^2. [*Hint:* For any rv Y, $E(Y^2) = V(Y) + [E(Y)]^2$. Apply this with $Y = \overline{X}$.]

b. For what value of k is the estimator $\overline{X}^2 - kS^2$ unbiased for μ^2? [*Hint:* Compute $E(\overline{X}^2 - kS^2)$.]

11. Of n_1 randomly selected male smokers, X_1 smoked filter cigarettes, whereas of n_2 randomly selected female smokers, X_2 smoked filter cigarettes. Let p_1 and p_2 denote the probabilities that a randomly selected male and female, respectively, smoke filter cigarettes.

a. Show that $(X_1/n_1) - (X_2/n_2)$ is an unbiased estimator for $p_1 - p_2$. [*Hint:* $E(X_i) = n_i p_i$ for $i = 1, 2$.]

b. What is the standard error of the estimator in part (a)?

c. How would you use the observed values x_1 and x_2 to estimate the standard error of your estimator?

d. If $n_1 = n_2 = 200$, $x_1 = 127$, and $x_2 = 176$, use the estimator of part (a) to obtain an estimate of $p_1 - p_2$.

e. Use the result of part (c) and the data of part (d) to estimate the standard error of the estimator.

12. Suppose a certain type of fertilizer has an expected yield per acre of μ_1 with variance σ^2, whereas the expected yield for a second type of fertilizer is μ_2 with the same variance σ^2. Let S_1^2 and S_2^2 denote the sample variances of yields based on sample sizes n_1 and n_2, respectively, of the two fertilizers. Show that the pooled (combined) estimator

$$\hat{\sigma}^2 = \frac{(n_1 - 1)S_1^2 + (n_2 - 1)S_2^2}{n_1 + n_2 - 2}$$

is an unbiased estimator of σ^2.

13. Consider a random sample $X_1, \ldots, X_n$ from the pdf

$$f(x; \theta) = .5(1 + \theta x) \quad -1 \le x \le 1$$

where $-1 \le \theta \le 1$ (this distribution arises in particle physics). Show that $\hat{\theta} = 3\overline{X}$ is an unbiased estimator of θ. [*Hint:* First determine $\mu = E(X) = E(\overline{X})$.]

14. A sample of n captured Pandemonium jet fighters results in serial numbers $x_1, x_2, x_3, \ldots, x_n$. The CIA knows that the aircraft were numbered consecutively at the factory starting with α and ending with β, so that the total number of planes manufactured is $\beta - \alpha + 1$ (e.g., if $\alpha = 17$ and $\beta = 29$, then $29 - 17 + 1 = 13$ planes having serial numbers 17, 18, 19, ..., 28, 29 were manufactured). However, the CIA does not know the values of α or β. A CIA statistician suggests using the estimator $\max(X_i) - \min(X_i) + 1$ to estimate the total number of planes manufactured.

a. If $n = 5$, $x_1 = 237$, $x_2 = 375$, $x_3 = 202$, $x_4 = 525$, and $x_5 = 418$, what is the corresponding estimate?

b. Under what conditions on the sample will the value of the estimate be exactly equal to the true total number of planes? Will the estimate ever be larger than the true total? Do you think the estimator is unbiased for estimating $\beta - \alpha + 1$? Explain in one or two sentences.

15. Let $X_1, X_2, \ldots, X_n$ represent a random sample from a Rayleigh distribution with pdf

$$f(x; \theta) = \frac{x}{\theta} e^{-x^2/(2\theta)} \quad x > 0$$

a. It can be shown that $E(X^2) = 2\theta$. Use this fact to construct an unbiased estimator of θ based on ΣX_i^2 (and use rules of expected value to show that it is unbiased).

b. Estimate θ from the following $n = 10$ observations on vibratory stress of a turbine blade under specified conditions:

16.88	10.23	4.59	6.66	13.68
14.23	19.87	9.40	6.51	10.95

16. Suppose the true average growth μ of one type of plant during a 1-year period is identical to that of a second type, but the variance of growth for the first type is σ^2, whereas for the second type the variance is $4\sigma^2$. Let $X_1, \ldots, X_m$ be m independent growth observations on the first type [so $E(X_i) = \mu$, $V(X_i) = \sigma^2$], and let $Y_1, \ldots, Y_n$ be n independent growth observations on the second type [$E(Y_i) = \mu$, $V(Y_i) = 4\sigma^2$].

a. Show that the estimator $\hat{\mu} = \delta\overline{X} + (1 - \delta)\overline{Y}$ is unbiased for μ (for $0 < \delta < 1$, the estimator is a weighted average of the two individual sample means).

b. For fixed m and n, compute $V(\hat{\mu})$, and then find the value of δ that minimizes $V(\hat{\mu})$. [*Hint:* Differentiate $V(\hat{\mu})$ with respect to δ.]

17. In Chapter 3, we defined a negative binomial rv as the number of failures that occur before the rth success in a sequence of independent and identical success/failure trials. The probability mass function (pmf) of X is

$$nb(x; r, p) =$$

$$\binom{x + r - 1}{x} p^r (1 - p)^x \quad x = 0, 1, 2, \ldots$$

a. Suppose that $r \geq 2$. Show that

$$\hat{p} = (r - 1)/(X + r - 1)$$

is an unbiased estimator for p. [*Hint*: Write out $E(\hat{p})$ and cancel $x + r - 1$ inside the sum.]

b. A reporter wishing to interview five individuals who support a certain candidate begins asking people whether (S) or not (F) they support the candidate. If the sequence of responses is *SFFSFFFSSS*, estimate p = the true proportion who support the candidate.

18. Let $X_1, X_2, \ldots, X_n$ be a random sample from a pdf $f(x)$ that is symmetric about μ, so that $\tilde{X}$ is an unbiased estimator of μ. If n is large, it can be shown that $V(\tilde{X}) \approx 1/(4n[f(\mu)]^2)$.

a. Compare $V(\tilde{X})$ to $V(\overline{X})$ when the underlying distribution is normal.

b. When the underlying pdf is Cauchy (see Example 6.7), $V(\overline{X}) = \infty$, so $\overline{X}$ is a terrible estimator. What is $V(\tilde{X})$ in this case when n is large?

19. An investigator wishes to estimate the proportion of students at a certain university who have violated the honor code. Having obtained a random sample of n students, she realizes that asking each, "Have you violated the honor code?" will probably result in some untruthful responses. Consider the following scheme, called a **randomized response** technique. The investigator makes up a deck of 100 cards, of which 50 are of type I and 50 are of type II.

Type I: Have you violated the honor code (yes or no)?

Type II: Is the last digit of your telephone number a 0, 1, or 2 (yes or no)?

Each student in the random sample is asked to mix the deck, draw a card, and answer the resulting question truthfully. Because of the irrelevant question on type II cards, a yes response no longer stigmatizes the respondent, so we assume that responses are truthful. Let p denote the proportion of honor-code violators (i.e., the probability of a randomly selected student being a violator), and let $\lambda = P(\text{yes response})$. Then λ and p are related by $\lambda = .5p + (.5)(.3)$.

a. Let Y denote the number of yes responses, so $Y \sim \text{Bin}(n, \lambda)$. Thus Y/n is an unbiased estimator of λ. Derive an estimator for p based on Y. If $n = 80$ and $y = 20$, what is your estimate? [*Hint*: Solve $\lambda = .5p + .15$ for p and then substitute Y/n for λ.]

b. Use the fact that $E(Y/n) = \lambda$ to show that your estimator $\hat{p}$ is unbiased.

c. If there were 70 type I and 30 type II cards, what would be your estimator for p?

6.2 Methods of Point Estimation

We now introduce two "constructive" methods for obtaining point estimators: the method of moments and the method of maximum likelihood. By *constructive* we mean that the general definition of each type of estimator suggests explicitly how to obtain the estimator in any specific problem. Although maximum likelihood estimators are generally preferable to moment estimators because of certain efficiency properties, they often require significantly more computation than do moment estimators. It is sometimes the case that these methods yield unbiased estimators.

The Method of Moments

The basic idea of this method is to equate certain sample characteristics, such as the mean, to the corresponding population expected values. Then solving these equations for unknown parameter values yields the estimators.

DEFINITION

Let $X_1, \ldots, X_n$ be a random sample from a pmf or pdf $f(x)$. For $k = 1, 2, 3, \ldots$, the **kth population moment**, or **kth moment of the distribution $f(x)$**, is $E(X^k)$. The **kth sample moment** is $(1/n)\sum_{i=1}^{n} X_i^k$.

Thus the first population moment is $E(X) = \mu$, and the first sample moment is $\Sigma X_i/n = \overline{X}$. The second population and sample moments are $E(X^2)$ and $\Sigma X_i^2/n$, respectively. The population moments will be functions of any unknown parameters $\theta_1, \theta_2, \ldots$.

DEFINITION

> Let $X_1, X_2, \ldots, X_n$ be a random sample from a distribution with pmf or pdf $f(x; \theta_1, \ldots, \theta_m)$, where $\theta_1, \ldots, \theta_m$ are parameters whose values are unknown. Then the **moment estimators** $\hat{\theta}_1, \ldots, \hat{\theta}_m$ are obtained by equating the first m sample moments to the corresponding first m population moments and solving for $\theta_1, \ldots, \theta_m$.

If, for example, $m = 2$, $E(X)$ and $E(X^2)$ will be functions of θ_1 and θ_2. Setting $E(X) = (1/n)\Sigma X_i \, (= \overline{X})$ and $E(X^2) = (1/n)\Sigma X_i^2$ gives two equations in θ_1 and θ_2. The solution then defines the estimators.

EXAMPLE 6.12 Let $X_1, X_2, \ldots, X_n$ represent a random sample of service times of n customers at a certain facility, where the underlying distribution is assumed exponential with parameter λ. Since there is only one parameter to be estimated, the estimator is obtained by equating $E(X)$ to $\overline{X}$. Since $E(X) = 1/\lambda$ for an exponential distribution, this gives $1/\lambda = \overline{X}$ or $\lambda = 1/\overline{X}$. The moment estimator of λ is then $\hat{\lambda} = 1/\overline{X}$. ∎

EXAMPLE 6.13 Let $X_1, \ldots, X_n$ be a random sample from a gamma distribution with parameters α and β. From Section 4.4, $E(X) = \alpha\beta$ and $E(X^2) = \beta^2\Gamma(\alpha + 2)/\Gamma(\alpha) = \beta^2(\alpha + 1)\alpha$. The moment estimators of α and β are obtained by solving

$$\overline{X} = \alpha\beta \qquad \frac{1}{n}\sum X_i^2 = \alpha(\alpha + 1)\beta^2$$

Since $\alpha(\alpha + 1)\beta^2 = \alpha^2\beta^2 + \alpha\beta^2$ and the first equation implies $\alpha^2\beta^2 = \overline{X}^2$, the second equation becomes

$$\frac{1}{n}\sum X_i^2 = \overline{X}^2 + \alpha\beta^2$$

Now dividing each side of this second equation by the corresponding side of the first equation and substituting back gives the estimators

$$\hat{\alpha} = \frac{\overline{X}^2}{(1/n)\sum X_i^2 - \overline{X}^2} \qquad \hat{\beta} = \frac{(1/n)\sum X_i^2 - \overline{X}^2}{\overline{X}}$$

To illustrate, the survival-time data mentioned in Example 4.24 is

152	115	109	94	88	137	152	77	160	165
125	40	128	123	136	101	62	153	83	69

from which $\bar{x} = 113.5$ and $(1/20)\Sigma x_i^2 = 14{,}087.8$. The parameter estimates are

$$\hat{\alpha} = \frac{(113.5)^2}{14{,}087.8 - (113.5)^2} = 10.7 \qquad \hat{\beta} = \frac{14{,}087.8 - (113.5)^2}{113.5} = 10.6$$

These estimates of α and β differ from the values suggested by Gross and Clark because they used a different estimation technique. ∎

EXAMPLE 6.14 Let $X_1, \ldots, X_n$ be a random sample from a generalized negative binomial distribution with parameters r and p (see Section 3.5). Since $E(X) = r(1 - p)/p$ and $V(X) = r(1 - p)/p^2$, $E(X^2) = V(X) + [E(X)]^2 = r(1 - p)(r - rp + 1)/p^2$. Equating $E(X)$ to $\overline{X}$ and $E(X^2)$ to $(1/n)\sum X_i^2$ eventually gives

$$\hat{p} = \frac{\overline{X}}{(1/n)\sum X_i^2 - \overline{X}^2} \qquad \hat{r} = \frac{\overline{X}^2}{(1/n)\sum X_i^2 - \overline{X}^2 - \overline{X}}$$

As an illustration, Reep, Pollard, and Benjamin **("Skill and Chance in Ball Games," *J. of Royal Stat. Soc.*, 1971: 623–629)** consider the negative binomial distribution as a model for the number of goals per game scored by National Hockey League teams. The data for 1966–1967 follows (420 games):

Goals	0	1	2	3	4	5	6	7	8	9	10
Frequency	29	71	82	89	65	45	24	7	4	1	3

Then,

$$\overline{x} = \sum x_i / 420 = [(0)(29) + (1)(71) + \cdots + (10)(3)]/420 = 2.98$$

and

$$\sum x_i^2 / 420 = [(0)^2(29) + (1)^2(71) + \cdots + (10)^2(3)]/420 = 12.40$$

Thus,

$$\hat{p} = \frac{2.98}{12.40 - (2.98)^2} = .85 \qquad \hat{r} = \frac{(2.98)^2}{12.40 - (2.98)^2 - 2.98} = 16.5$$

Although r by definition must be positive, the denominator of $\hat{r}$ could be negative, indicating that the negative binomial distribution is not appropriate (or that the moment estimator is flawed). ∎

Maximum Likelihood Estimation

The method of maximum likelihood was first introduced by R. A. Fisher, a geneticist and statistician, in the 1920s. Most statisticians recommend this method, at least when the sample size is large, since the resulting estimators have certain desirable efficiency properties (see the proposition on page 271).

EXAMPLE 6.15 The best protection against hacking into an online account is to use a password that has at least 8 characters consisting of upper- and lowercase letters, numerals, and special characters. [*Note:* The Jan. 2012 issue of ***Consumer Reports*** reported that only 25% of individuals surveyed used a strong password.] Suppose that 10 individuals who have email accounts with a certain provider are selected, and it is found that the first, third, and tenth individuals have such strong protection, whereas the others do not. Let $p = P(\text{strong protection})$, i.e., p is the proportion of all such account holders having strong protection. Define (Bernoulli) random variables $X_1, X_2, \ldots, X_{10}$ by

$$X_1 = \begin{cases} 1 \text{ if 1st has strong protection} \\ 0 \text{ if 1st does not have strong protection} \end{cases} \cdots X_{10} = \begin{cases} 1 \text{ if 10th has strong protection} \\ 0 \text{ if 10th does not have strong protection} \end{cases}$$

Then for the obtained sample, $X_1 = X_3 = X_{10} = 1$ and the other seven X_i's are all zero. The probability mass function of any particular X_i is $p^{x_i}(1 - p)^{1 - x_i}$, which becomes p if $x_i = 1$ and $1 - p$ when $x_i = 0$. Now suppose that the conditions of various passwords are independent of one another. This implies that the X_i's are independent, so their joint probability mass function is the product of the individual pmf's. Thus the joint pmf evaluated at the observed X_i's is

$$f(x_1, \ldots, x_{10}; p) = p(1-p)p \cdots p = p^3(1-p)^7 \qquad (6.4)$$

Suppose that $p = .25$. Then the probability of observing the sample that we actually obtained is $(.25)^3(.75)^7 = .002086$. If instead $p = .50$, then this probability is $(.50)^3(.50)^7 = .000977$. For what value of p is the obtained sample most likely to have occurred? That is, for what value of p is the joint pmf (6.4) as large as it can be? What value of p maximizes (6.4)? Figure 6.6(a) shows a graph of the *likelihood* (6.4) as a function of p. It appears that the graph reaches its peak above $p = .3 = $ the proportion of flawed helmets in the sample. Figure 6.6(b) shows a graph of the natural logarithm of (6.4); since $\ln[g(u)]$ is a strictly increasing function of $g(u)$, finding u to maximize the function $g(u)$ is the same as finding u to maximize $\ln[g(u)]$.

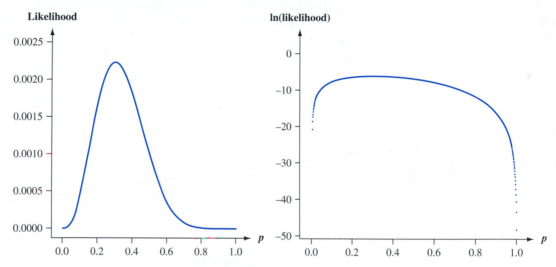

Figure 6.6 (a) Graph of the likelihood (joint pmf) (6.4) from Example 6.15 (b) Graph of the natural logarithm of the likelihood

We can verify our visual impression by using calculus to find the value of p that maximizes (6.4). Working with the natural log of the joint pmf is often easier than working with the joint pmf itself, since the joint pmf is typically a product so its logarithm will be a sum. Here

$$\ln[f(x_1, \ldots, x_{10}; p)] = \ln[p^3(1-p)^7] = 3\ln(p) + 7\ln(1-p) \qquad (6.5)$$

Thus

$$\frac{d}{dp}\{\ln[f(x_1, \ldots, x_{10}; p)]\} = \frac{d}{dp}\{3\ln(p) + 7\ln(1-p)\} = \frac{3}{p} + \frac{7}{1-p}(-1)$$

$$= \frac{3}{p} - \frac{7}{1-p}$$

[the (-1) comes from the chain rule in calculus]. Equating this derivative to 0 and solving for p gives $3(1-p) = 7p$, from which $3 = 10p$ and so $p = 3/10 = .30$ as conjectured. That is, our point estimate is $\hat{p} = .30$. It is called the *maximum likelihood estimate* because it is the parameter value that maximizes the likelihood (joint pmf) of the observed sample. In general, the second derivative should be examined to make sure a maximum has been obtained, but here this is obvious from Figure 6.5.

Suppose that rather than being told the condition of every password, we had only been informed that three of the ten were strong. Then we would have the observed value of a binomial random variable $X =$ the number with strong passwords. The pmf of X is $\binom{10}{x} p^x (1-p)^{10-x}$. For $x = 3$, this becomes $\binom{10}{3} p^3 (1-p)^7$. The binomial coefficient $\binom{10}{3}$ is irrelevant to the maximization, so again $\hat{p} = .30$. ∎

DEFINITION

> Let $X_1, X_2, \ldots, X_n$ have joint pmf or pdf
>
> $$f(x_1, x_2, \ldots, x_n; \theta_1, \ldots, \theta_m) \qquad (6.6)$$
>
> where the parameters $\theta_1, \ldots, \theta_m$ have unknown values. When $x_1, \ldots, x_n$ are the observed sample values and (6.6) is regarded as a function of $\theta_1, \ldots, \theta_m$, it is called the **likelihood function.** The maximum likelihood estimates (mle's) $\hat{\theta}_1, \ldots, \hat{\theta}_m$ are those values of the θ_i's that maximize the likelihood function, so that
>
> $$f(x_1, \ldots, x_n; \hat{\theta}_1, \ldots, \hat{\theta}_m) \geq f(x_1, \ldots, x_n; \theta_1, \ldots, \theta_m) \text{ for all } \theta_1, \ldots, \theta_m$$
>
> When the X_i's are substituted in place of the x_i's, the **maximum likelihood estimators** result.

The likelihood function tells us how likely the observed sample is as a function of the possible parameter values. Maximizing the likelihood gives the parameter values for which the observed sample is most likely to have been generated—that is, the parameter values that "agree most closely" with the observed data.

EXAMPLE 6.16 Suppose $X_1, X_2, \ldots, X_n$ is a random sample from an exponential distribution with parameter λ. Because of independence, the likelihood function is a product of the individual pdf's:

$$f(x_1, \ldots, x_n; \lambda) = (\lambda e^{-\lambda x_1}) \cdots (\lambda e^{-\lambda x_n}) = \lambda^n e^{-\lambda \Sigma x_i}$$

The natural logarithm of the likelihood function is

$$\ln[f(x_1, \ldots, x_n; \lambda)] = n \ln(\lambda) - \lambda \sum x_i$$

Equating $(d/d\lambda)[\ln(\text{likelihood})]$ to zero results in $n/\lambda - \Sigma x_i = 0$, or $\lambda = n/\Sigma x_i = 1/\bar{x}$. Thus the mle is $\hat{\lambda} = 1/\bar{X}$; it is identical to the method of moments estimator [but it is not an unbiased estimator, since $E(1/\bar{X}) \neq 1/E(\bar{X})$]. ∎

EXAMPLE 6.17 Let $X_1, \ldots, X_n$ be a random sample from a normal distribution. The likelihood function is

$$f(x_1, \ldots, x_n; \mu, \sigma^2) = \frac{1}{\sqrt{2\pi\sigma^2}} e^{-(x_1 - \mu)^2/(2\sigma^2)} \cdot \ldots \cdot \frac{1}{\sqrt{2\pi\sigma^2}} e^{-(x_n - \mu)^2/(2\sigma^2)}$$

$$= \left(\frac{1}{2\pi\sigma^2} \right)^{n/2} e^{-\Sigma(x_i - \mu)^2/(2\sigma^2)}$$

so

$$\ln[f(x_1, \ldots, x_n; \mu, \sigma^2)] = -\frac{n}{2} \ln(2\pi\sigma^2) - \frac{1}{2\sigma^2} \sum (x_i - \mu)^2$$

To find the maximizing values of μ and σ^2, we must take the partial derivatives of $\ln(f)$ with respect to μ and σ^2, equate them to zero, and solve the resulting two equations. Omitting the details, the resulting mle's are

$$\hat{\mu} = \overline{X} \qquad \hat{\sigma}^2 = \frac{\sum(X_i - \overline{X})^2}{n}$$

The mle of σ^2 is not the unbiased estimator, so two different principles of estimation (unbiasedness and maximum likelihood) yield two different estimators. ∎

EXAMPLE 6.18 In Chapter 3, we mentioned the use of the Poisson distribution for modeling the number of "events" that occur in a two-dimensional region. Assume that when the region R being sampled has area $a(R)$, the number X of events occurring in R has a Poisson distribution with parameter $\lambda a(R)$ (where λ is the expected number of events per unit area) and that nonoverlapping regions yield independent X's.

Suppose an ecologist selects n nonoverlapping regions $R_1, \ldots, R_n$ and counts the number of plants of a certain species found in each region. The joint pmf (likelihood) is then

$$p(x_1, \ldots, x_n; \lambda) = \frac{[\lambda \cdot a(R_1)]^{x_1} e^{-\lambda \cdot a(R_1)}}{x_1!} \cdot \ldots \cdot \frac{[\lambda \cdot a(R_n)]^{x_n} e^{-\lambda \cdot a(R_n)}}{x_n!}$$

$$= \frac{[a(R_1)]^{x_1} \cdot \cdots \cdot [a(R_n)]^{x_n} \cdot \lambda^{\sum x_i} \cdot e^{-\lambda \sum a(R_i)}}{x_1! \cdot \cdots \cdot x_n!}$$

The log likelihood is

$$\ln[p(x_1, \ldots, x_n; \lambda)] = \sum x_i \cdot \ln[a(R_i)] + \ln(\lambda) \cdot \sum x_i - \lambda \sum a(R_i) - \sum \ln(x_i!)$$

Taking $d/d\lambda \, [\ln(p)]$ and equating it to zero yields

$$\frac{\sum x_i}{\lambda} - \sum a(R_i) = 0$$

from which

$$\lambda = \frac{\sum x_i}{\sum a(R_i)}$$

The mle is then $\hat{\lambda} = \sum X_i / \sum a(R_i)$. This is intuitively reasonable because λ is the true density (plants per unit area), whereas $\hat{\lambda}$ is the sample density since $\sum a(R_i)$ is just the total area sampled. Because $E(X_i) = \lambda \cdot a(R_i)$, the estimator is unbiased.

Sometimes an alternative sampling procedure is used. Instead of fixing regions to be sampled, the ecologist will select n points in the entire region of interest and let $y_i =$ the distance from the ith point to the nearest plant. The cumulative distribution function (cdf) of $Y =$ distance to the nearest plant is

$$F_Y(y) = P(Y \leq y) = 1 - P(Y > y) = 1 - P\left(\begin{array}{c}\text{no plants in a} \\ \text{circle of radius } y\end{array}\right)$$

$$= 1 - \frac{e^{-\lambda \pi y^2}(\lambda \pi y^2)^0}{0!} = 1 - e^{-\lambda \cdot \pi y^2}$$

Taking the derivative of $F_Y(y)$ with respect to y yields

$$f_Y(y; \lambda) = \begin{cases} 2\pi\lambda y e^{-\lambda \pi y^2} & y \geq 0 \\ 0 & \text{otherwise} \end{cases}$$

If we now form the likelihood $f_Y(y_1; \lambda) \cdot \ldots \cdot f_Y(y_n; \lambda)$, differentiate ln(likelihood), and so on, the resulting mle is

$$\hat{\lambda} = \frac{n}{\pi \sum Y_i^2} = \frac{\text{number of plants observed}}{\text{total area sampled}}$$

which is also a sample density. It can be shown that in a sparse environment (small λ), the distance method is in a certain sense better, whereas in a dense environment the first sampling method is better. ∎

EXAMPLE 6.19 Let $X_1, \ldots, X_n$ be a random sample from a Weibull pdf

$$f(x; \alpha, \beta) = \begin{cases} \dfrac{\alpha}{\beta^{\alpha}} \cdot x^{\alpha-1} \cdot e^{-(x/\beta)^{\alpha}} & x \geq 0 \\ \\ 0 & \text{otherwise} \end{cases}$$

Writing the likelihood and ln(likelihood), then setting both $(\partial/\partial\alpha)[\ln(f)] = 0$ and $(\partial/\partial\beta)[\ln(f)] = 0$, yields the equations

$$\alpha = \left[\frac{\sum x_i^{\alpha} \cdot \ln(x_i)}{\sum x_i^{\alpha}} - \frac{\sum \ln(x_i)}{n} \right]^{-1} \qquad \beta = \left(\frac{\sum x_i^{\alpha}}{n} \right)^{1/\alpha}$$

These two equations cannot be solved explicitly to give general formulas for the mle's $\hat{\alpha}$ and $\hat{\beta}$. Instead, for each sample $x_1, \ldots, x_n$, the equations must be solved using an iterative numerical procedure. The R, SAS and Minitab software packages can be used for this purpose. Even moment estimators of α and β are somewhat complicated (see Exercise 21). ∎

Estimating Functions of Parameters

Once the mle for a parameter θ is available, the mle for any function of θ, such as $1/\theta$ or $\sqrt{\theta}$, is easily obtained.

PROPOSITION

The Invariance Principle

Let $\hat{\theta}_1, \hat{\theta}_2, \ldots, \hat{\theta}_m$ be the mle's of the parameters $\theta_1, \theta_2, \ldots, \theta_m$. Then the mle of any function $h(\theta_1, \theta_2, \ldots, \theta_m)$ of these parameters is the function $h(\hat{\theta}_1, \hat{\theta}_2, \ldots, \hat{\theta}_m)$ of the mle's.

EXAMPLE 6.20
(Example 6.17 continued)

In the normal case, the mle's of μ and σ^2 are $\hat{\mu} = \overline{X}$ and $\hat{\sigma}^2 = \Sigma(X_i - \overline{X})^2/n$. To obtain the mle of the function $h(\mu, \sigma^2) = \sqrt{\sigma^2} = \sigma$, substitute the mle's into the function:

$$\hat{\sigma} = \sqrt{\hat{\sigma}^2} = \left[\frac{1}{n} \sum (X_i - \overline{X})^2 \right]^{1/2}$$

The mle of σ is not the sample standard deviation S, though they are close unless n is quite small. ∎

EXAMPLE 6.21
(Example 6.19 continued)

The mean value of an rv X that has a Weibull distribution is

$$\mu = \beta \cdot \Gamma(1 + 1/\alpha)$$

The mle of μ is therefore $\hat{\mu} = \hat{\beta}\Gamma(1 + 1/\hat{\alpha})$, where $\hat{\alpha}$ and $\hat{\beta}$ are the mle's of α and β. In particular, $\overline{X}$ is not the mle of μ, though it is an unbiased estimator. At least for large n, $\hat{\mu}$ is a better estimator than $\overline{X}$.

For the data given in Example 6.3, the mle's of the Weibull parameters are $\hat{\alpha} = 11.9731$ and $\hat{\beta} = 77.0153$, from which $\hat{\mu} = 73.80$. This estimate is quite close to the sample mean 73.88. ∎

Large Sample Behavior of the MLE

Although the principle of maximum likelihood estimation has considerable intuitive appeal, the following proposition provides additional rationale for the use of mle's.

PROPOSITION

> Under very general conditions on the joint distribution of the sample, when the sample size n is large, the maximum likelihood estimator of any parameter θ is at least approximately unbiased $[E(\hat{\theta}) \approx \theta]$ and has variance that is either as small as or nearly as small as can be achieved by any estimator. Stated another way, the mle $\hat{\theta}$ is either exactly or at least approximately the MVUE of θ.

Because of this result and the fact that calculus-based techniques can usually be used to derive the mle's (though often numerical methods, such as Newton's method, are necessary), maximum likelihood estimation is the most widely used estimation technique among statisticians. Many of the estimators used in the remainder of the book are mle's. Obtaining an mle, however, does require that the underlying distribution be specified.

Some Complications

Sometimes calculus cannot be used to obtain mle's.

EXAMPLE 6.22   Suppose waiting time for a bus is uniformly distributed on $[0, \theta]$ and the results $x_1, \ldots, x_n$ of a random sample from this distribution have been observed. Since $f(x; \theta) = 1/\theta$ for $0 \le x \le \theta$ and 0 otherwise,

$$f(x_1, \ldots, x_n; \theta) = \begin{cases} \dfrac{1}{\theta^n} & 0 \le x_1 \le \theta, \ldots, 0 \le x_n \le \theta \\ 0 & \text{otherwise} \end{cases}$$

As long as $\max(x_i) \le \theta$, the likelihood is $1/\theta^n$, which is positive, but as soon as $\theta < \max(x_i)$, the likelihood drops to 0. This is illustrated in Figure 6.7. Calculus will not work because the maximum of the likelihood occurs at a point of discontinuity, but the figure shows that $\hat{\theta} = \max(X_i)$. Thus if my waiting times are 2.3, 3.7, 1.5, .4, and 3.2, then the mle is $\hat{\theta} = 3.7$. From Example 6.4, the mle is not unbiased.

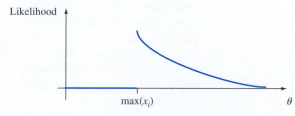

Figure 6.7   The likelihood function for Example 6.22   ∎

EXAMPLE 6.23 A method that is often used to estimate the size of a wildlife population involves performing a capture/recapture experiment. In this experiment, an initial sample of M animals is captured, each of these animals is tagged, and the animals are then returned to the population. After allowing enough time for the tagged individuals to mix into the population, another sample of size n is captured. With $X =$ the number of tagged animals in the second sample, the objective is to use the observed x to estimate the population size N.

The parameter of interest is $\theta = N$, which can assume only integer values, so even after determining the likelihood function (pmf of X here), using calculus to obtain N would present difficulties. If we think of a success as a previously tagged animal being recaptured, then sampling is without replacement from a population containing M successes and $N - M$ failures, so that X is a hypergeometric rv and the likelihood function is

$$p(x; N) = h(x; n, M, N) = \frac{\binom{M}{x} \cdot \binom{N - M}{n - x}}{\binom{N}{n}}$$

The integer-valued nature of N notwithstanding, it would be difficult to take the derivative of $p(x; N)$. However, if we consider the ratio of $p(x; N)$ to $p(x; N - 1)$, we have

$$\frac{p(x; N)}{p(x; N - 1)} = \frac{(N - M) \cdot (N - n)}{N(N - M - n + x)}$$

This ratio is larger than 1 if and only if (iff) $N < Mn/x$. The value of N for which $p(x; N)$ is maximized is therefore the largest integer less than Mn/x. If we use standard mathematical notation $[r]$ for the largest integer less than or equal to r, the mle of N is $\hat{N} = [Mn/x]$. As an illustration, if $M = 200$ fish are taken from a lake and tagged, and subsequently $n = 100$ fish are recaptured, and among the 100 there are $x = 11$ tagged fish, then $\hat{N} = [(200)(100)/11] = [1818.18] = 1818$. The estimate is actually rather intuitive; x/n is the proportion of the recaptured sample that is tagged, whereas M/N is the proportion of the entire population that is tagged. The estimate is obtained by equating these two proportions (estimating a population proportion by a sample proportion). ■

Suppose $X_1, X_2, \ldots, X_n$ is a random sample from a pdf $f(x; \theta)$ that is symmetric about θ but that the investigator is unsure of the form of the f function. It is then desirable to use an estimator $\hat{\theta}$ that is *robust*—that is, one that performs well for a wide variety of underlying pdf's. One such estimator is a trimmed mean. In recent years, statisticians have proposed another type of estimator, called an *M-estimator,* based on a generalization of maximum likelihood estimation. Instead of maximizing the log likelihood $\Sigma \ln[f(x; \theta)]$ for a specified f, one maximizes $\Sigma \rho(x_i; \theta)$. The "objective function" ρ is selected to yield an estimator with good robustness properties. The book by David Hoaglin et al. (see the bibliography) contains a good exposition of this topic.

EXERCISES Section 6.2 (20–30)

20. A diagnostic test for a certain disease is applied to n individuals known to not have the disease. Let $X =$ the number among the n test results that are positive (indicating presence of the disease, so X is the number of false positives) and $p =$ the probability that a disease-free individual's test result is positive (i.e., p is the true proportion of test results from disease-free individuals that are positive). Assume that only X is available rather than the actual sequence of test results.
    a. Derive the maximum likelihood estimator of p. If $n = 20$ and $x = 3$, what is the estimate?
    b. Is the estimator of part (a) unbiased?
    c. If $n = 20$ and $x = 3$, what is the mle of the probability $(1 - p)^5$ that none of the next five tests done on disease-free individuals are positive?

21. Let X have a Weibull distribution with parameters α and β, so

    $$E(X) = \beta \cdot \Gamma(1 + 1/\alpha)$$
    $$V(X) = \beta^2\{\Gamma(1 + 2/\alpha) - [\Gamma(1 + 1/\alpha)]^2\}$$

    a. Based on a random sample $X_1,\ldots, X_n$, write equations for the method of moments estimators of β and α. Show that, once the estimate of α has been obtained, the estimate of β can be found from a table of the gamma function and that the estimate of α is the solution to a complicated equation involving the gamma function.
    b. If $n = 20$, $\bar{x} = 28.0$, and $\Sigma x_i^2 = 16,500$, compute the estimates. [*Hint*: $[\Gamma(1.2)]^2/\Gamma(1.4) = .95$.]

22. Let X denote the proportion of allotted time that a randomly selected student spends working on a certain aptitude test. Suppose the pdf of X is

    $$f(x; \theta) = \begin{cases} (\theta + 1)x^\theta & 0 \le x \le 1 \\ 0 & \text{otherwise} \end{cases}$$

    where $-1 < \theta$. A random sample of ten students yields data $x_1 = .92$, $x_2 = .79$, $x_3 = .90$, $x_4 = .65$, $x_5 = .86$, $x_6 = .47$, $x_7 = .73$, $x_8 = .97$, $x_9 = .94$, $x_{10} = .77$.
    a. Use the method of moments to obtain an estimator of θ, and then compute the estimate for this data.
    b. Obtain the maximum likelihood estimator of θ, and then compute the estimate for the given data.

23. Let X represent the error in making a measurement of a physical characteristic or property (e.g., the boiling point of a particular liquid). It is often reasonable to assume that $E(X) = 0$ and that X has a normal distribution. Thus the pdf of any particular measurement error is

    $$f(x; \theta) = \frac{1}{\sqrt{2\pi\theta}} e^{-x^2/2\theta} \quad -\infty < x < \infty$$

(where we have used θ in place of σ^2). Now suppose that n independent measurements are made, resulting in measurement errors $X_1 = x_1, X_2 = x_2, \ldots, X_n = x_n$. Obtain the mle of θ.

24. A vehicle with a particular defect in its emission control system is taken to a succession of randomly selected mechanics until $r = 3$ of them have correctly diagnosed the problem. Suppose that this requires diagnoses by 20 different mechanics (so there were 17 incorrect diagnoses). Let $p = P(\text{correct diagnosis})$, so p is the proportion of all mechanics who would correctly diagnose the problem. What is the mle of p? Is it the same as the mle if a random sample of 20 mechanics results in 3 correct diagnoses? Explain. How does the mle compare to the estimate resulting from the use of the unbiased estimator given in Exercise 17?

25. The shear strength of each of ten test spot welds is determined, yielding the following data (psi):

    392 376 401 367 389 362 409 415 358 375
    a. Assuming that shear strength is normally distributed, estimate the true average shear strength and standard deviation of shear strength using the method of maximum likelihood.
    b. Again assuming a normal distribution, estimate the strength value below which 95% of all welds will have their strengths. [*Hint*: What is the 95th percentile in terms of μ and σ? Now use the invariance principle.]
    c. Suppose we decide to examine another test spot weld. Let $X =$ shear strength of the weld. Use the given data to obtain the mle of $P(X \le 400)$. [*Hint*: $P(X \le 400) = \Phi((400 - \mu)/\sigma)$.]

26. Consider randomly selecting n segments of pipe and determining the corrosion loss (mm) in the wall thickness for each one. Denote these corrosion losses by $Y_1, \ldots, Y_n$. The article **"A Probabilistic Model for a Gas Explosion Due to Leakages in the Grey Cast Iron Gas Mains"** (*Reliability Engr. and System Safety* (2013:270–279)) proposes a linear corrosion model: $Y_i = t_i R$, where t_i is the age of the pipe and R, the corrosion rate, is exponentially distributed with parameter λ. Obtain the maximum likelihood estimator of the exponential parameter (the resulting mle appears in the cited article). [*Hint*: If $c > 0$ and X has an exponential distribution, so does cX.]

27. Let $X_1, \ldots, X_n$ be a random sample from a gamma distribution with parameters α and β.
    a. Derive the equations whose solutions yield the maximum likelihood estimators of α and β. Do you think they can be solved explicitly?
    b. Show that the mle of $\mu = \alpha\beta$ is $\hat{\mu} = \bar{X}$.

28. Let $X_1, X_2,\ldots, X_n$ represent a random sample from the Rayleigh distribution with density function given in Exercise 15. Determine

a. The maximum likelihood estimator of θ, and then calculate the estimate for the vibratory stress data given in that exercise. Is this estimator the same as the unbiased estimator suggested in Exercise 15?

b. The mle of the median of the vibratory stress distribution. [*Hint*: First express the median in terms of θ.]

29. Consider a random sample $X_1, X_2,\ldots, X_n$ from the shifted exponential pdf

$$f(x; \lambda, \theta) = \begin{cases} \lambda e^{-\lambda(x-\theta)} & x \geq \theta \\ 0 & \text{otherwise} \end{cases}$$

Taking $\theta = 0$ gives the pdf of the exponential distribution considered previously (with positive density to the right of zero). An example of the shifted exponential distribution appeared in Example 4.5, in which the variable

of interest was time headway in traffic flow and $\theta = .5$ was the minimum possible time headway.

a. Obtain the maximum likelihood estimators of θ and λ.

b. If $n = 10$ time headway observations are made, resulting in the values 3.11, .64, 2.55, 2.20, 5.44, 3.42, 10.39, 8.93, 17.82, and 1.30, calculate the estimates of θ and λ.

30. At time $t = 0$, 20 identical components are tested. The lifetime distribution of each is exponential with parameter λ. The experimenter then leaves the test facility unmonitored. On his return 24 hours later, the experimenter immediately terminates the test after noticing that $y = 15$ of the 20 components are still in operation (so 5 have failed). Derive the mle of λ. [*Hint*: Let $Y =$ the number that survive 24 hours. Then $Y \sim \text{Bin}(n, p)$. What is the mle of p? Now notice that $p = P(X_i \geq 24)$, where X_i is exponentially distributed. This relates λ to p, so the former can be estimated once the latter has been.]

SUPPLEMENTARY EXERCISES (31–38)

31. An estimator $\hat{\theta}$ is said to be **consistent** if for any $\epsilon > 0$, $P(|\hat{\theta} - \theta| \geq \epsilon) \to 0$ as $n \to \infty$. That is, $\hat{\theta}$ is consistent if, as the sample size gets larger, it is less and less likely that $\hat{\theta}$ will be further than ϵ from the true value of θ. Show that $\overline{X}$ is a consistent estimator of μ when $\sigma^2 < \infty$ by using Chebyshev's inequality from Exercise 44 of Chapter 3. [*Hint*: The inequality can be rewritten in the form

$$P(|Y - \mu_Y| \geq \epsilon) \leq \sigma_Y^2/\epsilon$$

Now identify Y with $\overline{X}$.]

32. a. Let $X_1,\ldots, X_n$ be a random sample from a uniform distribution on $[0, \theta]$. Then the mle of θ is $\hat{\theta} = Y = \max(X_i)$. Use the fact that $Y \leq y$ iff each $X_i \leq y$ to derive the cdf of Y. Then show that the pdf of $Y = \max(X_i)$ is

$$f_Y(y) = \begin{cases} \dfrac{ny^{n-1}}{\theta^n} & 0 \leq y \leq \theta \\ 0 & \text{otherwise} \end{cases}$$

b. Use the result of part (a) to show that the mle is biased but that $(n + 1)\max(X_i)/n$ is unbiased.

33. At time $t = 0$, there is one individual alive in a certain population. A **pure birth process** then unfolds as follows. The time until the first birth is exponentially distributed with parameter λ. After the first birth, there are two individuals alive. The time until the first gives birth again is exponential with parameter λ, and similarly for the second individual. Therefore, the time until the next birth is the minimum of two exponential (λ) variables, which is

exponential with parameter 2λ. Similarly, once the second birth has occurred, there are three individuals alive, so the time until the next birth is an exponential rv with parameter 3λ, and so on (the memoryless property of the exponential distribution is being used here). Suppose the process is observed until the sixth birth has occurred and the successive birth times are 25.2, 41.7, 51.2, 55.5, 59.5, 61.8 (from which you should calculate the times between successive births). Derive the mle of λ. [*Hint*: The likelihood is a product of exponential terms.]

34. The **mean squared error** of an estimator $\hat{\theta}$ is MSE $(\hat{\theta}) = E(\hat{\theta} - \theta)^2$. If $\hat{\theta}$ is unbiased, then MSE $(\hat{\theta}) = V(\hat{\theta})$, but in general MSE $(\hat{\theta}) = V(\hat{\theta}) + (\text{bias})^2$. Consider the estimator $\hat{\sigma}^2 = KS^2$, where $S^2 =$ sample variance. What value of K minimizes the mean squared error of this estimator when the population distribution is normal? [*Hint*: It can be shown that

$$E[(S^2)^2] = (n + 1)\sigma^4/(n - 1)$$

In general, it is difficult to find $\hat{\theta}$ to minimize MSE $(\hat{\theta})$, which is why we look only at unbiased estimators and minimize $V(\hat{\theta})$.]

35. Let $X_1,\ldots,X_n$ be a random sample from a pdf that is symmetric about μ. An estimator for μ that has been found to perform well for a variety of underlying distributions is the *Hodges–Lehmann estimator*. To define it, first compute for each $i \leq j$ and each $j = 1, 2, \ldots, n$ the pairwise average $\overline{X}_{i,j} = (X_i + X_j)/2$. Then the estimator is $\hat{\mu} =$ the median of the $\overline{X}_{i,j}$'s. Compute the value of this estimate using the data of Exercise 44 of Chapter 1. [*Hint*: Construct a square table with the x_i's listed on the left margin and on top. Then compute averages on and above the diagonal.]

36. When the population distribution is normal, the statistic median $\{|X_1 - \tilde{X}|, \ldots, |X_n - \tilde{X}|\}/.6745$ can be used to estimate σ. This estimator is more resistant to the effects of outliers (observations far from the bulk of the data) than is the sample standard deviation. Compute both the corresponding point estimate and s for the data of Example 6.2.

37. When the sample standard deviation S is based on a random sample from a normal population distribution, it can be shown that

$$E(S) = \sqrt{2/(n-1)}\Gamma(n/2)\sigma/\Gamma((n-1)/2)$$

Use this to obtain an unbiased estimator for σ of the form cS. What is c when $n = 20$?

38. Each of n specimens is to be weighed twice on the same scale. Let X_i and Y_i denote the two observed weights for the ith specimen. Suppose X_i and Y_i are independent of one another, each normally distributed with mean value μ_i (the true weight of specimen i) and variance σ^2.

  a. Show that the maximum likelihood estimator of σ^2 is $\hat{\sigma}^2 = \Sigma(X_i - Y_i)^2/(4n)$. [*Hint*: If $\bar{z} = (z_1 + z_2)/2$, then $\Sigma(z_i - \bar{z})^2 = (z_1 - z_2)^2/2$.]

  b. Is the mle $\hat{\sigma}^2$ an unbiased estimator of σ^2? Find an unbiased estimator of σ^2. [*Hint*: For any rv Z, $E(Z^2) = V(Z) + [E(Z)]^2$. Apply this to $Z = X_i - Y_i$.]

BIBLIOGRAPHY

DeGroot, Morris, and Mark Schervish, *Probability and Statistics* (4th ed.), Addison-Wesley, Boston, MA, 2012. Includes an excellent discussion of both general properties and methods of point estimation; of particular interest are examples showing how general principles and methods can yield unsatisfactory estimators in particular situations.

Devore, Jay, and Kenneth Berk, *Modern Mathematical Statistics with Applications,* (2nd ed.), Springer, New York, 2012. The exposition is a bit more comprehensive and sophisticated than that of the current book.

Efron, Bradley, and Robert Tibshirani, *An Introduction to the Bootstrap,* Chapman and Hall, New York, 1993. The bible of the bootstrap.

Hoaglin, David, Frederick Mosteller, and John Tukey, *Understanding Robust and Exploratory Data Analysis,* Wiley, New York, 1983. Contains several good chapters on robust point estimation, including one on *M*-estimation.

Rice, John, *Mathematical Statistics and Data Analysis* (3rd ed.), Thomson-Brooks/Cole, Belmont, CA, 2007. A nice blending of statistical theory and data.

7 Statistical Intervals Based on a Single Sample

INTRODUCTION

A point estimate, because it is a single number, by itself provides no information about the precision and reliability of estimation. Consider, for example, using the statistic $\overline{X}$ to calculate a point estimate for the true average breaking strength (g) of paper towels of a certain brand, and suppose that $\bar{x} = 9322.7$. Because of sampling variability, it is virtually never the case that $\bar{x} = \mu$. The point estimate says nothing about how close it might be to μ. An alternative to reporting a single sensible value for the parameter being estimated is to calculate and report an entire interval of plausible values—an *interval estimate or confidence interval* (CI). A confidence interval is always calculated by first selecting a *confidence level*, which is a measure of the degree of reliability of the interval. A confidence interval with a 95% confidence level for the true average breaking strength might have a lower limit of 9162.5 and an upper limit of 9482.9. Then at the 95% confidence level, any value of μ between 9162.5 and 9482.9 is plausible. A confidence level of 95% implies that 95% of all samples would give an interval that includes μ, or whatever other parameter is being estimated, and only 5% of all samples would yield an erroneous interval. The most frequently used confidence levels are 95%, 99%, and 90%. The higher the confidence level, the more strongly we believe that the value of the parameter being estimated lies within the interval (an interpretation of any particular confidence level will be given shortly).

Information about the precision of an interval estimate is conveyed by the width of the interval. If the confidence level is high and the resulting interval is quite narrow, our knowledge of the value of the parameter is reasonably precise. A very wide confidence interval, however, gives the message that there is a great deal of uncertainty concerning the value of what we are estimating. Figure 7.1 shows 95% confidence intervals for true average breaking strengths of two different brands of paper towels. One of these intervals suggests precise knowledge about μ, whereas the other suggests a very wide range of plausible values.

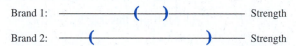

Figure 7.1 CIs indicating precise (brand 1) and imprecise (brand 2) information about μ

7.1 Basic Properties of Confidence Intervals

The basic concepts and properties of confidence intervals (CIs) are most easily introduced by first focusing on a simple, albeit somewhat unrealistic, problem situation. Suppose that the parameter of interest is a population mean μ and that

1. The population distribution is normal
2. The value of the population standard deviation σ is known

Normality of the population distribution is often a reasonable assumption. However, if the value of μ is unknown, it is typically implausible that the value of σ would be available (knowledge of a population's center typically precedes information concerning spread). We'll develop methods based on less restrictive assumptions in Sections 7.2 and 7.3.

EXAMPLE 7.1 Industrial engineers who specialize in ergonomics are concerned with designing workspace and worker-operated devices so as to achieve high productivity and comfort. The article **"Studies on Ergonomically Designed Alphanumeric Keyboards"** (***Human Factors,*** **1985: 175–187**) reports on a study of preferred height for an experimental keyboard with large forearm–wrist support. A sample of $n = 31$ trained typists was selected, and the preferred keyboard height was determined for each typist. The resulting sample average preferred height was $\bar{x} = 80.0$ cm. Assuming that the preferred height is normally distributed with $\sigma = 2.0$ cm (a value suggested by data in the article), obtain a confidence interval (interval of plausible values) for μ, the true average preferred height for the population of all experienced typists. ∎

The actual sample observations $x_1, x_2, \ldots, x_n$ are assumed to be the result of a random sample $X_1, \ldots, X_n$ from a normal distribution with mean value μ and standard deviation σ. The results described in Chapter 5 then imply that, irrespective of the sample size n, the sample mean $\bar{X}$ is normally distributed with expected value μ and standard deviation $\sigma/\sqrt{n}$. Standardizing $\bar{X}$ by first subtracting its expected value and then dividing by its standard deviation yields the standard normal variable

$$Z = \frac{\bar{X} - \mu}{\sigma/\sqrt{n}} \tag{7.1}$$

Because the area under the standard normal curve between -1.96 and 1.96 is $.95$,

$$P\left(-1.96 < \frac{\overline{X} - \mu}{\sigma/\sqrt{n}} < 1.96\right) = .95 \tag{7.2}$$

Now let's manipulate the inequalities inside the parentheses in (7.2) so that they appear in the equivalent form $l < \mu < u$, where the endpoints l and u involve $\overline{X}$ and $\sigma/\sqrt{n}$. This is achieved through the following sequence of operations, each yielding inequalities equivalent to the original ones.

1. Multiply through by $\sigma/\sqrt{n}$:

$$-1.96 \cdot \frac{\sigma}{\sqrt{n}} < \overline{X} - \mu < 1.96 \cdot \frac{\sigma}{\sqrt{n}}$$

2. Subtract $\overline{X}$ from each term:

$$-\overline{X} - 1.96 \cdot \frac{\sigma}{\sqrt{n}} < -\mu < -\overline{X} + 1.96 \cdot \frac{\sigma}{\sqrt{n}}$$

3. Multiply through by -1 to eliminate the minus sign in front of μ (which reverses the direction of each inequality):

$$\overline{X} + 1.96 \cdot \frac{\sigma}{\sqrt{n}} > \mu > \overline{X} - 1.96 \cdot \frac{\sigma}{\sqrt{n}}$$

that is,

$$\overline{X} - 1.96 \cdot \frac{\sigma}{\sqrt{n}} < \mu < \overline{X} + 1.96 \cdot \frac{\sigma}{\sqrt{n}}$$

The equivalence of each set of inequalities to the original set implies that

$$P\left(\overline{X} - 1.96 \frac{\sigma}{\sqrt{n}} < \mu < \overline{X} + 1.96 \frac{\sigma}{\sqrt{n}}\right) = .95 \tag{7.3}$$

The event inside the parentheses in (7.3) has a somewhat unfamiliar appearance; previously, the random quantity has appeared in the middle with constants on both ends, as in $a \le Y \le b$. In (7.3) the random quantity appears on the two ends, whereas the unknown constant μ appears in the middle. To interpret (7.3), think of a **random interval** having left endpoint $\overline{X} - 1.96 \cdot \sigma/\sqrt{n}$ and right endpoint $\overline{X} + 1.96 \cdot \sigma/\sqrt{n}$. In interval notation, this becomes

$$\left(\overline{X} - 1.96 \cdot \frac{\sigma}{\sqrt{n}}, \quad \overline{X} + 1.96 \cdot \frac{\sigma}{\sqrt{n}}\right) \tag{7.4}$$

The interval (7.4) is random because the two endpoints of the interval involve a random variable. It is centered at the sample mean $\overline{X}$ and extends $1.96\sigma/\sqrt{n}$ to each side of $\overline{X}$. Thus the interval's width is $2 \cdot (1.96) \cdot \sigma/\sqrt{n}$, a fixed number; only the location of the interval (its midpoint $\overline{X}$) is random (Figure 7.2). Now (7.3) can be paraphrased as "*the probability is .95 that the random interval (7.4) includes or covers the true value of* μ." Before any data is gathered, it is quite likely that μ will lie inside the interval (7.4).

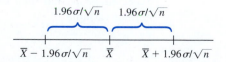

Figure 7.2 The random interval (7.4) centered at $\overline{X}$

DEFINITION

> If, after observing $X_1 = x_1, X_2 = x_2, \ldots, X_n = x_n$, we compute the observed sample mean $\bar{x}$ and then substitute $\bar{x}$ into (7.4) in place of $\overline{X}$, the resulting fixed interval is called a __95% confidence interval for μ__. This CI can be expressed either as
>
> $$\left(\bar{x} - 1.96 \cdot \frac{\sigma}{\sqrt{n}}, \bar{x} + 1.96 \cdot \frac{\sigma}{\sqrt{n}} \right) \text{ is a 95% CI for } \mu$$
>
> or as
>
> $$\bar{x} - 1.96 \cdot \frac{\sigma}{\sqrt{n}} < \mu < \bar{x} + 1.96 \cdot \frac{\sigma}{\sqrt{n}} \text{ with 95% confidence}$$
>
> A concise expression for the interval is $\bar{x} \pm 1.96 \cdot \sigma/\sqrt{n}$, where $-$ gives the left endpoint (lower limit) and $+$ gives the right endpoint (upper limit).

EXAMPLE 7.2
(Example 7.1
continued)

The quantities needed for computation of the 95% CI for true average preferred height are $\sigma = 2.0$, $n = 31$, and $\bar{x} = 80.0$. The resulting interval is

$$\bar{x} \pm 1.96 \cdot \frac{\sigma}{\sqrt{n}} = 80.0 \pm (1.96)\frac{2.0}{\sqrt{31}} = 80.0 \pm .7 = (79.3, 80.7)$$

That is, we can be highly confident, at the 95% confidence level, that $79.3 < \mu < 80.7$. This interval is relatively narrow, indicating that μ has been rather precisely estimated. ■

Interpreting a Confidence Level

The confidence level 95% for the interval just defined was inherited from the probability .95 for the random interval (7.4). Intervals having other levels of confidence will be introduced shortly. For now, though, consider how 95% confidence can be interpreted.

We started with an event whose probability was .95—that the random interval (7.4) would capture the true value of μ—and then used the data in Example 7.1 to compute the CI (79.3, 80.7). It is therefore tempting to conclude that μ is within this fixed interval with probability .95. But by substituting $\bar{x} = 80.0$ for $\overline{X}$, all randomness disappears; the interval (79.3, 80.7) is not a random interval, and μ is a constant (unfortunately unknown to us). Thus it is *incorrect* to write the statement $P(\mu$ lies in $(79.3, 80.7)) = .95$.

A correct interpretation of "95% confidence" relies on the long-run relative frequency interpretation of probability: To say that an event A has probability .95 is to say that if the experiment on which A is defined is performed over and over again, in the long run A will occur 95% of the time. Suppose we obtain another sample of typists' preferred heights and compute another 95% interval. Now consider repeating this for a third sample, a fourth sample, a fifth sample, and so on. Let A be the event that $\overline{X} - 1.96 \cdot \sigma/\sqrt{n} < \mu < \overline{X} + 1.96 \cdot \sigma/\sqrt{n}$. Since $P(A) = .95$, in the long run 95% of our computed CIs will contain μ. This is illustrated in Figure 7.3, where the vertical line cuts the measurement axis at the true (but unknown) value of μ. Notice that 7 of the 100 intervals shown fail to contain μ. In the long run, only 5% of the intervals so constructed would fail to contain μ.

According to this interpretation, the confidence level 95% is not so much a statement about any particular interval such as (79.3, 80.7). Instead it pertains to what would happen if a very large number of like intervals were to be constructed using the same CI formula. Although this may seem unsatisfactory, the root of the

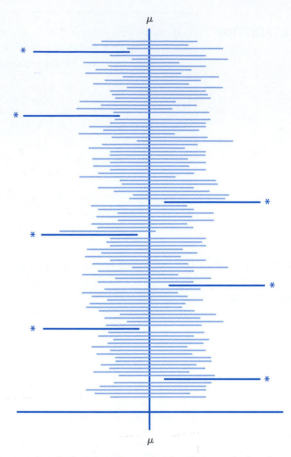

Figure 7.3 One hundred 95% CIs (asterisks identify intervals that do not include μ)

difficulty lies with our interpretation of probability—it applies to a long sequence of replications of an experiment rather than just a single replication. There is another approach to the construction and interpretation of CIs that uses the notion of subjective probability and Bayes' theorem, but the technical details are beyond the scope of this text; the book by DeGroot, et al. (see the Chapter 6 bibliography) is a good source. The interval presented here (as well as each interval presented subsequently) is called a "classical" CI because its interpretation rests on the classical notion of probability.

Other Levels of Confidence

The confidence level of 95% was inherited from the probability .95 for the initial inequalities in (7.2). If a confidence level of 99% is desired, the initial probability of .95 must be replaced by .99, which necessitates changing the z critical value from 1.96 to 2.58. A 99% CI then results from using 2.58 in place of 1.96 in the formula for the 95% CI.

In fact, any desired level of confidence can be achieved by replacing 1.96 or 2.58 with the appropriate standard normal critical value. Recall from Chapter 4 the notation for a z critical value: z_α is the number on the horizontal z scale that captures upper tail area α. As Figure 7.4 shows, a probability (i.e., central z curve area) of $1 - \alpha$ is achieved by using $z_{\alpha/2}$ in place of 1.96.

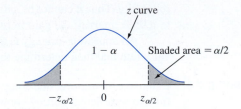

Figure 7.4 $P(-z_{\alpha/2} < Z < z_{\alpha/2}) = 1 - \alpha$

DEFINITION

> A **$100(1 - \alpha)\%$ confidence interval** for the mean μ of a normal population when the value of σ is known is given by
>
> $$\left(\bar{x} - z_{\alpha/2} \cdot \frac{\sigma}{\sqrt{n}}, \; \bar{x} + z_{\alpha/2} \cdot \frac{\sigma}{\sqrt{n}} \right) \qquad (7.5)$$
>
> or, equivalently, by $\bar{x} \pm z_{\alpha/2} \cdot \sigma/\sqrt{n}$.

The formula (7.5) for the CI can also be expressed in words as

point estimate of $\mu \pm$ (z critical value) (standard error of the mean).

EXAMPLE 7.3 The production process for engine control housing units of a particular type has recently been modified. Prior to this modification, historical data had suggested that the distribution of hole diameters for bushings on the housings was normal with a standard deviation of .100 mm. It is believed that the modification has not affected the shape of the distribution or the standard deviation, but that the value of the mean diameter may have changed. A sample of 40 housing units is selected and hole diameter is determined for each one, resulting in a sample mean diameter of 5.426 mm. Let's calculate a confidence interval for true average hole diameter using a confidence level of 90%. This requires that $100(1 - \alpha) = 90$, from which $\alpha = .10$ and $z_{\alpha/2} = z_{.05} = 1.645$ (corresponding to a cumulative z-curve area of .9500). The desired interval is then

$$5.426 \pm (1.645)\frac{.100}{\sqrt{40}} = 5.426 \pm .026 = (5.400, 5.452)$$

With a reasonably high degree of confidence, we can say that $5.400 < \mu < 5.452$. This interval is rather narrow because of the small amount of variability in hole diameter ($\sigma = .100$). ■

Confidence Level, Precision, and Sample Size

Why settle for a confidence level of 95% when a level of 99% is achievable? Because the price paid for the higher confidence level is a wider interval. Since the 95% interval extends $1.96 \cdot \sigma/\sqrt{n}$ to each side of $\bar{x}$, the width of the interval is $2(1.96) \cdot \sigma/\sqrt{n} = 3.92 \cdot \sigma/\sqrt{n}$. Similarly, the width of the 99% interval is $2(2.58) \cdot \sigma/\sqrt{n} = 5.16 \cdot \sigma/\sqrt{n}$. That is, we have more confidence in the 99% interval precisely because it is wider. The higher the desired degree of confidence, the wider the resulting interval will be.

If we think of the width of the interval as specifying its precision or accuracy, then the confidence level (or reliability) of the interval is inversely related to its precision. A highly reliable interval estimate may be imprecise in that the endpoints of the interval may be far apart, whereas a precise interval may entail relatively low reliability. Thus it cannot be said unequivocally that a 99% interval is to be preferred to a 95% interval; the gain in reliability entails a loss in precision.

An appealing strategy is to specify both the desired confidence level and interval width and then determine the necessary sample size.

EXAMPLE 7.4 Extensive monitoring of a computer time-sharing system has suggested that response time to a particular editing command is normally distributed with standard deviation 25 millisec. A new operating system has been installed, and we wish to estimate the true average response time μ for the new environment. Assuming that response times are still normally distributed with $\sigma = 25$, what sample size is necessary to ensure that the resulting 95% CI has a width of (at most) 10? The sample size n must satisfy

$$10 = 2 \cdot (1.96)(25/\sqrt{n})$$

Rearranging this equation gives

$$\sqrt{n} = 2 \cdot (1.96)(25)/10 = 9.80$$

so

$$n = (9.80)^2 = 96.04$$

Since n must be an integer, a sample size of 97 is required. ∎

A general formula for the sample size n necessary to ensure an interval width w is obtained from equating w to $2 \cdot z_{\alpha/2} \cdot \sigma/\sqrt{n}$ and solving for n.

The sample size necessary for the CI (7.5) to have a width w is

$$n = \left(2z_{\alpha/2} \cdot \frac{\sigma}{w}\right)^2$$

The smaller the desired width w, the larger n must be. In addition, n is an increasing function of σ (more population variability necessitates a larger sample size) and of the confidence level $100(1 - \alpha)\%$ (as α decreases, $z_{\alpha/2}$ increases).

The half-width $1.96\sigma/\sqrt{n}$ of the 95% CI is sometimes called the **bound on the error of estimation** associated with a 95% confidence level. That is, with 95% confidence, the point estimate $\bar{x}$ will be no farther than this from μ. Before obtaining data, an investigator may wish to determine a sample size for which a particular value of the bound is achieved. For example, with μ representing the average fuel efficiency (mpg) for all cars of a certain type, the objective of an investigation may be to estimate μ to within 1 mpg with 95% confidence. More generally, if we wish to estimate μ to within an amount B (the specified bound on the error of estimation) with $100(1 - \alpha)\%$ confidence, the necessary sample size results from replacing $2/w$ by $1/B$ in the formula in the preceding box.

Deriving a Confidence Interval

Let $X_1, X_2, \ldots, X_n$ denote the sample on which the CI for a parameter θ is to be based. Suppose a random variable satisfying the following two properties can be found:

1. The variable is a function of both $X_1, \ldots, X_n$ and θ.

2. The probability distribution of the variable does not depend on θ or on any other unknown parameters.

Let $h(X_1, X_2, \ldots, X_n; \theta)$ denote this random variable. For example, if the population distribution is normal with known σ and $\theta = \mu$, the variable $h(X_1, \ldots, X_n; \mu) = (\overline{X} - \mu)/(\sigma/\sqrt{n})$ satisfies both properties; it clearly depends functionally on μ, yet has the standard normal probability distribution irrespective of the value of μ. In general, the form of the h function is usually suggested by examining the distribution of an appropriate estimator $\hat{\theta}$.

For any α between 0 and 1, constants a and b can be found to satisfy

$$P(a < h(X_1, \ldots, X_n; \theta) < b) = 1 - \alpha \qquad (7.6)$$

Because of the second property, a and b do not depend on θ. In the normal example, $a = -z_{\alpha/2}$ and $b = z_{\alpha/2}$. Now suppose that the inequalities in (7.6) can be manipulated to isolate θ, giving the equivalent probability statement

$$P(l(X_1, X_2, \ldots, X_n) < \theta < u(X_1, X_2, \ldots, X_n)) = 1 - \alpha$$

Then $l(x_1, x_2, \ldots, x_n)$ and $u(x_1, \ldots, x_n)$ are the lower and upper confidence limits, respectively, for a $100(1 - \alpha)\%$ CI. In the normal example, we saw that $l(X_1, \ldots, X_n) = \overline{X} - z_{\alpha/2} \cdot \sigma/\sqrt{n}$ and $u(X_1, \ldots, X_n) = \overline{X} + z_{\alpha/2} \cdot \sigma/\sqrt{n}$.

EXAMPLE 7.5 A theoretical model suggests that the time to breakdown of an insulating fluid between electrodes at a particular voltage has an exponential distribution with parameter λ (see Section 4.4). A random sample of $n = 10$ breakdown times yields the following sample data (in min): $x_1 = 41.53$, $x_2 = 18.73$, $x_3 = 2.99$, $x_4 = 30.34$, $x_5 = 12.33$, $x_6 = 117.52$, $x_7 = 73.02$, $x_8 = 223.63$, $x_9 = 4.00$, $x_{10} = 26.78$. A 95% CI for λ and for the true average breakdown time are desired.

Let $h(X_1, X_2, \ldots, X_n; \lambda) = 2\lambda\Sigma X_i$. It can be shown that this random variable has a probability distribution called a chi-squared distribution with $2n$ degrees of freedom (df) ($\nu = 2n$, where ν is the parameter of a chi-squared distribution as mentioned in Section 4.4). Appendix Table A.7 pictures a typical chi-squared density curve and tabulates critical values that capture specified tail areas. The relevant number of df here is $2(10) = 20$. The $\nu = 20$ row of the table shows that 34.170 captures upper-tail area .025 and 9.591 captures lower-tail area .025 (upper-tail area .975). Thus for $n = 10$,

$$P\left(9.591 < 2\lambda\sum X_i < 34.170\right) = .95$$

Division by $2\Sigma X_i$ isolates λ, yielding

$$P\left(9.591/\left(2\sum X_i\right) < \lambda < \left(34.170/\left(2\sum X_i\right)\right)\right) = .95$$

The lower limit of the 95% CI for λ is $9.591/(2\Sigma x_i)$, and the upper limit is $34.170/(2\Sigma x_i)$. For the given data, $\Sigma x_i = 550.87$, giving the interval (.00871, .03101).

The expected value of an exponential rv is $\mu = 1/\lambda$. Since

$$P\left(2\sum X_i/34.170 < 1/\lambda < 2\sum X_i/9.591\right) = .95$$

the 95% CI for true average breakdown time is $(2\Sigma x_i/34.170, 2\Sigma x_i/9.591) = (32.24, 114.87)$. This interval is obviously quite wide, reflecting substantial variability in breakdown times and a small sample size. ■

In general, the upper and lower confidence limits result from replacing each $<$ in (7.6) by $=$ and solving for θ. In the insulating fluid example just considered, $2\lambda\Sigma x_i = 34.170$ gives $\lambda = 34.170/(2\Sigma x_i)$ as the upper confidence limit, and the lower limit is obtained from the other equation. Notice that the two interval limits are not equidistant from the point estimate, since the interval is not of the form $\hat{\theta} \pm c$.

Bootstrap Confidence Intervals

The bootstrap technique was introduced in Chapter 6 as a way of estimating $\sigma_{\hat{\theta}}$. It can also be applied to obtain a CI for θ. Consider again estimating the mean μ of a normal distribution when σ is known. Let's replace μ by θ and use $\hat{\theta} = \overline{X}$ as the point estimator. Notice that $1.96\sigma\sqrt{n}$ is the 97.5th percentile of the distribution of $\hat{\theta} - \theta$ [that is, $P(\overline{X} - \mu < 1.96\sigma/\sqrt{n}) = P(Z < 1.96) = .9750$]. Similarly, $-1.96\sigma/\sqrt{n}$ is the 2.5th percentile, so

$$.95 = P(2.5\text{th percentile} < \hat{\theta} - \theta < 97.5\text{th percentile})$$
$$= P(\hat{\theta} - 2.5\text{th percentile} > \theta > \hat{\theta} - 97.5\text{th percentile})$$

That is, with

$$l = \hat{\theta} - 97.5\text{th percentile of } \hat{\theta} - \theta$$
$$u = \hat{\theta} - 2.5\text{th percentile of } \hat{\theta} - \theta \tag{7.7}$$

the CI for θ is (l, u). In many cases, the percentiles in (7.7) cannot be calculated, but they *can* be estimated from bootstrap samples. Suppose we obtain $B = 1000$ bootstrap samples and calculate $\hat{\theta}_1^*, \ldots, \hat{\theta}_{1000}^*$, and $\overline{\theta}^*$ followed by the 1000 differences $\hat{\theta}_1^* - \overline{\theta}^*, \ldots, \hat{\theta}_{1000}^* - \overline{\theta}^*$. The 25th largest and 25th smallest of these differences are estimates of the unknown percentiles in (7.7). Consult the Devore and Berk or Efron books cited in Chapter 6 for more information.

EXERCISES Section 7.1 (1–11)

1. Consider a normal population distribution with the value of σ known.
   a. What is the confidence level for the interval $\overline{x} \pm 2.81\sigma/\sqrt{n}$?
   b. What is the confidence level for the interval $\overline{x} \pm 1.44\sigma/\sqrt{n}$?
   c. What value of $z_{\alpha/2}$ in the CI formula (7.5) results in a confidence level of 99.7%?
   d. Answer the question posed in part (c) for a confidence level of 75%.

2. Each of the following is a confidence interval for $\mu =$ true average (i.e., population mean) resonance frequency (Hz) for all tennis rackets of a certain type:

   (114.4, 115.6)   (114.1, 115.9)

   a. What is the value of the sample mean resonance frequency?
   b. Both intervals were calculated from the same sample data. The confidence level for one of these intervals is 90% and for the other is 99%. Which of the intervals has the 90% confidence level, and why?

3. Suppose that a random sample of 50 bottles of a particular brand of cough syrup is selected and the alcohol content of each bottle is determined. Let μ denote the average alcohol content for the population of all bottles of the brand under study. Suppose that the resulting 95% confidence interval is (7.8, 9.4).
   a. Would a 90% confidence interval calculated from this same sample have been narrower or wider than the given interval? Explain your reasoning.
   b. Consider the following statement: There is a 95% chance that μ is between 7.8 and 9.4. Is this statement correct? Why or why not?
   c. Consider the following statement: We can be highly confident that 95% of all bottles of this type of cough syrup have an alcohol content that is between 7.8 and 9.4. Is this statement correct? Why or why not?
   d. Consider the following statement: If the process of selecting a sample of size 50 and then computing the corresponding 95% interval is repeated 100 times, 95 of the resulting intervals will include μ. Is this statement correct? Why or why not?

4. A CI is desired for the true average stray-load loss μ (watts) for a certain type of induction motor when the line current is held at 10 amps for a speed of 1500 rpm. Assume that stray-load loss is normally distributed with $\sigma = 3.0$.
  a. Compute a 95% CI for μ when $n = 25$ and $\bar{x} = 58.3$.
  b. Compute a 95% CI for μ when $n = 100$ and $\bar{x} = 58.3$.
  c. Compute a 99% CI for μ when $n = 100$ and $\bar{x} = 58.3$.
  d. Compute an 82% CI for μ when $n = 100$ and $\bar{x} = 58.3$.
  e. How large must n be if the width of the 99% interval for μ is to be 1.0?

5. Assume that the helium porosity (in percentage) of coal samples taken from any particular seam is normally distributed with true standard deviation .75.
  a. Compute a 95% CI for the true average porosity of a certain seam if the average porosity for 20 specimens from the seam was 4.85.
  b. Compute a 98% CI for true average porosity of another seam based on 16 specimens with a sample average porosity of 4.56.
  c. How large a sample size is necessary if the width of the 95% interval is to be .40?
  d. What sample size is necessary to estimate true average porosity to within .2 with 99% confidence?

6. On the basis of extensive tests, the yield point of a particular type of mild steel-reinforcing bar is known to be normally distributed with $\sigma = 100$. The composition of bars has been slightly modified, but the modification is not believed to have affected either the normality or the value of σ.
  a. Assuming this to be the case, if a sample of 25 modified bars resulted in a sample average yield point of 8439 lb, compute a 90% CI for the true average yield point of the modified bar.
  b. How would you modify the interval in part (a) to obtain a confidence level of 92%?

7. By how much must the sample size n be increased if the width of the CI (7.5) is to be halved? If the sample size is increased by a factor of 25, what effect will this have on the width of the interval? Justify your assertions.

8. Let $\alpha_1 > 0$, $\alpha_2 > 0$, with $\alpha_1 + \alpha_2 = \alpha$. Then

$$P\left(-z_{\alpha_1} < \frac{\bar{X} - \mu}{\sigma/\sqrt{n}} < z_{\alpha_2}\right) = 1 - \alpha$$

  a. Use this equation to derive a more general expression for a $100(1 - \alpha)\%$ CI for μ of which the interval (7.5) is a special case.
  b. Let $\alpha = .05$ and $\alpha_1 = \alpha/4$, $\alpha_2 = 3\alpha/4$. Does this result in a narrower or wider interval than the interval (7.5)?

9. a. Under the same conditions as those leading to the interval (7.5), $P[(\bar{X} - \mu)/(\sigma/\sqrt{n}) < 1.645] = .95$. Use this to derive a one-sided interval for μ that has infinite width and provides a lower confidence bound on μ. What is this interval for the data in Exercise 5(a)?
  b. Generalize the result of part (a) to obtain a lower bound with confidence level $100(1 - \alpha)\%$.
  c. What is an analogous interval to that of part (b) that provides an upper bound on μ? Compute this 99% interval for the data of Exercise 4(a).

10. A random sample of $n = 15$ heat pumps of a certain type yielded the following observations on lifetime (in years):

| 2.0 | 1.3 | 6.0 | 1.9 | 5.1 | .4 | 1.0 | 5.3 |
| 15.7 | .7 | 4.8 | .9 | 12.2 | 5.3 | .6 | |

  a. Assume that the lifetime distribution is exponential and use an argument parallel to that of Example 7.5 to obtain a 95% CI for expected (true average) lifetime.
  b. How should the interval of part (a) be altered to achieve a confidence level of 99%?
  c. What is a 95% CI for the standard deviation of the lifetime distribution? [*Hint*: What is the standard deviation of an exponential random variable?]

11. Consider the next 1000 95% CIs for μ that a statistical consultant will obtain for various clients. Suppose the data sets on which the intervals are based are selected independently of one another. How many of these 1000 intervals do you expect to capture the corresponding value of μ? What is the probability that between 940 and 960 of these intervals contain the corresponding value of μ? [*Hint*: Let Y = the number among the 1000 intervals that contain μ. What kind of random variable is Y?]

7.2 Large-Sample Confidence Intervals for a Population Mean and Proportion

The CI for μ given in the previous section assumed that the population distribution is normal with the value of σ known. We now present a large-sample CI whose validity does not require these assumptions. After showing how the argument leading to this interval generalizes to yield other large-sample intervals, we focus on an interval for a population proportion p.

A Large-Sample Interval for μ

Let $X_1, X_2, \ldots, X_n$ be a random sample from a population having a mean μ and standard deviation σ. Provided that n is sufficiently large, the Central Limit Theorem (CLT) implies that $\overline{X}$ has approximately a normal distribution whatever the nature of the population distribution. It then follows that $Z = (\overline{X} - \mu)/(\sigma/\sqrt{n})$ has approximately a standard normal distribution, so that

$$P\left(-z_{\alpha/2} < \frac{\overline{X} - \mu}{\sigma/\sqrt{n}} < z_{\alpha/2}\right) \approx 1 - \alpha$$

An argument parallel to that given in Section 7.1 yields $\bar{x} \pm z_{\alpha/2} \cdot \sigma/\sqrt{n}$ as a large-sample CI for μ with a confidence level of *approximately* $100(1 - \alpha)\%$. That is, when n is large, the CI for μ given previously remains valid whatever the population distribution, provided that the qualifier "approximately" is inserted in front of the confidence level.

A practical difficulty with this development is that computation of the CI requires the value of σ, which will rarely be known. Consider replacing the population standard deviation σ in Z by the sample standard deviation to obtain the standardized variable

$$\frac{\overline{X} - \mu}{S/\sqrt{n}}$$

Previously, there was randomness only in the numerator of Z by virtue of $\overline{X}$. In the new standardized variable, both $\overline{X}$ *and* S vary in value from one sample to another. So it might seem that the distribution of the new variable should be more spread out than the z curve to reflect the extra variation in the denominator. This is indeed true when n is small. However, for large n the subsitituion of S for σ adds little extra variability, so this variable also has approximately a standard normal distribution. Manipulation of the variable in a probability statement, as in the case of known σ, gives a general large-sample CI for μ.

PROPOSITION

If n is sufficiently large, the standardized variable

$$Z = \frac{\overline{X} - \mu}{S/\sqrt{n}}$$

has approximately a standard normal distribution. This implies that

$$\bar{x} \pm z_{\alpha/2} \cdot \frac{s}{\sqrt{n}} \tag{7.8}$$

is a **large-sample confidence interval for μ** with confidence level approximately $100(1 - \alpha)\%$. This formula is valid regardless of the shape of the population distribution.

In words, the CI (7.8) is

point estimate of $\mu \pm$ (z critical value) (estimated standard error of the mean).

Generally speaking, $n > 40$ will be sufficient to justify the use of this interval. This is somewhat more conservative than the rule of thumb for the CLT because of the additional variability introduced by using S in place of σ.

EXAMPLE 7.6 Haven't you always wanted to own a Porsche? The author thought maybe he could afford a Boxster, the cheapest model. So he went to **www.cars.com** on Nov. 18, 2009, and found a total of 1113 such cars listed. Asking prices ranged from $3499 to $130,000 (the latter price was one of only two exceeding $70,000). The prices depressed him, so he focused instead on odometer readings (miles). Here are reported readings for a sample of 50 of these Boxsters:

2948	2996	7197	8338	8500	8759	12710	12925
15767	20000	23247	24863	26000	26210	30552	30600
35700	36466	40316	40596	41021	41234	43000	44607
45000	45027	45442	46963	47978	49518	52000	53334
54208	56062	57000	57365	60020	60265	60803	62851
64404	72140	74594	79308	79500	80000	80000	84000
113000	118634						

A boxplot of the data (Figure 7.5) shows that, except for the two outliers at the upper end, the distribution of values is reasonably symmetric (in fact, a normal probability plot exhibits a reasonably linear pattern, though the points corresponding to the two smallest and two largest observations are somewhat removed from a line fit through the remaining points).

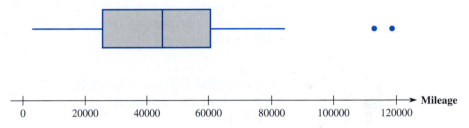

Figure 7.5 A boxplot of the odometer reading data from Example 7.6

Summary quantities include $n = 50$, $\bar{x} = 45{,}679.4$, $\tilde{x} = 45{,}013.5$, $s = 26{,}641.675$, $f_s = 34{,}265$. The mean and median are reasonably close (if the two largest values were each reduced by 30,000, the mean would fall to 44,479.4, while the median would be unaffected). The boxplot and the magnitudes of s and f_s relative to the mean and median both indicate a substantial amount of variability. A confidence level of about 95% requires $z_{.025} = 1.96$, and the interval is

$$45{,}679.4 \pm (1.96)\left(\frac{26{,}641.675}{\sqrt{50}}\right) = 45{,}679.4 \pm 7384.7$$

$$= (38{,}294.7, 53{,}064.1)$$

That is, $38{,}294.7 < \mu < 53{,}064.1$ with 95% confidence. This interval is rather wide because a sample size of 50, even though large by our rule of thumb, is not large enough to overcome the substantial variability in the sample. We do not have a very precise estimate of the population mean odometer reading.

Is the interval we've calculated one of the 95% that in the long run includes the parameter being estimated, or is it one of the "bad" 5% that does not do so? Without knowing the value of μ, we cannot tell. Remember that the confidence level refers to the long run capture percentage when the formula is used repeatedly on various samples; it cannot be interpreted for a single sample and the resulting interval. ∎

Unfortunately, the choice of sample size to yield a desired interval width is not as straightforward here as it was for the case of known σ. This is because the width of (7.8) is $2z_{\alpha/2}s/\sqrt{n}$. Since the value of s is not available before the data has been gathered, the width of the interval cannot be determined solely by the choice of n. The only option for an investigator who wishes to specify a desired width is to make an educated guess as to what the value of s might be. By being conservative and guessing a larger value of s, an n larger than necessary will be chosen. The investigator may be able to specify a reasonably accurate value of the population range (the difference between the largest and smallest values). Then if the population distribution is not too skewed, dividing the range by 4 gives a ballpark value of what s might be.

EXAMPLE 7.7 The charge-to-tap time (min) for carbon steel in one type of open hearth furnace is to be determined for each heat in a sample of size n. If the investigator believes that almost all times in the distribution are between 320 and 440, what sample size would be appropriate for estimating the true average time to within 5 min. with a confidence level of 95%?

A reasonable value for s is $(440 - 320)/4 = 30$. Thus

$$n = \left[\frac{(1.96)(30)}{5} \right]^2 = 138.3$$

Since the sample size must be an integer, $n = 139$ should be used. Note that estimating to within 5 min. with the specified confidence level is equivalent to a CI width of 10 min. ∎

A General Large-Sample Confidence Interval

The large-sample intervals $\bar{x} \pm z_{\alpha/2} \cdot \sigma/\sqrt{n}$ and $\bar{x} \pm z_{\alpha/2} \cdot s/\sqrt{n}$ are special cases of a general large-sample CI for a parameter θ. Suppose that $\hat{\theta}$ is an estimator satisfying the following properties: (1) It has approximately a normal distribution; (2) it is (at least approximately) unbiased; and (3) an expression for $\sigma_{\hat{\theta}}$, the standard deviation (standard error) of $\hat{\theta}$, is available. For example, in the case $\theta = \mu$, $\hat{\mu} = \bar{X}$ is an unbiased estimator whose distribution is approximately normal when n is large and $\sigma_{\hat{\mu}} = \sigma_{\bar{X}} = \sigma/\sqrt{n}$. Standardizing $\hat{\theta}$ yields the rv $Z = (\hat{\theta} - \theta)/\sigma_{\hat{\theta}}$, which has approximately a standard normal distribution. This justifies the probability statement

$$P\left(-z_{\alpha/2} < \frac{\hat{\theta} - \theta}{\sigma_{\hat{\theta}}} < z_{\alpha/2} \right) \approx 1 - \alpha \tag{7.9}$$

Assume first that $\sigma_{\hat{\theta}}$ does not involve any unknown parameters (e.g., known σ in the case $\theta = \mu$). Then replacing each $<$ in (7.9) by $=$ results in $\theta = \hat{\theta} \pm z_{\alpha/2} \cdot \sigma_{\hat{\theta}}$, so the lower and upper confidence limits are $\hat{\theta} - z_{\alpha/2} \cdot \sigma_{\hat{\theta}}$ and $\hat{\theta} + z_{\alpha/2} \cdot \sigma_{\hat{\theta}}$, respectively. Now suppose that $\sigma_{\hat{\theta}}$ does not involve θ but does involve at least one other unknown parameter. Let $s_{\hat{\theta}}$ be the estimate of $\sigma_{\hat{\theta}}$ obtained by using estimates in place of the unknown parameters (e.g., $s/\sqrt{n}$ estimates $\sigma/\sqrt{n}$). Under general conditions (essentially that $s_{\hat{\theta}}$ be close to $\sigma_{\hat{\theta}}$ for most samples), a valid CI is $\hat{\theta} \pm z_{\alpha/2} \cdot s_{\hat{\theta}}$. The large-sample interval $\bar{x} \pm z_{\alpha/2} \cdot s/\sqrt{n}$ is an example.

Finally, suppose that $\sigma_{\hat{\theta}}$ does involve the unknown θ. For example, we shall see momentarily that this is the case when $\theta = p$, a population proportion. Then $(\hat{\theta} - \theta)/\sigma_{\hat{\theta}} = z_{\alpha/2}$ can be difficult to solve. An approximate solution can often be obtained by replacing θ in $\sigma_{\hat{\theta}}$ by its estimate $\hat{\theta}$. This results in an estimated standard deviation $s_{\hat{\theta}}$, and the corresponding interval is again $\hat{\theta} \pm z_{\alpha/2} \cdot s_{\hat{\theta}}$.

In words, this CI is

point estimate of $\theta \pm$ (z critical value)(estimated standard error of the estimator)

A Confidence Interval for a Population Proportion

Let p denote the proportion of "successes" in a population, where *success* identifies an individual or object that has a specified property (e.g., individuals who graduated from college, computers that do not need warranty service, etc.). A random sample of n individuals or objects is to be selected, and X is the number of successes in the sample. Provided that n is small compared to the population size, X can be regarded as a binomial rv with $E(X) = np$ and $\sigma_X = \sqrt{np(1-p)}$. Furthermore, if both $np \geq 10$ and $nq \geq 10$, $(q = 1 - p)$, X has approximately a normal distribution.

The natural estimator of p is $\hat{p} = X/n$, the sample fraction of successes. Since $\hat{p}$ is just X multiplied by the constant $1/n$, $\hat{p}$ also has approximately a normal distribution. As shown in Section 6.1, $E(\hat{p}) = p$ (unbiasedness) and $\sigma_{\hat{p}} = \sqrt{p(1-p)/n}$. The standard deviation $\sigma_{\hat{p}}$ involves the unknown parameter p. Standardizing $\hat{p}$ by subtracting p and dividing by $\sigma_{\hat{p}}$ then implies that

$$P\left(-z_{\alpha/2} < \frac{\hat{p} - p}{\sqrt{p(1-p)/n}} < z_{\alpha/2}\right) \approx 1 - \alpha$$

Proceeding as suggested in the subsection "Deriving a Confidence Interval" (Section 7.1), the confidence limits result from replacing each $<$ by $=$ and solving the resulting equation for p. But whereas the equations $(\bar{x} - \mu)/(s/\sqrt{n}) = \pm z_{\alpha/2}$ employed in deriving the large-sample CI for μ are linear in μ, the equations here are quadratic (p^2 appears in the numerator when both sides of each equation are squared to eliminate the square root). The two roots are

$$p = \frac{\hat{p} + z_{\alpha/2}^2/2n}{1 + z_{\alpha/2}^2/n} \pm z_{\alpha/2} \frac{\sqrt{\hat{p}(1-\hat{p})/n + z_{\alpha/2}^2/4n^2}}{1 + z_{\alpha/2}^2/n}$$

$$= \tilde{p} \pm z_{\alpha/2} \frac{\sqrt{\hat{p}(1-\hat{p})/n + z_{\alpha/2}^2/4n^2}}{1 + z_{\alpha/2}^2/n}$$

PROPOSITION

Let $\tilde{p} = [\hat{p} + z_{\alpha/2}^2/2n]/[1 + z_{\alpha/2}^2/n]$. Then a **confidence interval for a population proportion** p with confidence level approximately $100(1 - \alpha)$% is

$$\tilde{p} \pm z_{\alpha/2} \frac{\sqrt{\hat{p}\hat{q}/n + z_{\alpha/2}^2/4n^2}}{1 + z_{\alpha/2}^2/n} \tag{7.10}$$

where $\hat{q} = 1 - \hat{p}$ and, as before, the $-$ in (7.10) corresponds to the lower confidence limit and the $+$ to the upper confidence limit.

This is often referred to as the *score CI* for p.

If the sample size n is very large, then $z^2/2n$ is generally quite negligible (small) compared to $\hat{p}$ and z^2/n is quite negligible compared to 1, from which $\tilde{p} \approx \hat{p}$. In this case $z^2/4n^2$ is also negligible compared to pq/n (n^2 is a much larger divisor than is n). As a result, the dominant term in the $\pm$ expression is $z_{\alpha/2}\sqrt{\hat{p}\hat{q}/n}$ and the score interval is approximately

$$\hat{p} \pm z_{\alpha/2}\sqrt{\hat{p}\hat{q}/n} \tag{7.11}$$

This latter interval has the general form $\hat{\theta} \pm z_{\alpha/2}\hat{\sigma}_{\hat{\theta}}$ of a large-sample interval suggested in the last subsection. The approximate CI (7.11) is the one that for decades has appeared in introductory statistics textbooks. It clearly has a much simpler and more appealing form than the score CI. So why bother with the latter?

First of all, suppose we use $z_{.025} = 1.96$ in the traditional formula (7.11). Then our *nominal* confidence level (the one we think we're buying by using that z critical value) is approximately 95%. So before a sample is selected, the probability that the random interval includes the actual value of p (i.e., the *coverage probability*) should be about .95. But as Figure 7.6 shows for the case $n = 100$, the actual coverage probability for this interval can differ considerably from the nominal probability .95, particularly when p is not close to .5 (the graph of coverage probability versus p is very jagged because the underlying binomial probability distribution is discrete rather than continuous). This is generally speaking a deficiency of the traditional interval—the actual confidence level can be quite different from the nominal level even for reasonably large sample sizes. Recent research has shown that the score interval rectifies this behavior—for virtually all sample sizes and values of p, its actual confidence level will be quite close to the nominal level specified by the choice of $z_{\alpha/2}$. This is due largely to the fact that the score interval is shifted a bit toward .5 compared to the traditional interval. In particular, the midpoint $\tilde{p}$ of the score interval is always a bit closer to .5 than is the midpoint $\hat{p}$ of the traditional interval. This is especially important when p is close to 0 or 1.

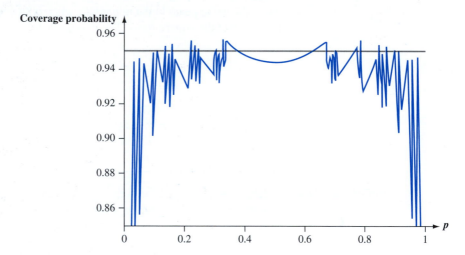

Figure 7.6 Actual coverage probability for the interval (7.11) for varying values of p when $n = 100$

In addition, the score interval can be used with nearly all sample sizes and parameter values. It is thus not necessary to check the conditions $n\hat{p} \geq 10$ and $n(1 - \hat{p}) \geq 10$ that would be required were the traditional interval employed. So rather than asking when n is large enough for (7.11) to yield a good approximation to (7.10), our recommendation is that the score CI should *always* be used. The slight additional tediousness of the computation is outweighed by the desirable properties of the interval.

EXAMPLE 7.8 The article **"Repeatability and Reproducibility for Pass/Fail Data" (*J. of Testing and Eval.*, 1997: 151–153)** reported that in $n = 48$ trials in a particular laboratory, 16 resulted in ignition of a particular type of substrate by a lighted cigarette. Let p denote the long-run proportion of all such trials that would result in ignition. A point estimate for p is $\hat{p} = 16/48 = .333$. A confidence interval for p with a confidence level of approximately 95% is

$$\frac{.333 + (1.96)^2/96}{1 + (1.96)^2/48} \pm (1.96)\frac{\sqrt{(.333)(.667)/48 + (1.96)^2/9216}}{1 + (1.96)^2/48}$$

$$= .345 \pm .129 = (.216, .474)$$

This interval is quite wide because a sample size of 48 is not at all large when estimating a proportion.

The traditional interval is

$$.333 \pm 1.96\sqrt{(.333)(.667)/48} = .333 \pm .133 = (.200, .466)$$

These two intervals would be in much closer agreement were the sample size substantially larger. ∎

Equating the width of the CI for p to a prespecified width w gives a quadratic equation for the sample size n necessary to give an interval with a desired degree of precision. Suppressing the subscript in $z_{\alpha/2}$, the solution is

$$n = \frac{2z^2\hat{p}\hat{q} - z^2w^2 \pm \sqrt{4z^4\hat{p}\hat{q}(\hat{p}\hat{q} - w^2) + w^2z^4}}{w^2} \tag{7.12}$$

Neglecting the terms in the numerator involving w^2 gives

$$n \approx \frac{4z^2\hat{p}\hat{q}}{w^2}$$

This latter expression is what results from equating the width of the traditional interval to w.

These formulas unfortunately involve the unknown $\hat{p}$. The most conservative approach is to take advantage of the fact that $\hat{p}\hat{q} [= \hat{p}(1 - \hat{p})]$ is maximized at $\hat{p} = .5$. Thus if $\hat{p} = \hat{q} = .5$ is used in (7.12), the width will be at most w regardless of what value of $\hat{p}$ results from the sample. Alternatively, if the investigator believes strongly, based on prior information, that $p \leq p_0 \leq .5$, then p_0 can be used in place of $\hat{p}$. A similar comment applies when $p \geq p_0 \geq .5$.

EXAMPLE 7.9 The width of the 95% CI in Example 7.8 is .258. The value of n necessary to ensure a width of .10 irrespective of the value of $\hat{p}$ is

$$n = \frac{2(1.96)^2(.25) - (1.96)^2(.01) \pm \sqrt{4(1.96)^4(.25)(.25 - .01) + (.01)(1.96)^4}}{.01} = 380.3$$

Thus a sample size of 381 should be used. The expression for n based on the traditional CI gives a slightly larger value of 385. ∎

One-Sided Confidence Intervals (Confidence Bounds)

The confidence intervals discussed thus far give both a lower confidence bound *and* an upper confidence bound for the parameter being estimated. In some circumstances, an investigator will want only one of these two types of bounds. For example, a psychologist may wish to calculate a 95% upper confidence bound for true average reaction time to a particular stimulus, or a reliability engineer may want only a lower confidence bound for true average lifetime of components of a certain type. Because the cumulative area under the standard normal curve to the left of 1.645 is .95,

$$P\left(\frac{\overline{X} - \mu}{S/\sqrt{n}} < 1.645\right) \approx .95$$

Manipulating the inequality inside the parentheses to isolate μ on one side and replacing rv's by calculated values gives the inequality $\mu > \overline{x} - 1.645s/\sqrt{n}$; the expression on the right is the desired lower confidence bound. Starting with $P(-1.645 < Z) \approx .95$ and manipulating the inequality results in the upper confidence bound. A similar argument gives a one-sided bound associated with any other confidence level.

PROPOSITION

A **large-sample upper confidence bound for μ** is

$$\mu < \overline{x} + z_\alpha \cdot \frac{s}{\sqrt{n}}$$

and a **large-sample lower confidence bound for μ** is

$$\mu > \overline{x} - z_\alpha \cdot \frac{s}{\sqrt{n}}$$

A **one-sided confidence bound for p** results from replacing $z_{\alpha/2}$ by z_α and $\pm$ by either $+$ or $-$ in the CI formula (7.10) for p. In all cases the confidence level is approximately $100(1 - \alpha)\%$.

EXAMPLE 7.10   Titanium and its alloys have found increasing use in aerospace and automotive applications because of durability and high strength-to-weight ratios. However, machining can be difficult because of low thermal conductivity. The article **"Modeling and Multi-Objective Optimization of Process Parameters of Wire Electrical Discharge Machining Using Non-Dominated Sorting Genetic Algorithm-II** (*J. of Engr. Manuf.*, 2012: 1186–2001) described an investigation into different settings that impact wire electrical discharge machining of titanium 6-2-4-2. One characteristic of interest was surface roughness (μg) of the metal after machining. A sample of 54 surface roughness observations resulted in a sample mean roughness of 1.9042 and a sample standard deviation of .1455. An upper confidence bound for true average roughness μ with confidence level 95% requires $z_{.05} = 1.645$ (not the value $z_{.025} = 1.96$ needed for a two-sided CI). The bound is

$$1.9042 + (1.645) \cdot \frac{(.1455)}{\sqrt{54}} = 1.9042 + .0326 = 1.9368$$

Thus we estimate with a confidence level of roughly 95% that $\mu < 1.9368$. ∎

EXERCISES Section 7.2 (12–27)

12. The following observations are lifetimes (days) subsequent to diagnosis for individuals suffering from blood cancer (**"A Goodness of Fit Approach to the Class of Life Distributions with Unknown Age,"** *Quality and Reliability Engr. Intl.*, 2012: 761–766):

115	181	255	418	441	461	516	739	743	789	807
865	924	983	1025	1062	1063	1165	1191	1222	1222	1251
1277	1290	1357	1369	1408	1455	1478	1519	1578	1578	1599
1603	1605	1696	1735	1799	1815	1852	1899	1925	1965	

  a. Can a confidence interval for true average lifetime be calculated without assuming anything about the nature of the lifetime distribution? Explain your reasoning. [*Note:* A normal probability plot of the data exhibits a reasonably linear pattern.]

b. Calculate and interpret a confidence interval with a 99% confidence level for true average lifetime. [*Hint:* $\bar{x} = 1191.6$ and $s = 506.6$.]

13. The article **"Gas Cooking, Kitchen Ventilation, and Exposure to Combustion Products"** (*Indoor Air,* **2006: 65–73**) reported that for a sample of 50 kitchens with gas cooking appliances monitored during a one-week period, the sample mean CO_2 level (ppm) was 654.16, and the sample standard deviation was 164.43.

a. Calculate and interpret a 95% (two-sided) confidence interval for true average CO_2 level in the population of all homes from which the sample was selected.

b. Suppose the investigators had made a rough guess of 175 for the value of s before collecting data. What sample size would be necessary to obtain an interval width of 50 ppm for a confidence level of 95%?

14. The negative effects of ambient air pollution on children's lung function has been well established, but less research is available about the impact of indoor air pollution. The authors of **"Indoor Air Pollution and Lung Function Growth Among Children in Four Chinese Cities"** (*Indoor Air,* **2012: 3–11**) investigated the relationship between indoor air-pollution metrics and lung function growth among children ages 6–13 years living in four Chinese cities. For each subject in the study, the authors measured an important lung-capacity index known as FEV_1, the forced volume (in ml) of air that is exhaled in 1 second. Higher FEV_1 values are associated with greater lung capacity. Among the children in the study, 514 came from households that used coal for cooking or heating or both. Their FEV_1 mean was 1427 with a standard deviation of 325. (A complex statistical procedure was used to show that burning coal had a clear negative effect on mean FEV_1 levels.)

a. Calculate and interpret a 95% (two-sided) confidence interval for true average FEV_1 level in the population of all children from which the sample was selected. Does it appear that the parameter of interest has been accurately estimated?

b. Suppose the investigators had made a rough guess of 320 for the value of s before collecting data. What sample size would be necessary to obtain an interval width of 50 ml for a confidence level of 95%?

15. Determine the confidence level for each of the following large-sample one-sided confidence bounds:
a. Upper bound: $\bar{x} + .84s/\sqrt{n}$
b. Lower bound: $\bar{x} - 2.05s/\sqrt{n}$
c. Upper bound: $\bar{x} + .67s/\sqrt{n}$

16. The alternating current (AC) breakdown voltage of an insulating liquid indicates its dielectric strength. The article **"Testing Practices for the AC Breakdown Voltage Testing of Insulation Liquids"** (*IEEE Electrical Insulation Magazine,* **1995: 21–26**) gave the accompanying sample observations on breakdown voltage (kV) of a particular circuit under certain conditions.

```
62 50 53 57 41 53 55 61 59 64 50 53 64 62 50 68
54 55 57 50 55 50 56 55 46 55 53 54 52 47 47 55
57 48 63 57 57 55 53 59 53 52 50 55 60 50 56 58
```

a. Construct a boxplot of the data and comment on interesting features.

b. Calculate and interpret a 95% CI for true average breakdown voltage μ. Does it appear that μ has been precisely estimated? Explain.

c. Suppose the investigator believes that virtually all values of breakdown voltage are between 40 and 70. What sample size would be appropriate for the 95% CI to have a width of 2 kV (so that μ is estimated to within 1 kV with 95% confidence)?

17. Exercise 1.13 gave a sample of ultimate tensile strength observations (ksi). Use the accompanying descriptive statistics output from Minitab to calculate a 99% lower confidence bound for true average ultimate tensile strength, and interpret the result.

N	Mean	Median	TrMean	StDev	SE Mean
153	135.39	135.40	135.41	4.59	0.37

Minimum	Maximum	Q1	Q3
122.20	147.70	132.95	138.25

18. The U.S. Army commissioned a study to assess how deeply a bullet penetrates ceramic body armor (**"Testing Body Armor Materials for Use by the U.S. Army– Phase III,"** **2012**). In the standard test, a cylindrical clay model is layered under the armor vest. A projectile is then fired, causing an indentation in the clay. The deepest impression in the clay is measured as an indication of survivability of someone wearing the armor. Here is data from one testing organization under particular experimental conditions; measurements (in mm) were made using a manually controlled digital caliper:

```
22.4   23.6   24.0   24.9   25.5   25.6
25.8   26.1   26.4   26.7   27.4   27.6
28.3   29.0   29.1   29.6   29.7   29.8
29.9   30.0   30.4   30.5   30.7   30.7
31.0   31.0   31.4   31.6   31.7   31.9
31.9   32.0   32.1   32.4   32.5   32.5
32.6   32.9   33.1   33.3   33.5   33.5
33.5   33.5   33.6   33.6   33.8   33.9
34.1   34.2   34.6   34.6   35.0   35.2
35.2   35.4   35.4   35.4   35.5   35.7
35.8   36.0   36.0   36.0   36.1   36.1
36.2   36.4   36.6   37.0   37.4   37.5
37.5   38.0   38.7   38.8   39.8   41.0
42.0   42.1   44.6   48.3   55.0
```

a. Construct a boxplot of the data and comment on interesting features.

b. Construct a normal probability plot. Is it plausible that impression depth is normally distributed? Is a normal distribution assumption needed in order to calculate a confidence interval or bound for the true average depth μ using the foregoing data? Explain.

c. Use the accompanying Minitab output as a basis for calculating and interpreting an upper confidence bound for μ with a confidence level of 99%.

Variable	Count	Mean	SE Mean	StDev
Depth	83	33.370	0.578	5.268

Q1	Median	Q3	IQR
30.400	33.500	36.000	5.600

19. The article **"Limited Yield Estimation for Visual Defect Sources"** (*IEEE Trans. on Semiconductor Manuf.*, 1997: 17–23) reported that, in a study of a particular wafer inspection process, 356 dies were examined by an inspection probe and 201 of these passed the probe. Assuming a stable process, calculate a 95% (two-sided) confidence interval for the proportion of all dies that pass the probe.

20. TV advertising agencies face increasing challenges in reaching audience members because viewing TV programs via digital streaming is gaining in popularity. The **Harris poll** reported on November 13, 2012, that 53% of 2343 American adults surveyed said they have watched digitally streamed TV programming on some type of device.

a. Calculate and interpret a confidence interval at the 99% confidence level for the proportion of all adult Americans who watched streamed programming up to that point in time.

b. What sample size would be required for the width of a 99% CI to be at most .05 irrespective of the value of $\hat{p}$?

21. In a sample of 1000 randomly selected consumers who had opportunities to send in a rebate claim form after purchasing a product, 250 of these people said they never did so (**"Rebates: Get What You Deserve,"** *Consumer Reports*, May 2009: 7). Reasons cited for their behavior included too many steps in the process, amount too small, missed deadline, fear of being placed on a mailing list, lost receipt, and doubts about receiving the money. Calculate an upper confidence bound at the 95% confidence level for the true proportion of such consumers who never apply for a rebate. Based on this bound, is there compelling evidence that the true proportion of such consumers is smaller than 1/3? Explain your reasoning.

22. The technology underlying hip replacements has changed as these operations have become more popular (over 250,000 in the United States in 2008). Starting in 2003, highly durable ceramic hips were marketed. Unfortunately, for too many patients the increased durability has been counterbalanced by an increased incidence of squeaking. The May 11, 2008, issue of the *New York Times* reported that in one study of 143 individuals who received ceramic hips between 2003 and 2005, 10 of the hips developed squeaking.

a. Calculate a lower confidence bound at the 95% confidence level for the true proportion of such hips that develop squeaking.

b. Interpret the 95% confidence level used in (a).

23. The **Pew Forum on Religion and Public Life** reported on Dec. 9, 2009, that in a survey of 2003 American adults, 25% said they believed in astrology.

a. Calculate and interpret a confidence interval at the 99% confidence level for the proportion of all adult Americans who believe in astrology.

b. What sample size would be required for the width of a 99% CI to be at most .05 irrespective of the value of $\hat{p}$?

24. A sample of 56 research cotton samples resulted in a sample average percentage elongation of 8.17 and a sample standard deviation of 1.42 (**"An Apparent Relation Between the Spiral Angle ϕ, the Percent Elongation E_1, and the Dimensions of the Cotton Fiber,"** *Textile Research J.*, 1978: 407–410). Calculate a 95% large-sample CI for the true average percentage elongation μ. What assumptions are you making about the distribution of percentage elongation?

25. A state legislator wishes to survey residents of her district to see what proportion of the electorate is aware of her position on using state funds to pay for abortions.

a. What sample size is necessary if the 95% CI for p is to have a width of at most .10 irrespective of p?

b. If the legislator has strong reason to believe that at least 2/3 of the electorate know of her position, how large a sample size would you recommend?

26. The superintendent of a large school district, having once had a course in probability and statistics, believes that the number of teachers absent on any given day has a Poisson distribution with parameter μ. Use the accompanying data on absences for 50 days to obtain a large-sample CI for μ. [*Hint*: The mean and variance of a Poisson variable both equal μ, so

$$Z = \frac{\overline{X} - \mu}{\sqrt{\mu/n}}$$

has approximately a standard normal distribution. Now proceed as in the derivation of the interval for p by making a probability statement (with probability $1 - \alpha$) and solving the resulting inequalities for μ — see the argument just after (7.10).]

Number of absences	0	1	2	3	4	5	6	7	8	9	10
Frequency	1	4	8	10	8	7	5	3	2	1	1

27. Reconsider the CI (7.10) for p, and focus on a confidence level of 95%. Show that the confidence limits agree quite well with those of the traditional interval (7.11) once two successes and two failures have been appended to the sample [i.e., (7.11) based on $x + 2$ S's in $n + 4$ trials]. [*Hint*: $1.96 \approx 2$. *Note:* Agresti and Coull showed that this adjustment of the traditional interval also has an actual confidence level close to the nominal level.]

7.3 Intervals Based on a Normal Population Distribution

The CI for μ presented in Section 7.2 is valid provided that n is large. The resulting interval can be used whatever the nature of the population distribution. The CLT cannot be invoked, however, when n is small. In this case, one way to proceed is to make a specific assumption about the form of the population distribution and then derive a CI tailored to that assumption. For example, we could develop a CI for μ when the population is described by a gamma distribution, another interval for the case of a Weibull distribution, and so on. Statisticians have indeed carried out this program for a number of different distributional families. Because the normal distribution is more frequently appropriate as a population model than is any other type of distribution, we will focus here on a CI for this situation.

ASSUMPTION

> The population of interest is normal, so that $X_1, \ldots, X_n$ constitutes a random sample from a normal distribution with both μ and σ unknown.

The key result underlying the interval in Section 7.2 was that for large n, the rv $Z = (\overline{X} - \mu)/(S/\sqrt{n})$ has approximately a standard normal distribution. When n is small, the additional variability in the denominator implies that the probability distribution of $(\overline{X} - \mu)/(S/\sqrt{n})$ will be more spread out than the standard normal distribution. The result on which inferences are based introduces a new family of probability distributions called t *distributions*.

THEOREM

> When $\overline{X}$ is the mean of a random sample of size n from a normal distribution with mean μ, the rv
>
> $$T = \frac{\overline{X} - \mu}{S/\sqrt{n}} \qquad (7.13)$$
>
> has a probability distribution called a t distribution with $n - 1$ degrees of freedom (df).

Properties of t Distributions

Before applying this theorem, a discussion of properties of t distributions is in order. Although the variable of interest is still $(\overline{X} - \mu)/(S/\sqrt{n})$, we now denote it by T to emphasize that it does not have a standard normal distribution when n is small. Recall that a normal distribution is governed by two parameters; each different choice of μ in combination with σ gives a particular normal distribution. Any particular t distribution results from specifying the value of a single parameter, called the **number of degrees of freedom**, abbreviated df. We'll denote this parameter by the Greek letter ν. Possible values of ν are the positive integers 1, 2, 3,.... So there is a t distribution with 1 df, another with 2 df, yet another with 3 df, and so on.

For any fixed value of ν, the density function that specifies the associated t curve is even more complicated than the normal density function. Fortunately, we need concern ourselves only with several of the more important features of these curves.

Properties of t Distributions

Let t_ν denote the t distribution with ν df.

1. Each t_ν curve is bell-shaped and centered at 0.
2. Each t_ν curve is more spread out than the standard normal (z) curve.
3. As ν increases, the spread of the corresponding t_ν curve decreases.
4. As $\nu \to \infty$, the sequence of t_ν curves approaches the standard normal curve (so the z curve is often called the t curve with df $= \infty$).

Figure 7.7 illustrates several of these properties for selected values of ν.

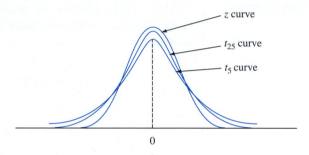

Figure 7.7 t_ν and z curves

The number of df for T in (7.13) is $n - 1$ because, although S is based on the n deviations $X_1 - \overline{X}, \ldots, X_n - \overline{X}$, $\Sigma(X_i - \overline{X}) = 0$ implies that only $n - 1$ of these are "freely determined." The number of df for a t variable is the number of freely determined deviations on which the estimated standard deviation in the denominator of T is based.

The use of t distribution in making inferences requires notation for capturing t-curve tail areas analogous to z_α for the z curve. You might think that t_α would do the trick. However, the desired value depends not only on the tail area captured but also on df.

NOTATION

Let $t_{\alpha,\nu} = $ the number on the measurement axis for which the area under the t curve with ν df to the right of $t_{\alpha,\nu}$ is α; $t_{\alpha,\nu}$ is called a **t critical value**.

For example, $t_{.05,6}$ is the t critical value that captures an upper-tail area of .05 under the t curve with 6 df. The general notation is illustrated in Figure 7.8. Because t curves are symmetric about zero, $-t_{\alpha,\nu}$ captures lower-tail area α. Appendix Table A.5 gives $t_{\alpha,\nu}$ for selected values of α and ν. This table also appears inside the back cover. The columns of the table correspond to different values of α. To obtain $t_{.05,15}$, go to the $\alpha = .05$ column, look down to the $\nu = 15$ row, and read $t_{.05,15} = 1.753$. Similarly, $t_{.05,22} = 1.717$ (.05 column, $\nu = 22$ row), and $t_{.01,22} = 2.508$.

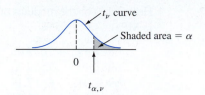

Figure 7.8 Illustration of a *t* critical value

The values of $t_{\alpha,\nu}$ exhibit regular behavior as we move across a row or down a column. For fixed ν, $t_{\alpha,\nu}$ increases as α decreases, since we must move farther to the right of zero to capture area α in the tail. For fixed α, as ν is increased (i.e., as we look down any particular column of the *t* table) the value of $t_{\alpha,\nu}$ decreases. This is because a larger value of ν implies a *t* distribution with smaller spread, so it is not necessary to go so far from zero to capture tail area α. Furthermore, $t_{\alpha,\nu}$ decreases more slowly as ν increases. Consequently, the table values are shown in increments of 2 between 30 df and 40 df and then jump to $\nu = 50, 60, 120$, and finally ∞. Because t_∞ is the standard normal curve, the familiar z_α values appear in the last row of the table. The rule of thumb suggested earlier for use of the large-sample CI (if $n > 40$) comes from the approximate equality of the standard normal and *t* distributions for $\nu \geq 40$.

The One-Sample *t* Confidence Interval

The standardized variable T has a *t* distribution with $n - 1$ df, and the area under the corresponding *t* density curve between $-t_{\alpha/2,\,n-1}$ and $t_{\alpha/2,\,n-1}$ is $1 - \alpha$ (area $\alpha/2$ lies in each tail), so

$$P\left(-t_{\alpha/2,\,n-1} < T < t_{\alpha/2,\,n-1}\right) = 1 - \alpha \tag{7.14}$$

Expression (7.14) differs from expressions in previous sections in that T and $t_{\alpha/2,n-1}$ are used in place of Z and $z_{\alpha/2}$, but it can be manipulated in the same manner to obtain a confidence interval for μ.

PROPOSITION

Let $\bar{x}$ and s be the sample mean and sample standard deviation computed from the results of a random sample from a normal population with mean μ. Then a **$100(1 - \alpha)\%$ confidence interval for μ** is

$$\left(\bar{x} - t_{\alpha/2,\,n-1} \cdot \frac{s}{\sqrt{n}}, \bar{x} + t_{\alpha/2,\,n-1} \cdot \frac{s}{\sqrt{n}}\right) \tag{7.15}$$

or, more compactly, $\bar{x} \pm t_{\alpha/2,\,n-1} \cdot s/\sqrt{n}$.

An **upper confidence bound for μ** is

$$\bar{x} + t_{\alpha,\,n-1} \cdot \frac{s}{\sqrt{n}}$$

and replacing $+$ by $-$ in this latter expression gives a **lower confidence bound for μ,** both with confidence level $100(1 - \alpha)\%$.

EXAMPLE 7.11

Even as traditional markets for sweetgum lumber have declined, large section solid timbers traditionally used for construction bridges and mats have become increasingly scarce. The article **"Development of Novel Industrial Laminated Planks from Sweetgum Lumber" (*J. of Bridge Engr.*, 2008: 64–66)** described the manufacturing and testing of composite beams designed to add value to low-grade sweetgum lumber.

Here is data on the modulus of rupture (psi; the article contained summary data expressed in MPa):

6807.99	7637.06	6663.28	6165.03	6991.41	6992.23
6981.46	7569.75	7437.88	6872.39	7663.18	6032.28
6906.04	6617.17	6984.12	7093.71	7659.50	7378.61
7295.54	6702.76	7440.17	8053.26	8284.75	7347.95
7422.69	7886.87	6316.67	7713.65	7503.33	7674.99

Figure 7.9 shows a normal probability plot from the R software. The straightness of the pattern in the plot provides strong support for assuming that the population distribution of MOR is at least approximately normal.

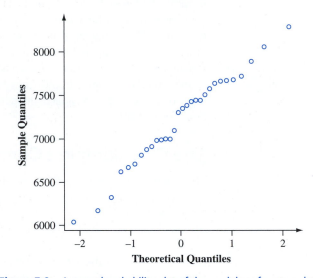

Figure 7.9 A normal probability plot of the modulus of rupture data

The sample mean and sample standard deviation are 7203.191 and 543.5400, respectively (for anyone bent on doing hand calculation, the computational burden is eased a bit by subtracting 6000 from each x value to obtain $y_i = x_i - 6000$; then $\Sigma y_i = 36{,}095.72$ and $\Sigma y_i^2 = 51{,}997{,}668.77$, from which $\bar{y} = 1203.191$ and $s_y = s_x$ as given).

Let's now calculate a confidence interval for true average MOR using a confidence level of 95%. The CI is based on $n - 1 = 29$ degrees of freedom, so the necessary t critical value is $t_{.025,29} = 2.045$. The interval estimate is now

$$\bar{x} \pm t_{.025,29} \cdot \frac{s}{\sqrt{n}} = 7203.191 \pm (2.045) \cdot \frac{543.5400}{\sqrt{30}}$$
$$= 7203.191 \pm 202.938 = (7000.253, 7406.129)$$

We estimate that $7000.253 < \mu < 7406.129$ with 95% confidence. If we use the same formula on sample after sample, in the long run 95% of the calculated intervals will contain μ. Since the value of μ is not available, we don't know whether the calculated interval is one of the "good" 95% or the "bad" 5%. Even with the moderately large sample size, our interval is rather wide. This is a consequence of the substantial amount of sample variability in MOR values.

A lower 95% confidence bound would result from retaining only the lower confidence limit (the one with $-$) and replacing 2.045 with $t_{.05,29} = 1.699$. ■

Unfortunately, it is not easy to select n to control the width of the t interval. This is because the width involves the unknown (before the data is collected) s and because n enters not only through $1/\sqrt{n}$ but also through $t_{\alpha/2,n-1}$. As a result, an appropriate n can be obtained only by trial and error.

In Chapter 15, we will discuss a small-sample CI for μ that is valid provided only that the population distribution is symmetric, a weaker assumption than normality. However, when the population distribution is normal, the t interval tends to be narrower than would be *any* other interval with the same confidence level.

A Prediction Interval for a Single Future Value

In many applications, the objective is to *predict* a single value of a variable to be observed at some future time, rather than to *estimate* the mean value of that variable.

EXAMPLE 7.12 Consider the following sample of fat content (in percentage) of $n = 10$ randomly selected hot dogs (**"Sensory and Mechanical Assessment of the Quality of Frankfurters," *J. of Texture Studies*, 1990: 395–409**):

| 25.2 | 21.3 | 22.8 | 17.0 | 29.8 | 21.0 | 25.5 | 16.0 | 20.9 | 19.5 |

Assuming that these were selected from a normal population distribution, a 95% CI for (interval estimate of) the population mean fat content is

$$\bar{x} \pm t_{.025,9} \cdot \frac{s}{\sqrt{n}} = 21.90 \pm 2.262 \cdot \frac{4.134}{\sqrt{10}} = 21.90 \pm 2.96$$
$$= (18.94, 24.86)$$

Suppose, however, you are going to eat a single hot dog of this type and want a *prediction* for the resulting fat content. A *point* prediction, analogous to a *point* estimate, is just $\bar{x} = 21.90$. This prediction unfortunately gives no information about reliability or precision. ∎

The general setup is as follows: We have available a random sample $X_1, X_2, \ldots, X_n$ from a normal population distribution, and wish to predict the value of X_{n+1}, a single future observation (e.g., the lifetime of a single lightbulb to be purchased or the fuel efficiency of a single vehicle to be rented). A point predictor is $\bar{X}$, and the resulting prediction error is $\bar{X} - X_{n+1}$. The expected value of the prediction error is

$$E(\bar{X} - X_{n+1}) = E(\bar{X}) - E(X_{n+1}) = \mu - \mu = 0$$

Since X_{n+1} is independent of $X_1, \ldots, X_n$, it is independent of $\bar{X}$, so the variance of the prediction error is

$$V(\bar{X} - X_{n+1}) = V(\bar{X}) + V(X_{n+1}) = \frac{\sigma^2}{n} + \sigma^2 = \sigma^2\left(1 + \frac{1}{n}\right)$$

The prediction error is normally distributed because it is a linear combination of independent, normally distributed rv's. Thus

$$Z = \frac{(\bar{X} - X_{n+1}) - 0}{\sqrt{\sigma^2\left(1 + \frac{1}{n}\right)}} = \frac{\bar{X} - X_{n+1}}{\sqrt{\sigma^2\left(1 + \frac{1}{n}\right)}}$$

has a standard normal distribution. It can be shown that replacing σ by the sample standard deviation S (of $X_1, \ldots, X_n$) results in

$$T = \frac{\overline{X} - X_{n+1}}{S\sqrt{1 + \frac{1}{n}}} \sim t \text{ distribution with } n - 1 \text{ df}$$

Manipulating this T variable as $T = (\overline{X} - \mu)/(S/\sqrt{n})$ was manipulated in the development of a CI gives the following result.

PROPOSITION

> A **prediction interval (PI)** for a single observation to be selected from a normal population distribution is
>
> $$\overline{x} \pm t_{\alpha/2, n-1} \cdot s\sqrt{1 + \frac{1}{n}} \qquad (7.16)$$
>
> The *prediction level* is $100(1 - \alpha)\%$. A lower prediction bound results from replacing $t_{\alpha/2}$ by t_α and discarding the $+$ part of (7.16); a similar modification gives an upper prediction bound.

The interpretation of a 95% prediction level is similar to that of a 95% confidence level. If the interval (7.16) is calculated for sample after sample and after each calculation X_{n+1} is observed, in the long run 95% of these intervals will include the corresponding future values.

EXAMPLE 7.13
(Example 7.12 continued)

With $n = 10$, $\overline{x} = 21.90$, $s = 4.134$, and $t_{.025,9} = 2.262$, a 95% PI for the fat content of a single hot dog is

$$21.90 \pm (2.262)(4.134)\sqrt{1 + \frac{1}{10}} = 21.90 \pm 9.81$$

$$= (12.09, 31.71)$$

This interval is quite wide, indicating substantial uncertainty about fat content. Notice that the width of the PI is more than three times that of the CI. ∎

The error of prediction is $\overline{X} - X_{n+1}$, a difference between two random variables, whereas the estimation error is $\overline{X} - \mu$, the difference between a random variable and a fixed (but unknown) value. The PI is wider than the CI because there is more variability in the prediction error (due to X_{n+1}) than in the estimation error. In fact, as n gets arbitrarily large, the CI shrinks to the single value μ, and the PI approaches $\mu \pm z_{\alpha/2} \cdot \sigma$. There is uncertainty about a single X value even when there is no need to estimate.

Tolerance Intervals

Consider a population of automobiles of a certain type, and suppose that under specified conditions, fuel efficiency (mpg) has a normal distribution with $\mu = 30$ and $\sigma = 2$. Then since the interval from -1.645 to 1.645 captures 90% of the area under the z curve, 90% of all these automobiles will have fuel efficiency values between $\mu - 1.645\sigma = 26.71$ and $\mu + 1.645\sigma = 33.29$. But what if the values of μ and σ are not known? We can take a sample of size n, determine the fuel efficiencies,

$\bar{x}$ and s, and form the interval whose lower limit is $\bar{x} - 1.645s$ and whose upper limit is $\bar{x} + 1.645s$. However, because of sampling variability in the estimates of μ and σ, there is a good chance that the resulting interval will include less than 90% of the population values. Intuitively, to have an *a priori* 95% chance of the resulting interval including at least 90% of the population values, when $\bar{x}$ and s are used in place of μ and σ we should also replace 1.645 by some larger number. For example, when $n = 20$, the value 2.310 is such that we can be 95% confident that the interval $\bar{x} \pm 2.310s$ will include at least 90% of the fuel efficiency values in the population.

Let k be a number between 0 and 100. A **tolerance interval** for capturing at least $k\%$ of the values in a normal population distribution with a confidence level 95% has the form

$$\bar{x} \pm \text{(tolerance critical value)} \cdot s$$

Tolerance critical values for $k = 90, 95$, and 99 in combination with various sample sizes are given in Appendix Table A.6. This table also includes critical values for a confidence level of 99% (these values are larger than the corresponding 95% values). Replacing $\pm$ by $+$ gives an upper tolerance bound, and using $-$ in place of $\pm$ results in a lower tolerance bound. Critical values for obtaining these one-sided bounds also appear in Appendix Table A.6.

EXAMPLE 7.14   As part of a larger project to study the behavior of stressed-skin panels, a structural component being used extensively in North America, the article **"Time-Dependent Bending Properties of Lumber"** (*J. of Testing and Eval.*, 1996: 187–193) reported on various mechanical properties of Scotch pine lumber specimens. Consider the following observations on modulus of elasticity (MPa) obtained 1 minute after loading in a certain configuration:

10,490	16,620	17,300	15,480	12,970	17,260	13,400	13,900
13,630	13,260	14,370	11,700	15,470	17,840	14,070	14,760

There is a pronounced linear pattern in a normal probability plot of the data. Relevant summary quantities are $n = 16$, $\bar{x} = 14{,}532.5$, $s = 2055.67$. For a confidence level of 95%, a two-sided tolerance interval for capturing at least 95% of the modulus of elasticity values for specimens of lumber in the population sampled uses the tolerance critical value of 2.903. The resulting interval is

$$14{,}532.5 \pm (2.903)(2055.67) = 14{,}532.5 \pm 5967.6 = (8{,}564.9, 20{,}500.1)$$

We can be highly confident that at least 95% of all lumber specimens have modulus of elasticity values between 8,564.9 and 20,500.1.

The 95% CI for μ is (13,437.3, 15,627.7), and the 95% prediction interval for the modulus of elasticity of a single lumber specimen is (10,017.0, 19,048.0). Both the prediction interval and the tolerance interval are substantially wider than the confidence interval. ∎

Intervals Based on Nonnormal Population Distributions

The one-sample t CI for μ is robust to small or even moderate departures from normality unless n is quite small. By this we mean that if a critical value for 95% confidence, for example, is used in calculating the interval, the actual confidence level will be reasonably close to the nominal 95% level. If, however, n is small and the population distribution is highly nonnormal, then the actual confidence level may be considerably different from the one you think you are using when you obtain a particular critical value from the t table. It would certainly be distressing to believe that your confidence level is about 95% when in fact it was really more like 88%! The bootstrap technique, introduced in Section 7.1, has been found to be quite successful at estimating parameters in a wide variety of nonnormal situations.

In contrast to the confidence interval, the validity of the prediction and tolerance intervals described in this section is closely tied to the normality assumption. These latter intervals should not be used in the absence of compelling evidence for normality. The excellent reference *Statistical Intervals,* cited in the bibliography at the end of this chapter, discusses alternative procedures of this sort for various other situations.

EXERCISES Section 7.3 (28–41)

28. Determine the values of the following quantities:

   a. $t_{.1,15}$ **b.** $t_{.05,15}$ **c.** $t_{.05,25}$ **d.** $t_{.05,40}$ **e.** $t_{.005,40}$

29. Determine the t critical value(s) that will capture the desired t-curve area in each of the following cases:

   a. Central area = .95, df = 10

   b. Central area = .95, df = 20

   c. Central area = .99, df = 20

   d. Central area = .99, df = 50

   e. Upper-tail area = .01, df = 25

   f. Lower-tail area = .025, df = 5

30. Determine the t critical value for a two-sided confidence interval in each of the following situations:

   a. Confidence level = 95%, df = 10

   b. Confidence level = 95%, df = 15

   c. Confidence level = 99%, df = 15

   d. Confidence level = 99%, n = 5

   e. Confidence level = 98%, df = 24

   f. Confidence level = 99%, n = 38

31. Determine the t critical value for a lower or an upper confidence bound for each of the situations described in Exercise 30.

32. According to the article **"Fatigue Testing of Condoms"** (***Polymer Testing*, 2009: 567–571),** "tests currently used for condoms are surrogates for the challenges they face in use," including a test for holes, an inflation test, a package seal test, and tests of dimensions and lubricant quality (all fertile territory for the use of statistical methodology!). The investigators developed a new test that adds cyclic strain to a level well below breakage and determines the number of cycles to break. A sample of 20 condoms of one particular type resulted in a sample mean number of 1584 and a sample standard deviation of 607. Calculate and interpret a

confidence interval at the 99% confidence level for the true average number of cycles to break. [*Note*: The article presented the results of hypothesis tests based on the t distribution; the validity of these depends on assuming normal population distributions.]

33. The article **"Measuring and Understanding the Aging of Kraft Insulating Paper in Power Transformers"** (***IEEE Electrical Insul. Mag.*, 1996: 28–34)** contained the following observations on degree of polymerization for paper specimens for which viscosity times concentration fell in a certain middle range:

418	421	421	422	425	427	431
434	437	439	446	447	448	453
454	463	465				

   a. Construct a boxplot of the data and comment on any interesting features.

   b. Is it plausible that the given sample observations were selected from a normal distribution?

   c. Calculate a two-sided 95% confidence interval for true average degree of polymerization (as did the authors of the article). Does the interval suggest that 440 is a plausible value for true average degree of polymerization? What about 450?

34. A sample of 14 joint specimens of a particular type gave a sample mean proportional limit stress of 8.48 MPa and a sample standard deviation of .79 MPa (**"Characterization of Bearing Strength Factors in Pegged Timber Connections,"** *J. of Structural Engr.*, 1997: 326–332).

   a. Calculate and interpret a 95% lower confidence bound for the true average proportional limit stress of all such joints. What, if any, assumptions did you make about the distribution of proportional limit stress?

b. Calculate and interpret a 95% lower prediction bound for the proportional limit stress of a single joint of this type.

35. Silicone implant augmentation rhinoplasty is used to correct congenital nose deformities. The success of the procedure depends on various biomechanical properties of the human nasal periosteum and fascia. The article **"Biomechanics in Augmentation Rhinoplasty"** (**J. of Med. Engr. and Tech.**, 2005: 14–17) reported that for a sample of 15 (newly deceased) adults, the mean failure strain (%) was 25.0, and the standard deviation was 3.5.

    a. Assuming a normal distribution for failure strain, estimate true average strain in a way that conveys information about precision and reliability.

    b. Predict the strain for a single adult in a way that conveys information about precision and reliability. How does the prediction compare to the estimate calculated in part (a)?

36. A normal probability plot of the $n = 26$ observations on escape time given in Exercise 36 of Chapter 1 shows a substantial linear pattern; the sample mean and sample standard deviation are 370.69 and 24.36, respectively.

    a. Calculate an upper confidence bound for population mean escape time using a confidence level of 95%.

    b. Calculate an upper prediction bound for the escape time of a single additional worker using a prediction level of 95%. How does this bound compare with the confidence bound of part (a)?

    c. Suppose that two additional workers will be chosen to participate in the simulated escape exercise. Denote their escape times by X_{27} and X_{28}, and let $\overline{X}_{new}$ denote the average of these two values. Modify the formula for a PI for a single x value to obtain a PI for $\overline{X}_{new}$, and calculate a 95% two-sided interval based on the given escape data.

37. A study of the ability of individuals to walk in a straight line (**"Can We Really Walk Straight?"** **Amer. J. of Physical Anthro.**, 1992: 19–27) reported the accompanying data on cadence (strides per second) for a sample of $n = 20$ randomly selected healthy men.

```
.95   .85   .92   .95   .93   .86   1.00   .92   .85   .81
.78   .93   .93   1.05  .93   1.06  1.06   .96   .81   .96
```

A normal probability plot gives substantial support to the assumption that the population distribution of cadence is approximately normal. A descriptive summary of the data from Minitab follows:

```
Variable N   Mean    Median  TrMean   StDev  SEMean
cadence  20  0.9255  0.9300  0.9261  0.0809 0.0181

Variable     Min     Max     Q1       Q3
cadence      0.7800  1.0600  0.8525   0.9600
```

    a. Calculate and interpret a 95% confidence interval for population mean cadence.

    b. Calculate and interpret a 95% prediction interval for the cadence of a single individual randomly selected from this population.

    c. Calculate an interval that includes at least 99% of the cadences in the population distribution using a confidence level of 95%.

38. Ultra high performance concrete (UHPC) is a relatively new construction material that is characterized by strong adhesive properties with other materials. The article **"Adhesive Power of Ultra High Performance Concrete from a Thermodynamic Point of View"** (**J. of Materials in Civil Engr.**, 2012: 1050–1058) described an investigation of the intermolecular forces for UHPC connected to various substrates. The following work of adhesion measurements (in mJ/m^2) for UHPC specimens adhered to steel appeared in the article:

```
107.1   109.5   107.4   106.8   108.1
```

    a. Is it plausible that the given sample observations were selected from a normal distribution?

    b. Calculate a two-sided 95% confidence interval for the true average work of adhesion for UHPC adhered to steel. Does the interval suggest that 107 is a plausible value for the true average work of adhesion for UHPC adhered to steel? What about 110?

    c. Predict the resulting work of adhesion value resulting from a single future replication of the experiment by calculating a 95% prediction interval, and compare the width of this interval to the width of the CI from (b).

    d. Calculate an interval for which you can have a high degree of confidence that at least 95% of all UHPC specimens adhered to steel will have work of adhesion values between the limits of the interval.

39. Exercise 72 of Chapter 1 gave the following observations on a receptor binding measure (adjusted distribution volume) for a sample of 13 healthy individuals: 23, 39, 40, 41, 43, 47, 51, 58, 63, 66, 67, 69, 72.

    a. Is it plausible that the population distribution from which this sample was selected is normal?

    b. Calculate an interval for which you can be 95% confident that at least 95% of all healthy individuals in the population have adjusted distribution volumes lying between the limits of the interval.

    c. Predict the adjusted distribution volume of a single healthy individual by calculating a 95% prediction interval. How does this interval's width compare to the width of the interval calculated in part (b)?

40. Exercise 13 of Chapter 1 presented a sample of $n = 153$ observations on ultimate tensile strength, and Exercise 17 of the previous section gave summary quantities and requested a large-sample confidence interval. Because the sample size is large, no assumptions about the population distribution are required for the validity of the CI.

    a. Is any assumption about the tensile-strength distribution required prior to calculating a lower prediction bound for the tensile strength of the next specimen selected using the method described in this section? Explain.

b. Use a statistical software package to investigate the plausibility of a normal population distribution.

c. Calculate a lower prediction bound with a prediction level of 95% for the ultimate tensile strength of the next specimen selected.

41. A more extensive tabulation of t critical values than what appears in this book shows that for the t distribution with 20 df, the areas to the right of the values .687, .860, and 1.064 are .25, .20, and .15, respectively. What is the confidence level for each of the following three confidence intervals for the mean μ of a normal population distribution? Which of the three intervals would you recommend be used, and why?

a. $(\bar{x} - .687s/\sqrt{21}, \bar{x} + 1.725s/\sqrt{21})$

b. $(\bar{x} - .860s/\sqrt{21}, \bar{x} + 1.325s/\sqrt{21})$

c. $(\bar{x} - 1.064s/\sqrt{21}, \bar{x} + 1.064s/\sqrt{21})$

7.4 Confidence Intervals for the Variance and Standard Deviation of a Normal Population

Although inferences concerning a population variance σ^2 or standard deviation σ are usually of less interest than those about a mean or proportion, there are occasions when such procedures are needed. In the case of a normal population distribution, inferences are based on the following result concerning the sample variance S^2.

THEOREM

Let $X_1, X_2, \ldots, X_n$ be a random sample from a normal distribution with parameters μ and σ^2. Then the rv

$$\frac{(n-1)S^2}{\sigma^2} = \frac{\Sigma(X_i - \overline{X})^2}{\sigma^2}$$

has a chi-squared (χ^2) probability distribution with $n - 1$ df.

As discussed in Sections 4.4 and 7.1, the chi-squared distribution is a continuous probability distribution with a single parameter ν, called the number of degrees of freedom, with possible values 1, 2, 3, ... The graphs of several χ^2 probability density functions (pdf's) are illustrated in Figure 7.10. Each pdf $f(x; \nu)$ is positive only for $x > 0$, and each has a positive skew (stretched out upper tail), though the distribution moves rightward and becomes more symmetric as ν increases. To specify inferential procedures that use the chi-squared distribution, we need notation analogous to that for a t critical value $t_{\alpha,\nu}$.

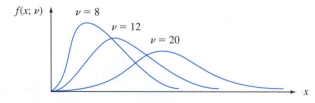

Figure 7.10 Graphs of chi-squared density functions

NOTATION

Let $\chi^2_{\alpha,\nu}$, called a **chi-squared critical value,** denote the number on the horizontal axis such that α of the area under the chi-squared curve with ν df lies to the right of $\chi^2_{\alpha,\nu}$.

Symmetry of t distributions made it necessary to tabulate only upper-tailed t critical values ($t_{\alpha,\nu}$ for small values of α). The chi-squared distribution is not symmetric, so Appendix Table A.7 contains values of $\chi^2_{\alpha,\nu}$ both for α near 0 and near 1, as illustrated in Figure 7.11(b). For example, $\chi^2_{.025,14} = 26.119$, and $\chi^2_{.95,20}$ (the 5th percentile) $= 10.851$.

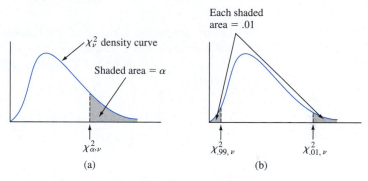

Figure 7.11 $\chi^2_{\alpha,\nu}$ notation illustrated

The rv $(n-1)S^2/\sigma^2$ satisfies the two properties on which the general method for obtaining a CI is based: It is a function of the parameter of interest σ^2, yet its probability distribution (chi-squared) does not depend on this parameter. The area under a chi-squared curve with ν df to the right of $\chi^2_{\alpha/2,\nu}$ is $\alpha/2$, as is the area to the left of $\chi^2_{1-\alpha/2,\nu}$. Thus the area captured between these two critical values is $1-\alpha$. As a consequence of this and the theorem just stated,

$$P\left(\chi^2_{1-\alpha/2,n-1} < \frac{(n-1)S^2}{\sigma^2} < \chi^2_{\alpha/2,n-1}\right) = 1 - \alpha \qquad (7.17)$$

The inequalities in (7.17) are equivalent to

$$\frac{(n-1)S^2}{\chi^2_{\alpha/2,n-1}} < \sigma^2 < \frac{(n-1)S^2}{\chi^2_{1-\alpha/2,n-1}}$$

Substituting the computed value s^2 into the limits gives a CI for σ^2, and taking square roots gives an interval for σ.

A **100$(1-\alpha)$% confidence interval for the variance σ^2 of a normal population** has lower limit

$$(n-1)s^2/\chi^2_{\alpha/2,n-1}$$

and upper limit

$$(n-1)s^2/\chi^2_{1-\alpha/2,n-1}$$

A **confidence interval for σ** has lower and upper limits that are the square roots of the corresponding limits in the interval for σ^2. An upper or a lower confidence bound results from replacing $\alpha/2$ with α in the corresponding limit of the CI.

EXAMPLE 7.15   The accompanying data on breakdown voltage of electrically stressed circuits was read from a normal probability plot that appeared in the article **"Damage of Flexible Printed Wiring Boards Associated with Lightning-Induced Voltage Surges"**

(IEEE Transactions on Components, Hybrids, and Manuf. Tech., 1985: 214–220). The straightness of the plot gave strong support to the assumption that breakdown voltage is approximately normally distributed.

1470	1510	1690	1740	1900	2000	2030	2100	2190
2200	2290	2380	2390	2480	2500	2580	2700	

Let σ^2 denote the variance of the breakdown voltage distribution. The computed value of the sample variance is $s^2 = 137,324.3$, the point estimate of σ^2. With df $= n - 1 = 16$, a 95% CI requires $\chi^2_{.975,16} = 6.908$ and $\chi^2_{.025,16} = 28.845$. The interval is

$$\left(\frac{16(137,324.3)}{28.845}, \frac{16(137,324.3)}{6.908} \right) = (76,172.3, 318,064.4)$$

Taking the square root of each endpoint yields (276.0, 564.0) as the 95% CI for σ. These intervals are quite wide, reflecting substantial variability in breakdown voltage in combination with a small sample size. ■

CIs for σ^2 and σ when the population distribution is not normal can be difficult to obtain. For such cases, consult a knowledgeable statistician.

EXERCISES Section 7.4 (42–46)

42. Determine the values of the following quantities:
   a. $\chi^2_{.1,15}$ **b.** $\chi^2_{.1,25}$
   c. $\chi^2_{.01,25}$ **d.** $\chi^2_{.005,25}$
   e. $\chi^2_{.99,25}$ **f.** $\chi^2_{.995,25}$

43. Determine the following:
   a. The 95th percentile of the chi-squared distribution with $\nu = 10$
   b. The 5th percentile of the chi-squared distribution with $\nu = 10$
   c. $P(10.98 \le \chi^2 \le 36.78)$, where χ^2 is a chi-squared rv with $\nu = 22$
   d. $P(\chi^2 < 14.611 \text{ or } \chi^2 > 37.652)$, where χ^2 is a chi-squared rv with $\nu = 25$

44. The amount of lateral expansion (mils) was determined for a sample of $n = 9$ pulsed-power gas metal arc welds used in LNG ship containment tanks. The resulting sample standard deviation was $s = 2.81$ mils. Assuming normality, derive a 95% CI for σ^2 and for σ.

45. Wire electrical-discharge machining (WEDM) is a process used to manufacture conductive hard metal components. It uses a continuously moving wire that serves as an electrode. Coating on the wire electrode allows for cooling of the wire electrode core and provides an improved cutting performance. The article **"High-Performance Wire Electrodes for Wire Electrical-Discharge Machining—A Review"** (*J. of Engr. Manuf.,* **2012: 1757–1773**) gave the following sample observations on total coating layer thickness (in μm) of eight wire electrodes used for WEDM:

21	16	29	35	42	24	24	25

Calculate a 99% CI for the standard deviation of the coating layer thickness distribution. Is this interval valid whatever the nature of the distribution? Explain.

46. The article **"Concrete Pressure on Formwork"** (*Mag. of Concrete Res.,* **2009: 407–417**) gave the following observations on maximum concrete pressure (kN/m²):

33.2	41.8	37.3	40.2	36.7	39.1	36.2	41.8
36.0	35.2	36.7	38.9	35.8	35.2	40.1	

   a. Is it plausible that this sample was selected from a normal population distribution?
   b. Calculate an upper confidence bound with confidence level 95% for the population standard deviation of maximum pressure.

SUPPLEMENTARY EXERCISES (47–62)

47. Example 1.11 introduced the accompanying observations on bond strength.

11.5	12.1	9.9	9.3	7.8	6.2	6.6	7.0
13.4	17.1	9.3	5.6	5.7	5.4	5.2	5.1
4.9	10.7	15.2	8.5	4.2	4.0	3.9	3.8
3.6	3.4	20.6	25.5	13.8	12.6	13.1	8.9
8.2	10.7	14.2	7.6	5.2	5.5	5.1	5.0
5.2	4.8	4.1	3.8	3.7	3.6	3.6	3.6

a. Estimate true average bond strength in a way that conveys information about precision and reliability. [*Hint*: $\Sigma x_i = 387.8$ and $\Sigma x_i^2 = 4247.08$.]

b. Calculate a 95% CI for the proportion of all such bonds whose strength values would exceed 10.

48. The article **"Distributions of Compressive Strength Obtained from Various Diameter Cores"** (*ACI Materials J.*, 2012: 597–606) described a study in which compressive strengths were determined for concrete specimens of various types, core diameters, and length-to-diameter ratios. For one particular type, diameter, and *l/d* ratio, the 18 tested specimens resulted in a sample mean compressive strength of 64.41 MPa and a sample standard deviation of 10.32 MPa. Normality of the compressive strength distribution was judged to be quite plausible.

a. Calculate a confidence interval with confidence level 98% for the true average compressive strength under these circumstances.

b. Calculate a 98% lower prediction bound for the compressive strength of a single future specimen tested under the given circumstances. [*Hint*: $t_{.02,17} = 2.224$.]

49. For those of you who don't already know, dragon boat racing is a competitive water sport that involves 20 paddlers propelling a boat across various race distances. It has become increasingly popular over the last few years. The article **"Physiological and Physical Characteristics of Elite Dragon Boat Paddlers"** (*J. of Strength and Conditioning*, 2013: 137–145) summarized an extensive statistical analysis of data obtained from a sample of 11 paddlers. It reported that a 95% confidence interval for true average force (N) during a simulated 200-m race was (60.2, 70.6). Obtain a 95% prediction interval for the force of a single randomly selected dragon boat paddler undergoing the simulated race.

50. A journal article reports that a sample of size 5 was used as a basis for calculating a 95% CI for the true average natural frequency (Hz) of delaminated beams of a certain type. The resulting interval was (229.764, 233.504). You decide that a confidence level of 99% is more appropriate than the 95% level used. What are the limits of the 99% interval? [*Hint*: Use the center of the interval and its width to determine $\bar{x}$ and s.]

51. Unexplained respiratory symptoms reported by athletes are often incorrectly considered secondary to exercise-induced asthma. The article **"High Prevalence of Exercise-Induced Laryngeal Obstruction in Athletes"** (*Medicine and Science in Sports and Exercise*, 2013: 2030–2035) suggested that many such cases could instead be explained by obstruction of the larynx. In a sample of 88 athletes referred for an asthma workup, 31 were found to have the EILO condition.

a. Calculate and interpret a confidence interval using a 95% confidence level for the true proportion of all athletes found to have the EILO condition under these circumstances.

b. What sample size is required if the desired width of the 95% CI is to be at most .04, irrespective of the sample results?

c. Does the upper limit of the interval in (a) specify a 95% upper confidence bound for the proportion being estimated? Explain.

52. High concentration of the toxic element arsenic is all too common in groundwater. The article **"Evaluation of Treatment Systems for the Removal of Arsenic from Groundwater"** (*Practice Periodical of Hazardous, Toxic, and Radioactive Waste Mgmt.*, 2005: 152–157) reported that for a sample of $n = 5$ water specimens selected for treatment by coagulation, the sample mean arsenic concentration was 24.3 μg/L, and the sample standard deviation was 4.1. The authors of the cited article used *t*-based methods to analyze their data, so hopefully had reason to believe that the distribution of arsenic concentration was normal.

a. Calculate and interpret a 95% CI for true average arsenic concentration in all such water specimens.

b. Calculate a 90% upper confidence bound for the standard deviation of the arsenic concentration distribution.

c. Predict the arsenic concentration for a single water specimen in a way that conveys information about precision and reliability.

53. Aphid infestation of fruit trees can be controlled either by spraying with pesticide or by inundation with ladybugs. In a particular area, four different groves of fruit trees are selected for experimentation. The first three groves are sprayed with pesticides 1, 2, and 3, respectively, and the fourth is treated with ladybugs, with the following results on yield:

Treatment	$n_i =$ Number of Trees	$\bar{x}_i$ (Bushels/Tree)	s_i
1	100	10.5	1.5
2	90	10.0	1.3
3	100	10.1	1.8
4	120	10.7	1.6

Let $\mu_i =$ the true average yield (bushels/tree) after receiving the ith treatment. Then

$$\theta = \frac{1}{3}(\mu_1 + \mu_2 + \mu_3) - \mu_4$$

measures the difference in true average yields between treatment with pesticides and treatment with ladybugs. When n_1, n_2, n_3, and n_4 are all large, the estimator $\hat{\theta}$ obtained by replacing each μ_i by $\bar{X}_i$ is approximately normal. Use this to derive a large-sample $100(1 - \alpha)\%$ CI for θ, and compute the 95% interval for the given data.

54. It is important that face masks used by firefighters be able to withstand high temperatures because firefighters commonly work in temperatures of 200–500°F. In a test of one type of mask, 11 of 55 masks had lenses pop out at 250°. Construct a 90% upper confidence bound for the true proportion of masks of this type whose lenses would pop out at 250°.

55. A manufacturer of college textbooks is interested in estimating the strength of the bindings produced by a particular binding machine. Strength can be measured by recording the force required to pull the pages from the binding. If this force is measured in pounds, how many books should be tested to estimate the average force required to break the binding to within .1 lb with 95% confidence? Assume that σ is known to be .8.

56. The accompanying data on crack initiation depth (μm) was read from a lognormal probability plot that appeared in the article **"Incorporating Small Fatigue Crack Growth in Probabilistic Life Prediction: Effect of Stress Ratio in Ti-6Al-2Sn-6Mo"** (*Intl. J. of Fatigue,* 2013: 83–95). Although the pattern in the plot was quite straight, a normal probability plot of the data also shows a reasonably linear pattern. And a boxplot indicates that the distribution is quite symmetric in the middle 50% of the data and only mildly skewed overall. It is therefore reasonable to estimate and predict using t intervals.

| 4.7 | 5.1 | 5.2 | 5.3 | 5.6 | 5.8 | 6.3 | 6.7 |
| 7.2 | 7.4 | 7.7 | 8.5 | 8.9 | 9.3 | 10.1 | 11.2 |

a. Estimate the true average crack initiation depth with a 99% CI and interpret the resulting interval.

b. Predict the value of a single crack initiation depth by constructing a 99% PI.

c. Interpret in context the meaning of 99% in (b).

57. In Example 6.8, we introduced the concept of a censored experiment in which n components are put on test and the experiment terminates as soon as r of the components have failed. Suppose component lifetimes are independent, each having an exponential distribution with parameter λ. Let Y_1 denote the time at which the first failure occurs, Y_2 the time at which the second failure occurs, and so on, so that $T_r = Y_1 + \cdots + Y_r + (n - r)Y_r$ is the total accumulated lifetime at termination. Then it can be shown that $2\lambda T_r$ has a chi-squared distribution with $2r$ df. Use this fact to develop a $100(1 - \alpha)\%$ CI formula for true average lifetime $1/\lambda$. Compute a 95% CI from the data in Example 6.8.

58. Let $X_1, X_2, \ldots, X_n$ be a random sample from a continuous probability distribution having median $\tilde{\mu}$ (so that $P(X_i \leq \tilde{\mu}) = P(X_i \geq \tilde{\mu}) = .5$).

a. Show that

$$P(\min (X_i) < \tilde{\mu} < \max (X_i)) = 1 - \left(\frac{1}{2}\right)^{n-1}$$

so that $(\min(x_i), \max(x_i))$ is a $100(1 - \alpha)\%$ confidence interval for $\tilde{\mu}$ with $\alpha = \left(\frac{1}{2}\right)^{n-1}$. [*Hint:* The complement of the event $\{\min (X_i) < \tilde{\mu} < \max (X_i)\}$ is $\{\max (X_i) \leq \tilde{\mu}\} \cup \{\min (X_i) \geq \tilde{\mu}\}$. But $\max (X_i) \leq \tilde{\mu}$ iff $X_i \leq \tilde{\mu}$ for all i.]

b. For each of six normal male infants, the amount of the amino acid alanine (mg/100 mL) was determined while the infants were on an isoleucine-free diet, resulting in the following data:

| 2.84 | 3.54 | 2.80 | 1.44 | 2.94 | 2.70 |

Compute a 97% CI for the true median amount of alanine for infants on such a diet (**"The Essential Amino Acid Requirements of Infants,"** *Amer. J. of Nutrition,* 1964: 322–330).

c. Let $x_{(2)}$ denote the second smallest of the x_i's and $x_{(n-1)}$ denote the second largest of the x_i's. What is the confidence level of the interval $(x_{(2)}, x_{(n-1)})$ for $\tilde{\mu}$?

59. Let $X_1, X_2, \ldots, X_n$ be a random sample from a uniform distribution on the interval $[0, \theta]$, so that

$$f(x) = \begin{cases} \dfrac{1}{\theta} & 0 \leq x \leq \theta \\ 0 & \text{otherwise} \end{cases}$$

Then if $Y = \max (X_i)$, it can be shown that the rv $U = Y/\theta$ has density function

$$f_U(u) = \begin{cases} nu^{n-1} & 0 \leq u \leq 1 \\ 0 & \text{otherwise} \end{cases}$$

a. Use $f_U(u)$ to verify that

$$P\left((\alpha/2)^{1/n} < \frac{Y}{\theta} \leq (1 - \alpha/2)^{1/n}\right) = 1 - \alpha$$

and use this to derive a $100(1 - \alpha)\%$ CI for θ.

b. Verify that $P(\alpha^{1/n} \le Y/\theta \le 1) = 1 - \alpha$, and derive a $100(1 - \alpha)\%$ CI for θ based on this probability statement.

c. Which of the two intervals derived previously is shorter? If my waiting time for a morning bus is uniformly distributed and observed waiting times are $x_1 = 4.2$, $x_2 = 3.5$, $x_3 = 1.7$, $x_4 = 1.2$, and $x_5 = 2.4$, derive a 95% CI for θ by using the shorter of the two intervals.

60. Let $0 \le \gamma \le \alpha$. Then a $100(1 - \alpha)\%$ CI for μ when n is large is

$$\left(\bar{x} - z_\gamma \cdot \frac{s}{\sqrt{n}}, \bar{x} + z_{\alpha-\gamma} \cdot \frac{s}{\sqrt{n}} \right)$$

The choice $\gamma = \alpha/2$ yields the usual interval derived in Section 7.2; if $\gamma \ne \alpha/2$, this interval is not symmetric about $\bar{x}$. The width of this interval is $w = s(z_\gamma + z_{\alpha-\gamma})/\sqrt{n}$. Show that w is minimized for the choice $\gamma = \alpha/2$, so that the symmetric interval is the shortest. [*Hints*: (a) By definition of z_α, $\Phi(z_\alpha) = 1 - \alpha$, so that $z_\alpha = \Phi^{-1}(1 - \alpha)$; (b) the relationship between the derivative of a function $y = f(x)$ and the inverse function $x = f^{-1}(y)$ is $(d/dy)f^{-1}(y) = 1/f'(x)$.]

61. Suppose $x_1, x_2, \ldots, x_n$ are observed values resulting from a random sample from a symmetric but possibly heavy-tailed distribution. Let $\tilde{x}$ and f_s denote the sample median and fourth spread, respectively. Chapter 11 of *Understanding Robust and Exploratory Data Analysis* (see the bibliography in Chapter 6) suggests the following robust 95% CI for the population mean (point of symmetry):

$$\tilde{x} \pm \left(\frac{\text{conservative } t \text{ critical value}}{1.075} \right) \cdot \frac{f_s}{\sqrt{n}}$$

The value of the quantity in parentheses is 2.10 for $n = 10$, 1.94 for $n = 20$, and 1.91 for $n = 30$. Compute this CI for the data of Exercise 45, and compare to the t CI appropriate for a normal population distribution.

62. a. Use the results of Example 7.5 to obtain a 95% lower confidence bound for the parameter λ of an exponential distribution, and calculate the bound based on the data given in the example.

b. If lifetime X has an exponential distribution, the probability that lifetime exceeds t is $P(X > t) = e^{-\lambda t}$. Use the result of part (a) to obtain a 95% lower confidence bound for the probability that breakdown time exceeds 100 min.

BIBLIOGRAPHY

DeGroot, Morris, and Mark Schervish, *Probability and Statistics* (4th ed.), Addison-Wesley, Upper Saddle River, NJ, 2012. A very good exposition of the general principles of statistical inference.

Devore, Jay, and Kenneth Berk, *Modern Mathematical Statistics with Applications,* Springer, New York, 2012. The exposition is a bit more comprehensive and sophisticated than that of the current book, and includes more material on bootstrapping.

Hahn, Gerald, and William Meeker, *Statistical Intervals,* Wiley, New York, 1991. Almost everything you ever wanted to know about statistical intervals (confidence, prediction, tolerance, and others).

8 Tests of Hypotheses Based on a Single Sample

INTRODUCTION

A parameter can be estimated from sample data either by a single number (a point estimate) or an entire interval of plausible values (a confidence interval). Frequently, however, the objective of an investigation is not to estimate a parameter but to decide which of two contradictory claims about the parameter is correct. Methods for accomplishing this comprise the part of statistical inference called *hypothesis testing*. In this chapter, we first discuss some of the basic concepts and terminology in hypothesis testing and then develop decision procedures for the most frequently encountered testing problems based on a sample from a single population.

8.1 Hypotheses and Test Procedures

A **statistical hypothesis**, or just *hypothesis,* is a claim or assertion either about the value of a single parameter (population characteristic or characteristic of a probability distribution), about the values of several parameters, or about the form of an entire probability distribution. One example of a hypothesis is the claim $\mu = .75$, where μ is the true average inside diameter of a certain type of PVC pipe. Another example is the statement $p < .10$, where p is the proportion of defective circuit boards among all circuit boards produced by a certain manufacturer. If μ_1 and μ_2 denote the true average breaking strengths of two different types of twine, one hypothesis is the assertion that $\mu_1 - \mu_2 = 0$, and another is the statement $\mu_1 - \mu_2 > 5$. Yet another example of a hypothesis is the assertion that vehicle braking distance under particular conditions has a normal distribution. Hypotheses of this latter sort will be considered in Chapter 14. In this and the next several chapters, we concentrate on hypotheses about parameters.

In any hypothesis-testing problem, there are two contradictory hypotheses under consideration. One hypothesis might be the claim $\mu = .75$ and the other $\mu \neq .75$, or the two contradictory statements might be $p \geq .10$ and $p < .10$. The objective is to decide, based on sample information, which of the two hypotheses is correct. There is a familiar analogy to this in a criminal trial. One claim is the assertion that the accused individual is innocent. In the U.S. judicial system, this is the claim that is initially believed to be true. Only in the face of strong evidence to the contrary should the jury reject this claim in favor of the alternative assertion that the accused is guilty. In this sense, the claim of innocence is the favored or protected hypothesis, and the burden of proof is placed on those who believe in the alternative claim.

Similarly, in testing statistical hypotheses, the problem will be formulated so that one of the claims is initially favored. This initially favored claim will not be rejected in favor of the alternative claim unless sample evidence contradicts it and provides strong support for the alternative assertion.

DEFINITION

> The **null hypothesis,** denoted by H_0, is the claim that is initially assumed to be true (the "prior belief" claim). The **alternative hypothesis,** denoted by H_a, is the assertion that is contradictory to H_0.
>
> The null hypothesis will be rejected in favor of the alternative hypothesis only if sample evidence suggests that H_0 is false. If the sample does not strongly contradict H_0, we will continue to believe in the plausibility of the null hypothesis. The two possible conclusions from a hypothesis-testing analysis are then *reject H_0* or *fail to reject H_0.*

A **test of hypotheses** is a method for using sample data to decide whether the null hypothesis should be rejected. Thus we might test $H_0: \mu = .75$ against the alternative $H_a: \mu \neq .75$. Only if sample data strongly suggests that μ is something other than .75 should the null hypothesis be rejected. In the absence of such evidence, H_0 should not be rejected, since it is still quite plausible.

Sometimes an investigator does not want to accept a particular assertion unless and until data can provide strong support for the assertion. As an example, suppose a company is considering putting a new type of coating on bearings that it produces.

The true average wear life with the current coating is known to be 1000 hours. With μ denoting the true average life for the new coating, the company would not want to make a change unless evidence strongly suggested that μ exceeds 1000. An appropriate problem formulation would involve testing $H_0: \mu = 1000$ against $H_a: \mu > 1000$. The conclusion that a change is justified is identified with H_a, and it would take conclusive evidence to justify rejecting H_0 and switching to the new coating.

Scientific research often involves trying to decide whether a current theory should be replaced by a more plausible and satisfactory explanation of the phenomenon under investigation. A conservative approach is to identify the current theory with H_0 and the researcher's alternative explanation with H_a. Rejection of the current theory will then occur only when evidence is much more consistent with the new theory. In many situations, H_a is referred to as the "researcher's hypothesis," since it is the claim that the researcher would really like to validate. The word *null* means "of no value, effect, or consequence," which suggests that H_0 should be identified with the hypothesis of no change (from current opinion), no difference, no improvement, and so on. Suppose, for example, that 10% of all circuit boards produced by a certain manufacturer during a recent period were defective. An engineer has suggested a change in the production process in the belief that it will result in a reduced defective rate. Let p denote the true proportion of defective boards resulting from the changed process. Then the research hypothesis, on which the burden of proof is placed, is the assertion that $p < .10$. Thus the alternative hypothesis is $H_a: p < .10$.

In our treatment of hypothesis testing, H_0 will generally be stated as an equality claim. If θ denotes the parameter of interest, the null hypothesis will have the form $H_0: \theta = \theta_0$, where θ_0 is a specified number called the **null value** of the parameter (value claimed for θ by the null hypothesis). As an example, consider the circuit board situation just discussed. The suggested alternative hypothesis was $H_a: p < .10$, the claim that the defective rate is reduced by the process modification. A natural choice of H_0 in this situation is the claim that $p \geq .10$, according to which the new process is either no better *or* worse than the one currently used. We will instead consider $H_0: p = .10$ versus $H_a: p < .10$. The rationale for using this simplified null hypothesis is that any reasonable decision procedure for deciding between $H_0: p = .10$ and $H_a: p < .10$ will also be reasonable for deciding between the claim that $p \geq .10$ and H_a. The use of a simplified H_0 is preferred because it has certain technical benefits, which will be apparent shortly.

The alternative to the null hypothesis $H_0: \theta = \theta_0$ will look like one of the following three assertions:

1. $H_a: \theta > \theta_0$ (in which case the implicit null hypothesis is $\theta \leq \theta_0$),

2. $H_a: \theta < \theta_0$ (in which case the implicit null hypothesis is $\theta \geq \theta_0$), or

3. $H_a: \theta \neq \theta_0$

For example, let σ denote the standard deviation of the distribution of inside diameters (inches) for a certain type of metal sleeve. If the decision was made to use the sleeve unless sample evidence conclusively demonstrated that $\sigma > .001$, the appropriate hypotheses would be $H_0: \sigma = .001$ versus $H_a: \sigma > .001$. The number θ_0 that appears in both H_0 and H_a (separates the alternative from the null) is the null value.

Test Procedures and *P*-Values

A test procedure is a rule, based on sample data, for deciding whether H_0 should be rejected. The key issue will be the following: Suppose that H_0 is in fact true. Then how likely is it that a (random) sample at least as contradictory to this hypothesis as our sample would result? Consider the following two scenarios:

1. There is only a .1% chance (a probability of .001) of getting a sample at least as contradictory to H_0 as what we obtained assuming that H_0 is true.

2. There is a 25% chance (a probability of .25) of getting a sample at least as contradictory to H_0 as what we obtained when H_0 is true.

In the first scenario, something as extreme as our sample is very unlikely to have occurred when H_0 is true—in the long run only 1 in 1000 samples would be at least as contradictory to the null hypothesis as the one we ended up selecting. In contrast, for the second scenario, in the long run 25 out of every 100 samples would be at least as contradictory to H_0 as what we obtained assuming that the null hypothesis is true. So our sample is quite consistent with H_0, and there is no reason to reject it.

We must now flesh out this reasoning by being more specific as to what is meant by "at least as contradictory to H_0 as the sample we obtained when H_0 is true." Before doing so in a general way, let's consider several examples.

EXAMPLE 8.1 The company that manufactures brand D Greek-style yogurt is anxious to increase its market share, and in particular persuade those who currently prefer brand C to switch brands. So the marketing department has devised the following blind taste experiment. Each of 100 brand C consumers will be asked to taste yogurt from two bowls, one containing brand C and the other brand D, and then say which one he or she prefers. The bowls are marked with a code so that the experimenters know which bowl contains which yogurt, but the experimental subjects do not have this information (Note: Such an experiment involving beers was actually carried out several decades ago, with the now defunct Schlitz beer playing the role of brand D and Michelob being the target beer).

Let p denote the proportion of all brand C consumers who would prefer C to D in such circumstances. Let's consider testing the hypotheses $H_0\colon p = .5$ versus $H_a\colon p < .5$. The alternative hypothesis says that a majority of brand C consumers actually prefer brand D. Of course the brand D company would like to have H_0 rejected so that H_a is judged the more plausible hypothesis. If the null hypothesis is true, then whether a single randomly selected brand C consumer prefers C or D is analogous to the result of flipping a fair coin.

The sample data will consist of a sequence of 100 preferences, each one a C or a D. Let X = the number among the 100 selected individuals who prefer C to D. This random variable will serve as our *test statistic*, the function of sample data on which we'll base our conclusion. Now X is a binomial random variable (the number of successes in an experiment with a fixed number of independent trials having constant success probability p). When H_0 is true, this test statistic has a binomial distribution with $p = .5$, in which case $E(X) = np = 100(.5) = 50$.

Intuitively, a value of X "considerably" smaller than 50 argues for rejection of H_0 in favor of H_a. Suppose the observed value of X is $x = 37$. How contradictory is this value to the null hypothesis? To answer this question, let's first identify values of X that are even more contradictory to H_0 than is 37 itself. Clearly 35 is one such value, and 30 is another; in fact, any number smaller than 37 is a value of X more contradictory to the null hypothesis than is the value we actually observed. Now consider the probability, computed assuming that the null hypothesis is true, of obtaining a value of X at least as contradictory to H_0 as is our observed value:

$$P(X \le 37 \text{ when } H_0 \text{ is true}) = P(X \le 37 \text{ when } X \sim \text{Bin}(100, .5))$$

$$= B(37; 100, .5) = .006$$

(from software). Thus if the null hypothesis is true, there is less than a 1% chance of seeing 37 or fewer successes amongst the 100 trials. This suggests that $x = 37$ is much more consistent with the alternative hypothesis than with the null, and that rejection of H_0 in favor of H_a is a sensible conclusion. In addition, note that $\sigma_X = \sqrt{npq} = \sqrt{100(.5)(.5)} = 5$ when H_0 is true. It follows that 37 is more than 2.5 standard deviations smaller than what we'd expect to see were H_0 true.

Now suppose that 45 of the 100 individuals in the experiment prefer C (45 successes). Let's again calculate the probability, assuming H_0 true, of getting a test statistic value at least as contradictory to H_0 as this:

$$P(X \le 45 \text{ when } H_0 \text{ is true}) = P(X \le 45 \text{ when } X \sim \text{Bin}(100, .5))$$
$$= B(45; 100, .5) = .184$$

So if in fact $p = .5$, it would not be surprising to see 45 or fewer successes. For this reason, the value 45 does not seem very contradictory to H_0 (it is only one standard deviation smaller than what we'd expect were H_0 true). Rejection of H_0 in this case does not seem sensible. ∎

EXAMPLE 8.2 According to the article **"Freshman 15: Fact or Fiction"** (*Obesity*, 2006: 1438– 1443), "A common belief among the lay public is that body weight increases after entry into college, and the phrase 'freshman 15' has been coined to describe the 15 pounds that students presumably gain over their freshman year." Let μ denote the true average weight gain of women over the course of their first year in college. The foregoing quote suggests that we should test the hypotheses H_0: $\mu = 15$ versus H_a: $\mu \ne 15$. For this purpose, suppose that a random sample of n such individuals is selected and the weight gain of each one is determined, resulting in a sample mean weight gain $\bar{x}$ and a sample standard deviation s (Note: The data here is actually *paired*, with each weight gain resulting from obtaining a (beginning, ending) weight pair and then subtracting to determine the difference; more will be said about such data in Section 9.3). Before data is obtained, the sample mean weight gain is a random variable $\bar{X}$ and the sample standard deviation is also a random variable S.

A natural test statistic (function of the data on which the decision will be based) is the sample mean $\bar{X}$ itself; if H_0 is true, then $E(\bar{X}) = \mu = 15$, whereas if μ differs considerably from 15, then the sample mean weight gain should do the same. But there is a more convenient test statistic that has appealing intuitive and technical properties: the sample mean standardized assuming that H_0 is true. Recall that the standard deviation (standard error) of $\bar{X}$ is $\sigma_{\bar{X}} = \sigma/\sqrt{n}$. Supposing that the population distribution of weight gains is normal, it follows that the sampling distribution of $\bar{X}$ itself is normal. Now standardizing a normally distributed variable gives a variable having a standard normal distribution (the z curve):

$$Z = \frac{\bar{X} - \mu}{\sigma/\sqrt{n}}$$

If the value of σ were known, we could obtain a test statistic simply by replacing μ by the null value $\mu_0 = 15$:

$$Z = \frac{\bar{X} - 15}{\sigma/\sqrt{n}}$$

If substitution of $\bar{x}$, σ, and n results in $z = 3$, the interpretation is that the observed value of the sample mean is three standard deviations larger than what we would have expected it to be were the null hypothesis true. Of course in "normal land" such an occurrence is exceedingly rare. Alternatively, if $z = -1$, then the sample mean is only one standard deviation less than what would be expected under H_0, a result not surprising enough to cast substantial doubt on H_0.

A practical glitch in the foregoing development is that the value of σ is virtually never available to an investigator. However, as discussed in the previous chapter, substitution of S for σ in Z typically introduces very little extra variability when n is large ($n > 40$ was our earlier rule of thumb). In this case the resulting variable still has *approximately* a standard normal distribution. The implied large-sample test statistic for our weight-gain scenario is

$$Z = \frac{\overline{X} - 15}{S/\sqrt{n}}$$

Thus when H_0 is true, Z has approximately a standard normal distribution.

Suppose that $\bar{x} = 13.7$, and that substitution of this along with s and n gives $z = -2.80$. Which values of the test statistic are at least as contradictory to H_0 as -2.80 itself? To answer this, let's first determine values of $\bar{x}$ that are at least as contradictory to H_0 as 13.7. One such value is 13.5, another is 13.0, and in fact *any* value smaller than 13.7 is more contradictory to H_0 than 13.7.

But that is not the whole story. Recall that the alternative hypothesis says that the value of μ is something other than 15. In light of this, the value 16.3 is just as contradictory to H_0 as is 13.7; it falls the same distance above the null value 15 as 13.7 does below 15—and the resulting z value is 3.0, just as extreme as -3.0. And any particular $\bar{x}$ that exceeds 16.3 is just as contradictory to H_0 as is a value the same distance below 15—e.g., 16.8 and 14.2, 17.0 and 13.0, and so on.

Just as values of $\bar{x}$ that are at most 13.7 correspond to $z \leq -2.80$, values of $\bar{x}$ that are at least 16.3 correspond to $z \geq 2.80$. Thus values of the test statistic that are at least as contradictory to H_0 as the value -2.80 actually obtained are $\{z: z \leq -2.80$ or $z \geq 2.80\}$. We can now calculate the probability, assuming H_0 true, of obtaining a test statistic value at least as contradictory to H_0 as what our sample yielded:

$$P(Z \leq -2.80 \text{ or } Z \geq 2.80 \text{ assuming } H_0 \text{ true})$$
$$\approx 2 \cdot (\text{area under the } z \text{ curve to the right of } 2.80)$$
$$= 2[1 - \Phi(2.80)] = 2[1 - .9974] = .0052$$

That is, if the null hypothesis is in fact true, only about one half of one percent of all samples would result in a test statistic value at least as contradictory to the null hypothesis as is our value. Clearly -2.80 is among the possible test statistic values that are most contradictory to H_0. It would therefore make sense to reject H_0 in favor of H_a.

Suppose we had instead obtained the test statistic value $z = .89$, which is less than one standard deviation larger than what we'd expect if H_0 were true. The foregoing probability would then be

$$P(Z \leq -0.89 \text{ or } Z \geq 0.89 \text{ assuming } H_0 \text{ true})$$
$$\approx 2 \cdot (\text{area under the } z \text{ curve to the right of } .89)$$
$$= 2[1 - \Phi(.89)] = 2[1 - .8133] = .3734$$

More than $1/3$ of all samples would give a test statistic value at least as contradictory to H_0 as is .89 when H_0 is true. So the data is quite consistent with the null hypothesis; it remains plausible that $\mu = 15$.

The article cited at the outset of this example reported that for a sample of 137 students, the sample mean weight gain was only 2.42 lb with a sample standard deviation of 5.72 lb (some students lost weight). This gives $z = (2.42 - 15)/(5.72/\sqrt{137}) = -25.7$! The probability of observing a value at least this extreme in either direction is essentially 0. The data very strongly contradicts the null hypothesis, and there is substantial evidence that true average weight gain is much closer to 0 than to 15. ∎

The type of probability calculated in Examples 8.1 and 8.2 will now provide the basis for obtaining general test procedures.

DEFINITIONS

A **test statistic** is a function of the sample data used as a basis for deciding whether H_0 should be rejected. The selected test statistic should discriminate effectively between the two hypotheses. That is, values of the statistic that tend to result when H_0 is true should be quite different from those typically observed when H_0 is not true.

The **P-value** is the probability, calculated assuming that the null hypothesis is true, of obtaining a value of the test statistic at least as contradictory to H_0 as the value calculated from the available sample data. A conclusion is reached in a hypothesis testing analysis by selecting a number α, called the **significance level** (alternatively, *level of significance*) of the test, that is reasonably close to 0. Then H_0 will be rejected in favor of H_a if P-value $\leq \alpha$, whereas H_0 will not be rejected (still considered to be plausible) if P-value $> \alpha$. The significance levels used most frequently in practice are (in order) $\alpha = .05, .01, .001,$ and $.10$.

For example, if we select a significance level of .05 and then compute P-value $=$ .0032, H_0 would be rejected because $.0032 \leq .05$. With this same P-value, the null hypothesis would also be rejected at the smaller significance level of .01 because $.0032 \leq .01$. However, at a significance level of .001 we would not be able to reject H_0 since $.0032 > .001$. Figure 8.1 illustrates the comparison of the P-value with the significance level in order to reach a conclusion.

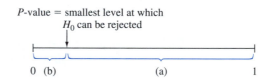

Figure 8.1 Comparing α and the P-value: (a) reject H_0 when α lies here; (b) do not reject H_0 when α lies here

We will shortly consider in some detail the consequences of selecting a smaller significance level rather than a larger one. For the moment, note that the smaller the significance level, the more protection is being given to the null hypothesis and the harder it is for that hypothesis to be rejected.

The definition of a P-value is obviously somewhat complicated, and it doesn't roll off the tongue very smoothly without a good deal of practice. In fact, many users of statistical methodology use the specified decision rule repeatedly to test hypotheses, but would be hard put to say what a P-value is! Here are some important points:

- The P-value is a probability.
- This probability is calculated assuming that the null hypothesis is true.
- To determine the P-value, we must first decide which values of the test statistic are at least as contradictory to H_0 as the value obtained from our sample.

- The smaller the P-value, the stronger is the evidence against H_0 and in favor of H_a.
- The P-value is not the probability that the null hypothesis is true or that it is false, nor is it the probability that an erroneous conclusion is reached.

EXAMPLE 8.3 Urban storm water can be contaminated by many sources, including discarded batteries. When ruptured, these batteries release metals of environmental significance. The article **"Urban Battery Litter" (*J. Environ. Engr.*, 2009: 46–57)** presented summary data for characteristics of a variety of batteries found in urban areas around Cleveland. A random sample of 51 Panasonic AAA batteries gave a sample mean zinc mass of 2.06 g. and a sample standard deviation of .141 g. Does this data provide compelling evidence for concluding that the population mean zinc mass exceeds 2.0 g.? Let's employ a significance level of .01 to reach a conclusion.

With μ denoting the true average zinc mass for such batteries, the relevant hypotheses are

$$H_0: \mu = 2.0 \text{ versus } H_a: \mu > 2.0.$$

The reasonably large sample size allows us to invoke the Central Limit Theorem, according to which the sample mean $\overline{X}$ has approximately a normal distribution. Furthermore, the standardized variable $Z = (\overline{X} - \mu)/(S/\sqrt{n})$ has approximately a standard normal distribution (the z curve). The test statistic results from standardizing $\overline{X}$ assuming that H_0 is true:

$$\text{Test statistic: } Z = \frac{\overline{X} - 2.0}{S/\sqrt{n}}$$

Substituting $n = 51$, $\bar{x} = 2.06$, and $s = .141$ gives $z = .06/.0197 = 3.04$. The sample mean here is roughly three (estimated) standard errors larger than would be expected were H_0 true (it does not appear to exceed 2 by very much, but there is only a small amount of variability in the 51 sample observations).

Any value of $\bar{x}$ larger than 2.06 is more contradictory to H_0 than 2.06 itself, and values of $\bar{x}$ that exceed 2.06 correspond to values of z that exceed 3.04. So any $z \geq 3.04$ is at least as contradictory to H_0. Since the test statistic has approximately a standard normal distribution when H_0 is true, we have

P-value $\approx P(\text{a standard normal rv is} \geq 3.04) = 1 - \Phi(3.04) = 1 - .9988 = .0012$

Because P-value $= .0012 \leq .01 = \alpha$, the null hypothesis should be rejected at the chosen significance level. It appears that true average zinc mass does indeed exceed 2. ■

Errors in Hypothesis Testing

The basis for choosing a particular significance level α lies in consideration of the errors that one might be faced with in drawing a conclusion. Recall the judicial scenario in which the null hypothesis is that the individual accused of committing a crime is in fact innocent. In rendering a verdict, the jury must consider the possibility of committing one of two different kinds of errors. One of these involves convicting an innocent person, and the other involves letting a guilty person go free. Similarly, there are two different types of errors that might be made in the course of a statistical hypothesis testing analysis.

DEFINITIONS

A **type I error** consists of rejecting the null hypothesis H_0 when it is true.

A **type II error** involves not rejecting H_0 when it is false.

As an example, a cereal manufacturer claims that a serving of one of its brands provides 100 calories (calorie content used to be determined by a destructive testing method, but the requirement that nutritional information appear on packages has led to more straightforward techniques). Of course the actual calorie content will vary somewhat from serving to serving (of the specified size), so 100 should be interpreted as an average. It could be distressing to consumers of this cereal if the true average calorie content exceeded the asserted value. So an appropriate formulation of hypotheses is to test H_0: $\mu = 100$ versus H_a: $\mu > 100$. The alternative hypothesis says that consumers are ingesting on average a greater amount of calories than what the company claims. A type I error here consists of rejecting the manufacturer's claim that $\mu = 100$ when it is actually true. A type II error results from not rejecting the manufacturer's claim when it is actually the case that $\mu > 100$.

Suppose μ_1 and μ_2 represent the true average lifetimes for two different brands of rollerball pen under controlled experimental conditions (utilizing a machine that writes continuously until a pen fails). It is natural to test the hypotheses H_0: $\mu_1 - \mu_2 = 0$ (i.e., $\mu_1 = \mu_2$) versus H_a: $\mu_1 - \mu_2 \neq 0$ (i.e., $\mu_1 \neq \mu_2$). A type I error would be to conclude that the true average lifetimes are different when in fact they are identical. A type II error involves deciding that the true average lifetimes may be the same when in fact they really differ from one another.

In the best of all possible worlds, we'd have a judicial system that never convicted an innocent person and never let a guilty person go free. This gold standard for judicial decisions has proven to be extremely elusive. Similarly, we would like to find test procedures for which neither type of error is ever committed. However, this ideal can be achieved only by basing a conclusion on an examination of the entire population. The difficulty with using a procedure based on sample data is that because of sampling variability, a sample unrepresentative of the population may result. In the calorie content scenario, even if the manufacturer's assertion is correct, an unusually large value of $\overline{X}$ may result in a P-value smaller than the chosen significance level and the consequent commission of a type I error. Alternatively, the true average calorie content may exceed what the manufacturer claims, yet a sample of servings may yield a relatively large P-value for which the null hypothesis cannot be rejected.

Instead of demanding error-free test procedures, we must seek procedures for which either type of error is unlikely to be committed. That is, a good procedure is one for which the probability of making a type I error is small and the probability of making a type II error is also small.

EXAMPLE 8.4 An automobile model is known to sustain no visible damage 25% of the time in 10-mph crash tests. A modified bumper design has been proposed in an effort to increase this percentage. Let p denote the proportion of all 10-mph crashes with this new bumper that result in no visible damage. The hypotheses to be tested are H_0: $p = .25$ (no improvement) versus H_a: $p > .25$. The test will be based on an experiment involving $n = 20$ independent crashes with prototypes of the new design. The natural test statistic here is $X = $ the number of crashes with no visible damage. If H_0 is true, $E(X) = np_0 = (20)(.25) = 5$. Intuition suggests that an observed value x much larger than this would provide strong evidence against H_0 and in support of H_a.

Consider using a significance level of .10. The P-value is $P(X \geq x$ when X has a binomial distribution with $n = 20$ and $p = .25) = 1 - B(x - 1; 20, .25)$ for $x > 0$.

Appendix Table A.1 shows that in this case,

$$P(X \geq 7) = 1 - B(6; 20, .25) = 1 - .786 = .214$$
$$P(X \geq 8) = 1 - .898 = .102 \approx .10, \; P(X \geq 9) = 1 - .959 = .041$$

Thus rejecting H_0 when P-value $\leq .10$ is equivalent to rejecting H_0 when $X \geq 8$. Therefore

P(committing a type I error) $= P$(rejecting H_0 when H_0 is true)

$$= P(X \geq 8 \text{ when } X \text{ has a binomial distribution with}$$
$$n = 20 \text{ and } p = .25)$$

$$= .102$$

$$\approx .10$$

That is, *the probability of a type I error is just the significance level* α. If the null hypothesis is true here and the test procedure is used over and over again, each time in conjunction with a group of 20 crashes, in the long run the null hypothesis will be incorrectly rejected in favor of the alternative hypothesis about 10% of the time. So our test procedure offers reasonably good protection against committing a type I error.

There is only one type I error probability because there is only one value of the parameter for which H_0 is true (this is one benefit of simplifying the null hypothesis to a claim of equality). Let β denote the probability of committing a type II error. Unfortunately there is not a single value of β, because there are a multitude of ways for H_0 to be false—it could be false because $p = .30$, because $p = .37$, because $p = .5$, and so on. There is in fact a different value of β for each different value of p that exceeds .25. At the chosen significance level .10, H_0 will be rejected if and only if $X \geq 8$, so H_0 will not be rejected if and only if $X \leq 7$. Thus

$$\beta(.3) = P(\text{type II error when } p = .3)$$

$$= P(H_0 \text{ is not rejected when } p = .3)$$

$$= P[X \leq 7 \text{ when } X \sim \text{Bin}(20, .3)]$$

$$= B(7; 20, .3) = .772$$

When p is actually .3 rather than .25 (a "small" departure from H_0), roughly 77% of all experiments of this type would result in H_0 being incorrectly not rejected!

The accompanying table displays β for selected values of p (each calculated as we just did for $\beta(.3)$). Clearly, β decreases as the value of p moves farther to the right of the null value .25. Intuitively, the greater the departure from H_0, the more likely it is that such a departure will be detected.

p	.3	.4	.5	.6	.7	.8
$\beta(p)$	.772	.416	.132	.021	.001	.000

The probability of committing a type II error here is quite large when $p = .3$ or .4. This is because those values are quite close to what H_0 asserts and the sample size of 20 is too small to permit accurate discrimination between .25 and those values of p.

The proposed test procedure is still reasonable for testing the more realistic null hypothesis that $p \leq .25$. In this case, there is no longer a single type I error probability α, but instead there is an α for each p that is at most .25: $\alpha(.25)$, $\alpha(.23)$, $\alpha(.20)$, $\alpha(.15)$, and so on. It is easily verified, though, that $\alpha(p) < \alpha(.25) = .102$ if $p < .25$. That is, the largest type I error probability occurs for the boundary value .25 between H_0 and H_a. Thus if α is small for the simplified null hypothesis, it will also be as small as or smaller for the more realistic H_0. ■

EXAMPLE 8.5   The drying time of a type of paint under specified test conditions is known to be normally distributed with mean value 75 min and standard deviation 9 min. Chemists have proposed a new additive designed to decrease average drying time. It is believed that drying times with this additive will remain normally distributed with $\sigma = 9$. Because of the expense associated with the additive, evidence should strongly suggest an improvement in average drying time before such a conclusion is adopted. Let μ denote the true average drying time when the additive is used. The appropriate hypotheses are H_0: $\mu = 75$ versus H_a: $\mu < 75$. Only if H_0 can be rejected will the additive be declared successful and used.

Experimental data is to consist of drying times from $n = 25$ test specimens. Let $X_1, \ldots, X_{25}$ denote the 25 drying times—a random sample of size 25 from a normal distribution with mean value μ and standard deviation $\sigma = 9$ (although the assumption of a known value of σ is generally unrealistic in practice, it considerably simplifies calculation of type II error probabilities). The sample mean drying time $\overline{X}$ then has a normal distribution with expected value $\mu_{\overline{X}} = \mu$ and standard deviation $\sigma_{\overline{X}} = \sigma/\sqrt{n} = 9/\sqrt{25} = 1.8$. When H_0 is true, we expect $\overline{X}$ to be 75; a sample mean much smaller than this would be contradictory to H_0 and supportive of H_a.

Our test statistic here will be $\overline{X}$ standardized assuming that H_0 is true:

$$Z = \frac{\overline{X} - \mu_0}{\sigma/\sqrt{n}} = \frac{\overline{X} - 75}{1.8}$$

The sampling distribution of $\overline{X}$ is normal because the population distribution is normal, which implies that Z has a standard normal distribution when H_0 is true (in contrast to Examples 8.2 and 8.3, we are assuming and using a known value of σ here).

Consider carrying out the test using a significance level of .01, i.e., H_0 will be rejected if P-value $\leq .01$. For a given value $\overline{x}$ of the sample mean and corresponding calculated value z, the form of the alternative hypothesis implies that values more contradictory to H_0 than this are values less than $\overline{x}$ and, correspondingly, values of the test statistic that are less than z. Thus the P-value is

$$
\begin{aligned}
P\text{-value} &= P(\text{obtaining a value of } Z \text{ at least as contradictory to} \\
&\quad\quad H_0 \text{ as } z \text{ when } H_0 \text{ is true}) \\
&= P(Z \leq z \text{ when } H_0 \text{ is true}) \\
&= \text{area under the standard normal curve to the left of } z \\
&= \Phi(z)
\end{aligned}
$$

So the P-value will equal .01 when z captures lower-tail area .01 under the z curve. From Appendix Table A.3, this happens when $z = -2.33$ [verify that $\Phi(-2.33) = .01$]. As illustrated in Figure 8.2, the P-value will therefore be at most .01 when $z \leq -2.33$. This in turn implies that

$$
\begin{aligned}
P(\text{type I error}) &= P(\text{rejecting } H_0 \text{ when } H_0 \text{ is true}) \\
&= P(P\text{-value} \leq .01 \text{ when } H_0 \text{ is true}) \\
&= P(Z \leq -2.33 \text{ when } Z \text{ has a standard normal distribution}) \\
&= .01
\end{aligned}
$$

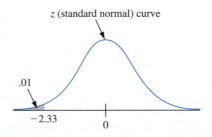

Figure 8.2 P-value $\leq .01$ if and only if $z \leq -2.33$

As in the previous example, the chosen significance level α is in fact the probability of committing a type I error. If the above test procedure [test statistic Z, reject H_0 if

P-value ≤ .01] is used repeatedly on sample after sample, in the long run the null hypothesis will be incorrectly rejected only 1% of the time. Our proposed test procedure offers excellent protection against the commission of a type I error. Note that if the more realistic null hypothesis $H_0: \mu \geq 75$ is considered, it can be shown that $P(\text{type I error}) \leq$ .01; the maximum occurs at the null value 75, which is the boundary between H_0 and H_a.

The calculation of $P(\text{type I error})$ in this example relied on the fact that *P*-value ≤ .01 is equivalent to $Z = (\overline{X} - 75)/1.8 \leq -2.33$. Multiplying both sides of this latter inequality by 1.8 and then adding 75 to both sides results in $\overline{X} \leq 70.8$. Thus rejecting H_0 at significance level .01 [if *P*-value ≤ .01] is equivalent to rejecting H_0 if $\overline{X} \leq 70.8$; H_0 will not be rejected if $\overline{X} > 70.8$. The probability of committing a type II error when $\mu = 72$ is now

$$\beta(72) = P(\text{not rejecting } H_0 \text{ when } \mu = 72)$$
$$= P(\overline{X} > 70.8 \text{ when } \overline{X} \sim \text{ normal with } \mu_{\overline{X}} = 72, \sigma_{\overline{X}} = 1.8)$$
$$= 1 - \Phi[(70.8 - 72)/1.8] = 1 - \Phi(-.67) = 1 - .2514 = .7486$$

This is an awfully large error probability. If the test with $\alpha = .01$ is used repeatedly on sample after sample and the actual value of μ is 72, almost 75% of the time the null hypothesis will not be rejected. The difficulty is that 72 is too close to the null value for a test with this sample size and value of α to have a good chance of detecting such a departure from H_0.

Similar calculations give

$$\beta(70) = 1 - \Phi[(70.8 - 70)/1.8] = .3300, \quad \beta(67) = .0174$$

These type II error probabilities are much smaller than $\beta(72)$ because 70 and 67 are both farther away from the null value than is 72. Figure 8.3 illustrates α and the first two type II error probabilities.

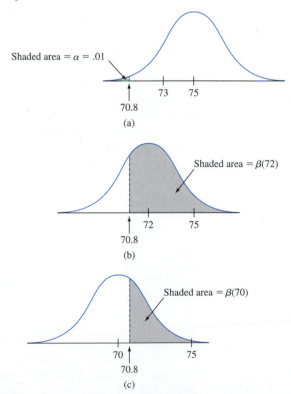

Shaded area = α = .01

70.8   73   75

(a)

Shaded area = $\beta(72)$

70.8   72   75

(b)

Shaded area = $\beta(70)$

70   70.8   75

(c)

Figure 8.3 α and β illustrated for Example 8.5: (a) the distribution of $\overline{X}$ when $\mu = 75$ (H_0 true); (b) the distribution of $\overline{X}$ when $\mu = 72$ (H_0 false); (c) the distribution of $\overline{X}$ when $\mu = 70$ (H_0 false)

The investigators might regard $\mu = 72$ as an important departure from the null hypothesis, in which case $\beta(72) = .7486$ is intolerably large. Consider changing the significance level (type I error probability from .01 to .05; that is, we now propose rejecting H_0 if P-value $\leq .05$. Appendix Table A.3 shows that the z critical value -1.645 captures a lower-tail z curve area of .05. Using the same reasoning that we previously applied when $\alpha = .01$, rejecting H_0 when P-value $\leq .05$ is equivalent to rejecting when $Z \leq -1.645$. This in turn is equivalent to rejecting when $\overline{X} \leq 72$ (notice that by increasing the significance level, we have made it easier for the null hypothesis to get rejected). Proceeding as in the previous calculations, we find that

$$\beta(72) = .5, \quad \beta(70) = .1335, \quad \beta(67) = .0027$$

These type II error probabilities are all smaller than their counterparts for the test with $\alpha = .01$. The important message here is that if a larger significance level (type I error probability) can be tolerated, then the resulting test will have better ability to detect when the null hypothesis is false. ∎

It is no accident that in the two foregoing examples, the significance level α turned out to be the probability of a type I error.

PROPOSITION

> The test procedure that rejects H_0 if P-value $\leq \alpha$ and otherwise does not reject H_0 has $P(\text{type I error}) = \alpha$. That is, the significance level employed in the test procedure is the probability of a type I error.

A partial proof of this proposition is sketched out at the end of the section.

The inverse relationship between the significance level α and type II error probabilities in Example 8.5 can be generalized in the following manner:

PROPOSITION

> Suppose an experiment or sampling procedure is selected, a sample size is specified, and a test statistic is chosen. Then increasing the significance level α, i.e., employing a larger type I error probability, results in a smaller value of β for any particular parameter value consistent with H_a.

This result is intuitively obvious because when α is increased, it becomes more likely that we'll have P-value $\leq \alpha$ and therefore less likely that P-value $> \alpha$.

The proposition implies that once the test statistic and n are fixed, it is not possible to make both α and any values of β that might be of interest arbitrarily small. Deciding on an appropriate significance level involves compromising between small α and small β's. In Example 8.5, the type II error probability for a test with $\alpha = .01$ was quite large for a value of μ close to the value in H_0. A strategy that is sometimes (but perhaps not often enough) used in practice is to specify α and also β for some alternative value of the parameter that is of particular importance to the investigator. Then the sample size n can be determined to satisfy these two conditions. For example, the article **"Cognitive Treatment of Illness Perceptions in Patients with Chronic Low Back Pain: A Randomized Controlled Trial" (*Physical Therapy*, 2013: 435–438)** contains the following passage: "A decrease of 18 to 24 mm on the PSC was determined as being a clinically relevant change in patients with low back pain. The sample size was calculated with a minimum change of 18 mm, a

2-sided α of .05, a $1 - \beta$ of .90, and a standard deviation of 26.01.... The sample size calculation resulted in a total of 135 participants." We'll consider such sample size determinations in subsequent sections and chapters.

In practice it is usually the case that the hypotheses of interest can be formulated so that a type I error is more serious than a type II error. The approach adhered to by most statistical practitioners is to reflect on the relative seriousness of a type I error compared to a type II error and then use the largest value of α that can be tolerated. This amounts to doing the best we can with respect to type II error probabilities while ensuring that the type I error probability is sufficiently small. For example, if $\alpha = .05$ is the largest significance level that can be tolerated, it would be better to use that rather than $\alpha = .01$, because all β's for the former α will be smaller than those for the latter one. As previously mentioned, the most frequently employed significance levels are $\alpha = .05, .01, .001,$ and .10. However, there are exceptions. Here is one example from particle physics: according to the article **"Discovery or Fluke: Statistics in Particle Physics"** (*Physics Today*, **July 2012: 45–50),** "the usual choice of alpha is 3×10^{-7}, corresponding to the 5σ of a Gaussian [i.e., normal] H_0 distribution.... Why so stringent? For one thing, recent history offers many cautionary examples of exciting 3σ and 4σ signals that went away when more data arrived."

If the distribution of the test statistic is continuous (e.g., if the test statistic has the standard normal distribution or a particular t distribution when H_0 is true), then any significance level α between 0 and 1 can be employed—for example, reject H_0 if P-value $\leq .035$. However, this is not necessarily the case if the distribution of the test statistic is discrete. As an example, consider again the bumper design scenario of Example 8.4 in which the hypotheses of interest were $H_0: p = .25$ versus $H_a: p > .25$. The test statistic X had a binomial distribution and

$$P\text{-value} = P(X \geq x \text{ when } n = 20 \text{ and } p = .25)$$

Appendix Table A.1 shows that in this case, $P(X \geq 8) = .102$ and $P(X \geq 9) = .041$. Thus if we want the significance level to be .05, the closest achievable level is actually .041: reject H_0 if P-value $\leq .041$.

Some Further Comments on the *P*-Value

Suppose that the P-value is calculated to be .038. The null hypothesis will then be rejected if $.038 \leq \alpha$ and not rejected otherwise. So H_0 can be rejected if $\alpha = .10$ or .05 but not if $\alpha = .01$ or .001. In fact, H_0 would be rejected for any significance level that is at least .038 but not for any level smaller than .038. For this reason, the P-value is often referred to as the **observed significance level** (OSL): it is the smallest value of α for which H_0 can be rejected.

One very appealing aspect of basing a conclusion from a hypothesis testing analysis on the P-value is that all widely used statistical software packages will calculate and output the P-value for any of the commonly used test procedures. Once the P-value is available, the investigator need only compare it to the selected significance level to decide whether H_0 should be rejected. This explains how an investigator can forget the definition of a P-value and still use it to reach a conclusion!

Sometimes a situation is encountered in which various individuals are interested in testing the same pair of hypotheses but may wish to use different significance levels. For example, suppose the true average time to pain relief for the current best-selling pain reliever is known to be 15 minutes. A new formulation has been developed that it is hoped will reduce this time. The relevant hypotheses are $H_0: \mu = 15$ versus $H_a: \mu < 15$, where μ is the true average time to relief using the new formulation. You may be quite satisfied with the current product and therefore wish to use a small significance level such as .01. I on the other hand may be less satisfied and thus more willing to switch, in which case a larger level such as .10 may be sensible. In using the larger α, I am giving less protection to H_0 than you are. Once the P-value is available, each of us can employ our own significance level irrespective of what the other person is using. Thus when medical journals report a P-value, a significance level is not mandated; instead it is left to the reader to select his or her own level and conclude accordingly. Furthermore, if someone else carried out the test and simply reported that H_0 was rejected at significance level .05 without revealing the P-value, then anyone wishing to use a smaller significance level would not know which conclusion is appropriate. That individual would have imposed his or her own significance level on other decision makers. Access to the P-value prevents such an imposition.

A final point concerning the utility of the P-value is that it allows one to distinguish between a close call and a very clear-cut conclusion at any particular significance level. For example, suppose you are told that H_0 was rejected at significance level .05. This conclusion is consistent with a P-value of .0498 and also with a P-value of .0003, since in each case P-value $\leq \alpha = .05$. But of course with a P-value of .0498, the null hypothesis is barely rejected, whereas with P-value $=$ .0003, the null hypothesis is rejected by a country mile. So it is always preferable to report the P-value rather than just stating the conclusion at a particular significance level.

Unfortunately most journal articles containing summaries of hypothesis testing analyses do not report exact P-values. Instead what typically appears is one of the following statements: "$P < .05$" if the P-value is between .05 and .01, "$P < .01$" if it is between .01 and .001, and "$P < .001$" if the P-value really is smaller than .001. In a tabular summary, you will often see *, **, and *** corresponding to these three cases.

Proof of the proposition stating that P(type I error) $=$ the significance level α:

Denote the test statistic by Y, and let $F(\cdot)$ be the cumulative distribution function of Y when H_0 is true (e.g., F might be the standard normal cdf Φ or the cdf of an rv having a t distribution with some specified number of df). Suppose the distribution of Y is continuous over some interval (often infinite in extent) so that F is a strictly increasing function over this interval. Then F has a well-defined inverse function F^{-1}. Consider the case in which only values of the test statistic smaller than the calculated value y are more contradictory to H_0 than y itself. This implies that

$$P\text{-value} = P(\text{obtaining a test statistic value at least}$$
$$\text{as contradictory to } H_0 \text{ when } H_0 \text{ is true}) = F(y)$$

Now before the sample data is available, the value of the test statistic is a random variable Y, and so the P-value itself is a random variable. Thus

$$P(\text{type I error}) = P(P\text{-value} \leq \alpha \text{ when } H_0 \text{ is true}) = P(F(Y) \leq \alpha)$$

Let's now apply F^{-1} to both sides of the inequality inside the last set of parentheses:

$$P(\text{type I error}) = P[F^{-1}(F(Y)) \leq F^{-1}(\alpha)] = P(Y \leq F^{-1}(\alpha)) = F(F^{-1}(\alpha)) = \alpha$$

The argument in the case in which only values of Y larger than y are more contradictory to H_0 than y itself is similar to what we have just shown. The case in which either large or small Y values are more contradictory to H_0 than y itself is a bit trickier. And when the test statistic has a discrete distribution, the inverse function F^{-1} is not uniquely defined, so extra care is needed to make the argument valid.

EXERCISES Section 8.1 (1–14)

1. For each of the following assertions, state whether it is a legitimate statistical hypothesis and why:
   a. $H: \sigma > 100$ b. $H: \tilde{x} = 45$
   c. $H: s \leq .20$ d. $H: \sigma_1/\sigma_2 < 1$
   e. $H: \overline{X} - \overline{Y} = 5$
   f. $H: \lambda \leq .01$, where λ is the parameter of an exponential distribution used to model component lifetime

2. For the following pairs of assertions, indicate which do not comply with our rules for setting up hypotheses and why (the subscripts 1 and 2 differentiate between quantities for two different populations or samples):
   a. $H_0: \mu = 100, H_a: \mu > 100$
   b. $H_0: \sigma = 20, H_a: \sigma \leq 20$
   c. $H_0: p \neq .25, H_a: p = .25$
   d. $H_0: \mu_1 - \mu_2 = 25, H_a: \mu_1 - \mu_2 > 100$
   e. $H_0: S_1^2 = S_2^2, H_a: S_1^2 \neq S_2^2$
   f. $H_0: \mu = 120, H_a: \mu = 150$
   g. $H_0: \sigma_1/\sigma_2 = 1, H_a: \sigma_1/\sigma_2 \neq 1$
   h. $H_0: p_1 - p_2 = -.1, H_a: p_1 - p_2 < -.1$

3. For which of the given P-values would the null hypothesis be rejected when performing a level .05 test?
   a. .001   b. .021   c. .078
   d. .047   e. .148

4. Pairs of P-values and significance levels, α, are given. For each pair, state whether the observed P-value would lead to rejection of H_0 at the given significance level.
   a. P-value = .084, α = .05
   b. P-value = .003, α = .001
   c. P-value = .498, α = .05
   d. P-value = .084, α = .10
   e. P-value = .039, α = .01
   f. P-value = .218, α = .10

5. To determine whether the pipe welds in a nuclear power plant meet specifications, a random sample of welds is selected, and tests are conducted on each weld in the sample. Weld strength is measured as the force required to break the weld. Suppose the specifications state that mean strength of welds should exceed 100 lb/in²; the inspection team decides to test $H_0: \mu = 100$ versus $H_a: \mu > 100$. Explain why it might be preferable to use this H_a rather than $\mu < 100$.

6. Let μ denote the true average radioactivity level (picocuries per liter). The value 5 pCi/L is considered the dividing line between safe and unsafe water. Would you recommend testing $H_0: \mu = 5$ versus $H_a: \mu > 5$ or $H_0: \mu = 5$ versus $H_a: \mu < 5$? Explain your reasoning. [*Hint:* Think about the consequences of a type I and type II error for each possibility.]

7. Before agreeing to purchase a large order of polyethylene sheaths for a particular type of high-pressure oil-filled submarine power cable, a company wants to see conclusive evidence that the true standard deviation of sheath thickness is less than .05 mm. What hypotheses should be tested, and why? In this context, what are the type I and type II errors?

8. Many older homes have electrical systems that use fuses rather than circuit breakers. A manufacturer of 40-amp fuses wants to make sure that the mean amperage at which its fuses burn out is in fact 40. If the mean amperage is lower than 40, customers will complain because the fuses require replacement too often. If the mean amperage is higher than 40, the manufacturer might be liable for damage to an electrical system due to fuse malfunction. To verify the amperage of the fuses, a sample of fuses is to be selected and inspected. If a hypothesis test were to be performed on the resulting data, what null and alternative hypotheses would be of interest to the manufacturer? Describe type I and type II errors in the context of this problem situation.

9. Water samples are taken from water used for cooling as it is being discharged from a power plant into a river. It has been determined that as long as the mean temperature of the discharged water is at most 150°F, there will be no negative effects on the river's ecosystem. To investigate whether the plant is in compliance with regulations that prohibit a mean discharge water temperature above 150°, 50 water samples will be taken at randomly selected times and the temperature of each sample recorded. The resulting data will be used to test the hypotheses $H_0: \mu = 150°$ versus $H_a: \mu > 150°$. In the context of this situation, describe type I and type II errors. Which type of error would you consider more serious? Explain.

10. A regular type of laminate is currently being used by a manufacturer of circuit boards. A special laminate has been developed to reduce warpage. The regular laminate will be used on one sample of specimens and the special laminate on another sample, and the amount of warpage will then be determined for each specimen. The manufacturer will then switch to the special laminate only if it can be demonstrated that the true average amount of warpage for that laminate is less than for the regular laminate. State the relevant hypotheses, and describe the type I and type II errors in the context of this situation.

11. Two different companies have applied to provide cable television service in a certain region. Let p denote the proportion of all potential subscribers who favor the first company over the second. Consider testing $H_0: p = .5$ versus $H_a: p \neq .5$ based on a random sample of 25 individuals. Let the test statistic X be the number in the

sample who favor the first company and x represent the observed value of X.

a. Describe type I and II errors in the context of this problem situation.

b. Suppose that $x = 6$. Which values of X are at least as contradictory to H_0 as this one?

c. What is the probability distribution of the test statistic X when H_0 is true? Use it to compute the P-value when $x = 6$.

d. If H_0 is to be rejected when P-value $\leq .044$, compute the probability of a type II error when $p = .4$, again when $p = .3$, and also when $p = .6$ and $p = .7$. [*Hint:* P-value $> .044$ is equivalent to what inequalities involving x (see Example 8.4)?]

e. Using the test procedure of (d), what would you conclude if 6 of the 25 queried favored company 1?

12. A mixture of pulverized fuel ash and Portland cement to be used for grouting should have a compressive strength of more than 1300 KN/m^2. The mixture will not be used unless experimental evidence indicates conclusively that the strength specification has been met. Suppose compressive strength for specimens of this mixture is normally distributed with $\sigma = 60$. Let μ denote the true average compressive strength.

a. What are the appropriate null and alternative hypotheses?

b. Let $\bar{X}$ denote the sample average compressive strength for $n = 10$ randomly selected specimens. Consider the test procedure with test statistic $\bar{X}$ itself (not standardized). If $\bar{x} = 1340$, should H_0 be rejected using a significance level of .01? [*Hint:* What is the probability distribution of the test statistic when H_0 is true?]

c. What is the probability distribution of the test statistic when $\mu = 1350$? For a test with $\alpha = .01$, what is the

probability that the mixture will be judged unsatisfactory when in fact $\mu = 1350$ (a type II error)?

13. The calibration of a scale is to be checked by weighing a 10-kg test specimen 25 times. Suppose that the results of different weighings are independent of one another and that the weight on each trial is normally distributed with $\sigma = .200$ kg. Let μ denote the true average weight reading on the scale.

a. What hypotheses should be tested?

b. With the sample mean itself as the test statistic, what is the P-value when $\bar{x} = 9.85$, and what would you conclude at significance level .01?

c. For a test with $\alpha = .01$, what is the probability that recalibration is judged unnecessary when in fact $\mu = 10.1$? When $\mu = 9.8$?

14. A new design for the braking system on a certain type of car has been proposed. For the current system, the true average braking distance at 40 mph under specified conditions is known to be 120 ft. It is proposed that the new design be implemented only if sample data strongly indicates a reduction in true average braking distance for the new design.

a. Define the parameter of interest and state the relevant hypotheses.

b. Suppose braking distance for the new system is normally distributed with $\sigma = 10$. Let $\bar{X}$ denote the sample average braking distance for a random sample of 36 observations. Which values of $\bar{x}$ are more contradictory to H_0 than 117.2, what is the P-value in this case, and what conclusion is appropriate if $\alpha = .10$?

c. What is the probability that the new design is not implemented when its true average braking distance is actually 115 ft and the test from part (b) is used?

8.2   *z* Tests for Hypotheses about a Population Mean

Recall from the previous section that a conclusion in a hypothesis testing analysis is reached by proceeding as follows:

i. Compute the value of an appropriate test statistic.

ii. Then determine the P-value—the probability, calculated assuming that the null hypothesis H_0 true, of observing a test statistic value at least as contradictory to H_0 as what resulted from the available data.

iii. Reject the null hypothesis if P-value $\leq \alpha$, where α is the specified or chosen significance level, i.e., the probability of a type I error (rejecting H_0 when it is true); if P-value $> \alpha$, there is not enough evidence to justify rejecting H_0 (it is still deemed plausible).

Determination of the P-value depends on the distribution of the test statistic when H_0 is true. In this section we describe z tests for testing hypotheses about a single population mean μ. By "z test," we mean that the test statistic has at least approximately a standard normal distribution when H_0 is true. The P-value will then be a z curve area which depends on whether the inequality in H_a is $>$, $<$, or $\neq$.

In the development of confidence intervals for μ in Chapter 7, we first considered the case in which the population distribution is normal with known σ, then relaxed the normality and known σ assumptions when the sample size n is large, and finally described the one-sample t CI for the mean of a normal population. In this section we discuss the first two cases, and then present the one-sample t test in Section 8.3.

A Normal Population Distribution with Known σ

Although the assumption that the value of σ is known is rarely met in practice, this case provides a good starting point because of the ease with which general procedures and their properties can be developed. The null hypothesis in all three cases will state that μ has a particular numerical value, the *null value*. We denote this value by the symbol μ_0, so the null hypothesis has the form H_0: $\mu = \mu_0$. Let $X_1, \dots, X_n$ represent a random sample of size n from the normal population. Then the sample mean $\overline{X}$ has a normal distribution with expected value $\mu_{\overline{X}} = \mu$ and standard deviation $\sigma_{\overline{X}} = \sigma/\sqrt{n}$. When H_0 is true, $\mu_{\overline{X}} = \mu_0$. Consider now the statistic Z obtained by standardizing $\overline{X}$ under the assumption that H_0 is true:

$$Z = \frac{\overline{X} - \mu_0}{\sigma/\sqrt{n}}$$

Substitution of the computed sample mean $\bar{x}$ gives z, the distance between $\bar{x}$ and μ_0 expressed in "standard deviation units." For example, if the null hypothesis is H_0: $\mu = 100$, $\sigma_{\overline{X}} = \sigma/\sqrt{n} = 10/\sqrt{25} = 2.0$, and $\bar{x} = 103$, then the test statistic value is $z = (103 - 100)/2.0 = 1.5$. That is, the observed value of $\bar{x}$ is 1.5 standard deviations (of $\overline{X}$) larger than what we expect it to be when H_0 is true. The statistic Z is a natural measure of the distance between $\overline{X}$, the estimator of μ, and its expected value when H_0 is true. If this distance is too great in a direction consistent with H_a, there is substantial evidence that H_0 is false.

Suppose first that the alternative hypothesis is of the form H_a: $\mu > \mu_0$. Then an $\bar{x}$ value that considerably exceeds μ_0 provides evidence against H_0. Such an $\bar{x}$ value corresponds to a large positive value z. This in turn implies that any value *exceeding* the calculated z is more contradictory to H_0 than is z itself. It follows that

$$P\text{-value} = P(Z \geq z \text{ when } H_0 \text{ is true})$$

Now here is the key point: when H_0 is true, the test statistic Z has a standard normal distribution—because we created Z by standardizing $\overline{X}$ assuming that H_0 is true (i.e., by subtracting μ_0). The implication is that in this case, the P-value is just the area under the standard normal curve to the right of z. Because of this, the test is referred to as *upper-tailed*. For example, in the previous paragraph we calculated $z = 1.5$. If in the alternative hypothesis there is H_a: $\mu > 100$, then P-value = area under the z curve to the right of $1.5 = 1 - \Phi(1.50) = .0668$. At significance level .05 we would not be able to reject the null hypothesis because the P-value exceeds α.

Now consider an alternative hypothesis of the form H_a: $\mu < \mu_0$. In this case any value of the sample mean smaller than our $\bar{x}$ is even more contradictory to the null hypothesis. Thus any test statistic value *smaller* than the calculated z is more contradictory to H_0 than is z itself. It follows that

$$P\text{-value} = P(Z \leq z \text{ when } H_0 \text{ is true})$$
$$= \text{area under the standard normal curve to the left of } z = \Phi(z)$$

The test in this case is customarily referred to as *lower-tailed*. If, for example, the alternative hypothesis is $H_a\colon \mu < 100$ and $z = -2.75$, then P-value $= \Phi(-2.75) = .0030$. This is small enough to justify rejection of H_0 at a significance level of either .05 or .01, but not .001.

The third possible alternative, $H_a\colon \mu \neq \mu_0$, requires a bit more careful thought. Suppose, for example, that the null value is 100 and that $\bar{x} = 103$ results in $z = 1.5$. Then any $\bar{x}$ value exceeding 103 is more contradictory to H_0 than is 103 itself. So any z exceeding 1.5 is likewise more contradictory to H_0 than is 1.5. However, 97 is just as contradictory to the null hypothesis as is 103, since it is the same distance below 100 as 103 is above 100. Thus $z = -1.5$ is just as contradictory to H_0 as is $z = 1.5$. Therefore any z smaller than -1.5 is more contradictory to H_0 than is 1.5 or -1.5. It follows that

$$P\text{-value} = P(Z \text{ either} \geq 1.5 \text{ or } \leq -1.5 \text{ when } H_0 \text{ is true})$$
$$= (\text{area under the } z \text{ curve to the right of } 1.5)$$
$$+ (\text{area under the } z \text{ curve to the left of } -1.5)$$
$$= 1 - \Phi(1.5) + \Phi(-1.5) = 2[1 - \Phi(1.5)]$$
$$= 2(.0668) = .1336$$

This would also be the P-value if $\bar{x} = 97$ results in $z = -1.5$. The important point is that because of the inequality $\neq$ in H_a, the P-value is the sum of an upper-tail area and a lower-tail area. By symmetry of the standard normal distribution, this becomes twice the area captured in the tail in which z falls. Equivalently, it is twice the area captured in the upper tail by $|z|$, i.e., $2[1 - \Phi(|z|)]$. It is natural to refer to this test as being *two-tailed* because z values far out in either tail of the z curve argue for rejection of H_0.

The test procedure is summarized in the accompanying box, and the P-value for each of the possible alternative hypotheses is illustrated in Figure 8.4.

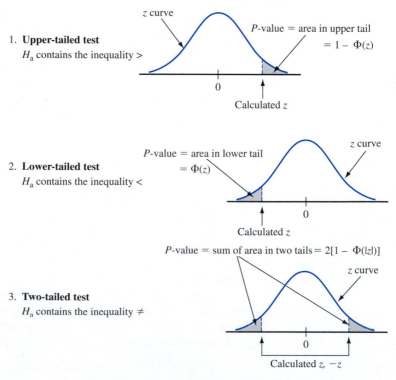

Figure 8.4 Determination of the P-value for a z test

Null hypothesis: $H_0: \mu = \mu_0$

Test statistic: $Z = \dfrac{\overline{X} - \mu_0}{\sigma/\sqrt{n}}$

Alternative Hypothesis	P-Value Determination
$H_a: \mu > \mu_0$	Area under the standard normal curve to the right of z
$H_a: \mu < \mu_0$	Area under the standard normal curve to the left of z
$H_a: \mu \neq \mu_0$	$2 \cdot$ (area under the standard normal curve to the right of $|z|$)

Assumptions: A normal population distribution with known value of σ.

Use of the following sequence of steps is recommended when testing hypotheses about a parameter. The plausibility of any assumptions underlying use of the selected test procedure should of course be checked before carrying out the test.

1. Identify the parameter of interest and describe it in the context of the problem situation.

2. Determine the null value and state the null hypothesis.

3. State the appropriate alternative hypothesis.

4. Give the formula for the computed value of the test statistic (substituting the null value and the known values of any other parameters, but *not* those of any sample-based quantities).

5. Compute any necessary sample quantities, substitute into the formula for the test statistic value, and compute that value.

6. Determine the *P*-value.

7. Compare the selected or specified significance level to the *P*-value to decide whether H_0 should be rejected, and state this conclusion in the problem context.

The formulation of hypotheses (Steps 2 and 3) should be done before examining the data, and the significance level α should be chosen prior to determination of the *P*-value.

EXAMPLE 8.6   A manufacturer of sprinkler systems used for fire protection in office buildings claims that the true average system-activation temperature is 130°. A sample of $n = 9$ systems, when tested, yields a sample average activation temperature of 131.08°F. If the distribution of activation times is normal with standard deviation 1.5°F, does the data contradict the manufacturer's claim at significance level $\alpha = .01$?

1. Parameter of interest: $\mu =$ true average activation temperature.

2. Null hypothesis: $H_0: \mu = 130$ (null value $= \mu_0 = 130$).

3. Alternative hypothesis: $H_a: \mu \neq 130$ (a departure from the claimed value in *either* direction is of concern).

4. Test statistic value:

$$z = \frac{\overline{x} - \mu_0}{\sigma/\sqrt{n}} = \frac{\overline{x} - 130}{1.5/\sqrt{n}}$$

5. Substituting $n = 9$ and $\overline{x} = 131.08$,

$$z = \frac{131.08 - 130}{1.5/\sqrt{9}} = \frac{1.08}{.5} = 2.16$$

That is, the observed sample mean is a bit more than 2 standard deviations above what would have been expected were H_0 true.

6. The inequality in H_a implies that the test is two-tailed, so the P-value results from doubling the captured tail area:

$$P\text{-value} = 2[1 - \Phi(2.16)] = 2(.0154) = .0308$$

7. Because P-value $= .0308 > .01 = \alpha$, H_0 cannot be rejected at significance level .01. The data does not give strong support to the claim that the true average differs from the design value of 130. ■

β and Sample Size Determination The z tests with known σ are among the few in statistics for which there are simple formulas available for β, the probability of a type II error. Consider first the alternative $H_a: \mu > \mu_0$. The null hypothesis is rejected if P-value $\le \alpha$, and the P-value is the area under the standard normal curve to the right of z. Suppose that $\alpha = .05$. The z critical value that captures an upper-tail area of .05 is $z_{.05} = 1.645$ (look for a cumulative area of .95 in Table A.3). Thus if the calculated test statistic value z is smaller than 1.645, the area to the right of z will be larger than .05 and the null hypothesis will then *not* be rejected. Now substitute $(\bar{x} - \mu_0)/(\sigma/\sqrt{n})$ in place of z in the inequality $z < 1.645$ and manipulate to isolate $\bar{x}$ on the left (multiply both sides by $\sigma/\sqrt{n}$ and then add μ_0 to both sides). This gives the equivalent inequality $\bar{x} < \mu_0 + z_\alpha \cdot \sigma/\sqrt{n}$. Now let μ' denote a particular value of μ that exceeds the null value μ_0. Then,

$$
\begin{aligned}
\beta(\mu') &= P(H_0 \text{ is not rejected when } \mu = \mu') \\
&= P(\bar{X} < \mu_0 + z_\alpha \cdot \sigma/\sqrt{n} \text{ when } \mu = \mu') \\
&= P\left(\frac{\bar{X} - \mu'}{\sigma/\sqrt{n}} < z_\alpha + \frac{\mu_0 - \mu'}{\sigma/\sqrt{n}} \text{ when } \mu = \mu'\right) \\
&= \Phi\left(z_\alpha + \frac{\mu_0 - \mu'}{\sigma/\sqrt{n}}\right)
\end{aligned}
$$

As μ' increases, $\mu_0 - \mu'$ becomes more negative, so $\beta(\mu')$ will be small when μ' greatly exceeds μ_0 (because the value at which Φ is evaluated will then be quite negative). Error probabilities for the lower-tailed and two-tailed tests are derived in an analogous manner.

If σ is large, the probability of a type II error can be large at an alternative value μ' that is of particular concern to an investigator. Suppose we fix α and also specify β for such an alternative value. In the sprinkler example, company officials might view $\mu' = 132$ as a very substantial departure from $H_0: \mu = 130$ and therefore wish $\beta(132) = .10$ in addition to $\alpha = .01$. More generally, consider the two restrictions $P(\text{type I error}) = \alpha$ and $\beta(\mu') = \beta$ for specified α, μ', and β. Then for an upper-tailed test, the sample size n should be chosen to satisfy

$$\Phi\left(z_\alpha + \frac{\mu_0 - \mu'}{\sigma/\sqrt{n}}\right) = \beta$$

This implies that

$$-z_\beta = \begin{array}{c} z \text{ critical value that} \\ \text{captures lower-tail area } \beta \end{array} = z_\alpha + \frac{\mu_0 - \mu'}{\sigma/\sqrt{n}}$$

This equation is easily solved for the desired n. A parallel argument yields the necessary sample size for lower- and two-tailed tests as summarized in the next box.

Alternative Hypothesis	Type II Error Probability $\beta(\mu')$ for a Level α Test
$H_a: \mu > \mu_0$	$\Phi\left(z_\alpha + \dfrac{\mu_0 - \mu'}{\sigma/\sqrt{n}}\right)$
$H_a: \mu < \mu_0$	$1 - \Phi\left(-z_\alpha + \dfrac{\mu_0 - \mu'}{\sigma/\sqrt{n}}\right)$
$H_a: \mu \neq \mu_0$	$\Phi\left(z_{\alpha/2} + \dfrac{\mu_0 - \mu'}{\sigma/\sqrt{n}}\right) - \Phi\left(-z_{\alpha/2} + \dfrac{\mu_0 - \mu'}{\sigma/\sqrt{n}}\right)$

where $\Phi(z) =$ the standard normal cdf.

The sample size n for which a level α test also has $\beta(\mu') = \beta$ at the alternative value μ' is

$$n = \begin{cases} \left[\dfrac{\sigma(z_\alpha + z_\beta)}{\mu_0 - \mu'}\right]^2 & \text{for a one-tailed} \\ & \text{(upper or lower) test} \\[2mm] \left[\dfrac{\sigma(z_{\alpha/2} + z_\beta)}{\mu_0 - \mu'}\right]^2 & \text{for a two-tailed test} \\ & \text{(an approximate solution)} \end{cases}$$

EXAMPLE 8.7 Let μ denote the true average tread life of a certain type of tire. Consider testing $H_0: \mu = 30{,}000$ versus $H_a: \mu > 30{,}000$ based on a sample of size $n = 16$ from a normal population distribution with $\sigma = 1500$. A test with $\alpha = .01$ requires $z_\alpha = z_{.01} = 2.33$. The probability of making a type II error when $\mu = 31{,}000$ is

$$\beta(31{,}000) = \Phi\left(2.33 + \frac{30{,}000 - 31{,}000}{1500/\sqrt{16}}\right) = \Phi(-.34) = .3669$$

Since $z_{.1} = 1.28$, the requirement that the level .01 test also have $\beta(31{,}000) = .1$ necessitates

$$n = \left[\frac{1500(2.33 + 1.28)}{30{,}000 - 31{,}000}\right]^2 = (-5.42)^2 = 29.32$$

The sample size must be an integer, so $n = 30$ tires should be used. ■

Large-Sample Tests

When the sample size is large, the foregoing z tests are easily modified to yield valid test procedures without requiring either a normal population distribution or known σ. The key result was used in Chapter 7 to justify large-sample confidence intervals: A large n implies that the standardized variable

$$Z = \frac{\overline{X} - \mu}{S/\sqrt{n}}$$

has *approximately* a standard normal distribution. Substitution of the null value μ_0 in place of μ yields the test statistic

$$Z = \frac{\overline{X} - \mu_0}{S/\sqrt{n}}$$

which has approximately a standard normal distribution when H_0 is true. The P-value is then determined exactly as was previously described in this section (e.g., $\Phi(z)$ when the alternative hypothesis is $H_a: \mu < \mu_0$). Rejecting H_0 when P-value $\leq \alpha$ gives a test with *approximate* significance level α. The rule of thumb $n > 40$ will again be used to characterize a large sample size.

EXAMPLE 8.8 A dynamic cone penetrometer (DCP) is used for measuring material resistance to penetration (mm/blow) as a cone is driven into pavement or subgrade. Suppose that for a particular application it is required that the true average DCP value for a certain type of pavement be less than 30. The pavement will not be used unless there is conclusive evidence that the specification has been met. Let's state and test the appropriate hypotheses using the following data (**"Probabilistic Model for the Analysis of Dynamic Cone Penetrometer Test Values in Pavement Structure Evaluation," *J. of Testing and Evaluation*, 1999: 7–14**):

14.1	14.5	15.5	16.0	16.0	16.7	16.9	17.1	17.5	17.8
17.8	18.1	18.2	18.3	18.3	19.0	19.2	19.4	20.0	20.0
20.8	20.8	21.0	21.5	23.5	27.5	27.5	28.0	28.3	30.0
30.0	31.6	31.7	31.7	32.5	33.5	33.9	35.0	35.0	35.0
36.7	40.0	40.0	41.3	41.7	47.5	50.0	51.0	51.8	54.4
55.0	57.0								

Figure 8.5 shows a descriptive summary obtained from Minitab. The sample mean DCP is less than 30. However, there is a substantial amount of variation in the data (sample coefficient of variation = $s/\bar{x}$ = .4265), so the fact that the mean is less than the design specification cutoff may be a consequence just of sampling variability. Notice that the histogram does not resemble at all a normal curve (and a normal probability plot does not exhibit a linear pattern). However, the large-sample z tests do not require a normal population distribution.

Descriptive Statistics

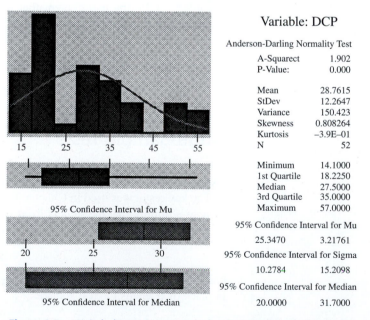

Variable: DCP

Anderson-Darling Normality Test

A-Squarect	1.902
P-Value:	0.000

Mean	28.7615
StDev	12.2647
Variance	150.423
Skewness	0.808264
Kurtosis	−3.9E–01
N	52

Minimum	14.1000
1st Quartile	18.2250
Median	27.5000
3rd Quartile	35.0000
Maximum	57.0000

95% Confidence Interval for Mu

25.3470	3.21761

95% Confidence Interval for Sigma

10.2784	15.2098

95% Confidence Interval for Median

20.0000	31.7000

95% Confidence Interval for Mu

95% Confidence Interval for Median

Figure 8.5 Minitab descriptive summary for the DCP data of Example 8.8

1. μ = true average DCP value

2. H_0: $\mu = 30$

3. H_a: $\mu < 30$ (so the pavement will not be used unless the null hypothesis is rejected)

4. $z = \dfrac{\bar{x} - 30}{s/\sqrt{n}}$

5. With $n = 52$, $\bar{x} = 28.76$, and $s = 12.2647$,

$$z = \frac{28.76 - 30}{12.2647/\sqrt{52}} = \frac{-1.24}{1.701} = -.73$$

6. The *P*-value for this lower-tailed *z* test is $\Phi(-.73) = .2327$.

7. Since $.2327 > .05$, H_0 cannot be rejected. We do not have compelling evidence for concluding that $\mu < 30$; use of the pavement is not justified. Note that in not rejecting H_0, we might possibly have committed a type II error. ∎

Determination of β and the necessary sample size for these large-sample tests can be based either on specifying a plausible value of σ and using the previous formulas (even though *s* is used in the test) or on using the methodology to be introduced in connection with the one-sample *t* tests discussed in Section 8.3.

EXERCISES Section 8.2 (15–28)

15. Let μ denote the true average reaction time to a certain stimulus. For a *z* test of H_0: $\mu = 5$ versus H_a: $\mu > 5$, determine the *P*-value for each of the following values of the *z* test statistic.
    a. 1.42 **b.** .90 **c.** 1.96 **d.** 2.48 **e.** −.11

16. Newly purchased tires of a particular type are supposed to be filled to a pressure of 30 psi. Let μ denote the true average pressure. A test is to be carried out to decide whether μ differs from the target value. Determine the *P*-value for each of the following *z* test statistic values.
    a. 2.10 **b.** −1.75 **c.** −.55 **d.** 1.41 **e.** −5.3

17. Answer the following questions for the tire problem in Example 8.7.
    a. If $\bar{x} = 30,960$ and a level $\alpha = .01$ test is used, what is the decision?
    b. If a level $.01$ test is used, what is $\beta(30,500)$?
    c. If a level $.01$ test is used and it is also required that $\beta(30,500) = .05$, what sample size *n* is necessary?
    d. If $\bar{x} = 30,960$, what is the smallest α at which H_0 can be rejected (based on $n = 16$)?

18. Reconsider the paint-drying situation of Example 8.5, in which drying time for a test specimen is normally distributed with $\sigma = 9$. The hypotheses H_0: $\mu = 75$ versus H_a: $\mu < 75$ are to be tested using a random sample of $n = 25$ observations.

    a. How many standard deviations (of $\overline{X}$) below the null value is $\bar{x} = 72.3$?
    b. If $\bar{x} = 72.3$, what is the conclusion using $\alpha = .002$?
    c. For the test procedure with $\alpha = .002$, what is $\beta(70)$?
    d. If the test procedure with $\alpha = .002$ is used, what *n* is necessary to ensure that $\beta(70) = .01$?
    e. If a level $.01$ test is used with $n = 100$, what is the probability of a type I error when $\mu = 76$?

19. The melting point of each of 16 samples of a certain brand of hydrogenated vegetable oil was determined, resulting in $\bar{x} = 94.32$. Assume that the distribution of the melting point is normal with $\sigma = 1.20$.
    a. Test H_0: $\mu = 95$ versus H_a: $\mu \neq 95$ using a two-tailed level $.01$ test.
    b. If a level $.01$ test is used, what is $\beta(94)$, the probability of a type II error when $\mu = 94$?
    c. What value of *n* is necessary to ensure that $\beta(94) = .1$ when $\alpha = .01$?

20. Lightbulbs of a certain type are advertised as having an average lifetime of 750 hours. The price of these bulbs is very favorable, so a potential customer has decided to go ahead with a purchase arrangement unless it can be conclusively demonstrated that the true average lifetime is smaller than what is advertised. A random sample of 50 bulbs was selected, the lifetime of each bulb determined,

and the appropriate hypotheses were tested using Minitab, resulting in the accompanying output.

```
Variable   N    Mean  StDev  SE Mean      Z  P-Value
lifetime  50  738.44  38.20     5.40  -2.14    0.016
```

What conclusion would be appropriate for a significance level of .05? A significance level of .01? What significance level and conclusion would you recommend?

21. The desired percentage of SiO_2 in a certain type of aluminous cement is 5.5. To test whether the true average percentage is 5.5 for a particular production facility, 16 independently obtained samples are analyzed. Suppose that the percentage of SiO_2 in a sample is normally distributed with $\sigma = .3$ and that $\bar{x} = 5.25$.
   a. Does this indicate conclusively that the true average percentage differs from 5.5?
   b. If the true average percentage is $\mu = 5.6$ and a level $\alpha = .01$ test based on $n = 16$ is used, what is the probability of detecting this departure from H_0?
   c. What value of n is required to satisfy $\alpha = .01$ and $\beta(5.6) = .01$?

22. To obtain information on the corrosion-resistance properties of a certain type of steel conduit, 45 specimens are buried in soil for a 2-year period. The maximum penetration (in mils) for each specimen is then measured, yielding a sample average penetration of $\bar{x} = 52.7$ and a sample standard deviation of $s = 4.8$. The conduits were manufactured with the specification that true average penetration be at most 50 mils. They will be used unless it can be demonstrated conclusively that the specification has not been met. What would you conclude?

23. Automatic identification of the boundaries of significant structures within a medical image is an area of ongoing research. The paper **"Automatic Segmentation of Medical Images Using Image Registration: Diagnostic and Simulation Applications"** (*J. of Medical Engr. and Tech.*, **2005: 53–63**) discussed a new technique for such identification. A measure of the accuracy of the automatic region is the average linear displacement (ALD). The paper gave the following ALD observations for a sample of 49 kidneys (units of pixel dimensions).

1.38	0.44	1.09	0.75	0.66	1.28	0.51
0.39	0.70	0.46	0.54	0.83	0.58	0.64
1.30	0.57	0.43	0.62	1.00	1.05	0.82
1.10	0.65	0.99	0.56	0.56	0.64	0.45
0.82	1.06	0.41	0.58	0.66	0.54	0.83
0.59	0.51	1.04	0.85	0.45	0.52	0.58
1.11	0.34	1.25	0.38	1.44	1.28	0.51

   a. Summarize/describe the data.
   b. Is it plausible that ALD is at least approximately normally distributed? Must normality be assumed prior to calculating a CI for true average ALD or testing hypotheses about true average ALD? Explain.

   c. The authors commented that in most cases the ALD is better than or of the order of 1.0. Does the data in fact provide strong evidence for concluding that true average ALD under these circumstances is less than 1.0? Carry out an appropriate test of hypotheses.
   d. Calculate an upper confidence bound for true average ALD using a confidence level of 95%, and interpret this bound.

24. Unlike most packaged food products, alcohol beverage container labels are not required to show calorie or nutrient content. The article **"What Am I Drinking? The Effects of Serving Facts Information on Alcohol Beverage Containers"** (*J. of Consumer Affairs*, **2008: 81–99**) reported on a pilot study in which each of 58 individuals in a sample was asked to estimate the calorie content of a 12-oz can of beer known to contain 153 calories. The resulting sample mean estimated calorie level was 191 and the sample standard deviation was 89. Does this data suggest that the true average estimated calorie content in the population sampled exceeds the actual content? Test the appropriate hypotheses at significance level .001.

25. Body armor provides critical protection for law enforcement personnel, but it does affect balance and mobility. The article **"Impact of Police Body Armour and Equipment on Mobility"** (*Applied Ergonomics*, **2013: 957–961**) reported that for a sample of 52 male enforcement officers who underwent an acceleration task that simulated exiting a vehicle while wearing armor, the sample mean was 1.95 sec, and the sample standard deviation was .20 sec. Does it appear that true average task time is less than 2 sec? Carry out a test of appropriate hypotheses using a significance level of .01.

26. The recommended daily dietary allowance for zinc among males older than age 50 years is 15 mg/day. The article **"Nutrient Intakes and Dietary Patterns of Older Americans: A National Study"** (*J. of Gerontology*, **1992: M145–150**) reports the following summary data on intake for a sample of males age 65–74 years: $n = 115$, $\bar{x} = 11.3$, and $s = 6.43$. Does this data indicate that average daily zinc intake in the population of all males ages 65–74 falls below the recommended allowance?

27. Show that for any $\Delta > 0$, when the population distribution is normal and σ is known, the two-tailed test satisfies $\beta(\mu_0 - \Delta) = \beta(\mu_0 + \Delta)$, so that $\beta(\mu')$ is symmetric about μ_0.

28. For a fixed alternative value μ', show that $\beta(\mu') \to 0$ as $n \to \infty$ for either a one-tailed or a two-tailed z test in the case of a normal population distribution with known σ.

8.3 The One-Sample *t* Test

When n is small, the Central Limit Theorem (CLT) can no longer be invoked to justify the use of a large-sample test. We faced this same difficulty in obtaining a small-sample confidence interval (CI) for μ in Chapter 7. Our approach here will be the same one used there: We will assume that the population distribution is at least approximately normal and describe test procedures whose validity rests on this assumption. If an investigator has good reason to believe that the population distribution is quite nonnormal, a distribution-free test from Chapter 15 may be appropriate. Alternatively, a statistician can be consulted regarding procedures valid for specific families of population distributions other than the normal family. Or a bootstrap procedure can be developed.

The key result on which tests for a normal population mean are based was used in Chapter 7 to derive the one-sample t CI: If X_1, X_2, ..., X_n is a random sample from a normal distribution, the standardized variable

$$T = \frac{\overline{X} - \mu}{S/\sqrt{n}}$$

has a t distribution with $n - 1$ degrees of freedom (df). Consider testing $H_0\colon \mu = \mu_0$ using the test statistic $T = (\overline{X} - \mu_0)/(S/\sqrt{n})$. That is, the test statistic results from standardizing $\overline{X}$ under the assumption that H_0 is true (using $S/\sqrt{n}$, the estimated standard deviation of $\overline{X}$, rather than $\sigma/\sqrt{n}$). When H_0 is true, this test statistic has a t distribution with $n - 1$ df. Knowledge of the test statistic's distribution when H_0 is true (the "null distribution") allows us to determine the P-value.

The test statistic is really the same here as in the large-sample case but is labeled T to emphasize that the reference distribution for P-value determination is a t distribution with $n - 1$ df rather than the standard normal (z) distribution. Instead of being a z curve area as was the case for large-sample tests, the P-value will now be an area under the t_{n-1} curve (see Figure 8.6).

The One-Sample t Test

Null hypothesis: $H_0\colon \mu = \mu_0$

Test statistic value: $t = \dfrac{\overline{x} - \mu_0}{s/\sqrt{n}}$

Alternative Hypothesis	P-Value Determination
$H_a\colon \ \mu > \mu_0$	Area under the t_{n-1} curve to the right of t
$H_a\colon \ \mu < \mu_0$	Area under the t_{n-1} curve to the left of t
$H_a\colon \ \mu \neq \mu_0$	$2 \cdot$ (Area under the t_{n-1} curve to the right of $|t|$)

Assumption: The data consists of a random sample from a normal population distribution.

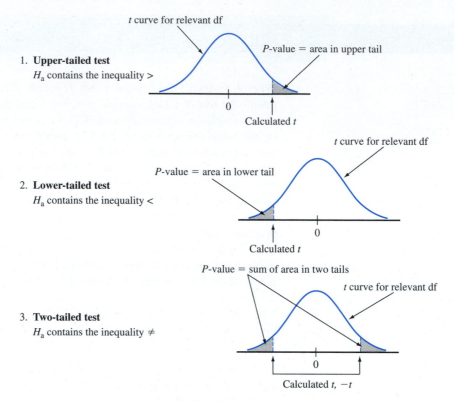

1. **Upper-tailed test**
   H_a contains the inequality >

2. **Lower-tailed test**
   H_a contains the inequality <

3. **Two-tailed test**
   H_a contains the inequality ≠

t curve for relevant df

P-value = area in upper tail

Calculated *t*

t curve for relevant df

P-value = area in lower tail

Calculated *t*

P-value = sum of area in two tails

t curve for relevant df

Calculated *t*, −*t*

Figure 8.6 *P*-values for *t* tests

Unfortunately the table of *t* critical values that we used for confidence and prediction interval calculations in Chapter 7 does not provide much information about *t* curve tail areas. This is because for each *t* distribution there are values for only the seven most commonly used tail areas: .10, .05, .025, .01, .005, .001, and .0005. *P*-value determination would be straightforward if we had a table of tail areas (or alternatively, cumulative areas) that resembled our *z* table: for each different *t* distribution, the area under the corresponding curve to the right (or the left) of values 0.00, 0.01, 0.02, 0.03, ... , 3.97, 3.98, 3.99, and finally 4.00. But this would necessitate an entire page of text for each different *t* distribution.

So we have included another *t* table in Appendix Table A.8. It contains a tabulation of upper-tail *t* curve areas but with less decimal accuracy than what the *z* table provides. Each different column of the table is for a different number of df, and the rows are for calculated values of the test statistic *t* ranging from 0.0 to 4.0 in increments of .1. For example, the number .074 appears at the intersection of the 1.6 row and the 8 df column. Thus the area under the 8 df curve to the right of 1.6 (an upper-tail area) is .074. Because *t* curves are symmetric about 0, .074 is also the area under the 8 df curve to the left of −1.6.

Suppose, for example, that a test of H_0: $\mu = 100$ versus H_a: $\mu > 100$ is based on the 8 df *t* distribution. If the calculated value of the test statistic is $t = 1.6$, then the *P*-value for this upper-tailed test is .074. Because .074 exceeds .05, we would not be able to reject H_0 at a significance level of .05. If the alternative hypothesis is H_a: $\mu < 100$ and a test based on 20 df yields $t = -3.2$, then Appendix Table A.7 shows that the *P*-value is the captured lower-tail area .002. The null hypothesis can be rejected at either level .05 or .01. In the next chapter, we will present a *t* test for hypotheses about a difference between two population means. Suppose the relevant hypotheses are H_0: $\mu_1 - \mu_2 = 0$ versus H_a: $\mu_1 - \mu_2 \neq 0$; the null hypothesis states that the means of the two populations are identical, whereas the alternative hypothesis states that they are different without specifying a direction of departure from H_0. If a *t* test is based on 20 df and $t = 3.2$, then the *P*-value for this two-tailed test is $2(.002) = .004$. This would also be the *P*-value for $t = -3.2$. The tail area is doubled because values both larger than 3.2 and smaller than -3.2 are more contradictory to H_0 than what was calculated (values farther out in *either* tail of the *t* curve).

EXAMPLE 8.9   Carbon nanofibers have potential application as heat-management materials, for composite reinforcement, and as components for nanoelectronics and photonics. The accompanying data on failure stress (MPa) of fiber specimens was read from a graph in the article **"Mechanical and Structural Characterization of Electrospun PAN-Derived Carbon Nanofibers"** (*Carbon*, 2005: 2175–2185).

300	312	327	368	400	425	470	556	573	575
580	589	626	637	690	715	757	891	900	

Summary quantities include $n = 19$, $\bar{x} = 562.68$, $s = 180.874$, $s/\sqrt{n} = 41.495$. Does the data provide compelling evidence for concluding that true average failure stress exceeds 500 MPa?

Figure 8.7 shows a normal probability plot of the data; the substantial linear pattern indicates that a normal population distribution of failure stress is quite plausible, giving us license to employ the one-sample *t* test (the box to the right of the plot gives information about a formal test of the hypothesis that the population distribution is normal; this will be discussed in Chapter 14).

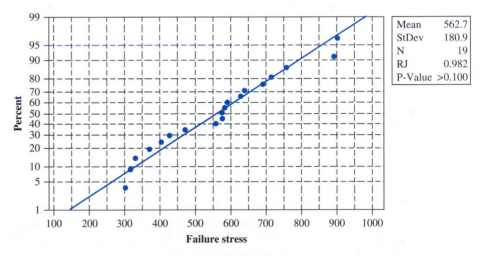

Figure 8.7   Normal probability plot of the failure stress data   ■

Let's carry out a test of the relevant hypotheses using a significance level of .05.

1. The parameter of interest is μ = the true average failure stress

2. The null hypothesis is $H_0: \mu = 500$

3. The appropriate alternative hypothesis is $H_a: \mu > 500$ (so we'll believe that true average failure stress exceeds 500 only if the null hypothesis can be rejected).

4. The one-sample *t* test statistic is $T = (\bar{X} - 500)/(S/\sqrt{n})$. Its value *t* for the given data results from replacing $\bar{X}$ by $\bar{x}$ and S by s.

5. The test-statistic value is $t = (562.68 - 500)/41.495 = 1.51$

6. The test is based on $19 - 1 = 18$ df. The entry in that column and the 1.5 row of Appendix Table A.8 is .075. Since the test is upper-tailed (because $>$ appears in H_a), it follows that P-value $\approx .075$ (Minitab says .074).

7. Because $.075 > .05$, there is not enough evidence to justify rejecting the null hypothesis at significance level .05. Rather than conclude that the true average failure stress exceeds 500, it appears that sampling variability provides a plausible explanation for the fact that the sample mean exceeds 500 by a rather substantial amount. ∎

EXAMPLE 8.10 Many deleterious effects of smoking on health have been well documented. The article **"Smoking Abstinence Impairs Time Estimation Accuracy in Cigarette Smokers"** (*Psychopharmacology Bull.*, **2003: 90–95**) described an investigation into whether time perception, an indicator of a person's ability to concentrate, is impaired during nicotine withdrawal. After a 24-hour smoking abstinence, each of 20 smokers was asked to estimate how much time had elapsed during a 45-second period. The following data on perceived elapsed time is consistent with summary quantities given in the cited article.

69	65	72	73	59	55	39	52	67	57
56	50	70	47	56	45	70	64	67	53

A normal probability plot of this data shows a very substantial linear pattern. Let's carry out a test of hypotheses at significance level .05 to decide whether true average perceived elapsed time differs from the known time 45.

1. μ = true average perceived elapsed time for all smokers exposed to the described experimental regimen

2. $H_0: \mu = 45$

3. $H_0: \mu \neq 45$

4. $t = (\bar{x} - 45)/(s/\sqrt{n})$

5. With $\bar{x} = 59.30$ and $s/\sqrt{n} = 9.84/\sqrt{20} = 2.200$, the test statistic value is $t = 14.3/2.200 = 6.50$.

6. The P-value for a two-tailed test is twice the area under the 19 df t curve to the right of 6.50. Since Table A.8 shows that the area under this t curve to the right of 4.0 is 0, the area to the right of 6.50 is certainly 0. The P-value is then $2(0) = 0$ (.00000 according to software).

7. A P-value as small as what we obtained argues very strongly for rejection of H_0 at any reasonable significance level, and in particular at significance level .05. The difference between the sample mean and its expected value when H_0 is true cannot plausibly be explained simply by chance variation. The true average perceived elapsed time is evidently something other than 45, so nicotine withdrawal does appear to impair perception of time. ∎

β and Sample Size Determination

The calculation of β at the alternative value μ' for a normal population distribution with known σ was carried out by converting the inequality P-value $> \alpha$ to a statement about $\bar{x}$ (e.g., $\bar{x} < \mu_0 + z_\alpha \cdot \sigma/\sqrt{n}$) and then subtracting μ' to standardize correctly. An equivalent approach involves noting that when $\mu = \mu'$, the test statistic $Z = (\bar{X} - \mu_0)/(\sigma/\sqrt{n})$ still has a normal distribution with variance 1, but now the mean value of Z is given by $(\mu' - \mu_0)/(\sigma/\sqrt{n})$. That is, when $\mu = \mu'$, the test statistic still has a normal distribution though not the standard normal distribution. Because of this, $\beta(\mu')$ is an area under the normal curve corresponding to mean value $(\mu' - \mu_0)/(\sigma/\sqrt{n})$ and variance 1. Both α and β involve working with normally distributed variables.

The calculation of $\beta(\mu')$ for the *t* test is much less straightforward. This is because the distribution of the test statistic $T = (\overline{X} - \mu_0)/(S/\sqrt{n})$ is quite complicated when H_0 is false and H_a is true. Thus, for an upper-tailed test, determining

$$\beta(\mu') = P(T < t_{\alpha, n-1} \text{ when } \mu = \mu' \text{ rather than } \mu_0)$$

involves integrating a very unpleasant density function. This must be done numerically. The results are summarized in graphs of β that appear in Appendix Table A.17. There are four sets of graphs, corresponding to one-tailed tests at level .05 and level .01 and two-tailed tests at the same levels.

To understand how these graphs are used, note first that both β and the necessary sample size n are as before functions not just of the absolute difference $|\mu_0 - \mu'|$ but of $d = |\mu_0 - \mu'|/\sigma$. Suppose, for example, that $|\mu_0 - \mu'| = 10$. This departure from H_0 will be much easier to detect (smaller β) when $\sigma = 2$, in which case μ_0 and μ' are 5 population standard deviations apart, than when $\sigma = 10$. The fact that β for the *t* test depends on d rather than just $|\mu_0 - \mu'|$ is unfortunate, since to use the graphs one must have some idea of the true value of σ. A conservative (large) guess for σ will yield a conservative (large) value of $\beta(\mu')$ and a conservative estimate of the sample size necessary for prescribed α and $\beta(\mu')$.

Once the alternative μ' and value of σ are selected, d is calculated and its value located on the horizontal axis of the relevant set of curves. The value of β is the height of the $n - 1$ df curve above the value of d (visual interpolation is necessary if $n - 1$ is not a value for which the corresponding curve appears), as illustrated in Figure 8.8.

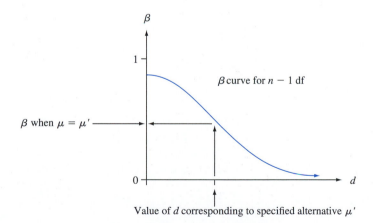

Figure 8.8 A typical β curve for the *t* test

Rather than fixing n (i.e., $n - 1$, and thus the particular curve from which β is read), one might prescribe both α (.05 or .01 here) and a value of β for the chosen μ' and σ. After computing d, the point (d, β) is located on the relevant set of graphs. The curve below and closest to this point gives $n - 1$ and thus n (again, interpolation is often necessary).

EXAMPLE 8.11 The true average voltage drop from collector to emitter of insulated gate bipolar transistors of a certain type is supposed to be at most 2.5 volts. An investigator selects a sample of $n = 10$ such transistors and uses the resulting voltages as a basis for testing $H_0: \mu = 2.5$ versus $H_a: \mu > 2.5$ using a t test with significance level $\alpha = .05$. If the standard deviation of the voltage distribution is $\sigma = .100$, how likely is it that H_0 will not be rejected when in fact $\mu = 2.6$? With $d = |2.5 - 2.6|/.100 = 1.0$, the point on the β curve at 9 df for a one-tailed test with $\alpha = .05$ above 1.0 has a height of approximately .1, so $\beta \approx .1$. The investigator might think that this is too large a value of β for such a substantial departure from H_0 and may wish to have $\beta = .05$ for this alternative value of μ. Since $d = 1.0$, the point $(d, \beta) = (1.0, .05)$ must be located. This point is very close to the 14 df curve, so using $n = 15$ will give both $\alpha = .05$ and $\beta = .05$ when the value of μ is 2.6 and $\sigma = .10$. A larger value of σ would give a larger β for this alternative, and an alternative value of μ closer to 2.5 would also result in an increased value of β. ∎

Most of the widely used statistical software packages are capable of calculating type II error probabilities. They generally work in terms of **power**, which is simply $1 - \beta$. A small value of β (close to 0) is equivalent to large power (near 1). A *powerful* test is one that has high power and therefore good ability to detect when the null hypothesis is false.

As an example, we asked Minitab to determine the power of the upper-tailed test in Example 8.11 for the three sample sizes 5, 10, and 15 when $\alpha = .05$, $\sigma = .10$, and the value of μ is actually 2.6 rather than the null value 2.5—a "difference" of. $2.6 - 2.5 = .1$. We also asked the software to determine the necessary sample size for a power of .9 ($\beta = .1$) and also .95. Here is the resulting output:

```
Power and Sample Size
Testing mean = null (versus > null)
Calculating power for mean = null + difference
Alpha = 0.05  Assumed standard deviation = 0.1
```

Difference	Sample Size	Power
0.1	5	0.579737
0.1	10	0.897517
0.1	15	0.978916

Difference	Sample Size	Target Power	Actual Power
0.1	11	0.90	0.924489
0.1	13	0.95	0.959703

The power for the sample size $n = 10$ is a bit smaller than .9. So if we insist that the power be at least .9, a sample size of 11 is required and the actual power for that n is roughly .92. The software says that for a target power of .95, a sample size of $n = 13$ is required, whereas eyeballing our β curves gave 15. When available, this type of software is more reliable than the curves. Finally, Minitab now also provides power curves for the specified sample sizes, as shown in Figure 8.9. Such curves illustrate how the power increases for each sample size as the actual value of μ moves farther and farther away from the null value.

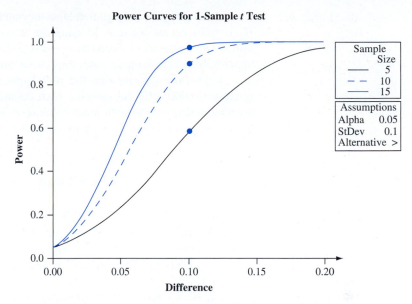

Figure 8.9   Power curves from Minitab for the *t* test of Example 8.11

Variation in *P*-values

The *P*-value resulting from carrying out a test on a selected sample is *not* the probability that H_0 is true, nor is it the probability of rejecting the null hypothesis. Once again, it is the probability, calculated assuming that H_0 is true, of obtaining a test statistic value at least as contradictory to the null hypothesis as the value that actually resulted. For example, consider testing H_0: $\mu = 50$ against H_0: $\mu < 50$ using a lower-tailed *t* test based on 20 df. If the calculated value of the test statistic is $t = -2.00$, then

$$P\text{-value} = P(T < -2.00 \text{ when } \mu = 50)$$
$$= \text{area under the } t_{20} \text{ to the left of } -2.00 = .030$$

But if a second sample is selected, the resulting value of *t* will almost surely be different from -2.00, so the corresponding *P*-value will also likely differ from .030. Because the test statistic value itself varies from one sample to another, the *P*-value will also vary from one sample to another. That is, the test statistic is a random variable, and so the *P*-value will also be a random variable. A first sample may give a *P*-value of .030, a second sample may result in a *P*-value of .117, a third may yield .061 as the *P*-value, and so on.

If H_0 is false, we hope the *P*-value will be close to 0 so that the null hypothesis can be rejected. On the other hand, when H_0 is true, we'd like the *P*-value to exceed the selected significance level so that the correct decision to not reject H_0 is made. The next example presents simulations to show how the *P*-value behaves both when the null hypothesis is true and when it is false.

EXAMPLE 8.12   The fuel efficiency (mpg) of any particular new vehicle under specified driving conditions may not be identical to the EPA figure that appears on the vehicle's sticker. Suppose that four different vehicles of a particular type are to be selected and driven over a certain course, after which the fuel efficiency of each one is to be determined.

Let μ denote the true average fuel efficiency under these conditions. Consider testing $H_0: \mu = 20$ versus $H_0: \mu > 20$ using the one-sample t test based on the resulting sample. Since the test is based on $n - 1 = 3$ degrees of freedom, the P-value for an upper-tailed test is the area under the t curve with 3 df to the right of the calculated t.

Let's first suppose that the null hypothesis is true. We asked Minitab to generate 10,000 different samples, each containing 4 observations, from a normal population distribution with mean value $\mu = 20$ and standard deviation $\sigma = 2$. The first sample and resulting summary quantities were

$$x_1 = 20.830, \, x_2 = 22.232, \, x_3 = 20.276, \, x_4 = 17.718$$

$$\bar{x} = 20.264 \quad s = 1.8864 \quad t = \frac{20.264 - 20}{.1.8864/\sqrt{4}} = .2799$$

The P-value is the area under the 3-df t curve to the right of .2799, which according to Minitab is .3989. Using a significance level of .05, the null hypothesis would of course not be rejected. The values of t for the next four samples were $-1.7591, .6082, -.7020,$ and $3.1053,$ with corresponding P-values .912, .293, .733, and .0265.

Figure 8.10(a) shows a histogram of the 10,000 P-values from this simulation experiment. About 4.5% of these P-values are in the first class interval from 0 to .05. Thus when using a significance level of .05, the null hypothesis is rejected in roughly 4.5% of these 10,000 tests. If we continued to generate samples and carry out the test for each sample at significance level .05, in the long run 5% of the P-values would be in the first class interval. This is because when H_0 is true and a test with significance level .05 is used, by definition the probability of rejecting H_0 is .05.

Looking at the histogram, it appears that the distribution of P-values is relatively flat. In fact, it can be shown that when H_0 is true, the probability distribution of the P-value is a uniform distribution on the interval from 0 to 1. That is, the density curve is completely flat on this interval, and thus must have a height of 1 if the total area under the curve is to be 1. Since the area under such a curve to the left of .05 is $(.05)(1) = .05$, we again have that the probability of rejecting H_0 when it is true that it is .05, the chosen significance level.

Now consider what happens when H_0 is false because $\mu = 21$. We again had Minitab generate 10,000 different samples of size 4 (each from a normal distribution with $\mu = 21$ and $\sigma = 2$), calculate $t = (\bar{x} - 20)/(s/\sqrt{4})$ for each one, and then determine the P-value. The first such sample resulted in $\bar{x} = 20.6411, \, s = .49637,$ $t = 2.5832, \, P$-value $= .0408$. Figure 8.10(b) gives a histogram of the resulting P-values. The shape of this histogram is quite different from that of Figure 8.10(a)— there is a much greater tendency for the P-value to be small (closer to 0) when $\mu = 21$ than when $\mu = 20$. Again H_0 is rejected at significance level .05 whenever the P-value is at most .05 (in the first class interval). Unfortunately, this is the case for only about 19% of the P-values. So only about 19% of the 10,000 tests correctly reject the null hypothesis; for the other 81%, a type II error is committed. The difficulty is that the sample size is quite small and 21 is not very different from the value asserted by the null hypothesis.

Figure 8.10(c) illustrates what happens to the P-value when H_0 is false because $\mu = 22$ (still with $n = 4$ and $\sigma = 2$). The histogram is even more concentrated toward values close to 0 than was the case when $\mu = 21$. In general, as μ moves farther to the right of the null value 20, the distribution of the P-value will become more and more concentrated on values close to 0. Even here a bit fewer than 50% of the P-values are smaller than .05. So it is still slightly more likely than

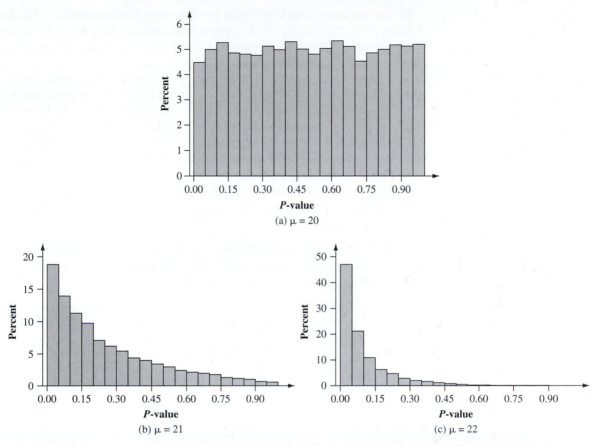

Figure 8.10 *P*-value simulation results for Example 8.12

not that the null hypothesis is incorrectly not rejected. Only for values of μ much larger than 20 (e.g., at least 24 or 25) is it highly likely that the *P*-value will be smaller than .05 and thus give the correct conclusion.

The big idea of this example is that because the value of any test statistic is random, the *P*-value will also be a random variable and thus have a distribution. The farther the actual value of the parameter is from the value specified by the null hypothesis, the more the distribution of the *P*-value will be concentrated on values close to 0 and the greater the chance that the test will correctly reject H_0 (corresponding to smaller β). ■

Whenever the observed value of a statistic such as $\overline{X}$ or $\hat{p}$ is reported, it is good statistical practice to include a quantitative measure of the statistic's precision, e.g., that the estimated standard error of $\overline{X}$ is $s/\sqrt{n}$. The *P*-value itself is a statistic—its value can be calculated once sample data is available and a particular test procedure is selected, and before such data is in hand, the *P*-value is subject to randomness. So it would be nice to have available σ_P or an estimate of this standard deviation.

Unfortunately the sampling distribution of a *P*-value is in general quite complicated. The simulation results of Example 8.12 suggest that the sampling distribution is quite skewed when H_0 is false (it is uniformly distributed on (0,1) when H_0 is true and the test statistic has a continuous distribution, e.g., a *t* distribution). A standard deviation is not as easy to interpret and use when there is substantial non-normality. The statisticians Dennis Boos and Leonard Stefanski investigated the random behavior of the P-value in their article **"P-Value Precision and Reproducibility"** (*The American Statistician, 2011: 213–221*). To address non-normality, they focused on the quantity $-\log(P\text{-value})$. The log-transformed P-value does for many test procedures have approximately a normal distribution when n is large.

Suppose application of a particular test procedure to sample data results in a P-value of .001. Then H_0 would be rejected using either a significance level of .05 or .01. If a new sample from the same population distribution is then selected, how likely is it that the P-value for this new data will lead to rejection of H_0 at a significance level of .05 or .01? This is what the authors of the foregoing article meant by "reproducibility": How likely is it that a new sample will lead to the same conclusion as that reached using the original sample? The answer to this question depends on the population distribution, the sample size, and the test procedure used. Nevertheless, based on their investigations, the authors suggested the following general guidelines:

If the P-value for the original data is .0001, then $P(\text{new } P\text{-value} \leq .05) \approx .97$, whereas this probability is roughly .91 if the original P-value is .001 and it is roughly .73 when the original P-value is .01.

Particularly when the original P-value is around .01, there is a reasonably good chance that a new sample will not lead to rejection of H_0 at the 5% significance level. Thus unless the original P-value is really small, it would not be surprising to have a new sample contradict the inference drawn from the original data. A P-value not too much smaller than a chosen significance level such as .05 or .01 should be viewed with some caution!

EXERCISES Section 8.3 (29–41)

29. The true average diameter of ball bearings of a certain type is supposed to be .5 in. A one-sample t test will be carried out to see whether this is the case. What conclusion is appropriate in each of the following situations?
   a. $n = 13, t = 1.6, \alpha = .05$
   b. $n = 13, t = -1.6, \alpha = .05$
   c. $n = 25, t = -2.6, \alpha = .01$
   d. $n = 25, t = -3.9$

30. A sample of n sludge specimens is selected and the pH of each one is determined. The one-sample t test will then be used to see if there is compelling evidence for concluding that true average pH is less than 7.0. What conclusion is appropriate in each of the following situations?
   a. $n = 6, t = -2.3, \alpha = .05$
   b. $n = 15, t = -3.1, \alpha = .01$
   c. $n = 12, t = -1.3, \alpha = .05$
   d. $n = 6, t = .7, \alpha = .05$
   e. $n = 6, \bar{x} = 6.68, s/\sqrt{n} = .0820$

31. The paint used to make lines on roads must reflect enough light to be clearly visible at night. Let μ denote the true average reflectometer reading for a new type of paint under consideration. A test of $H_0: \mu = 20$ versus $H_a: \mu > 20$ will be based on a random sample of size n from a normal population distribution. What conclusion is appropriate in each of the following situations?
   a. $n = 15, t = 3.2, \alpha = .05$
   b. $n = 9, t = 1.8, \alpha = .01$
   c. $n = 24, t = -.2$

32. The relative conductivity of a semiconductor device is determined by the amount of impurity "doped" into the device during its manufacture. A silicon diode to be used for a specific purpose requires an average cut-on voltage of .60 V, and if this is not achieved, the amount of impurity must be adjusted. A sample of diodes was selected and the cut-on voltage was determined. The accompanying SAS output resulted from a request to test the appropriate hypotheses.

N	Mean	Std Dev	T	Prob. > |T|
15	0.0453333	0.0899100	1.9527887	0.0711

[*Note:* SAS explicitly tests $H_0: \mu = 0$, so to test $H_0: \mu = .60$, the null value .60 must be subtracted from each x_i; the reported mean is then the average of the $(x_i - .60)$ values. Also, SAS's P-value is always for a two-tailed test.] What would be concluded for a significance level of .01? .05? .10?

33. The article **"The Foreman's View of Quality Control"** (*Quality Engr.*, **1990: 257–280**) described an investigation into the coating weights for large pipes resulting from a galvanized coating process. Production standards call for a true average weight of 200 lb per pipe. The accompanying descriptive summary and boxplot are from Minitab.

Variable	N	Mean	Median	TrMean	StDev	SEMean
ctg wt	30	206.73	206.00	206.81	6.35	1.16

Variable	Min	Max	Q1	Q3
ctg wt	193.00	218.00	202.75	212.00

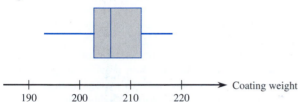

Coating weight

a. What does the boxplot suggest about the status of the specification for true average coating weight?

b. A normal probability plot of the data was quite straight. Use the descriptive output to test the appropriate hypotheses.

34. The following observations are on stopping distance (ft) of a particular truck at 20 mph under specified experimental conditions (**"Experimental Measurement of the Stopping Performance of a Tractor-Semitrailer from Multiple Speeds,"** **NHTSA, DOT HS 811 488, June 2011**):

32.1   30.6   31.4   30.4   31.0   31.9

The cited report states that under these conditions, the maximum allowable stopping distance is 30. A normal probability plot validates the assumption that stopping distance is normally distributed.

a. Does the data suggest that true average stopping distance exceeds this maximum value? Test the appropriate hypotheses using $\alpha = .01$.

b. Determine the probability of a type II error when $\alpha = .01$, $\sigma = .65$, and the actual value of μ is 31. Repeat this for $\mu = 32$ (use either statistical software or Table A.17).

c. Repeat (b) using $\sigma = .80$ and compare to the results of (b).

d. What sample size would be necessary to have $\alpha = .01$ and $\beta = .10$ when $\mu = 31$ and $\sigma = .65$?

35. The article **"Uncertainty Estimation in Railway Track Life-Cycle Cost"** (*J. of Rail and Rapid Transit*, **2009**) presented the following data on time to repair (min) a rail break in the high rail on a curved track of a certain railway line.

159 120 480 149 270 547 340 43 228 202 240 218

A normal probability plot of the data shows a reasonably linear pattern, so it is plausible that the population distribution of repair time is at least approximately normal. The sample mean and standard deviation are 249.7 and 145.1, respectively.

a. Is there compelling evidence for concluding that true average repair time exceeds 200 min? Carry out a test of hypotheses using a significance level of .05.

b. Using $\sigma = 150$, what is the type II error probability of the test used in (a) when true average repair time is actually 300 min? That is, what is $\beta(300)$?

36. Have you ever been frustrated because you could not get a container of some sort to release the last bit of its contents? The article **"Shake, Rattle, and Squeeze: How Much Is Left in That Container?"** (*Consumer Reports*, **May 2009: 8**) reported on an investigation of this issue for various consumer products. Suppose five 6.0 oz tubes of toothpaste of a particular brand are randomly selected and squeezed until no more toothpaste will come out. Then each tube is cut open and the amount remaining is weighed, resulting in the following data (consistent with what the cited article reported): .53, .65, .46, .50, .37. Does it appear that the true average amount left is less than 10% of the advertised net contents?

a. Check the validity of any assumptions necessary for testing the appropriate hypotheses.

b. Carry out a test of the appropriate hypotheses using a significance level of .05. Would your conclusion change if a significance level of .01 had been used?

c. Describe in context type I and II errors, and say which error might have been made in reaching a conclusion.

37. The accompanying data on cube compressive strength (MPa) of concrete specimens appeared in the article **"Experimental Study of Recycled Rubber-Filled High-Strength Concrete"** (*Magazine of Concrete Res.*, **2009: 549–556**):

112.3   97.0   92.7   86.0   102.0
99.2   95.8   103.5 89.0   86.7

a. Is it plausible that the compressive strength for this type of concrete is normally distributed?

b. Suppose the concrete will be used for a particular application unless there is strong evidence that true average strength is less than 100 MPa. Should the concrete be used? Carry out a test of appropriate hypotheses.

38. A random sample of soil specimens was obtained, and the amount of organic matter (%) in the soil was determined for each specimen, resulting in the accompanying data (from **"Engineering Properties of Soil,"** *Soil Science*, **1998: 93–102**).

1.10   5.09   0.97   1.59   4.60   0.32   0.55   1.45
0.14   4.47   1.20   3.50   5.02   4.67   5.22   2.69
3.98   3.17   3.03   2.21   0.69   4.47   3.31   1.17
0.76   1.17   1.57   2.62   1.66   2.05

The values of the sample mean, sample standard deviation, and (estimated) standard error of the mean are 2.481, 1.616, and .295, respectively. Does this data suggest that the true average percentage of organic matter in such soil is something other than 3%? Carry out a test of the appropriate hypotheses at significance level .10. Would your conclusion be different if $\alpha = .05$ had been used? [*Note:* A normal probability plot of the data shows an acceptable pattern in light of the reasonably large sample size.]

39. Reconsider the accompanying sample data on expense ratio (%) for large-cap growth mutual funds first introduced in Exercise 1.53.

0.52 1.06 1.26 2.17 1.55 0.99 1.10 1.07 1.81 2.05
0.91 0.79 1.39 0.62 1.52 1.02 1.10 1.78 1.01 1.15

A normal probability plot shows a reasonably linear pattern.

a. Is there compelling evidence for concluding that the population mean expense ratio exceeds 1%? Carry out a test of the relevant hypotheses using a significance level of .01.

b. Referring back to (a), describe in context type I and II errors and say which error you might have made in reaching your conclusion. The source from which the data was obtained reported that $\mu = 1.33$ for the population of all 762 such funds. So did you actually commit an error in reaching your conclusion?

c. Supposing that $\sigma = .5$, determine and interpret the power of the test in (a) for the actual value of μ stated in (b).

40. Polymer composite materials have gained popularity because they have high strength to weight ratios and are relatively easy and inexpensive to manufacture. However, their nondegradable nature has prompted development of environmentally friendly composites using natural materials. The article **"Properties of Waste Silk Short Fiber/Cellulose Green Composite Films" (*J. of Composite Materials*, 2012: 123–127)** reported that for a sample of 10 specimens with 2% fiber content, the sample mean tensile strength (MPa) was 51.3 and the sample standard deviation was 1.2. Suppose the true average strength for 0% fibers (pure cellulose) is known to be 48 MPa. Does the data provide compelling evidence for concluding that true average strength for the WSF/cellulose composite exceeds this value?

41. A spectrophotometer used for measuring CO concentration [ppm (parts per million) by volume] is checked for accuracy by taking readings on a manufactured gas (called span gas) in which the CO concentration is very precisely controlled at 70 ppm. If the readings suggest that the spectrophotometer is not working properly, it will have to be recalibrated. Assume that if it is properly calibrated, measured concentration for span gas samples is normally distributed. On the basis of the six readings—85, 77, 82, 68, 72, and 69—is recalibration necessary? Carry out a test of the relevant hypotheses using $\alpha = .05$.

8.4 Tests Concerning a Population Proportion

Let p denote the proportion of individuals or objects in a population who possess a specified property (e.g., college students who graduate without any debt, or computers that do not need service during the warranty period). If an individual or object with the property is labeled a success (*S*), then p is the population proportion of successes. Tests concerning p will be based on a random sample of size n from the population. Provided that n is small relative to the population size, X (the number of *S*'s in the sample) has (approximately) a binomial distribution. Furthermore, if n itself is large [$np \geq 10$ and $n(1 - p) \geq 10$], both X and the estimator $\hat{p} = X/n$ are approximately normally distributed. We first consider large-sample tests based on this latter fact and then turn to the small-sample case that directly uses the binomial distribution.

Large-Sample Tests

Large-sample tests concerning p are a special case of the more general large-sample procedures for a parameter θ. Let $\hat{\theta}$ be an estimator of θ that is (at least approximately) unbiased and has approximately a normal distribution. The null hypothesis has the form $H_0: \theta = \theta_0$ where θ_0 denotes a number (the null value) appropriate to the problem context. Suppose that when H_0 is true, the standard deviation of $\hat{\theta}$, $\sigma_{\hat{\theta}}$, involves no unknown parameters. For example, if $\theta = \mu$ and $\hat{\theta} = \overline{X}$, $\sigma_{\hat{\theta}} = \sigma_{\overline{X}} = \sigma/\sqrt{n}$, which involves no unknown parameters only if the value of σ is known. A large-sample test statistic results from standardizing $\hat{\theta}$ under the assumption that H_0 is true (so that $E(\hat{\theta}) = \theta_0$):

$$\text{Test statistic: } Z = \frac{\hat{\theta} - \theta_0}{\sigma_{\hat{\theta}}}$$

If the alternative hypothesis is $H_a: \theta > \theta_0$, an upper-tailed test whose significance level is approximately α has P-value $= 1 - \Phi(z)$. The other two alternatives, $H_a: \theta < \theta_0$ and $H_a: \theta \neq \theta_0$, are tested using a lower-tailed z test and a two-tailed z test, respectively.

In the case $\theta = p$, $\sigma_{\hat{\theta}}$ will not involve any unknown parameters when H_0 is true, but this is atypical. When $\sigma_{\hat{\theta}}$ does involve unknown parameters, it is often possible to use an estimated standard deviation $S_{\hat{\theta}}$ in place of $\sigma_{\hat{\theta}}$ and still have Z approximately normally distributed when H_0 is true (because this substitution does not increase variability in Z by very much). The large-sample test of the previous section furnishes an example of this: Because σ is usually unknown, we use $s_{\hat{\theta}} = s_{\overline{X}} = s/\sqrt{n}$ in place of $\sigma/\sqrt{n}$ in the denominator of z.

The estimator $\hat{p} = X/n$ is unbiased ($E(\hat{p}) = p$) and its standard deviation is $\sigma_{\hat{p}} = \sqrt{p(1-p)/n}$. These facts along with approximate normality were used in Section 7.2 to obtain a confidence interval for p. When H_0 is true, $E(\hat{p}) = p_0$ and $\sigma_{\hat{p}} = \sqrt{p_0(1-p_0)/n}$, so $\sigma_{\hat{p}}$ does not involve any unknown parameters. It then follows that when n is large and H_0 is true, the test statistic

$$Z = \frac{\hat{p} - p_0}{\sqrt{p_0(1-p_0)/n}}$$

has approximately a standard normal distribution. The P-value for the test is then a z curve area, just as it was in the case of large-sample z tests concerning μ. Its calculation depends on which of the three inequalities in H_a is under consideration.

Null hypothesis: $H_0: p = p_0$

Test statistic value: $z = \dfrac{\hat{p} - p_0}{\sqrt{p_0(1-p_0)/n}}$

Alternative Hypothesis	P-Value Determination
$H_a:\ p > p_0$	Area under the standard normal curve to the right of z
$H_a:\ p < p_0$	Area under the standard normal curve to the left of z
$H_a:\ p \neq p_0$	2·(Area under the standard normal curve to the right of $|z|$)

These test procedures are valid provided that $np_0 \geq 10$ and $n(1-p_0) \geq 10$.

They are referred to as *upper-tailed*, *lower-tailed*, and *two-tailed*, respectively, for the three different alternative hypotheses.

EXAMPLE 8.13 Student use of cell phones during class is perceived by many faculty to be an annoying but perhaps harmless distraction. However, the use of a phone to text during an exam is a serious breach of conduct. The article **"The Use and Abuse of Cell Phones and Text Messaging During Class: A Survey of College Students"** (*College Teaching*, 2012: 1–9) reported that 27 of the 267 students in a sample admitted to doing this. Can it be concluded at significance level .001 that more than 5% of all students in the population sampled had texted during an exam?

1. The parameter of interest is the proportion p of the sampled population that has texted during an exam.
2. The null hypothesis is H_0: $p = .05$
3. The alternative hypothesis is H_a: $p > .05$
4. Since $np_0 = 267(.05) = 13.35 \geq 10$ and $nq_0 = 267(.95) = 253.65 \geq 10$, the large-sample z test can be used. The test statistic value is $z = (\hat{p} - .05)/\sqrt{(.05)(.95)/n}$.
5. $\hat{p} = 27/267 = .1011$, from which $z = (.1011 - .05)/\sqrt{(.05)(.95)/267} = .0511/.0133 = 3.84$
6. The P-value for this upper-tailed z test is $1 - \Phi(3.84) < 1 - \Phi(3.49) = .0003$ (software gives .000062).
7. The null hypothesis is resoundingly rejected because P-value $< .0003 \leq .001 = \alpha$. The evidence for concluding that the population percentage of students who text during an exam exceeds 5% is very compelling. The cited article's abstract contained the following comment: "The majority of the students surveyed believe that instructors are largely unaware of the extent to which texting and other cell phone activities engage students in the classroom." Maybe it is time for instructors, administrators, and student leaders to become proactive about this issue ∎

β and Sample Size Determination   When H_0 is true, the test statistic Z has approximately a standard normal distribution. Now suppose that H_0 is *not* true and that $p = p'$. Then Z still has approximately a normal distribution (because it is a linear function of $\hat{p}$), but its mean value and variance are no longer 0 and 1, respectively. Instead,

$$E(Z) = \frac{p' - p_0}{\sqrt{p_0(1 - p_0)/n}} \qquad V(Z) = \frac{p'(1 - p')/n}{p_0(1 - p_0)/n}$$

The null hypothesis will not be rejected if P-value $> \alpha$. For an upper-tailed z test (inequality $>$ in H_a), we argued previously that this is equivalent to $z < z_\alpha$. The probability of a type II error (not rejecting H_0 when it is false) is $\beta(p') = P(Z < z_\alpha$ when $p = p')$. This can be computed by using the given mean and variance to standardize and then referring to the standard normal cdf. In addition, if it is desired that the level α test also have $\beta(p') = \beta$ for a specified value of β, this equation can be solved for the necessary n as in Section 8.2. General expressions for $\beta(p')$ and n are given in the accompanying box.

Alternative Hypothesis	$\beta(p')$
H_a: $p > p_0$	$\Phi\left[\dfrac{p_0 - p' + z_\alpha\sqrt{p_0(1 - p_0)/n}}{\sqrt{p'(1 - p')/n}}\right]$
H_a: $p < p_0$	$1 - \Phi\left[\dfrac{p_0 - p' - z_\alpha\sqrt{p_0(1 - p_0)/n}}{\sqrt{p'(1 - p')/n}}\right]$

$$H_a: \quad p \neq p_0 \qquad \Phi\left[\frac{p_0 - p' + z_{\alpha/2}\sqrt{p_0(1-p_0)/n}}{\sqrt{p'(1-p')/n}}\right]$$

$$-\Phi\left[\frac{p_0 - p' - z_{\alpha/2}\sqrt{p_0(1-p_0)/n}}{\sqrt{p'(1-p')/n}}\right]$$

The sample size n for which the level α test also satisfies $\beta(p') = \beta$ is

$$n = \begin{cases} \left[\dfrac{z_\alpha\sqrt{p_0(1-p_0)} + z_\beta\sqrt{p'(1-p')}}{p' - p_0}\right]^2 & \text{one-tailed test} \\[2em] \left[\dfrac{z_{\alpha/2}\sqrt{p_0(1-p_0)} + z_\beta\sqrt{p'(1-p')}}{p' - p_0}\right]^2 & \text{two-tailed test (an} \\ & \text{approximate solution)} \end{cases}$$

EXAMPLE 8.14   A package-delivery service advertises that at least 90% of all packages brought to its office by 9 A.M. for delivery in the same city are delivered by noon that day. Let p denote the true proportion of such packages that are delivered as advertised and consider the hypotheses $H_0: p = .9$ versus $H_a: p < .9$. If only 80% of the packages are delivered as advertised, how likely is it that a level .01 test based on $n = 225$ packages will detect such a departure from H_0? What should the sample size be to ensure that $\beta(.8) = .01$? With $\alpha = .01$, $p_0 = .9$, $p' = .8$, and $n = 225$,

$$\beta(.8) = 1 - \Phi\left(\frac{.9 - .8 - 2.33\sqrt{(.9)(.1)/225}}{\sqrt{(.8)(.2)/225}}\right)$$

$$= 1 - \Phi(2.00) = .0228$$

Thus the probability that H_0 will be rejected using the test when $p = .8$ is .9772; roughly 98% of all samples will result in correct rejection of H_0.

Using $z_\alpha = z_\beta = 2.33$ in the sample size formula yields

$$n = \left[\frac{2.33\sqrt{(.9)(.1)} + 2.33\sqrt{(.8)(.2)}}{.8 - .9}\right]^2 \approx 266 \qquad \blacksquare$$

Small-Sample Tests

Test procedures when the sample size n is small are based directly on the binomial distribution rather than the normal approximation. Consider the alternative hypothesis $H_a: p > p_0$ and again let X be the number of successes in the sample. Then X is the test statistic. When H_0 is true, X has a binomial distribution with parameters n and p_0, so

$$\begin{aligned} P\text{-value} &= P(X \geq x \text{ when } H_0 \text{ is true}) \\ &= P(X \geq x \text{ when } X \sim \text{Bin}(n, p_0)) \\ &= 1 - P(X \leq x - 1 \text{ when } X \sim \text{Bin}(n, p_0)) \\ &= 1 - B(x - 1; n, p_0) \end{aligned}$$

Because X has a discrete probability distribution, it is usually not possible to obtain a test for which $P(\text{type I error})$ is exactly the desired significance level α (e.g., .05 or .01; refer back to middle of page 323 for an example).

Let p' denote an alternative value of $p(p' > p_0)$. When $p = p'$, $X \sim \text{Bin}(n, p')$. The probability of a type II error is then calculated by expressing the condition P-value $> \alpha$ in the equivalent form $x < c_\alpha$. Then

$$\beta(p') = P(\text{type II error when } p = p')$$
$$= P(X < c_\alpha \text{ when } X \sim \text{Bin}(n, p')) = B(c_\alpha - 1; n, p')$$

That is, $\beta(p')$ is the result of a straightforward binomial probability calculation. The sample size n necessary to ensure that a level α test also has specified β at a particular alternative value p' must be determined by trial and error using the binomial cdf.

Test procedures for $H_a: p < p_0$ and for $H_a: p \neq p_0$ are constructed in a similar manner. In the former case, the P-value is $B(x; n, p_0)$. The P-value when the alternative hypothesis is $H_a: p \neq p_0$ is twice the smaller of the two probabilities $B(x; n, p_0)$ and $1 - B(x - 1; n, p_0)$.

EXAMPLE 8.15   A plastics manufacturer has developed a new type of plastic trash can and proposes to sell them with an unconditional 6-year warranty. To see whether this is economically feasible, 20 prototype cans are subjected to an accelerated life test to simulate 6 years of use. The proposed warranty will be modified only if the sample data strongly suggests that fewer than 90% of such cans would survive the 6-year period. Let p denote the proportion of all cans that survive the accelerated test. The relevant hypotheses are $H_0: p = .9$ versus $H_a: p < .9$. A decision will be based on the test statistic X, the number among the 20 that survive. Because of the inequality in H_a, any value smaller than the observed value x is more contradictory to H_0 than is x itself. Therefore

$$P\text{-value} = P(X \leq x \text{ when } H_0 \text{ is true}) = B(x; 20, .9)$$

From Appendix Table A.1, $B(15; 20, .9) = .043$, whereas $B(16; 20, .9) = .133$. The closest achievable significance level to .05 is therefore .043. Since $B(14; 20, .9) = .011$, H_0 would be rejected at this significance level if the accelerated test results in $x = 14$. It would then be appropriate to modify the proposed warranty. Because P-value $\leq .043$ is equivalent to $x \leq 15$, the probability of a type II error for the alternative value $p' = .8$ is

$$\beta(.8) = P(H_0 \text{ is not rejected when } X \sim \text{Bin}(20, .8))$$
$$= P(X \geq 16 \text{ when } X \sim \text{Bin}(20, .8))$$
$$= 1 - B(15; 20, .8) = 1 - .370 = .630$$

That is, when $p = .8$, 63% of all samples consisting of $n = 20$ cans would result in H_0 being incorrectly not rejected. This error probability is high because 20 is a small sample size and $p' = .8$ is close to the null value $p_0 = .9$. ■

EXERCISES Section 8.4 (42–52)

42. Consider using a z test to test $H_0: p = .6$. Determine the P-value in each of the following situations.
   a. $H_a: p > .6$, $z = 1.47$
   b. $H_a: p < .6$, $z = -2.70$
   c. $H_a: p \neq .6$, $z = -2.70$
   d. $H_a: p < .6$, $z = .25$

43. A common characterization of obese individuals is that their body mass index is at least 30 [BMI = weight/(height)^2, where height is in meters and weight is in kilograms]. The article **"The Impact of Obesity on Illness Absence and Productivity in an Industrial Population of Petrochemical Workers"** (*Annals of Epidemiology*, 2008: 8–14) reported that in a sample of female workers, 262 had BMIs of less than 25, 159 had BMIs that were at least 25 but less than 30, and 120 had BMIs exceeding 30. Is there compelling evidence for concluding that more than 20% of the individuals in the sampled population are obese?

   a. State and test appropriate hypotheses with a significance level of .05.

   b. Explain in the context of this scenario what constitutes type I and II errors.

   c. What is the probability of not concluding that more than 20% of the population is obese when the actual percentage of obese individuals is 25%?

44. A manufacturer of nickel-hydrogen batteries randomly selects 100 nickel plates for test cells, cycles them a specified number of times, and determines that 14 of the plates have blistered.

   a. Does this provide compelling evidence for concluding that more than 10% of all plates blister under such circumstances? State and test the appropriate hypotheses using a significance level of .05. In reaching your conclusion, what type of error might you have committed?

   b. If it is really the case that 15% of all plates blister under these circumstances and a sample size of 100 is used, how likely is it that the null hypothesis of part (a) will not be rejected by the level .05 test? Answer this question for a sample size of 200.

   c. How many plates would have to be tested to have $\beta(.15) = .10$ for the test of part (a)?

45. A random sample of 150 recent donations at a certain blood bank reveals that 82 were type A blood. Does this suggest that the actual percentage of type A donations differs from 40%, the percentage of the population having type A blood? Carry out a test of the appropriate hypotheses using a significance level of .01. Would your conclusion have been different if a significance level of .05 had been used?

46. It is known that roughly 2/3 of all human beings have a dominant right foot or eye. Is there also right-sided dominance in kissing behavior? The article **"Human Behavior: Adult Persistence of Head-Turning Asymmetry"** (*Nature*, 2003: 771) reported that in a random sample of 124 kissing couples, both people in 80 of the couples tended to lean more to the right than to the left.

   a. If 2/3 of all kissing couples exhibit this right-leaning behavior, what is the probability that the number in a sample of 124 who do so differs from the expected value by at least as much as what was actually observed?

   b. Does the result of the experiment suggest that the 2/3 figure is implausible for kissing behavior? State and test the appropriate hypotheses.

47. The article **"Effects of Bottle Closure Type on Consumer Perception of Wine Quality"** (*Amer. J. of Enology and Viticulture*, 2007: 182–191) reported that in a sample of 106 wine consumers, 22 (20.8%) thought that screw tops were an acceptable substitute for natural corks. Suppose a particular winery decided to use screw tops for one of its wines unless there was strong evidence to suggest that fewer than 25% of wine consumers found this acceptable.

   a. Using a significance level of .10, what would you recommend to the winery?

   b. For the hypotheses tested in (a), describe in context what the type I and II errors would be, and say which type of error might have been committed.

48. With domestic sources of building supplies running low several years ago, roughly 60,000 homes were built with imported Chinese drywall. According to the article **"Report Links Chinese Drywall to Home Problems"** (*New York Times*, Nov. 24, 2009), federal investigators identified a strong association between chemicals in the drywall and electrical problems, and there is also strong evidence of respiratory difficulties due to the emission of hydrogen sulfide gas. An extensive examination of 51 homes found that 41 had such problems. Suppose these 51 were randomly sampled from the population of all homes having Chinese drywall.

   a. Does the data provide strong evidence for concluding that more than 50% of all homes with Chinese drywall have electrical/environmental problems? Carry out a test of hypotheses using $\alpha = .01$.

   b. Calculate a lower confidence bound using a confidence level of 99% for the percentage of all such homes that have electrical/environmental problems.

   c. If it is actually the case that 80% of all such homes have problems, how likely is it that the test of (a) would not conclude that more than 50% do?

49. A plan for an executive travelers' club has been developed by an airline on the premise that 5% of its current customers would qualify for membership. A random sample of 500 customers yielded 40 who would qualify.

   a. Using this data, test at level .01 the null hypothesis that the company's premise is correct against the alternative that it is not correct.

   b. What is the probability that when the test of part (a) is used, the company's premise will be judged correct when in fact 10% of all current customers qualify?

50. Each of a group of 20 intermediate tennis players is given two rackets, one having nylon strings and the other synthetic gut strings. After several weeks of playing with the two rackets, each player will be asked to state a preference for one of the two types of strings. Let p denote the proportion of all such players who would prefer gut to nylon, and let X be the number of players in the sample who prefer gut. Because gut strings are more expensive, consider the null hypothesis that at most 50% of all such players prefer gut. We simplify this to $H_0: p = .5$, planning to reject H_0 only if sample evidence strongly favors gut strings.

a. Is a significance level of exactly .05 achievable? If not, what is the largest α smaller than .05 that is achievable?

b. If 60% of all enthusiasts prefer gut, calculate the probability of a type II error using the significance level from part (a). Repeat if 80% of all enthusiasts prefer gut.

c. If 13 out of the 20 players prefer gut, should H_0 be rejected using the significance level of (a)?

51. A manufacturer of plumbing fixtures has developed a new type of washerless faucet. Let $p = P$(a randomly selected faucet of this type will develop a leak within 2 years under normal use). The manufacturer has decided to proceed with production unless it can be determined that p is too large; the borderline acceptable value of p is specified as .10. The manufacturer decides to subject n of these faucets to accelerated testing (approximating 2 years of normal use). With $X =$ the number among the n faucets that leak before the test concludes, production will commence unless the observed X is too large. It is decided that if $p = .10$, the probability of not proceeding should be at most .10, whereas if $p = .30$ the probability of proceeding should be at most .10. Can $n = 10$ be used? $n = 20$? $n = 25$? What are the actual error probabilities for the chosen n?

52. In a sample of 171 students at an Australian university that introduced the use of plagiarism-detection software in a number of courses, 58 students indicated a belief that such software unfairly targets students (**"Student and Staff Perceptions of the Effectiveness of Plagiarism Detection Software,"** *Australian J. of Educ. Tech.,* **2008: 222–240).** Does this suggest that a majority of students at the university do not share this belief? Test appropriate hypotheses.

8.5 Further Aspects of Hypothesis Testing

We close this introductory chapter on hypothesis testing by briefly considering a variety of issues involving the use of test procedures: the distinction between statistical and practical significance, the relationship between tests and confidence intervals, the implications of multiple testing, and a general method for deriving test statistics.

Statistical Versus Practical Significance

Statistical significance means simply that the null hypothesis was rejected at the selected significance level. That is, in the judgment of the investigator, any observed discrepancy between the data and what would be expected were H_0 true cannot be explained solely by chance variation. However, a small P-value, which would ordinarily indicate statistical significance, may be the result of a large sample size in combination with a departure from H_0 that has little **practical significance**. In many experimental situations, only departures from H_0 of large magnitude would be worthy of detection, whereas a small departure from H_0 would have little practical significance.

As an example, let μ denote the true average IQ of all children in the very large city of Euphoria. Consider testing $H_0: \mu = 100$ versus $H_a: \mu > 100$ assuming a normal IQ distribution with $\sigma = 15$ (100 is conventionally believed to be the average IQ for all individuals, so parents of Euphorian children might be euphoric to have the null hypothesis rejected). But one IQ point is no big deal, so the value $\mu = 101$ certainly does not represent a departure from H_0 that has practical significance. For a reasonably large sample size $n,$ this μ would lead to an $\bar{x}$ value near 101, so we would not want this sample evidence to argue strongly for rejection of H_0 when $\bar{x} = 101$ is observed. For various sample sizes, Table 8.1 records both the P-value when $\bar{x} = 101$ and also the probability of not rejecting H_0 at level .01 when $\mu = 101$.

The second column in Table 8.1 shows that even for moderately large sample sizes, the P-value resulting from $\bar{x} = 101$ argues very strongly for rejection of H_0, whereas the observed $\bar{x}$ itself suggests that in practical terms the true value of μ differs

Table 8.1   An Illustration of the Effect of Sample Size on *P*-values and β

n	*P*-Value When $\bar{x} = 101$	$\beta(101)$ for Level .01 Test
25	.3707	.9772
100	.2514	.9525
400	.0918	.8413
900	.0228	.6293
1600	.0038	.3707
2500	.0004	.1587
5000	.0000012	.0087
10,000	.0000000	.0000075

little from the null value $\mu_0 = 100$. The third column points out that even when there is little practical difference between the true μ and the null value, for a fixed level of significance a large sample size will frequently lead to rejection of the null hypothesis at that level. To summarize, *one must be especially careful in interpreting evidence when the sample size is large, since any small departure from H_0 will almost surely be detected by a test, yet such a departure may have little practical significance.*

The Relationship between Confidence Intervals and Hypothesis Tests

Suppose the standardized variable $Z = (\hat{\theta} - \theta)/\hat{\sigma}_{\hat{\theta}}$ has (at least approximately) a standard normal distribution. The central z curve area captured between -1.96 and 1.96 is .95 (and the remaining area .05 is split equally between the two tails, giving area .025 in each one). This implies that a confidence interval for θ with confidence level 95% is $\hat{\theta} \pm 1.96\hat{\sigma}_{\hat{\theta}}$.

Now consider testing $H_0: \theta = \theta_0$ versus $H_a: \theta \neq \theta_0$ at significance level .05 using the test statistic $Z = (\hat{\theta} - \theta_0)/\hat{\sigma}_{\hat{\theta}}$. The phrase "$z$ test" implies that when the null hypothesis is true, Z has (at least approximately) a standard normal distribution. So the *P*-value will be twice the area under the z curve to the right of $|z|$. This *P*-value will be less than or equal to .05, allowing for rejection of the null hypothesis, if and only if either $z \geq 1.96$ or $z \leq -1.96$. The null hypothesis will therefore not be rejected if $-1.96 < z < 1.96$.

Substituting the formula for z into this latter system of inequalities and manipulating them to isolate θ_0 gives the equivalent system $\hat{\theta} - 1.96\hat{\sigma}_{\hat{\theta}} < \theta_0 < \hat{\theta} + 1.96\hat{\sigma}_{\hat{\theta}}$. The lower limit in this system is just the left endpoint of the 95% confidence interval, and the upper limit is the right endpoint of the interval. What this says is that the null hypothesis will not be rejected if and only if the null value θ_0 lies in the confidence interval. Suppose, for example, that sample data yields the 95% CI (68.6, 72.0). Then the null hypothesis $H_0: \theta = 70$ cannot be rejected at significance level .05 because 70 lies in the CI. But the null hypothesis $H_0: \theta = 65$ can be rejected because 65 does not lie in the CI. There is an analogous relationship between a 99% CI and a test with significance level .01—the null hypothesis cannot be rejected if the null value lies in the CI and should be rejected if the null value is outside the CI. There is a duality between a two-sided confidence interval with confidence level $100(1 - \alpha)\%$ and the conclusion from a two-tailed test with significance level α.

Now consider testing $H_0: \theta = \theta_0$ against the alternative $H_a: \theta > \theta_0$ at significance level .01. Because of the inequality in H_a, the *P*-value is the area under the z curve to the right of the calculated z. The z critical value 2.33 captures upper-tail area .01. Therefore the *P*-value (captured upper-tail area) will be at most .01 if and only if

$z \geq 2.33$; we will not be able to reject the null hypothesis if and only if $z < 2.33$. Again substituting the formula for z into this inequality and manipulating to isolate θ_0 gives the equivalent inequality $\hat{\theta} - 2.33\hat{\sigma}_{\hat{\theta}} < \theta_0$. The lower limit of this inequality is the lower confidence bound for θ with a confidence level of 99%. So the null hypothesis won't be rejected at significance level .01 if and only if the null value exceeds the lower confidence bound. Thus there is a duality between a lower confidence bound and the conclusion from an upper-tailed test. This is why the Minitab software package will output a lower confidence bound when an upper-tailed test is performed. If, for example, the 90% lower confidence bound is 25.3, i.e., $25.3 < \theta$ with confidence level 90%, then we would not be able to reject $H_0: \theta = 26$ versus $H_a: \theta > 26$ at significance level .10 but would be able to reject $H_0: \theta = 24$ in favor of $H_a: \theta > 24$. There is an analogous duality between an upper confidence bound and the conclusion from a lower-tailed test. And there are analogous relationships for t tests and t confidence intervals or bounds.

PROPOSITION

> Let $(\hat{\theta}_L, \hat{\theta}_U)$ be a confidence interval for θ with confidence level $100(1 - \alpha)\%$. Then a test of $H_0: \theta = \theta_0$ versus $H_a: \theta \neq \theta_0$ with significance level α rejects the null hypothesis if the null value θ_0 is not included in the CI and does not reject H_0 if the null value does lie in the CI. There is an analogous relationship between a lower confidence bound and an upper-tailed test, and also between an upper confidence bound and a lower-tailed test.

In light of these relationships, it is tempting to carry out a test of hypotheses by calculating the corresponding CI or CB. Don't yield to temptation! Instead carry out a more informative analysis by determining and reporting the P-value.

Simultaneous Testing of Several Hypotheses

Many published articles report the results of more than just a single test of hypotheses. For example, the article **"Distributions of Compressive Strength Obtained from Various Diameter Cores"** (*ACI Materials J.*, **2012: 597–606**) considered the plausibility of Weibull, normal, and lognormal distributions as models for compressive strength distributions under various experimental conditions. Table 3 of the cited article reported exact P-values for a total of 71 different tests.

Consider two different tests, one for a pair of hypotheses about a population mean and another for a pair of hypotheses about a population proportion—e.g., the mean wing length for adult Monarch butterflies and the proportion of schoolchildren in a particular state who are obese. Assume that the sample used to test the first pair of hypotheses is selected independently of that used to test the second pair. Then if each test is carried out at significance level .05 (type I error probability .05),

$P(\text{at least one type I error is committed}) = 1 - P(\text{no type I errors are committed})$

$= 1 - P(\text{no type I error in the 1st test}) \cdot P(\text{no type I error in the 2nd test})$

$= 1 - (.95)^2 = 1 - .9025 = .0975$

Thus the probability of committing at least one type I error when two independent tests are carried out is much higher than the probability that a type I error will result from a single test. If three tests are independently carried out, each at significance level .05, then the probability that at least one type I error is committed is $1 - (.95)^3 = .1426$. Clearly as the number of tests increases, the probability of committing at least one type I error gets larger and in fact will approach 1.

Suppose we want the probability of committing at least one type I error in two independent tests to be .05—an *experimentwise* error rate of .05. Then the significance level α for each test must be smaller than .05:

$$.05 = 1 - (1 - \alpha)^2 \Rightarrow 1 - \alpha = \sqrt{.95} = .975 \Rightarrow \alpha = .025$$

If the probability of committing at least one type I error in three independent tests is to be .05, the significance level for each one must be .017 (replace the square root by the cube root in the foregoing argument). As the number of tests increases, the significance level for each one must decrease to 0 in order to maintain an experimentwise error rate of .05.

Often it is not reasonable to assume that the various tests are independent of one another. In the example cited at the beginning of this subsection, four different tests were carried out based on the same sample involving one particular type of concrete in combination with a specified core diameter and length-to-diameter ratio. It is then no longer clear how the experimentwise error rate relates to the significance level for each individual test. Let A_i denote the event that the *ith* test results in a type I error. Then in the case of k tests,

$$P(\text{at least one type I error})$$
$$= P(A_1 \cup A_2 \cup \ldots \cup A_k) \le P(A_1) + \cdots + P(A_k) = k\alpha$$

(the inequality in the last line is called the *Bonferroni inequality*; it can be proved by induction on k). Thus a significance level of $.05/k$ for each test will ensure that the experimentwise significance level is at most .05.

Again, the central idea here is that in order for the probability of at least one type I error among k tests to be small, the significance level for each individual test must be quite small. If the significance level for each individual test is .05, for even a moderate number of tests it is rather likely that at least one type I error will be committed. That is, with $\alpha = .05$ for each test, when each null hypothesis is actually true, it is rather likely that at least one of the tests will yield a statistically significant result. This is why one should view a statistically significant result with skepticism when many tests are carried out using one of the traditional significance levels.

The Likelihood Ratio Principle

The test procedures presented in this and subsequent chapters will (at least for the most part) be intuitively sensible. But there are many situations that arise in practice where intuition is not a reliable guide to obtaining a test statistic. We now describe a general strategy for this purpose. Let $x_1, x_2, \ldots, x_n$ be the observations in a random sample of size n from a probability distribution $f(x; \theta)$. The joint distribution evaluated at these sample values is the product $f(x_1; \theta) \cdot f(x_2; \theta) \cdot \cdots \cdot f(x_n; \theta)$. As in the discussion of maximum likelihood estimation, the *likelihood function* is this joint distribution, regarded as a function of θ. Consider testing H_0: θ is in Ω_0 versus H_a: θ is in Ω_a, where Ω_0 and Ω_a are disjoint (for example, H_0: $\theta \le 100$ versus H_a: $\theta > 100$). The **likelihood ratio principle** for test construction proceeds as follows:

1. Find the largest value of the likelihood for any θ in Ω_0 (by finding the maximum likelihood estimate within Ω_0 and substituting back into the likelihood function).

2. Find the largest value of the likelihood for any θ in Ω_a.

3. Form the ratio

$$\lambda(x_1, \ldots, x_n) = \frac{\text{maximum likelihood for } \theta \text{ in } \Omega_0}{\text{maximum likelihood for } \theta \text{ in } \Omega_a}$$

The ratio $\lambda(x_1, \ldots, x_n)$ is called the *likelihood ratio statistic value*. Intuitively, the smaller the value of λ, the stronger is the evidence against H_0. It can, for example, be shown that for testing H_0: $\mu \leq \mu_0$ versus H_a: $\mu > \mu_0$ in the case of population normality, a small value of λ is equivalent to a large value of t. Thus the one-sample t test comes from applying the likelihood ratio principle. We emphasize that once a test statistic has been selected, its distribution when H_0 is true is required for P-value determination; statistical theory must again come to the rescue!

The likelihood ratio principle can also be applied when the X_i's have different distributions and even when they are dependent, though the likelihood function can be complicated in such cases. Many of the test procedures to be presented in subsequent chapters are obtained from the likelihood ratio principle. These tests often turn out to minimize β among all tests that have the desired α, so are truly best tests. For more details and some worked examples, refer to one of the references listed in the Chapter 6 bibliography.

A practical limitation is that, to construct the likelihood ratio test statistic, the form of the probability distribution from which the sample comes must be specified. Derivation of the t test from the likelihood ratio principle requires assuming a normal pdf. If an investigator is willing to assume that the distribution is symmetric but does not want to be specific about its exact form (such as normal, uniform, or Cauchy), then the principle fails because there is no way to write a joint pdf simultaneously valid for all symmetric distributions. In Chapter 15, we will present several **distribution-free** test procedures, so called because the probability of a type I error is controlled simultaneously for many different underlying distributions. These procedures are useful when the investigator has limited knowledge of the underlying distribution. We shall also consider criteria for selection of a test procedure when several sensible candidates are available, and comment on the performance of several procedures when an underlying assumption such as normality is violated.

EXERCISES Section 8.5 (53–56)

53. Reconsider the paint-drying problem discussed in Example 8.5. The hypotheses were H_0: $\mu = 75$ versus H_a: $\mu < 75$, with σ assumed to have value 9.0. Consider the alternative value $\mu = 74$, which in the context of the problem would presumably not be a practically significant departure from H_0.

  a. For a level .01 test, compute β at this alternative for sample sizes $n = 100$, 900, and 2500.

  b. If the observed value of $\overline{X}$ is $\bar{x} = 74$, what can you say about the resulting P-value when $n = 2500$? Is the data statistically significant at any of the standard values of α?

  c. Would you really want to use a sample size of 2500 along with a level .01 test (disregarding the cost of such an experiment)? Explain.

54. Consider the large-sample level .01 test in Section 8.4 for testing H_0: $p = .2$ against H_a: $p > .2$.

  a. For the alternative value $p = .21$, compute $\beta(.21)$ for sample sizes $n = 100$, 2500, 10,000, 40,000, and 90,000.

  b. For $\hat{p} = x/n = .21$, compute the P-value when $n = 100$, 2500, 10,000, and 40,000.

  c. In most situations, would it be reasonable to use a level .01 test in conjunction with a sample size of 40,000? Why or why not?

55. Consider carrying out m tests of hypotheses based on independent samples, each at significance level (exactly) .01.

  a. What is the probability of committing at least one type I error when $m = 5$? When $m = 10$?

  b. How many such tests would it take for the probability of committing at least one type I error to be at least .5?

56. A 95% CI for true average amount of warpage (mm) of laminate sheets under specified conditions was calculated as (1.81, 1.95), based on a sample size of $n = 15$ and the assumption that amount of warpage is normally distributed.

  a. Suppose you want to test H_0: $\mu = 2$ versus H_a: $\mu \neq 2$ using $\alpha = .05$. What conclusion would be appropriate, and why?

  b. If you wanted to use a significance level of .01 for the test in (a), what conclusion would be appropriate?

SUPPLEMENTARY EXERCISES (57–80)

57. A sample of 50 lenses used in eyeglasses yields a sample mean thickness of 3.05 mm and a sample standard deviation of .34 mm. The desired true average thickness of such lenses is 3.20 mm. Does the data strongly suggest that the true average thickness of such lenses is something other than what is desired? Test using $\alpha = .05$.

58. In Exercise 57, suppose the experimenter had believed before collecting the data that the value of σ was approximately .30. If the experimenter wished the probability of a type II error to be .05 when $\mu = 3.00$, was a sample size 50 unnecessarily large?

59. It is specified that a certain type of iron should contain .85 g of silicon per 100 g of iron (.85%). The silicon content of each of 25 randomly selected iron specimens was determined, and the accompanying Minitab output resulted from a test of the appropriate hypotheses.

```
Variable    N    Mean    StDev  SE Mean     T     P
sil cont   25  0.8880  0.1807   0.0361  1.05  0.30
```

   a. What hypotheses were tested?
   b. What conclusion would be reached for a significance level of .05, and why? Answer the same question for a significance level of .10.

60. One method for straightening wire before coiling it to make a spring is called "roller straightening." The article **"The Effect of Roller and Spinner Wire Straightening on Coiling Performance and Wire Properties"** (*Springs*, 1987: 27–28) reports on the tensile properties of wire. Suppose a sample of 16 wires is selected and each is tested to determine tensile strength (N/mm²). The resulting sample mean and standard deviation are 2160 and 30, respectively.

   a. The mean tensile strength for springs made using spinner straightening is 2150 N/mm². What hypotheses should be tested to determine whether the mean tensile strength for the roller method exceeds 2150?
   b. Assuming that the tensile strength distribution is approximately normal, what test statistic would you use to test the hypotheses in part (a)?
   c. What is the value of the test statistic for this data?
   d. What is the P-value for the value of the test statistic computed in part (c)?
   e. For a level .05 test, what conclusion would you reach?

61. Contamination of mine soils in China is a serious environmental problem. The article **"Heavy Metal Contamination in Soils and Phytoaccumulation in a Manganese Mine Wasteland, South China"** (*Air, Soil, and Water Res.*, 2008: 31–41) reported that, for a sample of 3 soil specimens from a certain restored mining area, the sample mean concentration of Total Cu was 45.31 mg/kg with a corresponding (estimated)

standard error of the mean of 5.26. It was also stated that the China background value for this concentration was 20. The results of various statistical tests described in the article were predicated on assuming normality.

   a. Does the data provide strong evidence for concluding that the true average concentration in the sampled region exceeds the stated background value? Carry out a test at significance level .01. Does the result surprise you? Explain.
   b. Referring back to the test of (a), how likely is it that the P-value would be at least .01 when the true average concentration is 50 and the true standard deviation of concentration is 10?

62. The article **"Orchard Floor Management Utilizing Soil-Applied Coal Dust for Frost Protection"** (*Agri. and Forest Meteorology*, 1988: 71–82) reports the following values for soil heat flux of eight plots covered with coal dust.

34.7   35.4   34.7   37.7   32.5   28.0   18.4   24.9

The mean soil heat flux for plots covered only with grass is 29.0. Assuming that the heat-flux distribution is approximately normal, does the data suggest that the coal dust is effective in increasing the mean heat flux over that for grass? Test the appropriate hypotheses using $\alpha = .05$.

63. The article **"Caffeine Knowledge, Attitudes, and Consumption in Adult Women"** (*J. of Nutrition Educ.*, 1992: 179–184) reports the following summary data on daily caffeine consumption for a sample of adult women: $n = 47$, $\bar{x} = 215$ mg, $s = 235$ mg, and range $= 5-1176$.

   a. Does it appear plausible that the population distribution of daily caffeine consumption is normal? Is it necessary to assume a normal population distribution to test hypotheses about the value of the population mean consumption? Explain your reasoning.
   b. Suppose it had previously been believed that mean consumption was at most 200 mg. Does the given data contradict this prior belief? Test the appropriate hypotheses at significance level .10.

64. Annual holdings turnover for a mutual fund is the percentage of a fund's assets that are sold during a particular year. Generally speaking, a fund with a low value of turnover is more stable and risk averse, whereas a high value of turnover indicates a substantial amount of buying and selling in an attempt to take advantage of short-term market fluctuations. Here are values of turnover for a sample of 20 large-cap blended funds (refer to Exercise 1.53 for a bit more information) extracted from **Morningstar.com:**

1.03   1.23   1.10   1.64   1.30   1.27   1.25   0.78   1.05   0.64
0.94   2.86   1.05   0.75   0.09   0.79   1.61   1.26   0.93   0.84

a. Would you use the one-sample t test to decide whether there is compelling evidence for concluding that the population mean turnover is less than 100%? Explain.

b. A normal probability plot of the 20 ln(turnover) values shows a very pronounced linear pattern, suggesting it is reasonable to assume that the turnover distribution is lognormal. Recall that X has a lognormal distribution if $\ln(X)$ is normally distributed with mean value μ and variance σ^2. Because μ is also the median of the $\ln(X)$ distribution, e^μ is the median of the X distribution. Use this information to decide whether there is compelling evidence for concluding that the median of the turnover population distribution is less than 100%.

65. The true average breaking strength of ceramic insulators of a certain type is supposed to be at least 10 psi. They will be used for a particular application unless sample data indicates conclusively that this specification has not been met. A test of hypotheses using $\alpha = .01$ is to be based on a random sample of ten insulators. Assume that the breaking-strength distribution is normal with unknown standard deviation.

a. If the true standard deviation is .80, how likely is it that insulators will be judged satisfactory when true average breaking strength is actually only 9.5? Only 9.0?

b. What sample size would be necessary to have a 75% chance of detecting that the true average breaking strength is 9.5 when the true standard deviation is .80?

66. The accompanying observations on residual flame time (sec) for strips of treated children's nightwear were given in the article **"An Introduction to Some Precision and Accuracy of Measurement Problems"** (**J. of Testing and Eval.**, **1982: 132–140).** Suppose a true average flame time of at most 9.75 had been mandated. Does the data suggest that this condition has not been met? Carry out an appropriate test after first investigating the plausibility of assumptions that underlie your method of inference.

9.85 9.93 9.75 9.77 9.67 9.87 9.67
9.94 9.85 9.75 9.83 9.92 9.74 9.99
9.88 9.95 9.95 9.93 9.92 9.89

67. The incidence of a certain type of chromosome defect in the U.S. adult male population is believed to be 1 in 75. A random sample of 800 individuals in U.S. penal institutions reveals 16 who have such defects. Can it be concluded that the incidence rate of this defect among prisoners differs from the presumed rate for the entire adult male population?

a. State and test the relevant hypotheses using $\alpha = .05$. What type of error might you have made in reaching a conclusion?

b. Based on the P-value calculated in (a), could H_0 be rejected at significance level .20?

68. In an investigation of the toxin produced by a certain poisonous snake, a researcher prepared 26 different vials, each containing 1 g of the toxin, and then determined the amount of antitoxin needed to neutralize the toxin. The sample average amount of antitoxin necessary was found to be 1.89 mg, and the sample standard deviation was .42. Previous research had indicated that the true average neutralizing amount was 1.75 mg/g of toxin. Does the new data contradict the value suggested by prior research? Test the relevant hypotheses. Does the validity of your analysis depend on any assumptions about the population distribution of neutralizing amount? Explain.

69. The sample average unrestrained compressive strength for 45 specimens of a particular type of brick was computed to be 3107 psi, and the sample standard deviation was 188. The distribution of unrestrained compressive strength may be somewhat skewed. Does the data strongly indicate that the true average unrestrained compressive strength is less than the design value of 3200? Test using $\alpha = .001$.

70. The Dec. 30, 2009, the **New York Times** reported that in a survey of 948 American adults who said they were at least somewhat interested in college football, 597 said the current Bowl Championship System should be replace by a playoff similar to that used in college basketball. Does this provide compelling evidence for concluding that a majority of all such individuals favor replacing the B.C.S. with a playoff? Test the appropriate hypotheses using a significant level of .001.

71. When $X_1, X_2, \ldots, X_n$ are independent Poisson variables, each with parameter μ, and n is large, the sample mean $\overline{X}$ has approximately a normal distribution with $\mu = E(\overline{X})$ and $V(\overline{X}) = \mu/n$. This implies that

$$Z = \frac{\overline{X} - \mu}{\sqrt{\mu/n}}$$

has approximately a standard normal distribution. For testing $H_0: \mu = \mu_0$, we can replace μ by μ_0 in the equation for Z to obtain a test statistic. This statistic is actually preferred to the large-sample statistic with denominator $S/\sqrt{n}$ (when the X_i's are Poisson) because it is tailored explicitly to the Poisson assumption. If the number of requests for consulting received by a certain statistician during a 5-day work week has a Poisson distribution and the total number of consulting requests during a 36-week period is 160, does this suggest that the true average number of weekly requests exceeds 4.0? Test using $\alpha = .02$.

72. An article in the Nov. 11, 2005, issue of the **San Luis Obispo Tribune** reported that researchers making random purchases at California Wal-Mart stores found scanners coming up with the wrong price 8.3% of the time. Suppose this was based on 200 purchases. The National Institute for Standards and Technology says that in the long run at most two out of every 100 items should have incorrectly scanned prices.

a. Develop a test procedure with a significance level of (approximately) .05, and then carry out the test to decide whether the NIST benchmark is not satisfied.

b. For the test procedure you employed in (a), what is the probability of deciding that the NIST benchmark has been satisfied when in fact the mistake rate is 5%?

73. The article **"Heavy Drinking and Polydrug Use Among College Students" (*J. of Drug Issues*, 2008: 445–466)** stated that 51 of the 462 college students in a sample had a lifetime abstinence from alcohol. Does this provide strong evidence for concluding that more than 10% of the population sampled had completely abstained from alcohol use? Test the appropriate hypotheses. [*Note:* The article used more advanced statistical methods to study the use of various drugs among students characterized as light, moderate, and heavy drinkers.]

74. The article **"Analysis of Reserve and Regular Bottlings: Why Pay for a Difference Only the Critics Claim to Notice?" (*Chance*, Summer 2005, pp. 9–15)** reported on an experiment to investigate whether wine tasters could distinguish between more expensive reserve wines and their regular counterparts. Wine was presented to tasters in four containers labeled A, B, C, and D, with two of these containing the reserve wine and the other two the regular wine. Each taster randomly selected three of the containers, tasted the selected wines, and indicated which of the three he/she believed was different from the other two. Of the $n = 855$ tasting trials, 346 resulted in correct distinctions (either the one reserve that differed from the two regular wines or the one regular wine that differed from the two reserves). Does this provide compelling evidence for concluding that tasters of this type have some ability to distinguish between reserve and regular wines? State and test the relevant hypotheses. Are you particularly impressed with the ability of tasters to distinguish between the two types of wine?

75. The American Academy of Pediatrics recommends a vitamin D level of at least 20 ng/ml for infants. The article **"Vitamin D and Parathormone Levels of Late-Preterm Formula Fed Infants During the First Year of Life" (*European J. of Clinical Nutr.*, 2012: 224–230)** reported that for a sample of 102 preterm infants judged to be of appropriate weight for their gestational age, the sample mean vitamin D level at 2 weeks was 21 with a sample standard deviation of 11. Does this provide convincing evidence that the population mean vitamin D level for such infants exceeds 20? Test the relevant hypotheses using a significance level of .10.

76. Chapter 7 presented a CI for the variance σ^2 of a normal population distribution. The key result there was that the rv $\chi^2 = (n-1)S^2/\sigma^2$ has a chi-squared distribution with $n-1$ df. Consider the null hypothesis $H_0: \sigma^2 = \sigma_0^2$ (equivalently, $\sigma = \sigma_0$). Then when H_0 is true, the test statistic $\chi^2 = (n-1)S^2/\sigma_0^2$ has a chi-squared distribution with $n-1$ df. If the relevant alternative is $H_a: \sigma^2 > \sigma_0^2$,

the P-value is the area under the χ^2 curve with $n-1$ df to the right of the calculated χ^2 value. To ensure reasonably uniform characteristics for a particular application, it is desired that the true standard deviation of the softening point of a certain type of petroleum pitch be at most .50°C. The softening points of ten different specimens were determined, yielding a sample standard deviation of .58°C. Does this strongly contradict the uniformity specification? Test the appropriate hypotheses using $\alpha = .01$. [*Hint:* Consult Table A.11.]

77. Referring to Exercise 76, suppose an investigator wishes to test $H_0: \sigma^2 = .04$ versus $H_a: \sigma^2 < .04$ based on a sample of 21 observations. The computed value of $20s^2/.04$ is 8.58. Place bounds on the P-value and then reach a conclusion at level .01. [*Hint:* Consult Table A.7.]

78. When the population distribution is normal and n is large, the sample standard deviation S has approximately a normal distribution with $E(S) \approx \sigma$ and $V(S) \approx \sigma^2/(2n)$. We already know that in this case, for any n, $\overline{X}$ is normal with $E(\overline{X}) = \mu$ and $V(\overline{X}) = \sigma^2/n$.

a. Assuming that the underlying distribution is normal, what is an approximately unbiased estimator of the 99th percentile $\theta = \mu + 2.33\sigma$?

b. When the X_i's are normal, it can be shown that $\overline{X}$ and S are independent rv's (one measures location whereas the other measures spread). Use this to compute $V(\hat{\theta})$ and $\sigma_{\hat{\theta}}$ for the estimator $\hat{\theta}$ of part (a). What is the estimated standard error $\hat{\sigma}_{\hat{\theta}}$?

c. Write a test statistic for testing $H_0: \theta = \theta_0$ that has approximately a standard normal distribution when H_0 is true. If soil pH is normally distributed in a certain region and 64 soil samples yield $\bar{x} = 6.33$, $s = .16$, does this provide strong evidence for concluding that at most 99% of all possible samples would have a pH of less than 6.75? Test using $\alpha = .01$.

79. Let $X_1, X_2, \ldots, X_n$ be a random sample from an exponential distribution with parameter λ. Then it can be shown that $2\lambda\Sigma X_i$ has a chi-squared distribution with $\nu = 2n$ (by first showing that $2\lambda X_i$ has a chi-squared distribution with $\nu = 2$).

a. Use this fact to obtain a test statistic for testing $H_0: \mu = \mu_0$. Then explain how you would determine the P-value when the alternative hypothesis is $H_a: \mu < \mu_0$. [*Hint:* $E(X_i) = \mu = 1/\lambda$, so $\mu = \mu_0$ is equivalent to $\lambda = 1/\mu_0$.]

b. Suppose that ten identical components, each having exponentially distributed time until failure, are tested. The resulting failure times are

95   16   11   3   42   71   225   64   87   123

Use the test procedure of part (a) to decide whether the data strongly suggests that the true average lifetime is less than the previously claimed value of 75. [*Hint:* Consult Table A.7.]

80. Because of variability in the manufacturing process, the actual yielding point of a sample of mild steel subjected to increasing stress will usually differ from the theoretical yielding point. Let p denote the true proportion of samples that yield before their theoretical yielding point. If on the basis of a sample it can be concluded that more than 20% of all specimens yield before the theoretical point, the production process will have to be modified.

a. If 15 of 60 specimens yield before the theoretical point, what is the P-value when the appropriate test is used, and what would you advise the company to do?

b. If the true percentage of "early yields" is actually 50% (so that the theoretical point is the median of the yield distribution) and a level .01 test is used, what is the probability that the company concludes a modification of the process is necessary?

BIBLIOGRAPHY

See the bibliographies at the ends of Chapter 6 and Chapter 7.

Inferences Based on Two Samples

9

INTRODUCTION

Chapters 7 and 8 presented confidence intervals (CI's) and hypothesis-testing procedures for a single mean μ, single proportion p, and a single variance σ^2. Here we extend these methods to situations involving the means, proportions, and variances of two different population distributions. For example, let μ_1 denote true average Rockwell hardness for heat-treated steel specimens and μ_2 denote true average hardness for cold-rolled specimens. Then an investigator might wish to use samples of hardness observations from each type of steel as a basis for calculating an interval estimate of $\mu_1 - \mu_2$, the difference between the two true average hardnesses. As another example, let p_1 denote the true proportion of nickel-cadmium cells produced under current operating conditions that are defective because of internal shorts, and let p_2 represent the true proportion of cells with internal shorts produced under modified operating conditions. If the rationale for the modified conditions is to reduce the proportion of defective cells, a quality engineer would want to use sample information to test the null hypothesis $H_0: p_1 - p_2 = 0$ (i.e., $p_1 = p_2$) versus the alternative hypothesis, $H_a: p_1 - p_2 > 0$ (i.e., $p_1 > p_2$).

Section 9.1 presents z intervals and tests for making inferences about a difference between two population means (i.e., procedures developed by starting with a standardized variable that has at least approximately a standard normal distribution). Two-sample t procedures for making inferences about $\mu_1 - \mu_2$ are the focus of Section 9.2. The validity of methods described in the first two sections depends on selecting samples from the two populations independently of one another. Often in practice data is gathered in pairs. For example, a sample of individuals might be selected, a measurement of some sort made before a treatment is applied, and then another measurement subsequent to application of the treatment. The analysis of such paired data is described in Section 9.3. Section 9.4 considers inferences about a difference between two population proportions, and Section 9.5 does the same thing for a ratio of population variances or standard deviations.

9.1 z Tests and Confidence Intervals for a Difference Between Two Population Means

The inferences discussed in this section concern a difference $\mu_1 - \mu_2$ between the means of two different population distributions. An investigator might, for example, wish to test hypotheses about the difference between true average breaking strengths of two different types of corrugated fiberboard. One such hypothesis would state that $\mu_1 - \mu_2 = 0$ that is, that $\mu_1 = \mu_2$. Alternatively, it may be appropriate to estimate $\mu_1 - \mu_2$ by computing a 95% CI. Such inferences necessitate obtaining a sample of strength observations for each type of fiberboard.

Basic Assumptions

1. $X_1, X_2, \ldots, X_m$ is a random sample from a distribution with mean μ_1 and variance σ_1^2.

2. $Y_1, Y_2, \ldots, Y_n$ is a random sample from a distribution with mean μ_2 and variance σ_2^2.

3. The X and Y samples are independent of one another.

The use of m for the number of observations in the first sample and n for the number of observations in the second sample allows for the two sample sizes to be different. Sometimes this is because it is more difficult or expensive to sample one population than another. In other situations, equal sample sizes may initially be specified, but for reasons beyond the scope of the experiment, the actual sample sizes may differ. For example, the abstract of the article **"A Randomized Controlled Trial Assessing the Effectiveness of Professional Oral Care by Dental Hygienists"** (***Intl. J. of Dental Hygiene*, 2008: 63–67**) states that "Forty patients were randomly assigned to either the POC group ($m = 20$) or the control group ($n = 20$). One patient in the POC group and three in the control group dropped out because of exacerbation of underlying disease or death." The data analysis was then based on $m = 19$ and $n = 16$.

The natural estimator of $\mu_1 - \mu_2$ is $\overline{X} - \overline{Y}$, the difference between the corresponding sample means. Inferential procedures are based on standardizing this estimator, so we need expressions for the expected value and standard deviation of $\overline{X} - \overline{Y}$.

PROPOSITION

The expected value of $\overline{X} - \overline{Y}$ is $\mu_1 - \mu_2$, so $\overline{X} - \overline{Y}$ is an unbiased estimator of $\mu_1 - \mu_2$. The standard deviation of $\overline{X} - \overline{Y}$ is

$$\sigma_{\overline{X} - \overline{Y}} = \sqrt{\frac{\sigma_1^2}{m} + \frac{\sigma_2^2}{n}}$$

Proof Both these results depend on the rules of expected value and variance presented in Chapter 5. Since the expected value of a difference is the difference of expected values,

$$E(\overline{X} - \overline{Y}) = E(\overline{X}) - E(\overline{Y}) = \mu_1 - \mu_2$$

Because the X and Y samples are independent, $\overline{X}$ and $\overline{Y}$ are independent quantities. Then the variance of the difference is the *sum* of $V(\overline{X})$ and $V(\overline{Y})$:

$$V(\overline{X} - \overline{Y}) = V(\overline{X}) + V(\overline{Y}) = \frac{\sigma_1^2}{m} + \frac{\sigma_2^2}{n}$$

The standard deviation of $\overline{X} - \overline{Y}$ is the square root of this expression. ∎

If we regard $\mu_1 - \mu_2$ as a parameter θ, then its estimator is $\hat{\theta} = \overline{X} - \overline{Y}$ with standard deviation $\sigma_{\hat{\theta}}$ given by the proposition. When σ_1^2 and σ_2^2 both have known values, the value of this standard deviation can be calculated. The sample variances must be used to estimate $\sigma_{\hat{\theta}}$ when σ_1^2 and σ_2^2 are unknown.

Test Procedures for Normal Populations with Known Variances

In Chapters 7 and 8, the first CI and test procedure for a population mean μ were based on the assumption that the population distribution was normal with the value of the population variance σ^2 known to the investigator. Similarly, we first assume here that *both* population distributions are normal and that the values of *both* σ_1^2 and σ_2^2 are known. Situations in which one or both of these assumptions can be dispensed with will be presented shortly.

Because the population distributions are normal, both $\overline{X}$ and $\overline{Y}$ have normal distributions. Furthermore, independence of the two samples implies that the two sample means are independent of one another. Thus the difference $\overline{X} - \overline{Y}$ is normally distributed, with expected value $\mu_1 - \mu_2$ and standard deviation $\sigma_{\overline{X}-\overline{Y}}$ given in the foregoing proposition. Standardizing $\overline{X} - \overline{Y}$ gives the standard normal variable

$$Z = \frac{\overline{X} - \overline{Y} - (\mu_1 - \mu_2)}{\sqrt{\dfrac{\sigma_1^2}{m} + \dfrac{\sigma_2^2}{n}}} \tag{9.1}$$

In a hypothesis-testing problem, the null hypothesis will state that $\mu_1 - \mu_2$ has a specified value. Denoting this null value by Δ_0, we have $H_0: \mu_1 - \mu_2 = \Delta_0$. Often $\Delta_0 = 0$, in which case H_0 says that $\mu_1 = \mu_2$. If μ_1 represents the true average fuel efficiency (mpg) for automobiles of a certain type equipped with a six-cylinder engine and μ_2 denotes true average efficiency for automobiles of the same type equipped with a four-cylinder engine, a sensible null hypothesis of interest might be $H_0: \mu_1 - \mu_2 = -3$. This is a fancy way of saying that on average the fuel efficiency for four-cylinder engines is 3 mpg higher than it is for six-cylinder engines.

Consider the alternative hypothesis $H_a: \mu_1 - \mu_2 > \Delta_0$. A value $\overline{x} - \overline{y}$ that considerably exceeds Δ_0 (the expected value of $\overline{X} - \overline{Y}$ when H_0 is true) provides evidence against H_0 and for H_a. Such a value of $\overline{x} - \overline{y}$ corresponds to a positive and large value of the test statistic. This implies that if the calculated sample means and sample sizes are substituted into the formula for Z and the resulting value is z, then values more contradictory to H_0 than z itself are those larger than z. Thus

$$P\text{-value} = P(\text{obtaining a test statistic value at least}$$
$$\text{as contradictory to } z \text{ when } H_0 \text{ is true})$$
$$= P(\text{a standard normal rv is } \geq z)$$
$$= \text{the area under the standard normal}$$
$$\text{curve to the right of } z$$
$$= 1 - \Phi(z)$$

The test procedure in this case is *upper-tailed* because the P-value is an upper-tail z curve area.

When the alternative hypothesis contains the inequality $<$, test statistic values more contradictory to H_0 than z itself are those smaller than z. The P-value is then the area under the standard normal curve to the left of z; the test is *lower-tailed*. Lastly, if the inequality $\neq$ appears in H_a, then values either larger than $|z|$ or smaller than $-|z|$ are more contradictory to H_0 than z itself (the absolute value around z takes care of both the z positive case and the z negative case). The implication is that the P-value is the sum of the area under the standard normal curve to the left of $-|z|$ and the area to the right of $|z|$—that is, a *two-tailed* test. This sum of two tail areas is the same as doubling the captured tail area.

Null hypothesis: $H_0: \mu_1 - \mu_2 = \Delta_0$

Test statistic value: $z = \dfrac{\bar{x} - \bar{y} - \Delta_0}{\sqrt{\dfrac{\sigma_1^2}{m} + \dfrac{\sigma_2^2}{n}}}$

Alternative Hypothesis	P-Value Determination
$H_a: \mu_1 - \mu_2 > \Delta_0$	Area under the standard normal curve to the right of z
$H_a: \mu_1 - \mu_2 < \Delta_0$	Area under the standard normal curve to the left of z
$H_a: \mu_1 - \mu_2 \neq \Delta_0$	$2 \cdot$ (Area under the standard normal curve to the right of $|z|$)

Assumptions: Two normal population distributions with known values of σ_1 and σ_2, two independent random samples.

EXAMPLE 9.1 Analysis of a random sample consisting of $m = 20$ specimens of cold-rolled steel to determine yield strengths resulted in a sample average strength of $\bar{x} = 29.8$ ksi. A second random sample of $n = 25$ two-sided galvanized steel specimens gave a sample average strength of $\bar{y} = 34.7$ ksi. Assuming that the two yield-strength distributions are normal with $\sigma_1 = 4.0$ and $\sigma_2 = 5.0$ (suggested by a graph in the article **"Zinc-Coated Sheet Steel: An Overview,"** *Automotive Engr.*, **Dec. 1984: 39–43**), does the data indicate that the corresponding true average yield strengths μ_1 and μ_2 are different? Let's carry out a test at significance level $\alpha = .01$.

1. The parameter of interest is $\mu_1 - \mu_2$, the difference between the true average strengths for the two types of steel.

2. The null hypothesis is $H_0: \mu_1 - \mu_2 = 0$.

3. The alternative hypothesis is $H_a: \mu_1 - \mu_2 \neq 0$; if H_a is true, then μ_1 and μ_2 are different.

4. With $\Delta_0 = 0$, the test statistic value is

$$z = \frac{\bar{x} - \bar{y}}{\sqrt{\dfrac{\sigma_1^2}{m} + \dfrac{\sigma_2^2}{n}}}$$

5. Substituting $m = 20, \bar{x} = 29.8, \sigma_1^2 = 16.0, n = 25, \bar{y} = 34.7$, and $\sigma_2^2 = 25.0$ into the formula for z yields

$$z = \frac{29.8 - 34.7}{\sqrt{\dfrac{16.0}{20} + \dfrac{25.0}{25}}} = \frac{-4.90}{1.34} = -3.66$$

That is, the observed value of $\bar{x} - \bar{y}$ is more than 3 standard deviations below what would be expected were H_0 true.

6. The $\neq$ inequality in H_a implies that a two-tailed test is appropriate. The P-value is

$$2[1 - \Phi(3.66)] \approx 2(0) = 0 \quad \text{(software gives .00025).}$$

7. Since $P\text{-value} \approx 0 \leq .01 = \alpha$, H_0 is therefore rejected at level .01 in favor of the conclusion that $\mu_1 \neq \mu_2$. In fact, with a P-value this small, the null hypothesis would be rejected at *any* sensible significance level. The sample data strongly suggests that the true average yield strength for cold-rolled steel differs from that for galvanized steel. ∎

Using a Comparison to Identify Causality

Investigators are often interested in comparing either the effects of two different treatments on a response or the response after treatment with the response after no treatment (treatment vs. control). If the individuals or objects to be used in the comparison are not assigned by the investigators to the two different conditions, the study is said to be **observational**. The difficulty with drawing conclusions based on an observational study is that although statistical analysis may indicate a significant difference in response between the two groups, the difference may be due to some underlying factors that had not been controlled rather than to any difference in treatments.

EXAMPLE 9.2   A letter in the ***Journal of the American Medical Association* (May 19, 1978)** reported that of 215 male physicians who were Harvard graduates and died between November 1974 and October 1977, the 125 in full-time practice lived an average of 48.9 years beyond graduation, whereas the 90 with academic affiliations lived an average of 43.2 years beyond graduation. Does the data suggest that the mean lifetime after graduation for doctors in full-time practice exceeds the mean lifetime for those who have an academic affiliation? (If so, those medical students who say that they are "dying to obtain an academic affiliation" may be closer to the truth than they realize; in other words, is "publish or perish" really "publish and perish"?)

Let μ_1 denote the true average number of years lived beyond graduation for physicians in full-time practice, and let μ_2 denote the same quantity for physicians with academic affiliations. Assume the 125 and 90 physicians to be random samples from populations 1 and 2, respectively (which may not be sensible if there is reason to believe that Harvard graduates have special characteristics that differentiate them from all other physicians—in this case inferences would be restricted just to the

"Harvard populations"). The letter from which the data was taken gave no information about variances, so for illustration assume that $\sigma_1 = 14.6$ and $\sigma_2 = 14.4$. The hypotheses are $H_0: \mu_1 - \mu_2 = 0$ versus $H_a: \mu_1 - \mu_2 > 0$, so Δ_0 is zero. The computed value of the test statistic is

$$z = \frac{48.9 - 43.2}{\sqrt{\dfrac{(14.6)^2}{125} + \dfrac{(14.4)^2}{90}}} = \frac{5.70}{\sqrt{1.70 + 2.30}} = 2.85$$

The P-value for an upper-tailed test is $1 - \Phi(2.85) = .0022$. At significance level .01, H_0 is rejected (because $\alpha > P$-value) in favor of the conclusion that $\mu_1 - \mu_2 > 0$ ($\mu_1 > \mu_2$). This is consistent with the information reported in the letter.

This data resulted from a **retrospective** observational study; the investigator did not start out by selecting a sample of doctors and assigning some to the "academic affiliation" treatment and the others to the "full-time practice" treatment, but instead identified members of the two groups by looking backward in time (through obituaries!) to past records. Can the statistically significant result here really be attributed to a difference in the type of medical practice after graduation, or is there some other underlying factor (e.g., age at graduation, exercise regimens, etc.) that might also furnish a plausible explanation for the difference? Observational studies have been used to argue for a causal link between smoking and lung cancer. There are many studies that show that the incidence of lung cancer is significantly higher among smokers than among nonsmokers. However, individuals had decided whether to become smokers long before investigators arrived on the scene, and factors in making this decision may have played a causal role in the contraction of lung cancer. ∎

A **randomized controlled experiment** results when investigators assign subjects to the two treatments in a random fashion. When statistical significance is observed in such an experiment, the investigator and other interested parties will have more confidence in the conclusion that the difference in response has been caused by a difference in treatments. A very famous example of this type of experiment and conclusion is the Salk polio vaccine experiment described in Section 9.4. Various aspects of experimental and sampling design are discussed at greater length in the (nonmathematical) books by Moore and by Freedman et al., listed in the Chapter 1 references.

β and the Choice of Sample Size

The probability of a type II error is easily calculated when both population distributions are normal with known values of σ_1 and σ_2. Consider the case in which the alternative hypothesis is $H_a: \mu_1 - \mu_2 > \Delta_0$. Let Δ' denote a value of $\mu_1 - \mu_2$ that exceeds Δ_0 (a value for which H_0 is false). As with the upper-tailed z tests of Chapter 8, the inequality P-value $\leq \alpha$ is equivalent to $z \geq z_\alpha$ (the area captured in the upper tail of the z curve will be at most α if and only if the calculated z is on or to the right of the z critical value that captures area α). This in turn is equivalent to $\bar{x} - \bar{y} \geq \Delta_0 + z_\alpha \sigma_{\bar{X}-\bar{Y}}$. Thus

$$\beta(\Delta') = P(\text{not rejecting } H_0 \text{ when } \mu_1 - \mu_2 = \Delta')$$
$$= P(\bar{X} - \bar{Y} < \Delta_0 + z_\alpha \sigma_{\bar{X}-\bar{Y}} \text{ when } \mu_1 - \mu_2 = \Delta')$$

When $\mu_1 - \mu_2 = \Delta'$, $\bar{X} - \bar{Y}$ is normally distributed with mean value Δ' and standard deviation $\sigma_{\bar{X}-\bar{Y}}$ (the same standard deviation as when H_0 is true); using these values to standardize the inequality in parentheses gives the desired probability.

Alternative Hypothesis	$\beta(\Delta') = P(\text{type II error when } \mu_1 - \mu_2 = \Delta')$
$H_a: \mu_1 - \mu_2 > \Delta_0$	$\Phi\left(z_\alpha - \dfrac{\Delta' - \Delta_0}{\sigma}\right)$
$H_a: \mu_1 - \mu_2 < \Delta_0$	$1 - \Phi\left(-z_\alpha - \dfrac{\Delta' - \Delta_0}{\sigma}\right)$
$H_a: \mu_1 - \mu_2 \neq \Delta_0$	$\Phi\left(z_{\alpha/2} - \dfrac{\Delta' - \Delta_0}{\sigma}\right) - \Phi\left(-z_{\sigma/2} - \dfrac{\Delta' - \Delta_0}{\sigma}\right)$

where $\sigma = \sigma_{\bar{X} - \bar{Y}} = \sqrt{(\sigma_1^2/m) + (\sigma_2^2/n)}$

EXAMPLE 9.3
(Example 9.1 continued)

Suppose that when μ_1 and μ_2 (the true average yield strengths for the two types of steel) differ by as much as 5, the probability of detecting such a departure from H_0 (the power of the test) should be .90. Does a level .01 test with sample sizes $m = 20$ and $n = 25$ satisfy this condition? The value of σ for these sample sizes (the denominator of z) was previously calculated as 1.34. The probability of a type II error for the two-tailed level .01 test when $\mu_1 - \mu_2 = \Delta' = 5$ is

$$\beta(5) = \Phi\left(2.58 - \frac{5-0}{1.34}\right) - \Phi\left(-2.58 - \frac{5-0}{1.34}\right)$$
$$= \Phi(-1.15) - \Phi(-6.31) = .1251$$

It is easy to verify that $\beta(-5) = .1251$ also. Thus the power is $1 - \beta(5) = .8749$. Because this is somewhat less than .9, slightly larger sample sizes should be used. ■

As in Chapter 8, sample sizes m and n can be determined that will satisfy both $P(\text{type I error}) = $ a specified α and $P(\text{type II error when } \mu_1 - \mu_2 = \Delta') = $ a specified β. For an upper-tailed test, equating the previous expression for $\beta(\Delta')$ to the specified value of β gives

$$\frac{\sigma_1^2}{m} + \frac{\sigma_2^2}{n} = \frac{(\Delta' - \Delta_0)^2}{(z_\alpha + z_\beta)^2}$$

When the two sample sizes are equal, this equation yields

$$m = n = \frac{(\sigma_1^2 + \sigma_2^2)(z_\alpha + z_\beta)^2}{(\Delta' - \Delta_0)^2}$$

These expressions are also correct for α lower-tailed test, whereas α is replaced by $\alpha/2$ for a two-tailed test.

Large-Sample Tests

The assumptions of normal population distributions and known values of σ_1 and σ_2 are fortunately unnecessary when both sample sizes are sufficiently large. In this case, the Central Limit Theorem guarantees that $\bar{X} - \bar{Y}$ has approximately a normal distribution regardless of the underlying population distributions. Furthermore, using S_1^2 and S_2^2 in place of σ_1^2 and σ_2^2 in Expression (9.1) gives a variable whose distribution is approximately standard normal:

$$Z = \frac{\bar{X} - \bar{Y} - (\mu_1 - \mu_2)}{\sqrt{\dfrac{S_1^2}{m} + \dfrac{S_2^2}{n}}}$$

A large-sample test statistic results from replacing $\mu_1 - \mu_2$ by Δ_0, the expected value of $\bar{X} - \bar{Y}$ when H_0 is true. This statistic Z then has approximately a standard normal distribution when H_0 is true, which allows for straightforward determination of a P-value as a z curve area.

Use of the test statistic value

$$z = \frac{\bar{x} - \bar{y} - \Delta_0}{\sqrt{\dfrac{s_1^2}{m} + \dfrac{s_2^2}{n}}}$$

along with the previously stated prescriptions for P-value determination gives large-sample tests whose significance levels are approximately α. These tests are usually appropriate if both $m > 40$ and $n > 40$.

EXAMPLE 9.4 What impact does fast-food consumption have on various dietary and health characteristics? The article **"Effects of Fast-Food Consumption on Energy Intake and Diet Quality Among Children in a National Household Study"** (*Pediatrics*, **2004: 112–118)** reported the accompanying summary data on daily calorie intake both for a sample of teens who said they did not typically eat fast food and another sample of teens who said they did usually eat fast food.

Eat Fast Food	Sample Size	Sample Mean	Sample SD
No	663	2258	1519
Yes	413	2637	1138

Does this data provide strong evidence for concluding that true average calorie intake for teens who typically eat fast food exceeds by more than 200 calories per day the true average intake for those who don't typically eat fast food? Let's investigate by carrying out a test of hypotheses at a significance level of approximately .05.

The parameter of interest is $\mu_1 - \mu_2$, where μ_1 is the true average calorie intake for teens who don't typically eat fast food and μ_2 is true average intake for teens who do typically eat fast food. The hypotheses of interest are

$$H_0: \mu_1 - \mu_2 = -200 \quad \text{versus} \quad H_a: \mu_1 - \mu_2 < -200$$

The alternative hypothesis asserts that true average daily intake for those who typically eat fast food exceeds that for those who don't by more than 200 calories. The test statistic value is

$$z = \frac{\bar{x} - \bar{y} - (-200)}{\sqrt{\dfrac{s_1^2}{m} + \dfrac{s_2^2}{n}}}$$

The calculated test statistic value is

$$z = \frac{2258 - 2637 + 200}{\sqrt{\dfrac{(1519)^2}{663} + \dfrac{(1138)^2}{413}}} = \frac{-179}{81.34} = -2.20$$

The inequality in H_a implies that *P*-value $= \Phi(-2.20) = .0139$ (a lower-tailed test). Since $.0139 \le .05$, the null hypothesis is rejected. At a significance level of .05, it does appear that true average daily calorie intake for teens who typically eat fast food exceeds by more than 200 the true average intake for those who don't typically eat such food. However, the *P*-value is not small enough to justify rejecting H_0 at significance level .01.

Notice that if the label 1 had instead been used for the fast-food condition and 2 had been used for the no-fast-food condition, then 200 would have replaced -200 in both hypotheses and H_a would have contained the inequality $>$, implying an upper-tailed test. The resulting test statistic value would have been 2.20, giving the same *P*-value as before. ∎

Confidence Intervals for $\mu_1 - \mu_2$

When both population distributions are normal, standardizing $\overline{X} - \overline{Y}$ gives a random variable Z with a standard normal distribution. Since the area under the z curve between $-z_{\alpha/2}$ and $z_{\alpha/2}$ is $1 - \alpha$, it follows that

$$P\left(-z_{\alpha/2} < \frac{\overline{X} - \overline{Y} - (\mu_1 - \mu_2)}{\sqrt{\dfrac{\sigma_1^2}{m} + \dfrac{\sigma_2^2}{n}}} < z_{\alpha/2} \right) = 1 - \alpha$$

Manipulation of the inequalities inside the parentheses to isolate $\mu_1 - \mu_2$ yields the equivalent probability statement

$$P\left(\overline{X} - \overline{Y} - z_{\alpha/2}\sqrt{\dfrac{\sigma_1^2}{m} + \dfrac{\sigma_2^2}{n}} < \mu_1 - \mu_2 < \overline{X} - \overline{Y} + z_{\alpha/2}\sqrt{\dfrac{\sigma_1^2}{m} + \dfrac{\sigma_2^2}{n}} \right) = 1 - \alpha$$

This implies that a $100(1 - \alpha)\%$ CI for $\mu_1 - \mu_2$ has lower limit $\bar{x} - \bar{y} - z_{\alpha/2} \cdot \sigma_{\overline{X}-\overline{Y}}$ and upper limit $\bar{x} - \bar{y} + z_{\alpha/2} \cdot \sigma_{\overline{X}-\overline{Y}}$, where $\sigma_{\overline{X}-\overline{Y}}$ is the square-root expression. This interval is a special case of the general formula $\hat{\theta} \pm z_{\alpha/2} \cdot \sigma_{\hat{\theta}}$.

If both m and n are large, the CLT implies that this interval is valid even without the assumption of normal populations; in this case, the confidence level is *approximately* $100(1 - \alpha)\%$. Furthermore, use of the sample variances S_1^2 and S_2^2 in the standardized variable Z yields a valid interval in which s_1^2 and s_2^2 replace σ_1^2 and σ_2^2.

Provided that m and n are both large, a CI for $\mu_1 - \mu_2$ with a confidence level of approximately $100(1 - \alpha)\%$ is

$$\bar{x} - \bar{y} \pm z_{\alpha/2}\sqrt{\frac{s_1^2}{m} + \frac{s_2^2}{n}}$$

where $-$ gives the lower limit and $+$ the upper limit of the interval. An upper or a lower confidence bound can also be calculated by retaining the appropriate sign ($+$ or $-$) and replacing $z_{\alpha/2}$ by z_α.

Our standard rule of thumb for characterizing sample sizes as large is $m > 40$ and $n > 40$.

EXAMPLE 9.5 Enhanced heavy oil recovery uses steam delivered to the production zone. The annulus between rock formation and the metal casing pipe is filled with cement. The article **"Thermal Stability of the Cement Sheath in Steam Treated Oil Wells"** (*J. of the Amer. Ceramic Soc.*, **2011: 4463–4470**) reported on a study of cement sheath performance when various thermal cements were cured at 35 °C and then heated to 230 °C. Here is summary data on Vicker's hardness (MPa) for both a control cement and an experimental cement:

Type	Sample Size	Sample Mean	Sample SD
Control	50	24.3	5.2
Experimental	50	27.0	5.8

Figure 9.1 shows a comparative boxplot of data consistent with these summary quantities. The main difference between the two samples appears to be where they are centered.

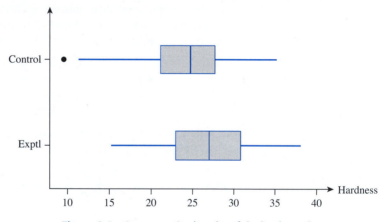

Figure 9.1 A comparative boxplot of the hardness data

Let's now calculate a confidence interval for the difference between true average hardness for the control cement (μ_1) and true average hardness for the experimental cement (μ_2) using a confidence level of 95%:

$$24.3 - 27.0 \pm (1.96)\sqrt{\frac{(5.2)^2}{50} + \frac{(5.8)^2}{50}} = -2.7 \pm (1.96)(1.1016)$$
$$= -2.7 \pm 2.2 = (-4.9, -.5)$$

That is, with 95% confidence, $-4.9 < \mu_1 - \mu_2 < -.5$. We can therefore be highly confident that true average hardness for the experimental cement exceeds that for the control cement by between .5 and 4.9 MPa. This CI does not include 0, so at the chosen confidence level, 0 is not a plausible value of $\mu_1 - \mu_2$. According to the relationship between CI's and HT's discussed in Section 8.5, the null hypothesis $H_0: \mu_1 - \mu_2 = 0$ should be rejected in favor of $H_a: \mu_1 - \mu_2 \neq 0$ at significance level .05 (the P-value for this test given in the cited article is not in agreement with other summary data).

Notice that if we relabel so that μ_1 refers to the experimental cement and μ_2 to the control cement, the CI becomes (.5, 4.9). The interpretation of the interval is exactly the same as was that of the first interval. ∎

If the variances σ_1^2 and σ_2^2 are at least approximately known and the investigator uses equal sample sizes, then the common sample size n that yields a $100(1 - \alpha)\%$ interval of width w is

$$n = \frac{4z_{\alpha/2}^2(\sigma_1^2 + \sigma_2^2)}{w^2}$$

which will generally have to be rounded up to an integer.

EXERCISES Section 9.1 (1–16)

1. An article in the November 1983 *Consumer Reports* compared various types of batteries. The average lifetimes of Duracell Alkaline AA batteries and Eveready Energizer Alkaline AA batteries were given as 4.1 hours and 4.5 hours, respectively. Suppose these are the population average lifetimes.

   a. Let $\overline{X}$ be the sample average lifetime of 100 Duracell batteries and $\overline{Y}$ be the sample average lifetime of 100 Eveready batteries. What is the mean value of $\overline{X} - \overline{Y}$ (i.e., where is the distribution of $\overline{X} - \overline{Y}$ centered)? How does your answer depend on the specified sample sizes?

   b. Suppose the population standard deviations of lifetime are 1.8 hours for Duracell batteries and 2.0 hours for Eveready batteries. With the sample sizes given in part (a), what is the variance of the statistic $\overline{X} - \overline{Y}$, and what is its standard deviation?

   c. For the sample sizes given in part (a), draw a picture of the approximate distribution curve of $\overline{X} - \overline{Y}$ (include a measurement scale on the horizontal axis). Would the shape of the curve necessarily be the same for sample sizes of 10 batteries of each type? Explain.

2. The **National Health Statistics Reports** dated Oct. 22, 2008, included the following information on the heights (in.) for non-Hispanic white females:

Age	Sample Size	Sample Mean	Std. Error Mean
20–39	866	64.9	.09
60 and older	934	63.1	.11

   a. Calculate and interpret a confidence interval at confidence level approximately 95% for the difference between population mean height for the younger women and that for the older women.

   b. Let μ_1 denote the population mean height for those aged 20–39 and μ_2 denote the population mean height for those aged 60 and older. Interpret the hypotheses $H_0: \mu_1 - \mu_2 = 1$ and $H_a: \mu_1 - \mu_2 > 1$, and then carry out a test of these hypotheses at significance level .001.

   c. Based on the *P*-value calculated in (b) would you reject the null hypothesis at any reasonable significance level? Explain your reasoning.

   d. What hypotheses would be appropriate if μ_1 referred to the older age group, μ_2 to the younger age group, and you wanted to see if there was compelling evidence for concluding that the population mean height for younger women exceeded that for older women by more than 1 in.?

3. Pilates is a popular set of exercises for the treatment of individuals with lower back pain. The method has six basic principles: centering, concentration, control, precision, flow, and breathing. The article **"Efficacy of the Addition of Modified Pilates Exercises to a Minimal Intervention in Patients with Chronic Low Back Pain: A Randomized Controlled Trial"** (*Physical Therapy*, 2013: 309–321) reported on an experiment involving 86 subjects with nonspecific low back pain. The participants were randomly divided into two groups of equal size. The first group received just educational materials, whereas the second group participated in 6 weeks of Pilates exercises. The sample mean level of pain (on a scale from 0 to 10) for the control group at a 6-week follow-up was 5.2 and the sample mean for the treatment group was 3.1; both sample standard deviations were 2.3.

   a. Does it appear that true average pain level for the control condition exceeds that for the treatment condition? Carry out a test of hypotheses using a significance level of .01 (the cited article reported statistical significance at this α, and a sample mean difference of 2.1 also suggests practical significance).

   b. Does it appear that true average pain level for the control condition exceeds that for the treatment condition by more than 1? Carry out a test of appropriate hypotheses.

4. Reliance on solid biomass fuel for cooking and heating exposes many children from developing countries to high levels of indoor air pollution. The article **"Domestic Fuels, Indoor Air Pollution, and Children's Health"** (*Annals of the N.Y. Academy of Sciences, 2008: 209–217*) presented information on various pulmonary characteristics in

samples of children whose households in India used either biomass fuel or liquefied petroleum gas (LPG). For the 755 children in biomass households, the sample mean peak expiratory flow (a person's maximum speed of expiration) was 3.30 L/s, and the sample standard deviation was 1.20. For the 750 children whose households used liquefied petroleum gas, the sample mean PEF was 4.25 and the sample standard deviation was 1.75.

a. Calculate a confidence interval at the 95% confidence level for the population mean PEF for children in biomass households and then do likewise for children in LPG households. What is the simultaneous confidence level for the two intervals?

b. Carry out a test of hypotheses at significance level .01 to decide whether true average PEF is lower for children in biomass households than it is for children in LPG households (the cited article included a P-value for this test).

c. FEV_1, the forced expiratory volume in 1 second, is another measure of pulmonary function. The cited article reported that for the biomass households the sample mean FEV_1 was 2.3 L/s and the sample standard deviation was .5 L/s. If this information is used to compute a 95% CI for population mean FEV_1, would the simultaneous confidence level for this interval and the first interval calculated in (a) be the same as the simultaneous confidence level determined there? Explain.

5. Persons having Reynaud's syndrome are apt to suffer a sudden impairment of blood circulation in fingers and toes. In an experiment to study the extent of this impairment, each subject immersed a forefinger in water and the resulting heat output (cal/cm²/min) was measured. For $m = 10$ subjects with the syndrome, the average heat output was $\bar{x} = .64$, and for $n = 10$ nonsufferers, the average output was 2.05. Let μ_1 and μ_2 denote the true average heat outputs for the two types of subjects. Assume that the two distributions of heat output are normal with $\sigma_1 = .2$ and $\sigma_2 = .4$.

a. Consider testing H_0: $\mu_1 - \mu_2 = -1.0$ versus H_a: $\mu_1 - \mu_2 < -1.0$ at level .01. Describe in words what H_a says, and then carry out the test.

b. What is the probability of a type II error when the actual difference between μ_1 and μ_2 is $\mu_1 - \mu_2 = -1.2$?

c. Assuming that $m = n$, what sample sizes are required to ensure that $\beta = .1$ when $\mu_1 - \mu_2 = -1.2$?

6. An experiment to compare the tension bond strength of polymer latex modified mortar (Portland cement mortar to which polymer latex emulsions have been added during mixing) to that of unmodified mortar resulted in $\bar{x} = 18.12$ kgf/cm² for the modified mortar ($m = 40$) and $\bar{y} = 16.87$ kgf/cm² for the unmodified mortar ($n = 32$). Let μ_1 and μ_2 be the true average tension bond strengths for the modified and unmodified mortars, respectively. Assume that the bond strength distributions are both normal.

a. Assuming that $\sigma_1 = 1.6$ and $\sigma_2 = 1.4$, test H_0: $\mu_1 - \mu_2 = 0$ versus H_a: $\mu_1 - \mu_2 > 0$ at level .01.

b. Compute the probability of a type II error for the test of part (a) when $\mu_1 - \mu_2 = 1$.

c. Suppose the investigator decided to use a level .05 test and wished $\beta = .10$ when $\mu_1 - \mu_2 = 1$. If $m = 40$, what value of n is necessary?

d. How would the analysis and conclusion of part (a) change if σ_1 and σ_2 were unknown but $s_1 = 1.6$ and $s_2 = 1.4$?

7. Is there any systematic tendency for part-time college faculty to hold their students to different standards than do full-time faculty? The article **"Are There Instructional Differences Between Full-Time and Part-Time Faculty?"** (*College Teaching*, **2009: 23–26**) reported that for a sample of 125 courses taught by full-time faculty, the mean course GPA was 2.7186 and the standard deviation was .63342, whereas for a sample of 88 courses taught by part-timers, the mean and standard deviation were 2.8639 and .49241, respectively. Does it appear that true average course GPA for part-time faculty differs from that for faculty teaching full-time? Test the appropriate hypotheses at significance level .01.

8. Tensile-strength tests were carried out on two different grades of wire rod (**"Fluidized Bed Patenting of Wire Rods,"** *Wire J.*, **June 1977: 56–61**), resulting in the accompanying data.

Grade	Sample Size	Sample Mean (kg/mm²)	Sample SD
AISI 1064	$m = 129$	$\bar{x} = 107.6$	$s_1 = 1.3$
AISI 1078	$n = 129$	$\bar{y} = 123.6$	$s_2 = 2.0$

a. Does the data provide compelling evidence for concluding that true average strength for the 1078 grade exceeds that for the 1064 grade by more than 10 kg/mm²? Test the appropriate hypotheses using a significance level of .01.

b. Estimate the difference between true average strengths for the two grades in a way that provides information about precision and reliability.

9. The article **"Evaluation of a Ventilation Strategy to Prevent Barotrauma in Patients at High Risk for Acute Respiratory Distress Syndrome"** (*New Engl. J. of Med.*, **1998: 355–358**) reported on an experiment in which 120 patients with similar clinical features were randomly divided into a control group and a treatment group, each consisting of 60 patients. The sample mean ICU stay (days) and sample standard deviation for the treatment group were 19.9 and 39.1, respectively, whereas these values for the control group were 13.7 and 15.8.

a. Calculate a point estimate for the difference between true average ICU stay for the treatment and control groups. Does this estimate suggest that there is a significant difference between true average stays under the two conditions?

b. Answer the question posed in part (a) by carrying out a formal test of hypotheses. Is the result different from what you conjectured in part (a)?

c. Does it appear that ICU stay for patients given the ventilation treatment is normally distributed? Explain your reasoning.

d. Estimate true average length of stay for patients given the ventilation treatment in a way that conveys information about precision and reliability.

10. An experiment was performed to compare the fracture toughness of high-purity 18 Ni maraging steel with commercial-purity steel of the same type (***Corrosion Science,*** **1971: 723–736**). For $m = 32$ specimens, the sample average toughness was $\bar{x} = 65.6$ for the high-purity steel, whereas for $n = 38$ specimens of commercial steel $\bar{y} = 59.8$. Because the high-purity steel is more expensive, its use for a certain application can be justified only if its fracture toughness exceeds that of commercial-purity steel by more than 5. Suppose that both toughness distributions are normal.

a. Assuming that $\sigma_1 = 1.2$ and $\sigma_2 = 1.1$, test the relevant hypotheses using $\alpha = .001$.

b. Compute β for the test conducted in part (a) when $\mu_1 - \mu_2 = 6$.

11. The level of lead in the blood was determined for a sample of 152 male hazardous-waste workers ages 20–30 and also for a sample of 86 female workers, resulting in a mean $\pm$ standard error of 5.5 ± 0.3 for the men and 3.8 ± 0.2 for the women (**"Temporal Changes in Blood Lead Levels of Hazardous Waste Workers in New Jersey, 1984–1987,"** ***Environ. Monitoring and Assessment,*** **1993: 99–107**). Calculate an estimate of the difference between true average blood lead levels for male and female workers in a way that provides information about reliability and precision.

12. The accompanying summary data on total cholesterol level (mmol/l) was obtained from a sample of Asian postmenopausal women who were vegans and another sample of such women who were omnivores (**"Vegetarianism, Bone Loss, and Vitamin D: A Longitudinal Study in Asian Vegans and Non-Vegans,"** ***European J. of Clinical Nutr.,*** **2012: 75–82**).

Diet	Sample Size	Sample Mean	Sample SD
Vegan	88	5.10	1.07
Omnivore	93	5.55	1.10

Calculate and interpret a 99% CI for the difference between population mean total cholesterol level for vegans and population mean total cholesterol level for omnivores (the cited article included a 95% CI). [*Note:* The article described a more sophisticated statistical analysis for investigating bone density loss taking into account other characteristics ("covariates") such as age,

body weight, and various nutritional factors; the resulting CI included 0, suggesting no diet effect.]

13. A mechanical engineer wishes to compare strength properties of steel beams with similar beams made with a particular alloy. The same number of beams, n, of each type will be tested. Each beam will be set in a horizontal position with a support on each end, a force of 2500 lb will be applied at the center, and the deflection will be measured. From past experience with such beams, the engineer is willing to assume that the true standard deviation of deflection for both types of beam is .05 in. Because the alloy is more expensive, the engineer wishes to test at level .01 whether it has smaller average deflection than the steel beam. What value of n is appropriate if the desired type II error probability is .05 when the difference in true average deflection favors the alloy by .04 in.?

14. The level of monoamine oxidase (MAO) activity in blood platelets (nm/mg protein/h) was determined for each individual in a sample of 43 chronic schizophrenics, resulting in $\bar{x} = 2.69$ and $s_1 = 2.30$, as well as for 45 normal subjects, resulting in $\bar{y} = 6.35$ and $s_2 = 4.03$. Does this data strongly suggest that true average MAO activity for normal subjects is more than twice the activity level for schizophrenics? Derive a test procedure and carry out the test using $\alpha = .01$. [*Hint:* H_0 and H_a here have a different form from the three standard cases. Let μ_1 and μ_2 refer to true average MAO activity for schizophrenics and normal subjects, respectively, and consider the parameter $\theta = 2\mu_1 - \mu_2$. Write H_0 and H_a in terms of θ, estimate θ, and derive $\hat{\sigma}_{\hat{\theta}}$ (**"Reduced Monoamine Oxidase Activity in Blood Platelets from Schizophrenic Patients,"** ***Nature,*** **July 28, 1972: 225–226**).]

15. a. Show for the upper-tailed test with σ_1 and σ_2 known that as either m or n increases, β decreases when $\mu_1 - \mu_2 > \Delta_0$.

b. For the case of equal sample sizes ($m = n$) and fixed α, what happens to the necessary sample size n as β is decreased, where β is the desired type II error probability at a fixed alternative?

16. To decide whether two different types of steel have the same true average fracture toughness values, n specimens of each type are tested, yielding the following results:

Type	Sample Average	Sample SD
1	60.1	1.0
2	59.9	1.0

Calculate the *P*-value for the appropriate two-sample z test, assuming that the data was based on $n = 100$. Then repeat the calculation for $n = 400$. Is the small *P*-value for $n = 400$ indicative of a difference that has practical significance? Would you have been satisfied with just a report of the *P*-value? Comment briefly.

9.2   The Two-Sample t Test and Confidence Interval

Values of the population variances will usually not be known to an investigator. In the previous section, we illustrated for large sample sizes the use of a z test and CI in which the sample variances were used in place of the population variances. In fact, for large samples, the CLT allows us to use these methods even when the two populations of interest are not normal.

In practice, though, it will often happen that at least one sample size is small and the population variances have unknown values. Without the CLT at our disposal, we proceed by making specific assumptions about the underlying population distributions. The use of inferential procedures that follow from these assumptions is then restricted to situations in which the assumptions are at least approximately satisfied. We could, for example, assume that both population distributions are members of the Weibull family or that they are both Poisson distributions. It shouldn't surprise you to learn that normality is often the most reasonable assumption.

ASSUMPTIONS

> Both population distributions are normal, so that $X_1, X_2,\ldots, X_m$ is a random sample from a normal distribution and so is $Y_1,\ldots, Y_n$ (with the X's and Y's independent of one another). The plausibility of these assumptions can be judged by constructing a normal probability plot of the x_i's and another of the y_i's.

The test statistic and confidence interval formula are based on the same standardized variable developed in Section 9.1, but the relevant distribution is now t rather than z.

THEOREM

> When the population distributions are both normal, the standardized variable
>
> $$T = \frac{\overline{X} - \overline{Y} - (\mu_1 - \mu_2)}{\sqrt{\dfrac{S_1^2}{m} + \dfrac{S_2^2}{n}}} \tag{9.2}$$
>
> has approximately a t distribution with df ν estimated from the data by
>
> $$\nu = \frac{\left(\dfrac{s_1^2}{m} + \dfrac{s_2^2}{n}\right)^2}{\dfrac{(s_1^2/m)^2}{m-1} + \dfrac{(s_2^2/n)^2}{n-1}} = \frac{[(se_1)^2 + (se_2)^2]^2}{\dfrac{(se_1)^4}{m-1} + \dfrac{(se_2)^4}{n-1}}$$
>
> where
>
> $$se_1 = \frac{s_1}{\sqrt{m}}, \quad se_2 = \frac{s_2}{\sqrt{n}}$$
>
> (round ν down to the nearest integer).

Manipulating T in a probability statement to isolate $\mu_1 - \mu_2$ gives a CI, whereas a test statistic results from replacing $\mu_1 - \mu_2$ by the null value Δ_0.

The **two-sample *t* confidence interval for $\mu_1 - \mu_2$** with confidence level $100(1 - \alpha)\%$ is then

$$\bar{x} - \bar{y} \pm t_{\alpha/2,\nu} \sqrt{\frac{s_1^2}{m} + \frac{s_2^2}{n}}$$

A one-sided confidence bound can be calculated as described earlier.

The **two-sample *t* test** for testing $H_0: \mu_1 - \mu_2 = \Delta_0$ is as follows:

$$\text{Test statistic value: } t = \frac{\bar{x} - \bar{y} - \Delta_0}{\sqrt{\frac{s_1^2}{m} + \frac{s_2^2}{n}}}$$

Alternative Hypothesis	P-Value Determination
$H_a: \mu_1 - \mu_2 > \Delta_0$	Area under the t_ν curve to the right of t
$H_a: \mu_1 - \mu_2 < \Delta_0$	Area under the t_ν curve to the left of t
$H_a: \mu_1 - \mu_2 \neq \Delta_0$	$2 \cdot$ (Area under the t_ν curve to the right of $\lvert t \rvert$)

Assumptions: Both population distributions are normal, and the two random samples are selected independently of one another.

EXAMPLE 9.6 The void volume within a textile fabric affects comfort, flammability, and insulation properties. Permeability of a fabric refers to the accessibility of void space to the flow of a gas or liquid. The article **"The Relationship Between Porosity and Air Permeability of Woven Textile Fabrics" (*J. of Testing and Eval.*, 1997: 108–114)** gave summary information on air permeability ($cm^3/cm^2/sec$) for a number of different fabric types. Consider the following data on two different types of plain-weave fabric:

Fabric Type	Sample Size	Sample Mean	Sample Standard Deviation
Cotton	10	51.71	.79
Triacetate	10	136.14	3.59

Assuming that the porosity distributions for both types of fabric are normal, let's calculate a confidence interval for the difference between true average porosity for the cotton fabric and that for the acetate fabric, using a 95% confidence level. Before the appropriate *t* critical value can be selected, df must be determined:

$$\text{df} = \frac{\left(\dfrac{.6241}{10} + \dfrac{12.8881}{10}\right)^2}{\dfrac{(.6241/10)^2}{9} + \dfrac{(12.8881/10)^2}{9}} = \frac{1.8258}{.1850} = 9.87$$

Thus we use $\nu = 9$; Appendix Table A.5 gives $t_{.025,9} = 2.262$. The resulting interval is

$$51.71 - 136.14 \pm (2.262)\sqrt{\frac{.6241}{10} + \frac{12.8881}{10}} = -84.43 \pm 2.63$$
$$= (-87.06, -81.80)$$

With a high degree of confidence, we can say that true average porosity for triacetate fabric specimens exceeds that for cotton specimens by between 81.80 and 87.06 cm^3/cm^2/sec. ∎

EXAMPLE 9.7 The deterioration of many municipal pipeline networks across the country is a growing concern. One technology proposed for pipeline rehabilitation uses a flexible liner threaded through existing pipe. The article **"Effect of Welding on a High-Density Polyethylene Liner"** (***J. of Materials in Civil Engr.*, 1996: 94–100**) reported the following data on tensile strength (psi) of liner specimens both when a certain fusion process was used and when this process was not used.

No fusion	2748	2700	2655	2822	2511			
	3149	3257	3213	3220	2753			
	$m = 10$	$\bar{x} = 2902.8$		$s_1 = 277.3$				
Fused	3027	3356	3359	3297	3125	2910	2889	2902
	$n = 8$	$\bar{y} = 3108.1$		$s_2 = 205.9$				

Figure 9.2 shows normal probability plots from Minitab. The linear pattern in each plot supports the assumption that the tensile strength distributions under the two conditions are both normal.

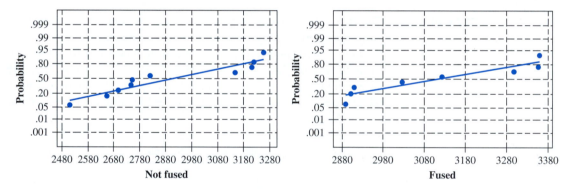

Figure 9.2 Normal probability plots from Minitab for the tensile strength data

The authors of the article stated that the fusion process increased the average tensile strength. The message from the comparative boxplot of Figure 9.3 is not all that clear. Let's carry out a test of hypotheses to see whether the data supports this conclusion.

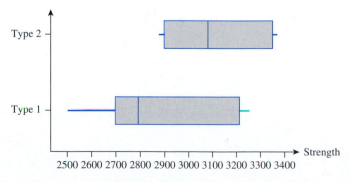

Figure 9.3 A comparative boxplot of the tensile-strength data

1. Let μ_1 be the true average tensile strength of specimens when the no-fusion treatment is used and μ_2 denote the true average tensile strength when the fusion treatment is used.

2. $H_0: \mu_1 - \mu_2 = 0$ (no difference in the true average tensile strengths for the two treatments)

3. $H_a: \mu_1 - \mu_2 < 0$ (true average tensile strength for the no-fusion treatment is less than that for the fusion treatment, so that the investigators' conclusion is correct)

4. The null value is $\Delta_0 = 0$, so the test statistic value is

$$t = \frac{\bar{x} - \bar{y}}{\sqrt{\dfrac{s_1^2}{m} + \dfrac{s_2^2}{n}}}$$

5. We now compute both the test statistic value and df for the test:

$$t = \frac{2902.8 - 3108.1}{\sqrt{\dfrac{(277.3)^2}{10} + \dfrac{(205.9)^2}{8}}} = \frac{-205.3}{113.97} = -1.8$$

Using $s_1^2/m = 7689.529$ and $s_2^2/n = 5299.351$,

$$\nu = \frac{(7689.529 + 5299.351)^2}{(7689.529)^2/9 + (5299.351)^2/7} = \frac{168{,}711{,}003.7}{10{,}581{,}747.35} = 15.94$$

so the test will be based on 15 df.

6. Appendix Table A.8 shows that the area under the 15 df t curve to the right of 1.8 is .046, so the P-value for a lower-tailed test is also .046. The following Minitab output summarizes all the computations:

```
Two-sample T for nofusion vs fused

               N     Mean    StDev    SE Mean
not fused     10     2903      277         88
fused          8     3108      206         73

95% C.I. for mu nofusion-mu fused: (-488, 38)
t-Test mu not fused = mu fused (vs <): T = -1.80  P = 0.046  DF = 15
```

7. Using a significance level of .05, we can barely reject the null hypothesis in favor of the alternative hypothesis, confirming the conclusion stated in the article. However, someone demanding more compelling evidence might select $\alpha = .01$, a level for which H_0 cannot be rejected.

If the question posed had been whether fusing increased true average strength by more than 100 psi, then the relevant hypotheses would have been $H_0: \mu_1 - \mu_2 = -100$ versus $H_a: \mu_1 - \mu_2 < -100$; that is, the null value would have been $\Delta_0 = -100$. ∎

Pooled t Procedures

Alternatives to the two-sample t procedures just described result from assuming not only that the two population distributions are normal but also that they have equal variances ($\sigma_1^2 = \sigma_2^2$). That is, the two population distribution curves are assumed normal with equal spreads, the only possible difference between them being where they are centered.

Let σ^2 denote the common population variance. Then standardizing $\overline{X} - \overline{Y}$ gives

$$Z = \frac{\overline{X} - \overline{Y} - (\mu_1 - \mu_2)}{\sqrt{\dfrac{\sigma^2}{m} + \dfrac{\sigma^2}{n}}} = \frac{\overline{X} - \overline{Y} - (\mu_1 - \mu_2)}{\sqrt{\sigma^2\left(\dfrac{1}{m} + \dfrac{1}{n}\right)}}$$

which has a standard normal distribution. Before this variable can be used as a basis for making inferences about $\mu_1 - \mu_2$, the common variance must be estimated from sample data. One estimator of σ^2 is S_1^2, the variance of the m observations in the first sample, and another is S_2^2, the variance of the second sample. Intuitively, a better estimator than either individual sample variance results from combining the two sample variances. A first thought might be to use $(S_1^2 + S_2^2)/2$. However, if $m > n$, then the first sample contains more information about σ^2 than does the second sample, and an analogous comment applies if $m < n$. The following *weighted* average of the two sample variances, called the **pooled** (i.e., combined) **estimator of σ^2**, adjusts for any difference between the two sample sizes:

$$S_p^2 = \frac{m-1}{m+n-2} \cdot S_1^2 + \frac{n-1}{m+n-2} \cdot S_2^2$$

The first sample contributes $m - 1$ degrees of freedom to the estimate of σ^2, and the second sample contributes $n - 1$ df, for a total of $m + n - 2$ df. Statistical theory says that if S_p^2 replaces σ^2 in the expression for Z, the resulting standardized variable has a t distribution based on $m + n - 2$ df. In the same way that earlier standardized variables were used as a basis for deriving confidence intervals and test procedures, this t variable immediately leads to the pooled t CI for estimating $\mu_1 - \mu_2$ and the pooled t test for testing hypotheses about a difference between means.

In the past, many statisticians recommended these pooled t procedures over the two-sample t procedures. The pooled t test, for example, can be derived from the likelihood ratio principle, whereas the two-sample t test is not a likelihood ratio test. Furthermore, the significance level for the pooled t test is exact, whereas it is only approximate for the two-sample t test. However, recent research has shown that although the pooled t test does outperform the two-sample t test by a bit (smaller β's for the same α) when $\sigma_1^2 = \sigma_2^2$, the former test can easily lead to erroneous conclusions if applied when the variances are different. Analogous comments apply to the behavior of the two confidence intervals. That is, the pooled t procedures are not robust to violations of the equal variance assumption.

It has been suggested that one could carry out a preliminary test of $H_0: \sigma_1^2 = \sigma_2^2$ and use a pooled t procedure if this null hypothesis is not rejected. Unfortunately, the usual "F test" of equal variances (Section 9.5) is quite sensitive to the assumption of normal population distributions—much more so than t procedures. We therefore recommend the conservative approach of using two-sample t procedures unless there is really compelling evidence for doing otherwise, particularly when the two sample sizes are different.

Type II Error Probabilities

Determining type II error probabilities (or equivalently, power $= 1 - \beta$) for the two-sample t test is complicated. There does not appear to be any simple way to use the β curves of Appendix Table A.17. The most recent version of Minitab (Version 16) will calculate power for the pooled t test but not for the two-sample t test. However, the UCLA Statistics Department homepage (http://www.stat.ucla.edu)

permits access to a power calculator that will do this. For example, we specified $m = 10$, $n = 8$, $\sigma_1 = 300$, $\sigma_2 = 225$ (these are the sample sizes for Example 9.7, whose sample standard deviations are somewhat smaller than these values of σ_1 and σ_2) and asked for the power of a two-tailed level .05 test of $H_0: \mu_1 - \mu_2 = 0$ when $\mu_1 - \mu_2 = 100$, 250, and 500. The resulting values of the power were .1089, .4609, and .9635 (corresponding to $\beta = .89$, .54, and .04), respectively. In general, β will decrease as the sample sizes increase, as α increases, and as $\mu_1 - \mu_2$ moves farther from 0. The software will also calculate sample sizes necessary to obtain a specified value of power for a particular value of $\mu_1 - \mu_2$.

EXERCISES Section 9.2 (17–35)

17. Determine the number of degrees of freedom for the two-sample t test or CI in each of the following situations:
  a. $m = 10$, $n = 10$, $s_1 = 5.0$, $s_2 = 6.0$
  b. $m = 10$, $n = 15$, $s_1 = 5.0$, $s_2 = 6.0$
  c. $m = 10$, $n = 15$, $s_1 = 2.0$, $s_2 = 6.0$
  d. $m = 12$, $n = 24$, $s_1 = 5.0$, $s_2 = 6.0$

18. Which way of dispensing champagne, the traditional vertical method or a tilted beer-like pour, preserves more of the tiny gas bubbles that improve flavor and aroma? The following data was reported in the article **"On the Losses of Dissolved CO_2 during Champagne Serving"** (*J. Agr. Food Chem.*, 2010: 8768–8775).

Temp (°C)	Type of Pour	n	Mean (g/L)	SD
18	Traditional	4	4.0	.5
18	Slanted	4	3.7	.3
12	Traditional	4	3.3	.2
12	Slanted	4	2.0	.3

Assume that the sampled distributions are normal.
  a. Carry out a test at significance level .01 to decide whether true average CO_2 loss at 18 °C for the traditional pour differs from that for the slanted pour.
  b. Repeat the test of hypotheses suggested in (a) for the 12° temperature. Is the conclusion different from that for the 18° temperature? *Note:* The 12° result was reported in the popular media.

19. Suppose μ_1 and μ_2 are true mean stopping distances at 50 mph for cars of a certain type equipped with two different types of braking systems. Use the two-sample t test at significance level .01 to test $H_0: \mu_1 - \mu_2 = -10$ versus $H_a: \mu_1 - \mu_2 < -10$ for the following data: $m = 6$, $\bar{x} = 115.7$, $s_1 = 5.03$, $n = 6$, $\bar{y} = 129.3$, and $s_2 = 5.38$.

20. Use the data of Exercise 19 to calculate a 95% CI for the difference between true average stopping distance for cars equipped with system 1 and cars equipped with system 2. Does the interval suggest that precise information about the value of this difference is available?

21. Quantitative noninvasive techniques are needed for routinely assessing symptoms of peripheral neuropathies, such as carpal tunnel syndrome (CTS). The article **"A Gap Detection Tactility Test for Sensory Deficits Associated with Carpal Tunnel Syndrome"** (*Ergonomics*, 1995: 2588–2601) reported on a test that involved sensing a tiny gap in an otherwise smooth surface by probing with a finger; this functionally resembles many work-related tactile activities, such as detecting scratches or surface defects. When finger probing was not allowed, the sample average gap detection threshold for $m = 8$ normal subjects was 1.71 mm, and the sample standard deviation was .53; for $n = 10$ CTS subjects, the sample mean and sample standard deviation were 2.53 and .87, respectively. Does this data suggest that the true average gap detection threshold for CTS subjects exceeds that for normal subjects? State and test the relevant hypotheses using a significance level of .01.

22. According to the article **"Modeling and Predicting the Effects of Submerged Arc Weldment Process Parameters on Weldment Characteristics and Shape Profiles"** (*J. of Engr. Manuf.*, 2012: 1230–1240), the submerged arc welding (SAW) process is commonly used for joining thick plates and pipes. The heat affected zone (HAZ), a band created within the base metal during welding, was of particular interest to the investigators. Here are observations on depth (mm) of the HAZ both when the current setting was high and when it was lower.

Non-high	1.04	1.15	1.23	1.69	1.92
	1.98	2.36	2.49	2.72	
	1.37	1.43	1.57	1.71	1.94
	2.06	2.55	2.64	2.82	
High	1.55	2.02	2.02	2.05	2.35
	2.57	2.93	2.94	2.97	

  a. Construct a comparative boxplot and comment on interesting features.
  b. Is it reasonable to use the two-sample t test to test hypotheses about the difference between true average HAZ depths for the two conditions?

c. Does it appear that true average HAZ depth is larger for the higher current condition than for the lower condition? Carry out a test of appropriate hypotheses using a significance level of .01.

23. Fusible interlinings are being used with increasing frequency to support outer fabrics and improve the shape and drape of various pieces of clothing. The article **"Compatibility of Outer and Fusible Interlining Fabrics in Tailored Garments" (*Textile Res. J.*, 1997: 137–142)** gave the accompanying data on extensibility (%) at 100 gm/cm for both high-quality (H) fabric and poor-quality (P) fabric specimens.

H	1.2	.9	.7	1.0	1.7	1.7	1.1	.9	1.7
	1.9	1.3	2.1	1.6	1.8	1.4	1.3	1.9	1.6
	.8	2.0	1.7	1.6	2.3	2.0			
P	1.6	1.5	1.1	2.1	1.5	1.3	1.0	2.6	

a. Construct normal probability plots to verify the plausibility of both samples having been selected from normal population distributions.

b. Construct a comparative boxplot. Does it suggest that there is a difference between true average extensibility for high-quality fabric specimens and that for poor-quality specimens?

c. The sample mean and standard deviation for the high-quality sample are 1.508 and .444, respectively, and those for the poor-quality sample are 1.588 and .530. Use the two-sample t test to decide whether true average extensibility differs for the two types of fabric.

24. Damage to grapes from bird predation is a serious problem for grape growers. The article **"Experimental Method to Investigate and Monitor Bird Behavior and Damage to Vineyards" (*Amer. J. of Enology and Viticulture*, 2004: 288–291)** reported on an experiment involving a bird-feeder table, time-lapse video, and artificial foods. Information was collected for two different bird species at both the experimental location and at a natural vineyard setting. Consider the following data on time (sec) spent on a single visit to the location.

Species	Location	n	$\bar{x}$	SE mean
Blackbirds	Exptl	65	13.4	2.05
Blackbirds	Natural	50	9.7	1.76
Silvereyes	Exptl	34	49.4	4.78
Silvereyes	Natural	46	38.4	5.06

a. Calculate an upper confidence bound for the true average time that blackbirds spend on a single visit at the experimental location.

b. Does it appear that true average time spent by blackbirds at the experimental location exceeds the true average time birds of this type spend at the natural location? Carry out a test of appropriate hypotheses.

c. Estimate the difference between the true average time blackbirds spend at the natural location and true average time that silvereyes spend at the natural

location, and do so in a way that conveys information about reliability and precision.

[*Note:* The sample medians reported in the article all seemed significantly smaller than the means, suggesting substantial population distribution skewness. The authors actually used the distribution-free test procedure presented in Section 2 of Chapter 15.]

25. The accompanying data consists of prices ($) for one sample of California cabernet sauvignon wines that received ratings of 93 or higher in the May 2013 issue of *Wine Spectator* and another sample of California cabernets that received ratings of 89 or lower in the same issue.

≥ 93:	100	100	60	135	195	195
	125	135	95	42	75	72

≤ 89:	80	75	75	85	75	35	85
	65	45	100	28	38	50	28

Assume that these are both random samples of prices from the population of all wines recently reviewed that received ratings of at least 93 and at most 89, respectively.

a. Investigate the plausibility of assuming that both sampled populations are normal.

b. Construct a comparative boxplot. What does it suggest about the difference in true average prices?

c. Calculate a confidence interval at the 95% confidence level to estimate the difference between μ_1, the mean price in the higher rating population, and μ_2, the mean price in the lower rating population. Is the interval consistent with the statement "Price rarely equates to quality" made by a columnist in the cited issue of the magazine?

26. The article **"The Influence of Corrosion Inhibitor and Surface Abrasion on the Failure of Aluminum-Wired Twist-On Connections" (*IEEE Trans. on Components, Hybrids, and Manuf. Tech.*, 1984: 20–25)** reported data on potential drop measurements for one sample of connectors wired with alloy aluminum and another sample wired with EC aluminum. Does the accompanying SAS output suggest that the true average potential drop for alloy connections (type 1) is higher than that for EC connections (as stated in the article)? Carry out the appropriate test using a significance level of .01. In reaching your conclusion, what type of error might you have committed? [*Note:* SAS reports the *P*-value for a two-tailed test.]

Type	N	Mean	Std Dev	Std Error
1	20	17.49900000	0.55012821	0.12301241
2	20	16.90000000	0.48998389	0.10956373

Variances	T	DF	Prob>|T|
Unequal	3.6362	37.5	0.0008
Equal	3.6362	38.0	0.0008

27. Anorexia Nervosa (AN) is a psychiatric condition leading to substantial weight loss among women who are fearful of becoming fat. The article **"Adipose Tissue Distribution After Weight Restoration and Weight Maintenance in Women with Anorexia Nervosa" (*Amer. J. of Clinical***

Nutr., 2009: 1132–1137) used whole-body magnetic resonance imagery to determine various tissue characteristics for both an AN sample of individuals who had undergone acute weight restoration and maintained their weight for a year and a comparable (at the outset of the study) control sample. Here is summary data on intermuscular adipose tissue (IAT; kg).

Condition	Sample Size	Sample Mean	Sample SD
AN	16	.52	.26
Control	8	.35	.15

Assume that both samples were selected from normal distributions.

a. Calculate an estimate for true average IAT under the described AN protocol, and do so in a way that conveys information about the reliability and precision of the estimation.

b. Calculate an estimate for the difference between true average AN IAT and true average control IAT, and do so in a way that conveys information about the reliability and precision of the estimation. What does your estimate suggest about true average AN IAT relative to true average control IAT?

28. As the population ages, there is increasing concern about accident-related injuries to the elderly. The article **"Age and Gender Differences in Single-Step Recovery from a Forward Fall"** (*J. of Gerontology*, 1999: M44–M50) reported on an experiment in which the maximum lean angle—the farthest a subject is able to lean and still recover in one step—was determined for both a sample of younger females (21–29 years) and a sample of older females (67–81 years). The following observations are consistent with summary data given in the article:

YF: 29, 34, 33, 27, 28, 32, 31, 34, 32, 27
OF: 18, 15, 23, 13, 12

Does the data suggest that true average maximum lean angle for older females is more than 10 degrees smaller than it is for younger females? State and test the relevant hypotheses at significance level .10.

29. The article **"Effect of Internal Gas Pressure on the Compression Strength of Beverage Cans and Plastic Bottles"** (*J. of Testing and Evaluation*, 1993: 129–131) includes the accompanying data on compression strength (lb) for a sample of 12-oz aluminum cans filled with strawberry drink and another sample filled with cola. Does the data suggest that the extra carbonation of cola results in a higher average compression strength? Base your answer on a *P*-value. What assumptions are necessary for your analysis?

Beverage	Sample Size	Sample Mean	Sample SD
Strawberry drink	15	540	21
Cola	15	554	15

30. The article **"Flexure of Concrete Beams Reinforced with Advanced Composite Orthogrids"** (*J. of Aerospace Engr.*, 1997: 7–15) gave the accompanying data on ultimate load (kN) for two different types of beams.

Type	Sample Size	Sample Mean	Sample SD
Fiberglass grid	26	33.4	2.2
Commercial carbon grid	26	42.8	4.3

a. Assuming that the underlying distributions are normal, calculate and interpret a 99% CI for the difference between true average load for the fiberglass beams and that for the carbon beams.

b. Does the upper limit of the interval you calculated in part (a) give a 99% upper confidence bound for the difference between the two μ's? If not, calculate such a bound. Does it strongly suggest that true average load for the carbon beams is more than that for the fiberglass beams? Explain.

31. Refer to Exercise 33 in Section 7.3. The cited article also gave the following observations on degree of polymerization for specimens having viscosity times concentration in a higher range:

429 430 430 431 436 437
440 441 445 446 447

a. Construct a comparative boxplot for the two samples, and comment on any interesting features.

b. Calculate a 95% confidence interval for the difference between true average degree of polymerization for the middle range and that for the high range. Does the interval suggest that μ_1 and μ_2 may in fact be different? Explain your reasoning.

32. The degenerative disease osteoarthritis most frequently affects weight-bearing joints such as the knee. The article **"Evidence of Mechanical Load Redistribution at the Knee Joint in the Elderly When Ascending Stairs and Ramps"** (*Annals of Biomed. Engr.*, 2008: 467–476) presented the following summary data on stance duration (ms) for samples of both older and younger adults.

Age	Sample Size	Sample Mean	Sample SD
Older	28	801	117
Younger	16	780	72

Assume that both stance duration distributions are normal.

a. Calculate and interpret a 99% CI for true average stance duration among elderly individuals.

b. Carry out a test of hypotheses at significance level .05 to decide whether true average stance duration is larger among elderly individuals than among younger individuals.

33. The article **"The Effects of a Low-Fat, Plant-Based Dietary Intervention on Body Weight, Metabolism,**

and Insulin Sensitivity in Postmenopausal Women" (*Amer. J. of Med.*, **2005: 991–997**) reported on the results of an experiment in which half of the individuals in a group of 64 postmenopausal overweight women were randomly assigned to a particular vegan diet, and the other half received a diet based on National Cholesterol Education Program guidelines. The sample mean decrease in body weight for those on the vegan diet was 5.8 kg, and the sample SD was 3.2, whereas for those on the control diet, the sample mean weight loss and standard deviation were 3.8 and 2.8, respectively. Does it appear the true average weight loss for the vegan diet exceeds that for the control diet by more than 1 kg? Carry out an appropriate test of hypotheses at significance level .05.

34. Consider the pooled t variable

$$T = \frac{(\overline{X} - \overline{Y}) - (\mu_1 - \mu_2)}{S_p \sqrt{\dfrac{1}{m} + \dfrac{1}{n}}}$$

which has a t distribution with $m + n - 2$ df when both population distributions are normal with $\sigma_1 = \sigma_2$

(see the Pooled t Procedures subsection for a description of S_p).
a. Use this t variable to obtain a pooled t confidence interval formula for $\mu_1 - \mu_2$.
b. A sample of ultrasonic humidifiers of one particular brand was selected for which the observations on maximum output of moisture (oz) in a controlled chamber were 14.0, 14.3, 12.2, and 15.1. A sample of the second brand gave output values 12.1, 13.6, 11.9, and 11.2 (**"Multiple Comparisons of Means Using Simultaneous Confidence Intervals,"** *J. of Quality Technology*, **1989: 232–241**). Use the pooled t formula from part (a) to estimate the difference between true average outputs for the two brands with a 95% confidence interval.
c. Estimate the difference between the two μ's using the two-sample t interval discussed in this section, and compare it to the interval of part (b).

35. Refer to Exercise 34. Describe the pooled t test for testing $H_0: \mu_1 - \mu_2 = \Delta_0$ when both population distributions are normal with $\sigma_1 = \sigma_2$. Then use this test procedure to test the hypotheses suggested in Exercise 33.

9.3 Analysis of Paired Data

In Sections 9.1 and 9.2, we considered making an inference about a difference between two means μ_1 and μ_2. This was done by utilizing the results of a random sample $X_1, X_2, \ldots X_m$ from the distribution with mean μ_1 and a completely independent (of the X's) sample $Y_1, \ldots, Y_n$ from the distribution with mean μ_2. That is, either m individuals were selected from population 1 and n different individuals from population 2, or m individuals (or experimental objects) were given one treatment and another set of n individuals were given the other treatment. In contrast, there are a number of experimental situations in which there is only one set of n individuals or experimental objects; making two observations on each one results in a natural pairing of values.

EXAMPLE 9.8 Trace metals in drinking water affect the flavor, and unusually high concentrations can pose a health hazard. The article **"Trace Metals of South Indian River"** (*Envir. Studies*, **1982: 62–66**) reported on a study in which six river locations were selected (six experimental objects) and the zinc concentration (mg/L) determined for both surface water and bottom water at each location. The six pairs of observations are displayed in the accompanying table. Does the data suggest that true average concentration in bottom water exceeds that of surface water?

	Location					
	1	2	3	4	5	6
Zinc concentration in bottom water (x)	.430	.266	.567	.531	.707	.716
Zinc concentration in surface water (y)	.415	.238	.390	.410	.605	.609
Difference	.015	.028	.177	.121	.102	.107

Figure 9.4(a) displays a plot of this data. At first glance, there appears to be little difference between the x and y samples. From location to location, there is a great deal of variability in each sample, and it looks as though any differences between the samples can be attributed to this variability. However, when the observations are identified by location, as in Figure 9.4(b), a different view emerges. At each location, bottom concentration exceeds surface concentration. This is confirmed by the fact that all $x - y$ differences displayed in the bottom row of the data table are positive. A correct analysis of this data focuses on these differences.

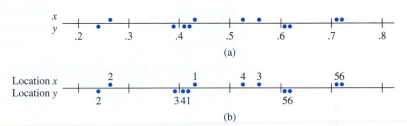

Figure 9.4 Plot of paired data from Example 9.8: (a) observations not identified by location; (b) observations identified by location ■

ASSUMPTIONS

The data consists of n independently selected pairs $(X_1, Y_1), (X_2, Y_2), \ldots (X_n, Y_n)$, with $E(X_i) = \mu_1$ and $E(Y_i) = \mu_2$. Let $D_1 = X_1 - Y_1, D_2 = X_2 - Y_2, \ldots, D_n = X_n - Y_n$ so the D_i's are the differences within pairs. The D_i's are assumed to be normally distributed with mean value μ_D and variance σ_D^2 (this is usually a consequence of the X_i's and Y_i's themselves being normally distributed).

We are again interested in making an inference about the difference $\mu_1 - \mu_2$. The two-sample t confidence interval and test statistic were obtained by assuming independent samples and applying the rule $V(\overline{X} - \overline{Y}) = V(\overline{X}) + V(\overline{Y})$. However, with paired data, the X and Y observations within each pair are often not independent. Then $\overline{X}$ and $\overline{Y}$ are not independent of one another. We must therefore abandon the two-sample t procedures and look for an alternative method of analysis.

The Paired t Test

Because different pairs are independent, the D_i's are independent of one another. Let $D = X - Y$, where X and Y are the first and second observations, respectively, within an arbitrary pair. Then the expected difference is

$$\mu_D = E(X - Y) = E(X) - E(Y) = \mu_1 - \mu_2$$

(the rule of expected values used here is valid even when X and Y are dependent). Thus any hypothesis about $\mu_1 - \mu_2$ can be phrased as a hypothesis about the mean difference μ_D. But since the D_i's constitute a normal random sample (of differences) with mean μ_D, hypotheses about μ_D can be tested using a one-sample t test. That is, *to test hypotheses about $\mu_1 - \mu_2$ when data is paired, form the differences $D_1, D_2, \ldots, D_n$ and carry out a one-sample t test (based on $n - 1$ df) on these differences.*

The Paired t Test

Null hypothesis: $H_0: \mu_D = \Delta_0$ (where $D = X - Y$ is the difference between the first and second observations within a pair, and $\mu_D = \mu_1 - \mu_2$)

Test statistic value: $t = \dfrac{\bar{d} - \Delta_0}{s_D/\sqrt{n}}$ (where $\bar{d}$ and s_D are the sample mean and standard deviation, respectively, of the d_i's)

Alternative Hypothesis	P-Value Determination
$H_a: \mu_D > \Delta_0$	Area under the t_{n-1} curve to the right of t
$H_a: \mu_D < \Delta_0$	Area under the t_{n-1} curve to the left of t
$H_a: \mu_D \neq \Delta_0$	$2 \cdot$ (Area under the t_{n-1} curve to the right of $|t|$)

Assumptions: The D_is constitute a random sample from a normal "difference" population.

EXAMPLE 9.9   Musculoskeletal neck-and-shoulder disorders are all too common among office staff who perform repetitive tasks using visual display units. The article **"Upper-Arm Elevation During Office Work" (*Ergonomics*, 1996: 1221–1230)** reported on a study to determine whether more varied work conditions would have any impact on arm movement. The accompanying data was obtained from a sample of $n = 16$ subjects. Each observation is the amount of time, expressed as a proportion of total time observed, during which arm elevation was below $30°$. The two measurements from each subject were obtained 18 months apart. During this period, work conditions were changed, and subjects were allowed to engage in a wider variety of work tasks. Does the data suggest that true average time during which elevation is below $30°$ differs after the change from what it was before the change?

Subject	1	2	3	4	5	6	7	8
Before	81	87	86	82	90	86	96	73
After	78	91	78	78	84	67	92	70
Difference	3	−4	8	4	6	19	4	3

Subject	9	10	11	12	13	14	15	16
Before	74	75	72	80	66	72	56	82
After	58	62	70	58	66	60	65	73
Difference	16	13	2	22	0	12	−9	9

Figure 9.5 shows a normal probability plot of the 16 differences; the pattern in the plot is quite straight, supporting the normality assumption. A boxplot of these differences appears in Figure 9.6; the boxplot is located considerably to the right of zero, suggesting that perhaps $\mu_D > 0$ (note also that 13 of the 16 differences are positive and only two are negative).

Let's now test the appropriate hypotheses.

1. Let μ_D denote the true average difference between elevation time before the change in work conditions and time after the change.

2. $H_0: \mu_D = 0$ (there is no difference between true average time before the change and true average time after the change)

3. $H_0: \mu_D \neq 0$

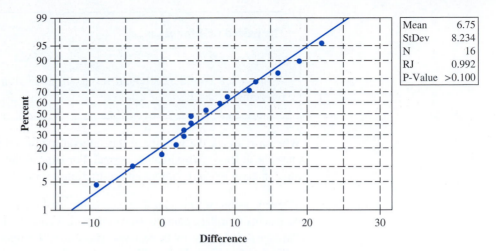

Mean	6.75
StDev	8.234
N	16
RJ	0.992
P-Value	>0.100

Figure 9.5 A normal probability plot from Minitab of the differences in Example 9.9

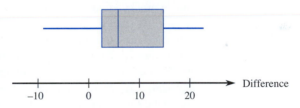

Figure 9.6 A boxplot of the differences in Example 9.9

4. $t = \dfrac{\bar{d} - 0}{s_D/\sqrt{n}} = \dfrac{\bar{d}}{s_D/\sqrt{n}}$

5. $n = 16$, $\Sigma d_i = 108$, and $\Sigma d_i^2 = 1746$, from which $\bar{d} = 6.75$, $s_D = 8.234$, and

$$t = \frac{6.75}{8.234/\sqrt{16}} = 3.28 \approx 3.3$$

6. Appendix Table A.8 shows that the area to the right of 3.3 under the t curve with 15 df is .002. The inequality in H_a implies that a two-tailed test is appropriate, so the P-value is approximately 2(.002) = .004 (Minitab gives .0051).

7. Since .004 < .01, the null hypothesis can be rejected at either significance level .05 or .01. It does appear that the true average difference between times is something other than zero; that is, true average time after the change is different from that before the change. ■

When the number of pairs is large, the assumption of a normal difference distribution is not necessary. The CLT validates the resulting z test.

The Paired t Confidence Interval

In the same way that the t CI for a single population mean μ is based on the t variable $T = (\bar{X} - \mu)/(S/\sqrt{n})$, a t confidence interval for $\mu_D (= \mu_1 - \mu_2)$ is based on the fact that

$$T = \frac{\bar{D} - \mu_D}{S_D/\sqrt{n}}$$

has a t distribution with $n - 1$ df. Manipulation of this t variable, as in previous derivations of CI's, yields the following $100(1 - \alpha)\%$ CI:

> The **paired t CI for μ_D** is
>
> $$\bar{d} \pm t_{\alpha/2, n-1} \cdot s_D/\sqrt{n}$$
>
> A one-sided confidence bound results from retaining the relevant sign and replacing $t_{\alpha/2}$ by t_α.

When n is small, the validity of this interval requires that the distribution of differences be at least approximately normal. For large n, the CLT ensures that the resulting z interval is valid without any restrictions on the distribution of differences.

EXAMPLE 9.10 Magnetic resonance imaging is a commonly used noninvasive technique for assessing the extent of cartilage damage. However, there is concern that the MRI sizing of articular cartilage defects may not be accurate. The article **"Preoperative MRI Underestimates Articular Cartilage Defect Size Compared with Findings at Arthroscopic Knee Surgery"** (*Amer. J. of Sports Med.*, 2013: 590–595) reported on a study involving a sample of 92 cartilage defects. For each one, the size of the lesion area was determined by an MRI analysis and also during arthroscopic surgery. Each MRI value was then subtracted from the corresponding arthroscopic value to obtain a difference value. The sample mean difference was calculated to be 1.04 cm^2, with a sample standard deviation of 1.67. Let's now calculate a confidence interval using a confidence level of (at least approximately) 95% for μ_D, the mean difference for the population of all such defects (as did the authors of the cited article). Because n is quite large here, we use the z critical value $z_{.025} = 1.96$ (an entry at the very bottom of our t table). The resulting CI is

$$1.04 \pm (1.96) \cdot \frac{1.67}{\sqrt{92}} = 1.04 \pm .34 = (.70, 1.38)$$

At the 95% confidence level, we believe that $.70 < \mu_D < 1.38$. Perhaps the most interesting aspect of this interval is that 0 is not included; only certain positive values of μ_D are plausible. It is this fact that led the investigators to conclude that MRIs tend to underestimate defect size. ■

Paired Data and Two-Sample t Procedures

Consider using the two-sample t test on paired data. The numerators of the two test statistics are identical, since $\bar{d} = \Sigma d_i/n = [\Sigma(x_i - y_i)]/n = (\Sigma x_i)/n - (\Sigma y_i)/n = \bar{x} - \bar{y}$. The difference between the statistics is due entirely to the denominators. Each test statistic is obtained by standardizing $\bar{X} - \bar{Y}(= \bar{D})$. But in the presence of dependence the two-sample t standardization is incorrect. To see this, recall from Section 5.5 that

$$V(X \pm Y) = V(X) + V(Y) \pm 2\,\text{Cov}(X, Y)$$

The correlation between X and Y is

$$\rho = \text{Corr}(X, Y) = \text{Cov}(X, Y)/[\sqrt{V(X)} \cdot \sqrt{V(Y)}]$$

It follows that

$$V(X - Y) = \sigma_1^2 + \sigma_2^2 - 2\rho\sigma_1\sigma_2$$

Applying this to $\bar{X} - \bar{Y}$ yields

$$V(\bar{X} - \bar{Y}) = V(\bar{D}) = V\left(\frac{1}{n}\Sigma D_i\right) = \frac{V(D_i)}{n} = \frac{\sigma_1^2 + \sigma_2^2 - 2\rho\sigma_1\sigma_2}{n}$$

The two-sample t test is based on the assumption of independence, in which case $\rho = 0$. But in many paired experiments, there will be a strong *positive* dependence between X and Y (large X associated with large Y), so that ρ will be positive and the variance of $\overline{X} - \overline{Y}$ will be smaller than $\sigma_1^2/n + \sigma_2^2/n$. Thus *whenever there is positive dependence within pairs, the denominator for the paired t statistic should be smaller than for t of the independent-samples test.* Often two-sample t will be much closer to zero than paired t, considerably understating the significance of the data.

Similarly, when data is paired, the paired t CI will usually be narrower than the (incorrect) two-sample t CI. This is because there is typically much less variability in the differences than in the x and y values.

Paired Versus Unpaired Experiments

In our examples, paired data resulted from two observations on the same subject (Example 9.9) or experimental object (location in Example 9.8). Even when this cannot be done, paired data with dependence within pairs can be obtained by matching individuals or objects on one or more characteristics thought to influence responses. For example, in a medical experiment to compare the efficacy of two drugs for lowering blood pressure, the experimenter's budget might allow for the treatment of 20 patients. If 10 patients are randomly selected for treatment with the first drug and another 10 independently selected for treatment with the second drug, an independent-samples experiment results.

However, the experimenter, knowing that blood pressure is influenced by age and weight, might decide to create pairs of patients so that within each of the resulting 10 pairs, age and weight were approximately equal (though there might be sizable differences between pairs). Then each drug would be given to a different patient within each pair for a total of 10 observations on each drug.

Without this matching (or "blocking"), one drug might appear to outperform the other just because patients in one sample were lighter and younger and thus more susceptible to a decrease in blood pressure than the heavier and older patients in the second sample. However, there is a price to be paid for pairing—a smaller number of degrees of freedom for the paired analysis—so we must ask when one type of experiment should be preferred to the other.

There is no straightforward and precise answer to this question, but there are some useful guidelines. If we have a choice between two t tests that are both valid (and carried out at the same level of significance α), we should prefer the test that has the larger number of degrees of freedom. The reason for this is that a larger number of degrees of freedom means smaller β for any fixed alternative value of the parameter or parameters. That is, for a fixed type I error probability, the probability of a type II error is decreased by increasing degrees of freedom.

However, if the experimental units are quite heterogeneous in their responses, it will be difficult to detect small but significant differences between two treatments. This is essentially what happened in the data set in Example 9.8; for both "treatments" (bottom water and surface water), there is great between-location variability, which tends to mask differences in treatments within locations. If there is a high positive correlation within experimental units or subjects, the variance of $\overline{D} = \overline{X} - \overline{Y}$ will be much smaller than the unpaired variance. Because of this reduced variance, it will be easier to detect a difference with paired samples than with independent samples. The pros and cons of pairing can now be summarized as follows.

1. If there is great heterogeneity between experimental units and a large correlation within experimental units (large positive ρ), then the loss in degrees of freedom will be compensated for by the increased precision associated with pairing, so a paired experiment is preferable to an independent-samples experiment.

2. If the experimental units are relatively homogeneous and the correlation within pairs is not large, the gain in precision due to pairing will be outweighed by the decrease in degrees of freedom, so an independent-samples experiment should be used.

Of course, values of σ_1^2, σ_2^2, and ρ will not usually be known very precisely, so an investigator will be required to make an educated guess as to whether Situation 1 or 2 obtains. In general, if the number of observations that can be obtained is large, then a loss in degrees of freedom (e.g., from 40 to 20) will not be serious; but if the number is small, then the loss (say, from 16 to 8) because of pairing may be serious if not compensated for by increased precision. Similar considerations apply when choosing between the two types of experiments to estimate $\mu_1 - \mu_2$ with a confidence interval.

EXERCISES Section 9.3 (36–48)

36. Consider the accompanying data on breaking load (kg/25 mm width) for various fabrics in both an unabraded condition and an abraded condition (**"The Effect of Wet Abrasive Wear on the Tensile Properties of Cotton and Polyester-Cotton Fabrics,"** *J. Testing and Evaluation*, **1993: 84–93**). Use the paired *t* test, as did the authors of the cited article, to test H_0: $\mu_D = 0$ versus H_a: $\mu_D > 0$ at significance level .01.

				Fabric				
	1	**2**	**3**	**4**	**5**	**6**	**7**	**8**
U	36.4	55.0	51.5	38.7	43.2	48.8	25.6	49.8
A	28.5	20.0	46.0	34.5	36.5	52.5	26.5	46.5

37. Hexavalent chromium has been identified as an inhalation carcinogen and an air toxin of concern in a number of different locales. The article **"Airborne Hexavalent Chromium in Southwestern Ontario"** (*J. of Air and Waste Mgmt. Assoc.*, **1997: 905–910**) gave the accompanying data on both indoor and outdoor concentration (nanograms/m³) for a sample of houses selected from a certain region.

				House					
	1	**2**	**3**	**4**	**5**	**6**	**7**	**8**	**9**
Indoor	.07	.08	.09	.12	.12	.12	.13	.14	.15
Outdoor	.29	.68	.47	.54	.97	.35	.49	.84	.86

				House				
	10	**11**	**12**	**13**	**14**	**15**	**16**	**17**
Indoor	.15	.17	.17	.18	.18	.18	.18	.19
Outdoor	.28	.32	.32	1.55	.66	.29	.21	1.02

				House				
	18	**19**	**20**	**21**	**22**	**23**	**24**	**25**
Indoor	.20	.22	.22	.23	.23	.25	.26	.28
Outdoor	1.59	.90	.52	.12	.54	.88	.49	1.24

				House				
	26	**27**	**28**	**29**	**30**	**31**	**32**	**33**
Indoor	.28	.29	.34	.39	.40	.45	.54	.62
Outdoor	.48	.27	.37	1.26	.70	.76	.99	.36

a. Calculate a confidence interval for the population mean difference between indoor and outdoor concentrations using a confidence level of 95%, and interpret the resulting interval.

b. If a 34th house were to be randomly selected from the population, between what values would you predict the difference in concentrations to lie?

38. Adding computerized medical images to a database promises to provide great resources for physicians.

However, there are other methods of obtaining such information, so the issue of efficiency of access needs to be investigated. The article **"The Comparative Effectiveness of Conventional and Digital Image Libraries"**(*J. of Audiovisual Media in Medicine*, 2001: 8–15) reported on an experiment in which 13 computer-proficient medical professionals were timed both while retrieving an image from a library of slides and while retrieving the same image from a computer database with a Web front end.

Subject	1	2	3	4	5	6	7
Slide	30	35	40	25	20	30	35
Digital	25	16	15	15	10	20	7
Difference	5	19	25	10	10	10	28

Subject	8	9	10	11	12	13
Slide	62	40	51	25	42	33
Digital	16	15	13	11	19	19
Difference	46	25	38	14	23	14

a. Construct a comparative boxplot of times for the two types of retrieval, and comment on any interesting features.

b. Estimate the difference between true average times for the two types of retrieval in a way that conveys information about precision and reliability. Be sure to check the plausibility of any assumptions needed in your analysis. Does it appear plausible that the true average times for the two types of retrieval are identical? Why or why not?

39. Scientists and engineers frequently wish to compare two different techniques for measuring or determining the value of a variable. In such situations, interest centers on testing whether the mean difference in measurements is zero. The article "**Evaluation of the Deuterium Dilution Technique Against the Test Weighing Procedure for the Determination of Breast Milk Intake**" (*Amer. J. of Clinical Nutr.*, 1983: 996–1003) reports the accompanying data on amount of milk ingested by each of 14 randomly selected infants.

	Infant				
	1	2	3	4	5
DD method	1509	1418	1561	1556	2169
TW method	1498	1254	1336	1565	2000
Difference	11	164	225	−9	169

	Infant				
	6	7	8	9	10
DD method	1760	1098	1198	1479	1281
TW method	1318	1410	1129	1342	1124
Difference	442	−312	69	137	157

	Infant			
	11	12	13	14
DD method	1414	1954	2174	2058
TW method	1468	1604	1722	1518
Difference	−54	350	452	540

a. Is it plausible that the population distribution of differences is normal?

b. Does it appear that the true average difference between intake values measured by the two methods is something other than zero? Determine the *P*-value of the test, and use it to reach a conclusion at significance level .05.

40. Lactation promotes a temporary loss of bone mass to provide adequate amounts of calcium for milk production. The paper **"Bone Mass Is Recovered from Lactation to Postweaning in Adolescent Mothers with Low Calcium Intakes"** (*Amer. J. of Clinical Nutr.*, 2004: 1322–1326) gave the following data on total body bone mineral content (TBBMC) (g) for a sample both during lactation (L) and in the postweaning period (P).

	Subject									
	1	2	3	4	5	6	7	8	9	10
L	1928	2549	2825	1924	1628	2175	2114	2621	1843	2541
P	2126	2885	2895	1942	1750	2184	2164	2626	2006	2627

a. Does the data suggest that true average total body bone mineral content during postweaning exceeds that during lactation by more than 25 g? State and test the appropriate hypotheses using a significance level of .05. [*Note:* The appropriate normal probability plot shows some curvature but not enough to cast substantial doubt on a normality assumption.]

b. Calculate an upper confidence bound using a 95% confidence level for the true average difference between TBBMC during postweaning and during lactation.

c. Does the (incorrect) use of the two-sample *t* test to test the hypotheses suggested in (a) lead to the same conclusion that you obtained there? Explain.

41. Antipsychotic drugs are widely prescribed for conditions such as schizophrenia and bipolar disease. The article **"Cardiometabolic Risk of Second-Generation Antipsychotic Medications During First-Time Use in Children and Adolescents"** (*J. of the Amer. Med. Assoc.*, 2009) reported on body composition and metabolic changes for individuals who had taken various antipsychotic drugs for short periods of time.

a. The sample of 41 individuals who had taken aripiprazole had a mean change in total cholesterol (mg/dL) of 3.75, and the estimated standard error $s_D/\sqrt{n}$ was

3.878. Calculate a confidence interval with confidence level approximately 95% for the true average increase in total cholesterol under these circumstances (the cited article included this CI).

b. The article also reported that for a sample of 36 individuals who had taken quetiapine, the sample mean cholesterol level change and estimated standard error were 9.05 and 4.256, respectively. Making any necessary assumptions about the distribution of change in cholesterol level, does the choice of significance level impact your conclusion as to whether true average cholesterol level increases? Explain. [*Note:* The article included a *P*-value.]

c. For the sample of 45 individuals who had taken olanzapine, the article reported (7.38, 9.69) as a 95% CI for true average weight gain (kg). What is a 99% CI?

42. Many freeways have service (or logo) signs that give information on attractions, camping, lodging, food, and gas services prior to off-ramps. These signs typically do not provide information on distances. The article **"Evaluation of Adding Distance Information to Freeway-Specific Service (Logo) Signs"** (*J. of Transp. Engr.*, **2011: 782–788**) reported that in one investigation, six sites along Virginia interstate highways where service signs are posted were selected. For each site, crash data was obtained for a three-year period before distance information was added to the service signs and for a one-year period afterward. The number of crashes per year before and after the sign changes were as follows:

Before:	15	26	66	115	62	64
After:	16	24	42	80	78	73

a. The cited article included the statement "A paired *t* test was performed to determine whether there was any change in the mean number of crashes before and after the addition of distance information on the signs." Carry out such a test. [*Note:* The relevant normal probability plot shows a substantial linear pattern.]

b. If a seventh site were to be randomly selected among locations bearing service signs, between what values would you predict the difference in number of crashes to lie?

43. Cushing's disease is characterized by muscular weakness due to adrenal or pituitary dysfunction. To provide effective treatment, it is important to detect childhood Cushing's disease as early as possible. Age at onset of symptoms and age at diagnosis (months) for 15 children suffering from the disease were given in the article **"Treatment of Cushing's Disease in Childhood and Adolescence by Transphenoidal Microadenomectomy"** (*New Engl. J. of Med.*, **1984: 889**). Here are the values of the differences between age at onset of symptoms and age at diagnosis:

−24	−12	−55	−15	−30	−60	−14	−21
−48	−12	−25	−53	−61	−69	−80	

a. Does the accompanying normal probability plot cast strong doubt on the approximate normality of the population distribution of differences?

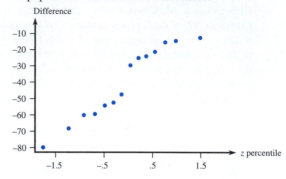

b. Calculate a lower 95% confidence bound for the population mean difference, and interpret the resulting bound.

c. Suppose the (age at diagnosis) – (age at onset) differences had been calculated. What would be a 95% upper confidence bound for the corresponding population mean difference?

44. Refer back to the previous exercise.

a. By far the most frequently tested null hypothesis when data is paired is $H_0: \mu_D = 0$. Is that a sensible hypothesis in this context? Explain.

b. Carry out a test of hypotheses to decide whether there is compelling evidence for concluding that on average diagnosis occurs more than 25 months after the onset of symptoms.

45. Torsion during hip external rotation (ER) and extension may be responsible for certain kinds of injuries in golfers and other athletes. The article **"Hip Rotational Velocities During the Full Golf Swing"** (*J. of Sports Science and Medicine*, **2009: 296–299**) reported on a study in which peak ER velocity and peak IR (internal rotation) velocity (both in deg·sec^{-1}) were determined for a sample of 15 female collegiate golfers during their swings. The following data was supplied by the article's authors.

Golfer	ER	IR	diff	z perc
1	−130.6	−98.9	−31.7	−1.28
2	−125.1	−115.9	−9.2	−0.97
3	−51.7	−161.6	109.9	0.34
4	−179.7	−196.9	17.2	−0.73
5	−130.5	−170.7	40.2	−0.34
6	−101.0	−274.9	173.9	0.97
7	−24.4	−275.0	250.6	1.83
8	−231.1	−275.7	44.6	−0.17
9	−186.8	−214.6	27.8	−0.52
10	−58.5	−117.8	59.3	0.00
11	−219.3	−326.7	107.4	0.17
12	−113.1	−272.9	159.8	0.73
13	−244.3	−429.1	184.8	1.28
14	−184.4	−140.6	−43.8	−1.83
15	−199.2	−345.6	146.4	0.52

a. Is it plausible that the differences came from a normally distributed population?

b. The article reported that mean ($\pm$ SD) $= -145.3(68.0)$ for ER velocity and $= -227.8(96.6)$ for IR velocity. Based just on this information, could a test of hypotheses about the difference between true average IR velocity and true average ER velocity be carried out? Explain.

c. The article stated that "The lead hip peak IR velocity was significantly greater than the trail hip ER velocity ($p = 0.003$, t value $= 3.65$)." (The phrasing suggests that an upper-tailed test was used.) Is that in fact the case? [*Note*: "$p = .033$" in Table 2 of the article is erroneous.]

46. Example 7.11 gave data on the modulus of elasticity obtained 1 minute after loading in a certain configuration. The cited article also gave the values of modulus of elasticity obtained 4 weeks after loading for the same lumber specimens. The data is presented here.

Observation	1 min	4 weeks	Difference
1	10,490	9,110	1380
2	16,620	13,250	3370
3	17,300	14,720	2580
4	15,480	12,740	2740
5	12,970	10,120	2850
6	17,260	14,570	2690
7	13,400	11,220	2180
8	13,900	11,100	2800
9	13,630	11,420	2210
10	13,260	10,910	2350
11	14,370	12,110	2260
12	11,700	8,620	3080
13	15,470	12,590	2880
14	17,840	15,090	2750
15	14,070	10,550	3520
16	14,760	12,230	2530

Calculate and interpret an upper confidence bound for the true average difference between 1-minute modulus and 4-week modulus; first check the plausibility of any necessary assumptions.

47. The article **"Slender High-Strength RC Columns Under Eccentric Compression"** (*Magazine of Concrete Res.*, **2005: 361–370**) gave the accompanying data on cylinder strength (MPa) for various types of columns cured under both moist conditions and laboratory drying conditions.

			Type			
	1	**2**	**3**	**4**	**5**	**6**
M:	82.6	87.1	89.5	88.8	94.3	80.0
LD:	86.9	87.3	92.0	89.3	91.4	85.9
	7	**8**	**9**	**10**	**11**	**12**
M:	86.7	92.5	97.8	90.4	94.6	91.6
LD:	89.4	91.8	94.3	92.0	93.1	91.3

a. Estimate the difference in true average strength under the two drying conditions in a way that conveys information about reliability and precision, and interpret the estimate. What does the estimate suggest about how true average strength under moist drying conditions compares to that under laboratory drying conditions?

b. Check the plausibility of any assumptions that underlie your analysis of (a).

48. Construct a paired data set for which $t = \infty$, so that the data is highly significant when the correct analysis is used, yet t for the two-sample t test is quite near zero, so the incorrect analysis yields an insignificant result.

9.4 Inferences Concerning a Difference Between Population Proportions

Having presented methods for comparing the means of two different populations, we now turn attention to the comparison of two population proportions. Regard an individual or object as a success S if he/she/it possesses some characteristic of interest (someone who graduated from college, a refrigerator with an icemaker, etc.). Let

$$p_1 = \text{the proportion of } S\text{'s in population \# 1}$$
$$p_2 = \text{the proportion of } S\text{'s in population \# 2}$$

Alternatively, $p_1(p_2)$ can be regarded as the probability that a randomly selected individual or object from the first (second) population is a success.

Suppose that a sample of size m is selected from the first population and independently a sample of size n is selected from the second one. Let X denote the number of S's in the first sample and Y be the number of S's in the second. Independence of the two samples implies that X and Y are independent. Provided that the two sample sizes are much smaller than the corresponding population sizes, X and Y can be regarded as having binomial distributions. The natural estimator for $p_1 - p_2$, the difference in population proportions, is the corresponding difference in sample proportions $X/m - Y/n$.

PROPOSITION

Let $\hat{p}_1 = X/m$ and $\hat{p}_2 = Y/n$, where $X \sim \text{Bin}(m, p_1)$ and $Y \sim \text{Bin}(n, p_2)$ with X and Y independent variables. Then

$$E(\hat{p}_1 - \hat{p}_2) = p_1 - p_2$$

so $\hat{p}_1 - \hat{p}_2$ is an unbiased estimator of $p_1 - p_2$, and

$$V(\hat{p}_1 - \hat{p}_2) = \frac{p_1 q_1}{m} + \frac{p_2 q_2}{n} \quad (\text{where } q_i = 1 - p_i) \tag{9.3}$$

Proof Since $E(X) = mp_1$ and $E(Y) = np_2$,

$$E\left(\frac{X}{m} - \frac{Y}{n}\right) = \frac{1}{m} E(X) - \frac{1}{n} E(Y) = \frac{1}{m} mp_1 - \frac{1}{n} np_2 = p_1 - p_2$$

Since $V(X) = mp_1 q_1$, $V(Y) = np_2 q_2$, and X and Y are independent,

$$V\left(\frac{X}{m} - \frac{Y}{n}\right) = V\left(\frac{X}{m}\right) + V\left(\frac{Y}{n}\right) = \frac{1}{m^2} V(X) + \frac{1}{n^2} V(Y) = \frac{p_1 q_1}{m} + \frac{p_2 q_2}{n} \quad \blacksquare$$

We will focus first on situations in which both m and n are large. Then because $\hat{p}_1$ and $\hat{p}_2$ individually have approximately normal distributions, the estimator $\hat{p}_1 - \hat{p}_2$ also has approximately a normal distribution. Standardizing $\hat{p}_1 - \hat{p}_2$ yields a variable Z whose distribution is approximately standard normal:

$$Z = \frac{\hat{p}_1 - \hat{p}_2 - (p_1 - p_2)}{\sqrt{\dfrac{p_1 q_1}{m} + \dfrac{p_2 q_2}{n}}}$$

A Large-Sample Test Procedure

The most general null hypothesis an investigator might consider would be of the form $H_0: p_1 - p_2 = \Delta_0$. Although for population means the case $\Delta_0 \neq 0$ presented no difficulties, for population proportions $\Delta_0 = 0$ and $\Delta_0 \neq 0$ must be considered separately. Since the vast majority of actual problems of this sort involve $\Delta_0 = 0$ (i.e., the null hypothesis $p_1 = p_2$), we'll concentrate on this case. When $H_0: p_1 - p_2 = 0$ is true, let p denote the common value of p_1 and p_2 (and similarly for q). Then the standardized variable

$$Z = \frac{\hat{p}_1 - \hat{p}_2 - 0}{\sqrt{pq\left(\dfrac{1}{m} + \dfrac{1}{n}\right)}} \tag{9.4}$$

has approximately a standard normal distribution when H_0 is true. However, this Z cannot serve as a test statistic because the value of p is unknown—H_0 asserts only that there is a common value of p, but does not say what that value is. A test statistic results from replacing p and q in (9.4) by appropriate estimators.

Assuming that $p_1 = p_2 = p$, instead of separate samples of size m and n from two different populations (two different binomial distributions), we really have a single sample of size $m + n$ from one population with proportion p. The total number of individuals in this combined sample having the characteristic of interest is $X + Y$. The natural estimator of p is then

$$\hat{p} = \frac{X + Y}{m + n} = \frac{m}{m + n} \cdot \hat{p}_1 + \frac{n}{m + n} \cdot \hat{p}_2 \qquad (9.5)$$

The second expression for $\hat{p}$ shows that it is actually a weighted average of estimators $\hat{p}_1$ and $\hat{p}_2$ obtained from the two samples. Using $\hat{p}$ and $\hat{q} = 1 - \hat{p}$ in place of p and q in (9.4) gives a test statistic having approximately a standard normal distribution when H_0 is true.

Null hypothesis: $H_0\colon p_1 - p_2 = 0$

Test statistic value (large samples): $z = \dfrac{\hat{p}_1 - \hat{p}_2}{\sqrt{\hat{p}\hat{q}\left(\dfrac{1}{m} + \dfrac{1}{n}\right)}}$

Alternative Hypothesis	P-Value Determination
$H_a\colon p_1 - p_2 > 0$	Area under the standard normal curve to the right of z
$H_a\colon p_1 - p_2 < 0$	Area under the standard normal curve to the left of z
$H_a\colon p_1 - p_2 \neq 0$	$2 \cdot$ (Area under the standard normal curve to the right of $|z|$)

The test can safely be used as long as $m\hat{p}_1$, $m\hat{q}_1$, $n\hat{p}_2$, and $n\hat{q}_2$ are all at least 10.

EXAMPLE 9.11   The article **"Aspirin Use and Survival After Diagnosis of Colorectal Cancer"** (***J. of the Amer. Med. Assoc.*, 2009: 649–658**) reported that of 549 study participants who regularly used aspirin after being diagnosed with colorectal cancer, there were 81 colorectal cancer-specific deaths, whereas among 730 similarly diagnosed individuals who did not subsequently use aspirin, there were 141 colorectal cancer-specific deaths. Does this data suggest that the regular use of aspirin after diagnosis will decrease the incidence rate of colorectal cancer-specific deaths? Let's test the appropriate hypotheses using a significance level of .05.

The parameter of interest is the difference $p_1 - p_2$, where p_1 is the true proportion of deaths for those who regularly used aspirin and p_2 is the true proportion of deaths for those who did not use aspirin. The use of aspirin is beneficial if $p_1 < p_2$, which corresponds to a negative difference between the two proportions. The relevant hypotheses are therefore

$$H_0\colon p_1 - p_2 = 0 \qquad \text{versus} \qquad H_a\colon p_1 - p_2 < 0$$

Parameter estimates are $\hat{p}_1 = 81/549 = .1475$, $\hat{p}_2 = 141/730 = .1932$, and $\hat{p} = (81 + 141)/(549 + 730) = .1736$. A z test is appropriate here because all of $m\hat{p}_1$, $m\hat{q}_1$, $n\hat{p}_2$, and $n\hat{q}_2$ are at least 10. The resulting test statistic value is

$$z = \frac{.1475 - .1932}{\sqrt{(.1736)(.8264)\left(\dfrac{1}{549} + \dfrac{1}{730}\right)}} = \frac{-.0457}{.021397} = -2.14$$

The corresponding P-value for a lower-tailed z test is $\Phi(-2.14) = .0162$. Because $.0162 \le .05$, the null hypothesis can be rejected at significance level .05. So anyone adopting this significance level would be convinced that the use of aspirin in these circumstances is beneficial. However, someone looking for more compelling evidence might select a significance level .01 and then not be persuaded. ∎

Type II Error Probabilities and Sample Sizes

Here the determination of β is a bit more cumbersome than it was for other large-sample tests. The reason is that the denominator of Z is an estimate of the standard deviation of $\hat{p} - \hat{p}_2$, assuming that $p_1 = p_2 = p$. When H_0 is false, $\hat{p}_1 - \hat{p}_2$ must be restandardized using

$$\sigma_{\hat{p}_1 - \hat{p}_2} = \sqrt{\frac{p_1 q_1}{m} + \frac{p_2 q_2}{n}} \tag{9.6}$$

The form of σ implies that β is not a function of just $p_1 - p_2$, so we denote it by $\beta(p_1, p_2)$.

Alternative Hypothesis	$\beta(p_1, p_2)$
$H_a:\ p_1 - p_2 > 0$	$\Phi\left[\dfrac{z_\alpha \sqrt{\bar{p}\,\bar{q}\left(\dfrac{1}{m} + \dfrac{1}{n}\right)} - (p_1 - p_2)}{\sigma}\right]$
$H_a:\ p_1 - p_2 < 0$	$1 - \Phi\left[\dfrac{-z_\alpha \sqrt{\bar{p}\,\bar{q}\left(\dfrac{1}{m} + \dfrac{1}{n}\right)} - (p_1 - p_2)}{\sigma}\right]$
$H_a:\ p_1 - p_2 \ne 0$	$\Phi\left[\dfrac{z_{\alpha/2} \sqrt{\bar{p}\,\bar{q}\left(\dfrac{1}{m} + \dfrac{1}{n}\right)} - (p_1 - p_2)}{\sigma}\right]$
	$-\Phi\left[\dfrac{-z_{\alpha/2} \sqrt{\bar{p}\,\bar{q}\left(\dfrac{1}{m} + \dfrac{1}{n}\right)} - (p_1 - p_2)}{\sigma}\right]$

where $\bar{p} = (mp_1 + np_2)/(m + n)$, $\bar{q} = (mq_1 + nq_2)/(m + n)$, and σ is given by (9.6).

Proof For the upper-tailed test ($H_a: p_1 - p_2 > 0$),

$$\beta(p_1, p_2) = P\left[\hat{p}_1 - \hat{p}_2 < z_\alpha \sqrt{\hat{p}\hat{q}\left(\frac{1}{m} + \frac{1}{n}\right)}\right]$$

$$= P\left[\frac{(\hat{p}_1 - \hat{p}_2 - (p_1 - p_2))}{\sigma} < \frac{z_a \sqrt{\hat{p}\hat{q}\left(\dfrac{1}{m} + \dfrac{1}{n}\right)} - (p_1 - p_2)}{\sigma}\right]$$

When m and n are both large,

$$\hat{p} = (m\hat{p}_1 + n\hat{p}_2)/(m + n) \approx (mp_1 + np_2)/(m + n) = \bar{p}$$

and $\hat{q} \approx \bar{q}$, which yields the previous (approximate) expression for $\beta(p_1, p_2)$. ■

Alternatively, for specified p_1, p_2 with $p_1 - p_2 = d$, the sample sizes necessary to achieve $\beta(p_1, p_2) = \beta$ can be determined. For example, for the upper-tailed test, we equate $-z_\beta$ to the argument of $\Phi(\cdot)$ (i.e., what's inside the parentheses) in the foregoing box. If $m = n$, there is a simple expression for the common value.

For the case $m = n$, the level α test has type II error probability β at the alternative values p_1, p_2 with $p_1 - p_2 = d$ when

$$n = \frac{\left[z_\alpha\sqrt{(p_1 + p_2)(q_1 + q_2)/2} + z_\beta\sqrt{p_1 q_1 + p_2 q_2}\right]^2}{d^2} \quad (9.7)$$

for an upper- or lower-tailed test, with $\alpha/2$ replacing α for a two-tailed test.

EXAMPLE 9.12 One of the truly impressive applications of statistics occurred in connection with the design of the 1954 Salk polio-vaccine experiment and analysis of the resulting data. Part of the experiment focused on the efficacy of the vaccine in combating paralytic polio. Because it was thought that without a control group of children, there would be no sound basis for assessment of the vaccine, it was decided to administer the vaccine to one group and a placebo injection (visually indistinguishable from the vaccine but known to have no effect) to a control group. For ethical reasons and also because it was thought that the knowledge of vaccine administration might have an effect on treatment and diagnosis, the experiment was conducted in a **double-blind** manner. That is, neither the individuals receiving injections nor those administering them actually knew who was receiving vaccine and who was receiving the placebo (samples were numerically coded). (Remember: at that point it was not at all clear whether the vaccine was beneficial.)

Let p_1 and p_2 be the probabilities of a child getting paralytic polio for the control and treatment conditions, respectively. The objective was to test $H_0: p_1 - p_2 = 0$ versus $H_a: p_1 - p_2 > 0$ (the alternative states that a vaccinated child is less likely to contract polio than an unvaccinated child). Supposing the true value of p_1 is .0003 (an incidence rate of 30 per 100,000), the vaccine would be a significant improvement if the incidence rate was halved—that is, $p_2 = .00015$. Using a level $\alpha = .05$ test, it would then be reasonable to ask for sample sizes for which $\beta = .1$ when $p_1 = .0003$ and $p_2 = .00015$. Assuming equal sample sizes, the required n is obtained from (9.7) as

$$n = \frac{\left[1.645\sqrt{(.5)(.00045)(1.99955)} + 1.28\sqrt{(.00015)(.99985) + (.0003)(.9997)}\right]^2}{(.0003 - .00015)^2}$$

$$= [(.0349 + .0271)/.00015]^2 \approx 171{,}000$$

The actual data for this experiment follows. Sample sizes of approximately 200,000 were used. The reader can easily verify that $z = 6.43$—a highly significant value. The vaccine was judged a resounding success!

Placebo: $m = 201{,}229$, $x =$ number of cases of paralytic polio $= 110$

Vaccine: $n = 200{,}745$, $y = 33$ ■

A Large-Sample Confidence Interval

As with means, many two-sample problems involve the objective of comparison through hypothesis testing, but sometimes an interval estimate for $p_1 - p_2$ is appropriate. Both $\hat{p}_1 = X/m$ and $\hat{p}_2 = Y/n$ have approximate normal distributions when m and n are both large. If we identify θ with $p_1 - p_2$, then $\hat{\theta} = \hat{p}_1 - \hat{p}_2$ satisfies the conditions necessary for obtaining a large-sample CI. In particular, the estimated standard deviation of $\hat{\theta}$ is $\sqrt{(\hat{p}_1\hat{q}_1/m) + (\hat{p}_2\hat{q}_2/n)}$. The general $100(1 - \alpha)\%$ interval $\hat{\theta} \pm z_{\alpha/2} \cdot \hat{\sigma}_{\hat{\theta}}$ then takes the following form.

> A CI for $p_1 - p_2$ with confidence level approximately $100(1 - \alpha)\%$ is
>
> $$\hat{p}_1 - \hat{p}_2 \pm z_{\alpha/2}\sqrt{\frac{\hat{p}_1\hat{q}_1}{m} + \frac{\hat{p}_2\hat{q}_2}{n}}$$
>
> This interval can safely be used as long as $m\hat{p}_1$, $m\hat{q}_1$, $n\hat{p}_2$, and $n\hat{q}_2$ are all at least 10.

Notice that the estimated standard deviation of $\hat{p}_1 - \hat{p}_2$ (the square-root expression) is different here from what it was for hypothesis testing when $\Delta_0 = 0$.

Recent research has shown that the actual confidence level for the traditional CI just given can sometimes deviate substantially from the nominal level (the level you think you are getting when you use a particular z critical value—e.g., 95% when $z_{\alpha/2} = 1.96$). The suggested improvement is to add one success and one failure to each of the two samples and then replace the $\hat{p}$'s and $\hat{q}$'s in the foregoing formula by $\tilde{p}$'s and $\tilde{q}$'s where $\tilde{p}_1 = (x + 1)/(m + 2)$, etc. This modified interval can also be used when sample sizes are quite small.

EXAMPLE 9.13 Do people who work long hours have more trouble sleeping? An investigation into this issue was described in the article **"Long Working Hours and Sleep Disturbances: The Whitehall II Prospective Cohort Study"** (*Sleep*, **2009: 737–745**). In one sample of 1501 British civil servants who worked more than 40 hours a week, 750 said they usually get less than 7 hours of sleep per night. In another sample of 958 British civil servants who worked between 35 and 40 hours per week, 407 said they usually get less than 7 hours of sleep per night. The investigators believed that these samples were representative of the populations to which they belong.

Let p_1 denote the proportion of British civil servants working more than 40 hours per week who usually get less than 7 hours of sleep per night, and let p_2 be the corresponding proportion for the 35–40 hours population. The point estimates of p_1 and p_2 are

$$\hat{p}_1 = \frac{750}{1501} = .500, \quad \hat{p}_2 = \frac{407}{958} = .425$$

from which $\hat{q}_1 = .500$, $\hat{q}_2 = .575$. All quantities $m\hat{p}_1$, $m\hat{q}_1$, $n\hat{p}_2$, $n\hat{q}_2$ are much larger than 10, so the large-sample CI for $p_1 - p_2$ can be used. The 99% interval is

$$.500 - .425 \pm 2.58\sqrt{\frac{(.500)(.500)}{1501} + \frac{(.425)(.575)}{958}} = .075 \pm (2.58)(.020534)$$

$$= 0.75 \pm 0.53 = (.022, .128)$$

At the 99% confidence level, we estimate that the proportion of those working longer hours who usually get less than 7 hours of sleep per night exceeds the corresponding proportion for those who work fewer hours by between .022 and .128. The fact that this interval includes only positive values suggests that those who work longer hours tend to get less sleep. But the study is observational rather than randomized controlled, so it would be dangerous to infer a causal relationship between work hours and amount of sleep. Because of the large sample sizes, the modified interval that uses $\tilde{p}_1, \tilde{q}_1, \tilde{p}_2,$ and $\tilde{q}_2$ is identical to the one we calculated. ∎

Small-Sample Inferences

On occasion an inference concerning $p_1 - p_2$ may have to be based on samples for which at least one sample size is small. Appropriate methods for such situations are not as straightforward as those for large samples, and there is more controversy among statisticians as to recommended procedures. One frequently used test, called the Fisher–Irwin test, is based on the hypergeometric distribution. Your friendly neighborhood statistician can be consulted for more information.

EXERCISES Section 9.4 (49–58)

49. Consider the following two questions designed to assess quantitative literacy:
  a. What is 15% of 1000?
  b. A store is offering a 15% off sale on all TVs. The most popular television is normally priced at $1000. How much money would a customer save on the television during this sale?

Suppose the first question is asked of 200 randomly selected college students, with 164 answering correctly; the second one is asked of a different random sample of 200 college students, resulting in 140 correct responses (the sample percentages agree with those given in the article **"Using the Right Yardstick: Assessing Financial Literacy Measures by Way of Financial Well-Being,"** *J. of Consumer Affairs*, **2013: 243–262**; the investigators found that those who answered such questions correctly, particularly questions with context, were significantly more successful in their investment decisions than those who did not answer correctly). Carry out a test of hypotheses at significance level .05 to decide if the true proportion of correct responses to the question without context exceeds that for the one with context.

50. Recent incidents of food contamination have caused great concern among consumers. The article **"How Safe Is That Chicken?"** (*Consumer Reports*, **Jan. 2010: 19–23**) reported that 35 of 80 randomly selected Perdue brand broilers tested positively for either campylobacter or salmonella (or both), the leading bacterial causes of food-borne disease, whereas 66 of 80 Tyson brand broilers tested positive.
  a. Does it appear that the true proportion of non-contaminated Perdue broilers differs from that for the Tyson brand? Carry out a test of hypotheses using a significance level .01.

  b. If the true proportions of non-contaminated chickens for the Perdue and Tyson brands are .50 and .25, respectively, how likely is it that the null hypothesis of equal proportions will be rejected when a .01 significance level is used and the sample sizes are both 80?

51. It is well known that a placebo, a fake medication or treatment, can sometimes have a positive effect just because patients often expect the medication or treatment to be helpful. The article **"Beware the Nocebo Effect"** (*New York Times*, **Aug. 12, 2012**) gave examples of a less familiar phenomenon, the tendency for patients informed of possible side effects to actually experience those side effects. The article cited a study reported in *The Journal of Sexual Medicine* in which a group of patients diagnosed with benign prostatic hyperplasia was randomly divided into two subgroups. One subgroup of size 55 received a compound of proven efficacy along with counseling that a potential side effect of the treatment was erectile dysfunction. The other subgroup of size 52 was given the same treatment without counseling. The percentage of the no-counseling subgroup that reported one or more sexual side effects was 15.3%, whereas 43.6% of the counseling subgroup reported at least one sexual side effect. State and test the appropriate hypotheses at significance level .05 to decide whether the nocebo effect is operating here. [*Note:* The estimated expected number of "successes" in the no-counseling sample is a bit shy of 10, but not by enough to be of great concern (some sources use a less conservative cutoff of 5 rather than 10).]

52. Do teachers find their work rewarding and satisfying? The article **"Work-Related Attitudes"** (*Psychological Reports*, **1991: 443–450**) reports the results of a survey of 395 elementary school teachers and 266 high school teachers. Of the elementary school teachers, 224 said they were very satisfied with their jobs, whereas 126 of the high school teachers were very satisfied with their work. Estimate the difference between the proportion of all elementary school teachers who are very satisfied and all high school teachers who are very satisfied by calculating and interpreting a CI.

53. Olestra is a fat substitute approved by the FDA for use in snack foods. Because there have been anecdotal reports of gastrointestinal problems associated with olestra consumption, a randomized, double-blind, placebo-controlled experiment was carried out to compare olestra potato chips to regular potato chips with respect to GI symptoms (**"Gastrointestinal Symptoms Following Consumption of Olestra or Regular Triglyceride Potato Chips,"** *J. of the Amer. Med. Assoc.*, **1998: 150–152).** Among 529 individuals in the TG control group, 17.6% experienced an adverse GI event, whereas among the 563 individuals in the olestra treatment group, 15.8% experienced such an event.

a. Carry out a test of hypotheses at the 5% significance level to decide whether the incidence rate of GI problems for those who consume olestra chips according to the experimental regimen differs from the incidence rate for the TG control treatment.

b. If the true percentages for the two treatments were 15% and 20%, respectively, what sample sizes ($m = n$) would be necessary to detect such a difference with probability .90?

54. Teen Court is a juvenile diversion program designed to circumvent the formal processing of first-time juvenile offenders within the juvenile justice system. The article **"An Experimental Evaluation of Teen Courts"** (*J. of Experimental Criminology*, **2008: 137–163**) reported on a study in which offenders were randomly assigned either to Teen Court or to the traditional Department of Juvenile Services method of processing. Of the 56 TC individuals, 18 subsequently recidivated (look it up!) during the 18-month follow-up period, whereas 12 of the 51 DJS individuals did so. Does the data suggest that the true proportion of TC individuals who recidivate during the specified follow-up period differs from the proportion of DJS individuals who do so? State and test the relevant hypotheses using a significance level of .10.

55. In medical investigations, the ratio $\theta = p_1/p_2$ is often of more interest than the difference $p_1 - p_2$ (e.g., individuals given treatment 1 are how many times as likely to recover as those given treatment 2?). Let $\hat{\theta} = \hat{p}_1/\hat{p}_2$. When m and n are both large, the statistic $\ln(\hat{\theta})$ has approximately a normal distribution with approximate mean value $\ln(\theta)$ and approximate standard deviation $[(m - x)/(mx) + (n - y)/(ny)]^{1/2}$.

a. Use these facts to obtain a large-sample 95% CI formula for estimating $\ln(\theta)$, and then a CI for θ itself.

b. Return to the heart-attack data of Example 1.3, and calculate an interval of plausible values for θ at the 95% confidence level. What does this interval suggest about the efficacy of the aspirin treatment?

56. Sometimes experiments involving success or failure responses are run in a paired or before/after manner. Suppose that before a major policy speech by a political candidate, n individuals are selected and asked whether (S) or not (F) they favor the candidate. Then after the speech the same n people are asked the same question. The responses can be entered in a table as follows:

		After	
		S	F
Before	S	x_1	x_2
	F	x_3	x_4

where $x_1 + x_2 + x_3 + x_4 = n$. Let $p_1, p_2, p_3,$ and p_4 denote the four cell probabilities, so that $p_1 = P(S$ before and S after), and so on. We wish to test the hypothesis that the true proportion of supporters (S) after the speech has not increased against the alternative that it has increased.

a. State the two hypotheses of interest in terms of $p_1, p_2, p_3,$ and p_4.

b. Construct an estimator for the after/before difference in success probabilities.

c. When n is large, it can be shown that the rv $(X_i - X_j)/n$ has approximately a normal distribution with variance given by $[p_i + p_j - (p_i - p_j)^2]/n$. Use this to construct a test statistic with approximately a standard normal distribution when H_0 is true (the result is called McNemar's test).

d. If $x_1 = 350$, $x_2 = 150$, $x_3 = 200$, and $x_4 = 300$, what do you conclude?

57. Two different types of alloy, A and B, have been used to manufacture experimental specimens of a small tension link to be used in a certain engineering application. The ultimate strength (ksi) of each specimen was determined, and the results are summarized in the accompanying frequency distribution.

	A	**B**
26 – < 30	6	4
30 – < 34	12	9
34 – < 38	15	19
38 – < 42	7	10
	$m = 40$	$m = 42$

Compute a 95% CI for the difference between the true proportions of all specimens of alloys A and B that have an ultimate strength of at least 34 ksi.

58. Using the traditional formula, a 95% CI for $p_1 - p_2$ is to be constructed based on equal sample sizes from the two populations. For what value of $n (= m)$ will the resulting interval have a width at most of .1, irrespective of the results of the sampling?

9.5 Inferences Concerning Two Population Variances

Methods for comparing two population variances (or standard deviations) are occasionally needed, though such problems arise much less frequently than those involving means or proportions. For the case in which the populations under investigation are normal, the procedures are based on a new family of probability distributions.

The *F* Distribution

The F probability distribution has two parameters, denoted by v_1 and v_2. The parameter v_1 is called the *number of numerator degrees of freedom,* and v_2 is the *number of denominator degrees of freedom;* here v_1 and v_2 are positive integers. A random variable that has an F distribution cannot assume a negative value. Since the density function is complicated and will not be used explicitly, we omit the formula. There is an important connection between an F variable and chi-squared variables. If X_1 and X_2 are independent chi-squared rv's with v_1 and v_2 df, respectively, then the rv

$$F = \frac{X_1/v_1}{X_2/v_2} \qquad (9.8)$$

(the ratio of the two chi-squared variables divided by their respective degrees of freedom), can be shown to have an F distribution.

Figure 9.7 illustrates the graph of a typical F density function. Analogous to the notation $t_{\alpha,v}$ and $\chi^2_{\alpha,v}$, we use F_{α,v_1,v_2} for the value on the horizontal axis that captures α of the area under the F density curve with v_1 and v_2 df in the upper tail. The density curve is not symmetric, so it would seem that both upper- and lower-tail critical values must be tabulated. This is not necessary, though, because of the fact that $F_{1-\alpha,v_1,v_2} = 1/F_{\alpha,v_2,v_1}$.

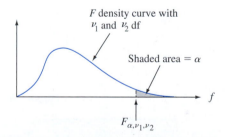

Figure 9.7 An *F* density curve and critical value

Appendix Table A.9 gives F_{α,v_1,v_2} for $\alpha = .10,\ .05,\ .01,$ and $.001$, and various values of v_1 (in different columns of the table) and v_2 (in different groups of rows of the table). For example, $F_{.05,6,10} = 3.22$ and $F_{.05,10,6} = 4.06$. The critical value $F_{.95,6,10}$, which captures .95 of the area to its right (and thus .05 to the left) under the F curve with $v_1 = 6$ and $v_2 = 10$, is $F_{.95,6,10} = 1/F_{.05,10,6} = 1/4.06 = .246$.

The *F* Test for Equality of Variances

A test procedure for hypotheses concerning the ratio σ_1^2/σ_2^2 is based on the following result.

THEOREM

Let $X_1, \ldots, X_m$ be a random sample from a normal distribution with variance σ_1^2, let $Y_1, \ldots, Y_n$ be another random sample (independent of the X_i's) from a normal distribution with variance σ_2^2, and let S_1^2 and S_2^2 denote the two sample variances. Then the rv

$$F = \frac{S_1^2/\sigma_1^2}{S_2^2/\sigma_2^2} \qquad (9.9)$$

has an F distribution with $v_1 = m - 1$ and $v_2 = n - 1$.

This theorem results from combining (9.8) with the fact that the variables $(m - 1)S_1^2/\sigma_1^2$ and $(n - 1)S_2^2/\sigma_2^2$ each have a chi-squared distribution with $m - 1$ and $n - 1$ df, respectively (see Section 7.4). Because F involves a ratio rather than a difference, the test statistic is the ratio of sample variances. The claim that $\sigma_1^2 = \sigma_2^2$ is implausible if the ratio differs by too much from 1.

Recall that the P-value for an upper-tailed t test is the area under an appropriate t curve to the right of the calculated t, whereas for a lower-tailed test the P-value is the area under the curve to the left of t. Analogously, the P-value for an upper-tailed F test is the area under an appropriate F curve (the one with specified numerator and denominator dfs) to the right of f, and the P-value for a lower-tailed test is the area under an F curve to the left of f. Because t curves are symmetric, the P-value for a two-tailed test is double the captured lower tail area if t is negative and double the captured upper tail area if t is positive. Although F curves are not symmetric, by analogy the P-value for a two-tailed F test is twice the captured lower tail area if f is below the median and twice the captured upper tail area if it is above the median. Figure 9.8 illustrates this for an upper-tailed test based on $v_1 = 4$ and $v_2 = 6$.

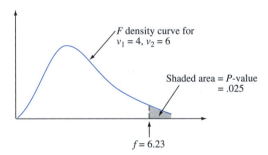

Figure 9.8 A P-value for an upper-tailed F test

Null hypothesis: $\quad H_0: \sigma_1^2 = \sigma_2^2$

Test statistic value: $\quad f = s_1^2/s_2^2$

Alternative Hypothesis	P-Value Determination
$H_a: \sigma_1^2 > \sigma_2^2$	A_R = Area under the $F_{m-1, n-1}$ curve to the right of f
$H_a: \sigma_1^2 < \sigma_2^2$	A_L = Area under the $F_{m-1, n-1}$ curve to the left of f
$H_a: \sigma_1^2 \neq \sigma_2^2$	$2 \cdot \min(A_R, A_L)$

Assumption: The population distributions are both normal, and the two random samples are independent of one another.

Tabulation of F-curve upper-tail areas is much more cumbersome than for t curves because two df's are involved. For each combination of ν_1 and ν_2, our F table gives only the four critical values that capture areas .10, .05, .01, and .001. Because of this, the table will generally provide only an upper or lower bound (or both) on the P-value. For example, suppose the test is upper-tailed and based on 4 numerator df and 6 denominator df. If $f = 5.82$, then the P-value is the area under the $F_{4,6}$ curve to the right of 5.82. Because $F_{.05,4,6} = 4.53$, the area to the right of 4.53 is by definition .05. Similarly, $F_{.01,4,6} = 9.15$ implies that the area under the curve to the right of this value is .01. Since 5.82 lies in between 4.53 and 9.15, the area to the right of 5.82 must be between .01 and .05. That is, $.01 < P\text{-value} < .05$. Figure 9.9 shows what can be said about the P-value depending on where f falls relative to the four relevant tabulated critical values.

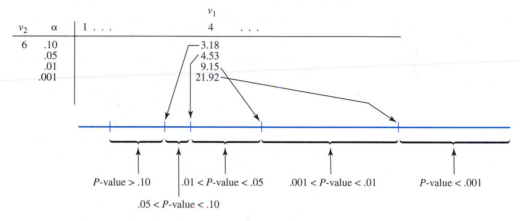

Figure 9.9   Obtaining *P*-value information from the *F* table for an upper-tailed *F* test

Again considering a test with $\nu_1 = 4$ and $\nu_2 = 6$,

$$f = 5.82 \implies .01 < P\text{-value} < .05$$
$$f = 2.16 \implies P\text{-value} > .10$$
$$f = 25.03 \implies P\text{-value} < .001$$

Only if f equals a tabulated value do we obtain an exact P-value (e.g., if $f = 4.53$, then P-value $= .05$). Once we know that $.01 < P\text{-value} < .05$, H_0 would be rejected at a significance level of .05 but not at a level of .01. When $P\text{-value} < .001$, H_0 should be rejected at any reasonable significance level.

The F tests discussed in succeeding chapters will all be upper-tailed. If, however, a lower-tailed F test is appropriate, then lower-tailed critical values should be obtained as described earlier so that a bound or bounds on the P-value can be established. In the case of a two-tailed test, the bound or bounds from a one-tailed test should be multiplied by 2. For example, if $f = 5.82$ when $\nu_1 = 4$ and $\nu_2 = 6$, then since 5.82 falls between the .05 and .01 critical values, $2(.01) < P\text{-value} < 2(.05)$, giving $.02 < P\text{-value} < .10$. H_0 would then be rejected if $\alpha = .10$ but not if $\alpha = .01$. In this case, we cannot say from our table what conclusion is appropriate when $\alpha = .05$ (since we don't know whether the P-value is smaller or larger than this). However, statistical software shows that the area to the right of 5.82 under this F curve is .029, so the P-value is .058 and the null hypothesis should therefore not be rejected at level .05. Various statistical software packages will, of course, provide an exact P-value for any F test.

EXAMPLE 9.14 A random sample of 200 vehicles traveling on gravel roads in a county with a posted speed limit of 35 mph on such roads resulted in a sample mean speed of 37.5 mph and a sample standard deviation of 8.6 mph, whereas another random sample of 200 vehicles in a county with a posted speed limit of 55 mph resulted in a sample mean and sample standard deviation of 35.8 mph and 9.2 mph, respectively (these means and standard deviations were reported in the article **"Evaluation of Criteria for Setting Speed Limits on Gravel Roads"** (*J. of Transp. Engr.*, **2011: 57–63**); the actual sample sizes result in dfs that exceed the largest of those in our F table). Let's carry out a test at significance level .10 to decide whether the two population distribution variances are identical.

1. σ_1^2 is the variance of the speed distribution on the 35 mph roads, and σ_2^2 is the variance of the speed distribution on 55 mph roads.

2. $H_0: \sigma_1^2 = \sigma_2^2$

3. $H_a: \sigma_1^2 \neq \sigma_2^2$

4. Test statistic value: $f = s_1^2/s_2^2$

5. Calculation: $f = (8.6)^2/(9.2)^2 = .87$

6. *P*-value determination: .87 lies in the lower tail of the F curve with 199 numerator df and 199 denominator df. A glance at the F table shows that $F_{.10,199,199} \approx F_{.10,200,200} \approx 1.20$ (consult the $v_1 = 120$ and $v_1 = 1000$ columns), implying $F_{.90,199,199} \approx 1/1.20 = .83$ (these values are confirmed by software). That is, the area under the relevant F curve to the left of .83 is .10. Thus the area under the curve to the left of .87 exceeds .10, and so *P*-value $> 2(.10) = .2$ (software gives .342).

7. The *P*-value clearly exceeds the mandated significance level. The null hypothesis therefore cannot be rejected; it is plausible that the two speed distribution variances are identical.

The sample sizes in the cited article were 2665 and 1868, respectively, and the *P*-value reported there was .0008. So for the actual data, the hypothesis of equal variances would be rejected not only at significance level .10—in contrast to our conclusion—but also at level .05, .01, and even .001. This illustrates again how quite large sample sizes can magnify a small difference in estimated values. Note also that the sample mean speed for the county with the lower posted speed limit was higher than for the county with the lower limit, a counterintuitive result that surprised the investigators; and because of the very large sample sizes, this difference in means is highly statistically significant. ∎

A Confidence Interval for σ_1/σ_2

The CI for σ_1^2/σ_2^2 is based on replacing F in the probability statement

$$P(F_{1-\alpha/2,v_1,v_2} < F < F_{\alpha/2,v_1,v_2}) = 1 - \alpha$$

by the F variable (9.9) and manipulating the inequalities to isolate σ_1^2/σ_2^2. An interval for σ_1/σ_2 results from taking the square root of each limit. The details are left for an exercise.

EXERCISES Section 9.5 (59–66)

59. Obtain or compute the following quantities:
   a. $F_{.05,5,8}$ b. $F_{.05,8,5}$ c. $F_{.95,5,8}$ d. $F_{.95,8,5}$
   e. The 99th percentile of the F distribution with $v_1 = 10$, $v_2 = 12$

   f. The 1st percentile of the F distribution with $v_1 = 10$, $v_2 = 12$
   g. $P(F \leq 6.16)$ for $v_1 = 6$, $v_2 = 4$
   h. $P(.177 \leq F \leq 4.74)$ for $v_1 = 10$, $v_2 = 5$

60. Give as much information as you can about the P-value of the F test in each of the following situations:
   a. $v_1 = 5$, $v_2 = 10$, upper-tailed test, $f = 4.75$
   b. $v_1 = 5$, $v_2 = 10$, upper-tailed test, $f = 2.00$
   c. $v_1 = 5$, $v_2 = 10$, two-tailed test, $f = 5.64$
   d. $v_1 = 5$, $v_2 = 10$, lower-tailed test, $f = .200$
   e. $v_1 = 35$, $v_2 = 20$, upper-tailed test, $f = 3.24$

61. Return to the data on maximum lean angle given in Exercise 28 of this chapter. Carry out a test at significance level .10 to see whether the population standard deviations for the two age groups are different (normal probability plots support the necessary normality assumption).

62. Refer to Example 9.7. Does the data suggest that the standard deviation of the strength distribution for fused specimens is smaller than that for not-fused specimens? Carry out a test at significance level .01.

63. Toxaphene is an insecticide that has been identified as a pollutant in the Great Lakes ecosystem. To investigate the effect of toxaphene exposure on animals, groups of rats were given toxaphene in their diet. The article **"Reproduction Study of Toxaphene in the Rat"** (*J. of Environ. Sci. Health*, **1988: 101–126**) reports weight gains (in grams) for rats given a low dose (4 ppm) and for control rats whose diet did not include the insecticide. The sample standard deviation for 23 female control rats was 32 g and for 20 female low-dose rats was 54 g. Does this data suggest that there is more variability in low-dose weight gains than in control weight gains? Assuming normality, carry out a test of hypotheses at significance level .05.

64. The following observations are on time (h) for a AA 1.5-volt alkaline battery to reach a 0.8 voltage (**"Comparing the Lifetimes of Two Brands of Batteries,"** *J. of Statistical Educ.*, **2013, online**):

Energizer:	8.65	8.74	8.91	8.72	8.85
Ultracell:	8.76	8.81	8.81	8.70	8.73
Energizer:	8.52	8.62	8.68	8.86	
Ultracell:	8.76	8.68	8.64	8.79	

Normal probability plots support the assumption that the population distributions are normal. Does the data suggest that the variance of the Energizer population distribution differs from that of the Ultracell population distribution? Test the relevant hypotheses using a significance level of .05. [*Note:* The two-sample t test for equality of population means gives a P-value of .763.] The Energizer batteries are much more expensive than the Ultracell batteries. Would you pay the extra money?

65. The article **"Enhancement of Compressive Properties of Failed Concrete Cylinders with Polymer Impregnation"** (*J. of Testing and Evaluation*, **1977: 333–337**) reports the following data on impregnated compressive modulus (psi $\times 10^6$) when two different polymers were used to repair cracks in failed concrete.

Epoxy	1.75	2.12	2.05	1.97
MMA prepolymer	1.77	1.59	1.70	1.69

Obtain a 90% CI for the ratio of variances by first using the method suggested in the text to obtain a general confidence interval formula.

66. Reconsider the data of Example 9.6, and calculate a 95% upper confidence bound for the ratio of the standard deviation of the triacetate porosity distribution to that of the cotton porosity distribution.

SUPPLEMENTARY EXERCISES (67–95)

67. The accompanying summary data on compression strength (lb) for $12 \times 10 \times 8$ in. boxes appeared in the article **"Compression of Single-Wall Corrugated Shipping Containers Using Fixed and Floating Test Platens"** (*J. Testing and Evaluation*, **1992: 318–320**). The authors stated that "the difference between the compression strength using fixed and floating platen method was found to be small compared to normal variation in compression strength between identical boxes." Do you agree? Is your analysis predicated on any assumptions?

Method	Sample Size	Sample Mean	Sample SD
Fixed	10	807	27
Floating	10	757	41

68. The article **"Supervised Exercise Versus Non-Supervised Exercise for Reducing Weight in Obese Adults"** (*The J. of Sports Med. and Physical Fitness*, **2009: 85–90**) reported on an investigation in which participants were randomly assigned to either a supervised exercise program or a control group. Those in the control group were told only that they should take measures to lose weight. After 4 months, the sample mean decrease in body fat for the 17 individuals in the experimental group was 6.2 kg with a sample standard deviation of 4.5 kg, whereas the sample mean and sample standard deviation for the 17 people in the control group were 1.7 kg and 3.1 kg, respectively. Assume normality of the two weight-loss distributions (as did the investigators).
   a. Calculate a 99% lower prediction bound for the weight loss of a single randomly selected individual subjected to the supervised exercise program. Can

you be highly confident that such an individual will actually lose weight?

b. Does it appear that true average decrease in body fat is more than two kg larger for the experimental condition than for the control condition? Use the accompanying Minitab output to reach a conclusion at significance level of .01. [*Note:* Minitab accepts such summary data as well as individual observations. Also, because the test is upper-tailed, the software provides a lower confidence bound rather than a conventional CI.]

```
Sample      N     Mean    StDev    SE Mean
Exptl.      17    6.20    4.50     1.1
Control     17    1.70    3.10     0.75

Difference = mu (1) − mu (2)
Estimate for difference:  4.50
95% lower bound for difference: 2.25
T-Test of difference = 2 (vs >):
T-Value = 1.89
P-Value = 0.035  DF = 28
```

69. Is the response rate for questionnaires affected by including some sort of incentive to respond along with the questionnaire? In one experiment, 110 questionnaires with no incentive resulted in 75 being returned, whereas 98 questionnaires that included a chance to win a lottery yielded 66 responses (**"Charities, No; Lotteries, No; Cash, Yes,"** *Public Opinion Quarterly*, 1996: 542–562). Does this data suggest that including an incentive increases the likelihood of a response? State and test the relevant hypotheses at significance level .10.

70. Shoveling is not exactly a high-tech activity, but it will continue to be a required task even in our information age. The article **"A Shovel with a Perforated Blade Reduces Energy Expenditure Required for Digging Wet Clay"** (*Human Factors*, 2010: 492–502) reported on an experiment in which 13 workers were each provided with both a conventional shovel and a shovel whose blade was perforated with small holes. The authors of the cited article provided the following data on stable energy expenditure [(kcal/kg(subject)/lb(clay)]:

Worker:	1	2	3	4
Conventional:	.0011	.0014	.0018	.0022
Perforated:	.0011	.0010	.0019	.0013

Worker:	5	6	7
Conventional	.0010	.0016	.0028
Perforated:	.0011	.0017	.0024

Worker:	8	9	10
Conventional:	.0020	.0015	.0014
Perforated:	.0020	.0013	.0013

Worker:	11	12	13
Conventional:	.0023	.0017	.0020
Perforated:	.0017	.0015	.0013

a. Calculate a confidence interval at the 95% confidence level for the true average difference between energy expenditure for the conventional shovel and the perforated shovel (the relevant normal probability plot shows a reasonably linear pattern). Based on this interval, does it appear that the shovels differ with respect to true average energy expenditure? Explain.

b. Carry out a test of hypotheses at significance level .05 to see if true average energy expenditure using the conventional shovel exceeds that using the perforated shovel.

71. The article **"Quantitative MRI and Electrophysiology of Preoperative Carpal Tunnel Syndrome in a Female Population"** (*Ergonomics*, 1997: 642–649) reported that $(-473.13, 1691.9)$ was a large-sample 95% confidence interval for the difference between true average thenar muscle volume (mm^3) for sufferers of carpal tunnel syndrome and true average volume for nonsufferers. Calculate and interpret a 90% confidence interval for this difference.

72. The following summary data on bending strength (lb-in/in) of joints is taken from the article **"Bending Strength of Corner Joints Constructed with Injection Molded Splines"** (*Forest Products J.*, April, 1997: 89–92).

Type	Sample Size	Sample Mean	Sample SD
Without side coating	10	80.95	9.59
With side coating	10	63.23	5.96

a. Calculate a 95% lower confidence bound for true average strength of joints with a side coating.

b. Calculate a 95% lower prediction bound for the strength of a single joint with a side coating.

c. Calculate an interval that, with 95% confidence, includes the strength values for at least 95% of the population of all joints with side coatings.

d. Calculate a 95% confidence interval for the difference between true average strengths for the two types of joints.

73. The article **"Urban Battery Litter"** cited in Example 8.14 gave the following summary data on zinc mass (g) for two different brands of size D batteries:

Brand	Sample Size	Sample Mean	Sample SD
Duracell	15	138.52	7.76
Energizer	20	149.07	1.52

Assuming that both zinc mass distributions are at least approximately normal, carry out a test at significance level .05 to decide whether true average zinc mass is different for the two types of batteries.

74. The derailment of a freight train due to the catastrophic failure of a traction motor armature bearing provided the impetus for a study reported in the article **"Locomotive Traction Motor Armature Bearing Life Study"** (*Lubrication Engr.*, Aug. 1997: 12–19). A sample of 17

high-mileage traction motors was selected, and the amount of cone penetration (mm/10) was determined both for the pinion bearing and for the commutator armature bearing, resulting in the following data:

	Motor					
	1	**2**	**3**	**4**	**5**	**6**
Commutator	211	273	305	258	270	209
Pinion	226	278	259	244	273	236

	Motor					
	7	**8**	**9**	**10**	**11**	**12**
Commutator	223	288	296	233	262	291
Pinion	290	287	315	242	288	242

	Motor				
	13	**14**	**15**	**16**	**17**
Commutator	278	275	210	272	264
Pinion	278	208	281	274	268

Calculate an estimate of the population mean difference between penetration for the commutator armature bearing and penetration for the pinion bearing, and do so in a way that conveys information about the reliability and precision of the estimate. [*Note:* A normal probability plot validates the necessary normality assumption.] Would you say that the population mean difference has been precisely estimated? Does it look as though population mean penetration differs for the two types of bearings? Explain.

75. *Headability* is the ability of a cylindrical piece of material to be shaped into the head of a bolt, screw, or other cold-formed part without cracking. The article **"New Methods for Assessing Cold Heading Quality"** (*Wire J. Intl.*, Oct. 1996: 66–72) described the result of a headability impact test applied to 30 specimens of aluminum killed steel and 30 specimens of silicon killed steel. The sample mean headability rating number for the steel specimens was 6.43, and the sample mean for aluminum specimens was 7.09. Suppose that the sample standard deviations were 1.08 and 1.19, respectively. Do you agree with the article's authors that the difference in headability ratings is significant at the 5% level (assuming that the two headability distributions are normal)?

76. The article **"Fatigue Testing of Condoms"** cited in Exercise 7.32 reported that for a sample of 20 natural latex condoms of a certain type, the sample mean and sample standard deviation of the number of cycles to break were 4358 and 2218, respectively, whereas a sample of 20 polyisoprene condoms gave a sample mean and sample standard deviation of 5805 and 3990, respectively. Is there strong evidence for concluding that true average number of cycles to break for the polyisoprene condom exceeds that for the natural latex condom by more than 1000 cycles? Carry out a test using a

significance level of .01. [*Note:* The cited paper reported *P*-values of *t* tests for comparing means of the various types considered.]

77. Information about hand posture and forces generated by the fingers during manipulation of various daily objects is needed for designing high-tech hand prosthetic devices. The article **"Grip Posture and Forces During Holding Cylindrical Objects with Circular Grips"** (*Ergonomics*, **1996: 1163–1176**) reported that for a sample of 11 females, the sample mean four-finger pinch strength (N) was 98.1 and the sample standard deviation was 14.2. For a sample of 15 males, the sample mean and sample standard deviation were 129.2 and 39.1, respectively.

   a. A test carried out to see whether true average strengths for the two genders were different resulted in $t = 2.51$ and *P*-value = .019. Does the appropriate test procedure described in this chapter yield this value of *t* and the stated *P*-value?

   b. Is there substantial evidence for concluding that true average strength for males exceeds that for females by more than 25 N? State and test the relevant hypotheses.

78. The article **"Pine Needles as Sensors of Atmospheric Pollution"** (*Environ. Monitoring*, **1982: 273–286**) reported on the use of neutron-activity analysis to determine pollutant concentration in pine needles. According to the article's authors, "These observations strongly indicated that for those elements which are determined well by the analytical procedures, the distribution of concentration is lognormal. Accordingly, in tests of significance the logarithms of concentrations will be used." The given data refers to bromine concentration in needles taken from a site near an oil-fired steam plant and from a relatively clean site. The summary values are means and standard deviations of the log-transformed observations.

Site	Sample Size	Mean Log Concentration	SD of Log Concentration
Steam plant	8	18.0	4.9
Clean	9	11.0	4.6

Let μ_1^* be the true average *log* concentration at the first site, and define μ_2^* analogously for the second site.

   a. Use the pooled *t* test (based on assuming normality *and* equal standard deviations) to decide at significance level .05 whether the two concentration distribution means are equal.

   b. If σ_1^* and σ_2^* (the standard deviations of the two log concentration distributions) are not equal, would μ_1 and μ_2 (the means of the concentration distributions) be the same if $\mu_1^* = \mu_2^*$? Explain your reasoning.

79. The article **"The Accuracy of Stated Energy Contents of Reduced-Energy, Commercially Prepared Foods"** (*J. of the Amer. Dietetic Assoc.*, **2010: 116–123**) presented the accompanying data on vendor-stated gross energy and measured value (both in kcal) for 10 different supermarket convenience meals):

Meal:	1	2	3	4	5	6	7	8	9	10
Stated:	180	220	190	230	200	370	250	240	80	180
Measured:	212	319	231	306	211	431	288	265	145	228

Carry out a test of hypotheses to decide whether the true average % difference from that stated differs from zero. [*Note*: The article stated "Although formal statistical methods do not apply to convenience samples, standard statistical tests were employed to summarize the data for exploratory purposes and to suggest directions for future studies."]

80. Arsenic is a known carcinogen and poison. The standard laboratory procedures for measuring arsenic concentration (μg/L) in water are expensive. Consider the accompanying summary data and Minitab output for comparing a laboratory method to a new relatively quick and inexpensive field method (from the article **"Evaluation of a New Field Measurement Method for Arsenic in Drinking Water Samples," *J. of Envir. Engr.*, 2008: 382–388**).

Two-Sample T-Test and CI

```
Sample    N      Mean    StDev    SE Mean
1         3      19.70   1.10     0.64
2         3      10.90   0.60     0.35
Estimate for difference: 8.800

95% CI for difference: (6.498, 11.102)
T-Test of difference = 0 (vs not =):
T-Value = 12.16 P-Value = 0.001 DF = 3
```

What conclusion do you draw about the two methods, and why? Interpret the given confidence interval. [*Note*: One of the article's authors indicated in private communication that they were unsure why the two methods disagreed.]

81. The accompanying data on response time appeared in the article **"The Extinguishment of Fires Using Low-Flow Water Hose Streams—Part II"** (*Fire Technology*, **1991: 291–320**).

Good visibility

      .43 1.17 .37 .47 .68 .58 .50 2.75

Poor visibility

    1.47 .80 1.58 1.53 4.33 4.23 3.25 3.22

The authors analyzed the data with the pooled *t* test. Does the use of this test appear justified? [*Hint*: Check for normality. The *z* percentiles for $n = 8$ are -1.53, $-.89$, $-.49$, $-.15$, .15, .49, .89, and 1.53.]

82. Acrylic bone cement is commonly used in total joint arthroplasty as a grout that allows for the smooth transfer of loads from a metal prosthesis to bone structure. The paper **"Validation of the Small-Punch Test as a Technique for Characterizing the Mechanical Properties of Acrylic Bone Cement"** (*J. of Engr. in Med.*, **2006: 11–21**) gave the following data on breaking force (N):

Temp	Medium	n	$\bar{x}$	s
22°	Dry	6	170.60	39.08
37°	Dry	6	325.73	34.97
22°	Wet	6	366.36	34.82
37°	Wet	6	306.09	41.97

Assume that all population distributions are normal.

a. Estimate true average breaking force in a dry medium at 37° in a way that conveys information about reliability and precision, and interpret your estimate.

b. Estimate the difference between true average breaking force in a dry medium at 37° and true average force at the same temperature in a wet medium, and do so in a way that conveys information about precision and reliability. Then interpret your estimate.

c. Is there strong evidence for concluding that true average force in a dry medium at the higher temperature exceeds that at the lower temperature by more than 100 N?

83. In an experiment to compare bearing strengths of pegs inserted in two different types of mounts, a sample of 14 observations on stress limit for red oak mounts resulted in a sample mean and sample standard deviation of 8.48 MPa and .79 MPa, respectively, whereas a sample of 12 observations when Douglas fir mounts were used gave a mean of 9.36 and a standard deviation of 1.52 (**"Bearing Strength of White Oak Pegs in Red Oak and Douglas Fir Timbers,"** *J. of Testing and Evaluation*, **1998, 109–114**). Consider testing whether or not true average stress limits are identical for the two types of mounts. Compare df's and *P*-values for the unpooled and pooled *t* tests.

84. How does energy intake compare to energy expenditure? One aspect of this issue was considered in the article **"Measurement of Total Energy Expenditure by the Doubly Labelled Water Method in Professional Soccer Players"** (*J. of Sports Sciences*, **2002: 391–397**), which contained the accompanying data (MJ/day).

	Player						
	1	**2**	**3**	**4**	**5**	**6**	**7**
Expenditure	14.4	12.1	14.3	14.2	15.2	15.5	17.8
Intake	14.6	9.2	11.8	11.6	12.7	15.0	16.3

Test to see whether there is a significant difference between intake and expenditure. Does the conclusion depend on whether a significance level of .05, .01, or .001 is used?

85. An experimenter wishes to obtain a CI for the difference between true average breaking strength for cables manufactured by company I and by company II. Suppose breaking strength is normally distributed for both types of cable with $\sigma_1 = 30$ psi and $\sigma_2 = 20$ psi.

a. If costs dictate that the sample size for the type I cable should be three times the sample size for the type II cable, how many observations are required if the 99% CI is to be no wider than 20 psi?

b. Suppose a total of 400 observations is to be made. How many of the observations should be made on type I cable samples if the width of the resulting interval is to be a minimum?

86. A study was carried out to compare two different methods, injection and nasal spray, for administering flu vaccine to children under the age of 5. All 8000 children in the study were given both an injection and a spray. However, the vaccine given to 4000 of the children actually contained just saltwater, and the spray given to the other 4000 children also contained just saltwater. At the end of the flu season, it was determined that 3.9% of the children who received the real vaccine via nasal spray contracted the flu, whereas 8.6% of the 4000 children receiving the real vaccine via injection contracted the flu.

a. Why do you think each child received both an injection and a spray?

b. Does one method for delivering the vaccine appear to be superior to the other? Test the appropriate hypotheses. [*Note:* The study was described in the article **"Spray Flu Vaccine May Work Better Than Injections for Tots,"** *San Luis Obispo Tribune,* **May 2, 2006**.]

87. Wait staff at restaurants have employed various strategies to increase tips. An article in the **Sept. 5, 2005,** *New Yorker* reported that "In one study a waitress received 50% more in tips when she introduced herself by name than when she didn't." Consider the following (fictitious) data on tip amount as a percentage of the bill:

Introduction: $m = 50$ $\bar{x} = 22.63$ $s_1 = 7.82$

No introduction: $n = 50$ $\bar{y} = 14.15$ $s_2 = 6.10$

Does this data suggest that an introduction increases tips on average by more than 50%? State and test the relevant hypotheses. [*Hint*: Consider the parameter $\theta = \mu_1 - 1.5\mu_2$.]

88. The paper **"Quantitative Assessment of Glenohumeral Translation in Baseball Players"** (*The Amer. J. of Sports Med.,* **2004: 1711–1715**) considered various aspects of shoulder motion for a sample of pitchers and another sample of position players [*glenohumeral* refers to the articulation between the humerus (ball) and the glenoid (socket)]. The authors kindly supplied the following data on anteroposterior translation (mm), a measure of the extent of anterior and posterior motion, both for the dominant arm and the nondominant arm.

	Pos Dom Tr	Pos ND Tr	Pit Dom Tr	Pit ND Tr
1	30.31	32.54	27.63	24.33
2	44.86	40.95	30.57	26.36
3	22.09	23.48	32.62	30.62
4	31.26	31.11	39.79	33.74
5	28.07	28.75	28.50	29.84
6	31.93	29.32	26.70	26.71
7	34.68	34.79	30.34	26.45
8	29.10	28.87	28.69	21.49
9	25.51	27.59	31.19	20.82

	Pos Dom Tr	Pos ND Tr	Pit Dom Tr	Pit ND Tr
10	22.49	21.01	36.00	21.75
11	28.74	30.31	31.58	28.32
12	27.89	27.92	32.55	27.22
13	28.48	27.85	29.56	28.86
14	25.60	24.95	28.64	28.58
15	20.21	21.59	28.58	27.15
16	33.77	32.48	31.99	29.46
17	32.59	32.48	27.16	21.26
18	32.60	31.61		
19	29.30	27.46		
mean	29.4463	29.2137	30.7112	26.6447
sd	5.4655	4.7013	3.3310	3.6679

a. Estimate the true average difference in translation between dominant and nondominant arms for pitchers in a way that conveys information about reliability and precision, and interpret the resulting estimate.

b. Repeat (a) for position players.

c. The authors asserted that "pitchers have greater difference in side-to-side anteroposterior translation of their shoulders compared with position players." Do you agree? Explain.

89. Suppose a level .05 test of $H_0: \mu_1 - \mu_2 = 0$ versus $H_a: \mu_1 - \mu_2 > 0$ is to be performed, assuming $\sigma_1 = \sigma_2 = 10$ and normality of both distributions, using equal sample sizes ($m = n$). Evaluate the probability of a type II error when $\mu_1 - \mu_2 = 1$ and $n = 25, 100, 2500$, and $10,000$. Can you think of real problems in which the difference $\mu_1 - \mu_2 = 1$ has little practical significance? Would sample sizes of $n = 10,000$ be desirable in such problems?

90. The invasive diatom species *Didymosphenia geminata* has the potential to inflict substantial ecological and economic damage in rivers. The article **"Substrate Characteristics Affect Colonization by the Bloom-Forming Diatom *Didymosphenia geminata"** (*Acquatic Ecology,* **2010: 33–40**) described an investigation of colonization behavior. One aspect of particular interest was whether the roughness of stones impacted the degree of colonization. The authors of the cited article kindly provided the accompanying data on roughness ratio (dimensionless) for specimens of sandstone and shale.

Sandstone:	5.74	2.07	3.29	0.75	1.23	
	2.95	1.58	1.83	1.61	1.12	
	2.91	3.22	2.84	1.97	2.48	
	3.45	2.17	0.77	1.44	3.79	

Shale:	.56	.84	.40	.55	.36	.72
	.29	.47	.66	.48	.28	
	.72	.31	.35	.32	.37	.43
	.60	.54	.43	.51		

Normal probability plots of both samples show a reasonably linear pattern. Estimate the difference between true average roughness for sandstone and that for shale in a way that provides information about reliability and precision, and interpret your estimate. Does it appear that true average roughness differs for the two types of rocks (a formal test of this was reported in the article)? [*Note:* The investigators concluded that more diatoms colonized the rougher surface than the smoother surface.]

91. Researchers sent 5000 resumes in response to job ads that appeared in the *Boston Globe* and *Chicago Tribune*. The resumes were identical except that 2500 of them had "white sounding" first names, such as Brett and Emily, whereas the other 2500 had "black sounding" names such as Tamika and Rasheed. The resumes of the first type elicited 250 responses and the resumes of the second type only 167 responses (these numbers are very consistent with information that appeared in a **Jan. 15, 2003**, report by the **Associated Press**). Does this data strongly suggest that a resume with a "black" name is less likely to result in a response than is a resume with a "white" name?

92. McNemar's test, developed in Exercise 56, can also be used when individuals are paired (matched) to yield n pairs and then one member of each pair is given treatment 1 and the other is given treatment 2. Then X_1 is the number of pairs in which both treatments were successful, and similarly for X_2, X_3, and X_4. The test statistic for testing equal efficacy of the two treatments is given by $(X_2 - X_3)/\sqrt{(X_2 + X_3)}$, which has approximately a standard normal distribution when H_0 is true. Use this to test whether the drug ergotamine is effective in the treatment of migraine headaches.

		Ergotamine	
		S	**F**
Placebo	**S**	44	34
	F	46	30

The data is fictitious, but the conclusion agrees with that in the article "**Controlled Clinical Trial of Ergotamine Tartrate**" (*British Med. J.*, 1970: 325–327).

93. The article **"Evaluating Variability in Filling Operations"** (*Food Tech.*, 1984: 51–55) describes two different filling operations used in a ground-beef packing plant. Both filling operations were set to fill packages with 1400 g of ground beef. In a random sample of size 30 taken from each filling operation, the resulting means and standard deviations were 1402.24 g and 10.97 g for operation 1 and 1419.63 g and 9.96 g for operation 2.

a. Using a .05 significance level, is there sufficient evidence to indicate that the true mean weight of the packages differs for the two operations?

b. Does the data from operation 1 suggest that the true mean weight of packages produced by operation 1 is higher than 1400 g? Use a .05 significance level.

94. Let $X_1, \ldots, X_m$ be a random sample from a Poisson distribution with parameter μ_1, and let $Y_1, \ldots, Y_n$ be a random sample from another Poisson distribution with parameter μ_2. We wish to test $H_0: \mu_1 - \mu_2 = 0$ against one of the three standard alternatives. When m and n are large, the large-sample z test of Section 9.1 can be used. However, the fact that $V(\overline{X}) = \mu/n$ suggests that a different denominator should be used in standardizing $\overline{X} - \overline{Y}$. Develop a large-sample test procedure appropriate to this problem, and then apply it to the following data to test whether the plant densities for a particular species are equal in two different regions (where each observation is the number of plants found in a randomly located square sampling quadrate having area 1 m², so for region 1 there were 40 quadrates in which one plant was observed, etc.):

				Frequency					
	0	**1**	**2**	**3**	**4**	**5**	**6**	**7**	
Region 1	28	40	28	17	8	2	1	1	$m = 125$
Region 2	14	25	30	18	49	2	1	1	$n = 140$

95. Referring to Exercise 94, develop a large-sample confidence interval formula for $\mu_1 - \mu_2$. Calculate the interval for the data given there using a confidence level of 95%.

BIBLIOGRAPHY

See the bibliography at the end of Chapter 7.

The Analysis of Variance

INTRODUCTION

In studying methods for the analysis of quantitative data, we first focused on problems involving a single sample of numbers and then turned to a comparative analysis of two such different samples. In one-sample problems, the data consisted of observations on or responses from individuals or experimental objects randomly selected from a single population. In two-sample problems, either the two samples were drawn from two different populations and the parameters of interest were the population means, or else two different treatments were applied to experimental units (individuals or objects) selected from a single population; in this latter case, the parameters of interest are referred to as true treatment means.

The **analysis of variance**, or more briefly **ANOVA**, refers broadly to a collection of experimental situations and statistical procedures for the analysis of quantitative responses from experimental units. The simplest ANOVA problem is referred to variously as a **single-factor, single-classification**, or **one-way ANOVA**. It involves the analysis either of data sampled from more than two numerical populations (distributions) or of data from experiments in which more than two treatments have been used. The characteristic that differentiates the treatments or populations from one another is called the **factor** under study, and the different treatments or populations are referred to as the **levels** of the factor. Examples of such situations include the following:

1. An experiment to study the effects of five different brands of gasoline on automobile engine operating efficiency (mpg)

2. An experiment to study the effects of the presence of four different sugar solutions (glucose, sucrose, fructose, and a mixture of the three) on bacterial growth

3. An experiment to investigate whether hardwood concentration in pulp (%) at three different levels impacts tensile strength of bags made from the pulp

4. An experiment to decide whether the color density of fabric specimens depends on which of four different dye amounts is used

In (1) the factor of interest is gasoline brand, and there are five different levels of the factor. In (2) the factor is sugar, with four levels (or five, if a control solution containing no sugar is used). In both (1) and (2), the factor is qualitative in nature, and the levels correspond to possible categories of the factor. In (3) and (4), the factors are concentration of hardwood and amount of dye, respectively; both these factors are quantitative in nature, so the levels identify different settings of the factor. When the factor of interest is quantitative, statistical techniques from regression analysis (discussed in Chapters 12 and 13) can also be used to analyze the data.

This chapter focuses on single-factor ANOVA. Section 10.1 presents the F test for testing the null hypothesis that the population or treatment means are identical. Section 10.2 considers further analysis of the data when H_0 has been rejected. Section 10.3 covers some other aspects of single-factor ANOVA. Chapter 11 introduces ANOVA experiments involving more than a single factor.

10.1 Single-Factor ANOVA

Single-factor ANOVA focuses on a comparison of more than two population or treatment means. Let

I = the number of populations or treatments being compared

μ_1 = the mean of population 1 or the true average response when treatment 1 is applied

$\vdots$

μ_I = the mean of population I or the true average response when treatment I is applied

The relevant hypotheses are

$H_0: \mu_1 = \mu_2 = \cdots = \mu_I$

versus

$H_a:$ at least two the of the μ_i's are different

If $I = 4$, H_0 is true only if all four μ_i's are identical. H_a would be true, for example, if $\mu_1 = \mu_2 \neq \mu_3 = \mu_4$, if $\mu_1 = \mu_3 = \mu_4 \neq \mu_2$, or if all four μ_i's differ from one another. A test of these hypotheses requires that we have available a random sample from each population or treatment.

EXAMPLE 10.1 The article **"Compression of Single-Wall Corrugated Shipping Containers Using Fixed and Floating Test Platens"** (*J. Testing and Evaluation*, 1992: 318–320) describes an experiment in which several different types of boxes were compared with respect to compression strength (lb). Table 10.1 presents the results of a single-factor ANOVA experiment involving $I = 4$ types of boxes (the sample means and standard deviations are in good agreement with values given in the article).

Table 10.1 The Data and Summary Quantities for Example 10.1

Type of Box	Compression Strength (lb)	Sample Mean	Sample SD
1	655.5 788.3 734.3 721.4 679.1 699.4	713.00	46.55
2	789.2 772.5 786.9 686.1 732.1 774.8	756.93	40.34
3	737.1 639.0 696.3 671.7 717.2 727.1	698.07	37.20
4	535.1 628.7 542.4 559.0 586.9 520.0	562.02	39.87
	Grand mean =	682.50	

With μ_i denoting the true average compression strength for boxes of type i ($i = 1, 2, 3, 4$), the null hypothesis is $H_0: \mu_1 = \mu_2 = \mu_3 = \mu_4$. Figure 10.1(a) shows a comparative boxplot for the four samples. There is a substantial amount of overlap among observations on the first three types of boxes, but compression strengths for the fourth type appear considerably smaller than for the other types. This suggests that H_0 is not true. The comparative boxplot in Figure 10.1(b) is based on adding 120 to each observation in the fourth sample (giving mean 682.02 and the same standard deviation) and leaving the other observations unaltered. It is no longer obvious whether H_0 is true or false. In situations such as this, we need a formal test procedure.

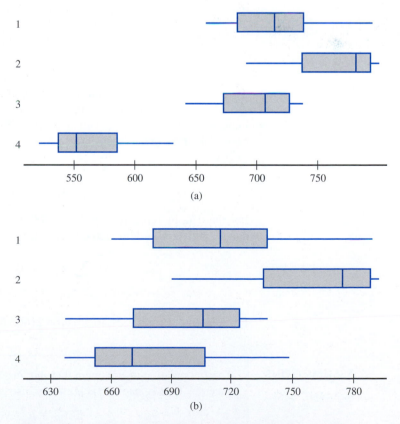

Figure 10.1 Boxplots for Example 10.1: (a) original data; (b) altered data

Notation and Assumptions

The letters X and Y were used in two-sample problems to differentiate the observations in one sample from those in the other. Because this is cumbersome for three or more samples, it is customary to use a single letter with two subscripts. The first subscript identifies the sample number, corresponding to the population or treatment being sampled, and the second subscript denotes the position of the observation within that sample. Let

$X_{i,j}$ = the random variable (rv) that denotes the jth measurement taken from the ith population, or the measurement taken on the jth experimental unit that receives the ith treatment

$x_{i,j}$ = the observed value of $X_{i,j}$ when the experiment is performed

The observed data is usually displayed in a rectangular table, such as Table 10.1. There samples from the different populations appear in different rows of the table, and $x_{i,j}$ is the jth number in the ith row. For example, $x_{2,3} = 786.9$ (the third observation from the second population), and $x_{4,1} = 535.1$. When there is no ambiguity, we will write x_{ij} rather than $x_{i,j}$ (e.g., if there were 15 observations on each of 12 treatments, x_{112} could mean $x_{1,12}$ or $x_{11,2}$). It is assumed that the X_{ij}'s within any particular sample are independent—a random sample from the ith population or treatment distribution—and that different samples are independent of one another.

In some experiments, different samples contain different numbers of observations. Here we'll focus on the case of equal sample sizes; the generalization to unequal sample sizes appears in Section 10.3. Let J denote the number of observations in each sample ($J = 6$ in Example 10.1). The data set consists of IJ observations. The individual sample means will be denoted by $\overline{X}_{1.}, \overline{X}_{2.}, \ldots, \overline{X}_{I.}$. That is,

$$\overline{X}_{i.} = \frac{\sum_{j=1}^{J} X_{ij}}{J} \quad i = 1, 2, \ldots, I$$

The dot in place of the second subscript signifies that we have added over all values of that subscript while holding the other subscript value fixed, and the horizontal bar indicates division by J to obtain an average. Similarly, the average of all IJ observations, called the **grand mean**, is

$$\overline{X}_{..} = \frac{\sum_{i=1}^{I} \sum_{j=1}^{J} X_{ij}}{IJ}$$

For the data in Table 10.1, $\overline{x}_{1.} = 713.00$, $\overline{x}_{2.} = 756.93$, $\overline{x}_{3.} = 698.07$, $\overline{x}_{4.} = 562.02$, and $\overline{x}_{..} = 682.50$. Additionally, let $S_1^2, S_2^2, \ldots, S_I^2$, denote the sample variances:

$$S_i^2 = \frac{\sum_{j=1}^{J} (X_{ij} - \overline{X}_{i.})^2}{J - 1} \quad i = 1, 2, \ldots, I$$

From Example 10.1, $s_1 = 46.55$, $s_1^2 = 2166.90$, and so on.

ASSUMPTIONS

The I population or treatment distributions are all normal with the same variance σ^2. That is, each X_{ij} is normally distributed with

$$E(X_{ij}) = \mu_i \quad V(X_{ij}) = \sigma^2$$

The I sample standard deviations will generally differ somewhat even when the corresponding σ's are identical. In Example 10.1, the largest among s_1, s_2, s_3, and s_4 is about 1.25 times the smallest. A rough rule of thumb is that if the largest s is not much more than two times the smallest, it is reasonable to assume equal σ^2's.

In previous chapters, a normal probability plot was suggested for checking normality. The individual sample sizes in ANOVA are often too small for I separate plots to be informative. A single plot can be constructed by subtracting $\bar{x}_{1\cdot}$ from each observation in the first sample, $\bar{x}_{2\cdot}$ from each observation in the second, and so on, and then plotting these IJ deviations against the z percentiles. Figure 10.2 gives such a plot for the data of Example 10.1. The straightness of the pattern gives strong support to the normality assumption.

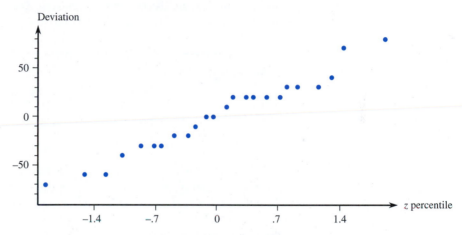

Figure 10.2 A normal probability plot based on the data of Example 10.1

If either the normality assumption or the assumption of equal variances is judged implausible, a method of analysis other than the usual F test must be employed. Please seek expert advice in such situations (one possibility, a data transformation, is suggested in Section 10.3, and another alternative is developed in Section 15.4).

The Test Statistic

If H_0 is true, the J observations in each sample come from a normal population distribution with the *same* mean value μ, in which case the sample means $\bar{x}_{1\cdot}, \ldots, \bar{x}_{I\cdot}$ should be reasonably close to one another. The test procedure is based on comparing a measure of differences among the $\bar{x}_{i\cdot}$'s ("between-samples" variation) to a measure of variation calculated from *within* each of the samples.

DEFINITION

Mean square for treatments is given by

$$\text{MSTr} = \frac{J}{I-1}[(\overline{X}_{1\cdot} - \overline{X}_{\cdot\cdot})^2 + (\overline{X}_{2\cdot} - \overline{X}_{\cdot\cdot})^2 + \cdots + (\overline{X}_{I\cdot} - \overline{X}_{\cdot\cdot})^2]$$

$$= \frac{J}{I-1}\sum_i (\overline{X}_{i\cdot} - \overline{X}_{\cdot\cdot})^2$$

and **mean square for error** is

$$\text{MSE} = \frac{S_1^2 + S_2^2 + \cdots + S_I^2}{I}$$

The test statistic for single-factor ANOVA is $F = \text{MSTr}/\text{MSE}$.

The terminology "mean square" will be explained shortly. Notice that uppercase $\overline{X}$'s and S^2's are used, so MSTr and MSE are defined as statistics. We will follow tradition and also use MSTr and MSE (rather than mstr and mse) to denote the calculated values of these statistics. Each S_i^2 assesses variation within a particular sample, so MSE is a measure of within-samples variation.

What kind of value of F provides evidence for or against H_0? If H_0 is true (all μ_i's are equal), the values of the individual sample means should be close to one another and therefore close to the grand mean, resulting in a relatively small value of MSTr. However, if the μ_i's are quite different, some $\bar{x}_{i\cdot}$'s should differ quite a bit from $\bar{x}..$. So the value of MSTr is affected by the status of H_0 (true or false). This is not the case with MSE, because the s_i^2's depend only on the underlying value of σ^2 and not on where the various distributions are centered. The following box presents an important property of $E(\text{MSTr})$ and $E(\text{MSE})$, the expected values of these two statistics.

PROPOSITION

When H_0 is true,

$$E(\text{MSTr}) = E(\text{MSE}) = \sigma^2$$

whereas when H_0 is false,

$$E(\text{MSTr}) > E(\text{MSE}) = \sigma^2$$

That is, both statistics are unbiased for estimating the common population variance σ^2 when H_0 is true, but MSTr tends to overestimate σ^2 when H_0 is false.

The unbiasedness of MSE is a consequence of $E(S_i^2) = \sigma^2$ whether H_0 is true or false. When H_0 is true, each $\overline{X}_{i\cdot}$ has the same mean value μ and variance σ^2/J, so $\Sigma(\overline{X}_{i\cdot} - \overline{X}..)^2/(I - 1)$, the "sample variance" of the $\overline{X}_{i\cdot}$'s, estimates σ^2/J unbiasedly; multiplying this by J gives MSTr as an unbiased estimator of σ^2 itself. The $\overline{X}_{i\cdot}$'s tend to spread out more when H_0 is false than when it is true, tending to inflate the value of MSTr in this case. Thus a value of F that greatly exceeds 1, corresponding to an MSTr much larger than MSE, casts considerable doubt on H_0. Determination of the P-value requires that the distribution of F when H_0 is true be known.

F Distributions and the F Test

In Section 9.5, we introduced a family of probability distributions called F distributions. An F distribution arises in connection with a ratio in which there is one number of degrees of freedom (df) associated with the numerator and another number of degrees of freedom associated with the denominator. Let ν_1 and ν_2 denote the number of numerator and denominator degrees of freedom, respectively, for a variable with an F distribution. Both ν_1 and ν_2 are positive integers. Figure 10.3 pictures

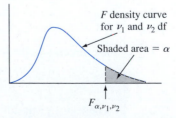

Figure 10.3   An F density curve and critical value F_{α,ν_1,ν_2}

an F density curve and the corresponding upper-tail critical value F_{α, ν_1, ν_2}. Appendix Table A.9 gives these critical values for $\alpha = .10, .05, .01,$ and $.001$. Values of ν_1 are identified with different columns of the table, and the rows are labeled with various values of ν_2. For example, the F critical value that captures upper-tail area $.05$ under the F curve with $\nu_1 = 4$ and $\nu_2 = 6$ is $F_{.05,4,6} = 4.53$, whereas $F_{.05,6,4} = 6.16$. The key theoretical result is that the test statistic F has an F distribution when H_0 is true.

THEOREM

Let $F = \text{MSTr}/\text{MSE}$ be the test statistic in a single-factor ANOVA problem involving I populations or treatments with a random sample of J observations from each one. When H_0 is true and the basic assumptions of this section are satisfied, F has an F distribution with $\nu_1 = I - 1$ and $\nu_2 = I(J - 1)$. Because a larger f is more contradictory to H_0 than a smaller f, the test is upper-tailed:

$$P\text{-value} = P(F \geq f \text{ when } H_0 \text{ is true})$$
$$= \text{area under the } F_{I-1, I(J-1)} \text{ curve to the right of } f$$

Statistical software will provide an exact P-value. Refer to Section 9.5 for a description of how our book's table of F critical values, Table A.9, can be used to obtain an upper or lower bound (or both) on the P-value.

The rationale for $\nu_1 = I - 1$ is that although MSTr is based on the I deviations $\overline{X}_{1.} - \overline{X}_{..}, \ldots, \overline{X}_{I.} - \overline{X}_{..}, \Sigma(\overline{X}_{i.} - \overline{X}_{..}) = 0$, so only $I - 1$ of these are freely determined. Because each sample contributes $J - 1$ df to MSE and these samples are independent, $\nu_2 = (J - 1) + \cdots + (J - 1) = I(J - 1)$.

EXAMPLE 10.2
(Example 10.1 continued)

The values of I and J for the strength data are 4 and 6, respectively, so numerator df $= I - 1 = 3$ and denominator df $= I(J - 1) = 20$. The grand mean is $\bar{x}_{..} = \Sigma\Sigma x_{ij}/(IJ) = 682.50$,

$$\text{MSTr} = \frac{6}{4-1}[(713.00 - 682.50)^2 + (756.93 - 682.50)^2$$
$$+ (698.07 - 682.50)^2 + (562.02 - 682.50)^2] = 42{,}455.86$$

$$\text{MSE} = \frac{1}{4}[(46.55)^2 + (40.34)^2 + (37.20)^2 + (39.87)^2] = 1691.92$$

$$f = \text{MSTr}/\text{MSE} = 42{,}455.86/1691.92 = 25.09$$

The largest F critical value in Table A.9 for $\nu_1 = 3$, $\nu_2 = 20$ is $F_{.001,3,20} = 8.10$. Since $f = 25.09 > 8.10$, the area under the $F_{3,20}$ curve to the right of 25.09 is smaller than $.001$. Therefore P-value $\leq .05$, so the null hypothesis $H_0: \mu_1 = \mu_2 = \mu_3 = \mu_4$ is resoundingly rejected at significance level $.05$. True average compression strength does appear to depend on box type. In fact, because the P-value is so small H_0 would be rejected at any reasonable significance level. ∎

EXAMPLE 10.3

The article **"Influence of Contamination and Cleaning on Bond Strength to Modified Zirconia"** (*Dental Materials*, **2009: 1541–1550**) reported on an experiment in which 50 zirconium-oxide disks were divided into five groups of 10 each. Then a different contamination/cleaning protocol was used for each group. The following summary data on shear bond strength (MPa) appeared in the article:

Treatment:	1	2	3	4	5	
Sample mean	10.5	14.8	15.7	16.0	21.6	Grand mean = 15.7
Sample sd	4.5	6.8	6.5	6.7	6.0	

The authors of the cited article used the F test, so hopefully examined a normal probability plot of the deviations (or a separate plot for each sample, since each sample size is 10) to check the plausibility of assuming normal treatment-response distributions. The five sample standard deviations are certainly close enough to one another to support the assumption of equal σ's.

1. μ_i = true average bond strength for protocol i (i = 1, 2, 3, 4, 5)
2. H_0: $\mu_1 = \mu_2 = \mu_3 = \mu_4 = \mu_5$ (true average bond strength does not depend on which protocol is used)
3. H_a: at least two of the μ_i's are different
4. The test statistic value is f = MSTr/MSE
5. Numerator and denominator dfs are $I - 1 = 4$ and $I(J - 1) = 5(9) = 45$. The mean squares are

$$\text{MSTr} = \frac{10}{5-1}[(10.5 - 15.7)^2 + (14.8 - 15.7)^2 + (15.7 - 15.7)^2$$
$$+ (16.0 - 15.7)^2 + (21.6 - 15.7)^2]$$
$$= 156.875$$
$$\text{MSE} = [(4.5)^2 + (6.8)^2 + (6.5)^2 + (6.7)^2 + (6.0)^2]/5 = 37.926$$

Thus the test statistic value is f = 156.875/37.926 = 4.14.

6. Table A.9 gives $F_{.01,4,40} = 3.83$, $F_{.01,4,50} = 3.72$, $F_{.001,4,40} = 5.70$, and $F_{.001,4,50} = 5.46$. Therefore $F_{.01,4,45} \approx 3.77$ and $F_{.001,4,45} \approx 5.56$. Because f = 4.14 falls between these latter two critical values, the area under the $F_{4,45}$ curve to the right of 4.14 (i.e., the P-value) is between .001 and .01 (software yields .0061).
7. Since P-value < .01, the null hypothesis should be rejected at this significance level. True average bond strength does appear to depend on which protocol is used. ∎

When the null hypothesis is rejected by the F test, as happened in both Examples 10.2 and 10.3, the experimenter will often be interested in further analysis of the data to decide which μ_i's differ from which others. Methods for doing this are called *multiple comparison procedures;* that is the topic of Section 10.2. The article cited in Example 10.3 summarizes the results of such an analysis.

Sums of Squares

The introduction of *sums of squares* facilitates developing an intuitive appreciation for the rationale underlying single-factor and multifactor ANOVAs. Let $x_{i.}$ represent the *sum* (not the average, since there is no bar) of the x_{ij}'s for i fixed (sum of the numbers in the ith row of the table) and $x..$ denote the sum of *all* the x_{ij}'s (the **grand total**).

DEFINITION

The **total sum of squares (SST), treatment sum of squares (SSTr),** and **error sum of squares (SSE)** are given by

$$\text{SST} = \sum_{i=1}^{I}\sum_{j=1}^{J}(x_{ij} - \bar{x}..)^2 = \sum_{i=1}^{I}\sum_{j=1}^{J}x_{ij}^2 - \frac{1}{IJ}x_{..}^2$$

$$\text{SSTr} = \sum_{i=1}^{I}\sum_{j=1}^{J}(\bar{x}_{i.} - \bar{x}..)^2 = \frac{1}{J}\sum_{i=1}^{I}x_{i.}^2 - \frac{1}{IJ}x_{..}^2$$

$$\text{SSE} = \sum_{i=1}^{I}\sum_{j=1}^{J}(x_{ij} - \bar{x}_{i.})^2 \quad \text{where } x_{i.} = \sum_{j=1}^{J}x_{ij} \quad x.. = \sum_{i=1}^{I}\sum_{j=1}^{J}x_{ij}$$

The sum of squares SSTr appears in the numerator of F, and SSE appears in the denominator of F; the reason for defining SST will be apparent shortly.

The expressions on the far right-hand side of SST and SSTr are convenient if ANOVA calculations will be done by hand, although the wide availability of statistical software makes this unnecessary. Both SST and SSTr involve $x_{..}^2/(IJ)$ (the square of the grand total divided by IJ), which is usually called the **correction factor for the mean** (CF). After the correction factor is computed, SST is obtained by squaring each number in the data table, adding these squares together, and subtracting the correction factor. SSTr results from squaring each row total, summing them, dividing by J, and subtracting the correction factor. SSE is then easily obtained by virtue of the following relationship.

> ### Fundamental Identity
>
> $$SST = SSTr + SSE \qquad (10.1)$$

Thus if any two of the sums of squares are computed, the third can be obtained through (10.1); SST and SSTr are easiest to compute, and then $SSE = SST - SSTr$. The proof follows from squaring both sides of the relationship

$$x_{ij} - \bar{x}.. = (\bar{x}_{ij} - \bar{x}_{i.}) + (\bar{x}_{i.} - \bar{x}..) \qquad (10.2)$$

and summing over all i and j. This gives SST on the left and SSTr and SSE as the two extreme terms on the right. The cross-product term is easily seen to be zero.

The interpretation of the fundamental identity is an important aid to an understanding of ANOVA. SST is a measure of the total variation in the data—the sum of all squared deviations about the grand mean. The identity says that this total variation can be partitioned into two pieces. SSE measures variation that would be present (within rows) whether H_0 is true or false, and is thus the part of total variation that is *unexplained* by the status of H_0. SSTr is the amount of variation (between rows) that *can be explained* by possible differences in the μ_i's. H_0 is rejected if the explained variation is large relative to unexplained variation.

Once SSTr and SSE are computed, each is divided by its associated df to obtain a mean square (*mean* in the sense of average). Then F is the ratio of the two mean squares.

> $$MSTr = \frac{SSTr}{I-1} \qquad MSE = \frac{SSE}{I(J-1)} \qquad F = \frac{MSTr}{MSE} \qquad (10.3)$$

The computations are often summarized in a tabular format, called an **ANOVA table**, as displayed in Table 10.2. Tables produced by statistical software customarily include a *P*-value column to the right of f.

Table 10.2 An ANOVA Table

Source of Variation	df	Sum of Squares	Mean Square	f
Treatments	$I-1$	SSTr	$MSTr = SSTr/(I-1)$	MSTr/MSE
Error	$I(J-1)$	SSE	$MSE = SSE/[I(J-1)]$	
Total	$IJ-1$	SST		

EXAMPLE 10.4 According to the article **"Evaluating Fracture Behavior of Brittle Polymeric Materials Using an IASCB Specimen"** (*J. of Engr. Manuf.*, 2013: 133–140), researchers have recently proposed an improved test for the investigation of fracture toughness of brittle polymeric materials. This new fracture test was applied to the brittle polymer polymethylmethacrylate (PMMA), more popularly known as Plexiglas, which is widely used in commercial products. The test was performed by applying asymmetric three-point bending loads on PMMA specimens. The location of one of the three loading points was then varied to determine its effect on fracture load. In one experiment, three loading point locations based on different distances from the center of the specimen's base were selected, resulting in the following fracture load data (kN):

						$x_{i.}$
	42 mm:	2.62	2.99	3.39	2.86	11.86
Distance	36 mm:	3.47	3.85	3.77	3.63	14.72
	31.2 mm:	4.78	4.41	4.91	5.06	19.16
					$x.. =$	45.74

Let μ_i denote true average fracture load when distance i is used ($i = 1, 2, 3$). The null hypothesis asserts that these three μ_i's are identical, whereas the alternative hypothesis says that not all the μ_i's are the same. Before using the F test at significance level .01, we should check the plausibility of underlying assumptions. The three sample standard deviations are .322, .167, and .278, respectively. Sure enough, the largest of these three is no more than twice the smallest. So the assumption of equal variances is plausible. Figure 10.4 shows a normal probability plot of the 12 residuals obtained by subtracting the mean of each sample from the four sample observations. They don't come much straighter than this! It is reasonable to assume that the three fracture load distributions are normal.

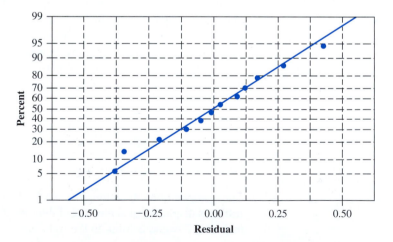

Figure 10.4 Normal probability plot of the residuals from Example 10.4

Squaring each of the 12 observations and adding gives $\Sigma\Sigma x_{ij}^2 = (2.62)^2 + \cdots + (5.06)^2 = 181.7376$. The values of the three sums of squares are

$$\text{SST} = 181.7376 - (45.74)^2/12 = 181.7376 - 174.3456 = 7.3920$$
$$\text{SSTr} = \tfrac{1}{4}[(11.86)^2 + (14.72)^2 + (19.16)^2] - 174.3456 = 6.7653$$
$$\text{SSE} = 7.3920 - 6.7653 = .6267$$

The accompanying ANOVA table from Minitab summarizes the computations. With a P-value of .000, the null hypothesis can be rejected at any sensible significance level, and in particular at the chosen level .01. There is compelling evidence for concluding that true average fracture load is not the same for all three distances.

Source	DF	SS	MS	F	P
Dist	2	6.7653	3.3826	48.58	0.000
Error	9	0.6267	0.0696		
Total	11	7.3920			

EXERCISES Section 10.1 (1–10)

1. In an experiment to compare the tensile strengths of $I = 5$ different types of copper wire, $J = 4$ samples of each type were used. The between-samples and within-samples estimates of σ^2 were computed as MSTr = 2673.3 and MSE = 1094.2, respectively. Use the F test at level .05 to test $H_0: \mu_1 = \mu_2 = \mu_3 = \mu_4 = \mu_5$ versus H_a: at least two μ_i's are unequal.

2. Suppose that the compression-strength observations on the fourth type of box in Example 10.1 had been 655.1, 748.7, 662.4, 679.0, 706.9, and 640.0 (obtained by adding 120 to each previous x_{4j}). Assuming no change in the remaining observations, carry out an F test with $\alpha = .05$.

3. The lumen output was determined for each of $I = 3$ different brands of lightbulbs having the same wattage, with $J = 8$ bulbs of each brand tested. The sums of squares were computed as SSE = 4773.3 and SSTr = 591.2. State the hypotheses of interest (including word definitions of parameters), and use the F test of ANOVA ($\alpha = .05$) to decide whether there are any differences in true average lumen outputs among the three brands for this type of bulb by obtaining as much information as possible about the P-value.

4. It is common practice in many countries to destroy (shred) refrigerators at the end of their useful lives. In this process material from insulating foam may be released into the atmosphere. The article **"Release of Fluorocarbons from Insulation Foam in Home Appliances During Shredding"** (*J. of the Air and Waste Mgmt. Assoc.*, **2007: 1452–1460**) gave the following data on foam density (g/L) for each of two refrigerators produced by four different manufacturers:

   1. 30.4, 29.2   2. 27.7, 27.1
   3. 27.1, 24.8   4. 25.5, 28.8

   Does it appear that true average foam density is not the same for all these manufacturers? Carry out an appropriate test of hypotheses by obtaining as much P-value information as possible, and summarize your analysis in an ANOVA table.

5. Consider the following summary data on the modulus of elasticity ($\times 10^6$ psi) for lumber of three different grades [in close agreement with values in the article **"Bending**

   Strength and Stiffness of Second-Growth Douglas-Fir Dimension Lumber" (*Forest Products J.*, **1991: 35–43**), except that the sample sizes there were larger]:

Grade	J	$\bar{x}_{i\cdot}$	s_i
1	10	1.63	.27
2	10	1.56	.24
3	10	1.42	.26

   Use this data and a significance level of .01 to test the null hypothesis of no difference in mean modulus of elasticity for the three grades.

6. The article **"Origin of Precambrian Iron Formations"** (*Econ. Geology*, **1964: 1025–1057**) reports the following data on total Fe for four types of iron formation (1 = carbonate, 2 = silicate, 3 = magnetite, 4 = hematite).

1:	20.5	28.1	27.8	27.0	28.0
	25.2	25.3	27.1	20.5	31.3
2:	26.3	24.0	26.2	20.2	23.7
	34.0	17.1	26.8	23.7	24.9
3:	29.5	34.0	27.5	29.4	27.9
	26.2	29.9	29.5	30.0	35.6
4:	36.5	44.2	34.1	30.3	31.4
	33.1	34.1	32.9	36.3	25.5

   Carry out an analysis of variance F test at significance level .01, and summarize the results in an ANOVA table.

7. An experiment was carried out to compare electrical resistivity for six different low-permeability concrete bridge deck mixtures. There were 26 measurements on concrete cylinders for each mixture; these were obtained 28 days after casting. The entries in the accompanying ANOVA table are based on information in the article **"In-Place Resistivity of Bridge Deck Concrete Mixtures"** (*ACI Materials J.*, **2009: 114–122**). Fill in the remaining entries and test appropriate hypotheses.

Source	df	Sum of Squares	Mean Square	f
Mixture				
Error			13.929	
Total		5664.415		

8. A study of the properties of metal plate-connected trusses used for roof support (**"Modeling Joints Made with Light-Gauge Metal Connector Plates,"** *Forest Products J.*, **1979: 39–44**) yielded the following observations on axial-stiffness index (kips/in.) for plate lengths 4, 6, 8, 10, and 12 in:

  4: 309.2 409.5 311.0 326.5 316.8 349.8 309.7
  6: 402.1 347.2 361.0 404.5 331.0 348.9 381.7
  8: 392.4 366.2 351.0 357.1 409.9 367.3 382.0
  10: 346.7 452.9 461.4 433.1 410.6 384.2 362.6
  12: 407.4 441.8 419.9 410.7 473.4 441.2 465.8

  Does variation in plate length have any effect on true average axial stiffness? State and test the relevant hypotheses using analysis of variance with $\alpha = .01$. Display your results in an ANOVA table. [*Hint:* $\Sigma\Sigma x_{ij}^2 = 5{,}241{,}420.79$.]

9. Six samples of each of four types of cereal grain grown in a certain region were analyzed to determine thiamin content, resulting in the following data ($\mu g/g$):

Wheat	5.2	4.5	6.0	6.1	6.7	5.8
Barley	6.5	8.0	6.1	7.5	5.9	5.6
Maize	5.8	4.7	6.4	4.9	6.0	5.2
Oats	8.3	6.1	7.8	7.0	5.5	7.2

  Does this data suggest that at least two of the grains differ with respect to true average thiamin content? Use a level $\alpha = .05$ test.

10. In single-factor ANOVA with I treatments and J observations per treatment, let $\mu = (1/I)\Sigma\mu_i$.
  a. Express $E(\overline{X}..)$ in terms of μ. [*Hint:* $\overline{X}.. = (1/I)\Sigma\overline{X}_{i.}$]
  b. Determine $E(\overline{X}_{i.}^2)$. [*Hint:* For any rv Y, $E(Y^2) = V(Y) + [E(Y)]^2$.]
  c. Determine $E(\overline{X}_{..}^2)$.
  d. Determine $E(SSTr)$ and then show that
  $$E(MSTr) = \sigma^2 + \frac{J}{I-1}\Sigma(\mu_i - \mu)^2$$
  e. Using the result of part (d), what is $E(MSTr)$ when H_0 is true? When H_0 is false, how does $E(MSTr)$ compare to σ^2?

10.2 Multiple Comparisons in ANOVA

When the computed value of the F statistic in single-factor ANOVA is not significant, the analysis is terminated because no differences among the μ_i's have been identified. But when H_0 is rejected, the investigator will usually want to know which of the μ_i's are different from one another. A method for carrying out this further analysis is called a **multiple comparisons procedure**.

Several of the most frequently used procedures are based on the following central idea. First calculate a confidence interval for each pairwise difference $\mu_i - \mu_j$ with $i < j$. Thus if $I = 4$, the six required CIs would be for $\mu_1 - \mu_2$ (but not also for $\mu_2 - \mu_1$), $\mu_1 - \mu_3$, $\mu_1 - \mu_4$, $\mu_2 - \mu_3$, $\mu_2 - \mu_4$, and $\mu_3 - \mu_4$. Then if the interval for $\mu_1 - \mu_2$ does not include 0, conclude that μ_1 and μ_2 *differ significantly* from one another; if the interval does include 0, the two μ's are judged not significantly different. Following the same line of reasoning for each of the other intervals, we end up being able to judge for each pair of μ's whether or not they differ significantly from one another.

The procedures based on this idea differ in how the various CIs are calculated. Here we present a popular method that controls the *simultaneous* confidence level for all $I(I-1)/2$ intervals.

Tukey's Procedure (the *T* Method)

Tukey's procedure involves the use of another probability distribution called the **Studentized range distribution.** The distribution depends on two parameters: a numerator df m and a denominator df ν. Let $Q_{\alpha,m,\nu}$ denote the upper-tail α critical value of the Studentized range distribution with m numerator df and ν denominator df (analogous to F_{α,ν_1,ν_2}). Values of $Q_{\alpha,m,\nu}$ are given in Appendix Table A.10.

PROPOSITION

With probability $1 - \alpha$,

$$\overline{X}_{i\cdot} - \overline{X}_{j\cdot} - Q_{\alpha, I, I(J-1)}\sqrt{MSE/J} \le \mu_i - \mu_j$$
$$\le \overline{X}_{i\cdot} - \overline{X}_{j\cdot} + Q_{\alpha, I, I(J-1)}\sqrt{MSE/J} \quad (10.4)$$

for *every* i and j ($i = 1, \ldots, I$ and $j = 1, \ldots, I$) with $i < j$.

Notice that numerator df for the appropriate Q_α critical value is I, the number of population or treatment means being compared, and not $I - 1$ as in the F test. When the computed $\bar{x}_{i\cdot}, \bar{x}_{j\cdot}$ and MSE are substituted into (10.4), the result is a collection of confidence intervals with *simultaneous* confidence level $100(1 - \alpha)\%$ for all pairwise differences of the form $\mu_i - \mu_j$ with $i < j$. Each interval that does not include 0 yields the conclusion that the corresponding values of μ_i and μ_j differ significantly from one another.

Since we are not really interested in the lower and upper limits of the various intervals but only in which include 0 and which do not, much of the arithmetic associated with (10.4) can be avoided. The following box gives details and describes how differences can be identified visually using an "underscoring pattern."

> ### The T Method for Identifying Significantly Different μ_is
>
> Select α, extract $Q_{\alpha, I, I(J-1)}$ from Appendix Table A.10, and calculate $w = Q_{\alpha, I, I(J-1)} \cdot \sqrt{MSE/J}$. Then list the sample means in increasing order and underline those pairs that differ by less than w. Any pair of sample means not underscored by the same line corresponds to a pair of population or treatment means that are judged significantly different.

Suppose, for example, that $I = 5$ and that

$$\bar{x}_{2\cdot} < \bar{x}_{5\cdot} < \bar{x}_{4\cdot} < \bar{x}_{1\cdot} < \bar{x}_{3\cdot}.$$

Then

1. Consider first the smallest mean $\bar{x}_{2\cdot}$. If $\bar{x}_{5\cdot} - \bar{x}_{2\cdot} \ge w$, proceed to Step 2. However, if $\bar{x}_{5\cdot} - \bar{x}_{2\cdot} < w$, connect these first two means with a line segment. Then if possible extend this line segment even further to the right to the largest $\bar{x}_{i\cdot}$ that differs from $\bar{x}_{2\cdot}$ by less than w (so the line may connect two, three, or even more means).

2. Now move to $\bar{x}_{5\cdot}$ and again extend a line segment to the largest $\bar{x}_{i\cdot}$ to its right that differs from $\bar{x}_{5\cdot}$ by less than w (it may not be possible to draw this line, or alternatively it may underscore just two means, or three, or even all four remaining means).

3. Continue by moving to $\bar{x}_{4\cdot}$ and repeating, and then finally move to $\bar{x}_{1\cdot}$.

To summarize, starting from each mean in the ordered list, a line segment is extended as far to the right as possible as long as the difference between the means is smaller than w. It is easily verified that a particular interval of the form (10.4) will contain 0 if and only if the corresponding pair of sample means is underscored by the same line segment.

EXAMPLE 10.5

An experiment was carried out to compare five different brands of automobile oil filters with respect to their ability to capture foreign material. Let μ_i denote the true average amount of material captured by brand i filters ($i = 1, \ldots, 5$) under controlled conditions. A sample of nine filters of each brand was used, resulting in the following sample mean amounts: $\bar{x}_1. = 14.5, \bar{x}_2. = 13.8, \bar{x}_3. = 13.3, \bar{x}_4. = 14.3$, and $\bar{x}_5. = 13.1$. Table 10.3 is the ANOVA table summarizing the first part of the analysis.

Table 10.3 ANOVA Table for Example 10.5

Source of Variation	df	Sum of Squares	Mean Square	f
Treatments (brands)	4	13.32	3.33	37.84
Error	40	3.53	.088	
Total	44	16.85		

Since $F_{.001,4,40} = 5.70$, the P-value is smaller than .001. Therefore, H_0 is rejected (decisively) at level .05. We now use Tukey's procedure to look for significant differences among the μ_i's. From Appendix Table A.10, $Q_{.05,5,40} = 4.04$ (the second subscript on Q is I and not $I - 1$ as in F), so $w = 4.04\sqrt{.088/9} = .4$. After arranging the five sample means in increasing order, the two smallest can be connected by a line segment because they differ by less than .4. However, this segment cannot be extended further to the right since $13.8 - 13.1 = .7 \geq .4$. Moving one mean to the right, the pair $\bar{x}_3.$ and $\bar{x}_2.$ cannot be underscored because these means differ by more than .4. Again moving to the right, the next mean, 13.8, cannot be connected to any further to the right. The last two means can be underscored with the same line segment.

$$\begin{array}{ccccc} \bar{x}_5. & \bar{x}_3. & \bar{x}_2. & \bar{x}_4. & \bar{x}_1. \\ \underline{13.1} & \underline{13.3} & 13.8 & \underline{14.3} & \underline{14.5} \end{array}$$

Thus brands 1 and 4 are not significantly different from one another, but are significantly higher than the other three brands in their true average contents. Brand 2 is significantly better than 3 and 5 but worse than 1 and 4, and brands 3 and 5 do not differ significantly.

If $\bar{x}_2. = 14.15$ rather than 13.8 with the same computed w, then the configuration of underscored means would be

$$\begin{array}{ccccc} \bar{x}_5. & \bar{x}_3. & \bar{x}_2. & \bar{x}_4. & \bar{x}_1. \\ \underline{13.1} & \underline{13.3} & \underline{14.15} & \underline{14.3} & \underline{14.5} \end{array}$$ ■

EXAMPLE 10.6 A biologist wished to study the effects of ethanol on sleep time. A sample of 20 rats, matched for age and other characteristics, was selected, and each rat was given an oral injection having a particular concentration of ethanol per body weight. The rapid eye movement (REM) sleep time for each rat was then recorded for a 24-hour period, with the following results:

Treatment (concentration of ethanol)					$x_i.$	$\bar{x}_i.$
0 (control)	88.6	73.2	91.4	68.0	75.2	396.4	79.28
1 g/kg	63.0	53.9	69.2	50.1	71.5	307.7	61.54
2 g/kg	44.9	59.5	40.2	56.3	38.7	239.6	47.92
4 g/kg	31.0	39.6	45.3	25.2	22.7	163.8	32.76

$x.. = 1107.5 \; \bar{x}.. = 55.375$

Does the data indicate that the true average REM sleep time depends on the concentration of ethanol? (This example is based on an experiment reported in **"Relationship of Ethanol Blood Level to REM and Non-REM Sleep Time and Distribution in the Rat,"** *Life Sciences*, 1978: 839–846.)

The $\bar{x}_{i.}$'s differ rather substantially from one another, but there is also a great deal of variability within each sample. To answer the question precisely we must carry out the ANOVA. The smallest and largest of the four sample standard deviations are 9.34 and 10.18, respectively, which supports the assumption of equal variances. A normal probability plot of the 20 residuals shows a reasonably linear pattern, justifying the assumption that the four REM sleep time distributions are normal. Thus it is legitimate to employ the F test.

Table 10.4 is a SAS ANOVA table. The last column gives the P-value as .0001. Using a significance level of .05, we reject the null hypothesis $H_0: \mu_1 = \mu_2 = \mu_3 = \mu_4$, since P-value $= .0001 < .05 = \alpha$. True average REM sleep time does appear to depend on concentration level.

Table 10.4 SAS ANOVA Table

```
                      Analysis of Variance Procedure
Dependent Variable:  TIME
                            Sum of            Mean
Source            DF       Squares           Square     F Value     Pr > F
Model              3     5882.35750       1960.78583      21.09     0.0001
Error             16     1487.40000         92.96250
Corrected
Total             19     7369.75750
```

There are $I = 4$ treatments and 16 df for error, from which $Q_{.05,4,16} = 4.05$ and $w = 4.05\sqrt{93.0/5} = 17.47$. Ordering the means and underscoring yields

$$\begin{array}{cccc} \bar{x}_{4.} & \bar{x}_{3.} & \bar{x}_{2.} & \bar{x}_{1.} \\ 32.76 & 47.92 & 61.54 & 79.28 \end{array}$$

The interpretation of this underscoring must be done with care, since we seem to have concluded that treatments 2 and 3 do not differ, 3 and 4 do not differ, yet 2 and 4 do differ. The suggested way of expressing this is to say that although evidence allows us to conclude that treatments 2 and 4 differ from one another, neither has been shown to be significantly different from 3. Treatment 1 has a significantly higher true average REM sleep time than any of the other treatments.

Figure 10.5 shows SAS output from the application of Tukey's procedure.

```
              Alpha =0.05 df =16 MSE = 92.9625
         Critical Value of Studentized Range = 4.046
            Minimum Significant Difference = 17.446

  Means with the same letter are not significantly different.

  Tukey Grouping            Mean        N     TREATMENT
              A           79.280        5     0(control)

              B           61.540        5     1 gm/kg
              B
       C      B           47.920        5     2 gm/kg
       C
       C                  32.760        5     4 gm/kg
```

Figure 10.5 Tukey's method using SAS

The Interpretation of α in Tukey's Method

We stated previously that the *simultaneous* confidence level is controlled by Tukey's method. So what does "simultaneous" mean here? Consider calculating a 95% CI for a population mean μ based on a sample from that population and then a 95% CI for a population proportion p based on another sample selected independently of the first one. Prior to obtaining data, the probability that the first interval will include μ is .95, and this is also the probability that the second interval will include p. Because the two samples are selected independently of one another, the probability that *both* intervals will include the values of the respective parameters is $(.95)(.95) = (.95)^2 \approx .90$. Thus the *simultaneous* or *joint* confidence level for the two intervals is roughly 90%—if pairs of intervals are calculated over and over again from independent samples, in the long run roughly 90% of the time the first interval will capture μ and the second will include p. Similarly, if three CI's are calculated based on independent samples, the simultaneous confidence level will be $100(.95)^3\% \approx 86\%$. Clearly, as the number of intervals increases, the simultaneous confidence level that all intervals capture their respective parameters will decrease.

Now suppose that we want to maintain the simultaneous confidence level at 95%. Then for two independent samples, the individual confidence level for each would have to be $100\sqrt{.95}\% \approx 97.5\%$. The larger the number of intervals, the higher the individual confidence level would have to be to maintain the 95% simultaneous level.

The tricky thing about the Tukey intervals is that they are not based on independent samples—MSE appears in every one, and various intervals share the same $\bar{x}_{i\cdot}$'s (e.g., in the case $I = 4$, three different intervals all use $\bar{x}_{1\cdot}$). This implies that there is no straightforward probability argument for ascertaining the simultaneous confidence level from the individual confidence levels. Nevertheless, it can be shown that if $Q_{.05}$ is used, the simultaneous confidence level is controlled at 95%, whereas using $Q_{.01}$ gives a simultaneous 99% level. To obtain a 95% simultaneous level, the individual level for each interval must be considerably larger than 95%. Said in a slightly different way, to obtain a 5% *experimentwise* or *family* error rate, the individual or per-comparison error rate for each interval must be considerably smaller than .05. Minitab asks the user to specify the family error rate (e.g., 5%) and then includes on output the individual error rate (see Exercise 16).

Confidence Intervals for Other Parametric Functions

In some situations, a CI is desired for a function of the μ_i's more complicated than a difference of $\mu_i - \mu_j$. Let $\theta = \Sigma c_i \mu_i$, where the c_i's are constants. One such function is $1/2(\mu_1 + \mu_2) - 1/3(\mu_3 + \mu_4 + \mu_5)$, which in the context of Example 10.5 measures the difference between the group consisting of the first two brands and that of the last three brands. Because the X_{ij}'s are normally distributed with $E(X_{ij}) = \mu_i$ and $V(X_{ij}) = \sigma^2$, $\hat{\theta} = \Sigma c_i \bar{X}_{i\cdot}$ is normally distributed, unbiased for θ, and

$$V(\hat{\theta}) = V\left(\sum_i c_i \bar{X}_{i\cdot}\right) = \sum_i c_i^2 V(\bar{X}_{i\cdot}) = \frac{\sigma^2}{J} \sum_i c_i^2$$

Estimating σ^2 by MSE and forming $\hat{\sigma}_{\hat{\theta}}$ results in a t variable $(\hat{\theta} - \theta)/\hat{\sigma}_{\hat{\theta}}$, which can be manipulated to obtain the following $100(1 - \alpha)\%$ confidence interval for $\Sigma c_i \mu_i$,

$$\sum c_i \bar{x}_{i\cdot} \pm t_{\alpha/2, I(J-1)} \sqrt{\frac{\text{MSE}\Sigma c_i^2}{J}} \tag{10.5}$$

EXAMPLE 10.7   The parametric function for comparing the first two (store) brands of oil filter with the last three (national) brands is $\theta = 1/2(\mu_1 + \mu_2) - 1/3(\mu_3 + \mu_4 + \mu_5)$, from which

$$\sum c_i^2 = \left(\frac{1}{2}\right)^2 + \left(\frac{1}{2}\right)^2 + \left(-\frac{1}{3}\right)^2 + \left(-\frac{1}{3}\right)^2 + \left(-\frac{1}{3}\right)^2 = \frac{5}{6}$$

With $\hat{\theta} = 1/2(\bar{x}_{1\cdot} + \bar{x}_{2\cdot}) - 1/3(\bar{x}_{3\cdot} + \bar{x}_{4\cdot} + \bar{x}_{5\cdot}) = .583$ and MSE $= .088$, a 95% interval is

$$.583 \pm 2.021\sqrt{5(.088)/[(6)(9)]} = .583 \pm .182 = (.401, .765)$$

Sometimes an experiment is carried out to compare each of several "new" treatments to a control treatment. In such situations, a multiple comparisons technique called Dunnett's method is appropriate.

EXERCISES Section 10.2 (11–21)

11. An experiment to compare the spreading rates of five different brands of yellow interior latex paint available in a particular area used 4 gallons ($J = 4$) of each paint. The sample average spreading rates (ft^2/gal) for the five brands were $\bar{x}_{1\cdot} = 462.0$, $\bar{x}_{2\cdot} = 512.8$, $\bar{x}_{3\cdot} = 437.5$, $\bar{x}_{4\cdot} = 469.3$, and $\bar{x}_{5\cdot} = 532.1$. The computed value of F was found to be significant at level $\alpha = .05$. With MSE $= 272.8$, use Tukey's procedure to investigate significant differences in the true average spreading rates between brands.

12. In Exercise 11, suppose $\bar{x}_{3\cdot} = 427.5$. Now which true average spreading rates differ significantly from one another? Be sure to use the method of underscoring to illustrate your conclusions, and write a paragraph summarizing your results.

13. Repeat Exercise 12 supposing that $\bar{x}_{2\cdot} = 502.8$ in addition to $\bar{x}_{3\cdot} = 427.5$.

14. Use Tukey's procedure on the data in Example 10.3 to identify differences in true average bond strengths among the five protocols.

15. Exercise 10.7 described an experiment in which 26 resistivity observations were made on each of six different concrete mixtures. The article cited there gave the following sample means: 14.18, 17.94, 18.00, 18.00, 25.74, 27.67. Apply Tukey's method with a simultaneous confidence level of 95% to identify significant differences, and describe your findings (use MSE $= 13.929$).

16. Reconsider the axial stiffness data given in Exercise 8. ANOVA output from Minitab follows:

Analysis of Variance for Stiffness

Source	DF	SS	MS	F	P
Length	4	43993	10998	10.48	0.000
Error	30	31475	1049		
Total	34	75468			

Level	N	Mean	StDev
4	7	333.21	36.59
6	7	368.06	28.57
8	7	375.13	20.83
10	7	407.36	44.51
12	7	437.17	26.00

Pooled StDev $= 32.39$

Tukey's pairwise comparisons

Family error rate = 0.0500
Individual error rate = 0.00693

Critical value = 4.10

Intervals for (column level mean) − (row level mean)

	4	6	8	10
6	−85.0			
	15.4			
8	−92.1	−57.3		
	8.3	43.1		
10	−124.3	−89.5	−82.4	
	−23.9	10.9	18.0	
12	−154.2	−119.3	−112.2	−80.0
	−53.8	−18.9	−11.8	20.4

a. Is it plausible that the variances of the five axial stiffness index distributions are identical? Explain.

b. Use the output (without reference to our F table) to test the relevant hypotheses.

c. Use the Tukey intervals given in the output to determine which means differ, and construct the corresponding underscoring pattern.

17. Refer to Exercise 5. Compute a 95% t CI for $\theta = 1/2(\mu_1 + \mu_2) - \mu_3$.

18. Consider the accompanying data on plant growth after the application of five different types of growth hormone.

1:	13	17	7	14
2:	21	13	20	17
3:	18	15	20	17
4:	7	11	18	10
5:	6	11	15	8

a. Perform an F test at level $\alpha = .05$.

b. What happens when Tukey's procedure is applied?

19. Consider a single-factor ANOVA experiment in which $I = 3$, $J = 5$, $\bar{x}_{1\cdot} = 10$, $\bar{x}_{2\cdot} = 12$, and $\bar{x}_{3\cdot} = 20$. Find a value of SSE for which $f > F_{.05,2,12}$, so that $H_0: \mu_1 = \mu_2 = \mu_3$ is rejected, yet when Tukey's procedure is applied none of the μ_i's can be said to differ significantly from one another.

20. Refer to Exercise 19 and suppose $\bar{x}_{1.} = 10, \bar{x}_{2.} = 15$, and $\bar{x}_{3.} = 20$. Can you now find a value of SSE that produces such a contradiction between the F test and Tukey's procedure?

21. The article **"The Effect of Enzyme Inducing Agents on the Survival Times of Rats Exposed to Lethal Levels of Nitrogen Dioxide"** (*Toxicology and Applied Pharmacology*, **1978: 169–174**) reports the following data on survival times for rats exposed to nitrogen dioxide (70 ppm) via different injection regimens. There were $J = 14$ rats in each group.

Regimen	$\bar{x}_{i.}$ (min)	s_i
1. Control	166	32
2. 3-Methylcholanthrene	303	53
3. Allylisopropylacetamide	266	54
4. Phenobarbital	212	35
5. Chlorpromazine	202	34
6. *p*-Aminobenzoic Acid	184	31

a. Test the null hypothesis that true average survival time does not depend on an injection regimen against the alternative that there is some dependence on an injection regimen using $\alpha = .01$.

b. Suppose that $100(1 - \alpha)\%$ CIs for k different parametric functions are computed from the same ANOVA data set. Then it is easily verified that the simultaneous confidence level is at least $100(1 - k\alpha)\%$. Compute CIs with a simultaneous confidence level of at least 98% for

$$\mu_1 - 1/5(\mu_2 + \mu_3 + \mu_4 + \mu_5 + \mu_6) \text{ and}$$
$$1/4 (\mu_2 + \mu_3 + \mu_4 + \mu_5) - \mu_6$$

10.3 More on Single-Factor ANOVA

We now briefly consider some additional issues relating to single-factor ANOVA. These include an alternative description of the model parameters, β for the F test, the relationship of the test to procedures previously considered, data transformation, a random effects model, and formulas for the case of unequal sample sizes.

The ANOVA Model

The assumptions of single-factor ANOVA can be described succinctly by means of the "model equation"

$$X_{ij} = \mu_i + \epsilon_{ij}$$

where ϵ_{ij} represents a random deviation from the population or true treatment mean μ_i. The ϵ_{ij}'s are assumed to be independent, normally distributed rv's (implying that the X_{ij}'s are also) with $E(\epsilon_{ij}) = 0$ [so that $E(X_{ij}) = \mu_i$] and $V(\epsilon_{ij}) = \sigma^2$ [from which $V(X_{ij}) = \sigma^2$ for every i and j]. An alternative description of single-factor ANOVA will give added insight and suggest appropriate generalizations to models involving more than one factor. Define a parameter μ by

$$\mu = \frac{1}{I}\sum_{i=1}^{I}\mu_i$$

and the parameters $\alpha_1, \ldots, \alpha_I$ by

$$\alpha_i = \mu_i - \mu \quad (i = 1, \ldots, I)$$

Then the treatment mean μ_i can be written as $\mu + \alpha_i$, where μ represents the true average overall response in the experiment, and α_i is the effect, measured as a departure from μ, due to the ith treatment. Whereas we initially had I parameters, we now have $I + 1$ $(\mu, \alpha_1, \ldots, \alpha_I)$. However, because $\Sigma\alpha_i = 0$ (the average departure from the overall mean response is zero), only I of these new parameters are independently determined, so there are as many independent parameters as there were before. In terms of μ and the α_i's, the model becomes

$$X_{ij} = \mu + \alpha_1 + \epsilon_{ij} \quad (i = 1, \ldots, I; \quad j = 1, \ldots, J)$$

In Chapter 11, we will develop analogous models for multifactor ANOVA. The claim that the μ_i's are identical is equivalent to the equality of the α_i's, and because $\Sigma \alpha_i = 0$, the null hypothesis becomes

$$H_0: \alpha_1 = \alpha_2 = \cdots = \alpha_I = 0$$

Recall that MSTr is an unbiased estimator of σ^2 when H_0 is true but otherwise tends to overestimate σ^2. Here is a more precise result:

$$E(\text{MSTr}) = \sigma^2 + \frac{J}{I - 1} \sum \alpha_i^2$$

When H_0 is true, $\Sigma \alpha_i^2 = 0$ so $E(\text{MSTr}) = \sigma^2$ (MSE is unbiased whether or not H_0 is true). If $\Sigma \alpha_i^2$ is used as a measure of the extent to which H_0 is false, then a larger value of $\Sigma \alpha_i^2$ will result in a greater tendency for MSTr to overestimate σ^2. In the next chapter, formulas for expected mean squares for multifactor models will be used to suggest how to form F ratios to test various hypotheses.

Proof of the Formula for $E(\text{MSTr})$ For any rv Y, $E(Y^2) = V(Y) + [E(Y)]^2$, so

$$E(\text{SSTr}) = E\left(\frac{1}{J} \sum_i X_{i\cdot}^2 - \frac{1}{IJ} X_{\cdot\cdot}^2 \right) = \frac{1}{J} \sum_i E(X_{i\cdot}^2) - \frac{1}{IJ} E(X_{\cdot\cdot}^2)$$

$$= \frac{1}{J} \sum_i \{V(X_{i\cdot}) + [E(X_{i\cdot})^2]\} - \frac{1}{IJ}\{V(X_{\cdot\cdot}) + [E(X_{\cdot\cdot})]^2\}$$

$$= \frac{1}{J} \sum_i \{J\sigma^2 + [J(\mu + \alpha_i)]^2\} - \frac{1}{IJ}[IJ\sigma^2 + (IJ\mu)^2]$$

$$= I\sigma^2 + IJ\mu^2 + 2\mu J \sum_i \alpha_i + J \sum_i \alpha_i^2 - \sigma^2 - IJ\mu^2$$

$$= (I - 1)\sigma^2 + J \sum_i \alpha_i^2 \quad (\text{since } \sum_i \alpha_i = 0)$$

The result then follows from the relationship $\text{MSTr} = \text{SSTr}/(I - 1)$. ■

β for the F Test

Consider a set of parameter values $\alpha_1, \alpha_2, \ldots, \alpha_I$ for which H_0 is not true. The probability of a type II error, β, is the probability that H_0 is not rejected when that set is the set of true values. One might think that β would have to be determined separately for each different configuration of α_i's. Fortunately, since β for the F test depends on the α_i's and σ^2 only through $\Sigma \alpha_i^2/\sigma^2$, it can be simultaneously evaluated for many different alternatives. For example, $\Sigma \alpha_i^2 = 4$ for each of the following sets of α_i's for which H_0 is false, so β is identical for all three alternatives:

1. $\alpha_1 = -1, \alpha_2 = -1, \alpha_3 = 1, \alpha_4 = 1$
2. $\alpha_1 = -\sqrt{2}, \alpha_2 = \sqrt{2}, \alpha_3 = 0, \alpha_4 = 0$
3. $\alpha_1 = -\sqrt{3}, \alpha_2 = \sqrt{1/3}, \alpha_3 = \sqrt{1/3}, \alpha_4 = \sqrt{1/3}$

The quantity $J\Sigma \alpha_i^2/\sigma^2$ is called the **noncentrality parameter** for one-way ANOVA (because when H_0 is false the test statistic has a *noncentral F distribution* with this as one of its parameters), and β is a decreasing function of the value of this parameter. Thus, for fixed values of σ^2 and J, the null hypothesis is more likely to be rejected for alternatives far from H_0 (large $\Sigma\alpha_i^2$) than for alternatives close to H_0. For a fixed value of $\Sigma\alpha_i^2$, β decreases as the sample size J on each treatment increases, and it increases as the variance σ^2 increases (since greater underlying variability makes it more difficult to detect any given departure from H_0).

Because hand computation of β and sample size determination for the F test are quite difficult (as in the case of t tests), statisticians have constructed sets of curves from which β can be obtained. Sets of curves for numerator df $\nu_1 = 3$ and $\nu_1 = 4$ are displayed in Figure 10.6* and Figure 10.7*, respectively. After the values of σ^2 and the α_i's for which β is desired are specified, these are used to compute the value of ϕ, where $\phi^2 = (J/I)\Sigma\alpha_i^2/\sigma^2$. We then enter the appropriate set of curves at

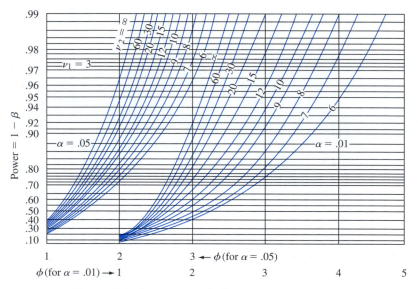

Figure 10.6   Power curves for the ANOVA F test ($\nu_1 = 3$)

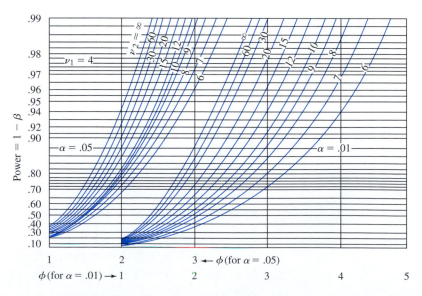

Figure 10.7   Power curves for the ANOVA F test ($\nu_1 = 4$)

* From E. S. Pearson and H. O. Hartley, "Charts of the Power Function for Analysis of Variance Tests, Derived from the Non-central F Distribution," *Biometrika*, vol. 38, 1951: 112.

the value of ϕ on the horizontal axis, move up to the curve associated with error df ν_2, and move over to the value of power on the vertical axis. Finally, $\beta = 1 -$ power.

EXAMPLE 10.8 The effects of four different heat treatments on yield point (tons/in^2) of steel ingots are to be investigated. A total of eight ingots will be cast using each treatment. Suppose the true standard deviation of yield point for any of the four treatments is $\sigma = 1$. How likely is it that H_0 will not be rejected at level .05 if three of the treatments have the same expected yield point and the other treatment has an expected yield point that is 1 ton/in^2 greater than the common value of the other three (i.e., the fourth yield is on average 1 standard deviation above those for the first three treatments)?

Suppose that $\mu_1 = \mu_2 = \mu_3$ and $\mu_4 = \mu_1 + 1$, $\mu = (\Sigma\mu_i)/4 = \mu_1 + 1/4$. Then $\alpha_1 = \mu_1 - \mu = -1/4$, $\alpha_2 = -1/4$, $\alpha_3 = -1/4$, $\alpha_4 = 3/4$, so

$$\phi^2 = \frac{8}{4}\left[\left(-\frac{1}{4}\right)^2 + \left(-\frac{1}{4}\right)^2 + \left(-\frac{1}{4}\right)^2 + \left(\frac{3}{4}\right)^2\right] = \frac{3}{2}$$

and $\phi = 1.22$. Degrees of freedom for the F test are $\nu_1 = I - 1 = 3$ and $\nu_2 = I(J - 1) = 28$, so interpolating visually between $\nu_2 = 20$ and $\nu_2 = 30$ gives power $\approx .47$ and $\beta \approx .53$. This β is rather large, so we might decide to increase the value of J. How many ingots of each type would be required to yield $\beta \approx .05$ for the alternative under consideration? By trying different values of J, it can be verified that $J = 24$ will meet the requirement, but any smaller J will not. ∎

As an alternative to the use of power curves, the SAS statistical software package has a function that calculates the cumulative area under a noncentral F curve (inputs F_α, numerator df, denominator df, and ϕ^2), and this area is β. Minitab does this and also something rather different. The user is asked to specify the maximum difference between μ_i's rather than the individual means. For example, we might wish to calculate the power of the test when $I = 4$, $\mu_1 = 100$, $\mu_2 = 101$, $\mu_3 = 102$, and $\mu_4 = 106$. Then the maximum difference is $106 - 100 = 6$. However, the power depends not only on this maximum difference but on the values of all the μ_i's. In this situation Minitab calculates the smallest possible value of power subject to $\mu_1 = 100$ and $\mu_4 = 106$, which occurs when the two other μ's are both halfway between 100 and 106. If this power is .85, then we can say that the power is at least .85 and β is at most .15 when the two most extreme μ's are separated by 6 (the common sample size, α, and σ must also be specified). The software will also determine the necessary common sample size if maximum difference and minimum power are specified.

Relationship of the *F* Test to the *t* Test

When the number of treatments or populations is $I = 2$, all formulas and results connected with the F test still make sense, so ANOVA can be used to test H_0: $\mu_1 = \mu_2$ versus H_a: $\mu_1 \neq \mu_2$. In this case, a two-tailed, two-sample t test can also be used. In Section 9.3, we mentioned the pooled t test, which requires equal variances, as an alternative to the two-sample t procedure. It can be shown that the single-factor ANOVA F test and the two-tailed pooled t test are equivalent; for any given data set, the P-values for the two tests will be identical, so the same conclusion will be reached by either test.

The two-sample t test is more flexible than the F test when $I = 2$ for two reasons. First, it is valid without the assumption that $\sigma_1 = \sigma_2$; second, it can be used to test H_a: $\mu_1 > \mu_2$ (an upper-tailed t test) or H_a: $\mu_1 < \mu_2$ as well as H_a: $\mu_1 \neq \mu_2$. In the case of $I \geq 3$, there is unfortunately no general test procedure known to have good properties without assuming equal variances.

Unequal Sample Sizes

When the sample sizes from each population or treatment are not equal, let $J_1, J_2, \ldots, J_I$ denote the I sample sizes, and let $n = \Sigma_i J_i$ denote the total number of observations. The accompanying box gives ANOVA formulas and the test procedure.

$$SST = \sum_{i=1}^{I} \sum_{j=1}^{J_i} (X_{ij} - \overline{X}..)^2 = \sum_{i=1}^{I} \sum_{j=1}^{J_i} X_{ij}^2 - \frac{1}{n} X_{..}^2 \qquad df = n - 1$$

$$SSTr = \sum_{i=1}^{I} \sum_{j=1}^{J_i} (\overline{X}_{i.} - \overline{X}..)^2 = \sum_{i=1}^{I} \frac{1}{J_i} X_{i.}^2 - \frac{1}{n} X_{..}^2 \quad df = I - 1$$

$$SSE = \sum_{i=1}^{I} \sum_{j=1}^{J_i} (X_{ij} - \overline{X}_{i.})^2 = \sum_{i=1}^{I} (J_i - 1)S_i^2 \qquad df = \sum(J_i - 1) = n - I$$

$$= SST - SSTr$$

Test statistic:

$$F = \frac{MSTr}{MSE} \quad \text{where MSTr} = \frac{SSTr}{I - 1} \quad MSE = \frac{SSE}{n - I}$$

Statistical theory says that the test statistic has an F distribution with numerator df $I - 1$ and denominator df $n - I$ when H_0 is true. As in the case of equal sample sizes, the larger the value of F, the stronger is the evidence against H_0. Therefore the test is upper-tailed; the P-value is the area under the $F_{I-1, n-I}$ curve to the right of f.

EXAMPLE 10.9 The article **"On the Development of a New Approach for the Determination of Yield Strength in Mg-based Alloys"** (*Light Metal Age*, Oct. 1998: 51–53) presented the following data on elastic modulus (GPa) obtained by a new ultrasonic method for specimens of a certain alloy produced using three different casting processes.

		J_i	$x_{i.}$	$\overline{x}_{i.}$
Permanent molding	45.5 45.3 45.4 44.4 44.6 43.9 44.6 44.0	8	357.7	44.71
Die casting	44.2 43.9 44.7 44.2 44.0 43.8 44.6 43.1	8	352.5	44.06
Plaster molding	46.0 45.9 44.8 46.2 45.1 45.5	6	273.5	45.58
		22	983.7	

Let μ_1, μ_2, and μ_3 denote the true average elastic moduli for the three different processes under the given circumstances. The relevant hypotheses are H_0: $\mu_1 = \mu_2 = \mu_3$ versus H_a: at least two of the μ_i's are different. The test statistic is, of course, $F = MSTr/MSE$, based on $I - 1 = 2$ numerator df and $n - I = 22 - 3 = 19$ denominator df. Relevant quantities include

$$\sum\sum x_{ij}^2 = 43{,}998.73 \quad CF = \frac{983.7^2}{22} = 43{,}984.80$$

$$SST = 43{,}998.73 - 43{,}984.80 = 13.93$$

$$SSTr = \frac{357.7^2}{8} + \frac{352.5^2}{8} + \frac{273.5^2}{6} - 43{,}984.84 = 7.93$$

$$SSE = 13.93 - 7.93 = 6.00$$

The remaining computations are displayed in the accompanying ANOVA table. Since $F_{.001,2,19} = 10.16 < 12.56 = f$, the P-value is smaller than .001. Thus the null hypothesis should be rejected at any reasonable significance level; there is compelling evidence for concluding that a true average elastic modulus somehow depends on which casting process is used.

Source of Variation	df	Sum of Squares	Mean Square	f
Treatments	2	7.93	3.965	12.56
Error	19	6.00	.3158	
Total	21	13.93		

There is more controversy among statisticians regarding which multiple comparisons procedure to use when sample sizes are unequal than there is in the case of equal sample sizes. The procedure that we present here is recommended in the excellent book *Beyond ANOVA: Basics of Applied Statistics* (see the chapter bibliography) for use when the I sample sizes $J_1, J_2, \ldots J_I$ are reasonably close to one another ("mild imbalance"). It modifies Tukey's method by using averages of pairs of $1/J_i$'s in place of $1/J$.

Let

$$w_{ij} = Q_{\alpha, I, n-I} \cdot \sqrt{\frac{\text{MSE}}{2}\left(\frac{1}{J_i} + \frac{1}{J_j}\right)}$$

Then the probability is *approximately* $1 - \alpha$ that

$$\overline{X}_{i.} - \overline{X}_{j.} - w_{ij} \le \mu_i - \mu_j \le \overline{X}_{i.} - \overline{X}_{j.} + w_{ij}$$

for every i and j ($i = 1, \ldots, I$ and $j = 1, \ldots, I$) with $i \ne j$.

The simultaneous confidence level $100(1 - \alpha)\%$ is only approximate rather than exact as it is with equal sample sizes. Underscoring can still be used, but now the w_{ij} factor used to decide whether $\bar{x}_{i.}$ and $\bar{x}_{j.}$ can be connected will depend on J_i and J_j.

EXAMPLE 10.10
(Example 10.9 continued)

The sample sizes for the elastic modulus data were $J_1 = 8$, $J_2 = 8$, $J_3 = 6$, and $I = 3$, $n - I = 19$, MSE $= .316$. A simultaneous confidence level of approximately 95% requires $Q_{.05,3,19} = 3.59$, from which

$$w_{12} = 3.59 \sqrt{\frac{.316}{2}\left(\frac{1}{8} + \frac{1}{8}\right)} = .713, \quad w_{13} = .771 \quad w_{23} = .771$$

Since $\bar{x}_{1.} - \bar{x}_{2.} = 44.71 - 44.06 = .65 < w_{12}$, μ_1 and μ_2 are judged not significantly different. The accompanying underscoring scheme shows that μ_1 and μ_3 appear to differ significantly, as do μ_2 and μ_3.

2. Die	1. Permanent	3. Plaster
44.06	44.71	45.58

Data Transformation

The use of ANOVA methods can be invalidated by substantial differences in the variances $\sigma_1^2, \ldots, \sigma_I^2$ (which until now have been assumed equal with common value σ^2). It sometimes happens that $V(X_{ij}) = \sigma_i^2 = g(\mu_i)$, a known function of μ_i (so that when

H_0 is false, the variances are not equal). For example, if X_{ij} has a Poisson distribution with parameter λ_i (approximately normal if $\lambda_i \geq 10$), then $\mu_i = \lambda_i$ and $\sigma_i^2 = \lambda_i$, so $g(\mu_i) = \mu_i$ is the known function. In such cases, one can often transform the X_{ij}'s to $h(X_{ij})$ so that they will have approximately equal variances (while leaving the transformed variables approximately normal), and then the F test can be used on the transformed observations. The key idea in choosing $h(\cdot)$ is that often $V[h(X_{ij})] \approx V(X_{ij}) \cdot [h'(\mu_i)]^2 = g(\mu_i) \cdot [h'(\mu_i)]^2$. We now wish to find the function $h(\cdot)$ for which $g(\mu_i) \cdot [h'(\mu_i)]^2 = c$ (a constant) for every i.

PROPOSITION

> If $V(X_{ij}) = g(\mu_i)$, a known function of μ_i, then a transformation $h(X_{ij})$ that "stabilizes the variance" so that $V[h(X_{ij})]$ is approximately the same for each i is given by $h(x) \propto \int [g(x)]^{-1/2} \, dx$.

In the Poisson case, $g(x) = x$, so $h(x)$ should be proportional to $\int x^{-1/2} \, dx = 2x^{1/2}$. Thus Poisson data should be transformed to $h(x_{ij}) = \sqrt{x_{ij}}$ before the analysis.

A Random Effects Model

The single-factor problems considered so far have all been assumed to be examples of a **fixed effects** ANOVA model. By this we mean that the chosen levels of the factor under study are the only ones considered relevant by the experimenter. The single-factor fixed effects model is

$$X_{ij} = \mu + \alpha_i + \epsilon_{ij} \qquad \sum \alpha_i = 0 \qquad (10.6)$$

where the ϵ_{ij}'s are random and both μ and the α_i's are fixed parameters.

In some single-factor problems, the particular levels studied by the experimenter are chosen, either by design or through sampling, from a large population of levels. For example, to study the effects on task performance time of using different operators on a particular machine, a sample of five operators might be chosen from a large pool of operators. Similarly, the effect of soil pH on the yield of maize plants might be studied by using soils with four specific pH values chosen from among the many possible pH levels. When the levels used are selected at random from a larger population of possible levels, the factor is said to be random rather than fixed, and the fixed effects model (10.6) is no longer appropriate. An analogous **random effects** model is obtained by replacing the fixed α_i's in (10.6) by random variables.

$$X_{ij} = \mu + A_i + \epsilon_{ij} \qquad \text{with } E(A_i) = E(\epsilon_{ij}) = 0$$

$$V(\epsilon_{ij}) = \sigma^2 \qquad V(A_i) = \sigma_A^2 \qquad (10.7)$$

all A_i's and ϵ_{ij}'s normally distributed and independent of one another.

The condition $E(A_i) = 0$ in (10.7) is similar to the condition $\Sigma \alpha_i = 0$ in (10.6); it states that the expected or average effect of the ith level measured as a departure from μ is zero.

For the random effects model (10.7), the hypothesis of no effects due to different levels is $H_0: \sigma_A^2 = 0$, which says that different levels of the factor contribute nothing to variability of the response. *Although the hypotheses in the single-factor fixed and random effects models are different, they are tested in exactly the same way*: P-value = area under the $F_{I-1, n-I}$ curve to the right of $f = $ MSTr/MSE, where $J_1, J_2, \ldots, J_I$ are the sample sizes and $n = \Sigma J_i$. This can be justified intuitively by noting that $E(\text{MSE}) = \sigma^2$ and

$$E(\text{MSTr}) = \sigma^2 + \frac{1}{I-1}\left(n - \frac{\sum J_i^2}{n}\right)\sigma_A^2 \tag{10.8}$$

The factor in parentheses on the right side of (10.8) is nonnegative, so again $E(\text{MSTr}) = \sigma^2$ if H_0 is true and $E(\text{MSTr}) > \sigma^2$ if H_0 is false.

EXAMPLE 10.11   The study of nondestructive forces and stresses in materials furnishes important information for efficient engineering design. The article **"Zero-Force Travel-Time Parameters for Ultrasonic Head-Waves in Railroad Rail"** (*Materials Evaluation,* **1985: 854–858**) reports on a study of travel time for a certain type of wave that results from longitudinal stress of rails used for railroad track. Three measurements were made on each of six rails randomly selected from a population of rails. The investigators used random effects ANOVA to decide whether some variation in travel time could be attributed to "between-rail variability." The data is given in the accompanying table (each value, in nanoseconds, resulted from subtracting 36.1 μ's from the original observation) along with the derived ANOVA table. The value f is highly significant; H_0: $\sigma_A^2 = 0$ is rejected in favor of the conclusion that differences between rails is a source of travel-time variability.

				x_i	Source of Variation	df	Sum of Squares	Mean Square	f
1:	55	53	54	162	Treatments	5	9310.5	1862.1	115.2
2:	26	37	32	95	Error	12	194.0	16.17	
3:	78	91	85	254	Total	17	9504.5		
4:	92	100	96	288					
5:	49	51	50	150					
6:	80	85	83	248					
				$x_{..} = 1197$					

EXERCISES Section 10.3 (22–34)

22. The following data refers to yield of tomatoes (kg/plot) for four different levels of salinity. *Salinity level* here refers to electrical conductivity (EC), where the chosen levels were EC = 1.6, 3.8, 6.0, and 10.2 nmhos/cm.

    1.6:   59.5   53.3   56.8   63.1   58.7
    3.8:   55.2   59.1   52.8   54.5
    6.0:   51.7   48.8   53.9   49.0
    10.2: 44.6   48.5   41.0   47.3   46.1

    Use the F test at level $\alpha = .05$ to test for any differences in true average yield due to the different salinity levels.

23. Apply the modified Tukey's method to the data in Exercise 22 to identify significant differences among the μ_i's.

24. The accompanying summary data on skeletal-muscle CS activity (nmol/min/mg) appeared in the article **"Impact of Lifelong Sedentary Behavior on Mitochondrial Function of Mice Skeletal Muscle"** (*J. of Gerontology,* **2009: 927–939**):

	Young	Old Sedentary	Old Active
Sample size	10	8	10
Sample mean	46.68	47.71	58.24
Sample sd	7.16	5.59	8.43

    Carry out a test to decide whether true average activity differs for the three groups. If appropriate, investigate differences amongst the means with a multiple comparisons method.

25. Lipids provide much of the dietary energy in the bodies of infants and young children. There is a growing interest in the quality of the dietary lipid supply during infancy as a major determinant of growth, visual and neural development, and long-term health. The article **"Essential Fat Requirements of Preterm Infants"** (*Amer. J. of Clinical Nutrition*, **2000: 245S–250S**) reported the following data on total polyunsaturated fats (%) for infants who were randomized to four different feeding regimens: breast milk, corn-oil-based formula, soy-oil-based formula, or soy-and-marine-oil-based formula:

Regimen	Sample Size	Sample Mean	Sample SD
Breast milk	8	43.0	1.5
CO	13	42.4	1.3
SO	17	43.1	1.2
SMO	14	43.5	1.2

a. What assumptions must be made about the four total polyunsaturated fat distributions before carrying out a single-factor ANOVA to decide whether there are any differences in true average fat content?

b. Carry out the test suggested in part (a). What can be said about the *P*-value?

26. Samples of six different brands of diet/imitation margarine were analyzed to determine the level of physiologically active polyunsaturated fatty acids (PAPFUA, in percentages), resulting in the following data:

Imperial	14.1	13.6	14.4	14.3	
Parkay	12.8	12.5	13.4	13.0	12.3
Blue Bonnet	13.5	13.4	14.1	14.3	
Chiffon	13.2	12.7	12.6	13.9	
Mazola	16.8	17.2	16.4	17.3	18.0
Fleischmann's	18.1	17.2	18.7	18.4	

(The preceding numbers are fictitious, but the sample means agree with data reported in the January 1975 issue of *Consumer Reports*.)

a. Use ANOVA to test for differences among the true average PAPFUA percentages for the different brands.

b. Compute CIs for all $(\mu_i - \mu_j)$'s.

c. Mazola and Fleischmann's are corn-based, whereas the others are soybean-based. Compute a CI for

$$\frac{(\mu_1 + \mu_2 + \mu_3 + \mu_4)}{4} - \frac{(\mu_5 + \mu_6)}{2}$$

[*Hint:* Modify the expression for $V(\hat{\theta})$ that led to (10.5) in the previous section.]

27. Although tea is the world's most widely consumed beverage after water, little is known about its nutritional value. Folacin is the only B vitamin present in any significant amount in tea, and recent advances in assay methods have made accurate determination of folacin content feasible. Consider the accompanying data on folacin content for randomly selected specimens of the four leading brands of green tea.

1:	7.9	6.2	6.6	8.6	8.9	10.1	9.6
2:	5.7	7.5	9.8	6.1	8.4		
3:	6.8	7.5	5.0	7.4	5.3	6.1	
4:	6.4	7.1	7.9	4.5	5.0	4.0	

(Data is based on **"Folacin Content of Tea,"** *J. of the Amer. Dietetic Assoc.*, **1983: 627–632**.) Does this data suggest that true average folacin content is the same for all brands?

a. Carry out a test using $\alpha = .05$.

b. Assess the plausibility of any assumptions required for your analysis in part (a).

c. Perform a multiple comparisons analysis to identify significant differences among brands.

28. For a single-factor ANOVA with sample sizes $J_i (i = 1, 2, \ldots I)$, show that SSTr $= \Sigma J_i (\bar{X}_{i\cdot} - \bar{X}_{\cdot\cdot})^2 = \Sigma J_i \bar{X}_{i\cdot}^2 - \bar{X}_{\cdot\cdot}^2$. where $n = \Sigma J_i$.

29. When sample sizes are equal ($J_i = J$), the parameters $\alpha_1, \alpha_2, \ldots \alpha_I$ of the alternative parameterization are restricted by $\Sigma \alpha_i = 0$. For unequal sample sizes, the most natural restriction is $\Sigma J_i \alpha_i = 0$. Use this to show that

$$E(\text{MSTr}) = \sigma^2 + \frac{1}{I-1} \Sigma J_i \alpha_i^2$$

What is $E(\text{MSTr})$ when H_0 is true? [This expectation is correct if $\Sigma J_i \alpha_i = 0$ is replaced by the restriction $\Sigma \alpha_i = 0$ (or any other single linear restriction on the α_i's used to reduce the model to I independent parameters), but $\Sigma J_i \alpha_i = 0$ simplifies the algebra and yields natural estimates for the model parameters (in particular, $\hat{\alpha}_i = \bar{x}_{i\cdot} - \bar{x}_{\cdot\cdot}$).]

30. Reconsider Example 10.8 involving an investigation of the effects of different heat treatments on the yield point of steel ingots.

a. If $J = 8$ and $\sigma = 1$, what is β for a level .05 F test when $\mu_1 = \mu_2, \mu_3 = \mu_1 - 1$, and $\mu_4 = \mu_1 + 1$?

b. For the alternative of part (a), what value of J is necessary to obtain $\beta = .05$?

c. If there are $I = 5$ heat treatments, $J = 10$, and $\sigma = 1$, what is β for the level .05 F test when four of the μ_i's are equal and the fifth differs by 1 from the other four?

31. When sample sizes are not equal, the noncentrality parameter is $\Sigma J_i \alpha_i^2 / \sigma^2$ and $\phi^2 = (1/I)\Sigma J_i \alpha_i^2 / \sigma^2$. Referring to Exercise 22, what is the power of the test when $\mu_2 = \mu_3, \mu_1 = \mu_2 - \sigma$, and $\mu_4 = \mu_2 + \sigma$?

32. In an experiment to compare the quality of four different brands of magnetic recording tape, five 2400-ft reels of each brand (A–D) were selected and the number of flaws in each reel was determined.

A: 10 5 12 14 8
B: 14 12 17 9 8
C: 13 18 10 15 18
D: 17 16 12 22 14

It is believed that the number of flaws has approximately a Poisson distribution for each brand. Analyze the data at level .01 to see whether the expected number of flaws per reel is the same for each brand.

33. Suppose that X_{ij} is a binomial variable with parameters n and p_i (so approximately normal when $np_i \geq 10$ and $nq_i \geq 10$). Then since $\mu_i = np_i$, $V(X_{ij}) = \sigma_i^2 = np_i(1 - p_i) = \mu_i(1 - \mu_i/n)$. How should the X_{ij}'s be transformed so as to stabilize the variance? [*Hint:* $g(\mu_i) = \mu_i(1 - \mu_i/n)$.]

34. Simplify $E(\text{MSTr})$ for the random effects model when $J_1 = J_2 = \cdots = J_I = J$.

SUPPLEMENTARY EXERCISES (35–46)

35. An experiment was carried out to compare flow rates for four different types of nozzle.
    a. Sample sizes were 5, 6, 7, and 6, respectively, and calculations gave $f = 3.68$. State and test the relevant hypotheses using $\alpha = .01$
    b. Analysis of the data using a statistical computer package yielded *P*-value $= .029$. At level .01, what would you conclude, and why?

36. Cortisol is a hormone that plays an important role in mediating stress. There is growing awareness that exposure of outdoor workers to pollutants may impact cortisol levels. The article **"Plasma Cortisol Concentration and Lifestyle in a Population of Outdoor Workers"** (*Intl. J. of Envir. Health Res.*, **2011: 62–71**) reported on a study involving three groups of police officers: (1) traffic police (TP), (2) drivers (D), and (3) other duties (O). Here is summary data on cortisol concentration (ng/ml) for a subset of the officers who neither drank nor smoked.

Group	Sample Size	Mean	SD
TP	47	174.7	50.9
D	36	160.2	37.2
O	50	153.5	45.9

Assuming that the standard assumptions for one-way ANOVA are satisfied, carry out a test at significance level .05 to decide whether true average cortisol concentration is different for the three groups. [*Note:* The investigators used more sophisticated statistical methodology (multiple regression) to assess the impact of age, length of employment, and drinking and smoking status on cortisol concentration; taking these factors into account, concentration appeared to be significantly higher in the TP group than in the other two groups.]

37. Numerous factors contribute to the smooth running of an electric motor (**"Increasing Market Share Through Improved Product and Process Design: An Experimental Approach,"** *Quality Engineering*, **1991: 361–369**). In particular, it is desirable to keep motor noise and vibration to a minimum. To study the effect that the brand of bearing has on motor vibration, five different motor bearing brands were examined by installing each type of bearing on different random samples of six motors. The amount of motor vibration (measured in microns) was recorded when each of the 30 motors was running. The data for this study follows. State and test the relevant hypotheses at significance level .05, and then carry out a multiple comparisons analysis if appropriate.

							Mean
1:	13.1	15.0	14.0	14.4	14.0	11.6	13.68
2:	16.3	15.7	17.2	14.9	14.4	17.2	15.95
3:	13.7	13.9	12.4	13.8	14.9	13.3	13.67
4:	15.7	13.7	14.4	16.0	13.9	14.7	14.73
5:	13.5	13.4	13.2	12.7	13.4	12.3	13.08

38. An article in the British scientific journal *Nature* (**"Sucrose Induction of Hepatic Hyperplasia in the Rat," August 25, 1972: 461**) reports on an experiment in which each of five groups consisting of six rats was put on a diet with a different carbohydrate. At the conclusion of the experiment, the DNA content of the liver of each rat was determined (mg/g liver), with the following results:

Carbohydrate	$\bar{x}_{i\cdot}$
Starch	2.58
Sucrose	2.63
Fructose	2.13
Glucose	2.41
Maltose	2.49

Assuming also that $\Sigma\Sigma x_{ij}^2 = 183.4$, does the data indicate that true average DNA content is affected by the type of carbohydrate in the diet? Construct an ANOVA table and use a .05 level of significance.

39. Referring to Exercise 38, construct a *t* CI for
$$\theta = \mu_1 - (\mu_2 + \mu_3 + \mu_4 + \mu_5)/4$$
which measures the difference between the average DNA content for the starch diet and the combined average for the four other diets. Does the resulting interval include zero?

40. Refer to Exercise 38. What is β for the test when true average DNA content is identical for three of the diets and falls below this common value by 1 standard deviation (σ) for the other two diets?

41. Four laboratories (1–4) are randomly selected from a large population, and each is asked to make three determinations of the percentage of methyl alcohol in specimens of a compound taken from a single batch. Based on the accompanying data, are differences among laboratories a source of variation in the percentage of methyl alcohol? State and test the relevant hypotheses using significance level .05.

1: 85.06 85.25 84.87
2: 84.99 84.28 84.88
3: 84.48 84.72 85.10
4: 84.10 84.55 84.05

42. The critical flicker frequency (cff) is the highest frequency (in cycles/sec) at which a person can detect the flicker in a flickering light source. At frequencies above the cff, the light source appears to be continuous even though it is actually flickering. An investigation carried out to see whether true average cff depends on iris color yielded the following data (based on the article **"The Effects of Iris Color on Critical Flicker Frequency,"** *J. of General Psych.*, 1973: 91–95):

Iris Color

	1. Brown	2. Green	3. Blue
	26.8	26.4	25.7
	27.9	24.2	27.2
	23.7	28.0	29.9
	25.0	26.9	28.5
	26.3	29.1	29.4
	24.8		28.3
	25.7		
	24.5		
J_i	8	5	6
$x_{i.}$	204.7	134.6	169.0
$\bar{x}_{i.}$	25.59	26.92	28.17

$n = 19$ $x.. = 508.3$

a. State and test the relevant hypotheses at significance level .05 [*Hint:* $\Sigma\Sigma x_{ij}^2 = 13,659.67$ and CF = 13,598.36.]
b. Investigate differences between iris colors with respect to mean cff.

43. Let $c_1, c_2, \ldots, c_I$ be numbers satisfying $\Sigma c_i = 0$. Then $\Sigma c_i \mu_i = c_1\mu_1 + \cdots + c_I\mu_I$ is called a *contrast* in the μ_i's. Notice that with $c_1 = 1$, $c_2 = -1, c_3 = \cdots = c_I = 0$, $\Sigma c_i \mu_i = \mu_1 - \mu_2$ which implies that every pairwise difference between μ_i's is a contrast (so is, e.g., $\mu_1 - .5\mu_2 - .5\mu_3$). A method attributed to Scheffé gives simultaneous CI's with simultaneous confidence level $100(1 - \alpha)\%$ for *all* possible contrasts (an infinite number of them!). The interval for $\Sigma c_i \mu_i$ is

$$\sum c_i \bar{x}_{i.} \pm \left(\sum c_i^2 / J_i \right)^{1/2} \cdot [(I - 1) \cdot \text{MSE} \cdot F_{\alpha, I-1, n-I}]^{1/2}$$

Using the critical flicker frequency data of Exercise 42, calculate the Scheffé intervals for the contrasts $\mu_1 - \mu_2, \mu_1 - \mu_3, \mu_2 - \mu_3$, and $.5\mu_1 + .5\mu_2 - \mu_3$ (this last contrast compares blue to the average of brown and green). Which contrasts appear to differ significantly from 0, and why?

44. Four types of mortars—ordinary cement mortar (OCM), polymer impregnated mortar (PIM), resin mortar (RM), and polymer cement mortar (PCM)—were subjected to a compression test to measure strength (MPa). Three strength observations for each mortar type are given in the article **"Polymer Mortar Composite Matrices for Maintenance-Free Highly Durable Ferrocement"** (*J. of Ferrocement*, 1984: 337–345) and are reproduced here. Construct an ANOVA table. Using a .05 significance level, determine whether the data suggests that the true mean strength is not the same for all four mortar types. If you determine that the true mean strengths are not all equal, use Tukey's method to identify the significant differences.

OCM	32.15	35.53	34.20
PIM	126.32	126.80	134.79
RM	117.91	115.02	114.58
PCM	29.09	30.87	29.80

45. Suppose the x_{ij}'s are "coded" by $y_{ij} = cx_{ij} + d$. How does the value of the F statistic computed from the y_{ij}'s compare to the value computed from the x_{ij}'s? Justify your assertion.

46. In Example 10.11, subtract $\bar{x}_{i.}$ from each observation in the ith sample ($i = 1, \ldots, 6$) to obtain a set of 18 residuals. Then construct a normal probability plot and comment on the plausibility of the normality assumption.

BIBLIOGRAPHY

Miller, Rupert, *Beyond ANOVA: The Basics of Applied Statistics,* Wiley, New York, 1986. An excellent source of information about assumption checking and alternative methods of analysis.

Montgomery, Douglas, *Design and Analysis of Experiments* (8th ed.), Wiley, New York, 2013. A very up-to-date presentation of ANOVA models and methodology.

Neter, John, William Wasserman, and Michael Kutner, *Applied Linear Statistical Models* (5th ed.), Irwin, Homewood, IL, 2004. The second half of this book contains a very well-presented survey of ANOVA; the level is comparable to that of the present text, but the discussion is more comprehensive, making the book an excellent reference.

Ott, R. Lyman and Michael Longnecker. *An Introduction to Statistical Methods and Data Analysis* (6th ed.), Cengage Learning, Boston, 2010. Includes several chapters on ANOVA methodology that can profitably be read by students desiring a very nonmathematical exposition; there is a good chapter on various multiple comparison methods.

Multifactor Analysis of Variance

11

INTRODUCTION

In the previous chapter, we used the analysis of variance (ANOVA) to test for equality of either I different population means or the true average responses associated with I different levels of a single factor (alternatively referred to as I different treatments). In many experimental situations, there are two or more factors that are of simultaneous interest. This chapter extends the methods of Chapter 10 to investigate such multifactor situations.

In the first two sections, we concentrate on the case of two factors. We will use I to denote the number of levels of the first factor (A) and J to denote the number of levels of the second factor (B). Then there are IJ possible combinations consisting of one level of factor A and one of factor B. Each such combination is called a treatment, so there are IJ different treatments. The number of observations made on treatment (i, j) will be denoted by K_{ij}. In Section 11.1, we consider $K_{ij} = 1$. An important special case of this type is a randomized block design, in which a single factor A is of primary interest but another factor, "blocks," is created to control for extraneous variability in experimental units or subjects. Section 11.2 focuses on the case $K_{ij} = K > 1$, with brief mention of the difficulties associated with unequal K_{ij}'s.

Section 11.3 considers experiments involving more than two factors. When the number of factors is large, an experiment consisting of at least one observation for each treatment would be expensive and time consuming. One frequently encountered situation, which we discuss in Section 11.4, is that in which there are p factors, each of which has two levels. There are then 2^p different treatments. We consider both the case in which observations are made on all these treatments (a complete design) and the case in which observations are made for only a selected subset of treatments (an incomplete design).

11.1 Two-Factor ANOVA with $K_{ij} = 1$

When factor A consists of I levels and factor B consists of J levels, there are IJ different combinations (pairs) of levels of the two factors, each called a treatment. With $K_{ij} =$ the number of observations on the treatment consisting of factor A at level i and factor B at level j, we restrict attention in this section to the case $K_{ij} = 1$, so that the data consists of IJ observations. Our focus is on the fixed effects model, in which the only levels of interest for the two factors are those actually represented in the experiment. Situations in which at least one factor is random are discussed briefly at the end of the section.

EXAMPLE 11.1 Is it really as easy to remove marks on fabrics from erasable pens as the word *erasable* might imply? Consider the following data from an experiment to compare three different brands of pens and four different wash treatments with respect to their ability to remove marks on a particular type of fabric (based on **"An Assessment of the Effects of Treatment, Time, and Heat on the Removal of Erasable Pen Marks from Cotton and Cotton/Polyester Blend Fabrics," *J. of Testing and Evaluation*, 1991: 394–397**). The response variable is a quantitative indicator of overall specimen color change; the lower this value, the more marks were removed.

		Washing Treatment					
		1	**2**	**3**	**4**	**Total**	**Average**
	1	.97	.48	.48	.46	2.39	.598
Brand of Pen	**2**	.77	.14	.22	.25	1.38	.345
	3	.67	.39	.57	.19	1.82	.455
	Total	2.41	1.01	1.27	.90	5.59	
	Average	.803	.337	.423	.300		.466

Is there any difference in the true average amount of color change due either to the different brands of pens or to the different washing treatments? ■

As in single-factor ANOVA, double subscripts are used to identify random variables and observed values. Let

X_{ij} = the random variable (rv) denoting the measurement when factor A is held at level i and factor B is held at level j

x_{ij} = the observed value of X_{ij}

The x_{ij}'s are usually presented in a rectangular table in which the various rows are identified with the levels of factor A and the various columns with the levels of factor B. In the erasable-pen experiment of Example 11.1, the number of levels of factor A is $I = 3$, the number of levels of factor B is $J = 4$, $x_{13} = .48$, $x_{22} = .14$, and so on.

Whereas in single-factor ANOVA we were interested only in row means and the grand mean, now we are interested also in column means. Let

$$\overline{X}_{i\cdot} = \text{the average of measurements obtained} \\ \text{when factor } A \text{ is held at level } i \quad\quad = \quad \frac{\displaystyle\sum_{j=1}^{J} X_{ij}}{J}$$

$$\overline{X}_{\cdot j} = \text{the average of measurements obtained} \\ \text{when factor } B \text{ is held at level } j \quad\quad = \quad \frac{\displaystyle\sum_{i=1}^{I} X_{ij}}{I}$$

$$\overline{X}_{\cdot\cdot} = \text{the grand mean} \quad\quad\quad\quad\quad\quad = \quad \frac{\displaystyle\sum_{i=1}^{I}\sum_{j=1}^{J} X_{ij}}{IJ}$$

with observed values $\bar{x}_{i\cdot}$, $\bar{x}_{\cdot j}$, and $\bar{x}_{\cdot\cdot}$. Totals rather than averages are denoted by omitting the horizontal bar (so $x_{\cdot j} = \Sigma_i x_{ij}$, etc.). Intuitively, to see whether there is any effect due to the levels of factor A, we should compare the observed $\bar{x}_{i\cdot}$'s with one another. Information about the different levels of factor B should come from the $\bar{x}_{\cdot j}$'s.

The Fixed Effects Model

Proceeding by analogy to single-factor ANOVA, one's first inclination in specifying a model is to let $\mu_{ij} = $ the true average response when factor A is at level i and factor B at level j. This results in IJ mean parameters. Then let

$$X_{ij} = \mu_{ij} + \epsilon_{ij}$$

where ϵ_{ij} is the random amount by which the observed value differs from its expectation. The ϵ_{ij}'s are assumed normal and independent with common variance σ^2. Unfortunately, there is no valid test procedure for this choice of parameters. This is because there are $IJ + 1$ parameters (the μ_{ij}'s and σ^2) but only IJ observations, so after using each x_{ij} as an estimate of μ_{ij}, there is no way to estimate σ^2.

The following alternative model is realistic yet involves relatively few parameters.

> Assume the existence of I parameters $\alpha_1, \alpha_2, \ldots, \alpha_I$ and J parameters $\beta_1, \beta_2, \ldots, \beta_J$, such that
>
> $$X_{ij} = \alpha_i + \beta_j + \epsilon_{ij} \quad (i = 1,\ldots,I, \quad j = 1,\ldots,J) \quad\quad (11.1)$$
>
> so that
>
> $$\mu_{ij} = \alpha_i + \beta_j \quad\quad\quad\quad\quad\quad\quad\quad (11.2)$$

Including σ^2, there are now $I + J + 1$ model parameters, so if $I \geq 3$ and $J \geq 3$, then there will be fewer parameters than observations (in fact, we will shortly modify (11.2) so that even $I = 2$ and/or $J = 2$ will be accommodated).

The model specified in (11.1) and (11.2) is called an **additive model** because each mean response μ_{ij} is the sum of an effect due to factor A at level i (α_i) and an effect due to factor B at level j (β_j). The difference between mean responses for factor A at level i and level i' when B is held at level j is $\mu_{ij} - \mu_{i'j}$. When the model is additive,

$$\mu_{ij} - \mu_{i'j} = (\alpha_i + \beta_j) - (\alpha_{i'} + \beta_j) = \alpha_i - \alpha_{i'}$$

which is independent of the level j of the second factor. A similar result holds for $\mu_{ij} - \mu_{ij'}$. Thus additivity means that the difference in mean responses for two levels of one of the factors is the same for all levels of the other factor. Figure 11.1(a) shows a set of mean responses that satisfy the condition of additivity. A nonadditive configuration is illustrated in Figure 11.1(b).

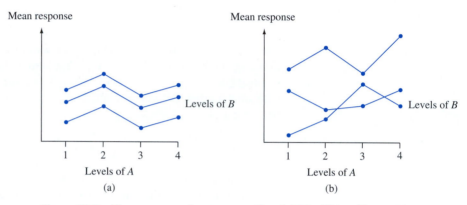

Figure 11.1 Mean responses for two types of model: (a) additive; (b) nonadditive

EXAMPLE 11.2
(Example 11.1 continued)

Plotting the observed x_{ij}'s in a manner analogous to that of Figure 11.1 results in Figure 11.2. Although there is some "crossing over" in the observed x_{ij}'s, the pattern is reasonably representative of what would be expected under additivity with just one observation per treatment.

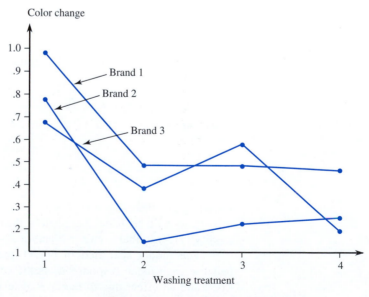

Figure 11.2 Plot of data from Example 11.1

Expression (11.2) is not quite the final model description because the α_i's and β_j's are not uniquely determined. Here are two different configurations of the α_i's and β_j's that yield the same additive μ_{ij}'s:

	$\beta_1 = 1$	$\beta_2 = 4$
$\alpha_1 = 1$	$\mu_{11} = 2$	$\mu_{12} = 5$
$\alpha_2 = 2$	$\mu_{21} = 3$	$\mu_{22} = 6$

	$\beta_1 = 2$	$\beta_2 = 5$
$\alpha_1 = 0$	$\mu_{11} = 2$	$\mu_{12} = 5$
$\alpha_2 = 1$	$\mu_{21} = 3$	$\mu_{22} = 6$

By subtracting any constant c from all α_i's and adding c to all β_j's, other configurations corresponding to the same additive model are obtained. This nonuniqueness is eliminated by use of the following model.

$$X_{ij} = \mu + \alpha_i + \beta_j + \epsilon_{ij} \tag{11.3}$$

where $\sum_{i=1}^{I} \alpha_i = 0$, $\sum_{j=1}^{J} \beta_j = 0$, and the ϵ_{ij}'s are assumed independent, normally distributed, with mean 0 and common variance σ^2.

This is analogous to the alternative choice of parameters for single-factor ANOVA discussed in Section 10.3. It is not difficult to verify that (11.3) is an additive model in which the parameters are uniquely determined (for example, for the μ_{ij}'s mentioned previously: $\mu = 4$, $\alpha_1 = -.5$, $\alpha_2 = .5$, $\beta_1 = -1.5$, and $\beta_2 = 1.5$). Notice that there are only $I - 1$ independently determined α_i's and $J - 1$ independently determined β_j's. Including μ, (11.3) specifies $I + J - 1$ mean parameters.

The interpretation of the parameters in (11.3) is straightforward: **μ is the true grand mean** (mean response averaged over all levels of both factors), **α_i is the effect of factor A at level** i (measured as a deviation from μ), and **β_j is the effect of factor B at level** j. Unbiased (and maximum likelihood) estimators for these parameters are

$$\hat{\mu} = \overline{X}.. \quad \hat{\alpha}_i = \overline{X}_{i.} - \overline{X}.. \quad \hat{\beta}_j = \overline{X}_{.j} - \overline{X}..$$

There are two different null hypotheses of interest in a two-factor experiment with $K_{ij} = 1$. The first, denoted by H_{0A}, states that the different levels of factor A have no effect on true average response. The second, denoted by H_{0B}, asserts that there is no factor B effect.

$$H_{0A}: \alpha_1 = \alpha_2 = \cdots = \alpha_I = 0$$
$$\text{versus } H_{aA}: \text{at least one } \alpha_i \neq 0$$
$$H_{0B}: \beta_1 = \beta_2 = \cdots = \beta_J = 0 \tag{11.4}$$
$$\text{versus } H_{aB}: \text{at least one } \beta_j \neq 0$$

(No factor A effect implies that all α_i's are equal, so they must all be 0 since they sum to 0, and similarly for the β_j's.)

Test Procedures

The description and analysis follow closely that for single-factor ANOVA. There are now four sums of squares, each with an associated number of df:

DEFINITION

$$\text{SST} = \sum_{i=1}^{I} \sum_{j=1}^{J} (X_{ij} - \overline{X}_{..})^2 \qquad\qquad \text{df} = IJ - 1$$

$$\text{SSA} = \sum_{i=1}^{I} \sum_{j=1}^{J} (\overline{X}_{i.} - \overline{X}_{..})^2 = J\sum_{i=1}^{I} (\overline{X}_{i.} - \overline{X}_{..})^2 \qquad \text{df} = I - 1$$

$$(11.5)$$

$$\text{SSB} = \sum_{i=1}^{I} \sum_{j=1}^{J} (\overline{X}_{.j} - \overline{X}_{..})^2 = I\sum_{j=1}^{J} (\overline{X}_{.j} - \overline{X}_{..})^2 \qquad \text{df} = J - 1$$

$$\text{SSE} = \sum_{i=1}^{I} \sum_{j=1}^{J} (X_{ij} - \overline{X}_{i.} - \overline{X}_{.j} + \overline{X}_{..})^2 \qquad \text{df} = (I-1)(J-1)$$

The fundamental identity is

$$\text{SST} = \text{SSA} + \text{SSB} + \text{SSE} \qquad\qquad (11.6)$$

There are computing formulas for SST, SSA, and SSB analogous to those given in Chapter 10 for single-factor ANOVA. But the wide availability of statistical software has rendered these formulas almost obsolete.

The expression for SSE results from replacing μ, α_i, and β_j by their estimators in $\Sigma[X_{ij} - (\mu + \alpha_i + \beta_j)]^2$. Error df is $IJ - $ number of mean parameters estimated $= IJ - [1 + (I-1) + (J-1)] = (I-1)(J-1)$. Total variation is split into a part (SSE) that is not explained by either the truth or the falsity of H_{0A} or H_{0B} and two parts that can be explained by possible falsity of the two null hypotheses.

Statistical theory now says that if we form F ratios as in single-factor ANOVA, when H_{0A} (H_{0B}) is true, the corresponding F ratio has an F distribution with numerator df $= I - 1$ $(J - 1)$ and denominator df $= (I-1)(J-1)$.

Hypotheses	Test Statistic Value	P-Value Determination
H_{0A} versus H_{aA}	$f_A = \dfrac{\text{MSA}}{\text{MSE}}$	Area under the $F_{I-1,(I-1)(J-1)}$ curve to the right of f_A
H_{0B} versus H_{aB}	$f_B = \dfrac{\text{MSB}}{\text{MSE}}$	Area under the $F_{J-1,(I-1)(J-1)}$ curve to the right of f_B

EXAMPLE 11.3
(Example 11.2 continued)

The $\overline{x}_{i.}$'s and $\overline{x}_{.j}$'s for the color-change data are displayed along the margins of the data table given previously. Table 11.1 summarizes the calculations.

Table 11.1 ANOVA Table for Example 11.3

Source of Variation	df	Sum of Squares	Mean Square	f
Factor A (brand)	$I - 1 = 2$	SSA = .1282	MSA = .0641	$f_A = 4.43$
Factor B (wash treatment)	$J - 1 = 3$	SSB = .4797	MSB = .1599	$f_B = 11.05$
Error	$(I-1)(J-1) = 6$	SSE = .0868	MSE = .01447	
Total	$IJ - 1 = 11$	SST = .6947		

Because $F_{.10,2,6} = 3.46 < 4.43 < 5.14 = F_{.05,2,6}$, the P-value for testing H_{0A} is between .05 and .10. Thus H_{0A} cannot be rejected at significance level .05. True average color change does not appear to depend on the brand of pen. Since $F_{.01,3,6} = 9.78$ and $F_{.001,3,6} = 23.70$, $.001 < P$-value $< .01$ for testing H_{0B}. Therefore this null hypothesis is rejected at significance level .05 in favor of the assertion that color change varies with washing treatment. A statistical computer package gives P-values of .066 and .007 for these two tests. ∎

Plausibility of the normality and constant variance assumptions can be investigated graphically. Define *predicted values* (also called *fitted values*) $\hat{x}_{ij} = \hat{\mu} + \hat{\alpha}_i + \hat{\beta}_j = \bar{x}_{..} + (\bar{x}_{i.} - \bar{x}_{..}) + (\bar{x}_{.j} - \bar{x}_{..}) = \bar{x}_{i.} + \bar{x}_{.j} - \bar{x}_{..}$, and the residuals (the differences between the observations and predicted values) $x_{ij} - \hat{x}_{ij} = x_{ij} - \bar{x}_{i.} - \bar{x}_{.j} + \bar{x}_{..}$. We can check the normality assumption with a normal probability plot of the residuals, and the constant variance assumption with a plot of the residuals against the fitted values. Figure 11.3 shows these plots for the data of Example 11.3.

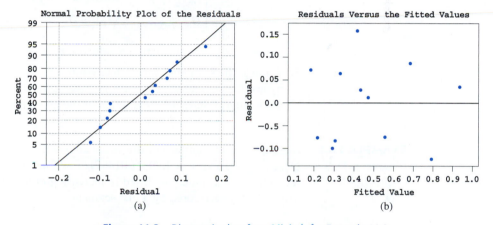

Figure 11.3 Diagnostic plots from Minitab for Example 11.3

The normal probability plot is reasonably straight, so there is no reason to question normality for this data set. On the plot of the residuals against the fitted values, look for substantial variation in vertical spread when moving from left to right. For example, a narrow range for small fitted values and a wide range for high fitted values would suggest that the variance is higher for larger responses (this happens often, and it can sometimes be cured by replacing each observation by its logarithm). Figure 11.3(b) shows no evidence against the constant variance assumption.

Expected Mean Squares

The plausibility of using the F tests just described is demonstrated by computing the expected mean squares. For the additive model,

$$E(\text{MSE}) = \sigma^2$$

$$E(\text{MSA}) = \sigma^2 + \frac{J}{I-1}\sum_{i=1}^{I}\alpha_i^2$$

$$E(\text{MSB}) = \sigma^2 + \frac{I}{J-1}\sum_{j=1}^{J}\beta_j^2$$

If H_{0A} is true, MSA is an unbiased estimator of σ^2, in which case F_A is a ratio of two unbiased estimators of σ^2. When H_{0A} is false, MSA tends to overestimate σ^2. Thus the larger the value of F_A, the more contradictory is the data to H_{0A}. This explains why the test is upper-tailed. Similar comments apply to MSB and H_{0B}.

Multiple Comparisons

After rejecting either H_{0A} or H_{0B}, Tukey's procedure can be used to identify significant differences between the levels of the factor under investigation.

1. For comparing levels of factor A, obtain $Q_{\alpha, I, (I-1)(J-1)}$.

   For comparing levels of factor B, obtain $Q_{\alpha, J, (I-1)(J-1)}$.

2. Compute

   $$w = Q \cdot (\text{estimated standard deviation of the sample}$$
   $$\text{means being compared})$$

   $$= \begin{cases} Q_{\alpha, I, (I-1)(J-1)} \cdot \sqrt{\text{MSE}/J} & \text{for factor A comparisons} \\ Q_{\alpha, J, (I-1)(J-1)} \cdot \sqrt{\text{MSE}/I} & \text{for factor B comparisons} \end{cases}$$

   (because, e.g., the standard deviation of $\overline{X}_{i\cdot}$ is $\sigma/\sqrt{J}$).

3. Arrange the sample means in increasing order, underscore those pairs differing by less than w, and identify pairs not underscored by the same line as corresponding to significantly different levels of the given factor.

EXAMPLE 11.4
(Example 11.3 continued)

Identification of significant differences among the four washing treatments requires $Q_{.05,4,6} = 4.90$ and $w = 4.90\sqrt{(.01447)/3} = .340$. The four factor B sample means (column averages) are now listed in increasing order, and any pair differing by less than .340 is underscored by a line segment:

$$\begin{array}{cccc} \overline{x}_{4\cdot} & \overline{x}_{2\cdot} & \overline{x}_{3\cdot} & \overline{x}_{1\cdot} \\ .300 & 337 & .423 & .803 \end{array}$$

Washing treatment 1 appears to differ significantly from the other three treatments, but no other significant differences are identified. In particular, it is not apparent which among treatments 2, 3, and 4 is best at removing marks. ∎

Randomized Block Experiments

In using single-factor ANOVA to test for the presence of effects due to the I different treatments under study, once the IJ subjects or experimental units have been chosen, treatments should be allocated in a completely random fashion. That is, J subjects should be chosen at random for the first treatment, then another sample of J chosen at random from the remaining $IJ - J$ subjects for the second treatment, and so on.

It frequently happens, though, that subjects or experimental units exhibit heterogeneity with respect to other characteristics that may affect the observed responses. Then, the presence or absence of a significant F value may be due to this extraneous variation rather than to the presence or absence of factor effects. This is why paired experiments were introduced in Chapter 9. The analogy to a paired experiment when $I > 2$ is called a **randomized block** experiment. An extraneous factor, "blocks," is constructed by dividing the IJ units into J groups with I units in each group. This grouping or blocking should be done so that within each block, the I units are homogeneous with respect to other factors thought to affect the responses. Then within each homogeneous block, the I treatments are randomly assigned to the I units or subjects.

EXAMPLE 11.5   A consumer product-testing organization wished to compare the annual power consumption for five different brands of dehumidifier. Because power consumption depends on the prevailing humidity level, it was decided to monitor each brand at four different levels ranging from moderate to heavy humidity (thus blocking on humidity level). Within each level, brands were randomly assigned to the five selected locations. The resulting observations (annual kWh) appear in Table 11.2, and the ANOVA calculations are summarized in Table 11.3.

Table 11.2   Power Consumption Data for Example 11.5

Treatments (brands)	Blocks (humidity level) 1	2	3	4	$x_{i.}$	$\bar{x}_{i.}$
1	685	792	838	875	3190	797.50
2	722	806	893	953	3374	843.50
3	733	802	880	941	3356	839.00
4	811	888	952	1005	3656	914.00
5	828	920	978	1023	3749	937.25
$x_{.j}$	3779	4208	4541	4797	17,325	
$\bar{x}_{.j}$	755.80	841.60	908.20	959.40		866.25

Table 11.3   ANOVA Table for Example 11.5

Source of Variation	df	Sum of Squares	Mean Square	f
Treatments (brands)	4	53,231.00	13,307.75	$f_A = 95.57$
Blocks	3	116,217.75	38,739.25	$f_B = 278.20$
Error	12	1671.00	139.25	
Total	19	171,119.75		

The F ratio for treatments considerably exceeds $F_{.001,4,12} = 9.63$, so P-value $<$ .001. Therefore at significance level .05, H_0 is rejected in favor of H_a. Power consumption appears to depend on the brand of humidifier. To identify significantly different brands, we use Tukey's procedure. $Q_{.05,5,12} = 4.51$ and $w = 4.51\sqrt{139.25/4} = 26.6$.

$\bar{x}_{1.}$	$\bar{x}_{3.}$	$\bar{x}_{2.}$	$\bar{x}_{4.}$	$\bar{x}_{5.}$
797.50	839.00	843.50	914.00	937.25

The underscoring indicates that the brands can be divided into three groups with respect to power consumption.

Because the block factor is of secondary interest, the corresponding P-value is not needed, though the computed value of F_B is clearly highly significant. Figure 11.4 shows SAS output for this data. At the top of the ANOVA table, the sums of squares (SS's) for treatments (brands) and blocks (humidity levels) are combined into a single "model" SS.

```
                         Analysis of Variance Procedure
Dependent Variable: POWERUSE
                                       Sum of              Mean
Source                      DF         Squares            Square       F Value       Pr > F
Model                        7      169448.750         24206.964        173.84       0.0001
Error                       12        1671.000           139.250
Corrected Total             19      171119.750

              R-Square            C.V.            Root MSE         POWERUSE Mean
              0.990235         1.362242            11.8004            866.25000

Source                      DF       Anova SS        Mean Square      F Value       PR > F
BRAND                        4      53231.000         13307.750         95.57       0.0001
HUMIDITY                     3     116217.750         38739.250        278.20       0.0001
```

```
              Alpha = 0.05  df = 12  MSE = 139.25
         Critical Value of Studentized Range = 4.508
           Minimum Significant Difference = 26.597
```

Means with the same letter are not significantly different.

Tukey Grouping	Mean	N	BRAND
A	937.250	4	5
A			
A	914.000	4	4
B	843.500	4	2
B			
B	839.000	4	3
C	797.500	4	1

Figure 11.4 SAS output for power consumption data ∎

In many experimental situations in which treatments are to be applied to subjects, a single subject can receive all I of the treatments. Blocking is then often done on the subjects themselves to control for variability between subjects; each subject is then said to act as its own control. Social scientists sometimes refer to such experiments as repeated-measures designs. The "units" within a block are then the different "instances" of treatment application. Similarly, blocks are often taken as different time periods, locations, or observers.

EXAMPLE 11.6 How does string tension in tennis rackets affect the speed of the ball coming off the racket? The article **"Elite Tennis Player Sensitivity to Changes in String Tension and the Effect on Resulting Ball Dynamics"** (*Sports Engr.*, **2008: 31–36**) described an experiment in which four different string tensions (N) were used, and balls projected from a machine were hit by 18 different players. The rebound speed (km/h) was then determined for each tension-player combination. Consider the following data in Table 11.4 from a similar experiment involving just six players (the resulting ANOVA is in good agreement with what was reported in the article).

The ANOVA calculations are summarized in Table 11.5. The P-value for testing to see whether true average rebound speed depends on string tension is .049. Thus $H_0: \alpha_1 = \alpha_2 = \alpha_3 = \alpha_4 = 0$ is barely rejected at significance level .05 in favor of the conclusion that true average speed does vary with tension ($F_{.05,3,15} = 3.29$). Application of Tukey's procedure to identify significant differences among tensions requires $Q_{.05,4,15} = 4.08$. Then $w = 7.464$. The difference between the largest and smallest sample mean tensions is 6.87. So although the F test is significant, Tukey's

Table 11.4   Rebound Speed Data for Example 11.6

Tension	Player 1	2	3	4	5	6	$\bar{x}_{i.}$
210	105.7	116.6	106.6	113.9	119.4	123.5	114.28
235	113.3	119.9	120.5	119.3	122.5	124.0	119.92
260	117.2	124.4	122.3	120.0	115.1	127.9	121.15
285	110.0	106.8	110.0	115.3	122.6	128.3	115.50
$\bar{x}_{.j}$	111.55	116.93	114.85	117.13	119.90	125.93	

Table 11.5   ANOVA Table for Example 11.6

Source	df	SS	MS	f	P
Tension	3	199.975	66.6582	3.32	0.049
Player	5	477.464	95.4928	4.76	0.008
Error	15	301.188	20.0792		
Total	23	978.626			

method does not identify any significant differences. This occasionally happens when the null hypothesis is just barely rejected. The configuration of sample means in the cited article is similar to ours. The authors commented that the results were contrary to previous laboratory-based tests, where higher rebound speeds are typically associated with low string tension. ■

In most randomized block experiments in which subjects serve as blocks, the subjects actually participating in the experiment are selected from a large population. The subjects then contribute random rather than fixed effects. This does not affect the procedure for comparing treatments when $K_{ij} = 1$ (one observation per "cell," as in this section), but the procedure is altered if $K_{ij} = K > 1$. We will shortly consider two-factor models in which effects are random.

More on Blocking   When $I = 2$, either the F test or the paired differences t test can be used to analyze the data. The resulting conclusion will not depend on which procedure is used, since $T^2 = F$ and $t^2_{\alpha/2,\nu} = F_{\alpha,1,\nu}$.

Just as with pairing, blocking entails both a potential gain and a potential loss in precision. If there is a great deal of heterogeneity in experimental units, the value of the variance parameter σ^2 in the one-way model will be large. The effect of blocking is to filter out the variation represented by σ^2 in the two-way model appropriate for a randomized block experiment. Other things being equal, a smaller value of σ^2 results in a test that is more likely to detect departures from H_0 (i.e., a test with greater power).

However, other things are not equal here, since the single-factor F test is based on $I(J - 1)$ degrees of freedom (df) for error, whereas the two-factor F test is based on $(I - 1)(J - 1)$ df for error. Fewer error df results in a decrease in power, essentially because the denominator estimator of σ^2 is not as precise. This loss in df can be especially serious if the experimenter can afford only a small number of observations. Nevertheless, if it appears that blocking will significantly reduce variability, the sacrifice of error df is sensible.

Models with Random and Mixed Effects

In many experiments, the actual levels of a factor used in the experiment, rather than being the only ones of interest to the experimenter, have been selected from a much larger population of possible levels of the factor. If this is true for both factors in a two-factor experiment, a **random effects model** is appropriate. The case in which the levels of one factor are the only ones of interest and the levels of the other factor are selected from a population of levels leads to a **mixed effects model**. The two-factor random effects model when $K_{ij} = 1$ is

$$X_{ij} = \mu + A_i + B_j + \epsilon_{ij} \quad (i = 1, \ldots, I, \quad j = 1, \ldots, J)$$

The A_i's, B_j's, and ϵ_{ij}'s are all independent, normally distributed rv's with mean 0 and variances σ_A^2, σ_B^2, and σ^2, respectively. The hypotheses of interest are then $H_{0A}: \sigma_A^2 = 0$ (level of factor A does not contribute to variation in the response) versus $H_{aA}: \sigma_A^2 > 0$ and $H_{0B}: \sigma_B^2 = 0$ versus $H_{aB}: \sigma_B^2 > 0$. Whereas $E(\text{MSE}) = \sigma^2$ as before, the expected mean squares for factors A and B are now

$$E(\text{MSA}) = \sigma^2 + J\sigma_A^2 \qquad E(\text{MSB}) = \sigma^2 + I\sigma_B^2$$

Thus when H_{0A} (H_{0B}) is true, F_A (F_B) is still a ratio of two unbiased estimators of σ^2. It can be shown that the P-value for testing H_{0A} versus H_{aA} is computed as for the case of fixed effects; an analogous comment applies for testing H_{0B} versus H_{aB}.

If factor A is fixed and factor B is random, the mixed model is

$$X_{ij} = \mu + \alpha_i + B_j + \epsilon_{ij} \quad (i = 1, \ldots, I, \quad j = 1, \ldots, J)$$

where $\Sigma\alpha_i = 0$ and the B_j's and ϵ_{ij}'s are normally distributed with mean 0 and variances σ_B^2 and σ^2, respectively. Now the two null hypotheses are

$$H_{0A}: \alpha_1 = \cdots = \alpha_I = 0 \quad \text{and} \quad H_{0B}: \sigma_B^2 = 0$$

with expected mean squares

$$E(\text{MSE}) = \sigma^2 \qquad E(\text{MSA}) = \sigma^2 + \frac{J}{I-1}\sum\alpha_i^2 \qquad E(\text{MSB}) = \sigma^2 + I\sigma_B^2$$

The test procedures for H_{0A} versus H_{aA} and H_{0B} versus H_{aB} are exactly as before. For example, in the analysis of the color-change data in Example 11.1, if the four wash treatments were randomly selected, then because $f_B = 11.05$ and $F_{.05,3,6} = 4.76$, $H_{0B}: \sigma_B^2 = 0$ is rejected in favor of $H_{aB}: \sigma_B^2 > 0$. An estimate of the "variance component" σ_B^2 is then given by $(\text{MSB} - \text{MSE})/I = .0485$.

Summarizing, when $K_{ij} = 1$, although the hypotheses and expected mean squares differ from the case of both effects fixed, the test procedures are identical.

EXERCISES Section 11.1 (1–15)

1. An experiment was carried out to investigate the effect of species (factor A, with $I = 4$) and grade (factor B, with $J = 3$) on breaking strength of wood specimens. One observation was made for each species—grade combination—resulting in SSA = 442.0, SSB = 428.6, and SSE = 123.4. Assume that an additive model is appropriate.
   a. Test $H_0: \alpha_1 = \alpha_2 = \alpha_3 = \alpha_4 = 0$ (no differences in true average strength due to species) versus H_a: at least one $\alpha_i \neq 0$ using a level .05 test.

   b. Test $H_0: \beta_1 = \beta_2 = \beta_3 = 0$ (no differences in true average strength due to grade) versus H_a: at least one $\beta_j \neq 0$ using a level .05 test.

2. Four different coatings are being considered for corrosion protection of metal pipe. The pipe will be buried in three different types of soil. To investigate whether the amount of corrosion depends either on the coating or on the type of soil, 12 pieces of pipe are selected. Each piece

is coated with one of the four coatings and buried in one of the three types of soil for a fixed time, after which the amount of corrosion (depth of maximum pits, in .0001 in.) is determined. The data appears in the table.

Soil Type (B)

		1	2	3
	1	64	49	50
Coating (A)	2	53	51	48
	3	47	45	50
	4	51	43	52

a. Assuming the validity of the additive model, carry out the ANOVA analysis using an ANOVA table to see whether the amount of corrosion depends on either the type of coating used or the type of soil. Use $\alpha = .05$.

b. Compute $\hat{\mu}$, $\hat{\alpha}_1$, $\hat{\alpha}_2$, $\hat{\alpha}_3$, $\hat{\alpha}_4$, $\hat{\beta}_1$, $\hat{\beta}_2$, and $\hat{\beta}_3$.

3. An investigation of the machinability of beryllium-copper alloy using two different dielectric mediums and four different working currents resulted in the following data on material removal rate (this is a subset of the data that appeared in the article **"Statistical Analysis and Optimization Study on the Machinability of Beryllium-Copper Alloy in Electro Discharge Machining,"** *J. of Engr. Manufacture*, **2012: 1847–1861**).

Working Current

		10	15	20	25
Medium	Oil	.2433	.3830	.5625	.7258
	Water	.1590	.2649	.3609	.4773

a. After constructing an ANOVA table, test at level .05 both the hypothesis of no medium effect against the appropriate alternative and the hypothesis of no working current effect against the appropriate alternative.

b. Use Tukey's procedure to investigate differences in expected material removal rate due to different working currents ($Q_{.05,4,3} = 6.825$).

4. In an experiment to see whether the amount of coverage of light-blue interior latex paint depends either on the brand of paint or on the brand of roller used, one gallon of each of four brands of paint was applied using each of three brands of roller, resulting in the following data (number of square feet covered).

Roller Brand

		1	2	3
	1	454	446	451
Paint	2	446	444	447
Brand	3	439	442	444
	4	444	437	443

a. Construct the ANOVA table. [*Hint*: The computations can be expedited by subtracting 400 (or any other convenient number) from each observation. This does not affect the final results.]

b. State and test hypotheses appropriate for deciding whether paint brand has any effect on coverage. Use $\alpha = .05$.

c. Repeat part (b) for brand of roller.

d. Use Tukey's method to identify significant differences among brands. Is there one brand that seems clearly preferable to the others?

5. In an experiment to assess the effect of the angle of pull on the force required to cause separation in electrical connectors, four different angles (factor A) were used, and each of a sample of five connectors (factor B) was pulled once at each angle (**"A Mixed Model Factorial Experiment in Testing Electrical Connectors,"** *Industrial Quality Control*, **1960: 12–16**). The data appears in the accompanying table.

		B				
		1	2	3	4	5
	0°	45.3	42.2	39.6	36.8	45.8
A	2°	44.1	44.1	38.4	38.0	47.2
	4°	42.7	42.7	42.6	42.2	48.9
	6°	43.5	45.8	47.9	37.9	56.4

Does the data suggest that true average separation force is affected by the angle of pull? State and test the appropriate hypotheses at level .01 by first constructing an ANOVA table (SST = 396.13, SSA = 58.16, and SSB = 246.97).

6. A particular county employs three assessors who are responsible for determining the value of residential property in the county. To see whether these assessors differ systematically in their assessments, 5 houses are selected, and each assessor is asked to determine the market value of each house. With factor A denoting assessors ($I = 3$) and factor B denoting houses ($J = 5$), suppose SSA = 11.7, SSB = 113.5, and SSE = 25.6.

a. Test $H_0: \alpha_1 = \alpha_2 = \alpha_3 = 0$ at level .05. (H_0 states that there are no systematic differences among assessors.)

b. Explain why a randomized block experiment with only 5 houses was used rather than a one-way ANOVA experiment involving a total of 15 different houses, with each assessor asked to assess 5 different houses (a different group of 5 for each assessor).

7. The accompanying data resulted from an experiment involving three different brands of lathe in combination with three different operators (the blocking factor). The response variable is the percentage of acceptable product produced during a full workday shift (from **"A Software-Based Resource Selection Process in Competitive Network Environment Using ANOVA,"** *Intl. J. of Computer Applications*, **2012: 17–21**).

Operator

		1	2	3
	1	86	85	82
Brand	2	86	86	83
	3	88	91	85

a. Construct the ANOVA table and test at level .05 to see whether brand of lathe has an effect on product acceptability.

b. Judging from the F ratio for operators (factor B), do you think that blocking on operators was effective in this experiment? Explain.

8. The paper **"Exercise Thermoregulation and Hyperprolac-tinaemia"** (*Ergonomics*, 2005: 1547–1557) discussed how various aspects of exercise capacity might depend on the temperature of the environment. The accompanying data on body mass loss (kg) after exercising on a semi-recumbent cycle ergometer in three different ambient temperatures (6°C, 18°C, and 30°C) was provided by the paper's authors.

		Cold	Neutral	Hot
	1	.4	1.2	1.6
	2	.4	1.5	1.9
	3	1.4	.8	1.0
	4	.2	.4	.7
Subject	5	1.1	1.8	2.4
	6	1.2	1.0	1.6
	7	.7	1.0	1.4
	8	.7	1.5	1.3
	9	.8	.8	1.1

a. Does temperature affect true average body mass loss? Carry out a test using a significance level of .01 (as did the authors of the cited paper).

b. Investigate significant differences among the temperatures.

c. The residuals are .20, .30, −.40, −.07, .30, .00, .03, −.20, −.14, .13, .23, −.27, −.04, .03, −.27, −.04, .33, −.10, −.33, −.53, .67, .11, −.33, .27, .01, −.13, .24. Use these as a basis for investigating the plausibility of the assumptions that underlie your analysis in (a).

9. The article **"The Effects of a Pneumatic Stool and a One-Legged Stool on Lower Limb Joint Load and Muscular Activity During Sitting and Rising"** (*Ergonomics*, 1993: 519–535) gives the accompanying data on the effort required of a subject to arise from four different types of stools (Borg scale). Perform an analysis of variance using $\alpha = .05$, and follow this with a multiple comparisons analysis if appropriate.

		\multicolumn{9}{c}{Subject}	

		1	2	3	4	5	6	7	8	9	$\bar{x}_{i.}$
Type of Stool	1	12	10	7	7	8	9	8	7	9	8.56
	2	15	14	14	11	11	11	12	11	13	12.44
	3	12	13	13	10	8	11	12	8	10	10.78
	4	10	12	9	9	7	10	11	7	8	9.22

10. The strength of concrete used in commercial construction tends to vary from one batch to another. Consequently, small test cylinders of concrete sampled from a batch are "cured" for periods up to about 28 days in temperature- and moisture-controlled environments before strength measurements are made. Concrete is then "bought and sold on the basis of strength test cylinders" (ASTM C 31 Standard Test Method for Making and Curing Concrete Test Specimens in the Field). The accompanying data resulted from an experiment carried out to compare three different curing methods with respect to compressive strength (MPa). Analyze this data.

Batch	Method A	Method B	Method C
1	30.7	33.7	30.5
2	29.1	30.6	32.6
3	30.0	32.2	30.5
4	31.9	34.6	33.5
5	30.5	33.0	32.4
6	26.9	29.3	27.8
7	28.2	28.4	30.7
8	32.4	32.4	33.6
9	26.6	29.5	29.2
10	28.6	29.4	33.2

11. For the data of Example 11.5, check the plausibility of assumptions by constructing a normal probability plot of the residuals and a plot of the residuals versus the predicted values, and comment on what you learn.

12. Suppose that in the experiment described in Exercise 6 the five houses had actually been selected at random from among those of a certain age and size, so that factor B is random rather than fixed. Test $H_0: \sigma_B^2 = 0$ versus $H_a: \sigma_B^2 > 0$ using a level .01 test.

13. a. Show that a constant d can be added to (or subtracted from) each x_{ij} without affecting any of the ANOVA sums of squares.

b. Suppose that each x_{ij} is multiplied by a nonzero constant c. How does this affect the ANOVA sums of squares? How does this affect the values of the F statistics F_A and F_B? What effect does "coding" the data by $y_{ij} = cx_{ij} + d$ have on the conclusions resulting from the ANOVA procedures?

14. Use the fact that $E(X_{ij}) = \mu + \alpha_i + \beta_j$ with $\Sigma\alpha_i = \Sigma\beta_j = 0$ to show that $E(\bar{X}_{i.} - \bar{X}_{..}) = \alpha_i$, so that $\hat{\alpha}_i = \bar{X}_{i.} - \bar{X}_{..}$ is an unbiased estimator for α_i.

15. The power curves of Figures 10.5 and 10.6 can be used to obtain $\beta = P(\text{type II error})$ for the F test in two-factor ANOVA. For fixed values of $\alpha_1, \alpha_2, \ldots, \alpha_I$, the quantity $\phi^2 = (J/I)\Sigma\alpha_i^2/\sigma^2$ is computed. Then the figure corresponding to $v_1 = I - 1$ is entered on the horizontal axis at the value ϕ, the power is read on the vertical axis from the curve labeled $v_2 = (I - 1)(J - 1)$, and $\beta = 1 - \text{power}$.

a. For the corrosion experiment described in Exercise 2, find β when $\alpha_1 = 4, \alpha_2 = 0, \alpha_3 = \alpha_4 = -2$, and $\sigma = 4$. Repeat for $\alpha_1 = 6, \alpha_2 = 0, \alpha_3 = \alpha_4 = -3$, and $\sigma = 4$.

b. By symmetry, what is β for the test of H_{0B} versus H_{aB} in Example 11.1 when $\beta_1 = .3, \beta_2 = \beta_3 = \beta_4 = -.1$, and $\sigma = .3$?

In Section 11.1, we analyzed data from a two-factor experiment in which there was one observation for each of the IJ combinations of factor levels. The μ_{ij}'s were assumed to have an additive structure with $\mu_{ij} = \mu + \alpha_i + \beta_j$, $\Sigma \alpha_i = \Sigma \beta_j = 0$. Additivity means that the difference in true average responses for any two levels of the factors is the same for each level of the other factor. For example, $\mu_{ij} - \mu_{i'j} = (\mu + \alpha_i + \beta_j) - (\mu + \alpha_{i'} + \beta_j) = \alpha_i - \alpha_{i'}$, independent of the level j of the second factor. This is shown in Figure 11.1(a) on p. 440, in which the lines connecting true average responses are parallel.

Figure 11.1(b) depicts a set of true average responses that does not have additive structure. The lines connectitng these μ_{ij}'s are not parallel, which means that the difference in true average responses for different levels of one factor does depend on the level of the other factor. When additivity does not hold, we say that there is **interaction** between the different levels of the factors. The assumption of additivity in Section 11.1 allowed us to obtain an estimator of the random error variance σ^2 (MSE) that was unbiased whether or not either null hypothesis of interest was true. When $K_{ij} > 1$ for at least one (i, j) pair, a valid estimator of σ^2 can be obtained without assuming additivity. Our focus here will be on the case $K_{ij} = K > 1$, so the number of observations per "cell" (for each combination of levels) is constant.

Fixed Effects Parameters and Hypotheses

Rather than use the μ_{ij}'s themselves as model parameters, it is customary to use an equivalent set that reveals more clearly the role of interaction.

NOTATION

$$\mu = \frac{1}{IJ} \sum_i \sum_j \mu_{ij} \qquad \mu_{i\cdot} = \frac{1}{J} \sum_j \mu_{ij} \qquad \mu_{\cdot j} = \frac{1}{I} \sum_i \mu_{ij} \qquad (11.7)$$

Thus μ is the expected response averaged over all levels of both factors (the true grand mean), $\mu_{i\cdot}$ is the expected response averaged over levels of the second factor when the first factor A is held at level i, and similarly for $\mu_{\cdot j}$.

DEFINITION

$\alpha_i = \mu_{i\cdot} - \mu = $ **the effect of factor A at level i**
$\beta_j = \mu_{\cdot j} - \mu = $ **the effect of factor B at level j**

$\gamma_{ij} = \mu_{ij} - (\mu + \alpha_i + \beta_j) = $ **interaction between factor A at level i and factor B at level j** $\qquad (11.8)$

from which

$$\mu_{ij} = \mu + \alpha_i + \beta_j + \gamma_{ij} \qquad (11.9)$$

The model is additive if and only if all γ_{ij}'s $= 0$. The γ_{ij}'s are referred to as the **interaction parameters**. The α_is are called the **main effects for factor A**, and the β_j's are the **main effects for factor B**. Although there are I α_i's, J β_j's, and IJ γ_{ij}'s in addition to μ, the conditions $\Sigma \alpha_i = 0$, $\Sigma \beta_j = 0$, $\Sigma_j \gamma_{ij} = 0$ for any i, and $\Sigma_i \gamma_{ij} = 0$ for any j [all by virtue of (11.7) and (11.8)] imply that only IJ of these new parameters are independently determined: $\mu, I - 1$ of the α_i's $J - 1$ of the β_j's, and $(I - 1)(J - 1)$ of the γ_{ij}'s.

There are now three sets of hypotheses to be considered:

H_{0AB}: $\gamma_{ij} = 0$ for all i, j	versus	H_{aAB}: at least one $\gamma_{ij} \neq 0$
H_{0A}: $\alpha_1 = \cdots = \alpha_I = 0$	versus	H_{aA}: at least one $\alpha_i \neq 0$
H_{0B}: $\beta_1 = \cdots = \beta_J = 0$	versus	H_{aB}: at least one $\beta_j \neq 0$

The no-interaction hypothesis H_{0AB} is usually tested first. If H_{0AB} is not rejected, then the other two hypotheses can be tested to see whether the main effects are significant. If H_{0AB} is rejected and H_{0A} is then tested and not rejected, the resulting model $\mu_{ij} = \mu + \beta_j + \gamma_{ij}$ does not lend itself to straightforward interpretation. In such a case, it is best to construct a picture similar to that of Figure 11.1(b) to try to visualize the way in which the factors interact.

The Model and Test Procedures

We now use triple subscripts for both random variables and observed values, with X_{ijk} and x_{ijk} referring to the kth observation (replication) when factor A is at level i and factor B is at level j.

The fixed effects model is

$$X_{ijk} = \mu + \alpha_i + \beta_j + \gamma_{ij} + \epsilon_{ijk} \qquad (11.10)$$
$$i = 1,\ldots, I, \quad j = 1,\ldots, J, \quad k = 1,\ldots, K$$

where the ϵ_{ijk}'s are independent and normally distributed, each with mean 0 and variance σ^2.

Again, a dot in place of a subscript denotes summation over all values of that subscript, and a horizontal bar indicates averaging. Thus $X_{ij\cdot}$ is the total of all K observations made for factor A at level i and factor B at level j [all observations in the (i, j)th cell], and $\overline{X}_{ij\cdot}$ is the average of these K observations. Test procedures are based on the following sums of squares:

DEFINITION

$$\text{SST} = \sum_i \sum_j \sum_k (X_{ijk} - \overline{X}_{\ldots})^2 \qquad \text{df} = IJK - 1$$

$$\text{SSE} = \sum_i \sum_j \sum_k (X_{ijk} - \overline{X}_{ij\cdot})^2 \qquad \text{df} = IJ(K - 1)$$

$$\text{SSA} = \sum_i \sum_j \sum_k (\overline{X}_{i\cdot\cdot} - \overline{X}_{\ldots})^2 \qquad \text{df} = I - 1$$

$$\text{SSB} = \sum_i \sum_j \sum_k (\overline{X}_{\cdot j\cdot} - \overline{X}_{\ldots})^2 \qquad \text{df} = J - 1$$

$$\text{SSAB} = \sum_i \sum_j \sum_k (\overline{X}_{ij\cdot} - \overline{X}_{i\cdot\cdot} - \overline{X}_{\cdot j\cdot} + \overline{X}_{\ldots})^2 \qquad \text{df} = (I - 1)(J - 1)$$

The fundamental identity is

$$\text{SST} = \text{SSA} + \text{SSB} + \text{SSAB} + \text{SSE}$$

SSAB is referred to as **interaction sum of squares.**

Total variation is thus partitioned into four pieces: unexplained (SSE—which would be present whether or not any of the three null hypotheses was true) and three pieces that may be attributed to the truth or falsity of the three H_0's. Each of four mean squares is defined by MS = SS/df. The expected mean squares suggest that each set of hypotheses should be tested using the appropriate ratio of mean squares with MSE in the denominator:

$$E(\text{MSE}) = \sigma^2$$

$$E(\text{MSA}) = \sigma^2 + \frac{JK}{I-1}\sum_{i=1}^{I}\alpha_i^2 \quad E(\text{MSB}) = \sigma^2 + \frac{IK}{J-1}\sum_{j=1}^{J}\beta_j^2$$

$$E(\text{MSAB}) = \sigma^2 + \frac{K}{(I-1)(J-1)}\sum_{i=1}^{I}\sum_{j=1}^{J}\gamma_{ij}^2$$

Each of the three mean square ratios can be shown to have an F distribution with appropriate dfs when the associated H_0 is true. If H_{0AB} is false, the expected value of the numerator mean square in F_{AB} exceeds that of the denominator mean square. The larger the value of this F ratio, the stronger is the evidence against the null hypothesis, again implying an upper-tailed test. Analogous comments apply to the tests for main effects.

Hypotheses	Test Statistic Value	P-Value Determination
H_{0A} versus H_{aA}	$f_A = \dfrac{\text{MSA}}{\text{MSE}}$	Area under the $F_{I-1,\,IJ(K-1)}$ curve to the right of f_A
H_{0B} versus H_{aB}	$f_B = \dfrac{\text{MSB}}{\text{MSE}}$	Area under the $F_{J-1,\,IJ(K-1)}$ curve to the right of f_B
H_{0AB} versus H_{aAB}	$f_{AB} = \dfrac{\text{MSAB}}{\text{MSE}}$	Area under the $F_{(I-1)(J-1),\,IJ(K-1)}$ curve to the right of f_{AB}

EXAMPLE 11.7 Lightweight aggregate asphalt mix has been found to have lower thermal conductivity than a conventional mix, which is desirable. The article **"Influence of Selected Mix Design Factors on the Thermal Behavior of Lightweight Aggregate Asphalt Mixes"** (*J. of Testing and Eval.*, 2008: 1–8) reported on an experiment in which various thermal properties of mixes were determined. Three different binder grades were used in combination with three different coarse aggregate contents (%), with two observations made for each such combination, resulting in the conductivity data (W/m·°K) that appears in Table 11.6.

Table 11.6 Conductivity Data for Example 11.7

Asphalt Binder Grade	Coarse Aggregate Content (%)			$\bar{x}_{i\cdot\cdot}$
	38	41	44	
PG58	.835, .845	.822, .826	.785, .795	.8180
PG64	.855, .865	.832, .836	.790, .800	.8297
PG70	.815, .825	.800, .820	.770, .790	.8033
$\bar{x}_{\cdot j\cdot}$	.8400	.8227	.7883	

Here $I = J = 3$ and $K = 2$ for a total of $IJK = 18$ observations. The results of the analysis are summarized in the ANOVA table which appears as Table 11.7 (a table with additional information appeared in the cited article).

Table 11.7 ANOVA Table for Example 11.7

Source	DF	SS	MS	f	P
AsphGr	2	.0020893	.0010447	14.12	0.002
AggCont	2	.0082973	.0041487	56.06	0.000
Interaction	4	.0003253	.0000813	1.10	0.414
Error	9	.0006660	.0000740		
Total	17	.0113780			

The *P*-value for testing for the presence of interaction effects is .414, which is clearly larger than any reasonable significance level, so the interaction null hypothesis cannot be rejected. Thus it appears that there is no interaction between the two factors. However, both main effects are significant at the 5% significance level (.002 ≤ .05 and .000 ≤ .05). So it appears that true average conductivity depends on which grade is used and also on the level of coarse-aggregate content.

Figure 11.5(a) shows an interaction plot for the conductivity data. Notice the nearly parallel sets of line segments for the three different asphalt grades, in agreement with the *F* test that shows no significant interaction effects. True average conductivity appears to decrease as aggregate content decreases. Figure 11.5(b) shows an interaction plot for the response variable *thermal diffusivity,* values of which appear in the cited article. The bottom two sets of line segments are close to parallel, but differ markedly from those for PG64; in fact, the *F* ratio for interaction effects is highly significant here.

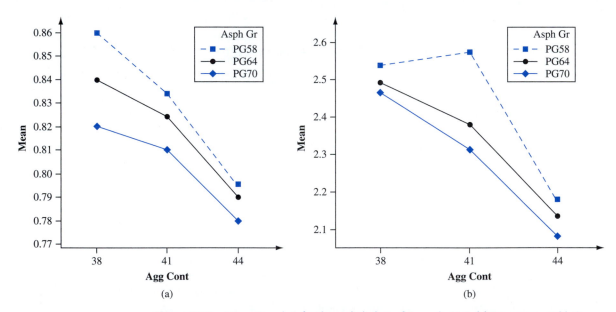

Figure 11.5 Interaction plots for the asphalt data of Example 11.7. (a) Response variable is conductivity. (b) Response variable is diffusivity

Plausibility of the normality and constant variance assumptions can be assessed by constructing plots similar to those of Section 11.1. Define the predicted (i.e., fitted) values to be the cell means: $\hat{x}_{ijk} = \bar{x}_{ij\cdot}$. For example, the predicted value for grade PG58 and aggregate content 38 is $\hat{x}_{11k} = (.835 + .845)/2 = .840$ for $k = 1, 2$. The residuals are the differences between the observations and corresponding predicted values: $x_{ijk} - \bar{x}_{ij\cdot}$. A normal probability plot of the residuals is shown in Figure 11.6(a).

The pattern is sufficiently linear that there should be no concern about lack of normality. The plot of residuals against predicted values in Figure 11.6(b) shows a bit less spread on the right than on the left, but not enough of a differential to be worrisome; constant variance seems to be a reasonable assumption.

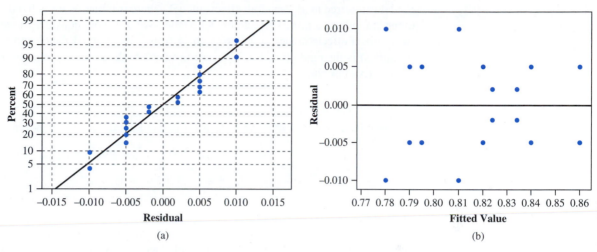

(a)

(b)

Figure 11.6   Plots for checking normality and constant variance assumptions in Example 11.7   ∎

Multiple Comparisons

When the no-interaction hypothesis H_{0AB} is not rejected and at least one of the two main effect null hypotheses is rejected, Tukey's method can be used to identify significant differences in levels. For identifying differences among the α_i's when H_{0A} is rejected,

1. Obtain $Q_{\alpha,I,IJ(K-1)}$, where the second subscript I identifies the number of levels being compared and the third subscript refers to the number of degrees of freedom for error.

2. Compute $w = Q\sqrt{\text{MSE}/(JK)}$, where JK is the number of observations averaged to obtain each of the $\bar{x}_{i\cdot\cdot}$'s compared in Step 3.

3. Order the $\bar{x}_{i\cdot\cdot}$'s from smallest to largest and, as before, underscore all pairs that differ by less than w. Pairs not underscored correspond to significantly different levels of factor A.

To identify different levels of factor B when H_{0B} is rejected, replace the second subscript in Q by J, replace JK by IK in w, and replace $\bar{x}_{i\cdot\cdot}$ by $\bar{x}_{\cdot j\cdot}$.

EXAMPLE 11.8
(Example 11.7 continued)

$I = J = 3$ for both factor A (grade) and factor B (aggregate content). With $\alpha = .05$ and error df $= IJ(K - 1) = 9$, $Q_{.05,3,9} = 3.95$. The yardstick for identifying significant differences is then $w = 3.95\sqrt{.0000740/6} = .00139$. The *grade* sample means in increasing order are .8033, .8180, and .8297. Only the difference between the two largest means is smaller than w. This gives the underscoring pattern

PG70   <u>PG58   PG64</u>

Grades PG58 and PG64 do not appear to differ significantly from one another in effect on true average conductivity, but both differ from the PG70 grade.

The ordered means for factor B are .7883, .8227, and .8400. All three pairs of means differ by more than .00139, so there are no underscoring lines. True average conductivity appears to be different for all three levels of aggregate content.   ∎

Models with Mixed and Random Effects

In some problems, the levels of either factor may have been chosen from a large population of possible levels, so that the effects contributed by the factor are random rather than fixed. As in Section 11.1, if both factors contribute random effects, the model is referred to as a random effects model, whereas if one factor is fixed and the other is random, a mixed effects model results. We will now consider the analysis for a mixed effects model in which factor A (rows) is the fixed factor and factor B (columns) is the random factor. The case in which both factors are random is dealt with in Exercise 26.

DEFINITION

> The **mixed effects model** when factor A is fixed and factor B is random is
>
> $$X_{ijk} = \mu + \alpha_i + B_j + G_{ij} + \epsilon_{ijk}$$
>
> $$i = 1,\ldots,I, \quad j = 1,\ldots,J, \quad k = 1,\ldots,K$$

Here μ and α_i's are constants with $\Sigma\alpha_i = 0$, and the B_j's, G_{ij}'s, and ϵ_{ijk}'s are independent, normally distributed random variables with expected value 0 and variances σ_B^2, σ_G^2, and σ^2, respectively.* The relevant hypotheses here are somewhat different from those for the fixed effects model.

> $H_{0A}: \alpha_1 = \alpha_2 = \cdots = \alpha_I = 0$ versus H_{aA}: at least one $\alpha_i \neq 0$
>
> $H_{0B}: \sigma_B^2 = 0$ versus $H_{aB}: \sigma_B^2 > 0$
>
> $H_{0G}: \sigma_G^2 = 0$ versus $H_{aG}: \sigma_G^2 > 0$

It is customary to test H_{0A} and H_{0B} only if the no-interaction hypothesis H_{0G} cannot be rejected.

Sums of squares and mean squares needed for the test procedures are defined and computed exactly as in the fixed effects case. The expected mean squares for the mixed model are

$$E(\text{MSE}) = \sigma^2$$

$$E(\text{MSA}) = \sigma^2 + K\sigma_G^2 + \frac{JK}{I-1}\sum\alpha_i^2$$

$$E(\text{MSB}) = \sigma^2 + K\sigma_G^2 + IK\sigma_B^2$$

$$E(\text{MSAB}) = \sigma^2 + K\sigma_G^2$$

The ratio $f_{AB} = \text{MSAB}/\text{MSE}$ is again appropriate for testing the no-interaction hypothesis, with the P-value determined as in the fixed effects case. However, for testing H_{0A} versus H_{aA}, the expected mean squares suggest that although the numerator of the F ratio should still be MSA, the denominator should be MSAB rather than MSE. MSAB is also the denominator of the F ratio for testing H_{0B}.

---

* This is referred to as an "unrestricted" model. An alternative "restricted" model requires that $\Sigma_i G_{ij} = 0$ for each j (so the G_{ij}'s are no longer independent). Expected mean squares and F ratios appropriate for testing certain hypotheses depend on the choice of model. Minitab's default option gives output for the unrestricted model.

For testing H_{0A} versus H_{aA} (A fixed, B random), the test statistic value is $f_A = \text{MSA}/\text{MSAB}$, and the P-value is the area under the $F_{I-1,(I-1)(J-1)}$ curve to the right of f_A. The test of H_{0B} versus H_{aB} utilizes $f_B = \text{MSB}/\text{MSAB}$; the P-value is the area under the $F_{J-1,(I-1)(J-1)}$ curve to the right of f_B.

EXAMPLE 11.9 A process engineer has identified two potential causes of electric motor vibration, the material used for the motor casing (factor A) and the supply source of bearings used in the motor (factor B). The accompanying data on the amount of vibration (microns) resulted from an experiment in which motors with casings made of steel, aluminum, and plastic were constructed using bearings supplied by five randomly selected sources.

		Supply Source				
		1	**2**	**3**	**4**	**5**
	Steel	13.1 13.2	16.3 15.8	13.7 14.3	15.7 15.8	13.5 12.5
Material	**Aluminum**	15.0 14.8	15.7 16.4	13.9 14.3	13.7 14.2	13.4 13.8
	Plastic	14.0 14.3	17.2 16.7	12.4 12.3	14.4 13.9	13.2 13.1

Only the three casing materials used in the experiment are under consideration for use in production, so factor A is fixed. However, the five supply sources were randomly selected from a much larger population, so factor B is random. The relevant null hypotheses are

$$H_{0A}: \alpha_1 = \alpha_2 = \alpha_3 = 0 \qquad H_{0B}: \sigma_B^2 = 0 \qquad H_{0AB}: \sigma_G^2 = 0$$

Minitab output appears in Figure 11.7. The P-value column in the ANOVA table indicates that the latter two null hypotheses should be rejected at significance level .05. Different casing materials by themselves do not appear to affect vibration, but interaction between material and supplier is a significant source of variation in vibration.

```
Factor       Type      Levels    Values
casmater     fixed          3        1    2    3
source       random         5        1    2    3    4    5

Source          DF           SS          MS          F          P
casmater         2       0.7047      0.3523       0.24      0.790
source           4      36.6747      9.1687       6.32      0.013
casmater*source  8      11.6053      1.4507      13.03      0.000
Error           15       1.6700      0.1113
Total           29      50.6547

Source              Variance  Error  Expected Mean Square for Each Term
                    component  term  (using unrestricted model)
1 casmater                        3  (4)+2(3)+Q[1]
2 source              1.2863      3  (4)+2(3)+6(2)
3 casmater*source     0.6697      4  (4)+2(3)
4 Error               0.1113         (4)
```

Figure 11.7 Output from Minitab's balanced ANOVA option for the data of Example 11.9 ■

When at least two of the K_{ij}'s are unequal, the ANOVA computations are much more complex than for the case $K_{ij} = K$. In addition, there is controversy as to which test procedures should be used. One of the chapter references can be consulted for more information.

EXERCISES Section 11.2 (16–26)

16. In an experiment to assess the effects of curing time (factor A) and type of mix (factor B) on the compressive strength of hardened cement cubes, three different curing times were used in combination with four different mixes, with three observations obtained for each of the 12 curing time–mix combinations. The resulting sums of squares were computed to be SSA = 30,763.0, SSB = 34,185.6, SSE = 97,436.8, and SST = 205,966.6.

   a. Construct an ANOVA table.

   b. Test at level .05 the null hypothesis H_{0AB}: all γ_{ij}'s = 0 (no interaction of factors) against H_{aAB}: at least one $\gamma_{ij} \neq 0$.

   c. Test at level .05 the null hypothesis H_{0A}: $\alpha_1 = \alpha_2 = \alpha_3 = 0$ (factor A main effects are absent) against H_{aA}: at least one $\alpha_i \neq 0$.

   d. Test H_{0B}: $\beta_1 = \beta_2 = \beta_3 = \beta_4 = 0$ versus H_{aB}: at least one $\beta_j \neq 0$ using a level .05 test.

   e. The values of the $\bar{x}_{i..}$'s were $\bar{x}_{1..} = 4010.88$, $\bar{x}_{2..} = 4029.10$, and $\bar{x}_{3..} = 3960.02$. Use Tukey's procedure to investigate significant differences among the three curing times.

17. The article **"Towards Improving the Properties of Plaster Moulds and Castings"** (*J. Engr. Manuf.*, **1991: 265–269**) describes several ANOVAs carried out to study how the amount of carbon fiber and sand additions affect various characteristics of the molding process. Here we give data on casting hardness and on wet-mold strength.

Sand Addition (%)	Carbon Fiber Addition (%)	Casting Hardness	Wet-Mold Strength
0	0	61.0	34.0
0	0	63.0	16.0
15	0	67.0	36.0
15	0	69.0	19.0
30	0	65.0	28.0
30	0	74.0	17.0
0	.25	69.0	49.0
0	.25	69.0	48.0
15	.25	69.0	43.0
15	.25	74.0	29.0
30	.25	74.0	31.0
30	.25	72.0	24.0
0	.50	67.0	55.0
0	.50	69.0	60.0
15	.50	69.0	45.0
15	.50	74.0	43.0
30	.50	74.0	22.0
30	.50	74.0	48.0

   a. An ANOVA for wet-mold strength gives SS Sand = 705, SSFiber = 1278, SSE = 843, and SST = 3105. Test for the presence of any effects using $\alpha = .05$.

   b. Carry out an ANOVA on the casting hardness observations using $\alpha = .05$.

   c. Plot sample mean hardness against sand percentage for different levels of carbon fiber. Is the plot consistent with your analysis in part (b)?

18. The accompanying data resulted from an experiment to investigate whether yield from a certain chemical process depended either on the formulation of a particular input or on mixer speed.

		Speed		
		60	**70**	**80**
		189.7	185.1	189.0
	1	188.6	179.4	193.0
		190.1	177.3	191.1
Formulation				
		165.1	161.7	163.3
	2	165.9	159.8	166.6
		167.6	161.6	170.3

A statistical computer package gave SS(Form) = 2253.44, SS(Speed) = 230.81, SS(Form*Speed) = 18.58, and SSE = 71.87.

   a. Does there appear to be interaction between the factors?

   b. Does yield appear to depend on either formulation or speed?

   c. Calculate estimates of the main effects.

   d. The *fitted values* are $\hat{x}_{ijk} = \hat{\mu} + \hat{\alpha}_i + \hat{\beta}_j + \hat{\gamma}_{ij}$, and the *residuals* are $x_{ijk} - \hat{x}_{ijk}$. Verify that the residuals are .23, −.87, .63, 4.50, −1.20, −3.30, −2.03, 1.97, .07, −1.10, −.30, 1.40, .67, −1.23, .57, −3.43, −.13, and 3.57.

   e. Construct a normal probability plot from the residuals given in part (d). Do the ϵ_{ijk}'s appear to be normally distributed?

19. A two-way ANOVA was carried out to assess the impact of type of farm (government agricultural settlement, established, individual) and tractor maintenance method (preventive, predictive, running, corrective, overhauling, breakdown) on the response variable *maintenance practice contribution*. There were two observations for each combination of factor levels. The resulting sums of squares were SSA = 35.75 (A = *type of farm*), SSB = 861.20, SSAB = 603.51, and SSE = 341.82 (**"Appraisal of Farm Practice Maintenance and Costs in Nigeria," *J. of Quality in Maintenance Engr.*, 2005: 152–168**). Assuming both factor effects to be fixed, construct an ANOVA table, test for the presence of interaction, and then test for the presence of main effects for each factor (all using level .01).

20. The article **"Fatigue Limits of Enamel Bonds with Moist and Dry Techniques"** (*Dental Materials*, 2009: 1527–1531) described an experiment to investigate the ability of adhesive systems to bond to mineralized tooth structures. The response variable is shear bond strength (MPa), and two different adhesives (Adper Single Bond Plus and OptiBond Solo Plus) were used in combination with two different surface conditions. The accompanying data was supplied by the authors of the article. The first 12 observations came from the SBP-dry treatment, the next 12 from the SBP-moist treatment, the next 12 from the OBP-dry treatment, and the last 12 from the OBP-moist treatment.

56.7	57.4	53.4	54.0	49.9	49.9
56.2	51.9	49.6	45.7	56.8	54.1
49.2	47.4	53.7	50.6	62.7	48.8
41.0	57.4	51.4	53.4	55.2	38.9
38.8	46.0	38.0	47.0	46.2	39.8
25.9	37.8	43.4	40.2	35.4	40.3
40.6	35.5	58.7	50.4	43.1	61.7
33.3	38.7	45.4	47.2	53.3	44.9

a. Construct a comparative boxplot of the data on the four different treatments and comment.

b. Carry out an appropriate analysis of variance and state your conclusions (use a significance level of .01 for any tests). Include any graphs that provide insight.

c. If a significance level of .05 is used for the two-way ANOVA, the interaction effect is significant (just as in general different glues work better with some materials than with others). So now it makes sense to carry out a one-way ANOVA on the four treatments SBP-D, SBP-M, OBP-D, and OBP-M. Do this, and identify significant differences among the treatments.

21. In an experiment to investigate the effect of "cement factor" (number of sacks of cement per cubic yard) on flexural strength of the resulting concrete (**"Studies of Flexural Strength of Concrete. Part 3: Effects of Variation in Testing Procedure,"** *Proceedings, ASTM,* 1957: 1127–1139), $I = 3$ different factor values were used, $J = 5$ different batches of cement were selected, and $K = 2$ beams were cast from each cement factor/batch combination. Sums of squares include SSA = 22,941.80, SSB = 22,765.53, SSE = 15,253.50, and SST = 64,954.70. Construct the ANOVA table. Then, assuming a mixed model with cement factor (A) fixed and batches (B) random, test the three pairs of hypotheses of interest at level .05.

22. A study was carried out to compare the writing lifetimes of four premium brands of pens. It was thought that the writing surface might affect lifetime, so three different surfaces were randomly selected. A writing machine was used to ensure that conditions were otherwise homogeneous (e.g., constant pressure and a fixed angle). The accompanying table shows the two lifetimes (min) obtained for each brand–surface combination.

		Writing Surface		
		1	2	3	$x_{i..}$
Brand	1	709, 659	713, 726	660, 645	4112
of Pen	2	668, 685	722, 740	692, 720	4227
	3	659, 685	666, 684	678, 750	4122
	4	698, 650	704, 666	686, 733	4137
	$x_{.j.}$	5413	5621	5564	16,598

Carry out an appropriate ANOVA, and state your conclusions.

23. The accompanying data was obtained in an experiment to investigate whether compressive strength of concrete cylinders depends on the type of capping material used or variability in different batches (**"The Effect of Type of Capping Material on the Compressive Strength of Concrete Cylinders,"** *Proceedings ASTM*, 1958: 1166–1186). Each number is a cell total ($x_{ij.}$) based on $K = 3$ observations.

		Batch			
		1	2	3	4	5
Capping Material	1	1847	1942	1935	1891	1795
	2	1779	1850	1795	1785	1626
	3	1806	1892	1889	1891	1756

In addition, $\Sigma\Sigma\Sigma x_{ijk}^2 = 16,815,853$ and $\Sigma\Sigma x_{ij.}^2 = 50,443,409$. Obtain the ANOVA table and then test at level .01 the hypotheses H_{0G} versus H_{aG}, H_{0A} versus H_{aA}, and H_{0B} versus H_{aB}, assuming that capping material is a fixed effects factor and batch is a random effects factor.

24. a. Show that $E(\overline{X}_{i..} - \overline{X}...) = \alpha_i$, so that $\overline{X}_{i..} - \overline{X}...$ is an unbiased estimator for α_i (in the fixed effects model).

b. With $\hat{\gamma}_{ij} = \overline{X}_{ij.} - \overline{X}_{i..} - \overline{X}_{.j.} + \overline{X}...$, show that $\hat{\gamma}_{ij}$ is an unbiased estimator for γ_{ij} (in the fixed effects model).

25. Show how a $100(1 - \alpha)\%$ t CI for $\alpha_i - \alpha_{i'}$ can be obtained. Then compute a 95% interval for $\alpha_2 - \alpha_3$ using the data from Exercise 19. [*Hint:* With $\theta = \alpha_2 - \alpha_3$, the result of Exercise 24(a) indicates how to obtain $\hat{\theta}$. Then compute $V(\hat{\theta})$ and $\sigma_{\hat{\theta}}$, and obtain an estimate of $\sigma_{\hat{\theta}}$ by using $\sqrt{\text{MSE}}$ to estimate σ (which identifies the appropriate number of df).]

26. When both factors are random in a two-way ANOVA experiment with K replications per combination of factor levels, the expected mean squares are $E(\text{MSE}) = \sigma^2$, $E(\text{MSA}) = \sigma^2 + K\sigma_G^2 + JK\sigma_A^2$, $E(\text{MSB}) = \sigma^2 + K\sigma_G^2 + IK\sigma_B^2$, and $E(\text{MSAB}) = \sigma^2 + K\sigma_G^2$.

a. What F ratio is appropriate for testing $H_{0G}: \sigma_G^2 = 0$ versus $H_{aG}: \sigma_G^2 > 0$?

b. Answer part (a) for testing $H_{0A}: \sigma_A^2 = 0$ versus $H_{aA}: \sigma_A^2 > 0$ and $H_{0B}: \sigma_B^2 = 0$ versus $H_{aB}: \sigma_B^2 > 0$.

11.3 Three-Factor ANOVA

To indicate the nature of models and analyses when ANOVA experiments involve more than two factors, we will focus here on the case of three fixed factors—A, B, and C. The numbers of levels of these factors will be denoted by I, J, and K, respectively, and L_{ijk} = the number of observations made with factor A at level i, factor B at level j, and factor C at level k. The analysis is quite complicated when the L_{ijk}'s are not all equal, so we further specialize to $L_{ijk} = L$. Then X_{ijkl} and x_{ijkl} denote the observed value, before and after the experiment is performed, of the lth replication ($l = 1, 2, \ldots, L$) when the three factors are fixed at levels i, j, and k.

To understand the parameters that will appear in the three-factor ANOVA model, first recall that in two-factor ANOVA with replications, $E(X_{ijk}) = \mu_{ij} = \mu + \alpha_i + \beta_j + \gamma_{ij}$, where the restrictions $\Sigma_i \alpha_i = \Sigma_j \beta_j = 0$, $\Sigma_i \gamma_{ij} = 0$ for every j, and $\Sigma_j \gamma_{ij} = 0$ for every i were necessary to obtain a unique set of parameters. If we use dot subscripts on the μ_{ij}'s to denote averaging (rather than summation), then

$$\mu_{i\cdot} - \mu_{\cdot\cdot} = \frac{1}{J} \sum_j \mu_{ij} - \frac{1}{IJ} \sum_i \sum_j \mu_{ij} = \alpha_i$$

is the effect of factor A at level i averaged over levels of factor B, whereas

$$\mu_{ij} - \mu_{\cdot j} = \mu_{ij} - \frac{1}{I} \sum_i \mu_{ij} = \alpha_i + \gamma_{ij}$$

is the effect of factor A at level i specific to factor B at level j. When the effect of A at level i depends on the level of B, there is interaction between the factors, and the γ_{ij}'s are not all zero. In particular,

$$\mu_{ij} - \mu_{\cdot j} - \mu_{i\cdot} + \mu_{\cdot\cdot} = \gamma_{ij} \tag{11.11}$$

The Fixed Effects Model and Test Procedures

> The **fixed effects model for three-factor ANOVA** with $L_{ijk} = L$ is
>
> $$X_{ijkl} = \mu_{ijk} + \varepsilon_{ijkl} \quad i = 1, \ldots, I, \quad j = 1, \ldots, J$$
> $$k = 1, \ldots, K, \quad l = 1, \ldots, L \tag{11.12}$$
>
> where the ϵ_{ijkl}'s are normally distributed with mean 0 and variance σ^2, and
>
> $$\mu_{ijk} = \mu + \alpha_i + \beta_i + \delta_k + \gamma_{ij}^{AB} + \gamma_{ik}^{AC} + \gamma_{jk}^{BC} + \gamma_{ijk} \tag{11.13}$$

The restrictions necessary to obtain uniquely defined parameters are that the sum over any subscript of any parameter on the right-hand side of (11.13) equal 0.

The parameters γ_{ij}^{AB}, γ_{ik}^{AC}, and γ_{jk}^{BC} are called two-factor interactions, and γ_{ijk} is called a three-factor interaction; the α_i's, β_j's, and δ_k's are the main effects parameters. For any fixed level k of the third factor, analogous to (11.11),

$$\mu_{ijk} - \mu_{i\cdot k} - \mu_{\cdot jk} + \mu_{\cdot\cdot k} = \gamma_{ij}^{AB} + \gamma_{ijk}$$

is the interaction of the ith level of A with the jth level of B specific to the kth level of C, whereas

$$\mu_{ij\cdot} - \mu_{i\cdot\cdot} - \mu_{\cdot j\cdot} + \mu_{\cdot\cdot\cdot} = \gamma_{ij}^{AB}$$

is the interaction between A at level i and B at level j averaged over levels of C. If the interaction of A at level i and B at level j does not depend on k, then all γ_{ijk}'s equal 0. Thus nonzero γ_{ijk}'s represent nonadditivity of the two-factor γ_{ij}^{AB}'s over the various levels of the third factor C. If the experiment included more than three factors, there would be corresponding higher-order interaction terms with analogous interpretations. Note that in the previous argument, if we had considered fixing the level of either A or B (rather than C, as was done) and examining the γ_{ijk}'s, their interpretation would be the same; if any of the interactions of two factors depend on the level of the third factor, then there are nonzero γ_{ijk}'s.

When $L > 1$, there is a sum of squares for each main effect, each two-factor interaction, and the three-factor interaction. To write these in a way that indicates how sums of squares are defined when there are more than three factors, note that any of the model parameters in (11.13) can be estimated unbiasedly by averaging X_{ijkl} over appropriate subscripts and taking differences. Thus

$$\hat{\mu} = \overline{X}_{....} \qquad \hat{\alpha}_i = \overline{X}_{i...} - \overline{X}_{....} \qquad \hat{\gamma}_{ij}^{AB} = \overline{X}_{ij..} - \overline{X}_{i...} - \overline{X}_{.j..} + \overline{X}_{....}$$

$$\hat{\gamma}_{ijk} = \overline{X}_{ijk.} - \overline{X}_{ij..} - \overline{X}_{i\cdot k} - \overline{X}_{.jk.} + \overline{X}_{i...} + \overline{X}_{.j..} + \overline{X}_{..k.} - \overline{X}_{....}$$

with other main effects and interaction estimators obtained by symmetry.

DEFINITION

Relevant **sums of squares** are

$$\text{SST} = \sum_i \sum_j \sum_k \sum_l (X_{ijkl} - \overline{X}_{....})^2 \qquad df = IJKL - 1$$

$$\text{SSA} = \sum_i \sum_j \sum_k \sum_l \hat{\alpha}_i^2 = JKL \sum_i (\overline{X}_{i...} - \overline{X}_{....})^2 \qquad df = I - 1$$

$$\text{SSAB} = \sum_i \sum_j \sum_k \sum_l (\hat{\gamma}_{ij}^{AB})^2 \qquad df = (I-1)(J-1)$$

$$= KL \sum_i \sum_j (\overline{X}_{ij..} - \overline{X}_{i...} - \overline{X}_{.j..} + \overline{X}_{....})^2$$

$$\text{SSABC} = \sum_i \sum_j \sum_k \sum_l \hat{\gamma}_{ijk}^2 = L \sum_i \sum_j \sum_k \hat{\gamma}_{ijk}^2 \qquad df = (I-1)(J-1)(K-1)$$

$$\text{SSE} = \sum_i \sum_j \sum_k \sum_l (X_{ijkl} - \overline{X}_{ijk.})^2 \qquad df = IJK(L-1)$$

with the remaining main effect and two-factor interaction sums of squares obtained by symmetry. SST is the sum of the other eight SSs.

Each sum of squares (excepting SST) when divided by its df gives a mean square. Expected mean squares are

$$E(\text{MSE}) = \sigma^2$$

$$E(\text{MSA}) = \sigma^2 + \frac{JKL}{I-1} \sum_i \alpha_i^2$$

$$E(\text{MSAB}) = \sigma^2 + \frac{KL}{(I-1)(J-1)} \sum_i \sum_j (\gamma_{ij}^{AB})^2$$

$$E(\text{MSABC}) = \sigma^2 + \frac{L}{(I-1)(J-1)(K-1)} \sum_i \sum_j \sum_k (\gamma_{ijk})^2$$

with similar expressions for the other expected mean squares. Main effect and interaction hypotheses are tested by forming F ratios with MSE in each denominator.

Null Hypothesis	Test Statistic Value	P-Value Determination
H_{0A}: all α_i's $= 0$	$f_A = \dfrac{\text{MSA}}{\text{MSE}}$	Area under the $F_{I-1,\,IJK(L-1)}$ curve to the right of f_A
H_{0AB}: all γ_{ij}^{AB}'s $= 0$	$f_{AB} = \dfrac{\text{MSAB}}{\text{MSE}}$	Area under the $F_{(I-1)(J-1),\,IJK(L-1)}$ curve to the right of f_{AB}
H_{0ABC}: all γ_{ijk}'s $= 0$	$f_{ABC} = \dfrac{\text{MSABC}}{\text{MSE}}$	Area under the $F_{(I-1)(J-1)(K-1),\,IJK(L-1)}$ curve to the right of f_{ABC}

Usually the main effect hypotheses are tested only if all interactions are judged not significant.

This analysis assumes that $L_{ijk} = L > 1$. If $L = 1$, then as in the two-factor case, the highest-order interactions must be assumed absent to obtain an MSE that estimates σ^2. Setting $L = 1$ and disregarding the fourth subscript summation over l, the foregoing formulas for sums of squares are still valid, and error sum of squares is SSE $= \Sigma_i\Sigma_j\Sigma_k\hat{\gamma}_{ijk}^2$ with $\overline{X}_{ijk\cdot} = X_{ijk}$ in the expression for $\hat{\gamma}_{ijk}$.

EXAMPLE 11.10   There has been increased interest in recent years in renewable fuels such as biodiesel, a form of diesel fuel derived from vegetable oils and animal fats. Advantages over petroleum diesel include nontoxicity, biodegradability, and lower greenhouse gas emissions. The article **"Application of the Full Factorial Design to Optimization of Base-Catalyzed Sunflower Oil Ethanolysis"** (*Fuel*, 2013: 433–442) reported on an investigation of three factors on the purity (%) of the biodiesel fuel *fatty acid ethyl ester* (FAEE). The factors and levels are as follows:

A:	Reaction temperature	25°C, 50°C, 75°C
B:	Ethanol-to-oil molar ratio	6:1, 9:1, 12:1
C:	Catalyst loading	.75 wt.%, 1.00 wt.%, 1.25 wt. %

The data appears in Table 11.8, where $I = J = K = 3$ and $L = 2$.

Table 11.8   Purity (%) data for Example 11.10

	B_1			B_2			B_3		
	C_1	C_2	C_3	C_1	C_2	C_3	C_1	C_2	C_3
A_1	81.07	88.71	95.42	81.54	89.12	96.32	86.07	92.05	97.02
	82.22	87.61	94.06	82.82	86.49	95.45	87.73	91.72	96.16
A_2	87.31	89.52	94.68	87.99	90.05	96.44	89.61	90.32	98.30
	87.94	88.75	95.45	88.98	90.42	96.47	89.02	90.61	96.62
A_3	90.66	91.60	93.65	92.14	92.55	97.41	92.88	96.12	97.66
	91.87	92.34	95.73	92.22	97.06	97.08	93.30	97.41	97.59

The resulting ANOVA table is shown in Table 11.9. The *P*-value for testing H_{0ABC} is .165, which is larger than any sensible significance level. This null hypothesis therefore cannot be rejected; it appears that the extent of interaction between any pair of factors is the same for each level of the remaining factor.

Figure 11.8 shows two-factor interaction plots. For example, the dots in the plot appearing in the *C* row and *B* column represent the $\overline{x}_{\cdot jk}$'s—that is, the observations averaged over the levels of the first factor for each combination of levels of the

Table 11.9 ANOVA Table for the Purity Data of Example 11.10

Source	DF	SS	MS	F	P
A	2	215.385	107.692	112.07	0.000
B	2	74.506	37.253	38.77	0.000
C	2	602.717	301.358	313.60	0.000
A*B	4	13.452	3.363	3.50	0.020
A*C	4	107.409	26.852	27.94	0.000
B*C	4	4.374	1.094	1.14	0.360
A*B*C	8	12.472	1.559	1.62	0.165

second and third factors. The bottom three dots connected by solid line segments represent the third level of factor C at each level of factor B. The fact that connected line segments are quite close to being parallel is evidence for the absence of BC interactions, and indeed the P-value in Table 11.9 for testing this null hypothesis is .360. However, the P-values for testing H_{0AB} and H_{0AC} are .020 and .000, respectively. So at significance level .05, we are forced to conclude that AB interactions and AC interactions are present. The line segments in the AC interaction plot are clearly not close to being parallel. It appears from the interaction plots that expected purity will be maximized when all factors are at their highest levels. As it happens, this is also the message from the main effects plots, but those cannot generally be trusted when interactions are present.

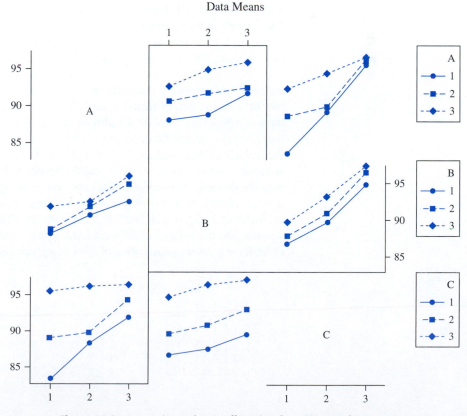

Interaction Plot for Purity

Data Means

Figure 11.8 Interaction and main effect plots from MINITAB for Example 11.10

Main Effects Plot for Purity

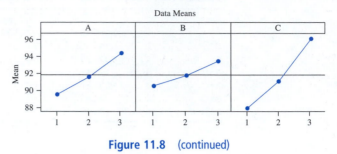

Data Means

Figure 11.8 (continued) ∎

Diagnostic plots for checking the normality and constant variance assumptions can be constructed as described in previous sections. Tukey's procedure can be used in three-factor (or more) ANOVA. The second subscript on Q is the number of sample means being compared, and the third is degrees of freedom for error.

Models with random and mixed effects are also sometimes appropriate. Sums of squares and degrees of freedom are identical to the fixed effects case, but expected mean squares are, of course, different for the random main effects or interactions. A good reference is the book by Douglas Montgomery listed in the chapter bibliography.

Latin Square Designs

When several factors are to be studied simultaneously, an experiment in which there is at least one observation for every possible combination of levels is referred to as a **complete layout**. If the factors are A, B, and C with I, J, and K levels, respectively, a complete layout requires at least IJK observations. Frequently an experiment of this size is either impracticable because of cost, time, or space constraints or literally impossible. For example, if the response variable is sales of a certain product and the factors are different display configurations, different stores, and different time periods, then only one display configuration can realistically be used in a given store during a given time period.

A three-factor experiment in which fewer than IJK observations are made is called an **incomplete layout**. There are some incomplete layouts in which the pattern of combinations of factors is such that the analysis is straightforward. One such three-factor design is called a **Latin square**. It is appropriate when $I = J = K$ (e.g., four display configurations, four stores, and four time periods) and all two- and three-factor interaction effects are assumed absent. If the levels of factor A are identified with the rows of a two-way table and the levels of B with the columns of the table, then the defining characteristic of a Latin square design is that *every level of factor C appears exactly once in each row and exactly once in each column*. Figure 11.9 shows examples of 3 × 3, 4 × 4, and 5 × 5 Latin squares. There are 12 different 3 × 3 Latin squares, and the number of different Latin squares increases rapidly with the number of levels (e.g., every permutation of rows of a given Latin square yields a Latin square,

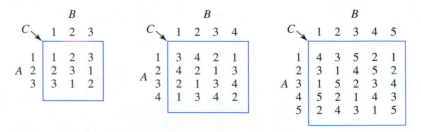

Figure 11.9 Examples of Latin squares

and similarly for column permutations). It is recommended that the square used in a an actual experiment be chosen at random from the set of all possible squares of the desired dimension; for further details, consult one of the chapter references.

The letter N will denote the common value of I, J, and K. Then a complete layout with one observation per combination would require N^3 observations, whereas a Latin square requires only N^2 observations. Once a particular square has been chosen, the value of k (the level of factor C) is completely determined by the values of i and j. To emphasize this, we use $x_{ij(k)}$ to denote the observed value when the three factors are at levels i, j, and k, respectively, with k taking on only one value for each i, j pair.

The model equation for a Latin square design is

$$X_{ij(k)} = \mu + \alpha_i + \beta_j + \delta_k + \epsilon_{ij(k)} \quad i, j, k = 1, \dots, N$$

where $\Sigma\alpha_i = \Sigma\beta_j = \Sigma\delta_k = 0$ and the $\epsilon_{ij(k)}$'s are independent and normally distributed with mean 0 and variance σ^2.

We employ the following notation for totals and averages:

$$X_{i\cdot\cdot} = \sum_j X_{ij(k)} \qquad X_{\cdot j\cdot} = \sum_i X_{ij(k)} \qquad X_{\cdot\cdot k} = \sum_{i,j} X_{ij(k)} \qquad X_{\cdots} = \sum_i \sum_j X_{ij(k)}$$

$$\overline{X}_{i\cdot\cdot} = \frac{X_{i\cdot\cdot}}{N} \qquad \overline{X}_{\cdot j\cdot} = \frac{X_{\cdot j\cdot}}{N} \qquad \overline{X}_{\cdot\cdot k} = \frac{X_{\cdot\cdot k}}{N} \qquad \overline{X}_{\cdots} = \frac{X_{\cdots}}{N^2}$$

Note that although $X_{i\cdot\cdot}$ previously suggested a double summation, now it corresponds to a single sum over all j (and the associated values of k).

DEFINITION

Sums of squares for a Latin square experiment are

$$\text{SST} = \sum_i \sum_j (X_{ij(k)} - \overline{X}_{\cdots})^2 \qquad\qquad \text{df} = N^2 - 1$$

$$\text{SSA} = \sum_i \sum_j (\overline{X}_{i\cdot\cdot} - \overline{X}_{\cdots})^2 \qquad\qquad \text{df} = N = 1$$

$$\text{SSB} = \sum_i \sum_j (\overline{X}_{\cdot j\cdot} - \overline{X}_{\cdots})^2 \qquad\qquad \text{df} = N - 1$$

$$\text{SSC} = \sum_i \sum_j (\overline{X}_{\cdot\cdot k} - \overline{X}_{\cdots})^2 \qquad\qquad \text{df} = N - 1$$

$$\text{SSE} = \sum_i \sum_j [X_{ij(k)} - (\hat{\mu} + \hat{\alpha}_i + \hat{\beta}_j + \hat{\delta}_k)]^2 \qquad \text{df} = N - 1$$

$$\quad = \sum_i \sum_j (X_{ij(k)} - \overline{X}_{i\cdot\cdot} - \overline{X}_{\cdot j\cdot} - \overline{X}_{\cdot\cdot k} + 2\overline{X}_{\cdots})^2 \qquad \text{df} = (N-1)(N-2)$$

$$\text{SST} = \text{SSA} + \text{SSB} + \text{SSC} + \text{SSE}$$

Each mean square is, of course, the ratio SS/df. For testing $H_{0C}: \delta_1 = \delta_2 = \cdots = \delta_N = 0$, the test statistic value is $f_C = \text{MSC}/\text{MSE}$, and the P-value is the area under the $F_{N-1, (N-1)(N-2)}$ curve to the right of f_C. The other two main effect null hypotheses are tested analogously.

If any of the null hypotheses is rejected, significant differences can be identified by using Tukey's procedure. After computing $w = Q_{\alpha,N,(N-1)(N-2)} \cdot \sqrt{\text{MSE}/N}$, pairs of sample means (the $\bar{x}_{i..}$'s, $\bar{x}_{.j.}$'s, or $\bar{x}_{..k}$'s) differing by more than w correspond to significant differences between associated factor effects (the α_i's, β_j's, or δ_k's).

The hypothesis H_{0C} is frequently the one of central interest. A Latin square design is used to control for extraneous variation in the A and B factors, as was done by a randomized block design for the case of a single extraneous factor. Thus in the product sales example mentioned previously, variation due to both stores and time periods is controlled by a Latin square design, enabling an investigator to test for the presence of effects due to different product-display configurations.

EXAMPLE 11.11 In an experiment to investigate the effect of relative humidity on abrasion resistance of leather cut from a rectangular pattern (**"The Abrasion of Leather," *J. Inter. Soc. Leather Trades' Chemists*, 1946: 287**), a 6×6 Latin square was used to control for possible variability due to row and column position in the pattern. The six levels of relative humidity studied were $1 = 25\%, 2 = 37\%, 3 = 50\%, 4 = 62\%, 5 = 75\%$, and $6 = 87\%$, with the following results:

		\multicolumn{6}{c}{B (columns)}					
		1	**2**	**3**	**4**	**5**	**6**	$x_{i..}$
	1	^{3}7.38	^{4}5.39	^{6}5.03	^{2}5.50	^{5}5.01	^{1}6.79	35.10
	2	^{2}7.15	^{1}8.16	^{5}4.96	^{4}5.78	^{3}6.24	^{6}5.06	37.35
	3	^{4}6.75	^{6}5.64	^{3}6.34	^{5}5.31	^{1}7.81	^{2}8.05	39.90
A (rows)	**4**	^{1}8.05	^{3}6.45	^{2}6.31	^{6}5.46	^{4}6.05	^{5}5.51	37.83
	5	^{6}5.65	^{5}5.44	^{1}7.27	^{3}6.54	^{2}7.03	^{4}5.96	37.89
	6	^{5}6.00	^{2}6.55	^{4}5.93	^{1}8.02	^{6}5.80	^{3}6.61	38.91
	$x_{.j.}$	40.98	37.63	35.84	36.61	37.94	37.98	

Also, $x_{..1} = 46.10$, $x_{..2} = 40.59$, $x_{..3} = 39.56$, $x_{..4} = 35.86$, $x_{..5} = 32.23$, $x_{..6} = 32.64$, $x_{...} = 226.98$. Further computations are summarized in Table 11.10.

Table 11.10 ANOVA Table for Example 11.11

Source of Variation	df	Sum of Squares	Mean Square	f
A (rows)	5	2.19	.438	2.50
B (columns)	5	2.57	.514	2.94
C (treatments)	5	23.53	4.706	26.89
Error	20	3.49	.175	
Total	35	31.78		

Since $F_{.001,5,20} = 6.46 < 26.89$, P-value $< .001$. Thus H_{0C} is rejected at any sensible significance level in favor of the hypothesis that relative humidity does on average affect abrasion resistance.

To apply Tukey's procedure, $w = Q_{.05,6,20} \cdot \sqrt{\text{MSE}/6} = 4.45 \sqrt{.175/6} = .76$. Ordering the $\bar{x}_{..k}$'s and underscoring yields

75%	87%	62%	50%	37%	25%
5.37	5.44	5.98	6.59	6.77	7.68

In particular, the lowest relative humidity appears to result in a true average abrasion resistance significantly higher than for any other relative humidity studied. ∎

27. The output of a continuous extruding machine that coats steel pipe with plastic was studied as a function of the thermostat temperature profile (A, at three levels), the type of plastic (B, at three levels), and the speed of the rotating screw that forces the plastic through a tube-forming die (C, at three levels). There were two replications ($L = 2$) at each combination of levels of the factors, yielding a total of 54 observations on output. The sums of squares were $SSA = 14{,}144.44$, $SSB = 5511.27$, $SSC = 244{,}696.39$, $SSAB = 1069.62$, $SSAC = 62.67$, $SSBC = 331.67$, $SSE = 3127.50$, and $SST = 270{,}024.33$.

   a. Construct the ANOVA table.

   b. Use appropriate F tests to show that none of the F ratios for two- or three-factor interactions is significant at level .05.

   c. Which main effects appear significant?

   d. With $x_{..1.} = 8242$, $x_{..2.} = 9732$, and $x_{..3.} = 11{,}210$, use Tukey's procedure to identify significant differences among the levels of factor C.

28. To see whether thrust force in drilling is affected by drilling speed (A), feed rate (B), or material used (C), an experiment using four speeds, three rates, and two materials was performed, with two samples ($L = 2$) drilled at each combination of levels of the three factors. Sums of squares were calculated as follows: $SSA = 19{,}149.73$, $SSB = 2{,}589{,}047.62$, $SSC = 157{,}437.52$, $SSAB = 53{,}238.21$, $SSAC = 9033.73$, $SSBC = 91{,}880.04$, $SSE = 56{,}819.50$, and $SST = 2{,}983{,}164.81$. Construct the ANOVA table and identify significant interactions using $\alpha = .01$. Is there any single factor that appears to have no effect on thrust force? (In other words, does any factor appear nonsignificant in every effect in which it appears?)

29. The article **"Effects of Household Fabric Softeners on Thermal Comfort of Cotton and Polyester Fabrics After Repeated Launderings"** (*Family and Consumer Science Research J.*, *2009: 535–549*) reported the results of a three-factor ANOVA carried out to investigate the impact of fabric softener treatment (A: no softener, rinse-cycle softener, dryer-sheet softener), fabric type (B: 100% cotton, 100% polyester), and number of laundering cycles (C: 1, 5, 25) on air permeability of fabric, which is an important determinant of thermal comfort.

   a. Five observations were made for each combination of factor levels. Resulting sums of squares were $SSA = 1043.27$, $SSB = 112{,}148.10$, $SSC = 3020.97$, $SSAB = 373.52$, $SSAC = 392.71$, $SSBC = 145.95$, $SSABC = 54.13$, and $SSE = 339.30$. Create an ANOVA table and carry out tests of all relevant hypotheses using a significance level of .01.

b. Because the test for the presence of three-factor interactions is insignificant, it makes sense to investigate two-factor interactions. Use the following values of various sample means to create interaction plots, and comment as to whether they are consistent with the test results of (a).

	A1	A2	A3
B1	67.10	56.50	65.93
B2	138.00	131.93	131.40
C1	110.25	105.55	103.30
C2	101.80	90.45	97.10
C3	95.60	86.65	95.60

(The cited article included a plot and commentary based on the AC means.)

30. The following summary quantities were computed from an experiment involving four levels of nitrogen (A), two times of planting (B), and two levels of potassium (C) (**"Use and Misuse of Multiple Comparison Procedures,"** *Agronomy J.*, *1977: 205–208*). Only one observation (N content, in percentage, of corn grain) was made for each of the 16 combinations of levels.

$SSA = .22625$ $SSB = .000025$ $SSC = .0036$
$SSAB = .004325$ $SSAC = .00065$
$SSBC = .000625$ $SST = .2384$.

   a. Construct the ANOVA table.

   b. Assume that there are no three-way interaction effects, so that MSABC is a valid estimate of σ^2, and test at level .05 for interaction and main effects.

   c. The nitrogen averages are $\bar{x}_{1..} = 1.1200$, $\bar{x}_{2..} = 1.3025$, $\bar{x}_{3..} = 1.3875$, and $\bar{x}_{4..} = 1.4300$. Use Tukey's method to examine differences in percentage N among the nitrogen levels ($Q_{.05,4,3} = 6.82$).

31. Nickel titanium (NiTi) shape memory alloy (SMA) has been widely used in medical devices. This is attributable largely to the alloy's shape memory effect (material returns to its original shape after heat deformation), superelasticity, and biocompatibility. An alloy element is usually coated on the surface of NiTi SMAs to prevent toxic Ni release.

   The article **"Parametrical Optimization of Laser Surface Alloyed NiTi Shape Memory Alloy with Co and Nb by the Taguchi Method"** (*J. of Engr. Manuf.*, *2012: 969–979*) described an investigation to see whether the percent by weight of nickel in the alloyed layer is affected by carbon monoxide powder paste thickness (C, at three levels), scanning speed (B, at three levels), and laser power (A, at three levels). One observation was made at each factor-level combination [*Note:* Thickness column headings were incorrect in the cited article]:

Power	Speed	Paste Thickness		
		.2	.3	.4
600	600	38.64	35.13	19.20
	900	38.16	34.24	26.23
	1200	37.54	33.46	30.44
700	600	36.56	35.91	34.62
	900	39.16	33.10	28.71
	1200	37.06	31.78	21.50
800	600	39.44	40.42	37.21
	900	39.34	37.64	35.65
	1200	39.30	34.97	32.50

a. Assuming the absence of three factor interactions (as did the investigators), SSE = SSABC can be used to obtain an estimate of σ^2. Construct an ANOVA table based on this data.

b. Use the appropriate F ratios to show that none of the two-factor interactions is significant at $\alpha = .05$.

c. Which main effects are significant at $\alpha = .05$?

d. Use Tukey's procedure with a simultaneous confidence level of 95% to identify significant differences between levels of paste thickness.

32. When factors A and B are fixed but factor C is random and the restricted model is used (see the footnote on page 438; there is a technical complication with the unrestricted model here), and $E(\text{MSE}) = \sigma^2$

$$E(\text{MSA}) = \sigma^2 + JL\sigma^2_{AC} + \frac{JKL}{I-1}\sum \alpha_i^2$$

$$E(\text{MSB}) = \sigma^2 + IL\sigma^2_{BC} + \frac{IKL}{J-1}\sum \beta_j^2$$

$$E(\text{MSC}) = \sigma^2 + IJL\sigma^2_C$$

$$E(\text{MSAB}) = \sigma^2 + L\sigma^2_{ABC}$$
$$+ \frac{KL}{(I-1)(J-1)}\sum_i \sum_j (\gamma_{ij}^{AB})^2$$

$$E(\text{MSAC}) = \sigma^2 + JL\sigma^2_{AC}$$

$$E(\text{MSBC}) = \sigma^2 + IL\sigma^2_{BC}$$

$$E(\text{MSABC}) = \sigma^2 + \sigma^2_{ABC}$$

a. Based on these expected mean squares, what F ratios would you use to test $H_0: \sigma^2_{ABC} = 0$; $H_0: \sigma^2_C = 0$; $H_0: \gamma_{ij}^{AB} = 0$ for all i, j; and $H_0: \alpha_1 = \cdots = \alpha_I = 0$?

b. In an experiment to assess the effects of age, type of soil, and day of production on compressive strength of cement/soil mixtures, two ages (A), four types of soil (B), and three days (C, assumed random) were used, with $L = 2$ observations made for each combination of factor levels. The resulting sums of squares were SSA = 14,318.24, SSB = 9656.40,

SSC = 2270.22, SSAB = 3408.93, SSAC = 1442.58, SSBC = 3096.21, SSABC = 2832.72, and SSE = 8655.60. Obtain the ANOVA table and carry out all tests using level .01.

33. Because of potential variability in aging due to different castings and segments on the castings, a Latin square design with $N = 7$ was used to investigate the effect of heat treatment on aging. With A = castings, B = segments, C = heat treatments, summary statistics include $x_{\cdots} = 3815.8$, $\Sigma x_{i\cdots}^2 = 297{,}216.90$, $\Sigma x_{\cdot j\cdot}^2 = 297{,}200.64$, $\Sigma x_{\cdots k}^2 = 297{,}155.01$, and $\Sigma\Sigma x_{ij(k)}^2 = 297{,}317.65$. Obtain the ANOVA table and test at level .05 the hypothesis that heat treatment has no effect on aging.

34. The article **"The Responsiveness of Food Sales to Shelf Space Requirements"** (*J. Marketing Research,* **1964: 63–67**) reports the use of a Latin square design to investigate the effect of shelf space on food sales. The experiment was carried out over a 6-week period using six different stores, resulting in the following data on sales of powdered coffee cream (with shelf space index in parentheses):

		Week		
		1	2	3
Store	1	27 (5)	14 (4)	18 (3)
	2	34 (6)	31 (5)	34 (4)
	3	39 (2)	67 (6)	31 (5)
	4	40 (3)	57 (1)	39 (2)
	5	15 (4)	15 (3)	11 (1)
	6	16 (1)	15 (2)	14 (6)

		Week		
		4	5	6
Store	1	35 (1)	28 (6)	22 (2)
	2	46 (3)	37 (2)	23 (1)
	3	49 (4)	38 (1)	48 (3)
	4	70 (6)	37 (4)	50 (5)
	5	9 (2)	18 (5)	17 (6)
	6	12 (5)	19 (3)	22 (4)

Construct the ANOVA table, and state and test at level .01 the hypothesis that shelf space does not affect sales against the appropriate alternative.

35. The article **"Variation in Moisture and Ascorbic Acid Content from Leaf to Leaf and Plant to Plant in Turnip Greens"** (*Southern Cooperative Services Bull.,* **1951: 13–17**) uses a Latin square design in which factor A is plant, factor B is leaf size (smallest to largest), factor C (in parentheses) is time of weighing, and the response variable is moisture content.

Leaf Size (B)

		1	2	3
	1	6.67 (5)	7.15 (4)	8.29 (1)
	2	5.40 (2)	4.77 (5)	5.40 (4)
Plant (A)	3	7.32 (3)	8.53 (2)	8.50 (5)
	4	4.92 (1)	5.00 (3)	7.29 (2)
	5	4.88 (4)	6.16 (1)	7.83 (3)

Leaf Size (B)

		4	5
	1	8.95 (3)	9.62 (2)
	2	7.54 (1)	6.93 (3)
Plant (A)	3	9.99 (4)	9.68 (1)
	4	7.85 (5)	7.08 (4)
	5	5.83 (2)	8.51 (5)

When all three factors are random, the expected mean squares are $E(\text{MSA}) = \sigma^2 + N\sigma_A^2$, $E(\text{MSB}) = \sigma^2 + N\sigma_B^2$, $E(\text{MSC}) = \sigma^2 + N\sigma_C^2$, and $E(\text{MSE}) = \sigma^2$. This implies that the F ratios for testing $H_{0A}: \sigma_A^2 = 0$, $H_{0B}: \sigma_B^2 = 0$, and $H_{0C}: \sigma_C^2 = 0$ are identical to those for fixed effects. Obtain the ANOVA table and test at level .05 to see whether there is any variation in moisture content due to the factors.

36. The article **"An Assessment of the Effects of Treatment, Time, and Heat on the Removal of Erasable Pen Marks** from Cotton and Cotton/Polyester Blend Fabrics (*J. of Testing and Eval.*, 1991: 394–397)** reports the following sums of squares for the response variable *degree of removal of marks*: SSA = 39.171, SSB = .665, SSC = 21.508, SSAB = 1.432, SSAC = 15.953, SSBC = 1.382, SSABC = 9.016, and SSE = 115.820. Four different laundry treatments, three different types of pen, and six different fabrics were used in the experiment, and there were three observations for each treatment-pen-fabric combination. Perform an analysis of variance using $\alpha = .01$ for each test, and state your conclusions (assume fixed effects for all three factors).

37. A four-factor ANOVA experiment was carried out to investigate the effects of fabric (*A*), type of exposure (*B*), level of exposure (*C*), and fabric direction (*D*) on extent of color change in exposed fabric as measured by a spectrocolorimeter. Two observations were made for each of the three fabrics, two types, three levels, and two directions, resulting in MSA = 2207.329, MSB = 47.255, MSC = 491.783, MSD = .044, MSAB = 15.303, MSAC = 275.446, MSAD = .470, MSBC = 2.141, MSBD = .273, MSCD = .247, MSABC = 3.714, MSABD = 4.072, MSABD = 4.072, MSACD = .767, MSBCD = .280, MSE = .977, and MST = 93.621 (**"Accelerated Weathering of Marine Fabrics," *J. Testing and Eval.*, 1992: 139–143**). Assuming fixed effects for all factors, carry out an analysis of variance using $\alpha = .01$ for all tests and summarize your conclusions.

11.4 2^p Factorial Experiments

If an experimenter wishes to study simultaneously the effect of p different factors on a response variable and the factors have $I_1, I_2, \ldots, I_p$ levels, respectively, then a complete experiment requires at least $I_1 \cdot I_2 \cdot \cdots \cdot I_p$ observations. In such situations, the experimenter can often perform a "screening experiment" with each factor at only two levels to obtain preliminary information about factor effects. An experiment in which there are p factors, each at two levels, is referred to as a **2^p factorial experiment**.

2^3 Experiments

As in Section 11.3, we let X_{ijkl} and x_{ijkl} refer to the observation from the lth replication, with factors A, B, and C at levels i, j, and k, respectively. The model for this situation is

$$X_{ijkl} = \mu + \alpha_i + \beta_j + \delta_k + \gamma_{ij}^{AB} + \gamma_{ik}^{AC} + \gamma_{jk}^{BC} + \gamma_{ijk} + \epsilon_{ijkl} \qquad (11.14)$$

for $i = 1, 2; j = 1, 2; k = 1, 2; l = 1, \ldots, n$. The ϵ_{ijkl}'s are assumed independent, normally distributed, with mean 0 and variance σ^2. Because there are only two levels of each factor, the side conditions on the parameters of (11.14) that uniquely specify the model are simply stated: $\alpha_1 + \alpha_2 = 0, \ldots, \gamma_{11}^{AB} + \gamma_{21}^{AB} = 0, \gamma_{12}^{AB} + \gamma_{22}^{AB} = 0$, $\gamma_{11}^{AB} + \gamma_{12}^{AB} = 0, \gamma_{21}^{AB} + \gamma_{22}^{AB} = 0$, and the like. These conditions imply that there is only one functionally independent parameter of each type (for each main effect and interaction). For example, $\alpha_2 = -\alpha_1$, whereas $\gamma_{21}^{AB} = -\gamma_{11}^{AB}$, $\gamma_{12}^{AB} = -\gamma_{11}^{AB}$, and $\gamma_{22}^{AB} = \gamma_{11}^{AB}$. Because of this, each sum of squares in the analysis will have 1 df.

The parameters of the model can be estimated by taking averages over various subscripts of the X_{ijkl}'s and then forming appropriate linear combinations of the averages. For example,

$$\hat{\alpha}_1 = \overline{X}_{1\cdots} - \overline{X}_{\cdots\cdots}$$
$$= \frac{(X_{111\cdot} + X_{121\cdot} + X_{112\cdot} + X_{122\cdot} - X_{211\cdot} - X_{212\cdot} - X_{221\cdot} - X_{222\cdot})}{8n}$$

and

$$\hat{\gamma}_{11}^{AB} = \overline{X}_{11\cdots} - \overline{X}_{1\cdots} - \overline{X}_{\cdot1\cdots} + \overline{X}_{\cdots\cdots}$$
$$= \frac{(X_{111\cdot} - X_{121\cdot} - X_{211\cdot} + X_{221\cdot} + X_{112\cdot} - X_{122\cdot} - X_{212\cdot} + X_{222\cdot})}{8n}$$

Each estimator is, except for the factor $1/(8n)$, a linear function of the cell totals ($X_{ijk\cdot}$'s) in which each coefficient is $+1$ or -1, with an equal number of each; such functions are called **contrasts** in the X_{ijk}'s. Furthermore, the estimators satisfy the same side conditions satisfied by the parameters themselves. For example,

$$\hat{\alpha}_1 + \hat{\alpha}_2 = \overline{X}_{1\cdots} - \overline{X}_{\cdots\cdots} + \overline{X}_{2\cdots} - \overline{X}_{\cdots\cdots} = \overline{X}_{1\cdots} + \overline{X}_{2\cdots} - 2\overline{X}_{\cdots\cdots}$$
$$= \frac{1}{4n}X_{1\cdots} + \frac{1}{4n}X_{2\cdots} - \frac{2}{8n}X_{\cdots\cdots} = \frac{1}{4n}X_{\cdots\cdots} - \frac{1}{4n}X_{\cdots\cdots} = 0$$

EXAMPLE 11.12 In an experiment to investigate the compressive strength properties of cement–soil mixtures, two different aging periods were used in combination with two different temperatures and two different soils. Two replications were made for each combination of levels of the three factors, resulting in the following data:

Age	Temperature	Soil 1	Soil 2
1	1	471, 413	385, 434
	2	485, 552	530, 593
2	1	712, 637	770, 705
	2	712, 789	741, 806

The computed cell totals are $x_{111\cdot} = 884$, $x_{211\cdot} = 1349$, $x_{121\cdot} = 1037$, $x_{221\cdot} = 1501$, $x_{112\cdot} = 819$, $x_{212\cdot} = 1475$, $x_{122\cdot} = 1123$, and $x_{222\cdot} = 1547$, so $x_{\cdots\cdots} = 9735$. Then

$$\hat{\alpha}_1 = (884 - 1349 + 1037 - 1501 + 819 - 1475 + 1123 - 1547)/16$$
$$= -125.5625 = -\hat{\alpha}_2$$
$$\hat{\gamma}_{11}^{AB} = (884 - 1349 - 1037 + 1501 + 819 - 1475 - 1123 + 1547)/16$$
$$= -14.5625 = -\hat{\gamma}_{12}^{AB} = -\hat{\gamma}_{21}^{AB} = \hat{\gamma}_{22}^{AB}$$

The other parameter estimates can be computed in the same manner. ∎

Analysis of a 2^3 Experiment Sums of squares for the various effects are easily obtained from the parameter estimates. For example,

$$\text{SSA} = \sum_i \sum_j \sum_k \sum_l \hat{\alpha}_i^2 = 4n \sum_{i=1}^{2} \hat{\alpha}_i^2 = 4n[\hat{\alpha}_1^2 + (-\hat{\alpha}_1)^2] = 8n\hat{\alpha}_1^2$$

and

$$SSAB = \sum_i \sum_j \sum_k \sum_l (\hat{\gamma}_{ij}^{AB})^2$$

$$= 2n \sum_{i=1}^{2} \sum_{j=1}^{2} (\hat{\gamma}_{ij}^{AB})^2 = 2n[(\hat{\gamma}_{11}^{AB})^2 + (-\hat{\gamma}_{11}^{AB})^2 + (-\hat{\gamma}_{11}^{AB})^2 + (\hat{\gamma}_{11}^{AB})^2]$$

$$= 8n(\hat{\gamma}_{11}^{AB})^2$$

Since each estimate is a contrast in the cell totals multiplied by $1/(8n)$, each sum of squares has the form $(\text{contrast})^2/(8n)$. Thus to compute the various sums of squares, we need to know the coefficients ($+1$ or -1) of the appropriate contrasts. The signs ($+$ or $-$) on each $x_{ijk\cdot}$ in each effect contrast are most conveniently displayed in a table. We will use the notation (1) for the experimental condition $i = 1, j = 1, k = 1$, a for $i = 2, j = 1, k = 1$, ab for $i = 2, j = 2, k = 1$, and so on. If level 1 is thought of as "low" and level 2 as "high," any letter that appears denotes a high level of the associated factor. Each column in Table 11.11 gives the signs for a particular effect contrast in the x_{ijk}'s associated with the different experimental conditions.

Table 11.11   Signs for Computing Effect Contrasts

Experimental Condition	Cell Total	\multicolumn{7}{c}{Factorial Effect}
		A	B	C	AB	AC	BC	ABC
(1)	$x_{111\cdot}$	$-$	$-$	$-$	$+$	$+$	$+$	$-$
a	$x_{211\cdot}$	$+$	$-$	$-$	$-$	$-$	$+$	$+$
b	$x_{121\cdot}$	$-$	$+$	$-$	$-$	$+$	$-$	$+$
ab	$x_{221\cdot}$	$+$	$+$	$-$	$+$	$-$	$-$	$-$
c	$x_{112\cdot}$	$-$	$-$	$+$	$+$	$-$	$-$	$+$
ac	$x_{212\cdot}$	$+$	$-$	$+$	$-$	$+$	$-$	$-$
bc	$x_{122\cdot}$	$-$	$+$	$+$	$-$	$-$	$+$	$-$
abc	$x_{222\cdot}$	$+$	$+$	$+$	$+$	$+$	$+$	$+$

In each of the first three columns, the sign is $+$ if the corresponding factor is at the high level and $-$ if it is at the low level. Every sign in the AB column is then the "product" of the signs in the A and B columns, with $(+)(+) = (-)(-) = +$ and $(+)(-) = (-)(+) = -$, and similarly for the AC and BC columns. Finally, the signs in the ABC column are the products of AB with C (or B with AC or A with BC). Thus, for example,

$$AC \text{ contrast} = +x_{111\cdot} - x_{211\cdot} + x_{121\cdot} - x_{221\cdot} - x_{112\cdot} + x_{212\cdot} - x_{122\cdot} + x_{222\cdot}$$

Once the seven effect contrasts are computed,

$$SS(\text{effect}) = \frac{(\text{effect contrast})^2}{8n}$$

Software for doing the calculations required to analyze data from factorial experiments is widely available (e.g., Minitab). Alternatively, here is an efficient method for hand computation due to Yates. Write in a column the eight cell totals in the **standard order**, as given in the table of signs, and establish three additional columns. In each of these three columns, the first four entries are the sums of entries 1 and 2, 3 and 4,

5 and 6, and 7 and 8 of the previous columns. The last four entries are the differences between entries 2 and 1, 4 and 3, 6 and 5, and 8 and 7 of the previous column. The last column then contains $x_{\ldots}$ and the seven effect contrasts in standard order. Squaring each contrast and dividing by $8n$ then gives the seven sums of squares.

EXAMPLE 11.13
(Example 11.12
continued)

Since $n = 2$, $8n = 16$. Yates's method is illustrated in Table 11.12.

Table 11.12 Yates's Method of Computation

Treatment Condition	$x_{ijk\cdot}$	1	2	Effect Contrast	$SS = (\text{contrast})^2/16$
$(1) = x_{111\cdot}$	884	2233	4771	9735	
$a = x_{211\cdot}$	1349	2538	4964	2009	252,255.06
$b = x_{121\cdot}$	1037	2294	929	681	28,985.06
$ab = x_{221\cdot}$	1501	2670	1080	-233	3,393.06
$c = x_{112\cdot}$	819	465	305	193	2,328.06
$ac = x_{212\cdot}$	1475	464	376	151	1,425.06
$bc = x_{122\cdot}$	1123	656	-1	71	315.06
$abc = x_{222\cdot}$	1547	424	-232	-231	3,335.06
					292,036.42

From the original data, $\sum_i\sum_j\sum_k\sum_l x_{ijkl}^2 = 6{,}232{,}289$, and

$$\frac{x_{\ldots}^2}{16} = 5{,}923{,}139.06$$

so

$$SST = 6{,}232{,}289 - 5{,}923{,}139.06 = 309{,}149.94$$

$$SSE = SST - [SSA + \cdots + SSABC] = 309{,}149.94 - 292{,}036.42$$

$$= 17{,}113.52$$

The ANOVA calculations are summarized in Table 11.13.

Table 11.13 ANOVA Table for Example 11.13

Source of Variation	df	Sum of Squares	Mean Square	f
A	1	252,255.06	252,255.06	117.92
B	1	28,985.06	28,985.06	13.55
C	1	2,328.06	2,328.06	1.09
AB	1	3,393.06	3,393.06	1.59
AC	1	1,425.06	1,425.06	.67
BC	1	315.06	315.06	.15
ABC	1	3,335.06	3,335.06	1.56
Error	8	17,113.52	2,139.19	
Total	15	309,149.94		

Figure 11.10 shows SAS output for this example. Only the P-values for age (A) and temperature (B) are less than .01, so only these effects are judged significant.

```
Analysis of Variance Procedure
Dependent Variable: STRENGTH

                            Sum of          Mean
Source              DF      Squares         Square      F Value   Pr > F
Model                7    292036.4375    41719.4911      19.50    0.0002
Error                8     17113.5000     2139.1875
Corrected Total     15    309149.9375

                 R-Square          C.V.      Root MSE        POWERUSE Mean

                 0.944643      7.601660      46.25135            608.437500

Source              DF     Anova SS    Mean Square    F Value   Pr > F

AGE                  1  252255.0625    252255.0625     117.92    0.0001
TEMP                 1   28985.0625     28985.0625      13.55    0.0062
AGE*TEMP             1    3393.0625      3393.0625       1.59    0.2434
SOIL                 1    2328.0625      2328.0625       1.09    0.3273
AGE*SOIL             1    1425.0625      1425.0625       0.67    0.4380
TEMP*SOIL            1     315.0625       315.0625       0.15    0.7111
AGE*TEMP*SOIL        1    3335.0625      3335.0625       1.56    0.2471
```

Figure 11.10 SAS output for strength data of Example 11.13 ∎

2^p Experiments for $p > 3$

The analysis of data from a 2^p experiment with $p > 3$ parallels that of the three-factor case. For example, if there are four factors A, B, C, and D, there are 16 different experimental conditions. The first 8 in standard order are exactly those already listed for a three-factor experiment. The second 8 are obtained by placing the letter d beside each condition in the first group. Yates's method is then initiated by computing totals across replications, listing these totals in standard order, and proceeding as before; with p factors, the pth column to the right of the treatment totals will give the effect contrasts.

For $p > 3$, there will often be no replications of the experiment (so only one complete replicate is available). One possible way to test hypotheses is to assume that certain higher-order effects are absent and then add the corresponding sums of squares to obtain an SSE. Such an assumption can, however, be misleading in the absence of prior knowledge (see the book by Montgomery listed in the chapter bibliography). An alternative approach involves working directly with the effect contrasts. Each contrast has a normal distribution with the same variance. When a particular effect is absent, the expected value of the corresponding contrast is 0, but this is not so when the effect is present. The suggested method of analysis is to construct a normal probability plot of the effect contrasts (or, equivalently, the effect parameter estimates, since estimate = contrast/2^p when $n = 1$). Points corresponding to absent effects will tend to fall close to a straight line, whereas points associated with substantial effects will typically be far from this line.

EXAMPLE 11.14 The accompanying data is from the article **"Quick and Easy Analysis of Unreplicated Factorials"** (*Technometrics*, 1989: 469–473). The four factors are A = acid strength, B = time, C = amount of acid, and D = temperature, and the response variable is the yield of isatin. The observations, in standard order, are .08, .04, .53, .43, .31, .09, .12, .36, .79, .68, .73, .08, .77, .38, .49, and .23. Table 11.14 displays the effect estimates as given in the article (which uses contrast/8 rather than contrast/16).

Table 11.14 Effect Estimates for Example 11.14

Effect	A	B	AB	C	AC	BC	ABC	D
Estimate	−.191	−.021	−.001	−.076	.034	−.066	.149	.274

Effect	AD	BD	ABD	CD	ACD	BCD	$ABCD$
Estimate	−.161	−.251	−.101	−.026	−.066	.124	.019

Figure 11.11 is a normal probability plot of the effect estimates. All points in the plot fall close to the same straight line, suggesting the complete absence of any effects (we will shortly give an example in which this is not the case).

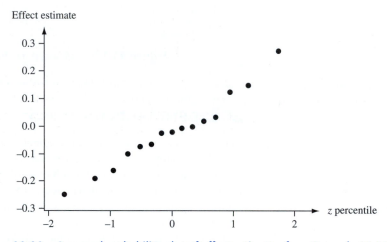

Figure 11.11 A normal probability plot of effect estimates from Example 11.14 ■

Visual judgments of deviation from straightness in a normal probability plot are rather subjective. The article cited in Example 11.14 describes a more objective technique for identifying significant effects in an unreplicated experiment.

Confounding

It is often not possible to carry out all 2^p experimental conditions of a 2^p factorial experiment in a homogeneous experimental environment. In such situations, it may be possible to separate the experimental conditions into 2^r homogeneous blocks ($r < p$), so that there are 2^{p-r} experimental conditions in each block. The blocks may, for example, correspond to different laboratories, different time periods, or different operators or work crews. In the simplest case, $p = 3$ and $r = 1$, so that there are two blocks, with each block consisting of four of the eight experimental conditions.

As always, blocking is effective in reducing variation associated with extraneous sources. However, when the 2^p experimental conditions are placed in 2^r blocks, the price paid for this blocking is that $2^r - 1$ of the factor effects cannot be estimated. This is because $2^r - 1$ factor effects (main effects and/or interactions) are mixed up, or **confounded**, with the block effects. The allocation of experimental conditions to blocks is then usually done so that only higher-level interactions are confounded, whereas main effects and low-order interactions remain estimable and hypotheses can be tested.

To see how allocation to blocks is accomplished, consider first a 2^3 experiment with two blocks ($r = 1$) and four treatments per block. Suppose we select ABC as the effect to be confounded with blocks. Then any experimental condition having an odd number of letters in common with ABC, such as b (one letter) or abc (three letters), is placed in one block, whereas any condition having an even number of letters in common with ABC (where 0 is even) goes in the other block. Figure 11.12 shows this allocation of treatments to the two blocks.

Block 1	Block 2
(1), ab, ac, bc	a, b, c, abc

Figure 11.12 Confounding ABC in a 2^3 experiment

In the absence of replications, the data from such an experiment would usually be analyzed by assuming that there were no two-factor interactions (additivity) and using SSE = SSAB + SSAC + SSBC with 3 df to test for the presence of main effects. Alternatively, a normal probability plot of effect contrasts or effect parameter estimates could be examined. Most frequently, though, there are replications when just three factors are being studied. Suppose there are u replicates, resulting in a total of $2^r \cdot u$ blocks in the experiment. Then after subtracting from SST all sums of squares associated with effects not confounded with blocks (computed using Yates's method), the block sum of squares is computed using the $2^r \cdot u$ block totals and then subtracted to yield SSE (so there are $2^r \cdot u - 1$ df for blocks).

EXAMPLE 11.15 The article **"Factorial Experiments in Pilot Plant Studies"** (*Industrial and Eng. Chemistry*, **1951: 1300–1306**) reports the results of an experiment to assess the effects of reactor temperature (A), gas throughput (B), and concentration of active constituent (C) on the strength of the product solution (measured in arbitrary units) in a recirculation unit. Two blocks were used, with the ABC effect confounded with blocks, and there were two replications, resulting in the data in Figure 11.13. The four block $\times$ replication totals are 288, 212, 88, and 220, with a grand total of 808, so

$$\text{SSB1} = \frac{(288)^2 + (212)^2 + (88)^2 + (220)^2}{4} - \frac{(808)^2}{16} = 5204.00$$

	Replication 1			Replication 2		
Block 1		Block 2		Block 1		Block 2
(1) 99		a 18		(1) 46		a 18
ab 52		b 51		ab −47		b 62
ac 42		c 108		ac 22		c 104
bc 95		abc 35		bc 67		abc 36

Figure 11.13 Data for Example 11.15

The other sums of squares are computed by Yates's method using the eight experimental condition totals, resulting in the ANOVA table given as Table 11.15. By comparison with $F_{.05,1,6} = 5.99$, we conclude that only the main effects for A and C differ significantly from zero (P-value $< .05$ for just f_A and f_C).

Table 11.15 ANOVA Table for Example 11.15

Source of Variation	df	Sum of Squares	Mean Square	f
A	1	12,996	12,996	39.82
B	1	702.25	702.25	2.15
C	1	2,756.25	2,756.25	8.45
AB	1	210.25	210.25	.64
AC	1	30.25	30.25	.093
BC	1	25	25	.077
Blocks	3	5,204	1,734.67	5.32
Error	6	1,958	326.33	
Total	15	23,882		

∎

Confounding Using More than Two Blocks

In the case $r = 2$ (four blocks), three effects are confounded with blocks. The experimenter first chooses two defining effects to be confounded. For example, in a five-factor experiment (A, B, C, D, and E), the two three-factor interactions BCD and CDE might be chosen for confounding. The third effect confounded is then the **generalized interaction** of the two, obtained by writing the two chosen effects side by side and then cancelling any letters common to both: $(BCD)(CDE) = BE$. Notice that if ABC and CDE are chosen for confounding, their generalized interaction is $(ABC)(CDE) = ABDE$, so that no main effects or two-factor interactions are confounded.

Once the two defining effects have been selected for confounding, one block consists of all treatment conditions having an even number of letters in common with both defining effects. The second block consists of all conditions having an even number of letters in common with the first defining contrast and an odd number of letters in common with the second contrast, and the third and fourth blocks consist of the "odd/even" and "odd/odd" contrasts. In a five-factor experiment with defining effects ABC and CDE, this results in the allocation to blocks as shown in Figure 11.14 (with the number of letters in common with each defining contrast appearing beside each experimental condition).

Block 1		Block 2		Block 3		Block 4	
(1)	(0, 0)	d	(0, 1)	a	(1, 0)	c	(1, 1)
ab	(2, 0)	e	(0, 1)	b	(1, 0)	ad	(1, 1)
de	(0, 2)	ac	(2, 1)	cd	(1, 2)	ae	(1, 1)
acd	(2, 2)	bc	(2, 1)	ce	(1, 2)	bd	(1, 1)
ace	(2, 2)	abd	(2, 1)	ade	(1, 2)	be	(1, 1)
bcd	(2, 2)	abe	(2, 1)	bde	(1, 2)	abc	(3, 1)
bce	(2, 2)	$acde$	(2, 3)	$abcd$	(3, 2)	cde	(1, 3)
$abde$	(2, 2)	$bcde$	(2, 3)	$abce$	(3, 2)	$abcde$	(3, 3)

Figure 11.14 Four blocks in a 2^5 factorial experiment with defining effects ABC and CDE

The block containing (1) is called the **principal block**. Once it has been constructed, a second block can be obtained by selecting any experimental condition not in the principal block and obtaining its generalized interaction with every condition in the principal block. The other blocks are then constructed in the same way by first selecting a condition not in a block already constructed and finding generalized interactions with the principal block.

For experimental situations with $p > 3$, there is often no replication, so sums of squares associated with nonconfounded higher-order interactions are usually pooled to obtain an error sum of squares that can be used in the denominators of the various F statistics. All computations can again be carried out using Yates's technique, with SSBl being the sum of sums of squares associated with confounded effects.

When $r > 2$, one first selects r defining effects to be confounded with blocks, making sure that no one of the effects chosen is the generalized interaction of any other two selected. The additional $2^r - r - 1$ effects confounded with the blocks are then the generalized interactions of all effects in the defining set (including not only generalized interactions of pairs of effects but also of sets of three, four, and so on).

Fractional Replication

When the number of factors p is large, even a single replicate of a 2^p experiment can be expensive and time consuming. For example, one replicate of a 2^6 factorial experiment involves an observation for each of the 64 different experimental conditions. An appealing strategy in such situations is to make observations for only a fraction of the 2^p conditions. Provided that care is exercised in the choice of conditions to be observed, much information about factor effects can still be obtained.

Suppose we decide to include only 2^{p-1} (half) of the 2^p possible conditions in our experiment; this is usually called a **half-replicate**. The price paid for this economy is twofold. First, information about a single effect (determined by the 2^{p-1} conditions selected for observation) is completely lost to the experimenter in the sense that no reasonable estimate of the effect is possible. Second, the remaining $2^p - 2$ main effects and interactions are paired up so that any one effect in a particular pair is confounded with the other effect in the same pair. For example, one such pair may be {A, BCD}, so that separate estimates of the A main effect and BCD interaction are not possible. It is desirable, then, to select a half-replicate for which main effects and low-order interactions are paired off (confounded) only with higher-order interactions rather than with one another.

The first step in specifying a half-replicate is to select a defining effect as the nonestimable effect. Suppose that in a five-factor experiment, ABCDE is chosen as the defining effect. Now the $2^5 = 32$ possible treatment conditions are divided into two groups with 16 conditions each, one group consisting of all conditions having an odd number of letters in common with ABCDE and the other containing an even number of letters in common with the defining contrast. Then either group of 16 conditions is used as the half-replicate. The "odd" group is

$$a, b, c, d, e, abc, abd, abe, acd, ace, ade, bcd, bce, bde, cde, abcde$$

Each main effect and interaction other than ABCDE is then confounded with (**aliased** with) its generalized interaction with ABCDE. Thus (AB)(ABCDE) = CDE, so the AB interaction and CDE interaction are confounded with each other. The resulting **alias pairs** are

{A, BCDE}	{B, ACDE}	{C, ABDE}	{D, ABCE}	{E, ABCD}
{AB, CDE}	{AC, BDE}	{AD, BCE}	{AE, BCD}	{BC, ADE}
{BD, ACE}	{BE, ACD}	{CD, ABE}	{CE, ABD}	{DE, ABC}

Note in particular that every main effect is aliased with a four-factor interaction. Assuming these interactions to be negligible allows us to test for the presence of main effects.

To specify a quarter-replicate of a 2^p factorial experiment (2^{p-2} of the 2^p possible treatment conditions), two defining effects must be selected. These two and their generalized interaction become the nonestimable effects. Instead of alias pairs as in the half-replicate, each remaining effect is now confounded with three other effects, each being its generalized interaction with one of the three nonestimable effects.

EXAMPLE 11.16   The article **"More on Planning Experiments to Increase Research Efficiency"** (*Industrial and Eng. Chemistry*, **1970: 60–65**) reports on the results of a quarter-replicate of a 2^5 experiment in which the five factors were A = condensation temperature, B = amount of material B, C = solvent volume, D = condensation time, and E = amount of material E. The response variable was the yield of the chemical process. The chosen defining contrasts were ACE and BDE, with generalized interaction $(ACE)(BDE) = ABCD$. The remaining 28 main effects and interactions can now be partitioned into seven groups of four effects each, such that the effects within a group cannot be assessed separately. For example, the generalized interactions of A with the nonestimable effects are $(A)(ACE) = CE$, $(A)(BDE) = ABDE$, and $(A)(ABCD) = BCD$, so one alias group is $\{A, CE, ABDE, BCD\}$. The complete set of alias groups is

$$\{A, CE, ABDE, BCD\} \qquad \{B, ABCE, DE, ACD\} \qquad \{C, AE, BCDE, ABD\}$$
$$\{D, ACDE, BE, ABC\} \qquad \{E, AC, BD, ABCDE\} \qquad \{AB, BCE, ADE, CD\}$$
$$\{AD, CDE, ABE, BC\}$$

■

Once the defining contrasts have been chosen for a quarter-replicate, they are used as in the discussion of confounding to divide the 2^p treatment conditions into four groups of 2^{p-2} conditions each. Then any one of the four groups is selected as the set of conditions for which data will be collected. Similar comments apply to a $1/2^r$ replicate of a 2^p factorial experiment.

Having made observations for the selected treatment combinations, a table of signs similar to Table 11.11 is constructed. The table contains a row only for each of the treatment combinations actually observed rather than the full 2^p rows, and there is a single column for each alias group (since each effect in the group would have the same set of signs for the treatment conditions selected for observation). The signs in each column indicate as usual how contrasts for the various sums of squares are computed. Yates's method can also be used, but the rule for arranging observed conditions in standard order must be modified.

The difficult part of a fractional replication analysis typically involves deciding what to use for error sum of squares. Since there will usually be no replication (though one could observe, e.g., two replicates of a quarter-replicate), some effect sums of squares must be pooled to obtain an error sum of squares. In a half-replicate of a 2^8 experiment, for example, an alias structure can be chosen so that the eight main effects and 28 two-factor interactions are each confounded only with higher-order interactions and that there are an additional 27 alias groups involving only higher-order interactions. Assuming the absence of higher-order interaction effects, the resulting 27 sums of squares can then be added to yield an error sum of squares, allowing 1 df tests for all main effects and two-factor interactions. However, in many cases tests for main effects can be obtained only by pooling some or all of the sums of squares associated with alias groups involving two-factor interactions, and the corresponding two-factor interactions cannot be investigated.

EXAMPLE 11.17
(Example 11.16
continued)

The set of treatment conditions chosen and resulting yields for the quarter-replicate of the 2^5 experiment were

e	ab	ad	bc	cd	ace	bde	$abcde$
23.2	15.5	16.9	16.2	23.8	23.4	16.8	18.1

The abbreviated table of signs is displayed in Table 11.16.

With SSA denoting the sum of squares for effects in the alias group $\{A, CE, ABDE, BCD\}$,

$$SSA = \frac{(-23.2 + 15.5 + 16.9 - 16.2 - 23.8 + 23.4 - 16.8 + 18.1)^2}{8} = 4.65$$

Table 11.16 Table of Signs for Example 11.17

	A	B	C	D	E	AB	AD
e	$-$	$-$	$-$	$-$	$+$	$+$	$+$
ab	$+$	$+$	$-$	$-$	$-$	$+$	$-$
ad	$+$	$-$	$-$	$+$	$-$	$-$	$+$
bc	$-$	$+$	$+$	$-$	$-$	$-$	$+$
cd	$-$	$-$	$+$	$+$	$-$	$+$	$-$
ace	$+$	$-$	$+$	$-$	$+$	$-$	$-$
bde	$-$	$+$	$-$	$+$	$+$	$-$	$-$
$abcde$	$+$	$+$	$+$	$+$	$+$	$+$	$+$

Similarly, SSB $= 53.56$, SSC $= 10.35$, SSD $= .91$ SSE$' = 10.35$ (the $'$ differentiates this quantity from error sum of squares SSE), SSAB $= 6.66$, and SSAD $= 3.25$, giving SST $= 4.65 + 53.56 + \cdots + 3.25 = 89.73$. To test for main effects, we use SSE $=$ SSAB $+$ SSAD $= 9.91$ with 2 df. The ANOVA table is in Table 11.17.

Table 11.17 ANOVA Table for Example 11.17

Source	df	Sum of Squares	Mean Square	f
A	1	4.65	4.65	.94
B	1	53.56	53.56	10.80
C	1	10.35	10.35	2.09
D	1	.91	.91	.18
E	1	10.35	10.35	2.09
Error	2	9.91	4.96	
Total	7	89.73		

Since $F_{.05,1,2} = 18.51$, none of the five main effects can be judged significant. Of course, with only 2 df for error, the test is not very powerful (i.e., it is quite likely to fail to detect the presence of effects). The article in *Industrial and Engineering Chemistry* from which the data came actually has an independent estimate of the standard error of the treatment effects based on prior experience, so it used a somewhat different analysis. Our analysis was done here only for illustrative purposes, since one would ordinarily want many more than 2 df for error. ∎

As an alternative to F tests based on pooling sums of squares to obtain SSE, a normal probability plot of effect contrasts can be examined.

EXAMPLE 11.18 An experiment was carried out to investigate shrinkage in the plastic casing material used for speedometer cables (**"An Explanation and Critique of Taguchi's Contribution to Quality Engineering,"** *Quality and Reliability Engr. Intl.*, **1988: 123–131**). The engineers started with 15 factors: liner outside diameter, liner die, liner material, liner line speed, wire braid type, braiding tension, wire diameter, liner tension, liner temperature, coating material, coating die type, melt temperature, screen pack, cooling method, and line speed. It was suspected that only a few of these factors were important, so a screening experiment in the form of a 2^{15-11} factorial (a $1/2^{11}$ fraction of a 2^{15} factorial experiment) was carried out. The resulting alias structure is quite complicated; in particular, every main effect is confounded with two-factor interactions. The response variable was the percentage of shrinkage for a cable specimen produced at designated levels of the factors.

Figure 11.15 displays a normal probability plot of the effect contrasts. All but two of the points fall quite close to a straight line. The discrepant points correspond to effects E = wire braid type and G = wire diameter, suggesting that these two factors are the only ones that affect the amount of shrinkage.

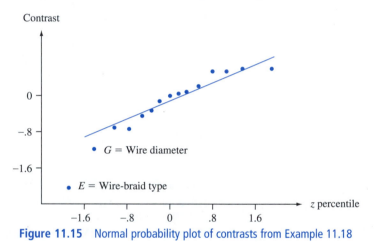

Figure 11.15 Normal probability plot of contrasts from Example 11.18 ■

The subjects of factorial experimentation, confounding, and fractional replication encompass many models and techniques we have not discussed. Please consult the chapter references for more information.

EXERCISES Section 11.4 (38–49)

38. The accompanying data resulted from an experiment to study the nature of dependence of welding current on three factors: welding voltage, wire feed speed, and tip-to-workpiece distance. There were two levels of each factor (a 2^3 experiment) with two replications per combination of levels (the averages across replications agree with values given in the article **"A Study on Prediction of Welding Current in Gas Metal Arc Welding,"** *J. Engr. Manuf.*, **1991: 64–69**). The first two given numbers are for the treatment (1), the next two for a, and so on in standard order: 200.0, 204.2, 215.5, 219.5, 272.7, 276.9, 299.5, 302.7, 166.6, 172.6, 186.4, 192.0, 232.6, 240.8, 253.4, 261.6.

a. Verify that the sums of squares are as given in the accompanying ANOVA table from Minitab.

b. Which effects appear to be important, and why?

Analysis of Variance for current

Source	DF	SS	MS	F	P
Volt	1	1685.1	1685.1	102.38	0.000
Speed	1	21272.2	21272.2	1292.37	0.000
Dist	1	5076.6	5076.6	308.42	0.000
Volt*speed	1	36.6	36.6	2.22	0.174
Volt*dist	1	0.4	0.4	0.03	0.877
Speed*dist	1	109.2	109.2	6.63	0.033
Volt*speed*dist	1	23.5	23.5	1.43	0.266
Error	8	131.7	16.5		
Total	15	28335.3			

39. The accompanying data resulted from a 2^3 experiment with three replications per combination of treatments designed to study the effects of concentration of detergent (A), concentration of sodium carbonate (B), and concentration of sodium carboxymethyl cellulose (C) on the cleaning ability of a solution in washing tests (a larger number indicates better cleaning ability than a smaller number).

Factor Levels

A	B	C	Condition	Observations
1	1	1	(1)	106, 93, 116
2	1	1	a	198, 200, 214
1	2	1	b	197, 202, 185
2	2	1	ab	329, 331, 307
1	1	2	c	149, 169, 135
2	1	2	ac	243, 247, 220
1	2	2	bc	255, 230, 252
2	2	2	abc	383, 360, 364

a. After obtaining cell totals $x_{ijk\cdot}$, compute estimates of β_1, γ_{11}^{AC}, and γ_{21}^{AC}.

b. Use the cell totals along with Yates's method to compute the effect contrasts and sums of squares. Then construct an ANOVA table and test all appropriate hypotheses using $\alpha = .05$.

c. Suppose a low water temperature has been used to obtain the data. The entire experiment is then repeated with a higher water temperature to obtain the following data. Use Yates's algorithm on the entire set of 48 observations to obtain the sums of squares and ANOVA table, and then test appropriate hypotheses at level .05.

Condition	Observations
d	144, 154, 158
ad	239, 227, 244
bd	232, 242, 246
abd	364, 362, 346
cd	194, 162, 203
acd	284, 295, 291
bcd	291, 287, 297
abcd	411, 406, 395

40. In a study of processes used to remove impurities from cellulose goods (**"Optimization of Rope-Range Bleaching of Cellulosic Fabrics," Textile Research J., 1976: 493–496**), the following data resulted from a 2^4 experiment involving the desizing process. The four factors were enzyme concentration (A), pH (B), temperature (C), and time (D).

Treatment	Enzyme (g/L)	pH	Temp. (°C)	Time (hr)	Starch % by Weight 1st Repl.	Starch % by Weight 2nd Repl.
(1)	.50	6.0	60.0	6	9.72	13.50
a	.75	6.0	60.0	6	9.80	14.04
b	.50	7.0	60.0	6	10.13	11.27
ab	.75	7.0	60.0	6	11.80	11.30
c	.50	6.0	70.0	6	12.70	11.37
ac	.75	6.0	70.0	6	11.96	12.05
bc	.50	7.0	70.0	6	11.38	9.92
abc	.75	7.0	70.0	6	11.80	11.10
d	.50	6.0	60.0	8	13.15	13.00
ad	.75	6.0	60.0	8	10.60	12.37
bd	.50	7.0	60.0	8	10.37	12.00
abd	.75	7.0	60.0	8	11.30	11.64
cd	.50	6.0	70.0	8	13.05	14.55
acd	.75	6.0	70.0	8	11.15	15.00
bcd	.50	7.0	70.0	8	12.70	14.10
abcd	.75	7.0	70.0	8	13.20	16.12

a. Use Yates's algorithm to obtain sums of squares and the ANOVA table.

b. Do there appear to be any second-, third-, or fourth-order interaction effects present? Explain your reasoning. Which main effects appear to be significant?

41. As with many dried products, sun-dried tomatoes can exhibit an undesirable discoloration during the drying and storage process. A replicated 2^3 experiment was conducted in an effort to relate color quality to the factors *storage time*, *temperature*, and *packaging type* (**"Use of Factorial Experimental Design for Analyzing the Effect of Storage Conditions on Color Quality of Sun-Dried Tomatoes," Sci. Res. and Essays, 2012: 477–489**). In the following table, higher values of the response variable (based on chromaticity measurements) are associated with higher color quality:

Storage time	Storage temp	Packaging	Color Quality Replication 1	Color Quality Replication 2
−	−	−	2.38	2.40
+	−	−	2.38	2.40
−	+	−	2.42	2.40
+	+	−	2.31	2.29
−	−	+	2.38	2.40
+	−	+	2.38	2.40
−	+	+	1.94	1.94
+	+	+	1.93	1.92

Construct an ANOVA table and use it as a basis for deciding which effects appear to be significant.

42. The following data on power consumption in electric-furnace heats (kW consumed per ton of melted product) resulted from a 2^4 factorial experiment with three replicates (**"Studies on a 10-cwt Arc Furnace,"** *J. of the Iron and Steel Institute, 1956: 22*). The factors were nature of roof (A, low/high), power setting (B, low/high), scrap used (C, tube/plate), and charge (D, 700 lb/1000 lb).

Treatment	x_{ijklm}	Treatment	x_{ijklm}
(1)	866, 862, 800	d	988, 808, 650
a	946, 800, 840	ad	966, 976, 876
b	774, 834, 746	bd	702, 658, 650
ab	709, 789, 646	abd	784, 700, 596
c	1017, 990, 954	cd	922, 808, 868
ac	1028, 906, 977	acd	1056, 870, 908
bc	817, 783, 771	bcd	798, 726, 700
abc	829, 806, 691	$abcd$	752, 714, 714

Construct the ANOVA table, and test all hypotheses of interest using $\alpha = .01$.

43. The article **"Statistical Design and Analysis of Qualification Test Program for a Small Rocket Engine"** (*Industrial Quality Control*, **1964: 14–18**) presents data from an experiment to assess the effects of vibration (A), temperature cycling (B), altitude cycling (C), and temperature for altitude cycling and firing (D) on thrust duration. A subset of the data is given here. (In the article, there were four levels of D rather than just two.) Use the Yates method to obtain sums of squares and the ANOVA table. Then assume that three- and four-factor interactions are absent, pool the corresponding sums of squares to obtain an estimate of σ^2, and test all appropriate hypotheses at level .05.

		D_1		D_2	
		C_1	C_2	C_1	C_2
A_1	B_1	21.60	21.60	11.54	11.50
	B_2	21.09	22.17	11.14	11.32
A_2	B_1	21.60	21.86	11.75	9.82
	B_2	19.57	21.85	11.69	11.18

44. a. In a 2^4 experiment, suppose two blocks are to be used, and it is decided to confound the $ABCD$ interaction with the block effect. Which treatments should be carried out in the first block [the one containing the treatment (1)], and which treatments are allocated to the second block?

b. In an experiment to investigate niacin retention in vegetables as a function of cooking temperature (A), sieve size (B), type of processing (C), and cooking time (D), each factor was held at two levels. Two blocks were used, with the allocation of blocks as given in part (a) to confound only the $ABCD$ interaction with blocks. Use Yates's procedure to obtain the ANOVA table for the accompanying data.

Treatment	x_{ijkl}	Treatment	x_{ijkl}
(1)	91	d	72
a	85	ad	78
b	92	bd	68
ab	94	abd	79
c	86	cd	69
ac	83	acd	75
bc	85	bcd	72
abc	90	$abcd$	71

c. Assume that all three-way interaction effects are absent, so that the associated sums of squares can be combined to yield an estimate of σ^2, and carry out all appropriate tests at level .05.

45. a. An experiment was carried out to investigate the effects on audio sensitivity of varying resistance (A), two capacitances (B, C), and inductance of a coil (D) in part of a television circuit. If four blocks were used with four treatments per block and the defining effects for confounding were AB and CD, which treatments appeared in each block?

b. Suppose two replications of the experiment described in part (a) were performed, resulting in the accompanying data. Obtain the ANOVA table, and test all relevant hypotheses at level .01.

Treatment	x_{ijkl1}	x_{ijkl2}	Treatment	x_{ijkl1}	x_{ijkl2}
(1)	618	598	d	598	585
a	583	560	ad	587	541
b	477	525	bd	480	508
ab	421	462	abd	462	449
c	601	595	cd	603	577
ac	550	589	acd	571	552
bc	505	484	bcd	502	508
abc	452	451	$abcd$	449	455

46. In an experiment involving four factors (A, B, C, and D) and four blocks, show that at least one main effect or two-factor interaction effect must be confounded with the block effect.

47. a. In a seven-factor experiment ($A, \ldots, G$), suppose a quarter-replicate is actually carried out. If the defining effects are $ABCDE$ and $CDEFG$, what is the third nonestimable effect, and what treatments are in the group containing (1)? What are the alias groups of the seven main effects?

b. If the quarter-replicate is to be carried out using four blocks (with eight treatments per block), what are the blocks if the chosen confounding effects are ACF and BDG?

48. The article **"Applying Design of Experiments to Improve a Laser Welding Process"** (*J. of Engr. Manufacture*, **2008: 1035–1042**) included the results of a half replicate of a 2^4 experiment. The four factors were: *A.* Power (2900 W, 3300 W), *B.* Current (2400 mV,

3600 mV), C. Laterals cleaning (No, Yes), and D. Roof cleaning (No, Yes).

a. If the effect $ABCD$ is chosen as the defining effect for the replicate and the group of eight treatments for which data is obtained includes treatment (1), what other treatments are in the observed group, and what are the alias pairs?

b. The cited article presented data on two different response variables, the percentage of defective joints for both the right laser welding cord and the left welding cord. Here we consider just the latter response. Observations are listed here in standard order after deleting the half not observed. Assuming that two- and three-factor interactions are negligible, test at level .05 for the presence of main effects. Also construct a normal probability plot.

8.936 9.130 4.314 7.692
.415   6.061 1.984 3.830

49. A half-replicate of a 2^5 experiment to investigate the effects of heating time (A), quenching time (B), drawing time (C), position of heating coils (D), and measurement position (E) on the hardness of steel castings resulted in the accompanying data. Construct the ANOVA table, and (assuming second and higher-order interactions to be negligible) test at level .01 for the presence of main effects. Also construct a normal probability plot.

Treatment	Observation	Treatment	Observation
a	70.4	acd	66.6
b	72.1	ace	67.5
c	70.4	ade	64.0
d	67.4	bcd	66.8
e	68.0	bce	70.3
abc	73.8	bde	67.9
abd	67.0	cde	65.9
abe	67.8	$abcde$	68.0

SUPPLEMENTARY EXERCISES (50–61)

50. The results of a study on the effectiveness of line drying on the smoothness of fabric were summarized in the article **"Line-Dried vs. Machine-Dried Fabrics: Comparison of Appearance, Hand, and Consumer Acceptance"** (*Home Econ. Research J.*, 1984: 27–35). Smoothness scores were given for nine different types of fabric and five different drying methods: (1) machine dry, (2) line dry, (3) line dry followed by 15-min tumble, (4) line dry with softener, and (5) line dry with air movement. Regarding the different types of fabric as blocks, construct an ANOVA table. Using a .05 significance level, test to see whether there is a difference in the true mean smoothness score for the drying methods.

		Drying Method				
		1	2	3	4	5
	Crepe	3.3	2.5	2.8	2.5	1.9
	Double knit	3.6	2.0	3.6	2.4	2.3
	Twill	4.2	3.4	3.8	3.1	3.1
	Twill mix	3.4	2.4	2.9	1.6	1.7
Fabric	**Terry**	3.8	1.3	2.8	2.0	1.6
	Broadcloth	2.2	1.5	2.7	1.5	1.9
	Sheeting	3.5	2.1	2.8	2.1	2.2
	Corduroy	3.6	1.3	2.8	1.7	1.8
	Denim	2.6	1.4	2.4	1.3	1.6

51. The water absorption of two types of mortar used to repair damaged cement was discussed in the article **"Polymer Mortar Composite Matrices for Maintenance-Free, Highly Durable Ferrocement"** (*J. of Ferrocement*, 1984: 337–345). Specimens of ordinary cement mortar (OCM) and polymer cement mortar (PCM) were submerged for varying lengths of time (5, 9, 24, or 48 hours) and water absorption (% by weight) was recorded. With mortar type as factor A (with two levels) and submersion period as factor B (with four levels), three observations were made for each factor level combination. Data included in the article was used to compute the sums of squares, which were SSA = 322.667, SSB = 35.623, SSAB = 8.557, and SST = 372.113. Use this information to construct an ANOVA table. Test the appropriate hypotheses at a .05 significance level.

52. Four plots were available for an experiment to compare clover accumulation for four different sowing rates (**"Performance of Overdrilled Red Clover with Different Sowing Rates and Initial Grazing Managements,"** *N. Zeal. J. of Exp. Ag.*, 1984: 71–81). Since the four plots had been grazed differently prior to the experiment and it was thought that this might affect clover accumulation, a randomized block experiment was used with all four sowing rates tried on a section of each plot. Use the given data to test the null hypothesis of no difference in true mean clover accumulation (kg DM/ha) for the different sowing rates.

		Sowing Rate (kg/ha)			
		3.6	6.6	10.2	13.5
	1	1155	2255	3505	4632
Plot	2	123	406	564	416
	3	68	416	662	379
	4	62	75	362	564

53. In an automated chemical coating process, the speed with which objects on a conveyor belt are passed through a chemical spray (belt speed), the amount of chemical sprayed (spray volume), and the brand of chemical used (brand) are factors that may affect the uniformity of the coating applied. A replicated 2^3 experiment was conducted in an effort to increase the coating uniformity. In the following table, higher values of the response variable are associated with higher surface uniformity:

				Surface Uniformity	
Run	Spray Volume	Belt Speed	Brand	Replication 1	Replication 2
1	−	−	−	40	36
2	+	−	−	25	28
3	−	+	−	30	32
4	+	+	−	50	48
5	−	−	+	45	43
6	+	−	+	25	30
7	−	+	+	30	29
8	+	+	+	52	49

Analyze this data and state your conclusions.

54. Coal-fired power plants used in the electrical industry have gained increased public attention because of the environmental problems associated with solid wastes generated by large-scale combustion (**"Fly Ash Binders in Stabilization of FGD Wastes,"** *J. of Environmental Engineering,* **1998: 43–49**). A study was conducted to analyze the influence of three factors—binder type (A), amount of water (B), and land disposal scenario (C)—that affect certain leaching characteristics of solid wastes from combustion. Each factor was studied at two levels. An unreplicated 2^3 experiment was run, and a response value EC50 (the effective concentration, in mg/L, that decreases 50% of the light in a luminescence bioassay) was measured for each combination of factor levels. The experimental data is given in the following table:

	Factor			Response
Run	A	B	C	EC50
1	−1	−1	−1	23,100
2	1	−1	−1	43,000
3	−1	1	−1	71,400
4	1	1	−1	76,000
5	−1	−1	1	37,000
6	1	−1	1	33,200
7	−1	1	1	17,000
8	1	1	1	16,500

Carry out an appropriate ANOVA, and state your conclusions.

55. Impurities in the form of iron oxides lower the economic value and usefulness of industrial minerals, such as kaolins, to ceramic and paper-processing industries. A 2^4 experiment was conducted to assess the effects of four factors on the percentage of iron removed from kaolin samples (**"Factorial Experiments in the Development of a Kaolin Bleaching Process Using Thiourea in Sulphuric Acid Solutions,"** *Hydrometallurgy,* **1997: 181–197**). The factors and their levels are listed in the following table:

Factor	Description	Units	Low Level	High Level
A	H_2SO_4	M	.10	.25
B	Thiourea	g/L	0.0	5.0
C	Temperature	°C	70	90
D	Time	min	30	150

The data from an unreplicated 2^4 experiment is listed in the next table.

Test Run	Iron Extraction (%)	Test Run	Iron Extraction (%)
(1)	7	d	28
a	11	ad	51
b	7	bd	33
ab	12	abd	57
c	21	cd	70
ac	41	acd	95
bc	27	bcd	77
abc	48	$abcd$	99

a. Calculate estimates of all main effects and two-factor interaction effects for this experiment.

b. Create a probability plot of the effects. Which effects appear to be important?

56. Factorial designs have been used in forestry to assess the effects of various factors on the growth behavior of trees. In one such experiment, researchers thought that healthy spruce seedlings should bud sooner than diseased spruce seedlings (**"Practical Analysis of Factorial Experiments in Forestry,"** *Canadian J. of Forestry,* **1995: 446–461**). In addition, before planting, seedlings were also exposed to three levels of pH to see whether this factor has an effect on virus uptake into the root system. The following table shows data from a 2×3 experiment to study both factors:

	pH		
Health Status	**3**	**5.5**	**7**
Diseased	1.2, 1.4, 1.0, 1.2, 1.4	.8, .6, .8, 1.0, .8	1.0, 1.0, 1.2, 1.4, 1.2
Healthy	1.4, 1.6, 1.6, 1.6, 1.4	1.0, 1.2, 1.2, 1.4, 1.4	1.2, 1.4, 1.2, 1.2, 1.4

The response variable is an average rating of five buds from a seedling. The ratings are 0 (bud not broken),

1 (bud partially expanded), and 2 (bud fully expanded). Analyze this data.

57. One property of automobile air bags that contributes to their ability to absorb energy is the permeability ($ft^3/ft^2/min$) of the woven material used to construct the air bags. Understanding how permeability is influenced by various factors is important for increasing the effectiveness of air bags. In one study, the effects of three factors, each at three levels, were studied (**"Analysis of Fabrics Used in Passive Restraint Systems—Airbags,"** *J. of the Textile Institute*, 1996: 554–571):

A (Temperature): 8°C, 50°C, 75°C
B (Fabric denier): 420-D, 630-D, 840-D
C (Air pressure): 17.2 kPa, 34.4 kPa, 103.4 kPa

Temperature 8°

Denier	Pressure 17.2	34.4	103.4
420-D	73	157	332
	80	155	322
630-D	35	91	288
	43	98	271
840-D	125	234	477
	111	233	464

Temperature 50°

Denier	Pressure 17.2	34.4	103.4
420-D	52	125	281
	51	118	264
630-D	16	72	169
	12	78	173
840-D	90	149	338
	100	155	350

Temperature 75°

Denier	Pressure 17.2	34.4	103.4
420-D	37	95	276
	31	106	281
630-D	30	91	213
	41	100	211
840-D	102	170	307
	98	160	311

Analyze this data and state your conclusions (assume that all factors are fixed).

58. A chemical engineer has carried out an experiment to study the effects of the fixed factors of vat pressure (A), cooking time of pulp (B), and hardwood concentration (C) on the strength of paper. The experiment involved two pressures, four cooking times, three concentrations, and two observations at each combination of these levels. Calculated sums of squares are SSA = 6.94, SSB = 5.61, SSC = 12.33, SSAB = 4.05, SSAC = 7.32, SSBC = 15.80, SSE = 14.40, and SST = 70.82. Construct the ANOVA table, and carry out appropriate tests at significance level .05.

59. The bond strength when mounting an integrated circuit on a metalized glass substrate was studied as a function of factor A = adhesive type, factor B = curve time, and factor C = conductor material (copper and nickel). The data follows, along with an ANOVA table from Minitab. What conclusions can you draw from the data?

Copper		Cure Time 1	2	3
	1	72.7	74.6	80.0
		80.0	77.5	82.7
Adhesive	2	77.8	78.5	84.6
		75.3	81.1	78.3
	3	77.3	80.9	83.9
		76.5	82.6	85.0

Nickel		1	2	3
	1	74.7	75.7	77.2
		77.4	78.2	74.6
Adhesive	2	79.3	78.8	83.0
		77.8	75.4	83.9
	3	77.2	84.5	89.4
		78.4	77.5	81.2

```
Analysis of Variance for strength
Source           DF       SS      MS     F      P
Adhesive          2  101.317  50.659  6.54  0.007
Curetime          2  151.317  75.659  9.76  0.001
Conmater          1    0.722   0.722  0.09  0.764
Adhes*curet       4   30.526   7.632  0.98  0.441
Adhes*conm        2    8.015   4.008  0.52  0.605
Curet*conm        2    5.952   2.976  0.38  0.687
Adh*curet*conm    4   33.298   8.325  1.07  0.398
Error            18  139.515   7.751
Total            35  470.663
```

60. The article **"Effect of Cutting Conditions on Tool Performance in CBN Hard Turning"** (*J. of Manuf. Processes*, 2005: 10–17) reported the accompanying data on cutting speed (m/s), feed (mm/rev), depth of cut (mm), and tool life (min). Carry out a three-factor ANOVA on tool life, assuming the absence of any factor interactions (as did the authors of the article).

Obs	Cut spd	Feed	Cut dpth	life
1	1.21	0.061	0.102	27.5
2	1.21	0.168	0.102	26.5
3	1.21	0.061	0.203	27.0
4	1.21	0.168	0.203	25.0
5	3.05	0.061	0.102	8.0
6	3.05	0.168	0.102	5.0
7	3.05	0.061	0.203	7.0
8	3.05	0.168	0.203	3.5

61. Analogous to a Latin square, a Greco-Latin square design can be used when it is suspected that three extraneous factors may affect the response variable and all four factors (the three extraneous ones and the one of interest) have the same number of levels. In a Latin square, each level of the factor of interest (C) appears once in each row (with each level of A) and once in each column (with each level of B). In a Greco-Latin square, each level of factor D appears once in each row, in each column, and also with each level of the third extraneous factor C. Alternatively, the design can be used when the four factors are all of equal interest, the number of levels of each is N, and resources are available only for N^2 observations. A 5×5 square is pictured in (a), with (k, l) in each cell denoting the kth level of C and lth level of D. In (b) we present data on weight loss in silicon bars used for semiconductor material as a function of volume of etch (A), color of nitric acid in the etch solution (B), size of bars (C), and time in the etch solution (D) (from **"Applications of Analytic Techniques to the Semiconductor Industry," Fourteenth Midwest Quality Control Conference, 1959**).

Let $x_{ij(kl)}$ denote the observed weight loss when factor A is at level i, B is at level j, C is at level k, and D is at level l. Assuming no interaction between factors, the total sum of squares SST (with $N^2 - 1$df) can be partitioned into SSA, SSB, SSC, SSD, and SSE. Give expressions for these sums of squares, including computing formulas,

obtain the ANOVA table for the given data, and test each of the four main effect hypotheses using $\alpha = .05$.

(C, D)		B			
	1	2	3	4	5
1	(1, 1)	(2, 3)	(3, 5)	(4, 2)	(5, 4)
2	(2, 2)	(3, 4)	(4, 1)	(5, 3)	(1, 5)
A **3**	(3, 3)	(4, 5)	(5, 2)	(1, 4)	(2, 1)
4	(4, 4)	(5, 1)	(1, 3)	(2, 5)	(3, 2)
5	(5, 5)	(1, 2)	(2, 4)	(3, 1)	(4, 3)

(a)

65	82	108	101	126
84	109	73	97	83
105	129	89	89	52
119	72	76	117	84
97	59	94	78	106

(b)

BIBLIOGRAPHY

Box, George, William Hunter, and Stuart Hunter, *Statistics for Experimenters* (2nd ed.), Wiley, New York, 2006. Contains a wealth of suggestions and insights on data analysis based on the authors' extensive consulting experience.

DeVor, R., T. Chang, and J. W. Sutherland, *Statistical Quality Design and Control,* (2nd ed.), Prentice-Hall, Englewood Cliffs, NJ, 2006. Includes a modern survey of factorial and fractional factorial experimentation with a minimum of mathematics.

Hocking, Ronald, *Methods and Applications of Linear Models* (2nd ed.), Wiley, New York, 2003. A very general treatment of analysis of variance written by one of the foremost authorities in this field.

Kleinbaum, David, Lawrence Kupper, Keith Muller, and Azhar Nizam, *Applied Regression Analysis and Other Multivariable Methods* (4th ed.), Duxbury Press, Boston, 2007. Contains an especially good discussion of problems associated with analysis of "unbalanced data"—that is, unequal K_{ij}'s.

Kuehl, Robert O., *Design of Experiments: Statistical Principles of Research Design and Analysis* (2nd ed.), Duxbury Press, Boston, 1999. A comprehensive treatment of designed experiments and analysis of the resulting data.

Montgomery, Douglas, *Design and Analysis of Experiments* (8th ed.), Wiley, New York, 2013. See the Chapter 10 bibliography.

Neter, John, William Wasserman, and Michael Kutner, *Applied Linear Statistical Models* (5th ed.), Irwin, Homewood, IL, 2004. See the Chapter 10 bibliography.

Vardeman, Stephen, *Statistics for Engineering Problem Solving,* PWS, Boston, 1994. A general introduction for engineers, with much descriptive and inferential methodology for data from designed experiments.

Simple Linear Regression and Correlation

12

INTRODUCTION

In the two-sample problems discussed in Chapter 9, we were interested in comparing values of parameters for the x distribution and the y distribution. Even when observations were paired, we did not try to use information about one of the variables in studying the other variable. This is precisely the objective of regression analysis: to exploit the relationship between two (or more) variables so that we can gain information about one of them through knowing values of the other(s).

Much of mathematics is devoted to studying variables that are *deterministically* related. Saying that x and y are related in this manner means that once we are told the value of x, the value of y is completely specified. For example, consider renting a van for a day, and suppose that the rental cost is $25.00 plus $.30 per mile driven. Letting $x =$ the number of miles driven and $y =$ the rental charge, then $y = 25 + .3x$. If the van is driven 100 miles ($x = 100$), then $y = 25 + .3(100) = 55$. As another example, if the initial velocity of a particle is v_0 and it undergoes constant acceleration a, then distance traveled $= y = v_0x + \frac{1}{2}ax^2$, where $x =$ time.

There are many variables x and y that would appear to be related to one another, but not in a deterministic fashion. A familiar example is given by variables $x =$ high school grade point average (GPA) and $y =$ college GPA. The value of y cannot be determined just from knowledge of x, and two different individuals could have the same x value but have very different y values. Yet there is a tendency for those who have high (low) high school GPAs also to have high (low) college GPAs. Knowledge of a student's high school GPA should be quite helpful in enabling us to predict how that person will do in college.

Other examples of variables related in a nondeterministic fashion include $x =$ age of a child and $y =$ size of that child's vocabulary, $x =$ size of an engine (cm³) and $y =$ fuel efficiency for an automobile equipped with that engine, and $x =$ applied tensile force and $y =$ amount of elongation in a metal strip.

Regression analysis is the part of statistics that investigates the relationship between two or more variables related in a nondeterministic fashion. In this chapter, we generalize the deterministic linear relation $y = \beta_0 + \beta_1 x$ to a linear probabilistic relationship, develop procedures for making various inferences based on the model, and obtain a quantitative measure (the correlation coefficient) of the extent to which the two variables are related. In Chapter 13, we will consider techniques for validating a particular model and investigate nonlinear relationships and relationships involving more than two variables.

12.1 The Simple Linear Regression Model

The simplest deterministic mathematical relationship between two variables x and y is a linear relationship $y = \beta_0 + \beta_1 x$. The set of pairs (x, y) for which $y = \beta_0 + \beta_1 x$ determines a straight line with slope β_1 and y-intercept β_0.* The objective of this section is to develop a linear probabilistic model.

If the two variables are not deterministically related, then for a fixed value of x, there is uncertainty in the value of the second variable. For example, if we are investigating the relationship between age of child and size of vocabulary and decide to select a child of age $x = 5.0$ years, then before the selection is made, vocabulary size is a random variable Y. After a particular 5-year-old child has been selected and tested, a vocabulary of 2000 words may result. We would then say that the observed value of Y associated with fixing $x = 5.0$ was $y = 2000$.

More generally, the variable whose value is fixed by the experimenter will be denoted by x and will be called the **independent, predictor,** or **explanatory variable**. For fixed x, the second variable will be random; we denote this random variable and its observed value by Y and y, respectively, and refer to it as the **dependent** or **response variable**.

Usually observations will be made for a number of settings of the independent variable. Let $x_1, x_2, \ldots, x_n$ denote values of the independent variable for which observations are made, and let Y_i and y_i, respectively, denote the random variable and observed value associated with x_i. The available bivariate data then consists of the n pairs $(x_1, y_1), (x_2, y_2), \ldots, (x_n, y_n)$. A picture of this data called a **scatterplot** gives preliminary impressions about the nature of any relationship. In such a plot, each (x_i, y_i) is represented as a point plotted on a two-dimensional coordinate system.

---

* The slope of a line is the change in y for a 1-unit increase in x. For example, if $y = -3x + 10$, then y decreases by 3 when x increases by 1, so the slope is -3. The y-intercept is the height at which the line crosses the vertical axis and is obtained by setting $x = 0$ in the equation.

EXAMPLE 12.1   Visual and musculoskeletal problems associated with the use of visual display terminals (VDTs) have become rather common in recent years. Some researchers have focused on vertical gaze direction as a source of eye strain and irritation. This direction is known to be closely related to ocular surface area (OSA), so a method of measuring OSA is needed. The accompanying representative data on y = OSA (cm^2) and x = width of the palprebal fissure (i.e., the horizontal width of the eye opening, in cm) is from the article **"Analysis of Ocular Surface Area for Comfortable VDT Workstation Layout"** (*Ergonomics*, **1996: 877–884**). The order in which observations were obtained was not given, so for convenience they are listed in increasing order of x values.

i	1	2	3	4	5	6	7	8	9	10	11	12	13	14	15
x_i	.40	.42	.48	.51	.57	.60	.70	.75	.75	.78	.84	.95	.99	1.03	1.12
y_i	1.02	1.21	.88	.98	1.52	1.83	1.50	1.80	1.74	1.63	2.00	2.80	2.48	2.47	3.05

i	16	17	18	19	20	21	22	23	24	25	26	27	28	29	30
x_i	1.15	1.20	1.25	1.25	1.28	1.30	1.34	1.37	1.40	1.43	1.46	1.49	1.55	1.58	1.60
y_i	3.18	3.76	3.68	3.82	3.21	4.27	3.12	3.99	3.75	4.10	4.18	3.77	4.34	4.21	4.92

Thus (x_1, y_1) = (.40, 1.02), (x_5, y_5) = (.57, 1.52), and so on. A Minitab scatterplot is shown in Figure 12.1; we used an option that produced a dotplot of both the x values and y values individually along the right and top margins of the plot, which makes it easier to visualize the distributions of the individual variables (histograms or boxplots are alternative options). Here are some things to notice about the data and plot:

- Several observations have identical x values yet different y values (e.g., $x_8 = x_9$ = .75, but y_8 = 1.80 and y_9 = 1.74). Thus the value of y is *not* determined solely by x but also by various other factors.

- There is a strong tendency for y to increase as x increases. That is, larger values of OSA tend to be associated with larger values of fissure width—a positive relationship between the variables.

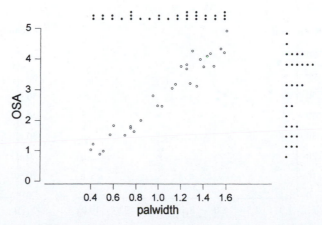

Figure 12.1   Scatterplot from Minitab for the data from Example 12.1, along with dotplots of x and y values

- It appears that the value of y could be predicted from x by finding a line that is reasonably close to the points in the plot (the authors of the cited article superimposed such a line on their plot). In other words, there is evidence of a substantial (though not perfect) linear relationship between the two variables. ∎

The horizontal and vertical axes in the scatterplot of Figure 12.1 intersect at the point $(0, 0)$. In many data sets, the values of x or y or the values of both variables differ considerably from zero relative to the range(s) of the values. For example, a study of how air conditioner efficiency is related to maximum daily outdoor temperature might involve observations for temperatures ranging from 80°F to 100°F. When this is the case, a more informative plot would show the appropriately labeled axes intersecting at some point other than $(0, 0)$.

EXAMPLE 12.2　Arsenic is found in many ground waters and some surface waters. Recent health effects research has prompted the Environmental Protection Agency to reduce allowable arsenic levels in drinking water so that many water systems are no longer compliant with standards. This has spurred interest in the development of methods to remove arsenic. The accompanying data on x = pH and y = arsenic removed (%) by a particular process was read from a scatterplot in the article **"Optimizing Arsenic Removal During Iron Removal: Theoretical and Practical Considerations"** (*J. of Water Supply Res. and Tech.*, 2005: 545–560).

x	7.01	7.11	7.12	7.24	7.94	7.94	8.04	8.05	8.07
y	60	67	66	52	50	45	52	48	40

x	8.90	8.94	8.95	8.97	8.98	9.85	9.86	9.86	9.87
y	23	20	40	31	26	9	22	13	7

Figure 12.2 shows two Minitab scatterplots of this data. In Figure 12.2(a), the software selected the scale for both axes. We obtained Figure 12.2(b) by specifying scaling for the axes so that they would intersect at roughly the point $(0, 0)$. The second plot is much more crowded than the first one; such crowding can make it difficult to ascertain the general nature of any relationship. For example, curvature can be overlooked in a crowded plot.

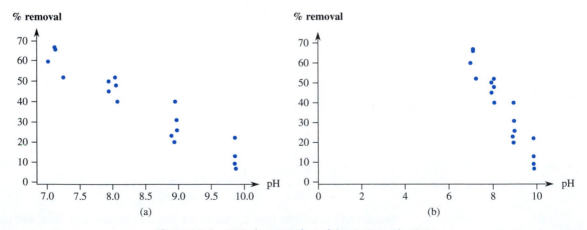

Figure 12.2　Minitab scatterplots of data in Example 12.2

Large values of arsenic removal tend to be associated with low pH, a negative or inverse relationship. Furthermore, the two variables appear to be at least approximately linearly related, although the points in the plot would spread out somewhat about any superimposed straight line (such a line appeared in the plot in the cited article). ■

A Linear Probabilistic Model

For the deterministic model $y = \beta_0 + \beta_1 x$, the actual observed value of y is a linear function of x. The appropriate generalization of this to a probabilistic model assumes that *the expected value of Y is a linear function of x*, but that for fixed x the variable Y differs from its expected value by a random amount.

DEFINITION

> ### The Simple Linear Regression Model
>
> There are parameters β_0, β_1, and σ^2, such that for any fixed value of the independent variable x, the dependent variable is a random variable related to x through the **model equation**
>
> $$Y = \beta_0 + \beta_1 x + \epsilon \tag{12.1}$$
>
> The quantity ϵ in the model equation is a random variable, assumed to be normally distributed with $E(\epsilon) = 0$ and $V(\epsilon) = \sigma^2$.

The variable ϵ is usually referred to as the **random deviation** or **random error term** in the model. Without ϵ, any observed pair (x, y) would correspond to a point falling exactly on the line $y = \beta_0 + \beta_1 x$, called the **true** (or **population**) **regression line**. The inclusion of the random error term allows (x, y) to fall either above the true regression line (when $\epsilon > 0$) or below the line (when $\epsilon < 0$). The points $(x_1, y_1), \ldots, (x_n, y_n)$ resulting from n independent observations will then be scattered about the true regression line, as illustrated in Figure 12.3. On occasion, the appropriateness of the simple linear regression model may be suggested by theoretical considerations (e.g., there is an exact linear relationship between the two variables, with ϵ representing measurement error). Much more frequently, though, the reasonableness of the model is indicated by a scatterplot exhibiting a substantial linear pattern (as in Figures 12.1 and 12.2).

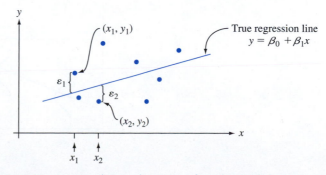

Figure 12.3 Points corresponding to observations from the simple linear regression model

Implications of the model equation (12.1) can best be understood with the aid of the following notation. Let x^* denote a particular value of the independent variable x and

$\mu_{Y \cdot x^*}$ = the expected (or mean) value of Y when x has value x^*

$\sigma_{Y \cdot x^*}^2$ = the variance of Y when x has value x^*

Alternative notation is $E(Y|x^*)$ and $V(Y|x^*)$. For example, if x = applied stress $(kg/mm)^2$ and y = time-to-fracture (hr), then $\mu_{Y \cdot 20}$ would denote the expected value of time-to-fracture when applied stress is 20 kg/mm^2. If we think of an entire population of (x, y) pairs, then $\mu_{Y \cdot x^*}$ is the mean of all y values for which $x = x^*$, and $\sigma_{Y \cdot x^*}^2$ is a measure of how much these values of y spread out about the mean value. If, for example, x = age of a child and y = vocabulary size, then $\mu_{Y \cdot 5}$ is the average vocabulary size for all 5-year-old children in the population, and $\sigma_{Y \cdot 5}^2$ describes the amount of variability in vocabulary size for this part of the population. Once x is fixed, the only randomness on the right-hand side of the model equation (12.1) is in the random error ϵ, and its mean value and variance are 0 and σ^2, respectively, whatever the value of x. This implies that

$$\mu_{Y \cdot x^*} = E(\beta_0 + \beta_1 x^* + \epsilon) = \beta_0 + \beta_1 x^* + E(\epsilon) = \beta_0 + \beta_1 x^*$$

$$\sigma_{Y \cdot x^*}^2 = V(\beta_0 + \beta_1 x^* + \epsilon) = V(\beta_0 + \beta_1 x^*) + V(\epsilon) = 0 + \sigma^2 = \sigma^2$$

Replacing x^* in $\mu_{Y \cdot x^*}$ by x gives the relation $\mu_{Y \cdot x} = \beta_0 + \beta_1 x$, which says that the *mean value* of Y, rather than Y itself, is a linear function of x. The true regression line $y = \beta_0 + \beta_1 x$ is thus the *line of mean values;* its height above any particular x value is the expected value of Y for that value of x. The slope β_1 of the true regression line is interpreted as the *expected* change in Y associated with a 1-unit increase in the value of x. The second relation states that the amount of variability in the distribution of Y values is the same at each different value of x (homogeneity of variance). If the independent variable is vehicle weight and the dependent variable is fuel efficiency (mpg), then the model implies that the average fuel efficiency changes linearly with weight (presumable β_1 is negative) and that the amount of variability in efficiency for any particular weight is the same as at any other weight. Finally, for fixed x, Y is the sum of a constant $\beta_0 + \beta_1 x$ and a normally distributed rv ϵ so itself has a normal distribution. These properties are illustrated in Figure 12.4. The variance parameter

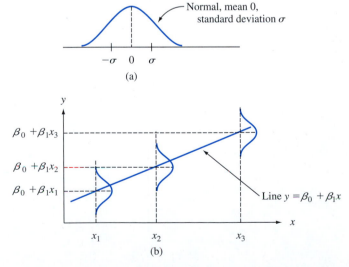

(a)

(b)

Figure 12.4   (a) Distribution of ϵ; (b) distribution of Y for different values of x

σ^2 determines the extent to which each normal curve spreads out about its mean value; roughly speaking, the value of σ is the size of a typical deviation from the true regression line. An observed point (x, y) will almost always fall quite close to the true regression line when σ is small, whereas observations may deviate considerably from their expected values (corresponding to points far from the line) when σ is large.

EXAMPLE 12.3   Suppose the relationship between applied stress x and time-to-failure y is described by the simple linear regression model with true regression line $y = 65 - 1.2x$ and $\sigma = 8$. Then for any fixed value x^* of stress, time-to-failure has a normal distribution with mean value $65 - 1.2x^*$ and standard deviation 8. In the population consisting of all (x, y) points, the magnitude of a typical deviation from the true regression line is about 8. For $x = 20$, Y has mean value $\mu_{Y \cdot 20} = 65 - 1.2(20) = 41$, so

$$P(Y > 50 \text{ when } x = 20) = P\left(Z > \frac{50 - 41}{8}\right) = 1 - \Phi(1.13) = .1292$$

Because $\mu_{Y \cdot 25} = 35$,

$$P(Y > 50 \text{ when } x = 25) = P\left(Z > \frac{50 - 35}{8}\right) = 1 - \Phi(1.88) = .0301$$

These probabilities are illustrated as the shaded areas in Figure 12.5.

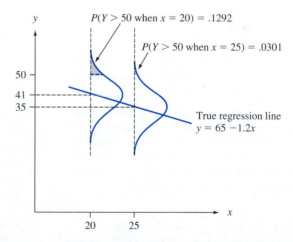

Figure 12.5   Probabilities based on the simple linear regression model

Suppose that Y_1 denotes an observation on time-to-failure made with $x = 25$ and Y_2 denotes an independent observation made with $x = 24$. Then $Y_1 - Y_2$ is normally distributed with mean value $E(Y_1 - Y_2) = \beta_1 = -1.2$, variance $V(Y_1 - Y_2) = \sigma^2 + \sigma^2 = 128$, and standard deviation $\sqrt{128} = 11.314$. The probability that Y_1 exceeds Y_2 is

$$P(Y_1 - Y_2 > 0) = P\left(Z > \frac{0 - (-1.2)}{11.314}\right) = P(Z > .11) = .4562$$

That is, even though we expected Y to decrease when x increases by 1 unit, it is not unlikely that the observed Y at $x + 1$ will be larger than the observed Y at x.   ■

EXERCISES Section 12.1 (1–11)

1. The efficiency ratio for a steel specimen immersed in a phosphating tank is the weight of the phosphate coating divided by the metal loss (both in mg/ft^2). The article **"Statistical Process Control of a Phosphate Coating Line" (*Wire J. Intl.*, May 1997: 78–81)** gave the accompanying data on tank temperature (x) and efficiency ratio (y).

Temp.	170	172	173	174	174	175	176
Ratio	.84	1.31	1.42	1.03	1.07	1.08	1.04

Temp.	177	180	180	180	180	180	181
Ratio	1.80	1.45	1.60	1.61	2.13	2.15	.84

Temp.	181	182	182	182	182	184	184
Ratio	1.43	.90	1.81	1.94	2.68	1.49	2.52

Temp.	185	186	188
Ratio	3.00	1.87	3.08

   a. Construct stem-and-leaf displays of both temperature and efficiency ratio, and comment on interesting features.
   b. Is the value of efficiency ratio completely and uniquely determined by tank temperature? Explain your reasoning.
   c. Construct a scatterplot of the data. Does it appear that efficiency ratio could be very well predicted by the value of temperature? Explain your reasoning.

2. The article **"Exhaust Emissions from Four-Stroke Lawn Mower Engines" (*J. of the Air and Water Mgmnt. Assoc.*, 1997: 945–952)** reported data from a study in which both a baseline gasoline mixture and a reformulated gasoline were used. Consider the following observations on age (yr) and NO$_x$ emissions (g/kWh):

Engine	1	2	3	4	5
Age	0	0	2	11	7
Baseline	1.72	4.38	4.06	1.26	5.31
Reformulated	1.88	5.93	5.54	2.67	6.53

Engine	6	7	8	9	10
Age	16	9	0	12	4
Baseline	.57	3.37	3.44	.74	1.24
Reformulated	.74	4.94	4.89	.69	1.42

   Construct scatterplots of NO$_x$ emissions versus age. What appears to be the nature of the relationship between these two variables? [*Note:* The authors of the cited article commented on the relationship.]

3. Bivariate data often arises from the use of two different techniques to measure the same quantity. As an example, the accompanying observations on x = hydrogen concentration (ppm) using a gas chromatography method and y = concentration using a new sensor method were read from a graph in the article **"A New Method to Measure the Diffusible Hydrogen Content in Steel Weldments Using a Polymer Electrolyte-Based Hydrogen Sensor" (*Welding Res.*, July 1997: 251s–256s)**.

x	47	62	65	70	70	78	95	100	114	118
y	38	62	53	67	84	79	93	106	117	116

x	124	127	140	140	140	150	152	164	198	221
y	127	114	134	139	142	170	149	154	200	215

   Construct a scatterplot. Does there appear to be a very strong relationship between the two types of concentration measurements? Do the two methods appear to be measuring roughly the same quantity? Explain your reasoning.

4. The accompanying data on y = ammonium concentration (mg/L) and x = transpiration (ml/h) was read from a graph in the article **"Response of Ammonium Removal to Growth and Transpiration of *Juncus effusus* During the Treatment of Artificial Sewage in Laboratory-Scale Wetlands" (*Water Research*, 2013: 4265–4273)**. The article's abstract stated "a linear correlation between the ammonium concentration inside the rhizosphere and the transpiration of the plant stocks implies that an influence of plant physiological activity on the efficiency of N-removal exists." (The rhizosphere is the narrow region of soil at the plant root–soil interface, and transpiration is the process of water movement through a plant and its evaporation.) The article reported summary quantities from a simple linear regression analysis. Based on a scatterplot, how would you describe the relationship between the variables, and does simple linear regression appear to be an appropriate modeling strategy?

x	5.8	8.8	11.0	13.6	18.5	21.0	23.7
y	7.8	8.2	6.9	5.3	4.7	4.9	4.3

x	26.0	28.3	31.9	36.5	38.2	40.4
y	2.7	2.8	1.8	1.9	1.1	.4

5. The article **"Objective Measurement of the Stretchability of Mozzarella Cheese" (*J. of Texture Studies*, 1992: 185–194)** reported on an experiment to investigate how the behavior of mozzarella cheese varied with temperature. Consider the accompanying data on x = temperature and y = elongation(%) at failure of the cheese. [*Note:* The

researchers were Italian and used *real* mozzarella cheese, not the poor cousin widely available in the United States.]

x	59	63	68	72	74	78	83
y	118	182	247	208	197	135	132

a. Construct a scatterplot in which the axes intersect at (0, 0). Mark 0, 20, 40, 60, 80, and 100 on the horizontal axis and 0, 50, 100, 150, 200, and 250 on the vertical axis.

b. Construct a scatterplot in which the axes intersect at (55, 100), as was done in the cited article. Does this plot seem preferable to the one in part (a)? Explain your reasoning.

c. What do the plots of parts (a) and (b) suggest about the nature of the relationship between the two variables?

6. One factor in the development of tennis elbow, a malady that strikes fear in the hearts of all serious tennis players, is the impact-induced vibration of the racket-and-arm system at ball contact. It is well known that the likelihood of getting tennis elbow depends on various properties of the racket used. Consider the scatterplot of $x =$ racket resonance frequency (Hz) and $y =$ sum of peak-to-peak acceleration (a characteristic of arm vibration, in m/sec/sec) for $n = 23$ different rackets (**"Transfer of Tennis Racket Vibrations into the Human Forearm," *Medicine and Science in Sports and Exercise*, 1992: 1134–1140**). Discuss interesting features of the data and scatterplot.

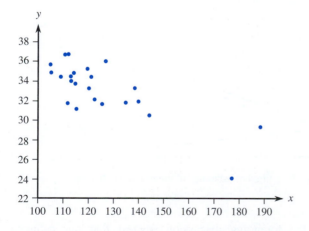

7. The article **"Some Field Experience in the Use of an Accelerated Method in Estimating 28-Day Strength of Concrete" (*J. of Amer. Concrete Institute*, 1969: 895)** considered regressing $y =$ 28-day standard-cured strength (psi) against $x =$ accelerated strength (psi). Suppose the equation of the true regression line is $y = 1800 + 1.3x$.

a. What is the expected value of 28-day strength when accelerated strength $= 2500$?

b. By how much can we expect 28-day strength to change when accelerated strength increases by 1 psi?

c. Answer part (b) for an increase of 100 psi.

d. Answer part (b) for a decrease of 100 psi.

8. Referring to Exercise 7, suppose that the standard deviation of the random deviation ϵ is 350 psi.

a. What is the probability that the observed value of 28-day strength will exceed 5000 psi when the value of accelerated strength is 2000?

b. Repeat part (a) with 2500 in place of 2000.

c. Consider making two independent observations on 28-day strength, the first for an accelerated strength of 2000 and the second for $x = 2500$. What is the probability that the second observation will exceed the first by more than 1000 psi?

d. Let Y_1 and Y_2 denote observations on 28-day strength when $x = x_1$ and $x = x_2$, respectively. By how much would x_2 have to exceed x_1 in order that $P(Y_2 > Y_1) = .95$?

9. The flow rate y (m³/min) in a device used for air-quality measurement depends on the pressure drop x (in. of water) across the device's filter. Suppose that for x values between 5 and 20, the two variables are related according to the simple linear regression model with true regression line $y = -.12 + .095x$.

a. What is the expected change in flow rate associated with a 1-in. increase in pressure drop? Explain.

b. What change in flow rate can be expected when pressure drop decreases by 5 in.?

c. What is the expected flow rate for a pressure drop of 10 in.? A drop of 15 in.?

d. Suppose $\sigma = .025$ and consider a pressure drop of 10 in. What is the probability that the observed value of flow rate will exceed .835? That observed flow rate will exceed .840?

e. What is the probability that an observation on flow rate when pressure drop is 10 in. will exceed an observation on flow rate made when pressure drop is 11 in.?

10. Suppose the expected cost of a production run is related to the size of the run by the equation $y = 4000 + 10x$. Let Y denote an observation on the cost of a run. If the variables' *size* and *cost* are related according to the simple linear regression model, could it be the case that $P(Y > 5500$ when $x = 100) = .05$ and $P(Y > 6500$ when $x = 200) = .10$? Explain.

11. Suppose that in a certain chemical process the reaction time y (hr) is related to the temperature (°F) in the chamber in which the reaction takes place according to the simple linear regression model with equation $y = 5.00 - .01x$ and $\sigma = .075$.

a. What is the expected change in reaction time for a 1°F increase in temperature? For a 10°F increase in temperature?

b. What is the expected reaction time when temperature is 200°F? When temperature is 250°F?

c. Suppose five observations are made independently on reaction time, each one for a temperature of 250°F.

What is the probability that all five times are between 2.4 and 2.6 hr?

d. What is the probability that two independently observed reaction times for temperatures 1° apart are such that the time at the higher temperature exceeds the time at the lower temperature?

12.2 Estimating Model Parameters

We will assume in this and the next several sections that the variables x and y are related according to the simple linear regression model. The values of β_0, β_1, and σ^2 will almost never be known to an investigator. Instead, sample data consisting of n observed pairs $(x_1, y_1), \ldots, (x_n, y_n)$ will be available, from which the model parameters and the true regression line itself can be estimated. These observations are assumed to have been obtained independently of one another. That is, y_i is the observed value of Y_i, where $Y_i = \beta_0 + \beta_1 x_i + \epsilon_i$ and the n deviations $\epsilon_1, \epsilon_2, \ldots, \epsilon_n$ are independent rv's. Independence of $Y_1, Y_2, \ldots, Y_n$ follows from independence of the ϵ_i's.

According to the model, the observed points will be distributed about the true regression line in a random manner. Figure 12.6 shows a typical plot of observed pairs along with two candidates for the estimated regression line. Intuitively, the line $y = a_0 + a_1 x$ is not a reasonable estimate of the true line $y = \beta_0 + \beta_1 x$ because, if $y = a_0 + a_1 x$ were the true line, the observed points would almost surely have been closer to this line. The line $y = b_0 + b_1 x$ is a more plausible estimate because the observed points are scattered rather closely about this line.

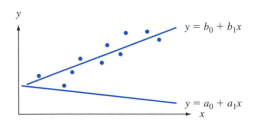

Figure 12.6 Two different estimates of the true regression line

Figure 12.6 and the foregoing discussion suggest that our estimate of $y = \beta_0 + \beta_1 x$ should be a line that provides in some sense a best fit to the observed data points. This is what motivates the principle of least squares, which can be traced back to the German mathematician Gauss (1777–1855). According to this principle, a line provides a good fit to the data if the vertical distances (deviations) from the observed points to the line are small (see Figure 12.7). The measure of the goodness of fit is the sum of the squares of these deviations. The best-fit line is then the one having the smallest possible sum of squared deviations.

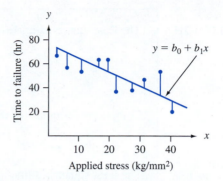

Figure 12.7 Deviations of observed data from line $y = b_0 + b_1x$

Principle of Least Squares

The vertical deviation of the point (x_i, y_i) from the line $y = b_0 + b_1x$ is

$$\text{height of point} - \text{height of line} = y_i - (b_0 + b_1x_i)$$

The sum of squared vertical deviations from the points $(x_1, y_1), \ldots, (x_n, y_n)$ to the line is then

$$f(b_0, b_1) = \sum_{i=1}^{n}[y_i - (b_0 + b_1x_i)]^2$$

The point estimates of β_0 and β_1, denoted by $\hat{\beta}_0$ and $\hat{\beta}_1$ and called the **least squares estimates**, are those values that minimize $f(b_0, b_1)$. That is, $\hat{\beta}_0$ and $\hat{\beta}_1$ are such that $f(\hat{\beta}_0, \hat{\beta}_1) \le f(b_0, b_1)$ for any b_0 and b_1. The **estimated regression line** or **least squares line** is then the line whose equation is $y = \hat{\beta}_0 + \hat{\beta}_1x$.

The minimizing values of b_0 and b_1 are found by taking partial derivatives of $f(b_0, b_1)$ with respect to both b_0 and b_1, equating them both to zero [analogously to $f'(b) = 0$ in univariate calculus], and solving the equations

$$\frac{\partial f(b_0, b_1)}{\partial b_0} = \sum 2(y_i - b_0 - b_1x_i)(-1) = 0$$

$$\frac{\partial f(b_0, b_1)}{\partial b_1} = \sum 2(y_i - b_0 - b_1x_i)(-x_i) = 0$$

Cancellation of the -2 factor and rearrangement gives the following system of equations, called the **normal equations**:

$$nb_0 + \left(\sum x_i\right)b_1 = \sum y_i$$

$$\left(\sum x_i\right)b_0 + \left(\sum x_i^2\right)b_1 = \sum x_iy_i$$

These equations are linear in the two unknowns b_0 and b_1. Provided that not all x_i's are identical, the least squares estimates are the unique solution to this system.

PROPOSITION

The least squares estimate of the slope coefficient β_1 of the true regression line is

$$b_1 = \hat{\beta}_1 = \frac{\sum(x_i - \bar{x})(y_i - \bar{y})}{\sum(x_i - \bar{x})^2} = \frac{S_{xy}}{S_{xx}} \qquad (12.2)$$

Computing formulas for the numerator and denominator of $\hat{\beta}_1$ are

$$S_{xy} = \sum x_i y_i - \left(\sum x_i\right)\left(\sum y_i\right)/n \qquad S_{xx} = \sum x_i^2 - \left(\sum x_i\right)^2/n$$

The least squares estimate of the intercept β_0 of the true regression line is

$$b_0 = \hat{\beta}_0 = \frac{\sum y_i - \hat{\beta}_1 \sum x_i}{n} = \bar{y} - \hat{\beta}_1\bar{x} \qquad (12.3)$$

The computational formulas for S_{xy} and S_{xx} require only the summary statistics Σx_i, Σy_i, Σx_i^2, and $\Sigma x_i y_i$ (Σy_i^2 will be needed shortly). In computing $\hat{\beta}_0$, use extra digits in $\hat{\beta}_1$ because, if $\bar{x}$ is large in magnitude, rounding will affect the final answer. In practice, the use of a statistical software package is preferable to hand calculation and hand-drawn plots. Once again, be sure that the scatterplot shows a linear pattern with relatively homogenous variation before fitting the simple linear regression model.

EXAMPLE 12.4

The cetane number is a critical property in specifying the ignition quality of a fuel used in a diesel engine. Determination of this number for a biodiesel fuel is expensive and time-consuming. The article **"Relating the Cetane Number of Biodiesel Fuels to Their Fatty Acid Composition: A Critical Study" (*J. of Automobile Engr.*, 2009: 565–583)** included the following data on $x =$ iodine value (g) and $y =$ cetane number for a sample of 14 biofuels. The iodine value is the amount of iodine necessary to saturate a sample of 100 g of oil. The article's authors fit the simple linear regression model to this data, so let's follow their lead.

x	132.0	129.0	120.0	113.2	105.0	92.0	84.0	83.2	88.4	59.0	80.0	81.5	71.0	69.2
y	46.0	48.0	51.0	52.1	54.0	52.0	59.0	58.7	61.6	64.0	61.4	54.6	58.8	58.0

The necessary summary quantities for hand calculation can be obtained by placing the x values in a column and the y values in another column and then creating columns for x^2, xy, and y^2 (these latter values are not needed at the moment but will be used shortly). Calculating the column sums gives $\Sigma x_i = 1307.5, \Sigma y_i = 779.2, \Sigma x_i^2 = 128{,}913.93, \Sigma x_i y_i = 71{,}347.30, \Sigma y_i^2 = 43{,}745.22$, from which

$$S_{xx} = 128{,}913.93 - (1307.5)^2/14 = 6802.7693$$
$$S_{xy} = 71{,}347.30 - (1307.5)(779.2)/14 = -1424.41429$$

The estimated slope of the true regression line (i.e., the slope of the least squares line) is

$$\hat{\beta}_1 = \frac{S_{xy}}{S_{xx}} = \frac{-1424.41429}{6802.7693} = -.20938742$$

We estimate that the expected change in true average cetane number associated with a 1 g increase in iodine value is $-.209$—i.e., a decrease of .209. Since $\bar{x} = 93.392857$ and $\bar{y} = 55.657143$, the estimated intercept of the true regression line (i.e., the intercept of the least squares line) is

$$\hat{\beta}_0 = \bar{y} - \hat{\beta}_1\bar{x} = 55.657143 - (-.20938742)(93.392857) = 75.212432$$

The equation of the estimated regression line (least squares line) is $y = 75.212 - .2094x$, exactly that reported in the cited article. Figure 12.8 displays a scatterplot of the data with the least squares line superimposed. This line provides a very good summary of the relationship between the two variables.

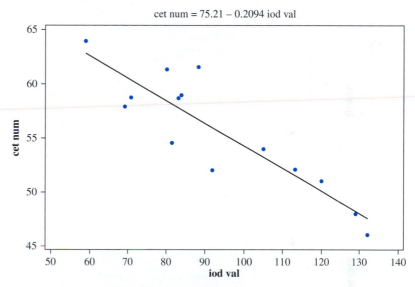

cet num = 75.21 − 0.2094 iod val

Figure 12.8 Scatterplot for Example 12.4 with least squares line superimposed, from Minitab

The estimated regression line can immediately be used for two different purposes. For a fixed x value x^*, $\hat{\beta}_0 + \hat{\beta}_1 x^*$ (the height of the line above x^*) gives either (1) a point estimate of the expected value of Y when $x = x^*$ or (2) a point prediction of the Y value that will result from a single new observation made at $x = x^*$.

EXAMPLE 12.5 Refer back to the iodine value–cetane number scenario described in the previous example. The estimated regression equation was $y = 75.212 - .2094x$. A point estimate of true average cetane number for all biofuels whose iodine value is 100 is

$$\hat{\mu}_{Y \cdot 100} = \hat{\beta}_0 + \hat{\beta}_1(100) = 75.212 - .2094(100) = 54.27$$

If a single biofuel sample whose iodine value is 100 is to be selected, 54.27 is also a point prediction for the resulting cetane number.

The least squares line should not be used to make a prediction for an x value much beyond the range of the data, such as $x = 40$ or $x = 150$ in Example 12.4. The **danger of extrapolation** is that the fitted relationship (a line here) may not be valid for such x values.

Estimating σ^2 and σ

The parameter σ^2 determines the amount of variability inherent in the regression model. A large value of σ^2 will lead to observed (x_i, y_i)'s that are typically quite spread out about the true regression line, whereas when σ^2 is small the observed points will tend to fall very close to the true line (see Figure 12.9). An estimate of σ^2 will be used in confidence interval (CI) formulas and hypothesis-testing procedures presented in the next two sections. Because the equation of the true line is unknown, the estimate is based on the extent to which the sample observations deviate from the *estimated* line. Many large deviations (residuals) suggest a large value of σ^2, whereas deviations all of which are small in magnitude suggest that σ^2 is small.

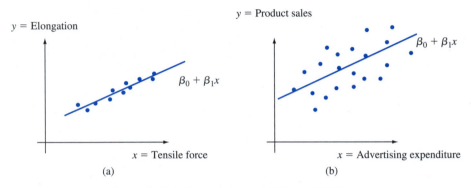

Figure 12.9   Typical sample for σ^2: (a) small; (b) large

DEFINITION

> The **fitted** (or **predicted**) values $\hat{y}_1, \hat{y}_2, \ldots, \hat{y}_n$ are obtained by successively substituting $x_1, \ldots, x_n$ into the equation of the estimated regression line: $\hat{y}_1 = \hat{\beta}_0 + \hat{\beta}_1 x_1, \hat{y}_2 = \hat{\beta}_0 + \hat{\beta}_1 x_2, \ldots, \hat{y}_n = \hat{\beta}_0 + \hat{\beta}_1 x_n$. The **residuals** are the differences $y_1 - \hat{y}_1, y_2 - \hat{y}_2, \ldots, y_n - \hat{y}_n$ between the observed and fitted y values.

In words, the predicted value $\hat{y}_i$ is the value of y that we would predict or expect when using the estimated regression line with $x = x_i$; $\hat{y}_i$ is the height of the estimated regression line above the value x_i for which the ith observation was made. The residual $y_i - \hat{y}_i$ is the vertical deviation between the point (x_i, y_i) and the least squares line—a positive number if the point lies above the line and a negative number if it lies below the line. If the residuals are all small in magnitude, then much of the variability in observed y values appears to be due to the linear relationship between x and y, whereas many large residuals suggest quite a bit of inherent variability in y relative to the amount due to the linear relation. Assuming that the line in Figure 12.7 is the least squares line, the residuals are identified by the vertical line segments from the observed points to the line. When the estimated regression line is obtained via the principle of least squares, the sum of the residuals should in theory be zero. In practice, the sum may deviate a bit from zero due to rounding.

EXAMPLE 12.6   Japan's high population density has resulted in a multitude of resource-usage problems. One especially serious difficulty concerns waste removal. The article **"Innovative Sludge Handling Through Pelletization Thickening"** (*Water Research*, **1999: 3245–3252**) reported the development of a new compression machine for processing sewage sludge. An important part of the investigation involved relating the moisture content of compressed pellets (y, in %) to the machine's filtration rate (x, in kg-DS/m/hr). The following data was read from a graph in the article:

x	125.3	98.2	201.4	147.3	145.9	124.7	112.2	120.2	161.2	178.9
y	77.9	76.8	81.5	79.8	78.2	78.3	77.5	77.0	80.1	80.2
x	159.5	145.8	75.1	151.4	144.2	125.0	198.8	132.5	159.6	110.7
y	79.9	79.0	76.7	78.2	79.5	78.1	81.5	77.0	79.0	78.6

Relevant summary quantities (*summary statistics*) are $\Sigma x_i = 2817.9$, $\Sigma y_i = 1574.8$, $\Sigma x_i^2 = 415{,}949.85$, $\Sigma x_i y_i = 222{,}657.88$, and $\Sigma y_i^2 = 124{,}039.58$, from which $\bar{x} = 140.895$, $\bar{y} = 78.74$, $S_{xx} = 18{,}921.8295$, and $S_{xy} = 776.434$. Thus

$$\hat{\beta}_1 = \frac{776.434}{18{,}921.8295} = .04103377 \approx .041$$

$$\hat{\beta}_0 = 78.74 - (.04103377)(140.895) = 72.958547 \approx 72.96$$

from which the equation of least squares line is $y = 72.96 + .041x$. For numerical accuracy, the fitted values are calculated from $\hat{y}_i = 72.958547 + .04103377 x_i$:

$$\hat{y}_1 = 72.958547 + .04103377(125.3) \approx 78.100, \quad y_1 - \hat{y}_1 \approx -.200, \text{ etc.}$$

Nine of the 20 residuals are negative, so the corresponding nine points in a scatterplot of the data lie below the estimated regression line. All predicted values (fits) and residuals appear in the accompanying table.

Obs	Filtrate	Moistcon	Fit	Residual
1	125.3	77.9	78.100	−0.200
2	98.2	76.8	76.988	−0.188
3	201.4	81.5	81.223	0.277
4	147.3	79.8	79.003	0.797
5	145.9	78.2	78.945	−0.745
6	124.7	78.3	78.075	0.225
7	112.2	77.5	77.563	−0.063
8	120.2	77.0	77.891	−0.891
9	161.2	80.1	79.573	0.527
10	178.9	80.2	80.299	−0.099
11	159.5	79.9	79.503	0.397
12	145.8	79.0	78.941	0.059
13	75.1	76.7	76.040	0.660
14	151.4	78.2	79.171	−0.971
15	144.2	79.5	78.876	0.624
16	125.0	78.1	78.088	0.012
17	198.8	81.5	81.116	0.384
18	132.5	77.0	78.396	−1.396
19	159.6	79.0	79.508	−0.508
20	110.7	78.6	77.501	1.099

In much the same way that the deviations from the mean in a one-sample situation were combined to obtain the estimate $s^2 = \Sigma(x_i - \bar{x})^2/(n - 1)$, the estimate of σ^2 in regression analysis is based on squaring and summing the residuals. We will continue to use the symbol s^2 for this estimated variance, so don't confuse it with our previous s^2.

<table>
<tr><td>**DEFINITION**</td><td>

The **error sum of squares** (equivalently, residual sum of squares), denoted by SSE, is

$$\text{SSE} = \sum (y_i - \hat{y}_i)^2 = \sum [y_i - (\hat{\beta}_0 + \hat{\beta}_1 x_i)]^2$$

and the estimate of σ^2 is

$$\hat{\sigma}^2 = s^2 = \frac{\text{SSE}}{n-2} = \frac{\sum (y_i - \hat{y}_i)^2}{n-2}$$

</td></tr>
</table>

The divisor $n - 2$ in s^2 is the number of degrees of freedom (df) associated with SSE and the estimate s^2. This is because to obtain s^2, the two parameters β_0 and β_1 must first be estimated, which results in a loss of 2 df (just as μ had to be estimated in one-sample problems, resulting in an estimated variance based on $n - 1$ df). Replacing each y_i in the formula for s^2 by the rv Y_i gives the estimator S^2. It can be shown that S^2 is an unbiased estimator for σ^2 (though the estimator S is not unbiased for σ). An interpretation of s here is similar to what we suggested earlier for the sample standard deviation: Very roughly, it is the size of a typical vertical deviation within the sample from the estimated regression line.

EXAMPLE 12.7 The residuals for the filtration rate–moisture content data were calculated previously. The corresponding error sum of squares is

$$\text{SSE} = (-.200)^2 + (-.188)^2 + \cdots + (1.099)^2 = 7.968$$

The estimate of σ^2 is then $\hat{\sigma}^2 = s^2 = 7.968/(20 - 2) = .4427$, and the estimated standard deviation is $\hat{\sigma} = s = \sqrt{.4427} = .665$. Roughly speaking, .665 is the magnitude of a typical deviation from the estimated regression line—some points are closer to the line than this and others are farther away. ∎

Computation of SSE from the defining formula involves much tedious arithmetic, because both the predicted values and residuals must first be calculated. Use of the following computational formula does not require these quantities.

$$\text{SSE} = \sum y_i^2 - \hat{\beta}_0 \sum y_i - \hat{\beta}_1 \sum x_i y_i = S_{yy} - \hat{\beta}_1 S_{xy}$$

The middle expression results from substituting $\hat{y}_i = \hat{\beta}_0 + \hat{\beta}_1 x_i$ into $\Sigma(y_i - \hat{y}_i)^2$, squaring the summand, carrying through the sum to the resulting three terms, and simplifying. These computational formulas are especially sensitive to the effects of rounding in $\hat{\beta}_0$ and $\hat{\beta}_1$, so carrying as many digits as possible in intermediate computations will protect against round-off error.

EXAMPLE 12.8 The article **"Promising Quantitative Nondestructive Evaluation Techniques for Composite Materials"** (*Materials Evaluation*, **1985: 561–565**) reports on a study to investigate how the propagation of an ultrasonic stress wave through a substance depends on the properties of the substance. The accompanying data on fracture strength (x, as a percentage of ultimate tensile strength) and attenuation (y, in neper/cm, the decrease in amplitude of the stress wave) in fiberglass-reinforced polyester composites was read from a graph that appeared in the article. The simple linear regression model is suggested by the substantial linear pattern in the scatterplot.

x	12	30	36	40	45	57	62	67	71	78	93	94	100	105
y	3.3	3.2	3.4	3.0	2.8	2.9	2.7	2.6	2.5	2.6	2.2	2.0	2.3	2.1

The necessary summary quantities are $n = 14$, $\Sigma x_i = 890$, $\Sigma x_i^2 = 67{,}182$, $\Sigma y_i = 37.6$, $\Sigma y_i^2 = 103.54$, and $\Sigma x_i y_i = 2234.30$, from which $S_{xx} = 10{,}603.4285714$, $S_{xy} = -155.98571429$, $\hat{\beta}_1 = -.0147109$, and $\hat{\beta}_0 = 3.6209072$. Then

$$\text{SSE} = 103.54 - (3.6209072)(37.6) - (-.0147109)(2234.30)$$
$$= .2624532$$

The same value results from

$$\text{SSE} = S_{yy} - \hat{\beta}_1 S_{xy} = 103.54 - (37.6)^2/14 - (-.0147109)(-155.98571429)$$

Thus $s^2 = .2624532/12 = .0218711$ and $s = .1479$. When $\hat{\beta}_0$ and $\hat{\beta}_1$ are rounded to three decimal places in the first computational formula for SSE, the result is

$$\text{SSE} = 103.54 - (3.621)(37.6) - (-.015)(2234.30) = .905$$

which is more than three times the correct value. ∎

The Coefficient of Determination

Figure 12.10 shows three different scatterplots of bivariate data. In all three plots, the heights of the different points vary substantially, indicating that there is much variability in observed y values. The points in the first plot all fall exactly on a straight line. In this case, all (100%) of the sample variation in y can be attributed to the fact that x and y are linearly related in combination with variation in x. The points in Figure 12.10(b) do not fall exactly on a line, but compared to overall y variability, the deviations from the least squares line are small. It is reasonable to conclude in this case that much of the observed y variation can be attributed to the approximate linear relationship between the variables postulated by the simple linear regression model. When the scatterplot looks like that of Figure 12.10(c), there is substantial variation about the least squares line relative to overall y variation, so the simple linear regression model fails to explain variation in y by relating y to x.

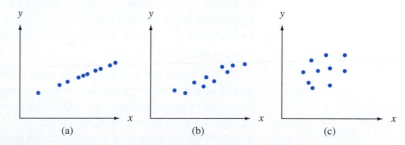

Figure 12.10 Using the model to explain y variation: (a) data for which all variation is explained; (b) data for which most variation is explained; (c) data for which little variation is explained

The error sum of squares SSE can be interpreted as a measure of how much variation in y is left unexplained by the model—that is, how much cannot be attributed to a linear relationship. In Figure 12.10(a), SSE = 0, and there is no unexplained variation, whereas unexplained variation is small for the data of Figure 12.10(b) and much larger in Figure 12.10(c). A quantitative measure of the total amount of variation in observed y values is given by the **total sum of squares**

$$\text{SST} = S_{yy} = \sum (y_i - \bar{y})^2 = \sum y_i^2 - \left(\sum y_i\right)^2 / n$$

Total sum of squares is the sum of squared deviations about the sample mean of the observed y values. Thus the same number $\bar{y}$ is subtracted from each y_i in SST, whereas SSE involves subtracting each different predicted value $\hat{y}_i$ from the corresponding observed y_i. Just as SSE is the sum of squared deviations about the least squares line $y = \hat{\beta}_0 + \hat{\beta}_1 x$, SST is the sum of squared deviations about the horizontal line at height $\bar{y}$ (since then vertical deviations are $y_i - \bar{y}$), as pictured in Figure 12.11. Furthermore, because the sum of squared deviations about the least squares line is smaller than the sum of squared deviations about *any* other line, SSE < SST unless the horizontal line itself is the least squares line. The ratio SSE/SST is the proportion of total variation that cannot be explained by the simple linear regression model, and 1 − SSE/SST (a number between 0 and 1) is the proportion of observed y variation explained by the model.

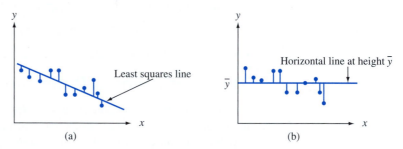

Figure 12.11 Sums of squares illustrated: (a) SSE = sum of squared deviations about the least squares line; (b) SST = sum of squared deviations about the horizontal line

DEFINITION

The **coefficient of determination,** denoted by r^2, is given by

$$r^2 = 1 - \frac{\text{SSE}}{\text{SST}}$$

It is interpreted as the proportion of observed y variation that can be explained by the simple linear regression model (attributed to an approximate linear relationship between y and x).

The higher the value of r^2, the more successful is the simple linear regression model in explaining y variation. When regression analysis is done by a statistical computer package, either r^2 or $100r^2$ (the percentage of variation explained by the model) is a prominent part of the output. If r^2 is small, an analyst will usually want to search for an alternative model (either a nonlinear model or a multiple regression model that involves more than a single independent variable) that can more effectively explain y variation.

EXAMPLE 12.9 The scatterplot of the iodine value–cetane number data in Figure 12.8 portends a reasonably high r^2 value. With

$$\hat{\beta}_0 = 75.212432 \qquad \hat{\beta}_1 = -.20938742 \qquad \Sigma y_i = 779.2$$

$$\Sigma x_i y_i = 71{,}347.30 \qquad \Sigma y_i^2 = 43{,}745.22$$

we have

$$\text{SST} = 43{,}745.22 - (779.2)^2/14 = 377.174$$

$$\text{SSE} = 43{,}745.22 - (75.212432)(779.2) - (-.20938742)(71{,}347.30) = 78.920$$

The coefficient of determination is then

$$r^2 = 1 - \text{SSE}/\text{SST} = 1 - (78.920)/(377.174) = .791$$

That is, 79.1% of the observed variation in cetane number is attributable to (can be explained by) the simple linear regression relationship between cetane number and iodine value (r^2 values are even higher than this in many scientific contexts, but social scientists would typically be ecstatic at a value anywhere near this large!).

Figure 12.12 shows partial Minitab output from the regression of cetane number on iodine value. The software will also provide predicted values, residuals, and other information upon request. The formats used by other packages differ slightly from that of Minitab, but the information content is very similar. Regression sum of squares will be introduced shortly. Other quantities in Figure 12.12 that have not yet been discussed will surface in Section 12.3 [excepting R-Sq(adj), which comes into play in Chapter 13 when multiple regression models are introduced].

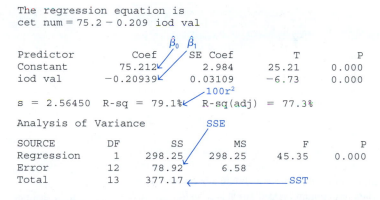

Figure 12.12 Minitab output for the regression of Examples 12.4 and 12.9 ■

The coefficient of determination can be written in a slightly different way by introducing a third sum of squares—**regression sum of squares**, SSR—given by $\text{SSR} = \Sigma(\hat{y}_i - \bar{y})^2 = \text{SST} - \text{SSE}$. Regression sum of squares is interpreted as the amount of total variation that *is* explained by the model. Then we have

$$r^2 = 1 - \text{SSE}/\text{SST} = (\text{SST} - \text{SSE})/\text{SST} = \text{SSR}/\text{SST}$$

the ratio of explained variation to total variation. The ANOVA table in Figure 12.12 shows that $\text{SSR} = 298.25$, from which $r^2 = 298.25/377.17 = .791$ as before.

Terminology and Scope of Regression Analysis

The term *regression analysis* was first used by Francis Galton in the late nineteenth century in connection with his work on the relationship between father's height x

and son's height y. After collecting a number of pairs (x_i, y_i), Galton used the principle of least squares to obtain the equation of the estimated regression line, with the objective of using it to predict son's height from father's height. In using the derived line, Galton found that if a father was above average in height, the son would also be expected to be above average in height, *but not by as much as the father was.* Similarly, the son of a shorter-than-average father would also be expected to be shorter than average, but not by as much as the father. Thus the predicted height of a son was "pulled back in" toward the mean; because *regression* means a coming or going back, Galton adopted the terminology *regression line.* This phenomenon of being pulled back in toward the mean has been observed in many other situations (e.g., batting averages from year to year in baseball) and is called the **regression effect**.

Our discussion thus far has presumed that the independent variable is under the control of the investigator, so that only the dependent variable Y is random. This was not, however, the case with Galton's experiment; fathers' heights were not preselected, but instead both X and Y were random. Methods and conclusions of regression analysis can be applied both when the values of the independent variable are fixed in advance and when they are random, but because the derivations and interpretations are more straightforward in the former case, we will continue to work explicitly with it. For more commentary, see the excellent book by John Neter et al. listed in the chapter bibliography.

EXERCISES Section 12.2 (12–29)

12. Refer back to the data in Exercise 4, in which y = ammonium concentration (mg/L) and x = transpiration (ml/h). Summary quantities include $n = 13$, $\Sigma x_i = 303.7$, $\Sigma y_i = 52.8$, $S_{xx} = 1585.230769$, $S_{xy} = -341.959231$, and $S_{yy} = 77.270769$.

   a. Obtain the equation of the estimated regression line and use it to calculate a point prediction of ammonium concentration for a future observation made when ammonium concentration is 25 ml/h.

   b. What happens if the estimated regression line is used to calculate a point estimate of true average concentration when transpiration is 45 ml/h? Why does it not make sense to calculate this point estimate?

   c. Calculate and interpret s.

   d. Do you think the simple linear regression model does a good job of explaining observed variation in concentration? Explain.

13. The accompanying data on x = current density (mA/cm^2) and y = rate of deposition (μm/min) appeared in the article **"Plating of 60/40 Tin/Lead Solder for Head Termination Metallurgy"** (*Plating and Surface Finishing*, Jan. 1997: 38–40). Do you agree with the claim by the article's author that "a linear relationship

was obtained from the tin-lead rate of deposition as a function of current density"? Explain your reasoning.

x	20	40	60	80
y	.24	1.20	1.71	2.22

14. Refer to the tank temperature–efficiency ratio data given in Exercise 1.

   a. Determine the equation of the estimated regression line.

   b. Calculate a point estimate for true average efficiency ratio when tank temperature is 182.

   c. Calculate the values of the residuals from the least squares line for the four observations for which temperature is 182. Why do they not all have the same sign?

   d. What proportion of the observed variation in efficiency ratio can be attributed to the simple linear regression relationship between the two variables?

15. Values of modulus of elasticity (MOE, the ratio of stress, i.e., force per unit area, to strain, i.e., deformation per unit length, in GPa) and flexural strength (a measure of the ability to resist failure in bending, in MPa) were determined for a sample of concrete beams of a certain type, resulting in the following data (read from a graph in the article **"Effects of Aggregates and Microfillers on the**

Flexural Properties of Concrete," *Magazine of Concrete Research*, **1997: 81–98)**:

MOE	29.8	33.2	33.7	35.3	35.5	36.1	36.2
Strength	5.9	7.2	7.3	6.3	8.1	6.8	7.0

MOE	36.3	37.5	37.7	38.7	38.8	39.6	41.0
Strength	7.6	6.8	6.5	7.0	6.3	7.9	9.0

MOE	42.8	42.8	43.5	45.6	46.0	46.9	48.0
Strength	8.2	8.7	7.8	9.7	7.4	7.7	9.7

MOE	49.3	51.7	62.6	69.8	79.5	80.0
Strength	7.8	7.7	11.6	11.3	11.8	10.7

a. Construct a stem-and-leaf display of the MOE values, and comment on any interesting features.

b. Is the value of strength completely and uniquely determined by the value of MOE? Explain.

c. Use the accompanying Minitab output to obtain the equation of the least squares line for predicting strength from modulus of elasticity, and then predict strength for a beam whose modulus of elasticity is 40. Would you feel comfortable using the least squares line to predict strength when modulus of elasticity is 100? Explain.

```
Predictor    Coef     Stdev    t-ratio      P
Constant    3.2925    0.6008      5.48   0.000
mod elas    0.10748   0.01280     8.40   0.000

s = 0.8657   R-sq = 73.8%   R-sq(adj) = 72.8%

Analysis of Variance

SOURCE        DF      SS       MS       F       P
Regression     1   52.870   52.870   70.55   0.000
Error         25   18.736    0.749
Total         26   71.605
```

d. What are the values of SSE, SST, and the coefficient of determination? Do these values suggest that the simple linear regression model effectively describes the relationship between the two variables? Explain.

16. The article **"Characterization of Highway Runoff in Austin, Texas, Area"** (*J. of Envir. Engr.*, **1998: 131–137**) gave a scatterplot, along with the least squares line, of x = rainfull volume (m³) and y = runoff volume (m³) for a particular location. The accompanying values were read from the plot.

x	5	12	14	17	23	30	40	47
y	4	10	13	15	15	25	27	46

x	55	67	72	81	96	112	127
y	38	46	53	70	82	99	100

a. Does a scatterplot of the data support the use of the simple linear regression model?

b. Calculate point estimates of the slope and intercept of the population regression line.

c. Calculate a point estimate of the true average runoff volume when rainfall volume is 50.

d. Calculate a point estimate of the standard deviation σ.

e. What proportion of the observed variation in runoff volume can be attributed to the simple linear regression relationship between runoff and rainfall?

17. No-fines concrete, made from a uniformly graded coarse aggregate and a cement-water paste, is beneficial in areas prone to excessive rainfall because of its excellent drainage properties. The article **"Pavement Thickness Design for No-Fines Concrete Parking Lots,"** *J. of Trans. Engr.*, **1995: 476–484)** employed a least squares analysis in studying how y = porosity (%) is related to x = unit weight (pcf) in concrete specimens. Consider the following representative data:

x	99.0	101.1	102.7	103.0	105.4	107.0	108.7	110.8
y	28.8	27.9	27.0	25.2	22.8	21.5	20.9	19.6

x	112.1	112.4	113.6	113.8	115.1	115.4	120.0
y	17.1	18.9	16.0	16.7	13.0	13.6	10.8

Relevant summary quantities are $\Sigma x_i = 1640.1$, $\Sigma y_i = 299.8$, $\Sigma x_i^2 = 179{,}849.73$, $\Sigma x_i y_i = 32{,}308.59$, $\Sigma y_i^2 = 6430.06$.

a. Obtain the equation of the estimated regression line. Then create a scatterplot of the data and graph the estimated line. Does it appear that the model relationship will explain a great deal of the observed variation in y?

b. Interpret the slope of the least squares line.

c. What happens if the estimated line is used to predict porosity when unit weight is 135? Why is this not a good idea?

d. Calculate the residuals corresponding to the first two observations.

e. Calculate and interpret a point estimate of σ.

f. What proportion of observed variation in porosity can be attributed to the approximate linear relationship between unit weight and porosity?

18. For the past decade, rubber powder has been used in asphalt cement to improve performance. The article **"Experimental Study of Recycled Rubber-Filled High-Strength Concrete"** (*Magazine of Concrete Res.*, **2009: 549–556**) includes a regression of y = axial strength (MPa) on x = cube strength (MPa) based on the following sample data:

x	112.3	97.0	92.7	86.0	102.0	99.2	95.8	103.5	89.0	86.7
y	75.0	71.0	57.7	48.7	74.3	73.3	68.0	59.3	57.8	48.5

a. Obtain the equation of the least squares line, and interpret its slope.

b. Calculate and interpret the coefficient of determination.

c. Calculate and interpret an estimate of the error standard deviation σ in the simple linear regression model.

19. The following data is representative of that reported in the article **"An Experimental Correlation of Oxides of Nitrogen Emissions from Power Boilers Based on Field Data"** (*J. of Engr. for Power*, **July 1973: 165–170**), with x = burner-area liberation rate (MBtu/hr-ft^2) and y = NO$_x$ emission rate (ppm):

x	100	125	125	150	150	200	200
y	150	140	180	210	190	320	280

x	250	250	300	300	350	400	400
y	400	430	440	390	600	610	670

a. Assuming that the simple linear regression model is valid, obtain the least squares estimate of the true regression line.

b. What is the estimate of expected NO$_x$ emission rate when burner area liberation rate equals 225?

c. Estimate the amount by which you expect NO$_x$ emission rate to change when burner area liberation rate is decreased by 50.

d. Would you use the estimated regression line to predict emission rate for a liberation rate of 500? Why or why not?

20. The bond behavior of reinforcing bars is an important determinant of strength and stability. The article **"Experimental Study on the Bond Behavior of Reinforcing Bars Embedded in Concrete Subjected to Lateral Pressure"** (*J. of Materials in Civil Engr.*, **2012: 125–133**) reported the results of one experiment in which varying levels of lateral pressure were applied to 21 concrete cube specimens, each with an embedded 16-mm plain steel round bar, and the corresponding bond capacity was determined. Due to differing concrete cube strengths (f_{cu}, in MPa), the applied lateral pressure was equivalent to a fixed proportion of the specimen's f_{cu} ($0, .1f_{cu}, \ldots, .6f_{cu}$). Also, since bond strength can be heavily influenced by the specimen's f_{cu}, bond capacity was expressed as the ratio of bond strength (MPa) to $\sqrt{f_{cu}}$.

Pressure	0	0	0	.1	.1	.1	.2
Ratio	0.123	0.100	0.101	0.172	0.133	0.107	0.217

Pressure	.2	.2	.3	.3	.3	.4	.4
Ratio	0.172	0.151	0.263	0.227	0.252	0.310	0.365

Pressure	.4	.5	.5	.5	.6	.6	.6
Ratio	0.239	0.365	0.319	0.312	0.394	0.386	0.320

a. Does a scatterplot of the data support the use of the simple linear regression model?

b. Use the accompanying Minitab output to give point estimates of the slope and intercept of the population regression line.

c. Calculate a point estimate of the true average bond capacity when lateral pressure is $.45f_{cu}$.

d. What is a point estimate of the error standard deviation σ, and how would you interpret it?

e. What is the value of total variation, and what proportion of it can be explained by the model relationship?

```
The regression equation is
Ratio = 0.101 + 0.461 Pressure

Predictor      Coef    SE Coef       T       P
Constant    0.10121    0.01308    7.74   0.000
Pressure    0.46071    0.03627   12.70   0.000

S=0.0332397   R-Sq=89.5%   R-Sq(adj)=88.9%

Analysis of Variance
Source           DF      SS       MS       F       P
Regression        1 0.17830  0.17830  161.37  0.000
Residual Error   19 0.02099  0.00110
Total            20 0.19929
```

21. Wrinkle recovery angle and tensile strength are the two most important characteristics for evaluating the performance of crosslinked cotton fabric. An increase in the degree of crosslinking, as determined by ester carboxyl band absorbence, improves the wrinkle resistance of the fabric (at the expense of reducing mechanical strength). The accompanying data on x = absorbance and y = wrinke resistance angle was read from a graph in the paper **"Predicting the Performance of Durable Press Finished Cotton Fabric with Infrared Spectroscopy"** (*Textile Res. J.*, **1999: 145–151**).

x	.115	.126	.183	.246	.282	.344	.355	.452	.491	.554	.651
y	334	342	355	363	365	372	381	392	400	412	420

Here is regression output from Minitab:

```
Predictor      Coef    SE Coef        T       P
Constant    321.878      2.483   129.64   0.000
absorb      156.711      6.464    24.24   0.000

S = 3.60498   R-Sq = 98.5%   R-Sq(adj) = 98.3%

Source           DF      SS       MS       F       P
Regression        1  7639.0   7639.0  587.81  0.000
Residual Error    9   117.0     13.0
Total            10  7756.0
```

a. Does the simple linear regression model appear to be appropriate? Explain.

b. What wrinkle resistance angle would you predict for a fabric specimen having an absorbance of .300?

c. What would be the estimate of expected wrinkle resistence angle when absorbance is .300?

22. Calcium phosphate cement is gaining increasing attention for use in bone repair applications. The article **"Short-Fibre Reinforcement of Calcium Phosphate Bone Cement"** (*J. of Engr. in Med.*, **2007: 203–211**) reported on a study in which polypropylene fibers were used in an attempt to improve fracture behavior. The following data on x = fiber weight (%) and y = compressive strength (MPa) was provided by the article's authors.

x	0.00	0.00	0.00	0.00	0.00	1.25	1.25	1.25	1.25
y	9.94	11.67	11.00	13.44	9.20	9.92	9.79	10.99	11.32

x	2.50	2.50	2.50	2.50	2.50	5.00	5.00	5.00	5.00
y	12.29	8.69	9.91	10.45	10.25	7.89	7.61	8.07	9.04

x	7.50	7.50	7.50	7.50	10.00	10.00	10.00	10.00
y	6.63	6.43	7.03	7.63	7.35	6.94	7.02	7.67

a. Fit the simple linear regression model to this data. Then determine the proportion of observed variation in strength that can be attributed to the model relationship between strength and fiber weight. Finally, obtain a point estimate of the standard deviation of ϵ, the random deviation in the model equation.

b. The average strength values for the six different levels of fiber weight are 11.05, 10.51, 10.32, 8.15, 6.93, and 7.24, respectively. The cited paper included a figure in which the average strength was regressed against fiber weight. Obtain the equation of this regression line and calculate the corresponding coefficient of determination. Explain the difference between the r^2 value for this regression and the r^2 value obtained in (a).

23. a. Obtain SSE for the data in Exercise 19 from the defining formula [$\text{SSE} = \Sigma(y_i - \hat{y}_i)^2$], and compare to the value calculated from the computational formula.

b. Calculate the value of total sum of squares. Does the simple linear regression model appear to do an effective job of explaining variation in emission rate? Justify your assertion.

24. The invasive diatom species *Didymosphenia geminata* has the potential to inflict substantial ecological and economic damage in rivers. The article **"Substrate Characteristics Affect Colonization by the Bloom-Forming Diatom *Didymosphenia geminata* (*Aquatic Ecology*, 2010: 33–40)** described an investigation of colonization behavior. One aspect of particular interest was whether y = colony density was related to x = rock surface area. The article contained a scatterplot and summary of a regression analysis. Here is representative data:

x	50	71	55	50	33	58	79	26
y	152	1929	48	22	2	5	35	7

x	69	44	37	70	20	45	49
y	269	38	171	13	43	185	25

a. Fit the simple linear regression model to this data, predict colony density when surface area = 70 and when surface area = 71, and calculate the corresponding residuals. How do they compare?

b. Calculate and interpret the coefficient of determination.

c. The second observation has a very extreme y value (in the full data set consisting of 72 observations, there were two of these). This observation may have had a substantial impact on the fit of the model and subsequent conclusions. Eliminate it and recalculate the equation of the estimated regression line. Does it appear to differ substantially from the equation before the deletion? What is the impact on r^2 and s?

25. Show that b_1 and b_0 of expressions (12.2) and (12.3) satisfy the normal equations.

26. Show that the "point of averages" $(\bar{x}, \bar{y})$ lies on the estimated regression line.

27. Suppose an investigator has data on the amount of shelf space x devoted to display of a particular product and sales revenue y for that product. The investigator may wish to fit a model for which the true regression line passes through $(0, 0)$. The appropriate model is $Y = \beta_1 x + \epsilon$. Assume that $(x_1, y_1), \ldots, (x_n, y_n)$ are observed pairs generated from this model, and derive the least squares estimator of β_1. [*Hint*: Write the sum of squared deviations as a function of b_1, a trial value, and use calculus to find the minimizing value of b_1.]

28. a. Consider the data in Exercise 20. Suppose that instead of the least squares line passing through the points $(x_1, y_1), \ldots, (x_n, y_n)$, we wish the least squares line passing through $(x_1 - \bar{x}, y_1), \ldots, (x_n - \bar{x}, y_n)$. Construct a scatterplot of the (x_i, y_i) points and then of the $(x_i - \bar{x}, y_i)$ points. Use the plots to explain intuitively how the two least squares lines are related to one another.

b. Suppose that instead of the model $Y_i = \beta_0 + \beta_1 x_i + \epsilon_i$ ($i = 1, \ldots, n$), we wish to fit a model of the form $Y_i = \beta_0^* + \beta_1^*(x_i - \bar{x}) + \epsilon_i$ ($i = 1, \ldots, n$). What are the least squares estimators of β_0^* and β_1^*, and how do they relate to $\hat{\beta}_0$ and $\hat{\beta}_1$?

29. Consider the following three data sets, in which the variables of interest are x = commuting distance and y = commuting time. Based on a scatterplot and the values of s and r^2, in which situation would simple linear regression be most (least) effective, and why?

Data Set	1		2		3	
	x	y	x	y	x	y
	15	42	5	16	5	8
	16	35	10	32	10	16
	17	45	15	44	15	22
	18	42	20	45	20	23
	19	49	25	63	25	31
	20	46	50	115	50	60
S_{xx}	17.50		1270.8333		1270.8333	
S_{xy}	29.50		2722.5		1431.6667	
$\hat{\beta}_1$	1.685714		2.142295		1.126557	
$\hat{\beta}_0$	13.666672		7.868852		3.196729	
SST	114.83		5897.5		1627.33	
SSE	65.10		65.10		14.48	

12.3 Inferences About the Slope Parameter β_1

In virtually all of our inferential work thus far, the notion of sampling variability has been pervasive. In particular, properties of sampling distributions of various statistics have been the basis for developing confidence interval formulas and hypothesis-testing methods. The key idea here is that the value of any quantity calculated from sample data—the value of any statistic—will vary from one sample to another.

EXAMPLE 12.10 Reconsider the data on x = burner area liberation rate and y = NO_x emission rate from Exercise 12.19 in the previous section. There are 14 observations, made at the x values 100, 125, 125, 150, 150, 200, 200, 250, 250, 300, 300, 350, 400, and 400, respectively. Suppose that the slope and intercept of the true regression line are $\beta_1 = 1.70$ and $\beta_0 = -50$, with $\sigma = 35$ (consistent with the values $\hat{\beta}_1 = 1.7114$, $\hat{\beta}_0 = -45.55$, $s = 36.75$). We proceeded to generate a sample of random deviations $\tilde{\epsilon}_1, \ldots, \tilde{\epsilon}_{14}$ from a normal distribution with mean 0 and standard deviation 35 and then added $\tilde{\epsilon}_i$ to $\beta_0 + \beta_1 x_i$ to obtain 14 corresponding y values. Regression calculations were then carried out to obtain the estimated slope, intercept, and standard deviation. This process was repeated a total of 20 times, resulting in the values given in Table 12.1.

Table 12.1 Simulation Results for Example 12.10

$\hat{\beta}_1$	$\hat{\beta}_0$	s	$\hat{\beta}_1$	$\hat{\beta}_0$	s
1. 1.7559	−60.62	43.23	11. 1.7843	−67.36	41.80
2. 1.6400	−49.40	30.69	12. 1.5822	−28.64	32.46
3. 1.4699	−4.80	36.26	13. 1.8194	−83.99	40.80
4. 1.6944	−41.95	22.89	14. 1.6469	−32.03	28.11
5. 1.4497	5.80	36.84	15. 1.7712	−52.66	33.04
6. 1.7309	−70.01	39.56	16. 1.7004	−58.06	43.44
7. 1.8890	−95.01	42.37	17. 1.6103	−27.89	25.60
8. 1.6471	−40.30	43.71	18. 1.6396	−24.89	40.78
9. 1.7216	−42.68	23.68	19. 1.7857	−77.31	32.38
10. 1.7058	−63.31	31.58	20. 1.6342	−17.00	30.93

There is clearly variation in values of the estimated slope and estimated intercept, as well as the estimated standard deviation. The equation of the least squares line thus varies from one sample to the next. Figure 12.13 on page 511 shows a dotplot of the estimated slopes as well as graphs of the true regression line and the 20 sample regression lines. ∎

The slope β_1 of the population regression line is the true average change in the dependent variable y associated with a 1-unit increase in the independent variable x. The slope of the least squares line, $\hat{\beta}_1$, gives a point estimate of β_1. In the same way that a confidence interval for μ and procedures for testing hypotheses about μ were based on properties of the sampling distribution of $\overline{X}$, further inferences about β_1 are based on thinking of $\hat{\beta}_1$ as a statistic and investigating its sampling distribution.

The values of the x_i's are assumed to be chosen before the experiment is performed, so only the Y_i's are random. The estimators (statistics, and thus random variables) for β_0 and β_1 are obtained by replacing y_i by Y_i in (12.2) and (12.3):

$$\hat{\beta}_1 = \frac{\sum(x_i - \bar{x})(Y_i - \overline{Y})}{\sum(x_i - \bar{x})^2} \qquad \hat{\beta}_0 = \frac{\sum Y_i - \hat{\beta}_1 \sum x_i}{n}$$

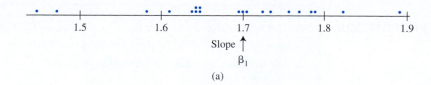

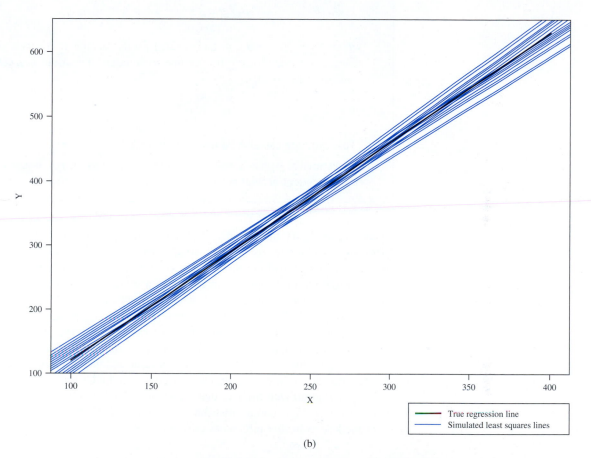

Figure 12.13   Simulation results from Example 12.10: (a) dotplot of estimated slopes; (b) graphs of the true regression line and 20 least squares lines (from S-Plus)

Similarly, the estimator for σ^2 results from replacing each y_i in the formula for s^2 by the rv Y_i:

$$\hat{\sigma}^2 = S^2 = \frac{\sum Y_i^2 - \hat{\beta}_0 \sum Y_i - \hat{\beta}_1 \sum x_i Y_i}{n - 2}$$

The denominator of $\hat{\beta}_1$, $S_{xx} = \Sigma(x_i - \bar{x})^2$, depends only on the x_i's and not on the Y_i's, so it is a constant. Then because $\Sigma(x_i - \bar{x})\bar{Y} = \bar{Y}\Sigma(x_i - \bar{x}) = \bar{Y} \cdot 0 = 0$, the slope estimator can be written as

$$\hat{\beta}_1 = \frac{\sum(x_i - \bar{x})Y_i}{S_{xx}} = \sum c_i Y_i \qquad \text{where } c_i = (x_i - \bar{x})/S_{xx}$$

That is, $\hat{\beta}_i$ is a linear function of the independent rv's $Y_1, Y_2, \ldots, Y_n$, each of which is normally distributed. Invoking properties of a linear function of random variables discussed in Section 5.5 leads to the following results.

PROPOSITION

> 1. The mean value of $\hat{\beta}_1$ is $E(\hat{\beta}_1) = \mu_{\hat{\beta}_1} = \beta_1$, so $\hat{\beta}_1$ is an unbiased estimator of β_1 (the distribution of $\hat{\beta}_1$ is always centered at the value of β_1).
> 2. The variance and standard deviation of $\hat{\beta}_1$ are
>
> $$V(\hat{\beta}_1) = \sigma_{\hat{\beta}_1}^2 = \frac{\sigma^2}{S_{xx}} \qquad \sigma_{\hat{\beta}_1} = \frac{\sigma}{\sqrt{S_{xx}}} \tag{12.4}$$
>
> where $S_{xx} = \Sigma(x_i - \bar{x})^2 = \Sigma x_i^2 - (\Sigma x_i)^2/n$. Replacing σ by its estimate s gives an estimate for $\sigma_{\hat{\beta}_1}$ (the estimated standard deviation, i.e., estimated standard error, of $\hat{\beta}_1$):
>
> $$s_{\hat{\beta}_1} = \frac{s}{\sqrt{S_{xx}}}$$
>
> (This estimate can also be denoted by $\hat{\sigma}_{\hat{\beta}_1}$.)
> 3. The estimator $\hat{\beta}_1$ has a normal distribution (because it is a linear function of independent normal rv's).

According to (12.4), the variance of $\hat{\beta}_1$ equals the variance σ^2 of the random error term—or, equivalently, of any Y_i, divided by $\Sigma(x_i - \bar{x})^2$. This denominator is a measure of how spread out the x_i's are about $\bar{x}$. Therefore making observations at x_i values that are quite spread out results in a more precise estimator of the slope parameter (smaller variance of $\hat{\beta}_1$), whereas values of x_i all close to one another imply a highly variable estimator. Of course, if the x_i's are spread out too far, a linear model may not be appropriate throughout the range of observation.

Many inferential procedures discussed previously were based on standardizing an estimator by first subtracting its mean value and then dividing by its estimated standard deviation. In particular, test procedures and a CI for the mean μ of a normal population utilized the fact that the standardized variable $(\bar{X} - \mu)/(S/\sqrt{n})$—that is, $(\bar{X} - \mu)/S_{\hat{\mu}}$—had a t distribution with $n - 1$ df. A similar result here provides the key to further inferences concerning β_1.

THEOREM

> The assumptions of the simple linear regression model imply that the standardized variable
>
> $$T = \frac{\hat{\beta}_1 - \beta_1}{S/\sqrt{S_{xx}}} = \frac{\hat{\beta}_1 - \beta_1}{S_{\hat{\beta}_1}}$$
>
> has a t distribution with $n - 2$ df.

A Confidence Interval for β_1

As in the derivation of previous CIs, we begin with a probability statement:

$$P\left(-t_{\alpha/2, n-2} < \frac{\hat{\beta}_1 - \beta_1}{S_{\hat{\beta}_1}} < t_{\alpha/2, n-2}\right) = 1 - \alpha$$

Manipulation of the inequalities inside the parentheses to isolate β_1 and substitution of estimates in place of the estimators gives the CI formula.

A 100$(1 - \alpha)$% **CI for the slope β_1** of the true regression line is

$$\hat{\beta}_1 \pm t_{\alpha/2, n-2} \cdot s_{\hat{\beta}_1}$$

This interval has the same general form as did many of our previous intervals. It is centered at the point estimate of the parameter, and the amount it extends out to each side depends on the desired confidence level (through the t critical value) and on the amount of variability in the estimator $\hat{\beta}_1$ (through $s_{\hat{\beta}_1}$, which will tend to be small when there is little variability in the distribution of $\hat{\beta}_1$ and large otherwise).

EXAMPLE 12.11   When damage to a timber structure occurs, it may be more economical to repair the damaged area rather than replace the entire structure. The article **"Simplified Model for Strength Assessment of Timber Beams Joined by Bonded Plates"** (***J. of Materials in Civil Engr.*, 2013: 980–990**) investigated a particular strategy for repair. The accompanying data was used by the authors of the article as a basis for fitting the simple linear regression model. The dependent variable is $y =$ rupture load (N) and the independent variable is anchorage length (the additional length of material used to bond at the junction, in mm).

x	50	50	80	80	110	110	140	140	170	170
y	17,052	14,063	26,264	19,600	21,952	26,362	26,362	26,754	31,654	32,928

Note that the relationship between anchorage length and rupture load is clearly not deterministic, since there are observations with identical x values but different y values. Figure 12.14 shows a scatterplot of the data (also displayed in the cited article); there appears to be a rather substantial positive linear relationship between the two variables.

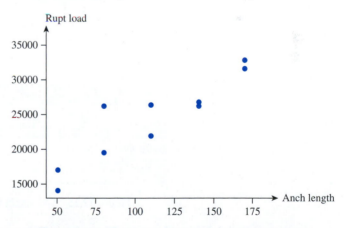

Figure 12.14   Scatterplot of the data from Example 12.11

Summary quantities include $S_{xx} = 18,000$, $S_{xy} = 2,225,579.40$, $S_{yy} = $ SST $= 331,839,568.9$, $\hat{\beta}_1 = 123.6433$, $\hat{\beta}_0 = 10,698.33$, SSE $= 56,661,439.1$, and $r^2 = .829$. Roughly 83% of the observed variation in rupture load can be attributed to the simple linear regression model relationship between rupture load and anchor length. Error df is $10 - 2 = 8$, from which $s^2 = 56,661,439.1/8 = 7,082,679.89$ and $s = 2661.33$. The estimated standard error of $\hat{\beta}_1$ is

$$s_{\hat{\beta}_1} = \frac{s}{\sqrt{S_{xx}}} = \frac{2661.33}{\sqrt{18,000}} = 19.836$$

A confidence level of 95% requires $t_{.025,8} = 2.306$. The CI is

$$123.64 \pm (2.306)(19.836) = 123.64 \pm 45.74 = (77.90, 169.38)$$

With a high degree of confidence, we estimate that an increase in true average rupture strength of between 77.90 N and 169.38 N is associated with an increase of 1 mm in anchorage length (at least for anchorage lengths between 50 mm and 170 mm). This interval is not overly narrow, a consequence of the small sample size and substantial variability about the estimated regression line. Notice that the interval includes only positive values, so we can be quite confident of the tendency for strength to increase as anchorage length increases.

Figure 12.15 displays regression output from the SAS package. The value of $s_{\hat{\beta}_1}$ is found under Parameter Estimates as the second number in the Standard Error column. There is also an estimated standard error for $\hat{\beta}_0$, from which a confidence interval for the intercept of the population regression line can be calculated. The last two columns of the Parameter Estimates table give information about testing certain hypotheses, our next topic of discussion.

Analysis of Variance					
Source	DF	Sum of Squares	Mean Square	F Value	Pr > F
Model	1	275178130	275178130	38.85	0.0003
Error	8	56661439	7082680		
Corrected Total	9	331839569			

Root MSE	2661.33047	R-Square	0.8293
Dependent Mean	24299	Adj R-Sq	0.8079
Coeff Var	10.95238		

Parameter Estimates					
Variable	DF	Parameter Estimate	Standard Error	t Value	Pr > |t|
Intercept	1	10698	2338.67544	4.57	0.0018
Anch Lngth	1	123.64333	19.83639	6.23	0.0003

Figure 12.15 SAS output for the data of Example 12.11 ■

Hypothesis-Testing Procedures

As before, the null hypothesis in a test about β_1 will be an equality statement. The null value (value of β_1 claimed true by the null hypothesis) is denoted by β_{10} (read "beta one nought," *not* "beta ten"). The test statistic results from replacing β_1 by the null value β_{10} in the standardized variable T—that is, from standardizing the estimator of β_1 under the assumption that H_0 is true. The test statistic thus has a t distribution with $n - 2$ df when H_0 is true, which allows for determination of a P-value as described for t tests in Chapters 8 and 9.

The most commonly encountered pair of hypotheses about β_1 is $H_0: \beta_1 = 0$ versus $H_a: \beta_1 \neq 0$. When this null hypothesis is true, $\mu_{Y \cdot x} = \beta_0$ independent of x. Then knowledge of x gives no information about the value of the dependent variable. A test of these two hypotheses is often referred to as the *model utility test* in simple linear regression. Unless n is quite small, H_0 will be rejected and the utility of the model confirmed precisely when r^2 is reasonably large. The simple linear regression model should not be used for further inferences (estimates of mean value or predictions of future values) unless the model utility test results in rejection of H_0 for a suitably small α.

Null hypothesis: H_0: $\beta_1 = \beta_{10}$

Test statistic value: $t = \dfrac{\hat{\beta}_1 - \beta_{10}}{s_{\hat{\beta}_1}}$

Alternative Hypothesis	P-Value Determination
H_a: $\beta_1 > \beta_{10}$	Area under the t_{n-2} curve to the right of t
H_a: $\beta_1 < \beta_{10}$	Area under the t_{n-2} curve to the left of t
H_a: $\beta_1 \neq \beta_{10}$	$2 \cdot$ (Area under the t_{n-2} curve to the right of $|t|$)

The **model utility test** is the test of H_0: $\beta_1 = 0$ versus H_a: $\beta_1 \neq 0$, in which case the test statistic value is the **t ratio** $t = \hat{\beta}_1 / s_{\hat{\beta}_1}$.

EXAMPLE 12.12 Mopeds are very popular in Europe because of cost and ease of operation. However, they can be dangerous if performance characteristics are modified. One of the features commonly manipulated is the maximum speed. The article **"Procedure to Verify the Maximum Speed of Automatic Transmission Mopeds in Periodic Motor Vehicle Inspections" (*J. of Automotive Engr.*, 2008: 1615–1623)** included a simple linear regression analysis of the variables x = test track speed (km/h) and y = rolling test speed. Here is data read from a graph in the article:

x	42.2	42.6	43.3	43.5	43.7	44.1	44.9	45.3	45.7
y	44	44	44	45	45	46	46	46	47

x	45.7	45.9	46.0	46.2	46.2	46.8	46.8	47.1	47.2
y	48	48	48	47	48	48	49	49	49

A scatterplot of the data shows a substantial linear pattern. The Minitab output in Figure 12.16 gives the coefficient of determination as $r^2 = .923$, which certainly portends a useful linear relationship. Let's carry out the model utility test at a significance level $\alpha = .01$.

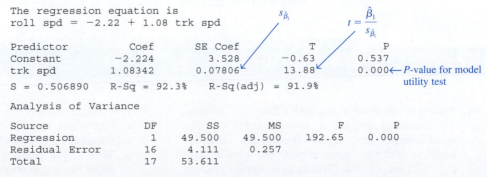

Figure 12.16 Minitab output for the moped data of Example 12.12

The parameter of interest is β_1, the expected change in rolling track speed associated with a 1 km/h increase in test speed. The null hypothesis $H_0: \beta_1 = 0$ will be rejected in favor of the alternative $H_0: \beta_1 \neq 0$ if the t ratio $t = \hat{\beta}_1 / s_{\hat{\beta}_1}$ falls too far into either tail of the t_{n-2} curve (resulting in a small P-value). From Figure 12.16, $\hat{\beta}_1 = 1.08342$, $s_{\hat{\beta}_1} = .07806$, and

$$t = \frac{1.08342}{.07806} = 13.88 \quad \text{(also on output)}$$

The P-value is twice the area captured under the 16 df t curve to the right of 13.88. Minitab gives P-value $= .000$. Thus the null hypothesis of no useful linear relationship can be rejected at any reasonable significance level. This confirms the utility of the model, and gives us license to calculate various estimates and predictions as described in Section 12.4. ∎

Regression and ANOVA

The decomposition of the total sum of squares $\Sigma(y_i - \bar{y})^2$ into a part SSE, which measures unexplained variation, and a part SSR, which measures variation explained by the linear relationship, is strongly reminiscent of one-way ANOVA. In fact, the null hypothesis $H_0: \beta_1 = 0$ can be tested against $H_a: \beta_1 \neq 0$ by constructing an ANOVA table (Table 12.2) and determining the P-value for the F test.

Table 12.2 ANOVA Table for Simple Linear Regression

Source of Variation	df	Sum of Squares	Mean Square	f
Regression	1	SSR	SSR	$\dfrac{\text{SSR}}{\text{SSE}/(n-2)}$
Error	$n-2$	SSE	$s^2 = \dfrac{\text{SSE}}{n-2}$	
Total	$n-1$	SST		

The F test gives exactly the same result as the model utility t test because $t^2 = f$ and $t^2_{\alpha/2, n-2} = F_{\alpha, 1, n-2}$. Virtually all computer packages that have regression options include such an ANOVA table in the output. For example, Figure 12.15 shows SAS output for the rupture load data of Example 12.11. The ANOVA table at the top of the output has $f = 38.85$ with a P-value of .0003 for the model utility test. The table of parameter estimates gives $t = 6.23$, again with $P = .0003$ and $38.85 \approx (6.23)^2$ (they would be identical if more decimal accuracy were shown).

EXERCISES Section 12.3 (30–43)

30. Reconsider the situation described in Exercise 7, in which $x = $ accelerated strength of concrete and $y = $ 28-day cured strength. Suppose the simple linear regression model is valid for x between 1000 and 4000 and that $\beta_1 = 1.25$ and $\sigma = 350$. Consider an experiment in which $n = 7$, and the x values at which observations are made are $x_1 = 1000$, $x_2 = 1500$, $x_3 = 2000$, $x_4 = 2500$, $x_5 = 3000$, $x_6 = 3500$, and $x_7 = 4000$.

a. Calculate $\sigma_{\hat{\beta}_1}$, the standard deviation of $\hat{\beta}_1$.

b. What is the probability that the estimated slope based on such observations will be between 1.00 and 1.50?

c. Suppose it is also possible to make a single observation at each of the $n = 11$ values $x_1 = 2000$, $x_2 = 2100,\ldots, x_{11} = 3000$. If a major objective is to estimate β_1 as accurately as possible, would the experiment with $n = 11$ be preferable to the one with $n = 7$?

31. During oil drilling operations, components of the drilling assembly may suffer from sulfide stress cracking. The article **"Composition Optimization of High-Strength Steels for Sulfide Cracking Resistance Improvement"** (*Corrosion Science*, 2009: 2878–2884) reported on a study in which the composition of a standard grade of steel was analyzed. The following data on y = threshold stress (% SMYS) and x = yield strength (MPa) was read from a graph in the article (which also included the equation of the least squares line).

x	635	644	711	708	836	820	810	870	856	923	878	937	948
y	100	93	88	84	77	75	74	63	57	55	47	43	38

$$\sum x_i = 10{,}576, \ \sum y_i = 894, \ \sum x_i^2 = 8{,}741{,}264,$$

$$\sum y_i^2 = 66{,}224, \ \sum x_i y_i = 703{,}192$$

a. What proportion of observed variation in stress can be attributed to the approximate linear relationship between the two variables?

b. Compute the estimated standard deviation $s_{\hat{\beta}_1}$.

c. Calculate a confidence interval using confidence level 95% for the expected change in stress associated with a 1 MPa increase in strength. Does it appear that this true average change has been precisely estimated?

32. Exercise 16 of Section 12.2 gave data on x = rainfall volume and y = runoff volume (both in m³). Use the accompanying Minitab output to decide whether there is a useful linear relationship between rainfall and runoff, and then calculate a confidence interval for the true average change in runoff volume associated with a 1 m³ increase in rainfall volume.

```
The regression equation is
runoff = −1.13 + 0.827 rainfall

Predictor    Coef     Stdev    t-ratio       P
Constant    −1.128    2.368     −0.48     0.642
rainfall    0.82697  0.03652    22.64     0.000

s = 5.240    R-sq = 97.5%    R-sq(adj) = 97.3%
```

33. Exercise 15 of Section 12.2 included Minitab output for a regression of flexural strength of concrete beams on modulus of elasticity.

a. Use the output to calculate a confidence interval with a confidence level of 95% for the slope β_1 of the population regression line, and interpret the resulting interval.

b. Suppose it had previously been believed that when modulus of elasticity increased by 1 GPa, the associated true average change in flexural strength would

be at most .1 MPa. Does the sample data contradict this belief? State and test the relevant hypotheses.

34. Electromagnetic technologies offer effective nondestructive sensing techniques for determining characteristics of pavement. The propagation of electromagnetic waves through the material depends on its dielectric properties. The following data, kindly provided by the authors of the article **"Dielectric Modeling of Asphalt Mixtures and Relationship with Density"** (*J. of Transp. Engr.*, 2011: 104–111), was used to relate y = dielectric constant to x = air void (%) for 18 samples having 5% asphalt content:

y	4.55	4.49	4.50	4.47	4.47	4.45	4.40	4.34	4.43
x	4.35	4.79	5.57	5.20	5.07	5.79	5.36	6.40	5.66
y	4.43	4.42	4.40	4.33	4.44	4.40	4.26	4.32	4.34
x	5.90	6.49	5.70	6.49	6.37	6.51	7.88	6.74	7.08

The following R output is from a simple linear regression of y on x:

```
               Estimate  Std. Error  t value  Pr(>|t|)
(Intercept)    4.858691    0.059768   81.283    <2e-16
AirVoid       −0.074676    0.009923   −7.526   1.21e-06

Residual standard error: 0.03551 on 16 DF Multiple
R-squared:    0.7797,   Adjusted   R-squared:   0.766
F-statistic: 56.63 on 1 and 16 DF, p-value: 1.214e-06

Analysis of Variance Table

Response: Dielectric
           Df    Sum Sq   Mean Sq   F value     Pr(>F)
Airvoid     1   0.071422  0.071422   56.635  1.214e-06
Residuals  16   0.20178   0.001261
```

a. Obtain the equation of the least squares line and interpret its slope.

b. What proportion of observed variation in dielectric constant can be attributed to the approximate linear relationship between dielectric constant and air void.

c. Does there appear to be a useful linear relationship between dielectric constant and air void? State and test the appropriate hypotheses.

d. Suppose it had previously been believed that when air void increased by 1 percent, the associated true average change in dielectric constant would be at least $-.05$. Does the sample data contradict this belief? Carry out a test of appropriate hypotheses using a significance level of .01.

35. How does lateral acceleration—side forces experienced in turns that are largely under driver control—affect nausea as perceived by bus passengers? The article **"Motion Sickness in Public Road Transport: The Effect of Driver, Route, and Vehicle"** (*Ergonomics*, 1999: 1646–1664) reported data on x = motion sickness dose (calculated in accordance with a British standard for evaluating similar motion at sea) and y = reported nausea (%). Relevant summary quantities are

$$n = 17, \ \sum x_i = 222.1, \ \sum y_i = 193, \ \sum x_i^2 = 3056.69,$$

$$\sum x_i y_i = 2759.6, \ \sum y_i^2 = 2975$$

Values of dose in the sample ranged from 6.0 to 17.6.

a. Assuming that the simple linear regression model is valid for relating these two variables (this is supported by the raw data), calculate and interpret an estimate of the slope parameter that conveys information about the precision and reliability of estimation.

b. Does it appear that there is a useful linear relationship between these two variables? Test appropriate hypotheses using $\alpha = .01$.

c. Would it be sensible to use the simple linear regression model as a basis for predicting % nausea when dose = 5.0? Explain your reasoning.

d. When Minitab was used to fit the simple linear regression model to the raw data, the observation (6.0, 2.50) was flagged as possibly having a substantial impact on the fit. Eliminate this observation from the sample and recalculate the estimate of part (a). Based on this, does the observation appear to be exerting an undue influence?

36. Mist (airborne droplets or aerosols) is generated when metal-removing fluids are used in machining operations to cool and lubricate the tool and workpiece. Mist generation is a concern to OSHA, which has recently lowered substantially the workplace standard. The article **"Variables Affecting Mist Generaton from Metal Removal Fluids"** (*Lubrication Engr.*, 2002: 10–17) gave the accompanying data on x = fluid-flow velocity for a 5% soluble oil (cm/sec) and y = the extent of mist droplets having diameters smaller than 10 μm (mg/m^3):

x	89	177	189	354	362	442	965
y	.40	.60	.48	.66	.61	.69	.99

a. The investigators performed a simple linear regression analysis to relate the two variables. Does a scatterplot of the data support this strategy?

b. What proportion of observed variation in mist can be attributed to the simple linear regression relationship between velocity and mist?

c. The investigators were particularly interested in the impact on mist of increasing velocity from 100 to 1000 (a factor of 10 corresponding to the difference between the smallest and largest x values in the sample). When x increases in this way, is there substantial evidence that the true average increase in y is less than .6?

d. Estimate the true average change in mist associated with a 1 cm/sec increase in velocity, and do so in a way that conveys information about precision and reliability.

37. Magnetic resonance imaging (MRI) is well established as a tool for measuring blood velocities and volume flows. The article **"Correlation Analysis of Stenotic Aortic Valve Flow Patterns Using Phase Contrast MRI,"** referenced in Exercise 1.67, proposed using this methodology for determination of valve area in patients with aortic stenosis. The accompanying data on peak velocity (m/s) from scans of 23 patients in two different planes was read from a graph in the cited paper.

Level-	.60	.82	.85	.89	.95	1.01	1.01	1.05
Level--	.50	.68	.76	.64	.68	.86	.79	1.03

Level-	1.08	1.11	1.18	1.17	1.22	1.29	1.28	1.32
Level--	.75	.90	.79	.86	.99	.80	1.10	1.15

Level-	1.37	1.53	1.55	1.85	1.93	1.93	2.14
Level--	1.04	1.16	1.28	1.39	1.57	1.39	1.32

a. Does there appear to be a difference between true average velocity in the two different planes? Carry out an appropriate test of hypotheses (as did the authors of the article).

b. The authors of the article also regressed level-- velocity against level- velocity. The resulting estimated intercept and slope are .14701 and .65393, with corresponding estimated standard errors .07877 and .05947, coefficient of determination .852, and $s = .110673$. The article included a comment that this regression showed evidence of a strong linear relationship but a regression slope well below 1. Do you agree?

38. Refer to the data on x = liberation rate and y = NO$_x$ emission rate given in Exercise 19.

a. Does the simple linear regression model specify a useful relationship between the two rates? Use the appropriate test procedure to obtain information about the P-value, and then reach a conclusion at significance level .01.

b. Compute a 95% CI for the expected change in emission rate associated with a 10 MBtu/hr-ft^2 increase in liberation rate.

39. Carry out the model utility test using the ANOVA approach for the filtration rate–moisture content data of Example 12.6. Verify that it gives a result equivalent to that of the t test.

40. Use the rules of expected value to show that $\hat{\beta}_0$ is an unbiased estimator for β_0 (assuming that $\hat{\beta}_1$ is unbiased for β_1).

41. a. Verify that $E(\hat{\beta}_1) = \beta_1$ by using the rules of expected value from Chapter 5.

b. Use the rules of variance from Chapter 5 to verify the expression for $V(\hat{\beta}_1)$ given in this section.

42. Verify that if each x_i is multiplied by a positive constant c and each y_i is multiplied by another positive constant d, the t statistic for testing $H_0: \beta_1 = 0$ versus $H_a: \beta_1 \neq 0$ is unchanged in value (the value of $\hat{\beta}_1$ will change, which shows that the magnitude of $\hat{\beta}_1$ is not by itself indicative of model utility).

43. The probability of a type II error for the t test for $H_0: \beta_1 = \beta_{10}$ can be computed in the same manner as it was computed for the t tests of Chapter 8. If the alternative value of β_1 is denoted by β_1', the value of

$$d = \frac{|\beta_{10} - \beta_1'|}{\sigma \sqrt{\dfrac{n-1}{S_{xx}}}}$$

is first calculated, then the appropriate set of curves in Appendix Table A.17 is entered on the horizontal axis at the value of d, and β is read from the curve for $n - 2$ df. An article in the **_Journal of Public Health Engineering_** reports the results of a regression analysis based on $n = 15$ observations in which x = filter application temerature (°C) and y = % efficiency of BOD removal. Calculated quantities include $\Sigma x_i = 402$, $\Sigma x_i^2 = 11{,}098$, $s = 3.725$, and $\hat{\beta}_1 = 1.7035$. Consider testing at level .01 $H_0: \beta_1 = 1$, which states that the expected increase in % BOD removal is 1 when filter application temperature increases by 1°C, against the alternative $H_a: \beta_1 > 1$. Determine P(type II error) when $\beta_1' = 2$, $\sigma = 4$.

12.4 Inferences Concerning $\mu_{Y \cdot x^*}$ and the Prediction of Future Y Values

Let x^* denote a specified value of the independent variable x. Once the estimates $\hat{\beta}_0$ and $\hat{\beta}_1$ have been calculated, $\hat{\beta}_0 + \hat{\beta}_1 x^*$ can be regarded either as a point estimate of $\mu_{Y \cdot x^*}$ (the expected or true average value of Y when $x = x^*$) or as a prediction of the Y value that will result from a single observation made when $x = x^*$. The point estimate or prediction by itself gives no information concerning how precisely $\mu_{Y \cdot x^*}$ has been estimated or Y has been predicted. This can be remedied by developing a CI for $\mu_{Y \cdot x^*}$ and a prediction interval (PI) for a single Y value.

Before we obtain sample data, both $\hat{\beta}_0$ and $\hat{\beta}_1$ are subject to sampling variability—that is, they are both statistics whose values will vary from sample to sample. Suppose, for example, that $\beta_0 = 50$ and $\beta_1 = 2$. Then a first sample of (x, y) pairs might give $\hat{\beta}_0 = 52.35$, $\hat{\beta}_1 = 1.895$; a second sample might result in $\hat{\beta}_0 = 46.52$, $\hat{\beta}_1 = 2.056$; and so on. It follows that $\hat{Y} = \hat{\beta}_0 + \hat{\beta}_1 x^*$ itself varies in value from sample to sample, so it is a statistic. If the intercept and slope of the population line are the aforementioned values 50 and 2, respectively, and $x^* = 10$, then this statistic is trying to estimate the value $50 + 2(10) = 70$. The estimate from a first sample might be $52.35 + 1.895(10) = 71.30$, from a second sample might be $46.52 + 2.056(10) = 67.08$, and so on.

This variation in the value of $\hat{\beta}_0 + \hat{\beta}_1 x^*$ can be visualized by returning to Figure 12.13 on page 511. Consider the value $x^* = 300$. The heights of the 20 pictured estimated regression lines above this value are all somewhat different from one another. The same is true of the heights of the lines above the value $x^* = 350$. In fact, there appears to be more variation in the value of $\hat{\beta}_0 + \hat{\beta}_1(350)$ than in the value of $\hat{\beta}_0 + \hat{\beta}_1(300)$. We shall see shortly that this is because 350 is further from $\bar{x} = 235.71$ (the "center of the data") than is 300.

Methods for making inferences about β_1 were based on properties of the sampling distribution of the statistic $\hat{\beta}_1$. In the same way, inferences about the mean Y value $\beta_0 + \beta_1 x^*$ are based on properties of the sampling distribution of the statistic $\hat{\beta}_0 + \hat{\beta}_1 x^*$. Substitution of the expressions for $\hat{\beta}_0$ and $\hat{\beta}_1$ into $\hat{\beta}_0 + \hat{\beta}_1 x^*$ followed by some algebraic manipulation leads to the representation of $\hat{\beta}_0 + \hat{\beta}_1 x^*$ as a linear function of the Y_i's:

$$\hat{\beta}_0 + \hat{\beta}_1 x^* = \sum_{i=1}^{n} \left[\frac{1}{n} + \frac{(x^* - \bar{x})(x_i - \bar{x})}{\sum (x_i - \bar{x})^2} \right] Y_i = \sum_{i=1}^{n} d_i Y_i$$

The coefficients $d_1, d_2, \ldots, d_n$ in this linear function involve the x_i's and x^*, all of which are fixed. Application of the rules of Section 5.5 to this linear function gives the following properties.

PROPOSITION

Let $\hat{Y} = \hat{\beta}_0 + \hat{\beta}_1 x^*$, where x^* is some fixed value of x. Then

1. The mean value of $\hat{Y}$ is

$$E(\hat{Y}) = E(\hat{\beta}_0 + \hat{\beta}_1 x^*) = \mu_{\hat{\beta}_0 + \hat{\beta}_1 x^*} = \beta_0 + \beta_1 x^*$$

Thus $\hat{\beta}_0 + \hat{\beta}_1 x^*$ is an unbiased estimator for $\beta_0 + \beta_1 x^*$ (i.e., for $\mu_{Y \cdot x^*}$).

2. The variance of $\hat{Y}$ is

$$V(\hat{Y}) = \sigma_{\hat{Y}}^2 = \sigma^2\left[\frac{1}{n} + \frac{(x^* - \bar{x})^2}{\sum x_i^2 - (\sum x_i)^2/n}\right] = \sigma^2\left[\frac{1}{n} + \frac{(x^* - \bar{x})^2}{S_{xx}}\right]$$

and the standard deviation $\sigma_{\hat{Y}}$ is the square root of this expression. The estimated standard deviation of $\hat{\beta}_0 + \hat{\beta}_1 x^*$, denoted by $s_{\hat{Y}}$ or $s_{\hat{\beta}_0 + \hat{\beta}_1 x^*}$, results from replacing σ by its estimate s:

$$s_{\hat{Y}} = s_{\hat{\beta}_0 + \hat{\beta}_1 x^*} = s\sqrt{\frac{1}{n} + \frac{(x^* - \bar{x})^2}{S_{xx}}}$$

3. $\hat{Y}$ has a normal distribution.

The variance of $\hat{\beta}_0 + \hat{\beta}_1 x^*$ is smallest when $x^* = \bar{x}$ and increases as x^* moves away from $\bar{x}$ in either direction. Thus the estimator of $\mu_{Y \cdot x^*}$ is more precise when x^* is near the center of the x_i's than when it is far from the x values at which observations have been made. This will imply that both the CI and PI are narrower for an x^* near $\bar{x}$ than for an x^* far from $\bar{x}$. Most statistical computer packages will provide both $\hat{\beta}_0 + \hat{\beta}_1 x^*$ and $s_{\hat{\beta}_0 + \hat{\beta}_1 x^*}$ for any specified x^* upon request.

Inferences Concerning $\mu_{Y \cdot x^*}$

Just as inferential procedures for β_1 were based on the t variable obtained by standardizing β_1, a t variable obtained by standardizing $\hat{\beta}_0 + \hat{\beta}_1 x^*$ leads to a CI and test procedures here.

THEOREM

The variable

$$T = \frac{\hat{\beta}_0 + \hat{\beta}_1 x^* - (\beta_0 + \beta_1 x^*)}{S_{\hat{\beta}_0 + \hat{\beta}_1 x^*}} = \frac{\hat{Y} - (\beta_0 + \beta_1 x^*)}{S_{\hat{Y}}} \tag{12.5}$$

has a t distribution with $n - 2$ df.

A probability statement involving this standardized variable can now be manipulated to yield a confidence interval for $\mu_{Y \cdot x^*}$.

A $100(1 - \alpha)\%$ **CI for** $\mu_{Y \cdot x^*}$, the expected value of Y when $x = x^*$, is

$$\hat{\beta}_0 + \hat{\beta}_1 x^* \pm t_{\alpha/2, n-2} \cdot s_{\hat{\beta}_0 + \hat{\beta}_1 x^*} = \hat{y} \pm t_{\alpha/2, n-2} \cdot s_{\hat{Y}} \tag{12.6}$$

This CI is centered at the point estimate for $\mu_{Y \cdot x^*}$ and extends out to each side by an amount that depends on the confidence level and on the extent of variability in the estimator on which the point estimate is based.

EXAMPLE 12.13   Corrosion of steel reinforcing bars is the most important durability problem for reinforced concrete structures. Carbonation of concrete results from a chemical reaction
that lowers the pH value by enough to initiate corrosion of the rebar. Representative
data on x = carbonation depth (mm) and y = strength (MPa) for a sample of core
specimens taken from a particular building follows (read from a plot in the article
**"The Carbonation of Concrete Structures in the Tropical Environment of
Singapore,"** *Magazine of Concrete Res.*, **1996: 293–300**).

x	8.0	15.0	16.5	20.0	20.0	27.5	30.0	30.0	35.0
y	22.8	27.2	23.7	17.1	21.5	18.6	16.1	23.4	13.4

x	38.0	40.0	45.0	50.0	50.0	55.0	55.0	59.0	65.0
y	19.5	12.4	13.2	11.4	10.3	14.1	9.7	12.0	6.8

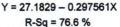
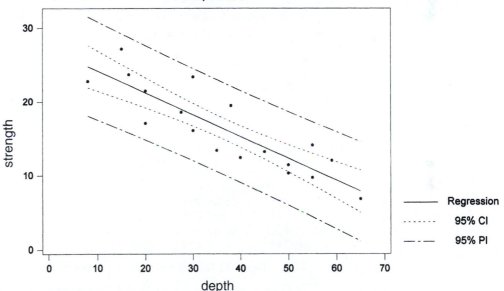

Figure 12.17   Minitab scatterplot with confidence intervals and prediction intervals for the data
of Example 12.13

A scatterplot of the data (see Figure 12.17) gives strong support for use of the simple
linear regression model. Relevant quantities are as follows:

$$\sum x_i = 659.0 \qquad \sum x_i^2 = 28{,}967.50 \qquad \bar{x} = 36.6111 \qquad S_{xx} = 4840.7778$$
$$\sum y_i = 293.2 \qquad \sum x_i y_i = 9293.95 \qquad \sum y_i^2 = 5335.76$$
$$\hat{\beta}_1 = -.297561 \qquad \hat{\beta}_0 = 27.182936 \qquad \text{SSE} = 131.2402$$
$$r^2 = .766 \qquad s = 2.8640$$

Let's now calculate a confidence interval, using a 95% confidence level, for the mean
strength for all core specimens having a carbonation depth of 45 mm—that is, a confidence interval for $\beta_0 + \beta_1(45)$. The interval is centered at

$$\hat{y} = \hat{\beta}_0 + \hat{\beta}_1(45) = 27.18 - .2976(45) = 13.79$$

The estimated standard deviation of the statistic $\hat{Y}$ is

$$s_{\hat{Y}} = 2.8640\sqrt{\frac{1}{18} + \frac{(45 - 36.6111)^2}{4840.7778}} = .7582$$

The 16 df t critical value for a 95% confidence level is 2.120, from which we determine the desired interval to be

$$13.79 \pm (2.120)(.7582) = 13.79 \pm 1.61 = (12.18, 15.40)$$

The narrowness of this interval suggests that we have reasonably precise information about the mean value being estimated. Remember that if we recalculated this interval for sample after sample, in the long run about 95% of the calculated intervals would include $\beta_0 + \beta_1(45)$. We can only hope that this mean value lies in the single interval that we have calculated.

Figure 12.18 shows Minitab output resulting from a request to fit the simple linear regression model and calculate confidence intervals for the mean value of strength at depths of 45 mm and 35 mm. The intervals are at the bottom of the output; note that the second interval is narrower than the first, because 35 is much closer to $\bar{x}$ than is 45. Figure 12.17 shows (1) curves corresponding to the confidence limits for each different x value and (2) prediction limits, to be discussed shortly. Notice how the curves get farther and farther apart as x moves away from $\bar{x}$.

```
The regression equation is strength = 27.2 − 0.298 depth

Predictor        Coef        Stdev      t-ratio          P
Constant       27.183        1.651        16.46      0.000
   depth      −0.29756      0.04116       −7.23      0.000

s = 2.864     R-sq = 76.6%       R-sq(adj) = 75.1%

Analysis of Variance

SOURCE           DF          SS          MS          F          P
Regression        1      428.62      428.62      52.25      0.000
    Error        16      131.24        8.20
    Total        17      559.86

  Fit      Stdev.Fit              95.0% C.I.              95.0% P.I.
13.793         0.758        (12.185, 15.401)        (7.510, 20.075)

  Fit      Stdev.Fit              95.0% C.I.              95.0% P.I.
16.768         0.678        (15.330, 18.207)       (10.527, 23.009)
```

Figure 12.18   Minitab regression output for the data of Example 12.13   ■

In some situations, a CI is desired not just for a single x value but for two or more x values. Suppose an investigator wishes a CI both for $\mu_{Y \cdot v}$ and for $\mu_{Y \cdot w}$, where v and w are two different values of the independent variable. It is tempting to compute the interval (12.6) first for $x = v$ and then for $x = w$. Suppose we use $\alpha = .05$ in each computation to get two 95% intervals. Then if the variables involved in computing the two intervals were independent of one another, the joint confidence coefficient would be $(.95) \cdot (.95) \approx .90$.

However, the intervals are not independent because the same $\hat{\beta}_0$, $\hat{\beta}_1$, and S are used in each. We therefore cannot assert that the joint confidence level for the two intervals is exactly 90%. It can be shown, though, that if the $100(1 - \alpha)$% CI (12.6) is computed both for $x = v$ and $x = w$ to obtain joint CIs for $\mu_{Y \cdot v}$ and $\mu_{Y \cdot w}$, then *the joint confidence level on the resulting pair of intervals is at least* $100(1 - 2\alpha)$%. In particular, using $\alpha = .05$ results in a joint confidence level of *at least* 90%, whereas using $\alpha = .01$ results in at least 98% confidence. For example, in Example 12.13 a 95% CI for $\mu_{Y \cdot 45}$ was (12.185, 15.401) and a 95% CI for $\mu_{Y \cdot 35}$ was (15.330, 18.207). The simultaneous or joint confidence level for the two statements $12.185 < \mu_{Y \cdot 45} < 15.401$ and $15.330 < \mu_{Y \cdot 35} < 18.207$ is at least 90%.

The validity of these joint or simultaneous CIs rests on a probability result called the **Bonferroni inequality**, so the joint CIs are referred to as **Bonferroni intervals**. The method is easily generalized to yield joint intervals for k different $\mu_{Y \cdot x}$'s. *Using the interval* (12.6) *separately first for* $x = x_1^*$, *then for* $x = x_2^*, \ldots$, *and finally for* $x = x_k^*$ *yields a set of k CIs for which the joint or simultaneous confidence level is guaranteed to be at least* $100(1 - k\alpha)\%$.

Tests of hypotheses about $\beta_0 + \beta_1 x^*$ are based on the test statistic T obtained by replacing $\beta_0 + \beta_1 x^*$ in the numerator of (12.5) by the null value μ_0. For example, $H_0: \beta_0 + \beta_1(45) = 15$ in Example 12.13 says that when carbonation depth is 45, expected (i.e., true average) strength is 15. The test statistic value is then $t = [\hat{\beta}_0 + \hat{\beta}_1(45) - 15]/s_{\hat{\beta}_0 + \hat{\beta}_1(45)}$, and the test is upper-, lower-, or two-tailed according to the inequality in H_a.

A Prediction Interval for a Future Value of Y

Rather than calculate an interval estimate for $\mu_{Y \cdot x^*}$, an investigator may wish to obtain an interval of plausible values for the value of Y associated with some future observation when the independent variable has value x^*. Consider, for example, relating vocabulary size y to age of a child x. The CI (12.6) with $x^* = 6$ would provide an estimate of true average vocabulary size for all 6-year-old children. Alternatively, we might wish an interval of plausible values for the vocabulary size of a particular 6-year-old child.

A CI refers to a parameter, or population characteristic, whose value is fixed but unknown to us. In contrast, a future value of Y is not a parameter but instead a random variable; for this reason we refer to an interval of plausible values for a future Y as a **prediction interval** rather than a confidence interval. The error of estimation is $\beta_0 + \beta_1 x^* - (\hat{\beta}_0 + \hat{\beta}_1 x^*)$, a difference between a fixed (but unknown) quantity and a random variable. The error of prediction is $Y - (\hat{\beta}_0 + \hat{\beta}_1 x^*)$, a difference between two random variables. There is thus more uncertainty in prediction than in estimation, so a PI will be wider than a CI. Because the future value Y is independent of the observed Y_i's,

$$
\begin{aligned}
V[Y - (\hat{\beta}_0 + \hat{\beta}_1 x^*)] &= \text{variance of prediction error} \\
&= V(Y) + V(\hat{\beta}_0 + \hat{\beta}_1 x^*) \\
&= \sigma^2 + \sigma^2 \left[\frac{1}{n} + \frac{(x^* - \bar{x})^2}{S_{xx}} \right] \\
&= \sigma^2 \left[1 + \frac{1}{n} + \frac{(x^* - \bar{x})^2}{S_{xx}} \right]
\end{aligned}
$$

Furthermore, because $E(Y) = \beta_0 + \beta_1 x^*$ and $E(\hat{\beta}_0 + \hat{\beta}_1 x^*) = \beta_0 + \beta_1 x^*$, the expected value of the prediction error is $E(Y - (\hat{\beta}_0 + \hat{\beta}_1 x^*)) = 0$. It can then be shown that the standardized variable

$$
T = \frac{Y - (\hat{\beta}_0 + \hat{\beta}_1 x^*)}{S \sqrt{1 + \dfrac{1}{n} + \dfrac{(x^* - \bar{x})^2}{S_{xx}}}}
$$

has a t distribution with $n - 2$ df. Substituting this T into the probability statement $P(-t_{\alpha/2, n-2} < T < t_{\alpha/2, n-1}) = 1 - \alpha$ and manipulating to isolate Y between the two inequalities yields the following interval.

A $100(1 - \alpha)\%$ **PI for a future Y observation to be made when $x = x^*$** is

$$\hat{\beta}_0 + \hat{\beta}_1 x^* \pm t_{\alpha/2, n-2} \cdot s \sqrt{1 + \frac{1}{n} + \frac{(x^* - \bar{x})^2}{S_{xx}}}$$

$$= \hat{\beta}_0 + \hat{\beta}_1 x^* \pm t_{\alpha/2, n-2} \cdot \sqrt{s^2 + s^2_{\hat{\beta}_0 + \hat{\beta}_1 x^*}} \qquad (12.7)$$

$$= \hat{y} \pm t_{\alpha/2, n-2} \cdot \sqrt{s^2 + s^2_{\hat{y}}}$$

The interpretation of the prediction level $100(1 - \alpha)\%$ is analogous to that of previous confidence levels—if (12.7) is used repeatedly, in the long run the resulting intervals will actually contain the observed y values $100(1 - \alpha)\%$ of the time. Notice that the 1 underneath the initial square root symbol makes the PI (12.7) wider than the CI (12.6), though the intervals are both centered at $\hat{\beta}_0 + \hat{\beta}_1 x^*$. Also, as $n \to \infty$, the width of the CI approaches 0, whereas the width of the PI does not (because even with perfect knowledge of β_0 and β_1, there will still be uncertainty in prediction).

EXAMPLE 12.14   Let's return to the carbonation depth-strength data of Example 12.13 and calculate a 95% PI for a strength value that would result from selecting a single core specimen whose depth is 45 mm. Relevant quantities from that example are

$$\hat{y} = 13.79 \qquad s_{\hat{y}} = .7582 \qquad s = 2.8640$$

For a prediction level of 95% based on $n - 2 = 16$ df, the t critical value is 2.120, exactly what we previously used for a 95% confidence level. The prediction interval is then

$$13.79 \pm (2.120)\sqrt{(2.8640)^2 + (.7582)^2} = 13.79 \pm (2.120)(2.963)$$

$$= 13.79 \pm 6.28 = (7.51, 20.07)$$

Plausible values for a single observation on strength when depth is 45 mm are (at the 95% prediction level) between 7.51 MPa and 20.07 MPa. The 95% confidence interval for mean strength when depth is 45 was (12.18, 15.40). The prediction interval is much wider than this because of the extra $(2.8640)^2$ under the square root. Figure 12.18, the Minitab output in Example 12.13, shows this interval as well as the confidence interval. ∎

The Bonferroni technique can be employed as in the case of confidence intervals. If a $100(1 - \alpha)\%$ PI is calculated for each of k different values of x, the simultaneous or joint prediction level for all k intervals is at least $100(1 - k\alpha)\%$.

EXERCISES Section 12.4 (44–56)

44. Fitting the simple linear regression model to the $n = 27$ observations on $x =$ modulus of elasticity and $y =$ flexural strength given in Exercise 15 of Section 12.2 resulted in $\hat{y} = 7.592$, $s_{\hat{y}} = .179$ when $x = 40$ and $\hat{y} = 9.741$, $s_{\hat{y}} = .253$ for $x = 60$.
   a. Explain why $s_{\hat{y}}$ is larger when $x = 60$ than when $x = 40$.

   b. Calculate a confidence interval with a confidence level of 95% for the true average strength of all beams whose modulus of elasticity is 40.
   c. Calculate a prediction interval with a prediction level of 95% for the strength of a single beam whose modulus of elasticity is 40.

d. If a 95% CI is calculated for true average strength when modulus of elasticity is 60, what will be the simultaneous confidence level for both this interval and the interval calculated in part (b)?

45. Reconsider the filtration rate–moisture content data introduced in Example 12.6 (see also Example 12.7).

   a. Compute a 90% CI for $\beta_0 + 125\beta_1$, true average moisture content when the filtration rate is 125.

   b. Predict the value of moisture content for a single experimental run in which the filtration rate is 125 using a 90% prediction level. How does this interval compare to the interval of part (a)? Why is this the case?

   c. How would the intervals of parts (a) and (b) compare to a CI and PI when filtration rate is 115? Answer without actually calculating these new intervals.

   d. Interpret the hypotheses $H_0: \beta_0 + 125\beta_1 = 80$ and $H_a: \beta_0 + 125\beta_1 < 80$, and then carry out a test at significance level .01.

46. Astringency is the quality in a wine that makes the wine drinker's mouth feel slightly rough, dry, and puckery. The paper **"Analysis of Tannins in Red Wine Using Multiple Methods: Correlation with Perceived Astringency"** (*Amer. J. of Enol. and Vitic.*, **2006:** 481–485) reported on an investigation to assess the relationship between perceived astringency and tannin concentration using various analytic methods. Here is data provided by the authors on x = tannin concentration by protein precipitation and y = perceived astringency as determined by a panel of tasters.

x	.718	.808	.924	1.000	.667	.529	.514	.559
y	.428	.480	.493	.978	.318	.298	−.224	.198
x	.766	.470	.726	.762	.666	.562	.378	.779
y	.326	−.336	.765	.190	.066	−.221	−.898	.836
x	.674	.858	.406	.927	.311	.319	.518	.687
y	.126	.305	−.577	.779	−.707	−.610	−.648	−.145
x	.907	.638	.234	.781	.326	.433	.319	.238
y	1.007	−.090	−1.132	.538	−1.098	−.581	−.862	−.551

Relevant summary quantities are as follows:

$$\sum x_i = 19.404, \sum y_i = -.549, \sum x_i^2 = 13.248032,$$
$$\sum y_i^2 = 11.835795, \sum x_i y_i = 3.497811$$
$$S_{xx} = 13.248032 - (19.404)^2/32 = 1.48193150,$$
$$S_{yy} = 11.82637622$$
$$S_{xy} = 3.497811 - (19.404)(-.549)/32$$
$$= 3.83071088$$

   a. Fit the simple linear regression model to this data. Then determine the proportion of observed variation in astringency that can be attributed to the model relationship between astringency and tannin concentration.

   b. Calculate and interpret a confidence interval for the slope of the true regression line.

   c. Estimate true average astringency when tannin concentration is .6, and do so in a way that conveys information about reliability and precision.

   d. Predict astringency for a single wine sample whose tannin concentration is .6, and do so in a way that conveys information about reliability and precision.

   e. Does it appear that true average astringency for a tannin concentration of .7 is something other than 0? State and test the appropriate hypotheses.

47. The simple linear regression model provides a very good fit to the data on rainfall and runoff volume given in Exercise 16 of Section 12.2. The equation of the least squares line is $y = -1.128 + .82697x$, $r^2 = .975$, and $s = 5.24$.

   a. Use the fact that $s_{\hat{y}} = 1.44$ when rainfall volume is 40 m^3 to predict runoff in a way that conveys information about reliability and precision. Does the resulting interval suggest that precise information about the value of runoff for this future observation is available? Explain your reasoning.

   b. Calculate a PI for runoff when rainfall is 50 using the same prediction level as in part (a). What can be said about the simultaneous prediction level for the two intervals you have calculated?

48. The catch basin in a storm-sewer system is the interface between surface runoff and the sewer. The catch-basin insert is a device for retrofitting catch basins to improve pollutant-removal properties. The article **"An Evaluation of the Urban Stormwater Pollutant Removal Efficiency of Catch Basin Inserts"** (*Water Envir. Res.*, **2005:** 500–510) reported on tests of various inserts under controlled conditions for which inflow is close to what can be expected in the field. Consider the following data, read from a graph in the article, for one particular type of insert on x = amount filtered (1000s of liters) and y = % total suspended solids removed.

x	23	45	68	91	114	136	159	182	205	228
y	53.3	26.9	54.8	33.8	29.9	8.2	17.2	12.2	3.2	11.1

Summary quantities are

$$\sum x_i = 1251, \sum x_i^2 = 199,365, \sum y_i = 250.6,$$
$$\sum y_i^2 = 9249.36, \sum x_i y_i = 21,904.4$$

   a. Does a scatterplot support the choice of the simple linear regression model? Explain.

   b. Obtain the equation of the least squares line.

   c. What proportion of observed variation in % removed can be attributed to the model relationship?

   d. Does the simple linear regression model specify a useful relationship? Carry out an appropriate test of hypotheses using a significance level of .05.

   e. Is there strong evidence for concluding that there is at least a 2% decrease in true average suspended solid removal associated with a 10,000 liter increase in the amount filtered? Test appropriate hypotheses using $\alpha = .05$.

f. Calculate and interpret a 95% CI for true average % removed when amount filtered is 100,000 liters. How does this interval compare in width to a CI when amount filtered is 200,000 liters?

g. Calculate and interpret a 95% PI for % removed when amount filtered is 100,000 liters. How does this interval compare in width to the CI calculated in (f) and to a PI when amount filtered is 200,000 liters?

49. You are told that a 95% CI for expected lead content when traffic flow is 15, based on a sample of $n = 10$ observations, is (462.1, 597.7). Calculate a CI with confidence level 99% for expected lead content when traffic flow is 15.

50. Silicon-germanium alloys have been used in certain types of solar cells. The paper **"Silicon-Germanium Films Deposited by Low-Frequency Plasma-Enhanced Chemical Vapor Deposition" (*J. of Material Res.*, 2006: 88–104)** reported on a study of various structural and electrical properties. Consider the accompanying data on $x = $ Ge concentration in solid phase (ranging from 0 to 1) and $y = $ Fermi level position (eV):

x	0	.42	.23	.33	.62	.60	.45	.87	.90	.79	1	1	1
y	.62	.53	.61	.59	.50	.55	.59	.31	.43	.46	.23	.22	.19

A scatterplot shows a substantial linear relationship. Here is Minitab output from a least squares fit. [*Note:* There are several inconsistencies between the data given in the paper, the plot that appears there, and the summary information about a regression analysis.]

```
The regression equation is
Fermi pos = 0.7217 - 0.4327 Ge conc

S = 0.0737573   R-Sq = 80.2%   R-Sq(adj) = 78.4%

Analysis of Variance

Source       DF      SS         MS        F      P
Regression    1   0.241728   0.241728   44.43  0.000
Error        11   0.059842   0.005440
Total        12   0.301569
```

a. Obtain an interval estimate of the expected change in Fermi-level position associated with an increase of .1 in Ge concentration, and interpret your estimate.

b. Obtain an interval estimate for mean Fermi-level position when concentration is .50, and interpret your estimate.

c. Obtain an interval of plausible values for position resulting from a single observation to be made when concentration is .50, interpret your interval, and compare to the interval of (b).

d. Obtain simultaneous CIs for expected position when concentration is .3, .5, and .7; the joint confidence level should be at least 97%.

51. Refer to Example 12.12 in which $x = $ test track speed and $y = $ rolling test speed.

a. Minitab gave $s_{\hat{\beta}_0 + \hat{\beta}_1(45)} = .120$ and $s_{\hat{\beta}_0 + \hat{\beta}_1(47)} = .186$. Why is the former estimated standard deviation smaller than the latter one?

b. Use the Minitab output from the example to calculate a 95% CI for expected rolling speed when test speed $= 45$.

c. Use the Minitab output to calculate a 95% PI for a single value of rolling speed when test speed $= 47$.

52. Plasma etching is essential to the fine-line pattern transfer in semiconductor processes. The article **"Ion Beam-Assisted Etching of Aluminum with Chlorine" (*J. of the Electrochem. Soc.*, 1985: 2010–2012)** gives the accompanying data (read from a graph) on chlorine flow (x, in SCCM) through a nozzle used in the etching mechanism and etch rate (y, in 100 A/min).

x	1.5	1.5	2.0	2.5	2.5	3.0	3.5	3.5	4.0
y	23.0	24.5	25.0	30.0	33.5	40.0	40.5	47.0	49.0

The summary statistics are $\Sigma x_i = 24.0$, $\Sigma y_i = 312.5$, $\Sigma x_i^2 = 70.50$, $\Sigma x_i y_i = 902.25$, $\Sigma y_i^2 = 11{,}626.75$, $\hat{\beta}_0 = 6.448718$, $\hat{\beta}_1 = 10.602564$.

a. Does the simple linear regression model specify a useful relationship between chlorine flow and etch rate?

b. Estimate the true average change in etch rate associated with a 1-SCCM increase in flow rate using a 95% confidence interval, and interpret the interval.

c. Calculate a 95% CI for $\mu_{Y \cdot 3.0}$, the true average etch rate when flow $= 3.0$. Has this average been precisely estimated?

d. Calculate a 95% PI for a single future observation on etch rate to be made when flow $= 3.0$. Is the prediction likely to be accurate?

e. Would the 95% CI and PI when flow $= 2.5$ be wider or narrower than the corresponding intervals of parts (c) and (d)? Answer without actually computing the intervals.

f. Would you recommend calculating a 95% PI for a flow of 6.0? Explain.

53. Consider the following four intervals based on the data of Exercise 12.17 (Section 12.2):

a. A 95% CI for mean porosity when unit weight is 110
b. A 95% PI for porosity when unit weight is 110
c. A 95% CI for mean porosity when unit weight is 115
d. A 95% PI for porosity when unit weight is 115

Without computing any of these intervals, what can be said about their widths relative to one another?

54. The height of a patient is useful for a variety of medical purposes, such as estimating tidal volume of someone in an intensive care who requires artificial ventilation. However, it can be difficult to make an accurate

measurement if the patient is confused, unconscious, or sedated. And measurement of height while an individual is lying down is also not straightforward. In contrast, ulna length measurements are generally quick and easy to obtain, even in chair- or bed-bound patients. The accompanying data on x = ulna length (cm) and y = height (cm) for males older than 65 was read from a graph in the article **"Ulna Length to Predict Height in English and Portuguese Patient Populations"** (*European J. of Clinical Nutr.*, 2012: 209–215).

x	22.5	22.8	22.8	23.3	23.3	24.4	25.0
y	158	155	156	160	161	162	164

x	25.0	25.0	25.0	26.0	26.0	26.8	28.2
y	166	167	170	166	173	178	174

Summary quantities include $\Sigma x_i = 346.1$, $\Sigma y_i = 2310$, $S_{xx} = 36.463571$, $S_{xy} = 137.60$, $S_{yy} = 626.00$.

a. Obtain the equation of the estimated regression line and interpret its slope.

b. Calculate and interpret the coefficient of determination.

c. Carry out a test of model utility.

d. Calculate prediction intervals for the heights of two individuals whose ulna lengths are 23 and 25, respectively; use a prediction level of 95% for each interval.

e. Based on the predictions of (d), would you agree with the statement in the cited article that "height can be predicted from ulna length with precision"?

55. Verify that $V(\hat{\beta}_0 + \hat{\beta}_1 x)$ is indeed given by the expression in the text. [*Hint*: $V(\Sigma d_i Y_i) = \Sigma d_i^2 \cdot V(Y_i)$.]

56. The article **"Bone Density and Insertion Torque as Predictors of Anterior Cruciate Ligament Graft Fixation Strength"** (*The Amer. J. of Sports Med.*, 2004: 1421–1429) gave the accompanying data on maximum insertion torque (N · m) and yield load (N), the latter being one measure of graft strength, for 15 different specimens.

Torque	1.8	2.2	1.9	1.3	2.1	2.2	1.6	2.1
Load	491	477	598	361	605	671	466	431

Torque	1.2	1.8	2.6	2.5	2.5	1.7	1.6
Load	384	422	554	577	642	348	446

a. Is it plausible that yield load is normally distributed?

b. Estimate true average yield load by calculating a confidence interval with a confidence level of 95%, and interpret the interval.

c. Here is output from Minitab for the regression of yield load on torque. Does the simple linear regression model specify a useful relationship between the variables?

```
Predictor       Coef     SE Coef        T        P
Constant       152.44       91.17     1.67    0.118
Torque         178.23       45.97     3.88    0.002

S = 73.2141   R-Sq = 53.6%   R-Sq(adj) = 50.0%

Source          DF          SS       MS       F        P
Regression       1       80554    80554   15.03    0.002
Residual Error  13       69684     5360
Total           14      150238
```

d. The authors of the cited paper state, "Consequently, we cannot but conclude that simple regression analysis-based methods are not clinically sufficient to predict individual fixation strength." Do you agree? [*Hint*: Consider predicting yield load when torque is 2.0.]

12.5 Correlation

There are many situations in which the objective in studying the joint behavior of two variables is to see whether they are related, rather than to use one to predict the value of the other. In this section, we first develop the sample correlation coefficient r as a measure of how strongly related two variables x and y are in a sample and then relate r to the correlation coefficient ρ defined in Chapter 5.

The Sample Correlation Coefficient r

Given n numerical pairs $(x_1, y_1), (x_2, y_2), \ldots, (x_n, y_n)$, it is natural to speak of x and y as having a positive relationship if large x's are paired with large y's and small x's with small y's. Similarly, if large x's are paired with small y's and small x's with large y's, then a negative relationship between the variables is implied. Consider the quantity

$$S_{xy} = \sum_{i=1}^{n}(x_i - \bar{x})(y_i - \bar{y}) = \sum_{i=1}^{n} x_i y_i - \frac{\left(\sum_{i=1}^{n} x_i\right)\left(\sum_{i=1}^{n} y_i\right)}{n}$$

Then if the relationship is strongly positive, an x_i above the mean $\bar{x}$ will tend to be paired with a y_i above the mean $\bar{y}$, so that $(x_i - \bar{x})(y_i - \bar{y}) > 0$, and this product will also be positive whenever both x_i and y_i are below their respective means. Thus a positive relationship implies that S_{xy} will be positive. An analogous argument shows that when the relationship is negative, S_{xy} will be negative, since most of the products $(x_i - \bar{x})(y_i - \bar{y})$ will be negative. This is illustrated in Figure 12.19.

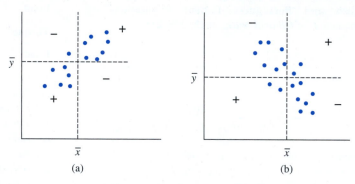

Figure 12.19 (a) Scatterplot with S_{xy} positive; (b) scatterplot with S_{xy} negative [+ means $(x_i - \bar{x})(y_i - \bar{y}) > 0$, and − means $(x_i - \bar{x})(y_i - \bar{y}) < 0$]

Although S_{xy} seems a plausible measure of the strength of a relationship, we do not yet have any idea of how positive or negative it can be. Unfortunately, S_{xy} has a serious defect: By changing the unit of measurement for either x or y, S_{xy} can be made either arbitrarily large in magnitude or arbitrarily close to zero. For example, if $S_{xy} = 25$ when x is measured in meters, then $S_{xy} = 25,000$ when x is measured in millimeters and .025 when x is expressed in kilometers. A reasonable condition to impose on any measure of how strongly x and y are related is that the calculated measure should not depend on the particular units used to measure them. This condition is achieved by modifying S_{xy} to obtain the sample correlation coefficient.

DEFINITION

The **sample correlation coefficient** for the n pairs $(x_1, y_1), \ldots, (x_n, y_n)$ is

$$r = \frac{S_{xy}}{\sqrt{\Sigma(x_i - \bar{x})^2}\sqrt{\Sigma(y_i - \bar{y})^2}} = \frac{S_{xy}}{\sqrt{S_{xx}}\sqrt{S_{yy}}} \qquad (12.8)$$

EXAMPLE 12.15 An accurate assessment of soil productivity is critical to rational land-use planning. Unfortunately, as the author of the article **"Productivity Ratings Based on Soil Series"** (*Prof. Geographer*, 1980: 158–163) argues, an acceptable soil productivity index is not so easy to come by. One difficulty is that productivity is determined partly by which crop is planted, and the relationship between the yield of two different crops planted in the same soil may not be very strong. To illustrate, the article presents the accompanying data on corn yield x and peanut yield y (mT/Ha) for eight different types of soil.

x	2.4	3.4	4.6	3.7	2.2	3.3	4.0	2.1
y	1.33	2.12	1.80	1.65	2.00	1.76	2.11	1.63

With $\Sigma x_i = 25.7$, $\Sigma y_i = 14.40$, $\Sigma x_i^2 = 88.31$, $\Sigma x_i y_i = 46.856$, and $\Sigma y_i^2 = 26.4324$,

$$S_{xx} = 88.31 - \frac{(25.7)^2}{8} = 5.75 \quad S_{yy} = 26.4324 - \frac{(14.40)^2}{8} = .5124$$

$$S_{xy} = 46.856 - \frac{(25.7)(14.40)}{8} = .5960$$

from which
$$r = \frac{.5960}{\sqrt{5.75}\sqrt{.5124}} = .347 \quad\blacksquare$$

Properties of r

The most important properties of r are as follows:

1. The value of r does not depend on which of the two variables under study is labeled x and which is labeled y.

2. The value of r is independent of the units in which x and y are measured.

3. $-1 \le r \le 1$

4. $r = 1$ if and only if (iff) all (x_i, y_i) pairs lie on a straight line with positive slope, and $r = -1$ iff all (x_i, y_i) pairs lie on a straight line with negative slope.

5. The square of the sample correlation coefficient gives the value of the coefficient of determination that would result from fitting the simple linear regression model—in symbols, $(r)^2 = r^2$.

Property 1 stands in marked contrast to what happens in regression analysis, where virtually all quantities of interest (the estimated slope, estimated y-intercept, s^2, etc.) depend on which of the two variables is treated as the dependent variable. However, Property 5 shows that the proportion of variation in the dependent variable explained by fitting the simple linear regression model does not depend on which variable plays this role.

Property 2 is equivalent to saying that r is unchanged if each x_i is replaced by cx_i and if each y_i is replaced by dy_i (a change in the scale of measurement), as well as if each x_i is replaced by $x_i - a$ and y_i by $y_i - b$ (which changes the location of zero on the measurement axis). This implies, for example, that r is the same whether temperature is measured in °F or °C.

Property 3 tells us that the maximum value of r, corresponding to the largest possible degree of positive relationship, is $r = 1$, whereas the most negative relationship is identified with $r = -1$. According to Property 4, the largest positive and largest negative correlations are achieved only when all points lie along a straight line. Any other configuration of points, even if the configuration suggests a deterministic relationship between variables, will yield an r value less than 1 in absolute magnitude. Thus *r measures the degree of linear relationship* among variables. A value of r near 0 is not evidence of the lack of a strong relationship, but only the absence of a linear relation, so that such a value of r must be interpreted with caution. Figure 12.20 illustrates several configurations of points associated with different values of r.

A frequently asked question is, "When can it be said that there is a strong correlation between the variables, and when is the correlation weak?" Here is an informal rule of thumb for characterizing the value of r:

Weak	Moderate	Strong
$-.5 \le r \le .5$	either $-.8 < r < -.5$ or $.5 < r < .8$	either $r \ge .8$ or $r \le -.8$

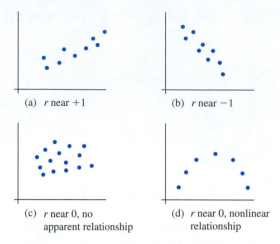

(a) r near $+1$

(b) r near -1

(c) r near 0, no
apparent relationship

(d) r near 0, nonlinear
relationship

Figure 12.20   Data plots for different values of r

It may surprise you that an r as substantial as .5 or $-.5$ goes in the weak category. The rationale is that if $r = .5$ or $-.5$, then $r^2 = .25$ in a regression with either variable playing the role of y. A regression model that explains at most 25% of observed variation is not in fact very impressive. In Example 12.15, the correlation between corn yield and peanut yield would be described as weak.

Inferences About the Population Correlation Coefficient

The correlation coefficient r is a measure of how strongly related x and y are in the observed sample. We can think of the pairs (x_i, y_i) as having been drawn from a bivariate population of pairs, with (X_i, Y_i) having some joint pmf or pdf. In Chapter 5, we defined the correlation coefficient $\rho(X, Y)$ by

$$\rho = \rho(X, Y) = \frac{\text{Cov}(X, Y)}{\sigma_X \cdot \sigma_Y}$$

where

$$\text{cov}(X, Y) = \begin{cases} \displaystyle\sum_x \sum_y (x - \mu_X)(y - \mu_Y)p(x, y) & (X, Y) \text{ discrete} \\[2ex] \displaystyle\int_{-\infty}^{\infty} \int_{-\infty}^{\infty} (x - \mu_X)(y - \mu_Y)f(x, y)\, dx\, dy & (X, Y) \text{ continuous} \end{cases}$$

If we think of $p(x, y)$ or $f(x, y)$ as describing the distribution of pairs of values within the entire population, ρ becomes a measure of how strongly related x and y are in that population. Properties of ρ analogous to those for r were given in Chapter 5.

The population correlation coefficient ρ is a parameter or population characteristic, just as μ_X, μ_Y, σ_X, and σ_Y are, so we can use the sample correlation coefficient to make various inferences about ρ. In particular, r is a point estimate for ρ, and the corresponding estimator is

$$\hat{\rho} = R = \frac{\sum(X_i - \overline{X})(Y_i - \overline{Y})}{\sqrt{\sum(X_i - \overline{X})^2}\sqrt{\sum(Y_i - \overline{Y})^2}}$$

EXAMPLE 12.16 Medical researchers have noted that adolescent females are much more likely to deliver low-birth-weight babies than are adult females. Because such babies have higher mortality rates, numerous investigations have focused on the relationship between mother's age and birth weight. One such study is described in the article **"Body Size and Intelligence in 6-Year-Olds: Are Offspring of Teenage Mothers at Risk? (*Maternal and Child Health J.*, 2009: 847–856)**. The following data on x = maternal age (yr) and y = baby's birth weight (g) is consistent with summary quantities given in the cited article as well as with data published by the National Center for Health Statistics.

x	15	17	18	15	16	19	17	16	18	19
y	2289	3393	3271	2648	2897	3327	2970	2535	3138	3573

A scatterplot of the data shows a rather substantial increasing linear pattern. Relevant summary quantities are $\Sigma x_i = 170$, $\Sigma y_i = 30{,}041$, $\Sigma x_i^2 = 3910$, $\Sigma y_i^2 = 91{,}785{,}351$, $\Sigma x_i y_i = 515{,}600$, from which $S_{xx} = 20$, $S_{yy} = 1{,}539{,}182.90$, $S_{xy} = 4903$. Then

$$r = \frac{4903}{\sqrt{20}\ \sqrt{1{,}539{,}182.90}} = .884$$

With ρ denoting the correlation between mother's age and baby's weight in the entire population of adolescent mothers who gave birth, the point estimate of ρ is $\hat{\rho} = r = .884$. ∎

The small-sample intervals and test procedures presented in Chapters 7–9 were based on an assumption of population normality. To test hypotheses about ρ, an analogous assumption about the distribution of pairs of (x, y) values in the population is required. We are now assuming that *both* X and Y are random (much of our regression work focused thus far on x fixed by the experimenter) with a bivariate normal probability distribution as described in Section 5.2. Recall that in this case, $\rho = 0$ implies that X and Y are independent rv's.

Assuming that the pairs are drawn from a bivariate normal distribution allows us to test hypotheses about ρ and to construct a CI. There is no completely satisfactory way to check the plausibility of the bivariate normality assumption. A partial check involves constructing two separate normal probability plots, one for the sample x_i's and another for the sample y_i's, since bivariate normality implies that the marginal distributions of both X and Y are normal. If either plot deviates substantially from a straight-line pattern, the following inferential procedures should not be used for small n.

Testing for the Absence of Correlation

Let R denote the sample correlation coefficient as a random variable (before data is obtained). When H_0: $\rho = 0$ is true, the test statistic

$$T = \frac{R\sqrt{n-2}}{\sqrt{1-R^2}}$$

has a t distribution with $n - 2$ df.

Alternative Hypothesis	P-Value Determination
H_a: $\rho > 0$	Area under the t_{n-2} curve to the right of t
H_a: $\rho < 0$	Area under the t_{n-2} curve to the left of t
H_a: $\rho \neq 0$	$2 \cdot$ (Area under the t_{n-2} curve to the right of $|t|$)

EXAMPLE 12.17 Neurotoxic effects of manganese are well known and are usually caused by high occupational exposure over long periods of time. In the fields of occupational hygiene and environmental hygiene, the relationship between lipid peroxidation (which is responsible for deterioration of foods and damage to live tissue) and occupational exposure has not been previously reported. The article **"Lipid Peroxidation in Workers Exposed to Manganese"** (*Scand. J. of Work and Environ. Health, 1996: 381–386*) gives data on x = manganese concentration in blood (ppb) and y = concentration (*m*mol/L) of malondialdehyde, which is a stable product of lipid peroxidation, both for a sample of 22 workers exposed to manganese and for a control sample of 45 individuals. The value of r for the control sample is .29, from which

$$t = \frac{(.29)\sqrt{45 - 2}}{\sqrt{1 - (.29)^2}} \approx 2.0$$

The corresponding P-value for a two-tailed t test based on 43 df is roughly .052 (the cited article reported only that P-value > .05). We would not want to reject the assertion that $\rho = 0$ at either significance level .01 or .05. For the sample of exposed workers, $r = .83$ and $t \approx 6.7$, clear evidence that there is a linear association in the entire population of exposed workers from which the sample was selected. ∎

Because ρ measures the extent to which there is a linear relationship between the two variables in the population, the null hypothesis $H_0: \rho = 0$ states that there is no such population relationship. In Section 12.3, we used the t ratio $\hat{\beta}_1/s_{\hat{\beta}_1}$ to test for a linear relationship between the two variables in the context of regression analysis. It turns out that the two test procedures are completely equivalent because $r\sqrt{n - 2}/\sqrt{1 - r^2} = \hat{\beta}_1/s_{\hat{\beta}_1}$. When interest lies only in assessing the strength of any linear relationship rather than in fitting a model and using it to estimate or predict, the test statistic formula just presented requires fewer computations than does the t-ratio.

Other Inferences Concerning ρ

The procedure for testing $H_0: \rho = \rho_0$ when $\rho_0 \neq 0$ is not equivalent to any procedure from regression analysis. The test statistic as well as a confidence interval formula are based on a transformation of R developed by the famous statistician R.A. Fisher.

PROPOSITION

When $(X_1, Y_1), \ldots, (X_n, Y_n)$ is a sample from a bivariate normal distribution, the rv

$$V = \frac{1}{2}\ln\left(\frac{1 + R}{1 - R}\right) \tag{12.9}$$

has approximately a normal distribution with mean and variance

$$\mu_V = \frac{1}{2}\ln\left(\frac{1 + \rho}{1 - \rho}\right) \qquad \sigma_V^2 = \frac{1}{n - 3}$$

The rationale for the transformation is to obtain a function of R that has a variance independent of ρ; this would not be the case with R itself. Also, the transformation should not be used if n is quite small, since the approximation will not be valid.

The test statistic for testing H_0: $\rho = \rho_0$ is

$$Z = \frac{V - \frac{1}{2}\ln[(1 + \rho_0)/(1 - \rho_0)]}{1/\sqrt{n - 3}}$$

Alternative Hypothesis	P-Value Determination
H_a: $\rho > \rho_0$	Area under the standard normal curve to the right of z
H_a: $\rho < \rho_0$	Area under the standard normal curve to the left of z
H_a: $\rho \neq \rho_0$	$2 \cdot$ (Area under the standard normal curve to the right of $|z|$)

EXAMPLE 12.18 The article **"Size Effect in Shear Strength of Large Beams—Behavior and Finite Element Modelling"** (*Mag. of Concrete Res.*, **2005: 497–509**) reported on a study of various characteristics of large reinforced concrete deep and shallow beams tested until failure. Consider the following data on x = cube strength and y = cylinder strength (both in MPa):

x	55.10	44.83	46.32	51.10	49.89	45.20	48.18	46.70	54.31	41.50
y	49.10	31.20	32.80	42.60	42.50	32.70	36.21	40.40	37.42	30.80

x	47.50	52.00	52.25	50.86	51.66	54.77	57.06	57.84	55.22
y	35.34	44.80	41.75	39.35	44.07	43.40	45.30	39.08	41.89

Then $S_{xx} = 367.74$, $S_{yy} = 488.54$, and $S_{xy} = 322.37$, from which $r = .761$. Does this provide strong evidence for concluding that the two measures of strength are at least moderately positively correlated?

Our previous interpretation of moderate positive correlation was $.5 < \rho < .8$, so we wish to test H_0: $\rho = .5$ versus H_a: $\rho > .5$. The computed value of V is then

$$v = .5 \ln\left(\frac{1 + .761}{1 - .761}\right) = .999 \qquad .5 \ln\left(\frac{1 + .5}{1 - .5}\right) = .549$$

Thus $z = (.999 - .549)\sqrt{19 - 3} = 1.80$. The P-value for this upper-tailed test is $1 - \Phi(1.80) = .0359$. The null hypothesis can therefore be rejected at significance level .05 but not at level .01. This latter result is somewhat surprising in light of the magnitude of r, but when n is small, a reasonably large r may result even when ρ is not all that substantial. At significance level .01, the evidence for a moderately positive correlation is not compelling. ∎

To obtain a CI for ρ, we first derive an interval for $\mu_V = \frac{1}{2}\ln[(1 + \rho)/(1 - \rho)]$. Standardizing V, writing a probability statement, and manipulating the resulting inequalities yields

$$\left(v - \frac{z_{\alpha/2}}{\sqrt{n - 3}}, v + \frac{z_{\alpha/2}}{\sqrt{n - 3}}\right) \tag{12.10}$$

as a $100(1 - \alpha)\%$ interval for μ_V, where $v = \frac{1}{2} \ln[(1 + r)/(1 - r)]$. This interval can then be manipulated to yield the desired CI.

A $100(1 - \alpha)\%$ confidence interval for ρ is

$$\left(\frac{e^{2c_1} - 1}{e^{2c_1} + 1}, \frac{e^{2c_2} - 1}{e^{2c_2} + 1} \right)$$

where c_1 and c_2 are the left and right endpoints, respectively, of the interval (12.11).

EXAMPLE 12.19 As far back as Leonardo da Vinci, it was known that $x =$ height and $y =$ wingspan (measured fingertip to fingertip while arms are outstretched side to side) are closely related. Here are measurements from a random sample of students taking a statistics course:

x	63.0	63.0	65.0	64.0	68.0	69.0	71.0	68.0
y	62.0	62.0	64.0	64.5	67.0	69.0	70.0	72.0
x	68.0	72.0	73.0	73.5	70.0	70.0	72.0	74.0
y	70.0	72.0	73.0	75.0	71.0	70.0	76.0	76.5

A scatterplot shows an approximate linear pattern, and so do normal probability plots of x and y. The sample correlation coefficient is computed to be $r = .9422$. Its Fisher transformation is

$$v = .5 \ln \left(\frac{1 + .9422}{1 - .9422} \right) = 1.757$$

A 95% CI for μ_V is

$$1.757 \pm \frac{1.96}{\sqrt{16 - 3}} = (1.213, 2.301) = (c_1, c_2)$$

The CI for ρ with a confidence level of approximately 95% is therefore

$$\left(\frac{e^{2(1.213)} - 1}{e^{2(1.213)} + 1}, \frac{e^{2(2.301)} - 1}{e^{2(2.301)} + 1} \right) = (.838, .980)$$

Notice that the interval includes only values exceeding .8, so it appears that there is a strong linear association between the two variables in the sampled population. ■

In Chapter 5, we cautioned that a large value of the correlation coefficient (near 1 or -1) implies only association and not causation. This applies to both ρ and r.

EXERCISES Section 12.5 (57–67)

57. The article **"Behavioural Effects of Mobile Telephone Use During Simulated Driving"** (*Ergonomics*, 1995: 2536–2562) reported that for a sample of 20 experimental subjects, the sample correlation coefficient for $x =$ age and $y =$ time since the subject had acquired a driving license (yr) was .97. Why do you think the value of r is so close to 1? (The article's authors give an explanation.)

58. The Turbine Oil Oxidation Test (TOST) and the Rotating Bomb Oxidation Test (RBOT) are two different procedures for evaluating the oxidation stability of steam turbine oils. The article **"Dependence of Oxidation Stability of Steam Turbine Oil on Base Oil Composition"** (*J. of the Society of Tribologists and Lubrication Engrs.*, Oct. 1997: 19–24) reported the accompanying observations on

x = TOST time (hr) and y = RBOT time (min) for 12 oil specimens.

TOST	4200	3600	3750	3675	4050	2770
RBOT	370	340	375	310	350	200
TOST	4870	4500	3450	2700	3750	3300
RBOT	400	375	285	225	345	285

a. Calculate and interpret the value of the sample correlation coefficient (as do the article's authors).
b. How would the value of r be affected if we had let x = RBOT time and y = TOST time?
c. How would the value of r be affected if RBOT time were expressed in hours?
d. Construct normal probability plots and comment.
e. Carry out a test of hypotheses to decide whether RBOT time and TOST time are linearly related.

59. Toughness and fibrousness of asparagus are major determinants of quality. This was the focus of a study reported in **"Post-Harvest Glyphosate Application Reduces Toughening, Fiber Content, and Lignification of Stored Asparagus Spears" (*J. of the Amer. Soc. of Hort. Science*, 1988: 569–572)**. The article reported the accompanying data (read from a graph) on x = shear force (kg) and y = percent fiber dry weight.

x	46	48	55	57	60	72	81	85	94
y	2.18	2.10	2.13	2.28	2.34	2.53	2.28	2.62	2.63
x	109	121	132	137	148	149	184	185	187
y	2.50	2.66	2.79	2.80	3.01	2.98	3.34	3.49	3.26

$n = 18$, $\Sigma x_i = 1950$, $\Sigma x_i^2 = 251{,}970$,
$\Sigma y_i = 47.92$, $\Sigma y_i^2 = 130.6074$, $\Sigma x_i y_i = 5530.92$

a. Calculate the value of the sample correlation coefficient. Based on this value, how would you describe the nature of the relationship between the two variables?
b. If a first specimen has a larger value of shear force than does a second specimen, what tends to be true of percent dry fiber weight for the two specimens?
c. If shear force is expressed in pounds, what happens to the value of r? Why?
d. If the simple linear regression model were fit to this data, what proportion of observed variation in percent fiber dry weight could be explained by the model relationship?
e. Carry out a test at significance level .01 to decide whether there is a positive linear association between the two variables.

60. Head movement evaluations are important because individuals, especially those who are disabled, may be able to operate communications aids in this manner. The article **"Constancy of Head Turning Recorded in Healthy Young Humans" (*J. of Biomed. Engr.*, 2008: 428–436)** reported data on ranges in maximum inclination angles of the head in the clockwise anterior, posterior, right, and left directions for 14 randomly selected subjects. Consider the accompanying data on average anterior maximum inclination angle (AMIA) both in the clockwise direction and in the counterclockwise direction.

Subj:	1	2	3	4	5	6	7
Cl:	57.9	35.7	54.5	56.8	51.1	70.8	77.3
Co:	44.2	52.1	60.2	52.7	47.2	65.6	71.4

Subj:	8	9	10	11	12	13	14
Cl:	51.6	54.7	63.6	59.2	59.2	55.8	38.5
Co:	48.8	53.1	66.3	59.8	47.5	64.5	34.5

a. Calculate a point estimate of the population correlation coefficient between Cl AMIA and Co AMIA ($\Sigma Cl = 786.7$, $\Sigma Co = 767.9$, $\Sigma Cl^2 = 45{,}727.31$, $\Sigma Co^2 = 43{,}478.07$, $\Sigma ClCo = 44{,}187.87$).
b. Assuming bivariate normality (normal probability plots of the Cl and Co samples are reasonably straight), carry out a test at significance level .01 to decide whether there is a linear association between the two variables in the population (as do the authors of the cited paper). Would the conclusion have been the same if a significance level of .001 had been used?

61. The authors of the paper **"Objective Effects of a Six Months' Endurance and Strength Training Program in Outpatients with Congestive Heart Failure" (*Medicine and Science in Sports and Exercise*, 1999: 1102–1107)** presented a correlation analysis to investigate the relationship between maximal lactate level x and muscular endurance y. The accompanying data was read from a plot in the paper.

x	400	750	770	800	850	1025	1200
y	3.80	4.00	4.90	5.20	4.00	3.50	6.30
x	1250	1300	1400	1475	1480	1505	2200
y	6.88	7.55	4.95	7.80	4.45	6.60	8.90

$S_{xx} = 36.9839$, $S_{yy} = 2{,}628{,}930.357$, $S_{xy} = 7377.704$. A scatterplot shows a linear pattern.

a. Test to see whether there is a positive correlation between maximal lactate level and muscular endurance in the population from which this data was selected.
b. If a regression analysis were to be carried out to predict endurance from lactate level, what proportion of observed variation in endurance could be attributed to the approximate linear relationship? Answer the analogous question if regression is used to predict lactate level from endurance—and answer both questions without doing any regression calculations.

62. The article **"Quantitative Estimation of Clay Mineralogy in Fine-Grained Soils"** (**J. of Geotechnical and Geoenvironmental Engr., 2011: 997–1008**) reported on various chemical properties of natural and artificial soils. Here are observations on x = cation exchange capacity (CEC, in meq/100 g) and y = specific surface area (SSA, in m^2/g) of 20 natural soils.

x	66	121	134	101	77	89	63	57	117	118
y	175	324	460	288	205	210	295	161	314	265
x	76	125	75	71	133	104	76	96	58	109
y	236	355	240	133	431	306	132	269	158	303

Minitab gave the following output in response to a request for r:

```
correlation of x and y = 0.853
```

Normal probability plots of x and y are quite straight.

a. Carry out a test of hypotheses to see if there is a positive linear association in the population from which the sample data was selected.

b. With $n = 20$, how small would the value of r have to be in order for the null hypothesis in the test of (a) to not be rejected at significance level .01?

c. Calculate a confidence interval for ρ using a 95% confidence level.

63. Physical properties of six flame-retardant fabric samples were investigated in the article **"Sensory and Physical Properties of Inherently Flame-Retardant Fabrics"** (**Textile Research, 1984: 61–68**). Use the accompanying data and a .05 significance level to determine whether a linear relationship exists between stiffness x (mg-cm) and thickness y (mm). Is the result of the test surprising in light of the value of r?

x	7.98	24.52	12.47	6.92	24.11	35.71
y	.28	.65	.32	.27	.81	.57

64. The accompanying data on x = UV transparency index and y = maximum prevalence of infection was read from a graph in the article **"Solar Radiation Decreases Parasitism in Daphnia"** (**Ecology Letters, 2012: 47–54**):

x	1.3	1.4	1.5	2.0	2.2	2.7	2.7	2.7	2.8
y	16	3	32	1	13	0	8	16	2
x	2.9	3.0	3.6	3.8	3.8	4.6	5.1	5.7	
y	1	7	36	25	10	35	58	56	

Summary quantities include $S_{xx} = 25.5224$, $S_{yy} = 5593.0588$, and $S_{xy} = 264.4882$.

a. Calculate and interpret the value of the sample correlation coefficient.

b. If you decided to fit the simple linear regression model to this data, what proportion of observed

variation in maximum prevalence could be explained by the model relationship?

c. If you decided to regress UV transparency index on maximum prevalence (i.e., interchange the roles of x and y), what proportion of observed variation could be attributed to the model relationship?

d. Carry out a test of $H_0: \rho = .5$ versus $H_a: \rho > .5$ using a significance level of .05. [*Note:* The cited article reported the P-value for testing $H_0: \rho = 0$ versus $H_0: \rho \neq 0$.]

65. Torsion during hip external rotation and extension may explain why acetabular labral tears occur in professional athletes. The article **"Hip Rotational Velocities During the Full Golf Swing"** (**J. of Sports Science and Med., 2009: 296–299**) reported on an investigation in which lead hip internal peak rotational velocity (x) and trailing hip peak external rotational velocity (y) were determined for a sample of 15 golfers. Data provided by the article's authors was used to calculate the following summary quantities:

$$\sum(x_i - \bar{x})^2 = 64,732.83, \sum(y_i - \bar{y})^2 = 130,566.96,$$
$$\sum(x_i - \bar{x})(y_i - \bar{y}) = 44,185.87$$

Separate normal probability plots showed very substantial linear patterns.

a. Calculate a point estimate for the population correlation coefficient.

b. Carry out a test at significance level .01 to decide whether there is a linear relationship between the two velocities in the sampled population.

c. Would the conclusion of (b) have changed if you had tested appropriate hypotheses to decide whether there is a positive linear association in the population? What if a significance level of .05 rather than .01 had been used?

66. Consider a time series—that is, a sequence of observations $X_1, X_2, \ldots$ obtained over time—with observed values $x_1, x_2, \ldots, x_n$. Suppose that the series shows no upward or downward trend over time. An investigator will frequently want to know just how strongly values in the series separated by a specified number of time units are related. The *lag-one sample autocorrelation coefficient* r_1 is just the value of the sample correlation coefficient r for the pairs $(x_1, x_2), (x_2, x_3), \ldots, (x_{n-1}, x_n)$, that is, pairs of values separated by one time unit. Similarly, the *lag-two sample autocorrelation coefficient* r_2 is r for the $n - 2$ pairs $(x_1, x_3), (x_2, x_4), \ldots, (x_{n-2}, x_n)$.

a. Calculate the values of r_1, r_2, and r_3 for the temperature data from Exercise 82 of Chapter 1, and comment.

b. Analogous to the population correlation coefficient ρ, let $\rho_1, \rho_2, \ldots$ denote the theoretical or long-run autocorrelation coefficients at the various lags. If all these ρ's are 0, there is no (linear) relationship at any lag. In this case, if n is large, each R_i has approximately a normal distribution with mean 0 and standard deviation $1/\sqrt{n}$, and different R_i's are almost independent.

Thus $H_0: \rho_i = 0$ can be tested using a z test with test statistic value $z_i = \sqrt{n} r_i$. If $n = 100$ and $r_1 = .16$, $r_2 = -.09$, and $r_3 = -.15$, at significance level .05 is there any evidence of theoretical autocorrelation at the first three lags?

c. If you are simultaneously testing the null hypothesis in part (b) for more than one lag, why might you want to increase the significance level for each test?

67. A sample of $n = 500$ (x, y) pairs was collected and a test of $H_0: \rho = 0$ versus $H_a: \rho \neq 0$ was carried out. The resulting P-value was computed to be .00032.

a. What conclusion would be appropriate at level of significance .001?

b. Does this small P-value indicate that there is a very strong linear relationship between x and y (a value of ρ that differs considerably from 0)? Explain.

c. Now suppose a sample of $n = 10,000$ (x, y) pairs resulted in $r = .022$. Test $H_0: \rho = 0$ versus $H_a: \rho \neq 0$ at level .05. Is the result statistically significant? Comment on the practical significance of your analysis.

SUPPLEMENTARY EXERCISES (68–87)

68. The appraisal of a warehouse can appear straightforward compared to other appraisal assignments. A warehouse appraisal involves comparing a building that is primarily an open shell to other such buildings. However, there are still a number of warehouse attributes that are plausibly related to appraised value. The article **"Challenges in Appraising 'Simple' Warehouse Properties" (Donald Sonneman**, *The Appraisal Journal*, **April 2001, 174–178)** gives the accompanying data on truss height (ft), which determines how high stored goods can be stacked, and sale price ($) per square foot.

Height	12	14	14	15	15	16	18	22	22	24
Price	35.53	37.82	36.90	40.00	38.00	37.50	41.00	48.50	47.00	47.50

Height	24	26	26	27	28	30	30	33	36
Price	46.20	50.35	49.13	48.07	50.90	54.78	54.32	57.17	57.45

a. Is it the case that truss height and sale price are "deterministically" related—i.e., that sale price is determined completely and uniquely by truss height? [*Hint*: Look at the data.]

b. Construct a scatterplot of the data. What does it suggest?

c. Determine the equation of the least squares line.

d. Give a point prediction of price when truss height is 27 ft, and calculate the corresponding residual.

e. What percentage of observed variation in sale price can be attributed to the approximate linear relationship between truss height and price?

69. Refer to the previous exercise, which gives data on truss heights for a sample of warehouses and the corresponding sale prices.

a. Estimate the true average change in sale price associated with a one-foot increase in truss height, and do so in a way that conveys information about the precision of estimation.

b. Estimate the true average sale price for all warehouses having a truss height of 25 ft, and do so in a

way that conveys information about the precision of estimation.

c. Predict the sale price for a single warehouse whose truss height is 25 ft, and do so in a way that conveys information about the precision of prediction. How does this prediction compare to the estimate of (b)?

d. Without calculating any intervals, how would the width of a 95% prediction interval for sale price when truss height is 25 ft compare to the width of a 95% interval when height is 30 ft? Explain your reasoning.

e. Calculate and interpret the sample correlation coefficient.

70. Forensic scientists are often interested in making a measurement of some sort on a body (alive or dead) and then using that as a basis for inferring something about the age of the body. Consider the accompanying data on age (yr) and % D-aspertic acid (hereafter %DAA) from a particular tooth **("An Improved Method for Age at Death Determination from the Measurements of D-Aspartic Acid in Dental Collagen," *Archaeometry*, 1990: 61–70.)**

Age	9	10	11	12	13	14	33	39	52	65	69
%DAA	1.13	1.10	1.11	1.10	1.24	1.31	2.25	2.54	2.93	3.40	4.55

Suppose a tooth from another individual has 2.01%DAA. Might it be the case that the individual is younger than 22? This question was relevant to whether or not the individual could receive a life sentence for murder.

A seemingly sensible strategy is to regress age on %DAA and then compute a PI for age when %DAA = 2.01. However, it is more natural here to regard age as the independent variable x and %DAA as the dependent variable y, so the regression model is %DAA $= \beta_0 + \beta_1 x + \epsilon$. After estimating the regression coefficients, we can substitute $y^* = 2.01$ into the estimated equation and then solve for a prediction of age $\hat{x}$. This "inverse" use of the regression line is called "calibration." A PI for age with prediction level approximately $100(1 - \alpha)\%$ is $\hat{x} \pm t_{\alpha/2, n-2} \cdot SE$ where

$$SE = \frac{s}{\hat{\beta}_1} \left\{ 1 + \frac{1}{n} + \frac{(\hat{x} - \bar{x})^2}{S_{xx}} \right\}^{1/2}$$

Calculate this PI for $y^* = 2.01$ and then address the question posed earlier.

71. Phenolic compounds are found in the effluents of coal conversion processes, petroleum refineries, herbicide manufacturing, and fiberglass manufacturing. These compounds are toxic, carcinogenic, and have contributed over the past decades to environmental pollution of aquatic environments. In one study reported in **"Photolysis, Biodegradation, and Sorption Behavior of Three Selected Phenolic Compounds on the Surface and Sediment of Rivers"** (*J. of Envir. Engr.*, 2011: 1114–1121), the authors examined the sorption characteristics of three selected phenolic compounds. The following data on y = sorbed concentration (μg/g) and x = equilibrium concentration (μg/mL) of 2, 4-Dinitrophenol (DNP) in a particular natural river sediment was read from a graph in the article.

x	.11	.13	.14	.18	.29	.44	.67	.78	.93
y	1.72	2.17	2.33	3.00	5.17	7.61	11.17	12.72	14.78

a. Calculate point estimates of the slope and intercept of the population regression line.
b. Does the simple linear regression model specify a useful relationship between y and x?
c. Confirm that $\hat{y} = 3.404$, $S_{\hat{Y}} = .107$ when $x = .2$, and $\hat{y} = 6.616$, $S_{\hat{Y}} = .088$ when $x = .4$. Explain why $s_{\hat{Y}}$ is larger when $x = .2$ than when $x = .4$.
d. Calculate a confidence interval with a confidence level of 95% for the true average DNP sorbed

concentration of all river sediment specimens having an equilibrium concentration of .4.
e. Calculate a prediction interval with a prediction level of 95% for the DNP sorbed concentration of a single river sediment specimen having an equilibrium concentration of .4.

72. The SAS output at the bottom of this page is based on data from the article **"Evidence for and the Rate of Denitrification in the Arabian Sea"** (*Deep Sea Research*, 1978: 431–435). The variables under study are x = salinity level (%) and y = nitrate level (μM/L).
a. What is the sample size n? [*Hint*: Look for degrees of freedom for SSE.]
b. Calculate a point estimate of expected nitrate level when salinity level is 35.5.
c. Does there appear to be a useful linear relationship between the two variables?
d. What is the value of the sample correlation coefficient?
e. Would you use the simple linear regression model to draw conclusions when the salinity level is 40?

73. The presence of hard alloy carbides in high chromium white iron alloys results in excellent abrasion resistance, making them suitable for materials handling in the mining and materials processing industries. The accompanying data on x = retained austenite content (%) and y = abrasive wear loss (mm^3) in pin wear tests with garnet as the abrasive was read from a plot in the article **"Microstructure-Property Relationships in High Chromium White Iron Alloys"** (*Intl. Materials Reviews*, 1996: 59–82).

x	4.6	17.0	17.4	18.0	18.5	22.4	26.5	30.0	34.0
y	.66	.92	1.45	1.03	.70	.73	1.20	.80	.91

x	38.8	48.2	63.5	65.8	73.9	77.2	79.8	84.0
y	1.19	1.15	1.12	1.37	1.45	1.50	1.36	1.29

SAS output for Exercise 72

Dependent Variable: NITRLVL

Analysis of Variance

Source	DF	Sum of Squares	Mean Square	F Value	Prob > F
Model	1	64.49622	64.49622	63.309	0.0002
Error	6	6.11253	1.01875		
C Total	7	70.60875			

Root MSE	1.00933	R-square	0.9134	
Dep Mean	26.91250	Adj R-sq	0.8990	
C.V.	3.75043			

Parameter Estimates

Variable	DF	Parameter Estimate	Standard Error	T for HO: Parameter = 0	Prob > |T|
INTERCEP	1	326.976038	37.71380243	8.670	0.0001
SALINITY	1	−8.403964	1.05621381	−7.957	0.0002

SAS output for Exercise 73

```
Dependent Variable: ABRLOSS
                              Analysis of Variance

Source         DF       Sum of Squares        Mean Square      F Value      Prob > F
Model           1              0.63690            0.63690       15.444        0.0013
Error          15              0.61860            0.04124
C Total        16              1.25551

          Root MSE       0.20308     R-square      0.5073
          Dep Mean       1.10765     Adj R-sq      0.4744
          C.V.          18.33410

                              Parameter Estimates

                        Parameter         Standard         T for H0:
Variable       DF        Estimate            Error      Parameter = 0      Prob > |T|
INTERCEP        1        0.787218       0.09525879             8.264          0.0001
AUSTCONT        1        0.007570       0.00192626             3.930          0.0013
```

Use the data and the SAS output above to answer the following questions.

a. What proportion of observed variation in wear loss can be attributed to the simple linear regression model relationship?

b. What is the value of the sample correlation coefficient?

c. Test the utility of the simple linear regression model using $\alpha = .01$.

d. Estimate the true average wear loss when content is 50% and do so in a way that conveys information about reliability and precision.

e. What value of wear loss would you predict when content is 30%, and what is the value of the corresponding residual?

74. The accompanying data was read from a scatterplot in the article **"Urban Emissions Measured with Aircraft"** (*J. of the Air and Waste Mgmt. Assoc.*, **1998: 16–25**). The response variable is ΔNO_y, and the explanatory variable is ΔCO.

ΔCO	50	60	95	108	135
ΔNO_y	2.3	4.5	4.0	3.7	8.2

ΔCO	210	214	315	720
ΔNO_y	5.4	7.2	13.8	32.1

a. Fit an appropriate model to the data and judge the utility of the model.

b. Predict the value of ΔNO_y that would result from making one more observation when ΔCO is 400, and do so in a way that conveys information about precision and reliability. Does it appear that ΔNO_y can be accurately predicted? Explain.

c. The largest value of ΔCO is much greater than the other values. Does this observation appear to have had a substantial impact on the fitted equation?

75. An investigation was carried out to study the relationship between speed (ft/sec) and stride rate (number of steps

taken/sec) among female marathon runners. Resulting summary quantities included $n = 11$, $\Sigma(\text{speed}) = 205.4$, $\Sigma(\text{speed})^2 = 3880.08$, $\Sigma(\text{rate}) = 35.16$, $\Sigma(\text{rate})^2 = 112.681$, and $\Sigma(\text{speed})(\text{rate}) = 660.130$.

a. Calculate the equation of the least squares line that you would use to predict stride rate from speed.

b. Calculate the equation of the least squares line that you would use to predict speed from stride rate.

c. Calculate the coefficient of determination for the regression of stride rate on speed of part (a) and for the regression of speed on stride rate of part (b). How are these related?

76. "Mode-mixity" refers to how much of crack propagation is attributable to the three conventional fracture modes of opening, sliding, and tearing. For plane problems, only the first two modes are present, and the mode-mixity angle is a measure of the extent to which propagation is due to sliding as opposed to opening. The article **"Increasing Allowable Flight Loads by Improved Structural Modeling"** (*AIAA J.*, **2006: 376–381**) gives the following data on $x = $ mode-mixity angle (degrees) and $y = $ fracture toughness (N/m) for sandwich panels use in aircraft construction.

x	16.52	17.53	18.05	18.50	22.39	23.89	25.50	24.89
y	609.4	443.1	577.9	628.7	565.7	711.0	863.4	956.2

x	23.48	24.98	25.55	25.90	22.65	23.69	24.15	24.54
y	679.5	707.5	767.1	817.8	702.3	903.7	964.9	1047.3

a. Obtain the equation of the estimated regression line, and discuss the extent to which the simple linear regression model is a reasonable way to relate fracture toughness to mode-mixity angle.

b. Does the data suggest that the average change in fracture toughness associated with a one-degree increase in mode-mixity angle exceeds 50 N/m? Carry out an appropriate test of hypotheses.

c. For purposes of precisely estimating the slope of the population regression line, would it have been preferable to make observations at the angles 16, 16, 18, 18, 20, 20, 20, 20, 22, 22, 22, 22, 24, 24, 26, and 26 (again a sample size of 16)? Explain your reasoning.

d. Calculate an estimate of true average fracture toughness and also a prediction of fracture toughness both for an angle of 18 degrees and for an angle of 22 degrees, do so in a manner that conveys information about reliability and precision, and then interpret and compare the estimates and predictions.

77. Open water oil spills can wreak terrible consequences on the environment and be expensive to clean up. Many physical and biological methods have been developed to recover oil from water surfaces. The article **"Capacity of Straw for Repeated Binding of Crude Oil from Salt Water and Its Effect on Biodegradation"** (*J. of Hazardous Toxic and Radioactive Waste*, 2012: 75–78) discussed how wheat straw could be used to extract crude oil from a water surface. An experiment was conducted in which crude oil (0 to 16.9g) was added to 100mL of salt-water in separate Petri dishes. Wheat straw (2g) was then added to each dish and all dishes were shaken at 70 rpm overnight. The following data, read from a graph, is based on the x = amount of oil added (in g) and y = the corresponding amount of oil recovered (in g) from wheat straw.

x	1.0	1.5	2.1	2.8	3.6	4.5	5.5
y	0.610	0.840	1.512	1.792	2.952	2.880	4.400

x	6.6	7.8	9.1	10.5	12.0	13.6	15.2	16.9
y	5.346	6.396	7.189	8.085	9.840	11.696	13.224	14.365

a. Construct a scatterplot of the data. Does it appear that recovered oil could be very well predicted by the value of added oil? Explain your reasoning.

b. Calculate and interpret the coefficient of determination.

c. Does the simple linear regression model appear to specify a useful relationship between these two variables? State and test the relevant hypotheses.

d. Predict the value of oil recovered when amount of oil added is 5.0, and do so in a way that conveys information about precision and reliability.

e. Without any further calculation, carry out a test of hypotheses to decide whether the value of ρ is something other than 0.

78. In Section 12.4, we presented a formula for $V(\hat{\beta}_0 + \hat{\beta}_1 x^*)$ and a CI for $\beta_0 + \beta_1 x^*$. Taking $x^* = 0$ gives $\sigma_{\hat{\beta}_0}^2$ and a CI for β_0. Use the data of Example 12.11 to calculate the estimated standard deviation of $\hat{\beta}_0$ and a 95% CI for the y-intercept of the true regression line.

79. Show that SSE $= S_{yy} - \hat{\beta}_1 S_{xy}$, which gives an alternative computational formula for SSE.

80. Suppose that x and y are positive variables and that a sample of n pairs results in $r \approx 1$. If the sample correlation coefficient is computed for the (x, y^2) pairs, will the resulting value also be approximately 1? Explain.

81. Let s_x and s_y denote the sample standard deviations of the observed x's and y's, respectively [that is, $s_x^2 = \Sigma(x_i - \bar{x})^2/(n - 1)$ and similarly for s_y^2].

a. Show that an alternative expression for the estimated regression line $y = \hat{\beta}_0 + \hat{\beta}_1 x$ is

$$y = \bar{y} + r \cdot \frac{s_y}{s_x}(x - \bar{x})$$

b. This expression for the regression line can be interpreted as follows. Suppose $r = .5$. What then is the predicted y for an x that lies 1 SD (s_x units) above the mean of the x_i's? If r were 1, the prediction would be for y to lie 1 SD above its mean $\bar{y}$, but since $r = .5$, we predict a y that is only .5 SD (.5s_y unit) above $\bar{y}$. Using the data in Exercise 64, when UV transparency index is 1 SD below the average in the sample, by how many standard deviations is the predicted maximum prevalence above or below its average for the sample?

82. Verify that the t statistic for testing $H_0: \beta_1 = 0$ in Section 12.3 is identical to the t statistic in Section 12.5 for testing $H_0: \rho = 0$.

83. Use the formula for computing SSE to verify that $r^2 = 1 - $ SSE/SST.

84. In biofiltration of wastewater, air discharged from a treatment facility is passed through a damp porous membrane that causes contaminants to dissolve in water and be transformed into harmless products. The accompanying data on x = inlet temperature (°C) and y = removal efficiency (%) was the basis for a scatterplot that appeared in the article **"Treatment of Mixed Hydrogen Sulfide and Organic Vapors in a Rock Medium Biofilter"** (*Water Environment Research*, 2001: 426–435).

Obs	Temp	Removal %	Obs	Temp	Removal %
1	7.68	98.09	17	8.55	98.27
2	6.51	98.25	18	7.57	98.00
3	6.43	97.82	19	6.94	98.09
4	5.48	97.82	20	8.32	98.25
5	6.57	97.82	21	10.50	98.41
6	10.22	97.93	22	16.02	98.51
7	15.69	98.38	23	17.83	98.71
8	16.77	98.89	24	17.03	98.79
9	17.13	98.96	25	16.18	98.87
10	17.63	98.90	26	16.26	98.76
11	16.72	98.68	27	14.44	98.58
12	15.45	98.69	28	12.78	98.73
13	12.06	98.51	29	12.25	98.45
14	11.44	98.09	30	11.69	98.37
15	10.17	98.25	31	11.34	98.36
16	9.64	98.36	32	10.97	98.45

Calculated summary quantities are $\Sigma x_i = 384.26$, $\Sigma y_i = 3149.04$, $\Sigma x_i^2 = 5099.2412$, $\Sigma x_i y_i = 37,850.7762$, and $\Sigma y_i^2 = 309,892.6548$.

a. Does a scatterplot of the data suggest appropriateness of the simple linear regression model?

b. Fit the simple linear regression model, obtain a point prediction of removal efficiency when temperature = 10.50, and calculate the value of the corresponding residual.

c. Roughly what is the size of a typical deviation of points in the scatterplot from the least squares line?

d. What proportion of observed variation in removal efficiency can be attributed to the model relationship?

e. Estimate the slope coefficient in a way that conveys information about reliability and precision, and interpret your estimate.

f. Personal communication with the authors of the article revealed that there was one additional observation that was not included in their scatterplot: (6.53, 96.55). What impact does this additional observation have on the equation of the least squares line and the values of s and r^2?

85. Normal hatchery processes in aquaculture inevitably produce stress in fish, which may negatively impact growth, reproduction, flesh quality, and susceptibility to disease. Such stress manifests itself in elevated and sustained corticosteroid levels. The article **"Evaluation of Simple Instruments for the Measurement of Blood Glucose and Lactate, and Plasma Protein as Stress Indicators in Fish"** (*J. of the World Aquaculture Society*, **1999: 276–284**) described an experiment in which fish were subjected to a stress protocol and then removed and tested at various times after the protocol had been applied. The accompanying data on x = time (min) and y = blood glucose level (mmol/L) was read from a plot.

x	2	2	5	7	12	13	17	18	23	24	26	28
y	4.0	3.6	3.7	4.0	3.8	4.0	5.1	3.9	4.4	4.3	4.3	4.4

x	29	30	34	36	40	41	44	56	56	57	60	60
y	5.8	4.3	5.5	5.6	5.1	5.7	6.1	5.1	5.9	6.8	4.9	5.7

Use the methods developed in this chapter to analyze the data, and write a brief report summarizing your conclusions (assume that the investigators are particularly interested in glucose level 30 min after stress).

86. The article **"Evaluating the BOD POD for Assessing Body Fat in Collegiate Football Players"** (*Medicine and Science in Sports and Exercise*, **1999: 1350–1356**) reports on a new air displacement device for measuring body fat. The customary procedure utilizes the hydrostatic weighing device, which measures the percentage of body fat by means of water displacement. Here is representative data read from a graph in the paper.

BOD	2.5	4.0	4.1	6.2	7.1	7.0	8.3	9.2	9.3	12.0	12.2
HW	8.0	6.2	9.2	6.4	8.6	12.2	7.2	12.0	14.9	12.1	15.3
BOD	12.6	14.2	14.4	15.1	15.2	16.3	17.1	17.9	17.9		
HW	14.8	14.3	16.3	17.9	19.5	17.5	14.3	18.3	16.2		

a. Use various methods to decide whether it is plausible that the two techniques measure on average the same amount of fat.

b. Use the data to develop a way of predicting an HW measurement from a BOD POD measurement, and investigate the effectiveness of such predictions.

87. Reconsider the situation of Exercise 73, in which x = retained austenite content using a garnet abrasive and y = abrasive wear loss were related via the simple linear regression model $Y = \beta_0 + \beta_1 x + \varepsilon$. Suppose that for a second type of abrasive, these variables are also related via the simple linear regression model $Y = \gamma_0 + \gamma_1 x + \varepsilon$ and that $V(\epsilon) = \sigma^2$ for both types of abrasive. If the data set consists of n_1 observations on the first abrasive and n_2 on the second and if SSE_1 and SSE_2 denote the two error sums of squares, then a pooled estimate of σ^2 is $\hat{\sigma}^2 = (SSE_1 + SSE_2)/(n_1 + n_2 - 4)$. Let SS_{x1} and SS_{x2} denote $\Sigma(x_i - \bar{x})^2$ for the data on the first and second abrasives, respectively. A test of $H_0: \beta_1 - \gamma_1 = 0$ (equal slopes) is based on the statistic

$$T = \frac{\hat{\beta}_1 - \hat{\gamma}_1}{\hat{\sigma}\sqrt{\dfrac{1}{SS_{x1}} + \dfrac{1}{SS_{x2}}}}$$

When H_0 is true, T has a t distribution with $n_1 + n_2 - 4$ df. Suppose the 15 observations using the alternative abrasive give $SS_{x2} = 7152.5578$, $\hat{\gamma}_1 = .006845$, and $SSE_2 = .51350$. Using this along with the data of Exercise 73, carry out a test at level .05 to see whether expected change in wear loss associated with a 1% increase in austenite content is identical for the two types of abrasive.

BIBLIOGRAPHY

Draper, Norman, and Harry Smith, *Applied Regression Analysis* (3rd ed.), Wiley, New York, 1999. A very comprehensive and authoritative book on regression analysis.

Neter, John, Michael Kutner, Christopher Nachtsheim, and William Wasserman, *Applied Linear Statistical Models* (5th ed.), Irwin, Homewood, IL, 2005. The first 14 chapters constitute an extremely readable and informative survey of regression analysis.

13 Nonlinear and Multiple Regression

INTRODUCTION

The probabilistic model studied in Chapter 12 specified that the observed value of the dependent variable Y deviated from the linear regression function $\mu_{Y \cdot x} = \beta_0 + \beta_1 x$ by a random amount. Here we consider two ways of generalizing the simple linear regression model. The first way is to replace $\beta_0 + \beta_1 x$ by a nonlinear function of x, and the second is to use a regression function involving more than a single independent variable. After fitting a regression function of the chosen form to the given data, it is of course important to have methods available for making inferences about the parameters of the chosen model. Before these methods are used, though, the data analyst should first assess the adequacy of the chosen model. In Section 13.1, we discuss methods, based primarily on a graphical analysis of the residuals (observed minus predicted y's), for checking model adequacy.

In Section 13.2, we consider nonlinear regression functions of a single independent variable x that are "intrinsically linear." By this we mean that it is possible to transform one or both of the variables so that the relationship between the resulting variables is linear. An alternative class of nonlinear relations is obtained by using polynomial regression functions of the form $\mu_{Y \cdot x} = \beta_0 + \beta_1 x + \beta_2 x^2 + \cdots + \beta_k x^k$; these polynomial models are the subject of Section 13.3. Multiple regression analysis involves building models for relating y to two or more independent variables. The focus in Section 13.4 is on interpretation of various multiple regression models and on understanding and using the regression output from various statistical computer packages. The last section of the chapter surveys some extensions and pitfalls of multiple regression modeling.

A plot of the observed pairs (x_i, y_i) is a necessary first step in deciding on the form of a mathematical relationship between x and y. It is possible to fit many functions other than a linear one ($y = b_0 + b_1 x$) to the data, using either the principle of least squares or another fitting method. Once a function of the chosen form has been fitted, it is important to check the fit of the model to see whether it is in fact appropriate. One way to study the fit is to superimpose a graph of the best-fit function on the scatterplot of the data. However, any tilt or curvature of the best-fit function may obscure some aspects of the fit that should be investigated. Furthermore, the scale on the vertical axis may make it difficult to assess the extent to which observed values deviate from the best-fit function.

Residuals and Standardized Residuals

A more effective approach to assessment of model adequacy is to compute the fitted or predicted values $\hat{y}_i$ and the residuals $e_i = y_i - \hat{y}_i$, and then plot various functions of these computed quantities. We then examine the plots either to confirm our choice of model or for indications that the model is not appropriate. Suppose the simple linear regression model is correct, and let $y = \hat{\beta}_0 + \hat{\beta}_1 x$ be the equation of the estimated regression line. Then the ith residual is $e_i = y_i - (\hat{\beta}_0 + \hat{\beta}_1 x_i)$. To derive properties of the residuals, let $e_i = Y_i - \hat{Y}_i$, represent the ith residual as a random variable (rv) before observations are actually made. Then

$$E(Y_i - \hat{Y}_i) = E(Y_i) - E(\hat{\beta}_0 + \hat{\beta}_1 x_i) = \beta_0 + \beta_1 x_i - (\beta_0 + \beta_1 x_i) = 0 \quad (13.1)$$

Because $\hat{Y}_i (= \hat{\beta}_0 + \hat{\beta}_1 x_i)$ is a linear function of the Y_j's, so is $Y_i - \hat{Y}_i$ (the coefficients depend on the x_j's). Thus the normality of the Y_j's implies that each residual is normally distributed. It can also be shown that

$$V(Y_i - \hat{Y}_i) = \sigma^2 \cdot \left[1 - \frac{1}{n} - \frac{(x_i - \bar{x})^2}{S_{xx}} \right] \quad (13.2)$$

Replacing σ^2 by s^2 and taking the square root of Equation (13.2) gives the estimated standard deviation of a residual.

Let's now standardize each residual by subtracting the mean value (zero) and then dividing by the estimated standard deviation.

The **standardized residuals** are given by

$$e_i^* = \frac{y_i - \hat{y}_i}{s \sqrt{1 - \frac{1}{n} - \frac{(x_i - \bar{x})^2}{S_{xx}}}} \qquad i = 1, \ldots, n \qquad (13.3)$$

If, for example, a particular standardized residual is 1.5, then the residual itself is 1.5 (estimated) standard deviations larger than what would be expected from fitting the correct model. Notice that the variances of the residuals differ from one another. In fact, because there is a $-$ sign in front of $(x_i - \bar{x})^2$, the variance of a residual decreases as x_i moves further away from the center of the data $\bar{x}$. Intuitively, this is because the least squares line is pulled toward an observation whose x_i value lies far to the right or left of other observations in the sample. Computation of the e_i^*'s can be tedious, but the most widely used statistical computer packages will provide these values and construct various plots involving them.

EXAMPLE 13.1 Exercise 19 in Chapter 12 presented data on $x =$ burner area liberation rate and $y = NO_x$ emissions. Here we reproduce the data and give the fitted values, residuals, and standardized residuals. The estimated regression line is $y = -45.55 + 1.71x$, and $r^2 = .961$. The standardized residuals are not a constant multiple of the residuals because the residual variances differ somewhat from one another.

x_i	y_i	$\hat{y}_i$	e_i	e_i^*
100	150	125.6	24.4	.75
125	140	168.4	−28.4	−.84
125	180	168.4	11.6	.35
150	210	211.1	−1.1	−.03
150	190	211.1	−21.1	−.62
200	320	296.7	23.3	.66
200	280	296.7	−16.7	−.47
250	400	382.3	17.7	.50
250	430	382.3	47.7	1.35
300	440	467.9	−27.9	−.80
300	390	467.9	−77.9	−2.24
350	600	553.4	46.6	1.39
400	610	639.0	−29.0	−.92
400	670	639.0	31.0	.99

■

Diagnostic Plots

The basic plots that many statisticians recommend for an assessment of model validity and usefulness are the following:

1. e^* (or e) on the vertical axis versus x on the horizontal axis—that is, a plot of the (x_i, e_i^*) pairs [or the (x_i, e_i) pairs]

2. e^* (or e) on the vertical axis versus $\hat{y}$ on the horizontal axis—that is, a plot of the $(\hat{y}_i, e_i^*)$ pairs [or the $(\hat{y}_i, e_i)$ pairs]

3. $\hat{y}$ on the vertical versus y on the horizontal—that is, a plot of the $(y_i, \hat{y}_i)$ pairs

4. A normal probability plot of the standardized residuals

Plots 1 and 2 are called **residual plots** (against the independent variable and fitted values, respectively). Since $\hat{y} = \hat{\beta}_0 + \hat{\beta}_1 x$ is a linear function of x, the general pattern of points in Plot 2 should be identical to that in Plot 1, though the horizontal scales will differ (in multiple regression, there is a Plot 1 for each predictor, and Plot 2 is a single omnibus picture that combines information from all of those). Provided that the chosen model is correct, neither residual plot should exhibit any discernible pattern. The residuals should be randomly distributed about 0 according to a normal distribution, so all or almost all e^*'s should lie between −2 and +2.

We hope that the fitted model will give predicted y values that are close to their observed counterparts. This would manifest itself in Plot 3 by plotted points falling close to a 45° line. Thus this plot provides a visual assessment of model effectiveness in making predictions. Plot 4 allows the analyst to assess the plausibility of assuming that the random deviation ε in the model equation has a normal distribution. If the pattern in the plot departs substantially from linearity, then the inferential procedures from Chapter 12 based on the t_{n-2} distribution should not be used as a basis for drawing conclusions.

EXAMPLE 13.2
(Example 13.1
continued)

Figure 13.1 presents a scatterplot of the data and the four plots just recommended. The plot of $\hat{y}$ versus y confirms the impression given by r^2 that x is effective in predicting y and also indicates that there is no observed y for which the predicted value is terribly far off the mark. Both residual plots show no unusual pattern or discrepant values. There is one standardized residual slightly outside the interval $(-2, 2)$, but this is not surprising in a sample of size 14. The normal probability plot of the standardized residuals is reasonably straight. In summary, the plots leave us with no qualms about either the appropriateness of a simple linear relationship or the fit to the given data.

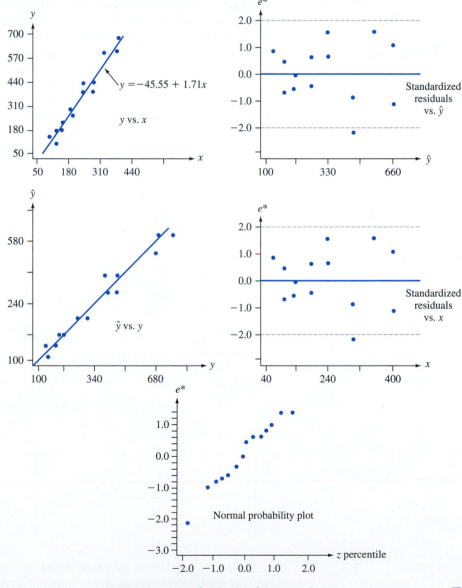

Figure 13.1 Plots for the data from Example 13.1

Difficulties and Remedies

Although we hope that our analysis will yield plots like those of Figure 13.1, quite frequently the plots will suggest one or more of the following difficulties:

1. A nonlinear probabilistic relationship between x and y is appropriate.
2. The variance of ϵ (and of Y) is not a constant σ^2, but instead depends somehow on x.
3. The selected model fits the data well except for a very few discrepant or outlying data values, which may have greatly influenced the choice of the best-fit function.
4. The error variable ϵ does not have a normal distribution.
5. When the subscript i indicates the time order of the observations, the ϵ_i's exhibit dependence over time.
6. One or more relevant independent variables have been omitted from the model.

Figure 13.2 presents residual plots corresponding to items 1–3, 5, and 6. In Chapter 4, we discussed patterns in normal probability plots that cast doubt on the assumption of an underlying normal distribution. Notice that the residuals from the data in Figure 13.2(d) with the circled point included would not by themselves necessarily suggest further analysis, yet when a new line is fit with that point deleted, the new line differs considerably from the original line. This type of behavior is more difficult to identify in multiple regression. It is most likely to arise when there is a single (or very few) data point(s) with independent variable value(s) far removed from the remainder of the data.

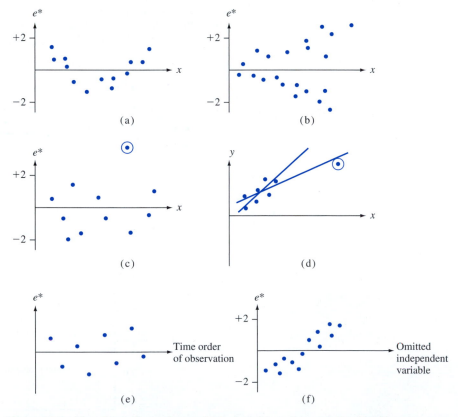

Figure 13.2 Plots that indicate abnormality in data: (a) nonlinear relationship; (b) nonconstant variance; (c) discrepant observation; (d) observation with large influence; (e) dependence in errors; (f) variable omitted

We now indicate briefly what remedies are available for the types of difficulties. For a more comprehensive discussion, one or more of the references on regression analysis should be consulted. If the residual plot looks something like that of Figure 13.2(a), exhibiting a curved pattern, then a nonlinear function of x may be fit.

The residual plot of Figure 13.2(b) suggests that, although a straight-line relationship may be reasonable, the assumption that $V(Y_i) = \sigma^2$ for each i is of doubtful validity. When the assumptions of Chapter 12 are valid, it can be shown that among all unbiased estimators of β_0 and β_1, the ordinary least squares estimators have minimum variance. These estimators give equal weight to each (x_i, Y_i). If the variance of Y increases with x, then Y_i's for large x_i should be given less weight than those with small x_i. This suggests that β_0 and β_1 should be estimated by minimizing

$$f_w(b_0, b_1) = \sum w_i[y_i - (b_0 + b_1x_i)]^2 \tag{13.4}$$

where the w_i's are weights that decrease with increasing x_i. Minimization of Expression (13.4) yields **weighted least squares** estimates. For example, if the standard deviation of Y is proportional to x (for $x > 0$)—that is, $V(Y) = kx^2$—then it can be shown that the weights $w_i = 1/x_i^2$ yield best estimators of β_0 and β_1. Weighted least squares is used quite frequently by econometricians (economists who use statistical methods) to estimate parameters.

When plots or other evidence suggest that the data set contains outliers or points having large influence on the resulting fit, one possible approach is to omit these outlying points and recompute the estimated regression equation. This would certainly be correct if it were found that the outliers resulted from errors in recording data values or experimental errors. If no assignable cause can be found for the outliers, it is still desirable to report the estimated equation both with and without outliers omitted. Yet another approach is to retain possible outliers but to use an estimation principle that puts relatively less weight on outlying values than does the principle of least squares. One such principle is MAD (minimize absolute deviations), which selects $\hat{\beta}_0$ and $\hat{\beta}_1$ to minimize $\Sigma|y_i - (b_0 + b_1x_i)|$. Unlike the estimates of least squares, there are no nice formulas for the MAD estimates; their values must be found by using an iterative computational procedure. Such procedures are also used when it is suspected that the ϵ_i's have a distribution that is not normal but instead have "heavy tails" (making it much more likely than for the normal distribution that discrepant values will enter the sample); robust regression procedures are those that produce reliable estimates for a wide variety of underlying error distributions. Least squares estimators are not robust in the same way that the sample mean $\overline{X}$ is not a robust estimator for μ.

When a plot suggests time dependence in the error terms, an appropriate analysis may involve a transformation of the y's or else a model explicitly including a time variable. Lastly, a plot such as that of Figure 13.2(f), which shows a pattern in the residuals when plotted against an omitted variable, suggests that a multiple regression model that includes the previously omitted variable should be considered.

EXERCISES Section 13.1 (1–14)

1. Suppose the variables x = commuting distance and y = comuting time are related according to the simple linear regression model with $\sigma = 10$.

   a. If $n = 5$ observations are made at the x values $x_1 = 5$, $x_2 = 10$, $x_3 = 15$, $x_4 = 20$, and $x_5 = 25$, calculate the standard deviations of the five corresponding residuals.

   b. Repeat part (a) for $x_1 = 5$, $x_2 = 10$, $x_3 = 15$, $x_4 = 20$, and $x_5 = 50$.

   c. What do the results of parts (a) and (b) imply about the deviation of the estimated line from the observation made at the largest sampled x value?

2. The x values and standardized residuals for the chlorine flow/etch rate data of Exercise 52 (Section 12.4) are displayed in the accompanying table. Construct a standardized residual plot and comment on its appearance.

x	1.50	1.50	2.00	2.50	2.50
$e*$	.31	1.02	−1.15	−1.23	.23

x	3.00	3.50	3.50	4.00
$e*$	.73	−1.36	1.53	.07

3. Example 12.6 presented the residuals from a simple linear regression of moisture content y on filtration rate x.
   a. Plot the residuals against x. Does the resulting plot suggest that a straight-line regression function is a reasonable choice of model? Explain your reasoning.
   b. Using $s = .665$, compute the values of the standardized residuals. Is $e_i^* \approx e_i$'s for $i = 1, \ldots, n$, or are the e_i^*'s not close to being proportional to the e_i's?
   c. Plot the standardized residuals against x. Does the plot differ significantly in general appearance from the plot of part (a)?

4. The accompanying data on y = normalized energy (J/m^2) and x = intraocular pressure (mmHg) appeared in a scatterplot in the article **"Evaluating the Risk of Eye Injuries: Intraocular Pressure During High Speed Projectile Impacts"** (*Current Eye Research*, **2012: 43–49**); an estimated regression function was superimposed on the plot.

x	2761	19764	25713	3980	12782	19008
y	1553	14999	32813	1667	8741	16526

x	20782	19028	14397	9606	3905	25731
y	26770	16526	9868	6640	1220	30730

   a. Here is Minitab output from fitting the simple linear regression model. Does the model appear to specify a useful relationship between the two variables?

```
Predictor      Coef      SE Coef       T       P
Constant      −5090         2257   −2.26   0.048
Pressure     1.2912       0.1347    9.59   0.000
S = 3679.36   R-Sq = 90.2%   R-Sq(adj) = 89.2%
```

   b. The standardized residuals resulting from fitting the simple linear regression model (in the same order as the observations) are .98, −1.57, 1.47, .50, −.76, −.84, 1.47, −.85, −1.03, −.20, .40, and .81. Construct a plot of $e*$ versus x and comment. [*Note:* The model fit in the cited article was not linear.]

5. As the air temperature drops, river water becomes supercooled and ice crystals form. Such ice can significantly affect the hydraulics of a river. The article **"Laboratory Study of Anchor Ice Growth"** (*J. of Cold Regions Engr.*, **2001: 60–66**) described an experiment in which ice thickness (mm) was studied as a function of elapsed time (hr) under specified conditions. The following data was read from a graph in the article: $n = 33$; $x = .17, .33, .50, .67, \ldots, 5.50$; $y = .50, 1.25, 1.50, 2.75, 3.50, 4.75, 5.75, 5.60, 7.00, 8.00, 8.25, 9.50, 10.50,$ 11.00, 10.75, 12.50, 12.25, 13.25, 15.50, 15.00, 15.25, 16.25, 17.25, 18.00, 18.25, 18.15, 20.25, 19.50, 20.00, 20.50, 20.60, 20.50, 19.80.

   a. The r^2 value resulting from a least squares fit is .977. Interpret this value and comment on the appropriateness of assuming an approximate linear relationship.
   b. The residuals, listed in the same order as the x values, are

−1.03	−0.92	−1.35	−0.78	−0.68	−0.11	0.21
−0.59	0.13	0.45	0.06	0.62	0.94	0.80
−0.14	0.93	0.04	0.36	1.92	0.78	0.35
0.67	1.02	1.09	0.66	−0.09	1.33	−0.10
−0.24	−0.43	−1.01	−1.75	−3.14		

Plot the residuals against elapsed time. What does the plot suggest?

6. The accompanying scatterplot is based on data provided by authors of the article **"Spurious Correlation in the USEPA Rating Curve Method for Estimating Pollutant Loads"** (*J. of Envir. Engr.*, **2008: 610–618**); here discharge is in ft³/s as opposed to m³/s used in the article. The point on the far right of the plot corresponds to the observation (140, 1529.35). The resulting standardized residual is 3.10. Minitab flags the observation with an R for large residual and an X for potentially influential observation. Here is some information on the estimated slope:

	Full sample	**(140, 1529.35) deleted**
$\hat{\beta}_1$	9.9050	8.8241
$s_{\hat{\beta}_1}$	.3806	.4734

Does this observation appear to have had a substantial impact on the estimated slope? Explain.

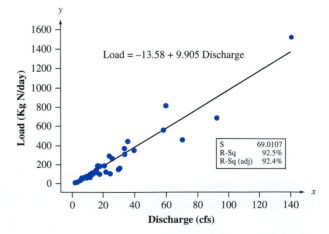

7. Composite honeycomb sandwich panels are widely used in various aerospace structural applications such as ribs, flaps, and rudders. The article **"Core Crush Problem in Manufacturing of Composite Sandwich Structures: Mechanisms and Solutions"** (*Amer. Inst. of Aeronautics and Astronautics J.*, **2006: 901–907**) fit

a line to the following data on x = prepreg thickness (mm) and y = core crush (%):

x	.246	.250	.251	.251	.254	.262	.264	.270
y	16.0	11.0	15.0	10.5	13.5	7.5	6.1	1.7

x	.272	.277	.281	.289	.290	.292	.293
y	3.6	0.7	0.9	1.0	0.7	3.0	3.1

a. Fit the simple linear regression model. What proportion of the observed variation in core crush can be attributed to the model relationship?

b. Construct a scatterplot. Does the plot suggest that a linear probabilistic relationship is appropriate?

c. Obtain the residuals and standardized residuals, and then construct residual plots. What do these plots suggest? What type of function should provide a better fit to the data than does a straight line?

8. Continuous recording of heart rate can be used to obtain information about the level of exercise intensity or physical strain during sports participation, work, or other daily activities. The article **"The Relationship Between Heart Rate and Oxygen Uptake During Non-Steady State Exercise"** (*Ergonomics*, **2000: 1578–1592**) reported on a study to investigate using heart rate response (x, as a percentage of the maximum rate) to predict oxygen uptake (y, as a percentage of maximum uptake) during exercise. The accompanying data was read from a graph in the article.

HR	43.5	44.0	44.0	44.5	44.0	45.0	48.0	49.0
VO_2	22.0	21.0	22.0	21.5	25.5	24.5	30.0	28.0

HR	49.5	51.0	54.5	57.5	57.7	61.0	63.0	72.0
VO_2	32.0	29.0	38.5	30.5	57.0	40.0	58.0	72.0

Use a statistical software package to perform a simple linear regression analysis, paying particular attention to the presence of any unusual or influential observations.

9. Consider the following four (x, y) data sets; the first three have the same x values, so these values are listed only once (**Frank Anscombe, "Graphs in Statistical Analysis," *Amer. Statistician*, 1973: 17–21**):

Data Set	1–3	1	2	3	4	4
Variable	x	y	y	y	x	y
	10.0	8.04	9.14	7.46	8.0	6.58
	8.0	6.95	8.14	6.77	8.0	5.76
	13.0	7.58	8.74	12.74	8.0	7.71
	9.0	8.81	8.77	7.11	8.0	8.84
	11.0	8.33	9.26	7.81	8.0	8.47
	14.0	9.96	8.10	8.84	8.0	7.04
	6.0	7.24	6.13	6.08	8.0	5.25
	4.0	4.26	3.10	5.39	19.0	12.50
	12.0	10.84	9.13	8.15	8.0	5.56
	7.0	4.82	7.26	6.42	8.0	7.91
	5.0	5.68	4.74	5.73	8.0	6.89

For each of these four data sets, the values of the summary statistics Σx_i, Σx_i^2, Σy_i, Σy_i^2, and $\Sigma x_i y_i$ are virtually identical, so all quantities computed from these five will be essentially identical for the four sets—the least squares line ($y = 3 + .5x$), SSE, s^2, r^2, t intervals, t statistics, and so on. The summary statistics provide no way of distinguishing among the four data sets. Based on a scatterplot and a residual plot for each set, comment on the appropriateness or inappropriateness of fitting a straight-line model; include in your comments any specific suggestions for how a "straight-line analysis" might be modified or qualified.

10. a. Show that $\Sigma_{i=1}^{n} e_i = 0$ when the e_i's are the residuals from a simple linear regression.

b. Are the residuals from a simple linear regression independent of one another, positively correlated, or negatively correlated? Explain.

c. Show that $\Sigma_{i=1}^{n} x_i e_i = 0$ for the residuals from a simple linear regression. (This result along with part (a) shows that there are two linear restrictions on the e_i's, resulting in a loss of 2 df when the squared residuals are used to estimate σ^2.)

d. Is it true that $\Sigma_{i=1}^{n} e_i^* = 0$? Give a proof or a counterexample.

11. a. Express the ith residual $Y_i - \hat{Y}_i$ (where $\hat{Y}_i = \hat{\beta}_0 + \hat{\beta}_1 x_i$) in the form $\Sigma c_j Y_j$, a linear function of the Y_j's. Then use rules of variance to verify that $V(Y_i - \hat{Y}_i)$ is given by Expression (13.2).

b. It can be shown that $\hat{Y}_i$ and $Y_i - \hat{Y}_i$ (the ith predicted value and residual) are independent of one another. Use this fact, the relation $Y_i = \hat{Y}_i + (Y_i - \hat{Y}_i)$, and the expression for $V(\hat{Y})$ from Section 12.4 to again verify Expression (13.2).

c. As x_i moves farther away from $\bar{x}$, what happens to $V(\hat{Y}_i)$ and to $V(Y_i - \hat{Y}_i)$?

12. a. Could a linear regression result in residuals 23, -27, 5, 17, -8, 9, and 15? Why or why not?

b. Could a linear regression result in residuals 23, -27, 5, 17, -8, -12, and 2 corresponding to x values 3, -4, 8, 12, -14, -20, and 25? Why or why not? [*Hint*: See Exercise 10.]

13. Recall that $\hat{\beta}_0 + \hat{\beta}_1 x$ has a normal distribution with expected value $\beta_0 + \beta_1 x$ and variance

$$\sigma^2 \left\{ \frac{1}{n} + \frac{(x - \bar{x})^2}{\Sigma(x_i - \bar{x})^2} \right\}$$

so that

$$Z = \frac{\hat{\beta}_0 + \hat{\beta}_1 x - (\beta_0 + \beta_1 x)}{\sigma \left(\frac{1}{n} + \frac{(x - \bar{x})^2}{\Sigma(x_i - \bar{x})^2} \right)^{1/2}}$$

has a standard normal distribution. If $S = \sqrt{SSE/(n - 2)}$ is substituted for σ, the resulting variable has a t distribution with $n - 2$ df. By analogy, what is the distribution of any particular standardized residual? If $n = 25$, what is the probability that a particular standardized residual falls outside the interval $(-2.50, 2.50)$?

14. If there is at least one x value at which more than one observation has been made, there is a formal test procedure for testing $H_0: \mu_{Y \cdot x} = \beta_0 + \beta_1 x$ for some values β_0, β_1 (the true regression function is linear)

versus

H_a: H_0 is not true (the true regression function is not linear)

Suppose observations are made at $x_1, x_2, \ldots, x_c$. Let $Y_{11}, Y_{12}, \ldots, Y_{1n_1}$ denote the n_1 observations when $x = x_1; \ldots; Y_{c1}, Y_{c2}, \ldots, Y_{cn_c}$ denote the n_c observations when $x = x_c$. With $n = \Sigma n_i$ (the total number of observations), SSE has $n - 2$ df. We break SSE into two pieces, SSPE (pure error) and SSLF (lack of fit), as follows:

$$\text{SSPE} = \sum_i \sum_j (Y_{ij} - \overline{Y}_{i \cdot})^2$$
$$= \sum \sum Y_{ij}^2 - \sum n_i \overline{Y}_{i \cdot}^2$$

$$\text{SSLF} = \text{SSE} - \text{SSPE}$$

The n_i observations at x_i contribute $n_i - 1$ df to SSPE, so the number of degrees of freedom for SSPE is $\Sigma_i(n_i - 1) = n - c$, and the degrees of freedom for SSLF is $n - 2 - (n - c) = c - 2$. Let $\text{MSPE} = \text{SSPE}/(n - c)$ and $\text{MSLF} = \text{SSLF}/(c - 2)$. Then it can be shown that

whereas $E(\text{MSPE}) = \sigma^2$ whether or not H_0 is true, $E(\text{MSLF}) = \sigma^2$ if H_0 is true and $E(\text{MSLF}) > \sigma^2$ if H_0 is false.

The test statistic is $F = \text{MSLF}/\text{MSPE}$, and the corresponding P-value is the area under the $F_{c-2, n-c}$ curve to the right of f.

The following data comes from the article **"Changes in Growth Hormone Status Related to Body Weight of Growing Cattle"** (*Growth*, **1977: 241–247**), with x = body weight and y = metabolic clearance rate/body weight.

x	110	110	110	230	230	230	360
y	235	198	173	174	149	124	115

x	360	360	360	505	505	505	505
y	130	102	95	122	112	98	96

(So $c = 4$, $n_1 = n_2 = 3$, $n_3 = n_4 = 4$.)

a. Test H_0 versus H_a at level .05 using the lack-of-fit test just described.

b. Does a scatterplot of the data suggest that the relationship between x and y is linear? How does this compare with the result of part (a)? (A nonlinear regression function was used in the article.)

13.2 Regression with Transformed Variables

The necessity for an alternative to the linear model $Y = \beta_0 + \beta_1 x + \epsilon$ may be suggested either by a theoretical argument or else by examining diagnostic plots from a linear regression analysis. In either case, settling on a model whose parameters can be easily estimated is desirable. An important class of such models is specified by means of functions that are "intrinsically linear."

DEFINITION

A function relating y to x is **intrinsically linear** if, by means of a transformation on x and/or y, the function can be expressed as $y' = \beta_0 + \beta_1 x'$, where x' = the transformed independent variable and y' = the transformed dependent variable.

Four of the most useful intrinsically linear functions are given in Table 13.1. In each case, the appropriate transformation is either a log transformation—either base 10 or natural logarithm (base e)—or a reciprocal transformation. Representative graphs of the four functions appear in Figure 13.3.

For an exponential function relationship, only y is transformed to achieve linearity, whereas for a power function relationship, both x and y are transformed. Because the variable x is in the exponent in an exponential relationship, y increases

Table 13.1 Useful Intrinsically Linear Functions*

Function	Transformation(s) to Linearize	Linear Form
a. Exponential: $y = \alpha e^{\beta x}$	$y' = \ln(y)$	$y' = \ln(\alpha) + \beta x$
b. Power: $y = \alpha x^{\beta}$	$y' = \log(y),\, x' = \log(x)$	$y' = \log(\alpha) + \beta x'$
c. $y = \alpha + \beta \cdot \log(x)$	$x' = \log(x)$	$y = \alpha + \beta x'$
d. Reciprocal: $y = \alpha + \beta \cdot \dfrac{1}{x}$	$x' = \dfrac{1}{x}$	$y = \alpha + \beta x'$

*When $\log(\,\cdot\,)$ appears, either a base 10 or a base e logarithm can be used.

(if $\beta > 0$) or decreases (if $\beta < 0$) much more rapidly as x increases than is the case for the power function, though over a short interval of x values it can be difficult to differentiate between the two functions. Examples of functions that are not intrinsically linear are $y = \alpha + \gamma e^{\beta x}$ and $y = \alpha + \gamma x^{\beta}$.

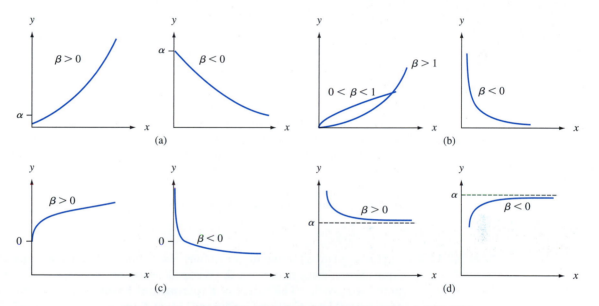

Figure 13.3 Graphs of the intrinsically linear functions given in Table 13.1

Intrinsically linear functions lead directly to probabilistic models that, though not linear in x as a function, have parameters whose values are easily estimated using ordinary least squares.

DEFINITION

> A probabilistic model relating Y to x is **intrinsically linear** if, by means of a transformation on Y and/or x, it can be reduced to a linear probabilistic model $Y' = \beta_0 + \beta_1 x' + \epsilon'$.

The intrinsically linear probabilistic models that correspond to the four functions of Table 13.1 are as follows:

a. $Y = \alpha e^{\beta x} \cdot \epsilon$, a multiplicative exponential model, from which $\ln(Y) = Y' = \beta_0 + \beta_1 x' + \epsilon'$ with $x' = x$, $\beta_0 = \ln(\alpha)$, $\beta_1 = \beta$, and $\epsilon' = \ln(\epsilon)$.

b. $Y = \alpha x^{\beta} \cdot \epsilon$, a multiplicative power model, so that $\log(Y) = Y' = \beta_0 + \beta_1 x' + \epsilon'$ with $x' = \log(x)$, $\beta_0 = \log(x) + \epsilon$, and $\epsilon' = \log(\epsilon)$.

c. $Y = \alpha + \beta \log(x) + \epsilon$, so that $x' = \log(x)$ immediately linearizes the model.

d. $Y = \alpha + \beta \cdot 1/x + \epsilon$, so that $x' = 1/x$ yields a linear model.

The additive exponential and power models, $Y = \alpha e^{\beta x} + \epsilon$ and $Y = \alpha x^{\beta} + \epsilon$, are not intrinsically linear. Notice that both (a) and (b) require a transformation on Y and, as a result, a transformation on the error variable ϵ. In fact, if ϵ has a lognormal distribution (see Chapter 4) with $E(\epsilon) = e^{\sigma^2/2}$ and $V(\epsilon) = \tau^2$ independent of x, then the transformed models for both (a) and (b) will satisfy all the assumptions of Chapter 12 regarding the linear probabilistic model; this in turn implies that all inferences for the parameters of the transformed model based on these assumptions will be valid. If σ^2 is small, $\mu_{Y \cdot x} \approx \alpha e^{\beta x}$ in (a) or αx^{β} in (b).

The major advantage of an intrinsically linear model is that the parameters β_0 and β_1 of the transformed model can be immediately estimated using the principle of least squares simply by substituting x' and y' into the estimating formulas:

$$\hat{\beta}_1 = \frac{\Sigma x_i' y_i' - \Sigma x_i' \Sigma y_i'/n}{\Sigma(x_i')^2 - (\Sigma x_i')^2/n}$$

$$\hat{\beta}_0 = \frac{\Sigma y_i' - \hat{\beta}_1 \Sigma x_i'}{n} = \bar{y}' - \hat{\beta}_1 \bar{x}' \tag{13.5}$$

Parameters of the original nonlinear model can then be estimated by transforming back $\hat{\beta}_0$ and/or $\hat{\beta}_1$ if necessary. Once a prediction interval for y' when $x' = x'^*$ has been calculated, reversing the transformation gives a PI for y itself. In cases (a) and (b), when σ^2 is small, an approximate CI for $\mu_{Y \cdot x^*}$ results from taking antilogs of the limits in the CI for $\beta_0 + \beta_1 x'^*$. (Strictly speaking, taking antilogs gives a CI for the *median* of the Y distribution, i.e., for $\tilde{\mu}_{Y \cdot x^*}$. Because the lognormal distribution is positively skewed, $\mu > \tilde{\mu}$; the two are approximately equal if σ^2 is close to 0.)

EXAMPLE 13.3 Taylor's equation for tool life y as a function of cutting time x states that $xy^c = k$ or, equivalently, that $y = \alpha x^{\beta}$ (see the Wikipedia entry on Tool wear for more information). The article **"The Effect of Experimental Error on the Determination of Optimum Metal Cutting Conditions"** (**J. of Engr. for Industry, 1967: 315–322**) observes that the relationship is not exact (deterministic) and that the parameters α and β must be estimated from data. Thus an appropriate model is the multiplicative power model $Y = \alpha \cdot x^{\beta} \cdot \epsilon$, which the author fit to the accompanying data consisting of 12 carbide tool life observations (Table 13.2). In addition to the x, y, x', and y' values, the predicted transformed values ($\hat{y}'$) and the predicted values on the original scale ($\hat{y}$, after transforming back) are given.

The summary statistics for fitting a straight line to the transformed data are $\Sigma x_i' = 74.41200$, $\Sigma y_i' = 26.22601$, $\Sigma x_i'^2 = 461.75874$, $\Sigma y_i'^2 = 67.74609$, and $\Sigma x_i' y_i' = 160.84601$, so

$$\hat{\beta}_1 = \frac{160.84601 - (74.41200)(26.22601)/12}{461.75874 - (74.41200)^2/12} = -5.3996$$

$$\hat{\beta}_0 = \frac{26.22601 - (-5.3996)(74.41200)}{12} = 35.6684$$

The estimated values of α and β, the parameters of the power function model, are $\hat{\beta} = \hat{\beta}_1 = -5.3996$ and $\hat{\alpha} = e^{\hat{\beta}_0} = 3.094491530 \cdot 10^{15}$. Thus the estimated

Table 13.2 Data for Example 13.3

	x	y	$x' = \ln(x)$	$y' = \ln(y)$	$\hat{y}'$	$\hat{y} = e^{\hat{y}'}$
1	600	2.35	6.39693	.85442	1.12754	3.0881
2	600	2.65	6.39693	.97456	1.12754	3.0881
3	600	3.00	6.39693	1.09861	1.12754	3.0881
4	600	3.60	6.39693	1.28093	1.12754	3.0881
5	500	6.40	6.21461	1.85630	2.11203	8.2650
6	500	7.80	6.21461	2.05412	2.11203	8.2650
7	500	9.80	6.21461	2.28238	2.11203	8.2650
8	500	16.50	6.21461	2.80336	2.11203	8.2650
9	400	21.50	5.99146	3.06805	3.31694	27.5760
10	400	24.50	5.99146	3.19867	3.31694	27.5760
11	400	26.00	5.99146	3.25810	3.31694	27.5760
12	400	33.00	5.99146	3.49651	3.31694	27.5760

regression function is $\hat{\mu}_{Y \cdot x} \approx 3.094491530 \cdot 10^{15} \cdot x^{-5.3996}$. To recapture Taylor's estimated) equation, set $y = 3.094491530 \cdot 10^{15} \cdot x^{-5.3996}$, whence $xy^{.185} = 740$.

Figure 13.4(a) gives a plot of the standardized residuals from the linear regression using transformed variables (for which $r^2 = .922$); there is no apparent pattern in the plot, though one standardized residual is a bit large, and the residuals look as they should for a simple linear regression. Figure 13.4(b) pictures a plot of $\hat{y}$ versus y, which indicates satisfactory predictions on the original scale.

To obtain a confidence interval for median tool life when cutting time is 500, we transform $x = 500$ to $x' = 6.21461$. Then $\hat{\beta}_0 + \hat{\beta}_1 x' = 2.1120$, and a 95% CI for $\beta_0 + \beta_1(6.21461)$ is (from Section 12.4) $2.1120 \pm (2.228)(.0824) = (1.928, 2.296)$. The 95% CI for $\tilde{\mu}_{Y \cdot 500}$ is then obtained by taking antilogs: $(e^{1.928}, e^{2.296}) = (6.876, 9.930)$. It is easily checked that for the transformed data $s^2 = \hat{\sigma}^2 \approx .081$. Because this is quite small, $(6.876, 9.930)$ is an approximate interval for $\mu_{Y \cdot 500}$.

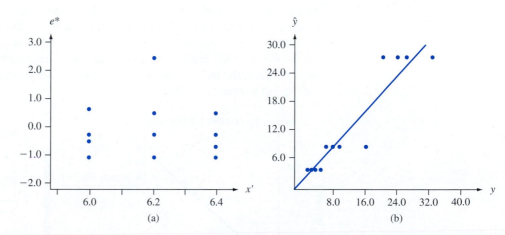

Figure 13.4 (a) Standardized residuals versus x' from Example 13.3; (b) $\hat{y}$ versus y from Example 13.3

EXAMPLE 13.4 The accompanying data on $x =$ length of a scamp (mm) and $y =$ mercury content (mg/kg) was extracted from a graph in the article **"Mercury in Groupers and Sea Basses from the Gulf of Mexico: Relationships with Size, Age, and Feeding Ecology"** (*Transactions of the Amer. Fisheries Soc.*, 2012: 1274–1286).

x	285	297	334	362	407	438	455	486	512
y	.076	.045	.064	.080	.061	.071	.131	.198	.084

x	520	535	560	573	590	598	612	667	690
y	.247	.223	.223	.278	.368	.497	.281	.257	.577

Figure 13.5 displays a scatterplot of the data and a plot of the standardized residuals from a linear regression of y on x, both from Minitab. In the latter plot, most points on the far left and right are above the zero line whereas most points in the middle are below the line. This indication of curvature is more apparent in the residual plot than in the scatterplot.

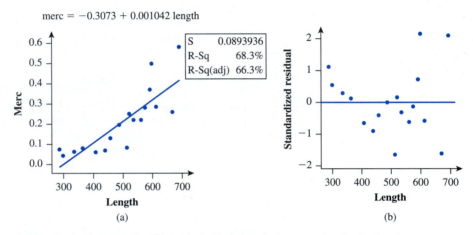

Figure 13.5 (a) Scatterplot; (b) residual plot from a linear regression for the data in Example 13.4

Figure 13.6 shows a scatterplot of $\ln(y)$ versus x and a plot of the standardized residuals resulting from a linear regression of $\ln(y)$ on x. There is no discernible pattern in the latter plot other than pure randomness, and a normal probability plot of the standardized residuals (not shown) has a very substantial linear pattern. The r^2 value for this transformed regression is reasonably impressive, and the P-value for the model utility test is .000. In fact, the cited article actually contained a plot of $\ln(y)$ versus x; the included regression equation is very close to ours, with $r^2 = .792$ for the full sample of 49 observations.

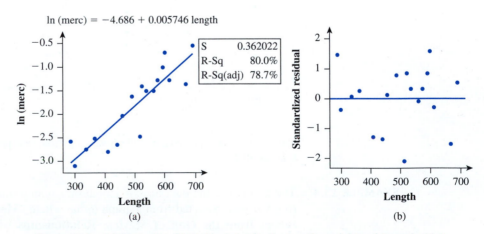

Figure 13.6 (a) Scatterplot; (b) Residual plot for the transformed data of Example 13.4

The estimated untransformed exponential regression function is $y = e^{-4.686 + .005746x} = .009224e^{.005746x}$, from which a point prediction for a future Y corresponding to any particular x can be obtained. A prediction interval for the mercury content of a single fish having length x^* is obtained by first using the simple linear regression to obtain a prediction interval for $\ln(Y)$ and then taking the antilog of the two endpoints. For example, a 95% PI when $x = 500$ is $(e^{-2.6026}, e^{-1.0246}) = (.074, .359)$. ■

In analyzing transformed data, one should keep in mind the following points:

1. Estimating β_1 and β_0 as in (13.5) and then transforming back to obtain estimates of the original parameters is not equivalent to using the principle of least squares directly on the original model. Thus, for the exponential model, we could estimate α and β by minimizing $\Sigma(y_i - \alpha e^{\beta x_i})^2$. Iterative computation would be necessary. In general, $\hat{\alpha} \neq e^{\hat{\beta}_0}$ and $\hat{\beta} \neq \hat{\beta}_1$.

2. If the chosen model is not intrinsically linear, the approach summarized in (13.5) cannot be used. Instead, least squares (or some other fitting procedure) would have to be applied to the untransformed model. Thus, for the additive exponential model $Y = \alpha e^{\beta x} + \epsilon$, least squares would involve minimizing $\Sigma(y_i - \alpha e^{\beta x_i})^2$. Taking partial derivatives with respect to α and β results in two nonlinear normal equations in α and β; these equations must then be solved using an iterative procedure.

3. When the transformed linear model satisfies all the assumptions described in Chapter 12, the method of least squares yields best estimates of the transformed parameters. However, estimates of the original parameters may not be best in any sense, though they will be reasonable. For example, in the exponential model, the estimator $\hat{\alpha} = e^{\hat{\beta}_0}$ will not be unbiased, though it will be the maximum likelihood estimator of α if the error variable ϵ' is normally distributed. Using least squares directly (without transforming) could yield better estimates.

4. If a transformation on y has been made and one wishes to use the standard formulas to test hypotheses or construct CIs, ϵ' should be at least approximately normally distributed. To check this, a normal probability plot of the standardized residuals from the transformed regression should be examined.

5. When y is transformed, the r^2 value from the resulting regression refers to variation in the y_i''s, explained by the transformed regression model. Although a high value of r^2 here indicates a good fit of the estimated original nonlinear model to the observed y_i's, r^2 does not refer to these original observations. Perhaps the best way to assess the quality of the fit is to compute the predicted values $\hat{y}_i'$ using the transformed model, transform them back to the original y scale to obtain $\hat{y}_i$, and then plot $\hat{y}$ versus y. A good fit is then evidenced by points close to the 45° line. One could compute SSE $= \Sigma(y_i - \hat{y}_i)^2$ as a numerical measure of the goodness of fit. When the model was linear, we compared this to SST $= \Sigma(y_i - \bar{y})^2$, the total variation about the horizontal line at height $\bar{y}$; this led to r^2. In the nonlinear case, though, it is not necessarily informative to measure total variation in this way, so an r^2 value is not as useful as in the linear case.

More General Regression Methods

Thus far we have assumed that either $Y = f(x) + \epsilon$ (an additive model) or that $Y = f(x) \cdot \epsilon$ (a multiplicative model). In the case of an additive model, $\mu_{Y \cdot x} = f(x)$, so estimating the regression function $f(x)$ amounts to estimating the curve of mean y values. On occasion, a scatterplot of the data suggests that there is no simple mathematical expression for $f(x)$. Statisticians have recently developed some more flexible methods that permit a wide variety of patterns to be modeled using the same fitting procedure. One such method is **LOWESS** (or LOESS), short for *locally weighted scatterplot smoother*. Let (x^*, y^*) denote a particular one of the n (x, y) pairs in the sample. The $\hat{y}$ value corresponding to (x^*, y^*) is obtained by fitting a straight line using only a specified percentage of the data (e.g., 25%) whose x values are closest to x^*. Furthermore, rather than use "ordinary" least squares, which gives equal weight to all points, those with x values closer to x^* are more heavily weighted than those whose x values are farther away. The height of the resulting line above x^* is the fitted value $\hat{y}^*$. This process is repeated for each of the n points, so n different lines are fit (you surely wouldn't want to do all this by hand). Finally, the fitted points are connected to produce a LOWESS curve.

EXAMPLE 13.5   Weighing large deceased animals found in wilderness areas is usually not feasible, so it is desirable to have a method for estimating weight from various characteristics of an animal that can be easily determined. Minitab has a stored data set consisting of various characteristics for a sample of $n = 143$ wild bears. Figure 13.7(a) displays a scatterplot of $y =$ weight versus $x =$ distance around the chest (chest girth). At first glance, it looks as though a single line obtained from ordinary least squares would effectively summarize the pattern. Figure 13.7(b) shows the LOWESS curve produced by Minitab using a span of 50% [the fit at (x^*, y^*) is determined by the closest 50% of the sample]. The curve appears to consist of two straight line segments joined together above approximately $x = 38$. The steeper line is to the right of 38, indicating that weight tends to increase more rapidly as girth does for girths exceeding 38 in.

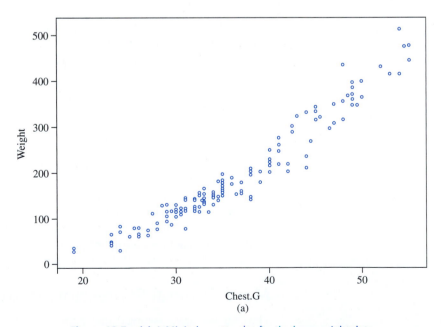

Figure 13.7   **(a)** A Minitab scatterplot for the bear weight data

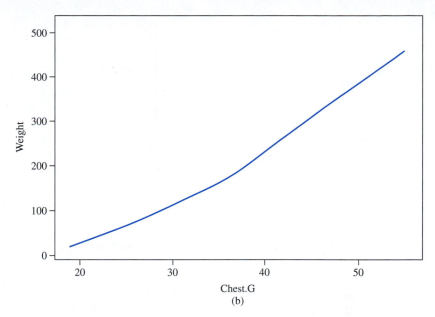

Figure 13.7 (b) A Minitab LOWESS curve for the bear weight data ■

It is complicated to make other inferences (e.g., obtain a CI for a mean y value) based on this general type of regression model. The bootstrap technique mentioned earlier can be used for this purpose.

Logistic Regression

The simple linear regression model is appropriate for relating a quantitative response variable to a quantitative predictor x. Consider now a dichotomous response variable with possible values 1 and 0 corresponding to success and failure. Let $p = P(S) = P(Y = 1)$. Frequently, the value of p will depend on the value of some quantitative variable x. For example, the probability that a car needs warranty service of a certain kind might well depend on the car's mileage, or the probability of avoiding an infection of a certain type might depend on the dosage in an inoculation. Instead of using just the symbol p for the success probability, we now use $p(x)$ to emphasize the dependence of this probability on the value of x. The simple linear regression equation $Y = \beta_0 + \beta_1 x + \epsilon$ is no longer appropriate, for taking the mean value on each side of the equation gives

$$\mu_{Y \cdot x} = 1 \cdot p(x) + 0 \cdot (1 - p(x)) = p(x) = \beta_0 + \beta_1 x$$

Whereas $p(x)$ is a probability and therefore must be between 0 and 1, $\beta_0 + \beta_1 x$ need not be in this range.

Instead of letting the mean value of Y be a linear function of x, we now consider a model in which some function of the mean value of Y is a linear function of x. In other words, we allow $p(x)$ to be a function of $\beta_0 + \beta_1 x$ rather than $\beta_0 + \beta_1 x$ itself. A function that has been found quite useful in many applications is the **logit function**

$$p(x) = \frac{e^{\beta_0 + \beta_1 x}}{1 + e^{\beta_0 + \beta_1 x}}$$

Figure 13.8 shows a graph of $p(x)$ for particular values of β_0 and β_1 with $\beta_1 > 0$. As x increases, the probability of success increases. For β_1 negative, the success probability would be a decreasing function of x.

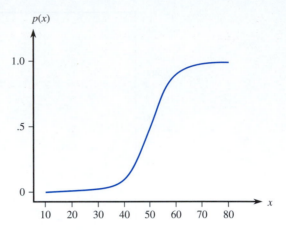

Figure 13.8 A graph of a logit function

Logistic regression means assuming that $p(x)$ is related to x by the logit function. Straightforward algebra shows that

$$\frac{p(x)}{1 - p(x)} = e^{\beta_0 + \beta_1 x}$$

The expression on the left-hand side is called the *odds*. If, for example, $\dfrac{p(60)}{1 - p(60)} = 3$, then when $x = 60$ a success is three times as likely as a failure. We now see that the logarithm of the odds is a linear function of the predictor. In particular, the slope parameter β_1 is the change in the log odds associated with a one-unit increase in x. This implies that the odds itself changes by the multiplicative factor e^{β_1} when x increases by 1 unit.

Fitting the logistic regression to sample data requires that the parameters β_0 and β_1 be estimated. This is usually done using the maximum likelihood technique described in Chapter 6. The details are quite involved, but fortunately the most popular statistical computer packages will do this on request and provide quantitative and pictorial indications of how well the model fits.

EXAMPLE 13.6 Here is data, in the form of a comparative stem-and-leaf display, on launch temperature and the incidence of failure of O-rings in 23 space shuttle launches prior to the Challenger disaster of 1986 (Y = yes, failed; N = no, did not fail). Observations on the left side of the display tend to be smaller than those on the right side.

Y		N	
873	5		
3	6	677789	Stem: Tens digit
500	7	002356689	Leaf : Ones digit
	8	1	

Figure 13.9 shows Minitab output for a logistic regression analysis and a graph of the estimated logit function from the R software. We have chosen to let p denote the probability of failure. The graph of $\hat{p}$ decreases as temperature increases because failures tended to occur at lower temperatures than did successes. The estimate of β_1 and its estimated standard deviation are $\hat{\beta}_1 = -.232$ and $s_{\hat{\beta}_1} = .1082$, respectively.

We assume that the sample size n is large enough here so that $\hat{\beta}_1$ has approximately a normal distribution. If $\beta_1 = 0$ (i.e., temperature does not affect the likelihood of O-ring failure), the test statistic $Z = \hat{\beta}_1/s_{\hat{\beta}}$ has approximately a standard normal distribution. The reported value of this ratio is $z = -2.14$, with a corresponding two-tailed P-value of .032 (some packages report a chi-square value which is just z^2, with the same P-value). At significance level .05, we reject the null hypothesis of no temperature effect.

Binary Logistic Regression: failure versus temp

```
Logistic Regression Table
                                            Odds    95%    CI
Predictor      Coef   SE Coef        Z      P  Ratio  Lower  Upper
Constant    15.0429   7.37862     2.04  0.041
temp        -0.232163 0.108236   -2.14  0.032   0.79   0.64   0.98

Goodness-of-Fit Tests
Method           Chi-Square   DF      P
Pearson            11.1303    14   0.676
Deviance           11.9974    14   0.607
Hosmer-Lemeshow     9.7119     8   0.286
```

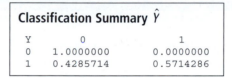

```
Classification Summary Ŷ

Y          0                  1
0     1.0000000          0.0000000
1     0.4285714          0.5714286
```

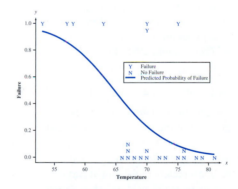

Figure 13.9   *Logistic regression output from Minitab for Example 13.6, and graph of estimated logistic function and classification probabilities from R*

The estimated odds of failure for any particular temperature value x is

$$\frac{p(x)}{1 - p(x)} = e^{15.0429 - .232163x}$$

This implies that the *odds ratio*—the odds of failure at a temperature of $x + 1$ divided by the odds of failure at a temperature of x—is

$$\frac{p(x + 1)/[1 - p(x + 1)]}{p(x)/[1 - p(x)]} = e^{-.232163} = .7928$$

The interpretation is that for each additional degree of temperature, we estimate that the odds of failure will decrease by a factor of .79 (21%). A 95% CI for the true odds ratio also appears on output. In addition, Minitab provides three different ways of assessing model lack-of-fit: the Pearson, deviance, and Hosmer-Lemeshow tests. Large P-values are consistent with a good model. These tests are useful in multiple logistic regression, where there is more than one predictor in the model relationship so there is no single graph like that of Figure 13.9(b). Various diagnostic plots are also available.

The R output provides information based on classifying an observation as a failure if the estimated $p(x)$ is at least .5 and as a non-failure otherwise. Since $p(x) = .5$ when $x = 64.80$, three of the seven failures (Ys in the graph) would be misclassified as non-failures (a misclassification proportion of .429), whereas none of the non-failure observations would be misclassified. A better way to assess the likelihood of misclassification is to use *cross-validation:* Remove the first observation from the sample, estimate the relationship, then classify the first observation based on this estimated relationship, and repeat this process with each of the other sample observations (so a sample observation does not affect its own classification).

The launch temperature for the *Challenger* mission was only 31°F. This temperature is much smaller than any value in the sample, so it is dangerous to extrapolate the estimated relationship. Nevertheless, it appears that O-ring failure is virtually a sure thing for a temperature this small. ∎

EXERCISES Section 13.2 (15–25)

15. No tortilla chip aficionado likes soggy chips, so it is important to find characteristics of the production process that produce chips with an appealing texture. The following data on x = frying time (sec) and y = moisture content (%) appeared in the article **"Thermal and Physical Properties of Tortilla Chips as a Function of Frying Time"** (*J. of Food Processing and Preservation*, 1995: 175–189).

x	5	10	15	20	25	30	45	60
y	16.3	9.7	8.1	4.2	3.4	2.9	1.9	1.3

    a. Construct a scatterplot of y versus x and comment.
    b. Construct a scatterplot of the $(\ln(x), \ln(y))$ pairs and comment.
    c. What probabilistic relationship between x and y is suggested by the linear pattern in the plot of part (b)?
    d. Predict the value of moisture content when frying time is 20, in a way that conveys information about reliability and precision.
    e. Analyze the residuals from fitting the simple linear regression model to the transformed data and comment.

16. Polyester fiber ropes are increasingly being used as components of mooring lines for offshore structures in deep water. The authors of the paper **"Quantifying the Residual Creep Life of Polyester Mooring Ropes"** (*Intl. J. of Offshore and Polar Exploration*, 2005: 223–228) used the accompanying data as a basis for studying how time to failure (hr) depended on load (% of breaking load):

x	77.7	77.8	77.9	77.8	85.5	85.5
y	5.067	552.056	127.809	7.611	.124	.077

x	89.2	89.3	73.1	85.5	89.2	85.5
y	.008	.013	49.439	.503	.362	9.930

x	89.2	85.5	89.2	82.3	82.0	82.3
y	.677	5.322	.289	53.079	7.625	155.299

A linear regression of log(time) versus load was fit. The investigators were particularly interested in estimating the slope of the true regression line relating these variables. Investigate the quality of the fit, estimate the slope, and predict time to failure when load is 80, in a way that conveys information about reliability and precision.

17. The following data on mass rate of burning x and flame length y is representative of that which appeared in the article **"Some Burning Characteristics of Filter Paper"** (*Combustion Science and Technology*, 1971: 103–120):

x	1.7	2.2	2.3	2.6	2.7	3.0	3.2
y	1.3	1.8	1.6	2.0	2.1	2.2	3.0

x	3.3	4.1	4.3	4.6	5.7	6.1
y	2.6	4.1	3.7	5.0	5.8	5.3

    a. Estimate the parameters of a power function model.
    b. Construct diagnostic plots to check whether a power function is an appropriate model choice.
    c. Test $H_0: \beta = 4/3$ versus $H_a: \beta < 4/3$, using a level .05 test.
    d. Test the null hypothesis that states that the median flame length when burning rate is 5.0 is twice the median flame length when burning rate is 2.5 against the alternative that this is not the case.

18. Failures in aircraft gas turbine engines due to high cycle fatigue is a pervasive problem. The article **"Effect of Crystal Orientation on Fatigue Failure of Single Crystal Nickel Base Turbine Blade Superalloys"** (*J. of Engineering for Gas Turbines and Power*, 2002: 161–176) gave the accompanying data and fit a nonlinear regression model in order to predict strain amplitude from cycles to failure. Fit an appropriate model, investigate the quality of the fit, and predict amplitude when cycles to failure = 5000.

Obs	Cycfail	Strampl	Obs	Cycfail	Strampl
1	1326	.01495	11	7356	.00576
2	1593	.01470	12	7904	.00580
3	4414	.01100	13	79	.01212
4	5673	.01190	14	4175	.00782
5	29516	.00873	15	34676	.00596
6	26	.01819	16	114789	.00600
7	843	.00810	17	2672	.00880
8	1016	.00801	18	7532	.00883
9	3410	.00600	19	30220	.00676
10	7101	.00575			

19. Thermal endurance tests were performed to study the relationship between temperature and lifetime of polyester enameled wire (**"Thermal Endurance of Polyester Enameled Wires Using Twisted Wire Specimens," *IEEE Trans. Insulation*, 1965: 38–44**), resulting in the following data.

Temp.	200	200	200	200	200	200
Lifetime	5933	5404	4947	4963	3358	3878

Temp.	220	220	220	220	220	220
Lifetime	1561	1494	747	768	609	777

Temp.	240	240	240	240	240	240
Lifetime	258	299	209	144	180	184

a. Does a scatterplot of the data suggest a linear probabilistic relationship between lifetime and temperature?

b. What model is implied by a linear relationship between expected ln(lifetime) and 1/temperature? Does a scatterplot of the transformed data appear consistent with this relationship?

c. Estimate the parameters of the model suggested in part (b). What lifetime would you predict for a temperature of 220?

d. Because there are multiple observations at each x value, the method in Exercise 14 can be used to test the null hypothesis that states that the model suggested in part (b) is correct. Carry out the test at level .01.

20. Exercise 14 presented data on body weight x and metabolic clearance rate/body weight y. Consider the following intrinsically linear functions for specifying the relationship between the two variables: (a) ln(y) versus x, (b) ln(y) versus ln(x), (c) y versus ln(x), (d) y versus $1/x$, and (e) ln(y) versus $1/x$. Use any appropriate diagnostic plots and analyses to decide which of these functions you would select to specify a probabilistic model. Explain your reasoning.

21. Mineral mining is one of the most important economic activities in Chile. Mineral products are frequently found in saline systems composed largely of natural nitrates. Freshwater is often used as a leaching agent for the extraction of nitrate, but the Chilean mining regions have scarce freshwater resources. An alternative leaching agent

is seawater. The authors of **"Recovery of Nitrates from Leaching Solutions Using Seawater" (*Hydrometallurgy*, 2013: 100–105)** evaluated the recovery of nitrate ions from discarded salts using freshwater and seawater leaching agents. Here is data on x = leaching time (h) and y = nitrate extraction percentage (seawater):

x	25.5	31.5	37.5	43.5	49.5	55.5
y	26.4	40.1	50.2	57.4	62.7	67.3
x	61.5	67.5	73.5	79.5	85.5	91.5
y	71.4	74.7	77.8	80.3	82.3	84.1
x	97.5	103.5	109.5	115.5	121.5	127.5
y	85.5	86.6	87.9	89.0	89.9	90.6
x	133.5	139.5	145.5	151.5	157.5	
y	91.2	91.8	92.3	92.8	93.3	

a. Construct a scatterplot. If the simple linear regression model were fit to this data, what would a plot of the (x, e^*) pairs look like?

b. Construct a scatterplot of y versus $x' = 1/x$ and speculate on the value of r^2 after fitting the simple linear regression model to the transformed data.

c. Obtain a 95% prediction interval for nitrogen extraction percentage when leaching time = 100 h.

22. In each of the following cases, decide whether the given function is intrinsically linear. If so, identify x' and y', and then explain how a random error term ϵ can be introduced to yield an intrinsically linear probabilistic model.

a. $y = 1/(\alpha + \beta x)$

b. $y = 1/(1 + e^{\alpha + \beta x})$

c. $y = e^{e^{\alpha + \beta x}}$ (a Gompertz curve)

d. $y = \alpha + \beta e^{\lambda x}$

23. Suppose x and y are related according to a probabilistic exponential model $Y = \alpha e^{\beta x} \cdot \epsilon$, with $V(\epsilon)$ a constant independent of x (as was the case in the simple linear model $Y = \beta_0 + \beta_1 x + \epsilon$). Is $V(Y)$ a constant independent of x [as was the case for $Y = \beta_0 + \beta_1 x + \epsilon$, where $V(Y) = \sigma^2$]? Explain your reasoning. Draw a picture of a prototype scatterplot resulting from this model. Answer the same questions for the power model $Y = \alpha x^\beta \cdot \epsilon$.

24. Kyphosis refers to severe forward flexion of the spine following corrective spinal surgery. A study carried out to determine risk factors for kyphosis reported the accompanying ages (months) for 40 subjects at the time of the operation; the first 18 subjects did have kyphosis and the remaining 22 did not.

Kyphosis	12	15	42	52	59	73
	82	91	96	105	114	120
	121	128	130	139	139	157

No kyphosis	1	1	2	8	11	18
	22	31	37	61	72	81
	97	112	118	127	131	140
	151	159	177	206		

Use the Minitab logistic regression output on the next page to decide whether age appears to have a significant impact on the presence of kyphosis.

Logistic regression table for Exercise 24

Predictor	Coef	StDev	Z	P	Odds Ratio	95% CI Lower	Upper
Constant	−0.5727	0.6024	−0.95	0.342			
age	0.004296	0.005849	0.73	0.463	1.00	0.99	1.02

25. The article **"Acceptable Noise Levels for Construction Site Offices"** (*Building Serv. Engr. Res. Tech.*, 2009: 87–94) analyzed responses from a sample of 77 individuals, each of whom was asked to say whether a particular noise level (dBA) to which he/she had been exposed was acceptable or unacceptable. Here is data provided by the article's authors:

Acceptable:

```
55.3  55.3  55.3  55.9  55.9  55.9  55.9  56.1  56.1  56.1  56.1
56.1  56.1  56.8  56.8  57.0  57.0  57.0  57.8  57.8  57.8  57.9
57.9  57.9  58.8  58.8  58.8  59.8  59.8  59.8  62.2  62.2  65.3
65.3  65.3  65.3  68.7  69.0  73.0  73.0
```

Unacceptable:

```
63.8  63.8  63.8  63.9  63.9  63.9  64.7  64.7  64.7  65.1  65.1
65.1  67.4  67.4  67.4  67.4  68.7  68.7  68.7  70.4  70.4  71.2
71.2  73.1  73.1  74.6  74.6  74.6  74.6  79.3  79.3  79.3  79.3
79.3  83.0  83.0  83.0
```

Interpret the accompanying Minitab logistic regression output, and sketch a graph of the estimated probability of a noise level being acceptable as a function of the level.

Logistic regression table for Exercise 25

Predictor	Coef	SE Coef	Z	P	Odds Ratio	95% CI Lower	Upper
Constant	23.2124	5.05095	4.60	0.000			
noise level	−0.359441	0.0785031	−4.58	0.000	0.70	0.60	0.81

13.3 Polynomial Regression

The nonlinear yet intrinsically linear models of Section 13.2 involved functions of the independent variable x that were either strictly increasing or strictly decreasing. In many situations, either theoretical reasoning or else a scatterplot of the data suggests that the true regression function $\mu_{Y \cdot x}$ has one or more peaks or valleys—that is, at least one relative minimum or maximum. In such cases, a polynomial function $y = \beta_0 + \beta_1 x + \cdots + \beta_k x^k$ may provide a satisfactory approximation to the true regression function.

DEFINITION

> The **kth-degree polynomial regression model equation** is
>
> $$Y = \beta_0 + \beta_1 x + \beta_2 x^2 + \cdots + \beta_k x^k + \epsilon \qquad (13.6)$$
>
> where ϵ is a normally distributed random variable with
>
> $$\mu_\epsilon = 0 \quad \sigma_\epsilon^2 = \sigma^2 \qquad (13.7)$$

From (13.6) and (13.7), it follows immediately that

$$\mu_{Y \cdot x} = \beta_0 + \beta_1 x + \cdots + \beta_k x^k \qquad \sigma_{Y \cdot x}^2 = \sigma^2 \qquad (13.8)$$

In words, the expected value of Y is a kth-degree polynomial function of x, whereas the variance of Y, which controls the spread of observed values about the regression function, is the same for each value of x. The observed pairs $(x_1, y_1), \ldots, (x_n, y_n)$ are assumed to have been generated independently from the model (13.6). Figure 13.10

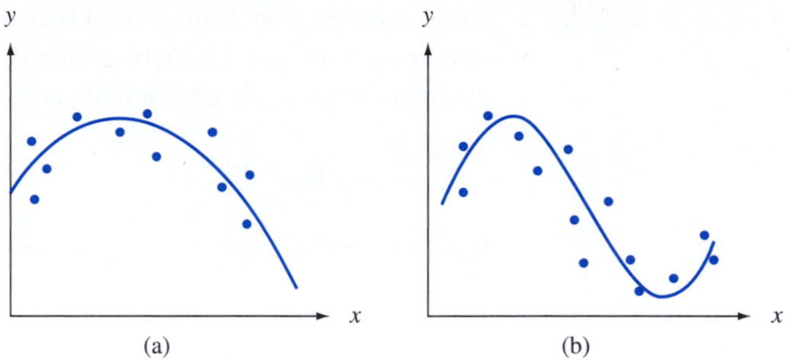

Figure 13.10 (a) Quadratic regression model; (b) cubic regression model

illustrates both a quadratic and cubic model; very rarely in practice is it necessary to go beyond $k = 3$.

Estimating Parameters

To estimate the β's, consider a trial regression function $y = b_0 + b_1x + \cdots + b_kx^k$. Then the goodness of fit of this function to the observed data can be assessed by computing the sum of squared deviations

$$f(b_0, b_1, \ldots, b_k) = \sum_{i=1}^{n} [y_i - (b_0 + b_1x_i + b_2x_i^2 + \cdots + b_kx_i^k)]^2 \qquad (13.9)$$

According to the principle of least squares, the estimates $\hat{\beta}_0, \hat{\beta}_1, \ldots, \hat{\beta}_k$ are those values of $b_0, b_1, \ldots, b_k$ that minimize Expression (13.9). It should be noted that when $x_1, x_2, \ldots, x_n$ are all different, there is a polynomial of degree $n - 1$ that fits the data perfectly, so that the minimizing value of (13.9) is 0 when $k = n - 1$. However, in virtually all applications, the polynomial model (13.6) with large k is quite unrealistic.

To find the minimizing values in (13.9), take the $k + 1$ partial derivatives $\partial f/\partial b_0, \partial f/\partial b_1, \ldots, \partial f/\partial b_k$ and equate them to 0. This gives a system of normal equations for the estimates. Because the trial function $b_0 + b_1x + \cdots + b_kx^k$ is linear in $b_0, \ldots, b_k$ (though not in x), the $k + 1$ normal equations are linear in these unknowns:

$$\begin{aligned}
b_0n + b_1\Sigma x_i + b_2\Sigma x_i^2 + \cdots + b_k\Sigma x_i^k &= \Sigma y_i \\
b_0\Sigma x_i + b_1\Sigma x_i^2 + b_2\Sigma x_i^3 + \cdots + b_k\Sigma x_i^{k+1} &= \Sigma x_iy_i \\
\vdots \qquad \vdots \qquad \vdots & \\
b_0\Sigma x_i^k + b_1\Sigma x_i^{k+1} + \cdots + b_k\Sigma x_i^{2k} &= \Sigma x_i^k y_i
\end{aligned} \qquad (13.10)$$

All standard statistical computer packages will automatically solve the equations in (13.10) and provide the estimates as well as much other information.*

EXAMPLE 13.7 The article **"Residual Stresses and Adhesion of Thermal Spray Coatings"** (**Surface Engineering, 2005: 35–40**) considered the relationship between the thickness (μm) of NiCrAl coatings deposited on stainless steel substrate and corresponding bond strength (MPa). The following data was read from a plot in the paper:

Thickness	220	220	220	220	370	370	370	370	440	440
Strength	24.0	22.0	19.1	15.5	26.3	24.6	23.1	21.2	25.2	24.0

Thickness	440	440	680	680	680	680	860	860	860	860
Strength	21.7	19.2	17.0	14.9	13.0	11.8	12.2	11.2	6.6	2.8

* We will see in Section 13.4 that polynomial regression is a special case of multiple regression, so a command appropriate for this latter task is generally used.

The scatterplot in Figure 13.11(a) supports the choice of the quadratic regression model. Figure 13.11(b) contains Minitab output from a fit of this model. The estimated regression coefficients are

$$\hat{\beta}_0 = 14.521 \qquad \hat{\beta}_1 = .04323 \qquad \hat{\beta}_2 = -.00006001$$

from which the estimated regression function is

$$y = 14.521 + .04323x - .00006001x^2$$

Substitution of the successive x values $220, 220, \ldots, 860$, and 860 into this function gives the predicted values $\hat{y}_1 = 21.128, \ldots, \hat{y}_{20} = 7.321$, and the residuals $y_1 - \hat{y}_1 = 2.872, \ldots, y_{20} - \hat{y}_{20} = -4.521$ result from subtraction. Figure 13.12 shows a plot of the standardized residuals versus $\hat{y}$ and also a normal probability plot of the standardized residuals, both of which validate the quadratic model.

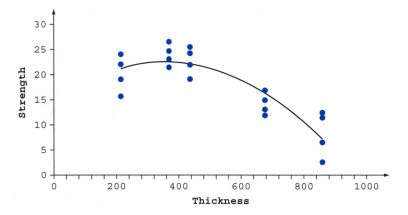

```
The regression equation is
strength = 14.5 + 0.0432 thickness − 0.000060 thicksqd

Predictor                 Coef          SE Coef          T           P
Constant                14.521            4.754        3.05       0.007
thickness              0.04323          0.01981        2.18       0.043
thicksqd            −0.00006001       0.00001786       −3.36       0.004

S = 3.26937      R-Sq = 78.0%          R-Sq(adj) = 75.4%

Analysis of Variance

Source                DF             SS             MS          F           P
Regression             2         643.29         321.65      30.09       0.000
Residual Error        17         181.71          10.69
Total                 19         825.00

Predicted Values for New Observations

New
Obs          Fit       SE Fit               95% CI                   95% PI
  1       21.136        1.167      (18.674, 23.598)        (13.812, 28.460)
  2       10.704        1.189      ( 8.195, 13.212)        ( 3.364, 18.043)

Values of Predictors for New Observations

New
Obs           thickness               thicksqd
  1                 500                 250000
  2                 800                 640000
```

Figure 13.11 Scatterplot of data from Example 13.7 and Minitab output from fit of quadratic model

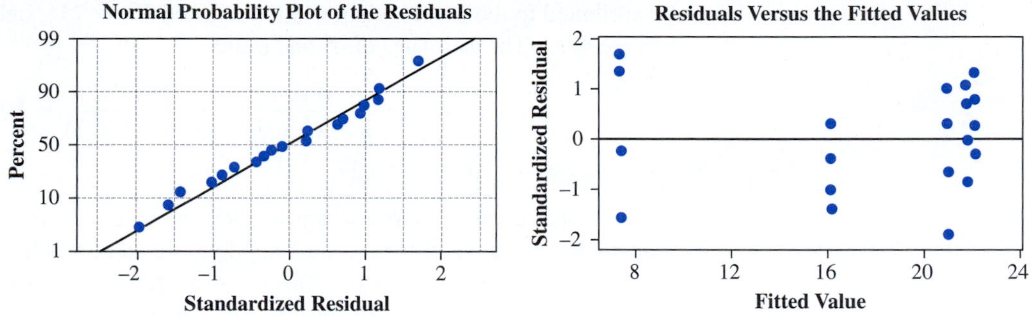

Normal Probability Plot of the Residuals

Residuals Versus the Fitted Values

Figure 13.12 Diagnostic plots for quadratic model fit to data of Example 13.7 ∎

$\hat{\sigma}^2$ and R^2

To make further inferences, the error variance σ^2 must be estimated. With $\hat{y}_i = \hat{\beta}_0 + \hat{\beta}_1 x_i + \cdots + \hat{\beta}_k x_i^k$, the ith residual is $y_i - \hat{y}_i$, and the sum of squared residuals (error sum of squares) is $\text{SSE} = \Sigma(y_i - \hat{y}_i)^2$. The estimate of σ^2 is then

$$\hat{\sigma}^2 = s^2 = \frac{\text{SSE}}{n - (k + 1)} = \text{MSE} \tag{13.11}$$

where the denominator $n - (k + 1)$ is used because $k + 1$ df are lost in estimating $\beta_0, \beta_1, \ldots, \beta_k$.

If we again let $\text{SST} = \Sigma(y_i - \bar{y})^2$, then SSE/SST is the proportion of the total variation in the observed y_i's that is not explained by the polynomial model. The quantity $1 - \text{SSE/SST}$, the proportion of variation explained by the model, is called the **coefficient of multiple determination** and is denoted by R^2.

Consider fitting a cubic model to the data in Example 13.7. Because this model includes the quadratic as a special case, the fit will be at least as good as the fit to a quadratic. More generally, with SSE_k = the error sum of squares from a kth-degree polynomial, $\text{SSE}_{k'} \leq \text{SSE}_k$ and $R^2_{k'} \geq R^2_k$ whenever $k' > k$. Because the objective of regression analysis is to find a model that is both simple (relatively few parameters) and provides a good fit to the data, a higher-degree polynomial may not specify a better model than a lower-degree model despite its higher R^2 value. To balance the cost of using more parameters against the gain in R^2, many statisticians use the **adjusted coefficient of multiple determination**

$$\text{adjusted } R^2 = 1 - \frac{n - 1}{n - (k + 1)} \cdot \frac{\text{SSE}}{\text{SST}} = \frac{(n - 1)R^2 - k}{n - 1 - k} \tag{13.12}$$

Adjusted R^2 adjusts the proportion of unexplained variation upward [since the ratio $(n - 1)/(n - k - 1)$ exceeds 1], which results in *adjusted $R^2 < R^2$*. For example, if $R^2_2 = .66$, $R^2_3 = .70$, and $n = 10$, then

$$\text{adjusted } R^2_2 = \frac{9(.66) - 2}{10 - 3} = .563 \qquad \text{adjusted } R^2_3 = \frac{9(.70) - 3}{10 - 4} = .550$$

Thus the small gain in R^2 in going from a quadratic to a cubic model is not enough to offset the cost of adding an extra parameter to the model.

EXAMPLE 13.8
(Example 13.7 continued)

SSE and SST are typically found on computer output in an ANOVA table. Figure 13.11(b) gives $\text{SSE} = 181.71$ and $\text{SST} = 825.00$ for the bond strength data, from which $R^2 = 1 - 181.71/825.00 = .780$ (alternatively, $R^2 = \text{SSR/SST} = 643.29/825.00 = .780$). Thus 78.0% of the observed variation in bond strength can be attributed to the model relationship. Adjusted $R^2 = .754$, only a small downward change in R^2. The estimates of σ^2 and σ are

$$\hat{\sigma}^2 = s^2 = \frac{SSE}{n - (k + 1)} = \frac{181.71}{20 - (2 + 1)} = 10.69$$

$$\hat{\sigma} = s = 3.27 \qquad \blacksquare$$

Besides computing R^2 and adjusted R^2, one should examine the usual diagnostic plots to determine whether model assumptions are valid or whether modification may be appropriate (see Figure 13.12). There is also a formal test of model utility, an F test based on the ANOVA sums of squares. Since polynomial regression is a special case of multiple regression, we defer discussion of this test to the next section.

Statistical Intervals and Test Procedures

Because the y_i's appear in the normal equations (13.10) only on the right-hand side and in a linear fashion, the resulting estimates $\hat{\beta}_0, \ldots, \hat{\beta}_k$ are themselves linear functions of the y_i's. Thus the estimators are linear functions of the Y_i's, so each $\hat{\beta}_i$ has a normal distribution. It can also be shown that each $\hat{\beta}_i$ is an unbiased estimator of β_i.

Let $\sigma_{\hat{\beta}_i}$ denote the standard deviation of the estimator $\hat{\beta}_i$. This standard deviation has the form

$$\sigma_{\hat{\beta}_i} = \sigma \cdot \left\{ \begin{array}{l} \text{a complicated expression involving } all \\ x_j\text{'s}, x_j^2\text{'s}, \ldots, \text{ and } x_j^k\text{'s} \end{array} \right\}$$

Fortunately, the expression in braces has been programmed into all of the most frequently used statistical software packages. The estimated standard deviation of $\hat{\beta}_i$ results from substituting s in place of σ in the expression for $\sigma_{\hat{\beta}_i}$. These estimated standard deviations $s_{\hat{\beta}_0}, s_{\hat{\beta}_1}, \ldots$, and $s_{\hat{\beta}_k}$ appear in output from all the aforementioned statistical packages. Let $S_{\hat{\beta}_i}$ denote the estimator of $\sigma_{\hat{\beta}_i}$—that is, the random variable whose observed value is $s_{\hat{\beta}_i}$. Then it can be shown that the standardized variable

$$T = \frac{\hat{\beta}_i - \beta_i}{S_{\hat{\beta}_i}} \qquad (13.13)$$

has a t distribution based on $n - (k + 1)$ df. This leads to the following inferential procedures.

A $100(1 - \alpha)\%$ CI for β_i, the coefficient of x^i in the polynomial regression function, is

$$\hat{\beta}_i \pm t_{\alpha/2, n-(k+1)} \cdot s_{\hat{\beta}_i}$$

A test of $H_0: \beta_i = \beta_{i0}$ is based on the t statistic value

$$t = \frac{\hat{\beta}_i - \beta_{i0}}{s_{\hat{\beta}_i}}$$

The test is based on $n - (k + 1)$ df and is upper-, lower-, or two-tailed according to whether the inequality in H_a is $>$, $<$, or $\neq$.

A point estimate of $\mu_{Y \cdot x}$—that is, of $\beta_0 + \beta_1 x + \cdots + \beta_k x^k$—is $\hat{\mu}_{Y \cdot x} = \hat{\beta}_0 + \hat{\beta}_1 x + \cdots + \hat{\beta}_k x^k$. The estimated standard deviation of the corresponding estimator is rather complicated. Many computer packages will give this estimated standard deviation for *any* x value upon request. This, along with an appropriate standardized t variable, can be used to justify the following procedures.

Let x^* denote a specified value of x. A $100(1 - \alpha)\%$ CI for $\mu_{Y \cdot x^*}$ is

$$\hat{\mu}_{Y \cdot x^*} \pm t_{\alpha/2, n-(k+1)} \cdot \begin{Bmatrix} \text{estimated SD of} \\ \hat{\mu}_{Y \cdot x^*} \end{Bmatrix}$$

With $\hat{Y} = \hat{\beta}_0 + \hat{\beta}_1 x^* + \cdots + \hat{\beta}_k (x^*)^k$, $\hat{y}$ denoting the calculated value of $\hat{Y}$ for the given data, and $s_{\hat{Y}}$ denoting the estimated standard deviation of the statistic $\hat{Y}$, the formula for the CI is much like the one in the case of simple linear regression:

$$\hat{y} \pm t_{\alpha/2, n-(k+1)} \cdot s_{\hat{Y}}$$

A $100(1 - \alpha)\%$ PI for a future y value to be observed when $x = x^*$ is

$$\hat{\mu}_{Y \cdot x^*} \pm t_{\alpha/2, n-(k+1)} \cdot \left\{ s^2 + \left(\begin{array}{c} \text{estimated SD} \\ \text{of } \hat{\mu}_{Y \cdot x^*} \end{array} \right)^2 \right\}^{1/2} = \hat{y} \pm t_{\alpha/2, n-(k+1)} \cdot \sqrt{s^2 + s_{\hat{Y}}^2}$$

EXAMPLE 13.9
(Example 13.8 continued)

Figure 13.11(b) shows that $\hat{\beta}_2 = -.00006001$ and $s_{\hat{\beta}_2} = .00001786$ (from the SE Coef column at the top of the output). The null hypothesis $H_0: \beta_2 = 0$ says that as long as the linear predictor x is retained in the model, the quadratic predictor x^2 provides no additional useful information. The relevant alternative is $H_a: \beta_2 \neq 0$, and the test statistic is $T = \hat{\beta}_2 / S_{\hat{\beta}_2}$, with computed value -3.36. The test is based on $n - (k + 1) = 17$ df. At significance level .05, the null hypothesis is rejected because the reported P-value is .004 (double the area under the t_{17} curve to the left of -3.36). Thus inclusion of the quadratic predictor in the model equation is justified.

The output in Figure 13.11(b) also contains estimation and prediction information both for $x = 500$ and for $x = 800$. In particular, for $x = 500$,

$$\hat{y} = \hat{\beta}_0 + \hat{\beta}_1(500) + \hat{\beta}_2(500)^2 = \text{Fit} = 21.136$$
$$s_{\hat{y}} = \text{estimated SD of } \hat{y} = \text{SE Fit} = 1.167$$

from which a 95% CI for mean strength when thickness = 500 is $21.136 \pm (2.110) \times (1.167) = (18.67, 23.60)$. A 95% PI for the strength resulting from a single bond when thickness = 500 is $21.136 \pm (2.110)[(3.27)^2 + (1.167)^2]^{1/2} = (13.81, 28.46)$. As before, the PI is substantially wider than the CI because s is large compared to SE Fit. ■

Centering x Values

For the quadratic model with regression function $\mu_{Y \cdot x} = \beta_0 + \beta_1 x + \beta_2 x^2$, the parameters β_0, β_1, and β_2 characterize the behavior of the function near $x = 0$. For example, β_0 is the height at which the regression function crosses the vertical axis $x = 0$, whereas β_1 is the first derivative of the function at $x = 0$ (instantaneous rate of change of $\mu_{Y \cdot x}$ at $x = 0$). If the x_i's all lie far from 0, we may not have precise information about the values of these parameters. Let $\bar{x}$ = the average of the x_i's for which observations are to be taken, and consider the model

$$Y = \beta_0^* + \beta_1^*(x - \bar{x}) + \beta_2^*(x - \bar{x})^2 + \epsilon \qquad (13.14)$$

In the model (13.14), $\mu_{Yx} = \beta_0^* + \beta_1^*(x - \bar{x}) + \beta_2^*(x - \bar{x})^2$, and the parameters now describe the behavior of the regression function near the center $\bar{x}$ of the data.

To estimate the parameters of (13.14), we simply subtract $\bar{x}$ from each x_i to obtain $x_i' = x_i - \bar{x}$ and then use the x_i''s in place of the x_i's. An important benefit of this is that the coefficients of $b_0, \ldots, b_k$ in the normal equations (13.10) will be of much smaller magnitude than would be the case were the original x_i's used. When the system is solved by computer, this centering protects against any round-off error that may result.

EXAMPLE 13.10   The article **"A Method for Improving the Accuracy of Polynomial Regression Analysis"** (**J. of Quality Tech., 1971: 149–155**) reports the following data on $x =$ cure temperature (°F) and $y =$ ultimate shear strength of a rubber compound (psi), with $\bar{x} = 297.13$:

x	280	284	292	295	298	305	308	315
x'	-17.13	-13.13	-5.13	-2.13	.87	7.87	10.87	17.87
y	770	800	840	810	735	640	590	560

A computer analysis yielded the results shown in Table 13.3.

Table 13.3   Estimated Coefficients and Standard Deviations for Example 13.10

Parameter	Estimate	Estimated SD	Parameter	Estimate	Estimated SD
β_0	$-26{,}219.64$	11,912.78	β_0^*	759.36	23.20
β_1	189.21	80.25	β_1^*	-7.61	1.43
β_2	$-.3312$	.1350	β_2^*	$-.3312$	.1350

The estimated regression function using the original model is $y = -26{,}219.64 + 189.21x - .3312x^2$, whereas for the centered model the function is $y = 759.36 - 7.61(x - 297.13) - .3312(x - 297.13)^2$. These estimated functions are identical; the only difference is that different parameters have been estimated for the two models. The estimated standard deviations indicate clearly that β_0^* and β_1^* have been more accurately estimated than β_0 and β_1. The quadratic parameters are identical ($\beta_2 = \beta_2^*$), as can be seen by comparing the x^2 term in (13.14) with the original model. We emphasize again that a major benefit of centering is the gain in computational accuracy, not only in quadratic but also in higher-degree models.   ∎

The book by Neter et al., listed in the chapter bibliography, is a good source for more information about polynomial regression.

EXERCISES Section 13.3 (26–35)

26. The article **"Physical Properties of Cumin Seed"** (**J. of Agric. Engr. Res., 1996: 93–98**) considered a quadratic regression of $y =$ bulk density on $x =$ moisture content.

Data from a graph in the article follows, along with Minitab output from the quadratic fit.

```
The regression equation is
bulkdens = 403 + 16.2  moiscont − 0.706  contsqd

Predictor          Coef      StDev        T         P
Constant         403.24      36.45     11.06     0.002
moiscont         16.164       5.451     2.97     0.059
contsqd          −0.7063      0.1852   −3.81     0.032

S = 10.15      R-Sq = 93.8%       R-Sq(adj) = 89.6%

Analysis of Variance

Source             DF        SS        MS        F        P
Regression          2    4637.7    2318.9    22.51    0.016
Residual Error      3     309.1     103.0
Total               5    4946.8

                                        StDev
                                                     St
Obs moiscont bulkdens    Fit  Fit Residual Resid
 1       7.0    479.00  481.78 9.35   −2.78 −0.70
 2      10.3    503.00  494.79 5.78    8.21  0.98
 3      13.7    487.00  492.12 6.49   −5.12 −0.66
 4      16.6    470.00  476.93 6.10   −6.93 −0.85
 5      19.8    458.00  446.39 5.69   11.61  1.38
 6      22.0    412.00  416.99 8.75   −4.99 −0.97

         StDev
Fit       Fit           95.0% CI            95.0% PI
491.10    6.52     (470.36, 511.83)   (452.71, 529.48)
```

a. Does a scatterplot of the data appear consistent with the quadratic regression model?

b. What proportion of observed variation in density can be attributed to the model relationship?

c. Calculate a 95% CI for true average density when moisture content is 13.7.

d. The last line of output is from a request for estimation and prediction information when moisture content is 14. Calculate a 99% PI for density when moisture content is 14.

e. Does the quadratic predictor appear to provide useful information? Test the appropriate hypotheses at significance level .05.

27. The following data on y = glucose concentration (g/L) and x = fermentation time (days) for a particular blend of malt liquor was read from a scatterplot in the article **"Improving Fermentation Productivity with Reverse Osmosis"** (*Food Tech.*, 1984: 92–96):

x	1	2	3	4	5	6	7	8
y	74	54	52	51	52	53	58	71

a. Verify that a scatterplot of the data is consistent with the choice of a quadratic regression model.

b. The estimated quadratic regression equation is $y = 84.482 − 15.875x + 1.7679x^2$. Predict the value of glucose concentration for a fermentation time of 6 days, and compute the corresponding residual.

c. Using SSE = 61.77, what proportion of observed variation can be attributed to the quadratic regression relationship?

d. The $n = 8$ standardized residuals based on the quadratic model are 1.91, −1.95, −.25, .58, .90, .04,

−.66, and .20. Construct a plot of the standardized residuals versus x and a normal probability plot. Do the plots exhibit any troublesome features?

e. The estimated standard deviation of $\hat{\mu}_{Y \cdot 6}$—that is, $\hat{\beta}_0 + \hat{\beta}_1(6) + \hat{\beta}_2(36)$—is 1.69. Compute a 95% CI for $\mu_{Y \cdot 6}$.

f. Compute a 95% PI for a glucose concentration observation made after 6 days of fermentation time.

28. The viscosity (y) of an oil was measured by a cone and plate viscometer at six different cone speeds (x). It was assumed that a quadratic regression model was appropriate, and the estimated regression function resulting from the $n = 6$ observations was

$$y = −113.0937 + 3.3684x − .01780x^2$$

a. Estimate $\mu_{Y \cdot 75}$, the expected viscosity when speed is 75 rpm.

b. What viscosity would you predict for a cone speed of 60 rpm?

c. If $\Sigma y_i^2 = 8386.43$, $\Sigma y_i = 210.70$, $\Sigma x_i y_i = 17,002.00$, and $\Sigma x_i^2 y_i = 1,419,780$, compute SSE $[= \Sigma y_i^2 − \hat{\beta}_0 \Sigma y_i − \hat{\beta}_1 \Sigma x_i y_i − \hat{\beta}_2 \Sigma x_i^2 y_i]$ and s.

d. From part (c), SST = $8386.43 − (210.70)^2/6 = 987.35$. Using SSE computed in part (c), what is the computed value of R^2?

e. If the estimated standard deviation of $\hat{\beta}_2$ is $s_{\hat{\beta}_2} = .00226$, test $H_0: \beta_2 = 0$ versus $H_a: \beta_2 \neq 0$ at level .01, and interpret the result.

29. High-alumina refractory castables have been extensively investigated in recent years because of their significant advantages over other refractory brick of the same class—lower production and application costs, versatility, and performance at high temperatures. The accompanying data on x = viscosity (MPa · s) and y = free-flow (%) was read from a graph in the article **"Processing of Zero-Cement Self-Flow Alumina Castables"** (*The Amer. Ceramic Soc. Bull.*, 1998: 60–66):

x	351	367	373	400	402	456	484
y	81	83	79	75	70	43	22

The authors of the cited paper related these two variables using a quadratic regression model. The estimated regression function is $y = −295.96 + 2.1885x − .0031662x^2$.

a. Compute the predicted values and residuals, and then SSE and s^2.

b. Compute and interpret the coefficient of multiple determination.

c. The estimated SD of $\hat{\beta}_2$ is $s_{\hat{\beta}_2} = .0004835$. Does the quadratic predictor belong in the regression model?

d. The estimated SD of $\hat{\beta}_1$ is .4050. Use this and the information in (c) to obtain joint CIs for the linear and quadratic regression coefficients with a joint confidence level of (at least) 95%.

e. The estimated SD of $\hat{\mu}_{Y \cdot 400}$ is 1.198. Calculate a 95% CI for true average free-flow when viscosity = 400 and also a 95% PI for free-flow resulting from a single observation made when viscosity = 400, and compare the intervals.

30. The accompanying data was extracted from the article **"Effects of Cold and Warm Temperatures on Springback of Aluminum-Magnesium Alloy 5083-H111"** (*J. of Engr. Manuf.*, 2009: 427–431). The response variable is yield strength (MPa), and the predictor is temperature (°C).

x	−50	25	100	200	300
y	91.0	120.5	136.0	133.1	120.8

Here is Minitab output from fitting the quadratic regression model (a graph in the cited paper suggests that the authors did this):

```
Predictor        Coef    SE Coef       T       P
Constant      111.277      2.100   52.98   0.000
temp          0.32845    0.03303    9.94   0.010
tempsqd    -0.0010050  0.0001213   -8.29   0.014

S = 3.44398    R-Sq = 98.1%    R-Sq(adj) = 96.3%

Analysis of Variance

Source        DF       SS      MS      F      P
Regression     2  1245.39  622.69  52.50  0.019
Residual Error 2    23.72   11.86
Total          4  1269.11
```

a. What proportion of observed variation in strength can be attributed to the model relationship?

b. Carry out a test of hypotheses at significance level .05 to decide if the quadratic predictor provides useful information over and above that provided by the linear predictor.

c. For a strength value of 100, $\hat{y} = 134.07$, $s_{\hat{y}} = 2.38$. Estimate true average strength when temperature is 100, in a way that conveys information about precision and reliability.

d. Use the information in (c) to predict strength for a single observation to be made when temperature is 100, and do so in a way that conveys information about precision and reliability. Then compare this prediction to the estimate obtained in (c).

31. The accompanying data on y = energy output (W) and x = temperature difference (°K) was provided by the authors of the article **"Comparison of Energy and Exergy Efficiency for Solar Box and Parabolic Cookers"** (*J. of Energy Engr.*, 2007: 53–62).

The article's authors fit a cubic regression model to the data. Here is Minitab output from such a fit.

The regression equation is

$$y = -134 + 12.7\ x - 0.377\ x^{**}2 + 0.00359\ x^{**}3$$

```
Predictor      Coef     SE Coef        T       P
Constant   -133.787       8.048   -16.62   0.000
x           12.7423      0.7750    16.44   0.000
x**2       -0.37652     0.02444   -15.41   0.000
x**3      0.0035861   0.0002529    14.18   0.000

S = 0.168354   R-Sq = 98.0%   R-Sq(adj) = 97.7%

Analysis of Variance

Source           DF       SS      MS       F       P
Regression        3  27.9744  9.3248  329.00   0.000
Residual Error   20   0.5669  0.0283
Total            23  28.5413
```

a. What proportion of observed variation in energy output can be attributed to the model relationship?

b. Fitting a quadratic model to the data results in $R^2 = .780$. Calculate adjusted R^2 for this model and compare to adjusted R^2 for the cubic model.

c. Does the cubic predictor appear to provide useful information about y over and above that provided by the linear and quadratic predictors? State and test the appropriate hypotheses.

d. When $x = 30$, $s_{\hat{y}} = .0611$. Obtain a 95% CI for true average energy output in this case, and also a 95% PI for a single energy output to be observed when temperature difference is 30. [*Hint*: $s_{\hat{y}} = .0611$.]

e. Interpret the hypotheses $H_0: \mu_{Y \cdot 35} = 5$ versus $H_a: \mu_{Y \cdot 35} \neq 5$, and then carry out a test at significance level .05 using the fact that when $x = 35$, $s_{\hat{y}} = .0523$.

32. The following data is a subset of data obtained in an experiment to study the relationship between x = soil pH and y = A1 Concentration/EC (**"Root Responses of Three *Gramineae* Species to Soil Acidity in an Oxisol and an Ultisol,"** *Soil Science*, 1973: 295–302):

x	4.01	4.07	4.08	4.10	4.18	
y	1.20	.78	.83	.98	.65	

x	4.20	4.23	4.27	4.30	4.41	
y	.76	.40	.45	.39	.30	

x	4.45	4.50	4.58	4.68	4.70	4.77
y	.20	.24	.10	.13	.07	.04

A cubic model was proposed in the article, but the version of Minitab used by the author of the present text

x	23.20	23.50	23.52	24.30	25.10	26.20	27.40	28.10	29.30	30.60	31.50	32.01
y	3.78	4.12	4.24	5.35	5.87	6.02	6.12	6.41	6.62	6.43	6.13	5.92

x	32.63	33.23	33.62	34.18	35.43	35.62	36.16	36.23	36.89	37.90	39.10	41.66
y	5.64	5.45	5.21	4.98	4.65	4.50	4.34	4.03	3.92	3.65	3.02	2.89

refused to include the x^3 term in the model, stating that "x^3 is highly correlated with other predictor variables." To remedy this, $\bar{x} = 4.3456$ was subtracted from each x value to yield $x' = x - \bar{x}$. A cubic regression was then requested to fit the model having regression function

$$y = \beta_0^* + \beta_1^* x' + \beta_2^*(x')^2 + \beta_3^*(x')^3$$

The following computer output resulted:

Parameter	Estimate	Estimated SD
β_0^*	.3463	.0366
β_1^*	−1.2933	.2535
β_2^*	2.3964	.5699
β_3^*	−2.3968	2.4590

a. What is the estimated regression function for the "centered" model?

b. What is the estimated value of the coefficient β_3 in the "uncentered" model with regression function $y = \beta_0 + \beta_1 x + \beta_2 x^2 + \beta_3 x^3$? What is the estimate of β_2?

c. Using the cubic model, what value of y would you predict when soil pH is 4.5?

d. Carry out a test to decide whether the cubic term should be retained in the model.

33. In many polynomial regression problems, rather than fitting a "centered" regression function using $x' = x - \bar{x}$, computational accuracy can be improved by using a function of the standardized independent variable $x' = (x - \bar{x})/s_x$, where s_x is the standard deviation of the x_i's. Consider fitting the cubic regression function $y = \beta_0^* + \beta_1^* x' + \beta_2^*(x')^2 + \beta_3^*(x')^3$ to the following data resulting from a study of the relation between thrust efficiency y of supersonic propelling rockets and the half-divergence angle x of the rocket nozzle (**"More on Correlating Data,"** *CHEMTECH*, **1976: 266–270**):

x	5	10	15	20	25	30	35
y	.985	.996	.988	.962	.940	.915	.878

Parameter	Estimate	Estimated SD
β_0^*	.9671	.0026
β_1^*	−.0502	.0051
β_2^*	−.0176	.0023
β_3^*	.0062	.0031

a. What value of y would you predict when the half-divergence angle is 20? When $x = 25$?

b. What is the estimated regression function $\hat{\beta}_0 + \hat{\beta}_1 x + \hat{\beta}_2 x^2 + \hat{\beta}_3 x^3$ for the "unstandardized" model?

c. Use a level .05 test to decide whether the cubic term should be deleted from the model.

d. What can you say about the relationship between SSEs and R^2's for the standardized and unstandardized models? Explain.

e. SSE for the cubic model is .00006300, whereas for a quadratic model SSE is .00014367. Compute R^2 for each model. Does the difference between the two suggest that the cubic term can be deleted?

34. The following data resulted from an experiment to assess the potential of unburnt colliery spoil as a medium for plant growth. The variables are x = acid extractable cations and y = exchangeable acidity/total cation exchange capacity (**"Exchangeable Acidity in Unburnt Colliery Spoil," *Nature*, 1969: 161**):

x	−23	−5	16	26	30	38	52
y	1.50	1.46	1.32	1.17	.96	.78	.77

x	58	67	81	96	100	113
y	.91	.78	.69	.52	.48	.55

Standardizing the independent variable x to obtain $x' = (x - \bar{x})/s_x$ and fitting the regression function $y = \beta_0^* + \beta_1^* x' + \beta_2^*(x')^2$ yielded the accompanying computer output.

Parameter	Estimate	Estimated SD
β_0^*	.8733	.0421
β_1^*	−.3255	.0316
β_2^*	.0448	.0319

a. Estimate $\mu_{Y \cdot 50}$.

b. Compute the value of the coefficient of multiple determination. (See Exercise 28(c).)

c. What is the estimated regression function $\hat{\beta}_0 + \hat{\beta}_1 x + \hat{\beta}_2 x^2$ using the unstandardized variable x?

d. What is the estimated standard deviation of $\hat{\beta}_2$ computed in part (c)?

e. Carry out a test using the standardized estimates to decide whether the quadratic term should be retained in the model. Repeat using the unstandardized estimates. Do your conclusions differ?

35. The article **"The Respiration in Air and in Water of the Limpets *Patella caerulea* and *Patella lusitanica*"** (*Comp. Biochemistry and Physiology*, **1975: 407–411**) proposed a simple power model for the relationship between respiration rate y and temperature x for *P. caerulea* in air. However, a plot of $\ln(y)$ versus x exhibits a curved pattern. Fit the quadratic power model $Y = \alpha e^{\beta x + \gamma x^2} \cdot \epsilon$ to the accompanying data.

x	10	15	20	25	30
y	37.1	70.1	109.7	177.2	222.6

13.4 Multiple Regression Analysis

In multiple regression, the objective is to build a probabilistic model that relates a dependent variable y to more than one independent or predictor variable. Let k represent the number of predictor variables ($k \geq 2$) and denote these predictors by $x_1, x_2, \ldots, x_k$. For example, in attempting to predict the selling price of a house, we might have $k = 3$ with $x_1 =$ size (ft^2), $x_2 =$ age (years), and $x_3 =$ number of rooms.

DEFINITION

> The **general additive multiple regression model equation** is
>
> $$Y = \beta_0 + \beta_1 x_1 + \beta_2 x_2 + \ldots + \beta_k x_k + \epsilon \qquad (13.15)$$
>
> where $E(\epsilon) = 0$ and $V(\epsilon) = \sigma^2$. In addition, for purposes of testing hypotheses and calculating CIs or PIs, it is assumed that ϵ is normally distributed.

Let $x_1^*, x_2^*, \ldots, x_k^*$ be particular values of $x_1, \ldots, x_k$. Then (13.15) implies that

$$\mu_{Y \cdot x_1^*, \ldots, x_k^*} = \beta_0 + \beta_1 x_1^* + \cdots + \beta_k x_k^* \qquad (13.16)$$

Thus just as $\beta_0 + \beta_1 x$ describes the mean Y value as a function of x in simple linear regression, the **true** (or **population**) **regression function** $\beta_0 + \beta_1 x_1 + \cdots + \beta_k x_k$ gives the expected value of Y as a function of $x_1, \ldots, x_k$. The β_i's are the **true** (or **population**) **regression coefficients.** The regression coefficient β_1 is interpreted as the expected change in Y associated with a 1-unit increase in x_1 *while $x_2, \ldots, x_k$ are held fixed.* Analogous interpretations hold for $\beta_2, \ldots, \beta_k$.

Models with Interaction and Quadratic Predictors

If an investigator has obtained observations on y, x_1, and x_2, one possible model is $Y = \beta_0 + \beta_1 x_1 + \beta_2 x_2 + \epsilon$. However, other models can be constructed by forming predictors that are mathematical functions of x_1 and/or x_2. For example, with $x_3 = x_1^2$ and $x_4 = x_1 x_2$, the model

$$Y = \beta_0 + \beta_1 x_1 + \beta_2 x_2 + \beta_3 x_3 + \beta_4 x_4 + \epsilon$$

has the general form of (13.15). In general, it is not only permissible for some predictors to be mathematical functions of others but also often highly desirable in the sense that the resulting model may be much more successful in explaining variation in y than any model without such predictors. This discussion also shows that polynomial regression is indeed a special case of multiple regression. For example, the quadratic model $Y = \beta_0 + \beta_1 x + \beta_2 x^2 + \epsilon$ has the form of (13.15) with $k = 2$, $x_1 = x$, and $x_2 = x^2$.

For the case of two independent variables, x_1 and x_2, consider the following four derived models.

1. The first-order model:

$$Y = \beta_0 + \beta_1 x_1 + \beta_2 x_2 + \epsilon$$

2. The second-order no-interaction model:

$$Y = \beta_0 + \beta_1 x_1 + \beta_2 x_2 + \beta_3 x_1^2 + \beta_4 x_2^2 + \epsilon$$

3. The model with first-order predictors and interaction:

$$Y = \beta_0 + \beta_1 x_1 + \beta_2 x_2 + \beta_3 x_1 x_2 + \epsilon$$

4. The complete second-order or full quadratic model:

$$Y = \beta_0 + \beta_1 x_1 + \beta_2 x_2 + \beta_3 x_1^2 + \beta_4 x_2^2 + \beta_5 x_1 x_2 + \epsilon$$

Understanding the differences among these models is an important first step in building realistic regression models from the independent variables under study.

The first-order model is the most straightforward generalization of simple linear regression. It states that for a fixed value of either variable, the expected value of Y is a linear function of the other variable and that the expected change in Y associated with a unit increase in x_1 (x_2) is β_1 (β_2) independent of the level of x_2 (x_1). Thus if we graph the regression function as a function of x_1 for several different values of x_2, we obtain as contours of the regression function a collection of parallel lines, as pictured in Figure 13.13(a). The function $y = \beta_0 + \beta_1 x_1 + \beta_2 x_2$ specifies a plane in three-dimensional space; the first-order model says that each observed value of the dependent variable corresponds to a point which deviates vertically from this plane by a random amount ϵ.

According to the second-order no-interaction model, if x_2 is fixed, the expected change in Y for a 1-unit increase in x_1 is

$$\beta_0 + \beta_1(x_1 + 1) + \beta_2 x_2 + \beta_3(x_1 + 1)^2 + \beta_4 x_2^2$$
$$- (\beta_0 + \beta_1 x_1 + \beta_2 x_2 + \beta_3 x_1^2 + \beta_4 x_2^2) = \beta_1 + \beta_3 + 2\beta_3 x_1$$

Because this expected change does not depend on x_2, the contours of the regression function for different values of x_2 are still parallel to one another. However, the dependence of the expected change on the value of x_1 means that the contours are now curves rather than straight lines. This is pictured in Figure 13.13(b). In this case, the regression surface is no longer a plane in three-dimensional space but is instead a curved surface.

The contours of the regression function for the first-order interaction model are nonparallel straight lines. This is because the expected change in Y when x_1 is increased by 1 is

$$\beta_0 + \beta_1(x_1 + 1) + \beta_2 x_2 + \beta_3(x_1 + 1)x_2$$
$$- (\beta_0 + \beta_1 x_1 + \beta_2 x_2 + \beta_3 x_1 x_2) = \beta_1 + \beta_3 x_2$$

This expected change depends on the value of x_2, so each contour line must have a different slope, as in Figure 13.13(c). The word *interaction* reflects the fact that an expected change in Y when one variable increases in value depends on the value of the other variable.

Finally, for the complete second-order model, the expected change in Y when x_2 is held fixed while x_1 is increased by 1 unit is $\beta_1 + \beta_3 + 2\beta_3 x_1 + \beta_5 x_2$, which is a function of both x_1 and x_2. This implies that the contours of the regression function are both curved and not parallel to one another, as illustrated in Figure 13.13(d).

Similar considerations apply to models constructed from more than two independent variables. In general, the presence of interaction terms in the model implies that the expected change in Y depends not only on the variable being increased or decreased but also on the values of some of the fixed variables. As in ANOVA, it is possible to have higher-way interaction terms (e.g., $x_1 x_2 x_3$), making model interpretation more difficult.

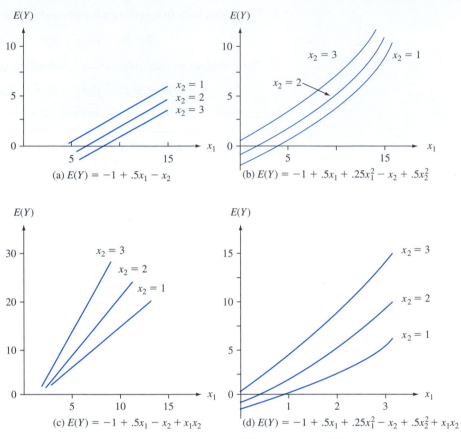

(a) $E(Y) = -1 + .5x_1 - x_2$

(b) $E(Y) = -1 + .5x_1 + .25x_1^2 - x_2 + .5x_2^2$

(c) $E(Y) = -1 + .5x_1 - x_2 + x_1x_2$

(d) $E(Y) = -1 + .5x_1 + .25x_1^2 - x_2 + .5x_2^2 + x_1x_2$

Figure 13.13 Contours of four different regression functions

Note that if the model contains interaction or quadratic predictors, the generic interpretation of a β_i given previously will not usually apply. This is because it is not then possible to increase x_i by 1 unit and hold the values of all other predictors fixed.

Models with Predictors for Categorical Variables

Thus far we have explicitly considered the inclusion of only quantitative (numerical) predictor variables in a multiple regression model. Using simple numerical coding, qualitative (categorical) variables, such as bearing material (aluminum or copper/lead) or type of wood (pine, oak, or walnut), can also be incorporated into a model. Let's first focus on the case of a dichotomous variable, one with just two possible categories—male or female, U.S. or foreign manufacture, and so on. With any such variable, we associate a **dummy** or **indicator variable** x whose possible values 0 and 1 indicate which category is relevant for any particular observation.

EXAMPLE 13.11 The article **"Estimating Urban Travel Times: A Comparative Study"** (**Trans. Res., 1980: 173–175**) described a study relating the dependent variable $y =$ travel time between locations in a certain city and the independent variable $x_2 =$ distance between locations. Two types of vehicles, passenger cars and trucks, were used in the study. Let

$$x_1 = \begin{cases} 1 & \text{if the vehicle is a truck} \\ 0 & \text{if the vehicle is a passenger car} \end{cases}$$

One possible multiple regression model is

$$Y = \beta_0 + \beta_1 x_1 + \beta_2 x_2 + \epsilon$$

The mean value of travel time depends on whether a vehicle is a car or a truck:

$$\text{mean time} = \beta_0 + \beta_2 x_2 \qquad \text{when } x_1 = 0 \ \ (\text{cars})$$
$$\text{mean time} = \beta_0 + \beta_1 + \beta_2 x_2 \quad \text{when } x_1 = 1 \ \ (\text{trucks})$$

The coefficient β_1 is the difference in mean times between trucks and cars with distance held fixed; if $\beta_1 > 0$, on average it will take trucks longer to traverse any particular distance than it will for cars.

A second possibility is a model with an interaction predictor:

$$Y = \beta_0 + \beta_1 x_1 + \beta_2 x_2 + \beta_3 x_1 x_2 + \epsilon$$

Now the mean times for the two types of vehicles are

$$\text{mean time} = \beta_0 + \beta_2 x_2 \qquad\qquad \text{when } x_1 = 0$$
$$\text{mean time} = \beta_0 + \beta_1 + (\beta_2 + \beta_3) x_2 \quad \text{when } x_1 = 1$$

For each model, the graph of the mean time versus distance is a straight line for either type of vehicle, as illustrated in Figure 13.14. The two lines are parallel for the first (no-interaction) model, but in general they will have different slopes when the second model is correct. For this latter model, the change in mean travel time associated with a 1-mile increase in distance depends on which type of vehicle is involved—the two variables "vehicle type" and "travel time" interact. Indeed, data collected by the authors of the cited article suggested the presence of interaction.

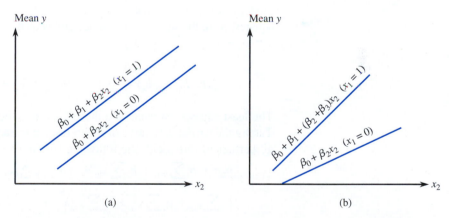

Figure 13.14 Regression functions for models with one dummy variable (x_1) and one quantitative variable x_2: (a) no interaction; (b) interaction ■

You might think that the way to handle a three-category situation is to define a single numerical variable with coded values such as 0, 1, and 2 corresponding to the three categories. This is incorrect, because it imposes an ordering on the categories that is not necessarily implied by the problem context. The correct approach to incorporating three categories is to define *two* different dummy variables. Suppose, for example, that y is the lifetime of a certain cutting tool, x_1 is cutting speed, and that there are three brands of tool being investigated. Then let

$$x_2 = \begin{cases} 1 & \text{if a brand A tool is used} \\ 0 & \text{otherwise} \end{cases} \qquad x_3 = \begin{cases} 1 & \text{if a brand B tool is used} \\ 0 & \text{otherwise} \end{cases}$$

When an observation on a brand A tool is made, $x_2 = 1$ and $x_3 = 0$, whereas for a brand B tool, $x_2 = 0$ and $x_3 = 1$. An observation made on a brand C tool has $x_2 = x_3 = 0$, and it is not possible that $x_2 = x_3 = 1$ because a tool cannot simultaneously be both brand A and brand B. The no-interaction model would have only the predictors x_1, x_2, and x_3. The following interaction model allows the mean change in lifetime associated with a 1-unit increase in speed to depend on the brand of tool:

$$Y = \beta_0 + \beta_1 x_1 + \beta_2 x_2 + \beta_3 x_3 + \beta_4 x_1 x_2 + \beta_5 x_1 x_3 + \epsilon$$

Construction of a picture like Figure 13.14 with a graph for each of the three possible (x_2, x_3) pairs gives three nonparallel lines (unless $\beta_4 = \beta_5 = 0$).

More generally, incorporating a categorical variable with c possible categories into a multiple regression model requires the use of $c - 1$ indicator variables (e.g., five brands of tools would necessitate using four indicator variables). Thus even one categorical variable can add many predictors to a model.

Estimating Parameters

The data in simple linear regression consists of n pairs $(x_1, y_1), \ldots, (x_n, y_n)$. Suppose that a multiple regression model contains two predictor variables, x_1 and x_2. Then the data set will consist of n triples $(x_{11}, x_{21}, y_1), (x_{12}, x_{22}, y_2), \ldots, (x_{1n}, x_{2n}, y_n)$. Here the first subscript on x refers to the predictor and the second to the observation number. More generally, with k predictors, the data consists of n $(k + 1)$-tuples $(x_{11}, x_{21}, \ldots, x_{k1}, y_1), (x_{12}, x_{22}, \ldots, x_{k2}, y_2), \ldots, (x_{1n}, x_{2n}, \ldots, x_{kn}, y_n)$, where x_{ij} is the value of the ith predictor x_i associated with the observed value y_j. The observations are assumed to have been obtained independently of one another according to the model (13.15). To estimate the parameters $\beta_0, \beta_1, \ldots, \beta_k$ using the principle of least squares, form the sum of squared deviations of the observed y_j's from a trial function $y = b_0 + b_1 x_1 + \cdots + b_k x_k$:

$$f(b_0, b_1, \ldots, b_k) = \sum_j [y_j - (b_0 + b_1 x_{1j} + b_2 x_{2j} + \cdots + b_k x_{kj})]^2 \quad (13.17)$$

The least squares estimates are those values of the b_i's that minimize $f(b_0, \ldots, b_k)$. Taking the partial derivative of f with respect to each $b_i (i = 0, 1, \ldots, k)$ and equating all partials to zero yields the following system of **normal equations:**

$$b_0 n + b_1 \sum x_{1j} + b_2 \sum x_{2j} + \cdots + b_k \sum x_{kj} = \sum y_j$$

$$b_0 \sum x_{1j} + b_1 \sum x_{1j}^2 + b_2 \sum x_{1j} x_{2j} + \cdots + b_k \sum x_{1j} x_{kj} = \sum x_{1j} y_j$$

$$\vdots \qquad \vdots \qquad \qquad \vdots \qquad \qquad (13.18)$$

$$b_0 \sum x_{kj} + b_1 \sum x_{1j} x_{kj} + \cdots + b_{k-1} \sum x_{k-1,j} x_{kj} + b_k \sum x_{kj}^2 = \sum x_{kj} y_j$$

These equations are linear in the unknowns $b_0, b_1, \ldots, b_k$. Solving (13.18) yields the least squares estimates $\hat{\beta}_0, \hat{\beta}_1, \ldots, \hat{\beta}_k$. This is best done by utilizing a statistical software package.

EXAMPLE 13.12   The article **"How to Optimize and Control the Wire Bonding Process: Part II"** (*Solid State Technology*, Jan. 1991: 67–72) described an experiment carried out to assess the impact of the variables $x_1 = $ force (gm), $x_2 = $ power (mW), $x_3 = $ temperature (°C), and x_4 = time (msec) on y = ball bond shear strength (gm). The following data* was generated to be consistent with the information given in the article:

Observation	Force	Power	Temperature	Time	Strength
1	30	60	175	15	26.2
2	40	60	175	15	26.3
3	30	90	175	15	39.8
4	40	90	175	15	39.7
5	30	60	225	15	38.6
6	40	60	225	15	35.5
7	30	90	225	15	48.8
8	40	90	225	15	37.8
9	30	60	175	25	26.6
10	40	60	175	25	23.4
11	30	90	175	25	38.6
12	40	90	175	25	52.1
13	30	60	225	25	39.5
14	40	60	225	25	32.3
15	30	90	225	25	43.0
16	40	90	225	25	56.0
17	25	75	200	20	35.2
18	45	75	200	20	46.9
19	35	45	200	20	22.7
20	35	105	200	20	58.7
21	35	75	150	20	34.5
22	35	75	250	20	44.0
23	35	75	200	10	35.7
24	35	75	200	30	41.8
25	35	75	200	20	36.5
26	35	75	200	20	37.6
27	35	75	200	20	40.3
28	35	75	200	20	46.0
29	35	75	200	20	27.8
30	35	75	200	20	40.3

A statistical computer package gave the following least squares estimates:

$$\hat{\beta}_0 = -37.48 \quad \hat{\beta}_1 = .2117 \quad \hat{\beta}_2 = .4983 \quad \hat{\beta}_3 = .1297 \quad \hat{\beta}_4 = .2583$$

Thus we estimate that .1297 gm is the average change in strength associated with a 1-degree increase in temperature when the other three predictors are held fixed; the other estimated coefficients are interpreted in a similar manner.

The estimated regression equation is

$$y = -37.48 + .2117x_1 + .4983x_2 + .1297x_3 + .2583x_4$$

A point prediction of strength resulting from a force of 35 gm, power of 75 mW, temperature of 200° degrees, and time of 20 msec is

$$\hat{y} = -37.48 + (.2117)(35) + (.4983)(75) + (.1297)(200) + (.2583)(20)$$
$$= 38.41 \, \text{gm}$$

* From the book *Statistics Engineering Problem Solving* by Stephen Vardeman, an excellent exposition of the territory covered by our book, albeit at a somewhat higher level.

This is also a point estimate of the mean value of strength for the specified values of force, power, temperature, and time. ∎

R^2 and $\hat{\sigma}^2$

Predicted or fitted values, residuals, and the various sums of squares are calculated as in simple linear and polynomial regression. The predicted value $\hat{y}_1$ results from substituting the values of the various predictors from the first observation into the estimated regression function:

$$\hat{y}_1 = \hat{\beta}_0 + \hat{\beta}_1 x_{11} + \hat{\beta}_2 x_{21} + \cdots + \hat{\beta}_k x_{k1}$$

The remaining predicted values $\hat{y}_2, \ldots, \hat{y}_n$ come from substituting values of the predictors from the 2nd, 3rd, …, and finally nth observations into the estimated function. For example, the values of the 4 predictors for the last observation in Example 13.12 are $x_{1,30} = 35$, $x_{2,30} = 75$, $x_{3,30} = 200$, and $x_{4,30} = 20$, so

$$\hat{y}_{30} = -37.48 + .2117(35) + .4983(75) + .1297(200) + .2583(20) = 38.41$$

The residuals $y_1 - \hat{y}_1, \ldots, y_n - \hat{y}_n$ are the differences between the observed and predicted values. The last residual in Example 13.12 is $40.3 - 38.41 = 1.89$. The closer the residuals are to 0, the better the job our estimated regression function is doing in making predictions corresponding to observations in the sample.

Error or residual sum of squares is $\text{SSE} = \Sigma(y_i - \hat{y}_i)^2$. It is again interpreted as a measure of how much variation in the observed y values is not explained by (not attributed to) the model relationship. The number of df associated with SSE is $n - (k + 1)$ because $k + 1$ df are lost in estimating the $k + 1$ β coefficients. Total sum of squares, a measure of total variation in the observed y values, is $\text{SST} = \Sigma(y_i - \bar{y})^2$. Regression sum of squares $\text{SSR} = \Sigma(\hat{y}_i - \bar{y})^2 = \text{SST} - \text{SSE}$ is a measure of explained variation. Then the **coefficient of multiple determination** R^2 is

$$R^2 = 1 - \text{SSE}/\text{SST} = \text{SSR}/\text{SST}$$

It is interpreted as the proportion of observed y variation that can be explained by the multiple regression model fit to the data.

Because there is no preliminary picture of multiple regression data analogous to a scatterplot for bivariate data, the coefficient of multiple determination is our first indication of whether the chosen model is successful in explaining y variation. Unfortunately, there is a problem with R^2: Its value can be inflated by adding lots of predictors into the model even if most of these predictors are rather frivolous. For example, suppose y is the sale price of a house. Then sensible predictors include $x_1 = $ the interior size of the house, $x_2 = $ the size of the lot on which the house sits, $x_3 = $ the number of bedrooms, $x_4 = $ the number of bathrooms, and $x_5 = $ the house's age. Now suppose we add in $x_6 = $ the diameter of the doorknob on the coat closet, $x_7 = $ the thickness of the cutting board in the kitchen, $x_8 = $ the thickness of the patio slab, and so on. Unless we are very unlucky in our choice of predictors, using $n - 1$ predictors (one fewer than the sample size) will yield $R^2 = 1$. So the objective in multiple regression is not simply to explain most of the observed y variation, but to do so using a model with relatively few predictors that are easily interpreted. It is thus desirable to adjust R^2, as was done in polynomial regression, to take account of the size of the model:

$$R_a^2 = 1 - \frac{\text{SSE}/(n - (k + 1))}{\text{SST}/(n - 1)} = 1 - \frac{n - 1}{n - (k + 1)} \cdot \frac{\text{SSE}}{\text{SST}}$$

Because the ratio in front of SSE/SST exceeds 1, R_a^2 is smaller than R^2. Furthermore, the larger the number of predictors k relative to the sample size n, the smaller R_a^2 will be relative to R^2. Adjusted R^2 can even be negative, whereas R^2 itself must be between 0 and 1. A value of R_a^2 that is substantially smaller than R^2 itself is a warning that the model may contain too many predictors.

The positive square root of R^2 is called the *multiple correlation coefficient* and is denoted by R. It can be shown that R is the sample correlation coefficient calculated from the $(\hat{y}_i, y_i)$ pairs (that is, use $\hat{y}_i$ in place of x_i in the formula for r from Section 12.5).

SSE is also the basis for estimating the remaining model parameter:

$$\hat{\sigma}^2 = s^2 = \frac{\text{SSE}}{n - (k + 1)} = \text{MSE}$$

EXAMPLE 13.13 Investigators carried out a study to see how various characteristics of concrete are influenced by $x_1 = \%$ limestone powder and $x_2 = $ water-cement ratio, resulting in the accompanying data (**"Durability of Concrete with Addition of Limestone Powder,"** *Magazine of Concrete Research*, **1996: 131–137**).

x_1	x_2	$x_1 x_2$	28-day Comp Str. (MPa)	Adsorbability (%)
21	.65	13.65	33.55	8.42
21	.55	11.55	47.55	6.26
7	.65	4.55	35.00	6.74
7	.55	3.85	35.90	6.59
28	.60	16.80	40.90	7.28
0	.60	0.00	39.10	6.90
14	.70	9.80	31.55	10.80
14	.50	7.00	48.00	5.63
14	.60	8.40	42.30	7.43

$\bar{y} = 39.317$, SST $= 278.52$ $\bar{y} = 7.339$, SST $= 18.356$

Consider first compressive strength as the dependent variable y. Fitting the first-order model results in

$$y = 84.82 + .1643x_1 - 79.67x_2, \quad \text{SSE} = 72.52 \ (\text{df} = 6), \quad R^2 = .741, \quad R_a^2 = .654$$

whereas including an interaction predictor gives

$$y = 6.22 + 5.779x_1 + 51.33x_2 - 9.357x_1 x_2$$
$$\text{SSE} = 29.35 \ (\text{df} = 5) \qquad R^2 = .895 \qquad R_a^2 = .831$$

Based on this latter fit, a prediction for compressive strength when % limestone $= 14$ and water–cement ratio $= .60$ is

$$\hat{y} = 6.22 + 5.779(14) + 51.33(.60) - 9.357(8.4) = 39.32$$

Fitting the full quadratic relationship results in virtually no change in the R^2 value. However, when the dependent variable is adsorbability, the following results are obtained: $R^2 = .747$ when just two predictors are used, .802 when the interaction predictor is added, and .889 when the five predictors for the full quadratic relationship are used. ■

In general, $\hat{\beta}_i$ can be interpreted as an estimate of the average change in Y associated with a 1-unit increase in x_i while values of all other predictors are held fixed. Sometimes, though, it is difficult or even impossible to increase the value of one predictor while holding all others fixed. In such situations, there is an alternative interpretation of the estimated regression coefficients. For concreteness, suppose that $k = 2$, and let $\hat{\beta}_1$ denote the estimate of β_1 in the regression of y on the two predictors x_1 and x_2. Then

1. Regress y against just x_2 (a simple linear regression) and denote the resulting residuals by $g_1, g_2, \ldots, g_n$. These residuals represent variation in y after removing or adjusting for the effects of x_2.

2. Regress x_1 against x_2 (that is, regard x_1 as the dependent variable and x_2 as the independent variable in this simple linear regression), and denote the residuals by $f_1, \ldots, f_n$. These residuals represent variation in x_1 after removing or adjusting for the effects of x_2.

Now consider plotting the residuals from the first regression against those from the second; that is, plot the pairs $(f_1, g_1), \ldots, (f_n, g_n)$. The result is called a *partial residual plot* or *adjusted residual plot.* If a regression line is fit to the points in this plot, the slope turns out to be exactly $\hat{\beta}_1$ (furthermore, the residuals from this line are exactly the residuals $e_1, \ldots, e_n$ from the multiple regression of y on x_1 and x_2). Thus $\hat{\beta}_1$ can be interpreted as the estimated change in y associated with a 1-unit increase in x_1 after removing or adjusting for the effects of any other model predictors. The same interpretation holds for other estimated coefficients regardless of the number of predictors in the model (there is nothing special about $k = 2$; the foregoing argument remains valid if y is regressed against all predictors other than x_1 in Step 1 and x_1 is regressed against the other $k - 1$ predictors in Step 2).

As an example, suppose that y is the sale price of an apartment building and that the predictors are number of apartments, age, lot size, number of parking spaces, and gross building area (ft^2). It may not be reasonable to increase the number of apartments without also increasing gross area. However, if $\hat{\beta}_5 = 16.00$, then we estimate that a \$16 increase in sale price is associated with each extra square foot of gross area after adjusting for the effects of the other four predictors.

A Model Utility Test

The absence of an informative picture of multivariate data and the aforementioned difficulty with R^2 provide compelling reasons for seeking a formal test of model utility. The model utility test in simple linear regression involved the null hypothesis $H_0: \beta_1 = 0$, according to which there is no useful relation between y and the single predictor x. Here we consider the assertion that $\beta_1 = 0, \beta_2 = 0, \ldots, \beta_k = 0$, which says that there is no useful relationship between y and *any* of the k predictors. If at least one of these β's is not 0, the corresponding predictor(s) is (are) useful. The test is based on a statistic that has a particular F distribution when H_0 is true.

Null hypothesis: $H_0: \beta_1 = \beta_2 = \cdots = \beta_k = 0$

Alternative hypothesis: H_a: at least one $\beta_i \neq 0$ $\quad (i = 1, \ldots, k)$

Test statistic value: $f = \dfrac{R^2/k}{(1 - R^2)/[n - (k + 1)]}$

$$= \dfrac{SSR/k}{SSE/[n - (k + 1)]} = \dfrac{MSR}{MSE} \qquad (13.19)$$

where SSR = regression sum of squares = SST − SSE

When H_0 is true, the test statistic F has an F distribution with k numerator df and $n - (k + 1)$ denominator df. The test is upper-tailed, so the P-value is the area under the $F_{k,\, n-(k+1)}$ curve to the right of f.

Except for a constant multiple, the test statistic here is $R^2/(1 - R^2)$, the ratio of explained to unexplained variation. If the proportion of explained variation is high relative to unexplained, we would naturally want to reject H_0 and confirm the utility of the model; this explains why the test is upper-tailed (only large values of f argue against H_0). However, if k is large relative to n, the factor $[(n - (k + 1))/k]$ will decrease f considerably.

EXAMPLE 13.14   Returning to the bond shear strength data of Example 13.12, a model with $k = 4$ predictors was fit, so the relevant hypotheses are

$$H_0: \beta_1 = \beta_2 = \beta_3 = \beta_4 = 0$$

$$H_a: \text{at least one of these four } \beta\text{'s is not } 0$$

Figure 13.15 shows output from the JMP statistical package. The values of s (Root Mean Square Error), R^2, and adjusted R^2 certainly suggest a useful model. The value of the model utility F ratio is

$$f = \frac{R^2/k}{(1 - R^2)/[n - (k + 1)]} = \frac{.713959/4}{.286041/(30 - 5)} = 15.60$$

Response: strength

Summary of Fit

RSquare	0.713959
RSquare Adj	0.668193
Root Mean Square Error	5.157979
Mean of Response	38.40667
Observations (or Sum Wgts)	30

Parameter Estimates

Term	Estimate	Std Error	t Ratio	Prob>|t|
Intercept	-37.47667	13.09964	-2.86	0.0084
force	0.2116667	0.210574	1.01	0.3244
power	0.4983333	0.070191	7.10	<.0001
temp	0.1296667	0.042115	3.08	0.0050
time	0.2583333	0.210574	1.23	0.2313

Whole-Model Test

Analysis of Variance

Source	DF	Sum of Squares	Mean Square	F Ratio
Model	4	1660.1400	415.035	15.6000
Error	25	665.1187	26.605	Prob>F
C Total	29	2325.2587		<.0001

Figure 13.15   Multiple regression output from JMP for the data of Example 13.14

This value also appears in the F Ratio column of the ANOVA table in Figure 13.15. The largest F critical value for 4 numerator and 25 denominator df in Appendix Table A.9 is 6.49, which captures an upper-tail area of .001. Thus P-value $<$ .001. The ANOVA table in the JMP output shows that P-value $<$ .0001. This is a highly significant result. The null hypothesis should be rejected at any reasonable significance level. We conclude that there *is* a useful linear relationship between y and *at least one* of the four predictors in the model. This does not mean that all four predictors are useful; we will say more about this subsequently.   ■

Inferences in Multiple Regression

Before testing hypotheses, constructing CI's, and making predictions, the adequacy of the model should be assessed and the impact of any unusual observations investigated. Methods for doing this are described at the end of the present section and in the next section.

Because each $\hat{\beta}_i$ is a linear function of the y_i's, the standard deviation (standard error) of each $\hat{\beta}_i$ is the product of σ and a function of the x_{ij}'s. An estimate $s_{\hat{\beta}_i}$ of this SD is obtained by substituting s for σ. The function of the x_{ij}'s is quite complicated, but all widely used statistical software packages compute and show the $s_{\hat{\beta}_i}$'s. Inferences concerning a single β_i are based on the standardized variable

$$T = \frac{\hat{\beta}_i - \beta_i}{S_{\hat{\beta}_i}}$$

which has a t distribution with $n - (k + 1)$ df.

The point estimate of $\mu_{Y \cdot x_1^*, \ldots, x_k^*}$, the expected value of Y when $x_1 = x_1^*, \ldots, x_k = x_k^*$, is $\hat{\mu}_{Y \cdot x_1^*, \ldots, x_k^*} = \hat{\beta}_0 + \hat{\beta}_1 x_1^* + \cdots + \hat{\beta}_k x_k^*$. The estimated standard deviation of the corresponding estimator is again a complicated expression involving the sample x_{ij}'s. However, appropriate software will calculate it on request. Inferences about $\hat{\mu}_{Y \cdot x_1^*, \ldots, x_k^*}$ are based on standardizing its estimator to obtain a t variable having $n - (k + 1)$ df.

---

1. A $100(1 - \alpha)\%$ CI for β_i, the coefficient of x_i in the regression function, is

$$\hat{\beta}_i \pm t_{\alpha/2, n-(k+1)} \cdot s_{\hat{\beta}_i}$$

2. A test for $H_0: \beta_i = \beta_{i0}$ uses the t statistic value $t = (\hat{\beta}_i - \beta_{i0})/s_{\hat{\beta}_i}$ based on $n - (k + 1)$ df. The test is upper-, lower-, or two-tailed according to whether H_a contains the inequality $>$, $<$, or $\neq$. The most frequently tested null hypothesis in practice is $H_0: \beta_i = 0$. The interpretation is that as long as all the other predictors $x_1, \ldots, x_i - 1, x_i + 1, \ldots, x_k$ remain in the model, the predictor x_i provides no additional useful information about y. The customary alternative is $H_a: \beta_i \neq 0$, according to which x_i does provide useful information over and above what is contained in the other $i - 1$ predictors. The test statistic value is the ***t* ratio** $\hat{\beta}_i/s_{\hat{\beta}_i}$, the ratio of the estimated coefficient to its estimated standard error.

3. A $100(1 - \alpha)\%$ CI for $\mu_{Y \cdot x_1^*, \ldots, x_k^*}$ is

$$\hat{\mu}_{Y \cdot x_1^*, \ldots, x_k^*} \pm t_{\alpha/2, n-(k+1)} \cdot \{\text{estimated SD of } \hat{\mu}_{Y \cdot x_1^*, \ldots, x_k^*}\} = \hat{y} \pm t_{\alpha/2, n-(k+1)} \cdot s_{\hat{Y}}$$

where $\hat{Y}$ is the statistic $\hat{\beta}_0 + \hat{\beta}_1 x_1^* + \cdots + \hat{\beta}_k x_k^*$ and $\hat{y}$ is the calculated value of $\hat{Y}$.

> **4.** A $100(1 - \alpha)\%$ PI for a future y value is
>
> $$\hat{\mu}_{Y \cdot x_1^*, \ldots, \, x_k^*} \pm t_{\alpha/2, n-(k+1)} \cdot \{s^2 + (\text{estimated SD of } \hat{\mu}_{Y \cdot x_1^*, \ldots, x_k^*})^2\}^{1/2}$$
>
> $$= \hat{y} \pm t_{\alpha/2, n-(k+1)} \cdot \sqrt{s^2 + s_{\hat{Y}}^2}$$
>
> Simultaneous intervals for which the simultaneous confidence or prediction level is controlled can be obtained by applying the Bonferroni technique.

EXAMPLE 13.15   The article **"Independent but Additive Effects of Fluorine and Nitrogen Substitution on Properties of a Calcium Aluminosilicate Glass"** (*J. of the Amer. Ceramic Soc.,* **2012: 600–606**) used multiple regression analyses to investigate various properties of glasses in the Ca-Si-Al-O-N-F system. The following data on microhardness (GPa) resulted from various compositions of 28Ca:57:Si :15Al:$(100 - x - y)$O:xN:yF glasses:

Obs	N	F	microhardness
1	0	0	6.1
2	20	0	7.5
3	0	1	6.2
4	20	1	7.6
5	40	1	8.6
6	0	5	6.1
7	5	5	6.4
8	10	5	6.7
9	15	5	6.9
10	20	5	7.2
11	0	0	6.1
12	0	1	6.2
13	0	3	6.1
14	0	5	6.1
15	20	0	7.5
16	20	1	7.6
17	20	3	7.2
18	20	5	7.2

The model fit by the investigators was $Y = \beta_0 + \beta_1 N + \beta_2 F + \epsilon$. Figure 13.16 shows output from Minitab:

```
MicroHard = 6.23 + 0.0618N − 0.0387F


Predictor        Coef     SE Coef        T        P
Constant      6.22769     0.04615   134.93    0.000
N            0.061823    0.002099    29.46    0.000
F            −0.03872     0.01122    −3.45    0.004


S = 0.100051   R−Sq = 98.4%   R−Sq(adj) = 98.2%


Source           DF         SS        MS        F       P
Regression        2     9.1348    4.5674   456.28   0.000
Residual Error   15     0.1502    0.0100
Total            17     9.2850
```

Figure 13.16   Minitab output for Example 13.15

In addition, when $N = 20$ and $F = 1$,

$$\hat{\mu}_{Y \cdot 20, 1} = \hat{y} = 6.22769 + (.061823)(20) - (.03872)(1) = 7.4254$$

and the estimated standard deviation of $\hat{Y}$ for these values of the predictors is $s_{\hat{y}} = .0332$.

The very high R^2 indicates that almost all of the observed variation in microhardness can be attributed to the model relationship and the fact that both nitrogen %

and fluorine % are varying. And the F ratio of 456.28 with a corresponding P-value of .000 in the ANOVA table resoundingly confirms the utility of the fitted model.

Inferences about the individual regression coefficients are based on the t distribution with $18 - (2 + 1) = 15$ df (degrees of freedom for error in the ANOVA table), and $t_{.025,15} = 2.131$. A 95% CI for β_1 is

$$\hat{\beta}_1 \pm t_{.025,15}s_{\hat{\beta}_1} = .061823 \pm (2.131)(.002099)$$
$$= .061823 \pm .004473 \approx (.0573, .0663)$$

We estimate that the expected change in microhardness associated with an increase of 1% in N while holding F fixed is between .0573 GPa and .0663 GPa. A similar calculation gives $(-.0626, -.0148)$ as a 95% CI for β_2. The Bonferroni technique implies that the simultaneous confidence level for both intervals is at least 90%.

The t ratio for testing $H_0: \beta_1 = 0$ vs. $H_a: \beta_1 \neq 0$ is $\hat{\beta}_1/s_{\hat{\beta}_1} = .061823/.002099 = 29.46$. The corresponding P-value is twice the area under the t_{15} curve to the right of 29.46, which according to Minitab output is .000. Thus even with F remaining in the model, the predictor N provides additional useful information about microhardness. The evidence for testing $H_0: \beta_2 = 0$ versus $H_a: \beta_2 \neq 0$ is not quite so compelling; Figure 13.16 shows the P-value to be .004. So at significance level .05 or .01, H_0 would be rejected; it appears that F also provides useful information over and above what is contained in N. There is no reason to delete either predictor from the model. Since neither 95% CI contains 0, it is no surprise that both null hypotheses are rejected at significance level .05.

A 95% CI for true average hardness when $N = 20$ and $F = 1$ is

$$7.4254 \pm (2.131)(.0332) = 7.4254 \pm .0707 \approx (7.35, 7.50)$$

A 95% prediction interval for the hardness resulting from a single observation when $N = 20$ and $F = 1$ is

$$7.4254 \pm (2.131)\sqrt{(.100051)^2 + (.0332)^2} = 7.4254 \pm .2246 \approx (7.20, 7.65)$$

The PI is about three times as wide as the CI, reflecting the extra uncertainty in prediction. ∎

An F Test for a Group of Predictors The model utility F test was appropriate for testing whether there is useful information about the dependent variable in *any* of the k predictors (i.e., whether $\beta_1 = \cdots = \beta_k = 0$). In many situations, one first builds a model containing k predictors and then wishes to know whether any of the predictors in a particular subset provide useful information about Y. For example, a model to be used to predict students' test scores might include a group of background variables such as family income and education levels and also some school characteristic variables such as class size and spending per pupil. One interesting hypothesis is that the school characteristic predictors can be dropped from the model.

Let's label the predictors as $x_1, x_2, \ldots, x_l, x_{l+1}, \ldots, x_k$, so that it is the last $k - l$ that we are considering deleting. The relevant hypotheses are as follows:

$H_0: \beta_{l+1} = \beta_{l+2} = \cdots = \beta_k = 0$

(so the "reduced" model $Y = \beta_0 + \beta_1 x_1 + \cdots + \beta_l x_l + \epsilon$ is correct)

versus

H_a: at least one among $\beta_{l+1}, \ldots, \beta_k$ is not 0

(so in the "full" model $Y = \beta_0 + \beta_1 x_1 + \cdots + \beta_k x_k + \epsilon$, at least one of the last $k - l$ predictors provides useful information)

The test is carried out by fitting both the full and reduced models. Because the full model contains not only the predictors of the reduced model but also some extra predictors, it should fit the data at least as well as the reduced model. That is, if we let SSE_k be the sum of squared residuals for the full model and SSE_l be the corresponding sum for the reduced model, then $SSE_k \leq SSE_l$. Intuitively, if SSE_k is a great deal smaller than SSE_l, the full model provides a much better fit than the reduced model; the appropriate test statistic should then depend on the reduction $SSE_l - SSE_k$ in unexplained variation.

SSE_k = unexplained variation for the full model

SSE_l = unexplained variation for the reduced model

$$\text{Test statistic value: } f = \frac{(SSE_l - SSE_k)/(k - l)}{SSE_k/[n - (k + 1)]}$$

(13.20)

The test is upper-tailed; the P-value is the area under the $F_{k-l,n-(k+1)}$ curve to the right of f.

EXAMPLE 13.16   Soluble dietary fiber (SDF) can provide health benefits by lowering blood cholesterol and glucose levels. The article **"Effects of Twin-Screw Extrusion on Soluble Dietary Fiber and Physicochemical Properties of Soybean Residue"** (*Food Chemistry*, **2013: 884–889)** reported the following data on y = SDF content (%) in soybean residue and the three predictors extrusion temperature (x_1, in °C), feed moisture (x_2, in %), and screw speed (x_3, in rpm) of a twin-screw extrusion process.

obs	x_1	x_2	x_3	y
1	35	110	160	11.13
2	25	130	180	10.98
3	30	110	180	12.56
4	30	130	200	11.46
5	30	110	180	12.38
6	30	110	180	12.43
7	30	110	180	12.55
8	25	110	160	10.59
9	30	130	160	11.15
10	30	90	200	10.55
11	30	90	160	9.25
12	25	90	180	9.58
13	35	110	200	11.59
14	35	90	180	10.68
15	35	130	180	11.73
16	25	110	200	10.81
17	30	110	180	12.68

The authors of the cited article fit the complete second-order model with predictors x_1, x_2, x_3, x_1^2, x_2^2, x_3^2, x_1x_2, x_1x_3, and x_2x_3. Figure 13.17 shows Minitab output resulting from fitting this model. Note that the $R^2 = .987$, so almost all of the observed variation in y can be explained by the model relationship, and adjusted R^2 is only slightly smaller than R^2 itself. Furthermore, the F ratio for model utility is 59.93 with a corresponding P-value of .000 (the area under the $F_{9,7}$ curve to the right of 59.93). So the null hypothesis that all nine β_i's corresponding to predictors have value 0 is resoundingly rejected. There appears to be a useful relationship between the dependent variable and at least one of the predictors.

```
The regression equation is
y = - 132 + 1.69 x1 + 0.777 x2 + 0.798 x3 - 0.0270 x1sqd - 0.00276 x2sqd
     - 0.00204 x3sqd - 0.000875 x1x2 + 0.000600 x1x3 - 0.000619 x2x3

Predictor          Coef       SE Coef        T       P
Constant        -131.61         10.41   -12.64   0.000
x1                1.6875       0.2764     6.10   0.000
x2               0.77688      0.06683    11.62   0.000
x3               0.79788      0.08484     9.40   0.000
x1sqd          -0.027000     0.003418    -7.90   0.000
x2sqd         -0.0027563    0.0002136   -12.90   0.000
x3sqd         -0.0020375    0.0002136    -9.54   0.000
x1x2          -0.0008750    0.0008767    -1.00   0.352
x1x3           0.0006000    0.0008767     0.68   0.516
x2x3          -0.0006188    0.0002192    -2.82   0.026

S = 0.175347    R-Sq = 98.7%    R-Sq(adj) = 97.1%

Analysis of Variance

Source            DF          SS        MS        F       P
Regression         9     16.5830    1.8426    59.93   0.000
Residual Error     7      0.2152    0.0307
Total             16     16.7982

Source    DF    Seq SS
x1         1    1.2561
x2         1    3.4585
x3         1    0.6555
x1sqd      1    2.5869
x2sqd      1    5.5393
x3sqd      1    2.7967
x1x2       1    0.0306
x1x3       1    0.0144
x2x3       1    0.2450
```

Figure 13.17 Minitab output from fitting the complete second-order model to the data of Example 13.16

Is the inclusion of the second-order predictors justified? That is, should the reduced model consisting of just the predictors x_1, x_2, and x_3 $(l = 3)$ be used? The hypotheses to be tested are

$$H_0: \beta_4 = \beta_5 = \cdots = \beta_9 = 0$$

versus

$$H_a: \text{at least one among } \beta_4, \ldots, \beta_9 \text{ is not 0}$$

SSE = .2152 for the full model (from the ANOVA table of Figure 13.17). Now we need SSE for the reduced model that contains only the three first-order predictors x_1, x_2, and x_3. It is actually not necessary to fit this model because of the "Sequential Sums of Squares" information at the bottom of Figure 13.17. Each number in the last column gives the *increase* in SSR (explained variation) when another predictor is entered into the model. So SSR for the reduced model is 1.2561 + 3.4585 + .6555 = 5.3701. Subtracting this from SST = 16.7982 (which is the same for both models) gives SSE = 11.4281. The value of the F statistic is then

$$f = \frac{(11.4281 - .2152)/3}{.2152/[17 - (9 + 1)]} = \frac{3.7376}{.03074} = 121.6$$

The P-value is the area under the $F_{3,7}$ curve to the right of this value, which unsurprisingly is 0. So the null hypothesis is resoundingly rejected. There is very convincing evidence for concluding that at least one of the second-order predictors is providing useful information over and above what is provided by the three first-order predictors. This conclusion makes intuitive sense because the full model leaves very little variation unexplained (SSE quite close to 0), whereas the reduced model has a rather substantial amount of unexplained variation relative to SST.

The t ratios of Figure 13.17 suggest that perhaps only the three quadratic predictors are useful and that the three interaction predictors can be eliminated. So let's now consider testing $H_0: \beta_7 = \beta_8 = \beta_9 = 0$ against the alternative that at least one of these three β_is is not 0. Again the sequential sums of squares information in Figure 13.17 allows us to obtain SSE for the reduced model (containing just the six predictors $x_1, x_2, x_3, x_1^2, x_2^2, x_3^2$) without actually fitting that model: SSR = the sum of the first six numbers in the Seq SS column = 16.2930, whence SSE = 16.7982 − 16.2930 = .5052. Then

$$f = [(.5052 - .2152)/3]/[.2152/(17 - (6 + 1))] = 4.49$$

Table A.9 gives $F_{.05,3,10} = 3.71$ and $F_{.01,3,10} = 6.55$, implying that the P-value is between .01 and .05. In particular, at a significance level of .01, the null hypothesis would not be rejected. The conclusion at that level is that none of the three interaction predictors provides additional useful information. ■

Assessing Model Adequacy

The standardized residuals in multiple regression result from dividing each residual by its estimated standard deviation; the formula for these standard deviations is substantially more complicated than in the case of simple linear regression. We recommend a normal probability plot of the standardized residuals as a basis for validating the normality assumption. If the pattern in this plot departs substantially from linearity, the t and F procedures developed in this section should not be used to make inferences. Plots of the standardized residuals versus each predictor and versus $\hat{y}$ should show no discernible pattern. Adjusted residual plots can also be helpful in this endeavor. The book by Neter et al. is an extremely useful reference.

EXAMPLE 13.17   Figure 13.18 shows a normal probability plot of the standardized residuals for the microhardness data and fitted model given in Example 13.15. The straightness of the plot casts little doubt on the assumption that the random deviation ϵ is normally distributed.

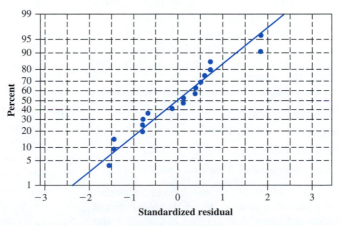

Figure 13.18   A normal probability plot of the standardized residuals for the data and model of Example 13.15

Figure 13.19 shows the other suggested plots for the microhardness data (fewer than 18 points appear because various observed and calculated values are duplicated).

Given the rather small sample size, there is not much evidence of a pattern in any of the first three plots other than randomness.

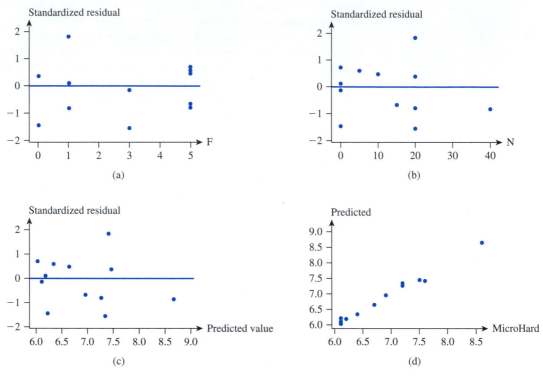

Figure 13.19 Diagnostic plots for the microhardness data: (a) standardized residual versus x_1; (b) standardized residual versus x_2; (c) standardized residual versus $\hat{y}$; (d) $\hat{y}$ versus y ∎

EXERCISES Section 13.4 (36–54)

36. Cardiorespiratory fitness is widely recognized as a major component of overall physical well-being. Direct measurement of maximal oxygen uptake (VO_2max) is the single best measure of such fitness, but direct measurement is time-consuming and expensive. It is therefore desirable to have a prediction equation for VO_2max in terms of easily obtained quantities. Consider the variables

$y = VO_2$max (L/min) $x_1 =$ weight (kg)

$x_2 =$ age (yr)

$x_3 =$ time necessary to walk 1 mile (min)

$x_4 =$ heart rate at the end of the walk (beats/min)

Here is one possible model, for male students, consistent with the information given in the article **"Validation of the Rockport Fitness Walking Test in College Males and Females"** (*Research Quarterly for Exercise and Sport*, **1994: 152–158**):

$Y = 5.0 + .01x_1 - .05x_2 - .13x_3 - .01x_4 + \epsilon$
$\sigma = .4$

a. Interpret β_1 and β_3.

b. What is the expected value of VO_2max when weight is 76 kg, age is 20 yr, walk time is 12 min, and heart rate is 140 b/m?

c. What is the probability that VO_2max will be between 1.00 and 2.60 for a single observation made when the values of the predictors are as stated in part (b)?

37. A trucking company considered a multiple regression model for relating the dependent variable $y =$ total daily travel time for one of its drivers (hours) to the predictors $x_1 =$ distance traveled (miles) and $x_2 =$ the number of deliveries made. Suppose that the model equation is

$$Y = -.800 + .060x_1 + .900x_2 + \epsilon$$

a. What is the mean value of travel time when distance traveled is 50 miles and three deliveries are made?

b. How would you interpret $\beta_1 = .060$, the coefficient of the predictor x_1? What is the interpretation of $\beta_2 = .900$?

c. If $\sigma = .5$ hour, what is the probability that travel time will be at most 6 hours when three deliveries are made and the distance traveled is 50 miles?

38. Let $y =$ wear life of a bearing, $x_1 =$ oil viscosity, and $x_2 =$ load. Suppose that the multiple regression model relating life to viscosity and load is

$$Y = 125.0 + 7.75x_1 + .0950x_2 - .0090x_1x_2 + \epsilon$$

a. What is the mean value of life when viscosity is 40 and load is 1100?

b. When viscosity is 30, what is the change in mean life associated with an increase of 1 in load? When viscosity is 40, what is the change in mean life associated with an increase of 1 in load?

39. Let $y =$ sales at a fast-food outlet (1000s of \$), $x_1 =$ number of competing outlets within a 1-mile radius, $x_2 =$ population within a 1-mile radius (1000s of people), and x_3 be an indicator variable that equals 1 if the outlet has a drive-up window and 0 otherwise. Suppose that the true regression model is

$$Y = 10.00 - 1.2x_1 + 6.8x_2 + 15.3x_3 + \epsilon$$

a. What is the mean value of sales when the number of competing outlets is 2, there are 8000 people within a 1-mile radius, and the outlet has a drive-up window?

b. What is the mean value of sales for an outlet without a drive-up window that has three competing outlets and 5000 people within a 1-mile radius?

c. Interpret β_3.

40. The article cited in Exercise 49 of Chapter 7 gave summary information on a regression in which the dependent variable was power output (W) in a simulated 200-m race and the predictors were $x_1 =$ arm girth (cm), $x_2 =$ excess post-exercise oxygen consumption (ml/kg), and $x_3 =$ immediate posttest lactate (mmol/L). The estimated regression equation was reported as

$$y = -408.20 + 14.06x_1 + .76x_2 - 3.64x_3$$
$$(n = 11, R^2 = .91)$$

a. Carry out the model utility test using a significance level of .01. [*Note:* All three predictors were judged to be important.]

b. Interpret the estimate 14.06.

c. Predict power output when arm girth is 36 cm, excess oxygen consumption is 120 ml/kg, and lactate is 10.0.

d. Calculate a point estimate for true average power output when values of the predictors are as given in (c).

e. Obtain a point estimate for the true average change in power output associated with a 1 mmol/L increase in lactate while arm girth and oxygen consumption remain fixed.

41. The article **"A Study of Factors Affecting the Human Cone Photoreceptor Density Measured by Adaptive Optics Scanning Laser Opthalmoscope"** (*Exptl. Eye Research*, **2013: 1–9**) included a summary of a multiple

regression analysis based on a sample of $n = 192$ eyes; the dependent variable was cone cell packing density (cells/mm^2), and the two independent variables were $x_1 =$ eccentricity (mm) and $x_2 =$ axial length (mm).

a. The reported coefficient of multiple determination was .834. Interpret this value, and carry out a test of model utility.

b. The estimated regression function was $y = 35,821.792 - 6294.729x_1 - 348.037x_2$. Calculate a point prediction for packing density when eccentricity is 1 mm and axial length is 25 mm.

c. Interpret the coefficient on x_1 in the estimated regression function in (b).

d. The estimated standard error of $\hat{\beta}_1$ was 203.702. Calculate and interpret a confidence interval with confidence level 95% for β_1.

e. The estimated standard error of the estimated coefficient on axial length was 134.350. Test the null hypothesis $H_0: \beta_2 = 0$ against the alternative $H_a: \beta_2 \neq 0$ using a significance level of .05, and interpret the result.

42. An investigation of a die-casting process resulted in the accompanying data on $x_1 =$ furnace temperature, $x_2 =$ die close time, and $y =$ temperature difference on the die surface (**"A Multiple-Objective Decision-Making Approach for Assessing Simultaneous Improvement in Die Life and Casting Quality in a Die Casting Process,"** *Quality Engineering*, **1994: 371–383**).

x_1	1250	1300	1350	1250	1300
x_2	6	7	6	7	6
y	80	95	101	85	92

x_1	1250	1300	1350	1350
x_2	8	8	7	8
y	87	96	106	108

Minitab output from fitting the multiple regression model with predictors x_1 and x_2 is given here.

```
The regression equation is
tempdiff = -200 + 0.210 furntemp
              + 3.00 clostime

Predictor      Coef     Stdev  t-ratio      p
Constant    -199.56     11.64   -17.14  0.000
furntemp   0.210000  0.008642    24.30  0.000
clostime     3.0000    0.4321     6.94  0.000

s = 1.058  R-sq = 99.1%     R-sq(adj) = 98.8%


Analysis of Variance
SOURCE       DF       SS      MS       F      p
Regression    2   715.50  357.75  319.31  0.000
Error         6     6.72    1.12
Total         8   722.22
```

a. Carry out the model utility test.

b. Calculate and interpret a 95% confidence interval for β_2, the population regression coefficient of x_2.

c. When $x_1 = 1300$ and $x_2 = 7$, the estimated standard deviation of $\hat{Y}$ is $s_{\hat{Y}} = .353$. Calculate a 95% confidence interval for true average temperature difference when furnace temperature is 1300 and die close time is 7.

d. Calculate a 95% prediction interval for the temperature difference resulting from a single experimental run with a furnace temperature of 1300 and a die close time of 7.

43. An experiment carried out to study the effect of the mole contents of cobalt (x_1) and the calcination temperature (x_2) on the surface area of an iron-cobalt hydroxide catalyst (y) resulted in the accompanying data (**"Structural Changes and Surface Properties of $Co_xFe_{3-x}O_4$ Spinels," *J. of Chemical Tech. and Biotech.*, 1994: 161–170**). A request to the SAS package to fit $\beta_0 + \beta_1 x_1 + \beta_2 x_2 + \beta_3 x_3$, where $x_3 = x_1 x_2$ (an interaction predictor) yielded the output below.

x_1	.6	.6	.6	.6	.6	1.0	1.0
x_2	200	250	400	500	600	200	250
y	90.6	82.7	58.7	43.2	25.0	127.1	112.3

x_1	1.0	1.0	1.0	2.6	2.6	2.6	2.6
x_2	400	500	600	200	250	400	500
y	19.6	17.8	9.1	53.1	52.0	43.4	42.4

x_1	2.6	2.8	2.8	2.8	2.8	2.8
x_2	600	200	250	400	500	600
y	31.6	40.9	37.9	27.5	27.3	19.0

a. Predict the value of surface area when cobalt content is 2.6 and temperature is 250, and calculate the value of the corresponding residual.

b. Since $\hat{\beta}_1 = -46.0$, is it legitimate to conclude that if cobalt content increases by 1 unit while the values of the other predictors remain fixed, surface area can be expected to decrease by roughly 46 units? Explain your reasoning.

c. Does there appear to be a useful linear relationship between y and the predictors?

d. Given that mole contents and calcination temperature remain in the model, does the interaction predictor x_3 provide useful information about y? State and test the appropriate hypotheses using a significance level of .01.

e. The estimated standard deviation of $\hat{Y}$ when mole contents is 2.0 and calcination temperature is 500 is $s_{\hat{Y}} = 4.69$. Calculate a 95% confidence interval for the mean value of surface area under these circumstances.

44. The accompanying Minitab regression output is based on data that appeared in the article **"Application of Design of Experiments for Modeling Surface Roughness in Ultrasonic Vibration Turning"** (*J. of Engr. Manuf.*, 2009: 641–652). The response variable is surface roughness (μm), and the independent variables are vibration amplitude (μm), depth of cut (mm), feed rate (mm/rev), and cutting speed (m/min), respectively.

a. How many observations were there in the data set?

b. Interpret the coefficient of multiple determination.

c. Carry out a test of hypotheses to decide if the model specifies a useful relationship between the response variable and at least one of the predictors.

d. Interpret the number 18.2602 that appears in the Coef column.

e. At significance level .10, can any single one of the predictors be eliminated from the model provided that all of the other predictors are retained?

f. The estimated SD of $\hat{Y}$ when the values of the four predictors are 10, .5, .25, and 50, respectively, is

SAS output for Exercise 43

```
Dependent Variable: SURFAREA

                    Analysis of Variance

   Source      DF    Sum of Squares    Mean Square    F Value    Prob>F
   Model        3       15223.52829     5074.50943     18.924    0.0001
   Error       16        4290.53971      268.15873
   C Total     19       19514.06800

   Root MSE      16.37555      R-square     0.7801
   Dep Mean      48.06000      Adj R-sq     0.7389
   C.V.          34.07314

                    Parameter Estimates

                     Parameter     Standard    T for H0:        Prob
   Variable   DF      Estimate        Error    Parameter = 0    > |T|
   INTERCEP    1     185.485740    21.19747682      8.750       0.0001
   COBCON      1     -45.969466    10.61201173     -4.332       0.0005
   TEMP        1      -0.301503     0.05074421     -5.942       0.0001
   CONTEMP     1       0.088801     0.02540388      3.496       0.0030
```

.1178. Calculate both a CI for true average roughness and a PI for the roughness of a single specimen, and compare these two intervals.

```
The regression equation is
Ra = −0.972 − 0.0312 a + 0.557 d + 18.3 f + 0.00282 v

Predictor        Coef    SE Coef       T       P
Constant       −0.9723    0.3923   −2.48   0.015
   a           −0.03117   0.01864  −1.67   0.099
   d            0.5568    0.3185    1.75   0.084
   f           18.2602    0.7536   24.23   0.000
   v            0.002822  0.003977  0.71   0.480

S = 0.822059   R-Sq = 88.6%   R-Sq(adj) = 88.0%

Source             DF      SS      MS       F       P
Regression          4   401.02  100.25  148.35  0.000
Residual Error     76    51.36    0.68
Total              80   452.38
```

45. The article **"Analysis of the Modeling Methodologies for Predicting the Strength of Air-Jet Spun Yarns"** (*Textile Res. J.*, 1997: 39–44) reported on a study carried out to relate yarn tenacity (y, in g/tex) to yarn count (x_1, in tex), percentage polyester (x_2), first nozzle pressure (x_3, in kg/cm^2), and second nozzle pressure (x_4, in kg/cm^2). The estimate of the constant term in the corresponding multiple regression equation was 6.121. The estimated coefficients for the four predictors were −.082, .113, .256, and −.219, respectively, and the coefficient of multiple determination was .946.

   a. Assuming that the sample size was $n = 25$, state and test the appropriate hypotheses to decide whether the fitted model specifies a useful linear relationship between the dependent variable and at least one of the four model predictors.

   b. Again using $n = 25$, calculate the value of adjusted R^2.

   c. Calculate a 99% confidence interval for true mean yarn tenacity when yarn count is 16.5, yarn contains 50% polyester, first nozzle pressure is 3, and second nozzle pressure is 5 if the estimated standard deviation of predicted tenacity under these circumstances is .350.

46. A regression analysis carried out to relate $y =$ repair time for a water filtration system (hr) to $x_1 =$ elapsed time since the previous service (months) and $x_2 =$ type of repair (1 if electrical and 0 if mechanical) yielded the following model based on $n = 12$ observations: $y = .950 + .400x_1 + 1.250x_2$. In addition, SST = 12.72, SSE = 2.09, and $s_{\hat{\beta}_2} = .312$.

   a. Does there appear to be a useful linear relationship between repair time and the two model predictors? Carry out a test of the appropriate hypotheses using a significance level of .05.

   b. Given that elapsed time since the last service remains in the model, does type of repair provide useful information about repair time? State and test the appropriate hypotheses using a significance level of .01.

   c. Calculate and interpret a 95% CI for β_2.

   d. The estimated standard deviation of a prediction for repair time when elapsed time is 6 months and the

repair is electrical is .192. Predict repair time under these circumstances by calculating a 99% prediction interval. Does the interval suggest that the estimated model will give an accurate prediction? Why or why not?

47. Efficient design of certain types of municipal waste incinerators requires that information about energy content of the waste be available. The authors of the article **"Modeling the Energy Content of Municipal Solid Waste Using Multiple Regression Analysis"** (*J. of the Air and Waste Mgmnt. Assoc.*, 1996: 650–656) kindly provided us with the accompanying data on $y =$ energy content (kcal/kg), the three physical composition variables $x_1 = \%$ plastics by weight, $x_2 = \%$ paper by weight, and $x_3 = \%$ garbage by weight, and the proximate analysis variable $x_4 = \%$ moisture by weight for waste specimens obtained from a certain region.

Obs	Plastics	Paper	Garbage	Water	Energy Content
1	18.69	15.65	45.01	58.21	947
2	19.43	23.51	39.69	46.31	1407
3	19.24	24.23	43.16	46.63	1452
4	22.64	22.20	35.76	45.85	1553
5	16.54	23.56	41.20	55.14	989
6	21.44	23.65	35.56	54.24	1162
7	19.53	24.45	40.18	47.20	1466
8	23.97	19.39	44.11	43.82	1656
9	21.45	23.84	35.41	51.01	1254
10	20.34	26.50	34.21	49.06	1336
11	17.03	23.46	32.45	53.23	1097
12	21.03	26.99	38.19	51.78	1266
13	20.49	19.87	41.35	46.69	1401
14	20.45	23.03	43.59	53.57	1223
15	18.81	22.62	42.20	52.98	1216
16	18.28	21.87	41.50	47.44	1334
17	21.41	20.47	41.20	54.68	1155
18	25.11	22.59	37.02	48.74	1453
19	21.04	26.27	38.66	53.22	1278
20	17.99	28.22	44.18	53.37	1153
21	18.73	29.39	34.77	51.06	1225
22	18.49	26.58	37.55	50.66	1237
23	22.08	24.88	37.07	50.72	1327
24	14.28	26.27	35.80	48.24	1229
25	17.74	23.61	37.36	49.92	1205
26	20.54	26.58	35.40	53.58	1221
27	18.25	13.77	51.32	51.38	1138
28	19.09	25.62	39.54	50.13	1295
29	21.25	20.63	40.72	48.67	1391
30	21.62	22.71	36.22	48.19	1372

Using Minitab to fit a multiple regression model with the four aforementioned variables as predictors of energy content resulted in the following output:

```
The regression equation is
enercont = 2245 + 28.9 plastics
                    + 7.64 paper + 4.30 garbage
                    −37.4 water

Predictor       Coef      StDev        T        P
Constant       2244.9     177.9     12.62    0.000
plastics       28.925     2.824     10.24    0.000
paper           7.644     2.314      3.30    0.003
garbage         4.297     1.916      2.24    0.034
water         -37.354     1.834    -20.36    0.000

s = 31.48       R-Sq = 96.4%     R-Sq(adj) = 95.8%

Analysis of Variance

Source      DF       SS       MS       F        P
Regression   4   664931   166233   167.71   0.000
Error       25    24779      991
Total       29   689710
```

a. Interpret the values of the estimated regression coefficients $\hat{\beta}_1$ and $\hat{\beta}_4$.

b. State and test the appropriate hypotheses to decide whether the model fit to the data specifies a useful linear relationship between energy content and at least one of the four predictors.

c. Given that % plastics, % paper, and % water remain in the model, does % garbage provide useful information about energy content? State and test the appropriate hypotheses using a significance level of .05.

d. Use the fact that $s_{\hat{Y}} = .7.46$ when $x_1 = 20$, $x_2 = 25$, $x_3 = 40$, and $x_4 = 45$ to calculate a 95% confidence interval for true average energy content under these circumstances. Does the resulting interval suggest that mean energy content has been precisely estimated?

e. Use the information given in part (d) to predict energy content for a waste sample having the specified characteristics, in a way that conveys information about precision and reliability.

48. An experiment to investigate the effects of a new technique for degumming of silk yarn was described in the article **"Some Studies in Degumming of Silk with Organic Acids"** (*J. Society of Dyers and Colourists*, 1992: 79–86). One response variable of interest was y = weight loss (%). The experimenters made observations on weight loss for various values of three independent variables: x_1 = temperature (°C) = 90, 100, 110; x_2 = time of teatment (min) = 30, 75, 120; x_3 = tartaric acid concentration (g/L) = 0, 8, 16. In the regression analyses, the three values of each variable were coded as −1, 0, and 1, respectively, giving the accompanying data (the value $y_8 = 19.3$ was reported, but our value $y_8 = 20.3$ results in regression output identical to that appearing in the article).

Obs	1	2	3	4	5	6	7	8
x_1	−1	−1	1	1	−1	−1	1	1
x_2	−1	1	−1	1	0	0	0	0
x_3	0	0	0	0	−1	1	−1	1
y	18.3	22.2	23.0	23.0	3.3	19.3	19.3	20.3

Obs	9	10	11	12	13	14	15
x_1	0	0	0	0	0	0	0
x_2	−1	−1	1	1	0	0	0
x_3	−1	1	−1	1	0	0	0
y	13.1	23.0	20.9	21.5	22.0	21.3	22.6

A multiple regression model with $k = 9$ predictors—x_1, x_2, x_3, $x_4 = x_1^2$, $x_5 = x_2^2$, $x_6 = x_3^2$, $x_7 = x_1x_2$, $x_8 = x_1x_3$, and $x_9 = x_2x_3$—was fit to the data, resulting in $\hat{\beta}_0 = 21.967$, $\hat{\beta}_1 = 2.8125$, $\hat{\beta}_2 = 1.2750$, $\hat{\beta}_3 = 3.4375$, $\hat{\beta}_4 = -2.208$, $\hat{\beta}_5 = 1.867$, $\hat{\beta}_6 = -4.208$, $\hat{\beta}_7 = -.975$, $\hat{\beta}_8 = -3.750, \hat{\beta}_9 = -2.325$, SSE $= 23.379$, and $R^2 = .938$.

a. Does this model specify a useful relationship? State and test the appropriate hypotheses using a significance level of .01.

b. The estimated standard deviation of $\hat{\mu}_Y$ when $x_1 = \cdots = x_9 = 0$ (i.e., when temperature = 100, time = 75, and concentration = 8) is 1.248. Calculate a 95% CI for expected weight loss when temperature, time, and concentration have the specified values.

c. Calculate a 95% PI for a single weight-loss value to be observed when temperature, time, and concentration have values 100, 75, and 8, respectively.

d. Fitting the model with only x_1, x_2, and x_3 as predictors gave $R^2 = .456$ and SSE = 203.82. Does at least one of the second-order predictors provide additional useful information? State and test the appropriate hypotheses.

49. Researchers carried out a study to see how y = ultimate deflection (mm), of reinforced ultrahigh toughness cementitious composite beams were influenced by x_1 = shear span ratio and x_2 = splitting tensile strength (MPa), resulting in the accompanying data (**"Shear Behavior of Reinforced Ultrahigh Toughness Cementitious Composite Beams without Transverse Reinforcement,"** *J. of Materials in Civil Engr.*, 2012: 1283–1294):

x_1	x_2	y	x_1	x_2	y
2.04	3.55	3.11	3.08	3.62	3.36
2.04	6.07	3.26	3.08	5.89	6.49
3.06	3.55	3.89	4.11	3.62	2.72
3.06	6.07	10.25	4.11	5.89	12.48
4.08	3.55	3.11	2.01	6.18	2.82
4.08	6.16	13.48	3.02	6.18	5.19
2.06	3.62	3.94	4.03	6.18	8.04
2.06	6.16	3.53			

a. Here is Minitab output from fitting the model with predictors x_1, x_2, and $x_3 = x_1x_2$:

```
The regression equation is
y = 17.3 − 6.37 x1 − 3.66 x2 + 1.71 x1x2

Predictor      Coef     SE Coef       T        P
Constant     17.279       7.167    2.41    0.035
x1           -6.368       2.260   22.82    0.017
x2           -3.658       1.364   22.68    0.021
x1x2          1.7067      0.4314    3.96    0.002
```

S=1.72225 R-Sq=82.5% R-Sq(adj)=77.8%

Analysis of Variance

Source	DF	SS	MS	F	P
Regression	3	154.033	51.344	17.31	0.000
Residual Error	11	32.627	2.966		
Total	14	186.660			

Carry out a test of model utility.

b. Should the interaction predictor be retained in the model? Carry out a test of hypotheses using a significance level of .05.

c. The estimated standard deviation of $\hat{Y}$ when $x_1 = 3$ and $x_2 = 6$ is $s_{\hat{y}} = .555$. Calculate and interpret a confidence interval with a 95% confidence level for true average deflection under these circumstances.

d. Using the information in (c), calculate and interpret a prediction interval using a 95% confidence level for a future value of ultimate deflection to be observed when $x_1 = 3$ and $x_2 = 6$.

50. When the model $Y = \beta_0 + \beta_1 x_1 + \beta_2 x_2 + \beta_3 x_1^2 + \beta_4 x_2^2 + \beta_5 x_1 x_2 + \epsilon$ is fit to the data of Exercise 49, the resulting value of SSE is 28.947. Given that the predictors x_1, x_2, and $x_1 x_2$ remain in the model, does either of the quadratic predictors x_1^2 or x_2^2 provide additional useful information? State and test the appropriate hypotheses.

51. The article **"Optimization of Surface Roughness in Drilling Using Vegetable-Based Cutting Oils Developed from Sunflower Oil"** (*Industrial Lubrication and Tribology*, 2011: 271–276) gave the following data on x_1 = spindle speed (rpm), x_2 = feed rate (mm/rev), x_3 = drilling depth (mm), and y = surface roughness (μm) when a semisynthetic cutting fluid was used:

x_1	x_2	x_3	y	e^*
320	.10	15	2.27	−1.32
320	.12	20	4.14	1.08
320	.14	25	4.69	0.26
420	.10	20	1.92	−0.40
420	.12	25	2.63	−0.79
420	.14	15	4.34	0.99
520	.10	25	2.03	1.64
520	.12	15	2.34	0.03
520	.14	20	2.67	−1.52

a. Here is partial Minitab output from fitting the model with x_1, x_2, and x_3 as predictors (authors of the cited article used Minitab for this purpose):

Predictor	Coef	SE Coef	T	P
Constant	0.099	1.871	0.05	0.960
x_1	−0.006767	0.002231	−3.03	0.029
x_2	45.67	11.16	4.09	0.009
x_3	0.01333	0.04463	0.30	0.777

S=0.546589 R-Sq = 83.9% R-Sq(adj)=74.2%

Does drilling depth provide useful information about roughness given that spindle speed and feed rate remain in the model?

b. Here is Minitab output from fitting the model with just x_1 and x_2 as predictors (the cited article made no mention of this model):

Predictor	Coef	SE Coef	T	P
Constant	0.365	1.514	0.24	0.817
x_1	−0.006767	0.002055	−3.29	0.017
x_2	45.67	10.28	4.44	0.004

S=0.503400 R-Sq=83.6% R-Sq(adj)=78.1%

Carry out a test of model utility using $\alpha = .05$.

c. Calculate and interpret a 95% CI for the population regression coefficient on x_1.

d. The estimated standard deviation of the predicted Y when $x_1 = 400$ and $x_2 = .125$ is .180. Calculate a 95% CI for true average roughness under these circumstances.

e. The e^* values that appear along with the data are from the regression of (b). Investigate model adequacy.

52. Utilization of sucrose as a carbon source for the production of chemicals is uneconomical. Beet molasses is a readily available and low-priced substitute. The article **"Optimization of the Production of β-Carotene from Molasses by Blakeslea Trispora"** (*J. of Chem. Tech. and Biotech.*, 2002: 933–943) carried out a multiple regression analysis to relate the dependent variable y = amount of β-carotene (g/dm^3) to the three predictors amount of linoleic acid, amount of kerosene, and amount of antioxidant (all g/dm^3).

Obs	Linoleic	Kerosene	Antiox	Betacaro
1	30.00	30.00	10.00	0.7000
2	30.00	30.00	10.00	0.6300
3	30.00	30.00	18.41	0.0130
4	40.00	40.00	5.00	0.0490
5	30.00	30.00	10.00	0.7000
6	13.18	30.00	10.00	0.1000
7	20.00	40.00	5.00	0.0400
8	20.00	40.00	15.00	0.0065
9	40.00	20.00	5.00	0.2020
10	30.00	30.00	10.00	0.6300
11	30.00	30.00	1.59	0.0400
12	40.00	20.00	15.00	0.1320
13	40.00	40.00	15.00	0.1500
14	30.00	30.00	10.00	0.7000
15	30.00	46.82	10.00	0.3460
16	30.00	30.00	10.00	0.6300
17	30.00	13.18	10.00	0.3970
18	20.00	20.00	5.00	0.2690
19	20.00	20.00	15.00	0.0054
20	46.82	30.00	10.00	0.0640

a. Fitting the complete second-order model in the three predictors resulted in $R^2 = .987$ and adjusted $R^2 = .974$, whereas fitting the first-order model gave $R^2 = .016$. What would you conclude about the two models?

b. For $x_1 = x_2 = 30, x_3 = 10$, a statistical software package reported that $\hat{y} = .66573, s_{\hat{y}} = .01785$, based on the complete second-order model. Predict the amount of β-carotene that would result from a single experimental run with the designated values of the independent variables, and do so in a way that conveys information about precision and reliability.

53. Snowpacks contain a wide spectrum of pollutants that may represent environmental hazards. The article **"Atmospheric PAH Deposition: Deposition Velocities and Washout Ratios"** (*J. of Environmental Engineering*, 2002: 186–195) focused on the deposition of polyaromatic hydrocarbons. The authors proposed a multiple regression model for relating deposition over a specified time period (y, in $\mu g/m^2$) to two rather complicated predictors x_1 (μg-sec/m³) and x_2 ($\mu g/m^2$), defined in terms of PAH air concentrations for various species, total time, and total amount of precipitation. Here is data on the species fluoranthene and corresponding Minitab output:

obs	x_1	x_2	flth
1	92017	.0026900	278.78
2	51830	.0030000	124.53
3	17236	.0000196	22.65
4	15776	.0000360	28.68
5	33462	.0004960	32.66
6	243500	.0038900	604.70
7	67793	.0011200	27.69
8	23471	.0006400	14.18
9	13948	.0004850	20.64
10	8824	.0003660	20.60
11	7699	.0002290	16.61
12	15791	.0014100	15.08
13	10239	.0004100	18.05
14	43835	.0000960	99.71
15	49793	.0000896	58.97
16	40656	.0026000	172.58
17	50774	.0009530	44.25

The regression equation is

flth = −33.5 + 0.00205 x1 + 29836 x2

Predictor	Coef	SE Coef	T	P
Constant	−33.46	14.90	−2.25	0.041
x1	0.0020548	0.0002945	6.98	0.000
x2	29836	13654	2.19	0.046

S = 44.28 R-Sq = 92.3% R-Sq(adj) = 91.2%

Analysis of Variance

Source	DF	SS	MS	F	P
Regression	2	330989	165495	84.39	0.000
Residual Error	14	27454	1961		
Total	16	358443			

Formulate questions and perform appropriate analyses to draw conclusions.

54. The use of high-strength steels (HSS) rather than aluminum and magnesium alloys in automotive body structures reduces vehicle weight. However, HSS use is still problematic because of difficulties with limited formability, increased springback, difficulties in joining, and reduced die life. The article **"Experimental Investigation of Springback Variation in Forming of High Strength Steels"** (*J. of Manuf. Sci. and Engr.*, 2008: 1–9) included data on y = springback from the wall opening angle and x_1 = blank holder pressure. Three different material suppliers and three different lubrication regimens (no lubrication, lubricant #1, and lubricant #2) were also utilized.

a. What predictors would you use in a model to incorporate supplier and lubrication information in addition to BHP?

b. The accompanying Minitab output resulted from fitting the model of (a) (the article's authors also used Minitab; amusingly, they employed a significance level of .06 in various tests of hypotheses). Does there appear to be a useful relationship between the response variable and at least one of the predictors? Carry out a formal test of hypotheses.

c. When BHP is 1000, material is from supplier 1, and no lubrication is used, $s_{\hat{y}} = .524$. Calculate a 95% PI for the spingback that would result from making an additional observation under these conditions.

d. From the output, it appears that lubrication regimen may not be providing useful information. A regression with the corresponding predictors removed resulted in SSE = 48.426. What is the coefficient of multiple determination for this model, and what would you conclude about the importance of the lubrication regimen?

e. A model with predictors for BHP, supplier, and lubrication regimen, as well as predictors for interactions between BHP and both supplier and lubrication regiment, resulted in SSE = 28.216 and $R^2 = .849$. Does this model appear to improve on the model with just BHP and predictors for supplier?

Predictor	Coef	SE Coef	T	P
Constant	21.5322	0.6782	31.75	0.000
BHP	−0.0033680	0.0003919	−8.59	0.000
Suppl_1	−1.7181	0.5977	−2.87	0.007
Suppl_2	−1.4840	0.6010	−2.47	0.019
Lub_1	−0.3036	0.5754	−0.53	0.602
Lub_2	0.8931	0.5779	1.55	0.133

S = 1.18413 R-Sq = 77.5% R-Sq(adj) = 73.8%

Source	DF	SS	MS	F	P
Regression	5	144.915	28.983	20.67	0.000
Residual Error	30	42.065	1.402		
Total	35	186.980			

In this section, we touch upon a number of issues that may arise when a multiple regression analysis is carried out. Consult the chapter references for a more extensive treatment of any particular topic.

Transformations

Sometimes, theoretical considerations suggest a nonlinear relation between a dependent variable and two or more independent variables, whereas on other occasions diagnostic plots indicate that some type of nonlinear function should be used. Frequently a transformation will linearize the model.

EXAMPLE 13.18 Natural single crystal diamond has been widely used in ultraprecision machining. However, its application to the cutting of ferrous metals has been problematic due to significant tool wear. The article **"Investigation on Frictional Wear of Single Crystal Diamond Against Ferrous Metals"** (*Intl. J. of Refractory Metals and Hard Materials*, **2013: 174–179**) presented the accompanying data on $x_1 =$ mechanical force (N), $x_2 =$ sliding velocity (m/s), $x_3 =$ carbon content (%), and $y =$ graphitized degree, a measure of diamond wear.

Obs	x_1	x_2	x_3	y
1	10	.84	.07	.18
2	10	1.05	.27	.19
3	10	1.26	.45	.22
4	20	.84	.27	.21
5	20	1.05	.45	.24
6	20	1.26	.07	.28
7	30	.84	.45	.26
8	30	1.05	.07	.30
9	30	1.26	.27	.33

The investigators proposed and fit the multiplicative power regression model $Y = \alpha x_1^{\beta_1} x_2^{\beta_2} x_3^{\beta_3} \epsilon$. Taking the natural logarithm of both sides of this equation gives

$$\ln(Y) = \ln(\alpha) + \beta_1 \ln(x_1) + \beta_2 \ln(x_2) + \beta_3 \ln(x_3) + \ln(\epsilon) \quad (13.21)$$

which is our general additive multiple regression equation with the dependent variable being the natural log of graphitized degree and predictors $\ln(x_1)$, $\ln(x_2)$, and $\ln(x_3)$. Presuming that ϵ in the original model equation has a lognormal distribution, the random error in our transformed model will be normally distributed. The plausibility of this assumption can be checked with a normal probability plot of the standardized residuals resulting from fitting the transformed model.

Table 13.4 shows Minitab output from fitting (13.21). The R^2 value is quite impressive—about 98% of the observed variation in $\ln(y)$ can be attributed to the model relationship—and adjusted R^2 is only slightly smaller than R^2 itself. Furthermore, the P-value for the model utility F test is .000 (the area under the $F_{3,5}$ curve to the right of 81.16), implying a useful relationship between $\ln(y)$ and at least one of the three predictors. The point estimates of β_1, β_2, and β_3 are .36557, .59366, and −.02074, respectively. The point estimate of $\ln(\alpha)$ is −2.53727, so the point estimate of α itself is $e^{-2.53727} = .079082$. The estimated original regression function is then $.079 x_1^{.366} x_2^{.594} x_3^{-.021}$; this appears in the cited article.

Table 13.4 Minitab output for the transformed regression in Example 13.18

```
The regression equation is
ln(y) = - 2.54 + 0.366 ln(x1) + 0.594 ln(x2) - 0.0207 ln(x3)
```

Predictor	Coef	SE Coef	T	P
Constant	−2.53727	0.08413	−30.16	0.000
ln(x1)	0.36557	0.02734	13.37	0.000
ln(x2)	0.59366	0.07480	7.94	0.001
ln(x3)	−0.02074	0.01580	−1.31	0.246

```
S = 0.0372066  R-Sq = 98.0%  R-Sq(adj) = 96.8%
```

Analysis of Variance

Source	DF	SS	MS	F	P
Regression	3	0.33706	0.11235	81.16	0.000
Residual Error	5	0.00692	0.00138		
Total	8	0.34399			

```
Predicted Values for New Observations
```

New Obs	Fit	SE Fit	95% CI	95% PI
1	−1.4134	0.0133	(−1.4477, −1.3791)	(−1.5150, −1.3118)

A point prediction of the value of graphitized degree when force = 20, velocity = 1, and carbon content = .25 requires that we first obtain a point prediction of $\ln(Y)$ by substituting $\ln(20)$, $\ln(0)$, and $\ln(.25)$ into the estimated regression equation in Table 13.4. The result is $\ln(\hat{y}) = -1.4134$, which appears in the last line of Minitab output. Then $\hat{y} = e^{-1.4134} = .243$. Similarly, the output gives a 95% PI for $\ln(Y)$, so a PI for Y itself is $(e^{-1.5150}, e^{-1.3118}) = (.220, .269)$.

The normal probability plot of Figure 13.20 exhibits a substantial linear pattern, validating the normality assumption for $\ln(\epsilon)$. And the plot of standardized residuals versus predicted values [of $\ln(y)$] does not show any pattern other than pure randomness, indicating no violation of model assumptions. However, looking back at Table 13.4, the *P*-value for testing $H_0: \beta_3 = 0$ is .246. Thus it appears that as long as $\ln(x_1)$ and $\ln(x_2)$ remain in the model, there is no useful information about the response variable contained in the natural log of carbon

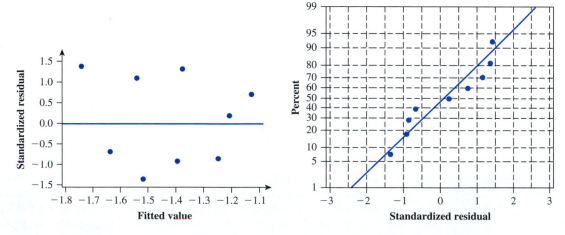

Figure 13.20 Standardized residual plot and normal probability plot for Example 13.18

content. Deleting that predictor and refitting gives $R^2 = .973$ and a model utility F ratio of 107.87. The estimates of β_1 and β_2 are almost identical to those for the three-predictor model. Also, the multiple exponential regression model $Y = \alpha e^{\beta_1 x_1 + \beta_2 x_2} \varepsilon$ [for which $\ln(Y)$ is regressed against x_1 and x_2 rather than against $\ln(x_1)$ and $\ln(x_2)$] fits the data about as well as does the power model. None of this was mentioned in the cited article. ∎

The logistic regression model was introduced in Section 13.2 to relate a dichotomous variable y to a single predictor. This model can be extended in an obvious way to incorporate more than one predictor. The probability of success p is now a function of the predictors $x_1, x_2, \ldots, x_k$:

$$p(x_1, \ldots, x_k) = \frac{e^{\beta_0 + \beta_1 x_1 + \cdots + \beta_k x_k}}{1 + e^{\beta_0 + \beta_1 x_1 + \cdots + \beta_k x_k}}$$

Simple algebra yields an expression for the odds:

$$\frac{p(x_1, \ldots, x_k)}{1 - p(x_1, \ldots, x_k)} = e^{\alpha + \beta_1 x_1 + \cdots + \beta_k x_k}$$

The interpretation of β_i $(i = 1, \ldots, k)$ is analogous to the interpretation for β_1 given in the logit function containing only a single predictor x. That is, the following argument shows that the odds change by the multiplicative factor e^{β_i} when x_i increases by 1 unit and all other predictors remain fixed.

$$\frac{p(x_1, \ldots, x_i + 1, \ldots, x_k)}{1 - p(x_1, \ldots, x_i + 1, \ldots, x_k)} = e^{\alpha + \beta_1 x_1 + \cdots \beta_i(x_i + 1) + \cdots + \beta_k x_k}$$

$$= e^{\alpha + \beta_1 x_1 + \cdots \beta_i x_i + \cdots + \beta_k x_k + \beta_i}$$

$$= \frac{p(x_1, \ldots, x_k)}{1 - p(x_1, \ldots, x_k)} e^{\beta_i}$$

Again, statistical software must be used to estimate parameters, calculate relevant standard deviations, and provide other inferential information.

EXAMPLE 13.19   Data was obtained from 189 women who gave birth during a particular period at the Bayside Medical Center in Springfield, MA, in order to identify factors associated with low birth weight. The accompanying Minitab output resulted from a logistic regression in which the dependent variable indicated whether (1) or not (0) a child had low birth weight ($<2500\,$g), and predictors were weight of the mother at her last menstrual period, age of the mother, and an indicator variable for whether (1) or not (0) the mother had smoked during pregnancy.

```
Logistic Regression Table
```

Predictor	Coef	SE Coef	Z	P	Odds Ratio	95% CI Lower	Upper
Constant	2.06239	1.09516	1.88	0.060			
Wt	−0.01701	0.00686	−2.48	0.013	0.98	0.97	1.00
Age	−0.04478	0.03391	−1.32	0.187	0.96	0.89	1.02
Smoke	0.65480	0.33297	1.97	0.049	1.92	1.00	3.70

It appears that age is not an important predictor of LBW, provided that the two other predictors are retained. The other two predictors do appear to be informative. The point estimate of the odds ratio associated with smoking status is 1.92 [ratio of the odds of LBW for a smoker to the odds for a nonsmoker, where $odds = P(Y = 1)/P(Y = 0)$];

at the 95% confidence level, the odds of a low-birth-weight child could be as much as 3.7 times higher for a smoker what it is for a nonsmoker. ∎

Please see one of the chapter references for more information on logistic regression, including methods for assessing model effectiveness and adequacy.

Standardizing Variables

In Section 13.3, we considered transforming x to $x' = x - \bar{x}$ before fitting a polynomial. For multiple regression, especially when values of variables are large in magnitude, it is advantageous to carry this coding one step further. Let $\bar{x}_i$ and s_i be the sample average and sample standard deviation of the x_{ij}'s ($j = 1, \ldots, n$). Now code each variable x_i by $x_i' = (x_i - \bar{x}_i)/s_i$. The coded variable x_i' simply reexpresses any x_i value in units of standard deviation above or below the mean. Thus if $\bar{x}_i = 100$ and $s_i = 20$, $x_i = 130$ becomes $x_i' = 1.5$, because 130 is 1.5, standard deviations above the mean of the values of x_i. For example, the coded full second-order model with two independent variables has regression function

$$E(Y) = \beta_0 + \beta_1\left(\frac{x_1 - \bar{x}_1}{s_1}\right) + \beta_2\left(\frac{x_2 - \bar{x}_2}{s_2}\right) + \beta_2\left(\frac{x_1 - \bar{x}_1}{s_1}\right)^2$$

$$+ \beta_4\left(\frac{x_2 - \bar{x}_2}{s_2}\right)^2 + \beta_5\left(\frac{x_1 - \bar{x}_1}{s_1}\right)\left(\frac{x_2 - \bar{x}_2}{s_2}\right)$$

$$= \beta_0 + \beta_1 x_1' + \beta_2 x_2' + \beta_3 x_3' + \beta_4 x_4' + \beta_5 x_5'$$

The benefits of coding are (1) increased numerical accuracy in all computations and (2) more accurate estimation than for the parameters of the uncoded model, because the individual parameters of the coded model characterize the behavior of the regression function near the center of the data rather than near the origin.

EXAMPLE 13.20 The article **"The Value and the Limitations of High-Speed Turbo-Exhausters for the Removal of Tar-Fog from Carburetted Water-Gas"** (*J. of the Chemical Industry Society*, **1946: 166–168**) presents the data (in Table 13.5) on y = tar content (grains/100 ft^3) of a gas stream as a function of x_1 = rotor speed (rpm) and x_2 = gas inlet temperature (°F). The data is also considered in the article **"Some Aspects of Nonorthogonal Data Analysis"** (*J. of Quality Tech.* **1973: 67–79**), which suggests using the coded model described previously.

The means and standard deviations are $\bar{x}_1 = 2991.13$, $s_1 = 387.81$, $\bar{x}_2 = 58.468$, and $s_2 = 6.944$, so $x_1' = (x_1 - 2991.13)/387.81$ and $x_2' = (x_2 - 58.468)/6.944$. With $x_3' = (x_1')^2$, $x_4' = (x_2')^2$, $x_5' = x_1' \cdot x_2'$, fitting the full second-order model yielded $\hat{\beta}_0 = 40.2660$, $\hat{\beta}_1 = -13.4041$, $\hat{\beta}_2 = 10.2553$, $\hat{\beta}_3 = 2.3313$, $\hat{\beta}_4 = -2.3405$, and $\hat{\beta}_5 = 2.5978$. The estimated regression equation is then

$$\hat{y} = 40.27 - 13.40x_1' + 10.26x_2' + 2.33x_3' - 2.34x_4' + 2.60x_5'$$

Thus if $x_1 = 3200$ and $x_2 = 57.0$, $x_1' = .539$, $x_2' = -.211$, $x_3' = (.539)^2 = .2901$, $x_4' = (-.211)^2 = .0447$, and $x_5' = (.539)(-.211) = -.1139$, so

$$\hat{y} = 40.27 - (13.40)(.539) + (10.26)(-.211) + (2.33)(.2901)$$

$$-(2.34)(.0447) + (2.60)(-.1139) = 31.16$$ ∎

Table 13.5   Data for Example 13.20

Run	y	x_1	x_2	x_1'	x_2'
1	60.0	2400	54.5	−1.52428	−.57145
2	61.0	2450	56.0	−1.39535	−.35543
3	65.0	2450	58.5	−1.39535	.00461
4	30.5	2500	43.0	−1.26642	−2.22763
5	63.5	2500	58.0	−1.26642	−.06740
6	65.0	2500	59.0	−1.26642	.07662
7	44.0	2700	52.5	−.75070	−.85948
8	52.0	2700	65.5	−.75070	1.01272
9	54.5	2700	68.0	−.75070	1.37276
10	30.0	2750	45.0	−.62177	−1.93960
11	26.0	2775	45.5	−.55731	−1.86759
12	23.0	2800	48.0	−.49284	−1.50755
13	54.0	2800	63.0	−.49284	.65268
14	36.0	2900	58.5	−.23499	.00461
15	53.5	2900	64.5	−.23499	.86870
16	57.0	3000	66.0	.02287	1.08472
17	33.5	3075	57.0	.21627	−.21141
18	34.0	3100	57.5	.28073	−.13941
19	44.0	3150	64.0	.40966	.79669
20	33.0	3200	57.0	.53859	−.21141
21	39.0	3200	64.0	.53859	.79669
22	53.0	3200	69.0	.53859	1.51677
23	38.5	3225	68.0	.60305	1.37276
24	39.5	3250	62.0	.66752	.50866
25	36.0	3250	64.5	.66752	.86870
26	8.5	3250	48.0	.66752	−1.50755
27	30.0	3500	60.0	1.31216	.22063
28	29.0	3500	59.0	1.31216	.07662
29	26.5	3500	58.0	1.31216	−.06740
30	24.5	3600	58.0	1.57002	−.06740
31	26.5	3900	61.0	2.34360	.36465

Variable Selection

Suppose an experimenter has obtained data on a response variable y as well as on p candidate predictors $x_1, \ldots, x_p$. How can a best (in some sense) model involving a subset of these predictors be selected? Recall that as predictors are added one by one into a model, SSE cannot increase (a larger model cannot explain less variation than a smaller one) and will usually decrease, albeit perhaps by a small amount. So there is no mystery as to which model gives the largest R^2 value—it must be the one containing all p predictors. What we'd really like is a model involving relatively few predictors that is easy to interpret and use yet explains a relatively large amount of observed y variation.

For any fixed number of predictors (e.g., 5), it is reasonable to identify the best model of that size as the one with the largest R^2 value—equivalently, the smallest value of SSE. The more difficult issue concerns selection of a criterion that will allow for comparison of models of different sizes. Let's use a subscript k to denote a quantity computed from a model containing k predictors (e.g., SSE_k). Three different criteria, each one a simple function of SSE_k, are popular.

1. R_k^2, the coefficient of multiple determination for a k-predictor model. Because R_k^2 will virtually always increase as k does (and can never decrease), we are not interested in the k that maximizes R_k^2. Instead, we wish to identify a small k for which R_k^2 is nearly as large as R^2 for all predictors in the model.

2. $\text{MSE}_k = \text{SSE}_k/(n - k - 1)$, the mean squared error for a k-predictor model. This is often used in place of R_k^2, because although R_k^2 never decreases with increasing k, a small decrease in SSE_k obtained with one extra predictor can be more than offset by a decrease of 1 in the denominator of MSE_k. The objective is then to find the model having minimum MSE_k. Since *adjusted* $R_k^2 = 1 - \text{MSE}_k/\text{MST}$, where $\text{MST} = \text{SST}/(n - 1)$ is constant in k, examination of *adjusted* R_k^2 is equivalent to consideration of MSE_k.

3. The rationale for the third criterion, C_k, is more difficult to understand, but the criterion is widely used by data analysts. Suppose the true regression model is specified by m predictors—that is,

$$Y = \beta_0 + \beta_1 x_1 + \cdots + \beta_m x_m + \epsilon \qquad V(\epsilon) = \sigma^2$$

so that

$$E(Y) = \beta_0 + \beta_1 x_1 + \cdots + \beta_m x_m$$

Consider fitting a model by using a subset of k of these m predictors; for simplicity, suppose we use $x_1, x_2, \ldots, x_k$. Then by solving the system of normal equations, estimates $\hat{\beta}_0, \hat{\beta}_1, \ldots, \hat{\beta}_k$ are obtained (but not, of course, estimates of any β's corresponding to predictors not in the fitted model). The true expected value $E(Y)$ can then be estimated by $\hat{Y} = \hat{\beta}_0 + \hat{\beta}_1 x_1 + \cdots + \hat{\beta}_k x_k$. Now consider the **normalized expected total error of estimation**

$$\Gamma_k = \frac{E\left(\sum_{i=1}^{n} [\hat{Y}_i - E(Y_i)]^2\right)}{\sigma^2} = \frac{E(\text{SSE}_k)}{\sigma^2} + 2(k + 1) - n \qquad (13.21)$$

The second equality in (13.21) must be taken on faith because it requires a tricky expected-value argument. A particular subset is then appealing if its Γ_k value is small. Unfortunately, though, $E(\text{SSE}_k)$ and σ^2 are not known. To remedy this, let s^2 denote the estimate of σ^2 based on the model that includes all predictors for which data is available, and define

$$C_k = \frac{\text{SSE}_k}{s^2} + 2(k + 1) - n$$

A desirable model is then specified by a subset of predictors for which C_k is small.

The total number of models that can be created from predictors in the candidate pool is 2^p (because each predictor can be included in or left out of any particular model—one of these is the model that contains no predictors). If $p \le 5$, then it would not be too tedious to examine all possible regression models involving these predictors using any good statistical software package. But the computational effort required to fit all possible models becomes prohibitive as the size of the candidate pool increases. Several software packages have incorporated algorithms which will sift through models of various sizes in order to identify the best one or more models of each particular size. Minitab, for example, will do this for $p \le 31$ and allows the user to specify the number of models of each size (1, 2, 3, 4, or 5) that will be identified as having best criterion values. You might wonder why we'd want to go beyond the best single model of each size. The answer is that the 2nd or 3rd best model may be just about as good as the best model and easier to interpret and use, or may be more satisfactory from a model-adequacy perspective. For example, suppose the candidate pool includes all predictors from a full quadratic model based on five independent variables. Then the best 3-predictor model might have predictors x_2, x_4^2, and $x_3 x_5$, whereas the second-best such model could be the one with predictors x_2, x_3, and $x_2 x_3$.

EXAMPLE 13.21   The review article by Ron Hocking listed in the chapter bibliography reports on an analysis of data taken from the 1974 issues of *Motor Trend* magazine. The dependent variable y was gas mileage, there were $n = 32$ observations, and the predictors for which data was obtained were x_1 = engine shape (1 = straight and 0 = V), x_2 = number of cylinders, x_3 = transmission type (1 = manual and 0 = auto), x_4 = number of transmission speeds, x_5 = engine size, x_6 = horsepower, x_7 = number of carburetor barrels, x_8 = final drive ratio, x_9 = weight, and x_{10} = quarter-mile time. In Table 13.6, we present summary information from the analysis. The table describes for each k the subset having minimum SSE_k; reading down the variables column indicates which variable is added in going from k to $k + 1$ (going from $k = 2$ to $k = 3$, both x_3 and x_{10} are added, and x_2 is deleted). Figure 13.21 contains plots of R_k^2, adjusted R_k^2, and C_k against k; these plots are an important visual aid in selecting a subset. The estimate of σ^2 is $s^2 = 6.24$, which is MSE_{10}. A simple model that rates highly according to all criteria is the one containing predictors x_3, x_9, and x_{10}.

Table 13.6   Best Subsets for Gas Mileage Data of Example 13.21

k = Number of Predictors	Variables	SSE_k	R_k^2	Adjusted R_k^2	C_k
1	9	247.2	.756	.748	11.6
2	2	169.7	.833	.821	1.2
3	3, 10, −2	150.4	.852	.836	.1
4	6	142.3	.860	.839	.8
5	5	136.2	.866	.840	1.8
6	8	133.3	.869	.837	3.4
7	4	132.0	.870	.832	5.2
8	7	131.3	.871	.826	7.1
9	1	131.1	.871	.818	9.0
10	2	131.0	.871	.809	11.0

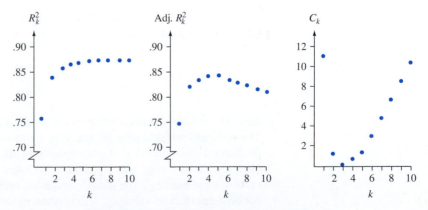

Figure 13.21 R_k^2 and C_k plots for the gas mileage data

Generally speaking, when a subset of k predictors $(k < m)$ is used to fit a model, the estimators $\hat{\beta}_0, \hat{\beta}_1, \ldots, \hat{\beta}_k$ will be biased for $\beta_0, \beta_1, \ldots, \beta_k$ and $\hat{Y}$ will also be a biased estimator for the true $E(Y)$ (all this because $m - k$ predictors are missing from the fitted model). However, as measured by the total normalized expected error Γ_k, estimates based on a subset can provide more precision than would be obtained using all possible predictors; essentially, this greater precision is obtained at the price of introducing a bias in the estimators. A value of k for which $C_k \approx k + 1$ indicates that the bias associated with this k-predictor model would be small.

EXAMPLE 13.22   The bond shear strength data introduced in Example 13.12 contains values of four different independent variables $x_1 - x_4$. We found that the model with only these four variables as predictors was useful, and there is no compelling reason to consider the inclusion of second-order predictors. Figure 13.22 is the Minitab output that results from a request to identify the two best models of each given size.

The best two-predictor model, with predictors power and temperature, seems to be a very good choice on all counts: R^2 is significantly higher than for models with fewer predictors yet almost as large as for any larger models, adjusted R^2 is almost at its maximum for this data, and C_2 is small and close to $2 + 1 = 3$.

```
Response is strength                              f    p
                                                  o    o    t    t
                                                  r    w    e    i
                          Adj.                    c    e    m    m
Vars    R-sq    R-sq      C-p        s            e    r    p    e
 1      57.7    56.2     11.0      5.9289         X
 1      10.8     7.7     51.9      8.6045              X
 2      68.5    66.2      3.5      5.2070         X    X
 2      59.4    56.4     11.5      5.9136         X              X
 3      70.2    66.8      4.0      5.1590         X    X         X
 3      69.7    66.2      4.5      5.2078    X    X    X
 4      71.4    66.8      5.0      5.1580    X    X    X         X
```

Figure 13.22   Output from Minitab's Best Subsets option ■

Stepwise Regression When the number of predictors is too large to allow for explicit or implicit examination of all possible subsets, several alternative selection procedures will generally identify good models. The simplest such procedure is the **backward elimination** (BE) method. This method starts with the model in which all predictors under consideration are used. Let the set of all such predictors be $x_1, \ldots, x_m$. Then each t ratio $\hat{\beta}_i / s_{\hat{\beta}_i} (i = 1, \ldots, m)$ appropriate for testing $H_0: \beta_i = 0$ versus $H_a: \beta_i \neq 0$ is examined. If the t ratio with the smallest absolute value is less than a prespecified constant t_{out}, that is, if

$$\min_{i=1,\ldots,m} \left| \frac{\hat{\beta}_i}{s_{\hat{\beta}_i}} \right| < t_{out}$$

then the predictor corresponding to the smallest ratio is eliminated from the model. The reduced model is now fit, the $m - 1$ t ratios are again examined, and another predictor is eliminated if it corresponds to the smallest absolute t ratio smaller than t_{out}. In this way, the algorithm continues until, at some stage, all absolute t ratios are at least t_{out}. The model used is the one containing all predictors that were not eliminated. The value $t_{out} = 2$ is often recommended since most $t_{.05}$ values are near 2. Some computer packages focus on P-values rather than t ratios.

EXAMPLE 13.23
(Example 13.20
continued)

For the coded full quadratic model in which $y = $ tar content, the five potential predictors are $x_1', x_2', x_3' = x_1'^2, x_4' = x_2'^2$, and $x_5' = x_1'x_2'$ (so $m = 5$). Without specifying t_{out}, the predictor with the smallest absolute t ratio (asterisked) was eliminated at each stage, resulting in the sequence of models shown in Table 13.7.

Table 13.7   Backward Elimination Results for the Data of Example 13.20

Step	Predictors	$|t\text{-ratio}|$ 1	2	3	4	5
1	1, 2, 3, 4, 5	16.0	10.8	2.9	2.8	1.8*
2	1, 2, 3, 4	15.4	10.2	3.7	2.0*	—
3	1, 2, 3	14.5	12.2	4.3*	—	—
4	1, 2	10.9	9.1*	—	—	—
5	1	4.4*	—	—	—	—

Using $t_{out} = 2$, the resulting model would be based on x_1', x_2', and x_3', since at Step 3 no predictor could be eliminated. It can be verified that each subset is actually the best subset of its size, though this is by no means always the case.   ■

An alternative to the BE procedure is **forward selection** (FS). FS starts with no predictors in the model and considers fitting in turn the model with only x_1, only $x_2, \ldots$, and finally only x_m. The variable that, when fit, yields the largest absolute t ratio enters the model provided that the ratio exceeds the specified constant t_{in}. Suppose x_1 enters the model. Then models with $(x_1, x_2), (x_1, x_3), \ldots (x_1, x_m)$ are considered in turn. The largest $|\hat{\beta}_j / s_{\hat{\beta}_j}| (j = 2, \ldots, m)$ then specifies the entering predictor provided that this maximum also exceeds t_{in}. This continues until at some step no absolute t ratios exceed t_{in}. The entered predictors then specify the model. The value $t_{in} = 2$ is often used for the same reason that $t_{out} = 2$ is used in BE. For the tarcontent data, FS resulted in the sequence of models given in Steps 5, 4, ..., 1 in Table 13.7 and thus is in agreement with BE. This will not always be the case.

The stepwise procedure most widely used is a combination of FS and BE, denoted by FB. This procedure starts as does forward selection, by adding variables to the model, but after each addition it examines those variables previously entered to see whether any is a candidate for elimination. For example, if there are eight predictors under consideration and the current set consists of x_2, x_3, x_5, and x_6 with x_5 having just been added, the t ratios $\hat{\beta}_2 / s_{\hat{\beta}_2}, \hat{\beta}_3 / s_{\hat{\beta}_3}$, and $\hat{\beta}_6 / s_{\hat{\beta}_6}$ are examined. If the smallest absolute ratio is less than t_{out}, then the corresponding variable is eliminated from the model (some software packages base decisions on $f = t^2$). The idea behind FB is that, with forward selection, a single variable may be more strongly related to y than to either of two or more other variables individually, but the combination of these variables may make the single variable subsequently redundant. This actually happened with the gas-mileage data discussed in Example 13.21, with x_2 entering and subsequently leaving the model.

Although in most situations these automatic selection procedures will identify a good model, there is no guarantee that the best or even a nearly best model will result. Close scrutiny should be given to data sets for which there appear to be strong relationships among some of the potential predictors; we will say more about this shortly.

Identification of Influential Observations

In simple linear regression, it is easy to spot an observation whose x value is much larger or much smaller than other x values in the sample. Such an observation may have a great impact on the estimated regression equation (whether it actually does depends on how far the point (x, y) falls from the line determined by the other points in the scatterplot). In multiple regression, it is also desirable to know whether the values of the predictors for a particular observation are such that it has the potential for exerting great influence on the estimated equation. One method for identifying potentially influential observations relies on the fact that because each $\hat{\beta}_i$ is a linear function of $y_1, y_2, \ldots, y_n$, each predicted y value of the form $\hat{y} = \hat{\beta}_0 + \hat{\beta}_1 x_1 + \cdots + \hat{\beta}_k x_k$ is also a linear function of the y_j's. In particular, the predicted values corresponding to sample observations can be written as follows:

$$\hat{y}_1 = h_{11}y_1 + h_{12}y_2 + \cdots + h_{1n}y_n$$
$$\hat{y}_2 = h_{21}y_1 + h_{22}y_2 + \cdots + h_{2n}y_n$$
$$\vdots \qquad \vdots \qquad \vdots \qquad \vdots$$
$$\hat{y}_n = h_{n1}y_1 + h_{n2}y_2 + \cdots + h_{nn}y_n$$

Each coefficient h_{ij} is a function only of the x_{ij}'s in the sample and not of the y_j's. It can be shown that $h_{ij} = h_{ji}$ and that $0 \le h_{jj} \le 1$.

Let's focus on the "diagonal" coefficients $h_{11}, h_{22}, \ldots, h_{nn}$. The coefficient h_{jj} is the weight given to y_j in computing the corresponding predicted value $\hat{y}_j$. This quantity can also be expressed as a measure of the distance between the point $(x_{1j}, \ldots, x_{kj})$ in k-dimensional space and the center of the data $(\bar{x}_1, \ldots, \bar{x}_k)$. It is therefore natural to characterize an observation whose h_{jj} is relatively large as one that has potentially large influence. Unless there is a perfect linear relationship among the k predictors, $\sum_{j=1}^{n} h_{jj} = k + 1$, so the average of the h_{jj}'s is $(k+1)/n$. Some statisticians suggest that if $h_{jj} > 2(k+1)/n$, the jth observation be cited as being potentially influential; others use $3(k+1)/n$ as the dividing line.

EXAMPLE 13.24   The accompanying data appeared in the article **"Testing for the Inclusion of Variables in Linear Regression by a Randomization Technique"** (*Technometrics,* **1966: 695–699**) and was reanalyzed in Hoaglin and Welsch, **"The Hat Matrix in Regression and ANOVA"** (*Amer. Statistician,* **1978: 17–23**). The h_{ij}'s (with elements below the diagonal omitted by symmetry) follow the data.

Beam Number	Specific Gravity (x_1)	Moisture Content (x_2)	Strength (y)
1	.499	11.1	11.14
2	.558	8.9	12.74
3	.604	8.8	13.13
4	.441	8.9	11.51
5	.550	8.8	12.38
6	.528	9.9	12.60
7	.418	10.7	11.13
8	.480	10.5	11.70
9	.406	10.5	11.02
10	.467	10.7	11.41

	1	2	3	4	5	6	7	8	9	10
1	.418	−.002	.079	−.274	−.046	.181	.128	.222	.050	.242
2		.242	.292	.136	.243	.128	−.041	.033	−.035	.004
3			.417	−.019	.273	.187	−.126	.044	−.153	.004
4				.604	.197	−.038	.168	−.022	.275	−.028
5					.252	.111	−.030	.019	−.010	−.010
6						.148	.042	.117	.012	.111
7							.262	.145	.277	.174
8								.154	.120	.168
9									.315	.148
10										.187

Here $k = 2$, so $(k + 1)/n = 3/10 = .3$; since $h_{44} = .604 > 2(.3)$, the fourth data point is identified as potentially influential. ∎

Another technique for assessing the influence of the jth observation that takes into account y_j as well as the predictor values involves deleting the jth observation from the data set and performing a regression based on the remaining observations. If the estimated coefficients from the "deleted observation" regression differ greatly from the estimates based on the full data, the jth observation has clearly had a substantial impact on the fit. One way to judge whether estimated coefficients change greatly is to express each change relative to the estimated standard deviation of the coefficient:

$$\frac{(\hat{\beta}_i \text{ before deletion}) - (\hat{\beta}_i \text{ after deletion})}{s_{\hat{\beta}_i}} = \frac{\text{change in } \hat{\beta}_i}{s_{\hat{\beta}_i}}$$

There exist efficient computational formulas that allow all this information to be obtained from the "no-deletion" regression, so that the additional n regressions are unnecessary.

EXAMPLE 13.25
(Example 13.24 continued)

Consider separately deleting observations 1 and 6, whose residuals are the largest, and observation 4, where h_{jj} is large. Table 13.8 contains the relevant information.

Table 13.8 Changes in Estimated Coefficients for Example 13.25

			Change When Point j Is Deleted		
Parameter	No-Deletions Estimates	Estimated SD	$j = 1$	$j = 4$	$j = 6$
β_0	10.302	1.896	2.710	−2.109	−.642
β_1	8.495	1.784	−1.772	1.695	.748
β_2	.2663	.1273	−.1932	.1242	.0329
		e_j:	−3.25	−.96	2.20
		h_{jj}:	.418	.604	.148

For deletion of both point 1 and point 4, the change in each estimate is in the range 1–1.5 standard deviations, which is reasonably substantial (this does not tell us what would happen if both points were simultaneously omitted). For point 6, however, the change is roughly .25 standard deviation. Thus points 1 and 4, but not 6, might well be omitted in calculating a regression equation. ∎

Multicollinearity

In many multiple regression data sets, the predictors $x_1, x_2, \ldots, x_k$ are highly interdependent. Consider the usual model

$$Y = \beta_0 + \beta_1 x_1 + \cdots + \beta_k x_k + \epsilon$$

with data $(x_{1j}, \ldots, x_{kj}, y_j)$ $(j = 1, \ldots, n)$ available for fitting. Suppose the principle of least squares is used to regress x_i on the other predictors $x_1, \ldots, x_{i-1}, x_{i+1}, \ldots, x_k$, resulting in

$$\hat{x}_i = a_0 + a_1 x_1 + \cdots + a_{i-1} x_{i-1} + a_{i+1} x_{i+1} + \cdots + a_k x_k$$

It can then be shown that

$$V(\hat{\beta}_i) = \frac{\sigma^2}{\displaystyle\sum_{j=1}^{n} (x_{ij} - \hat{x}_{ij})^2} \tag{13.22}$$

When the sample x_i values can be predicted very well from the other predictor values, the denominator of (13.22) will be small, so $V(\hat{\beta}_i)$ will be quite large. If this is the case for at least one predictor, the data is said to exhibit **multicollinearity.** Multicollinearity is often suggested by a regression computer output in which R^2 is large but some of the t ratios $\hat{\beta}_i / s_{\hat{\beta}_i}$ are small for predictors that, based on prior information and intuition, seem important. Another clue to the presence of multicollinearity lies in a $\hat{\beta}_i$ value that has the opposite sign from that which intuition would suggest, indicating that another predictor or collection of predictors is serving as a "proxy" for x_i.

An assessment of the extent of multicollinearity can be obtained by regressing each predictor in turn on the remaining $k - 1$ predictors. Let R_i^2 denote the value of R^2 in the regression with dependent variable x_i and predictors $x_1, \ldots, x_{i-1}, x_{i+1}, \ldots, x_k$. It has been suggested that severe multicollinearity is present if $R_i^2 > .9$ for any i. Some statistical software packages will refuse to include a predictor in the model when its R_i^2 value is quite close to 1.

There is no consensus among statisticians as to what remedies are appropriate when severe multicollinearity is present. One possibility involves continuing to use a model that includes all the predictors but estimating parameters by using something other than least squares. Consult a chapter reference for more details.

EXERCISES Section 13.5 (55–64)

55. The article **"The Influence of Honing Process Parameters on Surface Quality, Productivity, Cutting Angle, and Coefficient of Friction"** (*Industrial Lubrication and Tribology*, 2012: 77–83) included the following data on x_1 = cutting speed (m/s), x_2 = specific pressure of pre-honing process (N/mm^2), x_3 = specific pressure of finishing honing process, and y = productivity in the honing process (mm^3/s for a particular tool; productivity is the volume of the material cut in a second.

x_1	x_2	x_3	y	x_1	x_2	x_3	y
0.93	1.00	0.20	32.95	0.93	1.40	0.50	33.67
1.11	1.00	0.20	38.72	1.11	1.40	0.50	38.72
0.93	1.00	0.50	35.20	1.02	1.18	0.31	35.20
1.11	1.00	0.50	38.72	1.02	1.18	0.31	33.67
0.93	1.40	0.20	32.27	1.02	1.18	0.31	36.02
1.11	1.40	0.20	39.71	1.02	1.18	0.31	32.27

a. The article proposed a multivariate power model $Y = \alpha x_1^{\beta_1} x_2^{\beta_2} x_3^{\beta_3} \epsilon$. The implied linear regression model involves regressing $\ln(y)$ against the three predictors $\ln(x_1)$, $\ln(x_2)$, and $\ln(x_3)$. Partial Minitab output from fitting this latter model is as follows (the corresponding estimated power regression function appeared in the cited article).

Predictor	Coef	SE Coef	T	P
Constant	3.58797	0.04909	73.10	0.000
lnx1	0.8439	0.1952	4.32	0.003
lnx2	20.0280	0.1027	20.27	0.792
lnx3	0.02449	0.03768	0.65	0.534

S = 0.048848 R-Sq = 70.6% R-Sq(adj) = 59.5%

Carry out the model utility test at significance level .05.

b. The large P-value corresponding to the t ratio for $\ln(x_2)$ suggests that this predictor can be eliminated from the model. Doing so and refitting yields the following Minitab output.

Predictor	Coef	SE Coef	T	P
Constant	3.58329	0.04355	82.28	0.000
lnx1	0.8440	0.1849	4.57	0.001
lnx3	0.02449	0.03569	0.69	0.510

S = 0.0462680 R-Sq = 70.3% R-Sq(adj) = 63.7%

Given that $\ln(x_1)$ remains in the model, should $\ln(x_3)$ be retained?

c. Fit the simple linear regression model implied by your conclusion in (b) to the transformed data, and carry out a test of model utility.

d. The standardized residuals from the fit referred to in (c) are .03, .33. 1.69, .33, −.49, .96, .57, .33, −,25, −1.28, .29, −2.26. Plot these against $\ln(x_1)$. What does the pattern suggest?

e. Fitting a quadratic regression model to relate $\ln(y)$ to $\ln(x_1)$ gave the following Minitab output. Carry out a test of model utility at significance level .05 (the pattern in residual plots is satisfactory). Then use the fact that $s_{\ln(\hat{Y}')} = .0178 \ [Y' = \ln(Y)]$ when $x_1 = 1$ to obtain a 95% prediction interval for productivity.

Predictor	Coef	SE Coef	T	P
Constant	3.51879	0.01775	198.22	0.000
lnx1	0.6231	0.1683	3.70	0.005
lnx1 sqd	7.240	2.834	2.55	0.031

S = 0.0361358 R-Sq = 81.9% R-Sq(adj) = 77.9%

56. In an experiment to study factors influencing wood specific gravity (**"Anatomical Factors Influencing Wood Specific Gravity of Slash Pines and the Implications for the Development of a High-Quality Pulpwood,"** *TAPPI*, **1964: 401–404**), a sample of 20 mature wood samples was obtained, and measurements were taken on the number of fibers/mm^2 in springwood (x_1), number of fibers/mm^2 in summerwood (x_2), % springwood (x_3),

light absorption in springwood (x_4), and light absorption in summerwood (x_5).

a. Fitting the regression function $\mu_{Y \cdot x_1, x_2, x_3, x_4, x_5} = \beta_0 + \beta_1 x_1 + \cdots + \beta_5 x_5$ resulted in $R^2 = .769$. Does the data indicate that there is a linear relationship between specific gravity and at least one of the predictors? Test using $\alpha = .01$.

b. When x_2 is dropped from the model, the value of R^2 remains at .769. Compute adjusted R^2 for both the full model and the model with x_2 deleted.

c. When x_1, x_2, and x_4 are all deleted, the resulting value of R^2 is .654. The total sum of squares is SST = .0196610. Does the data suggest that all of x_1, x_2, and x_4 have zero coefficients in the true regression model? Test the relevant hypotheses at level .05.

d. The mean and standard deviation of x_3 were 52.540 and 5.4447, respectively, whereas those of x_5 were 89.195 and 3.6660, respectively. When the model involving these two standardized variables was fit, the estimated regression equation was $y = .5255 − .0236x_3' + .0097x_5'$. What value of specific gravity would you predict for a wood sample with % springwood = 50 and % light absorption in summerwood = 90?

e. The estimated standard deviation of the estimated coefficient $\hat{\beta}_3$ of x_3' (i.e., for $\hat{\beta}_3$ of the standardized model) was .0046. Obtain a 95% CI for β_3.

f. Using the information in parts (d) and (e), what is the estimated coefficient of x_3 in the unstandardized model (using only predictors x_3 and x_5), and what is the estimated standard deviation of the coefficient estimator (i.e., $s_{\hat{\beta}_3}$ for $\hat{\beta}_3$ in the unstandardized model)?

g. The estimate of σ for the two-predictor model is $s = .02001$, whereas the estimated standard deviation of $\hat{\beta}_0 + \hat{\beta}_3 x_3' + \hat{\beta}_5 x_5'$ when $x_3' = −.3747$ and $x_5' = −.2769$ (i.e., when $x_3 = 50.5$ and $x_5 = 88.9$) is .00482. Compute a 95% PI for specific gravity when % springwood = 50.5 and % light absorption in summerwood = 88.9.

57. In the accompanying table, we give the smallest SSE for each number of predictors k ($k = 1, 2, 3, 4$) for a regression problem in which y = cumulative heat of hardening in cement, x_1 = % tricalcium aluminate, x_2 = % tricalcium silicate, x_3 = % aluminum ferrate, and x_4 = % dicalcium silicate.

Number of Predictors k	Predictor(s)	SSE
1	x_4	880.85
2	x_1, x_2	58.01
3	x_1, x_2, x_3	49.20
4	x_1, x_2, x_3, x_4	47.86

In addition, $n = 13$ and SST = 2715.76.

a. Use the criteria discussed in the text to recommend the use of a particular regression model.

b. Would forward selection result in the best two-predictor model? Explain.

58. The article **"Response Surface Methodology for Protein Extraction Optimization of Red Pepper Seed"** (*Food Sci. and Tech.*, 2010: 226–231) gave data on the response variable y = protein yield (%) and the independent variables x_1 = temperature (°C), x_2 = pH, x_3 = extraction time (min), and x_4 = solvent/meal ratio.

   a. Fitting the model with the four x_i's as predictors gave the following output:

Predictor	Coef	SE Coef	T	P
Constant	−4.586	2.542	−1.80	0.084
x1	0.01317	0.02707	0.49	0.631
x2	1.6350	0.2707	6.04	0.000
x3	0.02883	0.01353	2.13	0.044
x4	0.05400	0.02707	1.99	0.058

Source	DF	SS	MS	F	P
Regression	4	19.8882	4.9721	11.31	0.000
Residual Error	24	10.5513	0.4396		
Total	28	30.4395			

Calculate and interpret the values of R^2 and adjusted R^2. Does the model appear to be useful?

   b. Fitting the complete second-order model gave the following results:

Predictor	Coef	SE Coef	T	P
Constant	−119.49	18.53	−6.45	0.000
x1	−0.1047	0.2839	−0.37	0.718
x2	28.678	3.625	7.91	0.000
x3	0.4074	0.1303	3.13	0.007
x4	0.2711	0.2606	1.04	0.316
x1sqd	−0.000752	0.002110	−0.36	0.727
x2sqd	−1.6452	0.2110	−7.80	0.000
x3sqd	0.0002121	0.0005275	0.40	0.694
x4sqd	−0.015152	0.002110	−7.18	0.000
x1x2	0.02150	0.02687	0.80	0.437
x1x3	0.000550	0.001344	0.41	0.688
x1x4	−0.000800	0.002687	−0.30	0.770
x2x3	−0.05900	0.01344	−4.39	0.001
x2x4	0.03900	0.02687	1.45	0.169
x3x4	0.002725	0.001344	2.03	0.062

$S = 0.268703$ R-Sq = 96.7%   R-Sq(adj) = 93.4%

Source	DF	SS	MS	F	P
Regression	14	29.4287	2.1020	29.11	0.000
Residual Error	14	1.0108	0.0722		
Total	28	30.4395			

Does at least one of the second-order predictors appear to be useful? Carry out an appropriate test of hypotheses.

   c. From the output in (b), a reasonable conjecture is that none of the predictors involving x_1 are providing useful information. When these predictors are eliminated, the value of SSE for the reduced regression model is 1.1887. Does this support the conjecture?

   d. Here is output from Minitab's best subsets option, with just the single best subset of each size identified. Which model(s) would you consider using (subject to checking model adequacy)?

Minitab output for Exercise 58d

Vars	R-Sq	R-Sq(adj)	Mallows Cp	S	x1	x2	x3	x4	x1sqd	x2sqd	x3sqd	x4sqd	x1x2	x1x3	x1x4	x2x3	x2x4	x3x4
1	52.7	50.9	174.4	0.73030	X													
2	67.9	65.4	112.5	0.61349						X						X		
3	77.7	75.0	73.1	0.52124	X				X									X
4	83.4	80.7	50.8	0.45835	X	X		X	X									
5	90.9	88.9	21.4	0.34731	X			X	X							X	X	
6	94.6	93.1	7.9	0.27422	X	X	X	X	X				X					
7	95.8	94.4	4.7	0.24683	X	X		X	X						X	X	X	
8	96.2	94.6	5.1	0.24137	X	X		X	X	X			X	X	X			
9	96.4	94.7	6.1	0.23962	X	X	X	X	X	X			X	X	X			
10	96.6	94.6	7.5	0.24132	X	X	X	X		X	X		X	X	X			
11	96.6	94.4	9.4	0.24716	X	X	X	X	X	X	X	X	X	X	X			
12	96.6	94.1	11.2	0.25328	X	X	X	X	X	X	X	X	X	X	X	X		
13	96.7	93.8	13.1	0.26041	X	X	X	X	X	X	X	X	X	X	X	X		
14	96.7	93.4	15.0	0.26870	X	X	X	X	X	X	X	X	X	X	X	X	X	X

59. Reconsider the wood specific gravity data referred to in Exercise 56.

   (a) Minitab's Best Regression option was used resulting in the accompanying output. Which model(s) would you recommend for investigating in more detail?

```
Response is spgrav
                                  s      %      s
                                  p  s   s   s  u
                                  r  u   p   p  m
                                  n  m   r   l  l
                                  g  r   w   t  t
                                  f  f   o   a  a
           R-Sq                   i  i   o   b  b
Vars R-Sq  (adj)  C-p       s     b  b   b   d  s   s
  1 56.4   53.9   10.6  0.021832              X
  1 10.6    5.7   38.5  0.031245                       X
  1  5.3    0.1   41.7  0.032155  X
  2 65.5   61.4    7.0  0.019975              X         X
  2 62.1   57.6    9.1  0.020950           X  X
  2 60.3   55.6   10.2  0.021439  X           X
  3 72.3   67.1    4.9  0.018461  X           X         X
  3 71.2   65.8    5.6  0.018807           X  X         X
  3 71.1   65.7    5.6  0.018846  X  X                  X
  4 77.0   70.9    4.0  0.017353  X  X   X              X
  4 74.8   68.1    5.4  0.018179  X  X   X              X
  4 72.7   65.4    6.7  0.018919  X  X   X              X
  5 77.0   68.9    6.0  0.017953  X  X   X   X  X
```

b. The accompanying Minitab output resulted from applying both the backward elimination method and the forward selection method. For each method, explain what occurred at every iteration of the algorithm.

```
Response is spgrav on 5 predictors,
with N = 20

Step            1        2        3        4
Constant     0.4421   0.4384   0.4381   0.5179

sprngfib     0.00011  0.00011  0.00012
T-Value         1.17     1.95     1.98

sumrfib      0.00001
T-Value         0.12

%sprwood    -0.00531 -0.00526 -0.00498 -0.00438
T-Value        -5.70    -6.56    -5.96    -5.20

spltabs     -0.0018  -0.0019
T-Value        -1.63    -1.76

sumltabs     0.0044   0.0044   0.0031   0.0027
T-Value         3.01     3.31     2.63     2.12

S            0.0180   0.0174   0.0185   0.0200
R-Sq          77.05    77.03    72.27    65.50

Step            1        2
Constant     0.7585   0.5179

%sprwood    -0.00444 -0.00438
T-Value        -4.82    -5.20

sumltabs              0.0027
T-Value                 2.12

S            0.0218   0.0200
R-Sq          56.36    65.50
```

60. Pillar stability is a most important factor to ensure safe conditions in underground mines. The authors of **"Developing Coal Pillar Stability Chart Using Logistic Regression"** (*Intl. J. of Rock Mechanics & Mining Sci.*, 2013: 55–60) used a logistic regression model to predict stability. The article reported the following data on x_1 = pillar height to width ratio, x_2 = pillar strength to stress ratio, and stability status for 29 coal pillars.

ID	x_1	x_2	Stable?	ID	x_1	x_2	Stable?
1	1.80	2.40	Y	16	0.80	1.37	N
2	1.65	2.54	Y	17	0.60	1.27	N
3	2.70	0.84	Y	18	1.30	0.87	N
4	3.67	1.68	Y	19	0.83	0.97	N
5	1.41	2.41	Y	20	0.57	0.94	N
6	1.76	1.93	Y	21	1.44	1.00	N
7	2.10	1.77	Y	22	2.08	0.78	N
8	2.10	1.50	Y	23	1.50	1.03	N
9	4.57	2.43	Y	24	1.38	0.82	N
10	3.59	5.55	Y	25	0.94	1.30	N
11	8.33	2.58	Y	26	1.58	0.83	N
12	2.86	2.00	Y	27	1.67	1.05	N
13	2.58	3.68	Y	28	3.00	1.19	N
14	2.90	1.13	Y	29	2.21	0.86	N
15	3.89	2.49	Y				

The corresponding logistic regression output from R is given here:

```
Coefficients:

               Estimate  Std. Error  z value  Pr(>|z|)
(Intercept)    -13.146       5.184    -2.536    0.0112
x1               2.774       1.477     1.878    0.0604
x2               5.668       2.642     2.145    0.0319
```

a. Using the output with $\alpha = .1$ to determine whether the two predictor variables appear to have a significant impact on pillar stability.

b. Provide interpretations for $e^{2.774}$ and $e^{5.668}$.

61. Reconsider the wood specific gravity data referred to in Exercise 56. The following R^2 values resulted from regressing each predictor on the other four predictors (in the first regression, the dependent variable was x_1 and the predictors were x_2–x_5, etc.): .628, .711, .341, .403, and .403. Does multicollinearity appear to be a substantial problem? Explain.

62. A study carried out to investigate the relationship between a response variable relating to pressure drops in a screen-plate bubble column and the predictors x_1 = superficial fluid velocity, x_2 = liquid viscosity, and x_3 = opening mesh size resulted in the accompanying data (**"A Correlation of Two-Phase Pressure Drops in Screen-Plate Bubble Column,"** *Canad. J. of Chem. Engr.*, 1993: 460–463). The standardized residuals and h_{ii} values resulted from the model with just x_1, x_2, and x_3 as predictors. Are there any unusual observations?

Data for Exercise 62

Observation	Velocity	Viscosity	Mesh Size	Response	Standardized Residual	h_{ii}
1	2.14	10.00	.34	28.9	2.01721	.202242
2	4.14	10.00	.34	26.1	1.34706	.066929
3	8.15	10.00	.34	22.8	.96537	.274393
4	2.14	2.63	.34	24.2	1.29177	.224518
5	4.14	2.63	.34	15.7	−.68311	.079651
6	8.15	2.63	.34	18.3	.23785	.267959
7	5.60	1.25	.34	18.1	.06456	.076001
8	4.30	2.63	.34	19.1	.13131	.074927
9	4.30	2.63	.34	15.4	−.74091	.074927
10	5.60	10.10	.25	12.0	−1.38857	.152317
11	5.60	10.10	.34	19.8	−.03585	.068468
12	4.30	10.10	.34	18.6	−.40699	.062849
13	2.40	10.10	.34	13.2	−1.92274	.175421
14	5.60	10.10	.55	22.8	−1.07990	.712933
15	2.14	112.00	.34	41.8	−1.19311	.516298
16	4.14	112.00	.34	48.6	1.21302	.513214
17	5.60	10.10	.25	19.2	.38451	.152317
18	5.60	10.10	.25	18.4	.18750	.152317
19	5.60	10.10	.25	15.0	−.64979	.152317

63. Multiple regression output from Minitab for the PAH data of Exercise 53 in the previous section included the following information:

```
Unusual Observations

Obs   x1    flth   Fit   SE Fit Residual St Resid
 6  243500 604.7 582.9  40.7     21.8     1.25X
 7   67793  27.7 139.3  12.3   −111.6    −2.62R

R denotes an observation with a large standard-
ized residual

X denotes an observation whose X value gives it
large influence.
```

What does this suggest about the appropriateness of using the previously given fitted equation as a basis for inferences? The investigators actually eliminated observation #7 and re-regressed. Does this make sense?

64. The article **"Bank Full Discharge of Rivers"** (*Water Resources J.*, **1978: 1141–1154**) reports data on discharge amount (q, in m^3/sec), flow area (a, in m^2), and slope of the water surface (b, in m/m) obtained at a number of floodplain stations. A subset of the data follows. The article proposed a multiplicative power model $Q = \alpha a^\beta b^\gamma \epsilon$.

q	17.6	23.8	5.7	3.0	7.5
a	8.4	31.6	5.7	1.0	3.3
b	.0048	.0073	.0037	.0412	.0416
q	89.2	60.9	27.5	13.2	12.2
a	41.1	26.2	16.4	6.7	9.7
b	.0063	.0061	.0036	.0039	.0025

Let $y = \ln(q)$, $x_1 = \ln(a)$, and $x_2 = \ln(b)$. Consider fitting the model $Y = \beta_0 + \beta_1 x_1 + \beta_2 x_2 + \epsilon$.

a. The resulting h_{ii}'s are .138, .302, .266, .604, .464, .360, .215, .153, .214, and .284. Does any observation appear to be influential?

b. The estimated coefficients are $\hat{\beta}_0 = 1.5652$, $\hat{\beta}_1 = .9450$, and $\hat{\beta}_2 = .1815$, and the corresponding estimated standard deviations are $s_{\hat{\beta}_0} = .7328$, $s_{\hat{\beta}_1} = .1528$, and $s_{\hat{\beta}_2} = .1752$. The second standardized residual is $e_2^* = 2.19$. When the second observation is omitted from the data set, the resulting estimated coefficients are $\hat{\beta}_0 = 1.8982$, $\hat{\beta}_1 = 1.025$, and $\hat{\beta}_2 = .3085$. Do any of these changes indicate that the second observation is influential?

c. Deletion of the fourth observation (why?) yields $\hat{\beta}_0 = 1.4592$, $\hat{\beta}_1 = .9850$, and $\hat{\beta}_2 = .1515$. Is this observation influential?

SUPPLEMENTARY EXERCISES (65–83)

65. Curing concrete is known to be vulnerable to shock vibrations, which may cause cracking or hidden damage to the material. As part of a study of vibration phenomena, the paper **"Shock Vibration Test of Concrete"** (*ACI Materials J.*, **2002: 361–370**) reported the accompanying data on peak particle velocity (mm/sec) and ratio of ultrasonic pulse velocity after impact to that before impact in concrete prisms.

Obs	ppv	Ratio	Obs	ppv	Ratio
1	160	.996	16	708	.990
2	164	.996	17	806	.984
3	178	.999	18	884	.986
4	252	.997	19	526	.991
5	293	.993	20	490	.993
6	289	.997	21	598	.993
7	415	.999	22	505	.993
8	478	.997	23	525	.990
9	391	.992	24	675	.991
10	486	.985	25	1211	.981
11	604	.995	26	1036	.986
12	528	.995	27	1000	.984
13	749	.994	28	1151	.982
14	772	.994	29	1144	.962
15	532	.987	30	1068	.986

Transverse cracks appeared in the last 12 prisms, whereas there was no observed cracking in the first 18 prisms.

a. Construct a comparative boxplot of ppv for the cracked and uncracked prisms and comment. Then estimate the difference between true average ppv for cracked and uncracked prisms in a way that conveys information about precision and reliability.

b. The investigators fit the simple linear regression model to the entire data set consisting of 30 observations, with ppv as the independent variable and ratio as the dependent variable. Use a statistical software package to fit several different regression models, and draw appropriate inferences.

66. The article **"Applying Regression Analysis to Improve Dyeing Process Quality: A Case Study" (_Intl. J. of Advanced Manuf. Tech._, 2010: 357–368)** examined the practice of adjust pH of dye liquor at a large manufacturer of automotive carpets. The investigation was based on a data set consisting of 114 observations included in the article). The dependent variable is $y = $ pH _before_ addition of dyes, and the predictors are $x_1 = $ carpet density (oz/yd^2), $x_2 = $ carpet weight (lb), $x_3 = $ dye weight (g), $x_4 = $ dye weight as a percentage of carpet weight (%), and $x_5 = $ pH _after_ addition of dyes.

a. Here is output from Minitab's Best Subsets Regression option. Which model(s) would you recommend, and why?

			Mallows		x	x	x	x	x
Vars	R-Sq	R-Sq(adj)	Cp	S	1	2	3	4	5
1	63.7	63.4	16.6	0.34971					X
1	4.4	3.5	223.6	0.56773	X				
2	68.7	68.1	1.2	0.32630			X		X
2	68.6	68.0	1.6	0.32684				X	X
3	69.0	68.2	2.2	0.32616	X	X			X

			Mallows		x	x	x	x	x
Vars	R-Sq	R-Sq(adj)	Cp	S	1	2	3	4	5
3	68.9	68.0	2.5	0.32668			X	X	X
4	69.0	67.9	4.1	0.32754	X	X	X	X	
4	69.0	67.9	4.1	0.32759	X	X	X		X
5	69.0	67.6	6.0	0.32894	X	X	X	X	X

b. The cited article recommended the model with just x_3 and x_5 as predictors. The following Minitab output resulted from fitting that model.

Predictor	Coef	SE Coef	T	P
Constant	0.9402	0.2814	3.34	0.001
x3	−0.00004639	0.00001104	−4.20	0.000
x5	0.73710	0.04813	15.31	0.000

$S = 0.326304$ R-Sq $= 68.7\%$ R-Sq(adj) $= 68.1\%$

Analysis of Variance

Source	DF	SS	MS	F	P
Regression	2	25.925	12.962	121.74	0.000
Residual Error	111	11.819	0.106		
Total	113	37.744			

Does this model appear to specify a useful relationship between the response variable and the predictors? [_Note:_ The pattern in a normal probability plot of the standardized residuals is very linear. The plots of standardized residuals against both x_3 and x_5 show no discernible pattern. There is one observation whose x_3 value is more than twice as large as for any other observation, but with $n = 114$, this observation has very little influence on the fit.]

c. Should either one of the two predictors be eliminated from the model provided that the other predictor is retained? Explain your reasoning.

d. Calculate and interpret 95% CIs for the β coefficients of the two model predictors.

e. The estimated standard deviation of $\hat{Y}$ when $x_3 = 1000$ and $x_5 = 6$ is .0336. Obtain and interpret a 95% CI for true average pH before addition of dyes under these circumstances.

67. The article **"Validation of the Rockport Fitness Walking Test in College Males and Females" (_Research Quarterly for Exercise and Sport_, 1994: 152–158)** recommended the following estimated regression equation for relating $y = \text{VO}_2\text{max}$ (L/min, a measure of cardiorespiratory fitness) to the predictors $x_1 = $ gender (female $= 0$, male $= 1$), $x_2 = $ weight (lb), $x_3 = $ 1-mile walk time (min), and $x_4 = $ heart rate at the end of the walk (beats/min):

$$y = 3.5959 + .6566x_1 + .0096x_2$$
$$- .0996x_3 - .0080x_4$$

a. How would you interpret the estimated coefficient $\hat{\beta}_3 = -.0996$?

b. How would you interpret the estimated coefficient $\hat{\beta}_1 = .6566$?

c. Suppose that an observation made on a male whose weight was 170 lb, walk time was 11 min, and heart rate was 140 beats/min resulted in $VO_2max = 3.15$. What would you have predicted for VO_2max in this situation, and what is the value of the corresponding residual?

d. Using SSE = 30.1033 and SST = 102.3922, what proportion of observed variation in VO_2max can be attributed to the model relationship?

e. Assuming a sample size of $n = 20$, carry out a test of hypotheses to decide whether the chosen model specifies a useful relationship between VO_2max and at least one of the predictors.

68. Feature recognition from surface models of complicated parts is becoming increasingly important in the development of efficient computer-aided design (CAD) systems. The article **"A Computationally Efficient Approach to Feature Abstraction in Design-Manufacturing Integration"** (*J. of Engr. for Industry*, 1995: 16–27) contained a graph of $\log_{10}$(total recognition time), with time in sec, versus $\log_{10}$(number of edges of a part), from which the following representative values were read:

Log(edges)	1.1	1.5	1.7	1.9	2.0	2.1
Log(time)	.30	.50	.55	.52	.85	.98

Log(edges)	2.2	2.3	2.7	2.8	3.0	3.3
Log(time)	1.10	1.00	1.18	1.45	1.65	1.84

Log(edges)	3.5	3.8	4.2	4.3
Log(time)	2.05	2.46	2.50	2.76

a. Does a scatterplot of log(time) versus log(edges) suggest an approximate linear relationship between these two variables?

b. What probabilistic model for relating y = recognition time to x = number of edges is implied by the simple linear regression relationship between the transformed variables?

c. Summary quantities calculated from the data are

$$n = 16 \quad \Sigma x_i' = 42.4 \quad \Sigma y_i' = 21.69$$
$$\Sigma(x_i')^2 = 126.34 \quad \Sigma(y_i')^2 = 38.5305$$
$$\Sigma x_i' y_i' = 68.640$$

Calculate estimates of the parameters for the model in part (b), and then obtain a point prediction of time when the number of edges is 300.

69. Air pressure (psi) and temperature (°F) were measured for a compression process in a certain piston-cylinder device, resulting in the following data **(from *Introduction to Engineering Experimentation*, Prentice-Hall, Inc., 1996, p. 153):**

Pressure	20.0	40.4	60.8	80.2	100.4
Temperature	44.9	102.4	142.3	164.8	192.2
Pressure	120.3	141.1	161.4	181.9	201.4
Temperature	221.4	228.4	249.5	269.4	270.8

Pressure	220.8	241.8	261.1	280.4	300.1
Temperature	291.5	287.3	313.3	322.3	325.8

Pressure	320.6	341.1	360.8
Temperature	337.0	332.6	342.9

a. Would you fit the simple linear regression model to the data and use it as a basis for predicting temperature from pressure? Why or why not?

b. Find a suitable probabilistic model and use it as a basis for predicting the value of temperature that would result from a pressure of 200, in the most informative way possible.

70. An aeronautical engineering student carried out an experiment to study how y = lift/drag ratio related to the variables x_1 = position of a certain forward lifting surface relative to the main wing and x_2 = tail placement relative to the main wing, obtaining the following data (*Statistics for Engineering Problem Solving, p. 133*):

x_1 (in.)	x_2 (in.)	y
−1.2	−1.2	.858
−1.2	0	3.156
−1.2	1.2	3.644
0	−1.2	4.281
0	0	3.481
0	1.2	3.918
1.2	−1.2	4.136
1.2	0	3.364
1.2	1.2	4.018

$$\bar{y} = 3.428, \text{SST} = 8.55$$

a. Fitting the first-order model gives SSE = 5.18, whereas including $x_3 = x_1 x_2$ as a predictor results in SSE = 3.07. Calculate and interpret the coefficient of multiple determination for each model.

b. Carry out a test of model utility using $\alpha = .05$ for each of the models described in part (a). Does either result surprise you?

71. An ammonia bath is the one most widely used for depositing Pd-Ni alloy coatings. The article **"Modelling of Palladium and Nickel in an Ammonia Bath in a Rotary Device"** (*Plating and Surface Finishing*, 1997: 102–104) reported on an investigation into how bath-composition characteristics affect coating properties. Consider the following data on x_1 = Pd concentration

(g/dm^3), $x_2 = $ Ni concentration (g/dm^3), $x_3 = $ pH, $x_4 = $ temperature (°C), $x_5 = $ cathode current density (A/dm^2), and $y = $ palladium content (%) of the coating.

Obs	pdconc	niconc	pH	temp	currdens	pallcont
1	16	24	9.0	35	5	61.5
2	8	24	9.0	35	3	51.0
3	16	16	9.0	35	3	81.0
4	8	16	9.0	35	5	50.9
5	16	24	8.0	35	3	66.7
6	8	24	8.0	35	5	48.8
7	16	16	8.0	35	5	71.3
8	8	16	8.0	35	3	62.8
9	16	24	9.0	25	3	64.0
10	8	24	9.0	25	5	37.7
11	16	16	9.0	25	5	68.7
12	8	16	9.0	25	3	54.1
13	16	24	8.0	25	5	61.6
14	8	24	8.0	25	3	48.0
15	16	16	8.0	25	3	73.2
16	8	16	8.0	25	5	43.3
17	4	20	8.5	30	4	35.0
18	20	20	8.5	30	4	69.6
19	12	12	8.5	30	4	70.0
20	12	28	8.5	30	4	48.2
21	12	20	7.5	30	4	56.0
22	12	20	9.5	30	4	77.6
23	12	20	8.5	20	4	55.0
24	12	20	8.5	40	4	60.6
25	12	20	8.5	30	2	54.9
26	12	20	8.5	30	6	49.8
27	12	20	8.5	30	4	54.1
28	12	20	8.5	30	4	61.2
29	12	20	8.5	30	4	52.5
30	12	20	8.5	30	4	57.1
31	12	20	8.5	30	4	52.5
32	12	20	8.5	30	4	56.6

a. Fit the first-order model with five predictors and assess its utility. Do all the predictors appear to be important?

b. Fit the complete second-order model and assess its utility.

c. Does the group of second-order predictors (interaction and quadratic) appear to provide more useful information about y than is contributed by the first-order predictors? Carry out an appropriate test of hypotheses.

d. The authors of the cited article recommended the use of all five first-order predictors plus the additional predictor $x_6 = (pH)^2$. Fit this model. Do all six predictors appear to be important?

72. The article **"An Experimental Study of Resistance Spot Welding in 1 mm Thick Sheet of Low Carbon Steel"** (*J. of Engr. Manufacture*, 1996: 341–348) discussed a statistical analysis whose basic aim was to establish a relationship that could explain the variation in weld strength (y) by relating strength to the process characteristics weld current (wc), weld time (wt), and electrode force (ef).

a. SST = 16.18555, and fitting the complete second-order model gave SSE = .80017. Calculate and interpret the coefficient of multiple determination.

b. Assuming that $n = 37$, carry out a test of model utility [the ANOVA table in the article states that $n - (k + 1) = 1$, but other information given contradicts this and is consistent with the sample size we suggest].

c. The given F ratio for the current–time interaction was 2.32. If all other predictors are retained in the model, can this interaction predictor be eliminated? [*Hint*: As in simple linear regression, an F ratio for a coefficient is the square of its t ratio.]

d. The authors proposed eliminating two interaction predictors and a quadratic predictor and recommended the estimated equation $y = 3.352 + .098wc + .222wt + .297ef - .0102(wt)^2 - .037(et)^2 + .0128(wc)(wt)$. Consider a weld current of 10 kA, a weld time of 12 ac cycles, and an electrode force of 6 kN. Supposing that the estimated standard deviation of the predicted strength in this situation is .0750, calculate a 95% PI for strength. Does the interval suggest that the value of strength can be accurately predicted?

73. The accompanying data on $x = $ frequency (MHz) and $y = $ output power (W) for a certain laser configuration was read from a graph in the article **"Frequency Dependence in RF Discharge Excited Waveguide CO_2 Lasers"** (*IEEE J. of Quantum Electronics*, 1984: 509–514).

x	60	63	77	100	125	157	186	222
y	16	17	19	21	22	20	15	5

A computer analysis yielded the following information for a quadratic regression model: $\hat{\beta}_0 = -1.5127$, $\hat{\beta}_1 = .391901$, $\hat{\beta}_2 = -.00163141$, $s_{\hat{\beta}_2} = .00003391$, SSE = .29, SST = 202.88, and $s_{\hat{Y}} = .1141$ when $x = 100$.

a. Does the quadratic model appear to be suitable for explaining observed variation in output power by relating it to frequency?

b. Would the simple linear regression model be nearly as satisfactory as the quadratic model?

c. Do you think it would be worth considering a cubic model?

d. Compute a 95% CI for expected power output when frequency is 100.

e. Use a 95% PI to predict the power from a single experimental run when frequency is 100.

74. The accompanying data on $x_1 = $ card cylinder speed (rpm), card production rate (kg/h), $x_3 = $ number of draw frame doubling, and $y = $ tenacity (RKM) appeared in the article **"Impact of Carding Parameters and Draw Frame Doubling on the Properties of Ring Spun Yarn"** (*J. of Engineered Fibers and Fabrics*, 2013: 72–78).

Obs	x_1	x_2	x_3	tenacity	Obs	x_1	x_2	x_3	tenacity
1	700	80	6	18.29	9	800	80	5	16.59
2	900	80	6	17.97	10	800	120	5	17.10
3	700	120	6	17.12	11	800	80	7	15.64
4	900	120	6	17.46	12	800	120	7	16.54
5	700	100	5	17.29	13	800	100	6	17.81
6	900	100	5	16.35	14	800	100	6	17.68
7	700	100	7	16.77	15	800	100	6	18.11
8	900	100	7	16.11					

a. Minitab's Best Subsets Regression option gave the following output when applied to the complete second-order model. Notice that adjusted R^2 for the model containing all predictors is much smaller than R^2 itself, indicating that the model contains too many predictors relative to the sample size. Which model(s) would you recommend, and why? [*Note:* The cited article reported results only for the model with all first- and second-order predictors.]

Vars	R-Sq	R-Sq(adj)	Mallows Cp	S	x1	x2	x3	x1 sqd	x2 sqd	x3 sqd	x1 x2	x1 x3	x2 x3
1	10.9	4.0	11.6	0.76552									X
1	10.3	3.3	11.7	0.76826						X			
2	73.4	69.0	−2.3	0.43510			X			X			
2	13.9	0.0	12.8	0.78320	X					X			
3	77.1	70.8	−1.2	0.42208	X		X			X			
3	77.1	70.8	−1.2	0.42229		X	X			X			
4	77.3	68.2	0.7	0.44046	X		X			X			X
4	77.2	68.1	0.8	0.44115	X		X		X	X			
5	79.6	68.2	2.2	0.44073	X		X			X	X	X	
5	79.3	67.9	2.2	0.44302			X	X	X	X	X		
6	80.0	65.0	4.1	0.46247	X		X			X	X	X	X
6	79.8	64.7	4.1	0.46426			X	X	X	X	X		X
7	80.2	60.4	6.0	0.49156	X		X	X		X	X	X	X
7	80.0	60.0	6.1	0.49408	X	X	X			X	X	X	X
8	80.2	53.9	8.0	0.53059	X	X	X		X	X	X	X	X
8	80.2	53.8	8.0	0.53094	X		X	X	X	X	X	X	X
9	80.2	44.7	10.0	0.58123	X	X	X	X	X	X	X	X	X

b. When the model with predictors x_3, x_3^2, and x_1 was fit, the t ratio corresponding to the coefficient on x_1 was −1.32. If the first two predictors remain in the model, is inclusion of x_1 justified? Explain your reasoning.

c. Here is output from relating y to x_3 via the quadratic regression model. A normal probability plot of the standardized residuals is quite straight, and the plot of e^* versus $\hat{y}$ shows no discernible pattern. Does this model specify a useful relationship, and should the quadratic predictor be retained in the model?

Predictor	Coef	SE Coef	T	P
Constant	−24.743	8.040	−3.08	0.010
x3	14.457	2.707	5.34	0.000
x3 sqd	−1.2284	0.2252	−5.46	0.000

S = 0.435097   R-Sq = 73.4%   R-Sq(adj) = 69.0%

d. For the model of part (c), the standard deviation of a predicted Y value when $x_3 = 6$ is .164. Predict tenacity in this situation in a way that conveys information about precision and reliability.

75. The effect of manganese (Mn) on wheat growth is examined in the article "Manganese Deficiency and Toxicity Effects on Growth, Development and Nutrient Composition in Wheat" (*Agronomy J.*, 1984: 213–217). A quadratic regression model was used to relate $y =$ plant height (cm) to $x = \log_{10}(\text{added Mn})$, with μM as the units for added Mn. The accompanying data was read from a scatterplot appearing in the article.

x	−1.0	−.4	0	.2	1.0
y	32	37	44	45	46

x	2.0	2.8	3.2	3.4	4.0
y	42	42	40	37	30

In addition, $\hat{\beta}_0 = 41.7422$, $\hat{\beta}_1 = 6.581$, $\hat{\beta}_2 = -2.3621$, $s_{\hat{\beta}_0} = .8522$, $s_{\hat{\beta}_1} = 1.002$, $s_{\hat{\beta}_2} = .3073$, and SSE = 26.98.

a. Is the quadratic model useful for describing the relationship between x and y? [*Hint*: Quadratic regression is a special case of multiple regression with $k = 2$, $x_1 = x$, and $x_2 = x^2$.] Apply an appropriate procedure.

b. Should the quadratic predictor be eliminated?

c. Estimate expected height for wheat treated with 10 μM of Mn using a 90% CI. [*Hint*: The estimated standard deviation of $\hat{\beta}_0 + \hat{\beta}_1 + \hat{\beta}_2$ is 1.031.]

76. The article **"Chemithermomechanical Pulp from Mixed High Density Hardwoods"** (*TAPPI*, **July 1988: 145–146**) reports on a study in which the accompanying data was obtained to relate y = specific surface area (cm^2/g) to x_1 = % NaOH used as a pretreatment chemical and x_2 = treatment time (min) for a batch of pulp.

x_1	x_2	y
3	30	5.95
3	60	5.60
3	90	5.44
9	30	6.22
9	60	5.85
9	90	5.61
15	30	8.36
15	60	7.30
15	90	6.43

The accompanying Minitab output resulted from a request to fit the model $Y = \beta_0 + \beta_1 x_1 + \beta_2 x_2 + \epsilon$.

```
The regression equation is
AREA = 6.05 + 0.142 NAOH − 0.0169 TIME

Predictor      Coef     Stdev   t-ratio       p
Constant     6.0483    0.5208     11.61   0.000
NAOH        0.14167   0.03301      4.29   0.005
TIME       -0.016944  0.006601    -2.57   0.043

s = 0.4851      R-sq = 80.7%      R-sq(adj) = 74.2%

Analysis of Variance

SOURCE      DF       SS       MS       F       p
Regression   2   5.8854   2.9427   12.51   0.007
Error        6   1.4118   0.2353
Total        8   7.2972
```

a. What proportion of observed variation in specific surface area can be explained by the model relationship?

b. Does the chosen model appear to specify a useful relationship between the dependent variable and the predictors?

c. Provided that % NaOH remains in the model, would you suggest that the predictor *treatment time* be eliminated?

d. Calculate a 95% CI for the expected change in specific surface area associated with an increase of 1% in NaOH when treatment time is held fixed.

e. Minitab reported that the estimated standard deviation of $\hat{\beta}_0 + \hat{\beta}_1(9) + \hat{\beta}_2(60)$ is .162. Calculate a prediction interval for the value of specific surface area to be observed when % NaOH = 9 and treatment time = 60.

77. The article **"Sensitivity Analysis of a 2.5 kW Proton Exchange Membrane Fuel Cell Stack by Statistical Method"** (*J. of Fuel Cell Sci. and Tech.*, **2009: 1–6**) used regression analysis to investigate the relationship between fuel cell power (W) and the independent variables x_1 = H$_2$ pressure (psi), x_2 = H$_2$ flow (stoc), x_3 = air pressure (psi) and x_4 = airflow (stoc).

a. Here is Minitab output from fitting the model with the aforementioned independent variables as predictors (also fit by the authors of the cited article):

```
Predictor      Coef    SE Coef        T        P
Constant     1507.3      206.8     7.29    0.000
x1           -4.282      4.969    -0.86    0.407
x2             7.46      62.11     0.12    0.907
x3          -0.9162     0.6227    -1.47    0.169
x4            90.60      24.84     3.65    0.004

S = 4.6885  R-Sq = 59.6%  R-Sq(adj) = 44.9%

Source          DF       SS      MS       F       P
Regression       4    40048   10012    4.06   0.029
Residual Error  11    27158    2469
Total           15    67206
```

a. Does there appear to be a useful relationship between power and at least one of the predictors? Carry out a formal test of hypotheses.

b. Fitting the model with predictors x_3, x_4, and the interaction $x_3 x_4$ gave $R^2 = .834$. Does this model appear to be useful? Can an F test be used to compare this model to the model of (a)? Explain.

c. Fitting the model with predictors $x_1 - x_4$ as well as all second-order interactions gave $R^2 = .960$ (this model was also fit by the investigators). Does it appear that at least one of the interaction predictors provides useful information about power over and above what is provided by the first-order predictors? State and test the appropriate hypotheses using a significance level of .05.

78. Coir fiber, derived from coconut, is an eco-friendly material with great potential for use in construction. The article **"Seepage Velocity and Piping Resistance of Coir Fiber Mixed Soils"** (*J. of Irrig. and Drainage Engr.*, **2008: 485–492**) included several multiple regression analyses. The article's authors kindly provided the accompanying data on x_1 = fiber content (%), x_2 = fiber length (mm), x_3 = hydraulic gradient (no unit provided), and y = seepage velocity (*cm*/sec).

Obs	cont	lngth	grad	vel
1	0.0	0	0.400	0.027
2	0.0	0	0.716	0.050
3	0.0	0	0.925	0.080
4	0.0	0	1.098	0.099
5	0.0	0	1.226	0.107
6	0.0	0	1.427	0.140
7	0.0	0	1.709	0.178
8	0.0	0	1.872	0.200
9	0.5	50	0.380	0.022
10	0.5	50	0.774	0.040
11	0.5	50	1.056	0.060
12	0.5	50	1.329	0.111
13	0.5	50	1.598	0.158
14	0.5	50	1.799	0.188
15	1.0	50	0.410	0.026
16	1.0	50	0.577	0.038
17	1.0	50	0.748	0.049
18	1.0	50	0.927	0.060
19	1.0	50	1.090	0.070
20	1.0	50	1.239	0.088
21	1.0	50	1.496	0.111
22	1.0	50	1.744	0.134
23	1.0	50	1.915	0.145
24	1.5	50	0.444	0.014
25	1.5	50	0.821	0.037
26	1.5	50	1.141	0.058
27	1.5	50	1.474	0.082
28	1.5	50	1.581	0.112
29	1.5	50	1.983	0.144
30	1.0	25	0.462	0.028
31	1.0	25	0.705	0.059
32	1.0	25	0.987	0.084
33	1.0	25	1.154	0.101
34	1.0	25	1.479	0.150
35	1.0	25	1.786	0.194
36	1.0	25	1.957	0.218
37	1.0	40	0.419	0.030
38	1.0	40	0.705	0.050
39	1.0	40	0.979	0.068
40	1.0	40	1.226	0.091
41	1.0	40	1.470	0.126
42	1.0	40	1.744	0.168
43	1.0	60	0.436	0.034
44	1.0	60	0.650	0.051
45	1.0	60	0.889	0.068
46	1.0	60	1.222	0.093
47	1.0	60	1.477	0.112
48	1.0	60	1.726	0.139
49	1.0	60	1.983	0.173

a. Here is output from fitting the model with the three x_i's as predictors:

```
Predictor         Coef    SE Coef        T       P
Constant     -0.002997   0.007639    -0.39   0.697
fib cont     -0.012125   0.007454    -1.63   0.111
fib lngth   -0.0003020   0.0001676   -1.80   0.078
hyd grad      0.102489   0.004711    21.76   0.000

S = 0.0162355   R-Sq = 91.6%   R-Sq(adj) = 91.1%
```

```
Source           DF       SS        MS       F      P
Regression        3  0.129898  0.043299  164.27  0.000
Residual Error   45  0.011862  0.000264
Total            48  0.141760
```

How would you interpret the number $-.0003020$ in the Coef column on output?

b. Does fiber content appear to provide useful information about velocity provided that fiber length and hydraulic gradient remain in the model? Carry out a test of hypotheses.

c. Fitting the model with just fiber length and hydraulic gradient as predictors gave the estimated regression coefficients $\hat{\beta}_0 = -.005315$, $\hat{\beta}_1 = -.0004968$, and $\hat{\beta}_2 = .102204$ (the t ratios for these two predictors are both highly significant). In addition, $s_{\hat{y}} = .00286$ when fiber length $= 25$ and hydraulic gradient $= 1.2$. Is there convincing evidence that true average velocity is something other than .1 in this situation? Carry out a test using a significance level of .05.

d. Fitting the complete second-order model (as did the article's authors) resulted in SSE $= .003579$. Does it appear that at least one of the second-order predictors provides useful information over and above what is provided by the three first-order predictors? Test the relevant hypotheses.

79. The article **"A Statistical Analysis of the Notch Toughness of 9% Nickel Steels Obtained from Production Heats"** (**J. of Testing and Eval., 1987: 355–363**) reports on the results of a multiple regression analysis relating Charpy v-notch toughness y (joules) to the following variables: $x_1 =$ plate thickness (mm), $x_2 =$ carbon content (%), $x_3 =$ manganese content (%), $x_4 =$ phosphorus content (%) $x_5 =$ sulphur content (%), $x_6 =$ silicon content (%), $x_7 =$ nickel content (%), $x_8 =$ yield strength (Pa), and $x_9 = .$ tensile strength (Pa)

a. The best possible subsets involved adding variables in the order x_5, x_8, x_6, x_3, x_2, x_7, x_9, x_1, and x_4. The values of R_k^2, MSE_k, and C_k are as follows:

No. of Predictors	1	2	3	4
R_k^2	.354	.453	.511	.550
MSE_k	2295	1948	1742	1607
C_k	314	173	89.6	35.7

No. of Predictors	5	6	7	8	9
R_k^2	.562	.570	.572	.575	.575
MSE_k	1566	1541	1535	1530	1532
C_k	19.9	11.0	9.4	8.2	10.0

Which model would you recommend? Explain the rationale for your choice.

b. The authors also considered second-order models involving predictors x_j^2 and $x_i x_j$. Information on the best such models starting with the variables x_2, x_3, x_5,

x_6, x_7, and x_8 is as follows (in going from the best four-predictor model to the best five-predictor model, x_8 was deleted and both x_2x_6 and x_7x_8 were entered, and x_8 was reentered at a later stage):

No. of Predictors	1	2	3	4	5
R_k^2	.415	.541	.600	.629	.650
MSE_k	2079	1636	1427	1324	1251
C_k	433	109	104	52.4	16.5

No. of Predictors	6	7	8	9	10
R_k^2	.652	.655	.658	.659	.659
MSE_k	1246	1237	1229	1229	1230
C_k	14.9	11.2	8.5	9.2	11.0

Which of these models would you recommend, and why? [*Note:* Models based on eight of the original variables did not yield marked improvement on those under consideration here.]

80. A sample of $n = 20$ companies was selected, and the values of y = stock price and $k = 15$ variables (such as quarterly dividend, previous year's earnings, and debt ratio) were determined. When the multiple regression model using these 15 predictors was fit to the data, $R^2 = .90$ resulted.

   a. Does the model appear to specify a useful relationship between y and the predictor variables? Carry out a test using significance level .05. [*Hint:* The F critical value for 15 numerator and 4 denominator df is 5.86.]

   b. Based on the result of part (a), does a high R^2 value by itself imply that a model is useful? Under what circumstances might you be suspicious of a model with a high R^2 value?

   c. With n and k as given previously, how large would R^2 have to be for the model to be judged useful at the .05 level of significance?

81. Does exposure to air pollution result in decreased life expectancy? This question was examined in the article **"Does Air Pollution Shorten Lives?"** (*Statistics and Public Policy,* **Reading, MA, Addison-Wesley, 1977**). Data on

   y = total mortality rate (deaths per 10,000)

   x_1 = mean suspended particle reading (μg/m³)

   x_2 = smallest sulfate reading ([μg/m³] $\times$ 10)

   x_3 = population density (people/mi²)

   x_4 = (percent nonwhite) $\times$ 10

   x_5 = (percent over 65) $\times$ 10

for the year 1960 was recorded for $n = 117$ randomly selected standard metropolitan statistical areas. The estimated regression equation was

$$y = 19.607 + .041x_1 + .071x_2$$
$$+ .001x_3 + .041x_4 + .687x_5$$

   a. For this model, $R^2 = .827$. Using a .05 significance level, perform a model utility test.

   b. The estimated standard deviation of $\hat{\beta}_1$ was .016. Calculate and interpret a 90% CI for β_1.

   c. Given that the estimated standard deviation of $\hat{\beta}_4$ is .007, determine whether percent nonwhite is an important variable in the model. Use a .01 significance level.

   d. In 1960, the values of x_1, x_2, x_3, x_4, and x_5 for Pittsburgh were 166, 60, 788, 68, and 95, respectively. Use the given regression equation to predict Pittsburgh's mortality rate. How does your prediction compare with the actual 1960 value of 103 deaths per 10,000?

82. Given that $R^2 = .723$ for the model containing predictors x_1, x_4, x_5, and x_8 and $R^2 = .689$ for the model with predictors x_1, x_3, x_5, and x_6, what can you say about R^2 for the model containing predictors

   a. x_1, x_3, x_4, x_5, x_6, and x_8? Explain.

   b. x_1 and x_4? Explain.

83. An article in *Lubrication Engr.* (**"Accelerated Testing of Solid Film Lubricants,"** 1972: 365–372) reported on an investigation of wear life (y, in hr) for solid film lubricant. Three sets of journal bearing tests were run on a Mil-L-8937-type film at each combination of three speeds (x_1, in rpm), and three loads (x_2, in 1000s of hr). The values of x_1 for the resulting 27 observations were $20, 20, \ldots, 20, 60, \ldots, 60, 100, \ldots, 100$, and the values of x_2 were $3, 3, 3, 6, 6, 6, 10, 10, 10, 3, 3, 3, \ldots,$ $10, 10, 10$. The corresponding values of y = wear life (hr) were $300.2, 310.8, 333.0, 99.6, 136.2, 142.4, 20.2,$ $28.2, 102.7, 67.3, 77.9, 93.9, 43.0, 44.5, 65.9, 10.7,$ $34.1, 39.1, 26.5, 22.3, 34.8, 32.8, 25.6, 32.7, 2.3, 4.4,$ and 5.8.

   The investigators commented that a lognormal distribution is appropriate for Y because $\ln(Y)$ is known to follow a normal law, and then proposed the multiplicative power regression model $Y = \alpha x_1^{\beta_1} x_2^{\beta_2} \epsilon$.

   a. Estimate the model parameters.

   b. Interpret R^2 for the transformed model, and then carry out a model utility test.

   c. Does it appear that both predictors provide useful information about wear life?

   d. Predict wear life when speed is 50 and load is 5 in a way that conveys information about precision and reliability.

BIBLIOGRAPHY

Chatterjee, Samprit, and Ali Hadi, *Regression Analysis by Example* (4th ed.), Wiley, New York, 2006. A brief but informative discussion of selected topics, especially multicollinearity and the use of biased estimation methods.

Daniel, Cuthbert, and Fred Wood, *Fitting Equations to Data* (2nd ed.), Wiley, New York, 1980. Contains many insights and methods that evolved from the authors' extensive consulting experience.

Draper, Norman, and Harry Smith, *Applied Regression Analysis* (3rd ed.), Wiley, New York, 1999. See Chapter 12 bibliography.

Hoaglin, David, and Roy Welsch, "The Hat Matrix in Regression and ANOVA," *American Statistician,* 1978: 17–23. Describes methods for detecting influential observations in a regression data set.

Hocking, Ron, "The Analysis and Selection of Variables in Linear Regression," *Biometrics,* 1976: 1–49. An excellent survey of this topic.

Neter, John, Michael Kutner, Christopher Nachtsheim, and William Wasserman, *Applied Linear Statistical Models* (5th ed.), Irwin, Homewood, IL, 2004. See Chapter 12 bibliography.

Goodness-of-Fit Tests and Categorical Data Analysis

14

INTRODUCTION

In the simplest type of situation considered in this chapter, each observation in a sample is classified as belonging to one of a finite number of categories (e.g., blood type could be one of the four categories O, A, B, or AB). Let p_i denoting the probability that any particular observation belongs in category i (or the proportion of the population belonging to category i). We then wish to test a null hypothesis that completely specifies the values of all the p_i's (such as H_0: $p_1 = .45$, $p_2 = .35$, $p_3 = .15$, $p_4 = .05$, when there are four categories). The test statistic is based on how different the numbers of observations in the various categories are from the corresponding expected numbers when H_0 is true. Because the reference distribution for determining the P-value is a chi-squared distribution, the procedure is called a chi-squared goodness-of-fit test.

Sometimes the null hypothesis specifies that the p_i's depend on some smaller number of parameters without specifying the values of these parameters. For example, with three categories the null hypothesis might state that $p_1 = \theta^2$, $p_2 = 2\theta(1 - \theta)$, and $p_3 = (1 - \theta)^2$. For a chi-squared test to be performed, the values of any unspecified parameters must be estimated from the sample data. Section 14.2 develops methodology for doing this. The methods are then applied to test a null hypothesis that states that the sample comes from a particular family of distributions, such as the Poisson family (with μ estimated from the sample) or the normal family (with μ and σ estimated). In addition, a test based on a normal probability plot is presented for the null hypothesis of population normality.

Chi-squared tests for two different situations are considered in Section 14.3. In the first, the null hypothesis states that the p_i's are the same for several different populations. The second type of situation involves taking a sample from a single population and classifying each individual with respect to two different categorical factors (such as religious preference and political-party registration). The null hypothesis in this situation is that the two factors are independent within the population.

14.1 Goodness-of-Fit Tests When Category Probabilities Are Completely Specified

A binomial experiment consists of a sequence of independent trials in which each trial can result in one of two possible outcomes: S (for success) and F (for failure). The probability of success, denoted by p, is assumed to be constant from trial to trial, and the number n of trials is fixed at the outset of the experiment. In Chapter 8, we presented a large-sample z test for testing $H_0: p = p_0$. Notice that this null hypothesis specifies both $P(S)$ and $P(F)$, since if $P(S) = p_0$, then $P(F) = 1 - p_0$. Denoting $P(F)$ by q and $1 - p_0$ by q_0, the null hypothesis can alternatively be written as $H_0: p = p_0, q = q_0$. The z test is two-tailed when the alternative of interest is $p \neq p_0$.

A **multinomial experiment** generalizes a binomial experiment by allowing each trial to result in one of k possible outcomes, where $k > 2$. For example, suppose a store accepts three different types of credit cards. A multinomial experiment would result from observing the type of credit card used—type 1, type 2, or type 3—by each of the next n customers who pay with a credit card. In general, we will refer to the k possible outcomes on any given trial as categories, and p_i will denote the probability that a trial results in category i. If the experiment consists of selecting n individuals or objects from a population and categorizing each one, then p_i is the proportion of the population falling in the ith category (such an experiment will be approximately multinomial provided that n is much smaller than the population size).

The null hypothesis of interest will specify the value of each p_i. For example, in the case $k = 3$, we might have $H_0: p_1 = .5$, $p_2 = .3$, $p_3 = .2$. The alternative hypothesis will state that H_0 is not true—that is, that at least one of the p_i's has a value different from that asserted by H_0 (in which case at least two must be different, since they sum to 1). The symbol p_{i0} will represent the value of p_i claimed by the null hypothesis. In the example just given, $p_{10} = .5$, $p_{20} = .3$, and $p_{30} = .2$.

Before the multinomial experiment is performed, the number of trials that will result in category i ($i = 1, 2, \ldots$, or k) is a random variable—just as the number of successes and the number of failures in a binomial experiment are random variables. This random variable will be denoted by N_i and its observed value by n_i. Since each trial results in exactly one of the k categories, $\Sigma N_i = n$, and the same is true of the n_i's. As an example, an experiment with $n = 100$ and $k = 3$ might yield $N_1 = 46$, $N_2 = 35$, and $N_3 = 19$.

The expected number of successes and expected number of failures in a binomial experiment are np and nq, respectively. When $H_0: p = p_0, q = q_0$ is true, the expected numbers of successes and failures are np_0 and nq_0, respectively. Similarly, in a multinomial experiment the expected number of trials resulting in category i is $E(N_i) = np_i (i = 1, \ldots, k)$. When $H_0: p_1 = p_{10}, \ldots, p_k = p_{k0}$ is true, these expected values become $E(N_1) = np_{10}, E(N_2) = np_{20}, \ldots, E(N_k) = np_{k0}$. For the case $k=3$, $H_0: p_1 = .5, p_2 = .3, p_3 = .2$, and $n = 100$, the expected frequencies when H_0 is true are $E(N_1) = 100(.5) = 50$, $E(N_2) = 30$, and $E(N_3) = 20$. The n_i's and corresponding expected frequencies are often displayed in a tabular format as shown in Table 14.1. The expected values when H_0 is true are displayed just below the observed values.

The N_i's and n_i's are often referred to as *observed cell counts* (or *observed cell frequencies*), and $np_{10}, np_{20}, \ldots, np_{k0}$ are the corresponding *expected cell counts* under H_0.

Table 14.1   Observed and Expected Cell Counts

Category	$i = 1$	$i = 2$	$\ldots$	$i = k$	Row total
Observed	n_1	n_2	$\ldots$	n_k	n
Expected	np_{10}	np_{20}	$\ldots$	np_{k0}	n

The n_i's should all be reasonably close to the corresponding np_{i0}'s when H_0 is true. On the other hand, several of the observed counts should differ substantially from these expected counts when the actual values of the p_i's differ markedly from what the null hypothesis asserts. The test procedure involves assessing the discrepancy between the n_i's and the np_{i0}'s. It is natural to base a measure of discrepancy on the squared deviations $(n_1 - np_{10})^2, (n_2 - np_{20})^2, \ldots, (n_k - np_{k0})^2$. A seemingly sensible way to combine these into an overall measure is to add them together to obtain $\Sigma(n_i - np_{i0})^2$. However, suppose $np_{10} = 100$ and $np_{20} = 10$. Then if $n_1 = 95$ and $n_2 = 5$, the two categories contribute the same squared deviations to the proposed measure. Yet n_1 is only 5% less than what would be expected when H_0 is true, whereas n_2 is 50% less. To take relative magnitudes of the deviations into account, each squared deviation is divided by the corresponding expected count.

Before giving a more detailed description, we must reintroduce a type of probability distribution called the *chi-squared* (χ^2) *distribution*. This distribution was first encountered in Section 4.4 and was used in Chapter 7 to obtain a confidence interval for the variance σ^2 of a normal population. The chi-squared distribution has a single parameter ν, called the number of degrees of freedom (df) of the distribution, with possible values 1, 2, 3, ... If $Y \sim \chi^2$ with ν df, then $E(Y) = \nu$ and $V(Y) = 2\nu$. Figure 14.1 shows a typical χ^2 density curve; it is positively skewed, but moves rightward and becomes more symmetric and spread out as ν increases.

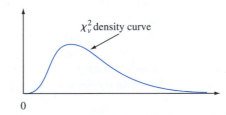

Figure 14.1   A typical chi-squared density curve (small ν).

THEOREM

Provided that $np_i \geq 5$ for every i ($i = 1, 2, \ldots, k$), the variable

$$\chi^2 = \sum_{i=1}^{k} \frac{(N_i - np_i)^2}{np_i} = \sum_{\text{all cells}} \frac{(\text{observed} - \text{expected})^2}{\text{expected}}$$

has approximately a chi-squared distribution with $k - 1$ df.

The fact that df $= k - 1$ is a consequence of the restriction $\Sigma N_i = n$. Although there are k observed cell counts, once any $k - 1$ are known, the remaining one is uniquely determined. That is, there are only $k - 1$ "freely determined" cell counts, and thus $k - 1$ df.

If np_{i0} is substituted for np_i in χ^2, the resulting test statistic has a chi-squared distribution when H_0 is true. The more the observed frequencies differ from expected frequencies, the larger the value of χ^2 will be. Since the test statistic utilizes expected frequencies assuming that H_0 is true, any test statistic value larger than the calculated χ^2 will be even more contradictory to H_0 than this calculated value. The implication is that the test is upper-tailed: The P-value will be the area under the relevant chi-squared curve to the right of the calculated χ^2 value.

Null hypothesis: H_0: $p_1 = p_{10}, p_2 = p_{20}, \ldots, p_k = p_{k0}$

Alternative hypothesis: H_a: at least one p_i does not equal p_{i0}

Test statistic value: $\chi^2 = \sum_{\text{all cells}} \dfrac{(\text{observed} - \text{expected})^2}{\text{expected}} = \sum_{i=1}^{k} \dfrac{(n_i - np_{i0})^2}{np_{i0}}$

Provided that $np_{i0} \geq 5$ for all i, the P-value is (approximately) the area under the χ^2_{k-1} curve to the right of the calculated value of χ^2. If $np_{i0} < 5$ for at least one i, categories should be combined in a sensible way to correct this deficiency.

Table A.7 gives chi-squared critical values $\chi^2_{\alpha,\nu}$ that capture specified areas α under various chi-squared curves (analogous to what $t_{\alpha,\nu}$ does for t curves). But because the tabulation is for only five small values of α, limited information about a P-value is available. We have therefore included another appendix table, similar to the t curve tail areas of Table A.8, that facilitates making more precise P-value statements.

The fact that t curves were all centered at zero allowed us to tabulate t-curve tail areas in a relatively compact way, with the left margin giving values ranging from 0.0 to 4.0 on the horizontal t scale and various columns displaying corresponding upper-tail areas for various df's. The rightward movement of chi-squared curves as df increases necessitates a somewhat different type of tabulation. The left margin of Appendix Table A.11 displays various upper-tail areas: .100, .095, .090, . . . , .005, and .001. Each column of the table is for a different value of df, and the entries are values on the horizontal chi-squared axis that capture these corresponding tail areas. For example, moving down to tail area .085 and across to the 4 df column, we see that the area to the right of 8.18 under the 4 df chi-squared curve is .085 (see Figure 14.2). Capturing the same upper-tail area under the 10 df curve requires going out to 16.54.

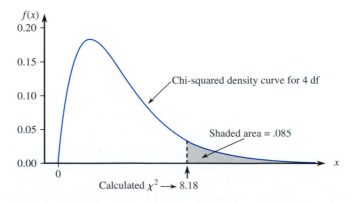

Figure 14.2 A P-value for an upper-tailed chi-squared test

Returning to the 4 df column of Table A.8, we see that the area under the χ_4^2 curve to the right of 8.33 is .08. Thus the calculated value $\chi^2 = 8.20$ implies that $.08 < P$-value $< .085$. In this case H_0 would be rejected at significance level .10 but not at levels .05 or .01. The top row of the 4 df column shows that if the calculated value of the chi-squared variable is smaller than 7.77, the captured tail area (the P-value) exceeds .10. Similarly, the bottom row in this column indicates that if the calculated value exceeds 18.46, the tail area is smaller than .001 (P-value $< .001$).

EXAMPLE 14.1 Genetics provides a rich area for application of chi-squared testing. Let's focus on two different characteristics of an organism, each controlled by a single gene, and consider crossing a pure strain having genotype *AABB* with a pure strain having genotype *aabb* (capital letters denoting dominant alleles and small letters recessive alleles). The resulting genotype will be *AaBb*. If these first-generation organisms are then crossed among themselves (a dihybrid cross), there will be four phenotypes depending on whether a dominant allele of either type is present. Mendel's laws of inheritance imply that these four phenotypes should have probabilities 9/16, 3/16, 3/16, and 1/16 of arising in any given dihybrid cross.

The article **"Linkage Studies of the Tomato"** (*Trans. Royal Canadian Institute*, **1931: 1–19**) reports the following data on phenotypes from a dihybrid cross of tall cut-leaf tomatoes with dwarf potato-leaf tomatoes. There are $k = 4$ categories corresponding to the four possible phenotypes, with the null hypothesis being

$$H_0: p_1 = \frac{9}{16}, \; p_2 = \frac{3}{16}, \; p_3 = \frac{3}{16}, \; p_4 = \frac{1}{16}$$

The expected cell counts are $9n/16$, $3n/16$, $3n/16$, and $n/16$, and the test is based on $k - 1 = 3$ df. The total sample size was $n = 1611$. Observed and expected counts are given in Table 14.2.

Table 14.2 Observed and Expected Cell Counts for Example 14.1

	$i = 1$ Tall, cut leaf	$i = 2$ Tall, potato leaf	$i = 3$ Dwarf, cut leaf	$i = 4$ Dwarf, potato leaf
n_i	926	288	293	104
np_{i0}	906.2	302.1	302.1	100.7

The contribution to χ^2 from the first cell is

$$\frac{(n_1 - np_{10})^2}{np_{10}} = \frac{(926 - 906.2)^2}{906.2} = .433$$

Cells 2, 3, and 4 contribute .658, .274, and .108, respectively, so $\chi^2 = .433 + .658 + .274 + .108 = 1.473$. Table A.11 shows that .10 is the area to the right of 6.25 under the chi-squared curve with 3 df. Therefore the area under this curve to the right of 1.473 considerably exceeds .10. That is, P-value $> .10$, so H_0 cannot be rejected even at this rather large level of significance. The data is quite consistent with Mendel's laws. ■

Although we have developed the chi-squared test for situations in which $k > 2$, it can also be used when $k = 2$. The null hypothesis in this case can be stated as $H_0: p_1 = p_{10}$, since the relations $p_2 = 1 - p_1$ and $p_{20} = 1 - p_{10}$ make the inclusion of

$p_2 = p_{20}$ in H_0 redundant. The alternative hypothesis is $H_a: p_1 \neq p_{10}$. These hypotheses can also be tested using a two-tailed z test with test statistic

$$Z = \frac{(N_1/n) - p_{10}}{\sqrt{\dfrac{p_{10}(1 - p_{10})}{n}}} = \frac{\hat{p}_1 - p_{10}}{\sqrt{\dfrac{p_{10}p_{20}}{n}}}$$

Surprisingly, the two test procedures are completely equivalent. This is because it can be shown that $Z^2 = \chi^2$ and $(z_{\alpha/2})^2 = \chi^2_{1,\alpha}$, so the relevant tail areas (*P*-values) are identical.* If the alternative is either $H_a: p_1 > p_{10}$ or $H_a: p_1 < p_{10}$, the chi-squared test cannot be used. One must then revert to an upper- or lower-tailed z test.

As is the case with all test procedures, one must be careful not to confuse statistical significance with practical significance. A computed χ^2 that exceeds $\chi^2_{\alpha,k-1}$ may be a result of a very large sample size rather than any practical differences between the hypothesized p_{i0}'s and true p_i's. Thus if $p_{10} = p_{20} = p_{30} = 1/3$, but the true p_i's have values .330, .340, and .330, a large value of χ^2 is sure to arise with a sufficiently large n. Before rejecting H_0, the $\hat{p}_i$'s should be examined to see whether they suggest a model different from that of H_0 from a practical point of view.

X^2 When the P_i's Are Functions of Other Parameters

Sometimes the p_i's are hypothesized to depend on a smaller number of parameters $\theta_1, \ldots, \theta_m$ ($m < k$). Then a specific hypothesis involving the θ_i's yields specific p_{i0}'s, which are then used in the χ^2 test.

EXAMPLE 14.2 In a well-known genetics article (**"The Progeny in Generations F_{12} to F_{17} of a Cross Between a Yellow-Wrinkled and a Green-Round Seeded Pea," *J. of Genetics*, 1923: 255–331**), the early statistician G. U. Yule analyzed data resulting from crossing garden peas. The dominant alleles in the experiment were Y = yellow color and R = round shape, resulting in the double dominant YR. Yule examined 269 four-seed pods resulting from a dihybrid cross and counted the number of YR seeds in each pod. Letting X denote the number of YRs in a randomly selected pod, possible X values are 0, 1, 2, 3, 4, which we identify with cells 1, 2, 3, 4, and 5 of a rectangular table (so, e.g., a pod with $X = 4$ yields an observed count in cell 5).

The hypothesis that the Mendelian laws are operative and that genotypes of individual seeds within a pod are independent of one another implies that X has a binomial distribution with $n = 4$ and $\theta = 9/16$. We thus wish to test $H_0: p_1 = p_{10}, \ldots, p_5 = p_{50}$, where

$$p_{i0} = P(i - 1 \text{ YRs among 4 seeds when } H_0 \text{ is true})$$

$$= \binom{4}{i - 1} \theta^{i-1}(1 - \theta)^{4-(i-1)} \quad i = 1, 2, 3, 4, 5; \theta = \frac{9}{16}$$

Yule's data and the computations are in Table 14.3, with expected cell counts $np_{i0} = 269p_{i0}$.

---

* The fact that $(z_{\alpha/2})^2 = \chi^2_{1,\alpha}$ is a consequence of the relationship between the standard normal distribution and the chi-squared distribution with 1 df; if $Z \sim N(0, 1)$, then Z^2 has a chi-squared distribution with $\nu = 1$.

Table 14.3 Observed and Expected Cell Counts for Example 14.2

Cell i	1	2	3	4	5
YR peas/pods	0	1	2	3	4
Observed	16	45	100	82	26
Expected	9.86	50.68	97.75	83.78	26.93
$\dfrac{(\text{observed} - \text{expected})^2}{\text{expected}}$	3.823	.637	.052	.038	.032

Thus $\chi^2 = 3.823 + \cdots + .032 = 4.582$. Appendix Table A.11 shows that because $4.582 < 7.77$, the P-value for the test exceeds .10. H_0 should not be rejected at any reasonable significance level. ■

χ^2 When the Underlying Distribution Is Continuous

We have so far assumed that the k categories are naturally defined in the context of the experiment under consideration. The χ^2 test can also be used to test whether a sample comes from a specific underlying continuous distribution. Let X denote the variable being sampled and suppose the hypothesized pdf of X is $f_0(x)$. As in the construction of a frequency distribution in Chapter 1, subdivide the measurement scale of X into k intervals $[a_0, a_1), [a_1, a_2), \ldots, [a_{k-1}, a_k)$, where the interval $[a_{i-1}, a_i)$ includes the value a_{i-1} but not a_i. The cell probabilities specified by H_0 are then

$$p_{i0} = P(a_{i-1} \le X < a_i) = \int_{a_{i-1}}^{a_i} f_0(x)\, dx$$

The cells should be chosen so that $np_{i0} \ge 5$ for $i = 1, \ldots, k$. Often they are selected so that the np_{i0}'s are equal.

EXAMPLE 14.3 To see whether the time of onset of labor among expectant mothers is uniformly distributed throughout a 24-hour day, we can divide a day into k periods, each of length $24/k$. The null hypothesis states that $f(x)$ is the uniform pdf on the interval $[0, 24]$, so that $p_{i0} = 1/k$. The article **"The Hour of Birth"** (***British J. of Preventive and Social Medicine*, 1953: 43–59)** reports on 1186 onset times, which were categorized into $k = 24$ 1-hour intervals beginning at midnight, resulting in cell counts of 52, 73, 89, 88, 68, 47, 58, 47, 48, 53, 47, 34, 21, 31, 40, 24, 37, 31, 47, 34, 36, 44, 78, and 59. Each expected cell count is $1186 \cdot 1/24 = 49.42$, and the resulting value of χ^2 is 162.77. Statistical software gives P-value = .000, so H_0 is resoundingly rejected at any sensible significance level. Generally speaking, it appears that labor is much more likely to commence very late at night than during normal waking hours. ■

For testing whether a sample comes from a specific normal distribution, the fundamental parameters are $\theta_1 = \mu$ and $\theta_2 = \sigma$, and each p_{i0} will be a function of these parameters.

EXAMPLE 14.4 At a certain university, final exams are supposed to last 2 hours. The psychology department constructed a departmental final for an elementary course that was believed to satisfy the following criteria: (1) actual time taken to complete the exam is normally distributed, (2) $\mu = 100$ min, and (3) exactly 90% of all students will finish within the 2-hour period. To see whether this is actually the case, 120 students were randomly selected, and their completion times recorded. It was decided that $k = 8$ intervals should be used. The criteria imply that the 90th percentile of the completion time distribution is $\mu + 1.28\sigma = 120$. Since $\mu = 100$, this implies that $\sigma = 15.63$.

The eight intervals that divide the standard normal scale into eight equally likely segments are $[0, .32)$, $[.32, .675)$, $[.675, 1.15)$, and $[1.15, \infty)$, and their four counterparts are on the other side of 0. For $\mu = 100$ and $\sigma = 15.63$, these intervals become $[100, 105)$, $[105, 110.55)$, $[110.55, 117.97)$, and $[117.97, \infty)$. Thus $p_{i0} = 1/8 = .125$ $(i = 1,\ldots, 8)$, so each expected cell count is $np_{i0} = 120(.125) = 15$. The observed cell counts were 21, 17, 12, 16, 10, 15, 19, and 10, resulting in a χ^2 of 7.73. The 8 df column of Table A.11 shows that P-value $> .10$, so there is no evidence for concluding that the criteria have not been met. ■

EXERCISES Section 14.1 (1–11)

1. What conclusion would be appropriate for an upper-tailed chi-squared test in each of the following situations?
   a. $\alpha = .05$, df = 4, $\chi^2 = 12.25$
   b. $\alpha = .01$, df = 3, $\chi^2 = 8.54$
   c. $\alpha = .10$, df = 2, $\chi^2 = 4.36$
   d. $\alpha = .01$, $k = 6$, $\chi^2 = 10.20$

2. The article **"Racial Stereotypes in Children's Television Commercials"** (**J. of Adver. Res., 2008: 80–93**) reported the following frequencies with which ethnic characters appeared in recorded commercials that aired on Philadelphia television stations.

Ethnicity:	African American	Asian	Caucasian	Hispanic
Frequency:	57	11	330	6

   The 2000 census proportions for these four ethnic groups are .177, .032, .734, and .057, respectively. Does the data suggest that the proportions in commercials are different from the census proportions? Carry out a test of appropriate hypotheses using a significance level of .01.

3. It is hypothesized that when homing pigeons are disoriented in a certain manner, they will exhibit no preference for any direction of flight after takeoff (so that the direction X should be uniformly distributed on the interval from 0° to 360°). To test this, 120 pigeons are disoriented, let loose, and the direction of flight of each is recorded; the resulting data follows. Use the chi-squared test at level .10 to see whether the data supports the hypothesis.

Direction	0–<45°	45–<90°	90–<135°
Frequency	12	16	17

Direction	135–<180°	180–<225°	225–<270°
Frequency	15	13	20

Direction	270–<315°	315–<360°
Frequency	17	10

4. The article **"Application of Methods for Central Statistical Monitoring in Clinical Trials"** (**Clinical Trials, 2013: 783–806**) made a strong case for central statistical monitoring as an alternative to more expensive onsite data verification. It suggested various methods for identifying data characteristics such as outliers, incorrect dates, anomalous data patterns, unusual correlation structures, and digit preference. Exercise 3.21 of this book introduced Benford's Law, which gives a probability model for the first significant digit in many large data sets: $p(x) = \log_{10}((x + 1)/x)$ for $x = 1, 2, \ldots, 9$. The cited article gave the following frequencies for the first significant digit in a variety of variables whose values were determined in one particular clinical trial:

Digit	1	2	3	4
Freq.	342	180	164	155

Digit	5	6	7	8	9
Freq.	86	65	54	47	56

   Carry out a test of hypotheses to see whether or not these frequencies are consistent with Benford's Law (the cited article gave P-value information).

5. An information-retrieval system has ten storage locations. Information has been stored with the expectation that the long-run proportion of requests for location i is given by $p_i = (5.5 - |i - 5.5|)/30$. A sample of 200 retrieval requests gave the following frequencies for locations 1–10, respectively: 4, 15, 23, 25, 38, 31, 32, 14, 10, and 8. Use a chi-squared test at significance level .10 to decide whether the data is consistent with the *a priori* proportions.

6. The article **"The Gap Between Wine Expert Ratings and Consumer Preferences"** (**Intl. J. of Wine Business**

Res., **2008: 335–351)** studied differences between expert and consumer ratings by considering medal ratings for wines, which could be gold (G), silver (S), or bronze (B). Three categories were then established: 1. Rating is the same [(G,G), (B,B), (S,S)]; 2. Rating differs by one medal [(G,S), (S,G), (S,B), (B,S)]; and 3. Rating differs by two medals [(G,B), (B,G)]. The observed frequencies for these three categories were 69, 102, and 45, respectively. On the hypothesis of equally likely expert ratings and consumer ratings being assigned completely by chance, each of the nine medal pairs has probability 1/9. Carry out an appropriate chi-squared test using a significance level of .10.

7. Criminologists have long debated whether there is a relationship between weather conditions and the incidence of violent crime. The author of the article **"Is There a Season for Homicide?"** (*Criminology,* **1988: 287–296)** classified 1361 homicides according to season, resulting in the accompanying data. Test the null hypothesis of equal proportions using $\alpha = .01$.

Winter	Spring	Summer	Fall
328	334	372	327

8. The article **"Psychiatric and Alcoholic Admissions Do Not Occur Disproportionately Close to Patients' Birthdays"** (*Psychological Reports,* **1992: 944–946)** focuses on the existence of any relationship between the date of patient admission for treatment of alcoholism and the patient's birthday. Assuming a 365-day year (i.e., excluding leap year), in the absence of any relation, a patient's admission date is equally likely to be any one of the 365 possible days. The investigators established four different admission categories: (1) within 7 days of birthday; (2) between 8 and 30 days, inclusive, from the birthday; (3) between 31 and 90 days, inclusive, from the birthday; and (4) more than 90 days from the birthday. A sample of 200 patients gave observed frequencies of 11, 24, 69, and 96 for categories 1, 2, 3, and 4, respectively. State and test the relevant hypotheses using a significance level of .01.

9. The response time of a computer system to a request for a certain type of information is hypothesized to have an exponential distribution with parameter $\lambda = 1$ sec (so if $X =$ response time, the pdf of X under H_0 is $f_0(x) = e^{-x}$ for $x \geq 0$).

a. If you had observed $X_1, X_2, \ldots, X_n$ and wanted to use the chi-squared test with five class intervals having equal probability under H_0, what would be the resulting class intervals?

b. Carry out the chi-squared test using the following data resulting from a random sample of 40 response times:

.10	.99	1.14	1.26	3.24	.12	.26	.80
.79	1.16	1.76	.41	.59	.27	2.22	.66
.71	2.21	.68	.43	.11	.46	.69	.38
.91	.55	.81	2.51	2.77	.16	1.11	.02
2.13	.19	1.21	1.13	2.93	2.14	.34	.44

10. a. Show that another expression for the chi-squared statistic is

$$\chi^2 = \sum_{i=1}^{k} \frac{N_i^2}{np_{i0}} - n$$

Why is it more efficient to compute χ^2 using this formula?

b. When the null hypothesis is $H_0: p_1 = p_2 = \cdots = p_k = 1/k$ (i.e., $p_{i0} = 1/k$ for all i), how does the formula of part (a) simplify? Use the simplified expression to calculate χ^2 for the pigeon/direction data in Exercise 4.

11. a. Having obtained a random sample from a population, you wish to use a chi-squared test to decide whether the population distribution is standard normal. If you base the test on six class intervals having equal probability under H_0, what should be the class intervals?

b. If you wish to use a chi-squared test to test H_0: the population distribution is normal with $\mu = .5$, $\sigma = .002$ and the test is to be based on six equiprobable (under H_0) class intervals, what should be these intervals?

c. Use the chi-squared test with the intervals of part (b) to decide, based on the following 45 bolt diameters, whether bolt diameter is a normally distributed variable with $\mu = .5$ in., $\sigma = .002$ in.

.4974	.4976	.4991	.5014	.5008	.4993
.4994	.5010	.4997	.4993	.5013	.5000
.5017	.4984	.4967	.5028	.4975	.5013
.4972	.5047	.5069	.4977	.4961	.4987
.4990	.4974	.5008	.5000	.4967	.4977
.4992	.5007	.4975	.4998	.5000	.5008
.5021	.4959	.5015	.5012	.5056	.4991
.5006	.4987	.4968			

14.2 Goodness-of-Fit Tests for Composite Hypotheses

In the previous section, we presented a goodness-of-fit test based on a χ^2 statistic for deciding between $H_0: p_1 = p_{10}, \ldots, p_k = p_{k0}$ and the alternative H_a stating that H_0 is not true. The null hypothesis was a **simple hypothesis** in the sense that each

p_{i0} was a specified number, so that the expected cell counts when H_0 was true were uniquely determined numbers.

In many situations, there are k naturally occurring categories, but H_0 states only that the p_i's are functions of other parameters $\theta_1, \ldots, \theta_m$ without specifying the values of these θ's. For example, a population may be in equilibrium with respect to proportions of the three genotypes AA, Aa, and aa. With $p_1, p_2,$ and p_3 denoting these proportions (probabilities), one may wish to test

$$H_0: p_1 = \theta^2, \ p_2 = 2\theta(1 - \theta), \ p_3 = (1 - \theta)^2 \tag{14.1}$$

where θ represents the proportion of gene A in the population. This hypothesis is **composite** because knowing that H_0 is true does not uniquely determine the cell probabilities and expected cell counts but only their general form. To carry out a χ^2 test, the unknown θ_i's must first be estimated.

Similarly, we may be interested in testing to see whether a sample came from a particular family of distributions without specifying any particular member of the family. To use the χ^2 test to see whether the distribution is Poisson, for example, the parameter μ must be estimated. In addition, because there are actually an infinite number of possible values of a Poisson variable, these values must be grouped so that there are a finite number of cells. If H_0 states that the underlying distribution is normal, use of a χ^2 test must be preceded by a choice of cells and estimation of μ and σ.

χ^2 When Parameters Are Estimated

As before, k will denote the number of categories or cells, and p_i will denote the probability of an observation falling in the ith cell. The null hypothesis now states that each p_i is a function of a small number of parameters $\theta_1, \ldots, \theta_m$ with the θ_i's otherwise unspecified:

$$H_0: p_1 = \pi_1(\boldsymbol{\theta}), \ldots, p_k = \pi_k(\boldsymbol{\theta}) \qquad \text{where } \boldsymbol{\theta} = (\theta_1, \ldots, \theta_m)$$

H_a: the hypothesis H_0 is not true $\tag{14.2}$

For example, for H_0 of (14.1), $m = 1$ (there is only one θ), $\pi_1(\theta) = \theta^2$, $\pi_2(\theta) = 2\theta(1 - \theta)$, and $\pi_3(\theta) = (1 - \theta)^2$.

In the case $k = 2$, there is really only a single rv, N_1 (since $N_1 + N_2 = n$), which has a binomial distribution. The joint probability that $N_1 = n_1$ and $N_2 = n_2$ is then

$$P(N_1 = n_1, N_2 = n_2) = \binom{n}{n_1} p_1^{n_1} \cdot p_2^{n_2} \propto p_1^{n_1} \cdot p_2^{n_2}$$

where $p_1 + p_2 = 1$ and $n_1 + n_2 = n$. For general k, the joint distribution of $N_1, \ldots, N_k$ is the multinomial distribution (Section 5.1) with

$$P(N_1 = n_1, \ldots, N_k = n_k) \propto p_1^{n_1} \cdot p_2^{n_2} \cdots p_k^{n_k} \tag{14.3}$$

When H_0 is true, (14.3) becomes

$$P(N_1 = n_1, \ldots, N_k = n_k) \propto [\pi_1(\boldsymbol{\theta})]^{n_1} \cdots [\pi_k(\boldsymbol{\theta})]^{n_k} \tag{14.4}$$

To apply a chi-squared test, $\boldsymbol{\theta} = (\theta_1, \ldots, \theta_m)$ must be estimated.

METHOD OF ESTIMATION

Let $n_1, n_2, \ldots, n_k$ denote the observed values of $N_1, \ldots, N_k$. Then $\hat{\theta}_1, \ldots, \hat{\theta}_m$ are those values of the θ_i's that maximize (14.4).

The resulting estimators $\hat{\theta}_1, \ldots, \hat{\theta}_m$ are the maximum likelihood estimators of $\theta_1, \ldots, \theta_m$; this principle of estimation was discussed in Section 6.2.

EXAMPLE 14.5   In humans there is a blood group, the MN group, that is composed of individuals having one of the three blood types M, MN, and N. Type is determined by two alleles, and there is no dominance, so the three possible genotypes give rise to three pheno-types. A population consisting of individuals in the MN group is in equilibrium if

$$P(M) = p_1 = \theta^2$$
$$P(MN) = p_2 = 2\theta(1 - \theta)$$
$$P(N) = p_3 = (1 - \theta)^2$$

for some θ. Suppose a sample from such a population yielded the results shown in Table 14.4.

Table 14.4 Observed Counts for Example 14.5

Type	M	MN	N	
Observed	125	225	150	$n = 500$

Then

$$[\pi_1(\theta)]^{n_1}[\pi_2(\theta)]^{n_2}[\pi_3(\theta)]^{n_3} = [(\theta^2)]^{n_1}[2\theta(1 - \theta)]^{n_2}[(1 - \theta)^2]^{n_3}$$
$$= 2^{n_2} \cdot \theta^{2n_1 + n_2} \cdot (1 - \theta)^{n_2 + 2n_3}$$

Maximizing this with respect to θ (or, equivalently, maximizing the natural logarithm of this quantity, which is easier to differentiate) yields

$$\hat{\theta} = \frac{2n_1 + n_2}{[(2n_1 + n_2) + (n_2 + 2n_3)]} = \frac{2n_1 + n_2}{2n}$$

With $n_1 = 125$ and $n_2 = 225$, $\hat{\theta} = 475/1000 = .475$. ∎

Once $\boldsymbol{\theta} = (\theta_1, \ldots, \theta_m)$ has been estimated by $\hat{\boldsymbol{\theta}} = (\hat{\theta}_1, \ldots, \hat{\theta}_m)$, the estimated expected cell counts are the $n\pi_i(\hat{\boldsymbol{\theta}})$'s. These are now used in place of the np_{i0}'s of Section 14.1 to specify a χ^2 statistic.

THEOREM   Under general "regularity" conditions on $\theta_1, \ldots, \theta_m$ and the $\pi_i(\boldsymbol{\theta})$'s, if $\theta_1, \ldots, \theta_m$ are estimated by the method of maximum likelihood as described previously and n is large,

$$\chi^2 = \sum_{\text{all cells}} \frac{(\text{observed} - \text{estimated expected})^2}{\text{estimated expected}} = \sum_{i=1}^{k} \frac{[N_i - n\pi_i(\hat{\boldsymbol{\theta}})]^2}{n\pi_i(\hat{\boldsymbol{\theta}})}$$

has approximately a chi-squared distribution with $k - 1 - m$ df when H_0 of (14.2) is true. The P-value is therefore (roughly) the area under the χ^2_{k-1-m} curve to the right of the calculated χ^2. In practice, the test can be used if $n\pi_i(\hat{\boldsymbol{\theta}}) \geq 5$ for every i.

Notice that *the number of degrees of freedom is reduced by the number of θ_i's estimated.*

EXAMPLE 14.6   With $\hat{\theta} = .475$ and $n = 500$, the estimated expected cell counts are $n\pi_1(\hat{\theta}) = 500(\hat{\theta})^2 =$
(Example 14.5 112.81, $n\pi_2(\hat{\theta}) = (500)(2)(.475)(1.475) = 249.38$, and $n\pi_3(\hat{\theta}) = 500 - 112.81 -$
continued) $249.38 = 137.81$. Then

$$\chi^2 = \frac{(125 - 112.81)^2}{112.81} + \frac{(225 - 249.38)^2}{249.38} + \frac{(150 - 137.81)^2}{137.81} = 4.78$$

Appendix Table A.11 shows that for df $= 3 - 1 - 1 = 1$, P-value $\approx .029$. Therefore H_0 is rejected at significance level .05 (but not at level .01). ∎

EXAMPLE 14.7 Consider a series of games between two teams, I and II, that terminates as soon as one team has won four games (with no possibility of a tie). A simple probability model for such a series assumes that outcomes of successive games are independent and that the probability of team I winning any particular game is a constant θ. We arbitrarily designate I the better team, so that $\theta \geq .5$. Any particular series can then terminate after 4, 5, 6, or 7 games. Let $\pi_1(\theta)$, $\pi_2(\theta)$, $\pi_3(\theta)$, $\pi_4(\theta)$ denote the probability of termination in 4, 5, 6, and 7 games, respectively. Then

$$\pi_1(\theta) = P(\text{I wins in 4 games}) + P(\text{II wins in 4 games})$$
$$= \theta^4 + (1 - \theta)^4$$
$$\pi_2(\theta) = P(\text{I wins 3 of the first 4 and the fifth})$$
$$+ P(\text{I loses 3 of the first 4 and the fifth})$$
$$= \binom{4}{3}\theta^3(1 - \theta) \cdot \theta + \binom{4}{1}\theta(1 - \theta)^3 \cdot (1 - \theta)$$
$$= 4\theta(1 - \theta)[\theta^3 + (1 - \theta)^3]$$
$$\pi_3(\theta) = 10\theta^2(1 - \theta)^2[\theta^2 + (1 - \theta)^2]$$
$$\pi_4(\theta) = 20\theta^3(1 - \theta)^3$$

The article **"Seven-Game Series in Sports" by Groeneveld and Meeden** (***Mathematics Magazine*, 1975: 187–192**) tested the fit of this model to results of National Hockey League playoffs during the period 1943–1967 (when league membership was stable). The data appears in Table 14.5.

Table 14.5 Observed and Expected Counts for the Simple Model

Cell	1	2	3	4	
Number of games played	**4**	**5**	**6**	**7**	
Observed frequency	15	26	24	18	$n = 83$
Estimated expected frequency	16.351	24.153	23.240	19.256	

The estimated expected cell counts are $83\pi_i(\hat{\theta})$, where $\hat{\theta}$ is the value of θ that maximizes

$$\{\theta^4 + (1 - \theta)^4\}^{15} \cdot \{4\theta(1 - \theta)[\theta^3 + (1 - \theta)^3]\}^{26}$$
$$\cdot \{10\theta^2(1 - \theta)^2[\theta^2 + (1 - \theta)^2]\}^{24} \cdot \{20\theta^3(1 - \theta)^3\}^{18} \quad (14.5)$$

Standard calculus methods fail to yield a nice formula for the maximizing value $\hat{\theta}$, so it must be computed using numerical methods. The result is $\hat{\theta} = .654$, from which $\pi_i(\hat{\theta})$ and the estimated expected cell counts are computed. The computed value of χ^2 is .360. According to the $k - 1 - m = 4 - 1 - 1 = 2$ df column of Table A.11, P-value $> .10$. There is thus no reason to reject the simple model as applied to the NHL playoff series.

The cited article also considered World Series data for the period 1903–1973. For the simple model, $\chi^2 = 5.97$; Table A.11 yields P-value $\approx .05$. At significance level .10, the model is of doubtful validity. The suggested reason for this is that

$$P(\text{series lasts six games}|\text{series lasts at least six games}) \geq .5 \quad (14.6)$$

whereas of the 38 series that actually lasted at least six games, only 13 lasted exactly six. The following alternative model is then introduced:

$$\pi_1(\theta_1, \theta_2) = \theta_1^4 + (1 - \theta_1)^4$$
$$\pi_2(\theta_1, \theta_2) = 4\theta_1(1 - \theta_1)[\theta_1^3 + (1 - \theta_1)^3]$$
$$\pi_3(\theta_1, \theta_2) = 10\theta_1^2(1 - \theta_1)^2\theta_2$$
$$\pi_4(\theta_1, \theta_2) = 10\theta_1^2(1 - \theta_1)^2(1 - \theta_2)$$

The first two π_i's are identical to the simple model, whereas θ_2 is the conditional probability of (14.6) (which can now be any number between 0 and 1). The values of $\hat{\theta}_1$ and $\hat{\theta}_2$ that maximize the expression analogous to expression (14.5) are determined numerically as $\hat{\theta}_1 = .614$, $\hat{\theta}_2 = .342$. A summary appears in Table 14.6, and $\chi^2 = .384$. Since two parameters are estimated, df $= k - 1 - m = 1$. The P-value considerably exceeds .10, indicating a good fit of the data to this new model.

Table 14.6 Observed and Expected Counts for the More Complex Model

Number of games played	4	5	6	7
Observed frequency	12	16	13	25
Estimated expected frequency	10.85	18.08	12.68	24.39

One of the conditions on the θ_i's in the theorem is that they be functionally independent of one another. That is, no single θ_i can be determined from the values of other θ_i's, so that m is the number of functionally independent parameters estimated. A general rule of thumb for degrees of freedom in a chi-squared test is the following.

$$\chi^2 \text{ df} = \left(\begin{array}{c} \text{number of freely} \\ \text{determined cell counts} \end{array}\right) - \left(\begin{array}{c} \text{number of independent} \\ \text{parameters estimated} \end{array}\right)$$

This rule will be used in connection with several different chi-squared tests in the next section.

Goodness of Fit for Discrete Distributions

Many experiments involve observing a random sample $X_1, X_2, \ldots, X_n$ from some discrete distribution. One may then wish to investigate whether the underlying distribution is a member of a particular family, such as the Poisson or negative binomial family. In the case of both a Poisson and a negative binomial distribution, the set of possible values is infinite, so the values must be grouped into k subsets before a chi-squared test can be used. The groupings should be done so that the expected frequency in each cell (group) is at least 5. The last cell will then correspond to X values of $c, c + 1, c + 2, \ldots$ for some value c.

This grouping can considerably complicate the computation of the $\hat{\theta}_i$'s and estimated expected cell counts. This is because the theorem requires that the $\hat{\theta}_i$'s be obtained from the cell counts $N_1, \ldots, N_k$ rather than the sample values $X_1, \ldots, X_n$.

EXAMPLE 14.8 Table 14.7 presents count data on the number of *Larrea divaricata* plants found in each of 48 sampling quadrats, as reported in the article **"Some Sampling Characteristics of Plants and Arthropods of the Arizona Desert"** (*Ecology*, 1962: 567–571).

Table 14.7 Observed Counts for Example 14.8

Cell	1	2	3	4	5
Number of plants	0	1	2	3	≥ 4
Frequency	9	9	10	14	6

The article's author fit a Poisson distribution to the data. Let μ denote the Poisson parameter and suppose for the moment that the six counts in cell 5 were actually 4, 4, 5, 5, 6, 6. Then denoting sample values by $x_1, \ldots, x_{48}$, nine of the x_i's were 0, nine were 1, and so on. The likelihood of the observed sample is

$$\frac{e^{-\mu}\mu^{x_1}}{x_1!} \cdots \cdots \frac{e^{-\mu}\mu^{x_{48}}}{x_{48}!} = \frac{e^{-48\mu}\mu^{\Sigma x_i}}{x_1! \cdots \cdots x_{48}!} = \frac{e^{-48\mu}\mu^{101}}{x_1! \cdots \cdots x_{48}!}$$

The value of μ for which this is maximized is $\hat{\mu} = \Sigma x_i/n = 101/48 = 2.10$ (the value reported in the article).

However, the $\hat{\mu}$ required for χ^2 is obtained by maximizing Expression (14.4) rather than the likelihood of the full sample. The cell probabilities are

$$\pi_i(\mu) = \frac{e^{-\mu}\mu^{i-1}}{(i-1)!} \quad i = 1, 2, 3, 4$$

$$\pi_5(\mu) = 1 - \sum_{i=0}^{3} \frac{e^{-\mu}\mu^i}{i!}$$

so the right-hand side of (14.4) becomes

$$\left[\frac{e^{-\mu}\mu^0}{0!} \right]^9 \left[\frac{e^{-\mu}\mu^1}{1!} \right]^9 \left[\frac{e^{-\mu}\mu^2}{2!} \right]^{10} \left[\frac{e^{-\mu}\mu^3}{3!} \right]^{14} \left[1 - \sum_{i=0}^{3} \frac{e^{-\mu}\mu^i}{i!} \right]^6$$

There is no nice formula for $\hat{\mu}$, the maximizing value of μ, in this latter expression, so it must be obtained numerically. ∎

Because the parameter estimates are usually more difficult to compute from the grouped data than from the full sample, they are typically computed using this latter method. If these "full" estimators are used in the chi-squared statistic, the distribution of the statistic when H_0 is true is quite complicated, so the actual P-value cannot be determined. However, the following result usually enables us to reach a conclusion at the desired significance level α.

THEOREM

Let $\hat{\theta}_1, \ldots, \hat{\theta}_m$ be the maximum likelihood estimators of $\theta_1, \ldots, \theta_m$ based on the full sample $X_1, \ldots, X_n$, and let χ^2 denote the statistic based on these estimators. Also let

P_1 = the P-value for an upper-tailed chi-squared test based on $k-1$ df
P_2 = the P-value for an upper-tailed chi-squared test based on $k-1-m$ df

Then it can be shown that

$$P_1 \leq P\text{-value} \leq P_2 \tag{14.7}$$

That is, the P-value for the test under consideration is sandwiched in between the P-values for two "pure" upper-tailed chi-squared tests based on different df's. The test procedure implied by (14.7) has the unusual feature that under some circumstances judgment must be withheld until more data is available.

> Select a significance level α. Then
> If $\alpha \le P_1$, do not reject H_0
> If $\alpha \ge P_2$, reject H_0 (14.8)
> If $P_1 < \alpha < P_2$, withhold judgment

Suppose, for example, that $k = 6$, $m = 2$, and $\alpha = .05$. The two relevant df's are $6 - 1 = 5$ and $6 - 1 - 2 = 3$. Then if $\chi^2 = 7.0$, Table A.11 shows that the P-value for a 3 df test is about .07 and the P-value for a 5 df test exceeds .10. Therefore we would not be able to reject H_0 because .05 is at most the smaller of the two pure chi-squared P-values. If, however, $\chi^2 = 15$, then the 3 df P-value is roughly .002 and the 5 df P-value is approximately .01. Because .05 is at least the larger of these pure P-values, we are given license to reject H_0. Only if .05 lies between the two pure chi-squared P-values would we not be able to reach a conclusion.

EXAMPLE 14.9
(Example 14.8 continued)

Using $\hat{\mu} = 2.10$, the estimated expected cell counts are computed from $n\pi_i(\hat{\mu})$, where $n = 48$. For example,

$$n\pi_1(\hat{\mu}) = 48 \cdot \frac{e^{-2.1}(2.1)^0}{0!} = (48)(e^{-2.1}) = 5.88$$

Similarly, $n\pi_2(\hat{\mu}) = 12.34$, $n\pi_3(\hat{\mu}) = 12.96$, $n\pi_4(\hat{\mu}) = 9.07$, and $n\pi_5(\mu) = 48 - 5.88 - \cdots - 9.07 = 7.75$. Then

$$\chi^2 = \frac{(9 - 5.88)^2}{5.88} + \cdots + \frac{(6 - 7.75)^2}{7.75} = 6.31$$

The relevant dfs are $5 - 1 = 4$ and $5 - 2 = 3$. Then Table A.11 shows that the P-value for a 3 df test is about .0955 and that for a 4 df test exceeds .10. Therefore at significance level .05, H_0 cannot be rejected because the P-exceeds .0955 and therefore certainly exceeds .05. At this level, it is plausible that the actual distribution is Poisson. However, if the selected significance level were instead .10, we'd be in the inconclusive situation because the P-value could be (slightly) smaller than .10 or larger than .10. ∎

Sometimes even the maximum likelihood estimates based on the full sample are quite difficult to compute. This is the case, for example, for the two-parameter (generalized) negative binomial distribution. In such situations, method-of-moments estimates are often used, though it is not known to what extent the use of moments estimators affects the null distribution of χ^2.

Goodness of Fit for Continuous Distributions

The chi-squared test can also be used to test whether the sample comes from a specified family of continuous distributions, such as the exponential family or the normal family. The choice of cells (class intervals) is even more arbitrary in the continuous case than in the discrete case. To ensure that the chi-squared test is valid, the cells should be chosen independently of the sample observations. Once the cells are chosen, it is almost always quite difficult to estimate unspecified parameters (such as μ and σ in the normal case) from the observed cell counts, so instead mle's based on the full sample are computed. The test procedure is again specified by (14.7) and (14.8).

EXAMPLE 14.10   The article **"Class Start Times, Sleep, and Academic Performance in College: A Path Analysis"** (*Chronobiology International*, **2012: 318–335**) reported on a study in which students were surveyed about various aspects of sleep behavior during a particular two-week period. Here is data on average sleep time per day (h) for 100 of the students:

5.75	5.93	5.96	6.00	6.19	6.36	6.36	6.43	6.50	6.51
6.51	6.57	6.69	6.78	6.93	7.04	7.05	7.05	7.11	7.18
7.21	7.25	7.26	7.30	7.32	7.39	7.40	7.43	7.43	7.50
7.50	7.52	7.53	7.54	7.57	7.60	7.61	7.63	7.64	7.64
7.64	7.67	7.71	7.73	7.75	7.79	7.81	7.83	7.83	7.84
7.86	7.86	7.87	7.88	7.89	7.93	7.96	7.98	7.99	8.00
8.00	8.04	8.07	8.11	8.17	8.18	8.18	8.20	8.21	8.21
8.29	8.29	8.43	8.49	8.49	8.52	8.54	8.59	8.61	8.68
8.71	8.71	8.75	8.79	8.81	8.82	8.88	8.89	8.93	9.00
9.05	9.15	9.19	9.25	9.32	9.80	9.85	9.87	9.96	10.62

Is it reasonable to assume that the population distribution of average sleep time is at least approximately normal? The histogram in Figure 14.3 is not persuasive. So let's carry out a chi-squared test of the null hypothesis that the distribution is normal.

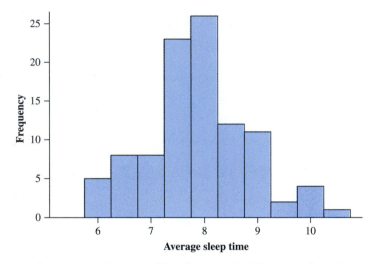

Figure 14.3   Histogram of the sleep time data from Example 14.10

Suppose that prior to sampling, it was believed that plausible values of μ and σ were 8 and 1, respectively. The eight equiprobable class intervals for the standard normal distribution (each with probability .125) are $(-\infty, -1.15)$, $[-1.15, -.67)$, $[-.67, -.32)$, $[-.32, 0)$, $[0, .32)$, $[.32, .67)$, $[.67, .1.15)$, and $[1.15, \infty)$, with each endpoint also giving the distance in standard deviations from the mean for any other normal distributions. For $\mu = 8$ and $\sigma = 1$, these intervals transform to $(-\infty, 6.85)$, $[6.85, 7.33)$, $[7.33, 7.68)$, $[7.68, 8.00)$, $[8.00, 8.32)$, $[8.32, 8.67)$, $[8.67, 9.15)$, and $[9.15, \infty)$.

To obtain the estimated cell probabilities $\pi_1(\hat{\mu}, \hat{\sigma}), \ldots, \pi_8(\hat{\mu}, \hat{\sigma})$, we first need the mle's $\hat{\mu}$ and $\hat{\sigma}$. In Chapter 6, the mle of σ was shown to be $[\Sigma(x_i - \bar{x})^2/n]^{1/2}$ (rather than s), so with $s = .9481$,

$$\hat{\mu} = \bar{x} = 7.876 \qquad \hat{\sigma} = \left[\frac{\Sigma(x_i - \bar{x})^2}{n}\right]^{1/2} = \left[\frac{(n-1)s^2}{n}\right]^{1/2} = .9433$$

Each $\pi_i(\hat{\mu}, \hat{\sigma})$ is then the probability that a normal rv X with mean 7.876 and standard deviation .9433 falls in the ith class interval. For example,

$$\pi_2(\hat{\mu}, \hat{\sigma}) = P(6.85 < X < 7.33) = P(-1.09 < Z < -.58) = .1431$$

so $n\pi_2(\hat{\mu}, \hat{\sigma}) = 100(.1431) = 14.31$. Observed and estimated expected cell counts are shown in Table 14.8.

Table 14.8 Observed and Expected Counts for Example 14.10

Cell	$(-\infty, 6.85)$	$[6.85, 7.33)$	$[7.33, 7.68)$	$[7.68, 8.00)$
Observed	14	11	17	9
Estimated expected	13.79	14.31	13.58	13.49

Cell	$[8.00, 8.32)$	$[8.32, 8.67)$	$[8.67, 9.15)$	$[9.15, \infty)$
Observed	11	8	12	8
Estimated expected	12.91	11.87	11.20	8.85

The computed value of χ^2 is 5.56. With $k = 8$ cells and $m = 2$ parameters estimated, the 7 df and 5 df columns of Table A.11 show that both pure chi-squared P-values exceed .10. Therefore our P-value certainly exceeds any reasonable α, indicating that the null hypothesis of population normality cannot be rejected. The evidence from the entire sample of $n = 253$ students is somewhat less supportive of H_0. The P-value from the special test for normality described in the next subsection is .086. ∎

EXAMPLE 14.11 The article **"Some Studies on Tuft Weight Distribution in the Opening Room"** (***Textile Research J.***, **1976: 567–573**) reports the accompanying data on the distribution of output tuft weight X (mg) of cotton fibers for the input weight $x_0 = 70$.

Interval	0–8	8–16	16–24	24–32	32–40	40–48	48–56	56–64	64–70
Observed frequency	20	8	7	1	2	1	0	1	0
Expected frequency	18.0	9.9	5.5	3.0	1.8	.9	.5	.3	.1

The authors postulated a truncated exponential distribution:

$$H_0: f(x) = \frac{\lambda e^{-\lambda x}}{1 - e^{-\lambda x_0}} \quad 0 \le x \le x_0$$

The mean of this distribution is

$$\mu = \int_0^{x_0} x f(x)\, dx = \frac{1}{\lambda} - \frac{x_0 e^{-\lambda x_0}}{1 - e^{-\lambda x_0}}$$

The parameter λ was estimated by replacing μ by $\bar{x} = 13.086$ and solving the resulting equation to obtain $\hat{\lambda} = .0742$ (so $\hat{\lambda}$ is a method-of-moments estimate and not an mle). Then with $\hat{\lambda}$ replacing λ in $f(x)$, the estimated expected cell frequencies as displayed previously are computed as

$$40\pi_i(\hat{\lambda}) = 40 P(a_{i-1} \le X < a_i) = 40 \int_{a_{i-1}}^{a_i} f(x)\, dx = \frac{40(e^{-\hat{\lambda} a_{i-1}} - e^{-\hat{\lambda} a_i})}{1 - e^{-\hat{\lambda} x_0}}$$

where $[a_{i-1}, a_i)$ is the ith class interval. To obtain expected cell counts of at least 5, the last six cells are combined to yield observed counts of 20, 8, 7, 5 and expected counts of 18.0, 9.9, 5.5, 6.6. The computed value of chi-squared is then $\chi^2 = 1.34$ with a corresponding P-value that exceeds .10. Therefore H_0 cannot be rejected at significance level .05, so the truncated exponential model provides a good fit. ∎

A Special Test for Normality

Probability plots were introduced in Section 4.6 as an informal method for assessing the plausibility of any specified population distribution as the one from which the given sample was selected. The straighter the probability plot, the more plausible is the distribution on which the plot is based. A normal probability plot is used for checking whether *any* member of the normal distribution family is plausible. Let's denote the sample x_i's when ordered from smallest to largest by $x_{(1)}, x_{(2)}, \ldots, x_{(n)}$. Then the plot suggested for checking normality was a plot of the points $(x_{(i)}, y_i)$, where $y_i = \Phi^{-1}((i - .5)/n)$.

A quantitative measure of the extent to which points cluster about a straight line is the sample correlation coefficient r introduced in Chapter 12. Consider calculating r for the n pairs $(x_{(1)}, y_1), \ldots, (x_{(n)}, y_n)$. The y_i's here are not observed values in a random sample from a y population, so properties of this r are quite different from those described in Section 12.5. However, it is true that the more r deviates from 1, the less the probability plot resembles a straight line (remember that a probability plot must slope upward). This implies that the test is lower-tailed: The P-value is the area under the density curve of R (the random variable whose computed value is r) when H_0 is true to the left of r. Unfortunately the distribution of R is very complicated. The developers of the Minitab software have provided critical values that capture lower-tail areas of .10, .05, and .01 for various sample sizes, which are included in our Table A.12. These critical values are based on a slightly different definition of the y_i's than that given previously.

Minitab will also construct a normal probability plot based on these y_i's. The plot will be almost identical in appearance to that based on the previous y_i's. When there are several tied $x_{(i)}$'s, Minitab computes r by using the average of the corresponding y_i's as the second number in each pair.

Let $y_i = \Phi^{-1}[(i - .375)/(n + .25)]$, and compute the sample correlation coefficient r for the n pairs $(x_{(1)}, y_1), \ldots, (x_{(n)}, y_n)$. The **Ryan-Joiner test** of

H_0: the population distribution is normal

versus

H_a: the population distribution is not normal

uses test statistic R (obtained by replacing the $x_{(i)}$'s in r by $X_{(i)}$'s). If r coincides with a critical value in Table A.12, we have an exact P-value (.10, .05, or .01). Otherwise we are able to say that either P-value $> .10$, $.05 < P$-value $< .10$, $.01 < P$-value $< .05$, or P-value $< .01$.

EXAMPLE 14.12   The following sample of $n = 20$ observations on dielectric breakdown voltage of a piece of epoxy resin first appeared in Exercise 4.89.

y_i	−1.871	−1.404	−1.127	−.917	−.742	−.587	−.446	−.313	−.186	−.062
$x_{(i)}$	24.46	25.61	26.25	26.42	26.66	27.15	27.31	27.54	27.74	27.94

y_i	.062	.186	.313	.446	.587	.742	.917	1.127	1.404	1.871
$x_{(i)}$	27.98	28.04	28.28	28.49	28.50	28.87	29.11	29.13	29.50	30.88

We asked Minitab to carry out the Ryan-Joiner test, and the result appears in Figure 14.4. The test statistic value is $r = .9881$, and Appendix Table A.12 gives .9600 as the critical value that captures lower-tail area .10 under the r sampling distribution curve when $n = 20$ and the underlying distribution is actually normal. Because $.9881 > .9600$, the P-value exceeds .10. Therefore the null hypothesis of normality cannot be rejected even for a significance level as large as .10.

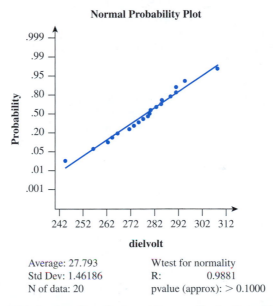

Normal Probability Plot

Average: 27.793 Wtest for normality
Std Dev: 1.46186 R: 0.9881
N of data: 20 pvalue (approx): > 0.1000

Figure 14.4 Minitab output from the Ryan-Joiner test for the data of Example 14.12 ■

EXERCISES Section 14.2 (12–23)

12. Consider a large population of families in which each family has exactly three children. If the genders of the three children in any family are independent of one another, the number of male children in a randomly selected family will have a binomial distribution based on three trials.

a. Suppose a random sample of 160 families yields the following results. Test the relevant hypotheses by proceeding as in Example 14.5.

Number of Male Children	0	1	2	3
Frequency	14	66	64	16

b. Suppose a random sample of families in a nonhuman population resulted in observed frequencies of 15, 20, 12, and 3, respectively. Would the chi-squared test be based on the same number of degrees of freedom as the test in part (a)? Explain.

13. A study of sterility in the fruit fly (**"Hybrid Dysgenesis in** *Drosophila melanogaster:* **The Biology of Female and Male Sterility,"** *Genetics*, **1979: 161–174**) reports the following data on the number of ovaries developed by each female fly in a sample of size 1388. One model for unilateral sterility states that each ovary develops with some probability p independently of the other ovary. Test the fit of this model using χ^2.

x = Number of Ovaries Developed	0	1	2
Observed Count	1212	118	58

14. The article **"Feeding Ecology of the Red-Eyed Vireo and Associated Foliage-Gleaning Birds"** (*Ecological Monographs*, **1971: 129–152**) presents the accompanying data on the variable $X =$ the number of hops before the first flight and preceded by a flight. The author then proposed and fit a geometric probability distribution

$[p(x) = P(X = x) = p^{x-1} \cdot q$ for $x = 1, 2, \ldots$, where $q = 1 - p]$ to the data. The total sample size was $n = 130$.

x	1	2	3	4	5	6	7	8	9	10	11	12
Number of Times x Observed	48	31	20	9	6	5	4	2	1	1	2	1

a. The likelihood is $(p^{x_1 - 1} \cdot q) \cdots (p^{x_n - 1} \cdot q) = p^{\Sigma x_i - n} \cdot q^n$. Show that the mle of p is given by $\hat{p} = (\Sigma x_i - n)/\Sigma x_i$, and compute $\hat{p}$ for the given data.

b. Estimate the expected cell counts using $\hat{p}$ of part (a) [expected cell counts $= n \cdot (\hat{p})^{x-1} \cdot \hat{q}$ for $x = 1, 2, \ldots$], and test the fit of the model using a χ^2 test by combining the counts for $x = 7, 8, \ldots$, and 12 into one cell $(x \geq 7)$.

15. A certain type of flashlight is sold with the four batteries included. A random sample of 150 flashlights is obtained, and the number of defective batteries in each is determined, resulting in the following data:

Number Defective	0	1	2	3	4
Frequency	26	51	47	16	10

Let X be the number of defective batteries in a randomly selected flashlight. Test the null hypothesis that the distribution of X is Bin$(4, \theta)$. That is, with $p_i = P(i$ defectives$)$, test

$$H_0: p_i = \binom{4}{i} \theta^i (1 - \theta)^{4-i} \qquad i = 0, 1, 2, 3, 4$$

[*Hint:* To obtain the mle of θ, write the likelihood (the function to be maximized) as $\theta^u (1 - \theta)^v$, where the exponents u and v are linear functions of the cell counts. Then take the natural log, differentiate with respect to θ, equate the result to 0, and solve for $\hat{\theta}$.]

16. Let $X =$ the number of adult police contacts for a randomly selected individual who previously had at least one such contact prior to age 18. The following frequencies were calculated from information given in the article **"Examining the Prevalence of Criminal Desistance"** (*Criminology*, **2003: 423–448**); our sample size differs slightly from what was reported because of rounding.

x	0	1	2	3	4	5	6	7
f	1627	421	219	130	107	51	15	22

x	8	9	10	11	12	13	14	15
f	8	14	5	8	5	0	3	2

a. Is it plausible that the population distribution of number of contacts is Poisson? Carry out a chi-squared test.

b. The cited article did not even entertain the possibility of a Poisson distribution. Instead several other models were proposed. One of these is based on the idea that each individual's number of contacts has a Poisson

distribution whose mean value μ is itself a random variable having a gamma distribution. This reasoning leads to a generalized negative binomial distribution having two parameters, which must then be estimated from the data. After doing so, the article reported the following estimated probabilities corresponding to the foregoing x values: .6099, .1657, .0838, .0489, .0305, .0197, .0130, .0088, .0060, .0041, .0029, .0020, .0014, .0010, .0007, and .0005. Test the plausibility of this model at significance level .05 by combining all x values exceeding 11 into a single category (this was done in the cited article, which included a *P*-value).

17. In a genetics experiment, investigators looked at 300 chromosomes of a particular type and counted the number of sister-chromatid exchanges on each (**"On the Nature of Sister-Chromatid Exchanges in 5-Bromodeoxyuridine-Substituted Chromosomes,"** *Genetics*, **1979: 1251–1264**). A Poisson model was hypothesized for the distribution of the number of exchanges. Test the fit of a Poisson distribution to the data by first estimating μ and then combining the counts for $x = 8$ and $x = 9$ into one cell.

$x =$ Number of Exchanges	0	1	2	3	4	5	6	7	8	9
Observed Counts	6	24	42	59	62	44	41	14	6	2

18. The article **"A Probabilistic Analysis of Dissolved Oxygen–Biochemical Oxygen Demand Relationship in Streams"** (*J. Water Resources Control Fed.*, **1969: 73–90**) reports data on the rate of oxygenation in streams at 20°C in a certain region. The sample mean and standard deviation were computed as $\bar{x} = .173$ and $s = .066$, respectively. Based on the accompanying frequency distribution, can it be concluded that oxygenation rate is a normally distributed variable? Use the chi-squared test with $\alpha = .05$.

Rate (per day)	Frequency
Below .100	12
.100–below .150	20
.150–below .200	23
.200–below .250	15
.250 or more	13

19. Each headlight on an automobile undergoing an annual vehicle inspection can be focused either too high (H), too low (L), or properly (N). Checking the two headlights simultaneously (and not distinguishing between left and right) results in the six possible outcomes HH, LL, NN, HL, HN, and LN. If the probabilities (population proportions) for the single headlight focus direction are $P(H) = \theta_1$, $P(L) = \theta_2$, and $P(N) = 1 - \theta_1 - \theta_2$ and the two headlights are focused independently of one another, the probabilities of the six outcomes for a randomly selected car are the following:

$$p_1 = \theta_1^2 \qquad p_2 = \theta_2^2 \qquad p_3 = (1 - \theta_1 - \theta_2)^2$$
$$p_4 = 2\theta_1\theta_2 \qquad p_5 = 2\theta_1(1 - \theta_1 - \theta_2)$$
$$p_6 = 2\theta_2(1 - \theta_1 - \theta_2)$$

Use the accompanying data to test the null hypothesis

$$H_0: p_1 = \pi_1(\theta_1, \theta_2), \ldots, p_6 = \pi_6(\theta_1, \theta_2)$$

where the $\pi_i(\theta_1, \theta_2)$s are given previously.

Outcome	HH	LL	NN	HL	HN	LN
Frequency	49	26	14	20	53	38

[*Hint:* Write the likelihood as a function of θ_1 and θ_2, take the natural log, then compute $\partial/\partial\theta_1$ and $\partial/\partial\theta_2$, equate them to 0, and solve for $\hat{\theta}_1, \hat{\theta}_2$.]

20. The article **"Compatibility of Outer and Fusible Interlining Fabrics in Tailored Garments (***Textile Res. J.***, 1997: 137–142)** gave the following observations on bending rigidity ($\mu N \cdot m$) for medium-quality fabric specimens, from which the accompanying Minitab output was obtained:

24.6	12.7	14.4	30.6	16.1	9.5	31.5	17.2
46.9	68.3	30.8	116.7	39.5	73.8	80.6	20.3
25.8	30.9	39.2	36.8	46.6	15.6	32.3	

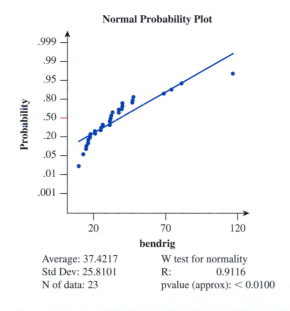

Normal Probability Plot

Average: 37.4217   W test for normality
Std Dev: 25.8101   R: 0.9116
N of data: 23   pvalue (approx): < 0.0100

Would you use a one-sample t confidence interval to estimate true average bending rigidity? Explain your reasoning.

21. The article from which the data in Exercise 20 was obtained also gave the accompanying data on the composite mass/outer fabric mass ratio for high-quality fabric specimens.

1.15	1.40	1.34	1.29	1.36	1.26	1.22	1.40
1.29	1.41	1.32	1.34	1.26	1.36	1.36	1.30
1.28	1.45	1.29	1.28	1.38	1.55	1.46	1.32

Minitab gave $r = .9852$ as the value of the Ryan-Joiner test statistic and reported that P-value $> .10$. Would you use the one-sample t test to test hypotheses about the value of the true average ratio? Why or why not?

22. The article **"A Method for the Estimation of Alcohol in Fortified Wines Using Hydrometer Baumé and Refractometer Brix" (***Amer. J. of Enol. and Vitic.***, 2006: 486–490)** gave duplicate measurements on distilled alcohol content (%) for a sample of 35 port wines. Here are averages of those duplicate measurements:

15.30	16.20	16.35	17.15	17.48	17.73	17.75
17.85	18.00	18.68	18.82	18.85	19.03	19.07
19.08	19.17	19.20	19.20	19.33	19.37	19.45
19.48	19.50	19.58	19.60	19.62	19.90	19.97
20.00	20.05	21.22	22.25	22.75	23.25	23.78

Use the Ryan-Joiner test to decide at significance level .05 whether a normal distribution provides a plausible model for alcohol content.

23. The article **"Nonbloated Burned Clay Aggregate Concrete" (***J. of Materials***, 1972: 555–563)** reports the following data on 7-day flexural strength of nonbloated burned clay aggregate concrete samples (psi):

257	327	317	300	340	340	343	374	377	386
383	393	407	407	434	427	440	407	450	440
456	460	456	476	480	490	497	526	546	700

Test at level .10 to decide whether flexural strength is a normally distributed variable.

14.3 Two-Way Contingency Tables

In the scenarios of Sections 14.1 and 14.2, the observed frequencies were displayed in a single row within a rectangular table. We now study problems in which the data also consists of counts or frequencies, but the data table will now have I rows ($I \geq 2$) and J columns, so IJ cells. There are two commonly encountered situations in which such data arises:

1. There are I populations of interest, each corresponding to a different row of the table, and each population is divided into the same J categories. A sample is taken from the ith population ($i = 1, \ldots, I$), and the counts are entered in the cells in the ith row of the table. For example, customers of each of

$I = 3$ department-store chains might have available the same $J = 5$ payment categories: cash, check, and credit cards from American Express, Visa, and MasterCard.

2. There is a single population of interest, with each individual in the population categorized with respect to two different factors. There are I categories associated with the first factor and J categories associated with the second factor. A single sample is taken, and the number of individuals belonging in both category i of factor 1 and category j of factor 2 is entered in the cell in row i, column j $(i = 1, \ldots, I; j = 1, \ldots, J)$. As an example, customers making a purchase might be classified according to both department in which the purchase was made, with $I = 6$ departments, and according to method of payment, with $J = 5$ as in (1) above.

Let n_{ij} denote the number of individuals in the sample(s) falling in the (i, j)th cell (row i, column j) of the table—that is, the (i, j)th cell count. The table displaying the n_{ij}'s is called a **two-way contingency table**; a prototype is shown in Table 14.9.

Table 14.9   A Two-Way Contingency Table

	1	2	...	j	...	J
1	n_{11}	n_{12}	...	n_{1j}	...	n_{1J}
2	n_{21}					$\vdots$
$\vdots$	$\vdots$					
i	n_{i1}	...		n_{ij}	...	
$\vdots$	$\vdots$					
I	n_{I1}	...				n_{IJ}

In situations of type 1, we want to investigate whether the proportions in the different categories are the same for all populations. The null hypothesis states that the populations are *homogeneous* with respect to these categories. In type 2 situations, we investigate whether the categories of the two factors occur independently of one another in the population.

Testing for Homogeneity

Suppose each individual in every one of the I populations belongs in exactly one of the same J categories. A sample of n_i individuals is taken from the ith population; let $n = \Sigma n_i$ and

$n_{ij} = $ the number of individuals in the ith sample who fall into category j

$$n_{\cdot j} = \sum_{i=1}^{I} n_{ij} = \text{the total number of individuals among the } n \text{ sample who fall into category } j$$

The n_{ij}'s are recorded in a two-way contingency table with I rows and J columns. The sum of the n_{ij}'s in the ith row is n_i, and the sum of entries in the jth column will be denoted by $n_{\cdot j}$.

Let

$$p_{ij} = \text{the proportion of the individuals in population } i \text{ who fall into category } j$$

Thus, for population 1, the J proportions are $p_{11}, p_{12}, \ldots, p_{1J}$ (which sum to 1) and similarly for the other populations. The **null hypothesis of homogeneity** states that the proportion of individuals in category j is the same for each population and that this is true for every category; that is, for every j, $p_{1j} = p_{2j} = \cdots = p_{Ij}$.

When H_0 is true, we can use $p_1, p_2, \ldots, p_J$ to denote the population proportions in the J different categories; these proportions are common to all I populations. The expected number of individuals in the ith sample who fall in the jth category when H_0 is true is then $E(N_{ij}) = n_i \cdot p_j$. To estimate $E(N_{ij})$, we must first estimate p_j, the proportion in category j. Among the total sample of n individuals, $N_{\cdot j}$'s fall into category j, so we use $\hat{p}_j = N_{\cdot j}/n$ as the estimator (this can be shown to be the maximum likelihood estimator of p_j). Substitution of the estimate $\hat{p}_j$ for p_j in $n_i p_j$ yields a simple formula for estimated expected counts under H_0:

$$\hat{e}_{ij} = \text{estimated expected count in cell } (i,j) = n_i \cdot \frac{n_{\cdot j}}{n}$$

$$= \frac{(i\text{th row total})(j\text{th column total})}{n} \tag{14.9}$$

The test statistic here has the same form as in Sections 14.1 and 14.2. The number of degrees of freedom comes from the general rule of thumb. In each row of Table 14.9 there are $J - 1$ freely determined cell counts (each sample size n_i is fixed), so there are a total of $I(J - 1)$ freely determined cells. Parameters $p_1, \ldots, p_J$ are estimated, but because $\Sigma p_i = 1$, only $J - 1$ of these are independent. Thus df $= I(J - 1) - (J - 1) = (J - 1)(I - 1)$.

Null hypothesis: $H_0: p_{1j} = p_{2j} = \cdots = p_{Ij}$ $j = 1, 2, \ldots, J$
Alternative hypothesis: $H_a: H_0$ is not true

Test statistic value:

$$\chi^2 = \sum_{\text{all cells}} \frac{(\text{observed} - \text{estimated expected})^2}{\text{estimated expected}} = \sum_{i=1}^{I} \sum_{j=1}^{J} \frac{(n_{ij} - \hat{e}_{ij})^2}{\hat{e}_{ij}}$$

When H_0 is true and $\hat{e}_{ij} \geq 5$ for all i, j, the test statistic has approximately a chi-squared distribution with $(I - 1)(J - 1)$ df. The test is again upper-tailed, so the P-value is the area under the $\chi^2_{(I-1)(J-1)}$ curve to the right of the calculated χ^2. Table A.11 can be used to obtain P-value information as described in Section 14.1.

EXAMPLE 14.13   A company packages a particular product in cans of three different sizes, each one using a different production line. Most cans conform to specifications, but a quality control engineer has identified the following reasons for nonconformance:

1. Blemish on can
2. Crack in can
3. Improper pull tab location
4. Pull tab missing
5. Other

A sample of nonconforming units is selected from each of the three lines, and each unit is categorized according to reason for nonconformity, resulting in the following contingency table data:

		Blemish	Crack	Location	Missing	Other	Sample Size
				Reason for Nonconformity			
	1	31	68	17	21	13	150
Production Line	**2**	19	47	30	19	10	125
	3	33	26	16	14	11	100
	Total	83	141	63	54	34	375

Does the data suggest that the proportions falling in the various nonconformance categories are not the same for the three lines? The parameters of interest are the various proportions, and the relevant hypotheses are

H_0: the production lines are homogeneous with respect to the five nonconformance categories; that is, $p_{1j} = p_{2j} = p_{3j}$ for $j = 1, \ldots, 5$

H_a: the production lines are not homogeneous with respect to the categories

The estimated expected frequencies (assuming homogeneity) must now be calculated. Consider the first nonconformance category for the first production line. When the lines are homogeneous,

estimated expected number among the 150 selected units that are blemished

$$= \frac{\text{(first row total)(first column total)}}{\text{total of sample sizes}} = \frac{(150)(83)}{375} = 33.20$$

The contribution of the cell in the upper-left corner to χ^2 is then

$$\frac{(\text{observed} - \text{estimated expected})^2}{\text{estimated expected}} = \frac{(31 - 33.20)^2}{33.20} = .146$$

The other contributions are calculated in a similar manner. Figure 14.5 shows Minitab output for the chi-squared test. The observed count is the top number in each cell, and directly below it is the estimated expected count. The contribution of

```
Expected counts are printed below observed counts
Chi-Square contributions are printed below expected counts

        Blemish    Crack    Location    Missing    Other    Total
   1         31       68          17         21       13      150
          33.20    56.40       25.20      21.60    13.60
          0.146    2.386       2.668      0.017    0.026

   2         19       47          30         19       10      125
          27.67    47.00       21.00      18.00    11.33
          2.715    0.000       3.857      0.056    0.157

   3         33       26          16         14       11      100
          22.13    37.60       16.80      14.40     9.07
          5.335    3.579       0.038      0.011    0.412

Total        83      141          63         54       34      375

Chi-Sq = 21.403,  DF = 8,  P-Value = 0.006
```

Figure 14.5 Minitab output for the chi-squared test of Example 14.13

each cell to χ^2 appears below the counts, and the test statistic value is $\chi^2 = 21.403$. All estimated expected counts are at least 5, so combining categories is unnecessary. The test is based on $(3 - 1)(5 - 1) = 8$ df. Appendix Table A.11 shows that the area under the 8 df chi-squared curve to the right of 20.09 is .010 and the area to the right of 21.95 is .005. Therefore we can say that $.005 < P\text{-value} < .01$; Minitab gives $P\text{-value} = .006$. Using a significance level of .01, the null hypothesis of homogeneity can be rejected in favor of the alternative that the distribution of reason for nonconformity is somehow different for the three production lines.

At this point it is desirable to seek an explanation for why the hypothesis of homogeneity is implausible. Figure 14.6 shows a stacked comparative bar chart of the data. It appears that the three lines are relatively homogenous with respect to the Other and Missing categories but not with respect to the Location, Crack, and Blemish categories. Line 1's incidence rate of crack nonconformities is much higher than for the other two lines, whereas location nonconformities appear to be more of a problem for line 2 than for the other two lines and blemish nonconformities occur much more frequently for line 3 than for the other two lines.

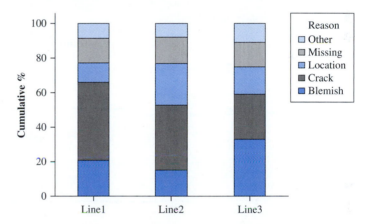

Figure 14.6 Stacked comparative bar chart for the data of Example 14.13 ■

Testing for Independence (Lack of Association)

We focus now on the relationship between two different factors in a single population. Each individual in the population is assumed to belong in exactly one of the I categories associated with the first factor and exactly one of the J categories associated with the second factor. For example, the population of interest might consist of all individuals who regularly watch the national news on television, with the first factor being preferred network (ABC, CBS, NBC, or PBS, so $I = 4$) and the second factor political philosophy (liberal, moderate, or conservative, giving $J = 3$).

For a sample of n individuals taken from the population, let n_{ij} denote the number among the n who fall both in category i of the first factor and category j of the second factor. The n_{ij}'s can be displayed in a two-way contingency table with I rows and J columns. In the case of homogeneity for I populations, the row totals were fixed in advance, and only the J column totals were random. Now only the total sample size is fixed, and both the n_i's and $n_{\cdot j}$'s are observed values of random variables. To state the hypotheses of interest, let

p_{ij} = the proportion of individuals in the population who belong in category i of factor 1 and category j of factor 2

$= P$(a randomly selected individual falls in both category i of factor 1 and category j of factor 2)

Then

$$p_{i\cdot} = \sum_j p_{ij} = P(\text{a randomly selected individual falls in category } i \text{ of factor 1})$$

$$p_{\cdot j} = \sum_i p_{ij} = P(\text{a randomly selected individual falls in category } j \text{ of factor 2})$$

Recall that two events, A and B, are independent if $P(A \cap B) = P(A) \cdot P(B)$. *The null hypothesis here says that an individual's category with respect to factor 1 is independent of the category with respect to factor 2.* In symbols, this becomes $p_{ij} = p_{i\cdot} \cdot p_{\cdot j}$ for every pair (i, j).

The expected count in cell (i, j) is $n \cdot p_{ij}$, so when the null hypothesis is true, $E(N_{ij}) = n \cdot p_{i\cdot} \cdot p_{\cdot j}$. To obtain a chi-squared statistic, we must therefore estimate the $p_{i\cdot}$'s $(i = 1, \ldots, I)$ and $p_{\cdot j}$'s $(j = 1, \ldots, J)$. The (maximum likelihood) estimates are

$$\hat{p}_{i\cdot} = \frac{n_{i\cdot}}{n} = \text{sample proportion for category } i \text{ of factor 1}$$

and

$$\hat{p}_{\cdot j} = \frac{n_{\cdot j}}{n} = \text{sample proportion for category } j \text{ of factor 2}$$

This gives estimated expected cell counts identical to those in the case of homogeneity.

$$\hat{e}_{ij} = n \cdot \hat{p}_{i\cdot} \cdot \hat{p}_{\cdot j} = n \cdot \frac{n_{i\cdot}}{n} \cdot \frac{n_{\cdot j}}{n} = \frac{n_{i\cdot} \cdot n_{\cdot j}}{n}$$

$$= \frac{(i\text{th row total})(j\text{th column total})}{n}$$

The test statistic is also identical to that used in testing for homogeneity, as is the number of degrees of freedom. This is because the number of freely determined cell counts is $IJ - 1$, since only the total n is fixed in advance. There are I estimated $p_{i\cdot}$'s, but only $I - 1$ are independently estimated since $\Sigma p_{i\cdot} = 1$; and similarly $J - 1$ $p_{\cdot j}$'s are independently estimated, so $I + J - 2$ parameters are independently estimated. The rule of thumb now yields df $= IJ - 1 - (I + J - 2) = IJ - I - J + 1 = (I - 1) \cdot (J - 1)$.

Null hypothesis: $H_0: p_{ij} = p_{i\cdot} \cdot p_{\cdot j}$ $\quad i = 1, \ldots, I; j = 1, \ldots, J$

Alternative hypothesis: $H_a: H_0$ is not true

Test statistic value:

$$\chi^2 = \sum_{\text{all cells}} \frac{(\text{observed} - \text{estimated expected})^2}{\text{estimated expected}} = \sum_{i=1}^{I} \sum_{j=1}^{J} \frac{(n_{ij} - \hat{e}_{ij})^2}{\hat{e}_{ij}}$$

When H_0 is true and $\hat{e}_{ij} \geq 5$ for all i, j, the test statistic has approximately a chi-squared distribution with $(I - 1)(J - 1)$ df. The test is again upper-tailed, so the P-value is the area under the $\chi^2_{(I-1)(J-1)}$ curve to the right of the calculated χ^2. Table A.11 can be used to obtain P-value information as described in Section 14.1.

EXAMPLE 14.14   The accompanying two-way table from Minitab (Table 14.10) gives a cross-classification in which the row factor is level of paternal education (completed university, partial university, secondary, partial secondary) and the column factor represents the quartile of neonatal (i.e., newborn) weight gain (Q1 = lowest 25%, Q2 = next lowest 25%, Q3, Q4); the data appeared in the article **"Impact of Neonatal Growth on IQ and Behavior at Early School Age"** (*Pediatrics*, **July 2013, e53–60**). Does it appear that educational level is independent of NWG in the sampled population?

Table 14.10   Observed and Estimated Expected Counts for Example 14.14

Expected counts are printed below observed counts
Chi-Square contributions are printed below expected counts

	Q1	Q2	Q3	Q4	Total
E1	422	433	429	414	1698
	411.63	444.79	422.64	418.93	
	0.261	0.313	0.096	0.058	
E2	1493	1655	1556	1605	6309
	1529.44	1652.65	1570.35	1556.56	
	0.868	0.003	0.131	1.508	
E3	1239	1276	1243	1179	4937
	1196.84	1293.25	1228.85	1218.06	
	1.485	0.230	0.163	1.252	
E4	61	110	73	74	318
	77.09	83.30	79.15	78.46	
	3.358	8.558	0.478	0.253	
Total	3215	3474	3301	3272	13262

The contribution to χ^2 from the cell in the upper-left corner is $(422 - 411.63)^2/411.63 = .261$. The 15 other contributions are calculated in the same way. Then $\chi^2 = .261 + \cdots + .253 = 19.016$. When H_0 is true, the test statistic has approximately a chi-squared distribution with $(4 - 1)(4 - 1) = 9$ df. The expected value of a chi-squared rv is just its number of degrees of freedom, so $E(\chi^2) = 9$ under the assumption of independence. Clearly the test statistic value exceeds what would be expected if the two factors were independent, but is it by enough to suggest implausibility of this null hypothesis? Table A.11 shows that .025 is the area to the right of 19.02 under the chi-squared curve with 9 df. Thus the *P*-value for the test is roughly .025 (which is the value calculated by Minitab; the cited article reported .03). At significance level .05, the null hypothesis of independence would be rejected since *P*-value $\approx$ .025 $\leq$ .05 = α. However, this conclusion would not be justified at a significance level of .01. The *P*-value is such that people might argue over what conclusion is appropriate.

Someone persuaded by our analysis to reject the assertion of independence would want to look more closely at the data to seek an explanation for that conclusion. Perhaps, for example, those in a higher quartile tend to have higher educational levels. Figure 14.7 shows histograms (bar graphs) of the percentages in the various educational level categories for each of the four different quartiles. The four histograms appear to be very similar; the visual impression is that the distribution over the four educational levels does not depend much on the NWG quartile. This seemingly contradicts the finding of statistical significance. Now note that the sample size here is extremely large, and this inflates the value of the chi-squared statistic. With the same percentages as in Figure 14.7 but a much more moderate sample size, the value of χ^2 would be much smaller and the *P*-value much larger. Our test result achieved statistical significance, but there does not seem to be any practical significance.

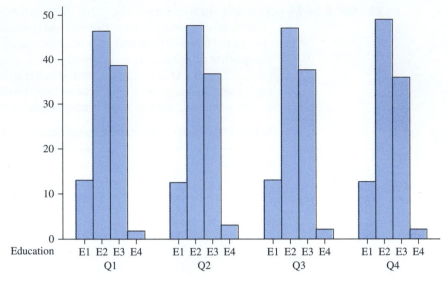

Figure 14.7 Histograms based on the data of Example 14.14 ■

Models and methods for analyzing data in which each individual is categorized with respect to three or more factors (multidimensional contingency tables) are discussed in several of the chapter references.

EXERCISES Section 14.3 (24–36)

24. The accompanying two-way table was constructed using data in the article **"Television Viewing and Physical Fitness in Adults"** (*Research Quarterly for Exercise and Sport*, **1990: 315–320**). The author hoped to determine whether time spent watching television is associated with cardiovascular fitness. Subjects were asked about their television-viewing habits and were classified as physically fit if they scored in the excellent or very good category on a step test. We include Minitab output from a chi-squared analysis. The four TV groups corresponded to different amounts of time per day spent watching TV (0, 1–2, 3–4, or 5 or more hours). The 168 individuals represented in the first column were those judged physically fit. Expected counts appear below observed counts, and Minitab displays the contribution to χ^2 from each cell. State and test the appropriate hypotheses using $\alpha = .05$.

```
                1         2      Total
1              35       147       182
            25.48    156.52
2             101       629       730
           102.20    627.80
3              28       222       250
            35.00    215.00
4               4        34        38
             5.32     32.68
Total         168      1032      1200
```

```
ChiSq = 3.557 + 0.579 +
        0.014 + 0.002 +
        1.400 + 0.228 +
        0.328 + 0.053 = 6.161
df = 3
```

25. In an investigation of alcohol use among college students, each male student in a sample was categorized both according to age group and according to the number of heavy drinking episodes during the previous 30 days (**"Alcohol Use in Students Seeking Primary Care Treatment at University Health Services,"** *J. of Amer. College Health*, **2012: 217–225**).

		Age Group		
		18–20	**21–23**	**≥24**
	None	357	293	592
	1–2	218	285	354
# Episodes	**3–4**	184	218	185
	≥5	328	331	147

Does there appear to be an association between extent of binge drinking and age group in the population from which the sample was selected? Carry out a test of hypotheses at significance level .01.

26. Contamination of various food products is an ongoing problem all over the world. The article **"Prevalence and Quantitative Detection of Salmonella in Retail Raw Chicken in Shaanxi, China"** (***J. of Food Production, 2013***) reported the following data on the occurrence of salmonella in chicken of three different types: (1) supermarket chilled, (2) supermarket frozen, and (3) wet market fresh slaughtered.

		Sample Size	# Salmonella Positive Samples
	1.	60	27
Type	**2.**	60	32
	3.	120	45

Does it appear that the incidence rate of salmonella occurrence depends on the type of chicken? State and test the appropriate hypotheses using a significance level of .05.

27. The article **"Human Lateralization from Head to Foot: Sex-Related Factors"** (***Science***, 1978: 1291–1292) reports for both a sample of right-handed men and a sample of right-handed women the number of individuals whose feet were the same size, had a bigger left than right foot (a difference of half a shoe size or more), or had a bigger right than left foot.

	L > R	L = R	L < R	Sample Size
Men	2	10	28	40
Women	55	18	14	87

Does the data indicate that gender has a strong effect on the development of foot asymmetry? State and test the appropriate hypotheses.

28. A random sample of 175 Cal Poly State University students was selected, and both the email service provider and cell phone provider were determined for each one, resulting in the accompanying data. State and test the appropriate hypotheses

		Cell Phone Provider		
		ATT	Verizon	Other
	gmail	28	17	7
Email Provider	**Yahoo**	31	26	10
	Other	26	19	11

29. The accompanying data on degree of spirituality for samples of natural and social scientists at research universities as well as for a sample of non-academics with graduate degrees appeared in the article **"Conflict Between Religion and Science Among Academic Scientists"** (***J. for the Scientific Study of Religion***, 2009: 276–292).

	Degree of Spirituality			
	Very	**Moderate**	**Slightly**	**Not at all**
N.S.	56	162	198	211
S.S.	56	223	243	239
G.D.	109	164	74	28

a. Is there substantial evidence for concluding that the three types of individuals are not homogenous with respect to their degree of spirituality? State and test the appropriate hypotheses.

b. Considering just the natural scientists and social scientists, is there evidence for non-homogeneity?

30. Three different design configurations are being considered for a particular component. There are four possible failure modes for the component. An engineer obtained the following data on number of failures in each mode for each of the three configurations. Does the configuration appear to have an effect on type of failure?

		Failure Mode			
		1	2	3	4
	1	20	44	17	9
Configuration	2	4	17	7	12
	3	10	31	14	5

31. A random sample of smokers was obtained, and each individual was classified both with respect to gender and with respect to the age at which he/she first started smoking. The data in the accompanying table is consistent with summary results reported in the article **"Cigarette Tar Yields in Relation to Mortality in the Cancer Prevention Study II Prospective Cohort"** (***British Med. J.***, 2004: 72–79).

		Gender	
		Male	**Female**
	<16	25	10
	16−17	24	32
Age	**18−20**	28	17
	>20	19	34

a. Calculate the proportion of males in each age category, and then do the same for females. Based on these proportions, does it appear that there might be an association between gender and the age at which an individual first smokes?

b. Carry out a test of hypotheses to decide whether there is an association between the two factors.

32. Eclosion refers to the emergence of an adult insect from an egg. The following data on eclosion rates when nymphs were exposed to heat for various durations was extracted from the article **"High Temperature Determines the Ups and Downs of Small Brown Planthopper *Laodelphax Striatellus* Population"** (***Insect Science***, 2012: 385–392).

Duration (d)	0	1	2	3	5	10	15
Sample size	120	41	47	44	46	42	10
# Emerged:	101	38	44	40	38	35	7

Carry out a chi-squared test to decide whether it is plausible that eclosion rate does not depend on exposure duration (the cited article included summary information from the test).

33. Show that the chi-squared statistic for the test of independence can be written in the form

$$\chi^2 = \sum_{i=1}^{I} \sum_{j=1}^{J} \left(\frac{N_{ij}^2}{\hat{E}_{ij}} \right) - n$$

Why is this formula more efficient computationally than the defining formula for χ^2?

34. Suppose that each student in a sample had been categorized with respect to political views, marijuana usage, and religious preference, with the categories of this latter factor being Protestant, Catholic, and other. The data could be displayed in three different two-way tables, one corresponding to each category of the third factor. With $p_{ijk} = P(\text{political category } i, \text{marijuana category } j, \text{and religious category } k)$, the null hypothesis of independence of all three factors states that $p_{ijk} = p_{i\cdot\cdot} \cdot p_{\cdot j\cdot} \cdot p_{\cdot\cdot k}$. Let n_{ijk} denote the observed frequency in cell (i, j, k). Show how to estimate the expected cell counts assuming that H_0 is true ($\hat{e}_{ijk} = n\hat{p}_{ijk}$, so the $\hat{p}_{ijk}$'s must be determined). Then use the general rule of thumb to determine the number of degrees of freedom for the chi-squared statistic.

35. Suppose that in a particular state consisting of four distinct regions, a random sample of n_k voters is obtained from the kth region for $k = 1, 2, 3, 4$. Each voter is then classified according to which candidate (1, 2, or 3) he or she prefers and according to voter registration (1 = Dem., 2 = Rep., 3 = Indep.). Let p_{ijk} denote the proportion of voters in region k who belong in candidate category i and registration category j. The null hypothesis of homogeneous regions is $H_0: p_{ij1} = p_{ij2} = p_{ij3} = p_{ij4}$ for all i, j (i.e., the proportion within each candidate/registration combination is the same for all four regions). Assuming that H_0 is true, determine $\hat{p}_{ijk}$ and $\hat{e}_{ijk}$ as functions of the observed n_{ijk}'s, and use the general rule of thumb to obtain the number of degrees of freedom for the chi-squared test.

36. Consider the accompanying 2×3 table displaying the sample proportions that fell in the various combinations of categories (e.g., 13% of those in the sample were in the first category of both factors).

	1	2	3
1	.13	.19	.28
2	.07	.11	.22

a. Suppose the sample consisted of $n = 100$ people. Use the chi-squared test for independence with significance level .10.

b. Repeat part (a), assuming that the sample size was $n = 1000$.

c. What is the smallest sample size n for which these observed proportions would result in rejection of the independence hypothesis?

SUPPLEMENTARY EXERCISES (37–49)

37. The article **"Birth Order and Political Success"** (*Psych. Reports*, 1971: 1239–1242) reports that among 31 randomly selected candidates for political office who came from families with four children, 12 were firstborn, 11 were middle born, and 8 were last born. Use this data to test the null hypothesis that a political candidate from such a family is equally likely to be in any one of the four ordinal positions.

38. Does the phase of the moon have any bearing on birthrate? Each of 222,784 births that occurred during a period encompassing 24 full lunar cycles was classified according to lunar phase. The following data is consistent with summary quantities that appeared in the article **"The Effect of the Lunar Cycle on Frequency of Births and Birth Complications"** (*Amer. J. of Obstetrics and Gynecology*, 2005: 1462–1464).

Lunar Phase	# Days in Phase	# Births
New moon	24	7680
Waxing crescent	152	48,442
First quarter	24	7579
Waxing gibbous	149	47,814
Full moon	24	7711
Waning gibbous	150	47,595
Last quarter	24	7733
Waning crescent	152	48,230

State and test the appropriate hypotheses to answer the question posed at the beginning of this exercise.

39. Each individual in a sample of nursing home patients was cross-classified both with respect to cognitive state (normal or mild impairment, moderate impairment, severe impairment) and with respect to drug status (psychotropic drug change, psychotropic user without a

change, no psychotropic medication). The following Minitab output resulted from a request to perform a chi-squared analysis.

```
          Drug change  No change  No med   Total
Normal             83         60      46     189
                90.06      64.11   34.83
                0.554      0.263   3.584
Moderate          237        151      70     458
               218.25     155.35   84.40
                1.611      0.122   2.456
Severe             86         78      41     205
                97.69      69.54   37.78
                1.398      1.030   0.275
Total             406        289     157     852
Chi-Sq = 11.294, DF = 4, P-Value = 0.023
```

(**"Psychotropic Drug Initiation or Increased Dosage and the Acute Risk of Falls," BMC Geriatrics, 2013: 13:19**).

a. Verify the expected frequency and contribution to χ^2 in the normal–drug change cell of the two-way table.

b. Does there appear to be an association between cognitive state and drug status? State and test the appropriate hypotheses using a significance level of .01. [*Note:* The cited article reported a *P*-value.]

40. The authors of the article **"Predicting Professional Sports Game Outcomes from Intermediate Game Scores"** (*Chance*, 1992: 18–22) used a chi-squared test to determine whether there was any merit to the idea that basketball games are not settled until the last quarter, whereas baseball games are over by the seventh inning. They also considered football and hockey. Data was collected for 189 basketball games, 92 baseball games, 80 hockey games, and 93 football games. The games analyzed were sampled randomly from all games played during the 1990 season for baseball and football and for the 1990–1991 season for basketball and hockey. For each game, the late-game leader was determined, and then it was noted whether the late-game leader actually ended up winning the game. The resulting data is summarized in the accompanying table.

Sport	Late-Game Leader Wins	Late-Game Leader Loses
Basketball	150	39
Baseball	86	6
Hockey	65	15
Football	72	21

The authors state that "*Late-game leader* is defined as the team that is ahead after three quarters in basketball and football, two periods in hockey, and seven innings in baseball. The chi-square value on three degrees of freedom is 10.52 ($P < .015$)."

a. State the relevant hypotheses and reach a conclusion using $\alpha = .05$.

b. Do you think that your conclusion in part (a) can be attributed to a single sport being an anomaly?

41. The accompanying two-way frequency table appears in the article **"Marijuana Use in College"** (*Youth and Society*, 1979: 323–334). Each of 445 college students was classified according to both frequency of marijuana use and parental use of alcohol and psychoactive drugs. Does the data suggest that parental usage and student usage are independent in the population from which the sample was drawn?

	Standard Level of Marijuana Use		
	Never	**Occasional**	**Regular**
Neither	141	54	40
Parental Use of Alcohol and Drugs **One**	68	44	51
Both	17	11	19

42. Much attention has recently focused on the incidence of concussions among athletes. Separate samples of soccer players, non-soccer athletes, and non-athletes were selected. The accompanying table then resulted from determining the number of concussions each individual reported on a medical history questionnaire (**"No Evidence of Impaired Neurocognitive Performance in Collegiate Soccer Players," Amer. J. of Sports Med., 2002: 157–162**).

	# of Concussions			
	0	**1**	**2**	**≥ 3**
Soccer	45	25	11	10
N-S Athletes	68	15	8	5
Non-athletes	45	5	3	0

Does the distribution of # of concussions appear to be different for the three types of individuals? Carry out a test of hypotheses.

43. In a study to investigate the extent to which individuals are aware of industrial odors in a certain region (**"Annoyance and Health Reactions to Odor from Refineries and Other Industries in Carson, California," Environmental Research, 1978: 119–132**), a sample of individuals was obtained from each of three different areas near industrial facilities. Each individual was asked whether he or she noticed odors (1) every day, (2) at least once/week, (3) at least once/month, (4) less often than once/month, or (5) not at all, resulting in the data and SPSS output at the bottom of the next page. State and test the appropriate hypotheses.

44. Many shoppers have expressed unhappiness because grocery stores have stopped putting prices on individual

grocery items. The article **"The Impact of Item Price Removal on Grocery Shopping Behavior"** (**J. of Marketing, 1980: 73–93**) reports on a study in which each shopper in a sample was classified by age and by whether he or she felt the need for item pricing. Based on the accompanying data, does the need for item pricing appear to be independent of age?

	Age				
	<30	**30–39**	**40–49**	**50–59**	**≥60**
Number in Sample	150	141	82	63	49
Number Who Want Item Pricing	127	118	77	61	41

45. Let p_1 denote the proportion of successes in a particular population. The test statistic value in Chapter 8 for testing $H_0: p_1 = p_{10}$ was $z = (\hat{p}_1 - p_{10})/\sqrt{p_{10}p_{20}/n}$, where $p_{20} = 1 - p_{10}$. Show that for the case $k = 2$, the chi-squared test statistic value of Section 14.1 satisfies $\chi^2 = z^2$. [*Hint:* First show that $(n_1 - np_{10})^2 = (n_2 - np_{20})^2$.]

46. The NCAA basketball tournament begins with 64 teams that are apportioned into four regional tournaments, each involving 16 teams. The 16 teams in each region are then ranked (seeded) from 1 to 16. During the 12-year period from 1991 to 2002, the top-ranked team

won its regional tournament 22 times, the second-ranked team won 10 times, the third-ranked team won 5 times, and the remaining 11 regional tournaments were won by teams ranked lower than 3. Let P_{ij} denote the probability that the team ranked i in its region is victorious in its game against the team ranked j. Once the P_{ij}'s are available, it is possible to compute the probability that any particular seed wins its regional tournament (a complicated calculation because the number of outcomes in the sample space is quite large). The paper **"Probability Models for the NCAA Regional Basketball Tournaments"** (**American Statistician, 1991: 35–38**) proposed several different models for the P_{ij}'s.

a. One model postulated $P_{ij} = .5 - \lambda(i - j)$ with $\lambda = 1/32$ (from which $P_{16,1} = \lambda$, $P_{16,2} = 2\lambda$, etc.). Based on this, P(seed # 1 wins) = .27477, P(seed #2 wins) = .20834, and P(seed # 3 wins) = .15429. Does this model appear to provide a good fit to the data?

b. A more sophisticated model has game probabilities $P_{ij} = .5 + .2813625 (z_i - z_j)$, where the z's are measures of relative strengths related to standard normal percentiles [percentiles for successive highly seeded teams are closer together than is the case for teams seeded lower, and .2813625 ensures that the range of probabilities is the same as for the model in part (a)]. The resulting probabilities of seeds 1, 2, or 3 winning their regional tournaments are .45883, .18813, and .11032, respectively. Assess the fit of this model.

```
SPSS output for Exercise 43

Crosstabulation: AREA BY CATEGORY

                  Count
                  Exp       Val
    CATEGORY ──→  Row       Pct
    AREA          Col       Pct    1.00    2.00    3.00    4.00    5.00     Row
                                                                           Total

                  1.00             20      28      23      14      12       97
                                   12.7    24.7    18.0    16.0    25.7     33.3%
                                   20.6%   28.9%   23.7%   14.4%   12.4%
                                   52.6%   37.8%   42.6%   29.2%   15.6%

                  2.00             14      34      21      14      12       95
                                   12.4    24.2    17.6    15.7    25.1     32.6%
                                   14.7%   35.8%   22.1%   14.7%   12.6%
                                   36.8%   45.9%   38.9%   29.2%   15.6%

                  3.00             4       12      10      20      53       99
                                   12.9    25.2    18.4    16.3    26.2     34.0%
                                   4.0%    12.1%   10.1%   20.2%   53.5%
                                   10.5%   16.2%   18.5%   41.7%   68.8%

                  Column          38      74      54      48      77       291
                  Total           13.1%   25.4%   18.6%   16.5%   26.5%    100.0%


    Chi-Square       D.F.    Significance    Min E.F.    Cells with E.F. < 5

     70.64156         8        .0000          12.405     None
```

47. Have you ever wondered whether soccer players suffer adverse effects from hitting "headers"? The authors of the article **"No Evidence of Impaired Neurocognitive Performance in Collegiate Soccer Players"** (*Amer. J. of Sports Med.*, **2002: 157–162**) investigated this issue from several perspectives.

    a. The paper reported that 45 of the 91 soccer players in their sample had suffered at least one concussion, 28 of 96 nonsoccer athletes had suffered at least one concussion, and only 8 of 53 student controls had suffered at least one concussion. Analyze this data and draw appropriate conclusions.

    b. For the soccer players, the sample correlation coefficient calculated from the values of $x =$ soccer exposure (total number of competitive seasons played prior to enrollment in the study) and $y =$ score on an immediate memory recall test was $r = -.220$. Interpret this result.

    c. Here is summary information on scores on a controlled oral word-association test for the soccer and nonsoccer athletes:

$$n_1 = 26, \bar{x}_1 = 37.50, s_1 = 9.13$$

$$n_2 = 56, \bar{x}_2 = 39.63, s_2 = 10.19$$

    Analyze this data and draw appropriate conclusions.

    d. Considering the number of prior nonsoccer concussions, the values of mean ± sd for the three groups were $.30 \pm .67$, $.49 \pm .87$, and $.19 \pm .48$. Analyze this data and draw appropriate conclusions.

48. Do the successive digits in the decimal expansion of π behave as though they were selected from a random number table (or came from a computer's random number generator)?

    a. Let p_0 denote the long run proportion of digits in the expansion that equal 0, and define $p_1, \ldots, p_9$ analogously. What hypotheses about these proportions should be tested, and what is df for the chi-squared test?

    b. H_0 of part (a) would not be rejected for the nonrandom sequence $012\ldots901\ldots901\ldots$. Consider nonoverlapping groups of two digits, and let p_{ij} denote the long run proportion of groups for which the first digit is i and the second digit is j. What hypotheses about these proportions should be tested, and what is df for the chi-squared test?

    c. Consider nonoverlapping groups of 5 digits. Could a chi-squared test of appropriate hypotheses about the p_{ijklm}'s be based on the first 100,000 digits? Explain.

    d. The article **"Are the Digits of π an Independent and Identically Distributed Sequence?"** (*The American Statistician*, **2000: 12–16**) considered the first 1,254,540 digits of π, and reported the following P-values for group sizes of $1, \ldots, 5$: .572, .078, .529, .691, .298. What would you conclude?

49. The Fibonacci sequence of numbers occurs in various scientific contexts. The first two numbers in the sequence are 1,1. Then every succeeding number is the sum of the two previous numbers: $1, 1, 1 + 1 = 2, 1 + 2 = 3, 2 + 3 = 5, 8, 13, 21, \ldots$. The first digit of any number in this sequence can be $1, 2, \ldots$, or 9. The frequencies of first digits for the first 85 numbers in the sequence are as follows: 25 (1's), 16 (2's), 11, 7, 7, 5, 4, 6, 4. Does the distribution of first digits in the Fibonacci sequence appear to be consistent with the Benford's Law distribution described in Exercise 21 of Chapter 3? State and test the relevant hypotheses.

BIBLIOGRAPHY

Agresti, Alan, *An Introduction to Categorical Data Analysis* (2nd ed.), Wiley, New York, 2007. An excellent treatment of various aspects of categorical data analysis by one of the most prominent researchers in this area.

Everitt, B. S., *The Analysis of Contingency Tables* (2nd ed.), Halsted Press, New York, 1992. A compact but informative survey of methods for analyzing categorical data, exposited with a minimum of mathematics.

Mosteller, Frederick, and Richard Rourke, *Sturdy Statistics,* Addison-Wesley, Reading, MA, 1973. Contains several very readable chapters on the varied uses of chi-square.

15

Distribution-Free Procedures

INTRODUCTION

When the underlying population or populations are nonnormal, the t and F tests and t confidence intervals of Chapters 7–13 will in general have actual levels of significance or confidence levels that differ from the nominal levels (those prescribed by the experimenter through the choice of, say, $t_{.025}$, $F_{.01}$, etc.) α and $100(1 - \alpha)\%$, although the difference between actual and nominal levels may not be large when the departure from normality is not too severe. Because the t and F procedures require the distributional assumption of normality, they are not "distribution-free" procedures—alternatively, because they are based on a particular parametric family of distributions (normal), they are not "nonparametric" procedures.

In this chapter, we describe procedures that are valid [actual significance level α or confidence level $100(1 - \alpha)\%$] simultaneously for many different types of underlying distributions. Such procedures are called **distribution-free** or **nonparametric**. One- and two-sample test procedures are presented in Sections 15.1 and 15.2, respectively. In Section 15.3, we develop distribution-free confidence intervals. Section 15.4 describes distribution-free ANOVA procedures. These procedures are all competitors of the parametric (t and F) procedures described in previous chapters, so it is important to compare the performance of the two types of procedures under both normal and nonnormal population models. Generally speaking, the distribution-free procedures perform almost as well as their t and F counterparts on the "home ground" of the normal distribution and will often yield a considerable improvement under nonnormal conditions.

15.1 The Wilcoxon Signed-Rank Test

A research chemist performed a particular chemical experiment a total of ten times under identical conditions, obtaining the following ordered values of reaction temperature:

$$-.57 \quad -.19 \quad -.05 \quad .76 \quad 1.30 \quad 2.02 \quad 2.17 \quad 2.46 \quad 2.68 \quad 3.02$$

The distribution of reaction temperature is of course continuous. Suppose the investigator is willing to assume that the reaction temperature distribution is symmetric; that is, there is a point of symmetry such that the density curve to the left of that point is the mirror image of the density curve to its right. This point of symmetry is the median $\tilde{\mu}$ of the distribution (and is also the mean value μ provided that the mean is finite). The assumption of symmetry may at first thought seem quite bold, but remember that any normal distribution is symmetric, so symmetry is actually a weaker assumption than normality.

Let's now consider testing $H_0: \tilde{\mu} = 0$ versus $H_a: \tilde{\mu} > 0$. The null hypothesis can be interpreted as saying that a temperature of any particular magnitude, for example, 1.50, is no more likely to be positive ($+1.50$) than it is to be negative (-1.50). A glance at the data suggests that this hypothesis is not very tenable; for example, the sample median is 1.66, which is far larger than the magnitude of any of the three negative observations.

Figure 15.1 shows two different symmetric pdf's, one for which H_0 is true and one for which H_a is true. When H_0 is true, we expect the magnitudes of the negative observations in the sample to be comparable to the magnitudes of the positive observations. If, however, H_0 is "grossly" untrue as in Figure 15.1(b), then observations of large absolute magnitude will tend to be positive rather than negative.

Figure 15.1 Distributions for which (a) $\tilde{\mu} = 0$; (b) $\tilde{\mu} \gg 0$

For the sample of ten reaction temperatures, let's for the moment disregard the signs of the observations and rank the absolute magnitudes from 1 to 10, with the smallest getting rank 1, the second smallest rank 2, and so on. Then apply the sign of each observation to the corresponding rank to obtain **signed ranks**. Typically some signed ranks will be negative (e.g., -3), whereas others will be positive (e.g., 8). The test statistic will be $S_+ =$ the sum of the positively signed ranks.

Absolute Magnitude	.05	.19	.57	.76	1.30	2.02	2.17	2.46	2.68	3.02
Rank	1	2	3	4	5	6	7	8	9	10
Signed Rank	-1	-2	-3	4	5	6	7	8	9	10

$$s_+ = 4 + 5 + 6 + 7 + 8 + 9 + 10 = 49$$

When the median of the distribution is much greater than 0, most of the observations with large absolute magnitudes should be positive, resulting in positively signed ranks and a large value of s_+. On the other hand, if the median is 0, magnitudes of positively signed observations should be intermingled with those of negatively signed observations, in which case s_+ will not be very large. So intuitively, a larger s_+ value provides more evidence against H_0 than does a smaller value. This implies that the test is upper-tailed: The P-value will be $P_0(S_+ \geq s_+)$, where P_0 represents the probability calculated assuming that H_0 is true. Thus we must determine the distribution of S_+ when the null hypothesis is true—that is, its *null distribution*.

Consider $n = 5$, in which case there are $2^5 = 32$ ways of applying signs to the five ranks 1, 2, 3, 4, and 5 (each rank could have a $-$ sign or a $+$ sign). The key point is that when H_0 is true, *any collection of five signed ranks has the same chance as does any other collection*. That is, the smallest observation in absolute magnitude is equally likely to be positive or negative, the same is true of the second smallest observation in absolute magnitude, and so on. Thus the collection $-1, 2, 3, -4, 5$ of signed ranks is just as likely as the collection $1, 2, 3, 4, -5$, and just as likely as any one of the other 30 possibilities.

Table 15.1 lists the 32 possible signed-rank sequences when $n = 5$, along with the value s_+, for each sequence. This immediately gives the null distribution of S_+ displayed in Table 15.2. For example, Table 15.1 shows that three of the 32 possible sequences have $s_+ = 8$, so $P_0(S_+ = 8) = 1/32 + 1/32 + 1/32 = 3/32$. Notice that the null distribution is symmetric about 7.5 [more generally, symmetrically distributed

Table 15.1 Possible Signed-Rank Sequences for $n = 5$

Sequence					s_+	Sequence					s_+
-1	-2	-3	-4	-5	0	-1	-2	-3	$+4$	-5	4
$+1$	-2	-3	-4	-5	1	$+1$	-2	-3	$+4$	-5	5
-1	$+2$	-3	-4	-5	2	-1	$+2$	-3	$+4$	-5	6
-1	-2	$+3$	-4	-5	3	-1	-2	$+3$	$+4$	-5	7
$+1$	$+2$	-3	-4	-5	3	$+1$	$+2$	-3	$+4$	-5	7
$+1$	-2	$+3$	-4	-5	4	$+1$	-2	$+3$	$+4$	-5	8
-1	$+2$	$+3$	-4	-5	5	-1	$+2$	$+3$	$+4$	-5	9
$+1$	$+2$	$+3$	-4	-5	6	$+1$	$+2$	$+3$	$+4$	-5	10
-1	-2	-3	-4	$+5$	5	-1	-2	-3	$+4$	$+5$	9
$+1$	-2	-3	-4	$+5$	6	$+1$	-2	-3	$+4$	$+5$	10
-1	$+2$	-3	-4	$+5$	7	-1	$+2$	-3	$+4$	$+5$	11
-1	-2	$+3$	-4	$+5$	8	-1	-2	$+3$	$+4$	$+5$	12
$+1$	$+2$	-3	-4	$+5$	8	$+1$	$+2$	-3	$+4$	$+5$	12
$+1$	-2	$+3$	-4	$+5$	9	$+1$	-2	$+3$	$+4$	$+5$	13
-1	$+2$	$+3$	-4	$+5$	10	-1	$+2$	$+3$	$+4$	$+5$	14
$+1$	$+2$	$+3$	-4	$+5$	11	$+1$	$+2$	$+3$	$+4$	$+5$	15

Table 15.2 Null Distribution of S_+ When $n = 5$

s_+	0	1	2	3	4	5	6	7
$p(s_+)$	$\frac{1}{32}$	$\frac{1}{32}$	$\frac{1}{32}$	$\frac{2}{32}$	$\frac{2}{32}$	$\frac{3}{32}$	$\frac{3}{32}$	$\frac{3}{32}$

s_+	8	9	10	11	12	13	14	15
$p(s_+)$	$\frac{3}{32}$	$\frac{3}{32}$	$\frac{3}{32}$	$\frac{2}{32}$	$\frac{2}{32}$	$\frac{1}{32}$	$\frac{1}{32}$	$\frac{1}{32}$

over the possible values $0, 1, 2,\ldots, n(n + 1)/2$]. This symmetry is important in relating the P-value for lower-tailed and two-tailed tests to that of an upper-tailed test.

For $n = 10$ there are $2^{10} = 1024$ possible signed-rank sequences, so a listing would involve much effort. Each sequence, though, would have probability $1/1024$ when H_0 is true, from which the distribution of S_+ when H_0 is true can be easily obtained.

We are now in a position to calculate a P-value for testing $H_0\text{: } \tilde{\mu} = 0$ versus $H_a\text{: } \tilde{\mu} > 0$ when $n = 5$. Suppose that $s_+ = 13$. Then

$$P\text{-value} = P(S_+ \geq 13 \text{ when } H_0 \text{ is true})$$
$$= P_0(S_+ = 13, 14, \text{ or } 15)$$
$$= \frac{1}{32} + \frac{1}{32} + \frac{1}{32} = \frac{3}{32}$$
$$= .094$$

If $s_+ = 14$, then P-value $= 2/32 = .063$. For the sample $x_1 = .58, x_2 = 2.50$, $x_3 = -.21, x_4 = 1.23, x_5 = .97$, the signed rank sequence is $-1, +2, +3, +4, +5$, so $s_+ = 14$. Thus H_0 would be rejected at significance level .10 because P-value $= .063 \leq .10 = \alpha$. However, at significance level .05 or .01, there would not be enough evidence to justify rejecting the null hypothesis.

General Description of the Test

Because the underlying distribution is assumed symmetric, $\mu = \tilde{\mu}$, so we will state the hypotheses of interest in terms of μ rather than $\tilde{\mu}$.*

ASSUMPTION

> $X_1, X_2,\ldots, X_n$ is a random sample from a continuous and symmetric probability distribution with mean (and median) μ.

When the hypothesized value of μ is μ_0, the absolute differences $|x_1 - \mu_0|,\ldots,$ $|x_n - \mu_0|$ must be ranked from smallest to largest.

> Null hypothesis: $H_0\text{: } \mu = \mu_0$
>
> Test statistic value: $s_+ =$ the sum of the ranks associated with positive $(x_i - \mu_0)$'s
>
> **Alternative Hypothesis**   **P-Value Determination**
>
> $H_a\text{: } \mu > \mu_0$ $P_0(S_+ \geq s_+)$
>
> $H_a\text{: } \mu < \mu_0$ $P_0(S_+ \leq s_+) = P_0(S_+ \geq n(n + 1)/2 - s_+)$
>
> $H_a\text{: } \mu \neq \mu_0$ $2P_0(S_+ \geq \max\{s_+, n(n + 1)/2 - s_+\})$
>
> Appendix Table A.13 gives $P_0(S_+ \geq c) = P(S_+ \geq c$ when H_0 is true) for values of c for which this probability is closest to .1, .05, .025, .01, and .005. This allows conclusions to be reached at significance levels that are at least approximately .10, .05, and .01.

Suppose, for example, that the test is upper-tailed and based on $n = 10$. Table A.13 shows that $P_0(S_+ \geq 41) = .097$ and $P_0(S_+ \geq 44) = .053$. So if $s_+ = 40$, then the

---

* If the tails of the distribution are "too heavy," as was the case with the Cauchy distribution mentioned in Chapter 6, then μ will not exist. In such cases, the Wilcoxon test will still be valid for tests concerning $\tilde{\mu}$.

P-value exceeds .10. The value $s_+ = 42$ implies that $.05 < P$-value $< .10$, allowing for rejection of the null hypothesis at significance level .10 but not at significance level .05. If $s_+ = 44$, it is a really close call at significance level .05.

In the case of a lower-tailed test based on $n = 10$, the value $s_+ = 13$ results in P-value $P_0(S_+ \le 13)$. By symmetry of the null distribution, this is identical to $P_0(S_+ \ge 10(11)/2 - 13) = P_0(S_+ \ge 42)$. The P-value is then between .05 and .10. If a two-tailed test results in $s_+ = 44$ when $n = 10$, then $\max\{44, 55 - 44\} = 44$. Thus the P-value is $2P_0(S_+ \ge 44) = 2(.053) = .106$. This would also be the P-value if $s_+ = 11$, since $\max\{11, 55 - 11\} = 44$; the value 11 is just as far out in the lower tail of the null distribution as 44 is in the upper tail.

EXAMPLE 15.1   A manufacturer of electric irons, wishing to test the accuracy of the thermostat control at the 500°F setting, instructs a test engineer to obtain actual temperatures at that setting for 15 irons using a thermocouple. The resulting measurements are as follows:

494.6	510.8	487.5	493.2	502.6	485.0	495.9	498.2
501.6	497.3	492.0	504.3	499.2	493.5	505.8	

The engineer believes it is reasonable to assume that a temperature deviation from 500° of any particular magnitude is just as likely to be positive as negative (the assumption of symmetry) but wants to protect against possible nonnormality of the actual temperature distribution, so she decides to use the Wilcoxon signed-rank test to see whether the data strongly suggests incorrect calibration of the iron.

The hypotheses are $H_0\colon \mu = 500$ versus $H_a\colon \mu \ne 500$, where $\mu =$ the true average actual temperature at the 500°F setting. Subtracting 500 from each x_i gives

−5.6	10.8	−12.5	−6.8	2.6	−15.0	−4.1	−1.8	1.6	−2.7
−8.0	4.3	−.8	−6.5	5.8					

The ranks are obtained by ordering these from smallest to largest without regard to sign.

Absolute Magnitude	.8	1.6	1.8	2.6	2.7	4.1	4.3	5.6	5.8	6.5	6.8	8.0	10.8	12.5	15.0
Rank	1	2	3	4	5	6	7	8	9	10	11	12	13	14	15
Sign	−	+	−	+	−	−	+	−	+	−	−	−	+	−	−

Thus $s_+ = 2 + 4 + 7 + 9 + 13 = 35$. With $n(n + 1)/2 = 120$, the P-value for a two-tailed test is $2P_0(S_+ \le 35) = 2P_0(S_+ \ge 85)$. Appendix Table A.13 shows that $P_0(S_+ \ge 89) = .053$, so P-value $> 2(.053) = .106$. Even at significance level .10, the null hypothesis cannot be rejected, so it certainly cannot be rejected at level .05. Software gives .164 as the P-value. There is no reason to question the plausibility of 500 as the value of the population mean and median.   ■

Although a theoretical implication of the continuity of the underlying distribution is that ties will not occur, in practice they often do because of the discreteness of measuring instruments. If there are several data values with the same absolute magnitude, then they would be assigned the average of the ranks they would receive if they differed very slightly from one another. For example, if in Example 15.1 $x_8 = 498.2$ is changed to 498.4, then two different values of $(x_i − 500)$ would have absolute magnitude 1.6. The ranks to be averaged would be 2 and 3, so each would be assigned rank 2.5.

Paired Observations

When the data consisted of pairs $(X_1, Y_1), \ldots, (X_n, Y_n)$ and the differences $D_1 = X_1 - Y_1, \ldots, D_n = X_n - Y_n$ were normally distributed, in Chapter 9 we used a paired t test to test hypotheses about the expected difference μ_D. If normality is not assumed, hypotheses about μ_D can be tested by using the Wilcoxon signed-rank test on the D_i's, provided that the distribution of the differences is continuous and symmetric. *If X_i and Y_i both have continuous distributions that differ only with respect to their means, then D_i will have a continuous symmetric distribution* (it is *not* necessary for the X and Y distributions to be symmetric individually). The null hypothesis is $H_0: \mu_D = \Delta_0$, and the test statistic S_+ is the sum of the ranks associated with the positive $(D_i - \Delta_0)$'s.

EXAMPLE 15.2   Intermittent fasting (IF) consists of repetitive bouts of short-term fasting. It is of potential interest because it may provide a simple tool to improve insulin sensitivity in individuals with insulin resistance (the latter increases the likelihood of type 2 diabetes and heart disease). The article **"Intermittent Fasting Does Not Affect Whole-Body Glucose, Lipid, or Protein Metabolism"** (*Amer. J. of Clinical Nutr.*, **2009: 1244–1251**) reported on a study in which resting energy expenditure (kcal/d) was determined for a sample of $n = 8$ subjects both while on an IF regimen and while on a standard diet. The authors of the article kindly provided the following data:

Subject	1	2	3	4	5	6	7	8
IF REE	1753.7	1604.4	1576.5	1279.7	1754.2	1695.5	1700.1	1717.0
Std REE	1755.0	1691.1	1697.1	1477.7	1785.2	1669.7	1901.3	1735.3
Difference	−1.3	−86.7	−120.6	−198.0	−31.0	25.8	−201.2	−18.3
Signed rank	−1	−5	−6	−7	−4	3	−8	−2

The article employed the Wilcoxon signed-rank test on the differences to decide whether there is any difference between true average REE for the IF diet and that for the standard diet. The relevant hypotheses are

$$H_0: \mu_D = 0 \text{ versus } H_a: \mu_D \neq 0$$

The test statistic value is clearly $s_+ = 3$ (only that signed rank is positive). For a two-tailed test, the P-value is $2P_0(S_+ \leq 3)$. In the case $n = 8$, there are $2^8 = 256$ possible sets of signed-ranks, all of which are equally likely when the null hypothesis is true. The signed-rank sets that result in test statistic values as small or smaller than the value 3 that came from the data are as follows (only positive signed ranks are displayed):

no positive signed ranks $(s_+ = 0)$; 1 $(s_+ = 1)$; 2 $(s_+ = 2)$; 1, 2 $(s_+ = 3)$; 3 $(s_+ = 3)$

So the P-value is $2(5/256) = .039$. The null hypothesis would thus be rejected at significance level .05 but not at significance level .01. The article reported only that P-value $< .05$.

Here is output from the R software package:

```
Wilcoxon signed rank test
data: y
V = 3, p-value = 0.03906
alternative hypothesis: true location is not equal to 0
Wilcoxon signed rank test with continuity correction
data: y
V = 3, p-value = 0.04232
alternative hypothesis: true location is not equal to 0
```

This latter *P*-value of .042, which Minitab also reports, is based on the normal approximation described in the next subsection along with a continuity correction. ∎

A Large-Sample Approximation

Appendix Table A.13 provides critical values for level α tests only when $n \le 20$. For $n > 20$, it can be shown that S_+ has approximately a normal distribution with

$$\mu_{S_+} = \frac{n(n+1)}{4} \qquad \sigma_{S_+}^2 = \frac{n(n+1)(2n+1)}{24}$$

when H_0 is true.

The mean and variance result from noting that when H_0 is true (the symmetric distribution is centered at μ_0), then the rank i is just as likely to receive a + sign as it is to receive a − sign. Thus

$$S_+ = W_1 + W_2 + W_3 + \cdots + W_n$$

where

$$W_1 = \begin{cases} 1 & \text{with probability .5} \\ 0 & \text{with probability .5} \end{cases} \cdots W_n = \begin{cases} n & \text{with probability .5} \\ 0 & \text{with probability .5} \end{cases}$$

($W_i = 0$ is equivalent to rank i being associated with a −, so i does not contribute to S_+.)

S_+ is then a sum of random variables, and when H_0 is true, these W_i's can be shown to be independent. Application of the rules of expected value and variance gives the mean and variance of S_+. Because the W_i's are not identically distributed, our version of the Central Limit Theorem cannot be applied, but there is a more general version of the theorem that can be used to justify the normality conclusion. Putting these results together gives the following large-sample test statistic.

$$Z = \frac{S_+ - n(n+1)/4}{\sqrt{n(n+1)(2n+1)/24}} \tag{15.1}$$

A *P*-value is computed using Appendix Table A.3 as it was for *z* tests in Chapters 8 and 9.

EXAMPLE 15.3 A particular type of steel beam has been designed to have a compressive strength (lb/in^2) of at least 50,000. For each beam in a sample of 25 beams, the compressive strength was determined and is given in Table 15.3. Assuming that actual compressive strength is distributed symmetrically about the true average value, let's use the Wilcoxon test with $a = .01$ to decide whether the true average compressive strength

Table 15.3 Data for Example 15.3

$x_i - 50{,}000$	Signed Rank	$x_i - 50{,}000$	Signed Rank	$x_i - 50{,}000$	Signed Rank
−10	−1	−99	−10	165	+18
−27	−2	113	+11	−178	−19
36	+3	−127	−12	−183	−20
−55	−4	−129	−13	−192	−21
73	+5	136	+14	−199	−22
−77	−6	−150	−15	−212	−23
−81	−7	−155	−16	−217	−24
90	+8	−159	−17	−229	−25
−95	−9				

is less than the specified value—that is, test $H_0: \mu = 50{,}000$ versus $H_a: \mu < 50{,}000$ (favoring the claim that average compressive strength is at least 50,000).

The sum of the positively signed ranks is $3 + 5 + 8 + 11 + 14 + 18 = 59$, $n(n + 1)/4 = 162.5$, and $n(n + 1)(2n + 1)/24 = 1381.25$, so

$$z = \frac{59 - 162.5}{\sqrt{1381.25}} = -2.78$$

The P-value for this lower-tailed test is $\Phi(-2.78) = .0027$. Since this is less than .01, H_0 is rejected in favor of the conclusion that true average compressive strength is less than 50,000.   ■

When there are ties in the absolute magnitudes, so that average ranks must be used, it is still correct to standardize S_+ by subtracting $n(n + 1)/4$, but the following corrected formula for variance should be used:

$$\sigma_{S_+}^2 = \frac{1}{24} n(n + 1)(2n + 1) - \frac{1}{48} \sum (\tau_i - 1)(\tau_i)(\tau_i + 1) \qquad (15.2)$$

where τ_i is the number of ties in the ith set of tied values and the sum is over all sets of tied values. If, for example, $n = 10$ and the signed ranks are 1, 2, -4, -4, 4, 6, 7, 8.5, 8.5, and 10, then there are two tied sets with $\tau_1 = 3$ and $\tau_2 = 2$, so the summation is $(2)(3)(4) + (1)(2)(3) = 30$ and $\sigma_{S_+}^2 = 96.25 - 30/48 = 95.62$. The denominator in (15.1) should be replaced by the square root of (15.2), though as this example shows, the correction is usually insignificant.

Efficiency of the Wilcoxon Signed-Rank Test

When the underlying distribution being sampled is normal, either the t test or the signed-rank test can be used to test a hypothesis about μ. The t test is the best test in such a situation because among all level α tests it is the one having minimum β. Since it is generally agreed that there are many experimental situations in which normality can be reasonably assumed, as well as some in which it should not be, there are two questions that must be addressed in an attempt to compare the two tests:

1. When the underlying distribution is normal (the "home ground" of the t test), how much is lost by using the signed-rank test?

2. When the underlying distribution is not normal, can a significant improvement be achieved by using the signed-rank test?

If the Wilcoxon test does not suffer much with respect to the t test on the "home ground" of the latter, and performs significantly better than the t test for a large number of other distributions, then there will be a strong case for using the Wilcoxon test.

Unfortunately, there are no simple answers to the two questions. Upon reflection, it is not surprising that the t test can perform poorly when the underlying distribution has "heavy tails" (i.e., when observed values lying far from μ are relatively more likely than they are when the distribution is normal). This is because the behavior of the t test depends on the sample mean, which can be very unstable in the presence of heavy tails. The difficulty in producing answers to the two questions is that β for the Wilcoxon test is very difficult to obtain and study for *any* underlying distribution, and the same can be said for the t test when the distribution is not normal. Even if β were easily obtained, any measure of efficiency would clearly depend on which underlying distribution was postulated. A number of different efficiency measures have been proposed by statisticians; one that many statisticians regard as credible is called **asymptotic relative efficiency** (ARE). The ARE of one test with respect to another is essentially the limiting ratio of sample sizes necessary to obtain identical error probabilities for the two tests. Thus if the ARE of one test with respect to a second equals .5, then when sample sizes are large, twice as large a sample size will be required of the first test to perform as well as the second test. Although the ARE does not characterize test performance for small sample sizes, the following results can be shown to hold:

1. When the underlying distribution is normal, the ARE of the Wilcoxon test with respect to the *t* test is approximately .95.

2. For any distribution, the ARE will be at least .86 and for many distributions will be much greater than 1.

We can summarize these results by saying that, in large-sample problems, the Wilcoxon test is never very much less efficient than the *t* test and may be much more efficient if the underlying distribution is far from normal. Though the issue is far from resolved in the case of sample sizes obtained in most practical problems, studies have shown that the Wilcoxon test performs reasonably and is thus a viable alternative to the *t* test.

EXERCISES Section 15.1 (1–9)

1. Give as much information as you can about the *P*-value for the Wilcoxon test in each of the following situations.
   a. $n = 12$, upper-tailed test, $s_+ = 56$
   b. $n = 12$, upper-tailed test, $s_+ = 62$
   c. $n = 12$, lower-tailed test, $s_+ = 20$
   d. $n = 14$, two-tailed test, $s_+ = 21$
   e. $n = 25$, two-tailed test, $s_+ = 300$

2. Here again is the data on expense ratio (%) for a sample of 20 large-cap blended mutual funds introduced in Exercise 1.53:

1.03	1.23	1.10	1.64	1.30	1.27	1.25
.78	1.05	.64	.94	.86	1.05	.75
.09	0.79	1.61	1.26	.93	.84	

   A normal probability plot shows a distinctly nonlinear pattern, primarily because of the single outlier on each end of the data. But a dotplot and boxplot exhibit a reasonable amount of symmetry. Assuming a symmetric population distribution, does the data provide compelling evidence for concluding that the population mean expense ratio exceeds 1%? Use the Wilcoxon test at significance level .1. [*Note*: The mean expense ratio for the population of all 825 such funds is actually 1.08.]

3. The accompanying data is a subset of the data reported in the article **"Synovial Fluid pH, Lactate, Oxygen and Carbon Dioxide Partial Pressure in Various Joint Diseases"** (*Arthritis and Rheumatism*, **1971: 476–477**). The observations are pH values of synovial fluid (which lubricates joints and tendons) taken from the knees of individuals suffering from arthritis. Assuming that true average pH for nonarthritic individuals is 7.39, test at level .05 to see whether the data indicates a difference between average pH values for arthritic and nonarthritic individuals.

7.02	7.35	7.34	7.17	7.28	7.77	7.09
7.22	7.45	6.95	7.40	7.10	7.32	7.14

4. A random sample of 15 automobile mechanics certified to work on a certain type of car was selected, and the time (in minutes) necessary for each one to diagnose a particular problem was determined, resulting in the following data:

30.6	30.1	15.6	26.7	27.1	25.4	35.0	30.8
31.9	53.2	12.5	23.2	8.8	24.9	30.2	

   Use the Wilcoxon test at significance level .10 to decide whether the data suggests that true average diagnostic time is less than 30 minutes.

5. Both a gravimetric and a spectrophotometric method are under consideration for determining phosphate content of a particular material. Twelve samples of the material are obtained, each is split in half, and a determination is made on each half using one of the two methods, resulting in the following data:

Sample	1	2	3	4
Gravimetric	54.7	58.5	66.8	46.1
Spectrophotometric	55.0	55.7	62.9	45.5

Sample	5	6	7	8
Gravimetric	52.3	74.3	92.5	40.2
Spectrophotometric	51.1	75.4	89.6	38.4
Sample	9	10	11	12
Gravimetric	87.3	74.8	63.2	68.5
Spectrophotometric	86.8	72.5	62.3	66.0

Use the Wilcoxon test to decide whether one technique gives on average a different value than the other technique for this type of material.

6. Reconsider the situation described in Exercise 39 of Section 9.3, and use the Wilcoxon test to test the appropriate hypotheses.

7. Use the large-sample version of the Wilcoxon test at significance level .05 on the data of Exercise 37 in Section 9.3 to decide whether the true mean difference between outdoor and indoor concentrations exceeds .20.

8. Reconsider the port alcohol content data from Exercise 14.22. A normal probability plot casts some doubt on the assumption of population normality. However, a dotplot shows a reasonable amount of symmetry, and the mean, median, and 5% trimmed mean are 19.257, 19.200, and 19.209, respectively. Use the Wilcoxon test at significance level .01 to decide whether there is substantial evidence for concluding that true average content exceeds 18.5.

9. Suppose that observations $X_1, X_2, \ldots, X_n$ are made on a process at times $1, 2, \ldots, n$. On the basis of this data, we wish to test

   H_0: the X_i's constitute an independent and identically distributed sequence

   versus

   H_a: X_{i+1} tends to be larger than X_i for $i = 1, \ldots, n$ (an increasing trend)

   Suppose the X_i's are ranked from 1 to n. Then when H_a is true, larger ranks tend to occur later in the sequence, whereas if H_0 is true, large and small ranks tend to be mixed together. Let R_i be the rank of X_i and consider the test statistic $D = \sum_{i=1}^{n}(R_i - i)^2$. Then small values of D give support to H_a (e.g., the smallest value is 0 for $R_1 = 10, R_2 = 2, \ldots, R_n = n$). When H_0 is true, any sequence of ranks has probability $1/n!$. Use this to determine the P-value in the case $n = 4$, $d = 2$. [*Hint:* List the 4! rank sequences, compute d for each one, and then obtain the null distribution of D. See the Lehmann book (in the chapter bibliography), p. 290, for more information.]

15.2 The Wilcoxon Rank-Sum Test

The two-sample t test is based on the assumption that both population distributions are normal. There are situations, though, in which an investigator would want to use a test that is valid even if the underlying distributions are quite nonnormal. We now describe such a test, called the **Wilcoxon rank-sum test**. An alternative name for the procedure is the Mann-Whitney test, though the Mann-Whitney test statistic is sometimes expressed in a slightly different form from that of the Wilcoxon test. The Wilcoxon test procedure is distribution-free because it will have the desired level of significance for a very large class of underlying distributions.

ASSUMPTIONS

$X_1, \ldots, X_m$ and $Y_1, \ldots, Y_n$ are two independent random samples from continuous distributions with means μ_1 and μ_2, respectively. The X and Y distributions have the same shape and spread, the only possible difference between the two being in the values of μ_1 and μ_2.

The null hypothesis $H_0: \mu_1 - \mu_2 = \Delta_0$ asserts that, the X distribution is shifted by the amount Δ_0 to the right of the Y distribution.

Development of the Test When $m = 3$, $n = 4$

Suppose the relevant hypotheses are $H_0: \mu_1 - \mu_2 = 0$ versus $H_a: \mu_1 - \mu_2 > 0$. If μ_1 is actually much larger than μ_2, then most of the observed x's will typically fall to the right of the observed y's. However, if H_0 is true, then the observed values from the two samples should be intermingled. The test statistic assesses how much intermingling there is in the two samples.

Consider the case $m = 3, n = 4$. Then if all three observed x's were to the right of all four observed y's, this would provide strong evidence for rejecting H_0 in favor of H_a. The test procedure involves pooling the X's and Y's into a combined sample of size $m + n = 7$ and ranking these observations from smallest to largest, with the smallest receiving rank 1 and the largest rank 7. If most of the largest ranks were associated with X observations, we would begin to doubt H_0. The test statistic that quantifies this reasoning is

$$W = \text{the sum of the ranks in the combined sample}$$
$$\text{associated with } X \text{ observations} \qquad (15.3)$$

For the values of m and n under consideration, the smallest possible value of W is $w = 1 + 2 + 3 = 6$ (if all three x's are smaller than all four y's), and the largest possible value is $w = 5 + 6 + 7 = 18$ (if all three x's are larger than all four y's).

As an example, suppose $x_1 = -3.10, x_2 = 1.67, x_3 = 2.01, y_1 = 5.27, y_2 = 1.89, y_3 = 3.86$, and $y_4 = .19$. Then the pooled ordered sample and corresponding ranks are as follows:

Ordered pooled sample:	−3.10	.19	1.67	1.89	2.01	3.86	5.27
Sample:	x	y	x	y	x	y	y
Rank:	1	2	3	4	5	6	7

Thus $w = 1 + 3 + 5 = 9$.

For the alternative under consideration, a larger value of W provides more evidence against H_0 than does a smaller value. This implies that the test is upper-tailed: P-value $= P_0(W \geq w)$, where again P_0 denotes the probability computed assuming that H_0 is true. So we need the "null distribution" of the test statistic. To this end, recall that when H_0 is true, all seven observations come from the same population. This means that under H_0, any possible triple of ranks associated with the three x's—such as $(1, 4, 5), (3, 5, 6)$, or $(5, 6, 7)$—has the same probability as any other possible rank triple. Since there are $\binom{7}{3} = 35$ possible rank triples, under H_0 each rank triple has probability $1/35$. From a list of all 35 rank triples and the w value associated with each, the probability distribution of W can immediately be determined. For example, there are four rank triples that have w value 11—$(1, 3, 7), (1, 4, 6), (2, 3, 6)$, and $(2, 4, 5)$—so $P(W = 11) = 4/35$. The computations are summarized in Table 15.4.

Table 15.4 Probability Distribution of $W(m = 3, n = 4)$ When H_0 Is True

w	6	7	8	9	10	11	12	13	14	15	16	17	18
$P_0(W = w)$	$\frac{1}{35}$	$\frac{1}{35}$	$\frac{2}{35}$	$\frac{3}{35}$	$\frac{4}{35}$	$\frac{4}{35}$	$\frac{5}{35}$	$\frac{4}{35}$	$\frac{4}{35}$	$\frac{3}{35}$	$\frac{2}{35}$	$\frac{1}{35}$	$\frac{1}{35}$

The null distribution of Table 15.4 is symmetric about $w = (6 + 18)/2 = 12$, which is the middle value in the ordered list of possible W values. This is because the two rank triples (r, s, t) (with $r < s < t$) and $(8 - t, 8 - s, 8 - r)$ have values of w symmetric about 12, so for each triple with w value below 12, there is a triple with w value above 12 by the same amount.

For the alternative under consideration, the test statistic value $w = 16$ results in P-value $= P_0(W \geq 16) = 2/35 + 1/35 + 1/35 = .114$. We would then not be able to reject the null hypothesis at significance level .05. The test is lower-tailed if the alternative hypothesis is $H_a: \mu_1 - \mu_2 < 0$. The test statistic value $w = 7$ gives P-value $= P_0(W \leq 7) = 1/35 + 1/35 = .057$. In the case of the alternative hypothesis $H_a: \mu_1 - \mu_2 \neq 0$, the test is two-tailed. Suppose, for example, that $w = 17$. Then 17 and 18 at least as contradictory to H_0 as the obtained value of W, and so also are 7 and 6; these latter two values are as far out in the lower tail of the null distribution as the former two are in the upper tail. The P-value is then $P_0(W \geq 17 \text{ or } \leq 7) = 1/35 + 1/35 + 1/35 + 1/35 = .114$.

General Description of the Test

The null hypothesis $H_0: \mu_1 - \mu_2 = \Delta_0$ is handled by subtracting Δ_0 from each X_i and using the $(X_i - \Delta_0)$'s as the X_i's were previously used. Note that for any positive integer K, the sum of the first K integers is $K(K + 1)/2$. This implies that the smallest possible value of the statistic W is $m(m + 1)/2$, which occurs when the $(X_i - \Delta_0)$s are all to the left of the Y sample. The largest possible value of W occurs when the $(X_i - \Delta_0)$'s lie entirely to the right of the Y's; in this case, $W = (n + 1) + \cdots + (m + n) = $ sum of first $m + n$ integers) $-$ (sum of first n integers), which gives $m(m + 2n + 1)/2$. As with the special case $m = 3, n = 4$, the distribution of W is symmetric about the value that is halfway between the smallest and largest values; this middle value is $m(m + n + 1)/2$. Because of this symmetry, P-values for lower-tailed and two-tailed tests are easily obtained from a tabulation of upper-tailed null distribution probabilities.

Null hypothesis: $H_0: \mu_1 - \mu_2 = \Delta_0$

Test statistic value: $w = \sum_{i=1}^{m} r_i$ where $r_i = $ rank of $(x_i - \Delta_0)$ in the combined sample of $m + n$ $(x - \Delta_0)$'s and y's

Alternative Hypothesis	P-Value Determination
$H_a: \mu_1 - \mu_2 > \Delta_0$	$P_0(W \geq w)$
$H_a: \mu_1 - \mu_2 < \Delta_0$	$P_0(W \leq w) = P_0(W \geq m(m + n + 1) - w)$
$H_a: \mu_1 - \mu_2 \neq \Delta_0$	$2P_0(W \geq \max\{w, m(m + n + 1) - w\})$

Appendix Table A.14 gives $P_0(W \geq c) = P(W \geq c$ when H_0 is true) for values of c for which this probability is closest to .05, .025, .01, and .005. This allows conclusions to be reached at significance levels that are at least approximately .05 and .01.

The table gives information only for $m = 3, 4, \ldots, 8$ and $n = m, m + 1, \ldots, 8$ (i.e., $3 \leq m \leq n \leq 8$). For values of m and n that exceed 8, a normal approximation can be used; of course statistical software will provide an exact P-value for any sample size. To use the table for small m and n, though, *the X and Y samples should be labeled so that $m \leq n$.*

EXAMPLE 15.4   Noroviruses are a leading cause of acute gastroenteritis, and as of December 2011, no vaccine was available to combat this virus. An experiment involved 15 patients, 8 of whom were randomly assigned to receive a new vaccine; the other 7 individuals received a placebo. The following data on duration (h) of Norwalk virus illness resulted:

Vaccine:	1.0	6.2	9.2	13.4	22.1	36.1	40.2	63.5
Placebo:	5.7	16.6	22.0	38.2	45.8	81.5	107.0	

Does it appear that true average duration is different for the two treatments?

Since use of Appendix Table A.14 requires $m \leq n$, let μ_1 denote true average duration when a placebo is given and μ_2 represent true average duration when the vaccine is administered.

We wish to test

$$H_0: \mu_1 - \mu_2 = 0 \text{ versus } H_a: \mu_1 - \mu_2 \neq 0$$

using the Wilcoxon rank-sum test at significance level .05. The x (placebo) ranks for the pooled sample consisting of all 15 observations are 2, 6, 7, 10, 12, 14, and 15, from which $w = 2 + \cdots + 15 = 66$. Since $\max\{w, m(m + n + 1) - w\} = \max\{66, 46\} = 66$,

$$P\text{-value} = 2P_0(W \geq 66) > 2P_0(W \geq 71) = 2(.047) = .094$$

($P_0(W \geq 71)$ comes from Table A.14). The P-value clearly exceeds .05, so at this significance level the null hypothesis cannot be rejected. There is not enough evidence to conclude that true average duration is different for the vaccine from what it is for the placebo. Software yields a P-value of .27, which is in excellent agreement with the value .28 based on larger sample sizes reported in the article **"Norovirus Vaccine against Experimental Human Norwalk Virus Illness"** (*New England J. of Medicine*, **2011: 2178–2187**). A large number of articles in this journal include summaries of data analyses using the Wilcoxon signed-rank test, rank-sum test, or both. ∎

Theoretically, the assumption of continuity of the two distributions ensures that all $m + n$ observed x's and y's will have different values. In practice, though, there will often be ties in the observed values. As with the Wilcoxon signed-rank test, the common practice in dealing with ties is to assign each of the tied observations in a particular set of ties the average of the ranks they would receive if they differed very slightly from one another.

A Normal Approximation for *W*

When both m and n exceed 8, the distribution of W can be approximated by an appropriate normal curve, and this approximation can be used in place of Appendix Table A.14. To obtain the approximation, we need μ_W and σ_W^2 when H_0 is true. In this case, the rank R_i of $X_i - \Delta_0$ is equally likely to be any one of the possible values 1, 2, 3, ..., $m + n$ (R_i has a discrete uniform distribution on the first $m + n$ positive integers), so $\mu_{R_i} = (m + n + 1)/2$. Since $W = \Sigma R_i$, this gives

$$\mu_W = \mu_{R_1} + \mu_{R_2} + \cdots + \mu_{R_m} = \frac{m(m + n + 1)}{2} \tag{15.4}$$

The variance of R_i is also easily computed to be $(m + n + 1)(m + n - 1)/12$. However, because the R_i's are not independent variables, $V(W) \neq mV(R_i)$. Using the fact that, for any two distinct integers a and b between 1 and $m + n$ inclusive, $P(R_i = a, R_j = b) = 1/[(m + n)(m + n - 1)]$ (two integers are being sampled without replacement), $\text{Cov}(R_i, R_j) = -(m + n + 1)/12$, which yields

$$\sigma_W^2 = \sum_{i=1}^{m} V(R_i) + \sum_{i \neq j} \sum \text{Cov}(R_i, R_j) = \frac{mn(m + n + 1)}{12} \tag{15.5}$$

A Central Limit Theorem can then be used to conclude that when H_0 is true, the *test statistic*

$$Z = \frac{W - m(m + n + 1)/2}{\sqrt{mn(m + n + 1)/12}}$$

has approximately a standard normal distribution. A P-value is computed using Appendix Table A.3 exactly as in the case of previous z tests.

EXAMPLE 15.5 The article **"A Lining for the Thermal Comfort of Trekking Boots–Experimental and Numerical Studies" (*Research J. of Textile and Apparel*, 2011: 50–61)** discussed the design and development of linings in footwear. The investigators used the rank-sum test to decide whether true average moisture accumulation (g) was different for two types of boots. The sample sizes and value of W were not provided, so suppose that $m = n = 9$ and $w = 37$. Then

$$\mu_W = 9(19)/2 = 85.5, \quad \sigma_W = \sqrt{9(9)(19)/12} = \sqrt{128.25} = 11.325$$

from which $z = (37 - 85.5)/11.325 = -4.28$ (this is the value of z given in the cited article). The P-value is $2\Phi(-4.28)$, which is less than $2\Phi(-3.49) = .0004$. At significance level .01 or even .001, we reject the null hypothesis and conclude that true average moisture accumulation is different for the two types of boots. ∎

If there are ties in the data, the numerator of Z is still appropriate, but the denominator should be replaced by the square root of the adjusted variance

$$\sigma_W^2 = \frac{mn(m + n + 1)}{12}$$

$$- \frac{mn}{12(m + n)(m + n - 1)} \sum (\tau_i - 1)(\tau_i)(\tau_i + 1) \tag{15.6}$$

where τ_i is the number of tied observations in the ith set of ties and the sum is over all sets of ties. Unless there are a great many ties, there is little difference between Equations (15.6) and (15.5).

Efficiency of the Wilcoxon Rank-Sum Test

When the distributions being sampled are both normal with $\sigma_1 = \sigma_2$, and therefore have the same shapes and spreads, either the pooled t test or the Wilcoxon test can be used (the two-sample t test assumes normality but not equal variances, so assumptions underlying its use are more restrictive in one sense and less in another than those for Wilcoxon's test). In this situation, the pooled t test is best among all possible tests in the sense of minimizing β for any fixed α. However, an investigator can never be absolutely certain that underlying assumptions are satisfied. It is therefore relevant to ask (1) how much is lost by using Wilcoxon's test rather than the pooled t test when the distributions are normal with equal variances and (2) how W compares to T in nonnormal situations.

The notion of test efficiency was discussed in the previous section in connection with the one-sample t test and Wilcoxon signed-rank test. The results for the two-sample tests are the same as those for the one-sample tests. When normality and equal variances both hold, the rank-sum test is approximately 95% as efficient as the pooled t test in large samples. That is, the t test will give the same error probabilities as the Wilcoxon test using slightly smaller sample sizes. On the other hand, the Wilcoxon test will always be at least 86% as efficient as the pooled t test and may be much more efficient if the underlying distributions are very nonnormal, especially with heavy tails. The comparison of the Wilcoxon test with the two-sample (unpooled) t test is less clear-cut. The t test is not known to be the best test in any sense, so it seems safe to conclude that as long as the population distributions have similar shapes and spreads, the behavior of the Wilcoxon test should compare quite favorably to the two-sample t test.

Lastly, we note that β calculations for the Wilcoxon test are quite difficult. This is because the distribution of W when H_0 is false depends not only on $\mu_1 - \mu_2$ but also on the shapes of the two distributions. For most underlying distributions, the nonnull distribution of W is virtually intractable. This is why statisticians have developed large-sample (asymptotic relative) efficiency as a means of comparing tests. With the capabilities of modern-day computer software, another approach to calculation of β is to carry out a simulation experiment.

EXERCISES Section 15.2 (10–16)

10. Say as much as you can about the *P*-value for the rank-sum test in each of the following situations.
    a. $m = 5$, $n = 6$, $w = 41$, upper-tailed test.
    b. $m = 5$, $n = 6$, $w = 22$, lower-tailed test.
    c. $m = 5$, $n = 6$, $w = 45$, two-tailed test.
    d. $m = n = 12$, upper-tailed test, *x* ranks = 4, 7, 8, 11, 12, 15, 17, 19, 20, 22, 23, 24.

11. In an experiment to compare the bond strength of two different adhesives, each adhesive was used in five bondings of two surfaces, and the force necessary to separate the surfaces was determined for each bonding. For adhesive 1, the resulting values were 229, 286, 245, 299, and 250, whereas the adhesive 2 observations were 213, 179, 163, 247, and 225. Let μ_i denote the true average bond strength of adhesive type *i*. Use the Wilcoxon rank-sum test at level .05 to test $H_0: \mu_1 = \mu_2$ versus $H_a: \mu_1 > \mu_2$.

12. The article **"A Study of Wood Stove Particulate Emissions"** (*J. of the Air Pollution Control Assoc.*, 1979: 724–728) reports the following data on burn time (hours) for samples of oak and pine. Test at level .05 to see whether there is any difference in true average burn time for the two types of wood.

    Oak   1.72   .67   1.55   1.56   1.42   1.23   1.77   .48
    Pine   .98   1.40   1.33   1.52   .73   1.20

13. The urinary fluoride concentration (parts per million) was measured both for a sample of livestock grazing in an area previously exposed to fluoride pollution and for a similar sample grazing in an unpolluted region:

Polluted	21.3 18.7 23.0 17.1 16.8 20.9 19.7
Unpolluted	14.2 18.3 17.2 18.4 20.0

    Does the data indicate strongly that the true average fluoride concentration for livestock grazing in the polluted region is larger than for the unpolluted region? Use the Wilcoxon rank-sum test at level $\alpha = .01$.

14. The article **"Multimodal Versus Unimodal Instruction in a Complex Learning Environment"** (*J. of Experimental Educ.*, 2002: 215–239) described an experiment carried out to compare students' mastery of certain software learned in two different ways. The first learning method (multimodal instruction) involved the use of a visual manual. The second technique (unimodal instruction) employed a textual manual. Here are exam scores for the two groups at the end of the experiment (assignment to the groups was random):

Method 1:	44.85	46.59	47.60	51.08	52.20
	56.87	57.03	57.07	60.35	60.82
	67.30	70.15	70.77	75.21	75.28
	76.60	80.30	81.23		

Method 2:	51.95	56.54	57.40	57.60	61.16
	39.91	42.01	43.58	48.83	49.07
	49.48	49.57	49.63	50.75	64.55
	65.31	68.59	72.40		

Does the data suggest that the true average score depends on which learning method is used?

15. The article **"Measuring the Exposure of Infants to Tobacco Smoke"** (*New England J. of Medicine*, 1984: 1075–1078) reports on a study in which various measurements were taken both from a random sample of infants who had been exposed to household smoke and from a sample of unexposed infants. The accompanying data consists of observations on urinary concentration of cotanine, a major metabolite of nicotine (the values constitute a subset of the original data and were read from a plot that appeared in the article). Does the data suggest that true average cotanine level is higher in exposed infants than in unexposed infants by more than 25? Carry out a test at significance level .05.

Unexposed	8	11	12	14	20	43	111
Exposed	35	56	83	92	128	150	176	208

16. Reconsider the situation described in Exercise 81 of Chapter 9 and the accompanying Minitab output (the Greek letter eta is used to denote a median).

```
Mann-Whitney Confidence Interval and Test
good                  N=8            Median=0.540
poor                  N=8            Median=2.400
Point estimate for ETA1-ETA2 is            -1.155
95.9 Percent CI for ETA1-ETA2 is (-3.160, -0.409)
W = 41.0

Test of ETA1 = ETA2 vs ETA1 < ETA2 is
significant at 0.0027
```

    a. Verify that the value of Minitab's test statistic is correct.
    b. Carry out an appropriate test of hypotheses using a significance level of .01.

15.3 Distribution-Free Confidence Intervals

The method we have used so far to construct a confidence interval (CI) can be described as follows: Start with a random variable (Z, T, χ^2, F, or the like) that depends on the parameter of interest and a probability statement involving the variable, manipulate the inequalities of the statement to isolate the parameter between random endpoints, and, finally, substitute computed values for random variables. Another general method for obtaining CIs takes advantage of the relationship between test procedures and CIs discussed in Section 8.5. A $100(1 - \alpha)\%$ CI for a parameter θ can be obtained from a level α test for $H_0: \theta = \theta_0$ versus $H_a: \theta \neq \theta_0$. This method will be used to derive intervals associated with the Wilcoxon signed-rank test and the Wilcoxon rank-sum test.

PROPOSITION

> Suppose we have a level α test procedure for testing $H_0: \theta = \theta_0$ versus $H_a: \theta \neq \theta_0$. For fixed sample values, let A denote the set of all values θ_0 for which H_0 is not rejected. Then A is a $100(1 - \alpha)\%$ CI for θ.

This makes intuitive sense because the CI consists of all values of the parameter that are plausible at the selected confidence level, and we do not want to reject H_0 in favor of H_a if θ_0 is a plausible value.

There are actually pathological examples in which the set A defined in the proposition is not an interval of θ values, but instead the complement of an interval or something even stranger. To be more precise, we should really replace the notion of a CI with that of a confidence set. In the cases of interest here, the set A does turn out to be an interval.

The Wilcoxon Signed-Rank Interval

To test $H_0: \mu = \mu_0$ versus $H_a: \mu \neq \mu_0$ using the Wilcoxon signed-rank test, where μ is the mean of a continuous symmetric distribution, the absolute values $|x_1 - \mu_0|, \ldots, |x_n - \mu_0|$ are ordered from smallest to largest, with the smallest receiving rank 1 and the largest rank n. Each rank is then given the sign of its associated $x_i - \mu_0$, and the test statistic is the sum of the positively signed ranks. The two-tailed test rejects H_0 if s_+ is either $\geq c$ or $\leq n(n + 1)/2 - c$, where c is obtained from Appendix Table A.13 once the desired level of significance α is specified. For fixed $x_1, \ldots, x_n$, the $100(1 - \alpha)\%$ signed-rank interval will consist of all μ_0 for which $H_0: \mu = \mu_0$ is not rejected at level α. To identify this interval, it is convenient to express the test statistic S_+ in another form.

> $S_+ =$ the number of pairwise averages $(X_i + X_j)/2$ with $i \leq j$ that are $\geq \mu_0$

That is, if we average each x_j in the list with each x_i to its left, including $(x_j + x_j)/2$ (which is just x_j), and count the number of these averages that are $\geq \mu_0$, s_+ results. In moving from left to right in the list of sample values, we are simply averaging every pair of observations in the sample [again including $(x_j + x_j)/2$] exactly once, so the order in which the observations are listed before averaging is not important. The equivalence of the two methods for computing s_+ is not difficult to verify. The number of pairwise averages is $\binom{n}{2} + n$ (the first term due to averaging of different observations and the second due to averaging each x_i with itself), which equals $n(n + 1)/2$. It can be shown that P-value $\leq \alpha$ if and only if either too many or too few of these pairwise averages are $\geq \mu_0$, in which case H_0 is rejected.

EXAMPLE 15.6   The following observations are values of cerebral metabolic rate for rhesus monkeys: $x_1 = 4.51$, $x_2 = 4.59$, $x_3 = 4.90$, $x_4 = 4.93$, $x_5 = 6.80$, $x_6 = 5.08$, $x_7 = 5.67$. The 28 pairwise averages are, in increasing order,

4.51	4.55	4.59	4.705	4.72	4.745	4.76	4.795	4.835	4.90
4.915	4.93	4.99	5.005	5.08	5.09	5.13	5.285	5.30	5.375
5.655	5.67	5.695	5.85	5.865	5.94	6.235	6.80		

The first few and the last few of these are pictured in Figure 15.2.

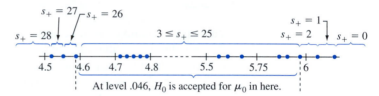

Figure 15.2   Plot of the data for Example 15.6

Because S_+ is a discrete rv, $\alpha = .05$ cannot be obtained exactly. Appendix Table A.13 shows that the P-value for a two-tailed test is $2(.023) = .046$ if either $s_+ = 26$ or 2. Thus H_0 will not be rejected at significance level .046 if $3 \leq s_+ \leq 25$. That is, if the number of pairwise averages $\geq \mu_0$ is between 3 and 25, inclusive, H_0 is not rejected. From Figure 15.2 the CI for μ with confidence level 95.4% (approximately 95%) is (4.59, 5.94). ■

In general, once the pairwise averages are ordered from smallest to largest, the endpoints of the Wilcoxon interval are two of the "extreme" averages. To express this precisely, let the smallest pairwise average be denoted by $\bar{x}_{(1)}$, the next smallest by $\bar{x}_{(2)}, \ldots$, and the largest by $\bar{x}_{(n(n+1)/2)}$.

PROPOSITION

> If the level α Wilcoxon signed-rank test for H_0: $\mu = \mu_0$ versus H_a: $\mu \neq \mu_0$ is to reject H_0 if either $s_+ \geq c$ or $s_+ \leq n(n + 1)/2 - c$, then a $100(1 - \alpha)$% CI for μ is
>
> $$(\bar{x}_{(n(n+1)/2-c+1)}, \bar{x}_{(c)}) \tag{15.7}$$

In words, the interval extends from the dth smallest pairwise average to the dth largest average, where $d = n(n + 1)/2 - c + 1$. Appendix Table A.15 gives the values of c that correspond to approximately the usual confidence levels for $n = 5, 6, \ldots, 25$.

EXAMPLE 15.7
(Example 15.6 continued)

For $n = 7$, the P-value for a two-tailed test is $2(.055) = .11$ if $s_+ = 24$ or $s_+ = 4$. Therefore the null hypothesis will be rejected at significance level .11 if $s_+ = 0, 1, 2, 3, 4, 24, 25, 26, 27,$ or 28. Thus an 89.0% interval (approximately 90%) is obtained by using $c = 24$. The interval is $(\bar{x}_{(28-24+1)}, \bar{x}_{(24)}) = (\bar{x}_{(5)}, \bar{x}_{(24)}) = (4.72, 5.85)$, which extends from the fifth smallest to the fifth largest pairwise average. ∎

The derivation of the interval depended on having a single sample from a continuous symmetric distribution with mean (median) μ. When the data is paired, the interval constructed from the differences $d_1, d_2, \ldots, d_n$ is a CI for the mean (median) difference μ_D. In this case, the symmetry of X and Y distributions need not be assumed; as long as the X and Y distributions have the same shape, the $X - Y$ distribution will be symmetric, so only continuity is required.

For $n > 20$, the large-sample approximation to the Wilcoxon test based on standardizing S_+ gives an approximation to c in (15.7). The result [for a $100(1 - \alpha)\%$ interval] is

$$c \approx \frac{n(n + 1)}{4} + z_{\alpha/2} \sqrt{\frac{n(n + 1)(2n + 1)}{24}}$$

The efficiency of the Wilcoxon interval relative to the t interval is roughly the same as that for the Wilcoxon test relative to the t test. In particular, for large samples when the underlying population is normal, the Wilcoxon interval will tend to be slightly wider than the t interval, but if the population is quite nonnormal (symmetric but with heavy tails), then the Wilcoxon interval will tend to be much narrower than the t interval.

The Wilcoxon Rank-Sum Interval

The Wilcoxon rank-sum test for testing H_0: $\mu_1 - \mu_2 = \Delta_0$ is carried out by first combining the $(X_i - \Delta_0)$'s and Y_j's into one sample of size $m + n$ and ranking them from smallest (rank 1) to largest (rank $m + n$). The test statistic W is then the sum of the ranks of the $(X_i - \Delta_0)$'s. For the two-sided alternative, H_0 is rejected if w is either too small or too large.

To obtain the associated CI for fixed x_i's and y_j's, we must determine the set of all Δ_0 values for which H_0 is not rejected. This is easiest to do if the test statistic is expressed in a slightly different form. The smallest possible value of W is $m(m + 1)/2$, corresponding to every $(X_i - \Delta_0)$ less than every Y_j, and there are mn differences of the form $(X_i - \Delta_0) - Y_j$. A bit of manipulation gives

$$W = [\text{number of} (X_i - Y_j - \Delta_0)\text{'s} \geq 0] + \frac{m(m + 1)}{2}$$

$$= [\text{number of} (X_i - Y_j)\text{'s} \geq \Delta_0] + \frac{m(m + 1)}{2}$$

(15.8)

The P-value will be at most α, leading to rejection of the null hypothesis, if w is relatively small (close to 0) or large (close to $m(m + 2n + 1)/2$). This is equivalent to rejecting H_0 if the number of $(x_i - y_j)$'s $\geq \Delta_0$ is either too small or too large.

Expression (15.8) suggests that we compute $x_i - y_j$ for each i and j and order these mn differences from smallest to largest. Then if the null value Δ_0 is neither smaller than most of the differences nor larger than most, H_0: $\mu_1 - \mu_2 = \Delta_0$ is not rejected. Varying Δ_0 now shows that a CI for $\mu_1 - \mu_2$ will have as its lower endpoint one of the ordered $(x_i - y_j)$'s, and similarly for the upper endpoint.

PROPOSITION

Let $x_1, \ldots, x_m$ and $y_1, \ldots, y_n$ be the observed values in two independent samples from continuous distributions that differ only in location (and not in shape). With $d_{ij} = x_i - y_j$ and the ordered differences denoted by $d_{ij(1)}, d_{ij(2)}, \ldots, d_{ij(mn)}$, the general form of a $100(1 - \alpha)\%$ CI for $\mu_1 - \mu_2$ is

$$(d_{ij(mn-c+1)}, d_{ij(c)}) \qquad (15.9)$$

where c is the critical constant for the two-tailed level α Wilcoxon rank-sum test.

Notice that the form of the Wilcoxon rank-sum interval (15.9) is very similar to the Wilcoxon signed-rank interval (15.7); that uses pairwise averages from a single sample, whereas (15.9) uses pairwise differences from two samples. Appendix Table A.16 gives values of c for selected values of m and n.

EXAMPLE 15.8

The article **"Some Mechanical Properties of Impregnated Bark Board"** (*Forest Products J.*, **1977: 31–38**) reports the following data on maximum crushing strength (psi) for a sample of epoxy-impregnated bark board and for a sample of bark board impregnated with another polymer:

Epoxy (x's)	10,860	11,120	11,340	12,130	14,380	13,070
Other (y's)	4590	4850	6510	5640	6390	

Let's obtain a 95% CI for the true average difference in crushing strength between the epoxy-impregnated board and the other type of board.

From Appendix Table A.16, since the smaller sample size is 5 and the larger sample size is 6, $c = 26$ for a confidence level of approximately 95%. The d_{ij}'s appear in Table 15.5. The five smallest d_{ij}'s $[d_{ij(1)}, \ldots, d_{ij(5)}]$ are 4350, 4470, 4610, 4730, and 4830; and the five largest d_{ij}'s are (in descending order) 9790, 9530, 8740, 8480, and 8220. Thus the CI is $(d_{ij(5)}, d_{ij(26)}) = (4830, 8220)$.

Table 15.5 Differences for the Rank-Sum Interval in Example 15.8

	d_{ij}	4590	4850	y_j 5640	6390	6510
	10,860	6270	6010	5220	4470	4350
	11,120	6530	6270	5480	4730	4610
x_i	**11,340**	6750	6490	5700	4950	4830
	12,130	7540	7280	6490	5740	5620
	13,070	8480	8220	7430	6680	6560
	14,380	9790	9530	8740	7990	7870

■

When m and n are both large, the Wilcoxon test statistic has approximately a normal distribution. This can be used to derive a large-sample approximation for the value c in interval (15.9). The result is

$$c \approx \frac{mn}{2} + z_{\alpha/2} \sqrt{\frac{mn(m + n + 1)}{12}}$$

As with the signed-rank interval, the rank-sum interval (15.9) is quite efficient with respect to the t interval; in large samples, it will tend to be only a bit wider than the t interval when the underlying populations are normal and may be considerably narrower than the t interval if the underlying populations have heavier tails than do normal populations.

EXERCISES Section 15.3 (17–22)

17. The article **"The Lead Content and Acidity of Christchurch Precipitation"** (*N. Zeal. J. of Science*, **1980: 311–312**) reports the accompanying data on lead concentration (μg/L) in samples gathered during eight different summer rainfalls: 17.0, 21.4, 30.6, 5.0, 12.2, 11.8, 17.3, and 18.8. Assuming that the lead-content distribution is symmetric, use the Wilcoxon signed-rank interval to obtain a 95% CI for μ.

18. Compute the 99% signed-rank interval for true average pH μ (assuming symmetry) using the data in Exercise 15.3. [*Hint:* Try to compute only those pairwise averages having relatively small or large values (rather than all 105 averages).]

19. An experiment was carried out to compare the abilities of two different solvents to extract creosote impregnated in test logs. Each of eight logs was divided into two segments, and then one segment was randomly selected for application of the first solvent, with the other segment receiving the second solvent.

Log	1	2	3	4	5	6	7	8
Solvent 1	3.92	3.79	3.70	4.08	3.87	3.95	3.55	3.76
Solvent 2	4.25	4.20	4.41	3.89	4.39	3.75	4.20	3.90

Calculate a CI using a confidence level of roughly 95% for the difference between the true average amount extracted using the first solvent and the true average amount extracted using the second solvent.

20. The following observations are amounts of hydrocarbon emissions resulting from road wear of bias-belted tires under a 522 kg load inflated at 228 kPa and driven at 64 km/hr for 6 hours (**"Characterization of Tire Emissions Using an Indoor Test Facility," *Rubber Chemistry and Technology*, 1978: 7–25**): .045, .117, .062, and .072. What confidence levels are achievable for this sample size using the signed-rank interval? Select an appropriate confidence level and compute the interval.

21. Compute the 90% rank-sum CI for $\mu_1 - \mu_2$ using the data in Exercise 11.

22. Compute a 99% CI for $\mu_1 - \mu_2$ using the data in Exercise 12.

15.4 Distribution-Free ANOVA

The single-factor ANOVA model of Chapter 10 for comparing I population or treatment means assumed that for $i = 1, 2, \ldots, I$, a random sample of size J_i was drawn from a normal population with mean μ_i and variance σ^2. This can be written as

$$X_{ij} = \mu_i + \epsilon_{ij} \quad j = 1, \ldots, J_i; i = 1, \ldots, I \qquad (15.10)$$

where the ϵ_{ij}'s are independent and normally distributed with mean zero and variance σ^2. Although the normality assumption was required for the validity of the F test described in Chapter 10, the next procedure for testing equality of the μ_i's requires only that the ϵ_{ij}'s have the same continuous distribution.

The Kruskal-Wallis Test

Let $N = \Sigma J_i$, the total number of observations in the data set, and suppose we rank all N observations from 1 (the smallest X_{ij}) to N (the largest X_{ij}). When

$H_0: \mu_1 = \mu_2 = \cdots = \mu_I$ is true, the N observations all come from the same distribution, in which case all possible assignments of the ranks $1, 2, \ldots, N$ to the I samples are equally likely and we expect ranks to be intermingled in these samples. If, however, H_0 is false, then some samples will consist mostly of observations having small ranks in the combined sample, whereas others will consist mostly of observations having large ranks. More specifically, if R_{ij} denotes the rank of X_{ij} among the N observations, and $R_{i.}$ and $\overline{R}_{i.}$ denote, respectively, the total and average of the ranks in the ith sample, then when H_0 is true,

$$E(R_{ij}) = \frac{N+1}{2} \qquad E(\overline{R}_{i.}) = \frac{1}{J_i}\sum_j E(R_{ij}) = \frac{N+1}{2}$$

The Kruskal-Wallis test statistic is a measure of the extent to which the $\overline{R}_{i.}$'s deviate from their common expected value $(N+1)/2$.

TEST STATISTIC

$$K = \frac{12}{N(N+1)}\sum_{j=1}^{I} J_i\left(\overline{R}_{i.} - \frac{N+1}{2}\right)^2$$

$$= \frac{12}{N(N+1)}\sum_{i=1}^{I}\frac{R_{i.}^2}{J_i} - 3(N+1)$$

(15.11)

The second expression for K is the computational formula; it involves the rank totals ($R_{i.}$'s) rather than the averages and requires only one subtraction.

Values of K at least as contradictory to H_0 as the calculated k are those that equal or exceed k. That is, the test is upper-tailed: P-value $= P_0(K \geq k)$. Under H_0, each possible assignment of the ranks to the I samples is equally likely, so in theory all such assignments can be enumerated, the value of K determined for each one, and the null distribution obtained by counting the number of times each value of K occurs. Clearly, this computation is tedious, so even though there are tables of the exact null distribution and critical values for small values of the J_i's, we will use the following "large-sample" approximation.

PROPOSITION

When H_0 is true and either

$$I = 3 \qquad J_i \geq 6 \qquad (i = 1, 2, 3)$$

or

$$I > 3 \qquad J_i \geq 5 \qquad (i = 1, \ldots, I)$$

then K has approximately a chi-squared distribution with $I - 1$ df. This implies that the approximate P-value is the area under the χ_{I-1}^2 curve to the right of k. Appendix Table A.11 gives a tabulation of chi-squared upper-tail curve areas.

EXAMPLE 15.9   The accompanying observations (Table 15.6) on axial stiffness index resulted from a study of metal-plate connected trusses in which five different plate lengths—4 in., 6 in., 8 in., 10 in., and 12 in.—were used (**"Modeling Joints Made with Light-Gauge Metal Connector Plates," *Forest Products J.*, 1979: 39–44**).

Table 15.6   Data and Ranks for Example 15.9

									$r_{i.}$	$\bar{r}_{i.}$
	$i = 1$ (4"):	309.2	309.7	311.0	316.8	326.5	349.8	409.5		
	$i = 2$ (6"):	331.0	347.2	348.9	361.0	381.7	402.1	404.5		
	$i = 3$ (8"):	351.0	357.1	366.2	367.3	382.0	392.4	409.9		
	$i = 4$ (10"):	346.7	362.6	384.2	410.6	433.1	452.9	461.4		
	$i = 5$ (12"):	407.4	410.7	419.9	441.2	441.8	465.8	473.4		

									$r_{i.}$	$\bar{r}_{i.}$
	$i = 1$:	1	2	3	4	5	10	24	49	7.00
	$i = 2$:	6	8	9	13	17	21	22	96	13.71
Ranks	$i = 3$:	11	12	15	16	18	20	25	117	16.71
	$i = 4$:	7	14	19	26	29	32	33	160	22.86
	$i = 5$:	23	27	28	30	31	34	35	208	29.71

The computed value of K is

$$k = \frac{12}{35(36)}\left[\frac{(49)^2}{7} + \frac{(96)^2}{7} + \frac{(117)^2}{7} + \frac{(160)^2}{7} + \frac{(208)^2}{7}\right] - 3(36)$$

$$= 20.21$$

Appendix Table A.11 shows that the area under the 4 df chi-squared curve to the right of 16.74 is .005 and the area under this curve to the right of 20.51 is .001. So the P-value for the test is slightly larger than .001 but much smaller than .005, and thus smaller than .01. Therefore H_0 is rejected at significance level .01, and we conclude that expected axial stiffness does depend on plate length.   ∎

Friedman's Test for a Randomized Block Experiment

Suppose $X_{ij} = \mu + \alpha_i + \beta_j + \epsilon_{ij}$, where α_i is the ith treatment effect, β_j is the jth block effect, and the ϵ_{ij}'s are drawn independently from the same continuous (but not necessarily normal) distribution. Then to test $H_0: \alpha_1 = \alpha_2 = \cdots = \alpha_I = 0$, the null hypothesis of no treatment effects, the observations are first ranked separately from 1 to I within each block, and then the rank average $\bar{r}_{i.}$ is computed for each of the I treatments. When H_0 is true, the $\bar{r}_{i.}$'s should be close to one another, since within each block all $I!$ assignments of ranks to treatments are equally likely. Friedman's test statistic measures the discrepancy between the expected value $(I + 1)/2$ of each rank average and the $\bar{r}_{i.}$'s.

TEST STATISTIC

$$F_r = \frac{12J}{I(I + 1)}\sum_{i=1}^{I}\left(\bar{R}_{i.} - \frac{I + 1}{2}\right)^2 = \frac{12}{IJ(I + 1)}\sum R_{i.}^2 - 3J(I + 1)$$

The test is again upper-tailed, because any value exceeding the calculated f_r is even more contradictory to H_0 than is f_r itself. For the cases $I = 3, J = 2,\ldots, 15$ and $I = 4, J = 2,\ldots, 8$, Lehmann's book (see the chapter bibliography) gives the upper-tail critical values from which P-value information can be obtained. Alternatively, for even moderate values of J, the test statistic F_r has approximately a chi-squared distribution with $I - 1$ df when H_0 is true, so the approximate P-value is the area under the χ_{I-1}^2 curve to the right of f_r.

EXAMPLE 15.10 The article **"Physiological Effects During Hypnotically Requested Emotions"** (***Psychosomatic Med.*, 1963: 334–343)** reports the following data (Table 15.7) on skin potential (mV) when the emotions of fear, happiness, depression, and calmness were requested from each of eight subjects.

Table 15.7 Data and Ranks for Example 15.10

x_{ij}	Blocks (Subjects)									
	1	2	3	4	5	6	7	8		
Fear	23.1	57.6	10.5	23.6	11.9	54.6	21.0	20.3		
Happiness	22.7	53.2	9.7	19.6	13.8	47.1	13.6	23.6		
Depression	22.5	53.7	10.8	21.1	13.7	39.2	13.7	16.3		
Calmness	22.6	53.1	8.3	21.6	13.3	37.0	14.8	14.8		

Ranks	1	2	3	4	5	6	7	8	$r_{i\cdot}$	$r_{i\cdot}^2$
Fear	4	4	3	4	1	4	4	3	27	729
Happiness	3	2	2	1	4	3	1	4	20	400
Depression	1	3	4	2	3	2	2	2	19	361
Calmness	2	1	1	3	2	1	3	1	14	196
										1686

Thus

$$f_r = \frac{12}{4(8)(5)}(1686) - 3(8)(5) = 6.45$$

The $v = 3$ column of Appendix Table A.11 shows that P-value $\approx .09$. Since this exceeds .05, H_0 cannot be rejected at that significance level. There is no evidence that average skin potential depends on which emotion is requested. ∎

The book by Myles Hollander et. al. (see the chapter bibliography) discusses multiple comparisons procedures associated with the Kruskal-Wallis and Friedman tests, as well as other aspects of distribution-free ANOVA.

EXERCISES Section 15.4 (23–27)

23. The accompanying data refers to concentration of the radioactive isotope strontium-90 in milk samples obtained from five randomly selected dairies in each of four different regions.

Region	1	6.4	5.8	6.5	7.7	6.1
	2	7.1	9.9	11.2	10.5	8.8
	3	5.7	5.9	8.2	6.6	5.1
	4	9.5	12.1	10.3	12.4	11.7

Test at level .10 to see whether true average strontium-90 concentration differs for at least two of the regions.

24. The article **"Production of Gaseous Nitrogen in Human Steady-State Conditions"** (***J. of Applied Physiology*, 1972: 155–159)** reports the following observations on the amount of nitrogen expired (in liters) under four dietary regimens: (1) fasting, (2) 23% protein, (3) 32% protein, and (4) 67% protein. Use the

Kruskal-Wallis test at level .05 to test equality of the corresponding μ_i's.

1.	4.079	4.859	3.540	5.047	3.298
2.	4.368	5.668	3.752	5.848	3.802
3.	4.169	5.709	4.416	5.666	4.123
4.	4.928	5.608	4.940	5.291	4.674
1.	4.679	2.870	4.648	3.847	
2.	4.844	3.578	5.393	4.374	
3.	5.059	4.403	4.496	4.688	
4.	5.038	4.905	5.208	4.806	

25. The accompanying data on cortisol level was reported in the article **"Cortisol, Cortisone, and 11-Deoxycortisol Levels in Human Umbilical and Maternal Plasma in Relation to the Onset of Labor"** (*J. of Obstetric Gynaecology of the British Commonwealth*, 1974: 737–745). Experimental subjects were pregnant women whose babies were delivered between 38 and 42 weeks gestation. Group 1 individuals elected to deliver by Caesarean section before labor onset, group 2 delivered by emergency Caesarean during induced labor, and group 3 individuals experienced spontaneous labor. Use the Kruskal-Wallis test at level .05 to test for equality of the three population means.

Group 1	262	307	211	323	454	339
	304	154	287	356		
Group 2	465	501	455	355	468	362
Group 3	343	772	207	1048	838	687

26. In a test to determine whether soil pretreated with small amounts of Basic-H makes the soil more permeable to water, soil samples were divided into blocks, and each block received each of the four treatments under study. The treatments were (*A*) water with .001% Basic-H flooded on control soil, (*B*) water without Basic-H on control soil, (*C*) water with Basic-H flooded on soil pretreated with Basic-H, and (*D*) water without Basic-H on soil pretreated with Basic-H. Test at level .01 to see whether there are any effects due to the different treatments.

	Blocks				
	1	**2**	**3**	**4**	**5**
A	37.1	31.8	28.0	25.9	25.5
B	33.2	25.3	20.2	20.3	18.3
C	58.9	54.2	49.2	47.9	38.2
D	56.7	49.6	46.4	40.9	39.4
	6	**7**	**8**	**9**	**10**
A	25.3	23.7	24.4	21.7	26.2
B	19.3	17.3	17.0	16.7	18.3
C	48.8	47.8	40.2	44.0	46.4
D	37.1	37.5	39.6	35.1	36.5

27. In an experiment to study the way in which different anesthetics affect plasma epinephrine concentration, ten dogs were selected and concentration was measured while they were under the influence of the anesthetics isoflurane, halothane, and cyclopropane (**"Sympathoadrenal and Hemodynamic Effects of Isoflurane, Halothane, and Cyclopropane in Dogs,"** *Anesthesiology*, 1974: 465–470). Test at level .05 to see whether there is an anesthetic effect on concentration.

	Dog				
	1	**2**	**3**	**4**	**5**
Isoflurane	.28	.51	1.00	.39	.29
Halothane	.30	.39	.63	.38	.21
Cyclopropane	1.07	1.35	.69	.28	1.24
	6	**7**	**8**	**9**	**10**
Isoflurane	.36	.32	.69	.17	.33
Halothane	.88	.39	.51	.32	.42
Cyclopropane	1.53	.49	.56	1.02	.30

SUPPLEMENTARY EXERCISES (28–36)

28. The article **"Effects of a Rice-Rich Versus Potato-Rich Diet on Glucose, Lipoprotein, and Cholesterol Metabolism in Noninsulin-Dependent Diabetics"** (*Amer. J. of Clinical Nutr.*, 1984: 598–606) gives the accompanying data on cholesterol-synthesis rate for eight diabetic subjects. Subjects were fed a standardized diet with potato or rice as the major carbohydrate source. Participants received both diets for specified periods of time, with cholesterol-synthesis rate (mmol/day) measured at the end of each dietary period. The analysis presented in this article used a distribution-free test. Use such a test with significance level .05 to determine whether the true mean cholesterol-synthesis rate differs significantly for the two sources of carbohydrates.

Cholesterol-Synthesis Rate							
Subject	**1**	**2**	**3**	**4**	**5**	**6**	**7**	**8**
Potato	1.88	2.60	1.38	4.41	1.87	2.89	3.96	2.31
Rice	1.70	3.84	1.13	4.97	.86	1.93	3.36	2.15

29. High-pressure sales tactics or door-to-door salespeople can be quite offensive. Many people succumb to such tactics, sign a purchase agreement, and later regret their actions. In the mid-1970s, the Federal Trade Commission implemented regulations clarifying and extending the rights of purchasers to cancel such agreements. The accompanying data is a subset of that given in the article **"Evaluating the FTC Cooling-Off Rule"** (*J. of Consumer Affairs*, **1977: 101–106**). Individual observations are cancellation rates for each of nine salespeople during each of 4 years. Use an appropriate test at level .05 to see whether true average cancellation rate depends on the year.

Salesperson

	1	2	3	4	5	6	7	8	9
1973	2.8	5.9	3.3	4.4	1.7	3.8	6.6	3.1	0.0
1974	3.6	1.7	5.1	2.2	2.1	4.1	4.7	2.7	1.3
1975	1.4	.9	1.1	3.2	.8	1.5	2.8	1.4	.5
1976	2.0	2.2	.9	1.1	.5	1.2	1.4	3.5	1.2

30. The given data on phosphorus concentration in topsoil for four different soil treatments appeared in the article **"Fertilisers for Lotus and Clover Establishment on a Sequence of Acid Soils on the East Otago Uplands"** (*N. Zeal. J. of Exptl. Ag.*, **1984: 119–129**). Use a distribution-free procedure to test the null hypothesis of no difference in true mean phosphorus concentration (mg/g) for the four soil treatments.

Treatment	I	8.1	5.9	7.0	8.0	9.0
	II	11.5	10.9	12.1	10.3	11.9
	III	15.3	17.4	16.4	15.8	16.0
	IV	23.0	33.0	28.4	24.6	27.7

31. Refer to the data of Exercise 30 and compute a 95% CI for the difference between true average concentrations for treatments II and III.

32. The study reported in **"Gait Patterns During Free Choice Ladder Ascents"** (*Human Movement Sci.*, **1983: 187–195**) was motivated by publicity concerning the increased accident rate for individuals climbing ladders. A number of different gait patterns were used by subjects climbing a portable straight ladder according to specified instructions. The ascent times for seven subjects who used a lateral gait and six subjects who used a four-beat diagonal gait are given.

Lateral	.86	1.31	1.64	1.51	1.53	1.39	1.09
Diagonal	1.27	1.82	1.66	.85	1.45	1.24	

  a. Carry out a test using $\alpha = .05$ to see whether the data suggests any difference in the true average ascent times for the two gaits.
  b. Compute a 95% CI for the difference between the true average gait times.

33. The **sign test** is a very simple procedure for testing hypotheses about a population median assuming only

that the underlying distribution is continuous. To illustrate, consider the following sample of 20 observations on component lifetime (hr):

1.7	3.3	5.1	6.9	12.6	14.4	16.4
24.6	26.0	26.5	32.1	37.4	40.1	40.5
41.5	72.4	80.1	86.4	87.5	100.2	

We wish to test $H_0: \tilde{\mu} = 25.0$ versus $H_a: \tilde{\mu} > 25.0$. The test statistic is $Y =$ the number of observations that exceed 25.

  a. Determine the P-value of the test when $Y = 15$. [*Hint:* Think of a "success" as a lifetime that exceeds 25.0. Then Y is the number of successes in the sample. What kind of a distribution does Y have when $\tilde{\mu} = 25.0$?]
  b. For the given data, should H_0 be rejected at significance level .05?

[*Note:* The test statistic is the number of differences $X_i - 25$ that have positive signs, hence the name *sign test.*]

34. Refer to Exercise 33, and consider a confidence interval associated with the sign test: the **sign interval.** The relevant hypotheses are now $H_0: \tilde{\mu} = \tilde{\mu}_0$ versus $H_a: \tilde{\mu} \neq \tilde{\mu}_0$.

  a. Suppose we decide to reject H_0 if either $Y \geq 15$ or $Y \leq 5$. What is the smallest α for which this equivalent to rejecting H_0 if P-value $\leq \alpha$?
  b. The confidence interval will consist of all values $\tilde{\mu}_0$ for which H_0 is not rejected. Determine the CI for the given data, and state the confidence level.

35. Suppose we wish to test.

H_0: the X and Y distributions are identical

versus

H_a: the X distribution is less spread out than the Y distribution

The accompanying figure pictures X and Y distributions for which H_a is true. The Wilcoxon rank-sum test is not appropriate in this situation because when H_a is true as pictured, the Y's will tend to be at the extreme ends of the combined sample (resulting in small and large Y ranks), so the sum of X ranks will result in a W value that is neither large nor small.

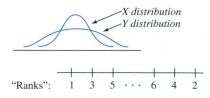

Consider modifying the procedure for assigning ranks as follows: After the combined sample of $m + n$ observations is ordered, the smallest observation is given rank 1, the largest observation is given rank 2, the second smallest is

given rank 3, the second largest is given rank 4, and so on. Then if H_a is true as pictured, the X values will tend to be in the middle of the sample and thus receive large ranks. Let W' denote the sum of the X ranks and consider an upper-tailed test based on this test statistic. When H_0 is true, every possible set of X ranks has the same probability, so W' *has the same distribution as does W when H_0 is true.* The accompanying data refers to medial muscle thickness for arterioles from the lungs of children who died from sudden infant death syndrome (x's) and a control group of children (y's). Carry out the test of H_0 versus H_a at level .05.

| **SIDS** | 4.0 | 4.4 | 4.8 | 4.9 | |
| **Control** | 3.7 | 4.1 | 4.3 | 5.1 | 5.6 |

Consult the Lehmann book (in the chapter bibliography) for more information on this test, called the *Siegel-Tukey test.*

36. The ranking procedure described in Exercise 35 is somewhat asymmetric, because the smallest observation receives rank 1, whereas the largest receives rank 2, and so on. Suppose both the smallest and the largest receive rank 1, the second smallest and second largest receive rank 2, and so on, and let W'' be the sum of the X ranks. The null distribution of W'' is not identical to the null distribution of W, so different tables are needed. Consider the case $m = 3, n = 4$. List all 35 possible orderings of the three X values among the seven observations (e.g., 1, 3, 7 or 4, 5, 6), assign ranks in the manner described, compute the value of W'' for each possibility, and then tabulate the null distribution of W''. What is the P-value if $w'' = 9$? This is the *Ansari-Bradley test;* for additional information, see the book by Hollander and Wolfe in the chapter bibliography.

BIBLIOGRAPHY

Hollander, Myles, Douglas Wolfe, and Eric Chicken, *Nonparametric Statistical Methods* (3rd ed.), Wiley, New York, 2013. A very good reference on distribution-free methods with an excellent collection of tables.

Lehmann, Erich, *Nonparametrics: Statistical Methods Based on Ranks,* Springer, New York, 2006. An excellent discussion of the most important distribution-free methods, presented with a great deal of insightful commentary.

16 Quality Control Methods

INTRODUCTION

Quality characteristics of manufactured products have received much attention from design engineers and production personnel as well as from those concerned with financial management. An article of faith over the years was that very high quality levels and economic well-being were incompatible goals. Recently, however, it has become increasingly apparent that raising quality levels can lead to decreased costs, a greater degree of consumer satisfaction, and thus increased profitability. This has resulted in renewed emphasis on statistical techniques for designing quality into products and for identifying quality problems at various stages of production and distribution.

Control charting is now used extensively as a diagnostic technique for monitoring production and service processes to identify instability and unusual circumstances. After an introduction to basic ideas in Section 16.1, a number of different control charts are presented in the next four sections. The basis for most of these lies in our previous work concerning probability distributions of various statistics such as the sample mean $\overline{X}$ and sample proportion $\hat{p} = X/n$.

Another commonly encountered situation in industrial settings involves a decision by a customer as to whether a batch of items offered by a supplier is of acceptable quality. In the last section of the chapter, we briefly survey some acceptance sampling methods for deciding, based on sample data, on the disposition of a batch.

Besides control charts and acceptance sampling plans, which were first developed in the 1920s and 1930s, statisticians and engineers have recently introduced many new statistical methods for identifying types and levels of production inputs that will ensure high-quality output. Japanese investigators, and in particular the engineer/statistician G. Taguchi and his disciples, have been very influential in this respect, and there is now a large body of material known as "Taguchi methods." The ideas of experimental design, and in particular fractional factorial experiments, are key ingredients. There is still much controversy in the statistical community as to which designs and methods of analysis are best suited to the task at hand. The expository article by George Box et al., cited in the chapter bibliography, gives an informative critique; the book by Thomas Ryan listed there is also a good source of information.

16.1 General Comments on Control Charts

A central message throughout this book has been the pervasiveness of naturally occurring variation associated with any characteristic or attribute of different individuals or objects. In a manufacturing context, no matter how carefully machines are calibrated, environmental factors are controlled, materials and other inputs are monitored, and workers are trained, diameter will vary from bolt to bolt, some plastic sheets will be stronger than others, some circuit boards will be defective whereas others are not, and so on. We might think of such natural random variation as uncontrollable background noise.

There are, however, other sources of variation that may have a pernicious impact on the quality of items produced by some process. Such variation may be attributable to contaminated material, incorrect machine settings, unusual tool wear, and the like. These sources of variation have been termed *assignable causes* in the quality control literature. **Control charts** provide a mechanism for recognizing situations where assignable causes may be adversely affecting product quality. Once a chart indicates an out-of-control situation, an investigation can be launched to identify causes and take corrective action.

A basic element of control charting is that samples have been selected from the process of interest at a sequence of time points. Depending on the aspect of the process under investigation, some statistic, such as the sample mean or sample proportion of defective items, is chosen. The value of this statistic is then calculated for each sample in turn. A traditional control chart then results from plotting these calculated values over time, as illustrated in Figure 16.1.

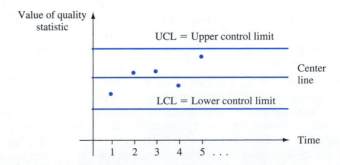

Figure 16.1 A prototypical control chart

Notice that in addition to the plotted points themselves, the chart has a center line and two control limits. The basis for the choice of a center line is sometimes a target value or design specification, for example, a desired value of the bearing diameter. In other cases, the height of the center line is estimated from the data. If the points on the chart all lie between the two control limits, the process is deemed to be in control. That is, the process is believed to be operating in a stable fashion reflecting only natural random variation. An out-of-control "signal" occurs whenever a plotted point falls outside the limits. This is assumed to be attributable to some assignable cause, and a search for such causes commences. The limits are designed so that an in-control process generates very few false alarms, whereas a process not in control quickly gives rise to a point outside the limits.

There is a strong analogy between the logic of control charting and our previous work in hypothesis testing. The null hypothesis here is that the process is in control. When an in-control process yields a point *outside* the control limits (an out-of-control signal), a type I error has occurred. On the other hand, a type II error results when an out-of-control process produces a point *inside* the control limits. Appropriate choice of sample size and control limits will make the associated error probabilities suitably small.

We emphasize that "in control" is not synonymous with "meets design specifications or tolerance." The extent of natural variation may be such that the percentage of items not conforming to specification is much higher than can be tolerated. In such cases, a major restructuring of the process will be necessary to improve process capability. An in-control process is simply one whose behavior with respect to variation is stable over time, showing no indications of unusual extraneous causes.

Software for control charting is now widely available. The journal *Quality Progress* contains many advertisements for statistical quality control computer packages. In addition, SAS and Minitab, among other general-purpose packages, have attractive quality control capabilities.

EXERCISES Section 16.1 (1–5)

1. A control chart for thickness of rolled-steel sheets is based on an upper control limit of .0520 in. and a lower limit of .0475 in. The first ten values of the quality statistic (in this case $\overline{X}$, the sample mean thickness of $n = 5$ sample sheets) are .0506, .0493, .0502, .0501, .0512, .0498, .0485, .0500, .0505, and .0483. Construct the initial part of the quality control chart, and comment on its appearance.

2. Refer to Exercise 1 and suppose the ten most recent values of the quality statistic are .0493, .0485, .0490, .0503, .0492, .0486, .0495, .0494, .0493, and .0488. Construct the relevant portion of the corresponding control chart, and comment on its appearance.

3. Suppose a control chart is constructed so that the probability of a point falling outside the control limits when the process is actually in control is .002. What is the probability that ten successive points (based on independently selected samples) will be within the control limits? What is the probability that 25 successive points will all lie within the control limits? What is the smallest number of successive

points plotted for which the probability of observing at least one outside the control limits exceeds .10?

4. A cork intended for use in a wine bottle is considered acceptable if its diameter is between 2.9 cm and 3.1 cm (so the *lower specification limit* is LSL = 2.9 and the *upper specification limit* is USL = 3.1).

   a. If cork diameter is a normally distributed variable with mean value 3.04 cm and standard deviation .02 cm, what is the probability that a randomly selected cork will conform to specification?

   b. If instead the mean value is 3.00 and the standard deviation is .05, is the probability of conforming to specification smaller or larger than it was in (a)?

5. If a process variable is normally distributed, in the long run virtually all observed values should be between $\mu - 3\sigma$ and $\mu + 3\sigma$, giving a process spread of 6σ.

   a. With LSL and USL denoting the lower and upper specification limits, one commonly used *process capability index* is $C_p = (\text{USL} - \text{LSL})/6\sigma$. The value

$C_p = 1$ indicates a process that is only marginally capable of meeting specifications. Ideally, C_p should exceed 1.33 (a "very good" process). Calculate the value of C_p for each of the cork production processes described in the previous exercise, and comment.

b. The C_p index described in (a) does not take into account process location. A capability measure that does involve the process mean is

$$C_{pk} = \min\{(USL - \mu)/3\sigma, (\mu - LSL)/3\sigma\}$$

Calculate the value of C_{pk} for each of the cork-production processes described in the previous exercise, and comment. [*Note:* In practice, μ and σ have to be estimated from process data; we show how to do this in Section 16.2]

c. How do C_p and C_{pk} compare, and when are they equal?

16.2 Control Charts for Process Location

Suppose the quality characteristic of interest is associated with a variable whose observed values result from making measurements. For example, the characteristic might be resistance of electrical wire (ohms), internal diameter of molded rubber expansion joints (cm), or hardness of a certain alloy (Brinell units). One important use of control charts is to see whether some measure of location of the variable's distribution remains stable over time. The most popular chart for this purpose is the $\overline{X}$ chart.

The $\overline{X}$ Chart Based on Known Parameter Values

Because there is uncertainty about the value of the variable for any particular item or specimen, we denote such a *random* variable (rv) by X. Assume that for an in-control process, X has a normal distribution with mean value μ and standard deviation σ. Then if $\overline{X}$ denotes the sample mean for a random sample of size n selected at a particular time point, we know that

1. $E(\overline{X}) = \mu$

2. $\sigma_{\overline{X}} = \sigma/\sqrt{n}$

3. $\overline{X}$ has a normal distribution.

It follows that

$$P(\mu - 3\sigma_{\overline{X}} \le \overline{X} \le \mu + 3\sigma_{\overline{X}}) = P(-3.00 \le Z \le 3.00) = .9974$$

where Z is a standard normal rv.* It is thus highly likely that for an in-control process, the sample mean will fall within 3 standard deviations ($3\sigma_{\overline{X}}$) of the process mean μ.

Consider first the case in which the values of both μ and σ are known. Suppose that at each of the time points $1, 2, 3, \ldots,$ a random sample of size n is available. Let $\overline{x}_1, \overline{x}_2, \overline{x}_3, \ldots$ denote the calculated values of the corresponding sample means. An $\overline{X}$ chart results from plotting these $\overline{x}_i$'s over time—that is, plotting points $(1, \overline{x}_1), (2, \overline{x}_2), (3, \overline{x}_3)$, and so on—and then drawing horizontal lines across the plot at

$$\text{LCL} = \text{lower control limit} = \mu - 3 \cdot \frac{\sigma}{\sqrt{n}}$$

$$\text{UCL} = \text{upper control limit} = \mu + 3 \cdot \frac{\sigma}{\sqrt{n}}$$

---

* The use of charts based on 3 SD limits is traditional, but tradition is certainly not inviolable.

Such a plot is often called a 3-sigma chart. Any point outside the control limits suggests that the process may have been out of control at that time, so a search for assignable causes should be initiated.

EXAMPLE 16.1   Once each day, three specimens of motor oil are randomly selected from the production process, and each is analyzed to determine viscosity. The accompanying data (Table 16.1) is for a 25-day period. Extensive experience with this process suggests that when the process is in control, viscosity of a specimen is normally distributed with mean 10.5 and standard deviation .18. Thus $\sigma_{\bar{X}} = \sigma/\sqrt{n} = .18/\sqrt{3} = .104$, so the 3 SD control limits are

$$\text{LCL} = \mu - 3 \cdot \frac{\sigma}{\sqrt{n}} = 10.5 - 3(.104) = 10.188$$

$$\text{UCL} = \mu + 3 \cdot \frac{\sigma}{\sqrt{n}} = 10.5 + 3(.104) = 10.812$$

Table 16.1   Viscosity Data for Example 16.1

Day	Viscosity Observations			$\bar{x}$	s	Range
1	10.37	10.19	10.36	10.307	.101	.18
2	10.48	10.24	10.58	10.433	.175	.34
3	10.77	10.22	10.54	10.510	.276	.55
4	10.47	10.26	10.31	10.347	.110	.21
5	10.84	10.75	10.53	10.707	.159	.31
6	10.48	10.53	10.50	10.503	.025	.05
7	10.41	10.52	10.46	10.463	.055	.11
8	10.40	10.38	10.69	10.490	.173	.31
9	10.33	10.35	10.49	10.390	.087	.16
10	10.73	10.45	10.30	10.493	.218	.43
11	10.41	10.68	10.25	10.447	.217	.43
12	10.00	10.60	10.71	10.437	.382	.71
13	10.37	10.50	10.34	10.403	.085	.16
14	10.47	10.60	10.75	10.607	.140	.28
15	10.46	10.46	10.56	10.493	.058	.10
16	10.44	10.68	10.32	10.480	.183	.36
17	10.65	10.42	10.26	10.443	.196	.39
18	10.73	10.72	10.83	10.760	.061	.11
19	10.39	10.75	10.27	10.470	.250	.48
20	10.59	10.23	10.35	10.390	.183	.36
21	10.47	10.67	10.64	10.593	.108	.20
22	10.40	10.55	10.38	10.443	.093	.17
23	10.24	10.71	10.27	10.407	.263	.47
24	10.37	10.69	10.40	10.487	.177	.32
25	10.46	10.35	10.37	10.393	.059	.11

All points on the control chart shown in Figure 16.2 are between the control limits, indicating stable behavior of the process mean over this time period (the standard deviation and range for each sample will be used in the next subsection).

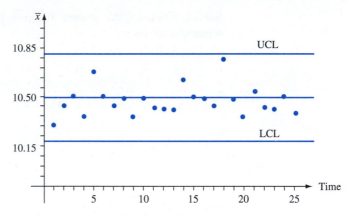

Figure 16.2 $\bar{X}$ chart for the viscosity data of Example 16.1 ■

$\bar{X}$ Charts Based on Estimated Parameters

In practice it frequently happens that values of μ and σ are unknown, so they must be estimated from sample data prior to determining the control limits. This is especially true when a process is first subjected to a quality control analysis. Denote the number of observations in each sample by n, and let k represent the number of samples available. Typical values of n are 3, 4, 5, or 6; it is recommended that k be at least 20. We assume that the k samples were gathered during a period when the process was believed to be in control. More will be said about this assumption shortly.

With $\bar{x}_1, \bar{x}_2, \ldots, \bar{x}_k$ denoting the k calculated sample means, the usual estimate of μ is simply the average of these means:

$$\hat{\mu} = \bar{\bar{x}} = \frac{\displaystyle\sum_{i=1}^{k} \bar{x}_i}{k}$$

There are two different commonly used methods for estimating σ: one based on the k sample standard deviations and the other on the k sample ranges (recall that the sample range is the difference between the largest and smallest sample observations). Prior to the wide availability of good calculators and statistical computer software, ease of hand calculation was of paramount consideration, so the range method predominated. However, in the case of a normal population distribution, the unbiased estimator of σ based on S is known to have smaller variance than that based on the sample range. Statisticians say that the former estimator is more *efficient* than the latter. The loss in efficiency for the estimator is slight when n is very small but becomes important for $n > 4$.

Recall that the sample standard deviation is not an unbiased estimator for σ. When $X_1, \ldots, X_n$ is a random sample from a normal distribution, it can be shown (cf. Exercise 6.37) that

$$E(S) = a_n \cdot \sigma$$

where

$$a_n = \frac{\sqrt{2}\,\Gamma(n/2)}{\sqrt{n-1}\,\Gamma[(n-1)/2]}$$

and $\Gamma(\cdot)$ denotes the gamma function (see Section 4.4). A tabulation of a_n for selected n follows:

n	3	4	5	6	7	8
a_n	.886	.921	.940	.952	.959	.965

Let

$$\overline{S} = \frac{\sum_{i=1}^{k} S_i}{k}$$

where $S_1, S_2, \ldots, S_k$ are the sample standard deviations for the k samples. Then

$$E(\overline{S}) = \frac{1}{K} E\left(\sum_{i=1}^{k} S_i\right) = \frac{1}{k} \sum_{i=1}^{k} E(S_i) = \frac{1}{k} \sum_{i=1}^{k} a_n \cdot \sigma = a_n \cdot \sigma$$

Thus

$$E\left(\frac{\overline{S}}{a_n}\right) = \frac{1}{a_n} E(\overline{S}) = \frac{1}{a_n} \cdot a_n \cdot \sigma = \sigma$$

so $\hat{\sigma} = \overline{S}/a_n$ is an unbiased estimator of σ.

Control Limits Based on the Sample Standard Deviations

$$\text{LCL} = \overline{\overline{x}} = 3 \cdot \frac{\overline{s}}{a_n\sqrt{n}}$$

$$\text{UCL} = \overline{\overline{x}} + 3 \cdot \frac{\overline{s}}{a_n\sqrt{n}}$$

where

$$\overline{\overline{x}} = \frac{\sum_{i=1}^{k} \overline{x}_i}{k} \qquad \overline{s} = \frac{\sum_{i=1}^{k} s_i}{k}$$

EXAMPLE 16.2 Referring to the viscosity data of Example 16.1, we had $n = 3$ and $k = 25$. The values of $\overline{x}_i$ and $s_i (i = 1, \ldots, 25)$ appear in Table 16.1, from which it follows that $\overline{\overline{x}} = 261.896/25 = 10.476$ and $\overline{s} = 3.834/25 = .153$. With $a_3 = .886$, we have

$$\text{LCL} = 10.476 - 3 \cdot \frac{.153}{.886\sqrt{3}} = 10.476 - .299 = 10.177$$

$$\text{UCL} = 10.476 + 3 \cdot \frac{.153}{.886\sqrt{3}} = 10.476 + .299 = 10.775$$

These limits differ a bit from previous limits based on $\mu = 10.5$ and $\sigma = .18$ because now $\hat{\mu} = 10.476$ and $\hat{\sigma} = \overline{s}/a_3 = .173$. Inspection of Table 16.1 shows that every $\overline{x}_i$ is between these new limits, so again no out-of-control situation is evident. ∎

To obtain an estimate of σ based on the sample range, note that if $X_1, \ldots, X_n$ form a random sample from a normal distribution, then

$$R = \text{range}(X_1, \ldots, X_n) = \max(X_1, \ldots, X_n) - \min(X_1, \ldots, X_n)$$

$$= \max(X_1 - \mu, \ldots, X_n - \mu) - \min(X_1 - \mu, \ldots, X_n - \mu)$$

$$= \sigma \left\{ \max\left(\frac{X_1 - \mu}{\sigma}, \cdots, \frac{X_n - \mu}{\sigma} \right) - \min\left(\frac{X_1 - \mu}{\sigma}, \cdots, \frac{X_n - \mu}{\sigma} \right) \right\}$$

$$= \sigma \cdot \{ \max(Z_1, \ldots, Z_n) - \min(Z_1, \ldots, Z_n) \}$$

where $Z_1, \ldots, Z_n$ are independent standard normal rv's. Thus

$$E(R) = \sigma \cdot E(\text{range of a standard normal sample})$$
$$= \sigma \cdot b_n$$

so that R/b_n is an unbiased estimator of σ.

Now denote the ranges for the k samples in the quality control data set by $r_1, r_2, \ldots, r_k$. The argument just given implies that the estimate

$$\hat{\sigma} = \frac{\dfrac{1}{k}\displaystyle\sum_{i=1}^{k} r_i}{b_n} = \frac{\bar{r}}{b_n}$$

comes from an unbiased estimator for σ. Selected values of b_n appear in the accompanying table [their computation is based on using statistical theory and numerical integration to determine $E(\min(Z_1, \ldots, Z_n))$ and $E(\max(Z_1, \ldots, Z_n))$].

n	3	4	5	6	7	8
b_n	1.693	2.058	2.325	2.536	2.706	2.844

Control Limits Based on the Sample Ranges

$$\text{LCL} = \bar{\bar{x}} - 3 \cdot \frac{\bar{r}}{b_n \sqrt{n}}$$

$$\text{UCL} = \bar{\bar{x}} + 3 \cdot \frac{\bar{r}}{b_n \sqrt{n}}$$

where $\bar{r} = \sum_{i=1}^{k} r_i / k$ and $r_1, \ldots, r_k$ are the k individual sample ranges.

EXAMPLE 16.3
(Example 16.2 continued)

Table 16.1 yields $\bar{r} = .292$, so $\hat{\sigma} = .292/b_3 = .292/1.693 = .172$ and

$$\text{LCL} = 10.476 - 3 \cdot \frac{.292}{1.693\sqrt{3}} = 10.476 - .299 = 10.177$$

$$\text{UCL} = 10.476 + 3 \cdot \frac{.292}{1.693\sqrt{3}} = 10.476 + .299 = 10.775$$

These limits are identical to those based on $\bar{s}$, and again every $\bar{x}_i$ lies between the limits. ∎

Recomputing Control Limits

We have assumed that the sample data used for estimating μ and σ was obtained from an in-control process. Suppose, though, that one of the points on the resulting control chart falls outside the control limits. Then if an assignable cause for this out-of-control situation can be found and verified, it is recommended that new control limits be calculated after deleting the corresponding sample from the data set. Similarly, if more than one point falls outside the original limits, new limits should be determined after eliminating any such point for which an assignable cause can be identified and dealt with. It may even happen that one or more points fall outside the new limits, in which case the deletion/recomputation process must be repeated.

Performance Characteristics of Control Charts

Generally speaking, a control chart will be effective if it gives very few out-of-control signals when the process is in control, but shows a point outside the control limits almost as soon as the process goes out of control. One assessment of a chart's effectiveness is based on the notion of "error probabilities." Suppose the variable of interest is normally distributed with known σ (the same value for an in-control or out-of-control process). In addition, consider a 3-sigma chart based on the target value μ_0, with $\mu = \mu_0$ when the process is in control. One error probability is

$$\alpha = P(\text{a single sample gives a point outside the control limits when } \mu = \mu_0)$$
$$= P(\overline{X} > \mu_0 + 3\sigma/\sqrt{n} \text{ or } \overline{X} < \mu_0 - 3\sigma/\sqrt{n} \text{ when } \mu = \mu_0)$$
$$= P\left(\frac{\overline{X} - \mu_0}{\sigma/\sqrt{n}} > 3 \text{ or } \frac{\overline{X} - \mu_0}{\sigma/\sqrt{n}} < -3 \text{ when } \mu = \mu_0\right)$$

The standardized variable $Z = (\overline{X} - \mu_0)/(\sigma/\sqrt{n})$ has a standard normal distribution when $\mu = \mu_0$, so

$$\alpha = P(Z > 3 \text{ or } Z < -3) = \Phi(-3.00) + 1 - \Phi(3.00) = .0026$$

If 3.09 rather than 3 had been used to determine the control limits (this is customary in Great Britain), then

$$\alpha = P(Z > 3.09 \text{ or } Z < -3.09) = .0020$$

The use of 3-sigma limits makes it highly unlikely that an out-of-control signal will result from an in-control process.

Now suppose the process goes out of control because μ has shifted to $\mu + \Delta\sigma$ (Δ might be positive or negative); Δ is the number of standard deviations by which μ has changed. A second error probability is

$$\beta = P\left(\begin{array}{c}\text{a single sample gives a point inside} \\ \text{the control limits when } \mu = \mu_0 + \Delta\sigma\end{array}\right)$$
$$= P(\mu_0 - 3\sigma/\sqrt{n} < \overline{X} < \mu_0 + 3\sigma/\sqrt{n} \text{ when } \mu = \mu_0 + \Delta\sigma)$$

We now standardize by first subtracting $\mu_0 + \Delta\sigma$ from each term inside the parentheses and then dividing by $\sigma/\sqrt{n}$:

$$\beta = P(-3 - \sqrt{n}\Delta < \text{standard normal rv} < 3 - \sqrt{n}\Delta)$$
$$= \Phi(3 - \sqrt{n}\Delta) - \Phi(-3 - \sqrt{n}\Delta)$$

This error probability depends on Δ, which determines the size of the shift, and on the sample size n. In particular, for fixed Δ, β will decrease as n increases (the larger the sample size, the more likely it is that an out-of-control signal will result), and for fixed n, β decreases as $|\Delta|$ increases (the larger the magnitude of a shift, the more likely it is that an out-of-control signal will result). The accompanying table gives β for selected values of Δ when $n = 4$.

Δ	.25	.50	.75	1.00	1.50	2.00	2.50	3.00
β **when** $n = 4$	.9936	.9772	.9332	.8413	.5000	.1587	.0668	.0013

It is clear that a small shift is quite likely to go undetected in a single sample.

If 3 is replaced by 3.09 in the control limits, then α decreases from .0026 to .002, but for any fixed n and σ, β will increase. This is just a manifestation of the inverse relationship between the two types of error probabilities in hypothesis testing. For example, changing 3 to 2.5 will increase α and decrease β.

The error probabilities discussed thus far are computed under the assumption that the variable of interest is normally distributed. If the distribution is only slightly nonnormal, the Central Limit Theorem effect implies that $\overline{X}$ will have approximately a normal distribution even when n is small, in which case the stated error probabilities will be approximately correct. This is, of course, no longer the case when the variable's distribution deviates considerably from normality.

A second performance assessment involves expected or average run length needed to observe an out-of-control signal. When the process is in control, we should expect to observe many samples before seeing one whose $\bar{x}$ lies outside the control limits. On the other hand, if a process goes out of control, the expected number of samples necessary to detect this should be small.

Let p denote the probability that a single sample yields an $\bar{x}$ value outside the control limits; that is,

$$p = P(\overline{X} < \mu_0 - 3\sigma/\sqrt{n} \quad \text{or} \quad \overline{X} > \mu_0 + 3\sigma/\sqrt{n})$$

Consider first an in-control process, so that $\overline{X}_1, \overline{X}_2, \overline{X}_3, \ldots$ are all normally distributed with mean value μ_0 and standard deviation $\sigma/\sqrt{n}$. Define an rv Y by

$$Y = \text{the first } i \text{ for which } \overline{X}_i \text{ falls outside the control limits}$$

If we think of each sample number as a trial and an out-of-control sample as a success, then Y is the number of (independent) trials necessary to observe a success. This Y has a geometric distribution, and we showed in Example 3.18 that $E(Y) = 1/p$. The acronym ARL (for *average run length*) is often used in place of $E(Y)$. Because $p = \alpha$ for an in-control process, we have

$$\text{ARL} = E(Y) = \frac{1}{p} = \frac{1}{\alpha} = \frac{1}{.0026} = 384.62$$

Replacing 3 in the control limits by 3.09 gives ARL $= 1/.002 = 500$.

Now suppose that, at a particular time point, the process mean shifts to $\mu = \mu_0 + \Delta\sigma$. If we define Y to be the first i subsequent to the shift for which a sample generates an out-of-control signal, it is again true that ARL $= E(Y) = 1/p$, but now $p = 1 - \beta$. The accompanying table gives selected ARLs for a 3-sigma chart when $n = 4$. These results again show the chart's effectiveness in detecting large shifts but also its inability to quickly identify small shifts. When sampling is done rather infrequently, a great many items are likely to be produced before a small shift in μ is detected. The CUSUM procedures discussed in Section 16.5 were developed to address this deficiency.

Δ	.25	.50	.75	1.00	1.50	2.00	2.50	3.00
ARL when $n = 4$	156.25	43.86	14.97	6.30	2.00	1.19	1.07	1.0013

Supplemental Rules for $\overline{X}$ Charts

The inability of $\overline{X}$ charts with 3-sigma limits to quickly detect small shifts in the process mean has prompted investigators to develop procedures that provide improved behavior in this respect. One approach involves introducing additional conditions that cause an out-of-control signal to be generated. The following conditions were recommended by Western Electric (then a subsidiary of AT&T). An intervention to take corrective action is appropriate whenever one of these conditions is satisfied:

1. Two out of three successive points fall outside 2-sigma limits on the same side of the center line.
2. Four out of five successive points fall outside 1-sigma limits on the same side of the center line.
3. Eight successive points fall on the same side of the center line.

A quality control text should be consulted for a discussion of these and other supplemental rules.

Robust Control Charts

The presence of outliers in the sample data tends to reduce the sensitivity of control-charting procedures when parameters must be estimated. This is because the control limits are moved outward from the center line, making the identification of unusual points more difficult. We do *not* want the statistic whose values are plotted to be resistant to outliers, because that would mask any out-of-control signal. For example, plotting sample medians would be less effective than plotting $\overline{x}_1, \overline{x}_2, \ldots$ as is done on an $\overline{X}$ chart.

The article **"Robust Control Charts" by David M. Rocke (*Technometrics, 1989: 173–184*)** presents a study of procedures for which control limits are based on statistics resistant to the effects of outliers. Rocke recommends control limits calculated from the *interquartile range* (IQR), which is very similar to the fourth spread introduced in Chapter 1. In particular,

$$\text{IQR} = \begin{cases} (\text{2nd largest } x_i) - (\text{2nd smallest } x_i) & n = 4, 5, 6, 7 \\ (\text{3rd largest } x_i) - (\text{3rd smallest } x_i) & n = 8, 9, 10, 11 \end{cases}$$

For a random sample from a normal distribution, $E(\text{IQR}) = k_n\sigma$; the values of k_n are given in the accompanying table.

n	4	5	6	7	8
k_n	.596	.990	1.282	1.512	.942

The suggested control limits are

$$\text{LCL} = \overline{\overline{x}} - 3 \cdot \frac{\text{IQR}}{k_n\sqrt{n}} \qquad \text{UCL} = \overline{\overline{x}} + 3 \cdot \frac{\text{IQR}}{k_n\sqrt{n}}$$

The values of $\overline{x}_1, \overline{x}_2, \overline{x}_3, \ldots$ are plotted. Simulations reported in the article indicated that the performance of the chart with these limits is superior to that of the traditional $\overline{X}$ chart.

EXERCISES Section 16.2 (6–15)

6. In the case of known μ and σ, what control limits are necessary for the probability of a single point being outside the limits for an in-control process to be .005?

7. Consider a 3-sigma control chart with a center line at μ_0 and based on $n = 5$. Assuming normality, calculate the probability that a single point will fall outside the control limits when the actual process mean is

   a. $\mu_0 + .5\sigma$

   b. $\mu_0 - \sigma$

   c. $\mu_0 + 2\sigma$

8. The table below gives data on moisture content for specimens of a certain type of fabric. Determine control limits for a chart with center line at height 13.00 based on $\sigma = .600$, construct the control chart, and comment on its appearance.

9. Refer to the data given in Exercise 8, and construct a control chart with an estimated center line and limits based on using the sample standard deviations to estimate σ. Is there any evidence that the process is out of control?

10. When installing a bath faucet, it is important to properly fasten the threaded end of the faucet stem to the water-supply line. The threaded stem dimensions must meet product specifications, otherwise malfunction and leakage may occur. Authors of **"Improving the Process Capability of a Boring Operation by the Application of Statistical Techniques"** (*Intl. J. Sci. Engr. Research, Vol. 3, Issue 5, May 2012*) investigated the production process of a particular bath faucet manufactured in India. The article reported the threaded stem diameter (target value being 13 mm) of each faucet in 25 samples of size 4 as shown here:

Subgroup	x_1	x_2	x_3	x_4
1	13.02	12.95	12.92	12.99
2	13.02	13.10	12.96	12.96
3	13.04	13.08	13.05	13.10
4	13.04	12.96	12.96	12.97
5	12.96	12.97	12.90	13.05
6	12.90	12.88	13.00	13.05
7	12.97	12.96	12.96	12.99
8	13.04	13.02	13.05	12.97
9	13.05	13.10	12.98	12.96
10	12.96	13.00	12.96	12.99
11	12.90	13.05	12.98	12.88
12	12.96	12.98	12.97	13.02

(continued)

Data for Exercise 8

Sample No.		Moisture-Content Observations				$\bar{x}$	s	Range
1	12.2	12.1	13.3	13.0	13.0	12.72	.536	1.2
2	12.4	13.3	12.8	12.6	12.9	12.80	.339	.9
3	12.9	12.7	14.2	12.5	12.9	13.04	.669	1.7
4	13.2	13.0	13.0	12.6	13.9	13.14	.477	1.3
5	12.8	12.3	12.2	13.3	12.0	12.52	.526	1.3
6	13.9	13.4	13.1	12.4	13.2	13.20	.543	1.5
7	12.2	14.4	12.4	12.4	12.5	12.78	.912	2.2
8	12.6	12.8	13.5	13.9	13.1	13.18	.526	1.3
9	14.6	13.4	12.2	13.7	12.5	13.28	.963	2.4
10	12.8	12.3	12.6	13.2	12.8	12.74	.329	.9
11	12.6	13.1	12.7	13.2	12.3	12.78	.370	.9
12	13.5	12.3	12.8	13.1	12.9	12.92	.438	1.2
13	13.4	13.3	12.0	12.9	13.1	12.94	.559	1.4
14	13.5	12.4	13.0	13.6	13.4	13.18	.492	1.2
15	12.3	12.8	13.0	12.8	13.5	12.88	.432	1.2
16	12.6	13.4	12.1	13.2	13.3	12.92	.554	1.3
17	12.1	12.7	13.4	13.0	13.9	13.02	.683	1.8
18	13.0	12.8	13.0	13.3	13.1	13.04	.182	.5
19	12.4	13.2	13.0	14.0	13.1	13.14	.573	1.6
20	12.7	12.4	12.4	13.9	12.8	12.84	.619	1.5
21	12.6	12.8	12.7	13.4	13.0	12.90	.316	.8
22	12.7	13.4	12.1	13.2	13.3	12.94	.541	1.3

Subgroup	x_1	x_2	x_3	x_4
13	13.00	12.96	12.99	12.90
14	12.88	12.94	13.05	13.00
15	12.96	12.96	13.04	12.98
16	12.99	12.94	13.00	13.05
17	13.05	13.02	12.88	12.96
18	13.08	13.06	13.10	13.05
19	13.02	13.05	13.04	12.97
20	12.96	12.90	12.97	13.05
21	12.98	12.99	12.96	13.00
22	12.97	13.02	12.96	12.99
23	13.04	13.00	12.98	13.10
24	13.02	12.90	13.05	12.97
25	12.93	12.88	12.91	12.90

Calculate control limits based on using the sample ranges to estimate σ. Does the process appear to be in control?

11. The accompanying table gives sample means and standard deviations, each based on $n = 6$ observations of the refractive index of fiber-optic cable. Construct a control chart, and comment on its appearance. [*Hint:* $\Sigma \bar{x}_i = 2317.07$ and $\Sigma s_i = 30.34$.]

Day	$\bar{x}$	s	Day	$\bar{x}$	s
1	95.47	1.30	13	97.02	1.28
2	97.38	.88	14	95.55	1.14
3	96.85	1.43	15	96.29	1.37
4	96.64	1.59	16	96.80	1.40
5	96.87	1.52	17	96.01	1.58
6	96.52	1.27	18	95.39	.98
7	96.08	1.16	19	96.58	1.21
8	96.48	.79	20	96.43	.75
9	96.63	1.48	21	97.06	1.34
10	96.50	.80	22	98.34	1.60
11	97.22	1.42	23	96.42	1.22
12	96.55	1.65	24	95.99	1.18

12. Refer to Exercise 11. An assignable cause was found for the unusually high sample average refractive index on day 22. Recompute control limits after deleting the data from this day. What do you conclude?

13. Consider the control chart based on control limits $\mu_0 \pm 2.81 \, \sigma/\sqrt{n}$.

a. What is the ARL when the process is in control?
b. What is the ARL when $n = 4$ and the process mean has shifted to $\mu = \mu_0 + \sigma$?
c. How do the values of parts (a) and (b) compare to the corresponding values for a 3-sigma chart?

14. Three-dimensional (3D) printing is a manufacturing technology that allows the production of three-dimensional solid objects through a meticulous layering process performed by a 3D printer. 3D printing has rapidly become a time-saving and economical way to create a wide variety of products such as medical implants, furniture, tools, and even jewelry. The article **"Process Capability Analysis of Cost Effective Rapid Casting Solution Based on Three Dimensional Printing"** (*MIT Intl. J. Mech. Engr.*, 2012: 31–38) considered the production process of metal castings by using a 3D printer. Data was collected on 16 batches (each having two castings), where the outer diameter of each casting (in mm) was recorded. The target diameter of each casting was 60 mm. The resulting data is given here:

Batch	x_1	x_2
1	59.664	59.675
2	59.661	59.648
3	59.679	59.652
4	59.665	59.654
5	59.667	59.678
6	59.673	59.657
7	59.676	59.661
8	59.648	59.651
9	59.681	59.675
10	59.655	59.672
11	59.691	59.676
12	59.682	59.651
13	59.651	59.682
14	59.668	59.685
15	59.691	59.682
16	59.661	59.673

Apply the supplemental rules suggested in the text to the data. Are there any out-of-control signals?

15. Calculate control limits for the data of Exercise 8 using the robust procedure presented in this section.

16.3 Control Charts for Process Variation

The control charts discussed in the previous section were designed to control the location (equivalently, central tendency) of a process, with particular attention to the mean as a measure of location. It is equally important to ensure that a process is under control with respect to variation. In fact, most practitioners recommend that control be established on variation *prior to* constructing an $\overline{X}$ chart or any other chart for controlling location. In this section, we consider charts for variation based on the sample standard deviation S and also charts based on the sample range R. The former are generally preferred because the standard deviation gives a more efficient assessment of variation than does the range, but R charts were used first and tradition dies hard.

The S Chart

We again suppose that k independently selected samples are available, each one consisting of n observations on a normally distributed variable. Denote the sample standard deviations by $s_1, s_2, \ldots, s_k$, with $\bar{s} = \Sigma s_i / k$. The values $s_1, s_2, s_3, \ldots$ are plotted in sequence on an S chart. The center line of the chart will be at height $\bar{s}$, and the 3-sigma limits necessitate determining $3\sigma_S$ (just as 3-sigma limits of an $\overline{X}$ chart required $3\sigma_{\overline{x}} = 3\sigma/\sqrt{n}$, with σ then estimated from the data).

Recall that for any rv Y, $V(Y) = E(Y^2) - [E(Y)]^2$, and that a sample variance S^2 is an unbiased estimator of σ^2, that is, $E(S^2) = \sigma^2$. Thus

$$V(S) = E(S^2) - [E(S)]^2 = \sigma^2 - (a_n\sigma)^2 = \sigma^2(1 - a_n^2)$$

where values of a_n for $n = 3, \ldots, 8$ are tabulated in the previous section. The standard deviation of S is then

$$\sigma_S = \sqrt{V(S)} = \sigma\sqrt{1 - a_n^2}$$

It is natural to estimate σ using $s_1, \ldots, s_k$, as was done in the previous section namely, $\hat{\sigma} = \bar{s}/a_n$. Substituting $\hat{\sigma}$ for σ in the expression for σ_S gives the quantity used to calculate 3-sigma limits.

The 3-sigma control limits for an S control chart are

$$\text{LCL} = \bar{s} - 3\bar{s}\sqrt{1 - a_n^2}/a_n$$

$$\text{UCL} = \bar{s} - 3\bar{s}\sqrt{1 - a_n^2}/a_n$$

The expression for LCL will be negative if $n \leq 5$, in which case it is customary to use LCL $= 0$.

EXAMPLE 16.4   Table 16.2 displays observations on stress resistance of plastic sheets (the force, in psi, necessary to crack a sheet). There are $k = 22$ samples, obtained at equally spaced time points, and $n = 4$ observations in each sample. It is easily verified that $\Sigma s_i = 51.10$ and $\bar{s} = 2.32$, so the center of the S chart will be at 2.32 (though because $n = 4$, LCL $= 0$ and the center line will not be midway between the control limits). From the previous section, $a_4 = .921$, from which the UCL is

$$\text{UCL} = 2.32 + 3(2.32)(\sqrt{1 - (.921)^2})/.921 = 5.26$$

Table 16.2 Stress-Resistance Data for Example 16.4

Sample No.	Observations				SD	Range
1	29.7	29.0	28.8	30.2	.64	1.4
2	32.2	29.3	32.2	32.9	1.60	3.6
3	35.9	29.1	32.1	31.3	2.83	6.8
4	28.8	27.2	28.5	35.7	3.83	8.5
5	30.9	32.6	28.3	28.3	2.11	4.3

(continued)

Table 16.2 Stress-Resistance Data for Example 16.4 (continued)

Sample No.	Observations				SD	Range
6	30.6	34.3	34.8	26.3	3.94	8.5
7	32.3	27.7	30.9	27.8	2.30	4.6
8	32.0	27.9	31.0	30.8	1.76	4.1
9	24.2	27.5	28.5	31.1	2.85	6.9
10	33.7	24.4	34.3	31.0	4.53	9.9
11	35.3	33.2	31.4	28.0	3.09	7.3
12	28.1	34.0	31.0	30.8	2.41	5.9
13	28.7	28.9	25.8	29.7	1.71	3.9
14	29.0	33.0	30.2	30.1	1.71	4.0
15	33.5	32.6	33.6	29.2	2.07	4.4
16	26.9	27.3	32.1	28.5	2.37	5.2
17	30.4	29.6	31.0	33.8	1.83	4.2
18	29.0	28.9	31.8	26.7	2.09	5.1
19	33.8	30.9	31.7	28.2	2.32	5.6
20	29.7	27.9	29.1	30.1	.96	2.2
21	27.9	27.7	30.2	32.9	2.43	5.2
22	30.0	31.4	27.7	28.1	1.72	3.7

The resulting control chart is shown in Figure 16.3. All plotted points are well within the control limits, suggesting stable process behavior with respect to variation.

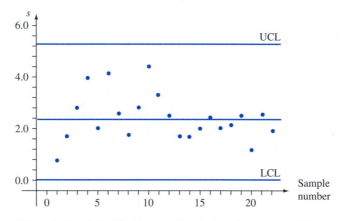

Figure 16.3 *S* chart for stress-resistance data for Example 16.4 ■

The *R* Chart

Let $r_1, r_2, \ldots r_k$ denote the k sample ranges and $\bar{r} = \Sigma r_i / k$. The center line of an R chart will be at height $\bar{r}$. Determination of the control limits requires σ_R, where R denotes the range (prior to making observations—as a random variable) of a random sample of size n from a normal distribution with mean value μ and standard deviation σ. Because

$$R = \max(X_1, \ldots, X_n) - \min(X_1, \ldots, X_n)$$
$$= \sigma\{\max(Z_1, \ldots, Z_n) - \min(Z_1, \ldots, Z_n)\}$$

where $Z_i = (X_i - \mu)/\sigma$, and the Z_i's are standard normal rv's, it follows that

$$\sigma_R = \sigma \cdot \left(\begin{array}{l}\text{standard deviation of the range of random sample} \\ \text{of size } n \text{ from a standard normal distribution}\end{array}\right)$$

$$= \sigma \cdot c_n$$

The values of c_n for $n = 3, \ldots, 8$ appear in the accompanying table.

n	3	4	5	6	7	8
c_n	.888	.880	.864	.848	.833	.820

It is customary to estimate σ by $\hat{\sigma} = \bar{r}/b_n$ as discussed in the previous section. This gives $\hat{\sigma}_R = c_n \bar{r}/b_n$ as the estimated standard deviation of R.

> The 3-sigma limits for an R chart are
>
> $$\text{LCL} = \bar{r} - 3c_n\bar{r}/b_n$$
>
> $$\text{UCL} = \bar{r} + 3c_n\bar{r}/b_n$$
>
> The expression for LCL will be negative if $n \leq 6$, in which case LCL = 0 should be used.

EXAMPLE 16.5 In tissue engineering, cells are seeded onto a scaffold that then guides the growth of new cells. The article **"On the Process Capability of the Solid Free-Form Fabrication: A Case Study of Scaffold Moulds for Tissue Engineering"** (*J. of Engr. in Med.,* **2008: 377–392**) used various quality control methods to study a method of producing such scaffolds. An unusual feature is that instead of sub-groups being observed over time, each subgroup resulted from a different design dimension (μm). Table 16.3 contains data from Table 2 of the cited article on the deviation from target in the perpendicular orientation (these deviations are indeed all positive—the printed beams exhibit larger dimensions than those designed).

Table 16.3 Deviation-from-Target Data for Example 16.5

des dim				mean	range	st dev
200	12	17	6	11.7	11	5.51
250	6	9	17	10.7	11	5.69
300	5	9	15	9.7	10	5.03
350	19	6	11	12.0	13	6.56
400	9	14	9	10.7	5	2.89
450	9	15	8	10.7	7	3.79
500	8	11	12	10.3	4	2.08
550	4	14	11	9.7	10	5.13
600	11	14	7	10.7	7	3.51
650	13	9	9	10.3	4	2.31
700	10	14	8	10.7	6	3.06
750	8	9	4	7.0	5	2.65
800	14	7	9	10.0	7	3.61
850	7	9	12	9.3	5	2.52
900	14	5	8	9.0	9	4.58
950	10	12	10	10.7	2	1.15
1000	7	11	15	11.0	8	4.00

Table 16.3 yields $\Sigma r_i = 124$, from which $\bar{r} = 7.29$. Since $n = 3$, LCL $= 0$. With $b_3 = 1.693$ and $c_3 = .888$,

$$\text{UCL} = 7.29 + 3 \cdot (.888)(7.29)/1.693 = 18.76$$

Figure 16.4 shows both an R chart and an $\bar{X}$ chart from the Minitab software package (the cited article also included these charts). All points are within the appropriate control limits, indicating an in-control process for both location and variation.

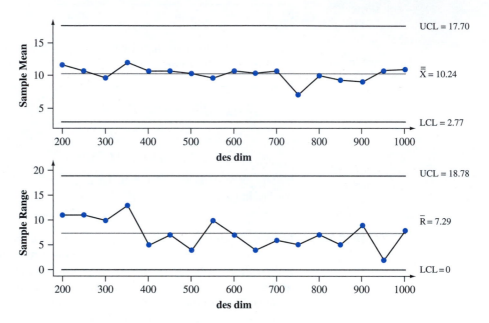

Figure 16.4 Control charts for the deviation-from-target data of Example 16.5 ■

Charts Based on Probability Limits

Consider an $\bar{X}$ chart based on the in-control (target) value μ_0 and known σ. When the variable of interest is normally distributed and the process is in control,

$$P(\bar{X}_i > \mu_0 + 3\sigma/\sqrt{n}) = .0013 = P(\bar{X}_i < \mu_0 - 3\sigma/\sqrt{n})$$

That is, the probability that a point on the chart falls above the UCL is .0013, as is the probability that the point falls below the LCL (using 3.09 in place of 3 gives .001 for each probability). When control limits are based on estimates of μ and σ, these probabilities will be approximately correct provided that n is not too small and k is at least 20.

By contrast, it is *not* the case for a 3-sigma S chart that $P(S_i > \text{UCL}) = P(S_i < \text{LCL}) = .0013$, nor is it true for a 3-sigma R chart that $P(R_i > \text{UCL}) = P(R_i < \text{LCL}) = .0013$. This is because neither S nor R has a normal distribution even when the population distribution is normal. Instead, both S and R have skewed distributions. The best that can be said for 3-sigma S and R charts is that an in-control process is quite unlikely to yield a point at any particular time that is outside the control limits. Some authors have advocated the use of control limits for which the "exceedance probability" for each limit is approximately .001. The book *Statistical Methods for Quality Improvement* (see the chapter bibliography) contains more information on this topic.

EXERCISES Section 16.3 (16–20)

16. A manufacturer of dustless chalk instituted a quality control program to monitor chalk density. The sample standard deviations of densities for 24 different subgroups, each consisting of $n = 8$ chalk specimens, were as follows:

.204 .315 .096 .184 .230 .212 .322 .287
.145 .211 .053 .145 .272 .351 .159 .214
.388 .187 .150 .229 .276 .118 .091 .056

Calculate limits for an S chart, construct the chart, and check for out-of-control points. If there is an out-of-control point, delete it and repeat the process.

17. Subgroups of power supply units are selected once each hour from an assembly line, and the high-voltage output of each unit is determined.
    a. Suppose the sum of the resulting sample ranges for 30 subgroups, each consisting of four units, is 85.2. Calculate control limits for an R chart.
    b. Repeat part (a) if each subgroup consists of eight units and the sum is 106.2.

18. The following data on the deviation from target in the parallel orientation is taken from Table 1 of the article cited in Example 16.5. Sometimes a transformation of the data is appropriate, either because of nonnormality or because subgroup variation changes systematically with the subgroup mean. The authors of the cited article suggested a square root transformation for this data (the *family of Box-Cox transformations* is $y = x^\lambda$, so $\lambda = .5$ here; Minitab will identify the best value of λ). Transform the data as suggested, calculate control limits for $\bar{X}$, R, and S charts, and check for the presence of any out-of-control signals.

des dim	observations		
200	14	31	12
250	22	13	9
300	12	22	16
350	11	28	1

des dim	observations		
400	15	12	36
450	6	31	14
500	13	24	9
550	21	18	16
600	6	16	20
650	8	17	23
700	3	26	17
750	17	12	22
800	41	17	3
850	18	11	21
900	9	15	22
950	25	4	17
1000	8	23	15

19. Calculate control limits for an S chart from the refractive index data of Exercise 11. Does the process appear to be in control with respect to variability? Why or why not?

20. When S^2 is the sample variance of a normal random sample, $(n - 1)S^2/\sigma^2$ has a chi-squared distribution with $n - 1$ df, so

$$P\left(\chi^2_{.999,n-1} < \frac{(n-1)S^2}{\sigma^2} < \chi^2_{.001,n-1}\right) = .998$$

from which

$$P\left(\frac{\sigma^2\chi^2_{.999,n-1}}{n-1} < S^2 < \frac{\sigma^2\chi^2_{.001,n-1}}{n-1}\right) = .998$$

This suggests that an alternative chart for controlling process variation involves plotting the sample variances and using the control limits

$$LCL = \bar{s^2}\chi^2_{.999,n-1}/(n-1)$$
$$UCL = \bar{s^2}\chi^2_{.001,n-1}/(n-1)$$

Construct the corresponding chart for the data of Exercise 11. [*Hint:* The lower- and upper-tailed chi-squared critical values for 5 df are .210 and 20.515, respectively.]

16.4 Control Charts for Attributes

The term *attribute data* is used in the quality control literature to describe two situations:

1. Each item produced is either defective or nondefective (conforms to specifications or does not).

2. A single item may have one or more defects, and the number of defects is determined.

In the former case, a control chart is based on the binomial distribution; in the latter case, the Poisson distribution is the basis for a chart.

The *p* Chart for Fraction Defective

Suppose that when a process is in control, the probability that any particular item is defective is p (equivalently, p is the long-run proportion of defective items for an in-control process) and that different items are independent of one another with respect to their conditions. Consider a sample of n items obtained at a particular time, and let X be the number of defectives and $\hat{p} = X/n$. Because X has a binomial distribution, $E(X) = np$ and $V(X) = np(1 - p)$, so

$$E(\hat{p}) = p \qquad V(\hat{p}) = \frac{p(1 - p)}{n}$$

Also, if $np \geq 10$ and $n(1 - p) \geq 10$, $\hat{p}$ has approximately a normal distribution.

In the case of known p (or a chart based on target value), the control limits are

$$\text{LCL} = p - 3\sqrt{\frac{p(1 - p)}{n}} \qquad \text{UCL} = p + 3\sqrt{\frac{p(1 - p)}{n}}$$

If each sample consists of n items, the number of defective items in the ith sample is x_i, and $\hat{p}_i = x_i/n$, then $\hat{p}_1, \hat{p}_2, \hat{p}_3, \ldots$ are plotted on the control chart.

Usually the value of p must be estimated from the data. Suppose that k samples from what is believed to be an in-control process are available, and let

$$\bar{p} = \frac{\sum_{i=1}^{k} \hat{p}_i}{k}$$

The estimate $\bar{p}$ is then used in place of p in the aforementioned control limits.

The p chart for the fraction of defective items has its center line at height $\bar{p}$ and control limits

$$\text{LCL} = \bar{p} - 3\sqrt{\frac{\bar{p}(1 - \bar{p})}{n}}$$

$$\text{UCL} = \bar{p} + 3\sqrt{\frac{\bar{p}(1 - \bar{p})}{n}}$$

If LCL is negative, it is replaced by 0.

EXAMPLE 16.6 A sample of 100 cups from a particular dinnerware pattern was selected on each of 25 successive days, and each was examined for defects. The resulting numbers of unacceptable cups and corresponding sample proportions are as follows:

Day (i)	1	2	3	4	5	6	7	8	9	10	11	12	13
x_i	7	4	3	6	4	9	6	7	5	3	7	8	4
$\hat{p}_i$	.07	.04	.03	.06	.04	.09	.06	.07	.05	.03	.07	.08	.04

Day (i)	14	15	16	17	18	19	20	21	22	23	24	25
x_i	6	2	9	7	6	7	11	6	7	4	8	6
$\hat{p}_i$	.06	.02	.09	.07	.06	.07	.11	.06	.07	.04	.08	.06

Assuming that the process was in control during this period, let's establish control limits and construct a p chart. Since $\Sigma \hat{p}_i = 1.52$, $\bar{p} = 1.52/25 = .0608$ and

$$\text{LCL} = .0608 - 3\sqrt{(.0608)(.9392)/100} = .0608 - .0717 = -.0109$$

$$\text{UCL} = .0608 + 3\sqrt{(.0608)(.9392)/100} = .0608 + .0717 = .1325$$

The LCL is therefore set at 0. The chart pictured in Figure 16.5 shows that all points are within the control limits. This is consistent with an in-control process.

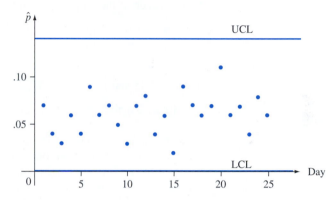

Figure 16.5 Control chart for fraction-defective data of Example 16.6

The c Chart for Number of Defectives

We now consider situations in which the observation at each time point is the number of defects in a unit of some sort. The unit may consist of a single item (e.g., one automobile) or a group of items (e.g., blemishes on a set of four tires). In the second case, the group size is assumed to be the same at each time point.

The control chart for number of defectives is based on the Poisson probability distribution. Recall that if Y is a Poisson random variable with parameter μ, then

$$E(Y) = \mu \qquad V(Y) = \mu \qquad \sigma_Y = \sqrt{\mu}$$

Also, Y has approximately a normal distribution when μ is large ($\mu \geq 10$ will suffice for most purposes). Furthermore, if $Y_1, Y_2, \ldots, Y_n$ are independent Poisson variables with parameters $\mu_1, \mu_2, \ldots, \mu_n$, it can be shown that $Y_1 + \cdots + Y_n$ has a Poisson distribution with parameter $\mu_1 + \cdots + \mu_n$. In particular, if $\mu_1 = \cdots = \mu_n = \mu$ (the distribution of the number of defects per item is the same for each item), then the Poisson parameter is $n\mu$.

Let μ denote the Poisson parameter for the number of defects in a unit (it is the expected number of defects per unit). In the case of known μ (or a chart based on a target value),

$$\text{LCL} = \mu - 3\sqrt{\mu} \qquad \text{UCL} = \mu + 3\sqrt{\mu}$$

With x_i denoting the total number of defects in the ith unit ($i = 1, 2, 3, \ldots$), then points at heights $x_1, x_2, x_3, \ldots$ are plotted on the chart. Usually the value of μ must be estimated from the data. Since $E(X_i) = \mu$, it is natural to use the estimate $\mu = \bar{x}$ (based on $x_1, x_2, \ldots, x_k$).

> The c chart for the number of defectives in a unit has center line at $\bar{x}$ and
>
> $$\text{LCL} = \bar{x} - 3\sqrt{\bar{x}}$$
>
> $$\text{UCL} = \bar{x} + 3\sqrt{\bar{x}}$$
>
> If LCL is negative, it is replaced by 0.

EXAMPLE 16.7   A company manufactures metal panels that are baked after first being coated with a slurry of powdered ceramic. Flaws sometimes appear in the finish of these panels, and the company wishes to establish a control chart for the number of flaws. The number of flaws in each of the 24 panels sampled at regular time intervals are as follows:

7	10	9	12	13	6	13	7	5	11	8	10
13	9	21	10	6	8	3	12	7	11	14	10

with $\Sigma x_i = 235$ and $\hat{\mu} = \bar{x} = 235/24 = 9.79$. The control limits are

$$\text{LCL} = 9.79 - 3\sqrt{9.79} = .40 \qquad \text{UCL} = 9.79 + 3\sqrt{9.79} = 19.18$$

The control chart is in Figure 16.6. The point corresponding to the fifteenth panel lies above the UCL. Upon investigation, the slurry used on that panel was discovered to be of unusually low viscosity (an assignable cause). Eliminating that observation gives $\bar{x} = 214/23 = 9.30$ and new control limits

$$\text{LCL} = 9.30 - 3\sqrt{9.30} = .15 \qquad \text{UCL} = 9.30 + 3\sqrt{9.30} = 18.45$$

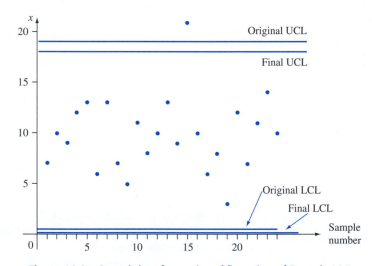

Figure 16.6   Control chart for number of flaws data of Example 16.7

The remaining 23 observations all lie between these limits, indicating an in-control process. ∎

Control Charts Based on Transformed Data

The use of 3-sigma control limits is presumed to result in $P(\text{statistic} < \text{LCL}) \approx P(\text{statistic} > \text{UCL}) \approx .0013$ when the process is in control. However, when p is small, the normal approximation to the distribution of $\hat{p} = X/n$ will often not be very accurate in the extreme tails. Table 16.3 gives evidence of this behavior for

Table 16.3 In-Control Probabilities for a *p* Chart

p	n	$P(\hat{p} < \text{LCL})$	$P(\hat{p} > \text{UCL})$	$P(\text{out-of-control point})$
.10	100	.00003	.00198	.00201
.10	200	.00048	.00299	.00347
.10	400	.00044	.00171	.00215
.05	200	.00004	.00266	.00270
.05	400	.00020	.00207	.00227
.05	600	.00031	.00189	.00220
.02	600	.00007	.00275	.00282
.02	800	.00036	.00374	.00410
.02	1000	.00023	.00243	.00266

selected values of *p* and *n* (the value of *p* is used to calculate the control limits). In many cases, the probability that a single point falls outside the control limits is very different from the nominal probability of .0026.

This problem can be remedied by applying a transformation to the data. Let $h(X)$ denote a function applied to transform the binomial variable *X*. Then $h(\cdot)$ should be chosen so that $h(X)$ has approximately a normal distribution *and* this approximation is accurate in the tails. A recommended transformation is based on the arcsin (i.e., $\sin^{-1}$) function:

$$Y = h(X) = \sin^{-1}(\sqrt{X/n})$$

Then *Y* is approximately normal with mean value $\sin^{-1}(\sqrt{p})$ and variance $1/(4n)$; note that the variance is independent of *p*. Let $y_i = \sin^{-1}(\sqrt{x_i/n})$. Then points on the control chart are at heights $y_1, y_2, \ldots$. For known *n*, the control limits are

$$\text{LCL} = \sin^{-1}(\sqrt{p}) - 3\sqrt{1/(4n)} \qquad \text{UCL} = \sin^{-1}(\sqrt{p}) + 3\sqrt{1/(4n)}$$

When *p* is not known, $\sin^{-1}(\sqrt{p})$ is replaced by $\bar{y}$.

Similar comments apply to the Poisson distribution when μ is small. The suggested transformation is $Y = h(X) = 2\sqrt{X}$, which has mean value $2\sqrt{\mu}$ and variance 1. Resulting control limits are $2\sqrt{\mu} \pm 3$ when μ is known and $\bar{y} \pm 3$ otherwise. The book *Statistical Methods for Quality Improvement* listed in the chapter bibliography discusses these issues in greater detail.

EXERCISES Section 16.4 (21–28)

21. On each of the previous 25 days, 100 electronic devices of a certain type were randomly selected and subjected to a severe heat stress test. The total number of items that failed to pass the test was 578.
   a. Determine control limits for a 3-sigma *p* chart.
   b. The highest number of failed items on a given day was 39, and the lowest number was 13. Does either of these correspond to an out-of-control point? Explain.

22. A sample of 200 ROM computer chips was selected on each of 30 consecutive days, and the number of nonconforming chips on each day was as follows: 10, 18, 24, 17,

37, 19, 7, 25, 11, 24, 29, 15, 16, 21, 18, 17, 15, 22, 12, 20, 17, 18, 12, 24, 30, 16, 11, 20, 14, 28. Construct a *p* chart and examine it for any out-of-control points.

23. When $n = 150$, what is the smallest value of $\bar{p}$ for which the LCL in a *p* chart is positive?

24. Refer to the data of Exercise 22, and construct a control chart using the $\sin^{-1}$ transformation as suggested in the text.

25. The accompanying observations are numbers of defects in 25 1-square-yard specimens of woven fabric of a certain type: 3, 7, 5, 3, 4, 2, 8, 4, 3, 3, 6, 7, 2, 3, 2, 4, 7, 3, 2, 4, 4, 1, 5, 4, 6. Construct a c chart for the number of defects.

26. For what $\bar{x}$ values will the LCL in a c chart be negative?

27. In some situations, the sizes of sampled specimens vary, and larger specimens are expected to have more defects than smaller ones. For example, sizes of fabric samples inspected for flaws might vary over time. Alternatively, the number of items inspected might change with time. Let

$$u_i = \frac{\text{the number of defects observed at time } i}{\text{size of entity inspected at time } i}$$

$$= \frac{x_i}{g_i}$$

where "size" might refer to area, length, volume, or simply the number of items inspected. Then a **u chart** plots $u_1, u_2, \ldots$, has center line $\bar{u}$, and the control limits for the ith observations are $\bar{u} \pm 3\sqrt{\bar{u}/g_i}$.

Painted panels were examined in time sequence, and for each one, the number of blemishes in a specified sampling region was determined. The surface area (ft^2) of the region examined varied from panel to panel. Results are given below. Construct a u chart.

Panel	Area Examined	No. of Blemishes
1	.8	3
2	.6	2
3	.8	3
4	.8	2
5	1.0	5
6	1.0	5
7	.8	10
8	1.0	12
9	.6	4
10	.6	2
11	.6	1
12	.8	3
13	.8	5
14	1.0	4
15	1.0	6
16	1.0	12
17	.8	3
18	.6	3
19	.6	5
20	.6	1

28. Construct a control chart for the data of Exercise 25 by using the transformation suggested in the text.

16.5   CUSUM Procedures

A defect of the traditional $\overline{X}$ chart is its inability to detect a relatively small change in a process mean. This is largely a consequence of the fact that whether a process is judged out of control at a particular time depends only on the sample at that time, and not on the past history of the process. **Cumulative sum** (**CUSUM**) control charts and procedures have been designed to remedy this defect.

There are two equivalent versions of a CUSUM procedure for a process mean, one graphical and the other computational. The computational version is used almost exclusively in practice, but the logic behind the procedure is most easily grasped by first considering the graphical form.

The V-Mask

Let μ_0 denote a target value or goal for the process mean, and define *cumulative sums* by

$$S_1 = \bar{x}_1 - \mu_0$$

$$S_2 = (\bar{x}_1 - \mu_0) + (\bar{x}_2 - \mu_0) = \sum_{i=1}^{2} (\bar{x}_i - \mu_0)$$

$$\vdots$$

$$S_l = (\bar{x}_1 - \mu_0) + \cdots + (\bar{x}_l - \mu_0) = \sum_{i=1}^{l} (\bar{x}_i - \mu_0)$$

(in the absence of a target value, $\bar{\bar{x}}$ is used in place of μ_0). These cumulative sums are plotted over time. That is, at time l, we plot a point at height S_l. At the current time point r, the plotted points are $(1, S_1), (2, S_2), (3, S_3), \ldots, (r, S_r)$.

Now a V-shaped "mask" is superimposed on the plot, as shown in Figure 16.7. The point 0, which lies a distance d behind the point at which the two arms of the mask intersect, is positioned at the current CUSUM point (r, S_r). At time r, the process is judged out of control if any of the plotted points lies outside the V-mask— either above the upper arm or below the lower arm. When the process is in control, the $\bar{x}_i$'s will vary around the target value μ_0, so successive S_i's should vary around 0. Suppose, however, that at a certain time, the process mean shifts to a value larger than the target. From that point on, differences $\bar{x}_i - \mu_0$ will tend to be positive, so that successive S_i's will increase and plotted points will drift upward. If a shift has occurred prior to the current time point r, there is a good chance that (r, S_r) will be substantially higher than some other points in the plot, in which case these other points will be below the lower arm of the mask. Similarly, a shift to a value smaller than the target will subsequently result in points above the upper arm of the mask.

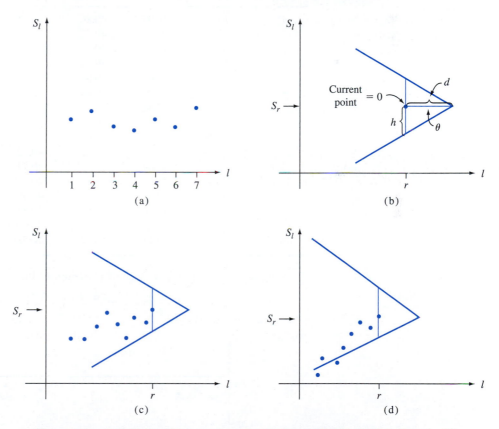

Figure 16.7 CUSUM plots: (a) successive points (l, S_l) in a CUSUM plot; (b) a V-mask with $0 = (r, S_r)$; (c) an in-control process; (d) an out-of-control process

Any particular V-mask is determined by specifying the "lead distance" d and "half-angle" θ, or, equivalently, by specifying d and the length h of the vertical line segment from 0 to the lower (or to the upper) arm of the mask. One method for deciding which mask to use involves specifying the size of a shift in the process mean that is of particular concern to an investigator. Then the parameters of the mask are chosen to give desired values of α and β, the false-alarm probability and the probability of not detecting the specified shift, respectively. An alternative method involves selecting the mask that yields specified values of the ARL (average run length) both for an in-control process and for a process in which the mean has shifted by a designated amount. After developing the computational form of the CUSUM procedure, we will illustrate the second method of construction.

EXAMPLE 16.8   A wood products company manufactures charcoal briquettes for barbecues. It packages these briquettes in bags of various sizes, the largest of which is supposed to contain 40 lbs. Table 16.4 displays the weights of bags from 16 different samples, each of size $n = 4$. The first 10 of these were drawn from a normal distribution with $\mu = \mu_0 = 40$ and $\sigma = .5$. Starting with the eleventh sample, the mean has shifted upward to $\mu = 40.3$.

Table 16.4   Observations, $\bar{x}$'s, and Cumulative Sums for Example 16.8

Sample Number		Observations			$\bar{x}$	$\Sigma(\bar{x}_i - 40)$
1	40.77	39.95	40.86	39.21	40.20	.20
2	38.94	39.70	40.37	39.88	39.72	−.08
3	40.43	40.27	40.91	40.05	40.42	.34
4	39.55	40.10	39.39	40.89	39.98	.32
5	41.01	39.07	39.85	40.32	40.06	.38
6	39.06	39.90	39.84	40.22	39.76	.14
7	39.63	39.42	40.04	39.50	39.65	−.21
8	41.05	40.74	40.43	39.40	40.41	.20
9	40.28	40.89	39.61	40.48	40.32	.52
10	39.28	40.49	38.88	40.72	39.84	.36
11	40.57	40.04	40.85	40.51	40.49	.85
12	39.90	40.67	40.51	40.53	40.40	1.25
13	40.70	40.54	40.73	40.45	40.61	1.86
14	39.58	40.90	39.62	39.83	39.98	1.84
15	40.16	40.69	40.37	39.69	40.23	2.07
16	40.46	40.21	40.09	40.58	40.34	2.41

Figure 16.8 displays an $\overline{X}$ chart with control limits

$$\mu \pm 3\sigma_{\overline{X}} = 40 \pm 3 \cdot (.5/\sqrt{4}) = 40 \pm .75$$

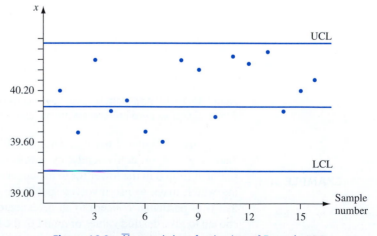

Figure 16.8 $\overline{X}$ control chart for the data of Example 16.8

No point on the chart lies outside the control limits. This chart suggests a stable process for which the mean has remained on target.

Figure 16.9 shows CUSUM plots with a particular V-mask superimposed. The plot in Figure 16.9(a) is for current time $r = 12$. All points in this plot lie inside the arms of the mask. However, the plot for $r = 13$ displayed in Figure 16.9(b) gives an out-of-control signal. The point falling below the lower arm of the mask suggests an increase in the value of the process mean. The mask at $r = 16$ is even more emphatic in its out-of-control message. This is in marked contrast to the $\overline{X}$ chart.

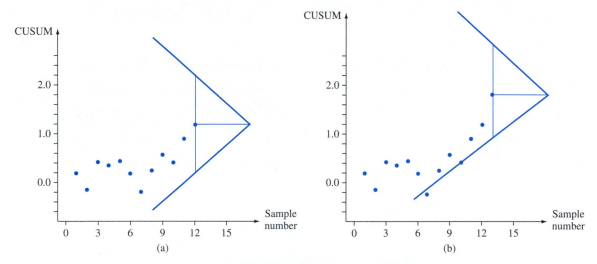

Figure 16.9   CUSUM plots and V-masks for data of Example 16.8: (a) V-mask at time $r = 12$, process in control; (b) V-mask at time $r = 13$, out-of-control signal

A Computational Version

The following computational form of the CUSUM procedure is equivalent to the previous graphical description.

Let $d_0 = e_0 = 0$, and calculate $d_1, d_2, d_3, \ldots$ and $e_1, e_2, e_3, \ldots$ recursively, using the relationships

$$d_l = \max[0, d_{l-1} + (\overline{x}_l - (\mu_0 + k))]$$
$$e_l = \max[0, e_{l-1} - (\overline{x}_l - (\mu_0 - k))] \qquad (l = 1, 2, 3, \ldots)$$

Here the symbol k denotes the slope of the lower arm of the V-mask, and its value is customarily taken as $\Delta/2$ (where Δ is the size of a shift in μ on which attention is focused).

If at current time r either $d_r > h$ or $e_r > h$, the process is judged to be out of control. The first inequality suggests that the process mean has shifted to a value greater than the target, whereas $e_r > h$ indicates a shift to a smaller value.

EXAMPLE 16.9   Reconsider the charcoal briquette data displayed in Table 16.4 of Example 16.8. The target value is $\mu_0 = 40$, and the size of a shift to be quickly detected is $\Delta = .3$. Thus

$$k = \frac{\Delta}{2} = .15 \qquad \mu_0 + k = 40.15 \qquad \mu_0 - k = 39.85$$

so

$$d_I = \max[0, d_{I-1} + (\bar{x}_I - 40.15)]$$

$$e_I = \max[0, e_{I-1} - (\bar{x}_I - 39.85)]$$

Calculations of the first few d_I's proceeds as follows:

$$d_0 = 0$$

$$d_1 = \max[0, d_0 + (\bar{x}_1 - 40.15)]$$

$$= \max[0, 0 + (40.20 - 40.15)]$$

$$= .05$$

$$d_2 = \max[0, d_1 + (\bar{x}_2 - 40.15)]$$

$$= \max[0, .05 + (39.72 - 40.15)]$$

$$= 0$$

$$d_3 = \max[0, d_2 + (\bar{x}_3 - 40.15)]$$

$$= \max[0, 0 + (40.42 - 40.15)]$$

$$= .27$$

The remaining calculations are summarized in Table 16.5.

Table 16.5 CUSUM Calculations for Example 16.9

Sample Number	$\bar{x}_I$	$\bar{x}_I - 40.15$	d_I	$\bar{x}_I - 39.85$	e_I
1	40.20	.05	.05	.35	0
2	39.72	−.43	0	−.13	.13
3	40.42	.27	.27	.57	0
4	39.98	−.17	.10	.13	0
5	40.06	−.09	.01	.21	0
6	39.76	−.39	0	−.09	.09
7	39.65	−.50	0	−.20	.29
8	40.41	.26	.26	.56	0
9	40.32	.17	.43	.47	0
10	39.84	−.31	.12	−.01	.01
11	40.49	.34	.46	.64	0
12	40.40	.25	.71	.55	0
13	40.61	.46	1.17	.76	0
14	39.98	−.17	1.00	.13	0
15	40.23	.08	1.08	.38	0
16	40.34	.19	1.27	.49	0

The value $h = .95$ gives a CUSUM procedure with desirable properties—false alarms (incorrect out-of-control signals) rarely occur, yet a shift of $\Delta = .3$ will usually be detected rather quickly. With this value of h, the first out-of-control signal comes after the 13th sample is available. Since $d_{13} = 1.17 > .95$, it appears that the mean has shifted to a value larger than the target. This is the same message as the one given by the V-mask in Figure 16.9(b). ■

To demonstrate equivalence, again let r denote the current time point, so that $\bar{x}_1, \bar{x}_2, \ldots, \bar{x}_r$ are available. Figure 16.10 displays a V-mask with the point labeled 0 at (r, S_r). The slope of the lower arm, which we denote by k, is h/d. Thus the points on the lower arm above $r, r-1, r-2, \ldots$ are at heights $S_r - h, S_r - h - k$, $S_r - h - 2k$, and so on.

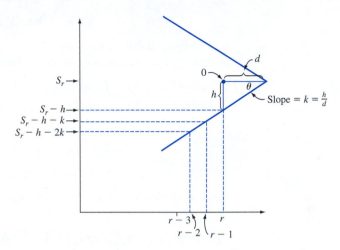

Figure 16.10   A V-mask with slope of lower arm = k

The process is in control if all points are on or between the arms of the mask. We wish to describe this condition algebraically. To do so, let

$$T_l = \sum_{i=1}^{l} [\bar{x}_i - (\mu_0 + k)] \qquad l = 1, 2, 3, \ldots, r$$

The conditions under which all points are on or above the lower arm are

$$S_r - h \leq S_r \qquad \text{(trivially satisfied) i.e., } S_r \leq S_r + h$$
$$S_r - h - k \leq S_{r-1} \qquad\qquad \text{i.e., } S_r \leq S_{r-1} + h + k$$
$$S_r - h - 2k \leq S_{r-2} \qquad\qquad \text{i.e., } S_r \leq S_{r-2} + h + 2k$$
$$\vdots \qquad\qquad\qquad\qquad \vdots$$

Now subtract rk from both sides of each inequality to obtain

$$S_r - rk \leq S_r - rk + h \qquad\qquad \text{i.e., } T_r \leq T_r + h$$
$$S_r - rk \leq S_{r-1} - (r-1)k + h \qquad\qquad \text{i.e., } T_r \leq T_{r-1} + h$$
$$S_r - rk \leq S_{r-2} - (r-2)k + h \qquad\qquad \text{i.e., } T_r \leq T_{r-2} + h$$
$$\vdots \qquad\qquad\qquad\qquad \vdots$$

Thus all plotted points lie on or above the lower arm if and only if (iff) $T_r - T_r \leq h$, $T_r - T_{r-1} \leq h$, $T_r - T_{r-2} \leq h$, and so on. This is equivalent to

$$T_r - \min(T_1, T_2, \ldots, T_r) \leq h$$

In a similar manner, if we let

$$V_r = \sum_{i=1}^{r} [\bar{x}_i - (\mu_0 - k)] = S_r + rk$$

it can be shown that all points lie on or below the upper arm iff

$$\max(V_1, \ldots, V_r) - V_r \leq h$$

If we now let

$$d_r = T_r - \min(T_1, \ldots, T_r)$$

$$e_r = \max(V_1, \ldots, V_r) - V_r$$

it is easily seen that $d_1, d_2, \ldots$ and $e_1, e_2, \ldots$ can be calculated recursively as illustrated previously. For example, the expression for d_r follows from consideration of two cases:

1. $\min(T_1, \ldots, T_r) = T_r$, whence $d_r = 0$

2. $\min(T_1, \ldots, T_r) = \min(T_1, \ldots, T_{r-1})$, so that

$$
\begin{aligned}
d_r &= T_r - \min(T_1, \ldots, T_{r-1}) \\
&= \bar{x}_r - (\mu_0 + k) + T_{r-1} - \min(T_1, \ldots, T_{r-1}) \\
&= \bar{x}_r - (\mu_0 + k) + d_{r-1}
\end{aligned}
$$

Since d_r cannot be negative, it is the larger of these two quantities.

Designing a CUSUM Procedure

Let Δ denote the size of a shift in μ that is to be quickly detected using a CUSUM procedure.* It is common practice to let $k = \Delta/2$. Now suppose a quality control practitioner specifies desired values of two average run lengths:

1. ARL when the process is in control ($\mu = \mu_0$)

2. ARL when the process is out of control because the mean has shifted by Δ ($\mu = \mu_0 + \Delta$ or $\mu = \mu_0 - \Delta$)

A chart developed by Kenneth Kemp (**"The Use of Cumulative Sums for Sampling Inspection Schemes," *Applied Statistics*, 1962: 23**), called a *nomogram*, can then be used to determine values of h and n that achieve the specified ARLs.[†] This chart is shown as Figure 16.11. The method for using the chart is described in the accompanying box. Either the value of σ must be known or an estimate is used in its place.

Using the Kemp Nomogram

1. Locate the desired ARLs on the in-control and out-of-control scales. Connect these two points with a line.

2. Note where the line crosses the k' scale, and solve for n using the equation

$$k' = \frac{\Delta/2}{\sigma/\sqrt{n}}$$

Then round n up to the nearest integer.

3. Connect the point on the k' scale with the point on the in-control ARL scale using a second line, and note where this line crosses the h' scale. Then $h = (\sigma/\sqrt{n}) \cdot h'$.

---

* This contrasts with previous notation, where Δ represented the number of standard deviations by which μ changed.

† The word *nomogram* is not specific to this chart; nomograms are used for many other purposes.

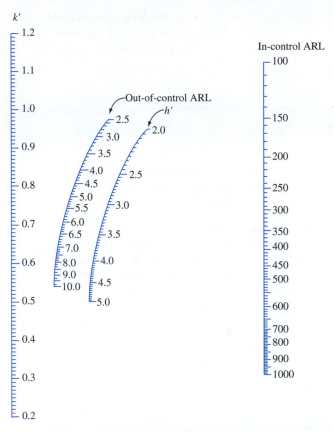

Figure 16.11 The Kemp nomogram*

The value $h = .95$ was used in Example 16.9. In that situation, it follows that the in-control ARL is 500 and the out-of-control ARL (for $\Delta = .3$) is 7.

EXAMPLE 16.10 The target value for the diameter of the interior core of a hydraulic pump is 2.250 in. If the standard deviation of the core diameter is $\sigma = .004$, what CUSUM procedure will yield an in-control ARL of 500 and an ARL of 5 when the mean core diameter shifts by the amount of .003 in.?

Connecting the point 500 on the in-control ARL scale to the point 5 on the out-of-control ARL scale and extending the line to the k' scale on the far left in Figure 16.11 gives $k' = .74$. Thus

$$k' = .74 = \frac{\Delta/2}{\sigma/\sqrt{n}} = \frac{.0015}{.004/\sqrt{n}} = .375\sqrt{n}$$

so

$$\sqrt{n} = \frac{.74}{.375} = 1.973 \qquad n = (1.973)^2 = 3.894$$

The CUSUM procedure should therefore be based on the sample size $n = 4$. Now connecting .74 on the k' scale to 500 on the in-control ARL scale gives $h' = 3.2$, from which

$$h = (\sigma/\sqrt{n}) \cdot (3.2) = (.004/\sqrt{4})(3.2) = .0064$$

An out-of-control signal results as soon as either $d_r > .0064$ or $e_r > .0064$. ■

* SOURCE: The Kemp nomogram—Kemp, Kenneth W., "The Use of Cumulative Sums for Sampling Inspection Schemes," *Applied Statistics,* Vol. XI, 1 962: 23.

We have discussed CUSUM procedures for controlling process location. There are also CUSUM procedures for controlling process variation and for attribute data. The chapter references should be consulted for information on these procedures.

EXERCISES Section 16.5 (29–32)

29. Containers of a certain treatment for septic tanks are supposed to contain 16 oz of liquid. A sample of five containers is selected from the production line once each hour, and the sample average content is determined. Consider the following results: 15.992, 16.051, 16.066, 15.912, 16.030, 16.060, 15.982, 15.899, 16.038, 16.074, 16.029, 15.935, 16.032, 15.960, 16.055. Using $\Delta = .10$ and $h = .20$, employ the computational form of the CUSUM procedure to investigate the behavior of this process.

30. The target value for the diameter of a certain type of driveshaft is .75 in. The size of the shift in the average diameter considered important to detect is .002 in. Sample average diameters for successive groups of $n = 4$ shafts are as follows: .7507, .7504, .7492, .7501, .7503,

.7510, .7490, .7497, .7488, .7504, .7516, .7472, .7489, .7483, .7471, .7498, .7460, .7482, .7470, .7493, .7462, .7481. Use the computational form of the CUSUM procedure with $h = .003$ to see whether the process mean remained on target throughout the time of observation.

31. The standard deviation of a certain dimension on an aircraft part is .005 cm. What CUSUM procedure will give an in-control ARL of 600 and an out-of-control ARL of 4 when the mean value of the dimension shifts by .004 cm?

32. When the out-of-control ARL corresponds to a shift of 1 standard deviation in the process mean, what are the characteristics of the CUSUM procedure that has ARLs of 250 and 4.8, respectively, for the in-control and out-of-control conditions?

16.6 Acceptance Sampling

Items coming from a production process are often sent in groups to another company or commercial establishment. A group might consist of all units from a particular production run or shift, in a shipping container of some sort, sent in response to a particular order, and so on. The group of items is usually called a *lot,* the sender is referred to as a *producer,* and the recipient of the lot is the *consumer.* Our focus will be on situations in which each item is either defective or nondefective, with p denoting the proportion of defective units in the lot. The consumer would naturally want to accept the lot only if the value of p is suitably small. Acceptance sampling is that part of applied statistics dealing with methods for deciding whether the consumer should accept or reject a lot.

Until quite recently, control chart procedures and acceptance sampling techniques were regarded by practitioners as equally important parts of quality control methodology. This is no longer the case. The reason is that the use of control charts and other recently developed strategies offers the opportunity to design quality into a product, whereas acceptance sampling deals with what has already been produced and thus does not provide for any direct control over process quality. This led the late American quality control expert W. E. Deming, a major force in persuading the Japanese to make substantial use of quality control methodology, to argue strongly against the use of acceptance sampling in many situations. In a similar vein, the recent book by Ryan (see the chapter bibliography) devotes several chapters to control charts and mentions acceptance sampling only in passing. As a reflection of this deemphasis, we content ourselves here with a brief introduction to basic concepts.

Single-Sampling Plans

The most straightforward type of acceptance sampling plan involves selecting a single random sample of size n and then rejecting the lot if the number of defectives in the sample exceeds a specified critical value c. Let the rv X denote the number of defective items in the lot and A denote the event that the lot is accepted. Then $P(A) = P(X \leq c)$ is a function of p; the larger the value of p, the smaller will be the probability of accepting the lot.

If the sample size n is large relative to N, $P(A)$ is calculated using the hypergeometric distribution (the number of defectives in the lot is Np):

$$P(X \leq c) = \sum_{x=0}^{c} h(x; n, Np, N) = \sum_{x=0}^{c} \frac{\binom{Np}{x} \cdot \binom{N(1-p)}{n-x}}{\binom{N}{n}}$$

When n is small relative to N (the rule of thumb suggested previously was $n \leq .05N$, but some authors employ the less conservative rule $n \leq .10N$), the binomial distribution can be used:

$$P(X \leq c) = \sum_{x=0}^{c} b(x; n, p) = \sum_{x=0}^{c} \binom{n}{x} p^x (1-p)^{n-x}$$

Finally, if $P(A)$ is large only when p is small (this depends on the value of c), the Poisson approximation to the binomial distribution is justified:

$$P(X \leq c) \approx \sum_{x=0}^{c} p(x; np) = \sum_{x=0}^{c} \frac{e^{-np}(np)^x}{x!}$$

The behavior of a sampling plan can be nicely summarized by graphing $P(A)$ as a function of p. Such a graph is called the **operating characteristic (OC) curve** for the plan.

EXAMPLE 16.11 Consider the sampling plan with $c = 2$ and $n = 50$. If the lot size N exceeds 1000, the binomial distribution can be used. This gives

$$P(A) = P(X \leq 2) = (1 - p)^{50} + 50p(1 - p)^{49} + 1255p^2(1 - p)^{48}$$

The accompanying table shows $P(A)$ for selected values of p, and the corresponding operating characteristic (OC) curve is shown in Figure 16.12.

p	.01	.02	.03	.04	.05	.06	.07	.08	.09	.10	.12	.15
$P(A)$	.986	.922	.811	.677	.541	.416	.311	.226	.161	.112	.051	.014

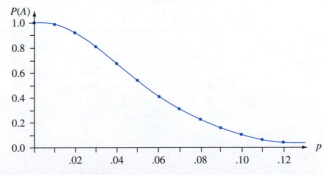

Figure 16.12 OC curve for sampling plan with $c = 2$, $n = 50$ ■

The OC curve for the plan of Example 16.11 has $P(A)$ near 1 for p very close to 0. However, in many applications a defective rate of 8% [for which $P(A) = .226$] or even just 5% [$P(A) = .541$] would be considered excessive, in which case the acceptance probabilities are too high. Increasing the critical value c while holding n fixed gives a plan for which $P(A)$ increases at each p (except 0 and 1), so the new OC curve lies above the old one. This is desirable for p near 0 but not for larger values of p. Holding c constant while increasing n gives a lower OC curve, which is fine for larger p but not for p close to 0. We want an OC curve that is higher for very small p and lower for larger p. This requires increasing n and adjusting c.

Designing a Single-Sample Plan

An effective sampling plan is one with the following characteristics:

1. It has a specified high probability of accepting lots that the producer considers to be of good quality.
2. It has a specified low probability of accepting lots that the consumer considers to be of poor quality.

A plan of this sort can be developed by proceeding as follows. Let's designate two different values of p, one for which $P(A)$ is a specified value close to 1 and the other for which $P(A)$ is a specified value near 0. These two values of p—say, p_1 and p_2—are often called the **acceptable quality level** (AQL) and the **lot tolerance percent defective** (LTPD). That is, we require a plan for which

1. $P(A) = 1 - \alpha$ when $p = p_1 = $ AQL (α small)
2. $P(A) = \beta$ when $p = p_2 = $ LTPD (β small)

This is analogous to seeking a hypothesis testing procedure with specified type I error probability α and specified type II error probability β. For example, we might have

$$\text{AQL} = .01 \quad \alpha = .05 \quad (P(A) = .95)$$

$$\text{LTPD} = .045 \quad \beta = .10 \quad (P(A) = .10)$$

Because X is discrete, we must typically be content with values of n and c that approximately satisfy these conditions.

Table 16.6 gives information from which n and c can be determined in the case $\alpha = .05, \beta = .10$.

Table 16.6 Factors for Determining n and c for a Single-Sample Plan with $\alpha = .05, \beta = .10$.

c	np_1	np_2	p_2/p_1	c	np_1	np_2	p_2/p_1
0	.051	2.30	45.10	8	4.695	12.99	2.77
1	.355	3.89	10.96	9	5.425	14.21	2.62
2	.818	5.32	6.50	10	6.169	15.41	2.50
3	1.366	6.68	4.89	11	6.924	16.60	2.40
4	1.970	7.99	4.06	12	7.690	17.78	2.31
5	2.613	9.28	3.55	13	8.464	18.86	2.24
6	3.285	10.53	3.21	14	9.246	20.13	2.18
7	3.981	11.77	2.96	15	10.040	21.29	2.12

EXAMPLE 16.12 Let's determine a plan for which AQL $= p_1 = .01$ and LTPD $= p_2 = .045$. The ratio of p_2 to p_1 is

$$\frac{\text{LTPD}}{\text{AQL}} = \frac{p_2}{p_1} = \frac{.045}{.01} = 4.50$$

This value lies between the ratio 4.89 given in Table 16.6, for which $c = 3$, and 4.06, for which $c = 4$. Once one of these values of c is chosen, n can be determined either by dividing the np_1 value in Table 16.6 by p_1 or via np_2/p_2. Thus four different plans (two values of c, and for each two values of n) give approximately the specified value of α and β. Consider, for example, using $c = 3$ and

$$n = \frac{np_1}{p_1} = \frac{1.366}{.01} = 136.6 \approx 137$$

Then

$$\alpha = 1 - P(X \le 3 \text{ when } p = p_1) = .050$$

(the Poisson approximation with $\mu = 1.37$ also gives .050) and

$$\beta = P(X \le 3 \text{ when } p = p_2) = .131$$

The plan with $c = 4$ and n determined from $np_2 = 7.99$ has $n = 178$, $\alpha = .034$, and $\beta = .094$. The larger sample size results in a plan with both α and β smaller than the corresponding specified values. ■

The book by Douglas Montgomery cited in the chapter bibliography contains a chart from which c and n can be determined for *any* specified α and β.

It may happen that the number of defective items in the sample reaches $c + 1$ before all items have been examined. For example, in the case $c = 3$ and $n = 137$, it may be that the 125th item examined is the fourth defective item, so that the remaining 12 items need not be examined. However, it is generally recommended that all items be examined even when this does occur, in order to provide a lot-by-lot quality history and estimates of p over time.

Double-Sampling Plans

In a double-sampling plan, the number of defective items x_1 in an initial sample of size n_1 is determined. There are then three possible courses of action: Immediately accept the lot, immediately reject the lot, or take a second sample of n_2 items and reject or accept the lot depending on the total number $x_1 + x_2$ of defective items in the two samples. Besides the two sample sizes, a specific plan is characterized by three further numbers—c_1, r_1, and c_2—as follows:

1. Reject the lot if $x_1 \ge r_1$.
2. Accept the lot if $x_1 \le c_1$.
3. If $c_1 < x_1 < r_1$, take a second sample; then accept the lot if $x_1 + x_2 \le c_2$ and reject it otherwise.

EXAMPLE 16.13 Consider the double-sampling plan with $n_1 = 80$, $n_2 = 80$, $c_1 = 2$, $r_1 = 5$, and $c_2 = 6$. Thus the lot will be accepted if (1) $x_1 = 0, 1,$ or 2; (2) $x_1 = 3$ and $x_2 = 0, 1, 2,$ or 3; or (3) $x_1 = 4$ and $x_2 = 0, 1,$ or 2.

Assuming that the lot size is large enough for the binomial approximation to apply, the probability $P(A)$ of accepting the lot is

$$P(A) = P(X_1 = 0, 1, \text{ or } 2) + P(X_1 = 3, \ X_2 = 0, 1, 2, \text{ or } 3)$$
$$+ \ P(X_1 = 4, X_2 = 0, 1, \text{ or } 2)$$
$$= \sum_{x_1=0}^{2} b(x_1; 80, p) + b(3; 80, p) \sum_{x_2=0}^{3} b(x_2; 80, p)$$
$$+ \ b(4; 80, p) \sum_{x_2=0}^{2} b(x_2; 80, p)$$

Again the graph of $P(A)$ versus p is the plan's OC curve. The OC curve for this plan appears in Figure 16.13.

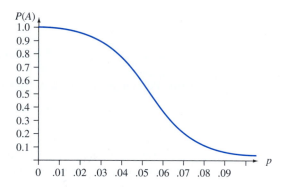

Figure 16.13 OC curve for the double-sampling plan of Example 16.13 ■

One standard method for designing a double-sampling plan involves proceeding as suggested earlier for single-sample plans. Specify values p_1 and p_2 along with corresponding acceptance probabilities $1 - \alpha$ and β. Then find a plan that satisfies these conditions. The book by Montgomery provides tables similar to Table 16.6 for this purpose in the cases $n_2 = n_1$ and $n_2 = 2n_1$ with $1 - \alpha = .95, \beta = .10$. Much more extensive tabulations of plans are available in other sources.

Analogous to standard practice with single-sample plans, it is recommended that all items in the first sample be examined even when the $(r_1 + 1)$st defective is discovered prior to inspection of the n_1th item. However, it is customary to terminate inspection of the second sample if the number of defectives is sufficient to justify rejection before all items have been examined. This is referred to as *curtailment* in the second sample. Under curtailment, it can be shown that the expected number of items inspected in a double-sampling plan is smaller than the number of items examined in a single-sampling plan when the OC curves of the two plans are close to being identical. This is the major virtue of double-sampling plans. For more on these matters as well as a discussion of multiple and sequential sampling plans (which involve selecting items for inspection one by one rather than in groups), a book on quality control should be consulted.

Rectifying Inspection and Other Design Criteria

In some situations, sampling inspection is carried out using *rectification*. For single-sample plans, this means that each defective item in the sample is replaced with a satisfactory one, and if the number of defectives in the sample exceeds the acceptance cutoff c, *all* items in the lot are examined and good items are substituted for any defectives. Let N denote the lot size. One important characteristic of a sampling plan with rectifying inspection is **average outgoing quality**, denoted by AOQ. This is the long-run proportion of defective items among those sent on after the sampling plan is employed. Now defectives will occur only among the $N - n$ items not inspected in a lot judged acceptable on the basis of a sample. Suppose, for example, that $P(A) = P(X \leq c) = .985$ when $p = .01$. Then, in the long run, 98.5% of the $N - n$ items not in the sample will not be inspected, of which we expect 1% to be defective. This implies that the expected number of defectives in a randomly selected batch is $(N - n) \cdot P(A) \cdot p = .00985(N - n)$. Dividing this by the number of items in a lot gives average outgoing quality:

$$\text{AOQ} = \frac{(N - n) \cdot P(A) \cdot p}{N}$$

$$\approx P(A) \cdot p \qquad \text{if } N \gg n$$

Because AOQ $= 0$ when either $p = 0$ or $p = 1$ [$P(A) = 0$ in the latter case], it follows that there is a value of p between 0 and 1 for which AOQ is a maximum. The maximum value of AOQ is called the **average outgoing quality limit, AOQL.** For example, for the plan with $n = 137$ and $c = 3$ discussed previously, AOQL $= .0142$, the value of AOQ at $p \approx .02$.

Proper choices of n and c will yield a sampling plan for which AOQL is a specified small number. Such a plan is not, however, unique, so another condition can be imposed. Frequently this second condition will involve the **average** (i.e., expected) **total number inspected**, denoted by **ATI**. The number of items inspected in a randomly chosen lot is a random variable that takes on the value n with probability $P(A)$ and N with probability $1 - P(A)$. Thus the expected number of items inspected in a randomly selected lot is

$$\text{ATI} = n \cdot P(A) + N \cdot (1 - P(A))$$

It is common practice to select a sampling plan that has a specified AOQL and, in addition, minimum ATI at a particular quality level p.

Standard Sampling Plans

It may seem as though the determination of a sampling plan that simultaneously satisfies several criteria would be quite difficult. Fortunately, others have already laid the groundwork in the form of extensive tabulations of such plans. MIL STD 105D, developed by the military after World War II, is the most widely used set of plans. A civilian version, ANSI/ASQC Z1.4, is quite similar to the military version. A third set of plans that is quite popular was developed at Bell Laboratories prior to World War II by two applied statisticians named Dodge and Romig. The book by Montgomery (see the chapter bibliography) contains a readable introduction to the use of these plans.

EXERCISES Section 16.6 (33–40)

33. Consider the single-sample plan with $c = 2$ and $n = 50$, as discussed in Example 16.11, but now suppose that the lot size is $N = 500$. Calculate $P(A)$, the probability of accepting the lot, for $p = .01, .02, ..., .10$, using the hypergeometric distribution. Does the binomial approximation give satisfactory results in this case?

34. A sample of 50 items is to be selected from a batch consisting of 5000 items. The batch will be accepted if the sample contains at most one defective item. Calculate the probability of lot acceptance for $p = .01, .02, ..., .10$, and sketch the OC curve.

35. Refer to Exercise 34 and consider the plan with $n = 100$ and $c = 2$. Calculate $P(A)$ for $p = .01, .02,..., .05$, and sketch the two OC curves on the same set of axes. Which of the two plans is preferable (leaving aside the cost of sampling) and why?

36. Develop a single-sample plan for which AQL = .02 and LTPD = .07 in the case $\alpha = .05, \beta = .10$. Once values of n and c have been determined, calculate the achieved values of α and β for the plan.

37. Consider the double-sampling plan for which both sample sizes are 50. The lot is accepted after the first sample if the number of defectives is at most 1, rejected if the number of defectives is at least 4, and rejected after the second sample if the total number of defectives is 6 or more. Calculate the probability of accepting the lot when $p = .01, .05$, and .10.

38. Some sources advocate a somewhat more restrictive type of doubling-sampling plan in which $r_1 = c_2 + 1$; that is, the lot is rejected if at either stage the (total)

number of defectives is at least r_1 (see the book by Montgomery). Consider this type of sampling plan with $n_1 = 50, n_2 = 100, c_1 = 1$, and $r_1 = 4$. Calculate the probability of lot acceptance when $p = .02, .05$, and .10.

39. Refer to Example 16.11, in which a single-sample plan with $n = 50$ and $c = 2$ was employed.
   a. Calculate AOQ for $p = .01, .02,..., .10$. What does this suggest about the value of p for which AOQ is a maximum and the corresponding AOQL?
   b. Determine the value of p for which AOQ is a maximum and the corresponding value of AOQL. [*Hint:* Use calculus.]
   c. For $N = 2000$, calculate ATI for the values of p given in part (a).

40. Consider the single-sample plan that utilizes $n = 50$ and $c = 1$ when $N = 2000$. Determine the values of AOQ and ATI for selected values of p, and graph each of these against p. Also determine the value of AOQL.

SUPPLEMENTARY EXERCISES (41–46)

41. Observations on shear strength for 26 subgroups of test spot welds, each consisting of six welds, yield $\Sigma \bar{x}_i = 10,980$, $\Sigma s_i = 402$, and $\Sigma r_i = 1074$. Calculate control limits for any relevant control charts.

42. The number of scratches on the surface of each of 24 rectangular metal plates is determined, yielding the following data: 8, 1, 7, 5, 2, 0, 2, 3, 4, 3, 1, 2, 5, 7, 3, 4, 6, 5, 2, 4, 0, 10, 2, 6. Construct an appropriate control chart, and comment.

43. The following numbers are observations on tensile strength of synthetic fabric specimens selected from a production process at equally spaced time intervals. Construct appropriate control charts, and comment (assume an assignable cause is identifiable for any out-of-control observations).

1.	51.3	51.7	49.5	12.	49.6	48.4	50.0
2.	51.0	50.0	49.3	13.	49.8	51.2	49.7
3.	50.8	51.1	49.0	14.	50.4	49.9	50.7
4.	50.6	51.1	49.0	15.	49.4	49.5	49.0
5.	49.6	50.5	50.9	16.	50.7	49.0	50.0
6.	51.3	52.0	50.3	17.	50.8	49.5	50.9
7.	49.7	50.5	50.3	18.	48.5	50.3	49.3
8.	51.8	50.3	50.0	19.	49.6	50.6	49.4
9.	48.6	50.5	50.7	20.	50.9	49.4	49.7
10.	49.6	49.8	50.5	21.	54.1	49.8	48.5
11.	49.9	50.7	49.8	22.	50.2	49.6	51.5

44. An alternative to the p chart for the fraction defective is the *np chart for number defective*. This chart has UCL $= n\bar{p} + 3\sqrt{n\bar{p}(1-\bar{p})}$, LCL $= n\bar{p} - 3\sqrt{n\bar{p}(1-\bar{p})}$,

and the *number* of defectives from each sample is plotted on the chart. Construct such a chart for the data of Example 16.6. Will the use of an np chart always give the same message as the use of a p chart (i.e., are the two charts equivalent)?

45. Resistance observations (ohms) for subgroups of a certain type of register gave the following summary quantities:

i	n_i	$\bar{x}_i$	s_i	i	n_i	$\bar{x}_i$	s_i
1	4	430.0	22.5	11	4	445.2	27.3
2	4	418.2	20.6	12	4	430.1	22.2
3	3	435.5	25.1	13	4	427.2	24.0
4	4	427.6	22.3	14	4	439.6	23.3
5	4	444.0	21.5	15	3	415.9	31.2
6	3	431.4	28.9	16	4	419.8	27.5
7	4	420.8	25.4	17	3	447.0	19.8
8	4	431.4	24.0	18	4	434.4	23.7
9	4	428.7	21.2	19	4	422.2	25.1
10	4	440.1	25.8	20	4	425.7	24.4

Construct appropriate control limits. [*Hint:* Use $\bar{x} = \Sigma n_i \bar{x}_i / \Sigma n_i$ and $s^2 = \Sigma (n_i - 1)s_i^2 / \Sigma (n_i - 1)$.]

46. Let α be a number between 0 and 1, and define a sequence $W_1, W_2, W_3,...$ by $W_0 = \mu$ and $W_t = \alpha \bar{X}_t + (1 - \alpha)W_{t-1}$ for $t = 1, 2,....$ Substituting for W_{t-1} its representation in terms of $\bar{X}_{t-1}$ and W_{t-2}, then substituting for W_{t-2}, and so on, results in

$$W_t = \alpha \bar{X}_t + \alpha(1 - \alpha)\bar{X}_{t-1} + \cdots$$
$$+ \alpha(1 - \alpha)^{t-1}\bar{X}_1 + (1 - \alpha)^t \mu$$

The fact that W_t depends not only on $\overline{X}_t$ but also on averages for past time points, albeit with (exponentially) decreasing weights, suggests that changes in the process mean will be more quickly reflected in the W_t's than in the individual $\overline{X}_t$'s.

a. Show that $E(W_t) = \mu$.

b. Let $\sigma_t^2 = V(W_t)$, and show that

$$\sigma_t^2 = \frac{\alpha[1 - (1 - \alpha)^{2t}]}{2 - \alpha} \cdot \frac{\sigma^2}{n}$$

c. An *exponentially weighted moving-average control chart* plots the W_t's and uses control limits $\mu_0 \pm 3\sigma_t$ (or $\overline{\overline{x}}$ in place of μ_0). Construct such a chart for the data of Example 16.9, using $\mu_0 = 40$.

BIBLIOGRAPHY

Box, George, Soren Bisgaard, and Conrad Fung, "An Explanation and Critique of Taguchi's Contributions to Quality Engineering," *Quality and Reliability Engineering International,* 1988: 123–131.

Montgomery, Douglas C., *Introduction to Statistical Quality Control* (7th ed.), Wiley, New York, 2012. This is a comprehensive introduction to many aspects of quality control at roughly the same level as this book.

Ryan, Thomas P., *Statistical Methods for Quality Improvement* (3rd ed.), Wiley, New York, 2011. Captures very nicely the modern flavor of quality control with minimal demands on the background of readers.

Vardeman, Stephen B., and J. Marcus Jobe, *Statistical Quality Assurance Methods for Engineers,* Wiley, New York, 1999. Includes traditional quality topics and also experimental design material germane to issues of quality; informal and authoritative.

Appendix Tables

Table A.1   Cumulative Binomial Probabilities

$$B(x; n, p) = \sum_{y=0}^{x} b(y; n, p)$$

a. $n = 5$

		0.01	0.05	0.10	0.20	0.25	0.30	0.40	0.50	0.60	0.70	0.75	0.80	0.90	0.95	0.99
	0	.951	.774	.590	.328	.237	.168	.078	.031	.010	.002	.001	.000	.000	.000	.000
	1	.999	.977	.919	.737	.633	.528	.337	.188	.087	.031	.016	.007	.000	.000	.000
x	2	1.000	.999	.991	.942	.896	.837	.683	.500	.317	.163	.104	.058	.009	.001	.000
	3	1.000	1.000	1.000	.993	.984	.969	.913	.812	.663	.472	.367	.263	.081	.023	.001
	4	1.000	1.000	1.000	1.000	.999	.998	.990	.969	.922	.832	.763	.672	.410	.226	.049

b. $n = 10$

		0.01	0.05	0.10	0.20	0.25	0.30	0.40	0.50	0.60	0.70	0.75	0.80	0.90	0.95	0.99
	0	.904	.599	.349	.107	.056	.028	.006	.001	.000	.000	.000	.000	.000	.000	.000
	1	.996	.914	.736	.376	.244	.149	.046	.011	.002	.000	.000	.000	.000	.000	.000
	2	1.000	.988	.930	.678	.526	.383	.167	.055	.012	.002	.000	.000	.000	.000	.000
	3	1.000	.999	.987	.879	.776	.650	.382	.172	.055	.011	.004	.001	.000	.000	.000
	4	1.000	1.000	.998	.967	.922	.850	.633	.377	.166	.047	.020	.006	.000	.000	.000
x	5	1.000	1.000	1.000	.994	.980	.953	.834	.623	.367	.150	.078	.033	.002	.000	.000
	6	1.000	1.000	1.000	.999	.996	.989	.945	.828	.618	.350	.224	.121	.013	.001	.000
	7	1.000	1.000	1.000	1.000	1.000	.998	.988	.945	.833	.617	.474	.322	.070	.012	.000
	8	1.000	1.000	1.000	1.000	1.000	1.000	.998	.989	.954	.851	.756	.624	.264	.086	.004
	9	1.000	1.000	1.000	1.000	1.000	1.000	1.000	.999	.994	.972	.944	.893	.651	.401	.096

c. $n = 15$

		0.01	0.05	0.10	0.20	0.25	0.30	0.40	0.50	0.60	0.70	0.75	0.80	0.90	0.95	0.99
	0	.860	.463	.206	.035	.013	.005	.000	.000	.000	.000	.000	.000	.000	.000	.000
	1	.990	.829	.549	.167	.080	.035	.005	.000	.000	.000	.000	.000	.000	.000	.000
	2	1.000	.964	.816	.398	.236	.127	.027	.004	.000	.000	.000	.000	.000	.000	.000
	3	1.000	.995	.944	.648	.461	.297	.091	.018	.002	.000	.000	.000	.000	.000	.000
	4	1.000	.999	.987	.836	.686	.515	.217	.059	.009	.001	.000	.000	.000	.000	.000
	5	1.000	1.000	.998	.939	.852	.722	.403	.151	.034	.004	.001	.000	.000	.000	.000
	6	1.000	1.000	1.000	.982	.943	.869	.610	.304	.095	.015	.004	.001	.000	.000	.000
x	7	1.000	1.000	1.000	.996	.983	.950	.787	.500	.213	.050	.017	.004	.000	.000	.000
	8	1.000	1.000	1.000	.999	.996	.985	.905	.696	.390	.131	.057	.018	.000	.000	.000
	9	1.000	1.000	1.000	1.000	.999	.996	.966	.849	.597	.278	.148	.061	.002	.000	.000
	10	1.000	1.000	1.000	1.000	1.000	.999	.991	.941	.783	.485	.314	.164	.013	.001	.000
	11	1.000	1.000	1.000	1.000	1.000	1.000	.998	.982	.909	.703	.539	.352	.056	.005	.000
	12	1.000	1.000	1.000	1.000	1.000	1.000	1.000	.996	.973	.873	.764	.602	.184	.036	.000
	13	1.000	1.000	1.000	1.000	1.000	1.000	1.000	1.000	.995	.965	.920	.833	.451	.171	.010
	14	1.000	1.000	1.000	1.000	1.000	1.000	1.000	1.000	1.000	.995	.987	.965	.794	.537	.140

(continued)

Table A.1   Cumulative Binomial Probabilities *(cont.)*

$$B(x; n, p) = \sum_{y=0}^{x} b(y; n, p)$$

d. $n = 20$

		p														
		0.01	0.05	0.10	0.20	0.25	0.30	0.40	0.50	0.60	0.70	0.75	0.80	0.90	0.95	0.99
	0	.818	.358	.122	.012	.003	.001	.000	.000	.000	.000	.000	.000	.000	.000	.000
	1	.983	.736	.392	.069	.024	.008	.001	.000	.000	.000	.000	.000	.000	.000	.000
	2	.999	.925	.677	.206	.091	.035	.004	.000	.000	.000	.000	.000	.000	.000	.000
	3	1.000	.984	.867	.411	.225	.107	.016	.001	.000	.000	.000	.000	.000	.000	.000
	4	1.000	.997	.957	.630	.415	.238	.051	.006	.000	.000	.000	.000	.000	.000	.000
	5	1.000	1.000	.989	.804	.617	.416	.126	.021	.002	.000	.000	.000	.000	.000	.000
	6	1.000	1.000	.998	.913	.786	.608	.250	.058	.006	.000	.000	.000	.000	.000	.000
	7	1.000	1.000	1.000	.968	.898	.772	.416	.132	.021	.001	.000	.000	.000	.000	.000
	8	1.000	1.000	1.000	.990	.959	.887	.596	.252	.057	.005	.001	.000	.000	.000	.000
	9	1.000	1.000	1.000	.997	.986	.952	.755	.412	.128	.017	.004	.001	.000	.000	.000
x	10	1.000	1.000	1.000	.999	.996	.983	.872	.588	.245	.048	.014	.003	.000	.000	.000
	11	1.000	1.000	1.000	1.000	.999	.995	.943	.748	.404	.113	.041	.010	.000	.000	.000
	12	1.000	1.000	1.000	1.000	1.000	.999	.979	.868	.584	.228	.102	.032	.000	.000	.000
	13	1.000	1.000	1.000	1.000	1.000	1.000	.994	.942	.750	.392	.214	.087	.002	.000	.000
	14	1.000	1.000	1.000	1.000	1.000	1.000	.998	.979	.874	.584	.383	.196	.011	.000	.000
	15	1.000	1.000	1.000	1.000	1.000	1.000	1.000	.994	.949	.762	.585	.370	.043	.003	.000
	16	1.000	1.000	1.000	1.000	1.000	1.000	1.000	.999	.984	.893	.775	.589	.133	.016	.000
	17	1.000	1.000	1.000	1.000	1.000	1.000	1.000	1.000	.996	.965	.909	.794	.323	.075	.001
	18	1.000	1.000	1.000	1.000	1.000	1.000	1.000	1.000	.999	.992	.976	.931	.608	.264	.017
	19	1.000	1.000	1.000	1.000	1.000	1.000	1.000	1.000	1.000	.999	.997	.988	.878	.642	.182

(continued)

Table A.1   Cumulative Binomial Probabilities *(cont.)*

$$B(x; n, p) = \sum_{y=0}^{x} b(y; n, p)$$

e. $n = 25$

		0.01	0.05	0.10	0.20	0.25	0.30	0.40	0.50	0.60	0.70	0.75	0.80	0.90	0.95	0.99
									p							
	0	.778	.277	.072	.004	.001	.000	.000	.000	.000	.000	.000	.000	.000	.000	.000
	1	.974	.642	.271	.027	.007	.002	.000	.000	.000	.000	.000	.000	.000	.000	.000
	2	.998	.873	.537	.098	.032	.009	.000	.000	.000	.000	.000	.000	.000	.000	.000
	3	1.000	.966	.764	.234	.096	.033	.002	.000	.000	.000	.000	.000	.000	.000	.000
	4	1.000	.993	.902	.421	.214	.090	.009	.000	.000	.000	.000	.000	.000	.000	.000
	5	1.000	.999	.967	.617	.378	.193	.029	.002	.000	.000	.000	.000	.000	.000	.000
	6	1.000	1.000	.991	.780	.561	.341	.074	.007	.000	.000	.000	.000	.000	.000	.000
	7	1.000	1.000	.998	.891	.727	.512	.154	.022	.001	.000	.000	.000	.000	.000	.000
	8	1.000	1.000	1.000	.953	.851	.677	.274	.054	.004	.000	.000	.000	.000	.000	.000
	9	1.000	1.000	1.000	.983	.929	.811	.425	.115	.013	.000	.000	.000	.000	.000	.000
	10	1.000	1.000	1.000	.994	.970	.902	.586	.212	.034	.002	.000	.000	.000	.000	.000
	11	1.000	1.000	1.000	.998	.980	.956	.732	.345	.078	.006	.001	.000	.000	.000	.000
x	12	1.000	1.000	1.000	1.000	.997	.983	.846	.500	.154	.017	.003	.000	.000	.000	.000
	13	1.000	1.000	1.000	1.000	.999	.994	.922	.655	.268	.044	.020	.002	.000	.000	.000
	14	1.000	1.000	1.000	1.000	1.000	.998	.966	.788	.414	.098	.030	.006	.000	.000	.000
	15	1.000	1.000	1.000	1.000	1.000	1.000	.987	.885	.575	.189	.071	.017	.000	.000	.000
	16	1.000	1.000	1.000	1.000	1.000	1.000	.996	.946	.726	.323	.149	.047	.000	.000	.000
	17	1.000	1.000	1.000	1.000	1.000	1.000	.999	.978	.846	.488	.273	.109	.002	.000	.000
	18	1.000	1.000	1.000	1.000	1.000	1.000	1.000	.993	.926	.659	.439	.220	.009	.000	.000
	19	1.000	1.000	1.000	1.000	1.000	1.000	1.000	.998	.971	.807	.622	.383	.033	.001	.000
	20	1.000	1.000	1.000	1.000	1.000	1.000	1.000	1.000	.991	.910	.786	.579	.098	.007	.000
	21	1.000	1.000	1.000	1.000	1.000	1.000	1.000	1.000	.998	.967	.904	.766	.236	.034	.000
	22	1.000	1.000	1.000	1.000	1.000	1.000	1.000	1.000	1.000	.991	.968	.902	.463	.127	.002
	23	1.000	1.000	1.000	1.000	1.000	1.000	1.000	1.000	1.000	.998	.993	.973	.729	.358	.026
	24	1.000	1.000	1.000	1.000	1.000	1.000	1.000	1.000	1.000	1.000	.999	.996	.928	.723	.222

Table A.2   Cumulative Poisson Probabilities

$$F(x; \mu) = \sum_{y=0}^{x} \frac{e^{-\mu} \mu^{y}}{y!}$$

		.1	.2	.3	.4	.5	.6	.7	.8	.9	1.0
						μ					
	0	.905	.819	.741	.670	.607	.549	.497	.449	.407	.368
	1	.995	.982	.963	.938	.910	.878	.844	.809	.772	.736
	2	1.000	.999	.996	.992	.986	.977	.966	.953	.937	.920
x	3		1.000	1.000	.999	.998	.997	.994	.991	.987	.981
	4				1.000	1.000	1.000	.999	.999	.998	.996
	5							1.000	1.000	1.000	.999
	6										1.000

(continued)

Table A.2 Cumulative Poisson Probabilities *(cont.)*

$$F(x; \mu) = \sum_{y=0}^{x} \frac{e^{-\mu}\mu^{y}}{y!}$$

	μ										
	2.0	**3.0**	**4.0**	**5.0**	**6.0**	**7.0**	**8.0**	**9.0**	**10.0**	**15.0**	**20.0**
0	.135	.050	.018	.007	.002	.001	.000	.000	.000	.000	.000
1	.406	.199	.092	.040	.017	.007	.003	.001	.000	.000	.000
2	.677	.423	.238	.125	.062	.030	.014	.006	.003	.000	.000
3	.857	.647	.433	.265	.151	.082	.042	.021	.010	.000	.000
4	.947	.815	.629	.440	.285	.173	.100	.055	.029	.001	.000
5	.983	.916	.785	.616	.446	.301	.191	.116	.067	.003	.000
6	.995	.966	.889	.762	.606	.450	.313	.207	.130	.008	.000
7	.999	.988	.949	.867	.744	.599	.453	.324	.220	.018	.001
8	1.000	.996	.979	.932	.847	.729	.593	.456	.333	.037	.002
9		.999	.992	.968	.916	.830	.717	.587	.458	.070	.005
10		1.000	.997	.986	.957	.901	.816	.706	.583	.118	.011
11			.999	.995	.980	.947	.888	.803	.697	.185	.021
12			1.000	.998	.991	.973	.936	.876	.792	.268	.039
13				.999	.996	.987	.966	.926	.864	.363	.066
14				1.000	.999	.994	.983	.959	.917	.466	.105
15					.999	.998	.992	.978	.951	.568	.157
16					1.000	.999	.996	.989	.973	.664	.221
17						1.000	.998	.995	.986	.749	.297
18							.999	.998	.993	.819	.381
19							1.000	.999	.997	.875	.470
20								1.000	.998	.917	.559
21									.999	.947	.644
22									1.000	.967	.721
23										.981	.787
24										.989	.843
25										.994	.888
26										.997	.922
27										.998	.948
28										.999	.966
29										1.000	.978
30											.987
31											.992
32											.995
33											.997
34											.999
35											.999
36											1.000

x

Table A.3   Standard Normal Curve Areas

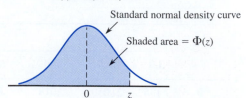

$\Phi(z) = P(Z \le z)$

z	.00	.01	.02	.03	.04	.05	.06	.07	.08	.09
−3.4	.0003	.0003	.0003	.0003	.0003	.0003	.0003	.0003	.0003	.0002
−3.3	.0005	.0005	.0005	.0004	.0004	.0004	.0004	.0004	.0004	.0003
−3.2	.0007	.0007	.0006	.0006	.0006	.0006	.0006	.0005	.0005	.0005
−3.1	.0010	.0009	.0009	.0009	.0008	.0008	.0008	.0008	.0007	.0007
−3.0	.0013	.0013	.0013	.0012	.0012	.0011	.0011	.0011	.0010	.0010
−2.9	.0019	.0018	.0017	.0017	.0016	.0016	.0015	.0015	.0014	.0014
−2.8	.0026	.0025	.0024	.0023	.0023	.0022	.0021	.0021	.0020	.0019
−2.7	.0035	.0034	.0033	.0032	.0031	.0030	.0029	.0028	.0027	.0026
−2.6	.0047	.0045	.0044	.0043	.0041	.0040	.0039	.0038	.0037	.0036
−2.5	.0062	.0060	.0059	.0057	.0055	.0054	.0052	.0051	.0049	.0038
−2.4	.0082	.0080	.0078	.0075	.0073	.0071	.0069	.0068	.0066	.0064
−2.3	.0107	.0104	.0102	.0099	.0096	.0094	.0091	.0089	.0087	.0084
−2.2	.0139	.0136	.0132	.0129	.0125	.0122	.0119	.0116	.0113	.0110
−2.1	.0179	.0174	.0170	.0166	.0162	.0158	.0154	.0150	.0146	.0143
−2.0	.0228	.0222	.0217	.0212	.0207	.0202	.0197	.0192	.0188	.0183
−1.9	.0287	.0281	.0274	.0268	.0262	.0256	.0250	.0244	.0239	.0233
−1.8	.0359	.0352	.0344	.0336	.0329	.0322	.0314	.0307	.0301	.0294
−1.7	.0446	.0436	.0427	.0418	.0409	.0401	.0392	.0384	.0375	.0367
−1.6	.0548	.0537	.0526	.0516	.0505	.0495	.0485	.0475	.0465	.0455
−1.5	.0668	.0655	.0643	.0630	.0618	.0606	.0594	.0582	.0571	.0559
−1.4	.0808	.0793	.0778	.0764	.0749	.0735	.0722	.0708	.0694	.0681
−1.3	.0968	.0951	.0934	.0918	.0901	.0885	.0869	.0853	.0838	.0823
−1.2	.1151	.1131	.1112	.1093	.1075	.1056	.1038	.1020	.1003	.0985
−1.1	.1357	.1335	.1314	.1292	.1271	.1251	.1230	.1210	.1190	.1170
−1.0	.1587	.1562	.1539	.1515	.1492	.1469	.1446	.1423	.1401	.1379
−0.9	.1841	.1814	.1788	.1762	.1736	.1711	.1685	.1660	.1635	.1611
−0.8	.2119	.2090	.2061	.2033	.2005	.1977	.1949	.1922	.1894	.1867
−0.7	.2420	.2389	.2358	.2327	.2296	.2266	.2236	.2206	.2177	.2148
−0.6	.2743	.2709	.2676	.2643	.2611	.2578	.2546	.2514	.2483	.2451
−0.5	.3085	.3050	.3015	.2981	.2946	.2912	.2877	.2843	.2810	.2776
−0.4	.3446	.3409	.3372	.3336	.3300	.3264	.3228	.3192	.3156	.3121
−0.3	.3821	.3783	.3745	.3707	.3669	.3632	.3594	.3557	.3520	.3482
−0.2	.4207	.4168	.4129	.4090	.4052	.4013	.3974	.3936	.3897	.3859
−0.1	.4602	.4562	.4522	.4483	.4443	.4404	.4364	.4325	.4286	.4247
−0.0	.5000	.4960	.4920	.4880	.4840	.4801	.4761	.4721	.4681	.4641

(continued)

Table A.3   Standard Normal Curve Areas *(cont.)* $\Phi(z) = P(Z \le z)$

z	.00	.01	.02	.03	.04	.05	.06	.07	.08	.09
0.0	.5000	.5040	.5080	.5120	.5160	.5199	.5239	.5279	.5319	.5359
0.1	.5398	.5438	.5478	.5517	.5557	.5596	.5636	.5675	.5714	.5753
0.2	.5793	.5832	.5871	.5910	.5948	.5987	.6026	.6064	.6103	.6141
0.3	.6179	.6217	.6255	.6293	.6331	.6368	.6406	.6443	.6480	.6517
0.4	.6554	.6591	.6628	.6664	.6700	.6736	.6772	.6808	.6844	.6879
0.5	.6915	.6950	.6985	.7019	.7054	.7088	.7123	.7157	.7190	.7224
0.6	.7257	.7291	.7324	.7357	.7389	.7422	.7454	.7486	.7517	.7549
0.7	.7580	.7611	.7642	.7673	.7704	.7734	.7764	.7794	.7823	.7852
0.8	.7881	.7910	.7939	.7967	.7995	.8023	.8051	.8078	.8106	.8133
0.9	.8159	.8186	.8212	.8238	.8264	.8289	.8315	.8340	.8365	.8389
1.0	.8413	.8438	.8461	.8485	.8508	.8531	.8554	.8577	.8599	.8621
1.1	.8643	.8665	.8686	.8708	.8729	.8749	.8770	.8790	.8810	.8830
1.2	.8849	.8869	.8888	.8907	.8925	.8944	.8962	.8980	.8997	.9015
1.3	.9032	.9049	.9066	.9082	.9099	.9115	.9131	.9147	.9162	.9177
1.4	.9192	.9207	.9222	.9236	.9251	.9265	.9278	.9292	.9306	.9319
1.5	.9332	.9345	.9357	.9370	.9382	.9394	.9406	.9418	.9429	.9441
1.6	.9452	.9463	.9474	.9484	.9495	.9505	.9515	.9525	.9535	.9545
1.7	.9554	.9564	.9573	.9582	.9591	.9599	.9608	.9616	.9625	.9633
1.8	.9641	.9649	.9656	.9664	.9671	.9678	.9686	.9693	.9699	.9706
1.9	.9713	.9719	.9726	.9732	.9738	.9744	.9750	.9756	.9761	.9767
2.0	.9772	.9778	.9783	.9788	.9793	.9798	.9803	.9808	.9812	.9817
2.1	.9821	.9826	.9830	.9834	.9838	.9842	.9846	.9850	.9854	.9857
2.2	.9861	.9864	.9868	.9871	.9875	.9878	.9881	.9884	.9887	.9890
2.3	.9893	.9896	.9898	.9901	.9904	.9906	.9909	.9911	.9913	.9916
2.4	.9918	.9920	.9922	.9925	.9927	.9929	.9931	.9932	.9934	.9936
2.5	.9938	.9940	.9941	.9943	.9945	.9946	.9948	.9949	.9951	.9952
2.6	.9953	.9955	.9956	.9957	.9959	.9960	.9961	.9962	.9963	.9964
2.7	.9965	.9966	.9967	.9968	.9969	.9970	.9971	.9972	.9973	.9974
2.8	.9974	.9975	.9976	.9977	.9977	.9978	.9979	.9979	.9980	.9981
2.9	.9981	.9982	.9982	.9983	.9984	.9984	.9985	.9985	.9986	.9986
3.0	.9987	.9987	.9987	.9988	.9988	.9989	.9989	.9989	.9990	.9990
3.1	.9990	.9991	.9991	.9991	.9992	.9992	.9992	.9992	.9993	.9993
3.2	.9993	.9993	.9994	.9994	.9994	.9994	.9994	.9995	.9995	.9995
3.3	.9995	.9995	.9995	.9996	.9996	.9996	.9996	.9996	.9996	.9997
3.4	.9997	.9997	.9997	.9997	.9997	.9997	.9997	.9997	.9997	.9998

Table A.4   The Incomplete Gamma Function

$$F(x; \alpha) = \int_0^x \frac{1}{\Gamma(\alpha)} y^{\alpha - 1} e^{-y} \, dy$$

x \ α	1	2	3	4	5	6	7	8	9	10
1	.632	.264	.080	.019	.004	.001	.000	.000	.000	.000
2	.865	.594	.323	.143	.053	.017	.005	.001	.000	.000
3	.950	.801	.577	.353	.185	.084	.034	.012	.004	.001
4	.982	.908	.762	.567	.371	.215	.111	.051	.021	.008
5	.993	.960	.875	.735	.560	.384	.238	.133	.068	.032
6	.998	.983	.938	.849	.715	.554	.394	.256	.153	.084
7	.999	.993	.970	.918	.827	.699	.550	.401	.271	.170
8	1.000	.997	.986	.958	.900	.809	.687	.547	.407	.283
9		.999	.994	.979	.945	.884	.793	.676	.544	.413
10		1.000	.997	.990	.971	.933	.870	.780	.667	.542
11			.999	.995	.985	.962	.921	.857	.768	.659
12			1.000	.998	.992	.980	.954	.911	.845	.758
13				.999	.996	.989	.974	.946	.900	.834
14				1.000	.998	.994	.986	.968	.938	.891
15					.999	.997	.992	.982	.963	.930

Table A.5 Critical Values for *t* Distributions

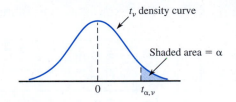

				α			
v	.10	.05	.025	.01	.005	.001	.0005
1	3.078	6.314	12.706	31.821	63.657	318.31	636.62
2	1.886	2.920	4.303	6.965	9.925	22.326	31.598
3	1.638	2.353	3.182	4.541	5.841	10.213	12.924
4	1.533	2.132	2.776	3.747	4.604	7.173	8.610
5	1.476	2.015	2.571	3.365	4.032	5.893	6.869
6	1.440	1.943	2.447	3.143	3.707	5.208	5.959
7	1.415	1.895	2.365	2.998	3.499	4.785	5.408
8	1.397	1.860	2.306	2.896	3.355	4.501	5.041
9	1.383	1.833	2.262	2.821	3.250	4.297	4.781
10	1.372	1.812	2.228	2.764	3.169	4.144	4.587
11	1.363	1.796	2.201	2.718	3.106	4.025	4.437
12	1.356	1.782	2.179	2.681	3.055	3.930	4.318
13	1.350	1.771	2.160	2.650	3.012	3.852	4.221
14	1.345	1.761	2.145	2.624	2.977	3.787	4.140
15	1.341	1.753	2.131	2.602	2.947	3.733	4.073
16	1.337	1.746	2.120	2.583	2.921	3.686	4.015
17	1.333	1.740	2.110	2.567	2.898	3.646	3.965
18	1.330	1.734	2.101	2.552	2.878	3.610	3.922
19	1.328	1.729	2.093	2.539	2.861	3.579	3.883
20	1.325	1.725	2.086	2.528	2.845	3.552	3.850
21	1.323	1.721	2.080	2.518	2.831	3.527	3.819
22	1.321	1.717	2.074	2.508	2.819	3.505	3.792
23	1.319	1.714	2.069	2.500	2.807	3.485	3.767
24	1.318	1.711	2.064	2.492	2.797	3.467	3.745
25	1.316	1.708	2.060	2.485	2.787	3.450	3.725
26	1.315	1.706	2.056	2.479	2.779	3.435	3.707
27	1.314	1.703	2.052	2.473	2.771	3.421	3.690
28	1.313	1.701	2.048	2.467	2.763	3.408	3.674
29	1.311	1.699	2.045	2.462	2.756	3.396	3.659
30	1.310	1.697	2.042	2.457	2.750	3.385	3.646
32	1.309	1.694	2.037	2.449	2.738	3.365	3.622
34	1.307	1.691	2.032	2.441	2.728	3.348	3.601
36	1.306	1.688	2.028	2.434	2.719	3.333	3.582
38	1.304	1.686	2.024	2.429	2.712	3.319	3.566
40	1.303	1.684	2.021	2.423	2.704	3.307	3.551
50	1.299	1.676	2.009	2.403	2.678	3.262	3.496
60	1.296	1.671	2.000	2.390	2.660	3.232	3.460
120	1.289	1.658	1.980	2.358	2.617	3.160	3.373
∞	1.282	1.645	1.960	2.326	2.576	3.090	3.291

Table A.6   Tolerance Critical Values for Normal Population Distributions

	Two-Sided Intervals						One-Sided Intervals					
Confidence Level	95%			99%			95%			99%		
Sample Size n / **% of Population Captured**	≥ 90%	≥ 95%	≥ 99%	≥ 90%	≥ 95%	≥ 99%	≥ 90%	≥ 95%	≥ 99%	≥ 90%	≥ 95%	≥ 99%
2	32.019	37.674	48.430	160.193	188.491	242.300	20.581	26.260	37.094	103.029	131.426	185.617
3	8.380	9.916	12.861	18.930	22.401	29.055	6.156	7.656	10.553	13.995	17.370	23.896
4	5.369	6.370	8.299	9.398	11.150	14.527	4.162	5.144	7.042	7.380	9.083	12.387
5	4.275	5.079	6.634	6.612	7.855	10.260	3.407	4.203	5.741	5.362	6.578	8.939
6	3.712	4.414	5.775	5.337	6.345	8.301	3.006	3.708	5.062	4.411	5.406	7.335
7	3.369	4.007	5.248	4.613	5.488	7.187	2.756	3.400	4.642	3.859	4.728	6.412
8	3.136	3.732	4.891	4.147	4.936	6.468	2.582	3.187	4.354	3.497	4.285	5.812
9	2.967	3.532	4.631	3.822	4.550	5.966	2.454	3.031	4.143	3.241	3.972	5.389
10	2.839	3.379	4.433	3.582	4.265	5.594	2.355	2.911	3.981	3.048	3.738	5.074
11	2.737	3.259	4.277	3.397	4.045	5.308	2.275	2.815	3.852	2.898	3.556	4.829
12	2.655	3.162	4.150	3.250	3.870	5.079	2.210	2.736	3.747	2.777	3.410	4.633
13	2.587	3.081	4.044	3.130	3.727	4.893	2.155	2.671	3.659	2.677	3.290	4.472
14	2.529	3.012	3.955	3.029	3.608	4.737	2.109	2.615	3.585	2.593	3.189	4.337
15	2.480	2.954	3.878	2.945	3.507	4.605	2.068	2.566	3.520	2.522	3.102	4.222
16	2.437	2.903	3.812	2.872	3.421	4.492	2.033	2.524	3.464	2.460	3.028	4.123
17	2.400	2.858	3.754	2.808	3.345	4.393	2.002	2.486	3.414	2.405	2.963	4.037
18	2.366	2.819	3.702	2.753	3.279	4.307	1.974	2.453	3.370	2.357	2.905	3.960
19	2.337	2.784	3.656	2.703	3.221	4.230	1.949	2.423	3.331	2.314	2.854	3.892
20	2.310	2.752	3.615	2.659	3.168	4.161	1.926	2.396	3.295	2.276	2.808	3.832
25	2.208	2.631	3.457	2.494	2.972	3.904	1.838	2.292	3.158	2.129	2.633	3.601
30	2.140	2.549	3.350	2.385	2.841	3.733	1.777	2.220	3.064	2.030	2.516	3.447
35	2.090	2.490	3.272	2.306	2.748	3.611	1.732	2.167	2.995	1.957	2.430	3.334
40	2.052	2.445	3.213	2.247	2.677	3.518	1.697	2.126	2.941	1.902	2.364	3.249
45	2.021	2.408	3.165	2.200	2.621	3.444	1.669	2.092	2.898	1.857	2.312	3.180
50	1.996	2.379	3.126	2.162	2.576	3.385	1.646	2.065	2.863	1.821	2.269	3.125
60	1.958	2.333	3.066	2.103	2.506	3.293	1.609	2.022	2.807	1.764	2.202	3.038
70	1.929	2.299	3.021	2.060	2.454	3.225	1.581	1.990	2.765	1.722	2.153	2.974
80	1.907	2.272	2.986	2.026	2.414	3.173	1.559	1.965	2.733	1.688	2.114	2.924
90	1.889	2.251	2.958	1.999	2.382	3.130	1.542	1.944	2.706	1.661	2.082	2.883
100	1.874	2.233	2.934	1.977	2.355	3.096	1.527	1.927	2.684	1.639	2.056	2.850
150	1.825	2.175	2.859	1.905	2.270	2.983	1.478	1.870	2.611	1.566	1.971	2.741
200	1.798	2.143	2.816	1.865	2.222	2.921	1.450	1.837	2.570	1.524	1.923	2.679
250	1.780	2.121	2.788	1.839	2.191	2.880	1.431	1.815	2.542	1.496	1.891	2.638
300	1.767	2.106	2.767	1.820	2.169	2.850	1.417	1.800	2.522	1.476	1.868	2.608
∞	1.645	1.960	2.576	1.645	1.960	2.576	1.282	1.645	2.326	1.282	1.645	2.326

Table A.7   Critical Values for Chi-Squared Distributions

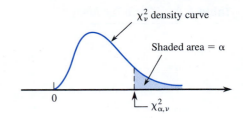

ν	.995	.99	.975	.95	.90	.10	.05	.025	.01	.005
1	0.000	0.000	0.001	0.004	0.016	2.706	3.843	5.025	6.637	7.882
2	0.010	0.020	0.051	0.103	0.211	4.605	5.992	7.378	9.210	10.597
3	0.072	0.115	0.216	0.352	0.584	6.251	7.815	9.348	11.344	12.837
4	0.207	0.297	0.484	0.711	1.064	7.779	9.488	11.143	13.277	14.860
5	0.412	0.554	0.831	1.145	1.610	9.236	11.070	12.832	15.085	16.748
6	0.676	0.872	1.237	1.635	2.204	10.645	12.592	14.440	16.812	18.548
7	0.989	1.239	1.690	2.167	2.833	12.017	14.067	16.012	18.474	20.276
8	1.344	1.646	2.180	2.733	3.490	13.362	15.507	17.534	20.090	21.954
9	1.735	2.088	2.700	3.325	4.168	14.684	16.919	19.022	21.665	23.587
10	2.156	2.558	3.247	3.940	4.865	15.987	18.307	20.483	23.209	25.188
11	2.603	3.053	3.816	4.575	5.578	17.275	19.675	21.920	24.724	26.755
12	3.074	3.571	4.404	5.226	6.304	18.549	21.026	23.337	26.217	28.300
13	3.565	4.107	5.009	5.892	7.041	19.812	22.362	24.735	27.687	29.817
14	4.075	4.660	5.629	6.571	7.790	21.064	23.685	26.119	29.141	31.319
15	4.600	5.229	6.262	7.261	8.547	22.307	24.996	27.488	30.577	32.799
16	5.142	5.812	6.908	7.962	9.312	23.542	26.296	28.845	32.000	34.267
17	5.697	6.407	7.564	8.682	10.085	24.769	27.587	30.190	33.408	35.716
18	6.265	7.015	8.231	9.390	10.865	25.989	28.869	31.526	34.805	37.156
19	6.843	7.632	8.906	10.117	11.651	27.203	30.143	32.852	36.190	38.580
20	7.434	8.260	9.591	10.851	12.443	28.412	31.410	34.170	37.566	39.997
21	8.033	8.897	10.283	11.591	13.240	29.615	32.670	35.478	38.930	41.399
22	8.643	9.542	10.982	12.338	14.042	30.813	33.924	36.781	40.289	42.796
23	9.260	10.195	11.688	13.090	14.848	32.007	35.172	38.075	41.637	44.179
24	9.886	10.856	12.401	13.848	15.659	33.196	36.415	39.364	42.980	45.558
25	10.519	11.523	13.120	14.611	16.473	34.381	37.652	40.646	44.313	46.925
26	11.160	12.198	13.844	15.379	17.292	35.563	38.885	41.923	45.642	48.290
27	11.807	12.878	14.573	16.151	18.114	36.741	40.113	43.194	46.962	49.642
28	12.461	13.565	15.308	16.928	18.939	37.916	41.337	44.461	48.278	50.993
29	13.120	14.256	16.147	17.708	19.768	39.087	42.557	45.772	49.586	52.333
30	13.787	14.954	16.791	18.493	20.599	40.256	43.773	46.979	50.892	53.672
31	14.457	15.655	17.538	19.280	21.433	41.422	44.985	48.231	52.190	55.000
32	15.134	16.362	18.291	20.072	22.271	42.585	46.194	49.480	53.486	56.328
33	15.814	17.073	19.046	20.866	23.110	43.745	47.400	50.724	54.774	57.646
34	16.501	17.789	19.806	21.664	23.952	44.903	48.602	51.966	56.061	58.964
35	17.191	18.508	20.569	22.465	24.796	46.059	49.802	53.203	57.340	60.272
36	17.887	19.233	21.336	23.269	25.643	47.212	50.998	54.437	58.619	61.581
37	18.584	19.960	22.105	24.075	26.492	48.363	52.192	55.667	59.891	62.880
38	19.289	20.691	22.878	24.884	27.343	49.513	53.384	56.896	61.162	64.181
39	19.994	21.425	23.654	25.695	28.196	50.660	54.572	58.119	62.426	65.473
40	20.706	22.164	24.433	26.509	29.050	51.805	55.758	59.342	63.691	66.766

For $\nu > 40$, $\chi^2_{a,\nu} \approx \nu\left(1 - \dfrac{2}{9\nu} + z_a \sqrt{\dfrac{2}{9\nu}}\right)^3$

Table A.8   *t* Curve Tail Areas

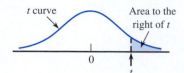

t curve   Area to the right of *t*

0

t

t \ ν	1	2	3	4	5	6	7	8	9	10	11	12	13	14	15	16	17	18
0.0	.500	.500	.500	.500	.500	.500	.500	.500	.500	.500	.500	.500	.500	.500	.500	.500	.500	.500
0.1	.468	.465	.463	.463	.462.	.462	.462	.461	.461	.461	.461	.461	.461	.461	.461	.461	.461	.461
0.2	.437	.430	.427	.426	.425	.424	.424	.423	.423	.423	.423	.422	.422	.422	.422	.422	.422	.422
0.3	.407	.396	.392	.390	.388	.387	.386	.386	.386	.385	.385	.385	.384	.384	.384	.384	.384	.384
0.4	.379	.364	.358	.355	.353	.352	.351	.350	.349	.349	.348	.348	.348	.347	.347	.347	.347	.347
0.5	.352	.333	.326	.322	.319	.317	.316	.315	.315	.314	.313	.313	.313	.312	.312	.312	.312	.312
0.6	.328	.305	.295	.290	.287	.285	.284	.283	.282	.281	.280	.280	.279	.279	.279	.278	.278	.278
0.7	.306	.278	.267	.261	.258	.255	.253	.252	.251	.250	.249	.249	.248	.247	.247	.247	.247	.246
0.8	.285	.254	.241	.234	.230	.227	.225	.223	.222	.221	.220	.220	.219	.218	.218	.218	.217	.217
0.9	.267	.232	.217	.210	.205	.201	.199	.197	.196	.195	.194	.193	.192	.191	.191	.191	.190	.190
1.0	.250	.211	.196	.187	.182	.178	.175	.173	.172	.170	.169	.169	.168	.167	.167	.166	.166	.165
1.1	.235	.193	.176	.167	.162	.157	.154	.152	.150	.149	.147	.146	.146	.144	.144	.144	.143	.143
1.2	.221	.177	.158	.148	.142	.138	.135	.132	.130	.129	.128	.127	.126	.124	.124	.124	.123	.123
1.3	.209	.162	.142	.132	.125	.121	.117	.115	.113	.111	.110	.109	.108	.107	.107	.106	.105	.105
1.4	.197	.148	.128	.117	.110	.106	.102	.100	.098	.096	.095	.093	.092	.091	.091	.090	.090	.089
1.5	.187	.136	.115	.104	.097	.092	.089	.086	.084	.082	.081	.080	.079	.077	.077	.077	.076	.075
1.6	.178	.125	.104	.092	.085	.080	.077	.074	.072	.070	.069	.068	.067	.065	.065	.065	.064	.064
1.7	.169	.116	.094	.082	.075	.070	.065	.064	.062	.060	.059	.057	.056	.055	.055	.054	.054	.053
1.8	.161	.107	.085	.073	.066	.061	.057	.055	.053	.051	.050	.049	.048	.046	.046	.045	.045	.044
1.9	.154	.099	.077	.065	.058	.053	.050	.047	.045	.043	.042	.041	.040	.038	.038	.038	.037	.037
2.0	.148	.092	.070	.058	.051	.046	.043	.040	.038	.037	.035	.034	.033	.032	.032	.031	.031	.030
2.1	.141	.085	.063	.052	.045	.040	.037	.034	.033	.031	.030	.029	.028	.027	.027	.026	.025	.025
2.2	.136	.079	.058	.046	.040	.035	.032	.029	.028	.026	.025	.024	.023	.022	.022	.021	.021	.021
2.3	.131	.074	.052	.041	.035	.031	.027	.025	.023	.022	.021	.020	.019	.018	.018	.018	.017	.017
2.4	.126	.069	.048	.037	.031	.027	.024	.022	.020	.019	.018	.017	.016	.015	.015	.014	.014	.014
2.5	.121	.065	.044	.033	.027	.023	.020	.018	.017	.016	.015	.014	.013	.012	.012	.012	.011	.011
2.6	.117	.061	.040	.030	.024	.020	.018	.016	.014	.013	.012	.012	.011	.010	.010	.010	.009	.009
2.7	.113	.057	.037	.027	.021	.018	.015	.014	.012	.011	.010	.010	.009	.008	.008	.008	.008	.007
2.8	.109	.054	.034	.024	.019	.016	.013	.012	.010	.009	.009	.008	.008	.007	.007	.006	.006	.006
2.9	.106	.051	.031	.022	.017	.014	.011	.010	.009	.008	.007	.007	.006	.005	.005	.005	.005	.005
3.0	.102	.048	.029	.020	.015	.012	.010	.009	.007	.007	.006	.006	.005	.004	.004	.004	.004	.004
3.1	.099	.045	.027	.018	.013	.011	.009	.007	.006	.006	.005	.005	.004	.004	.004	.003	.003	.003
3.2	.096	.043	.025	.016	.012	.009	.008	.006	.005	.005	.004	.004	.003	.003	.003	.003	.003	.002
3.3	.094	.040	.023	.015	.011	.008	.007	.005	.005	.004	.004	.003	.003	.002	.002	.002	.002	.002
3.4	.091	.038	.021	.014	.010	.007	.006	.005	.004	.003	.003	.003	.002	.002	.002	.002	.002	.002
3.5	.089	.036	.020	.012	.009	.006	.005	.004	.003	.003	.002	.002	.002	.002	.002	.001	.001	.001
3.6	.086	.035	.018	.011	.008	.006	.004	.004	.003	.002	.002	.002	.002	.001	.001	.001	.001	.001
3.7	.084	.033	.017	.010	.007	.005	.004	.003	.002	.002	.002	.002	.001	.001	.001	.001	.001	.001
3.8	.082	.031	.016	.010	.006	.004	.003	.003	.002	.002	.001	.001	.001	.001	.001	.001	.001	.001
3.9	.080	.030	.015	.009	.006	.004	.003	.002	.002	.001	.001	.001	.001	.001	.001	.001	.001	.001
4.0	.078	.029	.014	.008	.005	.004	.003	.002	.002	.001	.001	.001	.001	.001	.001	.001	.000	.000

(continued)

Table A.8   *t* Curve Tail Areas *(cont.)*

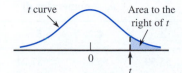

t \ ν	19	20	21	22	23	24	25	26	27	28	29	30	35	40	60	120	$\infty\,(=z)$
0.0	.500	.500	.500	.500	.500	.500	.500	.500	.500	.500	.500	.500	.500	.500	.500	.500	.500
0.1	.461	.461	.461	.461	.461	.461	.461	.461	.461	.461	.461	.461	.460	.460	.460	.460	.460
0.2	.422	.422	.422	.422	.422	.422	.422	.422	.421	.421	.421	.421	.421	.421	.421	.421	.421
0.3	.384	.384	.384	.383	.383	.383	.383	.383	.383	.383	.383	.383	.383	.383	.383	.382	.382
0.4	.347	.347	.347	.347	.346	.346	.346	.346	.346	.346	.346	.346	.346	.346	.345	.345	.345
0.5	.311	.311	.311	.311	.311	.311	.311	.311	.311	.310	.310	.310	.310	.310	.309	.309	.309
0.6	.278	.278	.278	.277	.277	.277	.277	.277	.277	.277	.277	.277	.276	.276	.275	.275	.274
0.7	.246	.246	.246	.246	.245	.245	.245	.245	.245	.245	.245	.245	.244	.244	.243	.243	.242
0.8	.217	.217	.216	.216	.216	.216	.216	.215	.215	.215	.215	.215	.215	.214	.213	.213	.212
0.9	.190	.189	.189	.189	.189	.189	.188	.188	.188	.188	.188	.188	.187	.187	.186	.185	.184
1.0	.165	.165	.164	.164	.164	.164	.163	.163	.163	.163	.163	.163	.162	.162	.161	.160	.159
1.1	.143	.142	.142	.142	.141	.141	.141	.141	.141	.140	.140	.140	.139	.139	.138	.137	.136
1.2	.122	.122	.122	.121	.121	.121	.121	.120	.120	.120	.120	.120	.119	.119	.117	.116	.115
1.3	.105	.104	.104	.104	.103	.103	.103	.103	.102	.102	.102	.102	.101	.101	.099	.098	.097
1.4	.089	.089	.088	.088	.087	.087	.087	.087	.086	.086	.086	.086	.085	.085	.083	.082	.081
1.5	.075	.075	.074	.074	.074	.073	.073	.073	.073	.072	.072	.072	.071	.071	.069	.068	.067
1.6	.063	.063	.062	.062	.062	.061	.061	.061	.061	.060	.060	.060	.059	.059	.057	.056	.055
1.7	.053	.052	.052	.052	.051	.051	.051	.051	.050	.050	.050	.050	.049	.048	.047	.046	.045
1.8	.044	.043	.043	.043	.042	.042	.042	.042	.042	.041	.041	.041	.040	.040	.038	.037	.036
1.9	.036	.036	.036	.035	.035	.035	.035	.034	.034	.034	.034	.034	.033	.032	.031	.030	.029
2.0	.030	.030	.029	.029	.029	.028	.028	.028	.028	.028	.027	.027	.027	.026	.025	.024	.023
2.1	.025	.024	.024	.024	.023	.023	.023	.023	.023	.022	.022	.022	.022	.021	.020	.019	.018
2.2	.020	.020	.020	.019	.019	.019	.019	.018	.018	.018	.018	.018	.017	.017	.016	.015	.014
2.3	.016	.016	.016	.016	.015	.015	.015	.015	.015	.015	.014	.014	.014	.013	.012	.012	.011
2.4	.013	.013	.013	.013	.012	.012	.012	.012	.012	.012	.012	.011	.011	.011	.010	.009	.008
2.5	.011	.011	.010	.010	.010	.010	.010	.010	.009	.009	.009	.009	.009	.008	.008	.007	.006
2.6	.009	.009	.008	.008	.008	.008	.008	.008	.007	.007	.007	.007	.007	.007	.006	.005	.005
2.7	.007	.007	.007	.007	.006	.006	.006	.006	.006	.006	.006	.006	.005	.005	.004	.004	.003
2.8	.006	.006	.005	.005	.005	.005	.005	.005	.005	.005	.005	.004	.004	.004	.003	.003	.003
2.9	.005	.004	.004	.004	.004	.004	.004	.004	.004	.004	.004	.003	.003	.003	.003	.002	.002
3.0	.004	.004	.003	.003	.003	.003	.003	.003	.003	.003	.003	.003	.002	.002	.002	.002	.001
3.1	.003	.003	.003	.003	.003	.002	.002	.002	.002	.002	.002	.002	.002	.002	.001	.001	.001
3.2	.002	.002	.002	.002	.002	.002	.002	.002	.002	.002	.002	.002	.001	.001	.001	.001	.001
3.3	.002	.002	.002	.002	.002	.001	.001	.001	.001	.001	.001	.001	.001	.001	.001	.001	.000
3.4	.002	.001	.001	.001	.001	.001	.001	.001	.001	.001	.001	.001	.001	.001	.001	.000	.000
3.5	.001	.001	.001	.001	.001	.001	.001	.001	.001	.001	.001	.001	.001	.001	.000	.000	.000
3.6	.001	.001	.001	.001	.001	.001	.001	.001	.001	.001	.001	.001	.000	.000	.000	.000	.000
3.7	.001	.001	.001	.001	.001	.001	.001	.001	.000	.000	.000	.000	.000	.000	.000	.000	.000
3.8	.001	.001	.001	.000	.000	.000	.000	.000	.000	.000	.000	.000	.000	.000	.000	.000	.000
3.9	.000	.000	.000	.000	.000	.000	.000	.000	.000	.000	.000	.000	.000	.000	.000	.000	.000
4.0	.000	.000	.000	.000	.000	.000	.000	.000	.000	.000	.000	.000	.000	.000	.000	.000	.000

Table A.9   Critical Values for *F* Distributions

		ν_1 = numerator df								
	α	1	2	3	4	5	6	7	8	9
	.100	39.86	49.50	53.59	55.83	57.24	58.20	58.91	59.44	59.86
1	.050	161.45	199.50	215.71	224.58	230.16	233.99	236.77	238.88	240.54
	.010	4052.20	4999.50	5403.40	5624.60	5763.60	5859.00	5928.40	5981.10	6022.50
	.001	405,284	500,000	540,379	562,500	576,405	585,937	592,873	598,144	602,284
	.100	8.53	9.00	9.16	9.24	9.29	9.33	9.35	9.37	9.38
2	.050	18.51	19.00	19.16	19.25	19.30	19.33	19.35	19.37	19.38
	.010	98.50	99.00	99.17	99.25	99.30	99.33	99.36	99.37	99.39
	.001	998.50	999.00	999.17	999.25	999.30	999.33	999.36	999.37	999.39
	.100	5.54	5.46	5.39	5.34	5.31	5.28	5.27	5.25	5.24
3	.050	10.13	9.55	9.28	9.12	9.01	8.94	8.89	8.85	8.81
	.010	34.12	30.82	29.46	28.71	28.24	27.91	27.67	27.49	27.35
	.001	167.03	148.50	141.11	137.10	134.58	132.85	131.58	130.62	129.86
	.100	4.54	4.32	4.19	4.11	4.05	4.01	3.98	3.95	3.94
4	.050	7.71	6.94	6.59	6.39	6.26	6.16	6.09	6.04	6.00
	.010	21.20	18.00	16.69	15.98	15.52	15.21	14.98	14.80	14.66
	.001	74.14	61.25	56.18	53.44	51.71	50.53	49.66	49.00	48.47
	.100	4.06	3.78	3.62	3.52	3.45	3.40	3.37	3.34	3.32
5	.050	6.61	5.79	5.41	5.19	5.05	4.95	4.88	4.82	4.77
	.010	16.26	13.27	12.06	11.39	10.97	10.67	10.46	10.29	10.16
	.001	47.18	37.12	33.20	31.09	29.75	28.83	28.16	27.65	27.24
	.100	3.78	3.46	3.29	3.18	3.11	3.05	3.01	2.98	2.96
6	.050	5.99	5.14	4.76	4.53	4.39	4.28	4.21	4.15	4.10
	.010	13.75	10.92	9.78	9.15	8.75	8.47	8.26	8.10	7.98
	.001	35.51	27.00	23.70	21.92	20.80	20.03	19.46	19.03	18.69
	.100	3.59	3.26	3.07	2.96	2.88	2.83	2.78	2.75	2.72
7	.050	5.59	4.74	4.35	4.12	3.97	3.87	3.79	3.73	3.68
	.010	12.25	9.55	8.45	7.85	7.46	7.19	6.99	6.84	6.72
	.001	29.25	21.69	18.77	17.20	16.21	15.52	15.02	14.63	14.33
	.100	3.46	3.11	2.92	2.81	2.73	2.67	2.62	2.59	2.56
8	.050	5.32	4.46	4.07	3.84	3.69	3.58	3.50	3.44	3.39
	.010	11.26	8.65	7.59	7.01	6.63	6.37	6.18	6.03	5.91
	.001	25.41	18.49	15.83	14.39	13.48	12.86	12.40	12.05	11.77
	.100	3.36	3.01	2.81	2.69	2.61	2.55	2.51	2.47	2.44
9	.050	5.12	4.26	3.86	3.63	3.48	3.37	3.29	3.23	3.18
	.010	10.56	8.02	6.99	6.42	6.06	5.80	5.61	5.47	5.35
	.001	22.86	16.39	13.90	12.56	11.71	11.13	10.70	10.37	10.11
	.100	3.29	2.92	2.73	2.61	2.52	2.46	2.41	2.38	2.35
10	.050	4.96	4.10	3.71	3.48	3.33	3.22	3.14	3.07	3.02
	.010	10.04	7.56	6.55	5.99	5.64	5.39	5.20	5.06	4.94
	.001	21.04	14.91	12.55	11.28	10.48	9.93	9.52	9.20	8.96
	.100	3.23	2.86	2.66	2.54	2.45	2.39	2.34	2.30	2.27
11	.050	4.84	3.98	3.59	3.36	3.20	3.09	3.01	2.95	2.90
	.010	9.65	7.21	6.22	5.67	5.32	5.07	4.89	4.74	4.63
	.001	19.69	13.81	11.56	10.35	9.58	9.05	8.66	8.35	8.12
	.100	3.18	2.81	2.61	2.48	2.39	2.33	2.28	2.24	2.21
12	.050	4.75	3.89	3.49	3.26	3.11	3.00	2.91	2.85	2.80
	.010	9.33	6.93	5.95	5.41	5.06	4.82	4.64	4.50	4.39
	.001	18.64	12.97	10.80	9.63	8.89	8.38	8.00	7.71	7.48

ν_2 = denominator df (left axis label)

(*continued*)

Table A.9   Critical Values for *F* Distributions *(cont.)*

				ν_1 = numerator df						
10	12	15	20	25	30	40	50	60	120	1000
60.19	60.71	61.22	61.74	62.05	62.26	62.53	62.69	62.79	63.06	63.30
241.88	243.91	245.95	248.01	249.26	250.10	251.14	251.77	252.20	253.25	254.19
6055.80	6106.30	6157.30	6208.70	6239.80	6260.60	6286.80	6302.50	6313.00	6339.40	6362.70
605,621	610,668	615,764	620,908	624,017	626,099	628,712	630,285	631,337	633,972	636,301
9.39	9.41	9.42	9.44	9.45	9.46	9.47	9.47	9.47	9.48	9.49
19.40	19.41	19.43	19.45	19.46	19.46	19.47	19.48	19.48	19.49	19.49
99.40	99.42	99.43	99.45	99.46	99.47	99.47	99.48	99.48	99.49	99.50
999.40	999.42	999.43	999.45	999.46	999.47	999.47	999.48	999.48	999.49	999.50
5.23	5.22	5.20	5.18	5.17	5.17	5.16	5.15	5.15	5.14	5.13
8.79	8.74	8.70	8.66	8.63	8.62	8.59	8.58	8.57	8.55	8.53
27.23	27.05	26.87	26.69	26.58	26.50	26.41	26.35	26.32	26.22	26.14
129.25	128.32	127.37	126.42	125.84	125.45	124.96	124.66	124.47	123.97	123.53
3.92	3.90	3.87	3.84	3.83	3.82	3.80	3.80	3.79	3.78	3.76
5.96	5.91	5.86	5.80	5.77	5.75	5.72	5.70	5.69	5.66	5.63
14.55	14.37	14.20	14.02	13.91	13.84	13.75	13.69	13.65	13.56	13.47
48.05	47.41	46.76	46.10	45.70	45.43	45.09	44.88	44.75	44.40	44.09
3.30	3.27	3.24	3.21	3.19	3.17	3.16	3.15	3.14	3.12	3.11
4.74	4.68	4.62	4.56	4.52	4.50	4.46	4.44	4.43	4.40	4.37
10.05	9.89	9.72	9.55	9.45	9.38	9.29	9.24	9.20	9.11	9.03
26.92	26.42	25.91	25.39	25.08	24.87	24.60	24.44	24.33	24.06	23.82
2.94	2.90	2.87	2.84	2.81	2.80	2.78	2.77	2.76	2.74	2.72
4.06	4.00	3.94	3.87	3.83	3.81	3.77	3.75	3.74	3.70	3.67
7.87	7.72	7.56	7.40	7.30	7.23	7.14	7.09	7.06	6.97	6.89
18.41	17.99	17.56	17.12	16.85	16.67	16.44	16.31	16.21	15.98	15.77
2.70	2.67	2.63	2.59	2.57	2.56	2.54	2.52	2.51	2.49	2.47
3.64	3.57	3.51	3.44	3.40	3.38	3.34	3.32	3.30	3.27	3.23
6.62	6.47	6.31	6.16	6.06	5.99	5.91	5.86	5.82	5.74	5.66
14.08	13.71	13.32	12.93	12.69	12.53	12.33	12.20	12.12	11.91	11.72
2.54	2.50	2.46	2.42	2.40	2.38	2.36	2.35	2.34	2.32	2.30
3.35	3.28	3.22	3.15	3.11	3.08	3.04	3.02	3.01	2.97	2.93
5.81	5.67	5.52	5.36	5.26	5.20	5.12	5.07	5.03	4.95	4.87
11.54	11.19	10.84	10.48	10.26	10.11	9.92	9.80	9.73	9.53	9.36
2.42	2.38	2.34	2.30	2.27	2.25	2.23	2.22	2.21	2.18	2.16
3.14	3.07	3.01	2.94	2.89	2.86	2.83	2.80	2.79	2.75	2.71
5.26	5.11	4.96	4.81	4.71	4.65	4.57	4.52	4.48	4.40	4.32
9.89	9.57	9.24	8.90	8.69	8.55	8.37	8.26	8.19	8.00	7.84
2.32	2.28	2.24	2.20	2.17	2.16	2.13	2.12	2.11	2.08	2.06
2.98	2.91	2.85	2.77	2.73	2.70	2.66	2.64	2.62	2.58	2.54
4.85	4.71	4.56	4.41	4.31	4.25	4.17	4.12	4.08	4.00	3.92
8.75	8.45	8.13	7.80	7.60	7.47	7.30	7.19	7.12	6.94	6.78
2.25	2.21	2.17	2.12	2.10	2.08	2.05	2.04	2.03	2.00	1.98
2.85	2.79	2.72	2.65	2.60	2.57	2.53	2.51	2.49	2.45	2.41
4.54	4.40	4.25	4.10	4.01	3.94	3.86	3.81	3.78	3.69	3.61
7.92	7.63	7.32	7.01	6.81	6.68	6.52	6.42	6.35	6.18	6.02
2.19	2.15	2.10	2.06	2.03	2.01	1.99	1.97	1.96	1.93	1.91
2.75	2.69	2.62	2.54	2.50	2.47	2.43	2.40	2.38	2.34	2.30
4.30	4.16	4.01	3.86	3.76	3.70	3.62	3.57	3.54	3.45	3.37
7.29	7.00	6.71	6.40	6.22	6.09	5.93	5.83	5.76	5.59	5.44

(continued)

Table A.9   Critical Values for *F* Distributions (*cont.*)

	α	1	2	3	4	5	6	7	8	9
						ν_1 = numerator df				
13	.100	3.14	2.76	2.56	2.43	2.35	2.28	2.23	2.20	2.16
	.050	4.67	3.81	3.41	3.18	3.03	2.92	2.83	2.77	2.71
	.010	9.07	6.70	5.74	5.21	4.86	4.62	4.44	4.30	4.19
	.001	17.82	12.31	10.21	9.07	8.35	7.86	7.49	7.21	6.98
14	.100	3.10	2.73	2.52	2.39	2.31	2.24	2.19	2.15	2.12
	.050	4.60	3.74	3.34	3.11	2.96	2.85	2.76	2.70	2.65
	.010	8.86	6.51	5.56	5.04	4.69	4.46	4.28	4.14	4.03
	.001	17.14	11.78	9.73	8.62	7.92	7.44	7.08	6.80	6.58
15	.100	3.07	2.70	2.49	2.36	2.27	2.21	2.16	2.12	2.09
	.050	4.54	3.68	3.29	3.06	2.90	2.79	2.71	2.64	2.59
	.010	8.68	6.36	5.42	4.89	4.56	4.32	4.14	4.00	3.89
	.001	16.59	11.34	9.34	8.25	7.57	7.09	6.74	6.47	6.26
16	.100	3.05	2.67	2.46	2.33	2.24	2.18	2.13	2.09	2.06
	.050	4.49	3.63	3.24	3.01	2.85	2.74	2.66	2.59	2.54
	.010	8.53	6.23	5.29	4.77	4.44	4.20	4.03	3.89	3.78
	.001	16.12	10.97	9.01	7.94	7.27	6.80	6.46	6.19	5.98
17	.100	3.03	2.64	2.44	2.31	2.22	2.15	2.10	2.06	2.03
	.050	4.45	3.59	3.20	2.96	2.81	2.70	2.61	2.55	2.49
	.010	8.40	6.11	5.19	4.67	4.34	4.10	3.93	3.79	3.68
	.001	15.72	10.66	8.73	7.68	7.02	6.56	6.22	5.96	5.75
18	.100	3.01	2.62	2.42	2.29	2.20	2.13	2.08	2.04	2.00
	.050	4.41	3.55	3.16	2.93	2.77	2.66	2.58	2.51	2.46
	.010	8.29	6.01	5.09	4.58	4.25	4.01	3.84	3.71	3.60
	.001	15.38	10.39	8.49	7.46	6.81	6.35	6.02	5.76	5.56
19	.100	2.99	2.61	2.40	2.27	2.18	2.11	2.06	2.02	1.98
	.050	4.38	3.52	3.13	2.90	2.74	2.63	2.54	2.48	2.42
	.010	8.18	5.93	5.01	4.50	4.17	3.94	3.77	3.63	3.52
	.001	15.08	10.16	8.28	7.27	6.62	6.18	5.85	5.59	5.39
20	.100	2.97	2.59	2.38	2.25	2.16	2.09	2.04	2.00	1.96
	.050	4.35	3.49	3.10	2.87	2.71	2.60	2.51	2.45	2.39
	.010	8.10	5.85	4.94	4.43	4.10	3.87	3.70	3.56	3.46
	.001	14.82	9.95	8.10	7.10	6.46	6.02	5.69	5.44	5.24
21	.100	2.96	2.57	2.36	2.23	2.14	2.08	2.02	1.98	1.95
	.050	4.32	3.47	3.07	2.84	2.68	2.57	2.49	2.42	2.37
	.010	8.02	5.78	4.87	4.37	4.04	3.81	3.64	3.51	3.40
	.001	14.59	9.77	7.94	6.95	6.32	5.88	5.56	5.31	5.11
22	.100	2.95	2.56	2.35	2.22	2.13	2.06	2.01	1.97	1.93
	.050	4.30	3.44	3.05	2.82	2.66	2.55	2.46	2.40	2.34
	.010	7.95	5.72	4.82	4.31	3.99	3.76	3.59	3.45	3.35
	.001	14.38	9.61	7.80	6.81	6.19	5.76	5.44	5.19	4.99
23	.100	2.94	2.55	2.34	2.21	2.11	2.05	1.99	1.95	1.92
	.050	4.28	3.42	3.03	2.80	2.64	2.53	2.44	2.37	2.32
	.010	7.88	5.66	4.76	4.26	3.94	3.71	3.54	3.41	3.30
	.001	14.20	9.47	7.67	6.70	6.08	5.65	5.33	5.09	4.89
24	.100	2.93	2.54	2.33	2.19	2.10	2.04	1.98	1.94	1.91
	.050	4.26	3.40	3.01	2.78	2.62	2.51	2.42	2.36	2.30
	.010	7.82	5.61	4.72	4.22	3.90	3.67	3.50	3.36	3.26
	.001	14.03	9.34	7.55	6.59	5.98	5.55	5.23	4.99	4.80

ν_2 = denominator df

(*continued*)

Table A.9   Critical Values for *F* Distributions *(cont.)*

					ν_1 = numerator df						
10	12	15	20	25	30	40	50	60	120	1000
2.14	2.10	2.05	2.01	1.98	1.96	1.93	1.92	1.90	1.88	1.85
2.67	2.60	2.53	2.46	2.41	2.38	2.34	2.31	2.30	2.25	2.21
4.10	3.96	3.82	3.66	3.57	3.51	3.43	3.38	3.34	3.25	3.18
6.80	6.52	6.23	5.93	5.75	5.63	5.47	5.37	5.30	5.14	4.99
2.10	2.05	2.01	1.96	1.93	1.91	1.89	1.87	1.86	1.83	1.80
2.60	2.53	2.46	2.39	2.34	2.31	2.27	2.24	2.22	2.18	2.14
3.94	3.80	3.66	3.51	3.41	3.35	3.27	3.22	3.18	3.09	3.02
6.40	6.13	5.85	5.56	5.38	5.25	5.10	5.00	4.94	4.77	4.62
2.06	2.02	1.97	1.92	1.89	1.87	1.85	1.83	1.82	1.79	1.76
2.54	2.48	2.40	2.33	2.28	2.25	2.20	2.18	2.16	2.11	2.07
3.80	3.67	3.52	3.37	3.28	3.21	3.13	3.08	3.05	2.96	2.88
6.08	5.81	5.54	5.25	5.07	4.95	4.80	4.70	4.64	4.47	4.33
2.03	1.99	1.94	1.89	1.86	1.84	1.81	1.79	1.78	1.75	1.72
2.49	2.42	2.35	2.28	2.23	2.19	2.15	2.12	2.11	2.06	2.02
3.69	3.55	3.41	3.26	3.16	3.10	3.02	2.97	2.93	2.84	2.76
5.81	5.55	5.27	4.99	4.82	4.70	4.54	4.45	4.39	4.23	4.08
2.00	1.96	1.91	1.86	1.83	1.81	1.78	1.76	1.75	1.72	1.69
2.45	2.38	2.31	2.23	2.18	2.15	2.10	2.08	2.06	2.01	1.97
3.59	3.46	3.31	3.16	3.07	3.00	2.92	2.87	2.83	2.75	2.66
5.58	5.32	5.05	4.78	4.60	4.48	4.33	4.24	4.18	4.02	3.87
1.98	1.93	1.89	1.84	1.80	1.78	1.75	1.74	1.72	1.69	1.66
2.41	2.34	2.27	2.19	2.14	2.11	2.06	2.04	2.02	1.97	1.92
3.51	3.37	3.23	3.08	2.98	2.92	2.84	2.78	2.75	2.66	2.58
5.39	5.13	4.87	4.59	4.42	4.30	4.15	4.06	4.00	3.84	3.69
1.96	1.91	1.86	1.81	1.78	1.76	1.73	1.71	1.70	1.67	1.64
2.38	2.31	2.23	2.16	2.11	2.07	2.03	2.00	1.98	1.93	1.88
3.43	3.30	3.15	3.00	2.91	2.84	2.76	2.71	2.67	2.58	2.50
5.22	4.97	4.70	4.43	4.26	4.14	3.99	3.90	3.84	3.68	3.53
1.94	1.89	1.84	1.79	1.76	1.74	1.71	1.69	1.68	1.64	1.61
2.35	2.28	2.20	2.12	2.07	2.04	1.99	1.97	1.95	1.90	1.85
3.37	3.23	3.09	2.94	2.84	2.78	2.69	2.64	2.61	2.52	2.43
5.08	4.82	4.56	4.29	4.12	4.00	3.86	3.77	3.70	3.54	3.40
1.92	1.87	1.83	1.78	1.74	1.72	1.69	1.67	1.66	1.62	1.59
2.32	2.25	2.18	2.10	2.05	2.01	1.96	1.94	1.92	1.87	1.82
3.31	3.17	3.03	2.88	2.79	2.72	2.64	2.58	2.55	2.46	2.37
4.95	4.70	4.44	4.17	4.00	3.88	3.74	3.64	3.58	3.42	3.28
1.90	1.86	1.81	1.76	1.73	1.70	1.67	1.65	1.64	1.60	1.57
2.30	2.23	2.15	2.07	2.02	1.98	1.94	1.91	1.89	1.84	1.79
3.26	3.12	2.98	2.83	2.73	2.67	2.58	2.53	2.50	2.40	2.32
4.83	4.58	4.33	4.06	3.89	3.78	3.63	3.54	3.48	3.32	3.17
1.89	1.84	1.80	1.74	1.71	1.69	1.66	1.64	1.62	1.59	1.55
2.27	2.20	2.13	2.05	2.00	1.96	1.91	1.88	1.86	1.81	1.76
3.21	3.07	2.93	2.78	2.69	2.62	2.54	2.48	2.45	2.35	2.27
4.73	4.48	4.23	3.96	3.79	3.68	3.53	3.44	3.38	3.22	3.08
1.88	1.83	1.78	1.73	1.70	1.67	1.64	1.62	1.61	1.57	1.54
2.25	2.18	2.11	2.03	1.97	1.94	1.89	1.86	1.84	1.79	1.74
3.17	3.03	2.89	2.74	2.64	2.58	2.49	2.44	2.40	2.31	2.22
4.64	4.39	4.14	3.87	3.71	3.59	3.45	3.36	3.29	3.14	2.99

(continued)

Table A.9 Critical Values for *F* Distributions (*cont.*)

	α	v_1 = numerator df								
		1	2	3	4	5	6	7	8	9
25	.100	2.92	2.53	2.32	2.18	2.09	2.02	1.97	1.93	1.89
	.050	4.24	3.39	2.99	2.76	2.60	2.49	2.40	2.34	2.28
	.010	7.77	5.57	4.68	4.18	3.85	3.63	3.46	3.32	3.22
	.001	13.88	9.22	7.45	6.49	5.89	5.46	5.15	4.91	4.71
26	.100	2.91	2.52	2.31	2.17	2.08	2.01	1.96	1.92	1.88
	.050	4.23	3.37	2.98	2.74	2.59	2.47	2.39	2.32	2.27
	.010	7.72	5.53	4.64	4.14	3.82	3.59	3.42	3.29	3.18
	.001	13.74	9.12	7.36	6.41	5.80	5.38	5.07	4.83	4.64
27	.100	2.90	2.51	2.30	2.17	2.07	2.00	1.95	1.91	1.87
	.050	4.21	3.35	2.96	2.73	2.57	2.46	2.37	2.31	2.25
	.010	7.68	5.49	4.60	4.11	3.78	3.56	3.39	3.26	3.15
	.001	13.61	9.02	7.27	6.33	5.73	5.31	5.00	4.76	4.57
28	.100	2.89	2.50	2.29	2.16	2.06	2.00	1.94	1.90	1.87
	.050	4.20	3.34	2.95	2.71	2.56	2.45	2.36	2.29	2.24
	.010	7.64	5.45	4.57	4.07	3.75	3.53	3.36	3.23	3.12
	.001	13.50	8.93	7.19	6.25	5.66	5.24	4.93	4.69	4.50
29	.100	2.89	2.50	2.28	2.15	2.06	1.99	1.93	1.89	1.86
	.050	4.18	3.33	2.93	2.70	2.55	2.43	2.35	2.28	2.22
	.010	7.60	5.42	4.54	4.04	3.73	3.50	3.33	3.20	3.09
	.001	13.39	8.85	7.12	6.19	5.59	5.18	4.87	4.64	4.45
30	.100	2.88	2.49	2.28	2.14	2.05	1.98	1.93	1.88	1.85
	.050	4.17	3.32	2.92	2.69	2.53	2.42	2.33	2.27	2.21
	.010	7.56	5.39	4.51	4.02	3.70	3.47	3.30	3.17	3.07
	.001	13.29	8.77	7.05	6.12	5.53	5.12	4.82	4.58	4.39
40	.100	2.84	2.44	2.23	2.09	2.00	1.93	1.87	1.83	1.79
	.050	4.08	3.23	2.84	2.61	2.45	2.34	2.25	2.18	2.12
	.010	7.31	5.18	4.31	3.83	3.51	3.29	3.12	2.99	2.89
	.001	12.61	8.25	6.59	5.70	5.13	4.73	4.44	4.21	4.02
50	.100	2.81	2.41	2.20	2.06	1.97	1.90	1.84	1.80	1.76
	.050	4.03	3.18	2.79	2.56	2.40	2.29	2.20	2.13	2.07
	.010	7.17	5.06	4.20	3.72	3.41	3.19	3.02	2.89	2.78
	.001	12.22	7.96	6.34	5.46	4.90	4.51	4.22	4.00	3.82
60	.100	2.79	2.39	2.18	2.04	1.95	1.87	1.82	1.77	1.74
	.050	4.00	3.15	2.76	2.53	2.37	2.25	2.17	2.10	2.04
	.010	7.08	4.98	4.13	3.65	3.34	3.12	2.95	2.82	2.72
	.001	11.97	7.77	6.17	5.31	4.76	4.37	4.09	3.86	3.69
100	.100	2.76	2.36	2.14	2.00	1.91	1.83	1.78	1.73	1.69
	.050	3.94	3.09	2.70	2.46	2.31	2.19	2.10	2.03	1.97
	.010	6.90	4.82	3.98	3.51	3.21	2.99	2.82	2.69	2.59
	.001	11.50	7.41	5.86	5.02	4.48	4.11	3.83	3.61	3.44
200	.100	2.73	2.33	2.11	1.97	1.88	1.80	1.75	1.70	1.66
	.050	3.89	3.04	2.65	2.42	2.26	2.14	2.06	1.98	1.93
	.010	6.76	4.71	3.88	3.41	3.11	2.89	2.73	2.60	2.50
	.001	11.15	7.15	5.63	4.81	4.29	3.92	3.65	3.43	3.26
1000	.100	2.71	2.31	2.09	1.95	1.85	1.78	1.72	1.68	1.64
	.050	3.85	3.00	2.61	2.38	2.22	2.11	2.02	1.95	1.89
	.010	6.66	4.63	3.80	3.34	3.04	2.82	2.66	2.53	2.43
	.001	10.89	6.96	5.46	4.65	4.14	3.78	3.51	3.30	3.13

v_2 = denominator df

(*continued*)

Table A.9   Critical Values for *F* Distributions *(cont.)*

ν_1 = numerator df

10	12	15	20	25	30	40	50	60	120	1000
1.87	1.82	1.77	1.72	1.68	1.66	1.63	1.61	1.59	1.56	1.52
2.24	2.16	2.09	2.01	1.96	1.92	1.87	1.84	1.82	1.77	1.72
3.13	2.99	2.85	2.70	2.60	2.54	2.45	2.40	2.36	2.27	2.18
4.56	4.31	4.06	3.79	3.63	3.52	3.37	3.28	3.22	3.06	2.91
1.86	1.81	1.76	1.71	1.67	1.65	1.61	1.59	1.58	1.54	1.51
2.22	2.15	2.07	1.99	1.94	1.90	1.85	1.82	1.80	1.75	1.70
3.09	2.96	2.81	2.66	2.57	2.50	2.42	2.36	2.33	2.23	2.14
4.48	4.24	3.99	3.72	3.56	3.44	3.30	3.21	3.15	2.99	2.84
1.85	1.80	1.75	1.70	1.66	1.64	1.60	1.58	1.57	1.53	1.50
2.20	2.13	2.06	1.97	1.92	1.88	1.84	1.81	1.79	1.73	1.68
3.06	2.93	2.78	2.63	2.54	2.47	2.38	2.33	2.29	2.20	2.11
4.41	4.17	3.92	3.66	3.49	3.38	3.23	3.14	3.08	2.92	2.78
1.84	1.79	1.74	1.69	1.65	1.63	1.59	1.57	1.56	1.52	1.48
2.19	2.12	2.04	1.96	1.91	1.87	1.82	1.79	1.77	1.71	1.66
3.03	2.90	2.75	2.60	2.51	2.44	2.35	2.30	2.26	2.17	2.08
4.35	4.11	3.86	3.60	3.43	3.32	3.18	3.09	3.02	2.86	2.72
1.83	1.78	1.73	1.68	1.64	1.62	1.58	1.56	1.55	1.51	1.47
2.18	2.10	2.03	1.94	1.89	1.85	1.81	1.77	1.75	1.70	1.65
3.00	2.87	2.73	2.57	2.48	2.41	2.33	2.27	2.23	2.14	2.05
4.29	4.05	3.80	3.54	3.38	3.27	3.12	3.03	2.97	2.81	2.66
1.82	1.77	1.72	1.67	1.63	1.61	1.57	1.55	1.54	1.50	1.46
2.16	2.09	2.01	1.93	1.88	1.84	1.79	1.76	1.74	1.68	1.63
2.98	2.84	2.70	2.55	2.45	2.39	2.30	2.25	2.21	2.11	2.02
4.24	4.00	3.75	3.49	3.33	3.22	3.07	2.98	2.92	2.76	2.61
1.76	1.71	1.66	1.61	1.57	1.54	1.51	1.48	1.47	1.42	1.38
2.08	2.00	1.92	1.84	1.78	1.74	1.69	1.66	1.64	1.58	1.52
2.80	2.66	2.52	2.37	2.27	2.20	2.11	2.06	2.02	1.92	1.82
3.87	3.64	3.40	3.14	2.98	2.87	2.73	2.64	2.57	2.41	2.25
1.73	1.68	1.63	1.57	1.53	1.50	1.46	1.44	1.42	1.38	1.33
2.03	1.95	1.87	1.78	1.73	1.69	1.63	1.60	1.58	1.51	1.45
2.70	2.56	2.42	2.27	2.17	2.10	2.01	1.95	1.91	1.80	1.70
3.67	3.44	3.20	2.95	2.79	2.68	2.53	2.44	2.38	2.21	2.05
1.71	1.66	1.60	1.54	1.50	1.48	1.44	1.41	1.40	1.35	1.30
1.99	1.92	1.84	1.75	1.69	1.65	1.59	1.56	1.53	1.47	1.40
2.63	2.50	2.35	2.20	2.10	2.03	1.94	1.88	1.84	1.73	1.62
3.54	3.32	3.08	2.83	2.67	2.55	2.41	2.32	2.25	2.08	1.92
1.66	1.61	1.56	1.49	1.45	1.42	1.38	1.35	1.34	1.28	1.22
1.93	1.85	1.77	1.68	1.62	1.57	1.52	1.48	1.45	1.38	1.30
2.50	2.37	2.22	2.07	1.97	1.89	1.80	1.74	1.69	1.57	1.45
3.30	3.07	2.84	2.59	2.43	2.32	2.17	2.08	2.01	1.83	1.64
1.63	1.58	1.52	1.46	1.41	1.38	1.34	1.31	1.29	1.23	1.16
1.88	1.80	1.72	1.62	1.56	1.52	1.46	1.41	1.39	1.30	1.21
2.41	2.27	2.13	1.97	1.87	1.79	1.69	1.63	1.58	1.45	1.30
3.12	2.90	2.67	2.42	2.26	2.15	2.00	1.90	1.83	1.64	1.43
1.61	1.55	1.49	1.43	1.38	1.35	1.30	1.27	1.25	1.18	1.08
1.84	1.76	1.68	1.58	1.52	1.47	1.41	1.36	1.33	1.24	1.11
2.34	2.20	2.06	1.90	1.79	1.72	1.61	1.54	1.50	1.35	1.16
2.99	2.77	2.54	2.30	2.14	2.02	1.87	1.77	1.69	1.49	1.22

Table A.10   Critical Values for Studentized Range Distributions

						m						
ν	α	2	3	4	5	6	7	8	9	10	11	12
5	.05	3.64	4.60	5.22	5.67	6.03	6.33	6.58	6.80	6.99	7.17	7.32
	.01	5.70	6.98	7.80	8.42	8.91	9.32	9.67	9.97	10.24	10.48	10.70
6	.05	3.46	4.34	4.90	5.30	5.63	5.90	6.12	6.32	6.49	6.65	6.79
	.01	5.24	6.33	7.03	7.56	7.97	8.32	8.61	8.87	9.10	9.30	9.48
7	.05	3.34	4.16	4.68	5.06	5.36	5.61	5.82	6.00	6.16	6.30	6.43
	.01	4.95	5.92	6.54	7.01	7.37	7.68	7.94	8.17	8.37	8.55	8.71
8	.05	3.26	4.04	4.53	4.89	5.17	5.40	5.60	5.77	5.92	6.05	6.18
	.01	4.75	5.64	6.20	6.62	6.96	7.24	7.47	7.68	7.86	8.03	8.18
9	.05	3.20	3.95	4.41	4.76	5.02	5.24	5.43	5.59	5.74	5.87	5.98
	.01	4.60	5.43	5.96	6.35	6.66	6.91	7.13	7.33	7.49	7.65	7.78
10	.05	3.15	3.88	4.33	4.65	4.91	5.12	5.30	5.46	5.60	5.72	5.83
	.01	4.48	5.27	5.77	6.14	6.43	6.67	6.87	7.05	7.21	7.36	7.49
11	.05	3.11	3.82	4.26	4.57	4.82	5.03	5.20	5.35	5.49	5.61	5.71
	.01	4.39	5.15	5.62	5.97	6.25	6.48	6.67	6.84	6.99	7.13	7.25
12	.05	3.08	3.77	4.20	4.51	4.75	4.95	5.12	5.27	5.39	5.51	5.61
	.01	4.32	5.05	5.50	5.84	6.10	6.32	6.51	6.67	6.81	6.94	7.06
13	.05	3.06	3.73	4.15	4.45	4.69	4.88	5.05	5.19	5.32	5.43	5.53
	.01	4.26	4.96	5.40	5.73	5.98	6.19	6.37	6.53	6.67	6.79	6.90
14	.05	3.03	3.70	4.11	4.41	4.64	4.83	4.99	5.13	5.25	5.36	5.46
	.01	4.21	4.89	5.32	5.63	5.88	6.08	6.26	6.41	6.54	6.66	6.77
15	.05	3.01	3.67	4.08	4.37	4.59	4.78	4.94	5.08	5.20	5.31	5.40
	.01	4.17	4.84	5.25	5.56	5.80	5.99	6.16	6.31	6.44	6.55	6.66
16	.05	3.00	3.65	4.05	4.33	4.56	4.74	4.90	5.03	5.15	5.26	5.35
	.01	4.13	4.79	5.19	5.49	5.72	5.92	6.08	6.22	6.35	6.46	6.56
17	.05	2.98	3.63	4.02	4.30	4.52	4.70	4.86	4.99	5.11	5.21	5.31
	.01	4.10	4.74	5.14	5.43	5.66	5.85	6.01	6.15	6.27	6.38	6.48
18	.05	2.97	3.61	4.00	4.28	4.49	4.67	4.82	4.96	5.07	5.17	5.27
	.01	4.07	4.70	5.09	5.38	5.60	5.79	5.94	6.08	6.20	6.31	6.41
19	.05	2.96	3.59	3.98	4.25	4.47	4.65	4.79	4.92	5.04	5.14	5.23
	.01	4.05	4.67	5.05	5.33	5.55	5.73	5.89	6.02	6.14	6.25	6.34
20	.05	2.95	3.58	3.96	4.23	4.45	4.62	4.77	4.90	5.01	5.11	5.20
	.01	4.02	4.64	5.02	5.29	5.51	5.69	5.84	5.97	6.09	6.19	6.28
24	.05	2.92	3.53	3.90	4.17	4.37	4.54	4.68	4.81	4.92	5.01	5.10
	.01	3.96	4.55	4.91	5.17	5.37	5.54	5.69	5.81	5.92	6.02	6.11
30	.05	2.89	3.49	3.85	4.10	4.30	4.46	4.60	4.72	4.82	4.92	5.00
	.01	3.89	4.45	4.80	5.05	5.24	5.40	5.54	5.65	5.76	5.85	5.93
40	.05	2.86	3.44	3.79	4.04	4.23	4.39	4.52	4.63	4.73	4.82	4.90
	.01	3.82	4.37	4.70	4.93	5.11	5.26	5.39	5.50	5.60	5.69	5.76
60	.05	2.83	3.40	3.74	3.98	4.16	4.31	4.44	4.55	4.65	4.73	4.81
	.01	3.76	4.28	4.59	4.82	4.99	5.13	5.25	5.36	5.45	5.53	5.60
120	.05	2.80	3.36	3.68	3.92	4.10	4.24	4.36	4.47	4.56	4.64	4.71
	.01	3.70	4.20	4.50	4.71	4.87	5.01	5.12	5.21	5.30	5.37	5.44
∞	.05	2.77	3.31	3.63	3.86	4.03	4.17	4.29	4.39	4.47	4.55	4.62
	.01	3.64	4.12	4.40	4.60	4.76	4.88	4.99	5.08	5.16	5.23	5.29

Table A.11   Chi-Squared Curve Tail Areas

Upper-Tail Area	$\nu = 1$	$\nu = 2$	$\nu = 3$	$\nu = 4$	$\nu = 5$
> .100	< 2.70	< 4.60	< 6.25	< 7.77	< 9.23
.100	2.70	4.60	6.25	7.77	9.23
.095	2.78	4.70	6.36	7.90	9.37
.090	2.87	4.81	6.49	8.04	9.52
.085	2.96	4.93	6.62	8.18	9.67
.080	3.06	5.05	6.75	8.33	9.83
.075	3.17	5.18	6.90	8.49	10.00
.070	3.28	5.31	7.06	8.66	10.19
.065	3.40	5.46	7.22	8.84	10.38
.060	3.53	5.62	7.40	9.04	10.59
.055	3.68	5.80	7.60	9.25	10.82
.050	3.84	5.99	7.81	9.48	11.07
.045	4.01	6.20	8.04	9.74	11.34
.040	4.21	6.43	8.31	10.02	11.64
.035	4.44	6.70	8.60	10.34	11.98
.030	4.70	7.01	8.94	10.71	12.37
.025	5.02	7.37	9.34	11.14	12.83
.020	5.41	7.82	9.83	11.66	13.38
.015	5.91	8.39	10.46	12.33	14.09
.010	6.63	9.21	11.34	13.27	15.08
.005	7.87	10.59	12.83	14.86	16.74
.001	10.82	13.81	16.26	18.46	20.51
< .001	> 10.82	> 13.81	> 16.26	> 18.46	> 20.51

Upper-Tail Area	$\nu = 6$	$\nu = 7$	$\nu = 8$	$\nu = 9$	$\nu = 10$
> .100	< 10.64	< 12.01	< 13.36	< 14.68	< 15.98
.100	10.64	12.01	13.36	14.68	15.98
.095	10.79	12.17	13.52	14.85	16.16
.090	10.94	12.33	13.69	15.03	16.35
.085	11.11	12.50	13.87	15.22	16.54
.080	11.28	12.69	14.06	15.42	16.75
.075	11.46	12.88	14.26	15.63	16.97
.070	11.65	13.08	14.48	15.85	17.20
.065	11.86	13.30	14.71	16.09	17.44
.060	12.08	13.53	14.95	16.34	17.71
.055	12.33	13.79	15.22	16.62	17.99
.050	12.59	14.06	15.50	16.91	18.30
.045	12.87	14.36	15.82	17.24	18.64
.040	13.19	14.70	16.17	17.60	19.02
.035	13.55	15.07	16.56	18.01	19.44
.030	13.96	15.50	17.01	18.47	19.92
.025	14.44	16.01	17.53	19.02	20.48
.020	15.03	16.62	18.16	19.67	21.16
.015	15.77	17.39	18.97	20.51	22.02
.010	16.81	18.47	20.09	21.66	23.20
.005	18.54	20.27	21.95	23.58	25.18
.001	22.45	24.32	26.12	27.87	29.58
< .001	> 22.45	> 24.32	> 26.12	> 27.87	> 29.58

(continued)

Table A.11   Chi-Squared Curve Tail Areas *(cont.)*

Upper-Tail Area	$\nu = 11$	$\nu = 12$	$\nu = 13$	$\nu = 14$	$\nu = 15$
> .100	< 17.27	< 18.54	< 19.81	< 21.06	< 22.30
.100	17.27	18.54	19.81	21.06	22.30
.095	17.45	18.74	20.00	21.26	22.51
.090	17.65	18.93	20.21	21.47	22.73
.085	17.85	19.14	20.42	21.69	22.95
.080	18.06	19.36	20.65	21.93	23.19
.075	18.29	19.60	20.89	22.17	23.45
.070	18.53	19.84	21.15	22.44	23.72
.065	18.78	20.11	21.42	22.71	24.00
.060	19.06	20.39	21.71	23.01	24.31
.055	19.35	20.69	22.02	23.33	24.63
.050	19.67	21.02	22.36	23.68	24.99
.045	20.02	21.38	22.73	24.06	25.38
.040	20.41	21.78	23.14	24.48	25.81
.035	20.84	22.23	23.60	24.95	26.29
.030	21.34	22.74	24.12	25.49	26.84
.025	21.92	23.33	24.73	26.11	27.48
.020	22.61	24.05	25.47	26.87	28.25
.015	23.50	24.96	26.40	27.82	29.23
.010	24.72	26.21	27.68	29.14	30.57
.005	26.75	28.29	29.81	31.31	32.80
.001	31.26	32.90	34.52	36.12	37.69
< .001	> 31.26	> 32.90	> 34.52	> 36.12	> 37.69

Upper-Tail Area	$\nu = 16$	$\nu = 17$	$\nu = 18$	$\nu = 19$	$\nu = 20$
> .100	< 23.54	< 24.77	< 25.98	< 27.20	< 28.41
.100	23.54	24.76	25.98	27.20	28.41
.095	23.75	24.98	26.21	27.43	28.64
.090	23.97	25.21	26.44	27.66	28.88
.085	24.21	25.45	26.68	27.91	29.14
.080	24.45	25.70	26.94	28.18	29.40
.075	24.71	25.97	27.21	28.45	29.69
.070	24.99	26.25	27.50	28.75	29.99
.065	25.28	26.55	27.81	29.06	30.30
.060	25.59	26.87	28.13	29.39	30.64
.055	25.93	27.21	28.48	29.75	31.01
.050	26.29	27.58	28.86	30.14	31.41
.045	26.69	27.99	29.28	30.56	31.84
.040	27.13	28.44	29.74	31.03	32.32
.035	27.62	28.94	30.25	31.56	32.85
.030	28.19	29.52	30.84	32.15	33.46
.025	28.84	30.19	31.52	32.85	34.16
.020	29.63	30.99	32.34	33.68	35.01
.015	30.62	32.01	33.38	34.74	36.09
.010	32.00	33.40	34.80	36.19	37.56
.005	34.26	35.71	37.15	38.58	39.99
.001	39.25	40.78	42.31	43.81	45.31
< .001	> 39.25	> 40.78	> 42.31	> 43.81	> 45.31

Table A.12   Approximate Critical Values for the
Ryan-Joiner Test of Normality

		α		
		.10	**.05**	**.01**
	4	.8951	.8734	.8318
	5	.9033	.8804	.8319
	6	.9114	.8893	.8409
	7	.9186	.8978	.8517
	8	.9248	.9054	.8622
	9	.9301	.9121	.8718
	10	.9347	.9179	.8804
	11	.9387	.9230	.8880
	12	.9422	.9275	.8947
	13	.9454	.9315	.9008
	14	.9481	.9351	.9061
n	15	.9506	.9383	.9109
	16	.9529	.9411	.9153
	17	.9549	.9437	.9192
	18	.9567	.9461	.9228
	19	.9584	.9483	.9260
	20	.9600	.9503	.9290
	25	.9662	.9582	.9407
	30	.9707	.9639	.9490
	40	.9767	.9715	.9597
	50	.9807	.9764	.9664
	60	.9835	.9799	.9709
	75	.9865	.9835	.9756

Source: Minitab Reference Manual.

Table A.13 Critical Values for the Wilcoxon Signed-Rank Test $\qquad$ $P_0(S_+ \geq c_1) = P(S_+ \geq c_1$ when H_0 is true$)$

n	c_1	$P_0(S_+ \geq c_1)$	n	c_1	$P_0(S_+ \geq c_1)$
3	6	.125		78	.011
4	9	.125		79	.009
	10	.062		81	.005
5	13	.094	14	73	.108
	14	.062		74	.097
	15	.031		79	.052
6	17	.109		84	.025
	19	.047		89	.010
	20	.031		92	.005
	21	.016	15	83	.104
7	22	.109		84	.094
	24	.055		89	.053
	26	.023		90	.047
	28	.008		95	.024
8	28	.098		100	.011
	30	.055		101	.009
	32	.027		104	.005
	34	.012	16	93	.106
	35	.008		94	.096
	36	.004		100	.052
9	34	.102		106	.025
	37	.049		112	.011
	39	.027		113	.009
	42	.010		116	.005
	44	.004	17	104	.103
10	41	.097		105	.095
	44	.053		112	.049
	47	.024		118	.025
	50	.010		125	.010
	52	.005		129	.005
11	48	.103	18	116	.098
	52	.051		124	.049
	55	.027		131	.024
	59	.009		138	.010
	61	.005		143	.005
12	56	.102	19	128	.098
	60	.055		136	.052
	61	.046		137	.048
	64	.026		144	.025
	68	.010		152	.010
	71	.005		157	.005
13	64	.108	20	140	.101
	65	.095		150	.049
	69	.055		158	.024
	70	.047		167	.010
	74	.024		172	.005

Table A.14   Critical Values for the Wilcoxon Rank-Sum Test $\qquad P_0(W \geq c) = P(W \geq c$ when H_0 is true$)$

m	n	c	$P_0(W \geq c)$	m	n	c	$P_0(W \geq c)$
3	3	15	.05			40	.004
	4	17	.057		6	40	.041
		18	.029			41	.026
	5	20	.036			43	.009
		21	.018			44	.004
	6	22	.048		7	43	.053
		23	.024			45	.024
		24	.012			47	.009
	7	24	.058			48	.005
		26	.017		8	47	.047
		27	.008			49	.023
	8	27	.042			51	.009
		28	.024			52	.005
		29	.012	6	6	50	.047
		30	.006			52	.021
4	4	24	.057			54	.008
		25	.029			55	.004
		26	.014		7	54	.051
	5	27	.056			56	.026
		28	.032			58	.011
		29	.016			60	.004
		30	.008		8	58	.054
	6	30	.057			61	.021
		32	.019			63	.01
		33	.010			65	.004
		34	.005	7	7	66	.049
	7	33	.055			68	.027
		35	.021			71	.009
		36	.012			72	.006
		37	.006		8	71	.047
	8	36	.055			73	.027
		38	.024			76	.01
		40	.008			78	.005
		41	.004	8	8	84	.052
5	5	36	.048			87	.025
		37	.028			90	.01
		39	.008			92	.005
		39	.008			92	.005

Table A.15   Critical Values for the Wilcoxon Signed-Rank Interval $(\bar{x}_{(n(n+1)/2-c+1)}, \bar{x}_{(c)})$

n	Confidence Level (%)	c	n	Confidence Level (%)	c	n	Confidence Level (%)	c
5	93.8	15	13	99.0	81	20	99.1	173
	87.5	14		95.2	74		95.2	158
6	96.9	21		90.6	70		90.3	150
	93.7	20	14	99.1	93	21	99.0	188
	90.6	19		95.1	84		95.0	172
7	98.4	28		89.6	79		89.7	163
	95.3	26	15	99.0	104	22	99.0	204
	89.1	24		95.2	95		95.0	187
8	99.2	36		90.5	90		90.2	178
	94.5	32	16	99.1	117	23	99.0	221
	89.1	30		94.9	106		95.2	203
9	99.2	44		89.5	100		90.2	193
	94.5	39	17	99.1	130	24	99.0	239
	90.2	37		94.9	118		95.1	219
10	99.0	52		90.2	112		89.9	208
	95.1	47	18	99.0	143	25	99.0	257
	89.5	44		95.2	131		95.2	236
11	99.0	61		90.1	124		89.9	224
	94.6	55	19	99.1	158			
	89.8	52		95.1	144			
12	99.1	71		90.4	137			
	94.8	64						
	90.8	61						

Table A.16   Critical Values for the Wilcoxon Rank-Sum Interval $(d_{ij(mn-c+1)}, d_{ij(c)})$

	Smaller Sample Size											
	5			**6**			**7**			**8**		
Larger Sample Size	**Confidence Level (%)**		*c*	**Confidence Level (%)**		*c*	**Confidence Level (%)**		*c*	**Confidence Level (%)**		*c*
5	99.2		25									
	94.4		22									
	90.5		21									
6	99.1		29	99.1		34						
	94.8		26	95.9		31						
	91.8		25	90.7		29						
7	99.0		33	99.2		39	98.9		44			
	95.2		30	94.9		35	94.7		40			
	89.4		28	89.9		33	90.3		38			
8	98.9		37	99.2		44	99.1		50	99.0		56
	95.5		34	95.7		40	94.6		45	95.0		51
	90.7		32	89.2		37	90.6		43	89.5		48
9	98.8		41	99.2		49	99.2		56	98.9		62
	95.8		38	95.0		44	94.5		50	95.4		57
	88.8		35	91.2		42	90.9		48	90.7		54
10	99.2		46	98.9		53	99.0		61	99.1		69
	94.5		41	94.4		48	94.5		55	94.5		62
	90.1		39	90.7		46	89.1		52	89.9		59
11	99.1		50	99.0		58	98.9		66	99.1		75
	94.8		45	95.2		53	95.6		61	94.9		68
	91.0		43	90.2		50	89.6		57	90.9		65
12	99.1		54	99.0		63	99.0		72	99.0		81
	95.2		49	94.7		57	95.5		66	95.3		74
	89.6		46	89.8		54	90.0		62	90.2		70

	Smaller Sample Size											
	9			**10**			**11**			**12**		
Larger Sample Size	**Confidence Level (%)**		*c*	**Confidence Level (%)**		*c*	**Confidence Level (%)**		*c*	**Confidence Level (%)**		*c*
9	98.9		69									
	95.0		63									
	90.6		60									
10	99.0		76	99.1		84						
	94.7		69	94.8		76						
	90.5		66	89.5		72						
11	99.0		83	99.0		91	98.9		99			
	95.4		76	94.9		83	95.3		91			
	90.5		72	90.1		79	89.9		86			
12	99.1		90	99.1		99	99.1		108	99.0		116
	95.1		82	95.0		90	94.9		98	94.8		106
	90.5		78	90.7		86	89.6		93	89.9		101

Table A.17 β Curves for t Tests

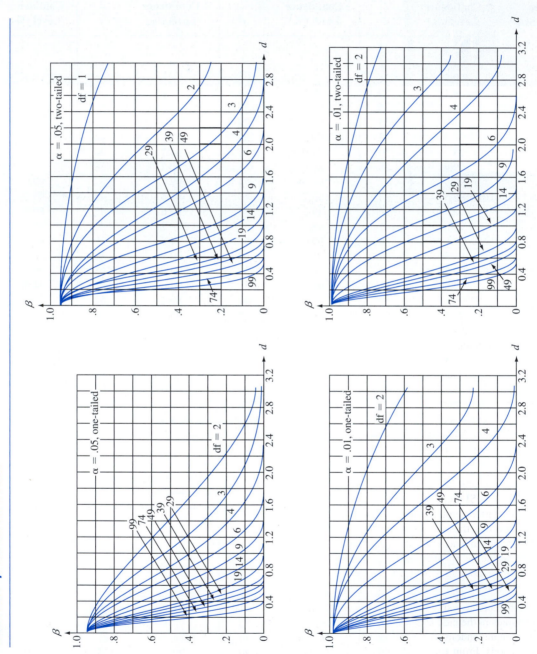

Answers to Selected Odd-Numbered Exercises

1. a. *Los Angeles Times, Oberlin Tribune, Gainesville Sun, Washington Post*
b. Duke Energy, Clorox, Seagate, Neiman Marcus
c. Vince Correa, Catherine Miller, Michael Cutler, Ken Lee
d. 2.97, 3.56, 2.20, 2.97

3. a. How likely is it that more than half of the sampled computers will need or have needed warranty service? What is the expected number among the 100 that need warranty service? How likely is it that the number needing warranty service will exceed the expected number by more than 10?
b. Suppose that 15 of the 100 sampled needed warranty service. How confident can we be that the proportion of *all* such computers needing warranty service is between .08 and .22? Does the sample provide compelling evidence for concluding that more than 10% of all such computers need warranty service?

5. a. No. All students taking a large statistics course who participate in an SI program of this sort.
b. Randomization protects against various biases and helps ensure that those in the SI group are as similar as possible to the students in the control group.
c. There would be no firm basis for assessing the effectiveness of SI (nothing to which the SI scores could reasonably be compared).

7. One could generate a simple random sample of all single-family homes in the city, or a stratified random sample by taking a simple random sample from each of the 10 district neighborhoods. From each of the selected homes, values of all desired variables would be determined. This would be an enumerative study because there exists a finite, identifiable population of objects from which to sample.

9. a. Possibly measurement error, recording error, differences in environmental conditions at the time of measurement, etc.

b. No. There is no sampling frame.

11.

3L	1
3H	56678
4L	000112222234
4H	5667888
5L	144
5H	58
6L	2
6H	6678
7L	
7H	5

stem: tenths digit

The stem-and-leaf display shows that .45 is a good representative value for the data. In addition, the display is not symmetric and appears to be positively skewed. The range of the data is $.75 - .31 = .44$, which is comparable to the typical value of .45. This constitutes a reasonably large amount of variation in the data. The data value .75 is a possible outlier.

13. a.

12	2
12	445
12	6667777
12	889999
13	00011111111
13	222222222233333333333333
13	44444444444444444445555555555555555555
13	666666666666777777777777
13	888888888888999999
14	0000001111
14	2333333
14	444
14	77

leaf: ones digit

symmetry

b. Close to bell-shaped, center ≈ 135, not insignificant dispersion, no gaps or outliers.

15.

	Am		Fr
		8	1
157020153504		9	00645632
9324		10	2563
6306		11	6913
Stem: Hundreds and tens 058		12	325528
Leaf: Ones	8	13	7
		14	
		15	8
	2	16	

Representative values: low 100s for Am and low 110's for Fr. Somewhat more variability in Fr times than in Am times. More extreme positive skew for Am than for Fr. 162 is an Am outlier, and 158 is perhaps a Fr outlier.

17. a. .639, .510   **b.** .491, .315
c. The relative frequencies are .065, .185, .241, .148, .102, .083, .056, .074, .028, and .019. The histogram is close to being unimodal (the peak at 9 would likely disappear with a significantly larger sample size) and positively skewed. A typical value of x here is 5, but there is substantial variability about that typical value. The data set contains no outliers.

19. a. .99 (99%), .71 (71%)   **b.** .64 (64%), .44 (44%)
c. Strictly speaking, the histogram is not unimodal, but is close to being so with a moderate positive skew. A much larger sample size would likely give a smoother picture.

21. a.

y	Freq.	Rel. freq.
0	17	.362
1	22	.468
2	6	.128
3	1	.021
4	0	.000
5	1	.021
	47	1.000

.362, .638

b.

z	Freq.	Rel. freq.
0	13	.277
1	11	.234
2	3	.064
3	7	.149
4	5	.106
5	3	.064
6	3	.064
7	0	.000
8	2	.043
	47	1.001

.894, .830

23. The class widths are not equal, so the density scale must be used. The densities for the six classes are .2030, .1373, .0303, .0086, .0021, and .0009, respectively. The resulting histogram is unimodal with a very substantial positive skew.

25.

Class	Freq.	Class	Freq.
10–<20	8	1.1–<1.2	2
20–<30	14	1.2–<1.3	6
30–<40	8	1.3–<1.4	7
40–<50	4	1.4–<1.5	9
50–<60	3	1.5–<1.6	6
60–<70	2	1.6–<1.7	4
70–<80	1	1.7–<1.8	5
	40	1.8–<1.9	1
			40

Original: positively skewed;
Transformed much more symmetric, not far from bell-shaped.

27. a. The observation 50 falls on a class boundary.
b.

Class	Freq.	Rel. freq.
0–<50	9	.18
50–<100	19	.38
100–<150	11	.22
150–<200	4	.08
200–<300	4	.08
300–<400	2	.04
400–<500	0	.00
500–<600	1	.02
	50	1.00

A representative (central) value is either a bit below or a bit above 100, depending on how one measures center. There is a great deal of variability in lifetimes, especially in values at the upper end of the data. There are several candidates for outliers.

c.

Class	Freq.	Rel. freq.
2.25–<2.75	2	.04
2.75–<3.25	2	.04
3.25–<3.75	3	.06
3.75–<4.25	8	.16
4.25–<4.75	18	.36
4.75–<5.25	10	.20
5.25–<5.75	4	.08
5.75–<6.25	3	.06
	50	1.00

There is much more symmetry in the distribution of the $\ln(x)$ values than in the x values themselves, and less variability. There are no longer gaps or obvious outliers.
d. .38, .14

29. A: .28 B: .19 C: .18
D: .17 E: .09 F: .09

31.

Class	Freq.	Cum. freq.	Cum. rel. freq.
0–<4	2	2	.050
4–<8	14	16	.400
8–<12	11	27	.675
12–<16	8	35	.875
16–<20	4	39	.975
20–<24	0	39	.975
24–<28	1	40	1.000

33. a. 640.5, 582.5
b. 610.5, 582.5
c. 591.2
d. 593.71

35. a. 1.237, .56; positive skew
b. 1.118; in between the two
c. .36

37. $\bar{x}_{tr(10)} = 11.46$

39. a. $\bar{x} = 1.0297, \tilde{x} = 1.009$ **b.** .383

41. a. .7   **b.** Also .7   **c.** 13

43. $\tilde{x} = 68.0, \bar{x}_{tr(20)} = 66.2, \bar{x}_{tr(30)} = 67.5$

45. a. $\bar{x} = 115.58$; the deviations are .82, .32, $-$.98, $-$.38, .22
   b. .482, .694   **c.** .482   **d.** .482

47. a. $\bar{x} = 14.88, \tilde{x} = 14.70$ **b.** .837   **c.** .837

49. a. 56.80, 197.8040   **b.** .5016, .708

51. a. 1264.766, 35.564   **b.** .351, .593

53. a. Bal: 1.121, 1.050, .536
   Gr: 1.244, 1.100, .448
   b. Typical ratios are quite similar for the two types. There is somewhat more variability in the Bal sample, due primarily to the two outliers (one mild, one extreme). For Bal, there is substantial symmetry in the middle 50% but positive skewness overall. For Gr, there is substantial positive skew in the middle 50% and mild positive skewness overall.

55. a. 33   **b.** No
   c. Slight positive skewness in the middle half, but rather symmetric overall. The extent of variability appears substantial.
   d. At most 32

57. a. Yes. 125.8 is an extreme outlier and 250.2 is a mild outlier.
   b. In addition to the presence of outliers, there is positive skewness both in the middle 50% of the data and, excepting the outliers, overall. Except for the two outliers, there appears to be a relatively small amount of variability in the data.

59. a. ED: .4, .10, 2.75, 2.65;
   Non-Ed: 1.60, .30, 7.90, 7.60
   b. ED: 8.9 and 9.2 are mild outliers, and 11.7 and 21.0 are extreme outliers.
   There are not outliers in the non-ED sample.
   c. Four outliers for ED, none for non-ED. Substantial positive skewness in both samples; less variability in ED (smaller f_s), and non-ED observations tend to be somewhat larger than ED observations.

61. Outliers, both mild and extreme, only at 6 A.M. Distributions at other times are quite symmetric. Variability increases somewhat until 2 P.M. and then decreases slightly, and the same is true of "typical" gasoline-vapor coefficient values.

63. $\bar{x} = 64.89, \tilde{x} = 64.70, s = 7.803$, lower $4^{th} = 57.8$, upper $4^{th} = 70.4, f_s = 12.6$. A histogram consisting of 8 classes starting at 52, each of width 4, is bimodal but close to unimodal with a positive skew. A boxplot shows no outliers,

there is a very mild negative skew in the middle 50%, and the upper whisker is much longer than the lower whisker.
   b. .9231, .9053
   c. .48

67. a. M: $\bar{x} = 3.64, \tilde{x} = 3.70, s = .269, f_s = .40$
   F: $\bar{x} = 3.28, \tilde{x} = 3.15, s = .478, f_s = .50$
   Female values are typically somewhat smaller than male values, and show somewhat more variability. An M boxplot shows negative skew whereas an F boxplot shows positive skew.
   b. F: $\bar{x}_{tr(10)} = 3.24$ M: $\bar{x}_{tr(10)} = 3.652 \approx 3.65$

69. a. $\bar{y} = a\bar{x} + b, s_y^2 = a^2 s_x^2$ **b.** 189.14, 1.87

71. a. The mean, median, and trimmed mean are virtually identical, suggesting a substantial amount of symmetry in the data; the fact that the quartiles are roughly the same distance from the median and that the smallest and largest observations are roughly equidistant from the center provides additional support for symmetry. The standard deviation is quite small relative to the mean and median.
   b. See the comments of (a). In addition, using $1.5(Q3 - Q1)$ as a yardstick, the two largest and three smallest observations are mild outliers.

73. $\bar{x} = .9255, s = .0809, \tilde{x} = .93$, small amount of variability, slight bit of skewness.

75. a. The "five-number summaries" ($\tilde{x}$, the two fourths, and the smallest and largest observations) are identical and there are no outliers, so the three individual boxplots are identical.
   b. Differences in variability, nature of gaps, and existence of clusters for the three samples.
   c. No. Detail is lost.

77. c. Representative depths are quite similar for the four types of soils—between 1.5 and 2. Data from the C and CL soils shows much more variability than for the other two types. The boxplots for the first three types show substantial positive skewness both in the middle 50% and overall. The boxplot for the SYCL soil shows negative skewness in the middle 50% and mild positive skewness overall. Finally, there are multiple outliers for the first three types of soils, including extreme outliers.

79. a. $\bar{x}_{n+1} = (n\bar{x}_n + x_{n+1})/(n + 1)$
   c. 12.53, .532

81. A substantial positive skew (assuming unimodality)

83. a. All points fall on a 45° line. Points fall below a 45° line.
   b. Points fall well below a 45° line, indicating a substantial positive skew.

Chapter 2

1. a. $\mathcal{S} = \{1324, 3124, 1342, 3142, 1423, 1432, 4123, 4132,$
   $2314, 2341, 3214, 3241, 2413, 2431, 4213, 4231\}$
   b. $A = \{1324, 1342, 1423, 1432\}$

   c. $B = \{2314, 2341, 3214, 3241, 2413, 2431, 4213, 4231\}$
   d. $A \cup B = \{1324, 1342, 1423, 1432, 2314, 2341, 3214,$
   $3241, 2413, 2431, 4213, 4231\},$

$A \cap B$ contains no outcomes (A and B are disjoint), $A' = \{3124, 3142, 4123, 4132, 2314, 2341, 3214, 3241, 2413, 2431, 4213, 4231\}$

3. a. $A = \{SSF, SFS, FSS\}$
b. $B = \{SSF, SFS, FSS, SSS\}$
c. $C = \{SFS, SSF, SSS\}$
d. $C' = \{FFF, FSF, FFS, FSS, SFF\}$,
$A \cup C = \{SSF, SFS, FSS, SSS\}$,
$A \cap C = \{SSF, SFS\}$,
$B \cup C = \{SSF, SFS, FSS, SSS\} = B$,
$B \cap C = \{SSF, SFS, SSS\} = C$

5. a. $\mathscr{S} = \{(1, 1, 1), (1, 1, 2), (1, 1, 3), (1, 2, 1), (1, 2, 2), (1, 2, 3),$
$(1, 3, 1), (1, 3, 2), (1, 3, 3), (2, 1, 1), (2, 1, 2), (2, 1, 3), (2, 2, 1),$
$(2, 2, 2), (2, 2, 3), (2, 3, 1), (2, 3, 2), (2, 3, 3), (3, 1, 1), (3, 1, 2),$
$(3, 1, 3), (3, 2, 1), (3, 2, 2), (3, 2, 3), (3, 3, 1), (3, 3, 2), (3, 3, 3)\}$
b. $\{(1, 1, 1), (2, 2, 2), (3, 3, 3)\}$ **c.** $\{(1, 2, 3), (1, 3, 2),$
$(2, 1, 3), (2, 3, 1), (3, 1, 2), (3, 2, 1)\}$ **d.** $\{(1, 1, 1), (1, 1, 3),$
$(1, 3, 1), (1, 3, 3), (3, 1, 1), (3, 1, 3), (3, 3, 1), (3, 3, 3)\}$

7. a. There are 35 outcomes in $\mathscr{S}$. **b.** {*AABABAB,*
AABAABB, AAABBAB, AAABABB, AAAABBB}

11. a. .07 **b.** .30 **c.** .57

13. a. .36 **b.** .64 **c.** .53
d. .47 **e.** .17 **f.** .75

15. a. .572 **b.** .879

17. a. There are statistical software packages other than SPSS and SAS.
b. .70 **c.** .80 **d.** .20

19. a. .8841 **b.** .0435

21. a. .10 **b.** .18, .19 **c.** .41 **d.** .59
e. .31 **f.** .69

23. a. .067 **b.** .400 **c.** .933 **d.** .533

25. a. .85 **b.** .15 **c.** .22 **d.** .35

27. a. .1 **b.** .7 **c.** .6

29. a. 676; 1296 **b.** 17,576; 46,656
c. 456,976; 1,679,616 **d.** .942

31. a. 45 **b.** 1440 days (almost 4 years)

33. a. 1,816,214,440 **b.** 659,067,881,572,000
c. 9,072,000

35. a. 38,760, .0048 **b.** .0054 **c.** .9946
d. .2885

37. a. 60 **b.** 10 **c.** .0456

39. a. .145 **b.** .075 **c.** .264 **d.** .154

41. a. 10,000 **b.** .9876 **c.** .03 **d.** .0337

43. .000394, .00394, .00001539

45. a. .447, .500, .200 **b.** .400, .447 **c.** .211

47. a. .73 **b.** .22 **c.** .50; among all those who have a Visa card, 50% have MasterCards; .75
d. .40 **e.** .85

49. a. .34, .40 **b.** .588 **c.** .50

51. a. .0312 **b.** .024

53. .083

55. .236

59. a. .21 **b.** .455 **c.** .264, .462, .274

61. a. .578, .278, .144 **b.** 0, .457, .543

63. b. .54 **c.** .68 **d.** .74 **e.** .7941

65. .087, .652, .261

67. .000329; very uneasy.

69. a. .126 **b.** .05 **c.** .1125
d. .2725 **e.** .5325 **f.** .2113

71. a. 300 **b.** .820 **c.** .146

75. .401, .722

77. a. .00648 **b.** .00421

79. .0059

81. a. .95

83. a. .10, .20 **b.** 0

85. a. $p(2 - p)$ **b.** $1 - (1 - p)^n$ **c.** $(1 - p)^3$
d. $.9 + (1 - p)^3(.1)$
e. $.1(1 - p)^3/[.9 + .1(1 - p)^3] = .0137$ for $p = .5$

87. a. .40 **b.** .571
c. No: $.571 \neq .65$, and also $.40 \neq (.65)(.7)$ **d.** .733

89. $[2\pi(1 - \pi)]/(1 - \pi^2)$

91. a. .333, .444 **b.** .150 **c.** .291

93. .45, .32

95. a. .0083 **b.** .2 **c.** .2 **d.** .1074

97. .905

99. a. .974 **b.** .9754

101. .926

103. a. .008 **b.** .018 **c.** .601

105. a. .883, .117 **b.** 23 **c.** .156

107. $1 - (1 - p_1)(1 - p_2) \cdots (1 - p_n)$

109. a. .0417 **b.** .375

111. $P(\text{hire \#1}) = 6/24$ for $s = 0$, $= 11/24$ for $s = 1$, $= 10/24$ for $s = 2$, and $= 6/24$ for $s = 3$, so $s = 1$ is best.

113. $1/4 = P(A_1 \cap A_2 \cap A_3)$
$\neq P(A_1) \cdot P(A_2) \cdot P(A_3) = 1/8$

Chapter 3

1. $x = 0$ for *FFF;* $x = 1$ for *SFF, FSF,* and *FFS;* $x = 2$ for *SSF, SFS,* and *FSS;* and $x = 3$ for *SSS*

3. Z = average of the two numbers, with possible values 2/2, 3/2, . . . , 12/2; W = absolute value of the difference, with possible values 0, 1, 2, 3, 4, 5

5. No. In Example 3.4, let $Y = 1$ if at most three batteries are examined and let $Y = 0$ otherwise. Then Y has only two values.

7. a. {0, 1, . . . , 12} ; discrete   **c.** {1, 2, 3, . . .}; discrete **e.** $\{z_1, z_2, . . . , z_N\}$, discrete because there are only a finite number N of different sales tax percentages across the entire country   **g.** $\{x : m \le x \le M\}$ where m (M) is the minimum (maximum) possible tension; continuous

9. a. {2, 4, 6, 8, . . .} , that is, {2(1), 2(2), 2(3), 2(4), . . .}, an infinite sequence; discrete
b. {2, 3, 4, 5, 6, . . .} , that is, {1 + 1, 1 + 2, 1 + 3, 1 + 4, . . .}, an infinite sequence; discrete

11. b. .55, .25   **c.** .70

13. a. .70   **b.** .45   **c.** .55
d. .71   **e.** .65   **f.** .45

15. a. (1, 2), (1, 3), (1, 4), (1, 5), (2, 3), (2, 4), (2, 5), (3, 4), (3, 5), (4, 5)   **b.** $p(0) = .3, p(1) = .6, p(2) = .1$
c. $F(x) = 0$ for $x < 0$, = .3 for $0 \le x < 1$, = .9 for $1 \le x < 2$, and = 1 for $2 \le x$

17. a. .81   **b.** .162   **c.** It is *A; AUUUA, UAUUA, UUAUA, UUUAA;* .00324

19. $p(0) = .09, p(1) = .40, p(2) = .32, p(3) = .19$

21. b. $p(x) = .301, .176, .125, .097, .079, .067, .058, .051,$ .046 for $x = 1, 2, . . . , 9$
c. $F(x) = 0$ for $x < 1$, = .301 for $1 \le x < 2$, = .477 for $2 \le x < 3$, . . . , = .954 for $8 \le x < 9$, = 1 for $x \ge 9$
d. .602, .301

23. a. .20   **b.** .33   **c.** .78   **d.** .53

25. a. $p(y) = (1 - p)^y \cdot p$ for $y = 0, 1, 2, 3, . . .$

27. a. 1234, 1243, 1324, . . . , 4321
b. $p(0) = 9/24, p(1) = 8/24, p(2) = 6/24, p(3) = 0,$ $p(4) = 1/24$

29. a. 6.45   **b.** 15.6475   **c.** 3.96   **d.** 15.6475

31. .4.49, 2.12, .68

33. a. p **b.** $p(1 - p)$ **c.** p

35. $E[h_3(X)] = \$4.93, E[h_4(X)] = \$5.33$, so 4 copies is better.

37. $E(X) = (n + 1)/2$, $E(X^2) = (n + 1)(2n + 1)/6, V(X) = (n^2 - 1)/12$

39. 2.3, .81, 88.5, 20.25

43. $E(X - c) = E(X) - c, E(X - \mu) = 0$

47. a. .001   **b.** .001   **c.** .147 **d.** .001
e. 1.000   **f.** .001

49. a. .354   **b.** .115   **c.** .918

51. a. 6.25   **b.** 2.17   **c.** .030

53. a. .403   **b.** .787   **c.** .774

55. .1478

57. .407, independence

59. a. .017   **b.** .811, .425   **c.** .006, .902, .586

61. When $p = .9$, the probability is .99 for A and .9963 for B. If $p = .5$, these probabilities are .75 and .6875, respectively.

63. The tabulation for $p > .5$ is unnecessary.

65. a. 20, 16   **b.** 70, 21

67. $P(|X - \mu| \ge 2\sigma) = .042$ when $p = .5$ and = .065 when $p = .75$, compared to the upper bound of .25. Using $k = 3$ in place of $k = 2$, these probabilities are .002 and .004, respectively, whereas the upper bound is .11.

69. .379, .879   **b.** .121   **c.** Use the binomial distribution with $n = 15, p = .10$

71. a. $h(x; 15, 10, 20)$ for $x = 5, . . . , 10$
b. .0325   **c.** .697

73. a. $h(x; 10, 10, 20)$ **b.** .033   **c.** $h(x; n, n, 2n)$

75. a. $nb(x; 2, .2)$ **b.** .0768   **c.** .1808
d. 8, 10

77. $nb(x; 6, .5)$, 6

79. a. .999   **b.** .184   **c.** .260   **d.** .080

81. a. .011   **b.** .441   **c.** .554, .459
d. .945

83. Poisson(5)   **a.** .492   **b.** .133

85. a. .122, .809, .283   **b.** 12, 3.464
c. .530, .011

87. a. .099   **b.** .135   **c.** 2

89. a. 4   **b.** .215   **c.** At least $-\ln(.1)/2 \approx$ 1.1513 years

91. a. .221   **b.** 6,800,000   **c.** $p(x; 20.106)$

95. b. 3.114, .405, .636

97. a. $b(x; 15, .75)$ **b.** .686
c. .313   **d.** 11.25, 2.81   **e.** .310

99. .991

101. a. $p(x; 2.5)$ **b.** .067   **c.** .109

103. 1.813, 3.05

105. $p(2) = p^2$, $p(3) = (1-p)p^2$, $p(4) = (1-p)p^2$, $p(x) = [1 - p(2) - \cdots - p(x-3)](1-p)p^2$ for $x = 5, 6, 7, \ldots$; .99950841

107. a. .0029   **b.** .0767, .9702

109. a. .135   **b.** .00144   **c.** $\sum_{x=0}^{\infty} [p(x; 2)]^5$

111. 3.590

113. a. No   **b.** .0273

115. b. $.5\mu_1 + .5\mu_2$ **c.** $.25(\mu_1 - \mu_2)^2 + .5(\mu_1 + \mu_2)$
d. .6 and .4 replace .5 and .5, respectively.

117. $\sum_{i=1}^{10} (p_{i+j+1} + p_{i-j-1})p_i$, where $p_k = 0$ if $k < 0$ or $k > 10$.

121. a. 2.50   **b.** 3.1

Chapter 4

1. b. .4625, same   **c.** .5, .278125

3. b. .5   **c.** .6875   **d.** .6328

5. a. .375   **b.** .125   **c.** .297   **d.** .578

7. b. .309   **c.** .494 ($\mu = 2.225$ by symmetry of the density curve)   **d.** .247

9. a. .451, .549   **b.** .312

11. a. .25   **b.** .1875   **c.** .4375   **d.** 1.4142
e. $f(x) = x/2$ for $0 < x < 2$ **f.** 1.33
g. .222, .471   **h.** 2

13. a. 3   **b.** 0 for $x \le 1$, $1 - x^{-3}$ for $x > 1$
c. .125, .088   **d.** 1.5, .866   **e.** .924

15. a. $F(x) = 0$ for $x \le 0$, $= 90\left[\frac{x^9}{9} - \frac{x^{10}}{10}\right]$ for $0 < x < 1$, $= 1$ for $x \ge 1$ **b.** .0107   **c.** .0107, .0107
d. .9036   **e.** .818, .111   **f.** .3137

17. a. $A + (B - A)p$ **b.** $E(X) = (A + B)/2$, $\sigma_X = (B - A)/\sqrt{12}$
c. $[B^{n+1} - A^{n+1}]/[(n+1)(B-A)]$

19. a. .597   **b.** .369
c. $f(x) = .3466 - .25 \ln(x)$ for $0 < x < 4$

21. 314.79

23. 248, 3.60

25. b. 1.8(90th percentile for X) + 32
c. $a(X \text{ percentile}) + b$

27. 0, 1.814

29. a. 2.14   **b.** .81   **c.** 1.17
d. .97   **e.** 2.41

31. a. 2.54   **b.** 1.34   **c.** $-.42$

33. a. .9664   **b.** .2451   **c.** .8664

35. a. .0455, .0455   **b.** Approximately 0
c. .6460   **d.** 2.13   **e.** .1700

37. a. 0, .5793, .5793   **b.** .3174, no
c. < 87.6 or > 120.4

39. a. .1003, .1003   **b.** 35.226
c. 21.888   **d.** 20.016, 39.984

41. .002

43. a. 58.31, 11.665   **b.** .4768   **c.** .1587

45. 7.3%

47. 21.155

49. a. .1190, .6969   **b.** .0021   **c.** .7019
d. > 5020 or < 1844 (using $z_{.0005} = 3.295$)
e. Normal, $\mu = 7.559$, $\sigma = 1.061$, .7019

51. .3174 for $k = 1$, .0456 for $k = 2$, .0026 for $k = 3$, as compared to the bounds of 1, .25, and .111, respectively.

53. a. Exact: .212, .577, .573; Approximate: .211, .567, .596
b. Exact: .885, .575, .017; Approximate: .885, .579, .012
c. Exact: .002, .029, .617; Approximate: .003, .033, .599

55. a. .9409   **b.** .9943

57. b. Normal, $\mu = 239$, $\sigma^2 = 12.96$

59. a. 1   **b.** 1   **c.** .982   **d.** .129

61. a. .480, .667, .187   **b.** .050, 0

63. a. short $\Rightarrow$ plan #1 better, whereas long $\Rightarrow$ plan #2 better
b. $1/\lambda = 10 \Rightarrow E[h_1(X)] = 100$, $E[h_2(X)] = 112.53$
$1/\lambda = 15 \Rightarrow E[h_1(X)] = 150$, $E[h_2(X)] = 138.51$

65. a. 3.01, 12.44   **b.** .238 (.237 using software)
c. .176

67. a. .424   **b.** .567, $\tilde{\mu} < 24$
c. 60   **d.** 66

69. a. $\cap A_i$ **b.** Exponential with $\lambda = .05$
c. Exponential with parameter $n\lambda$

73. a. .826, .826, .0636   **b.** .664   **c.** 172.727

77. a. 123.97, 117.373   **b.** .5517   **c.** .1587

79. a. 9.164, .385   **b.** .8790   **c.** .4247, skewness
d. No, since $P(X < 17,000) = .9332$

81. a. 3.962, 1.921   **b.** .0375   **c.** .2795
d. 7.77   **e.** 13.74   **f.** 4.52

83. $\alpha = \beta$

85. b. $[\Gamma(\alpha + \beta) \cdot \Gamma(m + \beta)]/[\Gamma(\alpha + \beta + m) \cdot \Gamma(\beta)]$, $\beta/(\alpha + \beta)$

87. Yes, since the pattern in the plot is quite linear.

89. Yes, because a normal probability plot shows a substantial linear pattern.

91. Yes

93. Plot $\ln(x)$ vs. z percentile. The pattern is straight, so a lognormal population distribution is plausible.

95. The pattern in the plot is quite linear; it is very plausible that strength is normally distributed.

97. There is substantial curvature in the plot. λ is a scale parameter (as is σ for the normal family).

99. a. $F(y) = \dfrac{1}{48}(y^2 - y^3/18)$ for $0 \le y \le 12$
   b. .259, .5, .241   **c.** 6, 43.2, 7.2
   d. .518   **e.** 3.75

101. a. $f(x) = x^2$ for $0 \le x < 1$ and $= \dfrac{7}{4} - \dfrac{3}{4}x$ for $1 \le x \le \dfrac{7}{3}$ **b.** .917   **c.** 1.213

103. a. .9162   **b.** .9549   **c.** 1.3374

105. .506

107. b. $F(x) = 0$ for $x < -1$, $= (4x - x^3/3)/9 + \dfrac{11}{27}$ for $-1 \le x \le 2$, and $= 1$ for $x > 2$
   c. No. $F(0) < .5 \Rightarrow \widetilde{\mu} > 0$
   d. $Y \sim \text{Bin}\left(10, \dfrac{5}{27}\right)$

109. a. .368, .828, .460   **b.** 352.53
   c. $1/\beta \cdot \exp\left[-\exp\left(-(x - \alpha)/\beta\right)\right] \cdot \exp\left(-(x - \alpha)/\beta\right)$
   d. α **e.** $\mu = 201.95$, mode $= 150$, $\widetilde{\mu} = 182.99$

111. a. μ **b.** No   **c.** 0   **d.** $(\alpha - 1)\beta$ **e.** $\nu - 2$

113. a. $\mu = p/\lambda_1 + (1 - p)/\lambda_2$
   $V(X) = 2p/\lambda_1^2 + 2(1 - p)/\lambda_2^2 - \mu^2$
   b. $p(1 - \exp(-\lambda_1 x)) + (1 - p)(1 - \exp(-\lambda_2 x))$
   for $x \ge 0$
   c. .403   **d.** .879
   e. 1, $CV > 1$ **f.** $CV < 1$

115. a. Lognormal   **b.** 1   **c.** 2.72, .0185

119. a. Exponential with $\lambda = 1$
   c. Gamma with parameters α and $c\beta$

121. a. $(1/365)^3$ **b.** $(1/365)^2$ **c.** .000002145

123. b. Let $u_1, u_2, u_3, \ldots$ be a sequence of observations from a Unif[0, 1] distribution (a sequence of random numbers). Then with $x_i = (-.1)\ln(1 - u_i)$, the x_i's are observations from an exponential distribution with $\lambda = 10$.

125. $g(E(X)) \le E(g(X))$

127. a. 710, 84.423, .684   **b.** .376

Chapter 5

1. a. .20   **b.** .42   **c.** At least one hose is in use at each pump; .70.   **d.** $p_X(x) = .16, .34, .50$ for $x = 0, 1, 2$, respectively; $p_Y(y) = .24, .38, .38$ for $y = 0, 1, 2$, respectively; .50   **e.** No; $p(0, 0) \neq p_X(0) \cdot p_Y(0)$

3. a. .15   **b.** .40   **c.** .22   **d.** .17, .46

5. a. .054   **b.** .00018

7. a. .030   **b.** .120   **c.** .300   **d.** .380   **e.** Yes

9. a. 3/380,000   **b.** .3024   **c.** .3593
   d. $10Kx^2 + .05$ for $20 \le x \le 30$ **e.** No

11. a. $e^{-\mu_1 - \mu_2} \cdot \mu_1^x \cdot \mu_2^y / x!y!$ **b.** $e^{-\mu_1 - \mu_2} \cdot [1 + \mu_1 + \mu_2]$
   c. $e^{-(\mu_1 + \mu_2)} \cdot (\mu_1 + \mu_2)^m / m!$; Poisson $(\mu_1 + \mu_2)$

13. a. $e^{-x - y}$ for $x \ge 0, y \ge 0$ **b.** .400   **c.** .594
   d. .330

15. a. $F(y) = 1 - e^{-\lambda y} + (1 - e^{-\lambda y})^2 - (1 - e^{-\lambda y})^3$ for $y \ge 0$
   b. $2/3\lambda$

17. a. .25   **b.** .318   **c.** .637
   d. $f_X(x) = 2\sqrt{R^2 - x^2}/\pi R^2$ for $-R \le x \le R$; no

19. a. $K(x^2 + y^2)/(10Kx^2 + .05)$; $K(x^2 + y^2)/(10Ky^2 + .05)$
   b. .556, .549   **c.** 25.37, 2.87

21. a. $f(x_1, x_2, x_3)/f_{X_1, X_2}(x_1, x_2)$ **b.** $f(x_1, x_2, x_3)/f_{X_1}(x_1)$

23. .15

25. L^2

27. .25 hr

29. $-\dfrac{2}{3}$

31. a. $-.1082$ **b.** $-.0131$

37. a.

$\bar{x}$	25	32.5	40	45	52.5	65
$p(\bar{x})$	.04	.20	.25	.12	.30	.09

,
   $E(\bar{X}) = \mu = 44.5$

   b.

s^2	0	112.5	312.5	800
$p(s^2)$	.38	.20	.30	.12

,
   $E(S^2) = 212.25 = \sigma^2$

39. This comes straight from the Bin(15, .8) distribution and Appendix Table A.1:

$x/15$	0	...	.333	.400	.467	...	.867	.933	1
$p(x/15)$	.000	...	.000	.001	.003	...	.231	.132	.035

41. a.

$\bar{x}$	1	1.5	2	2.5	3	3.5	4
$p(\bar{x})$	.16	.24	.25	.20	.10	.04	.01

   b. .85   **c.**

r	0	1	2	3
$p(r)$	.30	.40	.22	.08

47. a. .9876   **b.** .0009

49. a. .6026   **b.** .2981

51. .7720

53. a. .0062   **b.** 0

55. a. .9838   **b.** .8926

57. .9616

59. a. .9986, .9986   **b.** .9015, .3970
   c. .8357   **d.** .9525, .0003

61. a. 3.5, 2.27, 1.51   **b.** 15.4, 75.94, 8.71

63. a. .695   **b.** 4.0675 > 2.6775

65. a. .9232   **b.** .9660

67. .1588

69. a. 2400   **b.** 1205; independence   **c.** 2400, 41.77

71. a. 158, 430.25   **b.** .9788

73. a. Approximately normal with mean = 105, SD = 1.2649;
   Approximately normal with mean = 100, SD = 1.0142
   b. Approximately normal with mean = 5, SD = 1.6213
   c. .0068   **d.** .0010, yes

75. a. .2, .5, .3 for x = 12, 15, 20; .10, .35, .55 for y = 12, 15, 20
   b. .25   **c.** No   **d.** 33.35   **e.** 3.85

77. a. 3/81,250   **b.** $f_X(x) = k(250x - 10x^2)$ for $0 \leq x \leq 20$
   and $= k(450x - 30x^2 + \frac{1}{2}x^3)$ for $20 < x \leq 30$; $f_Y(y)$ results
   from substituting y for x in $f_X(x)$. They are not independent.
   c. .355   **d.** 25.969   **e.** 204.6154, −.894   **f.** 7.66

79. ≈ 1

81. a. 400 min   **b.** 70

83. 97

85. .9973

87. a. .2902   **b.** .8185
   c. The $\overline{X}$ distribution is much more concentrated about 13
   than is the population distribution.
   d. 0

91. b., c. Chi-squared with $\nu = n$.

93. a. $\sigma_W^2/(\sigma_W^2 + \sigma_E^2)$ **b.** .9999

95. 26, 1.64

97. a. .6   **b.** $U = \rho X + \sqrt{1 - \rho^2}\, Y$

Chapter 6

1. a. 8.14, $\overline{X}$ **b.** .77, $\widetilde{X}$ **c.** 1.66, S
   d. .148   **e.** .204, $S/\overline{X}$

3. a. 1.348, $\overline{X}$ **b.** 1.348, $\overline{X}$ **c.** 1.781, $\overline{X} + 1.28S$
   d. .6736   **e.** .0846

5. $N\overline{x} = 1,703,000$; $T - N\overline{d} = 1,591,300$; $T \cdot (\overline{x}/\overline{y}) = 1,601,438.281$

7. a. 120.6   **b.** 1,206,000   **c.** .80   **d.** 120.0

9. a. 2.11   **b.** .119

11. b. $\left[\dfrac{p_1 q_1}{n_1} + \dfrac{p_2 q_2}{n_2}\right]^{1/2}$ **c.** Use $\hat{p}_i = x_i/n_i$ and $\hat{q}_i = 1 - \hat{p}_i$
   in place of p_i and q_i in part (b) for $i = 1, 2$.
   d. −.245   **e.** .041

15. a. $\hat{\theta} = \Sigma X_i^2/2n$ **b.** 74.505

17. b. .444

19. a. $\hat{p} = 2\hat{\lambda} - .30 = .20$ **b.** $\hat{p} = (100\hat{\lambda} - 9)/70$

21. b. $\hat{\alpha} = 5, \hat{\beta} = 28.0/\Gamma(1.2)$

23. $\hat{\theta} = \Sigma X_i^2/n$

25. a. 384.4, 18.86   **b.** 415.42   **c.** .7967

29. a. $\hat{\theta} = \min(X_i), \hat{\lambda} = n/\Sigma[X_i - \min(X_i)]$
   b. .64, .202

33. With x_i = time between birth $i - 1$ and birth i,
   $\hat{\lambda} = 6/\sum_{i=1}^{6} ix_i = .0436$.

35. 29.5

37. 1.0132

Chapter 7

1. a. 99.5%   **b.** 85%   **c.** 2.96   **d.** 1.15

3. a. Narrower   **b.** No   **c.** No   **d.** No

5. a. (4.52, 5.18)   **b.** (4.12, 5.00)   **c.** .55   **d.** 94

7. By a factor of 4; the width is decreased by a factor of 5.

9. a. $(\overline{x} - 1.645\sigma/\sqrt{n}, \infty)$; (4.57, ∞)
   b. $(\overline{x} - z_\alpha \cdot \sigma/\sqrt{n}, \infty)$ **c.** $(-\infty, \overline{x} + z_\alpha \cdot \sigma/\sqrt{n})$;
   (−∞, 59.7)

11. 950, .8714

13. a. (608.58, 699.74)   **b.** 189

15. a. 80%   **b.** 98%   **c.** 75%

17. 134.53

19. (.513, .615)

21. $p < .273$ with 95% confidence; yes

23. a. (.225, .275)   **b.** 2655

25. a. 381   **b.** 339

29. a. 2.228   **b.** 2.086   **c.** 2.845   **d.** 2.680
e. 2.485   **f.** 2.571

31. a. 1.812   **b.** 1.753   **c.** 2.602   **d.** 3.747
e. 2.1716 (from Minitab)   **f.** Roughly 2.43

33. a. Reasonable amount of symmetry, no outliers
b. Yes (based on a normal probability plot)
c. (430.5, 446.1), yes, no

35. a. 95% CI: (23.1, 26.9)
b. 95% PI: (17.2, 32.8), roughly 4 times as wide

37. a. (.888, .964)   **b.** (.752, 1.100)   **c.** (.634, 1.218)

39. a. Yes   **b.** (6.45, 98.01)   **c.** (18.63, 85.83)

41. All 70%; (c), because it is shortest

43. a. 18.307   **b.** 3.940   **c.** .95   **d.** .10

45. (4.82, 21.85); no

47. a. 95% CI: (6.702, 9.456)   **b.** (.166, .410)

49. (47.4, 83.4)

51. a. (.260, .457)   **b.** 2398   **c.** No—97.5%

53. $(-.84, -.16)$

55. 246

57. $(2t_r/\chi^2_{1-\alpha/2,\,2r},\, 2t_r/\chi^2_{\alpha/2,\,2r}) = (65.3, 232.5)$

59. a. $(\max(x_i)/(1-\alpha/2)^{1/n},\ \max(x_i)/(\alpha/2)^{1/n})$
b. $(\max(x_i),\, \max(x_i)/\alpha^{1/n})$ **c.** (b); (4.2, 7.65)

61. (73.6, 78.8) versus (75.1, 79.6)

Chapter 8

1. a. Yes   **b.** No; $\bar{x}$ is the *sample* median, not a parameter
c. No; s is the *sample* standard deviation, not a parameter
d. Yes
e. No; $\overline{X}$ and $\overline{Y}$ are statistics, not parameters
f. Yes

3. a. Reject H_0 because $.001 \le .05 = \alpha$
b. Reject H_0 **c.** Don't reject H_0 because $.078 > .05$
d. Reject H_0 (a close call)
e. Don't reject H_0

5. Because this setup puts the burden of proof on the welds to show that they conform to specifications; only if there is compelling evidence for this will the welds be judged satisfactory.

7. $H_0\colon \sigma = .05$ versus $H_a\colon \sigma < .05$.
I: conclude variability in thickness is satisfactory when it is not.
II: conclude variability in thickness is not satisfactory when in fact it is.

9. I: concluding that the plant isn't in compliance when it is.
II: concluding that the plant is in compliance when it is not.

11. a. I: concluding that a majority favor one of the two companies when that is not the case.
II: concluding that potential subscribers are evenly split between the two companies when they aren't.
b. $x \le 6$ or $x \ge 19$

c. $X \sim \text{Bin}(25, .5)$, so P-value $= B(6{:}25,.5) + [1 - B(18{;}25, .5)] = .014$
d. Rejecting H_0 when P-value $\le .044$ is equivalent to rejecting when either $x \le 7$ or $x \ge 18$. Then $\beta(.3) = P(8 \le X \le 17 \text{ when } p = .3) = B(17{;}25, .3) - B(7{;}25, .3) = .488$, $\beta(.6) = .845$, $\beta(.7) = .488$.

13. a. $H_0\colon \mu = 10$ versus $H_a\colon \mu \ne 10$
b. P-value $= P(\overline{X} \le 9.85$ or ≥ 10.15 when H_0 is true$) = \Phi(-3.75) + [1 - \Phi(3.75)] \approx 0$ (software gives .00018). Since $0 \le .01$, H_0 should be rejected. The scale does not appear to be correctly calibrated.
c. .5319, .0078

15. a. .0778   **b.** .1841   **c.** .0250
d. .0066   **e.** .5438

17. a. P-value $= P(\overline{X} \ge 30{,}960$ when H_0 is true$) = 1 - \Phi(2.56) = .0052 \le .01$, so reject H_0
b. .8413   **c.** .143   **d.** .0052

19. a. $z = -2.27$, P-value $= .0232 > .01$, so don't reject H_0
b. .2266   **c.** 22

21. a. $z = -3.33$, P-value $= .0008 \le .01$, so reject H_0
b. .1056   **c.** 217

23. a. $\bar{x} = .750$, $\tilde{x} = .640$, $s = .3025$, $f_s = .480$. A boxplot shows substantial positive skew; there are no outliers.
b. No. A normal probability plot shows substantial curvature. No, since n is large.

c. $z = -5.79$, P-value ≈ 0, reject H_0 at any reasonable significance level; yes.
d. .821

25. No. H_0: $\mu = 2$ versus H_a: $\mu < 2$, $z = -1.80$, P-value $= .0359 > .01$, so don't reject H_0

29. a. P-value $= .136 > .05$, so don't reject H_0
b. $.136 > .05$, so don't reject H_0
c. P-value $= .016 > .01$, so don't reject H_0
d. P-value $\approx 0 \leq .05$, so reject H_0 in favor of H_a: $\mu \neq 5$

31. a. P-value $= .003 \leq .05$, so reject H_0
b. P-value $= .057 > .01$, so don't reject H_0
c. P-value $> .5 > \alpha$ for any reasonable α, so don't reject H_0

33. a. It appears that the specification has been violated; much of the boxplot lies to the right of 200.
b. $t = 5.8$, P-value ≈ 0, so conclude that $\mu \neq 200$

35. a. $t \approx 1.2$, P-value $\approx .128 > .05$, so H_0: $\mu = 200$ cannot be rejected
b. .30 (from software)

37. a. Yes, because the pattern in a normal probability plot is reasonably linear.
b. P-value $> .10$ (barely), so H_0: $\mu = 100$ should not be rejected at any sensible α, and the concrete should be used.

39. a. $t = 2.43$, P-value $\approx .013 > .01 = \alpha$, so there is not compelling evidence.
b. Yes, type II **c.** .66 (from software)

41. $t \approx 1.9$, so P-value $\approx .116$. H_0 should not be rejected.

43. a. H_0: $p = .2$ versus H_a: $p > .2$, $z = 1.27$, P-value $= .10 > .05$, so there is not compelling evidence for rejecting H_0.
b. I: say that more than 20% are obese when this is not the case; II: conclude that 20% are obese when the actual percentage exceeds 20%.
c. .121

45. $z = 3.67$, P-value ≈ 0, so reject H_0: $p = .40$. No

47. a. $z = -1.0$, so there is not enough evidence to conclude that $p < .25$; thus, use screwtops.

b. I: Don't use screwtops when their use is justified; II: Use screwtops when their use is not justified.

49. a. $z = 3.07$, P-value $= .0022 \leq .01$, so reject H_0 and the company's premise.
b. .0332

51. No, no, yes. $\alpha = .098$, $\beta = .090$

53. a. .8888, .1587, .0006 **b.** P-value ≈ 0. Yes
c. No

55. a. .049, .096 **b.** 69

57. $z = -3.12$, P-value $= .0018 \leq .05$, so conclude that $\mu \neq 3.20$

59. a. H_0: $\mu = .85$ versus H_a: $\mu \neq .85$
b. H_0 cannot be rejected at either α

61. a. No, because P-value $= .02 > .01$; yes, because 45.31 greatly exceeds 20, but n is very small.
b. $\beta = .57$ (software)

63. a. No, no
b. No, because $z = .44$ and P-value $= .33 > .10$.

65. a. Approximately .6; approximately .2 (from Appendix Table A.17)
b. $n = 28$

67. a. $z = 1.64$, P-value $\approx .1 > .05$, so H_0 cannot be rejected. Type II
b. Yes.

69. Yes, $z = -3.32$, P-value $= .0005 \leq .001$, so H_0 should be rejected.

71. No, since $z = 1.33$ and P-value $= .0918$

73. No, since P-value $= .2296$

75. $z = .92$, P-value $= .1788$, so it cannot be concluded that $\mu > 20$.

77. $.01 < P$-value $< .025$, so do not reject H_0; no contradiction

79. a. Test statistic is $\chi^2 = 2\Sigma X_i/\mu_0$, P-value $=$ area under the χ^2_{2n} curve to the left of the calculated χ^2.
b. $\chi^2 = 19.65$, P-value $> .10$, so H_0: $\mu = 75$ can't be rejected.

Chapter 9

1. a. $-.4$ hr; it doesn't **b.** .0724, .2691 **c.** No

3. a. Yes; P-value $\approx 0 \leq .01$, so reject H_0. **b.** P-value $= .0132$, so yes at significance level .05 but no at level .01.

5. a. $z = -2.90$, P-value $= .0019$, so reject H_0.
b. .8212 **c.** 66

7. No, since P-value for a 2-tailed test is .0602.

9. a. 6.2; yes **b.** $z = 1.14$, P-value $\approx .25$, no
c. No **d.** A 95% CI is (10.0, 21.8).

11. A 95% CI is (.99, 2.41).

13. 50

15. b. It increases.

17. a. 17 **b.** 21 **c.** 18 **d.** 26

19. $t = -1.20$, P-value $= .196$, so do not reject H_0.

21. No; $t = -2.46$, df $= 15$, P-value $\approx .013$, so do not reject H_0 (a close call).

23. b. No **c.** $t = -.38$, P-value $\approx .7$, so don't reject H_0.

25. a. Both normal probability plots exhibit substantial linear patterns.
b. Average price for the ≥ 93 wines appears to significantly exceed that for the ≤ 89 wines.
c. (16.1, 82.0); No, because this CI does not include 0.

27. a. 99% CI: (.33, .71)   **b.** 99% CI: $(-.07, .41)$, so 0 is a plausible value of the difference.

29. $t = -2.10$, df = 25, P-value = .023. At significance level .05, we would conclude that cola results in a higher average strength, but not at significance level .01.

31. a. Virtually identical centers, substantially more variability in medium range observations than in higher range observations
b. $(-7.9, 9.6)$, based on 23 df; no

33. No, $t = 1.33$, P-value = .094, don't reject H_0

35. $t = -2.2$, df = 16, P-value = $.021 > .01 = \alpha$, so don't reject H_0.

37. a. $(-.561, -.287)$ **b.** Between -1.224 and .376

39. a. Yes
b. $t = 2.7$, P-value = $.018 < .05 = \alpha$, so H_0 should be rejected.

41. a. $(-3.85, 11.35)$ **b.** Yes. Since P-value = .02, at level .05 there would appear to be an increase, but not at level .01.   **c.** (7.02, 10.06)

43. a. No   **b.** -49.1 **c.** 49.1

45. a. Yes, because of the linear pattern in a normal probability plot.   **b.** No, data is paired, not independent samples   **c.** $t = 3.66$, P-value = .001 (not .003), same conclusion.

47. a. 95% CI: $(-2.52, 1.05)$; plausible that they are identical
b. Linear pattern in npp implies normality of difference distribution is plausible.

49. $z = 2.84$, P-value = $.0023 \leq .05$, so H_0 can be rejected; the introduction of context appears to lower the correct response rate.

51. $z = 3.20$, P-value = .0007, so H_0 can be rejected at significance level .05, level .01, or even level .001. There does appear to be a nocebo effect.

53. a. $z = .80$, P-value $> .05$, so don't reject H_0.   **b.** $n = 1211$

55. a. The CI for $\ln(\theta)$ is $\ln(\hat{\theta}) \pm z_{\alpha/2}[(m - x)/(mx) + (n - y)/(ny)]^{1/2}$. Taking the antilogs of the lower and upper limits gives a CI for θ itself.
b. (1.43, 2.31); aspirin appears to be beneficial.

57. $(-.35, .07)$

59. a. 3.69   **b.** 4.82   **c.** .207   **d.** .271
e. 4.30   **f.** .212   **g.** .95   **h.** .94

61. $f = .384$, P-value $> .10$, so, don't reject H_0.

63. $f = 2.85 \geq F_{.05, 19, 22} \approx 2.08$. Thus P-value $< .05$, so reject H_0; there does appear to be more variability in low-dose weight gain.

65. $(s_2^2 F_{1-\alpha/2}/s_1^2, s_2^2 F_{\alpha/2}/s_1^2)$; (.023, 1.99)

67. No. $t = 3.2$, df = 15, P-value = .006, so reject H_0: $\mu_1 - \mu_2 = 0$ using either $\alpha = .05$ or .01.

69. $z > 0 \Rightarrow P$-value $> .5$, so H_0: $p_1 - p_2 = 0$ cannot be rejected.

71. $(-299.3, 1517.9)$

73. They appear to differ, since df = 14, $t = -5.19$, P-value = 0.

75. Yes, $t = -2.25$, df = 57, P-value $\approx .028$.

77. a. No. $t = -2.84$, df = 18, P-value $\approx .012$
b. No. $t = -.56$, P-value $\approx .29$

79. $t = 3.9$, P-value = .004, so H_0 is rejected at level .05 or .01.

81. No, nor should the two-sample t test be used, because a normal probability plot suggests that the good-visibility distribution is not normal.

83. Unpooled: df = 15, $t = -1.8$, P-value $\approx .092$
Pooled: df = 24, $t = -1.9$, P-value $\approx .070$

85. a. $m = 141, n = 47$ **b.** $m = 240, n = 160$

87. No, $z = .83$, P-value $\approx .20$

89. .9015, .8264, .0294, .0000; true average IQs; no

91. Yes; $z = 4.2$, P-value ≈ 0

93. a. Yes. $t = -6.4$, df = 57, and P-value ≈ 0
b. $t = 1.1$, P-value = .14, so don't reject H_0.

95. $(-1.29, -.59)$

Chapter 10

1. $f = 2.44$, $F_{.05,4,15} = 3.06$, $F_{.10,4,15} = 2.36$. Thus $.05 < P$-value $< .10$, so H_0 should not be rejected.

3. $f = 1.30 < 2.57 = F_{.10, 2, 21}$, so P-value $> .10$. H_0 cannot be rejected at any reasonable significance level.

5. $f = 1.73 < 2.51 = F_{.10,2,27}$, so P-value $> .10$ and the three grades don't appear to differ.

7. $f = 51.3$, P-value = 0, so H_0 can be rejected at any reasonable significance level.

9. $f = 3.96$ and $F_{.05, 3, 20} = 3.10 < 3.96 < 4.94 = F_{.01,3,20}$, so $.01 < P$-value $< .05$. Thus H_0 can be rejected at significance level .05; there appear to be differences among the grains.

11. $w = 36.09$

3	1	4	2	5
437.5	462.0	469.3	512.8	532.1

Brands 2 and 5 don't appear to differ, nor does there appear to be any difference between brands 1, 3, and 4, but each brand in the first group appears to differ significantly from all brands in the second group.

13.

3	1	4	2	5
427.5	462.0	469.3	502.8	532.1

15. 14.18 17.94 18.00 18.00 25.74 27.67

17. $(-.029, .379)$

19. Any value of SSE between 422.16 and 431.88 will work.

21. a. $f = 22.6$, $F_{.001, 5, 78} \approx 4.6$, P-value $< .001$, so reject H_0.
   b. $(-99.16, -35.64), (29.34, 94.16)$

23.

	1	2	3	4
1	–	2.88 ± 5.81	7.43 ± 5.81	12.78 ± 5.48
2	–	–	4.55 ± 6.13	9.90 ± 5.81
3	–	–	–	5.35 ± 5.81
4	–	–	–	–

4	3	2	1

25. a. Normal, equal variances
   b. SSTr $= 8.33$, SSE $= 77.79$, $f = 1.7$, H_0 should not be rejected (P-value $> .10$)

27. a. $f = 3.75$, $.01 < P$-value $< .05$, so at significance level .05 brands appear to differ.
   b. Normality is quite plausible (a normal probability plot of the residuals $x_{ij} - \bar{x}_{i.}$ shows a linear pattern).
   c. 4 3 2 1 Only brands 1 and 4 appear to differ significantly.

31. Approximately .62

33. $\arcsin (\sqrt{x/n})$

35. a. $.01 < P$-value $< .05$, so H_0 is not rejected.
   b. $.029 > .01$, so again H_0 is not rejected.

37. $f = 8.44 > 6.49 = F_{.001}$, so P-value $< .001$ and H_0 should be rejected.

5	3	1	4	2

39. The CI is $(-.144, .474)$, which does include 0.

41. $2.92 < f = 3.96 < 4.07$, so $.05 < P$-value $< .10$ and H_0: $\sigma_A^2 = 0$ cannot be rejected.

43. $(-3.70, 1.04), (-4.83, -.33), (-3.77, 1.27), (-3.99, .15)$. Only $\mu_1 - \mu_3$ among these four contrasts appears to differ significantly from zero.

45. They are identical.

Chapter 11

1. a. $f_A = 7.16$, $.01 < P$-value $< .05$, so reject H_{0A}.
   b. $f_B = 10.42$, $.01 < P$-value $< .05$, so reject H_{0B}.

3. a. $f_A = 18.77$, P-value $= .023$, $f_B = 21.10$, P-value $= .016$ (from software), so reject both H_{0A} and H_{0B} at significance level .05.
   b. $Q_{.05,4,3} = 6.825$, $w = .257$; .201 .324 .462 .602

5. $f_A = 2.56$, $F_{.01,3,12} = 5.95$, so there appears to be no effect due to angle of pull.

7. a. SSA $= 22.889$, SSB $= 27.556$, SSE $= 5.111$, $f_A = 8.96$, $.01 < P$-value $< .05$ (software gives .033), so at level .05 there does appear to be a brand effect.
   b. $f_B = 10.78$, P-value $= .024$, blocking does appear to have been effective.

9.

Source	df	SS	MS	f	$F_{.05}$
Treatments	3	81.19	27.06	22.4	3.01
Blocks	8	66.50	8.31		
Error	24	29.06	1.21		
Total	35	176.75			

1	4	3	2
8.56	9.22	10.78	12.44

11. A normal probability plot of the residuals shows a substantial linear pattern. There is no discernible pattern in a plot of the residuals versus the fitted values.

13. b. Each SS is multiplied by c^2, but f_A and f_B are unchanged.

15. a. Approximately .20, .43   **b.** Approximately .30

17. a. $f_A = 3.76$, $f_B = 6.82$, $f_{AB} = .74$, and $F_{.05,2,9} = 4.26$, so the amount of carbon fiber addition appears significant.
   b. $f_A = 6.54$, $f_B = 5.33$, $f_{AB} = .27$

19. $f_{AB} = 3.18$. Since $F_{.01,10,18} = 3.51$, P-value $> .01$ for testing H_{0AB}; hence the interaction effect is not significant. $f_A = .94$, P-value $> .10$; $f_B = 9.17$, P-value $< .001$. Thus type of farm doesn't seem to matter but maintenance method does.

21. a, b.

Source	df	SS	MS	f
A	2	22,941.80	11,470.90	22.98
B	4	22,765.53	5691.38	5.60
AB	8	3993.87	499.23	.49
Error	15	15,253.50	1016.90	
Total	29	64,954.70		

H_{0A} and H_{0B} are both rejected.

23.

Source	df	SS	MS	f
A	2	11,573.38	5786.69	$\frac{MSA}{MSAB} = 26.70$
B	4	17,930.09	4482.52	$\frac{MSB}{MSE} = 28.51$
AB	8	1734.17	216.77	$\frac{MSAB}{MSE} = 1.38$
Error	30	4716.67	157.22	
Total	44	35,954.31		

Since $F_{.01,8,30} = 3.17$, $F_{.01,2,8} = 8.65$, and $F_{.01,4,30} = 4.02$, H_{0G} is not rejected but both H_{0A} and H_{0B} are rejected.

25. $(-.373, -.033)$

27. a.

Source	df	SS	MS	f	$F_{.05}$
A	2	14,144.44	7072.22	61.06	3.35
B	2	5511.27	2755.64	23.79	3.35
C	2	244,696.39	122,348.20	1056.27	3.35
AB	4	1069.62	267.41	2.31	2.73
AC	4	62.67	15.67	.14	2.73
BC	4	331.67	82.92	.72	2.73
ABC	8	1080.77	135.10	1.17	2.31
Error	27	3127.50	115.83		
Total	53	270,024.33			

d. $Q_{.05,3,27} = 3.51$, $w = 8.90$, and all three of the levels differ significantly from one another.

29. a. $f_{ABC} = 2.87$, P-value $= .029$ for testing H_{0ABC}. However, all two-factor interaction F ratios are highly significant.

31.

Source	DF	SS	MS	F	P
A	2	124.60	62.30	4.85	0.042
B	2	20.61	10.30	0.80	0.481
C	2	356.95	178.47	13.89	0.002
A*B	4	57.49	14.37	1.12	0.412
A*C	4	61.39	15.35	1.19	0.383
B*C	4	11.06	2.76	0.22	0.923
Error	8	102.78	12.85		
Total	26	734.87			

a. The P-values in the foregoing ANOVA table for the AB, AC, and BC effects all considerably exceed .1, indicating that at any reasonable significance level, the hypotheses of no two-factor interactions cannot be rejected.
b. At significance level .05, the *Power* main effect is significant (somewhat of a close call) and the *Paste Thickness* main effect is highly significant.
c. $w = 4.83$. The sample means for the three levels of paste thickness are 29.562 (for .4), 35.183 (for .3), and 38.356 (for .2). So the level .4 can be judged significantly different from the other two levels.

33.

Source	df	SS	MS	f
A	6	67.32	11.02	
B	6	51.06	8.51	
C	6	5.43	.91	.61
Error	30	44.26	1.48	
Total	48	168.07		

$F_{.05,6,30} = 2.42$, $f_C = .61$, so H_{0C} is not rejected.

35.

Source	df	SS	MS	f
A	4	28.88	7.22	10.7
B	4	23.70	5.93	8.79
C	4	.62	.155	<1
Error	12	8.10	.675	
Total	24	61.30		

Since $F_{.05,4,12} = 3.26$, both A and B are significant.

37.

Source	df	MS	f
A	2	2207.329	2259*
B	1	47.255	48.4*
C	2	491.783	503*
D	1	.044	<1
AB	2	15.303	15.7*
AC	4	275.446	282*
AD	2	.470	<1
BC	2	2.141	2.19
BD	1	.273	<1
CD	2	.247	<1
ABC	4	3.714	3.80
ABD	2	4.072	4.17*
ACD	4	.767	<1
BCD	2	.280	<1
ABCD	4	.347	<1
Error	36	.977	
Total	71	93.621	

*Denotes a significant F ratio.

39. a. $\hat{\beta}_1 = 54.38$, $\hat{\gamma}_{11}^{AC} = -2.21$, $\hat{\gamma}_{21}^{AC} = 2.21$.

b.

Source	Effect Contrast	MS	f
A	1307	71,177.04	436.7
B	1305	70,959.34	435.4
C	529	11,660.04	71.54
AB	199	1650.04	10.12
AC	−53	117.04	<1
BC	57	135.38	<1
ABC	27	30.38	<1
Error		162.98	

41.

Source	DF	SS	MS	F	P
A	1	0.003906	0.003906	25.00	0.001
B	1	0.242556	0.242556	1552.36	0.000
C	1	0.178506	0.178506	1142.44	0.000
A*B	1	0.003906	0.003906	25.00	0.001
A*C	1	0.002256	0.002256	14.44	0.005
B*C	1	0.178506	0.178506	1142.44	0.000
A*B*C	1	0.002256	0.002256	14.44	0.005
Error	8	0.001250	0.000156		
Total	15	0.613144			

All effects are significant, and in particular the three-factor interaction effect is significant.

43.

Source	df	SS	f
A	1	.436	<1
B	1	.099	<1
C	1	.109	<1
D	1	414.12	851
AB	1	.003	<1
AC	1	.078	<1
AD	1	.017	<1
BC	1	1.404	3.62
BD	1	.456	<1
CD	1	2.190	4.50
Error	5	2.434	

$F_{.05,1,5} = 6.61$, so only the factor D main effect is judged significant.

45. a. 1: (1), *ab, cd, abcd*; 2: *a, b, acd, bcd*; 3: *c, d, abc, abd*; 4: *ac, bc, ad, bd.*

b.

Source	df	SS	f
A	1	12,403.125	27.18
B	1	92,235.125	202.13
C	1	3.125	0.01
D	1	60.500	0.13
AC	1	10.125	0.02
AD	1	91.125	0.20
BC	1	50.000	0.11
BD	1	420.500	0.92
ABC	1	3.125	0.01
ABD	1	0.500	0.00
ACD	1	200.000	0.44
BCD	1	2.000	0.00
Blocks	7	898.875	0.28
Error	12	5475.750	
Total	31	111,853.875	

$F_{.01,1,12} = 9.33$, so only the A and B main effects are significant.

47. a. *ABFG*; (1), *ab, cd, ce, de, fg, acf, adf, adg, aef, acg, aeg, bcg, bcf, bdf, bdg, bef, beg, abcd, abce, abde, abfg, cdfg, cefg, defg, acdef, acdeg, bcdef, bcdeg, abcdfg, abcefg, abdefg.* {*A, BCDE, ACDEFG, BFG*}, {*B, ACDE, BCDEFG, AFG*}, {*C, ABDE, DEFG, ABCFG*}, {*D, ABCE, CEFG, ABDFG*}, {*E, ABCD, CDFG, ABEFG*}, {*F, ABCDEF, CDEG, ABG*}, {*G, ABCDEG, CDEF, ABF*}. **b.** 1: (1), *aef, beg, abcd, abfg, cdfg, acdeg, bcdef*; 2: *ab, cd, fg, aeg, bef, acdef, bcdeg, abcdfg*; 3: *de, acg, adf, bcf, bdg, abce, cefg, abdefg*; 4: *ce, acf, adg, bcg, bdf, abde, defg, abcefg.*

49. SSA = 2.250, SSB = 7.840, SSC = .360, SSD = 52.563, SSE = 10.240, SSAB = 1.563, SSAC = 7.563, SSAD = .090, SSAE = 4.203, SSBC = 2.103, SSBD = .010, SSBE = .123, SSCD = .010, SSCE = .063, SSDE = 4.840. Error SS = sum of two-factor SS's = 20.568, Error MS = 2.057, $F_{.01,1,10} = 10.04$, so only the D main effect is significant.

51.

Source	df	SS	MS	f
A main effects	1	322.667	322.667	980.38
B main effects	3	35.623	11.874	36.08
Interaction	3	8.557	2.852	8.67
Error	16	5.266	0.329	
Total	23	372.113		

$F_{.05,3,16} = 3.24$, so interactions appear to be present.

53.

Source	df	SS	MS	f
A	1	30.25	30.25	6.72
B	1	144.00	144.00	32.00
C	1	12.25	12.25	2.72
AB	1	1122.25	1122.25	249.39
AC	1	1.00	1.00	.22
BC	1	12.25	12.25	2.72
ABC	1	16.00	16.00	3.56
Error	4	36.00	4.50	
Total	7			

Only the main effect for B and the AB interaction effect are significant at $\alpha = .01$.

55. a. $\hat{\alpha}_1 = 9.00, \hat{\beta}_1 = 2.25, \hat{\delta}_1 = 17.00, \hat{\gamma}_1 = 21.00,$ $(\hat{\alpha\beta})_{11} = 0, (\hat{\alpha\delta})_{11} = 2.00, (\hat{\alpha\gamma})_{11} = 2.75,$ $(\hat{\beta\delta})_{11} = .75, (\hat{\beta\gamma})_{11} = .50, (\hat{\delta\gamma})_{11} = 4.50$

b. A normal probability plot suggests that the A, C, and D main effects are quite important, and perhaps the CD interaction. In fact, pooling the 4 three-factor interaction SS's and the four-factor interaction SS to obtain an SSE based on 5 df and then constructing an ANOVA table suggests that these are the most important effects.

57.

Source	DF	SS	MS	F	P
A	2	34436	17218	436.92	0.000
B	2	105793	52897	1342.30	0.000
C	2	516398	258199	6552.04	0.000
A*B	4	6868	1717	43.57	0.000
A*C	4	10922	2731	69.29	0.000
B*C	4	10178	2545	64.57	0.000
A*B*C	8	6713	839	21.30	0.000
Error	27	1064	39		

Obviously all effects are highly significant.

59. Based on the P-values in the ANOVA table, statistically significant factors at the level $\alpha = .01$ are adhesive type and cure time. The conductor material does not have a statistically significant effect on bond strength. There are no significant interactions.

61.

Source	df	SS	MS	f
A	4	285.76	71.44	.594
B	4	227.76	56.94	.473
C	4	2867.76	716.94	5.958
D	4	5536.56	1384.14	11.502
Error	8	962.72	120.34	$F_{.05,4,8} = 3.84$
Total	24			

H_{0A} and H_{0B} cannot be rejected, while H_{0C} and H_{0D} are rejected.

Chapter 12

1. a. The accompanying displays are based on repeating each stem value five times (once for leaves 0 and 1, a second time for leaves 2 and 3, etc.).

17	0
17	2 3
17	4 4 5
17	6 7
17	
18	0 0 0 0 0 1 1
18	2 2 2 2
18	4 4 5
18	6
18	8

stem: hundreds and tens
leaf: ones

There are no outliers, no significant gaps, and the distribution is roughly bell-shaped with a reasonably high degree of concentration about its center at approximately 180.

0	8 8 9
1	0 0 0 0
1	3
1	4 4 4 4
1	6 6
1	8 8 8 9
2	1 1
2	
2	5
2	6
2	
3	0 0

stem: ones
leaf: tenths

A typical value is about 1.6, and there is a reasonable amount of dispersion about this value. The distribution is somewhat skewed toward large values, the two largest of which may be candidates for outliers.
b. No, because observations with identical x values have different y values.
c. No, because the points don't appear to fall at all close to a line or simple curve.

3. Yes. Yes.

5. b. Yes.
c. There appears to be an approximate quadratic relationship (points fall close to a parabola).

7. a. 5050   **b.** 1.3   **c.** 130   **d.** −130

9. a. .095   **b.** −.475   **c.** .830, 1.305
d. .4207, .3446   **e.** .0036

11. a. −.01, −.10   **b.** 3.00, 2.50
c. .3627   **d.** .4641

13. a. Yes, because $r^2 = .972$.

15. a.

2	9
3	3 3 5 5 6 6 6 7 7 8 8 9
4	1 2 2 3 5 6 6 8 9
5	1
6	2 9
7	9
8	0

Typical value in low 40s, reasonable amount of variability, positive skewness, two potential outliers.
b. No
c. $y = 3.2925 + .10748x = 7.59$. No; danger of extrapolation
d. 18.736, 71.605, .738, yes

17. a. $118.91 - .905x$; yes.   **b.** We estimate that the expected decrease in porosity associated with a 1-pcf increase in unit weight is .905%.   **c.** Negative prediction, but y can't be negative.   **d.** −.52, .49   **e.** .938, roughly the size of a typical deviation from the estimated regression line.   **f.** .974

19. a. $y = -45.5519 + 1.7114x$ **b.** 339.51
c. −85.57   **d.** The $\hat{y}_i$'s are 125.6, 168.4, 168.4, 211.1, 211.1, 296.7, 296.7, 382.3, 382.3, 467.9, 467.9, 553.4, 639.0, 639.0; a 45° line through (0, 0).

21. a. Yes; $r^2 = .985$ **b.** 368.89   **c.** 368.89

23. a. 16,213.64; 16,205.45
b. 414,235.71; yes, since $r^2 = .961$.

27. $\hat{\beta}_1 = \Sigma x_i Y_i / \Sigma x_i^2$

29.

Data set	r^2	s	
1	.43	4.03	Most effective: set 3
2	.99	4.03	Least effective: set 1
3	.99	1.90	

31. a. .893   **b.** .01837   **c.** $(-.216, -.136)$

33. a. (.081, .133)   **b.** $H_a: \beta_1 > .1$, P-value = .277, no

35. a. (.63, 2.44) is a 95% CI.
b. Yes. $t \approx 3.6$, P-value $\approx .004$
c. No; extrapolation
d. (.54, 2.82), no

37. a. Yes. $t = 7.99$, P-value ≈ 0. [*Note:* There is one mild outlier, so the resulting normal probability plot is not entirely satisfactory.]
b. Yes. $t = -5.8$, P-value ≈ 0, so reject $H_0: \beta_1 = 1$ in favor of $H_a: \beta_1 < 1$

39. $f = 71.97$, $s_{\hat{\beta}_1} = .004837$, $t = 8.48$, P-value = .000

43. $d = 1.20$, df = 13, and $\beta \approx .1$.

45. a. (77.80, 78.38)
b. (76.90, 79.28), same center but wider
c. wider, since 115 is farther from $\bar{x}$
d. $t = -11$, P-value = 0

47. a. 95% PI is (20.21, 43.69), no
b. (28.53, 51.92), at least 90%

49. (431.3, 628.5)

51. a. 45 is closer to $\bar{x} = 45.18$ **b.** (46.28, 46.78)
c. (47.56, 49.84)

53. (a) narrower than (b), (c) narrower than (d), (a) narrower than (c), (b) narrower than (d)

57. If, for example, 18 is the minimum age of eligibility, then for most people $y \approx x - 18$.

59. a. .966
b. The percent dry fiber weight for the first specimen tends to be larger than for the second.
c. No change   **d.** 93.3%
e. $t = 14.9$, P-value ≈ 0, so there does appear to be such a relationship.

61. a. $r = .748$, $t = 3.9$, P-value $= .001$. Using either $\alpha = .05$ or .01, yes.
b. .560 (56%), same

63. $r = .773$, yet $t = 2.44$, P-value $\approx .07$; so H_0: $\rho = 0$ cannot be rejected.

65. a. .481

b. $t = 1.98$, P-value $= .07$, so at level .01, no linear association.   **c.** At level .01, no positive linear association, but at level .05, there does appear to be positive linear association.

67. a. Reject H_0
b. No. P-value $= .00032 \Rightarrow z \approx 3.6 \Rightarrow r \approx .16$, which indicates only a weak relationship.
c. Yes, but very large $n \Rightarrow \rho \approx .022$, so no practical significance.

69. a. 95% CI: (.888, 1.086)
b. 95% CI: (47.730, 49.172)
c. 95% PI: (45.378, 51.524)
d. Narrower for $x = 25$, since 25 is closer to $\bar{x}$
e. .981

71. a. 16.0593, .1925 **b.** $t = 54.15$, P-value $= 0$
c. $\bar{x} = .408$, and .2 is farther from this than is .4.
d. (6.41, 6.82)   **e.** (5.96, 7.27)

73. a. .507   **b.** .712   **c.** P-value $= .0013 < .01 = \alpha$, so reject H_0: $\beta_1 = 0$ and conclude that there is a useful linear relationship.   **d.** A 95% CI is (1.056, 1.275).
e. 1.0143, .2143

75. a. $y = 1.69 + .0805x$ **b.** $y = -20.40 + 12.2254x$
c. .984 for both regressions.

77. a. Yes, the points fall very close to a straight line.
b. .996 **c.** Yes; $t = 54.6$, P-value $= 0$
d. 95% PI: (3.17, 4.57)
e. $t = 54.6$, P-value $= 0$

81. b. .573

87. $t = -1.14$, so it is plausible that $\beta_1 = \gamma_1$.

Chapter 13

1. a. 6.32, 8.37, 8.94, 8.37, and 6.32   **b.** 7.87, 8.49, 8.83, 8.94, and 2.83   **c.** The deviation is likely to be much smaller for the x values of part (b).

3. a. Yes.   **b.** $-.31, -.31, .48, 1.23, -1.15, .35, -.10,$ $-1.39, .82, -.16, .62, .09, 1.17, -1.50, .96, .02, .65,$ $-2.16, -.79, 1.74$. Here e/e^* ranges between .57 and .65, so e^* is close to e/s.   **c.** No.

5. a. About 98% of observed variation in thickness is explained by the relationship.
b. A nonlinear relationship

7. a. .776   **b.** Perhaps not, because of curvature.
c. Substantial curvature rather than a linear pattern, implying inadequacy of the linear model. A parabola (quadratic regression) provides a significantly better fit.

9. For set 1, simple linear regression is appropriate. A quadratic regression is reasonable for set 2. In set 3, (13, 12.74) appears very inconsistent with the remaining data. The estimated slope for set 4 depends largely on the single observation (19, 12.5), and evidence for a linear relationship is not compelling.

11. c. $V(\hat{Y}_i)$ increases, and $V(Y_i - \hat{Y}_i)$ decreases.

13. t with $n - 2$ df; .02

15. a. A curved pattern   **b.** A linear pattern
c. $Y = \alpha x^\beta \cdot \epsilon$ **d.** A 95% PI is (3.06, 6.50).
e. One standardized residual, corresponding to the third observation, is a bit large. There are only two positive standardized residuals, but two others are essentially 0. The patterns in a standardized residual plot and normal probability plot are marginally acceptable.

17. a. $\Sigma x'_i = 15.501$, $\Sigma y'_i = 13.352$, $\Sigma (x'_i)^2 = 20.228$,
$\Sigma x'_i y'_i = 18.109$, $\Sigma (y'_i)^2 = 16.572$, $\hat{\beta}_1 = 1.254$,
$\hat{\beta}_0 = -.468$, $\hat{\alpha} = .626$, $\hat{\beta} = 1.254$ **c.** $t = -1.02$, so don't reject H_0.   **d.** H_0: $\beta = 1$, $t = 3.28$, so reject H_0.

19. a. No **b.** $Y' = \beta_0 + \beta_1 \cdot (1/t) + \epsilon'$, where $Y' = \ln(Y)$,
so $Y = \alpha e^{\beta/t} \cdot \epsilon$. **c.** $\hat{\beta} = \hat{\beta}_1 = 3735.45$, $\hat{\beta}_0 = -10.2045$
$\hat{\alpha} = (3.70034) \cdot (10^{-5})$, $\hat{y}' = 6.7748$, $\hat{y} = 875.5$

d. SSE $= 1.39587$, SSPE $= 1.36594$ (using transformed values), $f = .33$, P-value $> .1$, so don't reject H_0.

21. a. Parabolic opening downward
   b. Very close to 1
   c. (83.89, 87.33)

23. For the exponential model, $V(Y|x) = \alpha^2 e^{2\beta x}\sigma^2$, which does depend on x. A similar result holds for the power model.

25. The z ratio for β_1 is highly significant, indicating that the likelihood of a level being acceptable does decrease as the level increases. We estimate that for each 1 dBA increase in noise level, the odds of acceptability decreases by a factor of .70.

27. b. 52.88, .12   **c.** .895   **d.** No
   e. (48.54, 57.22)   **f.** (42.85, 62.91)

29. a. SSE = 16.8, s = 2.048   **b.** R^2 = .995   **c.** Yes. $t = -6.55$, P-value = .003 (from Minitab)   **d.** 98% individual confidence levels $\Rightarrow$ joint confidence level $\geq$ 96%: (.671, 3.706), (−.00498, −.00135)
   e. (69.531, 76.186), (66.271, 79.446), using software

31. a. .980
   b. .747, much less than .977 for the cubic model.
   c. Yes, since $t = 14.18$, P-value = 0.
   d. (6.31, 6.57), (6.06, 6.81)
   e. $t = -5.6$, P-value = 0

33. a. .9671, .9407
   b. $.0000492x^3 - .000446058x^2 + .007290688x + .96034944$ **c.** $t = 2 < 3.182 = t_{.025,3}$, so the cubic term should be deleted.   **d.** Identical
   e. .987, .994, yes

35. $\hat{y} = 7.6883e^{.1799x - .0022x^2}$

37. a. 4.9   **b.** When number of deliveries is held fixed, the average change in travel time associated with a 1-mile increase in distance traveled is .060 hr. When distance traveled is held fixed, the average change in travel time associated with one extra delivery is .900 hr.   **c.** .9861

39. a. 77.3   **b.** 40.4

41. a. $f = 475$, P-value = 0   **b.** 20,826.14   **d.** (−6694.020, −5895.438)   **e.** $t = 2.59$, P-value = .01, retain x_2 in the model.

43. a. 48.31, 3.69   **b.** No. If x_1 increases, either x_3 or x_2 must change.   **c.** Yes, since $f = 18.924$, P-value = .001.
   d. Yes, using $\alpha = .01$, since $t = 3.496$ and P-value = .003.

45. a. $f = 87.6$, P-value = 0, so there does appear to be a useful linear relationship between y and at least one of the predictors.   **b.** .935   **c.** (9.095, 11.087)

47. b. P-value = .000, so conclude that the model is useful.
   c. P-value = $.034 \leq .05 = \alpha$, so reject H_0: $\beta_3 = 0$; % garbage does appear to provide additional useful information.   **d.** (1479.8, 1531.1), reasonable precision
   e. A 95% PI is (1435.7, 1575.2).

49. a. $f = 17.31$, P-value = .000, utility of the model is confirmed.   **b.** $t = 3.96$, P-value = .002, retain the interaction predictor.   **c.** (5.73, 8.17)   **d.** (2.97, 10.93).

51. a. $t = .30$, P-value = .777, so delete x_3 **b.** $f = 15.29$, $.001 < P$-value $< .01$, so confirm model utility

at significance level .05.   **c.** (−.01180, −.00174)
   d. (2.93, 3.81)   **e.** A normal probability plot of e^* is quite straight, and plots of e^* versus x_1 and e^* versus x_2 show no discernible pattern.

55. a. $f = 6.40$, $.01 < P$-value $< .05$, so at signficance level .05 model utility is confirmed.   **b.** No. Since P-value = .510, H_0: $\beta_3 = 0$ cannot be rejected.   **c.** $t = 4.69$, P-value = .001, so model utility is confirmed.   **d.** That a nonlinear model should be fit.   **e.** $f = 20.36$, P-value $< .001$, model utility is confirmed; (30.81, 36.97)

57.

k	R^2	adj. R^2	C_k
1	.676	.647	138.2
2	.979	.975	2.7
3	.9819	.976	3.2
4	.9824		4

   a. The model with $k = 2$ **b.** No

59. a. The model with predictors x_1, x_3, and x_5.

61. No. All R^2 values are much less than .9.

63. The impact of these two observations should be further investigated. Not entirely. The elimination of observation #6 followed by re-regressing should also be considered.

65. a. The two distributions have similar amounts of variability, are both reasonably symmetric, and contain no outliers. The main difference is that the median of the crack values is about 840, whereas it is about 480 for the no-crack values. A 95% t CI for the difference between means is (132, 557).
   b. $r^2 = .577$ for the simple linear regression model, P-value for model utility = 0, but one standardized residual is −4.11! Including an indicator for crack–no crack does not improve the fit, nor does including an indicator and interaction predictor.

67. a. When gender, weight, and heart rate are held fixed, we estimate that the average change in VO$_2$max associated with a 1-minute increase in walk time is −.0996.   **b.** When weight, walk time, and heart rate are held fixed, the estimate of average difference between VO$_2$max for males and females is .6566.   **c.** 3.669, −.519   **d.** .706
   e. $f = 9.0$, P-value $< .001$, so there does appear to be a useful relationship.

69. a. No. There is substantial curvature in the scatterplot.
   b. Cubic regression yields $R^2 = .998$ and a 95% PI of (261.98, 295.62), and the cubic predictor appears to be important (P-value = .001). A regression of y versus $\ln(x)$ has $r^2 = .991$, but there is a very large standardized residual and the standardized residual plot is not satisfactory.

71. a. $R^2 = .802$, $f = 21.03$, P-value = .000. pH is a candidate for deletion. Note that there is one extremely large standardized residual.
   b. $R^2 = .920$, adjusted $R^2 = .774$, $f = 6.29$, P-value = .002   **c.** $f = 1.08$, P-value $> .10$, don't reject H_0: $\beta_6 = \cdots = \beta_{20} = 0$. The group of second-order predictors does not appear to be useful.

d. $R^2 = .871$, $f = 28.50$, P-value $= .000$, and now all six predictors are judged important (the largest P-value for any t-ratio is .016); the importance of pH^2 was masked in the test of (c). Note that there are two rather large standardized residuals.

73. a. $f = 1783$, so the model appears useful.
b. $t = -48.1$, P-value $= 0$, so even at level .001 the quadratic predictor should be retained.
c. No **d.** (21.07, 21.65) **e.** (20.67, 22.05)

75. a. $f = 30.8$, P-value $< .001$, so the model appears useful.
b. $t = -7.69$ and P-value $< .001$, so retain the quadratic predictor. **c.** (44.01, 47.91)

77. a. At significance level .05, yes, since $f = 4.06$ and P-value $= .029$. **b.** Yes, because $f = 20.1$ and $F_{.05,3,12} = 3.49$. The full versus reduced F test cannot be used since the predictors in this model are not a subset of those in (a).

79. There are several reasonable choices in each case.

81. a. $f = 106$, P-value ≈ 0 **b.** (.014, .068)
c. $t = 5.9$, reject $H_0: \beta_4 = 0$, percent nonwhite appears to be important. **d.** 99.514, $y - \hat{y} = 3.486$

83. a. Estimates of α, β_1, and β_2 are 52,912.77, -1.2060, and -1.3988, respectively. **b.** $R^2 = .782$, $f = 42.95$, P-value $= 0$ **c.** P-values for testing $H_0: \beta_1 = 0$ and $H_0: \beta_2 = 0$ are both 0 **d.** A 95% PI is (14.18, 174.51)

Chapter 14

1. a. Reject H_0. **b.** Don't reject H_0.
c. Don't reject H_0. **d.** Don't reject H_0.

3. $\chi^2 = 4.80$, P-value $> .10$, so don't reject H_0.

5. $\chi^2 = 6.61$, P-value $> .10$, so don't reject H_0.

7. $\chi^2 = 4.03$ and P-value $> .10$, so don't reject H_0.

9. a. [0, .2231), [.2231, .5108), [.5108, .9163), [.9163, 1.6094), and [1.6094, ∞) **b.** $\chi^2 = 1.25$, P-value $> .10$, so the specified exponential distribution is quite plausible.

11. a. $(-\infty, -.97)$, $[-.97, -.43)$, $[-.43, 0)$, 0, .43), [.43, .97), and [.97, ∞) **b.** $(-\infty, .49806)$, [.49806, .49914), [.49914, .5), [.5, .50086), [.50086, .50194), and [.50194, ∞)
c. $\chi^2 = 5.53$, $\chi^2_{.10,5} = 9.236$, so P-value $> .10$, and the specified normal distribution is plausible.

13. $\hat{p} = .0843$, $\chi^2 = 280.3$, P-value $<.001$, so the model gives a poor fit.

15. The likelihood is proportional to $\theta^{233}(1 - \theta)^{367}$, from which $\hat{\theta} = .3883$. The estimated expected counts are 21.00, 53.33, 50.78, 21.50, and 3.41. Combining cells 4 and 5, $\chi^2 = 1.62$, so don't reject H_0.

17. $\hat{\mu} = 3.88$, estimated expected counts are 6.2, 24.0, 46.6, 60.3, 58.5, 45.4, 29.4, 16.3, and 13.3, from which $\chi^2 = 7.8$, P-value $> .10$, so the Poisson distribution provides a good fit.

19. $\hat{\theta}_1 = (2n_1 + n_3 + n_5)/2n = .4275$, $\hat{\theta}_2 = .2750$, $\chi^2 = 29.1$, P-value $< .001$, so reject H_0.

21. Yes. The null hypothesis of a normal population distribution cannot be rejected.

23. Minitab gives $r = .967$, and since $c_{.10} = .9707$ and $c_{.05} = .9639$, $.05 < $ P-value $< .10$. Using $\alpha = .05$, normality is judged plausible.

25. $\chi^2 = 212.9$, df $= 6$, P-value $= 0$, there appears to be an association due to younger people tending to drink more.

27. Yes. $\chi^2 = 44.98$ and P-value $< .001$.

29. a. Yes, since $\chi^2 = 213.2$, P-value $= 0$. **b.** Not at any reasonable significance level, since $\chi^2 = 3.1$, P-value $> .10$.

31. a. Yes. M: .26, .25, .29, .20: F: .11, .18, .34, .37
b. Reject H_0 at significance level .05 or .01, since $\chi^2 = 14.46$ so $.001 < $ P-value $< .005$.

35. $N_{ij.}/n$, $n_k N_{ij.}/n$, 24

37. $\chi^2 = 3.65$, P-value $> .10$, so H_0 cannot be rejected.

39. b. No. Since P-value $= .023$, the null hypothesis of no association can't be rejected at significance level .01 (but would be at level .05).

41. $\chi^2 = 22.4$ and P-value $< .001$, so the null hypothesis of independence is rejected.

43. P-value $= 0$, so the null hypothesis of homogeneity is rejected.

47. a. Test statistic value $= 19.2$, P-value $= 0$
b. Evidence of at best a weak relationship; test statistic value $= -2.13$
c. Test statistic value $= -.98$, P-value $> .10$
d. Test statistic value $= 3.3$, $.01 < $ P-value $< .05$

49. Combining 6 and 7 into one category and 8 and 9 into another gives a test based on 6 df for which $\chi^2 = .92$ and P-value $> .9$!

Chapter 15

1. a. P-value $= .102$ **b.** $.026 < P$-value $< .046$ **c.** $.055 < P$-value $< .102$ **d.** P-value $= .05$ **e.** $z = 3.7, P$-value ≈ 0.

3. $s_+ = 18, .02 < P$-value $< .05$, so H_0 is rejected.

5. $s_+ = 72, P$-value $< .01$, so H_0 is rejected.

7. $s_+ = 424, z = 2.56, P$-value $\approx .0052$,, so reject H_0.

9.

d	0	2	4	6	8	10	12	14	16	18	20
$p(d)$	$\frac{1}{24}$	$\frac{3}{24}$	$\frac{1}{24}$	$\frac{4}{24}$	$\frac{2}{24}$	$\frac{2}{24}$	$\frac{2}{24}$	$\frac{4}{24}$	$\frac{1}{24}$	$\frac{3}{24}$	$\frac{1}{24}$

P-value $= .167$

11. $w = 38, .008 < P$-value $< .028$, so H_0 is rejected.

13. $w = 25, P$-value $> .053$, so don't reject H_0.

15. $w = 39$ P-value $= .027$, so H_0 is rejected at significance level .05.

17. $(\bar{x}_{(5)}, \bar{x}_{(32)}) = (11.15, 23.80)$

19. $(-.585, .025)$

21. $(d_{ij(5)}, d_{ij(21)}) = (16, 87)$

23. $k = 14.06, .001 < P$-value $< .005$, so reject H_0.

25. $k = 9.23, P$-value $\approx .01$, so reject H_0.

27. $f_r = 2.60, P$-value $> .10$, so don't reject H_0.

29. $f_r = 9.62, .02 < P$-value $< .025$, so reject H_0 at significance level .05.

31. $(-5.9, -3.8)$

33. a. $.021$ **b.** $y = 12, P$-value $= .252$, so H_0 cannot be rejected.

35. $w' = 26, P$-value $> .056$, so don't reject H_0.

Chapter 16

1. All points on the chart fall between the control limits.

3. .9802, .9512, 53

5. a. 1.67, .67 **b.** 1, .67 **c.** $C_{pk} \leq C_p$, $=$ when $\mu = $ (USL + LSL)/2

7. a. .0301 **b.** .2236 **c.** .6808

9. LCL $= 12.20$, UCL $= 13.70$. No.

11. LCL $= 94.91$, UCL $= 98.17$. There appears to be a problem on the 22nd day.

13. a. 200 **b.** 4.78 **c.** 384.62 (larger), 6.30 (smaller)

15. LCL $= 12.37$, UCL $= 13.53$

17. a. LCL $= 0$, UCL $= 6.48$ **b.** LCL $= .48$, UCL $= 6.60$

19. LCL $= .045$, UCL $= 2.484$. Yes, since all points are inside the control limits.

21. a. LCL $= .105$, UCL $= .357$ **b.** Yes, since $.39 >$ UCL.

23. $\bar{p} > 3/53$

25. LCL $= 0$, UCL $= 10.1$

27. When area $= .6$, LCL $= 0$ and UCL $= 14.6$; when area $= .8$, LCL $= 0$ and UCL $= 13.4$; when area $= 1.0$, LCL $= 0$ and UCL $= 12.6$.

29.

l:	1	2	3	4	5	6	7	8
d_i:	0	.001	.017	0	0	.010	0	0
e_i:	0	0	0	.038	0	0	0	.054
l:	9	10	11	12	13	14	15	
d_i:	0	.024	.003	0	0	0	.005	
e_i:	0	0	0	.015	0	0	0	

There are no out-of-control signals.

31. $n = 5, h = .00626$

33. Hypergeometric probabilities (calculated on an HP21S calculator) are .9919, .9317, .8182, .6775, .5343, .4047, .2964, .2110, .1464, and .0994, whereas the corresponding binomial probabilities are .9862, .9216, .8108, .6767, .5405, .4162, .3108, .2260, .1605, and .1117. The approximation is satisfactory.

35. .9206, .6767, .4198, .2321, .1183; the plan with $n = 100$, $c = 2$ is preferable.

37. .9981, .5968, and .0688

39. a. .010, .018, .024, .027, .027, .025, .022, .018, .014, .011 **b.** .0477, .0274 **c.** 77.3, 202.1, 418.6, 679.9, 945.1, 1188.8, 1393.6, 1559.3, 1686.1, 1781.6

41. $\overline{X}$ chart based on sample standard deviations: LCL $= 402.42$, UCL $= 442.20$. $\overline{X}$ chart based on sample ranges: LCL $= 402.36$, UCL $= 442.26$. S chart: LCL $= .55$, UCL $= 30.37$. R chart: LCL $= 0$, UCL $= 82.75$.

43. S chart: LCL $= 0$, UCL $= 2.3020$; because $s_{21} = 2.931 >$ UCL, the process appears to be out of control at this time. Because an assignable cause is identified, recalculate limits after deletion: for an S chart, LCL $= 0$ and UCL $= 2.0529$; for an $\overline{X}$ chart, LCL $= 48.583$ and UCL $= 51.707$. All points on both charts lie between the control limits.

45. $\bar{\bar{x}} = 430.65$, $s = 24.2905$; for an S chart, UCL $= 62.43$ when $n = 3$ and UCL $= 55.11$ when $n = 4$; for an $\overline{X}$ chart, LCL $= 383.16$ and UCL $= 478.14$ when $n = 3$, and LCL $= 391.09$ and UCL $= 470.21$ when $n = 4$.

Glossary of Symbols/Abbreviations

Symbol/Abbreviation	Page	Description
n	13	sample size
x	13	variable on which observations are made
$x_1, x_2, \ldots, x_n$	13	sample observations on x
$\sum_{i=1}^{n} x_i$	30	sum of $x_1, x_2, \ldots, x_n$
$\bar{x}$	29	sample mean
μ	30	population mean
N	30	population size when the population is finite
$\tilde{x}$	31	sample median
$\tilde{\mu}$	32	population median
$\bar{x}_{tr}$	33	trimmed mean
x/n	34	sample proportion
s^2	37	sample variance
s	37	sample standard deviation
σ^2, σ	38	population variance and standard deviation
$n-1$	39	degrees of freedom for a single sample
S_{xx}	39	sum of squared deviations from the sample mean
f_s	40	sample fourth spread
$\mathcal{S}$	53	sample space of an experiment
$A, B, C_1, C_2, \ldots$	54	various events
A'	55	complement of the event A
$A \cup B$	55	union of the events A and B
$A \cap B$	55	intersection of the events A and B
$\varnothing$	56	the null event (event containing no outcomes)
$P(A)$	58	probability of the event A
N	63	number of equally likely outcomes
$N(A)$	64	number of outcomes in the event A
n_1, n_2	69	number of ways of selecting 1st (2nd) element of an ordered pair

Symbol/Abbreviation	Page	Description
$P_{k,n}$	70	number of permutations of size k from n distinct entities
$\binom{n}{k}$	71	number of combinations of size k from n distinct entities
$P(A\mid B)$	75	conditional probability of A given that B occurred
rv	96	random variable
X	96	a random variable
$X(s)$	96	value of the rv X associated with the outcome s
x	96	some particular value of the rv x
$p(x)$	99	probability distribution (mass function) of a discrete rv X
pmf	100	probability mass function
$p(x; \alpha)$	103	pmf with parameter α
$F(x)$	104	cumulative distribution function of an rv
cdf	104	cumulative distribution function
$a-$	106	largest possible X value smaller than a
$E(X), \mu_X, \mu$	110	mean or expected value of the rv X
$E[h(X)]$	112	expected value of the function $h(X)$
$V(X), \sigma_X^2, \sigma^2$	114	variance of the rv X
σ_X, σ	114	standard deviation of the rv X
S, F	118	success/failure on a single trial of a binomial experiment
n	118	number of trials in a binomial experiment
p	118	probability of success on a single trial of a binomial or negative binomial experiment
$X \sim \text{Bin}(n, p)$	120	the rv X has a binomial distribution with parameters n and p

Symbol/ Abbreviation	Page	Description
$b(x; n, p)$	120	binomial pmf with parameters n and p
$B(x; n, p)$	121	cumulative distribution function of a binomial rv
M	126	number of successes in a dichotomous population of size N
$h(x; n, M, N)$	126	hypergeometric pmf with parameters n, M, and N
r	129	number of desired successes in a negative binomial experiment
$nb(x; r, p)$	129	negative binomial pmf with parameters r and p
μ	131	parameter of a Poisson distribution
$p(x; \mu)$	131	Poisson pmf
$F(x; \mu)$	132	Poisson cdf
Δt	134	length of a short time interval
$o(\Delta t)$	134	quantity that approaches 0 faster than Δt does
α	134	rate parameter of a Poisson process
$\alpha(t)$	134	rate function of a variable-rate Poisson process
pdf	143	probability density function
$f(x)$	143	probability density function of a continuous rv X
$f(x; A, B)$	144	uniform pdf on the interval $[A, B]$
$F(x)$	148	cumulative distribution function
$\eta(p)$	151	$100p$th percentile of a continuous distribution
$\widetilde{\mu}$	152	median of a continuous distribution
$f(x; \mu, \sigma)$	157	pdf of a normally distributed rv
$N(\mu, \sigma^2)$	158	normal distribution with parameters μ and σ^2
Z	158	a standard normal rv
z curve	158	standard normal curve
$\Phi(z)$	158	cdf of a standard normal rv
z_α	160	value that captures upper-tail area α under the z curve
λ	170	parameter of an exponential distribution
$f(x; \lambda)$	170	exponential pdf
$\Gamma(\alpha)$	172	the gamma function
$f(x; \alpha, \beta)$	173	gamma pdf with parameters α and β
df	175	degrees of freedom
ν	175	number of df for a chi-squared distribution
$f(x; \alpha, \beta)$	177	Weibull pdf with parameters α and β
$f(x; \mu, \sigma)$	179	lognormal pdf with parameters μ and σ
$f(x; \alpha, \beta, A, B)$	181	beta pdf with parameters α, β, A, B

Symbol/ Abbreviation	Page	Description
θ_1, θ_2	190	location and scale parameters
$p(x, y)$	199	joint pmf of two discrete rv's X and Y
$p_X(x), p_Y(y)$	200	marginal pmf's of X and Y, respectively
$f_X(x), f_Y(y)$	202	marginal pdf's of X and Y, respectively
$p(x_1, \ldots, x_n)$	206	joint pmf of the n rv's $X_1, \ldots, X_n$
$f(x_1, \ldots, x_n)$	206	joint pdf of the n rv's $X_1, \ldots, X_n$
$f_{Y\|X}(y\|x)$	209	conditional pdf of Y given that $X = x$
$p_{Y\|X}(y\|x)$	209	conditional pmf of Y given that $X = x$
$E(Y\|X = x)$	209	expected value of Y given that $X = x$
$E[h(X, Y)]$	213	expected value of the function $h(X, Y)$
$\text{Cov}(X, Y)$	214	covariance between X and Y
$\text{Corr}(X, Y), \rho_{X,Y}, \rho$	216	correlation coefficient for X and Y
$\overline{X}$	221	the sample mean regarded as an rv
S^2	222	the sample variance regarded as an rv
CLT	232	Central Limit Theorem
θ	248	generic symbol for a parameter
$\hat{\theta}$	248	point estimate or estimator of θ
MVUE	256	minimum variance unbiased estimator (or estimate)
$\hat{\sigma}_{\hat{\theta}}, s_{\hat{\theta}}$	259	estimated standard deviation of $\hat{\theta}$
$x_1^*, \ldots, x_n^*$	259	bootstrap sample
$\hat{\theta}^*$	259	estimate of θ from a bootstrap sample
mle	270	maximum likelihood estimate (or estimator)
CI	277	confidence interval
$100(1 - \alpha)\%$	281	confidence level for a CI
T	295	variable having a t distribution
ν	296	degrees of freedom (df) parameter for a t distribution
t_ν	296	t distribution with ν df
$t_{\alpha, \nu}$	296	value that captures upper-tail area α under the t_ν density curve
PI	300	prediction interval
$\chi^2_{\alpha, \nu}$	304	value that captures upper-tail area α under the chi-squared density curve with ν df
H_0	311	null hypothesis
H_a	311	alternative hypothesis
α	319	significance level, probability of a type I error
β	319	probability of a type II error
μ_0	327	null value in a test concerning μ
Z	327	test statistic based on standard normal distribution
μ'	330	alternative value of μ in a β calculation

Symbol/ Abbreviation	Page	Description
$\beta(\mu')$	330	type II error probability when $\mu = \mu'$
T	335	test statistic based on t distribution
θ_0	346	null value in a test concerning θ
p_0	347	null value in a test concerning p
p'	347	alternative value of p in a β calculation
$\beta(p')$	348	type II error probability when $p = p'$
Ω_0, Ω_a	355	disjoint sets of parameter values in a likelihood ratio test
m, n	362	sample sizes in two-sample problems
Δ_0	363	null value in a test concerning $\mu_1 - \mu_2$
Δ'	366	alternative value of $\mu_1 - \mu_2$ in a β calculation
S_p^2	378	pooled estimator of σ^2
D_i	383	the difference $X_i - Y_i$ for the pair (X_i, Y_i)
$\bar{d}, s_D$	385	sample mean difference, sample standard deviation of differences for paired data
p	392	common value of p_1 and p_2 when $p_1 = p_2$
F	399	rv having an F distribution
ν_1, ν_2	399	numerator and denominator df for an F distribution
F_{α,ν_1,ν_2}	399	value capturing upper-tail area α under an F curve with ν_1, ν_2 df
ANOVA	409	analysis of variance
I	410	number of populations in a single-factor ANOVA
J	412	common sample size when sample sizes are equal
X_{ij}, x_{ij}	412	jth observation in a sample from the ith population
$\bar{X}_{i\cdot}$	412	mean of observations in sample from ith population
$\bar{X}_{\cdot\cdot}$	412	mean of all observations in a data set
MSTr	413	mean square for treatments
MSE	413	mean square for error
F	415	test statistic based on F distribution
$x_{i\cdot}$	415	total of observations in ith sample
$x_{\cdot\cdot}$	415	grand total of all observations
SST	416	total sum of squares
SSTr	416	treatment sum of squares
SSE	416	error sum of squares
m, ν	420	parameters for Studentized range distribution
$Q_{\alpha,m,\nu}$	420	value that captures upper-tail area α under the associated Studentized range density curve

Symbol/ Abbreviation	Page	Description
μ	426	average of population means in single-factor ANOVA
$\alpha_1, \ldots, \alpha_I$	426	treatment effects in a single-factor ANOVA
ϵ_{ij}	426	deviation of X_{ij} from its mean value
$J_1, \ldots, J_I$	430	individual sample sizes in a single-factor ANOVA
n	430	total number of observations in a single-factor ANOVA data set
$A_1, \ldots, A_I$	432	random effects in a single-factor ANOVA
A, B	437	factors in a two-factor ANOVA
K_{ij}	437	number of observations when factor A is at level i and factor B is at level j
I, J	437	number of levels of factors A and B, respectively
$\bar{X}_{i\cdot}, \bar{X}_{\cdot j}$	439	average of observations when A (B) is at level i (j)
μ_{ij}	439	expected response when A is at level i and B is at level j
α_i, β_j	441	effect of A (B) at level i (j)
f_A, f_B	442	F ratios for testing hypotheses about factor effects
A_i, B_j	448	factor effects in random effects model
σ_A^2, σ_B^2	448	variances of factor effects
K	451	sample size for each pair (i, j) of levels
γ_{ij}	451	interaction between A and B at levels i and j
A_i, B_j, G_{ij}	456	effects in mixed or random effects models
$\alpha_i, \beta_j, \delta_k$	456	main effects in a three-factor ANOVA
$\gamma_{ij}^{AB}, \gamma_{ik}^{AC}, \gamma_{jk}^{BC}$	460	two-factor interactions in a three-factor ANOVA
γ_{ijk}	460	three-factor interaction in a three-factor ANOVA
I, J, K	460–461	number of levels of A, B, C in a three-factor ANOVA
β_1, β_0	491	slope and intercept of population regression line
ϵ	491	deviation of Y from its mean value in simple linear regression
σ^2	491	variance of the random deviation ϵ
$\mu_{Y \cdot x^*}$	492	mean value of Y when $x = x^*$
$\sigma_{Y \cdot x^*}^2$	492	variance of Y when $x = x^*$
$\hat{\beta}_1, \hat{\beta}_0$	496	least squares estimates of β_1 and β_0
S_{xy}	498	$\Sigma(x_i - \bar{x})(y_i - \bar{y})$
$\hat{y}_i$	500	predicted value of y when $x = x_i$
SSE	502	error (residual) sum of squares

Symbol/ Abbreviation	Page	Description
SST	504	total sum of squares S_{yy}
r^2	504	coefficient of determination
$S_{\hat{\beta}_1}$	512	estimated standard deviation of $\hat{\beta}_1$
r, R	528	sample correlation coefficient
e_i^*	543	a standardized residual
$\beta_i (i = 1, \ldots, k)$	562	coefficient of x^i in polynomial regression
$\hat{\beta}_i$	563	least squares estimate of β_i
R^2	565	coefficient of multiple determination
β_i^*	567	coefficient in centered polynomial regression
β_i	572	population regression coefficient of predictor x_i
$\hat{\beta}_i$	576	least squares estimate of β_i
SSE_k, SSE_l	585	SSE for full and reduced models, respectively
Γ_k	600	normalized expected total estimation error
C_k	600	estimate of Γ_k
h_{ii}	604	coefficient of y_i in $\hat{y}_i$
$\chi^2_{\alpha,\nu}$	621	value that captures upper-tail area α under the χ^2 curve with ν df
χ^2	622	test statistic based on a chi-squared distribution
$p_{10}, \ldots, p_{k0}$	622	null values for a chi-squared test of a simple H_0
$\pi_i(\theta)$	628	category probability as a function of parameters $\theta_1, \ldots, \theta_m$
I, J	639	number of populations and categories in each population when testing for homogeneity
I, J	639	numbers of categories in each of two factors when testing for independence
n_{ij}	640	number of individuals in sample from population i who fall into category j

Symbol/ Abbreviation	Page	Description
$n_{\cdot j}$	640	total number of sampled individuals in category j
p_{ij}	640	proportion of population i in category j
$\hat{e}_{ij}$	641	estimated expected count in cell i, j
n_{ij}	643	number in sample falling into category i of 1st factor and category j of 2nd factor
p_{ij}	643	proportion of population in category i of 1st factor and category j of 2nd factor
S_+	654	signed-rank statistic
W	662	rank-sum statistic
K	672	Kruskal-Wallis test statistic
R_{ij}	672	rank of X_{ij} among all N observations in the data set
$\overline{R}_{i\cdot}$	672	average of ranks for observations in the sample from population or treatment i
F_r	674	Friedman's test statistic
UCL	679	upper control limit
LCL	679	lower control limit
C_p, C_{pk}	680–681	process capability indices
R	685	sample range
ARL	687	average run length
IQR	688	interquartile range
CUSUM	700	cumulative sum
OC	709	operating characteristic
AQL	710	acceptable quality level
LTPD	710	lot tolerance percent defective
AOQ	713	average outgoing quality
AOQL	713	average outgoing quality limit
ATI	713	average total number inspected

Index

Note: Page numbers preceded by the letter "A" indicate appendix table page numbers and the page numbers followed by a "f" indicate figures.